BIOLOGY OF ANIMAL BEHAVIOR

BIOLOGY OF ANIMAL BEHAVIOR

James W. Grier
Professor of Zoology
North Dakota State University
Fargo, North Dakota

Theodore Burk
Associate Professor of Biology
Creighton University
Omaha, Nebraska

Original artwork by
Barbara Bradley
Tucson, Arizona

SECOND EDITION

with 677 illustrations

Wm. C. Brown Publishers
Dubuque, Iowa • Melbourne, Australia • Oxford, England

To our own mates and closest kin

 Wm. C. Brown Communications, Inc.

President and Chief Executive Officer *G. Franklin Lewis*
Corporate Senior Vice President, President of WCB Manufacturing *Roger Meyer*
Corporate Senior Vice President and Chief Financial Officer *Robert Chesterman*

Cover photograph: *Animals, Animals.* © Robert Maier

A Times Mirror Company

Library of Congress Catalog Card Number: 91–33556

ISBN 0–697–23492–4

Printed in the United States of America by Wm. C. Brown Communications, Inc.,
2460 Kerper Boulevard, Dubuque, IA 52001

10 9 8 7 6 5 4 3 2 1

PREFACE

The subject of animal behavior is a large, diverse, and variably defined array of topics within the fields of biology and psychology. With time, some aspects become better defined and constrained, others change, and new interests, information, and theories tend to increase on most fronts. In this second edition of *Biology of Animal Behavior* we attempt to track the changing subject of animal behavior and continue to present the material in a manner that can be used flexibly by instructors with different needs and interests.

Changes from the First Edition

The writing of a book is rarely, if ever, finished. Usually the process is stopped so it can go to press. Subsequent editions allow the authors to spend additional time on the book and also enable them to take advantage of experience gained since the previous edition. Such has been the case with the second edition of *Biology of Animal Behavior*, and several changes have been made from the first edition.

One of the most significant changes is the addition of a coauthor, Theodore E. Burk. Dr. Burk contributes many new insights that greatly strengthen and enhance this edition, particularly in the area of social behavior of animals. At the same time, the authors worked together closely throughout the book to ensure a uniform manner of style and presentation.

Experience with the first edition indicated that the book was too large for an introductory course in animal behavior. One factor was the inclusion of material on basic biology. Although knowledge of basic biology is critical for an understanding of animal behavior it is best obtained before starting the study of behavior. Thus in this second edition the coverage of basic biology has been dropped. We strongly recommend that the reader have a background equivalent of at least one year of college-level biology, regardless of the student or instructor's major area. Alternatively and for review or reference whenever necessary, we recommend current and comprehensive introductory texts in biology at hand.

In addition, in this edition the encyclopedic aspects have been greatly reduced, and the text has been streamlined overall. Topics have been updated where appropriate, and there have been a number of cosmetic changes.

Intended Audience

The primary audience of this book remains unchanged from the first edition. The book should serve to present the subject both for those who want only an introduction and for those who may wish to delve deeper into the topic, take more advanced courses and seminars, or simply be able to read the original or more technical literature in animal behavior. We have willingly tackled difficult topics such as inclusive fitness and optimality theory, and we have continued to attempt to make them understandable although there has not been space to treat them at an advanced level. The only change is that the intended audience should have completed an introduction to biology at the level of a first year college course.

Approach and Organization

Although there have been numerous internal changes and reorganization, the overall design remains the same. The goal has been to integrate (1) the structure and function of behavior and (2) the topics of ethology, comparative psychology, and neurobiology. The presentation has been built on an underlying evolutionary framework. As in the first edition, before material was included in this book, ideas and conclusions were closely inspected to determine how well they fit the real rather than the purely academic world of animal life.

The book remains organized in four basic parts: (1) introductory topics, (2) behavior as observed output, (3) internal neural-endocrine aspects, and (4) miscellaneous, a chapter on the behavioral relationships between humans and other animals. Parts, chapters, and various subsections are essentially modular and are cross-referenced so that instructors may easily choose a different sequence of presentation or skip parts that they do not need or want.

Finally, like the first edition, this edition attempts to provide a broad perspective on the subject and focuses primarily on nonhuman behavior (although humans are included as appropriate). The text is as up-to-date as possible, and it can only point toward—not substitute for—the study of living animals. To appreciate and understand animals properly, one must spend a lot of time with the real thing. Good films on animal behavior are the next best route. We hope that all readers have access to both good films and live animals.

Distinctive Features

Each chapter begins with an overview and ends with a summary to help focus on the major points presented. A list of references for further reading, often categorized by subtopics, is provided for instructors preparing for lectures and instructors and students who are interested in delving deeper. An appendix offers suggestions for observing live animals.

A distinctive feature new to the second edition is the inclusion of "Guest Essays" in selected chapters. Some of these we have written ourselves. However, we also invited a number of people working with animal behavior to contribute their thoughts and insights on various subjects. We gave them wide latitude to address whatever they wanted and from whatever viewpoint and style the chose. Their contributions are short informational essays, editorials or differing viewpoints, interesting stories or anecdotes, insights from their experience, or additional points beyond what is covered in the chapter. We believe that these essays enhance certain topics of interest to students and provide a different perspective than what the authors provide throughout the text. Thus we have Frances Hamerstrom's reminiscences of Konrad

Lorenz in chapter 2, a description of the remarkable defense mechanism of the bombadier beetle by Thomas Eisner in chapter 9, and a discussion of the experiments conducted on *Aplysia* by William Beatty in chapter 20. These essays and others convey the fascinating nature of animal behavior in all its aspects.

Guest Essayists

Kenneth Able, Professor, State University of New York at Albany
William W. Beatty, Professor, University of Oklahoma
Charles R. Brown, Associate Professor, Yale University
Deborah Buitron, Adjunct Professor, North Dakota State University
Gordon M. Burghardt, Professor of Psychology, University of Tennessee-Knoxville
David Crews, Professor, University of Texas-Austin
Thomas Eisner, Professor of Biology, Cornell University
Frances Hamerstrom, Adjunct Professor, University of Wisconsin at Stevens Point
Jack Hailman, Professor, University of Wisconsin at Madison
Philip N. Lehner, Professor, Colorado State University
James W. Lloyd, Professor, University of Florida
Gary Nuechterlein, Associate Professor, North Dakota State University
Lewis W. Oring, Professor, University of Nevada at Reno
Asher Treat, Professor Emeritus, City University of New York

Reviewers

We wish to extend profound thanks to the following reviewers. Their helpful criticism was invaluable in improving and bringing this book to completion. In most cases we were able to address and incorporate their recommendations.

First Edition

Russell D. Fernald, University of Oregon
Merrill Frydendall, Mankato State University
Stephen J. Gaioni, Washington University
H. Carl Gerhardt, University of Missouri at Columbia
Robert Gibson, University of California at Los Angeles
Thomas C. Grubb, Jr., Ohio State University
Herbert E. Hays, Shippensberg State College
Stephen H. Jenkins, University of Nevada at Reno
Thomas Jenssen, Virginia Polytechnic Institute
John A. King, Michigan State University
Randall Lockwood, State University of New York at Stony Brook

Second Edition

Kenneth Able, State University of New York at Albany
Elizabeth Gardner, Pine Manor College
Neal Greenberg, University of Tennessee at Knoxville
Phil Lehner, Colorado State University
Laszlo Szijj, California State Polytechnic Institute at Pomona
George Waring, Southern Illinois State University at Carbondale
Fred Wasserman, Boston University

Valuable input and criticism were provided by many students and other instructors who used the first edition and drafts of this book, as well as our wives and other assistants who served as first critics and editors.

Request for Input

For the improvement of future editions, we welcome comments concerning the accuracy of facts and interpretations, topics and material that should be retained, added, or dropped, significant references or issues that need to be incorporated, new illustrations, etc. Our mailing addresses are:

James W. Grier
Zoology Department
North Dakota State University
Fargo, North Dakota 58105

Theodore Burk
Department of Biology
Creighton University
Omaha, Nebraska 68178-0103

Other Acknowledgments

Much of the credit for the content of this book must go to many persons who contributed greatly to our understanding of animals. The following were particularly significant and influential teachers and advisors.

For J.W. Grier: V. Sponseler, A. Potter, B. Eyestone, C. McCollum, R. Goss, L. Winier, V. Dowell, R. Tepaske, M. Grant, J.T. Emlen, W. Burns, J.F. Crow, J. Nees, M. Konishi, T.J. Cade, S.T. Emlen, A. van Tienhoven, H. Ambrose, and T. Eisner. Senior colleagues from whom I learned much are W. Dilger, D. Peakall, J. Wiens, H.C. Mueller, J.J. Hickey, F. and F. Hamerstrom, and J. Hailman. I particularly thank Fran Hamerstrom, who constantly reprimanded me as a student to write in "plain English." Influential peers who shared mutual interests in animal behavior include several personal friends in Ontario, Iowa, Wisconsin, Minnesota, North Dakota, and New York, as well as a large number of fellow undergraduate and graduate students at the University of Northern Iowa, University of Wisconsin at Madison, and Cornell University, my own present and past graduate and undergraduate students, many falconer friends and eagle/raptor enthusiasts, several faculty members of the Zoology and Psychology Departments at North Dakota State University and the University of North Dakota, and other biologists in the state and nearby, particularly at the Northern Prairie Wildlife Research Center, North Dakota Game and Fish Department, Minnesota Department of Natural Resources, Ontario Ministry of Natural Resources, Canadian Wildlife Service, and U.S. Fish and Wildlife Service.

For T.E. Burk: W.J. Bell, R. Jander, R. Dawkins, J.M. Cullen, J.R. Krebs, T.J. Walker, and J.E. Lloyd. I thank the Entomology faculty at the University of Kansas for treating a naive undergraduate like a colleague. All the members of the Animal Behaviour Research Group at the University of Oxford, faculty and students, provided a model of intellectual stimulation and camaraderie. I am indebted to friends and colleagues at the United States Department of Agriculture Insect Behavior Laboratory and the University of Florida Department of Entomology and Nematology in Gainesville, Florida, for providing new research opportunities and diverse role models of scientific creativity. My largest current debt is to my colleagues and especially my undergraduate students at Creighton University, who make every day a challenge and a pleasure.

In addition to W.W. Beatty, who helped write parts of the material on feeding and neural integration and who additionally reviewed psychological topics, we especially would like to thank the following for discussions and clarifications concerning various topics in this book: J.D. Brammer, D. Buitron, J.W. Gerst, B. Gladue, C.R. Gustavson, G. Nuechterlein, W. and R. Maki, and M. Sheridan at North Dakota State University, R. Crawford, J. Lang, L. Oring, and R. Seabloom at the University of North Dakota, G.M. Burghardt at the University of Tennessee, D. Crews at the University of Texas, J. Hailman at the University of Wisconsin at

Madison, D.W. Leger at the University of Nebraska at Lincoln, M.J. Ryan at the University of Texas at Austin, A.V. Hedrick of Simon Fraser University, S.B. Opp at California State University at Hayward, and a large number of persons who provided comments on the first edition.

Joyce and Dean Grier and Karlene Grier Froehling did much of the typing and entering of revisions into the word processing system and helped greatly with numerous other tasks. Barbara A. Bradley was responsible for most of the line drawings. Additional illustrations were done by Karen Schuler. Previously unpublished photographs were graciously donated by Tom Brakefield, Ed Bry, Gerald Holt, and Vladimir Beregovoy. Thomas P. Freeman, Kathy L.H. Iverson, Gary Fulton, and Jay Bjerke of the North Dakota State Univeristy Electron Microscope Laboratory prepared several electromicrographs and other photographs for the book. Several other persons, indicated in figure credits, also contributed photos. We thank Edward Murphy, Editor; Kathleen Scogna, Developmental Editor; Elisabeth Heitzeberg, Manuscript Editor; Peter Wold, and others at Mosby—Year Book for the preparation of this edition.

Finally, those in our immediate families who have both tolerated and shared our interests in animals have been of utmost importance. We particularly wish to thank our parents, Mr. and Mrs. P.H. Grier, and Mr. and Mrs. H.L. Burk; our wives, Joyce and Christine; our parents-in-law, Mr. and Mrs. E. Petersen and Mr. and Mrs. C.D. Spencer; and our children, Dean and Karlene, and Heather, Laurel, Rosemary, and Adelle. They have had to put up with opposums in clothes baskets, birds flying around the house, rattlesnakes in empty wash buckets, leaking aquaria full of invertebrates, pet praying mantises, basements full of cricket cages, numerous unmentionable items in the refrigerator and on the kitchen table, and our frequently being lost in thought or gone at work when they wanted to talk or spend time with us.

James W. Grier
Theodore Burk

RECOMMENDED GENERAL BIOLOGY/ZOOLOGY REFERENCES

Campbell, N.A. 1989. Biology, ed 2. Benjamin/Cummings, Menlo Park, Calif.
Hickman, C.P., Jr., L.S. Roberts, and F.M. Hickman. 1988. Integrated principles of zoology, ed 8. Times Mirror/Mosby College Publishing, St. Louis.
Keeton, W.T., and J.L. Gould. 1986. Biological science, ed 4. W.W. Norton.
Raven, P.H., and G.B. Johnson. 1986. Biology. Times Mirror/Mosby College Publishing, St. Louis.

ABOUT THE AUTHORS

James W. Grier is a professor in the Zoology Department at North Dakota State University at Fargo, North Dakota, where he has been since 1973. He received his B.A. degree in biology and science education from the University of Northern Iowa, an M.S. in zoology from the University of Wisconsin at Madison, and a Ph.D. in ecology and systematics from Cornell University in Ithaca, New York. His primary research focus has been on eagle behavior and population dynamics. He has taught a wide variety of courses in zoology and has a broad range of interests among different topics and groups of organisms. In addition to birds of prey, his favorite groups of animals include other birds, snakes, fishes, insects, several groups of aquatic and marine invertebrates, and a number of extinct, fossil groups. He is pictured here with Ithaca, a male golden eagle that he obtained from artificial insemination in 1972 during an extracurricular research project while he was a graduate student at Cornell. Photograph by Karen Maier.

Theodore Burk is an associate professor and chair of the Biology Department at Creighton University in Omaha, Nebraska, where he has been since 1982. He received his B.A. degree in biology from the University of Kansas in 1974 and his Ph.D. in zoology from the University of Oxford in 1979, where he was a Rhodes Scholar and a student in the Animal Behaviour Research Group. From 1979 to 1982 he was a post-doctoral associate at the University of Florida and the United States Department of Agriculture's Agricultural Research Service in Gainesville, Florida. He has taught courses on a variety of subjects, ranging from animal behavior to Charles Darwin. His academic interests are the behavioral ecology of insects and biology education; his outside interests are nature, history, and sports. He is shown examining a migrating Monarch butterfly.

CONTENTS
IN BRIEF

CONTENTS

PART TWO   **THE BEHAVIORAL REPERTOIRE**
OBSERVED FORMS AND INTERPRETED FUNCTIONS

PART THREE   INTERNAL CONTROL AND CHANGES IN BEHAVIOR

21 **INTERNAL INTEGRATION OF COMPLEX BEHAVIOR** 753
Feeding as a Case in Point

PART FOUR **ADDITIONAL TOPICS**

22 **BEHAVIORAL RELATIONSHIPS BETWEEN HUMANS AND OTHER
ANIMALS** 781

APPENDIX   **SUGGESTIONS FOR OBSERVING LIVE ANIMALS**   817

INTRODUCTION TO
BEHAVIOR AND ITS CAUSES

1

BEHAVIOR AND BIOLOGY

There are well over one million species of animals on Earth, including more than 40,000 species of vertebrates. All of them share some biological traits with other organisms such as plants and bacteria. Thus to understand animals, one must understand general biological principles. However, although animals respond to their environment in coordinated and functional ways, they do not all behave the same way. Different species display a vast array of behaviors; and within a species, males and females often behave quite differently.

The behavior of other animals is often subtle, complex, overlooked, and easily misinterpreted. The life histories, environments, capabilities, senses, and perspectives of other animals are usually quite different from ours. Furthermore, the behaviors of animals often are logistically difficult to study, which further challenges our attempts to understand them. Animal behavior is a field with many unanswered questions, although there has been much progress.

This chapter uses the fascinating predator-prey interaction between bats and moths as an example to introduce the subject of behavior and place it in a biological perspective, thus setting the stage for the remainder of the book. The interaction between bats and moths occurs at high speed under the cover of darkness. Attempts to unravel the interaction have involved an equally fascinating history of successes, failures, and the use of ingenuity and technology. The example illustrates several points and raises several questions about the biology of animal behavior. The causes of behavior, for example, are becoming better understood in some ways but remain far from clear in others, as we will see when we examine the issues of anthropomorphism and the mental experiences of animals.

ANIMAL BEHAVIOR: WHAT IS IT AND WHO CARES?

A botany student enrolled in an ornithology course once remarked that it was her most frustrating class. The birds kept moving, and she could not get close enough to them or keep them in sight long enough to study them. Several zoology students, on the other hand, have complained that botany is dull and boring because the plants do not "do anything." These statements certainly do not express the sentiments of all botanists and zoologists, but they do indicate the central difference between plants and animals. Most animals have nervous systems and muscles or analogous organelles and can actively move, whereas most plants cannot. Most animals are self-contained mobile organisms. Virtually all of an animal's anatomy and physiology is portable and supports or is otherwise affected by the animal's ability to move. Animal movements and other nervous system responses (i.e., what they do, how they do it, and why) are all parts of animal behavior.

Behavior is broadly defined as all observable or otherwise measurable muscular and secretory responses (or lack thereof in some cases) and related phenomena such as changes in blood flow and surface pigments in response to changes in an animal's internal or external environment. This definition is similar to Kandel's definition (1976), which is based on the tradition of Skinner (1938) and Hebb (1958). Our definition includes simple muscular contractions and glandular secretions, as well as higher-order categories such as courtship and communication. What can be observed about animals, however, must be viewed as only surface phenomena. Thus the study of behavior takes one far below the surface to things that cannot be sensed directly. One may consider and make inferences about relatively inaccessible matters such as molecular and evolutionary processes and perhaps even mental or cognitive states.

Movement, secretions, and related phenomena require responsive cells and mechanisms such as muscles, cilia, flagella, pseudopods, glands, and chromatophores (pigmented cells capable of redistributing the pigment and causing changes in appearance). Animals, however, do not just move and otherwise behave or change appearance willy-nilly. Rather, they usually respond in more or less functional ways to objects and events in the external environment or to changes in their internal environment.

Thus, in addition to muscles and other response mechanisms, animals and their behaviors also need the means to detect objects and events, both external and internal. They also must possess mechanisms, such as nervous and endocrine systems, to integrate and coordinate their perceptions and activities. These complex internal control systems operate silently behind the scenes and are unknown or taken for granted by most persons, but they are critical to behavior. All one has to do is cut some nerves, remove certain endocrine glands, or damage various parts of the brain or spinal cord to observe striking changes in behavior, including immobility or paralysis.

Nervous systems and endocrine tissues are soft, delicate, and often extremely complex. They have presented biologists with some of the greatest challenges in science. Neural and endocrine connections and pathways likewise are complicated and challenging. Many behavior patterns appear to be mostly fixed internally. They often are said to be innate or instinctive. Other patterns may be more open to modification or may be learned, as a result of events in the animal's life. A large portion of the study of behavior has been devoted to what goes on inside the animal.

However, there is more to understanding behavior than considering what an animal looks like and does or what is inside the animal. How an animal behaves is only part of the story. We must also ask why it behaves in certain ways and why there is so much diversity in the behavior of various species. Why are dogs frequently more sociable and alert to their masters than cats? Why do dogs eat meat, deer eat plants, and bears eat both meat and plants? Why are house flies active by day whereas most cockroaches hide by day and come out at night? If some flies need to see to fly, how and why do many mosquitoes (a type of fly) become active at night, and how do they keep from crashing into things? How do spiders or butterflies mate? What kinds of interactions and bodily contacts do porcupines have with each other?

Much animal behavior attracts us simply because it is so varied and seemingly mysterious. The botany student enrolled in ornithology, in spite of her frustration with trying to keep the birds in sight, nonetheless also became quite fascinated by birds and their behaviors. The rich diversity of bird behaviors, their migrations, vocalizations, courtship, nest building, feeding, and interactions with each other and other animals intrigue many people. The strange and unfamiliar behaviors of other animals, particularly exotic ones, are always popular subjects on television shows. Equally fascinating are the behaviors of common and familiar animals that surround us in our daily lives.

Beyond being merely interesting, behavior and an understanding of it are vital for understanding other aspects of animal biology, including morphology, physiology, ecology, genetics, and evolution. In fact, we will go so far as to suggest that one cannot truly appreciate and understand any other part of animal biology without considering behavior. This interaction between subjects is mutual and reciprocal: an understanding of behavior is inextricably interwoven with knowledge of the other fields of biology and even the social sciences.

In addition to its academic importance, animal behavior has many practical applications and is relevant to numerous human concerns. Many people are strongly attracted to dogs, cats, horses, and other animals as pets or for other captive uses, including the keeping of aquaria and zoos. Other people avidly participate in fishing, hunting, wildlife photography, birding, insect collecting, and similar activities. The better animal behavior is understood, the more rewarding becomes any of these activities. Much the same can be said of other human activities involving less direct use of animals, including farming of animals for food and other products; reintroductions to the wild; veterinary practice; and the use of animals in a variety of basic and applied research, including physiology and medicine.

A better understanding of animal behavior also advances contemporary human concerns over the ecological interactions between humans and other species—from those that are rare and endangered to game species and various pests. The study of animal behavior can even clarify some of the issues over animal rights, pain, and welfare (Burghardt and Herzog 1980).

Furthermore, scientific findings from the behavior of animals have begun spilling more into studies of human behavior. Psychologists have been at the forefront of behavioral biology for decades, and some sociologists are incorporating ideas and information from studies of social behavior in nonhuman species. The International Society of Human Ethology publishes a newsletter that includes current bibliographies.

Humans have had interest in and some understanding of animal behavior for several thousand years. Only during the last two centuries, however, has there been much scientific progress in understanding animal behavior. This progress has resulted in part from the growth and application of scientific methods in general, including carefully designed experiments. The understanding of behavior, along with many other facets of biology, also received a tremendous boost from work by Charles Darwin, which is discussed in more depth in the next chapter. In addition to advances in various individual areas of biology, the behavioral relationships between these areas have become clearer. Formerly distinct topics such as neurophysiology and the ecological aspects of behavior have begun to merge into a more unified subject. Animal behavior as a discipline has matured and taken on an academic status along with the other relatively recent developments of molecular biology and ecology.

When faced with a subject that pervades so many aspects of animal life, modern biology, and human interests and concerns, a textbook writer feels like a mosquito at a nudist camp: one does not know where to start. To illustrate how the study of animal behavior fits in with other biological topics and how it also raises a number of complex questions, we begin with the interaction between bats and moths. The example is a classic in animal behavior. The ecological and evolutionary aspects

James Grier

ESSAY

J.W. Grier

The Problem of Viewing Other Species of Animals Through Cultural Glasses

The particular culture and subculture that a person is raised and lives in greatly influences how one thinks about animals of other species in ways most people are not aware. We see other animals through "cultural glasses." Culture creates value judgments that can lead to prejudices and biases for or against other species just as it does toward different groups of people. These prejudices often do not become noticeable until we encounter contrasts between different cultures.

Most Americans of northern European descent, for example, consider chickens to be timid, fearful, and weak. But some Asian and Latin cultures with histories involving the sport of cock fighting consider chickens to be the embodiment of courage, virility, and toughness. Similarly with snakes, some cultures consider them to be evil, unclean, and disgusting, whereas other cultures revere and honor them.

Some cultural viewpoints toward particular animals are widespread. Nations, for example, may associate eagles or lions as national symbols, whereas few use spiders or shrews. Some views are solidly entrenched in culture, such as the idea that foxes are cunning. Other views, however, may depend on subculture or microculture. Among American ranchers, for example, sheep ranchers and cattle ranchers may have different views toward sheep and cattle. Similarly, many people who like to hunt or fish are favorably interested in the game species but not others.

Cultural views of animals arise from a variety of sources, including edibility, danger to humans, and even a scale of size where large animals are viewed differently than small ones. Some views have been in the culture for centuries. In the Old Testament (Leviticus, chapter 11), the early Israelites had a list of de-are fairly straightforward, and some of the anatomical and nervous system considerations are as simple, well studied, and as interesting as one is likely to find.

BATS AND MOTHS: A MICROCOSM IN THE STUDY OF ANIMAL BEHAVIOR

The animals involved in this frequent life-and-death drama are the little brown bat *(Myotis lucifugus)* and moths of the family Noctuidae. The setting might be an open field near a New England town on a warm summer evening. At first glance the behavior patterns of the bats and moths are not particularly remarkable. They seem to be merely flying about. Bats may be seen briefly overhead as dark forms silhouetted against the dusky sky. The moths may be noticed flitting against a window screen, attracted to the lights inside.

It is difficult to put oneself in these animals' positions, but try to do just that. To start breaking the habit of looking at the world only from the human viewpoint, stop reading and take a few minutes to think about the following questions. What did you do in the past hour? Where were you last night from 6:00 PM to 6:00 AM, and how did you spend the time? Why did you do all of those things? What did

testable animals, including creeping animals and some insects; anyone who even touched them was to be considered unclean. Other views are more recent. Mickey Mouse and Kermit the Frog have done wonders for the images of mice and frogs, respectively. Much of the source of cultural viewpoints is anthropomorphic (as discussed in this chapter). Some of the cultural bias is subtle and more difficult to recognize. This includes a general predisposition in Western culture to favor things that are uncommon and rare as opposed to those that are common.

Unfortunately, these cultural prejudices and biases influence our views as to which animals are valuable, noble, interesting, and worth studying as opposed to those which are disgusting (gross, icky, etc.) or otherwise not worthy of attention. This not only limits the scope of potential subjects, it seriously interferes with the ability to view animals biologically. In reality, except for domestication and a few other human evolutionary influences, animals are what they are for other reasons, shaped by millennia of evolution long before the appearance of our species on the planet.

Cultural views are arbitrary and artificial. Both eagles and spiders, for example, are interesting predators. Chickadees are not just cute little wimpy dickey birds; they are hardy animals that can withstand $-30°$ C or colder in the winter, fight vigorously (for example, if you try to hold one in your hands), have complex social behavior, and possess an incredible memory for different locations. The European starling in North America is no less interesting than one of the rare native species, although it is a common and introduced pest.

Cultural biases are unavoidable and we all have them, including the most objective biologist among us. But we should become aware of them and not let them hinder us. Understanding and appreciation of the biological world, including animal behavior, depend on the extent that one is capable of removing his or her cultural glasses. What cultural glasses are you wearing?

you eat for your most recent meal, and how did you get that food? Where did the food come from?

Now consider the lives of the bat and moth. These animals are largely inactive during the day and active at night. The little brown bat is a small mammal of the order Chiroptera (hand-wing). Its body is covered with soft fur. The bat's wings are largely composed of long, slender finger bones and naked skin. These bats sleep during the day by holding on with their back legs and hanging head down in the dark crevices of large trees, caves, mines, and buildings. They are highly social when not hunting and roost in groups, depending in part on their age, sex, and reproductive condition. They hibernate during the winter in more protected sites, hibernacula, which may be several kilometers from their roosts at other times of the year.

The moth family Noctuidae includes some large, soft-bodied insects with a wingspan of approximately 2 to 12 cm, depending on the species. Noctuids are beautiful and intriguing. If picked up they do not bite, but a dusty material, the fine scales that cover the wings, may rub off on one's hands. During the day noctuids hide in crevices or vegetation.

The New England summer evening can contain an overwhelming variety of objects, substances, and forms of energy, many of which are associated with humans. There are trees, shrubs, grasses, telephone poles, wires, houses, people, vehicles, cats, dogs, night-flying birds, mice, moths, a great variety and large number of other insects, bats, and much more. The air and environment contain innumerable odors, sounds, light, and other forms of energy. Consider the following sounds and mechanical vibrations: a slight breeze and many associated vibrations such as from rustling leaves, sounds of dogs barking and cats fighting, children crying, people talking, nighthawks buzzing and snipes winnowing overhead, high-pitched chirps of bats, humming of mosquitoes and droning of large flying beetles, chirping and whirring calls of many species of insects, twitters from perched birds, sounds of televisions and stereos coming from houses, roaring motorcycles, a police siren, various sounds of machines, vehicles and squeaking hinges, and a buzzing trans- former on a power pole. Energy in other forms includes light from the rising moon, starlight, radio and television signals from near and distant broadcasting stations, microwaves, cosmic rays and x rays plus many radio waves and other forms of radiation from space, the earth's magnetic field, streetlights, lights from houses, headlights from vehicles, light from fireflies, children with flashlights, and much more.

Out of this bewildering hodgepodge of substances and energies, a few are ex- tremely significant to the lives of bats and moths. These few may mean the difference between life and death. The remainder are irrelevant. The animals' sensory and nervous systems must somehow sort through all of this input, tune into the significant items, omit the rest, and then initiate the appropriate behavior. The means by which animals do this may not be conscious. The biological machinery that does the job has resulted from and been refined by millions of years of natural selection. The items associated with humans have not been around for long, but there have always been many objects and various vibrations, favorable, unfavorable, and neu- tral, in the environment.

The natural selective pressures resulting from the interactions between bats and moths are easy to understand. The bats and moths that continue living are those that, among other things, eat but are not eaten themselves. Bats that are better able to detect, capture, and eat insects in this noisy environment survive and produce new bats in greater numbers than those that are not good at obtaining food. On the other hand, the only moths that survive and reproduce are those that either do not encounter bats (among other hazards) or those that can detect and avoid the danger.

Both bats and moths are detecting and identifying each other on the basis of airborne mechanical vibrations or sounds. Sound waves are common in the environment, are sensed mechanically, and are used extensively by animals. The biological machinery, coming from two long, separated lines of evolution, is vastly different.

Echolocation and Hunting Behavior in Bats

Bats use a variety of sounds in many different ways (Griffin 1958, Simmons et al 1979). Some squeaks and chitters, for example, are uttered when the bats are disturbed. These sounds are often within the range of human hearing. The sounds of interest, however, are those that the bats use for detecting objects, such as obstacles or prey. These sounds are ultrasonic (above the range of normal human hearing). All of the calls are produced in the larynx, as in other mammals, but the bat's larynx is much more highly specialized. Bats also possess highly specialized ears (Figure 1-1).

The role of sound and hearing in the orientation of bats was first inferred by Lazzaro Spallanzani in Italy in 1793. He discovered that owls could not fly normally in complete darkness or with their eyes covered but that bats could. Furthermore, bats that he blinded, released, and then recovered later at their roost in the bell tower of the cathedral at Pavia had filled their stomachs with as much insect food as others that he had not blinded. But Spallanzani (in conjunction with a Swiss surgeon, Charles Jurine) found that tight plugging of the ears caused the bats to fly about randomly and collide with obstacles. Being ingenious and thorough and wishing to eliminate the possibility that the bats were disoriented by irritation of the ears, Spallanzani had some miniature brass tubes fitted to the bats' ears, which

Figure 1-1 External ear of little brown bat. The ears are cone-shaped funnels that serve to focus the sound, thus accentuating the sound intensity from a given direction and reducing the intensity from other directions. The tragus is believed to help focus sound in the vertical dimension.

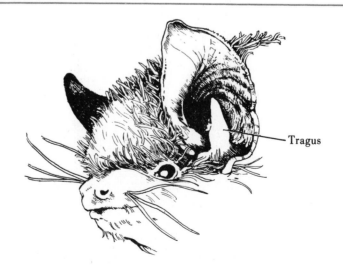

Tragus

he plugged and unplugged. With open tubes, bats could orient almost normally, but with closed tubes they could not.

No one knew of the ultrasonic calls of bats at that time, and Spallanzani was ridiculed. His results were nearly forgotten. The famous anatomist George Cuvier offered an alternate, armchair opinion that bats avoided obstacles by sense of touch (in spite of the fact that Spallanzani had already tested for that by covering bats' wings with varnish or flour paste and found that they still flew acceptably). Cuvier's opinion was accepted for 120 years until a Cambridge physiologist, H. Hartridge, contemplated a bat that strayed into his room one night. Hartridge knew of sonar, which had been developed for detecting submarines in World War I, and hypothesized that bats were using ultrasonic sounds.

The matter was not settled until 1938 when a curious Harvard undergraduate, Donald Griffin, checked some bats with an ultrasonic recorder. The equipment had been developed by a Harvard physicist, G.W. Pierce, for studying the high-pitched sounds of insects. Coincidentally and independently at about the same time, a Dutch zoologist with good hearing, Sven Dijkgraaf, listened to bat calls with his own hearing. He concluded that faint sounds were being used by the bats to detect objects.

More recent investigations have shown that different bats vary in their use of sound for echolocation or whether they even echolocate at all. There are around 800 known species of bats. Some of the larger old-world tropical bats of the suborder Megachiroptera, for example, have large eyes and relatively small ears and are thought to rely almost exclusively on vision while flying. Other species rely on both vision and echolocation.

The brief, loud, high-pitched calls used for echolocation by bats such as the little brown bat are given in rapid succession, and the frequency of each call is varied or modulated. Each call starts at a high pitch and rapidly drops (Figures 1-2 and 1-3). If the call is slowed down to bring it within the range of human hearing, it sounds like a chirp or wheeough. Because of the change in pitch, the call is said to be frequency modulated (FM), and the bats are called chirping or FM bats. Echoes from FM calls may contain information, such as the size of an object. (Objects reflect soundwaves of wavelengths only as large as or smaller than

Figure 1-2 Sound spectrograms of bat echolocation calls. See also Figure 1-3. FM calls (*left panel*) start at a high pitch (about 80 kHz here) and drop rapidly to a lower frequency (40 kHz). CF calls (*right panel*) remain at a constant frequency (around 83 kHz) but may contain an FM element at the end. The limit of human hearing is just below 20 kHz. Time spans are approximately 0.05 second in the left panel and 0.13 second in the right panel.

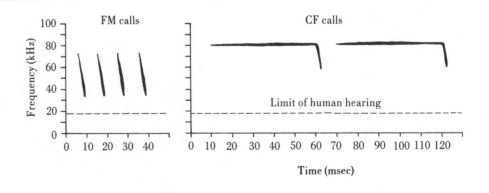

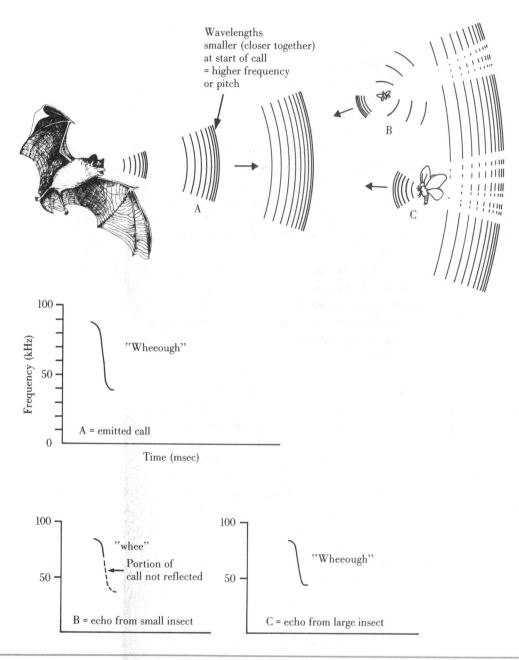

Wavelengths
smaller (closer together)
at start of call
= higher frequency
or pitch

A

B

C

Frequency (kHz)

100

"Wheeough"

50

A = emitted call

0

Time (msec)

100

"whee"
Portion of
call not reflected

50

B = echo from small insect

100

"Wheeough"

50

C = echo from large insect

Figure 1-3   **A flying little brown bat sends out bursts of high-pitched calls, chirps of descending frequency, which echo back from insects and other objects in the environment. As shown in the sound spectrograms, smaller objects reflect only the higher frequency portions of the calls, providing the bat with a means of determining their size.**
Modified from Griffin, D.R. 1959. Echoes of bats and men. Doubleday, New York.

their own dimensions. Larger wavelengths break up around the object, see Figure 1-3.) Furthermore, freshly reflected echoes retain their chirp quality and can be identified against a noisy background of reverberations from old echoes, other bats, and sounds from other sources.

Some species of bats, also mostly insectivorous, incorporate calls of a more constant frequency (CF). Echoes from CF calls do not permit recognition of the size of an object unless combined with an FM component, but they permit movement relative to the bat to be detected via Doppler shifts. These shifts represent changes in frequency from movement, such as in the commonly experienced sound of a passing vehicle. To use the Doppler effect, bats do not measure a change in the echo but rather adjust the frequency of their 'call to achieve an echo of constant pitch (Simmons et al 1979). The extent to which the call must be adjusted permits the bats to measure the magnitude of the Doppler shift.

A bat's sensory system operates to accentuate echoes, whereas the human system suppresses echoes. The echo suppression of the human system can be demonstrated by tape-recording a series of clicks or short bursts of sounds (such as those made by firing a cap pistol) in a large room or hallway with hard walls. When the sound is originally made and when the tape is played in the forward direction, the echoes are perceived only with difficulty or not at all. But if the tape is played backward, the echoes, which now precede the sound, are plainly audible.

The bat's hearing, on the other hand, is suppressed for the original emission of the call. The intensity at the ears is partly reduced by the structure of the mouth and nose, which direct the call forward. The hearing is further actively suppressed by the muscles in the middle ear, which are attached to the ear bones. During the brief emission of the bat's call the muscles clamp down and dampen the transmission of the sound to the inner ear. Then the muscles rapidly relax and permit full amplification of the echo before contracting again to suppress the next emission. Further suppression and amplification of calls and echoes, respectively, may occur by processing in the brain.

Echoes return earlier and louder from close objects than from distant ones. At any one time the incoming sound contains a mixture of echoes from close and distant objects, perhaps including echoes from earlier calls. Bats have an incredible ability to discriminate small distances on the order of 0.1 mm (Simmons 1979).

The information in echoes is not received continuously by bats because it must be interrupted to send new calls. However, in the bat's perception, the data contained in intermittent echoes may be fused for continuous interpretation, just as people are unaware of the flicker in a projected movie film or the normal blinking of the eyes.

When cruising, little brown bats emit calls that start at about 90 kHz and drop to 45 kHz and last about 1 or 2 msec (1 msec equals 1/1000 second). The calls are repeated at rates of 10 to 20 per second. Then when potential prey or small obstacles are encountered, the rate of calling may rise to as many as 250 calls per second, and each call lasts less than a msec (Figure 1-4, A). Bat detectors are available that render the calls audible to a human as clicks in an earphone or speaker. (For a circuit diagram of a simple bat detector, see Fenton 1983.) If one eavesdrops on bats with such a device, their cruising calls sound like the "putting" of a slow gasoline engine, and the final approach to prey turns into a rising tempo of whining rapid clicks.

To appreciate the significance of the numbers in the preceding paragraph, con-

Figure 1-4 A, Calls of a hunting little brown bat closing on prey **B,** contrasted with human speech. The bat calls occupy approximately 0.120 second (for 24 separate vocalizations) and span a frequency range of over 60 kHz. By comparison, human vocalization requires nearly 0.50 second per vocalization and spans a frequency range of only 2 to 3 kHz.

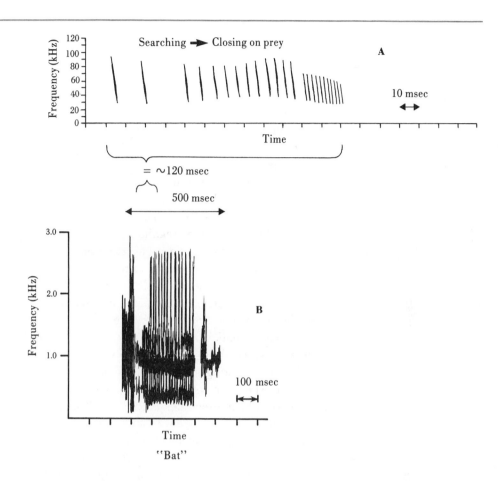

sider the vocal performances of a human. A human can make only about 10 distinct sounds per second. An average one-syllable human word takes about ½ second to produce and may span a range of 2 to 3 kHz (Figures 1-4, *B* and 1-5). In other words, a bat can search for and capture a flying insect, making over 20 separate calls, in less than one fourth the time it takes a human to say the word bat. Furthermore, between each call the bat listens for and interprets the returning echoes. Compared with the needle points of sound produced by bats, human vocalizations are low, slow, and slurred (see Figure 1-4).

Bats use these sounds to identify and capture insects. Some insects may be detected by the sounds they produce. The fact that bats take dead insects (Griffin et al 1965) and make characteristic call patterns on the final approach indicates quite clearly, however, that bats use their emitted calls as the primary detector. Little brown bats weighing 7 g have been shown to routinely catch 1 g of insects per hour of hunting. A smaller related species of bat studied in the laboratory caught at least 175 mosquitoes in 15 minutes or 1 every 6 seconds (Griffin 1959). The smallest prey recorded for a bat was a small gnat weighing 0.0002 g, which was found in the mouth of a bat that was killed while it was hunting. Noctuid moths are easily detected by bats, and many are captured and eaten.

The appearances of a flying moth's echoes were investigated by Kenneth Roeder

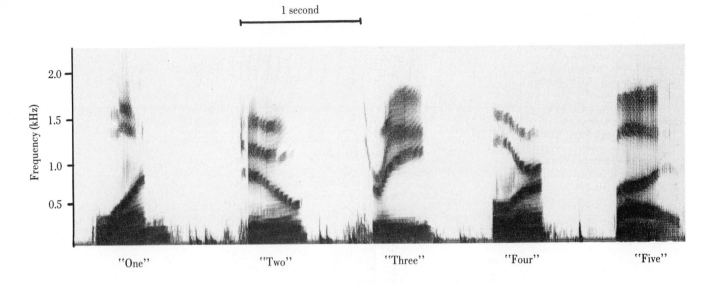

Figure 1-5   **Sound spectrograms of familiar human speech: a male counting.**
Sound spectrogram courtesy L.W. Oring.

(1963), who simulated a bat's calling and receiving system (Figure 1-6). The setup permitted him to simultaneously photograph the actual appearances of the moth with the companion echoes as they were displayed on an oscilloscope screen (Figure 1-7). He found that the echo was strongest when the moth was sideways to the bat and its wing surface was broadside to the sound. It was weakest when the moth was facing toward or away from the bat or when the moth's wings were down. Even when the wings were up and appeared similar visually, there were many variations in the echoes (see Figure 1-7). At best the moth would present a flickering sound return to the bat.

The perception of sound in the bat goes beyond the highly specialized ear. The vibrations from the air, having been transformed into nervous system impulses, are then transmitted via the auditory nerve to the bat's brain where they are further sorted, processed, and identified. After the detected object is identified as edible, the bat moves toward it and attempts to keep it in range and capture it. In addition to detecting and tracking the moth, the bat also needs information on its position in space, its speed and direction, air movement, and many other facts normally needed by pilots. Furthermore, all the detection, decisions, and movements have to be fast, often in small fractions of a second. If the whole process took as long as it does to write or read about, the bat not only would miss its meal but also would stall and crash into the ground.

Noctuid Moth Hearing and Behavior

Compared with bats, those moths that have ears (not all species do) have much simpler and different equipment for hearing. The moth's hearing is basically a detector for the bats' calls, somewhat of an organic radar detector. The moth's ears

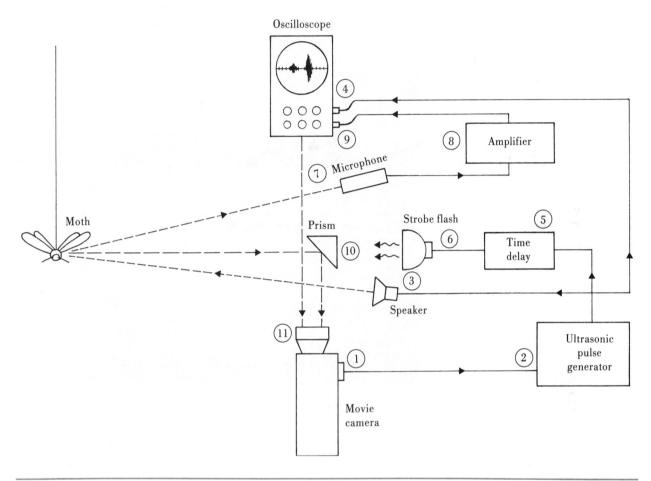

Figure 1-6   Artificial bat. This apparatus was used to simulate the calling (via ultrasonic speaker) and hearing (via microphone and amplifier) of a bat. A simultaneous photograph of a moth, reflected by a prism, and its echo are displayed on the oscilloscope, permitting a visual rendition and permanent record of the event. Numbers indicate the sequence of steps: the camera shutter opens *(1)*, and a contact at the camera triggers an ultrasonic pulse generator *(2)*, which produces an ultrasonic call at the speaker *(3)* and trace at the oscilloscope *(4)*. (The call is the left-most, small blip on the screen.) Sound travels more slowly than light so a time delay *(5)* is inserted, which flashes a strobe light *(6)* at the same time that the echo from the moth reaches the microphone *(7)* and shows up after amplification *(8)* at the oscilloscope *(9)* (right-most blip on the screen). The visual appearance of the moth and image on the oscilloscope reach the prism *(10)* and camera *(11)* at the same time. Photographs resulting from this setup are shown in Figure 1-7.
Modified from Roeder, K.D. 1967. Nerve cells and insect behavior. Harvard University Press, Cambridge, Mass.

Figure 1-7 Echo and visual appearances of moths with wings in different positions. Photographs obtained from the apparatus shown in Figure 1-6. (Only the echoes are shown, not the call traces.) **A,** The wings are in different positions and reflect quite different echoes. **B,** However, there is much variety among echoes even when the wing positions appear similar visually. **C,** Echoes from the rear view of a departing moth provide a minimum of broadside reflecting surface, hence a more constant and smaller echo, which would be more difficult for a bat to detect. From Roeder, K.D. 1967. Nerve cells and insect behavior, Harvard University Press, Cambridge, Mass.

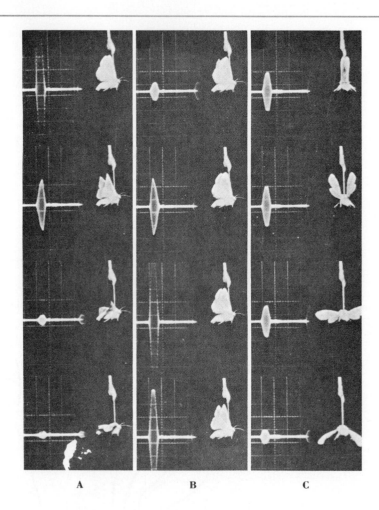

A   B   C

are located on the thorax (Figure 1-8). Each ear consists of a tympanic membrane connected to only two sensory cells (see Figure 17-1). The moth does not have several thousand sensory cells or an elaborate funnel-shaped ear, and it does not, so far as is known, use its eyes to detect bats. Yet moths not only detect bats but can determine the general location and distance of the bats and take appropriate action. Furthermore, they do it rapidly, in fractions of a second and generally in much less time than humans can respond to a stimulus.

Crashing is not as serious for moths as it is for bats because the moth is much lighter and possesses much less momentum. But slow or unlucky moths are eaten. Photographs of the tracks of a moth that won and another that lost in their encounters with bats are shown in Figure 1-9.

Depending on the distance of the approaching bat, responding moths have been observed to behave in one of two ways. If the bat is far enough away and the moth detects it, but the bat has not detected the moth yet, the moth makes a directed turn away from its present path. This frequently keeps it out of range of detection by the bat. If, however, the bat is closer, within the range where it may have detected the moth's echo and be closing in, then the moth displays a random,

Figure 1-8 Noctuid moth showing position of ear on thorax. See also Figure 17-1.

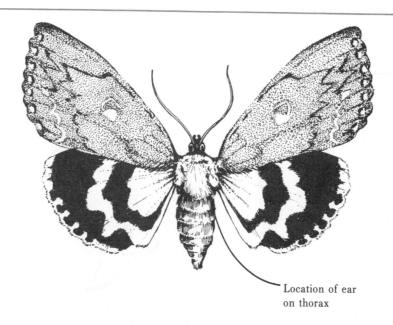

Location of ear on thorax

Figure 1-9 Flight paths of moths and bats recorded with camera shutter left open and live moths tossed into the air above an illuminated field. **A,** Moth (bright wavy track) and bat (smooth lower track) both enter from left. The moth dives and the paths approach each other, but then the moth turns sharply upward and escapes while the bat leaves picture to right. **B,** Bat enters from right side of photograph; moth enters in center from below in a looping flight. The paths intersect, indicating that the bat captured the moth.
From Roeder, K.D. 1967. Nerve cells and insect behavior, Harvard University Press, Cambridge, Mass.

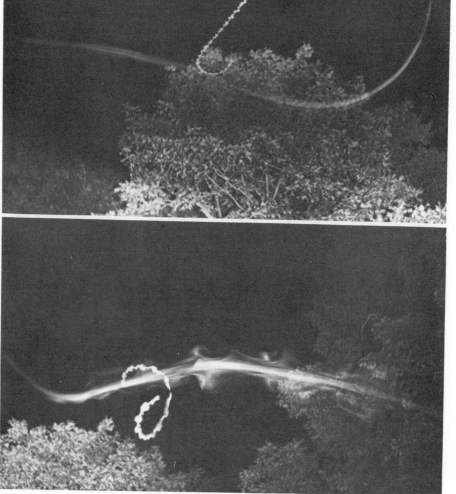

A

B

wildly gyrating, unpredictable dive. These rapidly changing moves are difficult for the bat to track and intercept, and there is a reduced chance that the moth will be caught.

The erratic behavior of noctuid moths in response to high-pitched sounds can be observed fairly easily, such as by jingling keys or making other high-pitched noises near moths at a streetlight. With patience the behavior and interactions of moths and bats can be observed at night with the aid of low illumination, such as from a 100-watt light bulb, or near yard lights, if a hand is held up to block the light of the lamp from one's eyes. Observations on moths' erratic behavior and information on moth ears were published as early as 1919. With later knowledge of bat sounds, correlations were made between moth behavior and the hunting of bats (Roeder 1967).

The directed turns and directional escapes, however, are more difficult to observe and were not even suspected before the 1960s. The possibility of such behavior was raised when neurophysiological studies (Chapter 17) showed that moths might be able to determine the direction of a sound source. Directional, rather than erratic, behavior was reasoned to be more advantageous if a bat was still far enough away.

Roeder (1962) confirmed the presence of two basic evasive behavior patterns in moths with a controlled source of artificial bat calls. An ultrasonic transmitter capable of producing a train of 70 kHz, 5 msec pulses at a rate of 30 per second was placed on a mast at the edge of a field. Intensity of the sound was adjusted to simulate bats at different distances. A floodlight and camera were used to observe and record the behavior of passing moths (Figure 1-10). The camera shutter was opened and kept open to photograph, as a continuous line, the path of a moth entering the field. After a sufficient length of flight had been recorded, the trans-

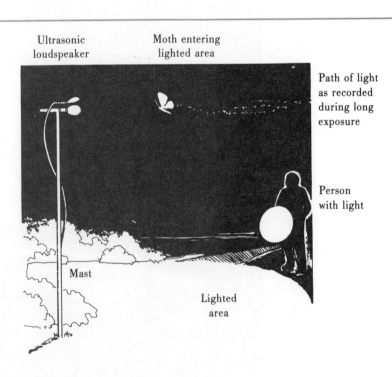

Figure 1-10 Method for observing moth responses to simulated bat calls. Calls are presented by an ultrasonic loudspeaker mounted on a mast. The area and any moths that enter the scene are lighted and photographed from the side. The camera shutter is left open to record moth flight paths as continuous tracks of light. Results of moth responses are shown in Figure 1-11.

Ultrasonic loudspeaker

Moth entering lighted area

Path of light as recorded during long exposure

Person with light

Mast

Lighted area

Figure 1-11   Flight responses of moths toward artificially presented, simulated bat calls. Arrows indicate position of moth when calls were initiated. Other objects and tracks in the pictures are from other insects at the edge of the field. **A,** Power dive. **B,** Passive dive interrupted by brief period of wing flapping. **C,** Looping dive. **D,** Directed upward movement away from sound source. **E,** and **F,** Directed horizontal moves away from sound.
From Roeder, K.D. 1967. Nerve cells and insect behavior, Harvard University Press, Cambridge, Mass.

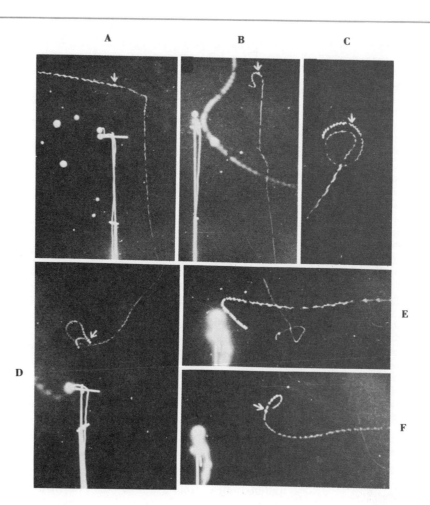

mitter was turned on. Many moths behaved as predicted and displayed one of the two behavior patterns, depending on the intensity of the ultrasonic sounds. Other moths did not react at all. After recording the behavior, Roeder attempted to capture each moth with a net to identify its family and species. Figure 1-11 shows a few of the photographs obtained during the course of over 1000 observations.

The tendency of noctuid moths to turn in an oriented manner away from an ultrasonic sound of low intensity was further confirmed in the laboratory by using flying moths fixed to an electronic device that measured their attempts to turn. The moth was attached to an insect pin with a drop of wax on the surface of the thorax in a manner that did not interfere with the flapping of the wings. The pin was inserted into a phonograph pickup for amplification and directional recording. A directional anemometer (a device for measuring wind speed) was placed behind the moth's wings to measure the direction of the wake produced by the movement of the wings. The moths in this setup showed clear directional responses to low-intensity ultrasonic sounds at one side or the other (Figure 1-12). If an ear was deafened, the moths always turned toward the silent side.

Some moths from another family (Arctiidae) display a different response to the calls of bats. These moths answer ultrasonic sounds with a series of rapid high-

Figure 1-12 Technique used to record directed flight behavior of moths in presence of artificial bat calls in laboratory. The ultrasonic loudspeaker *(1)* directs calls at one side of the moth, which is mounted by the thorax to a phonograph pickup *(2)*. A directional anemometer *(3)* uses thermistors to detect temperature changes created by differences in airflow as the moth attempts to turn while flying. Changes from the thermistors are balanced by a bridge circuit *(4)* and recorded as upward or downward deflections, depending on direction of turn, on the oscilloscope *(5)*. Thoracic vibrations and loudspeaker pulses also are recorded on the oscilloscope as indicated.
Modified from Roeder, K.D. 1967. Nerve cells and insect behavior. Harvard University Press, Cambridge, Mass.

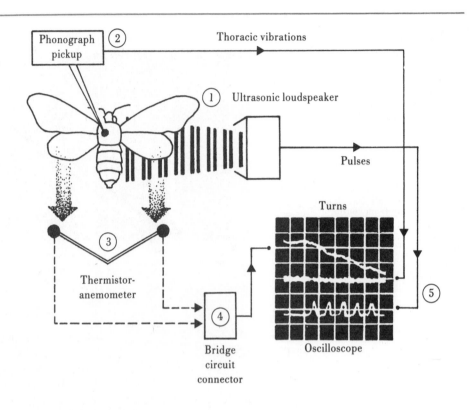

Figure 1-13 Silhouette of arctiid moth showing position of ear (*oval* to right) and timbal, a click-producing organ (*striated object* to left of ear).

pitched clicks from a timbal organ on their thoraxes (Figure 1-13). The moths of this group taste bad, and bats were observed to refuse capturing and eating them. The clicking behavior appears to warn bats of this distastefulness.

It may seem reasonable that moth hearing evolved at least partly under the selective pressure imposed by bats. Roeder and Asher Treat (1961) attempted to test this hypothesis by observing 402 encounters between moths and bats. Each encounter was scored for whether or not the moth took evasive action and whether or not it survived. For every 100 moths showing evasive behavior that survived, only 60 not showing evasive behavior survived. This demonstrates a clear selective advantage for the evasive behavior and gives some support to the hypothesis. (It also raises the question of how moths with nonevasive behavior continue to exist.)

A final twist in the evolution of moth hearing and their ability to detect and avoid bats involves yet another type of organism—parasitic mites that live in the ears of moths (Figure 1-14). Treat (1975) provided a description that would be difficult to improve on. In addition to describing the moth-mite-bat interaction, he further challenges our imaginations about the worlds in which these animals live:

> Let us shrink to the height of a moth ear mite (*Dicrocheles phalaenodectes*), creep under the wing of a sleeping noctuid, and roam for a little while through the sculptured caverns of the insect's ear. . . . Let us pierce the (tympanic) membrane and enter (the) inner chamber. The soft carpet yields to our feet at every step, and through the transparent inner wall we can see the moth's great muscle-engines resting

Figure 1-14 An adult
armyworm moth
(Pseudaletia unipuncta)
showing the right ear
tympanic recess (in center of
photograph) with one large
gravid and several immature
female moth ear mites.
Photograph by A.B. Klots,
courtesy of A.E. Treat.

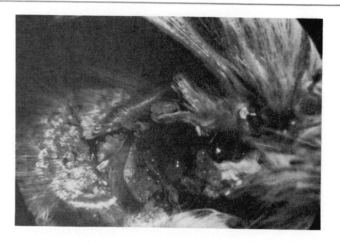

in readiness for the evening flight. Beside us, one above another, yawn the mouths of
four dark grottoes, the four pockets of the tympanic frame. Partitioned like the cells
of a cloister, but each of a singular shape, they form a hollow, four-storied pillar
between the thorax and the abdomen. . . . This, then, is the world of the moth ear
mite from the time of its emergence as a larva to its maturity as a fertile female,
perhaps to its death as a worn-out male, or to its destruction, with ill luck, in the
jaws of a hungry bat. The mite, of course, whatever its aesthetic sensibilities, does
not perceive this world as we have described it, because, if for no other reason, the
mite is blind. But the scene is there, the stage is set, the actors are ready, and
anyone with an entomological microscope can draw apart the curtain. . . .

Perhaps the most noteworthy aspect of a *D. phalaenodectes* colony is its unilaterality. The mites normally invade and occupy only one of their host's two ears,
leaving the other ear intact and fully functional as a detector of ultrasound. The
adaptive value of this behavior is obvious, for if an infested moth is caught by a
hungry bat, not only does the moth perish but the mites as well. A one-eared moth
may have its acoustic defenses somewhat impaired, but it is still a safer mite vehicle
than a moth that is deafened altogether. The chosen ear may be either the left one
or the right, but regardless of which it is, and regardless of how crowded it may
become, the other ear is almost invariably left wholly undisturbed. This seems even
more remarkable when one considers that most mites that travel in numbers on
insects distribute themselves quite symmetrically on their hosts. In some such
instances, where the mites become attached and only a small odd number are
present, the odd mite will occupy a median position. Mites that move about freely,
like the gamasine deutonymphs that commonly infest sexton beetles, manage nevertheless to maintain approximately equal numbers on the two sides of the host.
Even among the tympanicolous mites of moths, those that spare the ears from
damage use both ears equally or nearly so. It is only in certain species of *Dicrocheles*
that unilaterality is the rule and not the exception. In these species the restriction
of the colony to one ear appears to be the result of a highly evolved behavioral
pattern perhaps including pheromonal (chemical) and other types of communication.

SO WHAT?

This bat-moth story focuses on predator and antipredator behavior, sound and hearing, and a few species of animals. There are, however, many other kinds of behavior and other species of animals, and hearing plays only a small role in the broad, general consideration of behavior. If the bat-moth story is to be an example in the study of animal behavior, what has it illustrated? There are three major points, each with a set of subpoints.

The first point concerns human interests and scientific pursuit. Behaviors such as bats and moths flying around at night may seem fairly simple and uninteresting on the surface, but in fact they are extremely complex, fascinating, and puzzling. An attempt to satisfy our curiosity requires dedicated effort and special techniques used by many persons over long periods of time. One must consider not only the findings and current interpretations of science but also the historical and methodological substrate on which these findings rest. In this example we encounter curiosity, ingenuity, and persistence, along with other human emotions and interactions, including cooperation and jealousy. Recall, for example, Spallanzani, Jurine, Cuvier, Griffin, and Roeder.

The course of behavioral research is rarely smooth. Interpretations change, and the end is usually open. The discovery that bats use ultrasonic calls and echolocation took over 120 years, including some confusion and controversy, after Spallanzani first suggested the use of hearing. Armchair thinking is important, if not necessary, but it sometimes leads one astray, as with Cuvier's explanation of bat orientation. Methodology included controlled experiments, an ability to work with materials such as tiny brass tubes, and the use of ultrasonic, electronic sound-producing and detection equipment, bat-simulating devices, and directional anemometers.

Previous history and scientific methods are so important in behavioral biology that the next two chapters are devoted to these topics, and they also appear throughout the rest of the book.

The human interest points of this story are primarily academic, that is, related to the biology of the behavior along with the history and methodology that elucidated the biology. But a few applied points can also be made. Bats probably play a role in the control of some moths and other insects that we might regard as pests. An understanding of bat echolocation and the equipment used to study it has contributed to further development of radar, ultrasonic electronics, and related technology. An appreciation of bat behavior can also dispel long-held myths and misunderstandings, and reduce the fears that some people have of bats.

Our second major point is that it is clear that other animals and their behaviors are different from each other and from us, and there is much variability and diversity even under similar circumstances. This variability results from several factors.

One important cause of variability is that each animal has its own sensory and perceived world. This concept often is associated with the term *Umwelt* (German for own or self-world) introduced by Jakob von Uexküll in a classic book, *Umwelt und Innenwelt der Tiere* (1909). (For a translation of Uexküll's work, see Schiller, 1957. The matter also is discussed briefly in Chapter 16.) Humans usually take their own perceptual world for granted and assume that other animals see it the same way. But it is a serious mistake to think that other animals sense and perceive the world in the same way as humans. One of the pioneers and most highly respected scientists to experimentally investigate the senses of various species, particularly bees, was von Frisch. His contributions are discussed in more depth in Chapter 2.

The behavioral responses resulting from input to an animal are also quite variable. Some moths, for example, showed evasive behavior, and some did not; arctiids responded differently from noctuids. Different behavior may be shown by the same animal in different situations. Moths responded differently, for example, to close versus distant bats.

Much of an animal's behavior and the variability among different animals depend on the biological structures that the animals possess. The structures may be quite different in different organisms (e.g., the ears of bats and the ears of moths), but all are subject to the same principles, such as the physics of sound waves. The structures may be relatively simple or extremely elaborate and sophisticated. External structures tell only part of the story. How do the senses enter the information into the nervous system; how is it processed; and how are the appropriate movements brought about? Does experience alter subsequent behavior?

Further contemplation of the observations suggest that there is more to behavior and its variability than just the immediate conditions at hand. A moth, for example, must respond appropriately on its first encounter with a bat. It may not get a second chance, and it may not have time to invent a solution to the problem on the spot. Furthermore, although there is some variability in responses, many different moths may give similar responses in similar situations; such consistency would not be likely if each moth were acting completely independently. If an organism possesses particular biological structures and behavioral responses, what is their source?

Most behavioral responses are adaptive; that is, they are correlated with functions. The moth's responses (such as erratic versus directed flight) were appropriate to the particular situations (bats near or far). Thus it is reasonable to consider the long-term factors, events that occurred during the ancestry of the bats and moths, that led to their present-day adaptations.

All of this brings us to a third major point that can be gathered from the bat-moth example. There remains much that we do not understand and many questions that still can be asked at many levels of inquiry. In the case of the bats and moths, for example, we get an idea of what a flying moth might "look" like to a bat and vice versa, but our results, in spite of the ingenuity and technology, are still crude at best. We do not really know what goes on inside the mind of a bat or moth—or even if they have minds. Also, in spite of our reasonable guesses and speculations, we are not certain of the precise causes of their behaviors. Although we do not have all the answers, there are some generally accepted ways of approaching them from an evolutionary, biological viewpoint. In contrast, there are other ways, often more intuitive and popular among the general public, that are less accepted by those persons who have looked most closely at behavior. It is important to distinguish and clarify the different approaches as much as possible before proceeding with the subject. The following section discusses problems among different ways of asking about and interpreting animal behavior.

INTERPRETING THE CAUSES OF OBSERVED BEHAVIOR

Proximate versus Ultimate Causes

A cause-and-effect relationship usually involves a combination of factors—one that is immediately responsible for the effect plus a predisposing set of conditions. For a fire to occur, for example, there is a spark that ignites the flame, plus combustible materials; the fire will not occur without both. We can inquire further of the predisposing conditions and ask what caused the combustible materials to be present, or to be combustible in the first place. Similarly, when we ask what causes a particular behavior we can consider either the immediate factors and processes that triggered the observed behavior or the ecological and evolutionary factors that predisposed the animal to behave as it did under those circumstances; that is, how and why did it come to possess the behavioral traits and the underlying nervous system and related anatomical and physiological substrates? In contemporary parlance these two aspects of immediate causes and background causes are referred to, respectively, as proximate and ultimate causes.

The distinction between proximate and ultimate causes is particularly important for an understanding of behavior, both for conceptual reasons and as a basis for separating the entire subject of behavior into two major subdivisions. The historical development of interest in ultimate versus proximate causation is discussed in Chapter 2.

The proximate causes of particular behaviors are the more observable, immediate, and mechanistic or structural (anatomical and physiological) aspects associated with how certain behaviors occur. As an example, if one were to ask what causes a moth to flee from a bat, a proximate answer would be that the bat calls are sensed by the moth's ears, processed in the moth's nervous system, and cause a sequence of contractions in the moth's flight muscles that lead to fleeing behavior.

Ultimate causes, on the other hand, are the long-term, ecological, and evolutionary factors that lead to differential survival and reproduction, which, along with other factors such as genetic drift, produce the trends and differences that we see in organisms.

Thus an ultimate answer to why some moths flee from a bat would be that the ancestors of those present-day moths did it and as a result had higher rates of survival and reproduction and passed on the traits. Although we cannot turn back the calendar and directly observe the escape behavior and hearing of ancestral noctuid moths, it is reasonable to assume that these characteristics did not appear suddenly in present-day form. Rather, development probably occurred in steps. Initial changes in moth morphology resulting from genetic mutations or recombinations may have led to some moths becoming sensitive to airborne vibrations. If these coincidentally became linked to neural responses and altered flight patterns, they may have conferred a slight survival advantage by causing bats to miss their targets. Moths with these traits would live to successfully reproduce, whereas moths without these traits would be more likely to be captured and eaten. Subsequent genetic variation and differential survival in later generations could have further shaped the structures and behavioral responses into those that are observed today.

The distinction between proximate and ultimate factors can be applied to virtually all behavior. One could ask, for example, "Why does a male dog chase after a female in heat?" or, for that matter, "Why do humans engage in sex?" The answers to these questions are not just "to produce offspring."

A function is what a leg, wing, or other structure does. The current biological interpretation is that functions are the results of ultimate factors interacting with biological systems over long periods of time. A loose and less neutral term for function is *purpose*. Use of the word purpose should be avoided, however, because it may imply a larger design than many biologists are willing to accept or in other contexts may imply conscious purpose on the part of the animal. The question of consciousness and deliberate anticipation of the future in species other than humans has caused particular problems for discussions of behavior and has thus far resisted testing, although it generates much debate and is discussed further in the next section.

Anthropomorphism

Ultimate causes of behavior (or any other biological characteristic) emphasize the role of past and present events in survival and reproduction. Is there also a role, in the case of behavior, for future events or the anticipation of future events? This is a complex question that involves issues of both anthropomorphism and learning. Learning is given major attention in Chapters 2 and 19 to 21. Here we concentrate on the problem of anthropomorphism. The term is derived from the Greek *anthropomorphos*, which means "of human form." It means the attribution of human qualities to nonhuman objects, particularly other species of animals but also phenomena such as storms and objects like flags, vehicles, and buildings.

The general, uncritical anthropomorphic view is that animals are aware of what they do and of the consequences; hence they behave more or less purposefully. A turkey hen, for example, might be interpreted as taking care of her chicks "for the purpose (as understood by her) of raising them," "to give them love and care," "for the survival of the species," or for whatever other purpose one may wish to assert.

This type of anthropomorphism is based on the assumption that because much of human behavior is purposeful, based on reason, and often directed by the anticipation of future events, then it must be the same in other species. The resulting anthropomorphic view has been passed down as folklore among the general public, and it is fostered and reinforced by much of the entertainment industry. In its fullest expression, anthropomorphism views other animals, although they may not be smart or able to talk, as little people in fur, feathers, or chitinous shells.

But some careful observers noticed that not all animal behavior seems so reasonable or purposeful. The absence of purpose is seen most strikingly when something goes wrong with otherwise adaptive behaviors. Turkey hens normally care for their chicks, apparently as if they understand why they are doing it. If the hen is deafened, however, she will ignore or kill the chicks. If a speaker is placed on a kitten so that chick calls can be played back as if they were coming from the kitten, the turkey hen will attempt to mother the kitten (Schleidt et al 1960). Thus the nuturing behavior might be viewed simply as a more or less mechanical response to the chicks' vocalizations. Even among humans one often can see behavior that occurs without deliberate understanding or in spite of understanding and attempts to act otherwise. Examples include problems many people have with compulsive behavior (e.g., alcoholism, overeating, overwork, and a variety of sexual behaviors).

These observations and a reaction to unbridled anthropomorphism led to its rejection by many students of behavior. In the extreme alternate viewpoint, behavior is seen as being completely mechanical. Insects could be thought of, for example, simply as ganglia on legs. Most if not all animals could be considered as being no more aware of why, and perhaps even what, they do than are washing machines, automobiles, or battery-operated toys.

There has been some problem of where to fit humans into this picture. Most people who rejected anthropomorphism in other animals were nonetheless conscious of their own self-awareness. They considered that only the other animals were little living machines. Some persons (e.g., Skinner 1974), however, proposed that even humans, in spite of our capacity for learning and self-awareness, do not really understand why we behave as we do. According to this view, humans are not much different from the other animals, and people should be viewed simply as animals that are more advanced with respect to some aspects of the nervous system and behavior.

Bringing some moderation to these two extremes, Griffin (1976 a, b, 1977, 1981, 1984 a, b)—the same Griffin who worked with echolocation in bats—courageously questioned the view that nonhuman species lack self-awareness and purposeful behavior. He defined *awareness* on the basis of mental images of events that may be remote or near in time and space, a definition that he later (1981) described as compatible with accepted dictionary and philosophical usages of the term. His argument, in brief, is that if biologists are to take an evolutionary view of the structure of brains, nervous systems, and behavior, then all aspects, including awareness and purposefulness, should have some phylogenetic history. At least traces of these characteristics should be found in other species. If one is willing to use other animals such as white rats and pigeons (forms that are not direct ancestors of humans) as models of human learning, then why not also use them as models, at least to a degree, of awareness and purposefulness? In what he called *a possible window on the minds of animals*, Griffin proposed that humans could establish dialogues with other species through their species-specific channels of communication (see Chapter 15). Griffin, as might be anticipated, was criticized from several angles.

In a special Dahlem Workshop held in West Berlin, in 1981, nearly 50 selected participants of international scientific reputation in the field of behavior gathered to discuss the issue of animal mind versus human mind. The report of this workshop was edited by Griffin (1982). Several aspects of the issue were explored, including nervous system structures and capacity, verbal and nonverbal processes in humans, communication and social behavior in other animals, problem-solving and various other cognitive processes in animals, and possibly relevant ecological and evolutionary considerations. Bullock (1982), who provided a summary overview and some final opinions on the conference, stated that:

> Although our theme question has a large subjective element, it can be studied scientifically. . . . (It is reasonable) to conclude from the available body of evidence, meager but not inconsiderable, that there is a profound gradation, as well as a specialization for different styles of life among the taxa, be they invertebrate, vertebrate, mammalian, or primate.
>
> Although our conference report may disappoint the reader who looks for support for his favorite position or who expects a consensus of 50 experts as to which animals have a mind, we have achieved a modest goal. That is to point to a good many things that can be done to further understanding, by listing approaches and suggesting new research (as detailed elsewhere in the report).

Gordon M. Burghardt (1985, 1988 a) advocates a reasoned and middle-ground form of anthropomorphism that he calls *critical anthropomorphism*. In his words:

> My experience has been that unless challenged to separate description from interpretation, students readily use and defend the use of sloppy teleological and anthropomorphic *thinking*, not just the vocabulary. On the other hand, a studious, ideologically based opposition to using our own experience and intuitions, informed by a knowledge of natural history, in asking questions and designing studies in animal behavior, is ultimately sterile and dull.
>
> What I am calling for is a critical anthropomorphism and predictive inference that encourages the use of data from many sources (prior experiments, anecdotes,

publications, one's thoughts and feelings, neuroscience, imagining being the animal, naturalistic observations, insight from observing one's maiden aunt, etc.). But however eclectic in origin, the product must be an inference that can be tested or, failing that, can lead to predictions supportable by public (observable) data.

Persons wishing to further pursue the debate on this and closely related issues should refer to Corben et al (1974), Mason (1976), Hailman (1978 a), Premack and Woodruff (1978), Davidson and Davidson (1980), Griffin (1981, 1984), Burghardt (1985, 1988 a), Bitterman (1988), Lewin (1988), and references contained therein. Contemporary issues that are partially related to anthropomorphism but that we have not addressed here involve animal rights and the ethical treatment of animals; those topics are discussed in Chapters 3 and 22.

The Past and Future in Present Behavior: a Recap

Regardless of whether or how much an individual animal understands and guides its behavior by the anticipation of future events, such as the future welfare of its offspring, it appears that the reasons animals behave in specific ways to a considerable extent are rooted in evolutionary processes, that is, past and present events. Even the ability to think about the future depends mostly on the past.

Except for the possibility of some purposeful behavior in humans and occasionally other species, as discussed previously, and as stressed by Fisher (1958) and Hamilton (1963), the reasons most animals behave as they do are not "for the preservation of the species," "to survive," "or to reproduce." All of these statements project into the future. Rather the ultimate reasons lie in the past survival and reproduction of the animal's ancestors. They possessed suitable genotypes; that is, they were adapted to the particular ecological conditions under which they existed. Because those animals survived and reproduced, the characteristics were passed on to their descendants—the animals that are observed today.

Be wary of any causative statement that contains the words to or for. Ultimate factors should not be viewed as projecting into the future. They may lead to or result in improved survival and reproduction in the future (i.e., they are said to have a function), but that is a consequence and not a reason or cause. Skinner (1981) extended this principle (selection by consequence) to account for all learned behavior and the development of cultures. Most biologists are not prepared yet to go that far, but most might find Skinner's discussion thought-provoking.

If you remain confused by this, we recommend that you reread and continue to think about it, review the general subject of evolution in an introductory college-level biology textbook (see list at end of Preface), and follow continuing discussions among biologists (e.g., R. Dawkins 1987, S.J. Gould 1985).

SUMMARY

Animals, as generally self-contained, portable, mobile organisms, use various structures to move about and otherwise respond via their senses and their nervous and endocrine systems to both the external environment and internal conditions. Such responses are collectively referred to as *behavior*. An understanding of behavior is critical to a basic understanding of other aspects of biology, in turn requires an understanding of other biological disciplines, and also has numerous practical applications.

The interactions of bats and moths illustrate several points. First, behavior that appears to be simple and uninteresting may be found on closer inspection to be complex and fascinating. One can understand behavior by asking questions, by being patient and persistent, and by using proper methods. Second, the senses and perspectives of other animals give them a different Umwelt than that of humans. Variability in senses and perspectives, as well as in the morphological and physiological substrates of responses, leads to much diversity in behavior. Third, there is much about animal behavior that still is not understood. Behavior has structure and function, and it is important to distinguish between immediate or proximate and the longer-range ultimate causes of particular behavior. The proximate factors in behavior are mediated primarily (although not exclusively) through the neural and endocrine systems. Ultimate factors interact with living organisms through genetic, ecological, and evolutionary processes of differential survival and reproduction.

Interpretations of behavior in other animals may be complicated by subjective views of our own human behavior. This controversial problem, along with the related matter of animal consciousness, is considered under the topic of anthropomorphism, an issue that has not been resolved but toward which (in the minds of some persons) progress is being made.

FOR FURTHER READING   **More on bats, echolocation, and interactions with moths**

Fenton, M.B., and J.H. Fullard. 1979. The influence of moth hearing on bat echolocation strategies. Journal of Comparative Physiology 132:77-86.
Further investigates moth hearing and ecological aspects of the bat-moth interaction, and discusses countermeasures that various species of bats may take to foil the abilities of moths to detect them.

Fenton, M.B. 1982. Echolocation, insect hearing, and feeding ecology of insectivorous bats. In: T.H. Kunz, editor. Ecology of bats, pages 261-285. Plenum Press, New York.
A general, straightforward, and nontechnical review of bat echolocation among different species.

Fenton, M.B., P. Racey, and J.M.V. Rayner, editors. 1987. Recent advances in the study of bats. Cambridge University Press, New York.
Reviews of recent information on bat flight, echolocation, and other topics based on three symposia.

Griffin, D.R. 1958. Listening in the dark. Yale University Press, New Haven, Conn. (Reprinted 1974, Dover Publications, New York.)
A major, classic treatise on bats and echolocation.

Griffin, D.R. 1959. Echoes of bats and men. Doubleday, New York.
A concise discussion of bats and echolocation, somewhat similar to the previous reference but much briefer and with some different information.

Hill, J.E., and J.D. Smith. 1984. Bats, a natural history. University of Texas Press, Austin, Tex.
An excellent, well-illustrated, general overview of bat biology in a concise format and straightforward presentation.

Norberg, U.M., and J.M.V. Rayner. 1987. Ecological morphology and flight in bats (Mammalia; Chiroptera): wing adaptations, flight performance, foraging strategy and echolocation. Philosophical Transactions of the Royal Society of London. Series B 316:335-427.
A detailed, technical presentation.

Pollak, G., D. Marsh, R. Bodenhamer, and A. Souther. 1977. Echo-detecting characteristics of neurons in inferior colliculus of unanesthetized bats. Science 196:675-678.
A glimpse, albeit mostly technical, into the neural aspects of echo-detection by bats. The neural and related aspects of behavior are discussed in this book in Chapters 15 to 20.

Roeder, K.D. 1967. Nerve cells and insect behavior. Harvard University Press, Cambridge, Mass.
The classic book focusing on the moth-bat interaction, with emphasis on the moth's side of the story.

Simmons, J.A., M.B. Fenton, and M.J. O'Farrell. 1979. Echolocation and pursuit of prey by bats. Science 203:16-21.
An excellent journal review of the subject.

Simmons, J.A., and R.A. Stein. 1980. Acoustic imaging in bat sonar: echolocation signals and the evolution of echolocation. Journal of Comparative Physiology 135:61-84.

An in-depth, technical treatment of bat sonar and how echoes may be used for the perception of prey targets among different species of echolocating bats.

Wimsatt, W.A. 1970, 1977. Biology of bats. 3 vols. Academic Press, New York.

A standard reference on bat biology.

Discussions of problems related to anthropomorphism and the subjective aspects of behavior such as consciousness and emotion

Bateson, P.P.G., and P.H. Klopfer, editors. 1985. Perspectives in ethology. Vol. 6: mechanisms. Plenum Press, New York.

The first three chapters, by J.R. Durant, P.H. Klopfer, and H. Ursin, respectively, explore the neural and psychological substrates of emotions and provide a sense that at least some progress has been made in our understanding, although some problem concepts (such as the triune brain) persist and, overall, our understanding remains far from complete.

Bitterman, M.E. 1988. Creative deception. Science 239:1360.

A letter to the editor concerning a Research News report by R. Lewin. The content of this letter per se is perhaps of less interest than the tone and intent of the letter as they relate to the problem of understanding what goes on in other animals' minds. Also see Lewin (1988).

Burghardt, G.M. 1985. Animal awareness. American Psychologist 40:905-919.

A survey of attitudes by various groups of people toward animal feelings, emotions, and intellectual capacities; historical perspective; and an advocacy for critical anthropomorphism.

Griffin, D.R., editor. 1982. Animal mind - human mind. Springer-Verlag, New York.

A reasoned and objective discussion of the general issue of anthropomorphism and animals minds by a select group of animal behavior experts at several levels of inquiry, from neural to ecological and evolutionary.

Griffin, D.R. 1984 a. Animal thinking. Harvard University Press, Cambridge, Mass.

The most recent of a series of book-length essays by Griffin on the question of conscious mental experiences in animals.

Griffin, D.R. 1984 b. Animal thinking. American Scientist 72:456-464.

A short adaptation from Griffin's book of the same title.

Hinde, R.A. 1985. Was "the expression of the emotions" a misleading phrase? Animal Behavior 33:985-992.

Another attempt to deal with the complexity and difficulty of emotional states as they affect observed behavior. The discussion is placed into a context tracing back to Darwin (1872) and brought, unresolved, up to the present.

Lewin, R. 1988. Response to Bitterman. Science 239:1360.

Lewin consults his cat, Barbeque, on how to respond to Bitterman's interesting letter. The letter and response need to be read together to be appreciated.

2

HISTORY AND THE STUDY
OF BEHAVIOR

The historical paths leading to our current views of animal behavior have been many, varied, and intertwined. Humans undoubtedly have been interested in behavior since before recorded history, and from the start there have been a few good observers and critical thinkers. However, most of the progress in understanding behavior has occurred within the past 200 years. This was made possible by (1) the invention of the printing press, which permitted communication to a larger audience; (2) a cultural shift from relying on established authority to accepting and even valuing independent observation and thinking; (3) general scientific and technological advances; and (4) the development of three major lines of interest that effectively came together to focus on behavior. The three lines of interests were (1) a Darwinian interest in behavior from the standpoint of evolution, (2) an interest in our own human behavior, and (3) an interest in nervous systems from a primarily human medical and physiological viewpoint. From this base arose a division during the first half of the twentieth century into three disciplines: ethology, comparative psychology, and neurobiology. These were respectively concerned with the behavior of animals in their natural environments, animal and human psychology, and the biology of the nervous system. Since the middle of the twentieth century there has been a simultaneous merging of these different lines of inquiry and renewed focus on the ultimate and proximate aspects of behavior. Along the way, from before ancient Greece to today, many fascinating personalities have been involved. Some of these individuals are introduced in this chapter.

HISTORICAL INFLUENCES ON OUR UNDERSTANDING OF BEHAVIOR

Present understanding of behavior rests on a broad foundation of previous ideas, personalities, and techniques. One cannot truly understand today's ideas without also understanding the past. Although some past ideas are clearly wrong and even amusing, they may nonetheless contain elements of truth and remain at least heuristically useful when we learn what is wrong with them and why they no longer are accepted. It is useful, and sometimes humbling, to trace back the roots of modern ideas and discover that they often are not so modern after all. Frequently the idea, or an important kernel of it, was present some time ago, but there were insufficient data or a lack of appropriate methods to confirm its validity. Much of what is possessed today is not so much new ideas but the ability, based on improved information, to differentiate between past valid ideas and those that were less correct or actually wrong.

The history of understanding animal behavior is rich, complex, and fascinating. It also involves a large number of people, places, and events. This chapter only covers some of the highlights, essentially in summary fashion.* Persons interested in further details can refer to the references indicated in the footnote, other references in the chapter, and the additional readings given at the end of the chapter.

Behavioral study has experienced a recent exponential increase of interest in the subject and the number of persons involved. An estimate of the number of psychologists in the United States, for example, shows 228 in 1910, slightly over 1000 in 1930, and well over 45,000 in the late 1970s (Baron et al 1978). Comparable growth has taken place in other disciplines concerned with behavior, and a similar explosion has occurred in the literature. In terms of the contributions of individual persons, although there are a few important forefathers, it is no longer possible to give equal credit where credit is due. Today there are many highly qualified individuals who are making significant contributions on a broad front.

Contemporary interest in behavior can be roughly divided into two broad paths that follow the distinction between ultimate and proximate factors. Bullock (1981) described the difference, somewhat poetically, as "ecology over the eons" versus "physiology of the neurons." The two basic branches are quite distinct, but a complete understanding of behavior requires that they remain tied together and that both are studied.

The two current approaches to the study of behavior, ultimate and proximate, have largely developed from realignments of previous divisions and branches. These can be traced graphically as shown in Figure 2-1. The dimensions of the figure are time (on an exponential scale to represent increased growth and interest) and field of interest. Three of the major historical fields have been ethology, comparative psychology, and neurobiology. Each of these fields is discussed in more detail later in the chapter; here they are defined briefly.

Ethology is the attempt to understand the behavior of animals in their natural environment. Psychology is the study of mental processes and behavior. Comparative

*The historical perspective presented in this chapter was derived from a large number of sources. Although most of these are not cited specifically in the remainder of the chapter, they need to be acknowledged: Baron et al 1978, Beach 1955, Beer 1975, Bitterman 1975 and 1979, Burkhardt 1981, Boring 1957, Bullock et al 1977, Boakes 1984, Diamond 1974, Evans 1976, Flugel 1933, Hinde 1970, Hubel 1979, Jaynes 1969, Kandel 1976, Klein 1970, Klopfer 1974, Leahey 1987, Lorenz 1981, Manning 1967, McGrew 1985, Skinner 1976 and 1979, Smith 1983, Solso 1988, Sparks 1982, Thorpe 1979, Tinbergen 1951, and Watson 1971.

Figure 2-1 Descent of behavioral biology—a subjective impression of the history and relationships of the major disciplines contributing to the subject. This figure is an abstract, two-dimensional shadow of a historical structure that is better represented by three or more dimensions; see text for qualifications.

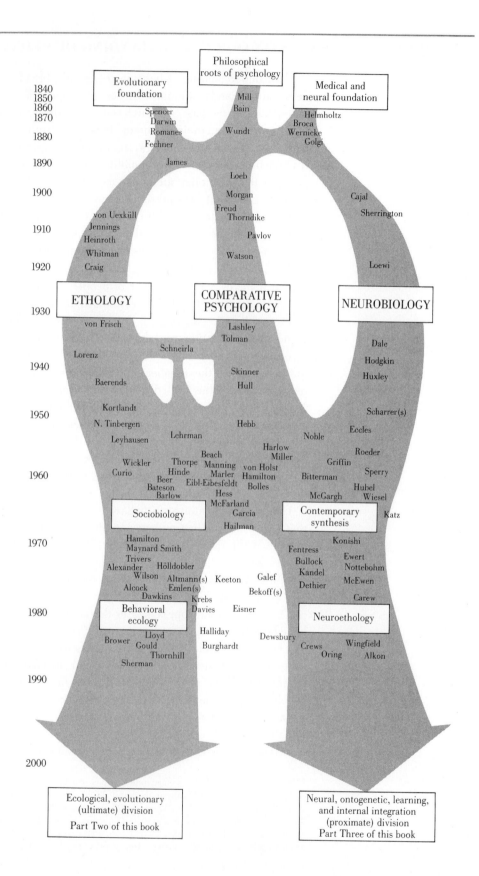

psychology initially was a fusion of nineteenth century human psychology and animal psychology. It leaned more toward laboratory experimentation than did ethology, placed more emphasis on using other animals as models of human behavior, and often focused on comparing learning processes under different conditions. Neurobiology is the biological study of the nervous system, including neuroendocrine and some endocrine secretions.

Although Figure 2-1 is in two dimensions, a better representation requires three or more dimensions. Neurobiology, for example, is as close to ethology as it is to comparative psychology. The three topics are more like the legs of a three-legged stool. Taxonomy could add another entire dimension to this picture. Interest in the behavior of particular groups of animals has been historically important in behavioral study (e.g., Van den Assem and Sevenster 1985, Pitcher 1986, and Shear 1986). Well-known examples include work by Schaller with gorillas and large cats, Geist with ungulates (hoofed mammals), Mech with wolves, several persons with particular groups of birds, fishes, and insects, and many people with primate studies.

Names are placed on the chart at single points, but many should be viewed in a flexible manner. Bitterman, for example, remains quite clearly in the traditional psychology viewpoint. On the other hand, von Holst appears to be under psychology in this chart, but his correct position is between ethology and neurobiology. Many of the persons have or have had multiple interests over time or for other reasons are difficult to pin to a single position.

The three branches of ethology, comparative psychology, and neurobiology developed from other lines of human interest. We therefore go back in time and work forward to the present.

BEGINNINGS OF BEHAVIORAL STUDY

It is difficult if not impossible to know where an understanding of behavior began. The earliest hunters and farmers undoubtedly had a practical understanding of the animals with which they were involved. Hunters learned when and where to find animals, how to hide from them, and how to attract them with decoys and baits. Similarly, persons involved in animal husbandry must have discovered much about behavior, (e.g., that males of many species cannot be confined together without serious fighting during the breeding season). Much information was probably relearned in subsequent generations or passed on by word of mouth until the advent of recorded communication. Animals figure prominently in early cave paintings. There is probably behavioral significance in many of the paintings, but most of the message has been lost. One of the earliest published items related to behavior is an Egyptian document (around 1700 BC; cited in Bullock et al 1977), which included descriptions of head injuries associated with brain functions. The ancient Greek anatomists and philosophers, including Alcmaeon (500 BC) and Aristotle (340 BC), became interested in more theoretical aspects of animal life, rather than just the practical. Furthermore, they published their observations and thoughts. Aristotle was a particularly accurate observer of some behavior. His *Historia Animalium*, which included scattered accounts of behavior, was published in several volumes.

Unfortunately, most of the early ideas and observations were either forgotten or solidified into authoritarian viewpoints not subject to revision and correction during the Middle Ages (from about the fifth to fifteenth centuries). To make matters worse, earlier information and writings were revised and muddled with superstition and inaccuracies. Pliny the Elder, for example, extended Aristotle's work by several volumes three and a half centuries after Aristotle by indiscriminately incorporating stories passed on to him from the general public. Pliny's work, *Historia Naturalis*, was further copied and reworked by others into the *Physiologus* to form a standard reference used for several centuries.

The long period of ignorance and misinformation continued with only a few isolated and mostly unimportant discoveries related to behavior. There were exceptions, one of the most notable being King Frederick II of the Middle European House of Hohenstaufen. He was the son of King Henry VI and became the Holy Roman Emperor of Germany and Italy in 1215 at age 21, reigning until his death in 1250. He also became King of Sicily and was crowned King of Jerusalem during the fifth Crusade. Frederick II was educated and enlightened. He was interested in politics, architecture, mathematics, and natural history, and founded the University of Naples. He was obsessed with the sport of falconry, even to the point of interrupting battles so he could practice the sport! He also was a true scientist, observing animals himself and double-checking other people's information. He was particularly interested in bird behavior, both of his beloved falcons and their prey species. Fortunately for us, Frederick II was also a writer, so his observations and thoughts survive to this day. His major writing, completed just before he died, is an immense work, *De Arte Venandi Cum Avibus (The Art of Falconry)*, which includes many descriptions of bird behavior. His potential contribution to science and the study of behavior, however, went largely unrealized and is only now appreciated in retrospect. He was philosophically ahead of his time, and he wrote before the invention of the printing press, so his scientific impact at the time was minimal.

A few other isolated accounts of animal behavior cropped up through the years, but they were mostly descriptive. An example uncovered by Hess (1973:68) and again noted by Kevan (1976) is the following from Thomas More's *Utopia* (1518; translated from Latin in 1551 by Robynson). The passage concerns the phenomenon of imprinting, a form of learning.

> They brynge vp a greate multytude of pulleyne, and that by a meruelous policie. For the hennes doo not syt vpon the egges: but by kepynge them in a certayne equall heate, they brynge lyfe into them, and hatche them. The chyckens, assone as they be come owte of the shell, followe men and women in steade of the hennes.

This account, like others of the time, can be viewed mostly as incidental; it did not in itself play a significant role in the history of behavioral science.

Taken together, however, these early accounts provide glimpses of both the independence of human curiosity and the renewal of interest in observing natural phenomena directly. That occasionally led to the questioning of established dogma and authority, and consequently gave rise to conflicts between different viewpoints. The early struggles and advances that were perhaps best known in physics with Copernicus, Kepler, Galileo, and Newton, and others could also be seen in other areas, such as biology. Recall, for example, the debate between Spallanzani and Cuvier in attempting to explain bat orientation as described in Chapter 1. The biggest conflict in biology, on the scale of the Copernican revolution in physics, however, did not come until later with the Darwin-Wallace theory of evolution. That controversy between observed facts and authoritarian dogma still lingers today, but it has waned, and we now hear mostly echoes of the past.

Thus, by fits and starts, from the fifteenth to the nineteenth centuries, scientific approaches to understanding nature in general developed and improved. That provided a background for understanding all aspects of nature, including animal behavior. If we focus specifically on behavior, three primary roots of the discipline can be identified: one concerned with anatomy and physiology; one based on evolution and behavior per se of all animals, including humans; and one that is mainly interested in human behavior.

The Contributions of Medical Anatomy and Physiology to the Study of Behavior

One of the significant beginnings of a scientific view of behavior along anatomical lines can be identified with Andreas Vesalius in 1543 with publication of *De Humani Corporis Fabrica*. This work included nothing on behavior per se, but it contained detailed figures of the human brain and clearly represents rekindled interest in original observation as opposed to citing ancient authorities. This was followed by scattered work on the brain and nerves, including investigations by Luigi Galvani in 1791 and Herman von Helmhotz in 1850 on conduction of nerve impulses in frogs. During the 1860s and 1870s, Paul Broca of France and Karl Wernicke of Germany demonstrated that particular speech problems in humans could be localized to specific brain areas. This path in neural anatomy and physiology formed the beginnings of neurobiology.

The Evolutionary Foundation

The second foundation for modern understanding of behavior arose with Charles Darwin and Alfred Russel Wallace's theory of evolution. In *The Origin of Species* (1859) Darwin established the general principles of evolution and devoted one chapter to the subject of instinct. In later books, particularly *Expressions of the Emotions in Man and Animals* (1872) and *The Descent of Man and Selection in Relation to Sex* (1871), Darwin concentrated more on behavior. He applied evolutionary arguments to behavior and to other biological traits and made many inferences concerning the internal mechanisms of behavior. The latter included his principle of serviceable associated habits, which closely resembled later notions of associative learning, his principle of antithesis, in which animals outwardly express their internal emotions, and his principle of the direct action of the excited nervous system of the body, independently of the will and in part of habit. Darwin provided numerous examples, mostly anecdotal, and proposed several ideas that, after translation to modern language, are still accepted. He also included a few ideas that have since been refuted. In a real sense, Darwin started the move toward progress in understanding behavior by providing the conceptual framework on which further development could take place.

Next in line was Darwin's student, George John Romanes. He published a comparative analysis of mental function and evolution, *Mental Evolution in Animals* (1884). He later extended similar arguments to humans in *Mental Evolution in Man* (1889). He included unpublished material given to him by Darwin in both books. One of the theses of these books was that behavior and psychological traits could be studied among different animals and, from that, one could infer phylogenetic relationships just as for morphological traits, a genuinely comparative viewpoint.

Romanes also discussed extensively conscious states and what he called *injective knowledge* (i.e., inferring what is going on inside another person [or animal] by observing how he, she, or it reacts to particular circumstances and by knowing how one feels in the same situation). For example, if one experiences fear in a particular dangerous situation and acts a certain way and if another animal is seen acting similarly in the same situation, then one may infer that it experiences fear also. This view is now considered anthropomorphic (as discussed in Chapter 1).

The Role of Human Psychology in Understanding Behavior

As with neural anatomy and physiology, psychology (Greek, *psyche*, soul or mind) and interest in human behavior can be traced back to the ancient civilizations. Increased attention to human psychology can be found in works from the sixteenth century onward. But most of the early psychology was more philosophically oriented and not on the same track as modern psychology. Generally, contemporary scientific psychology is considered to have started during the nineteenth century. It shares its basic beginnings with biological interests in behavior around the time of and after Darwin. Older, philosophical views of psychology were joined to physiological and other biological aspects during the latter half of the nineteenth century by such scholars as John Stuart Mill, Alexander Bain, William James, Wilhelm Wundt, and Herbert Spencer.

The person generally credited with shifting the emphasis in psychology from a philosophical to a scientific, experimental approach is Wilhelm Wundt. He was a tireless worker and writer who published an incredible amount of material over a period of nearly 70 years (1853-1920). Wundt entered the field of psychology and introduced the experimental emphasis from a background in anatomy and physiology as a medical student. The impact of Wundt's experimental approaches extended to all aspects of psychology, not just comparative psychology.

Another influential figure of this period was Herbert Spencer, who introduced the behavior of other animals into discussions of human behavior. In 1855, just before the publication in 1859 of Darwin's *The Origin of Species*, Spencer published *Principles of Psychology*. He anticipated some of Darwin's points of evolution but disagreed with or was wrong on several others; he believed, for example, in the inheritance of acquired characteristics. One of Spencer's important points, however, was his belief in the continuity of mental states (i.e., he thought there was some kind of continuity from the psychology of lower animals to that of higher animals). Later (1896) he expanded his viewpoint to rank psychic states from simple reflex behavior to volition (free will or choice) and suggested that they graded from one to the other and that the distinction among different animals was quantitative and not qualitative.

The Development of Different Disciplines: Ethology, Comparative Psychology, and Neurobiology

The period during and after the time of Romanes was exceptionally stimulating, and from about 1890 to 1910 many people became involved in the study of behavior. Numerous books and articles were published, and differences of opinion began to develop. These differences led to significant splits and to the development of three major disciplines that largely went their own ways. All were concerned in one way or another with the internal control of behavior, but they mostly developed independently of each other and also had other interests beyond the internal control of behavior. The three major branches were ethology, comparative psychology, and neurobiology.

The names that came to be associated with these branches, particularly *ethology* and *comparative psychology*, did not come into use in a precise and parallel fashion with the development of the disciplines themselves. Jaynes (1969) provides an excellent background history of the origins of the terms *ethology* and *comparative psychology*.

The word *ethology* (Greek, *ethos*, character or habit, from which we also get *ethics*) was in use by the seventeenth century, 200 years before Darwin, but it was used in the context of human stage actors or, later, relative to human ethics. That use continued well into the 1900s. During the middle of the nineteenth century and apparently out of ignorance of the other use of the term, *ethology* was used in a new context relative to living animals in a natural setting (as opposed to anatomical specimens in a laboratory). At that time, however, it was used more in an ecological rather than a behavioral sense. The term *ethology* gradually shifted from an ecological to a behavioral context, holding on by a thread of inconspicuous uses by a series of students of animal behavior through the end of the nineteenth and start of the twentieth centuries. In the late 1940s the term caught on and *ethology* finally came into popular use in relation to the study of animal behavior in the animal's natural surroundings.

The term *comparative psychology* was first used significantly in the title of a revised edition of a book by Pierre Flourens in 1864. Flourens represented a school of thought with emphasis on laboratory work that could be traced back to Cuvier and that was prejudiced against naturalistic observation. Flourens thus provided a name, *comparative psychology*, which quickly became popular and widely used by others, and he also provided a Cuverian philosophical orientation that would carry on for years to come and that is still present to some degree even today.

The divisions that developed among the three branches were often a result of ignorance or lack of interest in what the other branches were doing, and in some cases, they arose from sharp differences of opinion or outright hostility toward the other branches. Beer (1975) described the division by comparing ethology with a "broad river, fed by numerous tributaries, and braided in its lower reaches through the division and shifting of its channels in the loose gravel that is its bed" but is also "a single fluvial system that is separated from others flowing in its vicinity and through similar terrain."

The ethological viewpoint developed largely in Europe, and comparative psychology was concentrated in the United States. Ethology was devoted to the study of animal behavior under natural conditions and focused on consistent species-specific patterns of behavior with relatively little emphasis on learning, human behavior, or even mammals until the 1950s. Ethologists were primarily biologists-zoologists and maintained a fairly broad biological view of behavior, with particular attention to evolution and phylogenetic relationships. Much of their work was anecdotal, and only some of the early ethologists, such as Karl von Frisch, used good experimental design and statistics in their work.

Comparative psychology, on the other hand, minimized, then virtually lost, the evolutionary perspective. Comparative psychologists focused more narrowly on principles of associative learning, using laboratory and experimental studies with attention to statistics. Their basic underlying interest was the internal control (via learning) in humans. It was thought to resemble, if not be identical with (except for capacity), learning in other vertebrates. The work of the comparative psychologists concentrated on a few species, particularly white rats, pigeons, dogs, and occasionally other species of rodents and primates. It should be noted that white rats and other domesticated animals are many generations removed from their wild ancestors, and this can have important implications for their present day behaviors. (For further discussion of the relationship between domestication and behavior, see Chapter 22.)

Dewsbury (1984 a, b) hotly debated these (and other) characterizations of the differences between ethology and comparative psychology, referring to the distinctions as *myths*. The characterizations have been made by an impressive list of ethologists and comparative psychologists, however, and others (Demarest 1985, Burghardt 1986, Schleidt 1986) have either refuted or downplayed Dewsbury's contentions.

Neurobiologists came from a basic biological background but were initially interested in the immediate, proximate mechanisms of the nervous system rather than whole, functional units of behavior or ultimate, evolutionary considerations. Neurobiology proceeded as a mostly neutral, separate third party. It remained largely outside of the conflicts between ethology and comparative psychology.

One subdiscipline that was intermediate between psychology and biology, because its members generally were well versed in both areas, was physiological psychology. The physiological psychologists were concerned largely with the proximate causes of behavior and tended to lack interest in evolutionary aspects or adaptations of different animals living in different ecological settings. The underlying focus of the physiological psychologists, as with other types of psychologists, was human behavior. Other species were of interest mainly to the extent that they helped explain human behavior.

From among the three main branches and all of the individuals involved, only a few persons, such as G.K. Noble, Theodore C. Schneirla, and their students and associates, bridged the gulf between the psychological and ethological schools of thought. Noble was interested in amphibians, and Schneirla's primary focus was the learning and social behavior of ants (e.g., Schneirla 1971). In the midst of the period of sharpest division between the naturalist ethologists and the psychologists, Schneirla, his co-workers, and several productive students used information from both areas and presented a remarkably modern view of learning and the internal integration of behavior. A major publication along this line was *Principles of Animal Psychology* (Maier and Schneirla 1935).

Then and during the next two decades, Schneirla and a few others maintained an important comparative view by taking the best from both biology and psychology. He considered the internal mechanisms, such as the differences between the nervous systems of different taxonomic groups, and looked broadly at both invertebrates and vertebrates. He related differences among different organisms to their evolutionary backgrounds and to the different demands of surrounding, natural environmental factors. Particularly important for learning theory (discussed further in Chapter 20) was Schneirla's finding that maze-learning behavior in ants showed major differences from maze learning in rats. Today more details can be added and some of his ideas have been challenged, but most of his basic viewpoints are still considered valid.

As more people became involved in and aware of what the other disciplines were doing, ethology, comparative psychology, and neurobiology gradually began to reunite. There have been several heated exchanges. Ethologists accused the psychologists of being narrow minded and ignorant of evolutionary perspective, whereas psychologists often accused ethologists of being too experimentally unsophisticated and too ready to label behavior patterns as innate (discussed on page 47) or genetic

(Lehrman 1953). Some persons believed the old fields were dying or dead (Lockard 1971). The division between the two camps of ethology and comparative psychology was particularly deep, and much of it remains today, as evidenced by splits that still exist between respective departments on many university campuses. But generally there has been much merging of thought and information, and the two groups are at least speaking to each other and in many cases have genuinely joined, for example, as single departments at several universities.

Beer (1975) described the changed attitudes and atmosphere in a memorial to one of the major opponents, D.S. Lehrman. According to Beer, "The second half of the 1950s and early 1960s was a period of questioning and dismantling as far as ethological instinct theory was concerned." From the mid-1950s on there were several conferences that provided important interchanges of ideas. At one point Lorenz even helped hold a blackboard for Lehrman, an incident that drew loud applause from the audience. Beer remarked that the "conference was notable . . . for its spirit of harmony, enthusiasm, and optimism."

A modern synthesis seems to be gradually emerging, and the many remaining problems concerning internal control of behavior are being pursued jointly by persons with backgrounds in the three major branches. The following sections present a brief discussion of the history and major concepts of each branch.

ETHOLOGY

Continuing in what was already something of a tradition from earlier observers of animals under natural or seminatural conditions, two naturalists, Charles Whitman of the University of Chicago and Oskar Heinroth of the Berlin Zoo, are sometimes considered to be the founders of the discipline of ethology (Kandel 1976, Lorenz 1981). Whitman studied the reproductive behavior of closely related species of pigeons, and Heinroth worked with duck behavior. Among other findings and contributions, both came to essentially the same conclusion that the behavioral displays of different species are remarkably constant and can be used as taxonomic characters.

Early Ethological Concepts and Terms

One of Whitman's students, Wallace Craig, studied complex behaviors further and noted that they often consist of (1) a steering or appetitive component and (2) a consummatory part. During the appetitive part of the behavior, the animal actively searches and orients its movements to external stimuli. The appetitive part is variable, whereas the consummatory part is more fixed and stereotyped. A period of quiescence, during which the behavior will not occur again or in which the threshold for stimulation is high, often follows the completion of a consummatory act.

These characteristics can be observed in many behaviors, including feeding, sexual activity, and nest building. A classic example involves egg retrieval by birds such as geese (Figure 2-2). If an egg rolls out of the nest, the goose reaches out with its beak and brings it back. As the egg wobbles, the goose must correct for the displacement by moving its beak and steering the egg back. Once back in the nest, the egg is tucked under the bird's body in a stereotyped fashion that seems insensitive to external stimulation. In fact, if the egg is removed by an observer

Figure 2-2 Egg retrieval by grey lag goose. The goose uses its beak and neck movements to guide an egg back into the nest after it has accidentally rolled or been experimentally placed out of the nest.
From Lorenz, K., and N. Tinbergen. 1938. Zeitschrift für Tierpsychologie 2:1-29.

Figure 2-3 Konrad Lorenz.
Photograph by Thomas McAvoy, Life Magazine © 1955 Time.

before it is all the way under the goose, the goose continues to perform the full sequence.

Further work with different species, particularly in birds, fishes, and insects, was conducted by many naturalists, the most prominent being Konrad Lorenz (Figure 2-3), von Frisch (Figure 2-4), and Niko Tinbergen (Figure 2-5). Their methods are important in three aspects: (1) they observed animals in their natural habitat under conditions in which the behavior had evolved; (2) they studied both proximate and ultimate levels of causation; and (3) they worked with many different species of both invertebrates and vertebrates (Klopfer 1974). Tinbergen published a concise book, *The Study of Instinct*, in 1951. It summarized and synthesized the findings up to his time and has since become a classic. In 1973, Lorenz, von Frisch, and Tinbergen jointly received the Nobel Prize for their work.

Several important generalizations resulted from the work of these and several other naturalists. First, the expression of particular behaviors in many species is

Figure 2-4 Karl von Frisch.
From Hickman, C.P., Jr., L.S. Roberts, and F.M. Hickman. 1988. Biology of animals, ed 3. The C.V. Mosby Co., St. Louis. Photograph by W.S. Hoar.

Figure 2-5 Niko Tinbergen. From Hickman, C.P., Jr., L.S. Roberts, and F.M. Hickman. 1988. Biology of animals, ed 3. The C.V. Mosby Co., St. Louis. Photograph by L. Shaffer.

Dagmar, Konrad, and Gretl Lorenz

Frances Hamerstrom

Konrad Lorenz—Family Friend

"Please, send one woolen sock."

It was during the terrible post–World War II period that an ornithologist, writing from a prison camp in Russia, asked for only one sock. How bad can things be when you ask for only *one* sock? His name was Konrad Lorenz.

Konrad Lorenz? About 4 or 5 years earlier I had translated *Der Kumpan in der Umwelt des Vogels* for my professor, Aldo Leopold.

When, as chairman of the Wilson Ornithological Society's Committee for Relief for European Ornithologists, I sent one sock and then, a few weeks later, another wool sock, I did not dream that Konrad Lorenz would become a close family friend, that over the years we would argue an enormous variety of scientific questions, and that some day we would exchange daughters.

I never dreamt that some day I would kiss this man who seemed so hopelessly doomed by war, nor that I would swim with him in scummy, green water amongst his geese in Bavaria, nor that he would visit our primitive farmhouse in Wisconsin. It was his first visit to America, and I saw my own country with new eyes.

"Wolves," he asked, "you have wolves in Wisconsin? And bears?" It was in the early 1950s that Frederick and I got to worrying because our children were leading such uncultured lives. For example, my daughter Elva had never been remarkably stereotyped, (i.e., relatively constant, among all individuals of the same age and sex under similar circumstances). This led Lorenz to formulate the notion of the *fixed action pattern (FAP)*. More careful quantitative recent studies have shown that behavior is less constant than Lorenz's FAP suggested. (An early psychologist, William James, had been closer to the truth in 1890.) Accordingly, Barlow (1968, 1977) relaxed the requirement of stereotypy somewhat and referred to *modal action patterns (MAPs)*. Schleidt (1974) provides a further, detailed discussion of the concept of FAP and Gaioni and Evans (1986) present an alternate view of why FAPs may show variability.

A second important finding from the ethologists was that the stimulus required to trigger a response could be simple. Jakob von Uexküll (1934) emphasized this principle in a famous account of a tick. Females of this parasite, which require a meal of blood to produce their eggs, can live for over 18 years without a meal. They can wait in a sleeplike state as if time did not exist until triggered into action by a single odor, butyric acid, which is given off by the skin glands of mammals. Their waiting takes place on tips of twigs or grasses. When she senses butyric acid, the tick becomes alert and drops off her waiting post to either land on the mammal or miss her target—at which point she climbs back and resumes waiting. If she connects with a potential host, tactile and temperature cues direct her to move to an appropriate part of the skin, where she burrows in. The parasitic behavior of dropping from the perch, crawling onto the host, burrowing, and drinking the blood

taught how to come downstairs gracefully, nor had she ever been to dancing school. She went to spend most of a year with the Lorenz family when she was 12. The first dance she ever went to was at Buldern in the castle, and she danced with the baron.

Elva lived with Konrad, his charming wife, Gretl, and their daughter, Dagmar. In some ways she returned more sophisticated, but one day she did startle us by announcing, "I have two fathers."

"What do you mean?"

"Konrad is my father too."

Gretl gave large dinner parties, especially after they moved to Konrad's huge ancestral home (rather more like a castle) in Austria. Whenever conversation flagged, Konrad would shout, "Fran! Don't you think eagles are stupid?"

I'd put down my fork and yell back, "Not at all. It's geese that are stupid."

Visiting ethologists soon took sides. Some European gentlemen punctuate their arguments by brandishing steak knives.

These two families, Lorenz and Hamerstrom, loved each other. Usually, the Atlantic Ocean separated us, but we saw each other as often as possible. One day Konrad mentioned, "One has friends for two reasons: either because they are brilliant and useful, or, on the other hand, because they are such terribly nice people." Never did he spell out to which class the Hamerstroms belonged.

Figure 2-6 Courtship and fighting in three-spined sticklebacks. **A,** Two unrestrained males fighting. The fish on the left has assumed the threat posture, which elicits an attack from the fish on the right. **B,** Experimental manipulations of posture by restraining males in glass tubes. The vertical, threatening posture elicits an attack, whereas the horizontal, normal one does not. **C,** Experimental manipulations of stimuli (shape, color, realistic overall appearance) via models. All models with red bellies, regardless of overall shape and realistic appearance, elicited attack from male sticklebacks, whereas models without the red, including the most realistic fish, did not. The results shown in **B** and **C** suggest that a combination of two stimuli, posture and red belly color, are important in stimulating male fighting behavior. Male courting of female sticklebacks also is stimulated by limited features of the female. **D,** A female in normal posture and shape does not elicit vigorous courting, whereas a crude model with swollen belly or a fish of another species but in the proper posture (E) does.

A, B, and E from Pelkwijk, J.J. Ter, and N. Tinbergen. 1937. Zeitschrift für Tierpsychologie 1:193-204. C from Tinbergen, N. 1948. Wilson Bulletin 60:6-51. D from Tinbergen, N. 1942. Bibliotheca Biotheoretica 1:39-98.

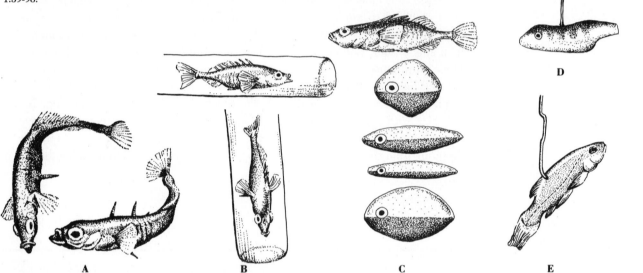

A   B   C   E

are all triggered by simple stimuli—the odor of butyric acid, tactile cues, and body temperature—and not by any other environmental cues. As described by von Uexküll (1934, translated by Schiller 1957:11-12), "Out of the vast world which surrounds the tick, three stimuli shine forth from the dark like beacons, and serve as guides to lead her unerringly to her goal."

This principle of simple stimuli for complex behavior was observed among a wide variety of species including vertebrates; classic examples are fighting responses of sticklebacks (Figure 2-6), territorial threat responses in European robins, begging responses in newly hatched gulls, and escape responses of birds triggered by silhouettes of hawks and other birds. (More is said about the hawk response later.) A live bird that lacks the appropriate signal might not trigger a response, whereas a properly colored bundle of feathers on a piece of wire will. Similarly, pieces of cardboard and other simple models shaped and painted properly can stimulate whole patterns of complex behavior. The critical stimulus properties, such as the odor of butyric acid for the tick, are called *sign stimuli*. Those that are used as social signals in communication by a species are referred to as *releasers*. The term *releaser* is sometimes used synonymously with sign stimulus, but technically releasers are special, intraspecific cases of sign stimuli.

Occasionally responses (FAPs) occur in the absence of the appropriate stimulus. These are called *vacuum behaviors*. Insectivorous birds deprived of flying insects, for example, may fly out and go through all the motions of catching, killing, and eating an imaginary insect. Occasionally vacuum behaviors are performed with inappropriate or minimal stimuli. Rats, for example, may carry and use their own tails in an attempt to build a nest in the absence of suitable materials (Eibl-Eibesfeldt 1958, cited in Lorenz 1981). One of the most remarkable cases of vacuum behavior that one of the authors (Grier) has personally observed involved a dog burying a bone on the linoleum in a corner of the kitchen. It first dug an imaginary hole and then dropped the bone into it. Next the dog carefully shoveled imaginary dirt up over the bone with its nose and tamped it in place. Then it walked away with the bone lying in full view on top of the linoleum.

Some artificial stimuli were found to elicit responses more effectively than the natural stimuli. A larger than normal egg, for example, may be incubated by some birds in preference over their own eggs (Figure 2-7). Such stimuli are called *supernormal releasers*.

Putting the sign stimulus together with the FAP led Lorenz to the concept of the *innate releasing mechanism* (IRM). The sign stimulus supposedly released the FAP via the IRM (Figure 2-8). The true neurological nature of the IRM was not known, but Lorenz in his famous hydraulic ("flush-toilet") model compared the process with the filling of a reservoir. The water could be released by a valve or might spill over the top. The hydraulic model was not meant to represent exactly what was going on inside the nervous system but only to serve as an analogy. Lorenz (1981)

Figure 2-7 Supernormal stimulus. The oyster catcher is attempting to incubate a large artificial egg chosen over a normal egg (*foreground*) and a herring gull egg (*left*).
From Tinbergen, N. 1951. The study of instinct. Oxford University Press, New York.

Figure 2-8 Relationship of the innate releasing mechanism (IRM) to the sign stimulus and fixed action pattern (FAP) in the classic ethological viewpoint.

Sign stimulus   IRM   FAP

Figure 2-9 **A,** Lorenz's old and **B,** new psycho-hydraulic models of motivation. The level of water represents the level of action-specific potential, which may be released by sign stimulus, or the weight at the end of the pulley (**A**) or by other input (in **B**). Modified from Lorenz, K.Z. 1981. The foundations of ethology. Springer-Verlag, New York.

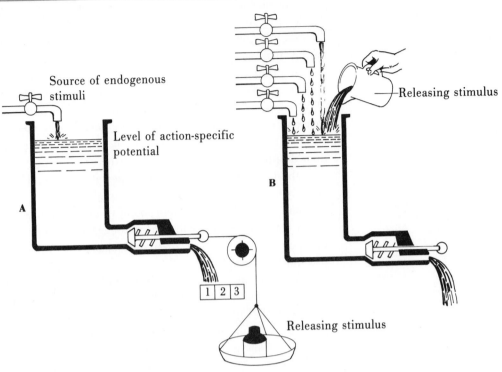

Source of endogenous stimuli

Level of action-specific potential

Releasing stimulus

Releasing stimulus

more recently resurrected and slightly modified his hydraulic analogy (Figure 2-9). In terms of the nervous system, Lorenz postulated an action-specific energy to describe a buildup of motivation for particular behavior patterns.

Problems and Arguments Concerning Ethological Concepts

The notions associated with action-specific energy, IRMs, and similar concepts such as drive have been substantially rejected as too vague and not based on neurological foundations (Lehrman 1953, Hinde 1956, 1970, McFarland and Sibly 1972, Andrew 1974). In discussing drives such as "thirst," Hinde (1970) pointed out the following problems that overlap to some extent:

1. Drives often are not "defined independently of the variations in behavior they are supposed to explain." For example, there is circularity in defining thirst as the tendency to drink.
2. How well drives correlate with behavior may depend on which aspects of the behavior are measured and how they are measured. Different measures of the same aspect of behavior often produce different results.
3. Drives have been defined in different ways by different authors; therefore the terminology is ambiguous.
4. It is not clear to which level of behavior the term *drive* should be applied. In nest-building behavior, for example, is there a reproductive drive, a nest-building drive, a stick-carrying drive, or all of these?
5. The use of drive can be an oversimplification and can obscure what really is occurring. One may think one has an answer when really there is only a new name for one's ignorance.
6. Drive can be misleading. One may look for or provide detailed explanations of things that do not even exist.

Hinde concludes "not that drive concepts are always useful or always useless, but that they have a limited range of usefulness and that they can be misleading and dangerous if misused." He notes further, "This, of course, is true of all explanatory concepts."

Dewsbury (1978) attempted to clarify the ambiguous definition of FAP. After comparing 10 sources on 11 attributes of the concept, he concluded, "It is apparent that there is little unanimity of opinion. The term *FAP* cannot presently be used in effective scientific communication." Many persons, however, would view that as an overstatement. One could make similar statements about other terms such as *learning*, *experience*, or *play*, yet few people would want to purge those words from our language.

The matters of innateness and instinct, particularly in opposition to learning, have been subjected to extensive and repeated review and scrutiny (Lehrman 1953, Beach 1955, Cassidy 1979, M. Dawkins 1986, and Burghardt 1988 a). The dictionary definition of innate (Latin, *innatus*) is "inborn," which superficially seems fairly straightforward. However, *innate* has been used ambiguously and in a variety of contexts including "genetic," "developmentally fixed," "unlearned," and "instinctive" (M. Dawkins 1986). The issue of "genetic" aspects of behavior is somewhat complicated and deserves further discussion (Chapter 4). Developmentally fixed and unlearned are roughly comparable concepts that refer to behaviors that develop and are expressed under appropriate conditions (e.g., by males during the breeding season) regardless of the environmental conditions under which an animal grows up. These two meanings, developmentally fixed and unlearned, are considered legitimate uses of the term *innate*, are the definitions usually implied or expressed in contemporary behavioral use by others, and are the meanings associated with the occasional use of *innate* in this text.

Instinct, often considered synonymous or otherwise associated with innate, has been used in numerous behavioral contexts and is more difficult to define. Its definition (Latin, *instinctus*) includes not only an inborn, native, or natural aspect but also a sense of internal motivation, impulse, or impelling. That tangles the idea with problems of motivation and drive. It also gives instinct a different meaning than innate, as illustrated in an example given by M. Dawkins (1986:65):

> . . . an animal may show 'innate' (meaning unlearned, developmentally fixed) anti-predator behaviour. It may respond quite appropriately with complex avoiding behaviour the first time it is exposed to a predator-like stimulus. And yet this would not imply that its anti-predator behaviour was necessarily driven from within and that the animal would search out predators to run away from if it had been deprived of the opportunity for doing so for some time.

Because of these problems, the term *instinct* is generally avoided (as it is in this text outside of the historical uses in this chapter).

Learning (i.e., the modification of behavior by experience or environmental influence) is itself an ambiguous and often confusing topic. It is discussed at length elsewhere in this book (later in this chapter plus Chapters 19 to 21). Learning appears to cover a multitude of phenomena. In addition, its opposite, that is, "unlearned" behavior, is not a simple, straightforward category. Many behaviors that are not considered learned may nonetheless change over time, often as a result of maturation, and many behaviors show elements of both innate and learned development. In a classic article titled "How an instinct is learned," Hailman (1969) described how the development of food-begging, pecking responses of gull chicks involve learning.

The escape responses of young birds to the appearance of hawks flying overhead have prompted several studies and differing interpretations. In one of the earliest investigations, Tinbergen (1948) demonstrated that chicks of several different species would respond to models of various hawks but not other birds such as geese (Figure 2-10). A general model that was flown over subjects in one direction resembled a hawk and elicited an escape response. When flown in the opposite direction, however, the same model resembled a goose and did not get the escape response. The important stimulus characteristic inferred from these results was that short neck was associated with hawk.

In a later study, however, Schleidt (1961) demonstrated that turkey chick escape reactions in response to hawk silhouettes involve learning; they respond fearfully

Figure 2-10 Models used to test the escape responses of young gallinaceous birds and waterfowl to flying birds of prey. Models that elicited responses are indicated with a plus sign.
From Tinbergen, N. 1948. Wilson Bulletin 60:6-51.

Figure 2-11 Hierarchical organization of instinct as envisioned by Tinbergen. **A,** Basic model of organization as shown for reproductive behavior in the male three-spined stickleback. **B,** Hypothesized application of the IRM concept to two levels of an instinct. **C,** Conjunction of **A** and **B** to form a hierarchical system of levels of instincts controlled by levels of IRMs.
From Tinbergen, N. 1951. The study of instinct. Oxford University Press, New York.

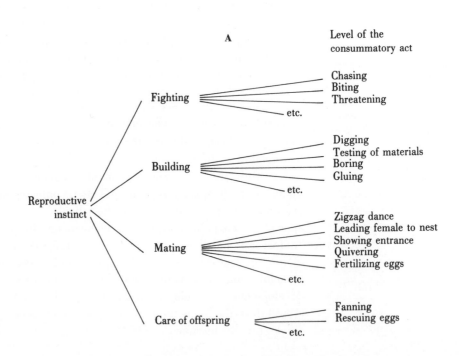

to any novel object passing overhead until they become habituated to familiar forms. In nature, hawks fly over rarely enough that the chicks do not habituate to them, but they do to forms of other birds that fly over more commonly. The hawk story, incidentally, is not finished. Mueller and Parker (1980) showed that naive ducklings do have some innate responses, such as more variable heart rates, to hawk versus nonhawk silhouettes. Some of the differences in responses may depend on the species studied. Cassidy (1979) provides a general review of the concept of innateness. Regardless of cases involving learning, many stimulus-response relationships have been shown to involve little learning. Even learned responses often may be triggered by simple cues.

Further observations by the ethologists revealed that during times when two or more stimuli occur simultaneously, the effect of one usually predominates. Occasionally, however, two stimuli are approximately equal, and this produces conflict behavior. An example is a male fighting another male at a territorial border where the impulse to flee may be balanced by the impulse to stay and fight. Two major categories of conflict behavior have been identified: redirected behavior, in which the proper response is directed at a different object (e.g., pecking the ground instead of the opponent), and displacement activity, in which an irrelevant response is suddenly given (e.g., preening or eating instead of either fleeing or fighting).

Several other behavioral responses also have been observed in conflict situations. These include inhibition of all but one response, intention movements (Chapter 5), alternation between behaviors, ambivalent behavior, immobility, compromise behavior, autonomic responses (e.g., defecation), and others (Hinde 1970).

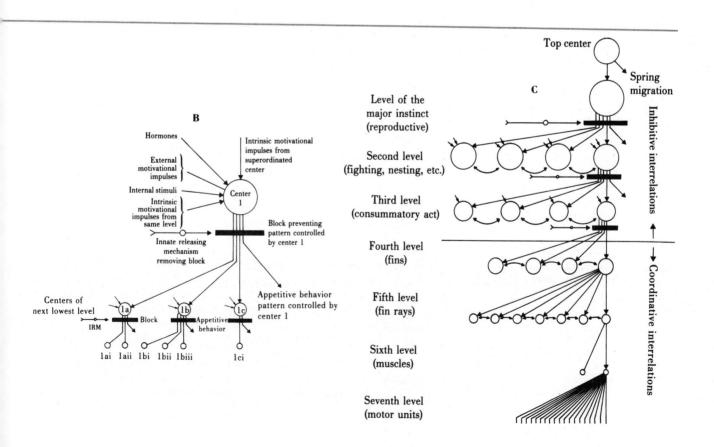

Some of the problems with FAPs were circumvented by using the concept and terminology of MAPs, as discussed previously, and MAPs are still being accepted and used in the literature (e.g., Brown and Colgan 1985). In an early attempt to bring Lorenz's IRMs more in line with the then-known neurological mechanisms, Tinbergen (1951) proposed hierarchical systems of centers underlying complex behavior. According to this view, there are major or higher levels of behavior, such as reproduction, which encompass lower, more specific behaviors such as territorial fighting, courtship, and nest building (Figure 2-11, *A*). The lower levels in turn are composed of yet lower levels such as particular courtship patterns, and these consist of specific muscular contractions and nervous system motor units. Various centers were thought to be stimulated by hormones and a variety of other internal and external factors, but higher levels were envisioned as being blocked or inhibited until released by the innate releasing mechanism in response to specific stimuli. Tinbergen's attempt to diagram all of this is shown in Figure 2-11, *B* and *C*.

The IRM in this scheme was viewed as a system of inhibitions that, when disinhibited, released lower centers. This view is closer to our modern understanding of internal neural mechanisms (see Part Three), but it concentrates too heavily on inhibitions that must be released (as opposed to centers that may be stimulated), and it conceives of an abstract organization, or hierarchy, that may have little real foundation in the nervous system. Furthermore, much of the stimulus filtering takes place peripherally at the sense organs rather than in the central nervous system. A few persons are still seeking support for the IRM concept (e.g., Baerends 1985).

The early ethological concepts and attempts to understand what goes on inside the animal, although now mostly replaced or revised by more recent information and ideas, nonetheless helped organize observations and led to progress in our current level of knowledge. The continued use of MAPs in the literature is evidence of the vitality still left in some of the classical concepts. The main contributions of ethologists have been to describe some of the general characteristics of input and output such as sign stimuli and MAPs, to catalog behavior in a wide variety of animals under diverse natural conditions, and to emphasize the ultimate roles of ecology and evolutionary forces in causing particular behaviors.

In addition to several persons in the classical ethological tradition, there have been a great number of other field-oriented zoologists such as Julian S. Huxley, Fraser Darling, and several amateurs who contributed significantly to the growing body of knowledge about behavior of animals under natural conditions. Margaret Nice, for example, was a housewife and amateur ornithologist who studied song sparrows around her home from 1928 to 1934 and contributed to the understanding of the importance of song and territorial behavior in birds' lives.

COMPARATIVE PSYCHOLOGY

Comparative psychology, which attempts to compare human psychology to that of other animals, initially developed in close association with the branch that eventually became ethology. One of the early workers who influenced both branches was William James (Figure 2-12). In 1890 he published *Principles of Psychology*, which dealt with instincts or bundles of reflexes. He noted that instincts were not invariable: "What is called an *instinct* is usually only a tendency to act in a way of which the average is pretty constant but which need not be mathematically 'true'" (1890:391). In addition, he proposed some general principles such as "inhibition by habit," by which the expression of particular behaviors is narrowed down, for example, to a particular mate or particular place to live. Another principle is the "law of transitoriness," which was an early recognition of what are called *critical periods* (Chapters 19 and 20).

The 30-year period after James's publication was active, and many people became involved, including Sigmund Freud. There was a profusion of interest in psychology, and numerous works were published at the time. During this period there was much controversy over consciousness and other internal states. Several schools of thought developed, including structuralism (concerned with the "structure" of mental states), introspectionism (which dealt with looking inward toward one's awareness to understand what is going on inside), and others, which are described later. Many of the problems leading to the controversies and different viewpoints are related to the issue of anthropomorphism (see Chapter 1). Efforts to deal with certain aspects of internal mental events were variously rejected and a shift was made toward

Figure 2-12 Development of the division between behaviorism and cognitive psychology.

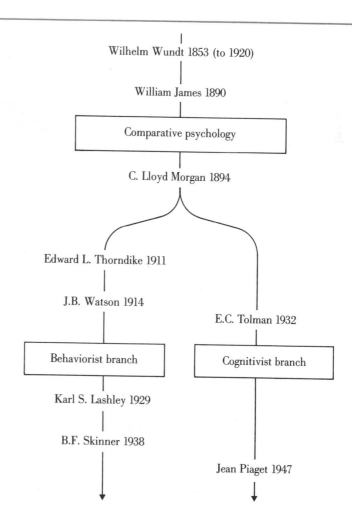

experimental approaches and observing outward expressions of behavior, the foundations of subsequent comparative psychology.

One important person in this movement was C. Lloyd Morgan, who published *Introduction to Comparative Psychology* in 1894. Among his other contributions, he rejected anecdotalism and undisciplined anthropomorphism in the interpretation of behavior in other animals. He called for a principle of theoretical parsimony, (i.e., the simplest explanation) which became known as *Morgan's canon*: "In no case may we interpret an action as the outcome of the exercise of a higher psychical faculty if it can be interpreted as the outcome of the exercise of one which stands lower in the psychological scale." A behavior should not be interpreted as resulting from thought, for example, if it can be ascribed to a conditioned reflex instead.

Reinforcement and Cognitive Theories

Comparative psychology after Morgan divided into two branches (see Figure 2-12). The first was strictly concerned with the outwardly observable relationships between the outputs and inputs of learned behavior. The theories that developed became known as *reinforcement theories*, and the associated school of thought eventually came to be known as *behaviorism*.

One of the initial architects of this line of thinking was Edward Lee Thorndike, who published a series of papers beginning in 1898. These papers later were collected into a single volume, *Animal Intelligence: Experimental Studies* (1911).

Figure 2-13 One of Thorndike's problem boxes (1911) from which animals had to learn to escape. Below the box is plotted the performance curve of a cat learning to escape from such a box during successive trials.
From Bitterman, M.E. 1979. Historical introduction. In: M.E. Bitterman, et al. Animal learning: survey and analysis. Plenum Publishing, New York.

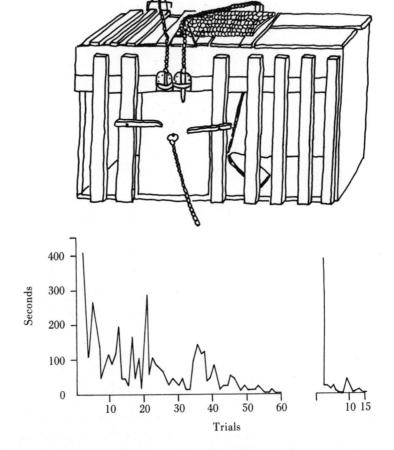

One of Thorndike's chief techniques was to put an animal into what he called a *problem box* (Figure 2-13). He plotted performance over subsequent trials. From this work Thorndike proposed his laws of exercise and effect. According to the law of exercise, performance of a behavior improves with practice. The law of effect states that the strength of a stimulus to evoke a response increases with pleasant consequences and decreases with unpleasant consequences. These laws are more commonly known as the *principles of reward and punishment*.

Comparisons of the performance of rats, sparrows, and monkeys in similar sorts of mazes suggested that learning was similar in all species. Information from elsewhere, such as Ivan Pavlov's classical conditioning studies of dog salivation in Russia (1906, translated into English in 1927), seemed to support these conclusions. In classical conditioning a normal response, such as salivation, becomes associated with a new stimulus, such as a ringing bell, rather than the normal stimulus, presence of food. The conditioning depends on pairing the unconditioned stimulus (US), in this case food, with the conditioned stimulus (CS), ringing of bell. As a result of the general consensus that learning is similar in all organisms, comparative psychology moved from comparing behavior in a variety of animals and concentrated on comparing the properties of learning under a variety of circumstances. We will return to the topic of behaviorism after a brief mention of the other branch, which remained concerned with the internal properties of learning and other behavioral phenomena.

Morgan, rejecting anecdotalism and anthropomorphism, did not deny internal events, but he attempted to clarify how we may investigate them scientifically. He distinguished between objective and subjective inferences, with only the objective category being amenable to science. Using an illustration given by Leahey (1987), imagine a dog greeting its master in a crowd of people. It is objectively legitimate to infer certain mental abilities in the dog: perception, memory, and recognition that permit it to identify its master from all of the other people. These characteristics can be investigated scientifically, for example, by sensory discrimination and memory capacity experiments, and thus are objective.

Other aspects, however, such as subjective emotions and the possibility that the dog can feel happiness when greeting its master, can be approached only through analogy to and projections of human feelings. As a result, they are subjective and beyond the scope of scientific inquiry. They may well exist, but they cannot be investigated.

Work with the objective aspects of internal processing in behavior led to an important class of theories known as *cognitive theories*, new attention to internal events, and development of a branch of psychology known as *cognitive psychology*. It encompasses several areas, including memory, attention, pattern recognition, perception, problem solving, and the general topic of intelligence. This branch of psychology expanded from the work of several persons including Edward C. Tolman (1932) and his students and Jean Piaget, a Swiss psychologist who studied the development of intelligence in children.

Behaviorism

The school of thought known as *behaviorism* was established after Thorndike by John Watson through a series of publications from 1913 to 1930. The last book in the series is titled *Behaviorism*. Watson's view was that one can infer the nature of the animal mind only from the outwardly observable behavior of the animal. The outward behavior has sometimes been referred to as *public events* and the internal mental processes as the *private events* beyond reach in an animal's life. Watson attempted to explain human behavior simply as a collection of stimulus-response reactions. Watson was active academically from 1905 to around 1920 and published *Comparative Psychology* in 1914. In later life he apparently did not spend much time in the laboratory or academic setting and was considered a bit eccentric. He was eventually fired because of an extramarital affair and went into the insurance business. Some of Watson's students, however, such as Lashley, carried on with his ideas.

Perhaps the most prominent behaviorist to follow Watson was B.F. Skinner (Figure 2-14). In 1938 he published *The Behavior of Organisms*, a book of major

Figure 2-14 B.F. Skinner. A modern automated Skinner box also is shown. When the rat pushes a lever, a food pellet or other form of reinforcement is delivered to the animal.
Photograph by Nina Leen, Life Magazine © 1964 Time.

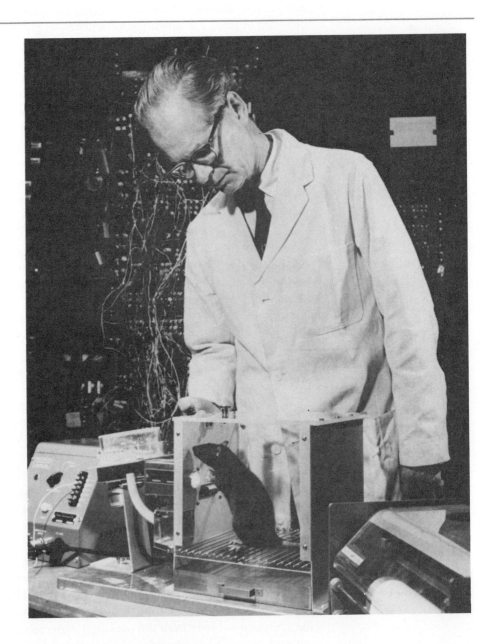

importance to psychology and studies of learning. Skinner pursued a class of conditioning that involved reinforcements. (Skinner preferred to avoid referring to these as *rewards* and *punishments*, because those words implied internal effects.) Responses of animals that may be reinforced are called *operants*, and this type of learning became known as *operant conditioning*. This category also has been called *instrumental learning*, a term that Skinner and his colleagues disliked. Operants include responses such as pressing bars in boxes, and a standard method of studying such learning involved the use of Skinner boxes (see Figure 2-14). Operant conditioning differs from classical conditioning in that it modifies the patterns of the response because reinforcement follows certain responses and not others. Conditioned responses are shaped through sequences of reward and punishment.

Skinner took an extreme operational view of learning. Reinforcement was viewed simply as that which alters the probability of behavior, and all that is needed is to discover how one variable (stimulus) affects another (response). According to this viewpoint, one does not need to become concerned with or develop theories about what is going on inside the animal, including neural or hormonal mechanisms. About 50 years after his 1938 book (in his typical provocative fashion) Skinner (1989) continued to emphasize this view in a paper on the vocabulary used to refer to feelings, states of mind, and cognitive psychology. He also proposed a linkage of behaviorists (behavior analysts) with ethologists and brain scientists in opposition to the cognitivists. He says (1989:18):

> No account of what is happening inside the human body, no matter how complete, will explain the origins of human behavior. . . . We can trace a small part of human behavior, and a much larger part of the behavior of other species, to natural selection and the evolution of the species, but the greater part of human behavior must be traced to contingencies of reinforcement, especially to the very complex social contingencies we call cultures. . . . Behavior analysts leave what is inside the black box to those who have the instruments and methods needed to study it properly. . . . Only brain science can fill those gaps. In doing so it completes the account; it does not give a different account of the same thing. Human behavior will eventually be explained (as it can only be explained) by the cooperative action of ethology, brain science, and behavior analysis.
>
> The analysis of behavior need not wait until brain scientists have done their part. The behavioral facts will not be changed. . . . Brain scientists may discover other kinds of variables affecting behavior, but they will turn to a behavioral analysis for the clearest account of the effects of these variables.

Much of the concern in comparative psychology with the properties of classical and operant conditioning continued up through the 1960s and is still present. The variations and refinements of the themes have been incredibly diverse. There has been much controversy as to whether the two categories of conditioning are really two or just one. The increased focus on learning studies using just a few species of subjects, particularly the white laboratory Norway rat, was plotted graphically by Beach in 1950 (Figure 2-15). Nearly 20 years later Hodos and Campbell (1969) observed that the situation had not changed much.

There have been a few notable exceptions to the mainstream focus on conditioning in comparative psychology. Schneirla, as discussed previously, was one. In addition, several other categories of learning, such as habituation and imprinting (Chapters

Figure 2-15 Numbers of published learning studies using laboratory rats compared with the use of other types of animals as subjects through 1948. Modified from Beach, F.A. 1950. American Psychologist 5:115-124.

■ Invertebrates
▦ Nonmammalian vertebrates
▨ Mammals except Norway rat
□ Norway rat

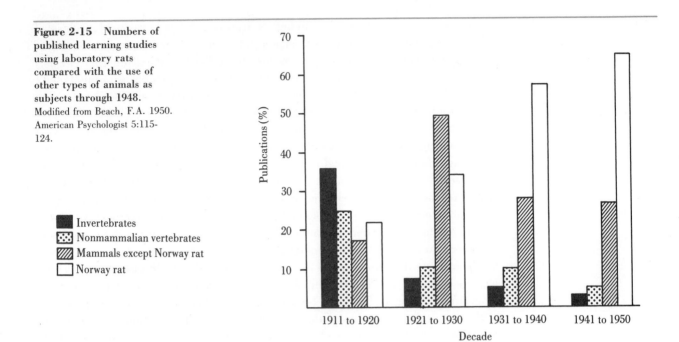

19 and 20), had to be taken into account and somehow fitted into or alongside conditioning paradigms. Cognitive viewpoints also demanded more attention. John Garcia and colleagues in the 1960s and 1970s (e.g., Garcia et al 1974) began quantifying taste aversion forms of learning and found that the results differed significantly from traditional forms of conditioning. It seemed that learning theory was developing cracks faster than they could be patched up.

Because of the major importance of learning and the contribution of comparative psychology to an understanding of the internal integration of behavior, and because this goes beyond the historical aspects, learning will be discussed further in Chapters 19 and 20.

Before leaving the psychologists, however, and to provide additional contrast with the ethologists, one can note a potential difference in the types of behavior that the two groups studied. Some persons (e.g., Denenberg 1972 a) have distinguished between species-specific behaviors, the stereotyped specific patterns of behavior that often are unique to each species, as denoted by their MAPs, and psychological behaviors, which are more general characteristics such as learning abilities, emotions, and memory traits. Ethologists generally have been interested in the former and psychologists in the latter. Whether or not the distinction is valid, however, remains to be seen. If it is, then some of the differences between ethological and psychological viewpoints may have resulted from the two groups working with different aspects of behavior. If the distinction is not valid, however, both it and many of the historical differences between ethology and psychology simply may be artifacts of different conceptual schemes. Whether the differences are in the subject or merely in the processes of study has not been settled.

NEUROBIOLOGY

Most of the basic concepts and terms associated with neurobiology, the study of nervous system anatomy and physiology, along with a few additional historical aspects, are placed in Chapters 16 to 18. In this section of the chapter we focus primarily on neurobiology's historical events and persons.

Neurobiology arose from the ranks of human medical sciences and the classic academic biological pursuits of anatomy and physiology. Because of these connections, neurobiology has benefited from larger numbers of workers and better funding than ethology or comparative psychology. The study of the nervous system also has tended to involve larger teams of workers and associated research institutes. Research institutes in neurobiology, often involving workers from different disciplines within physiology and medicine, began in Europe as early as the 1880s and subsequently grew and spread elsewhere.

The human medical sciences have been heavily and historically involved with the nervous system, particularly the eyes, ears, brain, spinal cord, and neuromuscular interactions. Although many nervous system disorders are congenital or disease related, the bulk of human problems have probably come from accidents, strokes, and injuries encountered in war, domestic quarrels, and other conflicts between people. The changes in behavior resulting from these externally caused problems, the findings associated with subsequent operations, and information from the extent and location of damage documented at autopsy have contributed to a gradually accumulating body of knowledge about the human nervous system that dates back to earliest recorded history in many cultures around the world.

One of the most striking medical history cases involving an accident is the story of Phineas Gage. Gage, the foreman of a crew of railroad construction workers, was the unfortunate victim of a blasting accident in 1884. While he was tamping the blasting powder with an iron rod into a hole drilled in a rock, a spark ignited the powder. The explosion blew the rod out of the hole and through Gage's head, from beneath his left eye, through his skull and brain, and out through a large hole in the top of his head. The rod, over 1 m long and 3 cm in diameter and weighing about 5 kg, landed 50 m away. Amazingly, Gage recovered consciousness within a few minutes and lived, but his behavior and personality were altered radically as a result of the brain damage. He regained his senses, speech, and memory but he changed over a period of days and weeks from a friendly, competent foreman to a cantankerous, unreliable, and unemployable drifter. His skull and the iron rod currently are on display in the Harvard Medical School museum.

Because of the difficulty of working with living nervous systems, most of the advances in our knowledge had to await the advent of modern technology. There have long been arguments over whether the control of behavior was in the brain or the heart; but all the way back to at least the Greek and Roman period there were some, including Alcmaeon, Hippocrates, and Galen, who recognized the importance of the brain and in some cases even proposed a connection between the brain, via the nerves, to the muscles. Much of the belief in the importance of the brain carried up into the sixteenth century, when Vesalius (1543) resumed original observations of the nervous system.

During the seventeenth and eighteenth centuries, with good anatomical descriptions available but still lacking the technology to adequately probe the physiology of the nervous system, philosophers could only guess how the system might actually work. The debates and arguments were numerous and noisy, with some sides being wrong and a few persons being essentially correct. René Descartes (1662), for example, speculated on the pathway from external stimuli being received, processed, and discriminated, to subsequent actions. A few decades later a clergyman and amateur biologist, Stephen Hales, began systematically studying frog reflexes by pricking the skin of decapitated specimens. He did not publish his findings but word spread and a Scottish physician and physiologist, Robert Whytt, repeated the work and published the results in 1751. These studies by Whytt were followed by Luigi Galvani speculating that nerve impulses were electrical in nature.

The pace gradually picked up, and during the nineteenth century the role of electricity in nerve impulses was confirmed by Alexander von Humboldt and Emil Du Bois-Reymond. Hermann von Helmholtz measured the speed of the nerve impulse in frogs at 35 to 40 m per second (a surprisingly slow speed compared to electrical speed in metallic conductors which is just under 300,000,000 meters per second). Important neuroanatomists of the nineteenth century included Paul Broca and Karl Wernicke, physicians who studied speech problems associated with brain dysfunction.

In addition to collecting information from human victims of accidents and other problems, medical researchers and animal physiologists have experimented with the nervous systems of nonhuman subjects in an attempt to obtain more insight into what is happening in human systems. Animals used in the latter work primarily have been cats, dogs, and a number of primates, particularly rhesus monkeys.

Sir Charles Sherrington conducted experiments on animals, particularly dogs and monkeys, from the 1890s up through World War II. His work included surgical removal of the cerebrum and isolation of specific connections in the spinal cord, leading to important information on reflex pathways and the connections between neurons (which he named *synapses*). Pavlov, who worked during the same period and after earlier work by another Russian, Ivan Sechenov, distinguished two types of reflex: innate and learned or conditioned. The latter involved functions in the cerebrum.

Other workers, in the meantime, were making progress at the cellular level. Much of the problem of working with neurons is their small size, fragility, and lack of pigmentation. The first significant progress in observing nervous tissue had to wait until the development of the microscope. But even under a good microscope neurons all blend together in an amorphous mass. With most stains they turn into a stained amorphous mass. In 1875 Camillo Golgi discovered a stain that (in a way still not understood) stains only a small fraction of the neurons present. This opened the door to studies of neuroanatomy. A contemporary of Golgi, Ramón y Cajal, used Golgi's new stain and devoted the rest of his life to studying of the nervous systems of humans and other vertebrates. In 1904 he published a monumental book, *Histologie du Systeme Nerveux de l'Homme et des Vertebres*. It is still considered the most important single work in neuroanatomy.

From the 1920s until the present, new techniques developed that led to numerous investigations of the operations of individual neurons. Significant technological advances include the electron microscope to better see cellular details, electronics for amplifying and measuring electrical events in the nervous system, and improved biochemical equipment and methodology for measuring chemical events. Most of the early emphasis in neurophysiology was on the components of the nervous system and physiological, molecular considerations, with less attention paid to the whole system or behavior per se. Invertebrates, particularly mollusks (such as squid and a marine snail *Aplysia*) and arthropods (such as crayfish and insects), have been important subjects of this work. Invertebrates have nervous systems that are much easier to reach than those of vertebrates. Where larger parts of the system were investigated, simpler movements and behaviors were considered for the most part. Attention eventually turned more toward more complex behavior, such as problems of learning and memory. There were a few early studies that focused on complex behavior, including work by Pavlov and work on moth hearing by Roeder and his colleagues (Chapters 1 and 17). With recent developments and the ability to record from several neurons simultaneously, there have been many advances in tracing neural pathways in simple animals and studying simple behaviors in more advanced organisms.

Along with research on the nervous system per se, there has been a significant contribution to the physiological understanding of behavior involving the roles of hormones (Chapter 18). Many persons have been involved including, to name only a few, Beach, Crews, Hinde, Lehrman, Rosenblatt, Wingfield, and E. and B. Sharrer.

Along the way, the field of neurobiology has been liberally sprinkled with Nobel Prizes. These include Golgi and y Cajal in 1906; Einthoven in 1924; Adrian and Sherrington in 1932; Dale and Loewi in 1936; Eccles, Hodgkin, and Huxley in 1963; and Sperry, Hubel, and Wiesel in 1981. The large number of prizes in this area partially reflects the fact that the Nobel Prizes include a section devoted to physiology or medicine. But it also reflects genuine progress and brilliant work with a complex and difficult topic.

SINCE THE SIXTIES

The classical history of behavioral science probably can be viewed as closing somewhere around the 1960s. The biological study of behavior since the 1960s has been developing along the two major lines indicated at the start of this chapter: the ultimate ecological and evolutionary aspects on one hand and proximate neurophysiology (including continued attention to learning and memory) on the other hand. The distinction between ethology and comparative psychology continues to fade (although there remains a major difference in emphasis on humans versus other species in the general fields of psychology versus biology.)

Along the ecological and evolutionary line there has been a notable shift from the classical approaches of Tinbergen, Lorenz, and von Frisch to evolutionary-ecological approaches reflecting the influence of Robert MacArthur, Edward O. Wilson, John Maynard Smith, George Williams, and Robert Trivers. Contemporary labels (extending to the names of professional societies and journals) for this branch of behavioral interest include *sociobiology* (from the title of a book by Wilson 1975) and *behavioral ecology* (from the title of a book by John Krebs and Nicholas B. Davies 1978). Studies by Jane Goodall (1986) and others show that the ecological and evolutionary approach has found a favorable response among psychologists and even anthropologists. In the area of neurobiology, one of the recent behavioral labels is *neuroethology*.

Stimulated in part by some of the work in neurophysiology, learning, and memory in combination with rapid and recent developments in computers, an almost wholly new subject has emerged involving artificial intelligence (AI), neural networking, connectionism, and related topics. That primarily involves machine (computer) behavior, however, and is mostly beyond the scope of this book, except where it is derived from information on biological systems or elucidates biological systems and as such will be treated briefly (Chapters 16 and 20).

One of the developments to affect the methods of studying animal behavior since the 1960s is an increase in concern for the welfare and rights of the animals being studied. The issue of animal welfare has generated much emotional debate and controversy. To prevent abuses of animals that occurred in some past studies, most organizations, institutions, and agencies now have strict guidelines and regulations, and there are even laws in some jurisdictions. Many institutions and agencies have animal welfare committees that must approve any proposed work with live animals before such work is permitted. These are discussed further in chapters 3 and 22 and in the appendix.

Most of the remainder of this book is organized along the two main lines of interest in behavior: Part Two covers the ecological and evolutionary aspects and Part Three covers the internal neurophysiological aspects. The recent histories of sociobiology and behavioral ecology, neuroethology, and the persons contributing to those areas are discussed where appropriate. History is not dead but continuously developing; it forms a bridge between the past and the present.

SUMMARY

Recorded observations and interpretations related to behavior go back well over 3000 years. Aristotle and other ancient scientists wrote about animals and their habits and considered the brain to be important to the control of human movement and behavior. Ideas became solidified, however, and for several hundred years society relied largely on quoting ancient authority, often muddled with myths and folk tales, rather than making progress through new observations and thinking. There were a few exceptions, such as King Frederick II during the thirteenth century, who was interested in falconry and became a keen observer of bird behavior. However, most of the independent observing and thinking of the time did not receive widespread attention because there were no mass media such as the printing press, and often because of the cultural bias, which favored a reliance on ancient authority.

This situation gradually changed with the appearance of printing, an enlightened cultural attitude, and the growth of science and technology in general. During the nineteenth century three lines of human interest came together to focus on the behavior of humans and other animals: (1) medical and physiological interests in the nervous system, (2) human psychology, and (3) evolutionary thinking about the origin and relationships of various species, including humans. Darwin contributed significantly not only because of his role in formulating the theory of evolution but also because he was interested in the behavior of humans and other species.

The merging of these three lines was followed by the development of three new disciplines, which at various times came to be known as (1) *ethology*, the study of animal behavior under natural or seminatural conditions; (2) *comparative psychology*, the comparison of animal and human psychology, usually under experimental and laboratory conditions and mostly aimed at a better understanding of human behavior; and (3) *neurobiology*, the study of the anatomy and physiology of the nervous system. Each discipline developed a body of concepts, terminology, and methodology. There were also subdivisions within the disciplines, such as between behaviorist and cognitivist schools of thought in comparative psychology.

In spite of a few persons who bridged the gaps between disciplines, the three disciplines differed in methodology, interpretations, and opinion. Since the middle of the twentieth century, however, the disciplines have reunited on a number of common grounds.

Current approaches to the study of behavior can be conceptually divided into two lines of interest based on either the ultimate or proximate aspects of behavior. New labels have cropped up, including *sociobiology*, *behavioral ecology*, and *neuroethology*.

A large number of people have contributed historically to our understanding of animal behavior. Only a few are highlighted in this chapter. Three, Aristotle, King Frederick II, and Darwin, have already been mentioned. We have arbitrarily chosen twelve more for this summary:

Wilhelm Wundt (published from 1853-1920)—helped move psychology from a philosophical subject to an experimental, laboratory science.

Jakob von Uexküll (early 1900s)—developed the concept of the Umwelt and contributed to early ethological ideas such as the sign stimulus.

Karl von Frisch (1910s to 1970s)—a thorough and meticulous scientist who experimented with animal senses and behavioral biology and was a key figure in ethology; among other important findings he discovered the dance language of honeybees.

Konrad Lorenz (1930s to 1980s)—a colorful figure who studied the phylogenetic relationships of waterfowl based on detailed analyses of stereotyped behavior, following a line of earlier similar studies; studied imprinting in geese; contributed to the concepts and terminology of ethology; and helped popularize the subject through his engaging writing style and lively personality.

Niko Tinbergen (1930s to 1980s)—studied the behavior of a variety of species, including wasps and gulls; developed models and experiments that were performed on animals in their natural settings; and contributed to, as well as helped synthesize, ethology's major concepts.

Ivan Pavlov (early 1900s)—a Russian physiologist who studied digestion and related nervous system reflexes, including conditioned reflexes or classical conditioning.

Burrhus F. Skinner (1930s to 1990)—an American psychologist who long spoke for and typified the behaviorist school of comparative psychology, which focuses on interpreting behavior from outwardly visible actions and their relationships to various reinforcing contingencies of the individual's environment.

Theodore C. Schneirla (1920s to 1960s)—a psychologist who was interested in the behavior of other animals, particularly army ants; discovered differences in learning phenomena between ants and mammals; helped bridge the gap between comparative psychology and ethology during the period of greatest differences.

Charles S. Sherrington (early 1900s)—studied the physiology, functions, and detailed anatomy of various parts of the mammalian nervous system; identified and named the synapse as the connection between adjacent neurons.

Alan L. Hodgkin and **Andrew F. Huxley** (1930s to 1950s)—physiologists who developed a technique (voltage clamping) to study the electrical and chemical processes of the neuron. The two worked together, which also typifies the team approach that has been so fruitful in neurobiology.

John C. Eccles (1950s to 1970s)—worked with others in the study of synaptic transmission and helped relate the detailed neural processes and pathways of the nervous system to the larger, integrated picture of movements and functional behavior.

The contributions of these and many others like them have greatly advanced our understanding of behavior and have helped make this book possible. We have inherited a rich legacy on a fascinating subject.

FOR FURTHER READING

Baars, B.J. 1986. The cognitive revolution in psychology. Guilford, New York.
An analysis, based in part on interviews of several scholars, of the shift in psychology from behaviorism to cognitive viewpoints.

Beach, F.A. 1955. The descent of instinct. Psychological Review 62:401-410.
A critical look at historical concepts.

Beer, C.G. 1975. Was Professor Lehrman an ethologist? Animal Behaviour 23:957-964.
A classic essay on the relationship between ethology and psychology.

Boakes, R. 1984. From Darwin to behaviourism. Cambridge University Press, Cambridge.
Analysis of the theories and theorists studying animal behavior and animal minds during the important period of 1870 to 1930.

Burghardt, G.M., editor. 1985. Foundations of comparative ethology. van Nostrand Rheinhold, New York.
A collection of reprinted excerpts from ethology's classical period, with a foreword by Lorenz.

Burghardt, G.M., 1986. Book review of Dewsbury. Ethology 73:78-88.
A lengthy and in many places spirited analysis of Dewsbury's book (1984 b) with rebuttal of many of his views.

Burkhardt, R.W. 1981. On the emergence of ethology as a scientific discipline. Conspectus of History 1(7):62-81.
A description of ethology's early development, with emphasis on Lorenz and his work. This presentation was sparked in part by disagreement over historical views and predictions presented by Wilson (1975).

Dethier, V.G. 1962. To know a fly. Holden-Day, Oakland, Calif.
A personal and engaging philosophical account of research on the feeding behavior of flies. This short and readable book provides insight into the nature of science and scientists by a neurobiologist who is able to communicate clearly with both fellow scientists and the general public.

Dewsbury, D.A. 1984 a. Comparative psychology in the twentieth century. Hutchinson Ross, Stroudsburg, Penn.
A lively and lucid, albeit defensive, view of comparative psychology. Includes several biographical sketches and what Dewsbury refers to as the ten myths about comparative psychology, which he attempts to dispel.

Dewsbury, D.A. 1984 b. Foundations of comparative psychology. van Nostrand Rheinhold, New York.
Reprints of classic writings from the field of comparative psychology plus editorial comments and an introduction in which Dewsbury continues his defense of comparative psychology and attempts to set the record straight.

Dewsbury, D.A. 1985. Leaders in the study of animal behavior: autobiographical perspectives. Bucknell University Press, Lewisburg, Penn.
Professional autobiographies of 19 leaders in the field of animal behavior.

Durant, J.R. 1986. The making of ethology: the Association for the Study of Animal Behaviour, 1936-1986.
Historical review of the ASAB on its fiftieth anniversary and discussion of how that history reflected ethology as a scientific discipline.

Jaynes, J. 1969. The historical origins of 'ethology' and 'comparative psychology.' Animal
     Behaviour. 17:601-606.
     Traces the origins of the terms and provides historical sketches of the disciplines themselves.
Kandel, E.R. 1976. Cellular basis of behavior. W.H. Freeman, San Francisco.
     *A neurobiology viewpoint with an excellent historical discussion of the relationship to other
     branches of behavioral science, pages 3 to 27.*
Klopfer, P.H. 1974. An introduction to animal behavior: ethology's first century. Prentice-Hall,
     Englewood Cliffs, N.J.
     *A brief textbook on animal behavior with a primarily historical viewpoint, covering the
     development of principal concepts.*
Leahey, T.H. 1987. A history of psychology. Prentice-Hall, Englewood Cliffs, N.J.
     *A comprehensive survey of the history of psychology from ancient Greek philosophers to the
     present.*
Lorenz, K.Z. 1952. King Solomon's ring. Thomas Y. Crowell, New York.
     *A delightful collection of essays, anecdotes, and comments from one of ethology's chief architects.
     This book conveys much of the flavor and excitement of early ethology.*
Lorenz, K.Z. 1981. The foundations of ethology. Springer-Verlag, New York.
     Lorenz's view of classical ethology.
Roitblat, T.G., B. and H.S. Terrace, editors. 1984. Animal cognition. Lawrence Erlbaum, New
     Jersey.
     *A review by several authors of contemporary cognitive theory applied primarily to nonhuman
     species.*
Solso, R.L. 1988. Cognitive psychology. Allyn and Bacon, Boston.
     A contemporary view of cognitive psychology with a historical introduction.
Sparks, J. 1982. The discovery of animal behaviour. Little, Brown and Co., Boston.
     *A companion to the BBC/Nature television series of the same name, which traces the early days
     of behavioral study and describes the relevance of various observers of behavior and their findings
     to our current views.*
Tinbergen, N. 1951. The study of instinct. Oxford University Press, New York.
     *The classic text on ethology, a relatively short (228 pages) overview of the field and concepts as
     of midtwentieth century.*
Tinbergen, N. 1958 (reprinted 1968). Curious naturalists. Natural History Library, Anchor Books,
     Doubleday, Garden City, New York.
     *A semiautobiography focused on research by Tinbergen and some of his colleagues and students.
     Provides historical insight into the man and his work, intended to encourage others to become
     involved in observing the behavior of animals. (Page 291: "If my stories of curious naturalists do
     not send out some readers to go and observe for themselves, this book will have missed its point.")
     This book, incidentally, started one of the authors (Burk) on the path to becoming a student of
     animal behavior, when he first read it at age 17.*

3

OBSERVATION AND MEASUREMENT OF BEHAVIOR

Awareness and understanding of the methods used to study animal behavior are important whether or not one personally works with the subject. By knowing the methods used to derive any given piece of knowledge, in conjunction with the historical background, we can better appreciate its limits, evaluate its current status, and understand where it might be headed in the future. Knowledge is not just a static entity obtained by past students and now merely waiting to be memorized by today's students. Rather, it is an ongoing, dynamic, and often exciting process of pursuing questions.

The study of animal behavior as a science uses many of the same approaches and methods as other branches of science. Some general aspects of methodology are particularly important for the study of behavior, and in addition, there are specific techniques that are required for answering particular behavioral questions. This chapter places the study of behavior into a general scientific context and uses an example of flocking behavior in birds to illustrate how questions about behavior are asked and pursued, how problems are a fact of scientific life, and why the answers usually remain elusive and alluring. After this example and the conclusions that can be drawn from it, the chapter provides a broad overview and introduction to the problems and methods of studying animal behavior.

SCIENCE AND THE STUDY OF BEHAVIOR

The common image of a scientist is a person who knows a lot of information, sort of like a walking encyclopedia. Unfortunately, that is also the way a lot of science is taught in schools and conveyed in the mass media. But what really drives most scientists is what they do not know and the desire to find out, like a detective searching for clues in a mystery or a hunter tracking elusive quarry. For a scientist, not knowing is an itch that refuses to go away, even with continuous scratching.

The process of answering scientific questions, however, is often difficult. There are some general guidelines, but the details usually get messy. Furthermore, it is not always clear whether or not one has found the right answers, good answers, or sometimes even what the question is. This applies not just to particular projects but to a general subject, such as behavior, and its subdisciplines as a whole. Thus, to understand the subject, one must also understand what brought it to where it is and what may take it further.

In addition to methodological aspects, attention must also be paid to associated hypotheses and theory. Behavior has been defined as the observable movements and other changes in the appearances of animals. What one observes and how one does it, however, are intimately connected with what one already thinks, hypothesizes, and knows. Theory without facts can only go so far, and the establishment of facts requires methods and techniques. The advancement of understanding is much like a person trying to move a large heavy object by himself. The facts are at one end and the theory at the other; progress is made by picking up one end, moving it forward and setting it down, then going back to the other end and walking it forward.

Progress in understanding behavior has resulted from the application of general biological and scientific methods, systematic observation and recording, critical but open attitudes, formulation of testable hypotheses, and experimentation. In addition, there have been advances in technology and the accumulation of specific knowledge about behavior. The study of behavior shares with other sciences a number of attributes, including general scientific methods and attitudes, much routine work of data gathering, and the need for quiet, uninterrupted contemplation. A major earmark of science as opposed to many other types of human endeavor is that many of the observations in science are or should be repeatable, except for certain unique historical aspects. As a result, scientists are prone to double-check themselves and each other. (For some recent interesting examples involving behavior in which the outcomes of earlier studies were not confirmed, see Immelmann et al 1982, Knight and Temple 1986 a, b, Ewing 1988, and Crossley 1988.)

This chapter discusses the study of behavior in a general scientific context and points to some aspects that are particularly important for understanding behavior. By considering behavior from this perspective, one can view it and the contents of this book not merely as a collection of interesting information but rather as a continuing pursuit of the unknown.

BEHAVIOR AND SCIENTIFIC METHODS

Science in general involves a complex body of techniques, approaches, and human activities. It includes not only observation and pure description but also various patterns of logic, assumptions, hunches and insights, discrimination and decision making about facts and ideas, experiments in some cases, plus a fair amount of luck (frequently known in the trade as serendipity). Although many textbooks make science appear clear-cut and purely objective, it is very much an art. What is beautiful in art is often said to be elegant in science, whether it is an outcome, method, hypothesis, or a whole piece of research.

Much science works indirectly by making inferences about the subject; these are educated guesses about what, how, or why things are. Initial guesses about cause and effect relationships are usually referred to as *hypotheses*, which may subsequently be confirmed, rejected, or further pursued. There are numerous familiar examples outside behavior. In astronomy, for example, the planet Pluto was inferred to exist through its effects on the planet Uranus before it was seen and known to actually exist. Atoms, the subject of chemistry, were seen only recently with powerful electron microscopes. Gregor Mendel never saw the units of inheritance (now called genes or cistrons); he inferred their existence from the way his garden peas grew and reproduced. Examples from behavior are found throughout most of this book.

Some techniques yield stronger inferences, that is, they permit more confidence in the interpretation, than do other techniques. We are almost never absolutely certain, however, and hypotheses are rarely proved. They are usually only tested. The strongest inferential technique involves the use of experiments, repeatable deliberate manipulations of a phenomenon where only one or a few factors of interest are permitted to vary at a time and then only in a controlled fashion. Other factors are either held constant or permitted to vary only in a restricted manner. Experimentation is not a guaranteed method, however, and even the best planned experiments can go awry or produce confusing or ambiguous results. Also, for a variety of reasons such as logistical problems, experimentation is not always possible.

There are a number of other, weaker inferential techniques, including correlations, comparative studies, historical inquiry, and modeling. Correlations involve relationships or patterns where changes in one factor may appear to be associated with changes in some other factor, such as an animal becoming more active with an increase in the average daily temperature. Comparative studies in biology consider similarities and differences among different species, ecological conditions, or other topics of interest. Historical studies in biology are usually long-term, involving the fossil record, or short-term, involving recent signs such as animal tracks or other evidence. Biological modeling usually involves mathematical simulations of phenomena under particular assumptions.

Many studies combine the correlational, comparative, historical, and modeling approaches in various ways. One can correlate, for example, certain anatomical structures (e.g., webbed feet) with particular environmental conditions (e.g., aquatic) and then extend that a step further to infer that if certain extinct species possessed the structure then they also may have lived in a similar environment. These four approaches (correlational, comparative, historical, and modeling) are usually considered weaker than experiments because they may overlook other factors

Philip N. Lehner

Philip N. Lehner

A Brief Look at Methods in Animal Behavior

As I write this essay my attention shifts from my computer screen to the scrub jays around the tray of birdseed in my backyard. I just saw the jay with orange leg bands chase away the one with green bands, and now orange has been chased off by the white-banded jay. Today, unfortunately, these behaviors are but alluring distractions I can only occasionally watch. To many people, the study of animal behavior appears to involve merely watching animals. However, animal behavior studies are based on many hours of detailed observation, a skill that requires concentration beyond mere watching.

The first step in studying behavior is the descriptive phase. For example, suppose I want to learn about the courtship behaviors of a particular species of bird. My search of the literature comes up empty so I make initial reconnaissance observations to determine where (habitat; microhabitat) and when (season; time-of-day) courtship occurs. I realize that I will have to hide in a blind and use a spotting scope to make observations at a distance and audio equipment to record sounds that accompany the visual displays. I also decide to videotape the rapid courtship bouts for repeated viewing in slow motion and analysis on a microcomputer. I then make the detailed observations that allow me to describe what behaviors are performed and who (which sex) performs them. When I have given each behavior a name and described its defining characteristics I have an ethogram of the courtship behaviors for this species. My ethogram will be more complete when I determine the frequency and duration of each behavior, as well as the sequence in which each sex performs its specific behaviors.

Suppose that analysis of the videotapes shows that one of the sounds made by males is correlated with both lifting of the head and wing fluttering. How is the sound made? I might decide to conduct an experiment and formulate the working hypothesis that the sound is a vocalization and not made with the wings. I then experimentally manipulate vocal structures in some males so that they cannot vocalize to test the null hypothesis that there is no difference in sound production between males that have been manipulated and those that I have handled but not operated on (controls).

I also notice variation between different males in where (treetops versus ground) and when (dawn versus midday) they perform courtship behaviors. Why do males show this variation? I suspect it affects their ability to obtain a mate (working hypothesis). I decide on an appropriate experimental design and sampling methods to obtain valid data for testing the null hypothesis that these variables have no effect. Data collection is followed by statistical analysis and I conclude that both where and when males perform their courtship behaviors have a significant effect on the males' ability to obtain a mate.

Procedures used to study animal behavior follow traditional scientific methods, but they vary greatly depending on the types of questions we are attempting to answer. The methods employed should be efficient, effective, and valid. As in any endeavor, the results of an animal behavior study are only as good as the methods used to obtain them.

or because other important pieces of information are missing. An animal may become more active on warmer days, for example, not because of an increase in temperature per se but because of an increase in biting insects or because of a decrease in humidity and the need to drink more water. Comparative studies of biological characteristics, including behavior, can be complicated by confusion or uncertainty between homologies versus analogies. Historical inferences are usually based on incomplete records and often have to be revised in the presence of new evidence. Modeling can be useful, but it depends on the correctness and precision of the relationships and assumptions that go into it; the relationships often are not well understood, and there usually are problems with assumptions. Thus these weaker approaches can be used to produce and test hypotheses, but they usually do not do a good job of it and cannot be trusted. However, they often are the only source

Figure 3-1   **A general picture of the scientific method.**

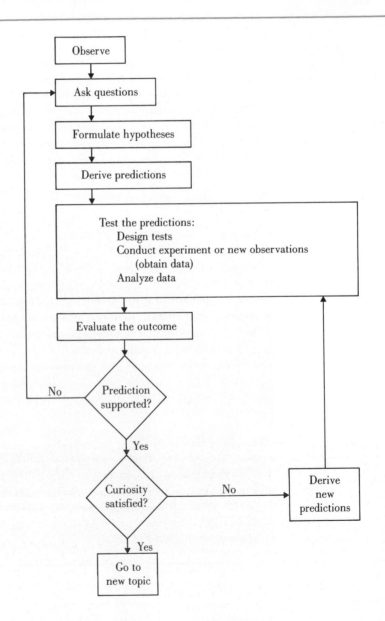

of inference available and frequently can be corroborated by other sources of information, including results of tests on related hypotheses.

Corroboration or confirmation and repeatability bring us to what the overall process of science is all about. Science is a large picture that is composed of smaller bits and pieces that overlap and merge with each other in stages, as illustrated in Figure 3-1. Depending on the outcomes at various stages, including whether or not outcomes match the predictions, the process may loop back to earlier stages. The observation and descriptive stages are usually followed by questions, hypotheses, and the next stages. If a hypothesis is correct, it should permit predictions that can be tested by experiments or, in weaker fashion, by the other techniques.

Working hypotheses are usually translated into null hypotheses, statements for testing purposes that usually say, in essence, that the hypothesized cause-and-effect relationship of interest is not true. The null hypothesis is something of a straw man, set up to be shot down.

Rejection of the null hypothesis produces a double negative ("reject a not-true statement"), which then supports the working hypothesis. That may seem like a confusing, backward approach, but it works statistically for reasons that are beyond the scope of this book.* Null hypotheses are rejected, that is, the results are said to be statistically significant, if the probability (P) of getting those results by chance alone under the null hypothesis is less than about 5% (or 0.05). To test the null hypothesis one designs an experiment or other approach, seeks money to cover the costs of conducting the work, collects the data, and analyzes the results statistically.

Rejecting null hypotheses and the complicated process of using statistics may seem unnecessary. But it is essential because of the variability and chance that exist in biology. Conducting experiments is much like playing a game of cards. If a deck of cards is shuffled and four cards are dealt, the chances for getting four aces are slim, but it still is possible. Just because a person plays cards day after day and never draws four aces in a row does not mean that it cannot happen. The experimental method is similar, and it is why even experiments—the strongest inferential technique known to humans—cannot always provide 100% certainty. One uses probability theory and an understanding of the "deck" to design the experiment and calculate the odds. Then if the observed outcome turns out to be unlikely, (e.g., $P < 0.05$) the results are said to be significant, the null hypothesis is rejected, and support is obtained for the hypothesis of interest. The importance of the technique is that it provides a standardized way of assessing inferences.

Skill in using and understanding experiments and the other scientific methods also depends on attitudes. Awareness that one might be making errors in inferences is one of these attitudes. Another attitude is to consider that important factors might have been overlooked in experiments. Skepticism of results is a golden quality in science. Skepticism not only helps detect errors, but it also leads to new questions and new hypotheses.

---

*Briefly, the reasons relate to the fact that the formal hypothesis that states that "a difference [effect] exists" (i.e., that the working hypothesis is true) usually involves a compound hypothesis. That is, there are many ways of being different, as opposed to a simple hypothesis of a specific difference. Because of that, it is difficult to determine the probability of a difference. It is much easier to assign probability to a statement of "no difference." As a result, for historical reasons with few people understanding why, it has become standard practice in the natural sciences to relegate the statement of a "difference" to the alternate hypothesis, H_a, and make the statement of "no difference" the null hypothesis, H_o, rather than the other way around. The α, or probability of a Type I error, can be assessed, presented, and discussed in the routine and familiar fashion whereas β, the probability of a Type II error, is just as routinely left to its own devices, out of sight and out of mind.

Basic scientific attitudes are acquired from the continued study of different sciences, (including behavioral research), from the study of history and philosophy of science, from reading original literature (including attention to methods), by associating with practicing scientists, and by engaging in science. Persons who want to become actively involved in conducting research on animal behavior can refer to the references at the end of this chapter or to the Appendix, which includes sections on how to read scientific papers and also how to write original scientific reports. The book, *Measuring Behavior* by Martin and Bateson (1986), is particularly recommended as an introduction to behavioral methodology.

We next analyze how the scientific method can be applied to a specific example of animal behavior: the question of why some animals flock or otherwise group together. We then present a number of general principles for the study of behavior.

AN EXAMPLE: DO BIRD GROUPINGS FUNCTION AS INFORMATION CENTERS?

The Anatomy of a Behavioral Hypothesis and its Testing

Almost all of the recent and some of the past study of behavior can be fitted somewhere into the general scientific scheme of observation, hypothesis formulation, and hypothesis testing. The first chapter used as an example the study of echolocation by bats. We now provide a recent example to illustrate the development and testing of a behavioral hypothesis. This example involves the question of whether bird groupings function as information centers, one of several hypotheses regarding why birds group together under certain circumstances. This example is relatively straightforward and typical and covers a relatively short history (approximately 40 years). Most of the literature is accessible so that further details of the story can be traced by anyone who is interested. This example is relevant not only for methodology, but also for the topics of evolution, foraging, social behavior, and learning, which are discussed in subsequent chapters. The developments are described mostly in chronological order, reflecting the sequence in which they occurred, except where some studies are more logically discussed together because of similar species or techniques.

Observation and Asking Questions

The fact that birds of many different species often group together in flocks, communal roosts, and colonies is a common and familiar observation. What has been less clear, however, is why they do it. What, if any, evolutionary advantages are there to such groupings? What ideas would you come up with to explain such behavior? Several hypotheses have been proposed through the years, including protection from predators, facilitation of reproduction, and improved efficiency in various aspects of life such as feeding and thermoregulation. Most of those hypotheses are considered in Chapter 13; for now we focus on one major hypothesis: that grouping can serve as a center from which individuals obtain information about food sources, similar to fishermen discovering good fishing places from each other.

Formulation of the Hypothesis

The general idea that grouping might somehow be advantageous for feeding can be traced to James Fisher (1954). He reviewed numerous observations going back much earlier, including an 1859 quote about soaring vultures observing distant neighbors' descents to quarry and following them down to the food. (Fisher [1954] described this as " . . . a 'stretched flock'—a great network of beaters spread to the limit of practical neighbor watching . . . ") Fisher did not attach a name to the idea and did not prove that there was indeed an advantage, remarking instead that " . . . the demonstration of this apparent truth seems . . . to derive from 'common sense' rather than from logical or formal proof."

A related concept called *local enhancement* had been around since at least the 1940s and early 1950s (Hinde and Fisher 1951, Thorpe 1951). The gist of this idea is that the presence of one or more birds feeding at a locality enhances the attractiveness of the site for other birds that see them, like fishermen on a lake congregating around other boats where fish are being caught.

Local enhancement, however, depends on direct observation of another animal. Ward (1965:211-212) clearly distinguished a difference between local enhancement and the new idea (which he did not name):

> "local enhancement". . . is practicable only over a limited area within which birds can see each other. It seems likely that the main function of the roost is to extend the benefits of this kind of feeding, so that social feeding may be practiced by a population together exploiting an area of hundreds of square kilometers. It seems reasonable to suppose that when the members of a roost fly out at dawn, their behaviour will depend partly on their success during the previous day. Those individuals which have left a good feeding place the evening before probably return to the same area, while those which have been less fortunate do not. It would obviously benefit the latter if, instead of going on a random search of new feeding grounds, they could simply join a group whose behaviour indicated that they were heading for an area where food was to be had. . . . It is difficult to prove this hypothesis, but there are a number of observations to support it.

Ward noted that the ecological nature of the food resource is important and must also be considered; social foraging involves food supplies that are unevenly distributed and sufficiently abundant to feed more than just the finder.

Ward was working in Nigeria with quelea weaverbirds (*Quelea quelea*, Figure 3-2), a serious agricultural pest and perhaps the most numerous bird in the world. Ward observed the behavior of the birds and their various assemblages (roosting, breeding, drinking), and analyzed their diets by collecting birds.

Henry Horn, conducting field work with Brewer's blackbird (*Euphagus cyano-*

Figure 3-2   Red-billed quelea *(Quelea quelea)*, a common and numerous bird in Africa, perhaps the most numerous bird of any species in the world, and a destructive pest of agricultural crops. **A,** Closeup of an individual bird. **B,** Large flock of many thousands of quelea leaving a roost at dawn on the Logone River in Chad, Africa. Photographs by J. Jackson (**A**) and M.T. Eliot (**B**). From Bruggers, R.L., and C.C.H. Elliott. 1989. *Quelea quelea: Africa's bird pest.* Oxford University Press, New York.

A

B

cephalus) in the state of Washington, referred (1968:690) to both Fisher and Ward, extended the ideas, and provided some direct but minor evidence:

> . . . I saw three cases in which a bird that had been foraging in the sagebrush followed a bird that had just come in from the water's edge with a bill full of food, waited until the latter had fed its young, then followed to forage near the edge of the water. Thus birds that are foraging successfully do communicate, albeit unintentionally, the location of better foraging to their less successful neighbors in the colony.

It seemed obvious to Horn that this was what the birds were doing, and he probably was correct, but three observations form a small sample size and there may have been other explanations, such as the movements merely being coincidence or the particular birds joining for some other unknown reason.

In a note published in England in the journal *Ibis*, Amotz Zahavi (1971) further reviewed published observations, added some of his own, contrasted antipredation versus feeding ideas, and referred (1971:108) to Ward's hypothesis, using the phrase *information centres* (English spelling), thus giving it a name.

Deriving Predictions and Attempts to Test Them

Knowing what to expect if a hypothesis is correct and thus how to test it is rarely easy. Usually there is a period of confusion in deciding what sorts of information are needed. A year after Zahavi named the hypothesis, Madhav Gadgil, in a note also published in *Ibis* and based on his observations of mixed flocks of birds in India, strongly disagreed with Zahavi. Gadgil argued for antipredator advantages to communal roosts. Gadgil (1972:531, 532) wrote, "If there were no positive advantage accruing to communally roosting birds through predator avoidance, it is difficult to see how such associations could have evolved." To make his case, he used mixed species groupings in which members of different species leave the flock independently and forage separately. He then extended his argument on theoretical grounds relating to reduced chances of being captured by a predator with increased numbers of individuals present (based on Hamilton 1971, discussed in Chapter 13).

John Krebs et al (1972) experimented with great tits (*Parus major*, birds that are closely related to North American chickadees) in captivity by using multiple hiding places for food on artificial trees in an aviary. The birds were more successful when searching for the hidden food in flocks of four than when restricted to searching singly or in pairs. The birds were in each other's presence while feeding, however, and thus the research involved local enhancement rather than the possibility of information centers separate from the site of feeding. Krebs subsequently (1973) extended the experiment to mixed (but closely related) species, with similar results to those in the 1972 paper.

In the next step of the story, Ward and Zahavi joined forces and published (1973) a major review on the hypothesis of information centers, using the phrase in the title of the paper. They summarized a mass of observations and proposed that the phenomenon was widespread and general (1973:517): " . . . we suggest that, with a few exceptions which we can explain, roosting and breeding assemblies of birds serve principally as information centres wherein knowledge of the location of food, or of good feeding sites, may be obtained by individuals temporarily lacking such knowledge." The hypothesis remained untested, however, merely serving as a means of interpreting the various observations. Different persons (e.g., Ward and Zahavi versus Gadgil) drew different interpretations, often from the same data, including each other's. Ward and Zahavi (1973), for example, discounted Gadgil's interpretations of Gadgil's own observations.

The search for a solution was on, and several more people became involved. At this point it might be worth stopping, thinking for a bit, and asking yourself, how would you go about proving or disproving the hypothesis of information centers? What kinds of observations would be needed involving wild birds, and can you think of logistically reasonable ways of getting such data? Can you imagine experiments that could be performed with either wild or captive birds?

Krebs continued working with the problem and directed his attention to the information-center hypothesis, not just local enhancement. He began working with a different species, the great blue heron *(Ardea herodias)*, in the wild in Vancouver, Canada. He observed (1974:99), "In spite of all this discussion in the literature, there is virtually no concrete evidence to show that sociality acts as a strategy for exploiting food resources. It was my aim in this study to collect field data which would test this hypothesis as directly as possible."

Krebs did several things. First he and assistants watched the flight paths of birds leaving the heronry. An observer in a blind inside the colony communicated by radio with another outside the colony, and they synchronized their records, by accurately recording times, with a third observer farther away to track individual birds. They analyzed 400 flights to different feeding areas, from different parts of the colony, using appropriate statistical tests, and concluded (1974:110) that birds were following neighboring birds. However, Krebs could only speculate whether the followers were more or less successful: "Some birds left the colony immediately after delivering food to the chicks while others flew to the edge of the colony and waited for a few minutes. It is tempting to speculate that the latter were unsuccessful birds." He also noted that, "In addition to following neighbors, the herons used at least two other methods of selecting feeding places. One was to fly over the feeding grounds and search for flocks (i.e., local enhancement) . . . and the other was to return to places where success had been high on previous trips."

Krebs attempted to determine whether local enhancement was occurring by experimentally placing model herons, made of molded and painted styrofoam, in different areas. He obtained a significant effect (Table 3-1, presented as an example of a good set of behavioral data). He also measured the rate of food intake of foraging adults under standardized conditions and observation times. Data were recorded

Table 3-1 Results of experiments in which model herons were placed in a tidepool. The figures are totals for 14 experiments, 7 with 1 model, and 7 with 5 models. Control data for the two groups did not differ significantly and are combined.	Number of Birds that Flew Over Model	Number of Birds that Landed in Pool	Percent Landing
Control	18	8	30.8
One model	6	11	64.7
Five models	4	31	88.6

Control versus one model, $X^2 = 4.8$, $P < 0.05$
Control versus five models, $X^2 = 21.6$, $P < 0.001$

One model versus five models, $X^2 = 4.19$, $P < 0.05$

From Krebs, J.R. 1974. Behaviour 51(1-2):99-131.

Figure 3-3 Rate of food intake of adult herons as a function of flock size. The data are grouped into blocks according to flock size purely to make the figure look neater; the analysis was done on the original points. The vertical bars represent standard errors of the means of grouped data.
From Krebs, J.R. 1974. Behaviour 51(1-2):99-131.

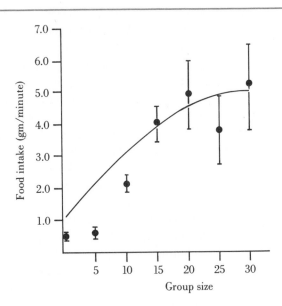

with a Dawkins behavior organ (Dawkins 1972), a small electronic organ that codes events as tones on a portable tape recorder. The tape later is decoded by a computer to give a record of events and the times at which they occurred. Food intake showed a highly significant relationship to flock size (Figure 3-3).

Krebs concluded that his results "suggest that birds follow each other from the colony to the feeding grounds, and particularly that neighbors tend to follow each other. Birds also locate good feeding places by looking for flocks on the feeding grounds and by returning to places where they have previously been successful. Thus the colony acts as an information centre."

Several other researchers followed with information from a variety of species and situations, with some supporting the information-center hypothesis, some not, and some being ambiguous. Helen Pratt (1980), for example, looked at the direction and timing of great blue heron foraging flights in a study that can be contrasted with Krebs'. Pratt observed local enhancement similar to Krebs, but she did not find evidence that herons were following each other from the rookery.

Erwin (1978) compared coloniality in different species of terns, correlating colony size with foraging conditions. He supported the information-center hypothesis but noted (1978:213), "While correlations may strongly suggest causal relationships, direct proof of the 'information-centre' hypothesis requires measures of food distribution patterns, evidence of communication, and comparisons of feeding success at varying group sizes. The first requirement presents great logistical difficulties for seabird researchers." Erwin later (1983) studied a mixed colony of six species of wading birds in North Carolina, observing flight directions, numbers of birds in arriving versus departing groups, and use of feeding sites away from the colony. He concluded that local enhancement was occurring but not information centers.

Emlen and DeMong (1975) proposed that colony members "pooling information" about food contributed importantly to the degree of synchrony of breeding in bank swallows *(Riparia riparia)*. Nestlings were more likely to become stunted or starved in nests that were later and less synchronized with the remainder of the colony, suggesting that they had less access to the social feeding information.

De Groot (1980) performed an experiment using captive birds, wild-caught quelea (the same species with which Ward originally worked and on which the hypothesis was based). Two adjacent aviaries were connected with four chambers and one-way tunnels and doors. A video system, light-sensitive diodes, timers, and a printer were used to record data and sequences of bird movements. Food of varying quality was placed in different locations, and naive birds were allowed to associate with knowledgeable birds. The somewhat confusing set of information that followed mostly supported the information-center hypothesis, but there were complications. In particular, naive birds followed the knowledgeable ones in only three of five trials. In the remaining two, the naive birds went first. De Groot discussed a number of possible explanations, including subtle cuing of information other than leading by knowledgable birds.

In an approach reminiscent of earlier studies on bee foraging (discussed in Chapters 7 and 15), some workers used feeding stations to investigate whether the first birds to use an area recruited others to the site. Loman and Tamm, working with hooded crows *(Corvus cornix))* and ravens *(C. corax)* in Sweden, put out dead pigs and piles of dead chickens. Their results were mixed. In some cases the numbers of birds increased as if information exchange were occurring. However, that did not occur all the time and, even when it did, they felt the results could also be explained by local enhancement or even by birds discovering the food piles independently. Andersson et al (1981) placed floating rafts with fish on them near a colony of black-headed gulls *(Larus ridibundus)* in a lake in Sweden. The gulls did not follow successful foragers from the colony, hence they refuted the information-center mechanism. Instead gulls left the colony and were then attracted to others that were already foraging, another example of local enhancement. Roger Evans (1983), studying black-billed gulls *(Larus bulleri)* in New Zealand, agreed with Andersson et al.

Theodore Fleming used supplemental feeding in a study of roosting pied wagtails *(Motacilla alba)* near Oxford, England. He hand-casted a gallon of maggots (about 30,000 maggots) at each site, then observed the wagtails and their behavior. He was plagued by other species taking the maggots but did get some information. He concluded (1981:475), "Results of the food supplementation experiments do not support the information centre hypothesis: birds exposed to maggots did not attract followers the next morning."

Along a somewhat parallel line, Gary Nuechterlein (1981), working with western grebes *(Aechmophorus occidentalis)* on the Delta Marsh in Manitoba, extended the information-center hypothesis to include information on more than food: he demonstrated with playbacks of recordings that Forster's tern *(Sterna forsteri)* aerial warning calls are used by grebes to detect danger (Table 3-2, presented as an example of the data). (Gadgil, 1972:533, disputed the use of roosts as sources of information on food and did not refer to warning signals as information; nonetheless he suggested that the mutual response of birds to warning signals would be advantageous.)

The information-center hypothesis was also considered experimentally in the laboratory in a different group, mammals. Bennett Galef and Stephen Wigmore (1983), in a psychology department at a Canadian university, experimented with demonstrator and observer rats. They showed a clear effect on subsequent food

Table 3-2 Responses of nesting western grebes to playbacks of Forster's tern alarm calls.

Responses	Stimulus Playback Call	
	Alarm Call	Control Basic Call
Leave nest	10	1
Alarm posture	11	2
No alarm reaction	5	23
$X^2 = 25.2$, 2 d.f., $P < 0.001$		

From Nuechterlein, G.L. 1981. Animal Behaviour 29:985-989.

preferences among observer rats from olfactory cues on demonstrators after they fed at a food source removed from the observers.

Range Bayer (1982) critically reviewed the status of the information-center hypothesis, the various evidence and methods used, and alternate explanations, and concluded (in the paper's abstract), "The evidence supporting this hypothesis . . . is indirect and could result from behaviors other than information exchange. Further, information exchange may not be as important as other mechanisms whereby birds may more effectively exploit their food resources by nesting colonially." The evidence was circumstantial at best or involved artificial conditions. Bayer's final remark (1982:38) was, "To confirm the Information Center hypothesis and its importance, proponents must document information exchange directly and show that information exchange at the colony is not rare."

There was further scuffling pro and con on the hypothesis (e.g., Weatherhead 1983, 1987, Waltz 1987, and Kiis and Moller 1986 [with a time lag in publication; the 1987 papers were submitted before the publication of Brown 1986, who is mentioned next]), and there were a few earlier studies that have not been included here. The hypothesis, however, clearly remained open.

Latest Evidence in Support of the Hypothesis

At this time, no direct experimental test of the information-center hypothesis is known to the authors. However, substantial supporting evidence of a correlational nature has been obtained by Charles Brown (1986). He and a number of assistants studied wild cliff swallows (Hirundo pyrrhonota) in Nebraska from 1982 through 1985. They observed 167 swallow colonies with 53,308 nests. The colonies were in treeless open terrain, and the observers could visually track the birds with binoculars. They recorded the birds arriving and departing, whether they carried food or not, where they foraged, which birds followed which others, and when. The results were clear (Table 3-3): unsuccessful birds followed successful ones to the food sources, as predicted if colonies of this species are serving as information centers for feeding.

In an expanded presentation of this work, including additional data, Brown (1988) further confirmed the existence of information exchange at the colonies and also discussed the additional presence of local enhancement, of which he identified three categories (network foraging, foraging from perches, and patch use near colonies). Brown concluded (1988:791), "Benefits of social foraging of one form or another probably account for the evolution of virtually all spacing behaviour in cliff swallows." Part of this research received the Animal Behavior Society's W.C. Allee

Table 3-3 Whether success on the previous foraging trip influenced whether birds followed or did not follow others on the subsequent trip, and whether success on the previous foraging trip influenced whether birds were followed by others or were not followed by others on the subsequent trip.

Subsequent Trip	Previous Trip	
	Successful	Unsuccessful
Followed	524	1355
Did not follow	2610	454
	$X^2 = 1647.6$, 1 d.f., $P < 0.001$	
Was followed	1378	172
Was not followed	1756	1637
	$X^2 = 632.8$, 1 d.f., $P < 0.001$	

From Brown C.R. 1986. Science 234:83-85.

Award in 1984. Thus the information-center hypothesis has been strongly supported, at least for nesting cliff swallows in Nebraska. It probably applies to several other cases as well. But it certainly does not apply to all instances of bird groupings, and it has not yet been clearly supported by direct experimental manipulation. The issue is still not closed.

Conclusions

Several important points about the course and methodology of behavior research can be drawn from this analysis of the information-center hypothesis story. First, there is a sequence of ideas and events that can be traced from the initial generation and identification of the hypothesis, through confusion, controversy, and numerous approaches, using different species and techniques—including both field and laboratory (aviary) and experimental and observational techniques—until at least a partial resolution is achieved.

Some of the evidence is indirect and circumstantial, whereas other data provide stronger inferences. Statistical techniques of various types are needed to objectively interpret the data. (Null hypotheses and statistical details were omitted from the information-center story except for a few examples, Tables 3-1, 3-2, and 3-3, for the sake of readers without a background in statistics.)

Eventually in the pursuit of scientific questions and hypotheses, direct manipulative experiments are required to obtain the strongest inferences and tests. An experiment that provides the most logical and least refutable evidence (whether for or against the hypothesis) is usually referred to as a *critical test*. To the best of our knowledge, such a test has not yet been performed for the information-center hypothesis in birds. In another, closely related example, the question of whether honeybees use dance language information in foraging, a critical test was performed (Gould 1975). That story is described in Chapter 15.

This example of the information-center hypothesis illustrates that the formulation of ideas and designing of tests require creativity and other thinking skills, and that logistics usually present major challenges. Research usually is not easy, although it often is fun and can be exciting.

In addition, this example illustrates that most questions remain open-ended and the story continues. In 1987, Erik Greene published the results of a study of information centers in ospreys that was as convincing as Brown's, but had three new twists. First, the birds ignored information about prey that was not concentrated but used information about clumped prey. Second, although Brown's cliff swallows did not actively alert other colony members that information was coming (observers merely "eavesdropped"), Greene's ospreys actually performed special displays when coming back with food. Finally, unlike Brown's swallows (but like von Frisch's bees), members of the osprey colonies are relatives. As Alice (in Wonderland) would say, nature gets ever "curiouser and curiouser."

There is one aspect of behavior study that is not well illustrated in this example. Aside from the special flight displays used by the ospreys, the example did not discuss unique, more-or-less stereotyped movement patterns and the need to break behavior into component behaviors or units. Flying and foraging, at least at the level of description discussed here, are straightforward and easy to work with. In many cases, however, it is necessary to look more closely at particular movement patterns. We consider that next, along with some general principles about behavioral methodology that can be drawn from the subject as a whole, not just specific examples such as the information-center hypothesis.

DESCRIPTION AND CATALOGING OF BEHAVIOR

The Units of Behavior

Before one can talk about or analyze parts of behavior, the behavior must be described and named. An animal's total behavior can be broken into a number of smaller components, much like dissecting the physical body into systems and parts. The process is largely arbitrary, however, similar to the classification of species in taxonomy. Dissection and naming of behavior are for the convenience of humans, not the animals', and there is not always agreement among observers of behavior on how best to divide a stream of behavior into parts.

Behavior involves various body parts such as limbs and individual muscles and neurons. From the most reductionist viewpoint, behavior is nothing more than sequences of movements or other changes of appearance in these individual parts. It is possible even in the extreme, however, to recognize and classify distinct patterns. The minimum identifiable units of behavior do not have a widely accepted name. The classical ethologists called them *action patterns* (Chapter 2). Ellis (1979) named the smallest units *ethons*, a name that has not yet gained widespread use. For simplicity and generality (including learned patterns) and to follow Lehner (1979), these units are referred to as *behavioral acts*. Sequences of acts usually are called *behavior patterns* or simply *behaviors*. Because of variability of behavior and a degree of arbitrariness in identifying specific behaviors, all persons do not necessarily recognize the same acts in the same animals. Nonetheless, consistency in recognizing units of behavior is generally quite high, particularly among persons familiar with the behavior. Hansen (1966, cited in Sackett 1968), for example, demonstrated high consistency between observers in a study of rhesus monkey behavior. Inexperienced observers required as little as 7 to 10 days of experience before they could reliably identify rhesus behavioral acts and properly record them.

Hinde (1970) lists two basic methods of naming individual acts: by physical description (such as low-pitched whistle, knee jerk, wing snap, head up, tip up) or by function or consequence (such as alarm call, sleep posture, food peck, aggressive jab). Naming by consequence may be more descriptive than its physical expression and may distinguish between similar motor patterns (such as food peck and aggressive jab), but the name includes an element of interpretation, which may inadvertently affect subsequent hypotheses and interpretations, blinding the observer to other alternatives. These names are also susceptible to overinterpretation. A learned peck response, for example, may or may not involve learning.

Once behavioral acts have been named, they can be classified, that is, grouped together with other behaviors. Hinde (1970) lists three ways of classifying: (1) immediate causation (e.g., agonistic fighting), (2) function (e.g., territorial threats), and (3) origin (e.g., those that originate from similar groups of muscles or have similar evolutionary origins). The first two categories may be quite similar when the cause and function are closely related.

Acts may be grouped into ever-larger collections. Several distinct acts may be involved in sexual behavior. Both sexual and parental categories can be grouped under the heading of reproductive behavior.

The list of an animal's entire behavioral repertoire, or at least a major segment of it, is called an *ethogram*. The number of acts that one observes in an animal's repertoire generally depends on three major factors: (1) the number the animal actually possesses, (2) how rare particular acts are, and (3) how much time one spends observing. The longer one watches, the more likely rare acts will be seen for the first time. Common, previously recorded acts will also be seen over and over again. The net result of this is seen in Figure 3-4. Fagen and Goldman (1977) provide a discussion and practical techniques for determining the amount of observation required for different species under different conditions.

Behaviors can be grouped, or the ethogram divided, into natural, logical categories based on function. A list of these general categories includes maintenance, feeding, orientation, and navigation, a number of interspecific (e.g., symbioses, predator-prey), and several categories of intraspecific social behaviors. The chapters in Part Two of this book are organized on the basis of these major categories. Figure 3-5 illustrates several behavioral acts involved in maintenance behavior of ducks. The entire duck repertoire is too large to illustrate here. A list indicating the size

Figure 3-4   Relationship between behavior observed and time spent observing.

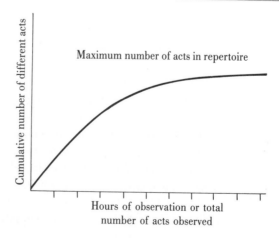

Figure 3-5 A few acts in the behavioral repertoire of the mallard duck. Each act depicted here has been given a specific name. Acts are associated with **A**, bathing **B**, shaking and **C**, nesting. A sequence of acts for oiling behavior is shown in **D**. A, C, and D from McKinney, F. 1965. Behaviour 25:120-220. Drawings by Peter Scott. B from McKinney, F. 1975. In: E.S.E. Hafez, editor. The behaviour of domestic animals, ed 3. The Williams & Wilkins, Baltimore.

of the ethogram for another group of animals, rodents, is presented on pages 82 and 83. For persons wanting other examples of ethograms, see Brown and Veltman (1987, magpie and crows), Jordan and Burghardt (1986, black bears), Leonard and Lukowiak (1984, 1986, sea-slug mollusks), and Schleidt et al (1984, quail).

There has been some discussion about standardizing ethogram methodology. Schleidt et al (1984) proposed coding observed behavior patterns into feature variables, each of which can assume only two or three states, and measuring orientation in reference to a 26-sided solid. They developed and refined the system for blue-breasted quail (*Coturnix chinensis*). Such a system could make ethograms more quantifiable and perhaps even permit their analysis by computer. However, it also would make them less understandable to other persons and less likely to be used than the more familiar verbal descriptions of behavior. Leonard and Lukowiak (1985)

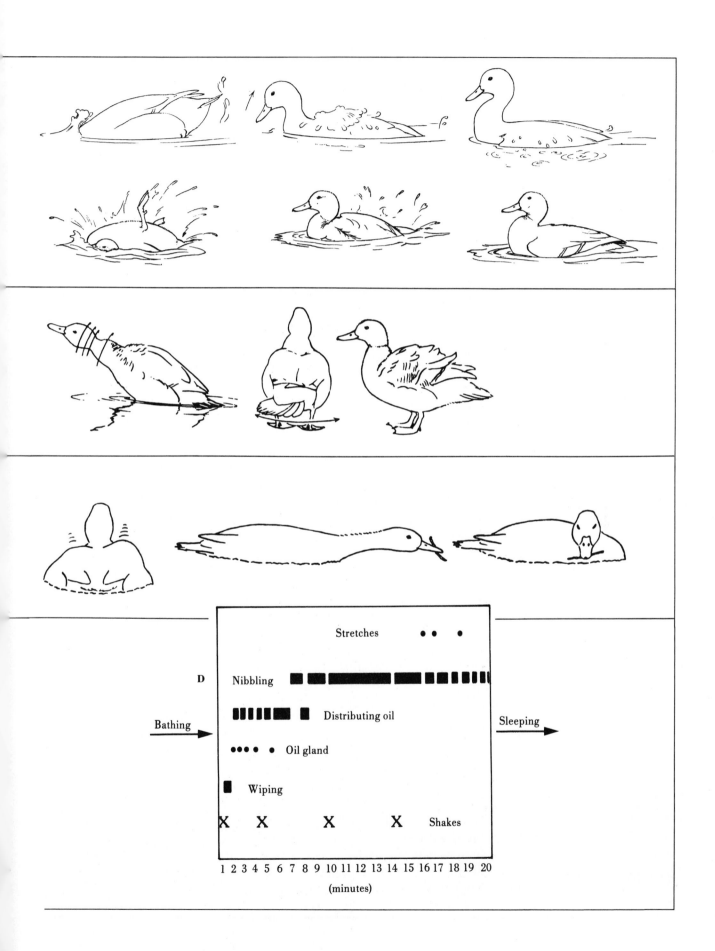

Rodent Ethogram

General Maintenance Behavior

Sleeping and resting
  Curled
  Stretched
    On ventrum
    On back
  Sitting

Locomotion
  On plane surface
    Diagonal
    Quadrupedal saltation
    Bipedal walk
    Bipedal saltation
    Jumping
  Climbing
    Diagonal coordination
    Fore and hind limb alteration
  Swimming

Care of the body surface and comfort
  movements
  Washing
    Mouthing the fur
      Licking
      Nibble
    Wiping with the forepaws
    Nibbling the toenails
  Scratching
  Sneezing
  Cough
  Sandbathing
    Ventrum rub
    Side rub
    Rolling over the back
    Writhing
  Stretch
  Yawn
  Shake
  Defecation
  Urination
  Marking
    Perineal drag
    Ventral rub
    Side rub

Ingestion
  Manipulation with forepaws
  Drinking (lapping)
  Gnawing (with incisors)
  Chewing (with molars)
  Swallowing
  Holding with the forepaws

Gathering foodstuffs and caching
  Sifting
  Dragging, carrying
  Picking up
    Forepaws
    Mouth
  Hauling in
  Chopping with incisors
  Digging
  Placing
  Pushing with forepaws
  Pushing with nose
  Covering
    Push
    Pat

Digging
  Forepaw movements
  Kick back
  Turn and push (forepaws and breast)
  Turn and push (nose)

Nest Building
  Gathering
  Stripping
    Biting
    Jerking
    Holding
  Pushing and patting
  Combing
  Molding
  Depositing

Isolated animal exploring
  Elongate, investigatory
  Upright
  Testing the air
  Rigid upright
  Freeze (on all fours)
  Escape leap
  Sniffing the substrate
  Whiskering

By permission of the Smithsonian Institution Press from *Proceedings of the U.S. National Museum*, Volume 22, No. 3597, "A comparative study in rodent ethology with emphasis on evolution of social behavior, I" by J.F. Eisenberg, Smithsonian Institution, Washington, D.C., 1967.

Rodent Ethogram—cont'd

Social Behavior

Initial contact and contact promoting
  Naso-nasal
  Naso-anal
  Grooming
  Head over-head under
  Crawling under and over
  Circling (mutual naso-anal)

Sexual
  Follow and driving
  Male patterns
    Mount
    Gripping with forelimbs
    Attempted mount
  Copulation
    Thrust
    Intromission
    Ejaculate
  Female Patterns
    Raising tail
    Lordosis
  Neck grip
  Postcopulatory wash

Approach
  Slow approach
  Turn toward
  Elongate

Agonistic
  Threat (proper) (remains on all four legs)
  Rush
  Flight
  Chase
  Turn away
  Move away
  Bite

Agonistic (continued)
  Locked fighting (mutual)
  Fight (single)
  Defense (on back)
  Side display
    Shouldering
    Sidling
    Rumping
  Uprights
    Upright threat
    Locked upright
  Striking, warding
  Sparring
  Tail flagging
  Kicking
  Attack leap
  Excape leap
  Submission posture
  Defeat posture
  Tooth chatter
  Drumming
  Pattering (with forepaws)
  Tail rattle

Miscellaneous patterns seen in a social context
  Sandbathing
  Digging and kick back
  Marking
    Ventral rub
    Side rub
    Perineal drag
  Pilo-erection
  Trembling

considered a standardized code to be a "two-edged sword." They tried the system of Schleidt et al on the mollusk *Aplysia californica* and found that it worked even for something as different from quail as this species. However, they wanted the information to be more accessible in human terms and subsequently published a typical descriptive ethogram (Leonard and Lukowiak 1986). They recommended (1985) that the proposed coded approach be considered only for cases where a more traditional ethogram exists first.

Behavior may be described not only via discrete acts but also by measuring continuous aspects of an animal's appearance, such as its posture, angles of one

Figure 3-6 Examples of descriptions and measurements of changes in animal behavior along a continuum, as opposed to discrete changes.
A, Darkening (from background color to black) in various regions of body of male guppy during courtship. CA, Copulatory attempts; S, sigmoid postures; S_i, sigmoid intention movements; 1, overall body darkening.
B, Measurement of body posture changes in hawks.
A from Baerends, G.P. et al. 1955. Behaviour 8:275. B from Grier, J.W. 1968. M.S. Thesis. University of Wisconsin, Madison.

Location on body

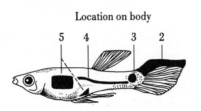

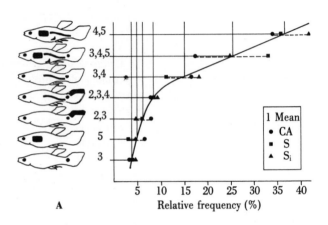

A

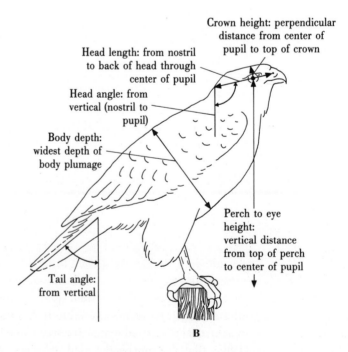

B

body part to another, positions of various limbs, or gradual changes in color (Figure 3-6). Furthermore, some action patterns may intergrade from one to another in a continuous rather than discrete fashion (Bond et al 1985). The time aspects of behaviors may also be considered, including the frequency and duration of particular behaviors.

COMMON PROBLEMS, NEEDS, AND PRINCIPLES IN THE STUDY OF BEHAVIOR

Any science is faced with various challenges and hindrances, such as logistical difficulties. The following five challenges are not unique to the study of behavior, but they are definitely of major importance. These need to be considered in virtually all behavioral studies.

Becoming Familiar with the Species

Perhaps the most important, basic consideration for studying animal behavior is how familiar a person is with the overall behavior and biology of the species being studied. Regardless of which aspect of methodology considered, the degree of familiarity with the species is critical. In fact, if there is a central dictum for the study of behavior, it is to "know thy animal."

Whether or not an ethogram has been compiled, general subjective understanding and familiarity with the species are required. Familiarity is gained by reading about the species and closely related species and by personal observation. Because of variability, the number of behaviors, interactions, and relationships involved and because many behaviors are shown only rarely, many hours of observation are generally required. Full-scale behavioral studies consume much more time than most realize. Wilson (1975) cites several examples, including a study by Dane et al of courtship displays in golden-eye ducks that used 22,000 feet of film, a study of Serengeti lions by Schaller that involved 2900 hours of observation, and a study of one troop of olive baboons by Ransom with 2555 hours of observation. Lindauer (1952), during one of several studies of bees, logged a total of 176 hours, 45 minutes watching a single worker!

Familiarity with a species is further enhanced by learning what is known from other sources such as publications, meetings, and correspondence or other contact with other persons working with the species. Most professional and serious amateur students of behavior spend much time reading, writing, attending meetings, and staying in contact with their colleagues.

Hence, whether involved in the study of behavior oneself or considering another's work, one must ask about the underlying familiarity with the species. The true experts generally have spent a considerable portion of their lives observing and working with particular animals, as well as reading and communicating with others on the subject.

Ethical and Legal Considerations

As indicated earlier (Chapter 2) and also discussed extensively in Chapter 22 and the Appendix, one is not at liberty to do whatever one wants with animals. For a variety of safety, ecological, ethical, regulatory, and legal reasons, there are numerous restraints on how, when, and where animals may be studied and who is qualified to work with the animals. Increasingly, these restraints have been restricting all aspects of animal study, including the study of behavior.

Whatever the laws and their reasons and origins, any researcher or educator working with any species of animal (and, for ecological, safety, and some ethical reasons, some species of plants) must (1) be aware of and adhere to the applicable, existing regulations and guidelines, (2) consider the subjects' welfare and potential pain and suffering, and (3) use common sense and discretion. This is an area of heated debate with proponents ranging from moderate to extreme on all sides of the issues. Philosophical viewpoints, conditions, regulations, and guidelines are likely to change and remain dynamic for some time to come.

Observing without Interfering

Observational or experimental artifacts and unintentional interference with the outcome plague most sciences. These problems are especially acute and pernicious in the study of behavior. The problem is that most animals, via their various senses and nervous systems, may be as aware of an observer as the observer is of them. Most vertebrates and many invertebrates are easily frightened, threatened, distracted, or otherwise interrupted from their normal activities. They generally respond to the presence of a human or a change in surroundings by attempting to escape or hide (Figure 3-7), becoming immobile, otherwise acting abnormally, or at least diverting a proportion of their attention to the intrusion.

Bruce Moore and Susan Stuttard (1979) uncovered and reported an example of observational interference that they called *tripping over the cat*. The case involved a study of cat behavior by E.R. Guthrie and G.P. Horton that was reported in *Cats*

Figure 3-7 A, Screech owl in normal undisturbed posture and **B,** in stiff-upright "freeze" posture, indicating that it is aware of presence of an observer. Photographs by James W. Grier.

Figure 3-8 Cats rubbing a vertical sensor rod in response to the presence of a human observer. During the 1940s similar behavior shown in a puzzle box was interpreted as evidence for learned responses. From Moore, B.R., and S. Stuttard. 1979. Science 205:1032. Photographs courtesy Bruce R. Moore.

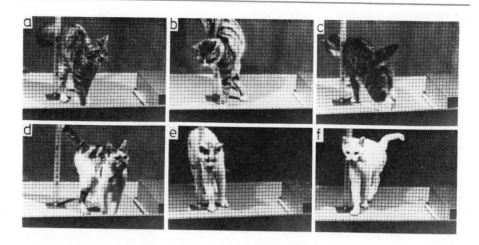

in a puzzle box. Guthrie and Horton described a box in which they supposedly conducted operant-learning experiments. However, the box contained a vertical rod that served as an ideal but unrecognized (by Guthrie and Horton) target for redirected (Chapter 2) rubbing. Many types of cats, including domestic cats, have a species-typical greeting reaction called *flank rubbing, head rubbing,* or *kopfchengeben*. The response is shown toward conspecifics and humans, which may be treated as con-specifics. The behavior involves arching the back, raising the tail, and brushing past or rubbing the head on the subject being greeted (Figure 3-8). If the subject cannot be reached easily, the cat will redirect the behavior toward a nearby object. The observers in the original experiments sat in full view of the cats, which triggered the rubbing. Moore and Stuttard repeated the work but controlled for the presence or absence of the observer and showed that the cat behavior resulted from observer presence rather than the supposed operant conditioning.

Two general techniques have been used to solve or reduce the problem of observer interference, both under natural conditions in the wild and for animals in captivity: (1) Hide or otherwise avoid detection by the subjects and (2) accustom the subject to the presence of the observer, equipment, or holding facilities.

Hiding can be accomplished in many different ways, such as by using blinds and barriers or observing from a considerable distance. Viewing can be aided with slits or other small openings, one-way glass or plastic, devices such as closed-circuit television, or, if at a distance, binoculars or telescopes. For sound there are many physical or electronic recording and "bugging" devices.

Working openly and often right in the midst of the animals is possible with many species that either ignore humans in the first place (an uncommon situation) or become accustomed to the presence of the observer. This type of observation has received much publicity, for example, Jane Goodall's work (1986) with chimpan-zees. Becoming accepted by the subjects to the point where the observer is ignored and the animals go about their routine activities generally requires much time, patience, and expertise.

Further, practical discussions of this issue are provided in the Appendix.

Gary Nuechterlein

Deborah Buitron

Gary Nuechterlein and Deborah Buitron

On Getting Close: Field Observation Methodology

Observing animals in the field under natural conditions can be a challenging endeavor. The act of observing is, in itself, an outside influence that easily causes a change in behavior.

There are several ways to deal with this problem. Spotting scopes and binoculars are essential in many studies. They enable the observer to remain at a distance that does not threaten the animal. Another way is to simply be persistently present in the study area. This time-consuming technique, made famous by Jane Goodall in her studies of wild chimpanzees, enabled her to eventually be accepted as just another "rather peculiar" troop member. We have used this technique in our studies of magpies, which soon learned to recognize our blue station wagon and showed only curiosity, not fear as we watched through scopes from the same spot, day after day. Betray that trust but once, however, and months of work may be lost. Only by borrowing a friend's car and dressing up with disguised face and a costume could we safely disturb the birds to check on eggs or nestlings without causing later impacts on our observation studies.

Perhaps the most common solution to overcoming the fear reactions of animals is the use of a blind that camouflages the presence of the researcher. Probably the biggest breakthrough in our 15 years of grebe studies came with the realization that these wary waterbirds could be approached easily by using small, mobile, floating blinds camouflaged to look like muskrat houses. These blinds consist of a cattail-covered wire dome constructed over a doughnut-shaped plywood platform and truck inner tube, which supports the observer and equipment. Clad in waders, one sits in a sling inside the "doughnut hole," waist-deep in water.

On our maiden voyage with these contraptions, we entered a breeding colony and, through slits in the sides of the dome, saw grebes on their floating nests eyeball to eyeball, barely at arm's length. Overnight, this blind freed us from the shackles of observing from a distance, which had been imposed by remote stationary blinds that somehow never seemed to be quite in the right position.

Identifying and Locating Individual Animals

The individual identity of animals is perhaps more important to the study of behavior than in any other biological discipline because of variability in behavior, the importance of recording data for the correct animal, and the need to know which animals are interacting with whom. It is easy to become confused over who is who, particularly in free-ranging situations where different individuals resemble each other.

Techniques to identify, follow, and relocate individual animals seem to be limited only by the imagination. The methods that have been used are numerous and often ingenious. Again, much usually depends on the investigator's personal experience and familiarity with the species and its natural life history. Species that are rare or otherwise difficult to find can often be found readily and routinely in their natural habitat by someone who knows how and where to look. But because of the differences

Forster's terns, which previously acted as sentinels for the wary grebes, now actually landed atop the blinds, only inches from our heads and, for hours, guarded their newly gained turf from intruders.

Suddenly we could see the world more from the animal's own point of view. We could see an incubating male cock his head as his mate called from a distance or see him bristle and, with a fixed stare, stop an intruder in his tracks, without so much as leaving the nest. The feeling of such close contact with our birds was exhilarating, and we realized the many subtle behaviors that were being missed by observing only from a distance.

One such mystery involved the grebe's unusual habit of swimming around with their young nestled beneath their back-feathers. Somehow the back-brooded young could always sense when a feeding parent was approaching, and the feathers would erupt with begging chicks even before he arrived. We learned their secret: the approaching parent gives a soft clucking call that the young had previously learned to associate with food. By experimentally playing the call at close range from the muskrat house blind we could get the chicks to pop out like little jack-in-the-boxes.

We had also always wondered why such back-brooding parents frequently would rear up and dump their chicks into the water, only to then allow them to immediately climb back on again. Were these their first swimming lessons? But if so, then why did they not last longer than a few seconds? The explanation turned out to be a simple matter of housekeeping: on hitting the water, the chicks instantly defecated. This is not the sort of observation that is easily seen through a spotting scope!

With each new field season, we develop a new muskrat house model as we incorporate the ideas of field assistants or graduate students using the blind. Our fleet now ranges from compact, streamlined, collapsible models to spacious "clunkers" that might be better termed *beaver-lodge blinds*. A wildlife photographer recently suggested replacing the puncture-prone inner tube with styrofoam. This innovation enabled us to enter via a hinged door, which certainly beats lifting a dripping cattail dome over your head. Now if we can only do something about the grebes, which each year seem to become more fond of stealing nest material from the sides.

between species and the length of time required for a person to gain experience, few people ever become intimately acquainted with more than a few species or groups of animals.

Individual animals are recognized by two primary means: their own unique individual characteristics and the use of artificial, applied marks or markers. Examples are numerous in either case; only a few will be given here. Natural individual differences in appearance have been used, for example, in porpoises *(Tursiops truncatus)*, dikdiks (Figure 3-9), and several primates. Individuals may be identified even through differences as subtle as vibrissae patterns in lions (Pennycuick and Rudnai 1970). Applied markers include bands, tags, paint, physical alterations—such as toe, tail, ear, scale, shell, hair, and feather clipping—and radio telemetry. Coding can be numerical, via color combinations, or through other, basically digital,

Figure 3-9 Individual markings in the ears and horns of dikdiks. From Hendrichs, H. 1975. Zeitschrift für Tierpsychologie 38:55-69.

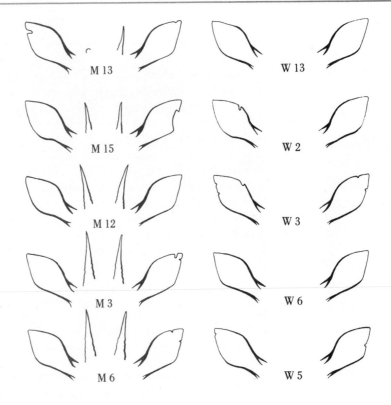

M 13   W 13

M 15   W 2

M 12   W 3

M 3   W 6

M 6   W 5

Figure 3-10 Examples of applying artificial markings to animals for purposes of individual identification. **A,** Clipping the edges of feathers on the wing of a martial eagle. **B,** Tagging the wing of a monarch butterfly with a numbered label. A courtesy John Snelling. B courtesy Fred A. Urquhart.

A   B

forms. Examples of artificial marking are shown in Figure 3-10. However, one must be sure that the markings are neutral to the animals. Burley et al (1982) found that leg bands of different colors increased or decreased the attractiveness of zebra finches to members of the opposite sex!

Radio telemetry deserves special mention. Modern electronic technology has provided access to methods for gathering information never before possible. Through radio telemetry (Figure 3-11) one can relocate animals where they otherwise might be impossible to find, and one can follow subjects over long distances and periods of time, often at a considerable distance. In addition to simple location information, sophisticated telemetry instrumentation can provide other data, such as on types and amounts of activity, heart rate, respiration and oxygen consumption, body temperature, and other behavioral and physiological characteristics. Overall, radio telemetry has proved quite useful in the study of animal behavior. Perhaps its biggest drawbacks are that it generally requires much time and effort, not to mention cost, to equip and follow the animals.

When animals with large ranges, such as birds, are marked, there is potential for overlap between different researchers and subsequent confusion and conflict. Bird banding on different continents has been largely standardized and coordinated through government agencies. Attempts to coordinate and control auxiliary markers, such as color markers on birds, depend on the cooperation of the researchers involved and have been partially successful. There is a limit to the information that can be coded reliably and read from applied markers. Researchers must coordinate about who should be coding for what and where, particularly in those instances where the animals may travel across the boundaries of research projects. None-

Figure 3-11 Transmitter weighing 3.0 g on the back of Swainson's thrush *(Catharus ustulatus).* From Demong, N.J., and S.T. Emlen. 1978. Bird Banding 49:342-359. Photograph by Stephen T. Emlen.

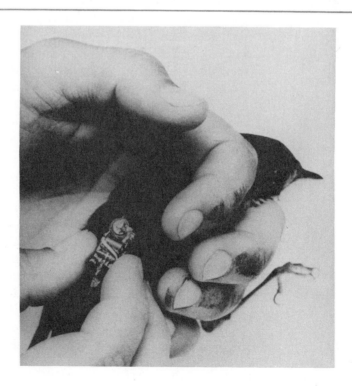

theless, applied individual markers provide the only means to get some types of information and remain a basic tool in much behavioral research; a good example involves monarch butterfly studies as discussed in Chapter 7.

Ingenuity and Technology

Science is a creative activity that relies on ingenuity and technological innovation. Both have played important roles in the study of behavior. The importance of technology means that there is a need for interdisciplinary approaches to the study of animal behavior. Persons from the traditional disciplines in biology and psychology frequently become involved with others from engineering and computer science. As a result, students and faculty from these various disciplines need to become more familiar with the other disciplines.

Techniques to locate and identify individual animals have already been mentioned. Other advances involve the expansion of one's senses and improved methods for recording data. The bat and moth interactions in Chapter 1 provide an introduction to these principles.

There are a great many devices that increase the sensitivity of one's seeing and hearing, shift sights and sounds into the range where one can detect them, render other sensory modes into visual or auditory format, and record sensory information for later or repeated analysis. Seeing is aided by binoculars and various telescopes and by a number of night-viewing devices and closed-circuit television. For hearing there are microphones or hydrophones (for underwater), amplifiers, quality headphones and speakers, and intercoms.

Events that occur too rapidly can be stopped by high-speed photographic techniques or by recording them at normal speed on film or magnetic tape and replaying them at slower speeds. At the other extreme, with events that occur slowly, the recordings are often speeded up, as in time-lapse photography or the speeded playback of whale songs. Special camera lenses, filters, and film or electronic photosensors allow one to obtain still or motion pictures or televised pictures in the infrared and ultraviolet ranges.

One of the most basic techniques for dealing with nonvisual items is to render them in visual format. Sounds can be converted to sound spectrograms and laid out in a two-dimensional format for the analysis of patterns, as was done in Chapter 1 for bat and human vocalizations. Electrical and magnetic fields can be plotted two or three dimensionally. Chemicals are analyzed with various sorts of chromatographs. Sensory data of many kinds are depicted in line, bar, and other types of graphs.

Not all persons have access to, funding for, or desire to use advanced technology. But nearly everyone has or can develop his or her imagination. In some instances the ingenuity in methods may be as interesting or more so than the actual results of the behavioral research. The ingenuity shown in behavioral research, with either simple or advanced equipment, has been remarkable in many cases.

Stephen T. Emlen, for example, devised several creative techniques to study bird migration. In one study where flying birds were to be followed at various altitudes, Demong and Emlen (1978) lifted birds in boxes suspended from balloons. The floors of the boxes were released by fuses cut to different lengths, depending on the time the balloon needed to rise to a particular height (Figure 3-12). After the birds were released at that height, they were tracked by specialized radar at

Figure 3-12 Diagram of box suspended from a balloon used to release experimental birds on migration under controlled conditions.
From Demong, N.J., and S.T. Emlen. 1978. Bird Banding 49:342-359.

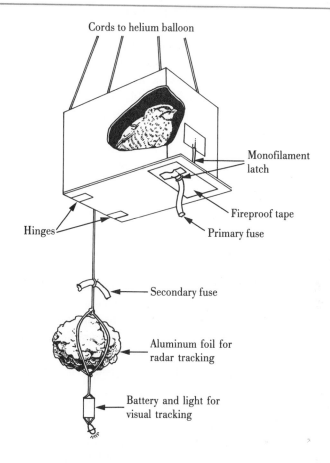

Cords to helium balloon

Monofilament latch

Fireproof tape

Primary fuse

Hinges

Secondary fuse

Aluminum foil for radar tracking

Battery and light for visual tracking

the Wallops Island NASA facility in Virginia. The arrangement included a small battery and light for visual tracking and a ball of aluminum foil for radar tracking until the bottom fell out of the box at the desired height. Then the radar picked up the bird and followed it. The radar was connected to computer facilities and permitted the flying birds to be tracked precisely in time and three-dimensional space. The radar equipment was available at times when it was not being used in NASA projects. Another classic example of ingenuity in technique is James L. Gould's (1975) manipulation of the sensitivity of bees to the presence of the sun to test their use of dance language information (see Chapter 15).

R. Stimson Wilcox (1979) used several ingenious techniques to study communication in water striders *(Gerris remigis)*. He used masks to block the striders' vision. Masks were created by painting black liquid-silicone rubber on the heads of dead striders, peeling off the cured masks, then using them on live individuals. Wilcox altered the frequency of water surface wave signals produced by the striders through the use of small magnets affixed to the striders' legs. By surrounding the setup with a wire coil, he could induce currents and force the striders to produce wave signals as desired for the experiments. From this, incidentally, he discovered that females and males produce different vibrations, and the male striders can use them to determine the sex of another individual, even when vision is blocked.

OBTAINING AND INTERPRETING DATA

Obtaining, analyzing, and interpreting data involves numerous routine and important details, but these will not be of concern to the beginning student or person reading about behavior. By the time information is published, it obviously has already been obtained, analyzed, and interpreted. However, those who publish information are not perfect, and even general readers need to be skeptical. Also, a general understanding of methodology, again, vastly improves one's appreciation of the information. Thus a broad overview is given here. Persons who want further detail can consult the references at the end of the chapter and the Appendix.

Sampling Behavior

How, when, where, and what behavior is observed depend in part on the underlying research questions being asked and the observer's familiarity with the species. When to observe presents additional problems. Even the most dedicated observer cannot spend 24 hours a day, 365 days a year watching all subjects.

Until recently the timing of behavioral observation could be described best either as "as much and often as possible" or as "hit and miss," depending on such factors as convenience and the amount of time an observer has available for observing. The name for such observation is *ad libitum* (Latin, at pleasure). This type of observation, however, presents a number of problems. Animals often display particular behaviors at regular intervals, which may or may not coincide with observation periods.

Depending on when observations are conducted, particular behaviors may be underrepresented or overrepresented, studies may not be repeatable, and it may be difficult or impossible to compare the results of one study with those of another. Most persons are familiar, for example, with the general springtime singing of songbirds. The birds are most active, vocal, and visible during the early morning hours around sunrise and again at dusk. An ornithologist who works only 9 AM to 5 PM will miss much of the birds' behavior. A study by someone who works for 2 hours at dawn and another 2 hours at dusk cannot be compared easily with a study by someone else who observes the species at midday, even if the total amount of time is the same.

Because of such problems as the lack of standardization with ad libitum sampling, improved statistical sampling techniques have been developed. They are described and compared by Altmann (1974), perhaps the most important methodological publication in the study of behavior.

The gist of statistical sampling of behavior is to use random sampling rather than subjective preferences to decide when and which subjects to observe. Within the overall time frame (e.g., 1 month or 3 years), one allots the amount of time available for observing into segments that are randomly dispersed over the total time the animals are to be observed. If there is more than one animal (or family, group, etc.) and all cannot be observed simultaneously, the animal or group to be observed at any one time is chosen at random. This procedure is called *focal animal sampling*.

The random allotments are chosen before observing begins; allotments may involve equal numbers in different cases to provide a balanced design; the initial design should include contingency plans in case problems develop and observations cannot proceed as desired. (Problems seem to be common when observing behavior.

Murphy's law applied to living organisms is that animals always do as they please.)

With random sampling, as with ad libitum observation, one does not obtain a complete record of every move that all animals make. However, random sampling should give a more representative quantitative picture than the ad libitum method. Furthermore, random sampling permits the application of standard statistical procedures. The time units are treated as sampling or experimental units, and one can tally frequencies of particular behavioral occurrences or measure proportions or lengths of times during the sampled time units for statistical purposes.

Data and Data Recording

The next methodological concern, whether one is making initial descriptive observations or seeking formal experimental answers to specific questions, involves the nature of the data and how to record them.

There are four main classes of data: (1) nominal (named) or categorical, such as male-female, adult-immature; (2) ordinal (from order) or ranked, such as position in a social hierarchy; (3) interval, where the size of the units is constant, such as degrees of temperature; and (4) ratio, where the units are of constant size and ratios are meaningful, such as in length or weight (e.g., 8 is twice the length of 4 m). The last two classes of data are often considered to be parametrical (with meaningful parameters such as mean and standard deviation) or metrical (strictly measurable), whereas the first two are usually considered nonparametrical.

Data in any particular class have all the properties of the lower classes but not vice versa. Something that is 8 m long, for example, not only is twice as long as something 4 m long, but the units are of constant size (hence of interval quality), 8 m is longer than 4 m (hence it is ordinal), and objects can be placed into named categories such as long-short (hence they can be nominal data). Data of lower levels, however, do not possess the properties of higher levels. An item that is male, for example, is not necessarily bigger, better, etc. (thus not ordinal), compared with female; and male is not a constant metric unit (hence not interval or ratio data). Different classes of data are measured, recorded, and analyzed in a number of different ways.

Good data recording can be done in many ways, depending on the particulars of the situation and the personal preferences of the observer. Some techniques are elegant in their simplicity. Emlen and Emlen (1966), for example, recorded the direction and amount of activity of restless indigo buntings by simply placing stamp pads in their cages and surrounding them with paper to catch the footprints as they jumped about (Figure 3-13). Contemporary methods are almost too numerous to mention. They range from the time-honored manual recording in data sheets to various magnetic tape recorders, mechanical chart recorders, computerized systems, and various high technology and automated techniques. Blough (1977), for example, devised an ingenious way to have pigeons record their own behavior directly into the computer (Figure 3-14). Computerized methods include using portable microcomputers (Hensler et al 1986) and even hand-held printing calculators (Ely 1987). One of the most recently developed techniques for recording data uses a portable bar-code reader to input universal bar codes (as used in product marketing) in real time from a selection of codes associated with particular behaviors. The stored information can then be directly loaded into a computer. (Interested readers can request information on the Time Wand from Videx, 1105 N.E. Circle Blvd., Corvallis, OR 97330-4285.)

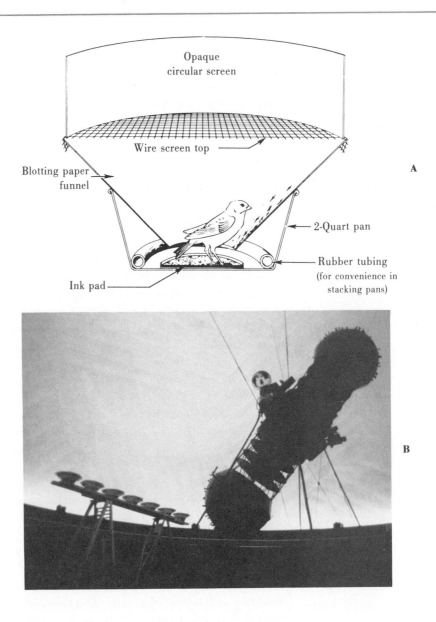

Interpreting and Presenting Behavioral Data

Once the behavioral observations have been made and recorded, what does one do with them? Ideally there was prior planning and understanding of why and how the data were gathered so that analyzing them will be almost mechanical and simply amount to plugging the facts into their proper places. In many cases observers proceed without design or planning and end up with answers for which there were no questions, that is, theory-free data; such information often presents an analytical quagmire that many statisticians refuse to touch.

Whether one has well-designed experimental results or just some kind of data hodgepodge, however, the methodology of data analysis and most of the statistical treatment are not unique to behavioral research but are more or less common to

Figure 3-14 Automated recording of pigeon pecking responses. In a study in which pigeons were to choose a small "o" amid an array of "x's" displayed on a television screen, responses were detected by a photocell attached to the bird's beak and sent directly to a computer. From Blough, D.S. 1977. Science 196:1013. Photo courtesy Donald S. Blough.

all branches of biology and other science; details are beyond the scope of this book, and only a few general comments will be made.

Because biological variability is a fact of life, an understanding of statistics is integral to a deeper understanding and pursuit of biology, including behavior. Most persons reading this introductory text probably will not yet have much if any exposure to formal statistics. Thus, in addition to keeping these comments general, we use this section to encourage all readers to learn more about statistics.

There are two general categories of statistics: descriptive and inferential. Descriptive statistics, such as the mean and variance, help describe the characteristics of a set of numbers. With inferential statistics, however, one attempts to infer, with the aid of various assumptions and particular tests (such as t tests, analyses of variance, chi square, and many others), how one group relates to another, whether there is any underlying natural structure or order, or perhaps whether the observed results can be accounted for by the inherent variability. If the latter is the case, then similar results might have been obtained by chance alone rather than because of some underlying effect of interest, as discussed previously.

A person rarely is interested in only one behavioral act at a time. Usually several must be considered. Three common categories of multiple-behavior analysis and data presentation are (1) comparisons of frequencies of different behavioral acts among different individuals (e.g., Tsingalia and Rowell 1984, monkeys), (2) analysis of behavioral interactions between two (or more) individuals (sociometric matrices) (e.g., Miura 1984, deer), and (3) sequences of behavior over time (Markov chains) (e.g., Robinson 1984, monkey vocalizations; Hailman et al 1987, chickadee calls). For introductions to Markov chains and related techniques and analysis, see Lehner (1979), Martin and Bateson (1986), and Bakeman and Gottman (1986).

Complex processes in biology, particularly in behavior, physiology, and ecology, usually include a host of interacting factors; that is, a particular result, such as how an animal responds to a predator or chooses where it will nest, may be an effect of several factors simultaneously and not just one or a few simple ones. Two basic statistical approaches for dealing with these processes are (1) experimental techniques whereby one or a few factors are deliberately varied in a known fashion while all others (it is hoped) are held constant and (2) a whole category of techniques known as multivariate statistics. Many things in nature occur together and are said to be correlated. They may occur together always, nearly always, or just a percentage of the time. When one occurs, the other occurs, and their possible single effects cannot be separated clearly. In a case used to illustrate this problem, there is a story about a person who studied crickets and noticed that they could be trained to jump every time he presented a certain sound. Then he removed all of the legs from a cricket, made the sound, and the cricket no longer jumped. He reasoned that it no longer jumped because it did not hear the noise; therefore the crickets' ears are located on their legs. (He was, in fact, correct but not because of his logic.)

As an illustration of more subtle but just as significant problem, imagine a pollster seeking to understand economic conditions in a community. He has a set of questions about such things as income and purchases. If he goes to the door of a house and asks the husband, he will obtain a certain amount of information. But what happens if he asks the wife the same questions? The answers may not be exactly the same, but it is reasonable to expect that at least some of the information will be the same and that little new information will be learned. One would expect the husband's and wife's answers to be highly correlated. If children in the family are asked the same questions, there probably would be less correlation because they know less about the family finances, but there might still be some correlation.

The problem of correlation is pernicious in science in general and behavioral studies in particular. As an example in behavior, consider a study that compares bird migration with measured weather variables. If only one variable, for example, wind direction, is considered, it may appear that migration relates to (depends on or itself correlates with) wind direction. But it could depend on a fall in air temperature instead, and a particular wind direction simply correlates with the decrease in air temperature. If several variables are measured, they all may simply yield similar correlated information. After the passage of a cold front in the northern hemisphere, for example, several things normally occur together: the wind changes to the northwest, temperature drops, cloud cover decreases, humidity drops, and barometric pressure increases. Depending on the amount of correlation and whether or not appropriate statistical techniques are used, one can have difficulty determining which, if any, single factor or set of factors is truly responsible for the bird migration.

The use of high-speed, modern computers has both helped and hindered the analysis and understanding of behavioral data. Computers have helped by permitting complex and advanced analysis in reasonable lengths of time and by making possible better graphical presentation. However, the speed with which the new techniques have proliferated and swept through biology and the ease with which data can be plugged into canned computer programs have led to two major problems. First, the complexity may become bewildering. Many persons have given up before trying to tackle such statistical situations. Second, it may become more difficult to interpret outcomes than to plug data into the computer, leading to questionable or illegitimate conclusions.

The bottom line is that one must always be cautious and skeptical, whether one produced a given set of information or only is reading about it, whether several potentially correlated factors are involved or not, and whether computers were used or not.

Figure 3-15 Graphic presentation of results from a multivariate (principal component) analysis of woodpecker nesting habitat. The positions of the five species relative to the first three components are shown in A and to the first two in B. The red-headed woodpecker *(R)* stands apart as unique and the most restricted in its use of habitat. The other species are downy *(D)*, hairy *(H)*, flicker *(F)*, and pileated *(P)*. From Conner, R.N., and C.S. Adkisson. 1977. Wilson Bulletin 89:122-129.

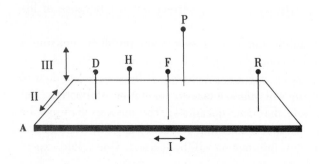

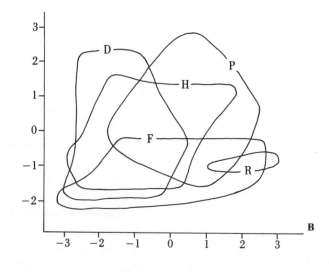

After data have been analyzed, the results normally are presented in a manner common to many areas of science. A variety of graphs and plots may be used, including three-dimensional forms. Some results of a multivariate study of woodpecker nesting habitat are shown in Figure 3-15.

SUMMARY

Information about animal behavior is intimately associated with the means by which it is derived. The contemporary study of behavior employs standard scientific approaches, including hypothesis formulation, testing, and inference by experimentation, comparative studies, historical analysis, and modeling. Verbal working hypotheses about cause-and-effect relationships are often translated into null hypotheses for the purpose of testing for statistical significance. Statistics are required because of natural variability in most phenomena of interest.

The hypothesis that social groupings of birds might function as centers for information about food sources illustrates the process and methodology of behavior research. This information-center hypothesis is difficult to pursue because of the logistical problems of working with the social behavior of birds foraging over large areas and also because of the large number of alternate explanations. This example illustrates the sequence of ideas, events, problems, and stages of behavior studies, including the fact that issues are rarely settled for certain.

Although not unique as a science, the study of behavior involves some important methodological considerations. These include identifying units of behavior (the list of which is called an *ethogram*), the need for basic familiarity with the species being studied, ethical and legal considerations, problems of observing or studying without interfering with the behavior, the importance of identifying and following individual animals, and the use of ingenuity and technology.

Other important aspects of behavioral study include a variety of sampling and data recording techniques, statistical analysis, and the interpretation of what it all means. As in any science, one must be cautious and skeptical of outcomes, whether the results are one's own or someone else's.

FOR FURTHER READING

General references for methodology

Altmann, J. 1974. Observational study of behavior: sampling methods. Behaviour 49(3,4):227-265.
The classic review of sampling techniques, a standard reference for all persons conducting behavioral observations.

Bakeman, R., and J.M. Gottman. 1986. Observing interaction: an introduction to sequential analysis. Cambridge University Press, New York.
A concise introduction to the collection and analysis of behavioral data, with emphasis on social interactions.

Brown, L., and J.F. Downhower. 1988. Analyses in behavioral ecology, a manual for lab and field. Sinauer Association, Sunderland, Mass.
This book includes 27 problems for laboratory and field inquiries, such as in behavior courses (but they are not classical labs because they have no "right" answer). Nearly half of the book, after the problems, is devoted to data analysis and statistical techniques.

Colgan, P.W., editor. 1978. Quantitative ethology. John Wiley & Sons, New York.
A collection of several papers on a wide range of topics related to methodology and analysis. The level of presentation ranges from introductory to moderately advanced.

Hazlett, B.A., editor. 1977. Quantitative methods in the study of animal behavior. Academic Press, New York.
A collection of papers similar in nature to Colgan's Quantitative Ethology *but with different authors and topics.*

Lehner, P.N. 1979. Handbook of ethological methods. Garland Publishing, New York.
An excellent, concise but thorough, and clear introduction to all steps of behavior research from planning through data recording, statistical analysis, and graphical presentation. Highly recommended for either an introduction to or a review of methods and techniques.

Martin, P., and P. Bateson. 1986. Measuring behavior. Cambridge University Press, New York.
Another excellent introduction to behavior research methods from design through data analysis. This book is accessible, paper bound, and also highly recommended, particularly for an introduction to methodology. It is more recent than Lehner but not as thorough and much less illustrated.

Zar, J.H. 1984. Biostatistical analysis. Prentice-Hall, Englewood Cliffs, N.J.
A recommended general introduction and reference for statistics. It is thorough, clearly written, and contains numerous examples.

A Videotape Program, with Accompanying Text/Workbook*

Hage, S.R., and J. Mellen. 1983. Research methods for studying animal behavior in a zoo setting. Minnesota and Washington Park Zoos. (Minnesota Zoo, Education Department, Apple Valley, MN 55124).
Two videocassettes and text covering behavioral observation and sampling techniques with practice examples for viewers. The content is for a zoo setting, but the material is useful for behavioral observation and research in general. This is an excellent and valuable resource for student and class use.

*Many of the videotapes and films on behavior, particularly from the PBS series *Nature* and *Nova* and syndicated series such as *National Geographic* and *Survival* specials, include segments on methods and techniques. A particularly good show for techniques is *Chemical Weapons*, on work by T. Eisner and his colleagues, from the *Nature* series.

Journals

Most of the numerous journals covering animal biology, particularly those on specific groups of animals such as birds, and in other areas such as ecology and physiology, now include many papers relevant to behavior. Methods sections are standard for virtually all such papers and provide specific examples of techniques and approaches to problems. The major journals in behavior are: Animal Behaviour, Behaviour, Behavioral Ecology and Sociobiology, and Ethology (formerly Zeitschrift für Tierpsychologie).

A selection of recommended readings involving the information center hypothesis (in chronological order)

Ward, P. 1965. Feeding ecology of the black-faced dioch *Quelea quelea* in Nigeria. Ibis 107:173-214.

Zahavi, A. 1971. The function of pre-roost gatherings and communal roosts. Ibis 113:106-109.

Ward, P., and A. Zahavi. 1973. The importance of certain assemblages of birds as "information-centres" for food-finding. Ibis 115:517-534.

Krebs, J.R. 1974. Colonial nesting and social feeding as strategies for exploiting food resources in the great blue heron *(Ardea herodias)*. Behaviour 51(1-2):99-131.

Pratt, H.M. 1980. Directions and timing of great blue heron foraging flights from a California colony: implications for social facilitation of food finding. Wilson Bulletin 92(4):489-496.

de Groot, P. 1980. Information transfer in a socially roosting weaver bird *(Quelea quelea;* Ploceinae): an experimental study. Animal Behaviour 28:1249-1254.

Bayer, R.D. 1982. How important are bird colonies as information centers? Auk 99:31-40.

Brown, C.R. 1986. Cliff swallow colonies as information centers. Science 234:83-85.

Brown, C.R. 1988. Social foraging in cliff swallows: local enhancement, risk sensitivity, competition and the avoidance of predators. Animal Behaviour 36:780-792.

4

THE BASIC GENETICS OF BEHAVIOR

An animal's behavior, like its anatomy and physiology, is a product of both its inherited genetic background and the environments in which it developed and now lives. Even the capacity to learn requires a genetic component. An appreciation of this is important for understanding the observed, proximate expression of behavior and the long-term, ultimate aspects. If there is any inherited component to behavior, then it is subject to natural selection and other evolutionary forces, as discussed in the next chapter.

This chapter describes the evidence for the role of genetics in behavior, establishes the basic principles of behavioral genetics, and explains how, in some cases within specific contexts, the genetic contribution to behavior can be measured. Other, more advanced topics in behavioral genetics and evolution, such as optimality, sexual selection, and evolutionarily stable strategies (ESSs), are treated in other chapters. Before focusing on behavior, however, we begin with some general background principles that are crucial to understanding the application of genetics to behavior.

ANIMALS, GENES, AND GENETICS: A BRIEF REVIEW AND BACKGROUND

Individual animals from the same species and even from the same parents (including identical twins) are different from each other. Variation is a rule and a fact of life in biology. This variation results partly from variation in the inherited instructions, located in DNA on the chromosomes received from the parents, and partly from variability in the environments surrounding different individuals. The environment contributes to the expression of DNA in complex ways through a variety of biochemical and developmental processes.

The instructions in DNA are basically analogous to the read-only memory of a computer: they may be copied for use, but they are not modified (except in abnormal situations such as with retroviruses). The original set of instructions stays protected on the shelf and is used only for making copies. The copies, molecules of messenger ribonucleic acid (mRNA), leave the nucleus and go out into the cytoplasm of the cell where they direct the assembly of proteins.

Proteins consist of various amino acids joined into specific sequences that determine the unique character of any particular protein. Different proteins form the basic structures of organelles, cells, tissues, and organs, including the brain and sensory organs. Proteins, in one way or another, also are responsible for the chemical activity within the nervous system. Enzymes, which catalyze and thus guide all of the body's metabolism, are largely protein.

Each cell contains the full set of instructions, similar to a complete library, for the entire organism. But any one cell at any one time uses only a small fraction of the available information in the process of manufacturing its protein; that is, different cells are doing different jobs at different times, and apparently they use only the relatively small pieces of information necessary for each job. Thus much of the DNA in an organism is being used for the synthesis of various proteins. But equally important for considering evolution, there are duplicate, unused sets of instructions that are set aside in the reproductive tissue (germinal epithelium, ovaries or testes). Depending on whether or not an animal survives until it can reproduce, and how often and how well it reproduces, copies of the DNA may or may not get passed on to the next generation.

The expression of information in DNA to achieve the final biochemical products is extremely complex, which is becoming more and more apparent with each new discovery. It involves a large variety of units and steps (i.e., atoms, molecules, and polymers, stages in electron transport, codons, introns, exons, nucleosomes, activator and repressor regulatory sites on the DNA, and much more). These interact with each other and with the surrounding environment.

Unfortunately, none of these units and steps correspond in any one-to-one manner with the phenotypical units (e.g., eye color, hair texture, behavioral traits) in which we humans are interested. The process produces whole organisms, not just collections of traits. Traits are arbitrarily defined entities that provide abstract handles for human understanding. Some of the units, particularly at the organ level such as eyes and hearts, might be relatively distinct, but even they are nonetheless arbitrary. An eye, for example, does not have absolute boundaries, depending on whether or not one includes the external muscles, optic nerve, and other associated tissues. The lack of one-to-one correspondence between any units or stages of the actual process of DNA expression and the arbitrary (human-defined) units of the outcome, that is, phenotypical traits, has led to much misunderstanding. The historical sequence by which we arrived at our present-day knowledge is largely responsible for this situation.

Early ideas about inheritance from at least Aristotle up through the nineteenth century included a theory known as *pangenesis*, which held that unknown particles called *pangenes* were distributed throughout the body and somehow came together to form the sperm and the egg. There were several other theories, including one that individuals contained an infinite series of miniature preformed offspring and another that parents' traits were "blended," like the mixing of paint, in their offspring. By the time Darwin proposed his theory of natural selection and Gregor Mendel worked with his garden peas, it was clear that something was inherited from one generation to the next. But whatever that something was, it was unknown and essentially unnamed. Neither Darwin nor Mendel knew about genes.

Mendel discovered some of the basic principles of heredity (segregation and independent assortment) during the 1860s, but they were not rediscovered and appreciated until the early 1900s. He referred to his units of heredity only in terms of factors. The name *gene* was first used by Wilhelm Johannsen in 1909. Then, during the first half of the twentieth century, there was a rapid proliferation of interest, information, and theory pertaining to genes and the field of genetics. The association between heredity and DNA became known during the 1940s and 1950s, culminating in the double-helix model of DNA by James Watson and Francis Crick in 1953.

This historical sequence of events created two major sources of confusion. First, terminology and concepts, including *genes* and *genetics,* were carried over from an earlier time when the subject seemed simpler. That contaminates and complicates the way we discuss the subject now. Second, the terms have accumulated a variety of meanings, which creates ambiguity. If one uses the term *genetic*, as in "the genetic control of behavior," what does that mean? Unfortunately, it can mean different things to different people. A lot of air and ink have been and continue to be spent in arguments resulting from these problems.

The general problem of outmoded terms and concepts is almost universal in science, but it varies in degree. The continued use of *genes*, for example, remains more widespread and accepted in genetics than does the use of *instinct* in the study of behavior. Because the gene and genetic terminology is still integral in the language of heredity, it is important to be aware and careful of their proper contemporary use, especially as they may apply to behavior. One must be alert to the following four particularly important points:

1. genes are best viewed as identifying the heritable differences between phenotypes, not as being responsible for constructing phenotypes;
2. genetic effects can be modified—they are not absolutely fixed and predetermined;
3. the "information" that is carried in DNA may be different than what we view as information in other contexts; and
4. DNA requires an environment for its expression—it does not operate in a vacuum.

We elaborate briefly on each of these points, then raise the issues again elsewhere in the chapter. The matter of the environment's role in the expression of DNA has many ramifications that come up frequently.

First, the issue of genes serving as labels for differences among phenotypes rather than being responsible for constructing phenotypes has been clarified nicely by

Marian Dawkins (1986). If someone says there is a single gene (or allele) for eye color in fruit flies *(Drosophila)*, for example, it means that the difference between flies that have red eyes and those that have white eyes can be accounted for by one gene, that is, one difference at one point (or locus) between the homologous chromosomes (in this case sex-linked on the X chromosome) of different individuals. A gene for eye color does not mean, as is so commonly misunderstood, that there is one gene or even one sequence of DNA consisting of a certain number of codons, that leads to the construction of red pigment in the eye.

A colored eye results from a host of processes that produce the pigment in certain locations, through the participation of numerous sequences of DNA and various environmental inputs at different times and developmental stages in many different ways. That entire process is responsible for the colored eye, nothing more and nothing less. However, a difference of one gene or allele, at only one point in one of the many sequences of DNA that are involved, may be responsible for a difference in whether the color turns out to be red or white. A gene for red eyes thus refers to the region of DNA where a difference exists, with that difference leading to an outcome of red eyes, compared with an alternate set of DNA that (in conjunction with various processes) produces white eyes.

The situation is analogous to a large, complex computer program or a complicated manufacturing process, say building an automobile. If only one change is made, even if only a single byte or character in the computer program or a single step in the assembly line, it may lead to a different final outcome. The focus is on the point of difference, not the whole process.

Differences are statistical properties that can be determined only by comparing two or more items. It does not make sense to talk about a difference in only one thing. Because of this, it is technically meaningless and incorrect to talk about genes and genetics in relation to individuals. However, because genetics is now associated with the process of DNA expression—which does occur in individuals— and also because we can turn our attention back to the individual after its differences from others have been identified, and further because this practice is so widespread, many biologists believe it is acceptable to talk about an individual's genes and genetics. Individuals are said, for example, to be homozygous or heterozygous for particular alleles. The danger is in thinking that the genes are responsible for the phenotypes. It is important to understand and remember that genes are responsible only for identifying heritable differences among the phenotypes.

The next important point is that genetic effects may be modified. Phenotypic differences among individuals can occur (or not occur) in many different ways in the myriad of paths, interactions, and processes that are involved in the expression of DNA. In some cases genetic differences among individuals are not observed as often as expected, as indicated by the genetic concepts of expressivity and incomplete penetrance. Some alleles produce multiple effects from a single allele at one locus (pleiotropy). Often there are multiple alleles (differences) that can occur at a given locus, as in the ABO blood types of humans. Also there are numerous interactions so that differences from some genetic sources may be counteracted by differences from other genetic sources (epistasis) or by environmental differences.

In some cases, such as with the various ABO blood types, which have or lack A and B antigens and in which, depending on which antigens are present, the anti-A and anti-B antibodies occur naturally in the blood without being stimulated by the introduction of foreign antigens, the differences between individuals are completely inherited and occur regardless of the environment, and it is difficult to change the outcome. But in other cases that also depend entirely or mostly on heredity, such as some human vision problems and some forms of diabetes, it is easy to change the situation by altering the environment, such as by adding eyeglasses or insulin, respectively.

Given that so much takes place on the way from DNA to its final phenotypic expression, it is amazing that phenotypes are as similar and constant as they are and that we have been able to trace the heredity connection to DNA at all.

Next, the concept of information and where it resides presents interesting and often overlooked implications for behavior. It arises not only in the present issue of genetic information but also with learning and phenotypic memory in the nervous system (Chapters 19 to 21), as well as in the subject of communication (Chapter 15). Information is an abstract and potentially problematic concept wherever it occurs. It can be defined as a coded reference to something. In its usual contexts, the meaning also carries an implicit or unstated sense that the information is important or relevant to some extent, in and of itself.

The nucleotide bases of DNA can be viewed as a coded sequence, legitimately considered as information (e.g., genetic information). However, one must be careful not to compare this information too closely with other sorts of information.

Because behavior is a sequence of muscular contractions, endocrine secretions, etc., that result from highly coordinated series of impulses in the nervous system, it resembles in its general properties (but certainly not in the mechanical details) other familiar forms of information processing. For example, computers use programs that consist of sequences of specific instructions (statements or commands). This resemblance can seduce one into asking, for example, whether the instructions and information are in the genes or being learned, as a result of experience and external, environmental effects.

The expression of DNA, however, involves far more than information. It also depends intensely on processes and developmental properties that we are only starting to comprehend. The term *information* may not be particularly appropriate for some of those processes and properties. For a simple analogy, imagine a ball rolling down a hill because of gravity. It may roll in different directions or at different speeds depending on the shape, orientation, and slope of the hill. But it may not be particularly useful to say that the hill contains information in the everyday sense of the word. Until we understand more completely both how behavioral phenotypes result from DNA and how the nervous system operates, one should be cautious about trying to compare the roles of information in DNA versus nervous systems. They may be different kinds of information or not really be information at all.

Finally, the expression of DNA requires environmental input, which has already been noted in some of the other points. The issue becomes a serious problem if one attempts to view the genetic, as opposed to the environmental, contributions to an animal's structure or behavior as being separate or simply additive. That is as absurd as saying that a cake is 60% recipe and 40% baking. The cake is a combination of both, and it is impossible to talk of only a 60% cake or a 40% cake. But there must be some way of separating the contributions of the two factors. Common sense and perhaps experience tell one that it makes a big difference in a cake whether or not the recipe includes sugar, for example. Also, the temperature of the oven and the length of baking time obviously are important. It thus also seems reasonable to consider heritable versus environmental factors in the phenotypic variation of behavior. There are some limited statistical ways to get at this, and we return to them later in the chapter (pages 113 to 117). For the moment, however, keep in mind that the expression of DNA requires environmental input.

With these points in mind, we next consider the basic genetics and evolution of behavior.

SINGLE GENES AND BEHAVIOR

Research since the time of Mendel has shown that most inheritance does not involve single-factor segregation, and different traits do not always segregate independently of each other. The traits in the garden peas that Mendel worked with, however, fortunately did show those characteristics. There also are a few traits in other organisms that show independent, single-factor segregation, including a few examples in behavior.

Probably the best known and most convincing example of single-gene behavior patterns involves honeybee nest cleaning. A bacterial disease, American foulbrood *(Bacillus larvae)*, kills honeybee larvae. If the bees remove dead larvae from the brood comb, further contamination and disease in the colony may be reduced.

The hygienic behavior has two components: first the caps are removed from brood cells that contain dead larvae; then the larvae are removed. Rather remarkably, each of these components was found to be associated with an independently segregating recessive allele (Rothenbuhler 1964). The alleles were named u for uncapping behavior and r for removal behavior. Both are recessive. Only worker bees that are homozygous recessive for both traits (uu rr) show the complete behavior. Workers that are homozygous recessive only for u (uu Rr or uu RR) uncap but do not remove dead larvae. Bees that are homozygous recessive for r but not for u (Uu rr or UU rr) do not uncap dead larval cells, but they will remove the larvae if another individual, such as another worker bee or a person, first uncaps them. Workers that are heterozygous (Uu Rr) or homozygous for the dominant allele do not perform the cleaning behavior (Figure 4-1). The behavior obviously is quite adaptive, and it clearly is affected by single alleles as classically defined and measured.

A more recent clear example of a single-gene effect involves frequency of activity in golden hamsters (Ralph and Menaker 1988). As discussed in more detail in Chapter 6, most animals show daily activity cycles with periods of about 24 hours. Normal or wild-type hamsters (Tn) show average activity cycle periods close to 24 hours. Those that are heterozygous for a short-period allele, Ts, have activity periods of about 22 hours; and animals that are homozygous for the allele, Tss, have activity periods of about 20 hours.

Several single-gene morphological traits, such as albinism in mice, are known to have behavioral correlations. Male albino mice, for example, were shown to mate much more successfully than black agouti males in one study of competitive mating (Levine 1958). The difference may have resulted from increased aggressiveness of the white males. Thiessen et al (1970) describe a large number of other behavioral correlations with single-gene traits in mice.

Figure 4-1 Single genes and bees.

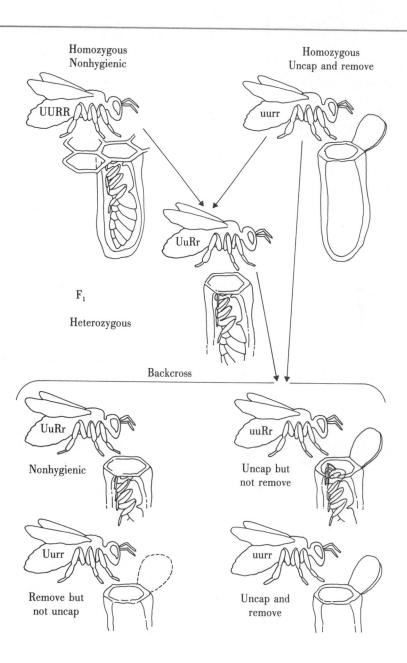

MULTIPLE GENES AND BEHAVIOR

Most behavior patterns that clearly show inheritance do not show simple segregation and independence. Many behavioral genetics examples involve crosses between subspecies or even different species, which would not be expected to show simple outcomes. The animals are too unrelated and the differences in traits involve too many genes to show simple, single-gene segregation.

One of the earliest and best examples of this was shown in research by David Bentley and Ronald R. Hoy with cricket song (Bentley and Hoy 1974). The familiar summer or tropical evening songs of crickets are produced by males that rapidly

Figure 4-2 Inheritance of
cricket song patterns as
indicated by hybridization
and backcrosses of two
different species. A
continuum of differences can
be seen, and control of the
pattern clearly is innate but
apparently is affected by a
large number of genes.
Modified from Bentley, D.R.
1971. Science 174:1139-1141.

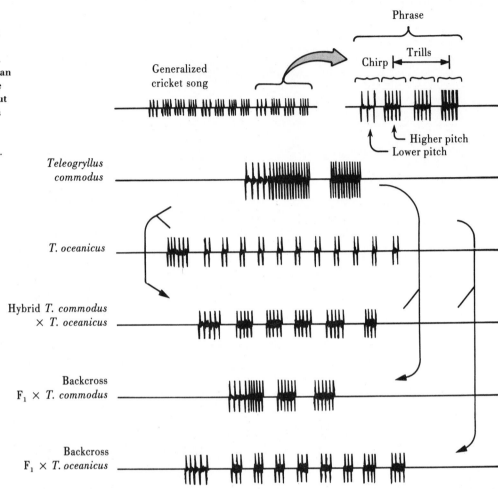

rub a scraper at the edge of one wing across a file on the other wing. Each species
has a repertoire of songs, including aggressive or rivalry songs that males sing
during fights with each other and calling songs that attract females to the singer.
The typical song consists of phrases that in turn are composed of various combinations of chirps and trills (Figure 4-2).

　　Different species have different, highly stereotyped calling song patterns. Hybrids
between different species show intermediate song patterns, and backcrosses—
between the hybrid and the wild type—are further intermediate, shifting back
toward the original pattern depending on which pattern it is, as can be seen by
studying Figure 4-2. Thus the shifts in song pattern depend on the proportion of
each wild-type genome present in a particular male. Bentley and Hoy analyzed 18
features of cricket song and found no single-gene effects. They concluded that
numerous genes were involved.

　　Direct neuromuscular studies in crickets have shown that neurons controlling
song are located in the thoracic ganglia of the central nervous system (Bentley
1971) but are inhibited by others in the head (Bentley and Hoy 1970). The researchers were even able to modify the songs artificially by altering the firing rates

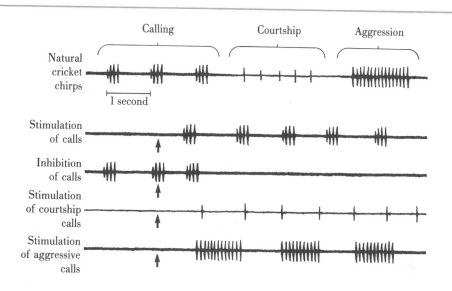

Figure 4-3 Stimulation and inhibition of natural cricket calls by artificial stimulation of different regions of the brain. Specific regions of the cricket brain control the pattern of chirping. Artificial stimulation of these regions triggers or inhibits the full, appropriate pattern as indicated. Arrows indicate start of brain stimulation. Modified from Bentley, D.R., and R.R. Hoy. 1970. Science 170:1409-1411.

of the command interneurons (Figure 4-3). Further work suggested that some genes regulating some song characteristics were located on the X chromosome. This was shown by a greater resemblance of F1s to their mother's species; male crickets have a single X sex chromosome that comes from the mother (there is no Y chromosome).

Perhaps even more fascinating is Bentley and Hoy's (1974) additional finding that female reception and response to male song is somehow apparently linked to the same set of genes that in the males leads to the production of the song. In experiments in which hybrid, backcross, or wild-type females were provided choices between different calls in a Y-maze situation, they chose calls from males belonging to the same genotype as themselves. Hybrid females, for example, preferred songs from hybrid males.

The phenomenon of hybridized animals showing intermediate or confused behavior has been demonstrated over a wide spectrum of species, including vertebrates. A classic example in vertebrates involves nest-building behavior in the small parrots known as lovebirds (Figure 4-4, Dilger 1962). Peach-faced lovebirds *(Agapornis roseicollis)* tear up strips of nesting material, reach back and tuck them under their rump feathers, and carry them to the nest. Fischer's lovebird *(A. fischeri)*, however, inherits a pattern of carrying behavior in which the nesting material is carried directly in the beak. Hybrids of these two species have difficulty carrying nesting materials. Initially the hybrids act as if completely confused and are unable to carry nesting material either way, although they try both. They eventually begin carrying material in the beak, but the seemingly simple behavior takes up to 3 years to perfect. Attempts to tuck material under the rump feathers diminish, but traces of the behavior such as incomplete movements persist indefinitely. Other examples of intermediate behavior in hybrids include courtship behavior of sunfish (Clarke et al 1984), choice of calling sites by tree frogs (Lamb 1987), cooing patterns by doves (Lade and Thorpe 1964), and exploration behavior in mice (Bateson and D'Udine 1986).

Figure 4-4 Confused carrying behavior in hybrid lovebirds (*Agapornis* sp.). A, Peach-faced lovebirds; B, Fischer's lovebird; C, hybrid. Figures based on information from Dilger, W.C. 1962. Scientific American 206(1):88-98.

OTHER KINDS OF EVIDENCE FOR GENETIC CONTROL OF BEHAVIOR

In addition to evidence such as that from Mendelian crosses and interspecific hybridization, there now are many other convincing lines of evidence. One is the presence of inherited behavioral mutations. The breeding of animals with such mutations, however, often is accomplished only with difficulty because many mutations are lethal, near lethal, or otherwise handicapping.

Kung et al (1975) have isolated over 300 behavioral mutations in the genus *Paramecium*. The paramecia studies were facilitated by the phenomenon of autogamy found in these protists. In autogamy the single-celled protist's chromosomes undergo meiotic replication and division followed by haploid mitosis. Subsequently there is self-fertilization, which produces greater levels of homozygosity. Some of the classes of mutants include paranoic, in which paramecia move backward for long periods, ts-paranoic, which are normal when grown at 23° C but paranoic at 35° C (illustrating some of the interplay between environment and genetics in determining behavior), pawns, which cannot swim backward at all, spinners, which spontaneously spin on hitting obstacles, "staccatos," which dash back and forth about 5 body lengths when stimulated, and sluggish, which move slowly or not at all.

Among fruit flies *(Drosophila melanogaster)*, as reviewed by Benzer (1973) with additional examples collected by Gould (1974), there are mutants that get stuck in copulation, others that go to the other extreme (coitus interruptus), some that do not climb, others that do not fly (although they appear to have normal wings and muscles), some that have seizures and faint when mechanically jolted, others that emerge from their pupal cases at abnormal times, some that respond abnormally to light or odors, and many more.

A large number of inherited mutations in behavioral traits have been noted also in vertebrates. Sidman et al (1965) cataloged 250 neurological-behavioral mutations in laboratory mice. Other examples include behavioral mutations (frequently behavioral defects) in domestic animals.

Closely associated with evidence from mutants is that from artificial selection of behavioral traits. A one-locus difference of locomotor activity in mice (Abeelen 1979) was identified through such selective breeding. Many examples come from selection for altered mating behavior (reviewed in Spieth 1974). Eoff (1977) was able to select artificially for behavior in two species of *Drosophila* that reduced the sexual isolation of the species. Recently Wallin (1988) was able to use artificial selection to alter the food choices of fruitflies, and Brandes et al (1988) produced lines with different learning ability in honeybees. Kovach (1980, Kovach and Wilson 1981) in a series of carefully controlled studies showed that color preferences in coturnix quail chicks *(Coturnix coturnix japonica)* can be modified by artificial selection. They produced lines of birds that preferred blue or red (Figure 4-5). Finally, domestic species provide numerous examples of unique behavioral traits that have been obtained via selection. These include pigeons (e.g., tumblers and pouters) (Nicolai 1976) and various breeds of dogs, cats, and livestock.

Another line of support for the genetic control of specific behavior includes cases where the neural regions controlling the behavior have been isolated and in some cases artificially modified. In fruit flies this has been accomplished with the aid of

Figure 4-5 Artificial selection for color preference in coturnix quail. Birds were selected over 14 generations for approach choices toward blue or red stimuli. Birds were tested by scoring color choice in 14 trials. Horizontal lines indicate mean number of choices, and vertical bars indicate standard deviations. From Kovach, J.K. 1980. Science 207:549-551.

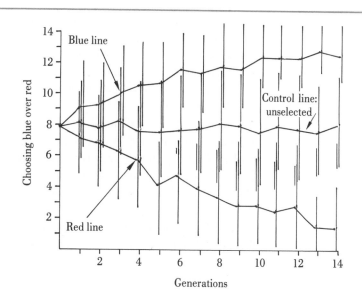

genetic mosaics, also referred to as *genetic chimeras*—individuals in which different parts of the body develop from different genotypes, even of different sex type (gynandromorphs) (Hotta and Benzer 1972)—and through direct neuromuscular studies (Levine and Wyman 1973).

Genetic mosaics have been produced in many species, including vertebrates. Rossler (1978), for example, was able to change feeding behavior to that of a donor species by transplanting the appropriate parts of the brain in developing amphibian embryos. Balaban et al (1988) transplanted parts of brains between developing embryos of domestic chickens and Japanese quail, which resulted in the exchange of species-typical crowing behavior. Unfortunately, newly hatched chickens that had received embryonic transplants of brain tissue from quails underwent tissue rejection after about 2 weeks and died. Birds less than 2 weeks old do not normally crow. Therefore, to stimulate crowing behavior in the chicks within 4 days of hatching, researchers used implanted capsules of the male hormone testosterone.

Sections of the brains of these chicks showed that donor and host brain cells had migrated and mixed together in certain regions of the brain during development. The developmental movement of brain cells is a complex and a poorly understood phenomenon that varies among different species of animals and in different parts of the brain even in a single animal (Walsh and Cepko 1988). This example not only demonstrates that inheritance plays a significant role, but along with information from other sources on the development of nervous systems, also provides a glimpse of just how circuitous and complex the route from DNA to behavior can be. The most recent line of evidence for behavioral genetics inheritance is recombinant DNA techniques. Scheller et al (1982) used these methods to identify specific genes and demonstrate that a family of genes control differences in egg-laying behavior in the marine snail *Aplysia*. They are mediated through different combinations of peptide hormones. (Hormones are discussed further in Chapter 18.)

Perhaps one of the least reliable, although commonly or casually accepted, lines of evidence for inherited behavioral traits involves the consistency of a particular behavior, i.e., FAP or MAP (Chapter 2, page 43) among different individuals of a species. The presence of consistent complex behavior is most convincing as evidence for inheritance only when demonstrated in individuals that develop in isolation from other members of the species and their normal environment. For example, if a newly laid bird egg is incubated artificially and the resulting chick is reared having no contact or experience with a nest or other members of its species, but at maturity and when provided with a variety of possible nesting materials it nonetheless builds a normal nest by choosing only those materials and using complex movements typical for the species, then it seems reasonable to conclude that the behavior is largely inherited.

Consistent behavior among members of species can also result from nongenetic sources. Hailman (1967) and others have warned that consistent behavior in a species may also indicate consistent environmental, developmental, and learning factors. As an obvious illustration, consider the behavior of automobile drivers, almost all of whom consistently raise their foot and step on the brake on seeing a red stoplight. It would be absurd to conclude that such species-specific and constant braking behavior is inherited. The same fallacy applies to many of the consistent behavior patterns seen in other species. This brings us to the problems of understanding and attempting to measure the roles of genes versus influences of the environment in behavior.

HERITABILITY AND BEHAVIOR

Genes are clearly involved in behavior. But it is equally clear that this is not the whole story. There can be no expression of DNA in isolation from the environment. Many of the complex interactions between the environment and protein synthesis are mediated via the endocrine and nervous systems. These processes are important in the development (ontogeny) and maintenance of all of an individual's characteristics, including body structures and physiology.

Behavior, however, is removed a step further from the structures. Behavior involves much more neural and endocrine input and processing, and it has the potential for much more lability and environmental influences than do morphological traits. Some behavior patterns, such as moving to a particular location to obtain food, may require regular modification and updating throughout an animal's life. Trying to understand and untangle the multitude of interactions and feedbacks involved in behavior has led to much debate, often under the banner of "nature versus nurture" or "instinct (innate) versus learning" (as discussed in Chapter 2, pages 47 to 49). Is it possible to say more than merely that both heredity and environment are involved? Can we quantitatively measure the roles of heredity in behavior?

Recall that genes involve differences among individuals, and we need to compare two or more individuals to identify the differences. With a sufficiently large number of individuals and proper attention to their environments and biological relationships in a given context, it is possible to statistically analyze the sources of variation in their phenotypes. One can then consider the contributions of genetic variation and environmental variation to the overall variation in phenotypes (in the particular context). Viewed this way, the contribution of genetic variation as a proportion (from 0 to 1) of the total variation is referred to as *heritability*.

Two extremely important points need to be emphasized: (1) heritability is a statistical feature of a group, not of single individuals, and (2) it is relative to a given context. Without the context the numbers may be meaningless, which we attempt to explain.

Some extreme cases are straightforward. Whether human blood has anti-A or anti-B antibodies always varies with the ABO blood type alleles regardless of normal environmental variation; thus it is reasonable to say that the presence of these particular antibodies has a heritability of 1. Whether humans have tattoos on their bodies, on the other hand (no pun intended), represents the other extreme. Tattooing is a trait in humans that varies only with the environment and appears to be entirely cultural. It is reasonable to guess that the presence of tattoos on humans has a heritability of 0. There are also some familiar intermediate cases. Unlike anti-A and anti-B antibodies, the presence of Rh antibodies in human blood, for example, depends on both genetic and environmental conditions in specific, well-known ways; the antibodies are not formed until after exposure to introduced Rh antigens, as in blood transfusions or Rh negative mothers with Rh positive (from the father) babies.

The observed variability in most behaviors involves important variation in both genes and environments. A complete discussion of heritability is far beyond the scope of this book; but because it is so relevant to the subject and interests of behavior, we present an introductory look at it. A full understanding of heritability requires good understandings of both statistics and genetics. Persons who are interested in more discussion, details of calculating heritabilities, and actual worked examples can refer to the appropriate sections of Crow (1986), Falconer (1981), and Ehrman and Parsons (1976).

The correct measure of variation for heritability is the statistical variance of some appropriately measured aspect of the subject of interest. Variance is extremely useful statistically because it can be broken into additive components. Variance, briefly, is defined as the sum of the squared differences between each of the measurements and the sample mean, divided by the number in the sample, as in the following formula:

$$S^2 = \frac{\Sigma(x - \bar{x})^2}{N}$$

It can be viewed as the average squared variation.

Because variance is additive, the total phenotypic variance (V_t or V_p) in a trait for a given population can be broken into the components of environmental variance (V_e) and genotypic variance (V_g). These are determined for each component respectively from measurements of traits for animals under a variety of environmental and genetic combinations.

Once measured (or, more properly, estimated), the proportion of the total variance resulting from genotypic variance can be calculated by the ratio $V_g : V_p$; this is referred to as *heritability in the broad sense* or H_B. However, H_B results from several underlying sources of genetic variation and as a consequence is not a useful measure. A more restricted category of heritability is desired: it is called *heritability in the narrow sense*, H_N or h^2. H_N gives a better measure of the genetic variability available for selection (artificial or natural) than H_B. H_N is a value of great interest and utility to breeders of crops and livestock who wish to develop certain traits such as milk production or weight gain under specified conditions such as diet. It is also potentially useful in studies of traits under natural conditions—if used and interpreted properly.

A straightforward interpretation of H_N has been provided by Crow (1986): heritability in the narrow sense is a measure of offspring deviation from the population mean as compared with the deviation of their parents (Figure 4-6). If a group of animals that are quite different from the population average were to be selected for breeding and their offspring were only one third as different from the population mean as their parents, then H_N would be one third. Note that we are still considering group or population values, not isolated individuals except as they contribute to the group statistics.

As an example of heritability measurements in behavior, Fulker (1966, cited in Ehrman and Parsons 1976) studied the mating speed of male *Drosophila melanogaster* using six inbred strains in various combinations and with several replications. Calculated heritabilities were 0.71 in the broad sense and 0.36 in the narrow sense. Heritability values have also been obtained for several other behaviors and in several different species.

In addition to being statistically complex, heritability measures require that one measure or experimentally control variability in the genotype or environment. This usually involves inferences based on known pedigrees and genetic lines or by reducing genetic variance to 0 or near 0 through clones (including identical twins) or lengthy inbreeding. Environments are notoriously difficult to deal with. Subtle differences can be important, and we often do not know which environmental factors are important in the first place. How does one know how much environmental

Figure 4-6 Heritability in the narrow sense, H_N. **A,** Illustrated as means of progeny compared to means of the parents. **B,** Change of means of a trait after two generations of selection for a trait with $H_N = 1/4$. From J.F. Crow. 1986. Basic concepts in population, quantitative, and evolutionary genetics. W.H. Freeman, New York.

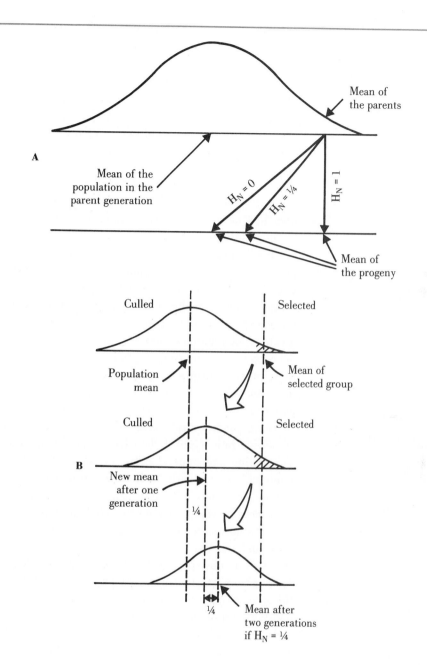

variability there is, or even when there is "no variability!?"

This is illustrated with a relatively simple hypothetical situation: running behavior of mice. Imagine a group of standard laboratory mice housed in simple cages. They can be provided with exercise wheels, and the amount of time they spend running in them can be measured. There will be variation in the running times of different mice, and the means and variances of those times can be calculated.

Next we can set up experiments to reduce variation either in genotypes or in the environment. If the genotypes are made uniform, for example by cloning or by an extensive inbreeding program, then the variance in running times as measured will reflect only environmental variation. That is, heritability (which is either broad or narrow) will be 0. That does not mean that the genotype played no part in activity but merely that in this context there was no contribution from genetic variability.

If, on the other hand, parent mice are selected at random from the stock population and all aspects of the environment (temperature, light, humidity, size of cage, food, cleaning, rearing conditions, etc.) are rigorously standardized from the moment of fertilization until the young are tested in the exercise wheels, then there should be no environmental variation, all observed variation should be genetic, and heritability would be 1. Notice that we have measured heritability ranging from 0 to 1 simply by specifying different contexts. Theoretically (although in reality it would be unlikely) we might reduce both genetic and environmental variation to 0 and then be unable to calculate heritability because there would be no variation in running times.

This example, hypothetically set to the two extremes, illustrates one of the main properties of heritability values: genetic variance is considered relative to the environmental variance under the specific set of genetic and environmental conditions being considered. Because heritability reflects genetic variance as a proportion of the total variance, it changes if the total variance is changed. Under a given set of genotypes, one can increase or decrease heritability values simply by decreasing or increasing environmental variability, respectively. In other words, heritability increases in a more restricted (less variable or more uniform) environment not necessarily because of a change or difference in the genetics but merely as a property of the way in which heritability is calculated.

The preceding discussion illustrates the situation with a single group of mice, a metric, ratio type of data (see Chapter 3), and a relatively simple behavior. Calculating and understanding heritability becomes much more difficult or even impossible in comparisons between unrelated groups of animals, with data that are ordinal or nominal (Chapter 3), or with complex, subjective behaviors. This is illustrated by expanding our hypothetical case with the mice. We could choose two different strains of mice that show different levels of activity, say active "jogger" mice and inactive "couch potato" mice. With each group we could perform the same kinds of experiments and get the same kinds of results (with heritabilities ranging from 0 to 1). But we would be unable to compare heritability values between the groups unless the environments of both groups were held strictly and completely constant and identical.

In reality we probably would not be able (or certain) to reduce either genetic or environmental variation to 0, and there would be other confounding sources of variation, including measurement error. That would force us to work with heritabilities somewhere between 0 and 1. Also, we might be interested in something more general than just simple* running time in mice; perhaps we want to study "athletic ability" of mice. Instead of just measuring running ability we might go for some kind of multiple or compound measure that attempts to simultaneously include measurements of the abilities of mice to climb, jump (length and height), dig holes, swim, lift weights, and fight with each other. Or imagine trying to compare results of these studies using simple cages with studies conducted in more variable environments, say enriched cages with numerous objects for the mice to play with,

---

*Running is simple only in a relative sense. It involves participation of numerous muscular, skeletal, neural, respiratory, cardiovascular, and metabolic factors, any or all of which could contribute to variation.

several levels and different partitions and rooms in the cage, highly variable diet, and so on. What would the resulting values for heritabilities mean? Good question! In fact, one might ask, why even bother to try measuring heritabilities? The numbers are useful in some situations, such as trying to artificially breed for specific traits under specific environmental conditions and perhaps in understanding genetics under specific, carefully interpreted natural conditions. But in many cases the heritability values are meaningless and unfortunately are frequently misunderstood and occasionally even misused.

Although this book is primarily concerned with nonhuman behavior, some of the most interesting and controversial potential applications of heritability to behavior involve human behavioral characteristics such as musical and athletic abilities, criminal behavior, and problem solving. We briefly consider one area that has received much attention, the issue of human intelligence. This is a prime example of the difficulties of measuring heritability in general—regardless of species. In addition, it is a matter of widespread continuing interest for the following reasons: it strikes close to all of us because it involves our own behavior; it is a story that is rich in detail; and it provides another fascinating slice of history in the subject of behavior.

HERITABILITY AND HUMAN INTELLIGENCE

Intelligence testing began quite innocently in the early 1900s with Alfred Binet, who attempted to diagnose and correct learning problems in school children. His focus was on developmental aspects, not heredity. It was Binet's view that intelligence, the ability to perform mental tasks, could be improved for students who were not performing up to their mental age. Binet worked with students who shared a common cultural background (the public schools of Paris).

Later workers modified the procedures to calculate the intelligence quotient (IQ), the ratio of mental age to chronological age. They also added test items to include adult intelligence. Many of these investigators differed sharply from Binet in philosophy and basic assumptions; they viewed intelligence as an inherited, innate, and unmodifiable trait. Thus began years of debate that have continued up to the present day and that have colored our collective understanding of heredity and behavior in general, including that of other species.

Measuring the heritability of intelligence is much like the hypothetical example involving the athletic ability of mice. One can replace references to mice with humans and athletic ability with intelligence. In addition, there are problems with ethical, emotional, and historical considerations and resistance by the subjects themselves, which do not permit the free selection and experimental breeding of humans. Because of this, neither the genotypes nor the environments can be controlled or restricted, a vital component in the equation of heritability. Random selection of subjects where called for is not readily possible.

There have been studies of natural experiments, involving identical twins raised in different homes and unrelated, adopted children raised in the same home. However, even here the placement of subjects has not been random. This raises two technical problems: interaction and covariance. *Interaction* involves the differences in the responses of particular genotypes to different environments. A possible genotype that improves problem-solving ability, for example, could do much better in a highly social environment than in a more isolated environment, whereas a different genotype might result in little difference in problem-solving ability whether in a social or nonsocial environment. *Covariance* refers to the possibility of factors varying together; children with certain genotypes (e.g., of certain races or specific phenotypic characteristics) may be more likely than others to be placed in certain home environments and cultures. All of this makes it logically impossible to make comparisons between groups, such as between different races or cultures.

Other problems are that intelligence is not a simple trait and the measurements are ordinal, not interval or metric data. It is not even completely clear just what intelligence is. In addition, standardized IQ tests are environmentally culture bound; that is, to take any given test the subject must have the necessary background for understanding and relating to the nature of the test questions. The environment is unavoidably tangled up in the test itself.

Not only is *intelligence* a complex concept, human environments also are exceedingly complex and the contributions from different factors are far from being understood. Furthermore, different factors have different effects at different times in the development of the child. To be investigated properly, individuals would need to be separated at the moment of fertilization (and even then there might be maternal effects in the ova). In utero effects, such as the position in the uterus, mother's diet, drugs taken, sound environment, sleep patterns, level of stress, etc., might influence the developing embryos. Many human environmental effects, such as a sudden traumatic experience, are unique, unpredictable, and uncontrollable. Humans (and most other species, for that matter) rarely live in cozy little restricted mouse cages.

Another complication with the heritability of IQ is the common misunderstanding that genetic means determined and unchangeable. As discussed previously, there is no necessary relationship between heritability and changeability. The whole problem of relating IQ to inheritance has proved to be like trying to put socks on an octopus.

If all of this were not enough, attempts to understand the inheritance of human IQ were thrown into disarray by the discovery that much of the data on which earlier studies were based were fraudulent. In 1976, after earlier suspicion, it was revealed that data from many years of twin studies by the famous and respected Sir Cyril Burt were fabricated. Two of his research associates in the work did not even exist. That caused researchers to reassess other data and required them virtually to start over in investigating the whole subject of IQ and inheritance.

Newer studies of twin data have been conducted, such as at the University of Minnesota Center for Twin and Adoption Research (e.g., Bouchard et al 1981, Segal 1985), and a critical reanalysis has been made of those portions of earlier data that are potentially valid (Bouchard and McGue 1981).

In addition to heritability values per se, as discussed previously, statistical analyses of twin data have included correlations of various types and, in more recent studies, a variety of sophisticated multivariate analyses and various methods designed both to test more specific models and to work better with twin and family data (as opposed to other methods that work better with experiments or artificial breeding).

Today some generalizations about heredity and intelligence can be made. One of the most consistent observations is that adopted children, regardless of race, sex, or any other correlation with their previous background, tend to have a higher average IQ than nonadopted children, apparently because the average "adoptive parents are not a random sample of households but tend to be older, richer, and more anxious to have children; and, of course, they have fewer children than the population at large. So the children they adopt receive the benefits of greater wealth, stability, and attention" (Lewontin et al 1984).

The Minnesota twin studies have considered the many assumptions and carefully sorted through the data for identical (monozygotic) twins reared together and apart, fraternal (dizygotic) twins reared together and apart, adopted nonrelated children raised together, various other types of family and nonfamily relationships, different kinds of IQ and cognitive tests, and a host of educational, social, and economic variables, including kinds of objects (tape recorders, power tools, magazines and books, pets, etc.) that are present in the home. These studies have shown, as might be expected, that both genetics and environment are important, and they work together. Variation in human intelligence does not result mostly from variation in either genetic inheritance or environmental factors. Identical twins are more similar in IQ and some (but not all) cognitive abilities than are fraternal twins; and fraternal twins show more similarities than are found in other types of possible comparisons, whether between other siblings or nonsiblings.

In response to the question "Do environmental similarities explain the similarity in intelligence of identical twins reared apart?" Bouchard (1983) offers a resounding "No!" At the same time, it is clear that the environment does play an important role (McGue and Bouchard 1987). Just what role the environment plays, however, remains unclear. McGue and Bouchard (1987, 1989) claim that 50% to 60% of the variance in IQ (in their studies) can be attributed to genetic variance and about 50% to environmental variance. Of the environmental variance, about 30% is associated with the family and about 20% with factors outside the family. From within the family, however, McGue and Bouchard (1989) were unable to find any relationship between the cognitive development of twins and the educational and socioeconomic status of their parents.

Similarly, aside from general findings, attempts to dissect out specific cognitive abilities have resulted in differing and sometimes confusing outcomes that depend on a host of specific conditions such as which ability and which measure is being considered. For example, twins reared apart are similar in cognitive measures involving general speed of response, but they are not similar in measures of speed in specific cognitive processes (McGue et al 1984). The pursuit and debate of these and a number of related issues is continuing.

If one attempts to compare intelligence or other behavioral traits among different species, the exercise becomes nonsensical. Humans and frogs, for example, obviously have vastly different mental abilities (and senses, as discussed in Chapter 16). These differences just as obviously come from a large genetic difference (and thus, in this context, few people would argue against the mental ability being mostly genetic). However, frogs, humans, and other species that one might want to compare on a genetic basis are just plain different in many ways, including their environments, and comparisons would be mostly meaningless. If an IQ test depended on one's ability to recognize a flying insect and rapidly flip out the tongue and catch it in midair, humans would rank pretty low.

GENES AND BEHAVIOR: CONCLUSIONS

Although heritability estimates using statistical variance among individuals are beset with problems, they are currently the best and almost only way to assess the contribution of inheritance to behavior. The few studies of heritability of behavior in various species indicate that few behaviors may be said to be entirely or mostly determined by inheritance or the environment in a given context. Instead behavioral variability appears to have both genetic and environmental components. Jacobs (1981) showed with the best variance data available that the proportions of variance accounted for by heritability are mostly unimodal and not bimodal (Figure 4-7); that is, there is not one peak for innate behavior patterns and another for learned patterns, as one would expect if most behavior were one or the other. This suggests, as has so often been argued, that the long-standing dichotomy is simply not useful. One must also keep in mind the relative nature of heritability values, which is a basic property of the technique itself, even in the "best" of heritability measures. Jacobs (1981) also notes that there is much random variation *(spontaneous variability*, as he calls it), that is, background "noise" in behavioral variability. Actual heritability may be low. Jacobs suggests that rather than being preoccupied with innate-acquired considerations, one should be looking at other relevant factors such as overall genetic diversity, environmental heterogeneity, and variable selection pressure (Jacobs 1981).

We return to some aspects of this problem, particularly as it relates to innateness versus learning, in Chapters 19 to 21. The means by which birds develop their specific calls and songs, for example, provide an excellent illustration of a range of inherited contributions. Some species have calls that seem impervious to normal environmental variation, other species learn their songs within narrow limits, and a few species of birds have an amazing plasticity in the sounds they can learn and mimic, including nonavian sounds.

The main point for our present purposes is that we may reasonably infer that nearly all behavior, even that involving learning ability, clearly is affected by heredity to at least some degree, and much of animal behavior is inherited to a large extent. If there is any heritability in behavior, then it is subject to the forces of evolution. If heritability is low, then evolutionary change may be more difficult or take longer, all other factors being equal, but it can still occur. The evolution of behavior is the subject of the next chapter.

Figure 4-7 Frequency distribution of heritability values measured for 82 instances of animal behavior. Measurements ranged over a variety of taxonomic groups and types of behavior, including reproductive, social, individual behaviors, orientation, learning capacity, and emotionality. Two distinctly separate peaks do not exist, suggesting that all behavior cannot be classified simply as innate or noninnate but span a wide range. A few can be viewed as largely innate, a few as largely noninnate, and most are somewhere between. Modified from Jacobs, J. 1981. Zeitschrift für Tierpsychologie 55:1-18.

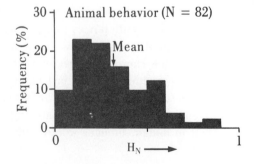

SUMMARY

Variation in animal behavior is derived in part from inherited factors. When considering the possible genetics of behavior, there are four basic points that must be kept in mind:

1. Genes identify the heritable differences among the phenotypes of two or more individuals; they are not responsible for constructing phenotypes.
2. Genetic effects may be modified by environmental effects; genetic does not mean fixed and determined.
3. The information in DNA may not be strictly comparable with our concept of information in other contexts. The expression of DNA is intensely dependent on processes and properties that may shape and influence the final outcome beyond the information aspects.
4. DNA requires an environment for its expression.

Single-gene effects in behavior can be observed in honeybee nest cleaning, hamster activity cycles, and a number of behaviors associated (pleiotropically) with single-gene morphological traits. Beyond single-gene effects, there are numerous other lines of evidence for connections between heredity and behavior: hybridization (e.g., in crickets and lovebirds), behavioral mutations, artificial selection for behavioral traits, genetic mosaics by transplants of nervous system tissue, and recombinant DNA techniques involving behavior. One approach for demonstrating genetic effects, the presence of consistent or stereotyped behavior per se in a species, is not reliable. Environmental circumstances also may lead to stereotyped behavior.

One measure of genetic influence on behavior is heritability, the proportion of variability of the behavior in a group of individuals, as measured statistically by variance, that is accounted for by genetic variance out of the total variance (genetic plus environmental). This is a relative measure that is highly dependent on the specific context of variabilities in the genetic relationships and environmental conditions. Many of the problems in attempts to measure the heritability of behavior are illustrated by the issue of human intelligence. The available information to date on human intelligence and data on behavior in other species suggests that variation in most behaviors arises from both genetic and environmental variation, not mostly one or the other; there is not a bimodal distribution in heritability values. Because most behavior has a genetic component in its variability, behavior is subject to the forces of evolution, the topic of the next chapter.

FOR FURTHER READING

General

Bateson, P., and B. D'Udine. 1986. Exploration in two inbred strains of mice and their hybrids: additive and interactive models of gene expression. Animal Behaviour 34:1026-1032.
An original paper involving hybrid behavior that also elucidates some of the complications and pitfalls of such work.

Crow, J.F. 1986. Basic concepts in population, quantitative, and evolutionary genetics. W.H. Freeman, New York.

Dawkins, M.S. 1986. Unravelling animal behaviour. Longman Group, Essex, England.
This concise book specifically addresses a number of the common points of misunderstanding in behavioral genetics. It provides further discussion of several of the issues raised in this chapter (and topics such as innateness, instinct, optimality, and inclusive fitness discussed elsewhere in the book).

Ehrman, L., and P.A. Parsons. 1976. The genetics of behavior. Sinauer Associates, Sunderland, Mass.

A general text on the subject of behavioral genetics.

Lamb, T. 1987. Call site selection in a hybrid population of treefrogs. Animal Behaviour 35:1140-1144.

An original research paper, clearly written and relatively short with straightforward outcome.

Nei, M. 1987. Molecular evolutionary genetics. Columbia University Press, New York.

Plomin, R., J.C. DeFries, and G.E. McClearn. 1990. Behavioral genetics: a primer. W.H. Freeman, New York.

An updated introduction to the genetics of behavior, with basic background genetics from molecular to statistical aspects. Although this book does not go into much depth on any topic, it provides a good overview and expands several topics beyond their level of coverage in a general biology text.

Heritability and Human Intelligence

Bouchard, T.J., Jr. 1983. Do environmental similarities explain the similarity in intelligence of identical twins reared apart? Intelligence 7:175-184.

An analysis and rebuttal of earlier claims.

Bouchard, T.J., Jr., and M. McGue. 1981. Familial studies of intelligence: a review. Science 212:1055-1059.

A thorough analysis of world literature up to that point.

Lewontin, R.C., S. Rose, and L.J. Kamin. 1984. Not in our genes. Pantheon Books, New York.

A strong rebuttal of genetic determinism—the belief that behavior is determined by genes— particularly human behavior. Devotes much attention to the IQ problem, including the history of intelligence testing.

McGue, M., and T.J. Bouchard, Jr. 1989. Genetic and environmental determinants of information processing and special mental abilities: a twin analysis. In: R.J. Sternberg, editor. Advances in the psychology of human intelligence. Vol. 5:7-45.

A thorough, detailed, technical survey of data from the Minnesota twin studies.

Montagu, A. 1975. Race and IQ. Oxford University Press, New York.

A somewhat dated but nevertheless valid collection of reprints on the issue, including extensive discussions of the problems and misuses. See particularly the articles by Gould, Lewontin, Layzer, and Bodmer.

Paul, D.B. 1985. Textbook treatments of the genetics of intelligence. Quarterly Review of Biology 60:317-326.

Further discussion of the issue and laments that, while dropping the name of Cyril Burt, many textbooks continue to repeat the misconceptions and misuses. (We hope that this textbook does not make that mistake!)

5

THE EVOLUTION
OF BEHAVIOR
AN INTRODUCTION

Evolution forms a central theme and foundation for the consideration of all other topics in behavior. As such, the evolution of behavior is more than the subject of a single chapter; it is what this entire book is about. Because of that, this short chapter should be viewed as an introduction for the remainder of the book. This chapter introduces some of the basic principles, with examples, that form the basis for understanding all subsequent material. Other, more advanced evolutionary topics, such as evolutionarily stable strategies (ESSs, Chapter 14), are presented in other chapters, where appropriate. The principles discussed in this chapter are presented primarily as they apply to behavior. For even more general and basic biological principles of evolution and additional background information, refer to a general biology text such as one of those listed at the end of the Preface.

THE EVOLUTION OF BEHAVIOR: CONSEQUENCES OF GENETIC AND ENVIRONMENTAL VARIABILITY

Not only are animals highly variable, with much of the variability being inherited, but the earth's environment is also highly complex, dynamic, and variable, both temporally and spatially. The heterogeneity and uncertainty are widespread and local, with unsuitable habitats often adjacent to suitable habitats. Much of an animal's environment consists of other animals, including others of the same species.

Those animals that survive and reproduce in the midst of all this variability and uncertainty are those with suitable structures and behavior patterns. Behavior can be viewed in a sense simply as putting certain structures into action. Structures (and behavior patterns) are said to be adaptive if they permit an animal to avoid or tolerate unfavorable conditions or if they permit the animal to obtain favorable or necessary surroundings and resources. It should also be noted that animals may also possess characteristics that are not necessarily explained as being directly or primarily adaptive. Many traits of organisms occur through genetic drift, founder and other phylogenetic historical effects, or as secondary consequences resulting from other traits. Secondary traits, such as perhaps many of the colored patterns of mollusk shells, are analogous to spandrels, the tapered triangular spaces in the ceilings of cathedrals that are formed as "architectural by-products of mounting a dome on rounded arches" (Gould and Lewontin 1979). Darwin, incidentally, took a pluralistic view of the sources of evolved traits in organisms.

Traits that are adaptive often represent an accumulation of compromises. Under two different conditions that an individual may have to face, a given structure or behavior can be advantageous in one and a handicap in the other. A heavy insulation of feathers, for example, is useful at low temperatures but may lead to heat prostration or even death under hotter conditions. The capture and killing of a moving object may be useful if it is potential food but not if it is a potential mate. How all of this gets sorted out biologically, via differential survival and reproduction, is the general subject matter of the topic of evolution. Fortunately the explanation for behavior appears to be basically the same as for the evolution of bones, body shapes, and other biological characteristics. The best way to understand the evolution of behavior is to first understand biological evolution in general. If one understands the basic principles, then the evolution of behavior becomes (almost) self-evident.

The view that behavior can be shaped and changed by evolution in a manner similar to morphological and physiological traits can be traced back to Darwin and other early workers. Elaboration of the principles as applied to behavior received much impetus from the work of Konrad Lorenz, Karl von Frisch, and Niko Tinbergen, for which they were awarded the Nobel Prize in 1973, and has received continued attention through the work and writings of J. Maynard Smith, George C. Williams, Robert L. Trivers, Edward O. Wilson, Stephen J. Gould, Richard and Marian S. Dawkins, Mark Ridley, and many others.

Evolution is studied through several approaches, including experimentation. Experiments that attempt to test the mechanism of evolution generally compare differences in survival or reproductive success among groups of animals that differ in a particular characteristic of interest under either natural or artificial (experimental) situations. An example using a natural situation is described on page 20

Male size	No. males	x̄ Matings	% of total matings
Large	21 (21) [21]	5.8 (3.9) [3.1]	60 (48) [43]
Medium	28 (28) [28]	1.9 (2.3) [2.1]	27 (37) [40]
Small	21 (21) [21]	1.2 (1.2) [1.2]	13 (15) [17]

Table 5-1 Male size in relation to mating success in *P. latipennis* under three resource (dead cricket) abundances. First number = 2 crickets/ enclosure (N = 201 matings). Number in ( ) = 4 crickets/enclosure (N = 170 matings). Number in [ ] = 6 crickets/enclosure (N = 150 matings).

From Thornhill, R. 1986. In: M.H. Nitecki and J.A. Kitchell, editors. Evolution of animal behavior: paleontological and field approaches. Oxford University Press, New York.

where moths with and without ears were compared in their survival against bat attacks. An example of deliberate experimental manipulation on selection (in this case sexual selection, see Chapter 11) involves studies by Randy Thornhill (1986) on differences of reproductive success in scorpionflies (genus *Panorpa*) in response to manipulated differences in mating circumstances. He conducted experiments in large screened enclosures placed in the field of the insects' natural habitat by exposing adults to various combinations of available food resources (dead crickets), sex ratios, and male body size. One example of Thornhill's findings was that larger males were more successful at mating than smaller males in all cases and the difference became greater when resources were more limited and had to be defended against other males (Table 5-1).

More commonly, however, evolution is investigated through two other major approaches: historical, using the fossil record, and comparative (Chapter 3, see also Nitecki and Kitchell 1986). The various approaches used to draw inferences about evolution are based on the premise known as *uniformitarianism*, which is the belief that physical and biological processes operate in a uniform manner throughout time (Simpson 1970). What we can observe in the present is the key to the past.

Animal behavior, being a step removed from the morphological structures, is not easily fossilized. The inferences are more difficult and invoke less confidence. Nonetheless, uniformitarianism can be applied to behavior. The role of head ornaments of dinosaurs, for example, has been inferred from the behavior of animals such as deer and certain beetles that have head ornaments today (Molnar 1977). Likewise, the behaviors that produced certain nest structures, patterns of footprints in the mud, and teeth marks on the bones of other animals in the fossil record are inferred from the behaviors that produce similar patterns and tracks today. Ostrom (1986) provides a concise, well-written perspective on what can be inferred about dinosaur behavior from the fossil record.

Although behavior may leave few fossil records, it is frequently more complex and richer in detail than most morphological structures (except at the molecular level). Species that appear almost identical morphologically may behave differently. This often permits the comparative approach to be quite productive, and these methods have been of utmost importance to understanding the evolution of behavior.

Jack Hailman

Jack Hailman
The Necessity of a "Show-me" Attitude in Science

In 1899, congressman Willard D. Vandiver told a dinner audience in Philadelphia that "frothy eloquence neither convinces nor satisfies me. I am from Missouri. You have got to show me." It is not necessary to have been born in St Louis (as I happened to have been) to follow suit. Behavioral studies are finally emerging from an era where the frothy eloquence of sociobiological theory asked us to believe without evidence in such things as genes for helping behavior and natural selection operating at the level of genes (instead of organisms or populations). Especially when historical processes are involved—as they are in cosmology, geology, and evolution—it is wise to adopt a show-me attitude.

Consider the myth of industrial melanism. A prominent biologist asserted that evolution by natural selection was happening before our eyes. The peppered moth *(Biston betularia)* has a high incidence of melanic individuals in an industrial area of England. Field experiments on avian predation showed that birds took more black than mottled forms released in a polution-free rural area in Dorset, whereas in Birmingham where soot blackened the tree-trunks birds captured more mottled forms than black. Therefore we were supposed to believe that avian predation was the force of natural selection promoting the evolution of melanic moths in industrial areas.

Show me the rest of the evidence. Has anyone actually seen the ratio of peppered to black forms change over time? Apparently not, at least not since 1937 when the Oxford biologist began his studies. The best he could do was point to midnineteenth century collections that contained few black individuals. Those collections were not scientific samples but were made by amateurs who probably favored pretty things. Perhaps they did not like ugly black moths. Do birds eat sufficient numbers of moths to shift the ratio? It is improbable, according to population estimates showing extraordinary densities of the little moths (up to 100,000 moths per square kilometer). Is there a clear correlation between black moths in industrial areas and peppered forms elsewhere? No. What happens if you feed the products of industrial pollution to the caterpillars? More experiments would be nice, but those that had been done seemed to show increased numbers of melanic individuals. At least half a dozen scientific papers have been published attacking the industrial melanism story, yet it still appears in textbooks as a prime example of evolution through natural selection. The will to believe seems stronger than the contrary evidence.

Evolutionary biology has suffered chronically from a confounding of mechanisms and results, making it difficult to live by the show-me dictum. An English

THE COMPARATIVE STUDY OF ANIMAL BEHAVIOR

The comparative method is really two different methods (or more, depending on how they are conceptualized, Lauder 1986). The first looks at evolutionary radiation among closely related organisms and attempts to reveal ancestral relationships and ultimate causes through homologous characteristics and environmental differences. The second considers the functions of behavior as potentially analogous traits among clergyman argued in a nineteenth century book that animals were at least as intricately constructed as pocketwatches. Since it is unimaginable, he continued, for a watch to have come into existence without a watchmaker, so animals must have been fashioned by a divine watchmaker. No one today accepts this argument as proving the existence of God, yet many biologists use a curiously similar argument to "prove" the existence of natural selection. The coloration, anatomy, physiology, and behavior of animals seems so well suited to the environments in which they live that natural selection "must" have adapted the animals to their environments. Come on, can't we do better than that?

To begin, we can at least tackle separate problems separately. The knotty problem of how evolution adapts animals to their environments is still unsolved. Darwin knew that natural selection among individuals was insufficient, so he proposed sexual selection and then realized that even the combination could not explain things like eusociality of ants, bees, wasps, and termites. Genetic drift, pleiotropism, and other evolutionary processes have been added to the list, and even the formerly stomped-on notion of group selection has been resurrected with new enthusiasm. Keep an open mind.

Identifying adaptations is fortunately an easier problem. If we ignore the recent naive attempts to prove that animals have "optimal" behavior and other characteristics, we still have the convincing direct evidence of convergent evolution. When unrelated species living in similar environments show similar traits, these traits can be called adaptations. It cannot be an accident that boobies, auks, gulls, terns, and other cliff-nesting birds behave similarly to one another but differently from close relatives that nest on the ground or in trees.

Turn the comparative method around to help trace phylogeny, the course of evolution. When related species behave similarly regardless of their ecological diversity, we must be dealing with evolutionarily conservative traits, telling us something about the behavior of common ancestors. Only those estrilidid finches living in the desert drink by sucking water through their bills, but all pigeons and doves do this. The trait is a typical adaptation (in this case to arid environments), but dove species that have spread to moist climes retain the trait of their desert ancestors.

It is fun to speculate about evolution, and speculation leads to new ideas. But ideas in science need to be testable through observable fact. Show me and I will believe. Students beginning their exploration of the world opened by science are all welcome to join me as honorary Missourians.

For additional readings see Bishop (1975) and Hailman (1982, 1988).

unrelated organisms and attempts to explain what caused those traits by searching for correlated environmental factors. The two different comparative approaches focus basically on divergence versus convergence, respectively. The two methods are often combined in a given investigation.

The study of ancestral relationships is based on the premise that closely related organisms share more traits than less-related organisms. The more traits that are considered, the more complete the picture. Homologous behaviors (i.e., those that are derived from a common ancestor) may show some divergence but still contain enough recognizable elements to identify them as homologous, just as the forelimbs of most terrestrial vertebrates retain two bones, the ulna and radius.

Studies of convergent, analogous behaviors are often more difficult to interpret than convergence in morphological or physiological traits. The strong morphological resemblance, for example, between dolphins and fish, two groups that are not closely related, is relatively easy to understand because both groups have evolved under the influence of aquatic environments (i.e., selection pressures for streamlining and means of propulsion in water). Both fish and dolphins also show group behavior, forming schools and pods, respectively, but the causes for that are much less obvious. Furthermore, it is not clear whether these grouping behaviors are truly even convergent from a functional standpoint or whether there is merely some superficial resemblance. Schooling does not necessarily have the same function among all the different kinds of fishes or even for different size or age groups in the same species (Chapter 13).

Polygamy is another example of behavior that may have similar appearances but different explanations, resulting from one set of factors in some species and a different set in other species (Chapter 12). The environmental conditions that result in convergence in behavior frequently involve interactions with other animals and the inanimate aspects of the environment. Much of this is discussed at greater length in chapters dealing with intraspecific (social) and interspecific behavior.

The comparative study of behavior differs from other aspects of biology in that most sequences of behavior are shown only infrequently and they may be of short duration. One can always see a leg or wing, for example, but a particular courtship behavior or act of predation may be seen only rarely. Also, the behavior may appear only under natural or undisturbed conditions. Furthermore, behavior may be less consistent than morphological and physiological traits. Thus the comparative study of behavior requires much time, often years of observation.

Another ramification of the dependence of behavior on neural output is that it is potentially more evolutionarily flexible than morphological characteristics. As a consequence, it becomes more difficult to discern between homologous and analogous behavioral traits. Just because two species have homologous neural pathways and other morphological features, for example, does not guarantee that their behaviors will also be homologous; in fact, just the opposite sometimes occurs (Lauder 1986). In other words, a behavior can be lost during the course of evolution so it no longer is considered homologous, but then a similar behavior can secondarily evolve and thus become analogous. In the meantime, however, the supporting muscular, skeletal, and basic neural structures might have undergone no change. To avoid spurious apparent homologies, one must consider several behaviors and search for congruence among them.

Comparative Study of Behavioral Homologies and Divergence: Inferences about Phylogenetic Origins and Ultimate Causes of Behavior

Detailed comparative studies involving homologous (or potentially homologous) behaviors have been conducted on a large number of taxonomic groups. A familiar example of homologous behavior involves head scratching in mammals and many birds (Figure 5-1). Quadrupedal animals generally scratch their heads by propping themselves on the front two and one of the back legs and using the other back leg for scratching. In its movement the back leg comes up and over the front leg on that side. Birds also use their back legs for scratching (the forelegs having become wings). Although it would seem logical to simply bring the leg forward to scratch

Figure 5-1 Conservative nature, or lack of evolutionary change, of vertebrate head scratching. Many birds retain the trait of scratching the head by reaching over the wing ("front leg") as shown by present quadrupeds and presumably inherited from quadrupedal ancestors. Modified from Lorenz, K.Z. 1958. Scientific American 199(6):67-68.

Figure 5-2 Courtship displays of surface-feeding ducks. Ten acts as illustrated by the mallard: *(1)* bill-shake, *(2)* head-flick, *(3)* tail-shake, *(4)* grunt-whistle, *(5)* head-up-tail-up, *(6)* turn-toward-female, *(7)* nod-swim, *(8)* turn-back-of-head, *(9)* bridling, and *(10)* down-up. Different species link these acts together in different sequences during courtship (top panels). Modified from Lorenz, K.Z. 1958. Scientific American 199(6):67-68.

| 3 | 2 | 3 | 1 | 4 | 3 | 5 | 6 | 7 | 8 | Mallard |

| 4 | 3 | 2 | 3 | 5 | 6 | 10 | 6 | Gadwall |

| 3 | 2 | 3 | 10 | 4 | 3 | 2 | 5 | 6 | 8 | European teal |

the head, only a few birds do it this way. Instead most species of birds have retained the primitive quadrupedal pattern and scratch by dropping the wing and bringing the leg, somewhat awkwardly, up and over the top.

Other homologous behavior patterns include many display postures, such as the courtship actions in waterfowl (Figure 5-2). Different species may use similar acts, but the sequences are different, as in sequences of numbers in combination padlocks.

Some studies of large numbers of behavior patterns in particular taxonomic groups permit the determination of the likely ancestral relationships through the apparent homologies. Comparative studies of different but closely related species and the environments in which they live sometimes provide insight into possible reasons for the divergence of the species and the ultimate causes of their behaviors.

Following are the general steps in a comparative study of ancestral relationships involving behavior:

1. Qualitatively describe the behavioral repertoire (i.e., construct an ethogram, Chapter 3).
2. Compare different species.
3. Formulate an evolutionary, phylogenetic hypothesis for the behavior.
4. Quantitatively test and refine the hypothesis.
5. Evaluate the outcome against other evidence.

Two of the classic examples of deriving phylogenetic relationships from behavior and demonstrating the evolution and divergence of behavior are waterfowl (Lorenz 1941) and the group of birds that includes pelicans (Tets 1965). Similar examples have been worked out for a large number of taxonomic groups. For illustrations of different points we have chosen three examples: vocalizations in cranes, nesting behavior of gulls, and courtship in empid flies.

Examples of Behavioral Homology and Divergence

Crane vocalizations and courtship dance behavior

The family of cranes (Gruidae) consists of 15 living species found in various parts of the world (see Figures 5-3 and 5-4 and boxed material on page 132). They are large, long-legged, spectacular birds with populations that vary from common to extremely endangered. Of the two species found in North America, the sandhill

Figure 5-3 **Different species of cranes found in the world. Although all species are shown together, they are widely distributed in nature, and only a few species can be found in the same location.**
Modified from Singer, A. 1978. Audubon 80(2):17-24.

Figure 5-4 **Distribution of crane species throughout the world. Names are located on the continent where the birds nest.**

crane is now relatively common, whereas the whooping crane is well known for its endangered status. For details on the natural history of the various species, see Johnsgard (1983).

Cranes are vocal and have a repertoire of about 10 distinct calls, including contact, stress, foodbegging, and others. The guard call and the closely related unison call are the subject of this example. The guard call is a single, short vocalization (approximately 1 second) given in threat or fear contexts when flight is not yet the course of action. It is independent of other vocalizations that may precede or follow.

The unison call is a series of calls lasting several seconds to over 1 minute and given in duet by both members of a pair of cranes. The unison call generally occurs in a courtship or sexual context. It may be accompanied by a variety of movements, particularly of the head, neck, and wings, commonly known as the "dance." Unison calls often incorporate guard calls as components in various manners, depending on the species. Figure 5-5 presents a sound spectrogram of the unison call of the white-naped crane. The unison calls and associated behaviors of 13 of the species were studied intensively by George Archibald (1976 a, b; Wood 1979).

For a behavior to be useful in elucidating the evolutionary relationships of a group, it should meet the following three criteria: (1) be homologous among the members of the group, (2) be largely innate (see Chapter 2), with little variation resulting from experience, and (3) not be used as a reproductive isolating mech-

Classification of Cranes*

Family	Gruidae	
Subfamily	Balearicinae	
Genus	*Balearica*	Crowned cranes
Species	1. *pavonina*	West crowned
	B.p. *pavonina*	Nigerian crowned
	B.p. *ceciliae*	Sudan crowned
	2. *regulorum*	East crowned
	B.r. *regulorum*	Southern crowned
	B.r. *gibberceps*	Kenyan crowned
Subfamily	Gruinae	
Genus	*Bugeranus*	
Species	1. *carunculatus*	Wattled
Genus	*Anthropoides*	
Species	1. *virgo*	Demoiselle
	2. *paradisea*	Stanley
Genus	*Grus*	
Species	1. *canadensis*	Sandhill
	G.c. *canadensis*	Lesser sandhill
	G.c. *rowani*	Canadian sandhill
	G.c. *tabida*	Greater sandhill
	G.c. *pratensis*	Florida sandhill
	G.c. *pulla*	Mississippi sandhilll
	G.c. *nesiotes*	Cuban sandhill
	2. *antigone*	Sarus
	G.a. *antigone*	Indian sarus
	G.a. *sharpii*	Eastern sarus
	3. *rubicunda*	Brolga
	4. *vipio*	White-naped
	5. *leucogeranus*	Siberian
	6. *monacha*	Hooded
	7. *grus*	Common
	G.g. *grus*	Europena
	G.g. *lilfordi*	Lilford
	8. *nigricollis*	Black-necked
	9. *americana*	Whooping
	10. *japonensis*	Manchurian

*In subsequent tables and figures scientific names are abbreviated to the first letters of the genus and species.

mechanism, which might create potentially confusing and excessive divergence of the display (Marler 1957). (In many species, behavior is important in reproductive isolation, see Chapter 11.)

The crane unison calls meet all of Marler's criteria. The unison calls are sexual pair displays in all species, and the sound spectrograms show varying degrees of structural homology in the calls, thus they may be reasonably considered as homologous.

Figure 5-5   Example of the unison call and associated dance movements for one species of crane, the white-naped crane *(Grus vipio).* The call lasts for 9 seconds, as shown in the 3-second panels. The sequence of calling by both male and female is indicated beneath the sound spectrogram along with diagrams of the associated postures.
Modified from Archibald, G.W. 1976. Ph.D. thesis. Cornell University, Ithaca, N.Y.

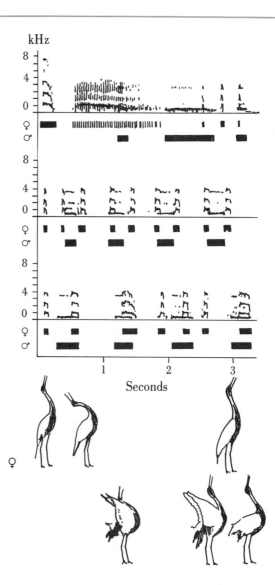

Chicks that are incubated, hatched, and reared under unnatural conditions, such as by humans or with foster species, and isolated from their own species develop unison calls that are indistinguishable from those of chicks hatched and raised by their own parents. Thus the calling appears to be innate.

Concerning reproductive isolation, various subspecies of sandhill cranes winter together and have indistinguishable unison calls but do not breed together; hooded and common cranes have similar unison calls and winter together in some areas and breed sympatrically in Siberia but have not been observed to hybridize. Eastern sarus cranes, which recently entered Australia, have different unison calls from the native brolga cranes and have been observed to hybridize with the brolgas. There appears to be little reproductive isolation associated with the unison calls. They function instead as a sexual display between members of a pair in all species and as a territorial threat display in some.

Table 5-2 Characteristics of the unison-call dance for 15 species of cranes*

	Bp	Br	Bc	Av	Ap	Gc	Ga	Gr	Gv	Gm	Gg	Gam	Gj	Gl	Gn
1. UC a series of GCs	X	X		X	X	X									
2. GCs often precede UC	X	X		X	X	X									
3. GCs rarely precede Uc							X	X	X	X	X	X	X		X
4. Gcs never precede UC			X											X	
5. Unique introductory calls absent	X	X				X				X	X	X	X		X
6. Unique introductory calls present			X	X	X	X	X	X	X	X	X	X	X	X	X
7. Either sex begins UC	X	X		X	X	X				X	X	X	X	X	X
8. Only ♀ begins UC			X				X	X	X						
9. Usually ♀ begins UC										X	X	X	X		X
10. Both broken and unbroken calls			X	X	X	X	X	X	X						
11. Broken to unbroken to broken to unbroken sequence				X	X										
12. Unbroken to broken to broken sequence			X			X	X	X	X						
13. Only unbroken calls			X			X				X	X	X	X	X	X
14. Glissando to regular calls			X							X	X	X	X	X	X
15. Several short calls to regular calls										X	X	X	X		X
16. Glissando and several short calls to regular calls										X	X	X	X		X
Regular Elements of the UC															
17. Guard calling mostly	X														
18. Boom calling mostly		X													
19. Boom	X	X													
20. Shrill calls			X	X	X	X	X	X	X	X	X	X	X	X	X
21. Calls generally uniform throughout			X				X	X	X	X	X	X	X	X	X
22. Two types of calls	X	X	X	X	X	X	X	X		X	X	X	X		X
23. Last call low, long, broken			X			X		X							
24. First few regular calls shorter							X	X	X	X	X	X	X		X
25. Guard calls and regular calls	X	X		X	X	X	X								
26. Monosyllabic and disyllabic calls										X	X	X	X		X
27. Usually monosyllabic												X	X		
28. Usually disyllabic										X	X				X
29. Glissandi present							X			X		X	X	X	X
30. Shrill call modified by gular sac								X							
31. Regular calls longer than regular ♂ calls	X	X	X				X								
32. Regular ♂ calls longer than regular ♀ calls				X	X	X	X	X	X	X	X	X	X		X
33. One ♂ call per ♀ call	X	X		X	X			X						X	

Data for the first 13 species are from Archibald, G.W. 1976. The unison call of cranes as a useful taxonomic tool. Ph.D. Thesis. Cornell University, Ithaca, N.Y. Details for the last 2 species were obtained later and provided by Archibald.

*X means the character is present in either both male or female. For full species names, see box on page 132.

Table 5-2 Characteristics of the unison-call dance for 15 species of cranes—cont'd

							Species								
	Bp	**Br**	**Bc**	**Av**	**Ap**	**Gc**	**Ga**	**Gr**	**Gv**	**Gm**	**Gg**	**Gam**	**Gj**	**Gl**	**Gn**
34. Two ♀ calls per ♂ call						X	X	X	X	X	X	X	X		X
35. More than two ♀ calls per ♂ call			X												
36. More than one ♂ call per ♀ call							X								
37. ♀ calls between 0.30 and 0.52 second							X			X		X	X	X	
38. ♀ calls less than 0.22 second			X	X	X	X	X	X	X		X				X
39. ♂ calls 0.11 second			X												
40. ♂ calls 0.21 to 0.30 second	X	X		X	X	X		X	X						
41. ♂ calls 0.38 to 0.65 second							X			X	X	X	X	X	X
42. UC determinate in length			X	X	X										
43. UC indeterminate	X	X					X	X	X	X	X	X	X	X	X
44. UC usually between 4 and 40 seconds							X	X	X	X	X	X	X	X	X
45. UC can last more than 1 minute	X	X												X	
46. Basal frequency below 0.70 kHz	X	X						X							
47. Basal frequency about 0.70 kHz to 0.90 kHz				X	X	X		X	X		X	X			X
48. Basal frequency about 0.90 kHz to 1.2 kHz				X	X	X			X	X	X		X		
49. Basal frequency about 1.4 kHz			X	X										X	
50. Only basal frequency with significant amounts of sound energy	X	X		X	X										
51. Much sound energy in several harmonics above the basal frequency			X			X	X	X	X	X	X	X	X	X	X
52. More than 80 dB at 10 feet							X	X							
53. Between 75 and 80 dB at 10 feet	X	X	X	X	X	X			X	X	X	X	X	X	X
54. Asynchronous UC	X	X													
55. Synchronous UC			X	X	X	X	X	X	X	X	X	X	X	X	X
56. After introduction, no apparent synchrony			X												
57. Depending on which sex starts UC the other synchronizes with it				X	X									X	
58. After the introduction, ♀ synchronizes her calls with ♂						X	X	X	X	X	X	X	X		X

Visual, Dance Characters Associated with UC

	Bp	**Br**	**Bc**	**Av**	**Ap**	**Gc**	**Ga**	**Gr**	**Gv**	**Gm**	**Gg**	**Gam**	**Gj**	**Gl**	**Gn**
59. Stand stationary	X	X	X	X	X	X	X	X	X	X	X	X	X	X	X
60. Usually walk										X	X	X	X	X	X
61. May or may not stand near each other	X	X												X	
62. Usually stand near each other			X	X	X	X	X	X	X	X	X	X	X		X

Continued.

Table 5-2 Characteristics of the unison-call dance for 15 species of cranes—cont'd

Species

	Bp	Br	Bc	Av	Ap	Gc	Ga	Gr	Gv	Gm	Gg	Gam	Gj	Gl	Gn
63. Sexes sometimes touch while UC					X		X	X	X						
64. Wings folded throughout	X	X	X	X	X	X	X	X	X	X	X	X	X	X	X
65. Always raise elbows			X		X		X	X	X						
66. Raise elbows with increased threat						X				X	X	X	X	X	X
67. Lower primaries with increased threat										X	X	X	X	X	X
68. Head lowered to shoulder level at start	X	X	X											X	
69. Head lowered to shoulder level throughout	X	X													
70. Wings pumped						X	X	X	X						
71. Upward neck thrust			X												
72. Neck back beyond vertical at start				X	X		X	X	X	X	X	X	X		X
73. Neck back beyond vertical throughout				X	X	X	X	X	X	X	X	X	X		X
74. Neck vertical throughout				X		X	X	X	X	X	X	X	X		X
75. Neck forward beyond vertical throughout			X							X	X	X	X		X
76. Head movement from side to side	X	X													
77. Head held in vertical plane throughout			X	X	X	X	X	X	X	X	X	X	X		X
78. Gular sac inflated	X	X						X							
79. Sexual context	X	X	X	X	X	X	X	X	X	X	X	X	X	X	X
80. Threat context				X	X	X	X	X	X	X	X	X	X	X	X

Table 5-3 Number of unison-call dance characteristics shared among different cranes*

	Bp	Br	Bc	Av	Ap	Gc	Ga	Gr	Gv	Gm	Gg	Gam	Gj
Bp	25												
Br	24	25											
Bc	6	6	29										
Av	12	12	13	29									
Ap	12	12	13	27	30								
Gc	12	12	15	22	22	34							
Ga	7	7	17	17	19	25	37						
Gr	9	7	17	18	19	25	30	37					
Gv	6	6	15	18	20	25	30	30	32				
Gm	8	8	14	17	17	20	27	22	23	40			
Gg	8	8	15	18	18	26	26	24	25	38	40		
Gam	8	8	14	18	17	20	26	23	23	38	37	40	
Gj	8	8	14	17	17	20	27	22	23	39	37	39	40

From Archibald, G.W. 1976. The unison call of cranes as a useful taxonomic tool. Ph.D. Thesis. Cornell University, Ithaca, N.Y.

*Total number of characters in the dance repertoire of each species is shown in the central diagonal. For species identification of symbols, see box on page 132.

Figure 5-6 A chart of crane unison-call dance characters arranged to reveal shared characteristics among 13 of the species of cranes.

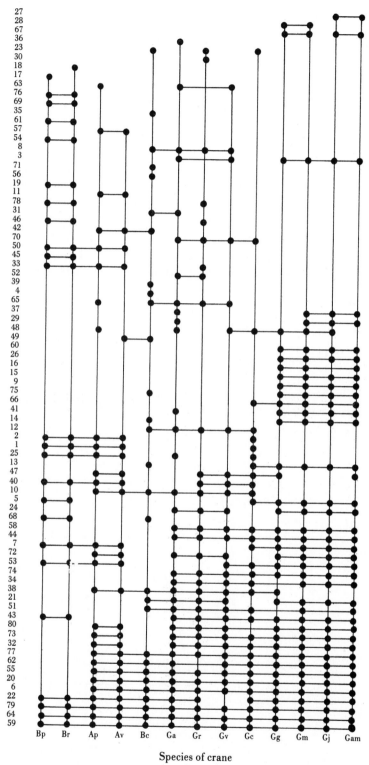

Unison call character

Species of crane

Archibald (1976 a) broke the unison calls and associated movements into 80 identifiable characters. The list is presented for all 15 species in Table 5-2 (two species were added after the initial study). The total number of characters in a given species' unison calls ranges from 25 to 40 (Table 5-3).

The proper analysis and construction of phylogenetic dendrograms ("trees") is complicated, depends on different approaches, and is best performed with the aid of computers (e.g., Sneath and Sokal 1973, Schnell 1970 a,b). For the sake of illustration, an alternate, manual technique is used, which was introduced by Lorenz (1941) for waterfowl. This method renders a visual pattern of all the species simultaneously; it is accomplished by inspecting a graph of species versus behavioral characters and rearranging the characters and species in different sequences until the maximum number of contiguous shared relationships is found. The job can be facilitated by starting with the characters that are found in all or most of the species, then working up to those that are least common. An outcome of one such effort, with the crane data, is shown in Figure 5-6 (see also Table 5-3).

Assuming that the most closely related species share the largest number of characters and the most distantly related share the fewest, one can use this infor-

Figure 5-7 Inferred family tree of crane species based on shared characteristics of the unison-call dance.

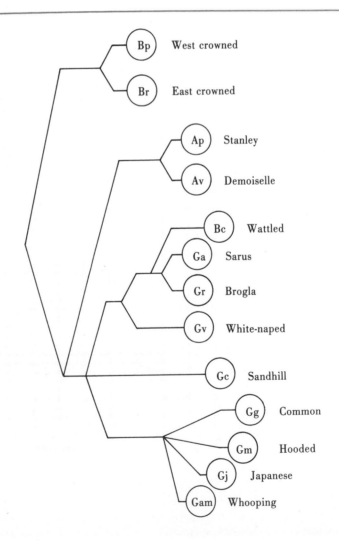

mation to infer the probable phylogenetic tree as shown in Figure 5-7. (The remaining two species of cranes, the Siberian and black-necked, were not included in the initial study because specimens could not be obtained for observation. These species have been observed since that time, and the results are shown in Table 5-2. The task of fitting these two species into the analysis performed for the other 13 will be left to interested readers as an exercise.)

The use of the unison calls to infer phylogenetic relationships corroborated the picture obtained from the fossil record, geographical distribution, physiological cold adaptedness, external morphology, and tracheal anatomy (Archibald 1976 a). For some of the comparisons it provided better resolution, comparable with that obtained from a more detailed morphological analysis (Wood 1979). The results of the behavioral comparisons were also nearly identical to a more recent phylogenetic analysis based on DNA hybridization (Krajewski 1989), differing only in that the DNA study showed the wattled crane to be more closely related to the Stanley and demoiselle cranes than the sarus group.

The reasons for divergence in the unison calls and dance behaviors of cranes are only partially understood at present (Archibald 1976 a). Some of the differences may be just accumulated genetic drift that developed after the species first separated geographically, when populations are typically small and genetic founder effects (resulting from the particular genotypes included by chance in the founding groups) and subsequent drift are quite likely. Some of the differences seem to be fairly clear adaptations to different ecological conditions. The two species with the loudest calls, sarus and brolga, are found in wide-open spaces where the birds are widely separated and loud calls more likely would be heard. The species with the quietest call, the wattled crane, is found under closer conditions where the birds are more grouped together. But the reasons for most of the differences and why the most recently evolved species, hooded, common, whooping, and Manchurian cranes, should be so specialized require further study and consideration. (Chapter 11 discusses current ideas on divergence in sexual displays in other species.)

Although the divergence and ultimate causes of crane courtship are only partially understood, there are other cases for which fuller interpretations have been proposed. Two of these are briefly described next.

Reproductive behavior of gulls, including nest building

One of the best examples of the evolution of behavior in which the selective forces seem apparent involves gull nesting behavior, with emphasis on cliff dwellers, the kittiwakes. Much of the research leading to an understanding of gull reproductive behavior was conducted by Tinbergen (1953, 1954, 1959, 1960) and his student Esther Cullen (1957).

The 35 or so species of gulls are a remarkably similar group of birds belonging to only a few genera; most are in the genus *Larus*. There are some groupings, such as large gulls and hooded gulls, and there are certainly differences in their behavior. But for the most part the differences are slight, and the similarities are perhaps more noticeable. There are several common components in their courtship and nesting displays (Figure 5-8), including long calls, mew calls, choking displays at the nest, upright threat postures, grass pulling threat displays, and facing away appeasement postures in which birds turn their heads and weapons (beaks) away from each other. Differences among the species primarily amount to variations on these common themes.

Figure 5-8 Common display postures of gulls, as shown by the herring gull. The oblique, mew, forward, and sometimes choking movements are accompanied by characteristic calls. These are incorporated in different ways into various displays by different species of gulls. Modified from Tinbergen, N. 1960. The evolution of behavior in gulls. Scientific American 203(6):118-130.

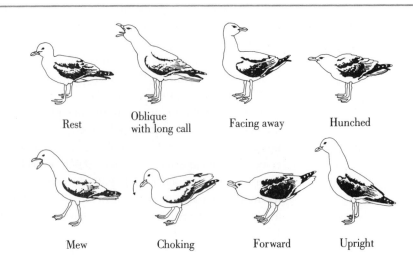

Rest

Oblique with long call

Facing away

Hunched

Mew

Choking

Forward

Upright

Gulls as a group also show much similarity in their life histories and ecology. Most are generalist scavengers, and the majority nest on the ground on lake or oceanic islands or along shorelines.

A couple of species, including the kittiwake, *Rissa tridactyla*, however, show a striking departure from the other gulls. Kittiwakes nest on narrow ledges high above the ground or water, presumably as an adaptation against ground predators (Cullen 1957). Along with the difference in nesting substrate, they show a whole constellation of behavioral differences from the other gulls. Visible elements of the common behavior patterns remain. Kittiwake eggs retain traces of cryptic coloration and blotching, and kittiwake chicks are able to run about under some circumstances. So it seems reasonable that kittiwakes evolved from a ground-dwelling ancestor. However, the behavioral differences are considerable. As an example, chicks of most gull species roam about from the nest starting a day or so after hatching. They may run for cover in the presence of danger, and they may flee with food from nest mates. Roaming about on small ledges could be fatal and, accordingly, chicks of kittiwakes stay put in the nests. In interactions with nest mates, rather than running away, they show head-flagging appeasement behavior, movements that do not occur in other species until the birds become older.

Cullen tallied about 30 behavioral or behavior-related differences between kittiwakes and ground-nesting gulls (see box on pages 142 and 143). The evolutionary interpretations of these differences, as they relate to ground versus cliff nesting, seem quite reasonable, if not almost obvious. The interpretations can be derived by inspection and thinking.

Imagine, for example, variation in the behavior of ancestral gulls incubating eggs in response to the appearance of a distant predator. Variation could range from birds that sit tight to those that flee from the nest, and from those that leave the vicinity completely to those that fly at and attack (mob) the predator. The selective advantages are different for ground versus cliff nesting. For ground nesters, fleeing hens would escape and mobbing would have the effect of driving away the predator, whereas hens that sat tight would be captured and eaten, obviously reducing both survival and reproduction. Ground predators would pose no threat to cliff nesters, however; fleeing would result in no selective advantage and could be a disadvantage by wasting energy and exposing eggs during inclement weather. Similar straight-forward arguments can be made for the other differences.

Courtship of empid flies

A classic example of evolutionary interpretation of behavior involves the courtship of a type of predatory dipteran commonly called *empid flies* (from the family Empididae); they are also called *dance* and *balloon flies*. Male empid flies of one species give empty silk balloons to the females during courtship. A comparative study (Kessel 1955) revealed that males of some species give nothing and occasionally get eaten themselves. In other species the males give prey wrapped in silk. In yet others the males present the females with prey only. If P = prey, W = wrapping, and presence or absence of the behavior is indicated by + or − respectively, the reconstructed hypothetical sequence of evolution of the behavior is

$$P - W - \text{ to } P + W - \text{ to } P + W + \text{ to } P - W +$$

Clearly the behavior has greatly changed from its ancestral form. The original function (distraction of the female with edible prey during the time the male is mating) has been replaced by a behavior that distracts without providing food. The cause of the change (i.e., why empty balloons rather than wrapped prey are given by the males) is not known. It may have resulted from genetic drift or some other nonadaptive source. It is also reasonable to infer, however, that empty balloons are adaptive; they require less energy expenditure by males than would the work of catching prey. Figure 5-9 arranges the different species by their behaviors in this order.

Figure 5-9 Some of the major categories of empid fly courtship behavior, presented in order of inferred evolutionary development. These flies are predatory, normally eating other insects, but they also may kill and eat conspecifics. A, Males of some species go directly to the female and may either be eaten by the female or be successful at copulation. B, Males in these groups of species carry prey when courting and present it to the female. The male copulates while the female eats the item. C, Males capture prey and wrap it in silk balloons before presenting it to the female. The male copulates while the female unwraps and eats the prey. D, Males of at least one species *(Hilara sartor)* do not capture prey first but simply present the female with an empty balloon. Other intermediate stages in the evolution of this behavior also have been identified.

Based on information in Kessel, E.L. 1955. Sys. Zoology 4:97-104.

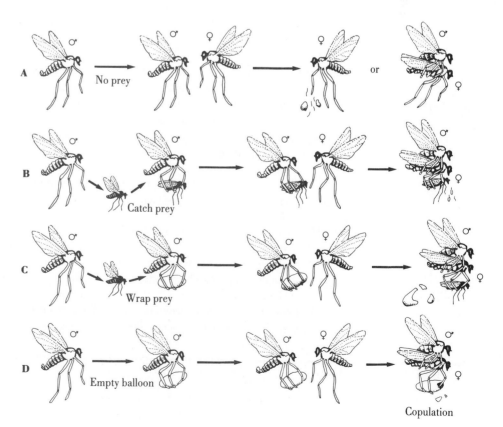

Behavioral differences between ground-nesting gulls and the cliff-nesting kittiwake

Ground-Nesting Gulls	Kittiwake
High predation rate in nesting colonies	**Predation pressure relaxed on cliffs**
Alarm call frequent	Alarm call rarer
Adults leave nest when predator some way distant	Remain on nest until predator close
Vigorously attacks predator intruding in colony	Weak attacks at intruding predator
Brooding birds disperse droppings and carry egg shells away from nest	Neither droppings nor egg shells dispersed
Young cryptic in appearance and behavior	Young not cryptic in appearance or behavior
Clutch size normally three eggs	Clutch size normally two eggs
Suited to Life in Colony on Ground	**Adapted to Life on Cliffs**
A. Several fighting methods	Specialized to fighting in one way (grabbing beak and twisting)
Upright threat posture occurs, derived from preparation to peck down at opponent	No upright threat
Beak does not especially direct attacks	Beak releases and directs attacks
Not known if it is as strong a releasing mechanism as in the kittiwakes	
Beak turned away in appeasement but not elaborately hidden	Beak turned away in appeasement and elaborately hidden
B. Young run away when attacked	Young do not run when attacked
No head flagging in young	Head turning and hiding of beak in young when pecked and appropriate behavior in attacker
No neck band	Possess black neck band
C. Number of nest sites less restricted and therefore probably less competition for nest sites	Number of nest sites restricted, probably more competition
Often first occupy pairing territories before nesting territories and pairs form away from nest	Occupy nesting ledges at arrival in breeding area and pairs form on the nest
Choking not normally used by unmated males as advertisement display	Choking normal advertisement display of unmated males
D. Copulation on the ground, female stands	Copulation on the tiny ledge or nest, female sits on tarsi
E. Nest material collected near nest, building not synchronised, individual collecting	Nest material collected in unfamiliar places, synchronization of building and social collecting
Little stealing of nest material	Birds readily steal nest material
Nests often unguarded before laying of first egg	Nests guarded

Modified from Cullen, E. 1957. Adaptations in the kittiwake to cliff-nesting. Ibis 99:275-301.

	Behavioral differences between ground-nesting gulls and the cliff-nesting kittiwake—cont'd.	
	Nest building technique relatively simple	Nest building technique elaborate
	Mud not used	Mud as nest material
	Only one or, at most, short series of depositing jerks	Prolonged jerking of head when depositing nest material
	Only traces of trampling on nest material	Prolonged trampling on nest material
	Nest has relatively shallow cup	Nest has deep cup
F.	Young leave nest a few days after hatching	Young stay in nest for long period
	Young fed by regurgitation on the ground	Young fed from throat
	Nest cleaning absent or less conspicuous	Young and adults pick up and throw away strange objects falling into nest
	Parents have feeding call, probably to attract young	Parents have no feeding call
	Hungry young make themselves conspicuous to parents by head pumping	Head pumping absent in young
	Parents learn to recognize own young in a few days	Parents do not recognize own chicks at least up to the age of 4 weeks
G.	Young face any direction. Vigorous wing flapping in young	Young face wall much of the time. Flight movements weak
H.	Weak claws, cannot hold on well	Strongly developed claws and toe musculature

Comparative Study of Behavioral Analogies and Convergence

Although some behaviors yield fairly logical and straightforward interpretations based on inspecting of and thinking about homologies among closely related species, such comparisons are not always possible or sufficient. An additional or alternate approach is to broaden the perspective and compare possible analogous, convergent behaviors among species that are only distantly related. An important, relatively recent approach involves the detailed analysis of environmental correlates of particular behaviors among a large number of species in which the behavioral traits could have arisen independently.

This approach developed from studies by several persons, including important work by Chance (1959) and Clutton-Brock and Harvey (1977) on primate sociality and Crook (1964) and Baker and Parker (1979) on bird coloration and behavior. All of these studies considered the environmental differences that may have led to the observed behavioral differences.

As the methodology improved, more statistics became involved, to the point where this approach is considered as strong or nearly as strong as the genuinely experimental method. In essence, one is merely taking advantage of experiments that have already taken place in nature where there is a sufficiently large sample size to reduce the impact of spurious correlations. The problems and dangers of correlation, nonindependent homologous traits, and small sample sizes, however, lurk just beneath the surface and almost never can be completely ruled out.

Figure 5-10 Precopulation guarding in amphipods. From Krebs, J.R., and N.B. Davies. 1987. An introduction to behavioural ecology. Blackwell Scientific Publications, Boston.

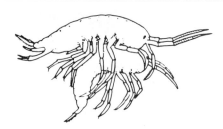

Mark Ridley (1983) discussed these problems in detail plus a number of other logical and logistical aspects of comparative study using convergence. He then applied it to two examples, one of which—precopulatory guarding—provided a relatively clear conclusion.

In precopulatory guarding the male literally grabs hold of the female and hangs on, staying with her for long periods of time, usually days to weeks. Fertilization occurs and the two separate and go their own ways. The phenomenon occurs widely among arthropods, some other invertebrates, and anurans (frogs and toads). Familiar examples include many of the crustaceans known as amphipods or sideswimmers (Figure 5-10), and some species of toads in which the males ride around on the females' backs for days. (Other forms of mate guarding also occur throughout the animal kingdom, see Chapter 11. Precopulatory guarding is considered here only when the male physically attaches himself to the female for long periods of time until fertilization occurs, after which they separate.)

Ridley reviewed the literature and compared observed versus predicted instances of precopulatory guarding. This behavior was predicted to evolve when females would become receptive for mating during predictable, brief periods of time such as after a particular molt or in a short breeding season but not if females are continuously receptive or if their periods of receptivity are unpredictable. Using taxonomic groupings above the species level to ensure independent trials, Ridley found a highly significant relationship between observed and predicted behavior. Precopulatory behavior was observed 19 of 20 times where it was predicted, and it was not found 10 out of 11 times where it was not predicted. At the species level it seems even clearer (only 2 out of 401 species not observed as predicted); however, many of the species were related, and might have inherited the behavior homologously from a common ancestor, and the cases could not be considered independent at the species level. Thus, even though quantitative, studies of analogous traits remain correlational and the outcomes speculative to a degree. The comparative method is a strong tool, but it is far from perfect.

EVOLUTIONARY CHARACTERISTICS OF BEHAVIOR

When behavior has been described and compared among various species, one finds virtually all of the evolutionary characteristics associated with morphology and physiology. Some examples of divergence and convergence have been discussed previously. Another example of divergent behavior involves the feeding behaviors that accompany the different beak shapes and body sizes of Darwin's finches in the Galapagos Islands. Familiar possibly convergent (at least in expression) behaviors include parental care, other social behaviors of many types, learning, and the gliding behavior of "flying" lizards, frogs, and squirrels. Parallel evolution, (similar changes in different groups) and clines (a range of variation over a geographical distance) have been demonstrated in a number of behavioral traits such as displays of jays (Brown 1963), feeding behavior of garter snakes (Burghardt 1970 a), and nesting behavior of some mice (King et al 1964).

Many behaviors show polymorphism (different, discrete phenotypes or "morphs") among different individuals of the same species. An example of behavioral polymorphism involves the migratory activity in old world blackcap warblers *(Sylvia atricapilla)*. The behavioral morphs are migratory (measured by migratory restlessness in captivity, discussed in Chapter 8) and sedentary (not showing migratory restlessness) (Berthold and Querner 1986). Perhaps one of the most polymorphic behaviors that has been documented involves army ants *(Eciton burchelli)* in which there are four castes of workers that differ in their physical structure and associated behaviors (Franks 1985).

In addition, behavior shows some new evolutionary twists such as ritualization. A particular behavior may evolve away from an original function to become ritualized in a new context, generally involving communication displays. The empid fly behavior described previously is an example of ritualized behavior. Once the principle of ritualization was recognized, it became easier to understand several confusing

Figure 5-11 Evolution of courtship in galliform birds from the food call.
A, domestic rooster;
B, ring-necked pheasant;
C, impeyan pheasant;
D, turkey; and **E,** peacock.
After Schenkel, R. 1956; modified from Eibl-Eibesfeldt, I. 1970. Ethology. Holt, Rinehart and Winston, New York.

behaviors. Another evolutionary feature seen in behavior, often in conjunction with ritualization, is exaggeration, for example, of coloration, structures, and movement.

A familiar example of both ritualization and exaggeration involves tail-spreading displays among galliform (poultrylike) birds (Schenkel 1956; Eibl-Eibesfeldt 1972) (Figure 5-11). Male jungle fowl and domestic chickens *(Gallus domesticus)* that are descended from jungle fowl have a calling-to-food behavior in which the cock attracts females by bending forward and pointing to a food item on the ground with its beak. As the bird's body tips forward, the tail is raised as a physical consequence of the action, but it does not seem overly accentuated (i.e., more than jungle fowl tails normally are in any other position). In ring-necked pheasants the long tail appears quite prominent when the bird tips forward during calling-to-food behavior. In various other species such as other pheasants, grouse, and turkey *(Meleagris gallopavo)* the tail shows various intermediate degrees of development, patterning, and coloration, and in many species, spreading or fanning movements. The tail morphology and accompanying behavior during courtship show the peak of ritualization and exaggeration in the peacock *(Pavo cristatus)*.

There are several sources for behaviors that subsequently undergo ritualization and exaggeration in evolution. Many of the sources involve maintenance or other routine behavior (see Chapter 6). Some of the commonly recognized sources include the following:

1. Thermoregulation. This includes hair and feather raising and the distribution of blood (as in blushing).
2. Intention movements for locomotion. Many birds, for example, lean forward and lower their bodies before launching themselves into flight. Jumping animals of other types similarly show a preliminary change in posture before leaping.
3. Protective movements. Some facial movements of dogs, cats, and primates and head turning in gulls are thought to be examples.
4. Conflict behavior. Examples of conflict behaviors as sources for the evolution of new behaviors include displacement or redirected body care (e.g., preening) and eating movements.

Other evolutionary characteristics of behavior include transfer of a display beyond the animal, such as in the bower bird displays (Chapter 12), and developmental acceleration of behavior for display purposes. With acceleration, young animals show behaviors that would otherwise be seen only in older animals and often in different contexts. An example of acceleration is head flagging in kittiwake chicks as described on pages 142 and 143.

Although most behavior might be subject to evolution, some categories of behavior seem to be more evolutionarily labile (i.e., they show greater change and diversity among species than other types of behavior). Why? The following are proposed reasons: (1) many behaviors, such as communication and courtship, may have a species-specific or species-isolating function; (2) they may permit more efficient use of resources; (3) competition may lead to increased specialization; and (4) other interactions between two or more parties can lead to rapid, even runaway, coevolution. The latter situation includes courtship interactions (see the topic of sexual selection, Chapter 11) and predator-prey, parasite-host, and other interspecific interactions (see in Chapters 9 and 10).

It also may be worth considering why some other categories of behavior are more stable and do not change.

1. Perhaps the variation has not occurred. (Imagine an empid fly not having a prey item when it went to court.) In other words, evolution along a particular line simply has not yet had the opportunity to start.
2. There may be no preadapted behavior (a reason closely related to the first). *Preadapted* (or *exapted*, Gould and Vrba 1982) refers to traits existing first in a different, even nonfunctional, context before they serve the particular role or selective advantage being discussed. Nonpredatory flies, for example, normally do not carry prey. The change needs a foundation or starting point.
3. There may be no advantage to changing. Some changes may not offer a selective advantage; they might even be disadvantageous. In the case of the empid flies, there was a clear selective advantage to males giving females something to eat other than themselves.

Dawkins (1982) provides an extensive discussion of these and other constraints on evolution, including the problem of time lags. As an example of the time-lag problem, Dawkins notes the response of European hedgehogs to roll up in balls when they encounter predators. That does not work when encountering automobiles, but there has not yet been enough time for the evolutionary development of an appropriate response toward automobiles in hedgehogs. A similar problem in North America exists for armadillos, which jump into the air when startled. They are frequently killed by jumping into the undersides of automobiles passing overhead. Many other species, however, have a more generalized response to predators— escape by running away—which does work for automobiles.

One aspect of evolution is noted mostly in passing: the point(s) at which selection operates, whether at the level of individuals, groups, or molecules (selfish genes). This is a debated subject that often involves behavioral issues. The matter is unresolved and open-ended. Because much space would be required to develop and address the arguments on various positions, we decided not to treat the topic at length in this edition, although it is alluded to in Chapter 14. We defer to discussions elsewhere. For students or instructors who want to explore this topic, we recommend starting with Dawkins 1982, 1989, Wilson 1980, and Brandon and Burian 1984.

In the remainder of this book, the validity of behavioral evolution is assumed as a given, underlying premise. This is not to say that all behavior is evolutionarily predetermined or fixed. As discussed in the previous chapter on the related topic of genetics, even behavior that has high heritability and patterns that are largely innate can be modified in some cases. Modification and changes of behavior during the life of an individual are discussed in Chapters 19 and 21.

SUMMARY

Because of the role of heredity in behavior, selection and other evolutionary forces affect behavior as in morphology and physiology, leading to evolutionary changes. Inferences on the evolution of behavior have been obtained through several approaches, including experimentation, historical (via the fossil record on such factors as tracks and morphological counterparts of behavior), and comparative techniques. The comparative approach has been particularly important for an evolutionary understanding of behavior involving both the divergence of homologous traits and convergence of analogous behavioral traits.

Evolutionary features that are familiar in morphology and physiology, such as the divergence and convergence already mentioned, plus parallel evolution, clines, and polymorphism, can also be seen in behavior. Some features, such as ritualization, exaggeration, transfer of display beyond the animal, and developmental acceleration, are of particular interest in behavior. Several examples illustrate these principles, including crane vocalizations and associated dance behavior, gull nesting, and the courtship of predatory empid flies, courtship in gallinaceous birds, and others. There are several sources of evolutionary change in behavior, and some categories of behavior appear to be more subject to change than others.

FOR FURTHER READING

Aronson, L.R., E. Tobach, D.S. Lehrman, and J.S. Rosenblatt, editors. 1970. Development and evolution of behavior. W.H. Freeman, San Francisco.
A collection of articles by various authors.

Dawkins, M.S. 1986. Unravelling animal behaviour. Longman Group Limited, Essex, England.
This concise book specifically addresses a number of the common points of misunderstanding in behavioral genetics and evolution. It provides further discussion of several of the issues raised in this chapter (as well as topics such as innateness, instinct, optimality, inclusive fitness, and evolutionarily stable strategies discussed elsewhere in the book).

Dawkins, R. 1982. The extended phenotype. W.H. Freeman, San Francisco.
Discusses evolution using numerous behavioral examples.

Dawkins, R. 1989. The selfish gene. Oxford University Press, New York.
The second edition of a popular book (first published in 1976) concerning the level (gene, individual, group) at which selection operates, with much discussion of behavior.

Futuyma, D.J. 1986. Evolutionary biology. ed. 2. Sinauer Association, Sunderland, Mass.
A further introduction to the general subject of evolution.

Greenwood, P.J., P.H. Harvey, and M. Slatkin, editors. 1985. Evolution: essays in honour of John Maynard Smith. Cambridge University Press, Cambridge.
For persons wishing to pursue somewhat more advanced topics, including many pertaining to behavior.

Johnsgard, P.A. 1983. Cranes of the World. Indiana University Press, Bloomington, Ind.
This book provides a further introduction and background to cranes and their natural history.

Nitecki, M.H., and J.A. Kitchell, editors. 1986. Evolution of animal behavior: paleontological and field approaches. Oxford University Press, New York.
A short collection of seven papers, partially review and partially original reports, divided into two sections, one on historical approaches and one on field and experimental approaches to the evolution of behavior.

Ridley, M. 1983. The explanation of organic diversity. Clarendon Press, Oxford.
This book begins with a discussion of the comparative method, with emphasis on using convergence to study adaptation, then applies the techniques to two examples of mating behavior: precopulatory mate guarding and homogamy (assortative mating).

PART TWO

THE BEHAVIORAL REPERTOIRE
OBSERVED FORMS AND INTERPRETED FUNCTIONS

6

MAINTENANCE BEHAVIOR AND TEMPORAL CONSIDERATIONS

Maintenance behaviors of animals permit individuals to meet the minimum requirements of living. Such behaviors are important both to the animals and to students observing them, for a host of reasons. This category of behavior includes those behavior patterns that afford the animal protection from routine adversities of the physical environment and that keep the body operational. More specifically, there are various types of grooming and toilet behavior, miscellaneous simple body movements, thermoregulatory behavior, related behaviors such as shelter seeking or construction, and feeding and drinking movements. Feeding is a broad topic that is considered separately (Chapter 9). The remainder of the maintenance behaviors are briefly surveyed in this chapter. Many of these activities occur on a rhythmic schedule alternating with inactivity, including sleep, which raises the issue of time in behavior. The chapter concludes with a consideration of biological rhythms and the enigmatic phenomenon of sleep.

THE IMPORTANCE OF MAINTENANCE BEHAVIOR

Any behavior that keeps an animal in proper operating condition and equips it for normal environmental conditions is obviously important to the animal itself. An awareness of this somewhat mundane category of animal behavior is also important to any person studying behavior, for four major reasons. First, basic behavior and inactivity occupy much of an animal's time and behavioral repertoire. Such behaviors, as a result, are observed most commonly.

Second, these common, relatively simple behavior patterns often serve as good subjects for understanding the internal machinery and control of behavior. To understand the role of sensory feedback, for example, one can remove a single leg from an insect and then analyze subsequent locomotion changes in the remaining five legs (Hughes 1957).

Third, except for feeding and shelter construction, the basic behavior patterns in their original function generally are conservative and least likely to be modified by either evolution (Chapter 5) or learning (Chapter 20). Hence they may provide insight into both of these phenomena. The relatively fixed nature of these behaviors can be seen in a variety of contexts. A classic example of the conservative nature of maintenance behaviors involves head scratching in birds (Chapter 5).

Finally, these behaviors serve as important sources for the evolution of other behaviors and displays (Chapter 5). Although the basic behaviors, in their normal context, show little change, they may become incorporated into other behavior patterns, which then show much change and variability, in ritualization, exaggeration, and divergence. The speculum-flashing displays of many ducks, for example, clearly seem to have been derived from preening movements. Another interesting and possibly derived display in waterfowl, and perhaps in other birds, is tail wagging. Hailman and Dzelzkalns (1974) suggested that tail wagging in waterfowl may have evolved from a simple tail-shaking movement, as in getting rid of water on the feathers, to a form of punctuation in their communication. The subjects of animal displays and communication, which so importantly incorporate and elaborate some of the maintenance movements, are discussed in Chapters 11 to 15. The remainder of this chapter is devoted to brief descriptions of maintenance behaviors with comments on evolutionary aspects where appropriate.

CATEGORIES OF MAINTENANCE AND RELATED BEHAVIOR

The activities of animals may be classified on a functional basis (Chapter 3). A functional category can involve behaviors that range from simple movements to complex behaviors. Thermoregulation, for example, may simply require raising or lowering the hair or feathers. In other instances, however, it may involve changing the orientation of the body or moving considerable distances to obtain a different microhabitat, such as into or out of sunlight, to warm rocks, or underground. Because of such variability, maintenance behaviors sometimes defy easy classification; any attempt to define categories is arbitrary.

Respiration and General Body Care

Many relatively simple behaviors are associated with respiration, elimination, stretching and yawning, grooming, scratching, washing and bathing, drying, and molting or ecdysis. Most species, including humans, show characteristic, stereo-typed patterns of movement in these basic behaviors.

Respiratory movements include the following: (1) gill movements among oste-ichthyes (bony fishes), (2) continuous swimming or pharyngeal pumping movements in chondrichthyes (cartilaginous fishes), (3) gill waving among animals with external gills, (4) pumping of the oral and pharyngeal regions in amphibians with positive pressure breathing, in which air is pumped into the lungs by force from the mouth region, and (5) bellowlike sucking movements of the chest or abdomen by amniotic vertebrates and various invertebrates that possess negative pressure (sucking) breathing. Some aquatic species with rudimentary lungs gulp air during periods of low oxygen concentration in the water. Other aquatic but air-breathing organisms, including both invertebrates and vertebrates, direct much of their movement toward the surface to breathe periodically. Some invertebrates and humans have acquired various means for carrying or storing air outside the body for use underwater. Other respiratory or respiratory-related movements include coughing, hiccuping, and sneezing.

Elimination, defecation, and urination (micturition), involve characteristic pos-tures (Figure 6-1) that also may differ between the sexes or even with social status. Male and female dogs, for example, urinate quite differently; the males cock one leg and urinate to the side, whereas females usually squat. Sprague and Anisko (1973) described in detail several characteristic elimination behavior patterns in the beagle dog. Cats (Felidae) are well known for their toilet behavior. They dig in a soft substrate, eliminate, then cover the waste by pawing and pulling material over it. Birds usually defecate by leaning forward, raising the tail, and spreading the feathers in the anal area. Some birds eject the defecated material great distances, whereas other species merely drop it. Young birds in the nest may back to the edge to defecate over the edge; they may defecate with droppings in a membrane, the package of which is picked up and removed by the parent; or they may simply defecate on the nest material, which, depending on the species and individual,

Figure 6-1 Pronghorn
antelope defecating.
Courtesy T. Brakefield.

may be picked up with the nest material and removed (Lorenz 1970), covered up with new nest material, or ignored. There is much variation even among individuals of the same species, and some nests may be clean, whereas others are excessively fouled.

Eliminative behavior has evolved in many different ways, depending in part on the species' place and manner of living. Many species that have relatively permanent dens, nests, or other living sites show a degree of voluntary control over when and where (but usually not how) they will defecate and urinate. The place usually is located away from normal living quarters. The selective advantages of this are obvious: indiscriminate defecation and urination could attract predators via the odor, could lead to disease and health problems, and could raise humidity to harmful levels. The waste products also could accumulate excessively. In species that normally move from place to place, such as ungulates, or that live above the substrate (as in arboreal species, including most primates, and aquatic species), the waste products simply drop away and pose little problem. There would be little selective advantage to voluntarily monitoring and controlling the process. Accordingly, the behavior remains under involuntary control. (This is probably the chief reason monkeys are not more popular as house pets.)

Sloths provide interesting, if not extreme, exceptions to the general patterns of eliminative behavior for both den-living and arboreal species. The extinct giant ground sloths defecated in their caves with the dung accumulating to great depths (Long and Martin 1974). The arboreal three-toed sloths (*Bradypus variegatus*) of Central and South America hold their urine and feces for about a week at a time, then slowly and laboriously descend to the ground where they form a depression at the base of a tree with their stubby tails, defecate and urinate, cover it with leaves, then climb back up into the tree. Montgomery and Sunquist (1975, also see Montogomery 1983) speculated that this strange behavior may function to recycle significant amounts of nutrients back to their food trees.

Vultures have an interesting behavior known as urohydrosis. They direct their defecation onto their own legs; the evaporation cools the legs and aids thermoregulation of the body during periods of high temperature.

Stretching and yawning are seen among osteichthyes, reptiles, birds, mammals, and perhaps some of the other vertebrates. The function of stretching appears to be physiological, including increased heart rate, blood pressure, circulation in muscles, and muscle and joint flexibility (McArdle et al 1981, Shephard 1982).

Luttenberger (1975) showed in a study of yawning and stretching in tortoises (*Testudo hermanni*) that yawning increases with muscular fatigue, an increase in CO_2, apparently with hunger, and also with some temperature changes. He showed that yawning was suppressed, but limb stretching was stimulated, by a shortage of O_2. Yawning and stretching may affect the muscles directly or simply increase or improve circulation. Hadidian (1980) compared yawning in black monkeys (*Macaca nigra*) with numerous other vertebrates and showed that structurally it was highly stereotyped and clearly homologous. In addition to the simple maintenance function associated with fatigue, however, it was seen in other contexts, as described also by other researchers for other primates (reviewed by Hadidian). The other contexts include stress and threat or canine (teeth) displays, hence communication. Yawning occurs most frequently in adult males; rates increase with age.

Yawning in humans has received some study (e.g., Provine 1986, Provine et al 1987). Yawning may occur at any time but is most frequent in humans after waking or before sleeping. Stretching is much more frequent after waking than before sleeping. Stretching is often accompanied by yawning, yawning is less often accompanied by stretching. In humans yawning appears to have little if any physiological function per se (there is no support for the popular idea that it is associated with CO_2 or O_2 levels in the blood). As an indication, increased breathing through the nose will not satisfy the urge to yawn. Instead human yawning may serve a largely communicative function as a paralinguistic signal (Provine et al 1987) for drowsiness and may help synchronize the behavioral state of a group. Provine (1986) considers human yawning to be a classic action pattern with stereotyped expression and identifiable releasing stimuli, which include seeing others yawn, thinking about yawning, or even reading about it.

Yawning may have originated in fish. Fish yawning seems homologous to yawning in terrestrial vertebrates but apparently does not occur under similar situations (Rasa 1971). In fish, it is associated with increased activity rather than sleepiness or the relaxation of tension; it is not associated with breathing and O_2 or CO_2 discrepancies; and it does not have the infectious quality found in other vertebrates (Rasa 1971).

Washing, bathing, and grooming are performed either with movements of the entire body or with sequences involving the head and various limbs. These movements are obvious and familiar to most people. Intensive studies of grooming behavior of hymenopteran insects (bees, wasps, ants) contributed to an understanding of the phylogeny of the order (Farish 1972). There are also detailed studies of grooming behavior in a wide variety of other organisms. They range, for example, from grooming in flies (Dawkins and Dawkins 1976) and mantids (Zack 1978) to kangaroo rats (Randall 1981) and herring gulls (Rhijn 1977). Dust bathing, seen in many birds and some mammals, apparently functions to reduce parasitism. Many animals, particularly mammals, roll or wallow in dust or mud. Canids (dogs, foxes, wolves) often rub their necks or roll in strong-smelling substances such as feces or rotten carcasses, much to the dismay of many dog owners. A unique apparent bathing or grooming behavior in many birds is called *anting*. They lie or roll in ant nests or pick up ants and wipe them about on their feathers. The function(s), if any, of these behaviors is not known.

Preening is a frequent behavior seen in birds and some insects such as flies. Animals pull their feathers and wings (in birds) or antennae (in insects) through or across the beak or legs, respectively, in a combing like motion. Avian preening helps spread oil obtained from the uropygial gland, near the base of the tail, onto the feathers for waterproofing. The movements also help reunite the hooks and hooklets of feather vanes that may have become separated. It seems amazing that some birds can preen, given the shapes of their beaks. Insects apparently use preening to keep body parts clean. Objects that fly, whether aircraft or living organisms, generally require more maintenance of parts than objects that do not fly.

Scratching to remove surface irritations can be seen among all vertebrates, including amphibians and fishes. If the irritation is located where it cannot be reached with an appendage, such as in fishes, the animal may rub itself on rocks, the ground, or some other structure in the environment. Scratching also involves species-specific movements. For any particular part of the body, most individuals, even humans, will use the same appendage and manner of scratching.

Drying behavior, also remarkably species-consistent, may occur after deliberate or accidental wetting. Drying may be accomplished by shaking, rubbing, or extending various appendages to increase the surface area.

Shedding of epidermis (including hair and feathers) or exoskeleton, as in arthropods, can be seen in many species; the process is called *molting* or *ecdysis* and involves characteristic movements to accomplish. Some insects shed larval or nymphal exoskeletons, emerge as adults, and assume different habitats and ways of life. Shedding organisms may move to specific locations, and they usually engage in characteristic and quite complex movements that remove the old skin or otherwise extricate the animal from it. Carlson (1977), for example, provides a detailed description of the neural and muscular sequences involved in cricket molt. There also may be characteristic postmolting behavior (Cloarec 1980). A somewhat analogous situation is seen in behavior associated with birth or hatching (Oppenheim 1972).

Thermoregulatory Behavior

Movements associated with the regulation of internal body temperature are shown by many if not most species. Some are highly complex, such as those involved with hibernation, discussed later in this chapter in conjunction with rhythms, or in long-distance migrations, which are discussed in the next two chapters.

Before we focus on behavioral aspects, however, a few terms need to be defined and clarified (Bligh and Johnson 1973, Ostrom 1980). Many of the terms are more-or-less, but not exactly, synonymous.

Animals are classically divided into two main groups on the basis of temperature regulation: the so-called warm-blooded mammals, birds, and possibly dinosaurs (Thomas and Olson 1980) and all others, which are cold blooded. Traditional technical names for the two groups are *homeothermic* and *poikilothermic*, respectively. The differences supposedly derive from the ability, or lack thereof, to generate body heat by carefully regulated, high rates of metabolism. The homeotherms stay warm and maintain a relatively constant deep or core body temperature (peripheral

Figure 6-2 Theoretical relationships between body and ambient temperatures in homeotherms and poikilotherms. In homeotherms the body temperature is largely independent of ambient temperature, within limits, whereas in poikilotherms body temperature varies with ambient temperature.

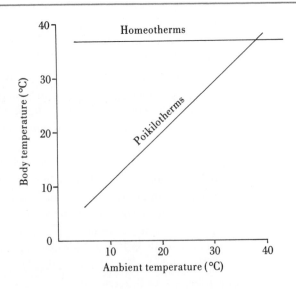

and limb temperatures may fluctuate more) regardless of ambient (environmental) temperature, whereas the body temperatures of poikilotherms fluctuate with ambient temperature (Figure 6-2).

The traditional view, however, is a bit oversimplified, and several other terms have been introduced to handle some of the differences. A term almost synonymous with poikilotherm is *temperature conformer*. In contrast, however, *temperature regulator* is not synonymous with homeotherm. Homeothermy technically refers to keeping the body temperature within relatively narrow limits. *Heterothermy* involves temperature regulation where body temperature is permitted to fluctuate more than in homeothermy, such as with torpor at night (e.g., in bats and hummingbirds) or during hibernation, but body temperature is nonetheless regulated. Furthermore, there are many ways of regulating body temperature. Endothermy, via controlled internal metabolic processes, is only one general way. Ectothermy is thermoregulation by acquiring body heat from external sources, usually via behavioral means. Heliothermy involves variation in exposure to solar radiation by behavioral means. The total heat budget involves heat absorption, generation, radiation, convection, and conduction. Lyman et al (1982) discussed the terminology further along with the topics of hibernation and torpor.

In the real world of animals, there is much diversity with little respect for arbitrary categories. Mammals in general have the most constant body temperatures. Human body temperature, for example, is regulated quite precisely around 37° C. Mammals that show more variation in body temperature include bats and hibernators. Body temperatures of birds generally are more variable than in mammals, commonly in the range of 38° to 40° C. Sometimes they range from 35° to 41° C or more,

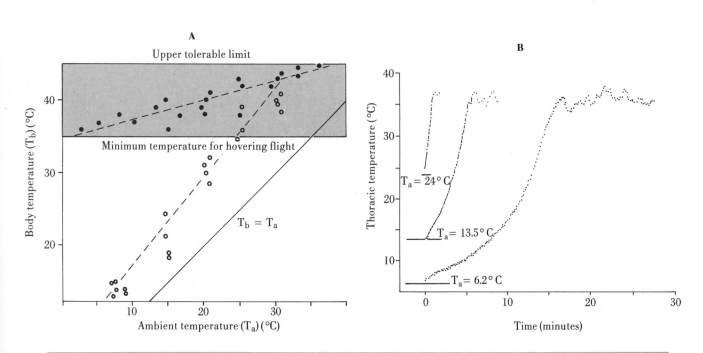

Figure 6-3   Bumblebee body temperatures during **A,** flight and **B,** warm-up at different ambient temperatures. Dark circles, thoracic temperatures; open circles, abdominal temperatures.

From Heinrich, B. 1975. Journal of Comparative Physiology 96:155-166.

depending on whether the bird is flying, active, excited, calm, or sleeping. The body temperatures of hummingbirds in torpor at night can drop almost to ambient temperature.

Outside the groups classically considered homeotherms are many organisms, particularly among flying insects, that should be considered true thermoregulators and even a few homeotherms and good endotherms. Figure 6-3, *A*, for example, illustrates almost classic homeothermy in the thoracic (but not abdominal) temperatures of a flying bumblebee. Bumblebees warm up (Figure 6-3, *B*), as do other bees, nocturnal moths, some katydids, beetles, flies, and dragonflies, by shivering and contracting flight muscles. Honeybees, some other hymenopterans, and many termites (Figure 6-4) achieve remarkably constant temperatures by heating and air conditioning their hives, nests, or mounds to within narrow, precise temperature ranges (see also Chapters 13 and 14). Among fish, the bluefin tuna maintains a relatively constant body temperature regardless of water temperature within normal limits (Carey and Lawson 1973).

Behavioral thermoregulation is well known for reptiles (Greenberg 1976). A thermoregulatory ethogram for the horned lizard (*Phrynosoma* sp.) is shown in Figure

Figure 6-4 Temperature control and air conditioning in giant mounds of African termites *(Macrotermes)*. Warm air rises in the mound, drawing cooler air from below. The pattern of air movements depends on the arrangement of passageways. Cooling is also achieved by evaporation of water droplets brought from underground tunnels and placed along the air passageways.

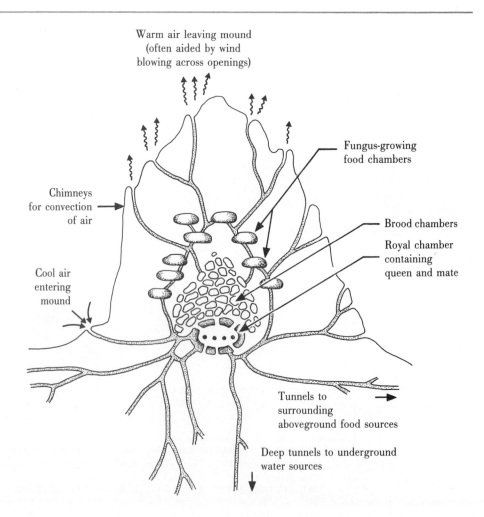

Warm air leaving mound (often aided by wind blowing across openings)

Fungus-growing food chambers

Chimneys for convection of air

Brood chambers

Royal chamber containing queen and mate

Cool air entering mound

Tunnels to surrounding aboveground food sources

Deep tunnels to underground water sources

Figure 6-5 Behavioral
temperature regulation in
the horned lizard
(*Phrynosoma* sp.).
From Heath, J.E. 1965. Zoology
64:97-136.

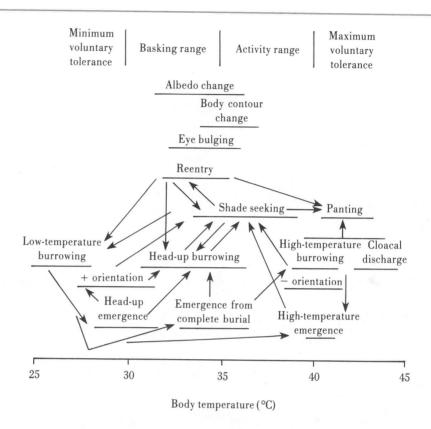

6-5. Many reptiles, such as crocodilians, select a variable temperature regime, moving to warm areas after eating and during digestion and food assimilation. They then move to lower temperatures, which reduce energy demands but still permit growth, thus maximizing net energy gain and utilization (Lang 1987). Crocodilians also select different temperatures depending on age and reproductive condition. Thermal conditions, including a variety or range of temperatures for behavioral selection, are important not only for animals under natural conditions but also for those held in captivity.

Numerous examples of ectothermal behavior are also well documented for many other organisms. Some arctic insects seek out the solar reflecting and focusing flowers of solar-tracking plants (Kevan 1975). Black desert grasshoppers (*Taeniopoda eques*) have four stereotyped postures used in thermoregulation (Figure 6-6, Whitman 1987). Many if not most fish (Beitinger et al 1975) have preferred temperatures and actively avoid temperature zones outside a particular range. Herring gulls orient their bodies to the sun (often to reduce absorption) and rotate with the sun during the day (Lustick et al 1978). The postures and locations of seals while out of the water have been shown to depend importantly on thermoregulatory factors (Gentry 1973).

Much thermoregulation is accomplished by basking and other behavioral means in virtually all temperature regulators, which brings us to the relevance of this discussion. Ectothermy and behavioral choice of the thermal environment are im-

Figure 6-6   Stereotyped postures associated with thermoregulation, under appropriate environmental conditions, in the grasshopper *Taeniopoda eques*. **A,** Flanking, a broadside exposure to the sun; **B,** ground flanking, flanking while crouching next to a warm surface under cool conditions; **C,** stilting, raising the body high above hot surfaces; and **D,** stem-shading, quiescence in shade.
From Whitman, D.W. 1987. Animal Behaviour 35:1814-1826.

portant to most organisms, including humans, regardless of the presence of endothermy. Ectothermy constitutes an important but often unrecognized component of the behavioral repertoire. Anyone who doubts that ectothermy is important even in humans has only to observe the species under the sun at the beach, on the lawn, on rooftops, standing over radiators and hot-air registers, and in saunas. When it comes to regulation of temperature in the home or work environment, humans in Western culture are almost in the same category of preferring ranges and precise control as African mound termites.

A historically puzzling behavior, or group of behaviors, is spread-winged postures of perched or reclining birds of different species (Figure 6-7). The behavior has received much attention and generated some controversy (Kennedy 1969). Ohmart and Lasiewski (1971) proposed that roadrunners sunned to acquire heat. However,

Figure 6-7   Subadult bateleur eagle *(Terathopius ecaudatus)* sunning under experimental conditions in captivity with artificial lights.
Photograph by James W. Grier.

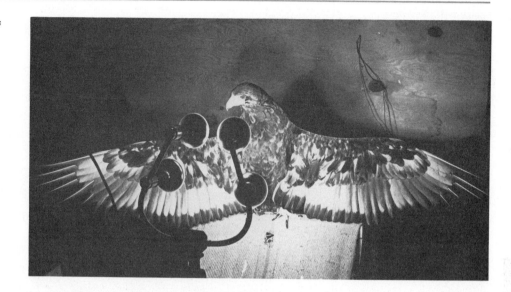

the researchers looked only at birds in sunning postures under light or in normal postures in the dark and not at birds using versus not using the postures under similar conditions of radiation—a serious defect in the experimental design leading to unavoidable correlations. Grier (1975) investigated spread-winged and non-spread-winged behavior of several species under natural and artificial radiation and found no thermal effects from the unique postures. Whether or not the birds' body temperatures or oxygen consumption changed depended on the presence or absence of radiation and not on the presence or absence of the spread-winged posture. This posture was influenced instead by the presence of water on the feathers in bright light, suggesting a drying function.

Drinking

The maintenance of body water is accomplished in most species with characteristic and familiar behavior patterns. Most birds scoop up a beak full of water, then tip back the head, letting the water run into the throat. A few birds, notably the Columbiformes (doves and pigeons), place the beak or head in the water and suck it up, as do most reptiles. The Columbiformes are considered relatively advanced and, since so few other birds show sucking behavior, it may have been secondarily derived. Many species obtain water physiologically from their food. Drinking habits and frequency depend, in addition, on a number of other body characteristics (such as kidney, respiratory, and epidermal water losses) and environmental factors such as temperature and relative humidity. Many animals never drink during their entire lives.

Some species have unusual drinking or other water-provisioning behaviors. The male desert Namaqua sandgrouse *(Pterocles namagua)* in Africa, for example, soaks up water from pools in specially adapted abdominal feathers (Figure 6-8). It then flies as far as 80 km back to the young, which drink by pecking at and stripping the water from the feathers (Cade and Maclean 1967). A few other birds also have

Figure 6-8 Namaqua sandgrouse soaking water into its feathers to provide water for the nestlings. From Cade, T.L., and G.L. Maclean. 1967. Condor 69:323-343.

been reported to carry water back to the young on feathers or via the mouth. Some snakes drink rain water that collects on their bodies. One desert species, the great basin rattlesnake (*Crotalus viridis lutosus*), has been observed coiling tightly during a rain, trapping the water, and subsequently drinking from it (Aird and Aird 1990). Chimpanzees use crushed or chewed leaves as sponges to get water from tree holes (Goodall 1986).

A different technique for obtaining water is used by three species of tenebrionid beetles that live in the Namib desert of Africa (Seely and Hamilton 1976). During foggy periods the beetles dig long trenches in dunes perpendicular to the wind. The trenches trap moisture, which is then consumed by the beetles (Figure 6-9). In one case where individuals from a population were sampled before and after drinking their fog water, body moisture increased nearly 14% during a single fog.

Shelter Seeking and Construction

A list of the behaviors needed to meet minimum living requirements should include whatever animals do to obtain or provide protective shelter. Many species remain exposed in whichever environment surrounds them; they have no shelter or home. There may be just a depression where they rest or give birth. At the other extreme are species that construct elaborate structures. At the peak of construction behavior, excluding humans, are some species of termites and hymenopterans, many birds, and a few rodents such as the beaver. In between the two extremes are many tube-, den-, and hole-dwelling species. Many species construct structures, including birds that build nests, many web- and cocoon-spinning arthropods, and miscellaneous shelter builders such as caddis flies. Several examples of animal constructions are pictured in Figure 6-10.

Construction behavior is a vast topic that has been significantly synthesized only

Figure 6-9 Namib Desert Tenebrionid beetle (*Lepidochora*) collecting water from its fog-catching sand trench.
From Seely, M.L., and W.J. Hamilton III. 1976. Science 195:485.

Figure 6-10 A potpourri of animal construction. As demonstrated by studies of the construction behavior of native animals and alterations caused by brain surgery (e.g., Van der Kloot 1956), most construction behavior is thought to result from innate motor programs in the brain.

Burrowing

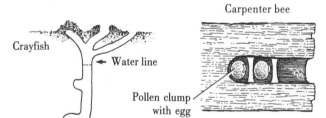

Crayfish

← Water line

Carpenter bee

Pollen clump with egg

Halicitine bees

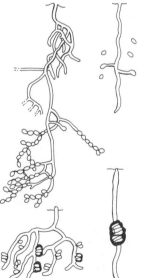

Mammal

Continued.

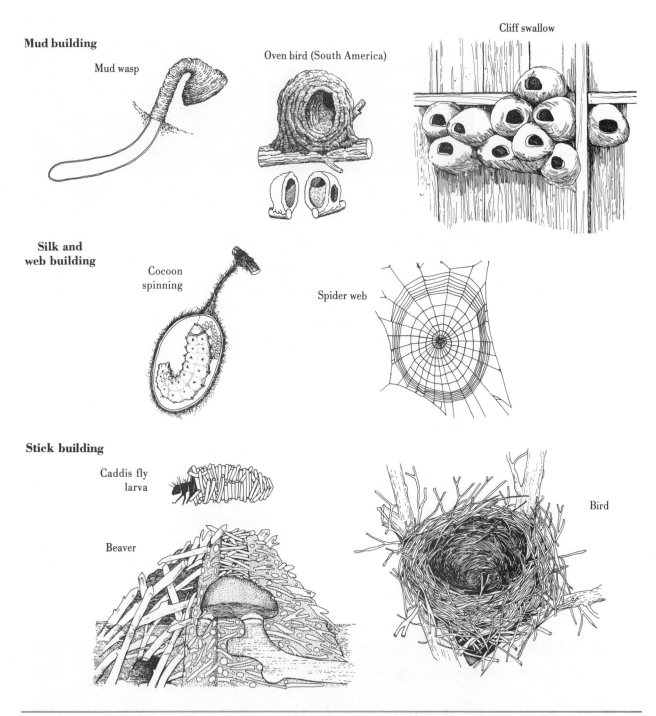

Mud building

Mud wasp

Oven bird (South America)

Cliff swallow

Silk and web building

Cocoon spinning

Spider web

Stick building

Caddis fly larva

Beaver

Bird

Figure 6-10 cont'd   A potpourri of animal construction. As demonstrated by studies of the construction behavior of native animals and alterations caused by brain surgery (e.g., Van der Kloot 1956), most construction behavior is thought to result from innate motor programs in the brain.

Figure 6-10 cont'd

Weaving and rolling

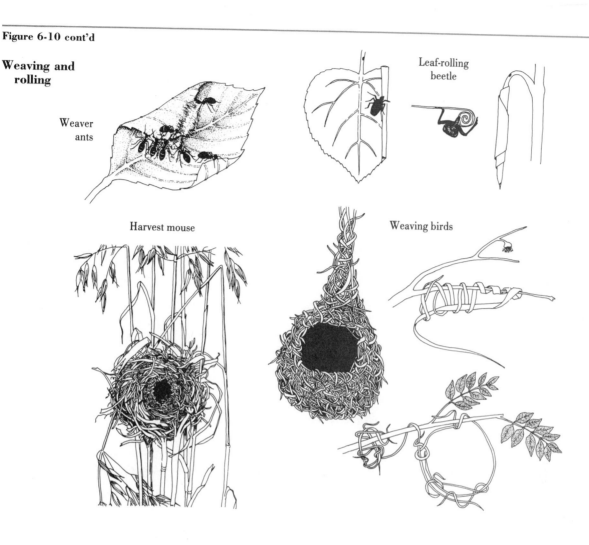

recently by Hansell (1984), after more modest earlier attempts by von Frisch (1974) and Collias and Collias (1976). Hansell (1984) provides such an excellent review of the subject that we merely indicate some of the salient points and defer to Hansell for anyone who would like more information.

Hansell (1984) provides a thorough phylogenetic survey of the animals that build structures, the materials they use, construction methods and types of structures that result, and discussions of the physiological, genetic, evolutionary, and control aspects. The functions of animal constructions include protection from the physical environment (extreme temperatures, water, and atmospheric aspects such as wind and low humidity), protection from predators, mechanisms for food gathering or storage, and communication.

Sequences of behavior involved in construction can be complex (Figure 6-11). Construction by most nonhuman species appears to be largely innate, and evolutionary relationships are often apparent (Figure 6-12). Many of the most elaborate

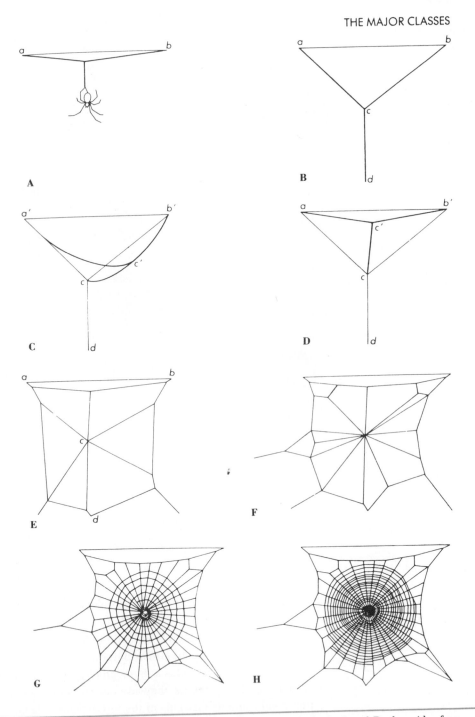

Figure 6-11 Sequence of construction of a typical orb spider web: **A** and **B**, the spider forms an inverted triangle below the initial bridge line, the apex (*c*) of which ultimately forms the hub of the completed web; **C** and **D**, the spider inserts another radius (*c-c*); **E** and **F**, more radii are added; **G**, the temporary scaffolding spiral is laid from the hub of the web out toward its periphery; **H**, finally, the sticky spiral is laid from the web periphery to the hub as the scaffoding spiral is gathered up.

From Hansell, M.H. 1984. Animal architecture and building behaviour. Longman, London.

Figure 6-12   Stages in the construction of the inverted-ladder web of *Scoloderus* sp. The web is built like an orb web except for the great exaggeration of the northerly sector.
From Hansell, M.H. 1984. Animal architecture and building behaviour. Longman, London.

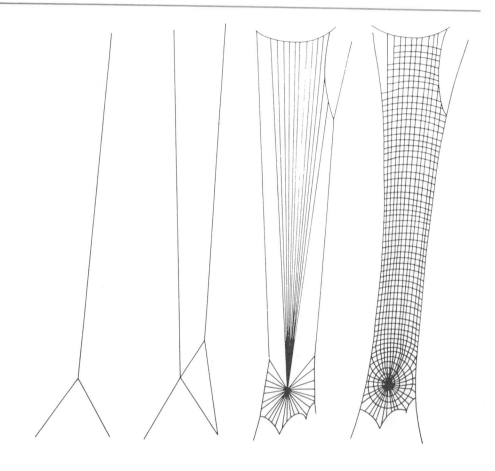

Figure 6-13   The web of the colonial ecribellate spider *Metabus gravidus* is composed of individual orb webs attached to one another.
From Hansell, M.H. 1984. Animal architecture and building behaviour. Longman, London.

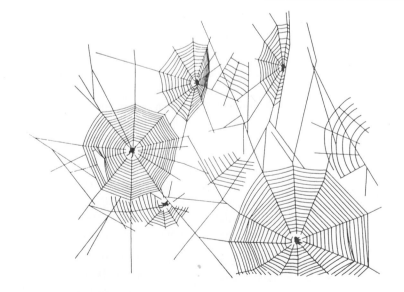

Figure 6-14   Evidence
suggestive of the learning of
building skills in the village
weaverbird *Textor
cucullatus*. The ragged nest
on the right is built by a
young, inexperienced male
and the nest on the left by
an experienced adult.
Figure and legend from Hansell,
M.H. 1984. *Animal architecture
and building behaviour.*
Longman, London.

forms of animal engineering are social endeavors (Figure 6-13), but in other cases
the construction is on an individual basis. In almost all cases, learning from other
animals appears to play little if any part in the specific patterns and choice of
building materials used in construction behavior. However, animals may learn or
improve with experience (Figure 6-14).

The proximate control of construction behavior is particularly fascinating. The
starting, continuation, progress, and stopping of various steps in the building process
may depend on external stimuli, disinhibition or inhibition of various neural centers,
interactions with hormones and other physiological conditions, the supply of resources (such as silk in a silk gland or external building materials), sensory feedback
from the structure itself, and combinations of these factors. The proximate control
of construction behavior can be experimentally manipulated relatively easily and
accordingly has received much attention. In some cases the behavior is clearly
affected by external stimuli, including such phenomena as gravity. In one example
described by Hansell (1984:183-184):

> If part of the anterior of the roof is removed from the house of the caddis larva *Silo
> pallipes* its absence is detected, probably by thoracic hairs; this stimulates searching.
> Contact with loose particles produces scratching of the substrate with the legs. If,
> during this behaviour, a large particle is contacted, the larva picks it up and handles
> it. If the size of the particle is suitable, the larva proceeds to attempt to fit it and, if
> the particle is of the appropriate size and, probably, shape, it is fitted into the gap in
> the roof. It is easy to imagine that the whole sequence of repair is brought to an end
> by the stimulus of a completed roof so that the entire sequence is controlled by a
> chain of stimulus and response. This is an especially well documented case of a
> sequence of behaviour under apparent stimulus-response control and ample evidence
> is available of the general importance of stimuli from the environment, or more
> specifically from the artefact itself, in controlling building behaviour.

However, in other situations, such as cocoon building by silkworm moths (*Hyalo-
phora* sp.), external cues may be of less importance (Hansell 1984:190):

> Van der Kloot and Williams [1953 b] . . . demonstrated that certain changes of
> internal state resulting from building could overrule the influence of external stimuli.

For example, a larva that was ready to start spinning a cocoon but was instead sewn into an outer cocoon layer built by another individual, none the less spun a complete cocoon of its own with outer, middle and inner layers. Similarly, a larva removed from its cocoon after completing only the outer layer and placed inside an inner layer still spun an inner layer of its own. Also, if an unfortunate larva was obliged to spin 60-70 per cent of its silk on the smooth inner surface of an inflated balloon (which only permits a flat sheet of silk to be produced), and was then transferred to a normal cocoon-spinning site, it omitted the first two stages of cocoon construction and proceeded directly to the building of the inner cocoon layer before pupating.

As would be expected from the diversity of organisms and their constructions, much diversity can be found in the control aspects. (The general subject of proximate control of behavior, including internal physiological mechanisms and learning, is covered in Part Three of this text, Chapters 16 to 21.) Overall, the same principles, whether at the proximate or ultimate level of inquiry, can be observed in construction behavior as in other behaviors.

TEMPORAL CONSIDERATIONS IN BEHAVIOR

Alternation Between Activity and Inactivity

The amount and timing of activity shown by different species and individuals are highly variable. Anyone who spends time watching animals soon discovers that most animals spend much time doing little or nothing, or engaging in seemingly insignificant activities (Herbers 1981). Even during the busy reproductive period animals may be inactive for long periods (Figure 6-15). Over the span of an entire year or lifetime, inactivity is much more common than most persons realize (Table 6-1). Even those species traditionally considered to be most active, such as shrews and bees, spend much time being inactive.

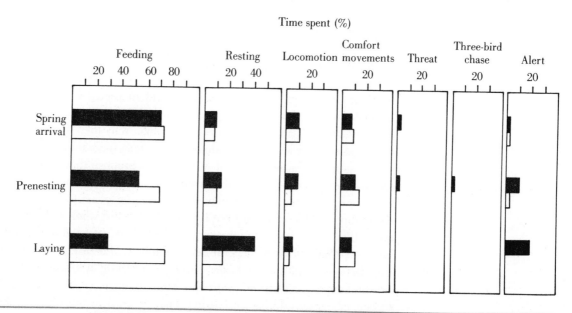

Figure 6-15 Time budgets for gadwall ducks during the breeding season. During this active time of the year most of the time is spent feeding, resting, and with basic locomotion and comfort movements. Males are represented by dark bars, females by open bars.
From Dwyer, T.J. 1975. Wilson Bulletin 87:335-343.

Table 6-1 Time budget of free-ranging shorthorn cattle in Australia during 24-hour periods throughout the year*	Period	Grazing	Walking	Standing/Ruminating	Standing/Resting	Lying/Ruminating	Lying/Resting	Drinking
	Day	46	9	8	12	15	11	—
	Night	38	6	7	9	22	15	—
	24-hour period	42	8	8	10	19	13	0.2

Modified from Low, W.A., et al. 1981. Applied Animal Ethology 7:27-38.
*Values are percentages based on minutes per day, night, or 24-hour period.

Inactivity is an important behavior that could be viewed as maintenance in some situations. It may be highly adaptive if it permits physiological recuperation, conserves energy by reducing expenditures, enhances survival by reducing exposure to potential hazards, or creates some combination of the above. In many instances inactivity may occur simply because the animal literally has nothing else to do.

Many birds of prey and other carnivores, when they are not actively involved in hunting and eating, reproduction, migration, or escaping from danger, show fairly discrete but irregular periods of inactivity alternating with activity. During active periods birds of prey, for example, appear restless and may fly or frequently change perches. Active periods occur whether the birds are satiated or not and whether they are in the wild or in captivity (Grier 1971). Just as noticeable are periods of inactivity. When inactive, birds of prey sit for long periods in characteristic relaxed postures on one foot, feathers slightly fluffed out, and with an almost hypnotized appearance, although they are still quite alert and move their heads frequently. Eagles, for example, if not disturbed may perch quietly for several hours at a time. But the activity varies. Hawks of the genus *Accipiter* are much more active on average than those of the genus *Buteo*, and for any one species there is much individual variation. Mueller and Berger (1973) showed that times during the day for activities such as migrating flight also may vary among the different genera of hawks and falcons (Figure 6-16).

Amounts of inactivity and the irregularity of inactivity appear to increase in predators compared with nonpredators, and in predators as prey size increases. Animals that are inactive (lying, perching, etc.) and not pursuing food appear to have enough food and perhaps even a surplus. Why then do so many predators die of starvation, and why don't predator populations increase in response to the available resources? Sutherland and Moss (1985) propose that it is not the amount of food per se but the unpredictable availability of the prey that makes the difference. Larger prey items are less common (than smaller ones) to begin with, and unpredictability may make them even less available. But when a predator of large prey does obtain food, it can become satiated and does not need to feed again for a long period of time. Other behavioral needs can be taken care of and there may still be time left over with nothing to do, forming something of a time shadow after eating.

The activity patterns of kestrels (*Falco tinnunculus*), for example, were shown to depend on hunting success: when unsuccessful in the morning they continued to hunt actively through the rest of the day, but when successful in the morning they became inactive in the afternoon (Rijnsdorp et al 1981). Beyond irregular activity directly related to success, sporadic periods of activity and moving may be

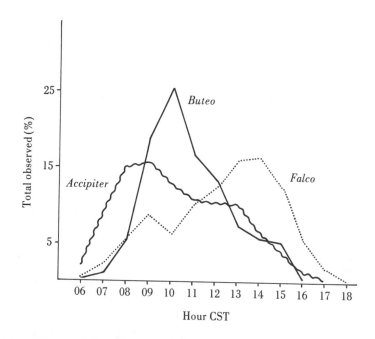

Figure 6-16 Differences in migration activity times among different genera of birds of prey passing along Lake Michigan at Cedar Grove, Wis. Data are based on 7,906 birds observed from 1958 to 1961.

Modified from Mueller, H.C., and D.D. Berger. 1973. Auk 90:591-596.

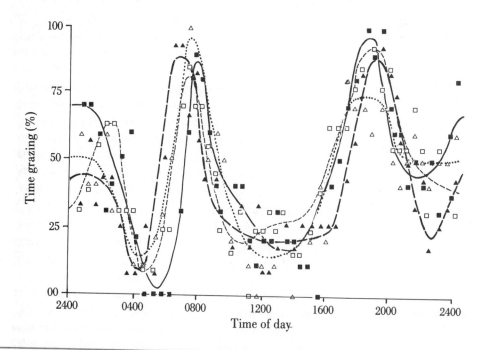

Figure 6-17 Grazing patterns of free-ranging cattle in central Australia. The percentage of time cows spent grazing during 30-minute intervals are remarkably consistent regardless of season. Two sharp peaks occur during morning and late afternoon; there is a moderate amount during the night, with a minimum just before sunrise and a lull during midday.

From Low, W.A., et al. 1981. Applied Animal Ethology 7:11-26.

a general adaptation that places the predators in different locations, thereby increasing opportunities for encountering prey. Either way, animals with irregular diets may have more irregular lives. Animals with more predictable food resources, such as horses (Duncan 1985), also show time budgets that are influenced by environmental factors, such as the presence of biting flies. But herbivore activity patterns in general are much more regular, sometimes remarkably so, as seen in Figure 6-17.

Rhythms

Regular patterns of alternating activity and inactivity often are associated with and regulated by an animal's internal, physiological biological rhythms. The most commonly recognized rhythms are those of about a day, that is, circadian (*circa*, about and *dies*, day), those of about a year, circannual, and those associated with the daily tides and monthly movement of the moon (e.g., literal meaning of menstrual).

That the rhythms are only approximate is well expressed by the prefix *circa*. They are kept regular, or entrained, by environmental cues called *Zeitgebern* (German, timegivers; *Zeitgeber*, singular), such as the regular rising and setting of the sun. If the normal Zeitgeber is blocked from the animal (as in a laboratory), it may readily switch to a substitute (such as the sound of a janitor coming in to clean the room next door every day), and this has created some problems for research on the subject. Not only may the Zeitgeber be switched, it can be used to reset the internal rhythms, a characteristic used to advantage in clock shifting (discussed in Chapter 7).

If the animal is isolated completely so that all of Zeitgebern are blocked, the approximate nature of the rhythm becomes apparent: the activity or other output cycles begin to drift and become out of phase with cycles of other individuals that remain in contact with their Zeitgebern. Examples of these drifting periods of activity are shown for both circadian (Figure 6-18) and circannual (Figure 6-19) rhythms. The continued presence of the rhythm after all external cues are eliminated demonstrates that they are derived in part from internal, physiological factors. Many of the internal clocks are now partially understood, as discussed later in the chapter.

Cole (1957), who has questioned the existence of many cycles in nature, demonstrated a diurnal cycle in the metabolic rate of unicorns based on purely random numbers. However, the accumulation of a massive amount of evidence since that time, including some details on the internal mechanisms, provide ample testimony to the presence of the rhythms. Few biologists today doubt that rhythms exist.

The length of the free-running circadian period tau (τ), is highly variable. It may range from 22 to 28 hours, depending on species, environmental conditions (particularly temperature and illumination level), and conditions under which an animal was raised (Aschoff 1979). Few clear correlations have been shown between τ and such factors as nocturnal versus diurnal species, body size, or taxonomic relationships (Aschoff 1979). For humans τ is approximately 25 hours, as has been shown in studies of persons who have been isolated from normal schedules (e.g., Aschoff study cited in Takahashi and Zatz 1982) or the case of a blind person trying to live under a normal schedule (Miles et al 1977). The 25-hour τ for humans, incidentally, coincides with the daily lunar period, which could either imply a relationship between the two or be purely coincidental.

There is some evidence in humans that there are at least two circadian pacemakers, one for the sleep-wake cycle and another for body temperature. Although

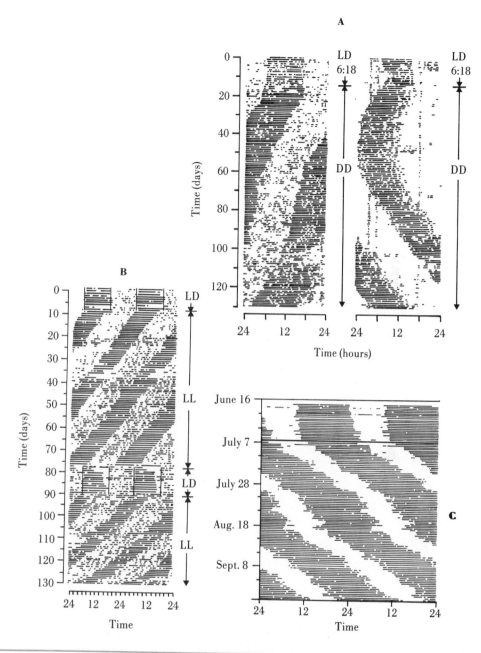

Figure 6-18   Circadian rhythms of animal activity during normal light-dark (*LD*) cycles versus periods of continuous light (*LL*) or continuous dark (*DD*). Activity is indicated by dark lines. Under constant conditions the free-running period (τ) is only approximately 24 hours. If less than 24 hours, activity commences earlier and earlier each day, with patterns shifting to the left on the graph. If longer than 24 hours, patterns shift to the right. **A,** Locomotor activity of two house sparrows. **B,** Activity of a pig-tailed macaque (*Macaca memestrina*). **C,** Awake activity (not constantly active) of a human living alone in a cave with a watch. At first the subject tried to maintain a constant schedule; then he gave up and slept as he felt inclined, after the indicated point. 12 = Noon; 24 = midnight.

A From Eskin, A. 1971. In: M. Menaker, editor. Biochronometry. National Academy of Science, Washington, D.C. B from Aschoff, J. 1979. Zeitschrift für Tierpsychologie 49:225-249. C modified from Halberg, F. 1973. In: J.N. Mills, editor. Biological aspects of circadian rhythms. Plenum Press, New York. Based on data in Mills, J.N. 1964. Journal of Physiology 174:217-231.

Figure 6-19   A, Natural circannual rhythms of ground squirrels and for those held under constant environmental conditions of 12 hours of light each day and constant temperatures of either **B,** 22° C or **C,** 0° C. Animals continued to show hibernation cycles in spite of constant conditions. Food was constantly available. Squirrels continued to eat, more during the higher temperature, and became occasionally active even while hibernating. Cycles of food consumption and weight change were quite marked. Modified from Pengelley, E.T., and S.J. Asmundon. 1971. Scientific American 224(4):72-79.

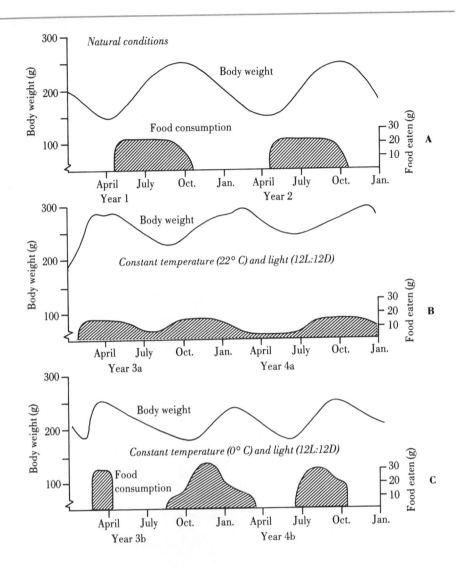

these normally coincide, they may uncouple if given the opportunity (Wever 1979, cited and illustrated in Johnson and Hastings 1986). The human period of 25 hours explains why it is easier to go to bed and get up later rather than earlier, rotate work to later shifts rather than to earlier shifts, and travel east to west rather than vice versa. Czeisler et al (1982) applied this principle to improving work conditions under rotating work-shift schedules.

Precise rhythmical phenomena are so widespread (although not universal, e.g., Stockman 1985, Gherardi 1988) among both plants and animals that two significant ramifications exist. The first is a practical aspect: rhythmicity and periodic differences must be recognized and taken into account or at least suspected in virtually all physiological and behavioral studies. This is discussed in Chapter 3 in reference to sampling techniques. Harcourt (1978) studied rhythms in gorilla behavior and demonstrated that effects extend even into social relationships.

The second aspect is more theoretical: the widespread presence of rhythmicity suggests that it actually is important, that is, highly adaptive. Is biological timing per se advantageous, or is it just a by-product or consequence of other processes? One can argue from a logical basis that rhythm and timing are advantageous. Rhythms could synchronize and coordinate internal, physiological events, coordinate the animal with the external environment, or do both. There is considerable temporal variation in the suitability of the environment for particular behaviors. Some of this variability is highly periodic and regular, such as with effects from the daily rotation and yearly orbit of the earth's movements. The optimal time (optimality is discussed in Chapter 9) to seek food is when it is most available and not at other times. The safest time to be active is when predators are not active and vice versa. Any biological mechanism that synchronizes the organism with these and other environmental variables should offer a huge selective advantage. An internal rhythm could assist this coordination during times when an animal is removed from direct external cues such as the sun; for example, when it is sleeping in a shelter. The reasonableness of the explanation is not proof, but it certainly seems difficult to refute.

Whether specific timing mechanisms are merely consequences of other processes or directly advantageous, there is rapidly accumulating evidence for the location and nature of such mechanisms. The diversity of mechanisms suggests strongly that rhythms are analogous rather than homologous. The hormonal changes associated with circannual and seasonal rhythms probably are the best understood (Chapter 18). Circadian pacemakers involve many varied endocrine and cellular or molecular mechanisms. Pacemakers appear to be located in places such as the base of the eye in some mollusks (Block and Wallace 1982), the optic lobes in the protocerebrum of the brain of cockroaches (Page 1982), a part of the vertebrate brain called the suprachiasmatic nucleus (SCN) in the hypothalamus (Rusak and Groos 1982, Albers et al 1984), and perhaps the pineal gland—long a favorite target of searches for vertebrate biological clocks (Underwood 1977). In rats the SCN of developing fetuses in the uterus can be entrained and coordinated somehow through the mother's system, depending on her visual ability (Reppert and Schwartz 1983).

The location of the cockroach circadian pacemakers was demonstrated convincingly and remarkably by first rearing cockroaches under different schedules of light and dark to achieve different free-running activity rhythms (22.7- and 24.2-hour periods). Then the optic lobes of different cockroach brains were transplanted to other individuals. After recovery the recipients assumed the free-running rhythm previously shown by the donor (Page 1982). Similar results have been obtained in house sparrows by transplanting pineal glands (Zimmerman and Menaker 1979).

How are free-running pacemaker rhythms coupled with the external Zeitgebern? The exact mechanisms are not yet understood. At least for rhythms involving light, however, the chemical serotonin, which is affected by light, has been implicated in both vertebrates and invertebrates. In the marine snail *Aplysia*, for example, treatment with serotonin, a chemical found naturally in the eye, shifted the phase of the circadian output of optic nerves (Corrent et al 1978). In birds the pineal gland inside the skull is light sensitive (Deguchi 1981), whereas in mammals, light sensitivity for the serotonin pathway uses the main light sensor, the retina of the eye (Moore 1978). The retinal pathway, however, appears to use different neural fibers directed to different parts of the brain than those used in vision. Light is also clearly involved in human circadian rhythms (e.g., Czeisler et al 1986), and the importance of light in setting these rhythms is useful even in clinical situations, such as in reducing psychological depression during the short day lengths of winter (Lewy et al 1987). For further details on the mechanisms of these biological clocks and how they are thought to be coupled with the Zeitgebern, see Aschoff et al (1982), Moore-Ede et al (1982), and Takahashi and Zatz (1982).

Time and Behavior

A number of recent findings raise intriguing questions about the whole issue of time in animals' lives. Schleidt (1988), after analyzing common, everyday human movements of the arms, legs, head, and trunk across four different cultures, proposed that our species makes repetitive movements with a time constant of about 3 seconds. Ralph and Menaker (1988) discovered a single gene mutation in hamsters that shortens their circadian τ to 22 hours in heterozygous and 20 hours in homozygous animals, as also discussed in Chapter 4.

Per (period) mutations in fruit flies have been known for some years and have received increased attention recently. There have been 4 per types identified, all of which map to the X chromosome: per^+, which is the wild type with a τ of about 24 hours, per^l for long with a τ of 28 to 30 hours, per^s for short with a τ of 18 to 20 hours, and per^0, which has no apparent sleep-wake cycle. (For a further general description of the per mutations, see Kolata, 1985. For discussion of some of the claimed effects on courtship songs, see Ewing 1988 and Crossley 1988.) Clock mutants have also been reported in other invertebrates (see references in Ralph and Menaker 1988).

Are there precise physiological timing mechanisms in animals that are roughly analogous to the crystals in quartz clocks? Do internal clocks do more than coordinate the animal with night and day, such as permitting synchrony in communication and other interactions among animals—somewhat akin to the baud rate in computers? What implications do environmental factors other than light, such as temperature, have on time in animals' lives? Are there effects of body size? Schmidt-Nielsen (1984) notes that a 30 g mouse breathing 150 times per minute, with a heart rate of 600 beats per minute, will have about the same number of breaths and heartbeats (200 million and 800 million, respectively) in a 3-year lifetime as a 5-ton elephant living 40 years. (He also notes that humans "live several times as long as our body size suggests we should.") Schmidt-Nielsen also considers real time versus metabolic time and suggests that the same principles, except in general terms, may not apply to both warm-blooded and cold-blooded animals.

Before leaving the subject of time and rhythms, we need to add a few comments about circannual rhythms. They are important for coordinating animals with long-term, annual events and are significant for many behavioral and physiological aspects of some animals' lives such as hibernation, food storage, migration, seasonal breeding, and related phenomena such as feather molt in birds. Although circannual rhythms are common, they are not universal. Some tropical birds breed on 10-month cycles, and some tropical desert and marine species show no long-term regular patterns in their lives. However, animals that live in temperate and arctic or annual wet-dry situations, which includes most of the species in the world, show marked circannual rhythms.

Circannual rhythms appear to have internal timing like circadian rhythms, with free-running periods observed to range from 6 to 16 months, and cycles may persist under an artificial constant environment for years (Gwinner 1986). There has been much research on circannual rhythms regarding differences among species, the nature of possible Zeitgebern, and underlying physiological mechanisms. However, circannual rhythms are much more poorly understood than circadian rhythms, partly because they take so much longer to study. The physiological relationships, if any, between circannual and circadian rhythms are not yet known. For more information on circannual rhythms refer to Gwinner (1986).

Sleep and Dreams

Sleep is the most extreme state of inactivity. It is associated with immobility and change of alertness thresholds (Figure 6-20), sometimes taxon-specific postures, secluded or hidden and protected locations, and characteristic physiological and neural states. Sleep is a fascinating topic that we are just starting to understand. The more it is investigated the more complex and interesting it proves to be.

Sleep has been studied and described most extensively in mammals, particularly in humans. There are several categories and stages of sleep, with some having more than one name. The basic mammalian pattern of sleep during a given time period consists of a series of sleep cycles with successive cycles becoming shallower and shallower (Figure 6-21). A typical cycle involves going from the active, awake state through deeper stages classified as stages I to IV, then returning to stage I. Each of the stages of sleep is associated with relatively distinct brain wave patterns (see Figure 6-21) and different arousal thresholds. It is more difficult, for example, to awaken from stage IV than from stage I. After the return to stage I, there is a period of rapid eye movement, occasional muscle twitches, and otherwise general muscular immobility. The period of rapid eye movement commonly is called *REM sleep*. This period of sleep also goes by several other names in the literature, including *active sleep (AS)*, *paradoxical sleep*, *rhombencephalic sleep*, and *fast* or *dreaming sleep*. The remainder of the cycle is called *non-REM (NREM)* or *quiet sleep (QS)*. The period of REM sleep normally marks the end of a sleep cycle.

Generally, a sleeping mammal is immobile, aside from the REM eye movements and muscle twitches, during both REM and most of NREM sleep. Yet between

Figure 6-20 Sleeping gray seal (*Halichoerus grypus*) floating in the open sea. From Ridgeway, S.D., and P.L. Romano. 1975. Science, Feb. 14, 1975.

Figure 6-21 A,
Generalized mammalian
(including human) sleep
patterns and **B,** associated
brain waves.
A modified from Hobson, J.A.,
et al. 1978. Science 201:1251-
1253. B modified from Dement,
W., and N. Kleitman. 1975.
Electroencephalography and
Clinical Neurophysiology 9:673.

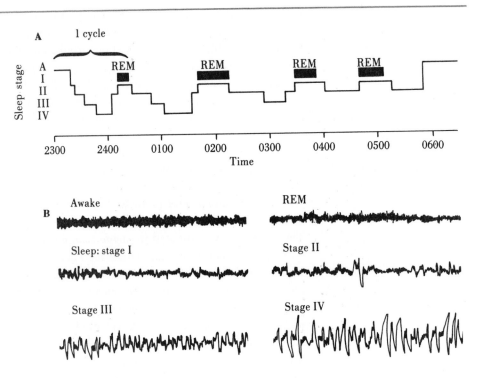

REM and NREM periods the individual may change posture, that is, turn over and otherwise move about, while remaining asleep. These periods of postural body movements show up strikingly on time-lapse photographs.

REM sleep is called paradoxical sleep because the brain is showing activity that during awake periods would lead to movement while at the same time showing inhibition and immobilization of the muscles that otherwise would be moving. The twitching that sometimes accompanies REM sleep apparently results when the brain's messages to move and not move do not quite cancel each other. Chase and Morales (1983), who studied the phenomenon at the neuronal level, concluded that the immobilization is adaptive for the organism to protect itself from the deleterious consequence of undirected and inappropriate movements when it is blind and unconscious. One might ask, however, why would the brain produce muscle-stimulating impulses in the first place?

Morrison (1983) and others (referenced in Morrison) have discovered that the immobilization or paralysis of the muscles that occurs during REM sleep can be turned off in cats by lesioning (damaging) appropriate places in the brain (e.g., in the pons). By doing this they could release various movements in the cats during REM sleep, including lifting the head, standing, walking, and even imaginary chasing and attacking behavior (as in vacuum activities).

At least in humans, who are able to talk about their experiences, REM sleep is the time of dreaming. For most persons the total amount of dreaming each night is about 90 to 120 minutes (about the duration of a feature-length movie).

Dreams vary among individuals and between the sexes and apparently depend on what goes on in an individual's life. Researchers who have systematically awakened subjects during dreams (Cartwright, cited in Kiester 1980) find a more-or-less standard format to the sequence of a night's dreams. The general sequence proceeds from first dreaming about concerns from the previous day, to generalizing about similar previous experiences, to a final imaginary and often tense and vivid conclusion. All components might be quite fanciful. If anything is remembered for long after waking, it usually is only the final scene, as if one saw only the end of a confusing movie.

In the classic Freudian view (reviewed in McCarley and Hobson 1977), dreams were considered to function as censors that disguised repressed wishes to prevent them from waking the sleeping individual. This view, labeled the *guardian-censorship* or *wish fulfillment-disguise theory*, ascribes a psychological purpose to dreaming. This led to much interest in dream analysis, in attempts to determine which wishes were being repressed and to use the results in psychotherapy.

Recent alternate interpretations, however, have focused more on the physiological aspects of dreaming. Hobson and McCarley (1977) proposed an activation-synthesis theory for dreaming based on neurophysiological evidence. According to this theory, parts of the brain periodically and endogenously generate impulses during sleep that activate the forebrain. The forebrain then attempts to make sense of, or synthesize, the hodgepodge of impulses, perhaps using past memories from the real world in an attempt to match the new input. As stated by Hobson and McCarley (1977), "The forebrain may be making the best of a bad job in producing even partially coherent dream imagery from the relatively noisy signals sent up to it from the brainstem."

Thus it may be that dreams are mostly literal representations rather than mystical, symbolic, or repressed thoughts. According to this view, a dream about a banana, for example, is just that and not a dream about a penis. Hobson and McCarley later commented (in Kiester 1980) that dreaming is essentially a consequence of neurological processes in which dreams "may be the signals made by the system as it steps through a built-in test pattern—a kind of brain tune-up crucial to prepare the organism for behavioral competence."

Sleep: Need or Compulsion?

Whether or not there is a need in mammals for dreams or whatever dreaming represents or, more generally, for REM sleep is not clear. The function of sleep, if any, has eluded biologists. Only mammals and birds, with some indications in reptiles, have been shown to possess the brain and muscular characteristics of REM sleep, suggesting that REM sleep is a relatively recent evolutionary phenomenon. However, representatives of virtually all taxonomic groups show periodic immobility and several other characteristics (Table 6-2) of what might be called sleep (Meddis 1975).

Some persons (e.g., Allison and Van Twyver 1970) have tried to correlate sleep habits with general life-history patterns, such as predators versus prey species. But more recent information suggests the picture is more complex (Table 6-3). Note that cats and mice sleep similar amounts of time; small animals with high metabolic rates (bats and shrews) are at opposite extremes; large and small animals (elephants and shrews) are different but more similar than bats and shrews; and related cetaceans (Dall's porpoise and bottle-nosed dolphins) are different in their sleeping habits.

Hypotheses on the function of sleep fall into three general categories: no function,

Table 6-2 Some characteristics of sleep and their phylogenetic distribution

Characteristic	Primates	Other Mammals	Mono-tremes	Birds	Reptiles	Amphibia	Fish	Molluscs	Insects
Prolonged period of inactivity	+	+	+	+	+	+	+	+	+
Circadian organization	+	+	+	+	+	+	+	+	+
Reduced alertness	+	+	+	+	+	+	+	+	+
Specific sleep sites/postures	+	+	+	+	+	+	+	+	+
High-voltage slow brain waves	+	+	+	+	−	−	−	−	−
REM sleep	+	+	−	+	−?	−	−	−	−

From Meddis, R. 1975. Animal Behaviour 23:676-691.

Table 6-3 Total sleep time per 24 hours for various mammals for which figures are available

Hours	Mammal
20	Two-toed sloth
19	Armadillo, opossum, bat
18	
17	
16	Lemur, tree shrew (Tupaia)
15	
14	Hamster, squirrel, mountain beaver
13	Rat, cat, mouse, pig, phalanger
12	Chinchilla, spiny anteater [echidna]
11	Jaguar
10	Hedgehog, chimpanzee, rabbit, mole rat
9	
8	Human, mole
7	Guinea pig, cow
6	Tapir, sheep
5	Okapi, horse, bottle-nosed dolphin, pilot whale
4	Giraffe, elephant
3	
2	
1	
0	Dall's porpoise, shrew

From Meddis, R. 1975. Animal Behaviour 23:676-691.

recuperation or restoration, and immobility. The first category suggests that there is no function and that the environmental rhythm (of night and day) is so regular and complete that over millions of years nervous systems have somehow locked into the periodicity and now show a proximate but no ultimate need for sleep. Note that the no-function hypothesis does not deny that sleep or cycles exist, as discussed earlier for circadian rhythms; the hypothesis simply suggests that sleep has no adaptive function.

Recuperation hypotheses propose that sleep may be necessary to somehow recharge or reorganize physiological or neurological processes. Sleep may permit more efficient and complete repair of other bodily systems, such as the resting of muscles. It is not clear in this case, however, why an animal should become so deeply and physiologically immobilized; why doesn't the animal just rest quietly?

Immobility hypotheses propose that sleep evolved to keep animals quiet and immobile when their full sensory abilities cannot operate (e.g., vision at night) and when they may be subject to predation or other dangers such as falling (Freemon 1972). It has even been suggested (Freemon 1972) that different parts of the brain take turns recharging, while other parts remain on guard for sudden emergencies requiring arousal, and that this leads to the two observed forms of sleep. Sagan (1977) speculated that the sleep condition in some modern mammals is a relic from the period when large and dangerous reptiles roamed the earth along with smaller primitive mammals.

In attempting to distinguish between these categories of hypotheses, however, there may be unavoidable correlations and sometimes the data can be interpreted to support any of the alternatives. Among animals that sleep, including humans, there seems to be a proximate need for sleep. Humans and other animals deprived of sleep, for example, seem to become increasingly unable to function normally until sleep is restored. Debilitation and drowsiness under sleep deprivation have been used as evidence for a recuperation function; that is, without repair and recuperation the system breaks down. But it has never been clear just what is breaking down; and because the lack of sleep is linked and correlated to the deprivation techniques, it may be the techniques that are causing the problems.

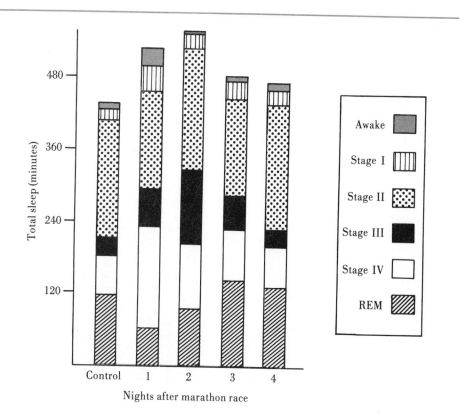

Figure 6-22   Amounts of sleep after a 92-km marathon race. Averages are shown for six subjects. Primary differences after the heavy exercise were increases in sleep of stages 3 and 4 and a reduction of REM sleep.
From Shapiro, C.M., et al. 1981. Science 214:1253-1254.

The other two theories state that sleep simply has become fixed in the behavioral repertoire of some species, complete with appetitive and consummatory components, and the apparent debilitation is nothing more than an increased attempt of the motivation system to find expression. The no-function hypothesis suggests that it just happens. Immobilization proponents suggest that the selective advantage is to enforce periods of inactivity to prevent wasting energy or avoid predators rather than for any physiological benefit.

There are supportive data and arguments on all sides. Shapiro et al (1981), for example, show significant correlations between amount of sleep and exercise (Figure 6-22). Meddis (1975), on the other hand, states:

> Recuperation theories would seem to imply that a healthy existence without sleep is impossible, at the very least among mammals. . . . [Those mammals] whose sleep requirements are negligible must therefore remain the Achilles' heel of recuperation theories. By contrast, the immobilization hypothesis not only tolerates the possibility but predicts that such conditions should occur among species who have little spare time or who have alternative procedures for obtaining the advantages normally resulting from the sleep instinct.

Meddis then notes several species (see Table 6-3) and individuals (including bona fide cases involving active healthy humans) that show little or no sleep.

Perhaps the most convincing data supporting a recuperative, physiological function of sleep have come from a study of paired rats in which only one of the pair was deprived of sleep (Rechtschaffen et al 1983). Previous studies, as mentioned previously, are suspect because it may have been the deprivation methods rather than the lack of sleep that caused problems. Rechtschaffen et al, however, used a mild deprivation technique. Their design (another example of ingenuity in behavior research) involved rats in neighboring cages joined by a rotatable disk suspended over shallow water on which they lived (Figure 6-23). Electrodes in the rats recorded their brain waves and muscle activity.

The disk the rats lived on was rotated slowly whenever the deprived rat started to sleep and stopped when it woke up. Starting and stopping of the disk was controlled by the deprived rat's brain waves and a microcomputer. Both rats were

Figure 6-23   Design of rotating-disk apparatus for depriving rats of sleep. From Rechtshaffen A., et al. 1983. Science 221:182-184.

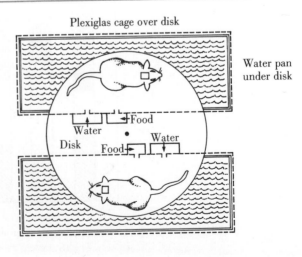

exposed to the same amount of movement, but the control rat could sleep whenever the deprived rat was spontaneously awake and the disk was not moving. None of the rats slept in the water. The deprivation was mild and nonfatiguing in that the rats were forced to walk only an estimated 0.9 mile per day; rats on exercise wheels run up to 30 miles per day.

The only external difference between the two rats in each pair was that the deprived rat could not sleep, whereas the control rat could. The control rats also showed some reduction of sleep over normal baseline levels, but they were still able to sleep much more than the deprived rats and, aside from two rats with minor problems, remained healthy and normal. The deprived rats, however, all showed one or more of a wide variety of physiological problems, including collapsed lungs, brain wave abnormalities, stomach ulcers, edema, and atrophied testicles. Three of the deprived rats died suddenly while being observed. The physiological problems were so varied that the precise nature of the deficit was not known; but it was clearly related to the lack of sleep. Stress from various sources, via the adrenal hormones and perhaps the nervous system, can lead to immunological and physiological dysfunction; the experimental rats may merely have been displaying a generalized stress syndrome in response to unfulfilled proximate urges to sleep, thus the problem remains unsolved.

The immobilization hypothesis permits much diversity among sleep patterns, depending on different organisms' natural histories. According to Meddis (1975), for example, the cat has few predators and "can sleep long and securely," whereas the mouse has numerous predators but obtains security in a protected nest. In the case of small-bodied, high-metabolism species, "The shrew tackles its energy problem with almost incessant food gathering. The bat adopts the alternative of almost total inactivity and spends a great deal of time asleep." Large herbivores, such as elephants, do not spend much time sleeping because they must be constantly eating. In short the immobilization hypothesis states that immobility occurs if it is needed or can be afforded but not if it is not needed and cannot be afforded.

None of the sleep hypotheses can yet be eliminated completely, and they are not mutually exclusive. The immobility hypothesis, for example, cannot deny that there still might be some recuperative selective advantages. Different advantages or even no advantages might occur in different situations among different species. The function of sleep remains an intriguing unknown.

The only safe generalizations about sleep are (1) that most animals show periods of inactivity, some of which may be fairly regular, and an animal may become less alert and therefore sleeps; and (2) that there are considerable differences in activity and sleep patterns among different species and, within a species, during different seasons. A striking example of this variability is shown by two sympatric, closely related species of desert rattlesnakes (Figure 6-24). The sidewinder (*Crotalus cerastes*) is consistently nocturnal in its activity patterns, whereas the speckled rattlesnake (*C. mitchelli*) is similar to the sidewinder during the summer months but shifts to diurnal patterns the remainder of the year.

Shaffery et al (1985) were able to manipulate daytime sleep in herring gulls by artificially providing increased food in their territories. Rather than increasing as predicted, sleep was actually reduced. The gulls with increased food spent more time in their territories than control gulls, but they slept less. The decrease in sleep was only of one type: front-sleep (or rest-sleep, a form of light sleep or napping). Another category of sleep, back-sleep, a deeper sleep with the head

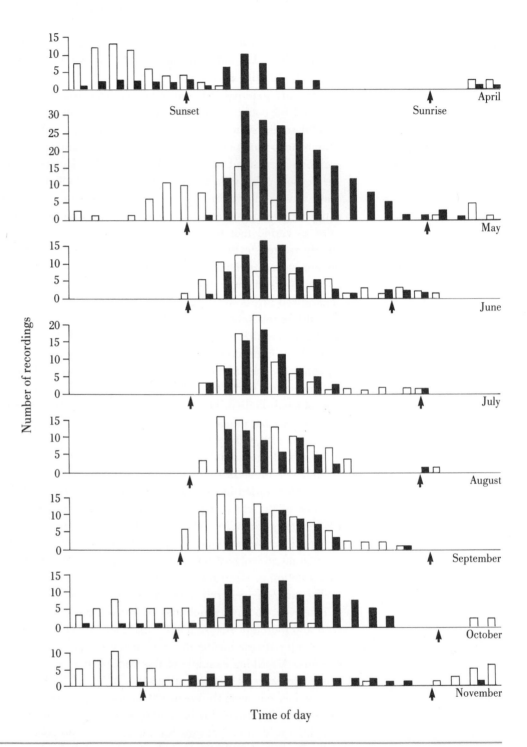

Figure 6-24   Species-specific differences in activity patterns of sympatric desert rattlesnakes. The two species are active at different times of the day or night during most of the year, but both become nocturnal during the hot summer months. Open bars represent *Crotalus mitchelli*, solid bars *C. cerastes*. Arrows indicate sunrise and sunset.
Modified from Moore, R.G. 1978. Copeia 3:439-442.

turned and resting on the back, was similar in both control and experimental gulls. This led the authors to suggest that gulls have different types of sleep with different functions.

All that is clear from the varied information about sleep in different species is that either we do not yet truly understand it or else it is really not a singular phenomenon. Perhaps it is simply a multitude of different phenomena that only appear similar but that cannot be usefully generalized.

SUMMARY

Knowledge of the maintenance behavior of animals is important to an overall understanding of animal behavior for several reasons. Maintenance movements are the most basic to the individual animal's survival, generally form the greatest portion of the total repertoire, and consume most of the animal's time on a daily basis. Relatively simple behaviors have served as useful subjects for studies of the internal control of behavior. These basic behaviors are frequently conservative in their original role; that is, they show little evolutionary change or lability, and they may be less subject to learning, individual variation, and voluntary control than many other forms of behavior. Finally, in new roles, maintenance behaviors often form the substrate for evolutionary development of other behaviors such as communicative and reproductive displays, rituals, and highly species-specific behaviors.

Included among the list of arbitrarily defined maintenance behaviors are feeding (covered separately in Chapter 9) and drinking, respiration and related movements, elimination, stretching, grooming, scratching, washing and bathing, drying, the periodic shedding of protective body coverings, thermoregulation, the construction of shelter, and possibly inactivity.

Inactivity, biological rhythms, and the enigmatic phenomenon of sleep are all components of animal behavior. Common rhythms include those of about a day (circadian), about a year (circannual), and about a month (e.g., menstrual). The internal, free-cycling period, tau (τ), of circadian rhythms varies from species to species and even among individuals. The internal rhythm is synchronized or entrained with external cycles such as the sun's by regular contact with various external cues or Zeitgebern (singular, Zeitgeber). The anatomical and physiological bases of the internal rhythms are becoming better understood; the details vary among different types of animals.

Sleep is widespread but highly variable among animals. It may represent different phenomena that are only superficially similar. Sleep has been researched most extensively in mammals, which show particular cycles of neural and muscular activity during sleep, including periods of rapid eye movement (REM). REM sleep is associated with dreaming, at least in humans, and may represent (e.g., according to the activation-synthesis theory) either functions or consequences of other nervous system activities that are not yet well understood. Previous interpretations, such as those espoused by Freud, have become less accepted. Functions proposed for sleep include no-function (it just happens), recuperation, and immobility. At the present time, sleep remains a fascinating topic of continuing research.

FOR FURTHER READING

Books and General Articles

Aschoff, S.D., and G.A. Groos, editors. 1980. Vertebrate circadian systems. Structure and physiology. Springer-Verlag, New York.
Papers from a symposium reviewing the status of knowledge on circadian rhythms, with much detail on the underlying neural aspects.

Borbely, A. 1986. Secrets of sleep. Basic Books, New York.
A concise, nontechnical book on sleep that expands on many of the points about sleep covered in this chapter.

Collias, N.E., and E.C. Collias, editors. 1976. External construction by animals: benchmark papers in animal behavior, vol. 4. Dowden, Hutchinson, and Ross, Stroudsburg, Pa.
A collection of excerpts and classic papers on the subject, including both vertebrates and invertebrates.

Collias, N.E., and E.C. Collias. 1984. Nest building and bird behavior. Princeton University Press, Princeton, New Jersey.
A comprehensive monograph on bird nests with emphasis on their evolutionary ecology. A widely acclaimed reference on the subject.

Frisch, K. von. 1974. Animal architecture. Harcourt Brace Jovanovich, New York.
A thorough review with many excellent illustrations.

Gwinner, E. 1986. Circannual rhythms. Springer-Verlag, New York.
A concise, thorough, and up-to-date review and presentation of summarized information on circannual rhythms.

Hansell, M.H. 1984. Animal architecture and building behaviour. Longman, London.

Lyman, C.P., J.S. Willis, A. Malan, and L.C.H. Wang. 1982. Hibernation and torpor in mammals and birds. Academic Press, New York.
Discusses terminology and provides information, mostly physiological, concerning hibernation and torpor. A basic reference for persons interested in this subject.

Morrison, A.R. 1983. A window on the sleeping brain. Scientific American 248:94-102.
Reports on work with lesions in cat brains that turn off the normal REM immobility of muscles, providing insight into the neural mechanisms of sleep and dreaming.

Recommended Original Research Papers

Carlson, J.R. 1977. The imaginal ecdysis of the cricket *Teleogryllus ocianicus*. 1. Organization of motor programs and roles of central and sensory control. Journal of Comparative Physiology 115:299-317.

Lewy, A.J., R.L. Sack, L.S. Miller, and T.M. Hoban. 1987. Antidepressant and circadian phase-shifting effects of light. Science 235:352-354.

Martin, R.R., and M. Menaker. 1988. A mutation of the circadian system in golden hamsters. Science 241:1225-1227.

Rechtschaffen, A., M.A. Gilliland, B.M. Bergmann, and J.B. Winter. 1983. Physiological correlates of prolonged sleep deprivation in rats. Science 221:182-184.

Sutherland, W.J., and D. Moss. 1985. The inactivity of animals: influence of stochasticity and prey size. Behavior 92:1-8.

Whitman, D.W. 1987. Thermoregulation and daily activity patterns in a black desert grasshopper, *Taeniopoda eques*. Animal Behaviour 35:1814-1826.

7

SPATIAL ASPECTS OF BEHAVIOR
INTRODUCTION, HABITAT SELECTION, AND LOCAL MOVEMENTS

The ability to move from one location to another is an important feature in the life and behavior of all but a few sessile animals (animals that are fixed to one location). Even many of those that are sessile during some part of their lives are capable of moving about in other, usually immature stages. Almost all animals travel locally to obtain necessary resources and avoid adverse conditions. Sometimes the movements are only to suitable conditions per se, or away from adverse ones. In other cases, however, the movements involve travels to and from specific locations, such as to and from a home, nest, shelter, or particular feeding location. This chapter concentrates on the local, relatively short-distance movements common to most animals, including how the animals find their ways about.

Some animals engage in extended travels, often of incredible distances, using methods that are only partially understood. The methods may involve either highly refined extensions of the means used for local travel or entirely different mechanisms, including celestial navigation or use of the Earth's magnetic field. These long-distance travels are considered in the next chapter.

One species that illustrates both local and long-distance movement is the monarch butterfly (*Danaus plexippus*) of North America. We begin with it.

INTRODUCTION TO THE SPATIAL ASPECTS OF BEHAVIOR: THE MONARCH BUTTERFLY

It seems incredible that fragile insects with big flappy wings such as adult monarch butterflies can fly at all. Yet they travel about superbly, spending much of their lives on the wing. They fly about during the summer searching for mates, occasional resting places, or their primary food source: a single genus of plants, *Asclepias*, the milkweeds. As far as is known, individual monarch butterflies do not home in on specific places from and to which they travel. Rather they travel about and concentrate on certain types of places, usually places with milkweed.

At the end of the summer they begin to travel longer distances, mostly in a southwesterly direction. Some monarchs travel all the way from Ontario, Canada, to one of a few small isolated locations in the mountains of Mexico, a distance of approximately 3000 km (1800 miles). A few are thought to return to Ontario the following spring. Others from elsewhere in North America join those from Ontario or travel to a second major location in California (Figure 7-1). The wing tagging and research that resulted in these findings were conducted over a 40-year period

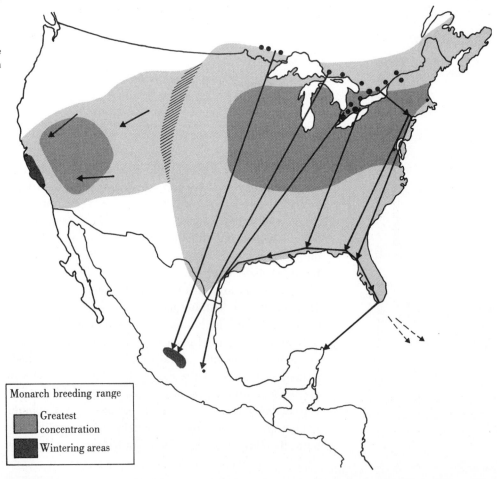

Figure 7-1 Major breeding ranges, migration routes, and wintering sites of the monarch butterfly. Dots and arrows from Canada indicate specific information based on wing-tagging studies. Arrows to the southeast from Florida indicate a possible route to the Antilles. Modified from Urquhart, F.A. 1976. National Geographic 150:160-173, and Urquhart, F.A., and N.R. Urquhart. 1979. Canadian Field-Naturalist 93:41-47.

Monarch breeding range

Greatest concentration

Wintering areas

by the Urquharts (1976, 1979) and thousands of volunteers and fellow enthusiasts in the Insect Migration Association. Reports of butterfly locations came from many persons. In one case a monarch landed on a golf ball in California just as the golfer started into his swing. It was too late to stop the swing, but the remains of the monarch and its tag were recovered and mailed to the researchers. After several frustrating years, the wintering locations of monarchs in Mexico were finally discovered by a dedicated couple, Ken and Cathy Brugger from Mexico City, after they became aware of the problem through newspaper advertisements. Monarch migrations have attracted increasing scientific and popular attention since then (Brower 1985, Urquhart 1987).

The monarch butterflies breed throughout much of the United States and in scattered locations along the southern edge of Canada (see Figure 7-1). Monarchs in the west migrate to and winter primarily along the California coast between Monterey and Los Angeles. Tagged butterflies from further east, as described previously, winter at a number of sites in the Neovolcanic Mountain region near Mexico City.

In these mountain sites, at approximate elevations of 3000 m (10,000 feet) and temperatures that continually hover around freezing, the butterflies mass together by the millions in dense clusters that cover the trees, bushes, and ground (Figure 7-2). In one of the first areas discovered, the clusters of butterflies completely covered the branches and trunks of more than 1000 trees. The butterflies pass the winter at these high altitudes and low temperatures in a state of semidormancy. Then, apparently with the lengthening daylight of spring, clouds of butterflies take to the air and begin the return trip to the northeast.

The spring migration begins in late February and March. Mating occurs when the masses begin to disperse and continues along the northward journey. Few males

Figure 7-2 Overwintering roosting monarch butterflies in California.

Photo from Animals, animals. © M.A. Chappell.

travel beyond central Texas, but some females may travel all the way back into Ontario and other northern areas, arriving throughout late May, June, and early July. They deposit eggs en route and next-generation adults, resulting from those eggs, begin appearing in June and early July. They are much brighter than the tattered, torn, and faded butterflies that traveled all the way to Mexico and back. First-generation adults from eggs laid in Canada show up from mid-July through August and yet another generation may appear through September, depending on the latitude. The adults from the previous year have all presumably died by this time. From mid-August throughout the fall the adults of the new generation begin migrating toward Mexico.

The butterflies form small clusters overnight at traditional trees and locations en route. The first butterfly at an overnight site flies around until a suitable location is found; then it lands and opens its brightly colored wings, which appears to attract others. The means by which they select the trees and why the same sites are chosen in subsequent years are not known. There is no evidence that a particular odor is involved. When trees are removed, the butterflies select other trees nearby (Urquhart and Urquhart 1979). The mechanism or mechanisms by which the monarchs navigate to the small isolated wintering sites are not known and have been the subject of much speculation. It may involve some kind of a species memory, fixed hereditarily in the nervous system, but it clearly does not involve individual memory because the migration each fall consists of new individuals. Do these small animals with a nervous system smaller than a grain of wheat have a simple but highly accurate means of navigating to specific locations, perhaps based on the Earth's magnetic field? Or instead is there some means other than a specific guidance system that results in monarch butterflies becoming concentrated in particular places during the winter? We will return to that issue in the next chapter on long-distance travels.

How do the local and long-distance travels of butterflies compare with those of other organisms, including other flying animals such as birds and bats and other animals that move about in other ways? Do humans fit the pattern of other species, and do we ourselves perhaps have some or all of the same navigational abilities and mechanisms as butterflies and birds? Although we usually take it for granted, we ourselves are capable of traveling without an external map to many places where we have been before. How do animals choose the best places to live or even know when they are in such a place? If they travel over long distances, beyond a distance over which they can maintain direct sensory contact with familiar landmarks, how do they keep from getting lost? How is something even as seemingly simple as familiar landmarks integrated in the brain? These questions involve several different aspects of spatial behavior, including the underlying proximate and ultimate causes. We will try to unravel these by starting with the most fundamental aspect: locomotion and its ramifications.

LOCOMOTION AND ITS OPPORTUNITIES AND HAZARDS

Animals move about in many different ways, and these travels accomplish many different functions. It should be noted, incidentally, that the movements and spacial orientation being considered in this and the next chapter involve only whole-body movements from one location to another. There is another category of movement and spatial orientation that involves balance, body posture, and the immediate three-dimensional space surrounding an individual's body; that is a somewhat separate matter that we are not considering here (see Schone 1984).

Some functions of traveling about are readily apparent, such as movements toward food, shelter, and mates or away from danger, whereas other activity appears to be largely exploratory. The apparent advantage of exploration is that it may expose an animal to new opportunities, increasing the probability of favorable encounters with food, mates, or other situations. It also may be important for building a spatial map of an animal's surrounding within the animal's nervous system.

Patterns of locomotion within a given species may be fairly consistent and recognizable. Many types of birds, for example, fly in recognizable ways. Some species fly in a straight line with a steady, specific wing-beat frequency. Many woodpeckers (Piciformes) fly in an undulating pattern, flapping and rising, then not flapping and falling. Hawks of the genus *Accipiter* typically fly by interspersing somewhat regular sequences of flapping and gliding. Different species of birds, in spite of considerable structural similarity, show many differences in terrestrial gaits (Clark 1975). Some walk on alternate legs, some hop on both legs, and some alternate between stepping and hopping. In many species, there are ontogenetic (developmental) changes in gait patterns. Although most species show stereotyped patterns, including differences in various age, sex, feeding, and social contexts, individuals can compensate and change gait patterns under conditions such as injury. Similar generalizations can be drawn about the gaits of mammals, swimming of fishes, flight and walking or hopping of insects, and locomotion movements of other animals. A person familiar with differences in locomotion patterns often can identify a species at a great distance by a moving silhouette or when other details of an animal's appearance cannot be seen.

The following is a simple list of different types of basic locomotion:
1. Movement by pseudopods, cilia, or flagella in protozoans
2. Pumping actions and other jet-propulsive techniques as in cnidarians, scallops, and cephalopods
3. Paddling, flipper, and rowing motions for swimming in many forms of invertebrates and vertebrates
4. Undulatory swimming and terrestrial movements as in swimming by fish and leeches or the terrestrial movements of animals such as snakes
5. Tunneling, burrowing, digging, peristalsis-like motions, and other underground movements as in many worms, amphibians, reptiles, some mammals, and others
6. Gliding, friction, and extension type of movements as in flatworms, gastropods, inchworms, earthworms, and snakes
7. Walking
8. Jumping
9. Running, trotting, pacing, and cantering
10. Climbing and related movements in trees and on other vertical surfaces
11. Gliding and various kinds of flight by different invertebrates and vertebrates

The cost of locomotion has received considerable attention (Denny 1980). It requires more energy, for a given weight and assuming the appropriate structure is present, to run than to fly and to fly than to swim (Figure 7-3). Under various circumstance and in various animals, however, the costs of different forms of

Figure 7-3 Costs of
locomotion.
From Denny, M. 1980. Science
208:1288-1290.

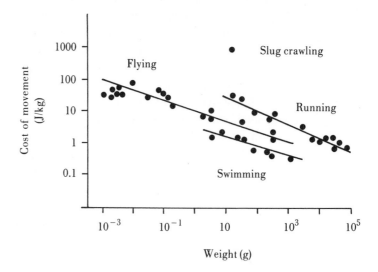

locomotion can vary widely so that whether flying or running costs more, for example, depends on many factors. The costliest form of locomotion, interestingly, appears to be crawling on a mucous track by mollusks because of the production of mucus that holds the mollusk to the substance (Denny 1980). Animals that are capable of moving by several means, such as walking and flying, or walking versus trotting or galloping, generally choose the most energetically efficient method in any given situation (e.g., Hoyt and Taylor 1981). This matter is of considerable importance to the subject of optimal foraging, which will be discussed in Chapter 9.

For some movements, particularly jumping and flying, there are preparatory, orienting movements. They usually involve a preliminary posture and a readiness to move, with limbs in the proper position for springing or taking off. They are referred to as intention movements. In birds, for example, the bird stands or perches on both legs, leans forward, and slightly lowers the body; it may also stretch the neck forward, extend the wings slightly, and gaze in the direction toward which it is ready to move. Intention movements are significant not only as preliminary actions for completed movements but also because they frequently are seen by themselves; that is, the subsequent movement is abandoned and not carried out. They also have played an important role in the evolution of many communication displays (Chapter 15).

Regardless of how animals move, however, they do not travel about purely at random, like so many molecules in a gas or like simple wind-up toys. Their movements are usually functional and often directed.

MOVEMENTS WITHIN AND BETWEEN LOCATIONS: GENERAL CONSIDERATIONS

Imagine a map that traces the entire path of an individual animal's movements throughout its life from birth to death, including all small, local, daily movements and any long-distance travels. This can be referred to as the *lifetime track* of the animal (Baker 1978, 1982). The memory of a lifetime track may be useful to some animals for developing the familiar areas and paths by which they navigate, as discussed later in this chapter and in the next chapter. The image of such a track can also be useful for us to help envision and tie together the numerous, diverse aspects of animal travel.

What is your own lifetime track to the present? What local daily movements do you now make, say from residence hall or apartment to classrooms, shopping, and places of relaxation and recreation; and what have been your recent longer travels, such as to a summer job, to visit relatives or friends, or to take a break from school? Could you trace and mark on a map all of your movements since birth? For most

Figure 7-4 Tracks of four radio-tagged freshwater crabs, *Potamon fluviatile*, showing differences among individuals. These four are examples from a larger number that were tracked. Locations were recorded every 4 hours for an average of 20 days for each crab. From Gherardi, F., et al. 1988. Ethology 77:300-316.

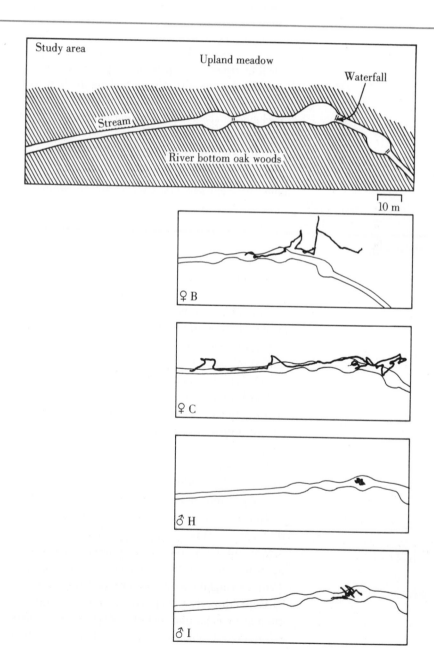

animals, including humans, the lifetime track is a mixture of lines heavily concentrated around one or a few local areas plus some occasional lines beyond those places and a few lines of much greater distance.

Few lifetime tracks have ever been followed and plotted for any normally mobile animal. There are numerous examples, however, of tracks for shorter periods of time. Figure 7-4 shows the daily tracks of different individual freshwater crabs *(Potamon fluviatile)* along a stream where they were followed by radio telemetry (Gherardi et al 1988).

The lifetime track of an animal's travels can be placed into an evolutionary and ecological context, as described by Baker (1982):

> One of the problems that faces all animals is where to live. Individuals have little control over where or at what stage of development they first appear. Parents decide these things according to their own best interests. Thereafter, the individual has to wend its way through time and space until it dies. The most successful individuals in terms of the number and quality of offspring they produce (i.e. reproductive success), will be those with a path that exploits the environment to the full, minimizing costs, maximizing benefits, always seeking the best trade-off between the two. The path that an animal produces is its lifetime track, the outward manifestation of the individual's solution to spatial and temporal problems, the playing off of inherited predispositions and acquired experience against the environmental backcloth, running the gauntlet of natural selection for yet another generation.

The accumulated pattern of travels by an animal during its lifetime involves many different aspects, each of which we consider in turn.

DISPERSAL

The vast majority of animals, regardless of species or phylogenetic affiliation, sooner or later move away from the immediate place of their birth or hatching. Some move considerable distances, depending on a scale relative to the animal's body size and means of locomotion. Many animals subsequently return to their original location or nearby, perhaps not having gone far in the first place. The movement away from the initial point of origin and some other movements during an animal's life have been referred to as *dispersal*.

The literal meaning of *dispersal* is "to break up and scatter or spread about in all directions." This is something that groups do but that individuals technically do not unless they were to explode or perhaps be torn apart and scattered about by predators or scavengers. In addition, dispersal remains poorly defined and ambiguous because it has been used to cover a wide variety of animal movements from initial departures from their point of origin, to long-distance movements with or without returning, to all movements that did not fit some other definition.

However, the predominant biological connotation of the term *dispersal* is "non-directed movement away from a particular location of interest" (Stenseth 1983:65). Movement away from a place can be important biologically, and it has received attention from several workers (e.g., Hamilton and May 1977, Lidicker and Caldwell 1982, Swingland and Greenwood 1983, and Haig and Oring 1988).

Depending on the species and a host of circumstances, dispersal as opposed to staying in the same location offers several theoretical advantages. Many of these relate to problems of competition that result from increased numbers of individuals in a particular place. For our purposes, we simply list some of the potential advantages; these are not all mutually exclusive:

1. Dispersing individuals may find and colonize new or open habitat, even if the present location remains suitable to some extent.
2. Dispersal may lead to improved tracking of changing environments.
3. Dispersal from a deteriorating environment (including situations with too high a population density) may increase the probability of finding more suitable conditions elsewhere. This category includes refugees fleeing from desperate situations and is thought to be the major cause of mass movements among lemmings and many other species. (Although many individuals may die under such circumstances, including drowning from exhaustion while swimming across bodies of water, the old popular view that the lemmings are bent on suicide is not founded on valid biological considerations.) When some animals disperse under deteriorating conditions and others remain behind, with differences in aggressiveness, for example, there may be important genetic implications for the population.
4. Even in a stable, nondeteriorating environment, forced dispersal of some (e.g., offspring) by others (e.g., parents) may result in an advantageous reduction of competition for those who stay, even if there is a cost of increased mortality among relatives who are forced out.
5. Dispersal results in a reduced potential for inbreeding.
6. Dispersal may keep potential prey a step ahead of their predators, increasing chances of survival.

Further discussion of these issues can be found in Horn (1983), Stenseth (1983), and the references contained therein. The main points of significance for this chapter are that dispersal is ecologically and evolutionarily important; it is widespread and common, if not almost universal, in the lives of virtually all animals; and it results in animals moving into new locations where they have not been before. They may then be exposed to both the potential of suitable, advantageous conditions and the risks of situations that are not only disadvantageous but often downright dangerous. That places a selective premium on any ability to identify and seek out suitable conditions while avoiding the unsuitable. Animals display a wide variety of behavior in accomplishing this.

SIMPLE FORMS OF CHOOSING A SUITABLE PLACE

The two simplest categories of behavior that functionally move animals into suitable locations and then keep them there are kinesis and taxis. Kinesis is a nondirected, apparently random movement in response to environmental factors such as light, sound, temperature, moisture, or chemical cues. When in an unfavorable location or when moving into an increasingly unfavorable place, the organism increases its speed of movement or frequency of turning away. These actions remove it from the present location and expose it to new conditions in new locations. When more suitable conditions are encountered, the animal simply moves less and remains in the new area (Figure 7-5).

Taxis, on the other hand, involves directed movements; that is, animals detect the source or differences in intensity of some factor, then move appropriately.

Figure 7-5 Kinesis in animal orientation. Track of a louse approaching a favorable diffuse chemical stimulus.
Modified from Fraenkel, G.S., and D.L. Gunn. 1961. The orientation of animals. Dover Publications, New York.

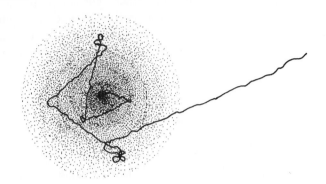

Figure 7-6 A, Track of a planarian toward a favorable diffuse stimulus. B, Taxis in a crawling maggot. The maggot turns away from a light that has been directed toward its side. Technically this would be a negative-photo-klino-taxis (away-from-light-turning-taxis).
A modified from Carthy, J.D. 1956. Animal navigation. Charles Scribner's Sons, New York. B modified from Fraenkel, G.S., and D.L. Gunn. 1961. The orientation of animals. Dover Publications, New York.

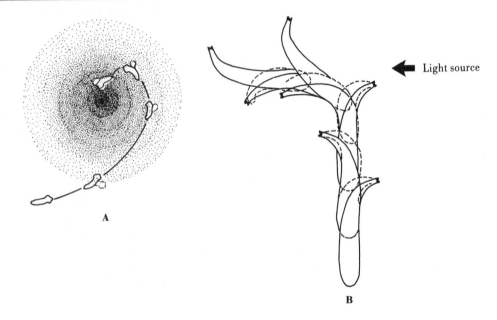

Light source

A

B

Detection of concentration or direction of a source of something may be accomplished by repeated testing of the environment. The best known examples of repeated testing involve planarians and fly maggots (Figure 7-6). An alternate method of sensing the source or intensity of something is shown in many species with directional sensory capabilities, often through the use of paired sense organs. Taxis is frequently identified with the environmental condition to which the organism is responding. For example, there are phototaxis, geotaxis, and hydrotaxis, to or from light, gravity, and water, respectively. An animal may also be attracted to or repelled from a concentration of some particular chemical, including those associated with food, nest sites, or potential mates. Orientation on the basis of odors and air movements has been referred to as anemotaxis. A positive or negative taxis means movement toward or away from the factor, respectively. A host of other terminology has been used as well.

HABITAT SELECTION AND USE

From the relatively simple level of kinesis and taxis, one can consider the problem of finding a place to live on a broader scale, involving organisms that have more complex sensory and nervous systems and that are able to move considerable distances under their own power. Various means of finding good locations to live are not necessarily exclusive or confined to certain groups of animals. Even the most mobile and complex animals may use taxis to some extent in guiding their movements. Chemical cues or odors are used not only in kinesis and taxis but also for some long-distance movements. Odor following is perhaps one of the most basic and general mechanisms for finding a place and, although not particularly important among humans, can be found in almost all taxonomic groups from protozoa to mollusks, insects, fishes, reptiles (e.g., turtles, snakes), mammals, and even birds (Grubb 1977).

The problem of how an animal ends up living in a certain area can be viewed in at least two ways: first, the general aspect of the type or kind of environment, and second, the specific location(s) within which a particular animal spends most of its life, including different locations at different seasons and the pathways between these places. The first topic usually is considered under the banner of habitat selection or use and the second under some name such as home range or territory.

Habitat selection assumes that animals are actively choosing the area, whereas habitat use simply notes the presence of the animals and hence implies much less. The difference between habitat selection and use can be temporal, original choice of an area versus subsequent activities in that area. Or the difference can simply refer to the animal's active (not necessarily conscious) choice of the area versus simply being there, without focusing on how the animal got there.

Similarly, territory is generally considered to involve some degree of active defense and interaction between animals (Chapters 11 to 15), whereas home range just refers to where the animals spend their time, whether that place is actively defended or not. For a study that considered both home range and territory (in a species of lizard), see Smith 1985. The different topics of habitat selection or use and home range are discussed separately here.

In distinguishing between habitat selection and use, it must be emphasized that the places animals end up living are not always determined by the animals themselves. Many species, particularly some aquatic organisms and small insects and a few others, are true drifters. They are simply carried about by water or wind currents, and may occasionally become concentrated by eddies of current. They have no sense of location, direction, or the past, but move or are carried about in an undirected manner in the environment in which they find themselves, more like the seeds, pollen, or spores of plants. They are often considered minor or insignificant from our perspective, but they can be diverse and abundant enough in the real world to fill the stomachs of whales. Some animals, such as many birds on migration, combine methods; they may drift with the wind or water, but they select which currents and when to ride so that there is a degree of active participation.

Another method by which animals may occur in particular locations without their active choice involves differential survival imposed by the environment, such as from predation or weather factors. This may lead to the false impression that the animals chose their location. Imagine, for example, a large population of animals that disperse, then settle randomly into several types of habitats. In some locations, however, the animals are more vulnerable to predators, perhaps because they are more conspicuous or there are fewer hiding places, and these individuals are eliminated. Others in more suitable habitats, however, survive and remain (Turner 1961). Their presence in just those habitats may give the impression that the animals chose the places when in fact it was determined by the absence of predation.

Competition, somewhat along the lines of dispersal, also may greatly influence where an animal lives. Studies of fish (Werner and Hall 1976), chipmunks (Chappell 1978), mice (Randall 1978), and many other organisms have shown that different species, when tested alone, may have similar preferences for habitat, but when together, one species may partially or completely exclude the other and push it into a less-preferred type of habitat.

In the mouse example (Randall 1978), two sympatric species of *Microtus* were given choice tests in outdoor enclosures. *M. montanus* is found normally in grass habitat and *M. longicaudus* in shrub habitat. The enclosures provided both types of habitat. When tested individually in the enclosures, *M. montanus* spent slightly more time in the grass than *M. longicaudus* (about 75% to 90% versus 65% to 70%, respectively), but both species showed a distinct preference for being in the grass, whether the test subjects were wild caught, reared in plastic boxes with wood shavings, or reared in the opposite type of habitat than where normally found. However, when animals from both species were placed together, *M. longicaudus* was observed to withdraw from the more aggressive *M. montanus*. Thus *M. montanus* is thought to occur naturally in grass habitat because of its preference for grass, whereas *M. longicaudus* probably occurs in shrub habitat because it is excluded from the grass by the more dominant *M. montanus*.

In the extreme case this relates to the ecological principle of competitive exclusion, by which one species may even completely eliminate another from an area. Even with increased density of a single species, some individuals may have to accept less-preferred places to live. Based on a study of reproductive output in herring gulls, Pierotti (1982) obtained results that supported a model proposed by Fretwell and Lucas (1970). "Increasing density in a preferred habitat can create a situation whereby fitness may actually be greater in less-preferred habitat."

In spite of these situations, however, it seems advantageous for animals to actively choose where they live, and there is at least circumstantial evidence that many if not most animals do choose within the range of options open to them. Any good birder, fisherman, or hunter knows that certain species will be found almost exclusively in certain habitats within a region, although the animals could easily move on their own to other places and different habitats. One of the first to demonstrate this quantitatively and employ the term *habitat selection* was David Lack (1933). He demonstrated that different species groupings of birds were found in vegetation stands of different ages. MacArthur (1958), in a much cited classic study, showed that species may be segregated even in their microhabitat use of individual trees (Figure 7-7).

One could state tautologically that the ultimate reason particular species select particular habitats is because those are the places to which the animals are best adapted. However, that involves potentially circular reasoning, and the argument is related to the semantically difficult topic of the niche. The niche has been defined in many ways, including an animal's position, what it does, its "profession" in nature, or, in a more contemporary view, the species' set of environmental requirements. (For references and a good discussion of the history and theoretical problems associated with the niche concept, see Krebs 1985.)

Figure 7-7 Preferred parts of trees for feeding habitat by different species of warblers. Diagram is based on percentage of total seconds of observation for three (of several) species involved in the study. Modified from MacArthur, R.H. 1958. Ecology 39:599-619.

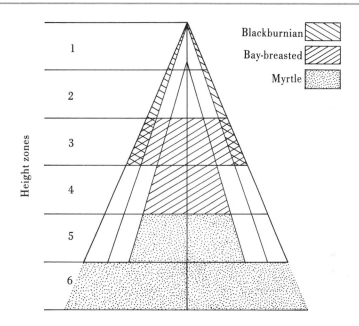

Beyond the semantic problems, however, the topic of habitats to which animals are adapted and how they select and use them is important for both basic and applied reasons. From an applied standpoint, for example, the activities of humans, including habitat changes, have threatened the existence of many species of animals, commonly classified as endangered species. To manage and correct the problems, one needs to better understand species' habitat requirements and the environmental factors to which their senses are tuned. The factors they need and those they select may not always be the same; they may simply occur together. Also, many persons are interested in maintaining various species, including domestic, under the best conditions in captivity. Although the natural habitat contains many objects and environmental conditions, only a few or particular subsets are really important to a given animal; there is no need to duplicate the entire set to satisfy the needs of the species. The trick is to find those factors that are most important.

Dealing with the proximate factors in habitat selection may be easier than the ultimate ones. However, even the proximate aspects may be elusive; measuring a species' habitat is difficult. Habitat, rather than being a small, isolated object on which the animal and a researcher can focus their attentions, encompasses and surrounds the animal, with no well-defined boundaries. It generally is variable and under the best of conditions consists of many items and factors. The animal may respond to different factors at different times and seasons and under different circumstances. Habitat is a fairly vague, complex, and abstract entity.

In the case of simple organisms and their simple orientation movements the animals often respond to specific, readily identifiable factors such as light or temperature. Habitat selection in these cases probably could be considered as being synonymous with kinesis or taxis. But as animals and their sensory systems become more complex, so does the sophistication of their habitat selection.

Habitat selection, like any other kind of selection, clearly has to be based on sensory cues. But, except for a few cases, little work has been done on what cues are used by which animals under what circumstances. Furthermore, there has not been much research on the heritability of specific habitat selection. The identity of the preferred habitat in some species appears to be learned in part (via a form of imprinting, discussed in Chapter 19). It is clear that species are variable in their specificity. Many species are known as *specialists* and others are known as *generalists* in their choice of habitat.

The apparent choosing or selecting of habitat by many animals can be observed outright. Some rotifers, for example, become sessile, but initially they are able to move about. The habitats that some species of these microscopic animals choose may be as restricted as the left side of certain types of snail shells. Wallace (1980) described how larvae settle once they have reached that period of their life: they swim to various surfaces, pause, and contact the surface with the corona (the wheel-like ciliated organ around the mouth from which they get their name); then, depending on whether it is the appropriate surface, they temporarily attach or swim on and try a new location. In suitable habitat, they may detach and reattach several times in different spots before becoming permanently fixed. At each initial attachment they twist around and bend over, touching the substrate with the corona as if trying out the site for preference.

Similar descriptive studies of habitat use have been published for many organisms, often showing much variability among different individuals, different seasons, and for different activities. Gottfried and Franks (1975), for example, showed differential use of habitat by two different flocks of juncos *(Junco hyemalis)* and for one of the flocks before and after a month-long period of heavy snow accumulation during which the flock disbanded (Figure 7-8). Hunter (1980) described micro-habitat selection in great tits *(Parus major)* in which the birds chose different

Figure 7-8 Habitat used by two different flocks, including two different times for one flock, of one species of junco. The percentages of nonroosting time spent in the different habitats show considerable variability. Modified from Gottfried, B.M., and E.C. Franks. 1975. Wilson Bulletin 87:374-383.

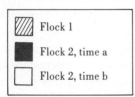

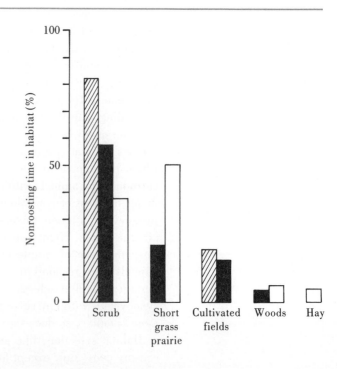

Figure 7-9 Microhabitat selection for different activities by great tits. From Hunter, M.L., Jr. 1980. Animal Behaviour 28:468-475.

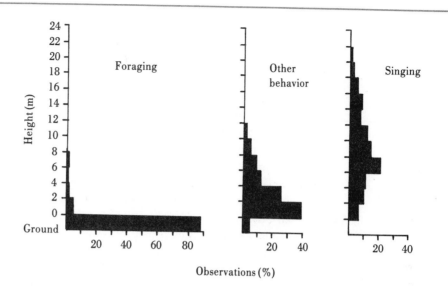

habitat for different activities (Figure 7-9). The most hidden, protected habitat was used when the birds were singing and might attract predators.

Laboratory experiments to investigate the role of prior experience in habitat selection have been conducted with several types of animals. The basic approach has been to provide captive animals with a choice of habitats. Depending on the particular study, individuals have been captured from the wild or reared in the laboratory. Laboratory or hand-reared animals may have come from wild parents or from several generations of captive parents. They have been reared in natural or unnatural types of habitat or in restricted conditions with no exposure to anything resembling a normal habitat. Klopfer (1963), for example, hand-raised birds in cloth-covered aviaries with no foliage and under more natural situations. Randall (1978) used similar procedures in the *Microtus* example, investigating competition.

The results of such experiments have generally demonstrated definite habitat preferences by individual animals, both species effects and experience effects, and differences among closely related species. In the study by Klopfer (1963), for example, white-throated sparrows *(Zonotrichia albicollis)*, when given choices between pine and oak branches on opposite sides of an observation chamber, showed a slight but not strong preference for the oak, whereas chipping sparrows *(Spizella passerina)*, whether wild-trapped, hand-reared in pine, hand-reared in oak, or reared without foliage, all preferred the pine. Those reared in oak, however, showed less of a preference for pine. In studies of blue tits *(Parus caeruleus)*, which normally are found in broad-leaved trees, and coal tits *(P. ater)*, which live in conifers, both wild-trapped and hand-reared birds tested in choice conditions preferred the habitat as predicted, but hand-reared birds showed less of a preference (Gibb 1957, Partridge 1974).

In a classic study of *Peromyscus* mice, Wecker (1963, 1964) tested a prairie subspecies, *P. maniculatus bairdi*, in a large observation pen that half enclosed a field and half enclosed the edge of a woodlot. The subspecies normally is found in fields. Wild-trapped mice and first-generation laboratory-raised offspring all showed a preference for the field side of the enclosure. After 12 to 20 generations of laboratory rearing, however, offspring showed no preference for either field or woods unless they were raised first in one or the other. In that case, field-reared offspring preferred the field and woods-reared offspring showed no preference, indicating that an inherent bias for the field still remained. Similar results have been obtained among fish (e.g., Sale 1971).

Studies of animals under natural conditions have been more indirect and often difficult to interpret. The usual plan has been to observe a sample of individuals from a species where they are found and measure as many aspects of their surroundings as possible, or at least as many variables as are deemed reasonable by the investigator. Comparable measures are made for other species in the general area or for a number of randomly chosen points. Then the mass of data is subjected to multivariate analyses to boil it into more comprehensible form. The approach is an extension of studies in which distributions of birds are plotted along continuums of plant communities (Bond 1957).

Two of the pioneering applications of multivariate analyses for animal habitat use were by Martin L. Cody (1968) and Frances C. James (1971). In James's study, for example, 15 vegetation variables were measured for 46 species of birds from 18 counties in Arkansas. From her results she was able to determine the positions of the different species in an abstract habitat space (as illustrated for woodpeckers in Figure 3-15). She also identified the characteristic habitat appearance, the niche-gestalt, from the viewpoint of the species. The characteristic appearance depended on the visible elements of the vegetational structure that were present consistently in the surroundings of animals for each species. Examples of James' niche-gestalts for four species of warblers are shown in Figure 7-10.

Subsequent multivariate analyses of animal habitat use proliferated, and various techniques were used to better illustrate differences among species while still allowing for variability and overlap (e.g., Figure 7-11). For a recent compilation of habitat selection information in birds, see Cody 1985.

The basic problem with the multivariate approach to habitat use is that it is correlational at best. One cannot determine which, if any, factors the animals themselves are actually choosing, and it is not clear that the most relevant variables have been measured. Furthermore, outcomes of analyses of preference may shift and yield different interpretations, depending on which particular variables are included or omitted. Johnson (1980) discussed the problem of shifts of inferred preferences and proposed a solution whereby ranks rather than actual measurements would be used for different elements. Johnson (1981) further noted that "a bird might find James's (1971) outline drawings of niche-gestalts to be meaningful, and would be willing to select its habitat based on those drawings. But the bird would be hard-pressed to plug the values of 15 or 20 variables into a number of linear combinations, compare the calculated values to one another, and select a habitat with values closest to its liking."

The problem of not including all variables was clearly recognized by James (1971). She stated, "The space also contains gradients in types of food, nest-sites, microclimate, etc. Although these variables are undefined in the present study, they would have to be included in a thorough analysis of the ecology of adaptation."

What really is needed to answer whether habitat selection is occurring and which factors are being chosen is to experimentally vary the components of animal habitat and measure the response of the animals. That, however, is a large order of business. The "test-tube" or experimental unit for such studies might be several hectares or

Yellowthroat

Redstart

Black-and-white warbler

Hooded warbler

Figure 7-10 Four examples of F.C. James's niche-gestalts for different species of birds. These outline drawings represent features of vegetation consistently measured where the birds were found. Modified from James, F.C. 1971. Wilson Bulletin 83:215-236.

Figure 7-11 Example of multivariate plotting by the first two principal components of the habitat, which indicates both species differences and species variability to a limited extent. (The areas plotted are 1% confidence ellipses.) From White, D.H., and D. James. 1978. Wilson Bulletin 90:99-111.

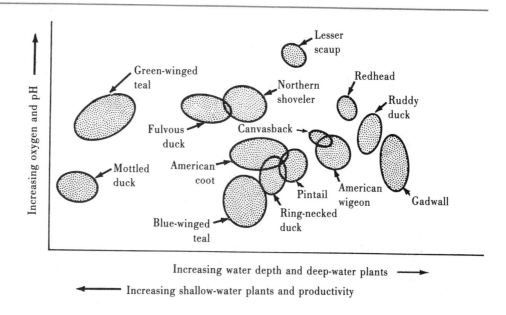

even square kilometers in size; sufficient natural habitat may no longer exist for experimental manipulation; and one may not have the foggiest idea how to identify those components to be manipulated.

Jacobs et al (1978), for example, considered color in the nest selection by female village weaverbirds *(Ploceus cucullatus)*. They painted nests brown or green and found that females showed only a slight, nonsignificant preference in whether they accepted or rejected nests. Males, however, destroyed brown nests much more often than green nests after particular nests had been rejected by females. These results are difficult to interpret but suggest that color is not the only or major element in the choice of something as discrete as a nest. That is, the correct component or components have not yet been identified. It is most likely that several factors are important together in habitat selection. But, unlike the ease with which simple models of animals can be varied for studying some behavior (Figure 2-6, page 44), habitat is not as easy to experiment with.

In a few instances the important factors in habitat selection may involve something as simple as presence or absence of nest holes in trees, cover (of almost any kind) from predators or inclement weather, or specific types of food. But more often than not, the situation is more complex, and the multivariate approach, used with caution and the best subjective insight from knowledge about the species, is the best tool available. At the same time one must be careful not to infer too much from the results of such studies.

Until it is shown beyond a reasonable doubt through experimental confirmation that an animal indeed is actively selecting its habitat, it must be considered that the animal may be where it is found by default, such as from predation and other sources of mortality or from competition with other animals. Not all humans live in the dwellings or geographical areas of first choice, and it is unlikely that many other animals do either. Often there is a multitude of contributing factors. Furthermore, as is amply clear from a small sample of experimental studies of mice and birds, the factors can be quite different even for closely related species.

HOME RANGE/FAMILIAR AREA

Okay, so we have placed the animal in a suitable location, often after dispersal of some sort from a previous place. Now it is time to consider the specific places where an animal lives and whether specific places are important or whether the animal simply wanders around within a generally suitable area.

Most individuals tend to remain within relatively confined areas most of the time. Animals of some species travel about more than others, and males commonly cover a larger area than females (Schoener and Schoener 1982). But even those individuals generally tend to return to places they have been before and stay more or less within a given area. Thus the lifetime track shows a heavy concentration of repeated paths and activity in certain places.

The ultimate advantages of being able to return to a familiar place, such as home, should be clear. Such places may afford protection, hence increasing the chances of survival, increased resources, and perhaps even stored resources. Simply being in a familiar area may aid survival. Mice allowed to become familiar with an area, for example, are less vulnerable to predation by owls than are mice newly introduced to the area (Metzgar 1967). Any biological mechanisms that permit an animal to nonrandomly return to a previous location should have high selective value.

The area within which an animal stays is called its *home range*. Home ranges are usually dynamic; that is, they tend to change with new paths and locations over time (Figure 7-12). Several techniques exist for estimating the size and boundaries of home ranges, and there are a number of associated terms and concepts. Lehner (1979) surveys and discusses the older techniques. For recent methods, comparisons of alternatives, and general reviews of the subject, see Schoener (1981) and Anderson (1982).

The most common method by which animals appear to maintain their spatial orientation and ability to stay within a given area is what can be called simply the

Figure 7-12 Shifts in home range over time for a female bank vole, *Cleithrionomys glareolus*, on Skomer Island, Wales.
From Baker, R.R. 1978. The evolutionary ecology of animal migration. Holmes & Meier Publishers, New York.

familiar area or *familiar path*. Jakob von Uexküll (1934) was one of the first persons to draw attention to the familiar path. The familiar area for an animal requires two important things: (1) the ability to learn and remember the area and (2) some form of at least minimal sensory contact with consistent features or landmarks. Landmarks in a generalized sense can be visual, auditory, tactile, olfactory, or involve different senses, including some less familiar ones such as electrical fields and magnetism, which are discussed in the next chapter. Frequently the marks of a path or area are made by the animal itself, such as odor trails or signposts of some kind left along the way.

A classic example of the use of landmarks involves digger wasps (Tinbergen 1951). The wasp first digs a hole for a nest. Emerging from it, the wasp flies around for a few seconds learning landmarks. Artificial landmarks, such as pinecones can be placed around the nest entrance while the wasp is inside (Figure 7-13). If these landmarks are moved to a new location while the wasp is away capturing prey, it

Figure 7-13 Landmarks in the location of nest holes by the digger wasp (*Philanthus trangulum*). **A,** This wasp first digs a hole; then, on emerging from it, it spends a few seconds flying around the area learning the landmarks. The landmark in this instance is a ring of pine cones placed around the nest entrance while the wasp was inside. **B,** If the cones are moved to a new location while the wasp is away, it will fly to the new location with prey for the nest and be unable to find the actual nest entrance. The characteristics of landmarks used by these wasps were investigated further by using different shapes of surrounding objects, objects other than pine cones, incomplete perimeters, and combinations of objects, such as cubes and twigs; Modified from Tinbergen, N. 1951. The study of instinct. Oxford University Press, Oxford.

Nest

will fly to the new location when it returns and not be able to find the actual nest entrance. The characteristics of landmark learning were investigated by using different kinds and patterns of surrounding objects, incomplete perimeters, and combinations of objects such as cubes and twigs (see references in Tinbergen 1951).

Spatial memory, whereby an animal forms some kind of mental image or map of its familiar area, may be one of the most basic and universal forms of memory and learning (Chapter 20, Konishi 1986). However, the nature of the internal representation in the nervous system is not known. It may be quite different in other species, perhaps consisting of series of directions more like a computer program than the apparent three-dimensional, external photograph-like image that humans seem to experience. Not only the way in which individuals mentally orient themselves relative to their spatial surroundings but also the means by which the landmarks themselves are recognized and remembered theoretically could be accomplished in different ways. There are at least two ways that landmarks could be remembered: (1) by a few basic characteristics (as parameters or in sectors) or (2) as a more complete image or picture. Gould (1987) described the difference and implications with a familiar example:

> To use a human analogy, the information content of an advertisement specifying the important parameters of, say, an automobile (make, model, year, body type, colour, accessories, mileage, etc.) is far lower than the information content of an actual picture of the item for sale; the list of important parameters is clearly an adequate and inexpensive way to store (or communicate) information in certain contexts.

Gould (1987) proceeded to experiment with honeybees and concluded that they remember landmarks pictorially. The resolution of the image as calculated by Gould was adequate for a pictorial representation, but it was not what would be considered high-resolution by modern photographic standards; hence the stored image would only be picturelike. Even humans do not have truly high-resolution visual images, so in reality what is mentally stored is probably a biological compromise between the extremes of parameter-list and true picture. Gould and others who have studied bees, incidentally, believe that landmark memory is a separate process from flower memory, although more research is needed to clarify that matter.

Regardless of how the landmarks and system of orientation are handled by the brain, the familiar area is clearly a common means by which many types of animals find their ways about. Baker (1982) refers to the familiar area as a *sense of location*. The system is so familiar to humans that most of us simply take it for granted. It operates by the individual starting from a given point and becoming better and better acquainted with the features of the area or path during travels from, to, and around that point. This process of exploration continues, and the boundaries of the familiar area are pushed outward. Perhaps the best way to envision the process is to recall the feelings associated with times when you first arrived at a new place, such as a college campus or city where you had never been before. One may feel lost and overwhelmed when starting with a new, unknown area. Humans commonly seek out maps, ask directions or accompany others to particular locations, and start exploring. It is time consuming, but eventually the sense of location returns and one can then travel efficiently to desired locations within the familiar area.

Exploration can be easily confused with dispersal if one is not careful to distinguish between them (Baker 1982). Exploration involves return movements, often frequent returns, and staying within a given area.

there is habitat selection. Habitat use implies less active choosing on the part of the animal and recognizes that animals may end up in a particular location that would not be chosen otherwise for a variety of reasons, such as differences in predation or competition.

Animals of many species establish and maintain behavioral connections with specific locations and establish a meaningful home range or familiar area. The process of becoming familiar with the area or path requires an ability to learn and remember and to maintain sensory contact with landmarks.

Under conditions in which landmarks fail or cannot be used, some animals employ other, nonlandmark methods of spatial orientation, including the use of a sun compass, which is linked to internal circadian rhythms. Examples include honeybees, reptiles, birds, and many other animals. This leads into the topic of the next chapter, the long-distance travels of animals.

FOR FURTHER READING

Baker, R.R. 1978. The evolutionary ecology of animal migration. Holmes & Meier Publishers, New York.
An extensive (massive!) treatment of spatial orientation in animals, with much emphasis on the familiar area aspects.

Brower, L. 1985. New perspectives on the migration biology of the monarch butterfly, *Danaus plexippus* L. In: M.A. Rankin, editor. Migration: mechanisms and adaptive significance. Contributions in Marine Science, vol. 27 (supplement):748-785.
The best overall reference for monarch migration.

Carthy, J.D. 1956. Animal navigation. Charles Scribner's Sons, New York.
An older but somewhat classic, concise treatment of the basics of animal orientation.

Cody, M.L., editor. 1985. Habitat selection in birds. Academic Press, Orlando, Fla.
A collection of papers by several authors providing information and some discussion on habitat selection in birds.

Fraenkel, G.S., and D.L. Gunn. 1961. The orientation of animals. Dover Publications, New York.
Similar to Carthy, this is another of the older but somewhat classic treatments of the topic.

Horn, H.S. 1983. Some theories about dispersal. In: I.R. Swingland and P.J. Greenwood, editors. The ecology of animal movement. Clarendon Press, Oxford.
Review of information and a theoretical synthesis of the topic of dispersal.

Klopfer, P.H. 1969. Habitats and territories. Basic Books, New York.
Discusses habitat selection and animal territories.

Klopfer, P.H., and J.P. Hailman. 1965. Habitat selection in birds. Advances in the Study of Behaviour 1:279-303.
A review of research on habitat selection in birds; somewhat outdated but still excellent information.

Schmidt-Koenig, K., and W.T. Keeton. 1978. Animal migration, navigation, and homing. Springer-Verlag New York, New York.
Another somewhat general discussion and review that is useful for both this and the next chapter.

Smith, D.C. 1985. Home range and territory in the striped plateau lizard (*Sceloporus virgatus*). Animal Behaviour 33:417-427.
A good example of data and discussion pertaining to the sister topics of home range and territory.

Stenseth, N.C. 1983. Causes and consequences of dispersal in small animals. In: I.R. Swingland and P.J. Greenwood, editors. The ecology of animal movement. Clarendon Press, Oxford.
Review of the topic concerning small mammals, including lemmings. Includes much excellent theoretical discussion of the topic of dispersal in general.

Urquhart, F.A. 1976. Found at last: the monarch's winter home. National Geographic 150:160-173.
A readily accessible and well-illustrated general report on the monarch's overwintering locations and their discovery.

Urquhart, F.A. 1987. The monarch butterfly: international traveler. Nelson-Hall, Chicago.
An updated and also well-illustrated account of the monarch butterfly and its travels.

8

SPATIAL ASPECTS OF BEHAVIOR LONG-DISTANCE MOVEMENTS

The previous chapter discussed the topic of spatial orientation in animals at the local level, which, in species that maintain contact with specific locations, relies heavily on learning familiar areas or familiar paths through sensory contacts with environmental cues (e.g., landmarks). It is clear, however, that a familiar area with landmarks is not always sufficient to explain how animals get from one location to another, such as after artificial or accidental displacement beyond their obvious familiar area or under certain adverse conditions, including fog. Also, the familiar area seems inadequate to account for many of the long-distance travels of animals. A number of nonlandmark techniques plus some extended systems of familiar areas may permit animals to travel long distances, well beyond the local level of movements and in some cases literally around the Earth.

As might be expected, the animals that travel the farthest are those that are most mobile and travel in the largest continuous expanses of habitat: the air and oceans, as well as large lakes, long rivers, and some terrestrial regions of the continents. Fishes and birds are the best-known long-distance travelers, but there are other animals who travel great distances as well, including ourselves and a few other mammals. This chapter considers long-distance travel by animals and how it might be accomplished. It is a subject that has many unknowns and several ongoing controversies.

TERMINOLOGY

Orientation

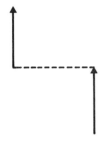

Navigation

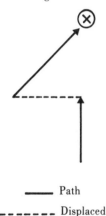

——— Path

------ Displaced

Figure 8-1   Difference between directional orientation and navigation, as indicated by experimentally displacing *(dashed line)* the animal from its path *(solid line)*. An animal that is merely orienting in a particular direction continues in the same direction but not toward a specific goal or target. An animal that is navigating, however, compensates for the displacement and travel toward the specific end point.
Modified from Baker, R.R. 1978. The evolutionary ecology of animal migration. Holmes & Meier Publishers, New York.

Before proceeding into this chapter, we need to clarify several terms and concepts. This chapter is primarily concerned with an animal's ability to move long distances to specific locations that cannot be sensed directly. The ability to move to only a particular kind of place, without emphasis on the specific location, is better viewed as habitat selection, as discussed in the previous chapter. Movement away from a place without a return or directed move to another specific location is simply dispersal or emigration.

Migration is considered in this book to involve a specific, directed long-distance move from one location to another with a subsequent return to the first. It involves travel, usually seasonal, back and forth between two distant points. A given individual, assuming that it otherwise survives, may make only one round trip in its lifetime (such as to return to a breeding location), or it may make several round trips or only part of a trip as one link in a sequence of different individuals—like a relay race and as occurs with monarch butterflies (Chapter 7). Migration as defined here is typified by some species of temperate or high-latitude birds that regularly travel long distances during the spring and fall.

Two other terms that sometimes are used interchangeably and that may be confused are *orientation* and *navigation*. These terms are inherently ambiguous because of their histories of popular usage. Furthermore, meanings have changed in some cases even during their uses by some persons (e.g., Baker 1978 and 1982 versus 1984).

In this book we use the generally accepted (dictionary) uses of the terms as follows: *orientation* refers to the general, generic meaning indicating relationships among things being discussed; *directional orientation* refers to travel or pointing in a particular compass direction; and *navigation* refers to travel or pointing toward a specific location, endpoint, or goal.

Directional orientation can be distinguished from navigation by interrupting and displacing an animal's movement (Figure 8-1). If the animal is merely orienting in a particular compass direction, it continues on in the same direction but not toward a specific goal or target. An animal that is navigating, however, compensates for the displacement, changes direction, and heads toward the specific endpoint. Navigation usually implies that the endpoint is not directly sensed itself but some other means is required for determining its location.

To maintain or use directional orientation, animals must have a sense (perception or awareness) of direction. The sense of direction has also been referred to as a compass sense (e.g., Waterman 1989). There are some animals at the extreme that basically just keep on moving forward through life. They stop only briefly at particular locations and never return to the same places. Individuals often maintain a consistent direction for long periods of time or even through their entire life. Different individuals in a population may travel in the same or different directions, but each individual seems to have an awareness of its own direction.

A good example of species that keep moving in a particular direction is the small white butterfly *(Pieris rapae)*, studied in England by Baker (e.g., 1978). (The species is also a common imported pest in the United States, where it is called the *cabbage butterfly*.) These butterflies stop in suitable habitat long enough to feed, mate, lay eggs, or spend the night, but then they move on. They are basically transients without a home and keep moving essentially in the same direction (Figure 8-2).

Figure 8-2   Track of the
small white butterfly, *Pieris
rapae*, an example of a
linear path and a sense of
direction. Visual tracking
was conducted starting at
Bristol University, England.
Wind direction and speed
are shown by arrows.
From Baker, R.R. 1978. The
evolutionary ecology of animal
migration. Holmes & Meier
Publishers, New York.

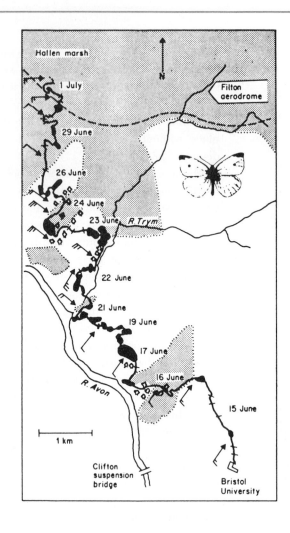

Figure 8-3   Transition
from summer to autumn
directions of movement by
the small white butterfly.
Length of arrows indicates
number of butterflies moving
in the principal compass
directions, with large arrows
indicating the mean
direction.
From Baker, R.R. 1978. The
evolutionary ecology of animal
migration. Holmes & Meier
Publishers, New York.

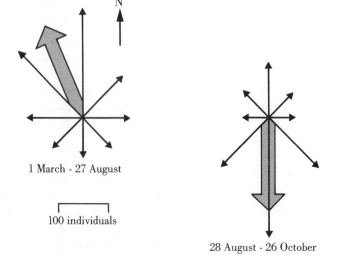

At different seasons, however, the directions may reverse (Figure 8-3) so that individuals or their offspring may end up fairly near where they started.

Navigation involves an awareness of location or place or, as it is sometimes referred to, a map sense (Waterman 1989). Maps may be one of four types: (1) linear, involving simply a path or gradient, (2) polar, involving radiating relationships to and from a central point, (3) grid, involving two-dimensional coordinates such as the system of latitude and longitude that humans use (and that becomes polar for longitudes at the north and south poles), and (4) mosaic, which involves a patchwork of irregular, known areas with landmarks. Location can be determined from each of these map types respectively by one's position along the path or gradient, distance from the central point, reference to the coordinates, or in relation to the landmarks.

In addition to an awareness of location, and depending on the type of map in use, navigation also generally requires information or an awareness of distance, direction, and time (usually in conjunction with measuring distance by length of time traveled, monitoring fuel consumption, or in tracking direction by time compensation for moving celestial objects). The need for information on time by humans navigating the open seas, incidentally, was primarily responsible for the origin and early development of chronometers (clocks) in Western society.

The familiar, common forms of human navigation are piloting, dead reckoning, celestial navigation, and geophysical navigation (Waterman 1989). Piloting involves following landmarks (including artificial markers such as buoys and electronic signal devices) with remembered or external mosaic maps. Dead reckoning uses vectoring distances and directions between points of interest. Celestial navigation uses reference to the direction of objects such as the sun and other stars. Geophysical navigation uses information from cues such as the Earth's electromagnetic field. During any particular travel, humans may use combinations of these techniques.

ULTIMATE FACTORS IN ANIMAL MIGRATION

Before discussing the possible proximate mechanisms of orientation and navigation, we should pause to contemplate the possible ultimate factors in long-distance migration. For example, why do animals migrate at all, or in certain ways, at certain times, and to particular places? At present we can only speculate on these questions. The different locations at the two ends of a migratory trip apparently are better suited to survival or reproduction at different times. The differential advantages, furthermore, must be sufficient to outweigh the high risks of getting lost, running out of fuel, or getting caught by predators or inclement weather in unfamiliar territory during the long journey.

How long-distance migration originates is not clear. It seems to be derived, as in other evolved characteristics, from the natural variation in the movements of the animals. Those individuals that move from areas that become inhospitable would, according to standard Darwinian reasoning, have better chances of survival than those that remain behind. Of those that do move away, some would then have to return later and have greater survival and reproduction than those that do not return.

Explanations for the origins of long-distance migration frequently invoke the roles of past glaciation and perhaps even continental drift, in addition to the pressures of present seasonal differences at different locations. Competition from other species already present in various locations may also be important. Otherwise it does not seem that there is any selective advantage, and in fact it appears disadvantageous, for example, for arctic terns to go all the way to antarctic regions or for other birds to go so far into the southern hemisphere just to escape the northern winters.

For many species of birds with a broad latitudinal range, those living at lower latitudes may show less tendency to migrate than others of the same species at higher latitudes; they even may be permanent residents. Birds from farther north may fly on beyond the middle latitudes, in somewhat of a leap-frog fashion, and winter further south than the nonmigrants. In many species of birds there is a northward movement in the late summer and early fall, before the major migration to the south. In some species, such as bald eagles, birds that nest far to the south, such as in Florida, may briefly leave and travel north after the nesting season.

Speculating on the origins of the monarch butterfly migration, Urquhart (1976) notes that most of the hundred species of milkweed in North America are native to Mexico. He considers that perhaps the monarch originated in Mexico:

> Now, in returning there each winter, the butterfly is "going home," after straying, perhaps over eras of a warming trend, farther and ever farther north. Anyway, I'm convinced that the monarch's selection of the Sierra Madre for overwintering is no random choice.

Waterman (1989) also discusses other global factors, including, for example, the possible role of drifting in the Atlantic Ocean on the movements of green turtles between Brazil and Ascension Island (see Chapter 7).

THE TRAVELERS AND NAVIGATORS

There are numerous animals capable of accurate, directed long-distance travel, including not only the familiar birds and others previously mentioned but also some crustaceans, many fish, various mammals, some amphibians, and perhaps some of the dinosaurs and other extinct reptiles. More complete phylogenetic treatments are given by Dingle (1980) and Schmidt-Koenig (1975). Here we concentrate primarily on fish and birds, with additional mention given to some mammals—including humans.

Fish

Movements and migrations of fish have attracted almost as much interest as those of birds, partially because of the economic importance of fish. However, migration research is often more difficult with fish than with birds. There are fewer recovery encounters between fish and humans, even with commercial and sport fishing, than between birds and humans, which has resulted in a smaller return of fish data. Nonetheless, enough is known to demonstrate a considerable diversity among fish homing abilities and mechanisms.

A few species of fish probably do not home but simply drift, which is also the case in many species of insects. A large number of species probably return via the basic familiar-area methods. Some, however, show remarkable migrational and navigational abilities. For further reading, see Harden Jones 1986, McKeown 1984, and Smith 1985. Details for two cases, salmon and eels, are discussed here.

Salmon are *anadromous* fish that live in the ocean as adults and move up into fresh water to spawn. They spend most of their lives, often several years, in the ocean or other large bodies of water; then they return to rivers and smaller streams to breed. Their migrational and homing abilities are legendary. They may travel 2000 to 4000 km out to sea; then when they return, they go back to the exact stream from which they hatched. Large numbers of tagging returns, for example, are found at the home stream and not at other nearby streams.

Understanding this homing behavior is important for economic and scientific reasons. Salmon provide a large commercial and sport fishing resource. At the same time, numerous dams have been and continue to be constructed across streams, inadvertently blocking the salmon's return. Also, water pollution in rivers and streams may make the home streams unsuitable for the fish or otherwise prevent their continued use. Fish ladders to aid the movement of returning fish around dams have been partially successful, but large numbers of fish continue to be lost.

A long series of studies, associated primarily with A.D. Hasler and a number of his students (e.g., Hasler et al 1978; also see comments in Hasler 1985, 1986), have established convincingly that the salmon home on the basis of stream odor. Different streams are believed to possess unique odors from the soil and vegetative characteristics of the drainage basin or from the fish themselves. Young salmon are believed to imprint to (Chapter 19) and memorize the odors of the streams where they hatch, then seek out those same odors at maturity.

These conclusions are based on a diversity of experiments that initially involved such manipulations as conditioning (Chapter 20) fish of various species to different odors using food rewards and shock punishment. Other work included cutting nerves from the nose to prevent any use of smell. Early objections were raised that interference with smell, for example, was interfering with other abilities, such as the ability to feed, and that one could not conclude that smell was responsible for homing (Peters 1971).

Results of several different approaches to the subject, however, leave little room for further doubt. In an important laboratory experiment, fish were tested in a four-armed maze with water cascading down a series of steps into the center, from which it drained. Different odors, with which fish had or had not been raised, were introduced to different arms. The fish accepted or rejected odor-treated water and went up the appropriate arms as predicted.

In field tests, researchers captured returning salmon and plugged the noses of half of them. Fish with plugged noses went to different streams at random, whereas fish with unplugged noses returned to their proper streams. In something of a hybrid laboratory-field experiment, juvenile salmon were exposed to artificial odors (morpholine or phenylethyl alcohol) or plain water (the control group) and then released into Lake Michigan. About 18 months later when returning fish would be expected to move into streams along the lake, 19 streams were monitored. Two of these were artificially scented with either morpholine or phenylethyl alcohol. Of the treated fish that were recovered, 94% to 98% of the morpholine-treated fish and 91% to 93% of the alcohol-treated fish were recovered in the respectively scented streams. Controls were recovered in a number of different streams. To further study the response of these fish, the researchers captured some fish and fitted them with telemetry transmitters. Fish homed to scented streams as would be expected if they were using the artificial odors. All of these results indicate clearly that the fish were homing to odors to which they initially had been imprinted.

The next case of fish migration involves eels of the genus *Anguilla*. They are *catadromous*, which means they live most of their lives in rivers and breed in the ocean, the opposite of salmon. The global patterns of anadromous versus catadromous fish migrations, incidentally, suggest that the direction depends on productivity and whether fresh waters or the ocean are more productive in the particular region (Gross et al 1988). Eel movements are only partially understood after nearly a century of investigation and debate. There are thought to be 14 to 18 species of eels worldwide, mostly in the western Pacific. The one or two species in the North Atlantic, however, have been the subject of a continuing controversy.

Atlantic eels are found along the coastal rivers of both eastern North America and western Europe. Early studies established that eels in the two areas have different growth rates and different numbers of vertebrae; therefore it was thought that they represented two different species. Initially it was not known where they spawned. Interest in the eels and their movements arose both from curiosity and from practical, economic concerns. Eels provide a considerable fishing resource. The catch along coasts of European nations in 1965 alone, for example, amounted to 17,000 metric tons.

From what is known of the species, adult eels migrate to deep (400 to 700 m), warm ocean waters, which overlie much deeper cold water. The breeding area for the Atlantic is thought to be the Sargasso Sea region (Figure 8-4), in or near the so-called Bermuda Triangle. The eggs are shed into the open, deep waters where they drift with the currents and hatch into flat, leaflike larvae called leptocephali (small-headed). They drift for 1 to 3 years or more, growing slowly, until they reach coastal regions, at which time they metamorphose into small eels called elvers or glass eels. They swim up into rivers where they spend several years growing and

Figure 8-4   Eel spawning area in the Sargasso Sea region of the Atlantic and known or hypothesized movements to and from the area.

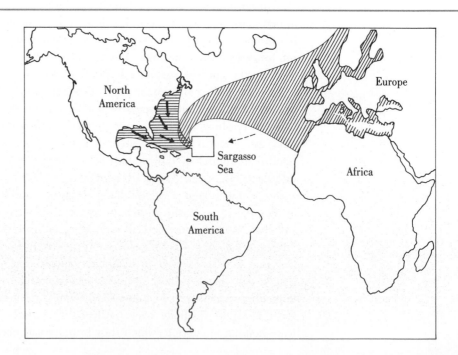

maturing and are called *yellow eels*, because of their color. They then change to a silver color and acquire the characteristics of deep-sea fish, including enlarged eyes and different retinal pigment molecules. They move back out to the ocean and are believed to return to their deep-water spawning areas.

A Danish biologist, Johannes Schmidt, began studying eels around the turn of the century, first with a tiny research vessel named Thor, then by hitching rides on commercial fishing ships. After being recognized as an outstanding authority on eels, he eventually obtained ships outfitted for systematic research. Part of the motivation for the work was concern that the species might be overharvested. Schmidt's research spanned more than 25 years. By following catches of progressively smaller and younger leptocephali, he tracked them backward to the Sargasso Sea. He also discovered that eels from both America and Europe were coming from the Sargasso Sea; apparently both species were spawning at the same time.

No adult eels were ever caught between the continental shelves and the Sargasso Sea, but based on the fairly complete picture from the leptocephali, Schmidt hypothesized that there were two species, one North American and one European, that returned to the Sargasso to spawn. Schmidt published at least 21 papers on his research (cited in Harden Jones 1968).

Research on development of fish of other species, including trout and a blenny, however, has shown that growth rates and numbers of vertebrae may depend on water temperature at a critical period of embryonic development; that is, differences can be environmental rather than genetic. This led Denys Tucker (1959) to propose that there was only one species of Atlantic eel, with spawning stock coming from North America. He proposed that the eels found in Europe were simply from leptocephali that drifted there, taking longer and growing more slowly and with more vertebrae, but originating from American eels. Under this hypothesis the European adults did not return to the Sargasso Sea to breed.

A third possibility is that the two forms of Atlantic eels do represent a single species, as Tucker proposed, but adults from both sides of the Atlantic return to the Sargasso. European adults could migrate directly across the Atlantic at great depths, backtrack over the route they had drifted earlier, or perhaps take some other indirect route. Eels are long-lived, and many routes seem possible.

Adult eels and eggs, however, have never been taken in the open ocean, so the issue remains unsettled. A few other anomalies in the details of the story, reviewed in Harden Jones (1968), suggest that perhaps the Sargasso Sea is not the only location where Atlantic eels spawn. From the comfort of one's chair, in restricted surroundings, and faced with impressive amounts of material published by other people, it is easy to lose (or not obtain in the first place) a proper perspective on how large the world is. It can be very difficult to find things out there.

Birds

General background

We now come to the most famous group of animal homers and navigators: birds. Birds of many species migrate between continents and return, often to the exact same nest site. The pied flycatcher, for example, breeds in northern Europe and winters in tropical Africa. Of 829 banding recoveries for that species during the breeding season, over half have been within 1 km of the nest where the birds were hatched (Berndt and Sternberg 1969). The tendency to return to the same area, faithfulness to a place, frequently is referred to by the German term *Ortstreue* or a related term, *philopatry* (German, love of fatherland). It should be noted that philopatry is not universal in birds; many species shift breeding or wintering locations, often depending on shifting food or habitat resources (such as water for ducks and food resources for some arctic species).

The best distance records are for arctic terns, which nest in the arctic and winter in the antarctic. An arctic tern banded in Russia was recovered 14,400 km (9000 miles) away. These birds are believed to make an annual round trip of over 35,000 km (22,000 miles). Not only the total length of trips but even the length of single nonstop flights by birds is impressive. A radio-equipped thrush was tracked 560 km (350 miles) in one night (Cochran et al 1967). A thrush tracked more recently by Cochran (1987) during a 7-day period averaged over 250 km per night during the 6 nights that it flew (it stopped during the day and one of the 7 nights). Golden plovers may leave Labrador, Canada, and fly 3800 km (2400 miles) nonstop to South America—on 2 ounces of fuel (fat). The path of a flock of blue geese was observed and recorded by various commercial aircraft in 1952; the geese made a nonstop trip from James Bay, Canada, to Louisiana. They traveled 2700 km (1700 miles) in 60 hours (average 30 MPH) at an altitude of 900 to 2400 m (3000 to 8000 feet).

The altitude at which migrants fly is generally between 300 and 900 m (1000 and 3000 feet) but commonly goes up to 3300 m (11,000 feet), and large birds have been recorded up to 6100 m (20,000 feet) on overseas flights or over mountain passes. A human would have difficulty with the low pressure and oxygen conditions at those heights even under resting conditions, let alone with heavy muscular exertion.

The proximate weather conditions under which birds initiate or continue migration probably have been investigated more thoroughly than in any other group of or-

Figure 8-5 A few of the saw-whet owls netted during a migration study to determine proximate weather factors. These birds were being retained for banding, after which they were released back to the wild. Courtesy Lance Parthe and David Evans.

ganisms. Because of the numerous factors involved in weather, multivariate analyses are preferred (Able 1974). In one such analysis of 1401 saw-whet owls netted in a banding operation (Figure 8-5) at Duluth, Minnesota, Evans (1980) showed that changes in migration volume correlated with wind speed, barometric pressure, and temperature. In an extensive review of the literature, Richardson (1978) concluded: "Causative and coincidental relationships remain difficult to separate and at least a few birds migrate in almost any weather conditions. However, maximum numbers migrate with fair weather, with tailwinds and with temperature, pressure and humidity conditions that accompany tailwinds." Beyond this, specific conditions vary among different groups of birds such as waterfowl, shorebirds, and hawks.

Berthold (1986) conducted a comparative review of several long-term studies of proximate factors controlling migration in European warblers of the genus *Sylvia*. The studies involved over 20,000 birds of 15 to 19 species from a wide geographic region, with populations that included varying proportions of migratory versus non-migratory, year-round residents. The birds studied included 1300 hand-raised individuals and selective breeding of some of the birds held in captivity. Factors investigated included: preparation for, onset, and termination of migration; migratory directions and endpoints; resting behavior en route; and the proportions of populations that migrated. As might be expected, there was considerable variation among species and among populations of any given species from different geographical areas. The variation in the proportions of birds that were migratory or not was confirmed through selective breeding experiments to have a large genetic component. Variation in several of the other factors also was thought to be largely genetic.

In addition to showing natural return behavior during migration, many birds can be artificially displaced, often by several hundred kilometers, released, and they successfully return. Homing pigeons are well-known for their abilities to return, but similar displacements have been conducted with many other species as well. One of the most famous was a Manx shearwater *(Puffinus puffinus)* flown by jet across the Atlantic from England to Boston. It appeared back at its nest 12 days later (Matthews 1953).

Present information on bird migration consists of a varied assortment of facts and hypotheses. Before we plunge into the variety of information, however, some of the techniques used to study bird migration are mentioned briefly.

Techniques of studying bird migration and homing

Because of their sizes, numbers, and differences in sensory and motor abilities relative to those of humans, birds are not particularly easy to follow. Early indications of their travels came from the familiar observations of the seasonal disappearances and returns of various species over many parts of the Earth. Long-distance human travelers noted that certain species were seen at some times only in one region and at other times only in other, often far removed, places. Banding (called *ringing* in England and Europe), color marking, and other means of individually identifying birds have produced a wealth of information concerning endpoints and occasionally intermediate points of journeys but not much information on actual paths, times of travel, and possible mechanisms. Solutions to some of these more difficult facets of the problem have been ingenious.

1. The direction and a few other aspects of migration have been studied with birds of some species held in circular cages. During the times when these birds would normally be flying, they show heightened activity or migratory

restlessness (German, Zugunruhe) in the cages. Interestingly, the direction in which they jump and attempt to escape from the cage often correlates with the direction in which they would otherwise be migrating. This behavior has made it possible to experiment with several aspects of migration, including external sensory cues and the importance of the birds' internal rhythms. The directions in which the birds jump have been recorded, for later statistical analysis, visually, by treadles and electronic counters, and by the ink-pad technique (Chapter 3).

2. Many facets of navigation have been investigated by transporting or displacing birds and then releasing them and determining the direction toward which they depart. Departure points have been plotted visually from the ground, with the aid of radio telemetry, which permits the birds to be followed beyond the horizon or nearby obstacles, or by tracking them via fixed-wing airplanes or helicopters. The data have generally been plotted and analyzed on circular graphs.

3. Night migrants have been tracked with the aid of lights directed at the sky (ceilometers) and surveillance and tracking radar. Demong and Emlen (1978) prompted a proportion of experimental birds to initiate migratory flights by lifting them to the desired altitude by balloon; then they followed them with the tracking radar at Wallops Island (Chapter 3). Birds have also been followed at night with radio telemetry and tracking from ground vehicles or aircraft.

4. The birds' senses have been impaired experimentally by cutting the appropriate sensory nerves, by placing the birds in artificial or altered sensory environments, and by physically blocking the sensory organs at or near the surface. Nostrils and ears have been blocked, and the eyes have been covered with contact lenses in various experiments. Homing pigeons with frosted contact eye lenses, for example, have been able to return remarkably close to their lofts. Attempts to impair potential magnetic senses have included fitting the birds with magnetic bars and electromagnetic Helmholtz coils.

5. One of the major techniques used in conjunction with the various methods listed here has been clock shifting, or altering an animal's internal diurnal rhythms relative to the external real time. Clock shifting is accomplished by artificially changing the photoperiod, the Zeitgeber (Chapter 6). If the natural photoperiod is 12 hours of light from 6:00 AM to 6:00 PM, for example, it can be moved forward 3 hours by turning lights on at 3:00 AM and off at 3:00 PM and by preventing the animal from seeing any light after 3:00 PM (Figure 8-6). After a period of time the animal's internal rhythm changes to the altered time period. Then if exposed to the external real photoperiod the animal in essence, subjectively perceives the real time now as being 3 hours later, as if the animal had been moved three time zones to the west. Shifts can be of any magnitude, either forward or backward. To understand this, imagine being placed under these conditions yourself and consider when and where the sun would appear at particular times.

These and other techniques have been conducted under a variety of environmental conditions (such as night versus day, clear versus overcast skies, different seasons), with a wide variety of species, ages, and experiences of individuals, and with variable experimental designs and sample sizes. The variety of conditions, tech-

Figure 8-6 Clock shifting. The animal's subjective impression of time of day can be altered by modifying the times that the surroundings become light and dark. Technically the shifted Zeitgebern are said to entrain the diurnal rhythm to new times.

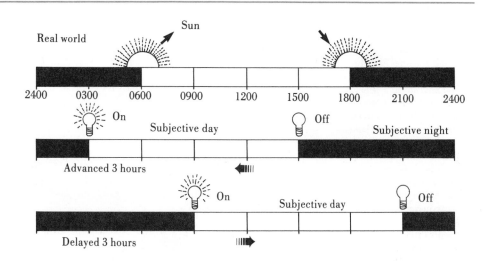

niques, and factors has produced a volume of data that, until fairly recently, seemed rather bewildering. Patience, keen insight, and new experiments, however, have clarified much of the picture.

Genetics versus experience in bird migration

The genetic contribution to variation in avian orientation and navigation has not been well quantified through analysis of variance (Chapter 4), but it is clear that there is an important inherited component in many species. In some species the adults leave before the departure of the young of the year. Yet the young travel to the same area, a trip that obviously could not have been learned individually or made by traveling with experienced individuals. In European storks the direction that the birds take to travel around the Mediterranean Sea appears to be inherited. Young storks migrate independently from their parents. Storks in western Europe travel to the southwest and cross the Mediterranean Sea at the Strait of Gibraltar, whereas those further east move toward the southeast and cross on the eastern side. Young from eastern eggs artificially relocated to nests in the west traveled in a southeasterly direction (Schuz 1971).

In many facets of avian migration and navigation there are also important roles for experience and learning. Learning commonly is involved in determining the specific locations that are occupied at different times of the year and how to navigate. In probably the majority of species the location of the initial home area is learned around the time when the young first begin flying. If they subsequently migrate or are artificially displaced, this is the area to which they later return. If young are moved before they are capable of flight, as in the translocations and reintroductions of endangered species, the birds later return to that new area.

If the birds leave their natal area naturally on migration, they may accompany the adults (e.g., cranes), move in a particular distance in a general direction (e.g., storks), or more or less simply wander (e.g., bald eagles). After finding suitable wintering locations and even some particular temporary sites en route, the birds apparently remember these places and return at those times of the year in future years, thus establishing learned locations for different seasons. It is conceivable

Figure 8-7 Interaction between innate orientation tendencies and experience as shown by displaced migrating starlings. Starlings from northern and eastern Europe migrate in a general southwesterly direction during the fall. Birds were captured in the Netherlands and transported to Switzerland, where they were released. Banding recoveries depended on whether the birds were adults or juveniles; adults traveled northwest and returned to their normal wintering areas, whereas juveniles, including those that had been released in the company of adults, continued toward the southwest, including areas outside the normal wintering range.

Based on information and figures in Perdeck, A.C. 1958. Ardea 46:1-37.

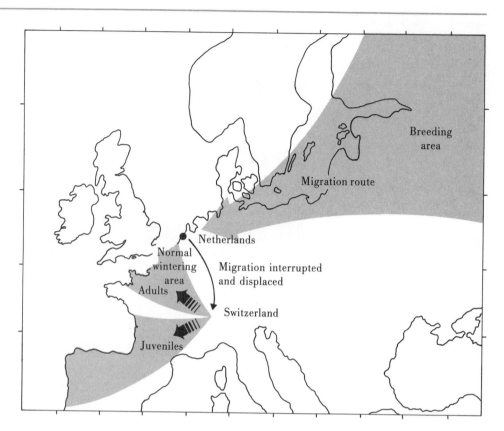

that the travels of some birds to specific locations are determined genetically but that does not seem plausible and there is no direct evidence for such.

Starlings in some parts of Europe have a tendency to migrate southwest in the fall and return to specific wintering locations. The difference between the inherited tendency to move west and the importance of learning a specific winter location was demonstrated by Perdeck (1958) by trapping and displacing birds of different ages during their fall movements. Immature birds simply continued southwestward from the new release sites, whereas adults corrected for the displacement and navigated to their normal specific wintering locations (Figure 8-7). The tendency of European starlings to travel west, incidentally, may be a partial reason for their successful introduction to the North American continent. Several small-scale introductions of starlings into North America were not successful. There were two known large-scale introductions, one in Portland, Oregon, in 1889, which was not successful, and another in New York City in 1890, which eventually led to the permanent establishment of the species in North America (Kessel 1953).

The mechanics of avian orientation and navigation

We now have arrived at the most intriguing part of the whole problem of avian migration and navigation: regardless of the extent of learning involved, how do birds manage to orient themselves and navigate so accurately? We must skim over many years' and dollars' worth of painstaking research for a few highlights and current interpretations. Several excellent general reviews of the subject are available

Figure 8-8 Wandering courses over increasingly larger area until familiar areas are encountered. This interpretation was suggested by following displaced gannets *(Morus bassanus)* with the aid of light aircraft at a distance. Based on information and figures from Griffin, D.R., and R.J. Hock. 1949. Ecology 30:176-198, and Griffin, D.R. 1955. Bird navigation. In: A. Wolfson, editor. Recent studies in avian biology. University of Illinois Press, Urbana.

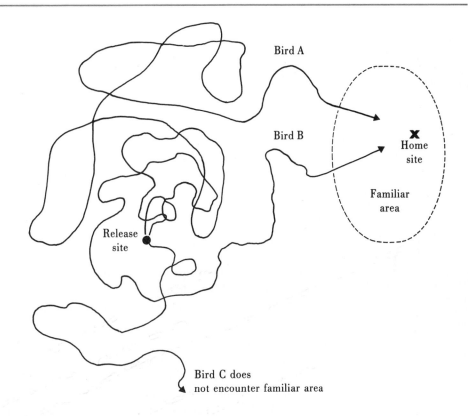

(e.g., Keeton 1974, Able and Bingman 1987, and references cited therein).

Where possible, some species, such as gannets, appear to use landmarks or some aspect of the familiar area (Figure 8-8). However, it is not clear to what extent birds use visual landmarks except at the end for landing. Many species head straight home, often without visual landmarks; birds often home accurately when vision is obscured by darkness, fog, or artificial impairment such as with frosted contact lenses. In many cases, particularly after experimental manipulation, birds with normal vision seem to ignore visual landmarks and bypass or become lost in what should be familiar areas. Pigeons that have been clock shifted, for example, may be disoriented and fly away from the home direction even when their loft is in plain sight (Keeton 1974).

Similarly (and as also reviewed by Keeton 1974), even major topographic features such as mountains and shorelines seem to have little influence on pigeons once they have established a course of flight, except where high mountain peaks form substantial obstacles. Local topography appears to affect only the initial departure directions of released pigeons, which then correct for these release site effects and turn toward their true homeward course.

Much of the story for many species involves one or more of various compasses. Diurnal migrants and homing species (e.g., starlings, mallards, pigeons) can use the sun as a compass. That they use the sun at least partially has been demonstrated (1) by clock shifting, in which case the orientation is altered by the predicted angle, (2) by altering the apparent position of the sun with mirrors, with the birds orienting to the new mirrored position as would be predicted if they were using the sun, and

Figure 8-9 Paths of mallards released at night. Birds were tracked visually with the aid of small flashlights attached to their feet. Release point is at the center of the circle. Dots at the edge of the graph indicate the direction from the release point of each departing duck. Under overcast sky the ducks wandered in several directions and appeared disoriented. Under clear sky, however, all individuals departed promptly in a consistent direction. These results suggest that the birds needed and were using celestial cues. Some other species of birds are not disoriented under overcast sky.
Modified from Bellrose, F.C. 1958. Bird Banding 29:75-90.

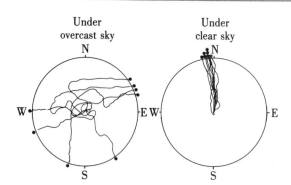

(3) by keeping the birds under an artificial light source in a constant position, in which case the birds change their orientation by an average of 15 degrees per hour, as if they were correcting for the normal movement of the sun. For many nocturnal migrants, the position of the sun when it sets is important in orienting the night's flight.

Some species of nocturnal migrants have been shown to use stars in a compass manner. These species include European warblers, indigo buntings, and mallards. Evidence for mallards includes observations that birds released at night under a clear sky depart in a consistent, predicted direction, whereas those that are released under overcast skies depart more randomly (Figure 8-9, Bellrose 1958).

Emlen (1975) conducted perhaps the most thorough analysis of star-compass navigation using indigo buntings in conjunction with the Zugunruhe and stamp-pad technique discussed previously (see Figure 3-13). By using naive young birds and artificial skies in a planetarium, Emlen showed that the star patterns (constellations) are important and that the young birds learn them at night as nestlings. The moon may cause disorientation, but the orientation to the stars is not affected by clock shifting. The orientation is based on learning the position of the pole star among the other stars by the extent of the Earth's rotation during the night. (Polaris is the current pole star in the northern hemisphere. The position of the pole star gradually changes over a 26,000-year period with the precession of the earth's rotation.) The general orientation of the birds, which depends on their circannual rhythm and hormonal conditions, is away from the pole star, that is, south during the fall and toward it or north during the spring movements.

Without being able to see the stars, mallards and indigo buntings are confused, but European robins, swans, and many others are not. Also, pigeons trained to fly under adverse conditions may home correctly under heavy overcast, with frosted contact lenses, and at night. Some birds, at least under some combinations of conditions, are clearly using something more for navigation. The Earth's magnetic field has long been suspected, but the topic has been beset with controversy, reminiscent of the Spallanzani bat-sonar story described in Chapter 1.

Evidence for the use of magnetism by birds for navigation includes the following:

1. Alternatives have been eliminated. Many birds orient correctly after all other apparent cues have been exhausted.

2. Caged birds of some species correctly orient without any visual cues.

3. Caged birds shifted their orientation in certain imposed magnetic fields (Figure 8-10).
4. Experimental free-flying homing pigeons (without other cues such as the sun or landmarks) are disoriented with magnetic bars attached to them, whereas control birds under similar conditions but with brass bars attached are not disoriented (Figure 8-11).
5. Released pigeons that depart correctly with Helmholtz coil caps and a given battery polarity depart in the opposite direction if the polarity of the battery connections is switched (Figure 8-12).

Figure 8-10   Example of induced, altered magnetic field effects on direction of migratory restlessness in caged birds. In this case European robins *(Erithacus rubecula)* were subjected to experimentally modified magnetic fields as indicated. Dashed lines indicate direction of normal migratory movements; solid arrows are mean direction of activity in altered fields. The birds' directions changed in response to the altered fields appropriately to the season (toward magnetic north in spring, away from magnetic north in autumn).
Modified from Wiltschko, W. 1972. NASA Special Publication 262:569-578.

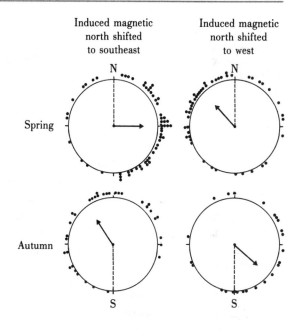

Figure 8-11   Effects of magnetic bars on homing pigeons. Arrows indicate mean direction of departure of pigeons from an unfamiliar release site. The length of the arrow indicates the consistency of departure direction among different birds. The dashed line represents the direction of the birds' loft from the release site. (Birds were released in several different locations; thus direction toward home loft is not any single compass heading.) Pigeons were disoriented and vanished in several directions if they wore a magnetic bar and were released under an overcast sky. Control birds with nonmagnetic brass bars or those released under a sunny sky (in which case the sun could be used for orientation) all departed in a homeward directed.
Modified from Keeton, W.T. 1974. Scientific American 231(6):96-107.

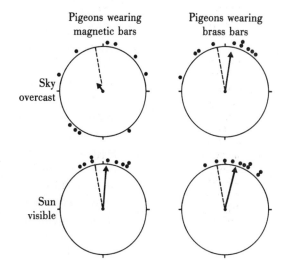

Figure 8-12 Effects of electromagnetic coil (Helmholtz coil) caps worn on the heads of homing pigeons. Magnetic fields were induced by connecting the coils by wires to small mercury batteries worn on the pigeons' backs. The effect depended not only on whether or not the birds could see the sun (see text and Figure 8-14) but also on the direction of current flow and magnetic field in the coils. Direction was reversed simply by reversing the connections to the battery. Details of the figure, such as arrow length, are as in previous figures.
Modified from Keeton, W.T. 1974. Scientific American 231(6):96-107. Based on material originally in Walcott, C., and R.P. Green. 1974. Science 184:180-182.

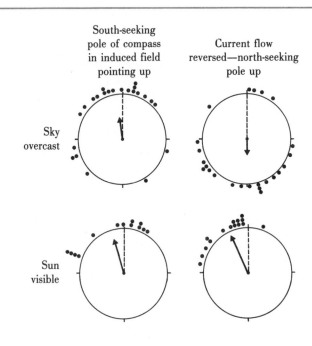

The intriguing questions at this point are: where are the sensory detectors in the birds' bodies and what are they? Such organs were never previously identified or recognized, in spite of years of anatomical study of birds. This seems comparable to the situation for bats before Griffin's discovery of bat sonar.

Walcott et al (1979) reported microscopic spicules of magnetite in a small region between the brain and skull of pigeons. Magnetite has been found also in numerous other species. Recent information (reviewed by Wiltschko and Wiltschko 1988), however, suggests that the Earth's magnetic forces are not sensed with the magnetite but rather within the visual system in conjunction with light detection, perhaps even using the same molecules (rhodopsin). The magnetic sense is still not well understood either in terms of where it is located or how it works. All that is clear is that it exists and it influences the orientation of many species of birds (and other organisms; see discussion and references in Wiltschko and Wiltschko 1988).

Redundancy and combinations of navigational input, including compasses and maps

Some of the previous confusion in understanding bird navigation appears to have been caused not only by the variability already discussed (among species, techniques, studies, etc.) but also by the fact that many species, particularly those with the most sophisticated abilities, possess much redundancy in their navigational systems. That is, many birds have more than one system, with the various systems serving as backups in case the others cannot be used. Also among different birds, even within the same species, there is variation depending on the individual's developmental (ontogenetic) experiences and specific environment (Walcott 1986, Able and Bingman 1987).

The presence of combinations and hierarchies of navigational cues has been demonstrated best in homing pigeons (Keeton 1974 a,b). Experienced pigeons use the sun first if it is visible; clock-shifted birds will be off in their orientation under clear daytime skies. If the sun is not visible, however, then the experienced pigeons use the magnetic field, and clock shifting will not alter the direction of their orientation under such conditions. Inexperienced homing pigeons need both the sun and magnetic fields for proper orientation. Experience permits the birds to use different systems alone or, in a slightly different interpretation, to get by with less information.

Wiltschko and Wiltschko (1985) have shown that young inexperienced pigeons require cues obtained during their outward journeys from home (perhaps by permitting a return by dead reckoning or through the formation of a kind of linear or mosaic map). With increased experience, they apparently switch to relying more on local map information obtained at the release site.

The information needed for navigation to specific locations requires one or more of the kinds of maps (page 215); a compass alone will not get one home. One must first know at least the direction and preferably also the distance to home; then the compass can be used to orient the animal in the correct direction.

Figure 8-13 Sun-arc hypothesis proposed by Matthews to explain bird homing. The bird is hypothesized to learn the path of the sun across the sky; then, if displaced as in homing experiments, it could compare the predicted path with an observed path and travel back home to correct the disparity. The concept is shown from two different angles to aid understanding. **A,** The view to an external observer, as originally diagrammed by Matthews. **B,** View as might be seen by the bird. The circle is the observed sun, and the solid circle is the predicted position. The observed sun, for the path it is taking, is too low and too far along in its arc. Therefore the bird must be north (for the sun to be lower) and east (for the sun to be ahead in its position) of home. To get home it should travel south and west.
A from Matthews, G.V.T. 1955. Bird navigation. Cambridge University Press, Cambridge. **B** modified from Keeton, W.T. 1974. Scientific American 231(6):96-107.

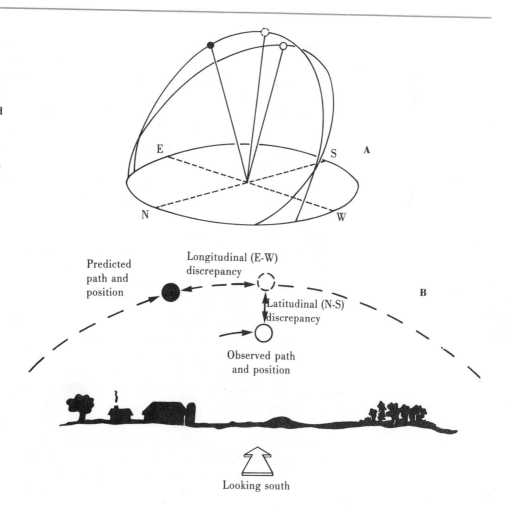

G.T.V. Matthews (1951 a,b, 1955) proposed a sun-arc hypothesis for pigeon homing, which states that coordinate information, as with a grid map, could be obtained from the path of the sun across the sky, resulting from the earth's rotation and inclination. This hypothesis has been discounted and is not worth learning just for itself. However, science, education, and the process of developing understanding—as stressed throughout this book—are more than just acquiring and memorizing the most recent views and information. They also involve an ongoing process of wrestling with the unknown and growing in one's appreciation of the historical and methodological context. The rejection of the sun-arc hypothesis illustrates those points, helps explain by contrast the most recent, alternate view, and additionally serves as a reminder that we must always remain open to new information and hypotheses. Even the latest view might have to be replaced as new information and thinking arise. Therefore the details of the sun-arc hypothesis are considered briefly.

According to the sun-arc hypothesis (Figure 8-13), if one knew the time of day and the path of the sun at home or the location where one was headed, then one could figure where one was from the differences in the sun's path. For example, imagine that one is in the northern hemisphere and located southwest of where one wanted to be. The sun would appear too high in altitude, the angular height above the horizon, because of one's being too far south. The sun also would not be far enough along its path (the azimuth, the position of a vertical line drawn from the sun to the horizon). That is, the sun would be too far east for that time because one was west of the described position. To correct the situation and position oneself so the sun appeared in the right place, one would have to travel north and east the appropriate distances, hence moving to home or the desired location. This hypothesis suffered from a serious problem, however. The apparent path of the sun changes only slowly and would have to be traced for a period of time to determine the relationships of its azimuth and altitude to the Earth's compass directions. Released birds, however, do not wait long enough to trace the path of the sun and obtain the necessary information before setting off in the proper direction.

Gustav Kramer (1953) proposed an alternate hypothesis, the map-compass hypothesis. According to this, the sun is used only for compass information and something else is serving as the map.

To distinguish between the two hypotheses, consider an experiment in which a bird is delayed (i.e., clock-shifted back) by 6 hours and then moved south 100 km into an unfamiliar location and released at noon. Because its internal clock has been delayed 6 hours, the bird perceives its real home time as being 6:00 AM rather than noon. If the sun-arc hypothesis were correct, the bird would recognize that the sun is at the height of its path, a condition that would occur only much farther east. To get home it must travel six time zones to the west, a fourth of the distance around the earth. It also would notice that the sun is slightly too high because of the southward displacement, and it should go 100 km toward the north. Thus the pigeon would depart in a primarily western orientation. If, on the other hand, Kramer's map-compass hypothesis were correct, then the unknown map or coordinate system would somehow tell the pigeon that it was 100 km south of home and that it needs to travel north. The clock-shifted pigeon, on the altered basis that it is 6:00 AM, would determine north to be 90 degrees to the left of the sun. (The sun at "6:00 AM" should be in the east, and north would be at a 90-degree angle to the left.) Thus the bird should depart at a right angle to the left of the sun. But the sun, rather than being in the east at 6:00 AM, is in the south because

Figure 8-14   The critical test of the sun-arc versus the map-compass hypothesis. The birds were clock shifted back 6 hours, as if they were living six time zones west, then released to the south of their loft at noon. Predicted paths under the two hypotheses are indicated. The clock-shifted birds actually traveled east. Modified from Keeton, W.T. 1974. Advances in the Study of Behavior 5:47-132.

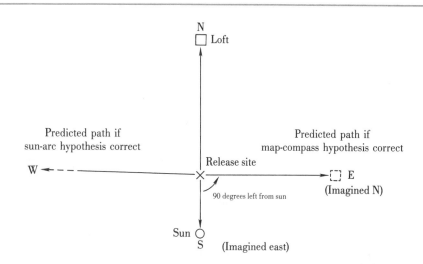

it is really noon. The bird, in traveling at a right angle to the left from the sun, would head east.

When the experiments were conducted, the pigeons headed east, as predicted by the map-compass hypothesis instead of west, as was expected from the sun-arc hypothesis (Figure 8-14, Schmidt-Koenig, Graue, and Keeton, reviewed in Keeton 1974). For the moment, until better data or interpretations are obtained, the map-compass theory appears to be the best interpretation. However, the map-compass hypothesis has not been explained fully, and the physical basis of the map is not yet known.

One possibility is that some animals have an innate grid map that is unknown to humans. Our system of latitude and longitude, or any similar grid system that we might devise, is arbitrary, requires a complex set of accompanying external references such as maps, globes, compasses, sextants and clocks, or electronic broadcast signal positioning equipment, and it has to be learned. If there is a natural grid map, what is its basis and how do other species have access to it? Could the Earth's magnetic field provide such a map?

As far as we know, the Earth's magnetic field does not provide full grid information. The lines of force run between the north and south magnetic poles, being horizontal and parallel to the surface of the Earth at the equator, then dipping or pointing down elsewhere at an increasing angle (the dip angle). They are vertical over the magnetic poles. Experiments with caged birds in imposed artificial electromagnetic fields show that they sense and respond to the strength and inclination of the magnetic lines of force, that is, the dip angle, which provide latitudinal gradients between the equator and the magnetic poles.

Interestingly, the magnetic strength and inclination are the only aspects the birds respond to and through those characteristics they can only determine poleward versus equatorward. They do not respond to specific polarity and hence do not distinguish between north and south, as do human compasses. This further suggests that something other than magnetite is being used to sense the magnetism.

In addition to latitudinal gradients in the strength and dip angle, there is local variation in the strength of the magnetic field that forms magnetic peaks, valleys, plains, and occasional anomalies. That could furnish unique landmarks, as with the visual topography familiar to humans, and possibly be used as a mosaic map. But there is no consistent lateral or east-west variation, and thus the magnetic field—as far as we know—cannot produce a complete grid in both directions.

One intriguing and controversial suggestion for the basis of the physical map for birds involves a learned mosaic or gradient map composed of odors or some other airborne factor. This is discussed next.

The enigma of pigeon olfaction

Pigeons have served as fruit flies for the study of bird navigation. They are common, easy to keep in captivity and experiment with, and navigate on demand so that one does not have to wait for seasonal migrations. Findings from pigeons, however, although shedding some light on mechanisms, continue to raise puzzling questions. Able and Bingman (1987) introduce the problem by noting the fact that pigeons without previous homing experience, when

> released under clear skies, usually orient homeward even at a considerable distance from their home loft [e.g., Wallraff 1970; summary in Keeton 1974], and even if their entire lives up to that point had been spent confined in an aviary [e.g., Wallraff 1966, 1970]. These experiments do not address the question of the basis of the navigational map, but they do indicate that exploration and flight experience in the vicinity of the loft are not necessary for homeward orientation at distant sites even on the first flight.

The unknowns are: (1) what environmental factors are pigeons exposed to while they are confined in a loft; (2) if they are transported away from their loft, what do they experience during the process of being transported; and (3) how are the factors at the release site related to conditions at the loft and/or during transportation? That animals can obtain inertial information on the outward journey via some form of path integration and return by dead reckoning, even with altered magnetic fields and with visual and other external cues blocked, has been clearly demonstrated in at least some cases (e.g., golden hamsters carrying food back to their nests, Etienne et al 1986), but that system is not considered to be sufficiently accurate at increased distances.

The only ways to totally prevent young, inexperienced pigeons from orienting toward home, depending on distance and a host of other confounding conditions, are to transport them in total darkness (recall that for the magnetic sense to work there must be at least a minimum level of light) or confine them in a loft with solid walls (including glass, through which they can see the sun). The latter observation suggests that there might be some "airborne" factor that is important to the birds as they are reared in their loft. Just what that "airborne" factor might be has been the subject of hot debate and is the current focus of much research.

Based initially on studies in Italy with pigeons that had their olfactory nerves cut or nostrils plugged with cotton and that subsequently were disoriented, Floriano Papi and colleagues suggested that pigeons formed a mosaic map of the odors surrounding their home loft (e.g., Papi et al 1971, 1972). According to this hypothesis, young pigeons learn the odors in the immediate loft area and in surrounding areas under different wind directions as the air moves past the loft. Then, by sensing the odors during their outward journey and at the release site, they determine their location relative to home and use one of the compasses (sun, magnetic field) to travel in the proper direction. Wallraff (1974) suggested that olfactory cues created a bicoordinate gradient map rather than just a mosaic map.

The proposal that pigeons might be using olfactory navigation, however, met stiff resistance (reviewed in Keeton 1974 and Schmidt-Koenig 1987). On theoretical grounds: (1) pigeons were not considered to have adequate olfactory capabilities, although some other birds have been shown to have a good sense of smell, use it for locating food or nests, and it could not be ruled out for pigeons; (2) the hypothesis is difficult to support meteorologically because of diffusion and mixing of air from different sources, particularly at distances from which young, first-flight pigeons are capable of homing; and (3) the techniques of nerve cutting and nostril plugging seemed likely to interfere with normal behavior. On the basis of actual data, researchers in other locations were unable to reliably duplicate the Italian results.

Papi and colleagues persisted, however, developed different techniques to test the hypothesis, and continued to provide data suggesting that pigeons were using olfaction (Papi 1982). A hodgepodge of confusing information from other sources supported or failed to support the olfactory hypothesis to varying degrees. For details and references, see the reviews cited above and the list at the end of the chapter. Some of the highlights are briefly described below.

Pigeons that were rendered unable to smell (anosmic) through the use of local anesthetics sprayed into the nostrils, rather than by nerve sectioning or nostril plugging, were significantly disoriented in Italy but not elsewhere as in the initial studies. Pigeons reared behind solid walls (including glass and plastic) were unable to home normally, as already mentioned, whereas birds exposed to air movement through openings (wire, baffles, etc.) could orient themselves on their first flights.

One technique used in several studies involved deflector lofts in which wind-deflecting walls are set up around an otherwise open loft. Pigeons raised in such lofts are confused as if they had been getting odor landmarks from the deflected positions. Based on results of Waldvogel et al (1988), however, the effect may instead be a result of a simultaneous deflection of celestial polarized light patterns.

One of the best, least disputed experiments involved the use of fans to duplicate or reverse the normal air flow at lofts, in which case released pigeons either homed normally or flew off in the wrong direction, respectively (Figure 8-15). In another

Figure 8-15 Vanishing bearings of pigeons raised in reversed winds at the home loft. **A,** Pigeons were housed from fledging in one of three types of corridors. In one, they were exposed to natural winds flowing parallel with the axis of the corridor *(N, open dots)*; in another a fan automatically blew in the same direction as the ambient wind whenever winds paralleled the axis of the corridor array *(N, solid dots)*; the third corridor had fans that automatically blew in the reversed direction *(R)*, opposite the ambient wind. **B,** When released at localities on the axis of the loft corridors, birds that had experienced the correct wind direction were homeward oriented *(N)*, whereas those that had lived in reversed winds tended to fly directly away from home *(R)*. From Able, K.P., and V.P. Bingman. 1987. Quarterly Review of Biology 62:1-29. Based on data from Ioale et al (1978) and Ioale (1980).

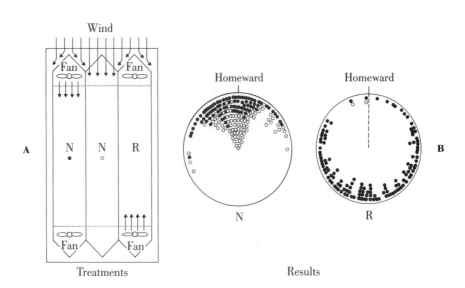

interesting experiment, large plastic bags were used to collect air from different release sites, then pigeons were transported from the loft to various sites in these air bags. Birds released at sites from which their air had been collected were significantly oriented toward home. Those released at sites after being carried in air from different sites, on the other hand, were disoriented (Schmidt-Koenig and Kiepenheuer 1986). (Birds received a local olfactory anesthetic before release so they could not detect odors at the site itself.)

In one attempt (of many) to explain the differences between Italy and elsewhere, it was suggested that there might be site-specific differences, perhaps related to prevailing winds and the consistency of local odors. Further work, however, demonstrated the difference to result from rearing procedures (Wiltschko and Wiltschko 1989). The birds in Italy were being raised in more open aviary-type lofts on building rooftops, whereas pigeons elsewhere were mostly being raised in lofts at ground level in the protection of nearby buildings and vegetation. When both types of lofts were constructed at Frankfurt, Germany, the roof loft produced similar results to those obtained in Italy, with anosmic birds being disoriented. Birds from the ground loft, however, oriented correctly whether anosmic or normal, as had been seen previously.

The meaning of all of this remains unclear. Rearing and local environmental differences definitely affect the acquisition of different navigational abilities and techniques in pigeons; and part of the process seems to depend on the birds being exposed to normal air movement at their lofts. But it is not known whether the airborne factors are really olfactory or just how they exert their influence. The results are perplexing to say the least.

We return to bird navigation, homing, and migration and attempt to integrate that information into a general picture of animal orientation and navigation after considering a few more long-distance navigators, some of the mammals.

Nonhuman Mammals

Mammals probably home and navigate mostly on the basis of familiar area, including many of those species that move long distances, such as bighorn sheep, caribou, bison, and wildebeest. These animals usually travel in large herds, the young accompany the adults, and the familiar area probably involves a spatial memory that is passed by experience from one generation to the next as they travel together. The only species for which this is well documented, however, is bighorn sheep; if traditional routes are interrupted or blocked or young are translocated artificially, the long-distance travels cease.

Other mechanisms may exist in marine mammals such as cetaceans, pinnipeds, and polar bears. However, the mechanisms of their long-distance travels have not been researched to the extent of similar travels in birds.

The most likely other candidates for sophisticated navigational systems in mammals would be those capable of strong flight, the bats. They have been shown to home after artificial displacement up to a distance of around 500 km (Griffin 1970). A few species migrate in a north-south direction up to a distance of around 1200 km. But most travel only shorter distances between seasons to traditional protected hibernating sites (*hibernacula*). An example of banding recoveries for little brown bats in the northeastern United States is shown in Figure 8-16. Bats may be affected by the earth's magnetic field in that they seem attracted to and hit radio towers under certain conditions of overcast and fog, similar to collisions by birds with

Figure 8-16 A few of the distant recoveries of little brown bats banded at a large hibernaculum in a mountain cave in Vermont. Bats were banded in the winter and recovered at other times of the year. In subsequent winters bats from this site would return to it. Note that to get to this location, bats from southeast would have to travel *north* to spend the winter. In spite of being further north, however, the cave provided advantageous protection for hibernation.
From Griffin, D.R. 1970. Migrations and homing of bats. In: W.A. Wimsatt, Biology of bats. Academic Press. Based in part on results of Davis and Hitchcock (1965), who banded over 70,000 of these bats.

towers. Bat navigation over long distances, however, has not yet received the extensive study given to birds, and the evidence is circumstantial at best. Further work is needed with individually tracked bats, using the diversity of approaches that have been applied to birds. The whole phenomenon of collisions with radio towers by both bats and birds needs further investigation. Are there differences in numbers of collisions when radio broadcasting is on or off? Are collisions caused by electromagnetic effects from large, grounded steel objects?

Humans

Humans clearly and successfully engage in long-distance travel even without the advantage of modern technological aids. Our species has long traveled great distances through forests and across deserts, plains, and large bodies of water. Micronesian and Polynesian natives in the South Pacific, for example, have extensively and successfully navigated between islands over large expanses of open ocean since around 6000 years ago (Waterman 1989).

How do we do it? In the case of other animals, we are faced with the old problem of not being able to ask them directly but only observing their outward responses to particular situations (Chapter 1 under anthropomorphism and Chapter 2 under behaviorism and cognition). Perhaps we can ask ourselves and each other to gain some insight. As it turns out, however, we have not been entirely successful yet even at self-inspection.

Our common, everyday knowledge tells us that we know where we are via the familiar area and we travel by piloting and the other methods described earlier in this chapter. It is known by sailors and the Polynesian travelers that even on the seemingly featureless ocean there can be landmarks such as clouds or flocks of birds over islands that are beyond the horizon, different appearances of sky over different colors of water, and recognizable water currents, waves, and various weather patterns from different directions. But is that enough, or all there is?

R. Robin Baker and his colleagues from the University of Manchester in England initiated a series of experiments on humans to parallel some of the work that has been done with pigeons. His results (Baker 1980 and several publications since then; see lists in Baker 1982, 1984, and 1987) supposedly provide evidence that humans also orient to "home" using a geomagnetic sense. Those claims touched off an intense controversy and spurred numerous attempts to replicate and validate or discredit the findings (see Gould and Able 1981 and the references listed in Baker 1987). One study (Bayliss et al 1985), apparently in satire, was a deliberate hoax. Some studies have included the use of magnets and Helmholtz coil helmets plus various control treatments.

The studies have involved three types of experiment: (1) bus, in which blindfolded subjects are driven around, then asked to point, draw an arrow, or write down the compass direction toward home, (2) chair, where blindfolded and earmuffed subjects are rotated on spinning chairs then asked to state what compass direction they are facing when the chair stops, and (3) walkabout, where subjects are led on a tortuous path through woods and periodically asked to indicate where home is. In addition to the basic conclusions on nonvisual orientation and a geomagnetic sense, Baker has purported to discover a number of other rather remarkable aspects, including different results that depend on which compass direction a subject sleeps at night and on whether subjects wear polyester clothes, cotton, or nothing during the experiment (referenced and summarized in Baker 1984).

Science is noted for skepticism, questioning, and even resistance to new ideas and findings, and those of Baker (as might be expected) have been no different. Baker is convinced of the validity of his results, and many others remain strongly unconvinced. To begin with, there are the usual uncertainties of methodological detail. In the chair experiments, for example, Baker states (1984:105-106), "the pattern of turning is critical. Too little rotation and the subject may follow by inner ear mechanisms. Too much rotation and the ability to judge compass direction disappears altogether. This and other evidence suggests that magnetic receptors dislike being spun as much as inertial receptors." (We might add that the last statement cannot be exactly true or there would be no way to separate inertial and magnetic senses, which further raises the question as to whether spinning is a good way to test magnetic reception in the first place.)

Another problem involves replication. Most other persons who tried to replicate Baker's findings have believed that their results did not confirm Baker's. Baker (1987), however, in a statistical attempt to assess all of the other work, excluding only his own, believes that the other work does confirm his. He discusses the problem extensively. In a summary statement in his abstract he states, "This stark contrast between positive results and negative interpretations remains an unexplained feature of the literature on human magnetoreception."

The lack of positive evidence for a magnetic or other nonlandmark homing sense in humans may result in part from using subjects without extensive experience under conditions lacking visual cues. At this point we insert a remarkable personal experience that involved one of us (Grier). During 1970, while engaged in peregrine falcon population research with the Canadian Wildlife Service, Grier and a friend were traveling along the southwest coast of Baffin Island in a boat guided by an Inuit (Eskimo). We came to a large bay, with Baffin Island on one side and the open Hudson straits on the other. We were returning to camp and, rather than waste time following the shoreline closely around the bay, we decided to save a few kilometers of travel by cutting across from one side of the bay to the other. As we were part way across, however, a heavy fog developed. The fog became so dense that we could barely see the other end of the boat, much less where we were going. The sun could not be seen; there was no wind blowing for direction; the water was calm; and no waves could be heard breaking on the distant shoreline. We had several kilometers to travel, and an inertial sense of direction seemed too inaccurate.

Our Inuit guide, however, proceeded forward at full speed. Two or three times we encountered seals coming to the surface for air, at which time our Inuit friend would stop the motor and allow the boat to drift, hoping for a shot at the animal for food. He would light up a cigarette and perhaps shoot at the seal a time or two when it appeared. In the meantime we drifted. My friend and I could not tell how much or in what direction, but our movements in the boat were enough to cause it to change direction slightly. At one point our guide shot one of the seals, paddled over to retrieve it, and hauled it over the side, all of which moved the boat considerably.

After each stop he would start the motor again, swing the boat around as if he knew exactly where we were, and head off at full speed. After a long time the fog began to lift slightly, and the rocky point toward which we had been headed suddenly appeared before us. We were right on course. It did not occur to us to inquire at the time, so we never found out how our Inuit friend knew where we were. But good navigation obviously is essential and adaptive in an environment frequently obscured by fog or snow and with wide expanses of featureless landscape or open ocean.

What about humans that live in agricultural, industrialized, or other relatively sedentary conditions? Are long-distance navigational abilities a natural human characteristic that can be developed, lost if not used, or perhaps lying dormant just beneath the surface of our consciousness? Perhaps the full set of senses, whether honed for long-distance movements or not, exist at the subconscious level as another autonomic response. Even when we know where we are at the local level, how do we know? Is it based completely on the familiar area and landmarks, or is there more?

The Inuit anecdote is just that, merely an anecdote, and may well have a different explanation than what has been considered. Additional and less ambiguous research is necessary. In the meantime, readers can only decide for themselves. We recommend further reading on both sides of the issue of possible magnetoreception in humans (references indicated previously and at end of chapter).

THE CURRENT STATE OF UNDERSTANDING ABOUT LONG-DISTANCE NAVIGATION

So where does all of this leave us? Do other animals orient their travels and navigate differently than humans, and is there anything that is truly mysterious and cannot be explained? Is there any relationship between the clear and remarkable homing abilities of artificially displaced alligators, homing pigeons, and Manx shearwaters? Do individual monarch butterflies each possess a navigational system in their tiny nervous systems that permits them to accurately travel to a specific place in Mexico all the way from Ontario, Canada, or is there some other explanation?

Baker (1982) believes that although there is much we do not know, including the internal mechanisms, we already know enough about animal senses (including the geomagnetic sense) and abilities to explain animal homing and long-distance movement, and in spite of the obvious superficial differences among different species and taxonomic groups, the underlying, fundamental principles are similar in all, and evolution has merely accentuated the details differently in different organisms. Baker, incidentally, includes the drifters in his overall scheme; he views them as simply searching for suitable environmental resources without keeping track of where they have been. Baker thinks an explanation does not require the use of a grid map or any ability that we humans do not naturally possess (i.e., "stripped naked" of our technological navigation aids). The feats of other species might still seem amazing but they are no longer mysterious, according to this view.

Baker's central theme involves animals searching for suitable habitat or environmental resources to match the needs imposed by their inherited, genetic makeup. Because of the large selective advantage of being able to return to particular places, many if not most species have evolved a sense of location and a homing ability. This involves a familiar area, the key to which, regardless of distance involved, according to Baker, is exploration. He does not say how exploration is integrated internally, including the specific roles of the various senses, but merely that exploration as a general process is required.

Baker (1982) extends the concept of familiar areas to familiar area maps. The familiar area is bounded by the most distant places where an individual has actually been, whereas the map extends to the most distant landmarks that an individual has perceived and remembered. A person living in a city all of his or her life, for example, could be familiar with the sight and location of a distant mountain peak without ever actually visiting it. Thus the familiar area map may be bigger, in some cases much bigger, than the familiar area itself. This may seem trivial, but according to Baker it has significant ramifications for longer-distance travel. A bird, for example, simply by flying straight up into the air a few hundred meters, would have visual access to an area of several hundred square kilometers. With extensive exploration, a bird could build up a familiar area map that is much larger than we might initially appreciate. The use of visual landmarks may or may not be important to the bird; whether or not landmarks are visual is not critical to the view. It may be using something else as landmarks, such as odors or the topography of the Earth's magnetic field. Landmarks in general may be of any sort, including not only visual, olfactory, and magnetic, but also auditory, tactile, and electromagnetic (in many electric field-sensing fishes), as long as they provide location-specific, map types of information.

Exploration may be generalized to include passive experiences, such as with airborne factors moving past a young pigeon's loft or being artificially carried on an outward journey to a release site. Some persons, however, might object that such a definition of exploration is too broad and useless because, unless there is inherited site-specific information (for which there is no evidence), experience of some kind is obviously required before an animal can relate to specific locations.

A map sense in the extreme case could occur without a sense of direction, except for the orientation toward familiar landmarks. At the other extreme, some animals may have a compass sense with little or no sense of location. Most species are presumably somewhere in between and possess an awareness of both location and direction. The lives of animals in some species may shift back and forth between being location-oriented part of the time and direction-oriented at other times.

According to this view, long-distance movements are extensions of local movements and do not require any new principles or mysterious accurate navigational abilities. The scales of magnitude, say meters versus kilometers, may differ and some species may be better equipped than others, such as having a larger memory to hold a larger familiar area map or having better senses for increased resolution of the landmarks, but the fundamental principles are the same. Concerning memory, Baker (1984) suggests that birds use a mosaic map composed of a few significant landmarks that may be spread far apart. Thus spatial memory for a larger area may not require more capacity but merely involve a larger scale. A bird seeing (or otherwise sensing) a large area at a scale of kilometers from the air may be comparable with, say, a mouse learning an area at a scale of meters from the ground.

How can this concept be applied to those species that travel the longest distances and under conditions of little or no contact with landmarks, except possibly the magnetic field? First, most animals may be doing much more exploring than we are aware of. It is apparent that the young of bird species that use a star compass must be doing more at night than quietly sleeping; they must have their eyes open and be watching the sky at least part of the time or they would not be able to learn the star patterns rotating around the pole star. Similarly, most birds may spend a major portion of their lives exploring, often over a large area.

We are aware of birds when they return in the spring, during the time they spend at their nests and breeding locations, and perhaps at a few other times during the year. Ornithologists conducting banding studies are aware of where the birds are when they are captured and marked and where they occasionally show up later. But what are the birds doing and where are they all the rest of the time? Chances are good that they are directly or indirectly (while searching for food, etc.) exploring.

We already noted that a bird could establish a large familiar area map simply by flying straight up into the air and looking around. Most birds, even permanent resident species, obviously do more than just fly straight up. Over time, particularly if they move around much after leaving the breeding area, typical birds can build up huge familiar area maps, involving hundreds or thousands of square kilometers and even significant segments of a continent. This certainly happens with human airplane pilots and even in humans who travel a lot on the ground or water.

The larger a familiar area map becomes, the bigger the target and the easier it becomes to encounter from a distance, including returns from migration and accidental or artificial displacement. All that is needed is a sense of direction that is accurate enough to get onto the familiar area map. To get to a specific point, such as a pigeon loft, from outside the familiar area map, the animal needs only to get to the familiar area map, then it can travel to the desired point, as in the study of gannets. The animal would not need a pinpoint navigational system that takes it straight from point A outside a familiar area map to point B inside the familiar area.

Birds on migration, in this view, would expand their familiar areas along the route. Exploration along the route, incidentally, might help explain some observed movements (such as lateral movements and reverse migration, in which birds are seen flying in the "wrong" direction) that previously have been puzzling. (Alternately, the birds might simply be responding to changed weather conditions or some other factor.) Birds on a return migration would not have to take same route as originally traveled, as long as they eventually came back onto some part of the familiar area map, including some part of the long previous route which they could encounter by traveling more or less perpendicular to it. Also, birds could travel when vision was obscured without knowing precisely where they were as long as they maintained a compass sense and were headed toward some part of their familiar area map. The magnetic field, if they are using that, would never be obscured except in the occasional places of magnetic anomalies.

This picture does include some sense of direction (of which inertial senses may roughly work in some cases), and it also needs to keep open the option of a magnetic sense. Magnetic effects have been demonstrated for many organisms beyond a reasonable doubt (aside from the questions that remain about humans) (Kirschvink et al 1985).

What about birds that locate small oceanic islands for the first time? That could be explained partially by birds seeing and being attracted to or following other birds in the distance, and partially as a matter of numbers and time: given enough birds and enough time, some encounter those locations purely by chance. Other birds that are not so lucky will be lost at sea and eventually die. Any sense of direction would greatly improve the odds. A human observer finding the species, even a marked bird, at an isolated island (or any other place for that matter) could easily but mistakenly conclude that the bird somehow traveled directly and specifically to just that precise place. After the first visit to a location, even a small island far from the mainland, the chances of subsequently finding and recognizing the place may be greatly increased.

If return movements can be explained as above for birds, then they should work for many other animals as well, particularly other vertebrates. Again, the familiar landmarks may be visual, olfactory, auditory, or another nature. But they are all familiar landmarks, and the process involves a sense of location.

A sense of direction played an important role in much of the above discussion, and it may be sufficient in combination with large population numbers and chance to explain a few remaining cases, such as the monarch butterfly story with which we started in the previous chapter. Baker (1978, 1982, 1984 b; also see discussion and references in Douglas 1986) provides an explanation along these lines based on an extension of the picture given for small white butterflies (Figure 8-17). In brief, monarchs and their offspring tend to move in a preferred northeastward direction during the spring and summer with many of them eventually reaching southern Canada. Along the way they are feeding, depositing eggs, and occasionally resting. With the approach of fall the butterflies stop reproducing and build up food reserves while moving shorter distances, perhaps 2 km or less a day.

Then the preferred direction reverses and they begin moving toward the sun's azimuth or perhaps using the sun otherwise as a compass with some time compensation. Most of the movement is during midday and early afternoon so that the average direction is toward the southwest. The daily movements may increase, perhaps to around 50 or more km per day with the assistance of suitable winds, so they end up as much as 2000 km south and west of where they originally emerged as adults. As they advance south the rate of movement then slows.

This process could theoretically funnel large numbers of butterflies into certain locations such as those observed in Mexico and California without requiring much navigational accuracy or that the entire species goes to such places. Those butterflies

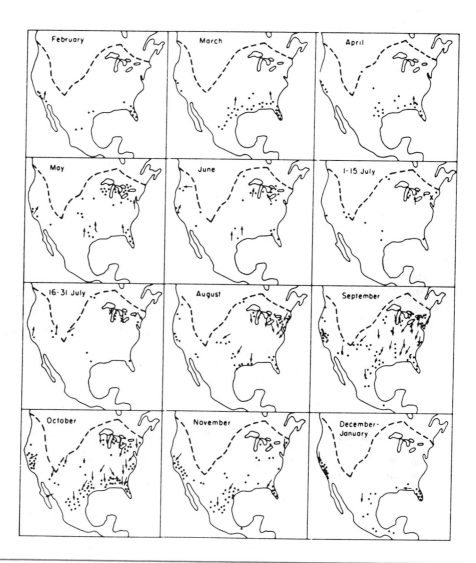

Figure 8-17   The seasonal return and remigration of the monarch butterfly in North America. Dashed line, northern limit to distribution in summer; dots, monarchs observed flying; arrows, peak migration direction.
From Baker, R.R. 1978. After data from Urquhart.

that do end up there would then just stop moving during the cool high-altitude winter conditions. Other butterlies could go on about their business elsewhere. As it turns out, during the winter monarch butterflies are also to be found in other places in Mexico and across the southern United States, including Florida, although not in the concentrations seen at the famous wintering sites. Urquhart (1987) refers to these as *aberrant migrants*. Some continue breeding, feeding, and local wandering throughout the winter. A few may even remain behind and survive in a dormant condition further north. Thus the incredible numbers seen at some wintering sites may be only a portion of an even larger and more incredible number to begin with. Even if not all monarchs are at or depend on the Mexican and California wintering sites, the sites nonetheless are spectacular, fascinating, serve as refuges for large numbers of monarchs, and deserve protection (cf. Norman 1986).

Ken Able

Ken Able

Maps, Compasses, and Birds

The study of bird orientation and navigation is based on a singular but unassailable observation. Individuals of many species of wild birds (and their laboratory counterpart, the homing pigeon) have the ability to return to specific pinpoints on the earth from hundreds and even thousands of kilometers away. The undeniability of that fact and our inability to explain in a step-by-step way how it is accomplished are the classic ingredients of all good scientific mysteries. At one level, we could explain the phenomenon by postulating familiar area maps expanded through exploration, coupled with a sense of direction, as described near the end of this chapter. Most scientists, however, will be unsatisfied until we know precisely what information constitutes a map and compass and which senses are used to perceive those cues and rigorously testable hypotheses about those mechanisms have been evaluated.

Scientific research, like all human enterprise, proceeds on the basis of many unstated assumptions and is subject to various biases of the researchers themselves. A number of the important recent advances in bird-orientation research illustrate how such constraints retard progress until some breakthrough forces a change in research philosophy.

In the 1950s and 1960s, it was generally assumed that we were searching for a single kind of compass for a given type of bird. Discoveries during those years made a neat story: a sun compass for diurnal migration and a star compass for nocturnal flights. Ever more persuasive evidence for the existence of a magnetic compass, even in species known to possess one of the visual compasses, threw this simple scenario into a cocked hat. The implicit assumption of a unitary mechanism impeded the discovery of multiple compass capabilities. Their demonstration forced the focus of research to shift from a search for

Much of the story for both birds and butterflies, however, is conjectural and many persons remain unconvinced by Baker's arguments. There are too many nagging, unsettling questions to accept his views at face value. Even after allowing for nonvisual landmarks, for example, pigeons that accurately travel to the immediate vicinity of their home lofts while wearing frosted lens over their eyes or that under other experimental conditions become lost even when their lofts are in plain view, and Manx Shearwaters that can return home quickly after being displaced far from familiar areas across several hundred miles of open ocean all seem to strain Baker's model too far.

Our present understanding of animal migration and homing, particularly in respect to the nature of the potential "map(s)," seems comparable with where human understanding was for bat echolocation before Griffin (Chapter 1), the information-center hypothesis before Brown (Chapter 3), or the bee-language controversy before Gould (material in Chapter 15). As described in Chapter 3, it is not knowing something that drives scientists; animal travels have certainly been no exception.

'the' compass to investigations of how the several abilities are related one to another in the behavior of the bird.

Although this may have brought us closer to the truth, it greatly complicated experimental design because one must take into account several stimuli simultaneously. If, for example, a night migrant was shown to orient in the expected direction in the absence of stars, one could no longer conclude that it does not use stars when they are present. In this case the bird may simply be relying on its magnetic compass. The more recent discovery that at least some night migrants also use information at sunset for orientation, including patterns of polarized light in the sky, has further complicated the task of sorting out the relationships among these compass capabilities.

In some ways, history seems to be repeating itself in the search for the map component of bird navigation. At present there are two viable hypotheses for the basis of the map of homing pigeons: the odor map and a map based on the earth's magnetic field. Much of the controversy that surrounds research on this problem stems from the assumption that the map is based on a single stimulus and the hypotheses therefore are mutually exclusive. That need not be the case and, as so often in science, the truth may lie between the positions taken by the proponents of one hypothesis or another. In this instance, both sides may be correct.

The ontogeny of orientation behavior involves both genetic constraints on what is learned and important influences of experience. Birds are not the automatons we sometimes assume them to be. Whereas developmental plasticity complicates our efforts to understand the mechanisms of orientation and navigation, it equips the birds with the flexibility that may enable them to learn to rely on those environmental cues that provide the most reliable and useful information at a given time or place. Although this may be inconvenient for us, it is no doubt adaptive for the birds.

SUMMARY

The long-distance travels of animals include homing and round-trip migrations. The subject involves several important terms and concepts, including the following:

Migration: long-distance travel, usually with a return, to specific locations.

Directional orientation: orientation only toward a particular compass direction, not a specific location.

Navigation: orientation toward a specific location.

Compass sense: a sense, perception, or awareness of direction.

Map sense: a sense, perception, or awareness of specific locations.

Linear map: a one-dimensional (but not necessarily straight) path or gradient.

Polar map: radiating relationships about a central point.

Grid map: bicoordinate system such as with human longitude and latitude.

Mosaic map: patchwork of irregular, known areas with landmarks (in the generic sense, not restricted to visual).

Piloting: navigation by following landmarks (including artificial).

Dead reckoning: navigation by vectoring distances and directions.

Celestial navigation: navigation with reference to celestial objects such as the Sun and other stars.

Geophysical navigation: navigation with the aid of the Earth's magnetic field.

Long-distance travel, including migration, involves a number of both ultimate and proximate factors. Examples of long-distance travelers are fish (e.g., salmon and eels), birds, and some mammals, including marine mammals, bats, humans, and others.

Techniques used in research on the phenomena of homing and migration include tagging, artificial displacement, radar, radio telemetry, and a host of experimental methods such as manipulation of senses, clock shifting, and the use of artificial magnetic fields. A natural response of migratory birds that has helped elucidate migratory phenomena is Zugunruhe or migratory restlessness.

As in most behavior, much of the variation in migratory and navigational behavior is genetic, whereas much clearly depends on experience and specific environmental aspects.

Compass senses may be based on time-compensated apparent motions of the sun, star patterns, or the inclination (dip angle) and strength of the Earth's magnetic field.

Animals that routinely navigate or home over long distances have been shown to possess redundancy in their navigation. The different backup systems possessed by such animals generally operate in a hierarchy. Depending on the conditions and distance, parts of the hierarchy in homing pigeons are (1) sun compass, (2) magnetic compass, and apparently (3) various map information. Different individuals may acquire different navigational abilities and techniques depending on their particular rearing experiences and local environments. The details of just how these systems are acquired and used, however, remain poorly understood.

Some facets of pigeon homing, such as what they experience and sense at their home loft, during transportation, and at unfamiliar release sites, have received much attention and generated continuing controversy. One proposal, for example, suggests that pigeons use olfaction in their navigation. Exposure to some airborne factor or factors appears to be important in pigeon navigation, but the nature of the factor(s) and how it (they) operates remain unclear.

One view of animal orientation and navigation is that, as animals travel around seeking suitable habitat and environmental resources, many if not most species have evolved a sense of location based on exploration, landmarks, and spatial memory. Compasses, such as the sun and stars, often are involved. In this scheme, long-distance movements may simply be an extension of local movements. Instead of a sense of location, a compass sense alone may guide the movements in some species of animals. Many of the views and interpretations of animal navigation, homing, and long distance movements are controversial and not accepted by all persons studying the issue.

Three of the major unsettled issues are: (1) the nature and location of the magnetic sense in animals, (2) whether humans possess a magnetic sense and the same fundamental, natural abilities for navigation as other animals, and (3) the nature of the navigational "map(s)" used by birds and other animals.

FOR FURTHER READING   Able, K.P., and V.P. Bingman. 1987. The development of orientation and navigation behavior in birds. Quarterly Review of Biology 62:1-29.
Reviews and discusses the numerous interacting factors involved in the ontogeny of homing and navigation.

Baker, R.R. 1982. Migration. Paths through time and space. Hodder and Stoughton, London.
This is a condensed and somewhat updated version of the much more voluminous Baker 1978. It provides an accessible summary of Baker's general thesis, with its strong flavor of advocacy, as well as much of the information on the subject.

Baker, R.R. 1984 a. Bird navigation. The solution of a mystery? Hodder and Stoughton, London.
More information and Baker's views as focused on bird migration and homing. Includes a summary of the work on human magnetoreception by Baker and his colleagues at Manchester University.

Baker, R.R. 1984 b. The dilemma: when and how to go or stay. In: R.I. Vane-Wright and P.R. Ackery, editors. The biology of butterflies. Symposium of the Royal Entomological Society of London, no. 11. Academic Press, New York.
This is a treatment of Baker's views, with further theoretical considerations, as applied primarily to butterflies.

Baker, R.R. 1987. Human navigation and magnetoreception: the Manchester experiments do replicate. Animal Behaviour 35:691-704.
A review, including references, of the several attempts to replicate Baker's magnetoreception studies in humans. This reference and the many other studies it cites can be used as the basis for further thinking and discussion about the topic.

Cochran, W.W. 1987. Orientation and other migratory behaviours of a Swainson's thrush followed for 1500 km. Animal Behaviour 35:927-929.
A good example of actual data plus discussion on possible orientation mechanisms (sun and geomagnetic aspects).

Etienne, A.S., R. Maurer, F. Saucy, and E. Teroni. 1986. Short-distance homing in the golden hamster after a passive outward journey. Animal Behaviour 34:696-715.
A good experimental, technical-level paper on route-based versus location-based cues and the subject of path integration during passive transport.

Gauthreaux, S.A., Jr., editor. 1980. Animal migration, orientation, and navigation. Academic Press, New York.
A collection of papers by several authorities working in this field of research. A fairly standard reference on the subject and a good balance as opposed to the more advocate-type position of Baker.

Gould, J.L. 1987. Landmark learning by honeybees. Animal Behaviour 35:26-34.
A technical paper and a good example of contemporary work and findings on landmark use by animals.

Keeton, W.T. 1974. The orientational and navigational basis of homing in birds. Advances in the Study of Behaviour 5:47-132.
A major, thorough, systematic review of the subject. Although slightly outdated, this continues to provide one of the best overviews and historical perspectives.

Lidicker, W.Z., Jr., and R.L. Caldwell. 1982. Dispersal and migration. Benchmark Papers in Ecology/11. Hutchinson Ross, Stroudsburg, Pennsylvania.
A collection of readings on both dispersal and migration.

Rodda, G.H. 1985. Navigation in juvenile alligators. Zeitschrift für Tierpsychologie 68:65-77.
A fascinating original paper on the artificial displacement of young alligators and their subsequent returns to the original sites.

Walcott, C. 1986. Homing in pigeons: are differences in experimental results due to different home-loft environments? In: H. Ouellet, editor. Acta XIX Congressus Internationalis Ornithologici (vol. 1:305-308). University of Ottawa Press.

Waterman, T.H. 1989. Animal navigation. Scientific American Library, New York.
A well-illustrated, up-to-date, general survey of the subject.

Wiltschko, R., and W. Wiltschko. 1985. Pigeon homing: change in navigational strategy during ontogeny. Animal Behaviour 33:583-590.
Discusses the changes from inexperienced pigeons requiring outward journey information to experienced birds being able to rely on release site cues, including the means and results by which this conclusion was reached.

Wiltschko, W., and R. Wiltschko. 1988. Magnetic orientation in birds. Current Ornithology 5:67-121.
A major, thorough review of the literature on the topic of magnetic field sensing and orientation in birds.

Recent Reviews and Technical Papers on the Possible Role of Olfaction in Pigeon Homing

Papi, F. 1982. Olfaction and homing in pigeons: ten years of experiments. In: F. Papi and H.G. Wallraff, editors. Avian Navigation. Springer-Verlag, Berlin.

Schmidt-Koenig, K. 1987. Bird navigation: has olfactory orientation solved the problem? Quarterly Review of Biology 62:31-47.

Spetch, M.L., and C.A. Edwards. 1988. Pigeons', *Columba livia*, use of global and local cues for spatial memory. Animal Behaviour 36:293-296.

Wallraff, H.G., and U. Sinsch. 1988. The role of "outward-journey information" in homing experiments with pigeons: new data on ontogeny of navigation and general survey. Ethology 77:10-27.

Waldvogel, J.A., J.B. Phillips, and A.I. Brown. 1988. Changes in the short-term deflector loft effect are linked to the sun compass of homing pigeons. Animal Behaviour 36:150-158.

Wiltschko, W., R. Wiltschko, and M. Jahnel. 1987. The orientation behaviour of anosmic pigeons in Frankfurt, Germany. Animal Behaviour 35:1324-1333.

Wiltschko, W., R. Wiltschko, W.T. Keeton, and A.I. Brown. 1987. Pigeon homing: the orientation of young birds that had been prevented from seeing the sun. Ethology 76:27-32.

Wiltschko, R., and W. Wiltschko. 1989. Pigeon homing: olfactory orientation—a paradox. Behavioral Ecology Sociobiology 24:163-173.

9

FORAGING AND ANTIPREDATOR BEHAVIOR

Foraging behavior is a complex and important subject that has received considerable attention. Just as surely as the function of sleep has been a puzzle, the function of eating is obvious. It is the means by which animals refuel, that is, replenish their bodies with energy and chemical building blocks. All animals at some times in their lives must confront the problems of finding adequate amounts of food and regulating intake to meet energy requirements and to provide protein and other nutrients essential for growth, maintenance, and reproduction. Most must also confront danger to themselves from other foraging animals.

The variety of foods consumed by animals is enormous, and scientists have devoted much effort to the description of foraging methods used in obtaining food. Much of this food consists of other living organisms whose interests are not served by being eaten. As a result, coevolution of improved foraging methods and superior antipredation methods has occurred, producing many extremely intriguing behaviors. The drama of the evolutionary conflict between predators and prey, in their broadest sense, has fascinated not only scientists but the general public as well.

This chapter covers the visible manifestations of feeding, along with some ecological and evolutionary aspects. The topic of feeding is also touched on in relation to other subjects elsewhere in the book. Recall, for example, the notions of appetitive, consummatory, and quiescent stages of behavior such as feeding in Chapter 2. Feeding is also discussed in relation to the internal control and integration of behavior in Chapter 21.

CLASSIFICATION OF FORAGING TYPES

In a broad sense the basic nutritional requirements of all animals are similar. Yet these basically similar needs are satisfied by the consumption of an extraordinary variety of foods in accordance with the particular biochemical resources of the environments the animals inhabit. Every conceivable kind of food is eaten, including such apparently unpalatable items as horn, carrion, poison ivy, and the crude tartar in the bottom of wine casks. The hawksbill turtle (*Eretmochelys imbricata*) has a specialized diet of sponges and consumes vast quantities of both glass and toxic chemicals in the process (Meylan 1988).

The meat-packing industry in the United States has always taken pride in its efficiency. ("Everything is used but the squeal.") The feeding behavior of animals is often no less opportunistic or efficient. Different species display a wide range of physiological and behavioral adaptations that enable them to use a variety of sources as foods.

Foraging types can be classified in a variety of ways. A familiar way is on the basis of the kind of food taken. Herbivores are animals that consume living plant tissues; carnivores consume the tissues of living animals; omnivores eat a mixture of plant and animal foods; detritovores consume dead organic matter. A variety of more specialized subtypes can be identified under each of these main types. Under herbivores, for example, one could list frugivores (fruit eaters) and folivores (leaf eaters). Under carnivores, one could include insectivores (insect eaters), piscivores (fish eaters), etc.

Another classification is based on the effects on the organism consumed. If the prey is immediately killed and mostly consumed, that is referred to as *predation* (Latin, *praedari*, to plunder or take by force). Note that this is a broader use of the term than is usually made by the general public, which uses it as equivalent to "a killing carnivore." If part of the prey organism is consumed but death is not caused, that is termed *parasitism*. Parasites can be internal (e.g., the *Plasmodium* protozoans that live in red blood cells, causing malaria) or external (ectoparasites), such as fleas or mosquitoes. Of course, parasitism can lead to death if an individual is parasitized by many individuals, with small individual hurts summing to a fatal decline. Entomologists speak of an intermediate type, *parasitoids*. Parasitoids, which include fly and wasp larvae, live inside their victim and consume it slowly—as do parasites; but they also routinely kill their victim in the end—as do predators. Note that a comprehensive use of this scheme leads to some classifications not usually made by the general public; for example, a cow is a parasite of grass.

One other way to classify animal foragers is by their mode of activity. Two major types are thus arrived at: searchers, which move actively through their environment while foraging, and sit-and-wait foragers (or ambushers), which establish a position in the environment and catch their prey in that place. Among searchers, a distinction may be made between grazers, which search for food that is widespread and abundant, and hunters, whose food is more patchily distributed.

This last classification system is useful because it helps us to focus on general kinds of adaptations. For example, efficient sit-and-wait foragers must evolve abilities to respond to habitat patches where many prey will arrive; traps, webs, nets, or filters with which to catch them; or lures to attract them (see discussion of "aggressive mimics" later in this chapter). Searchers' adaptations include keen prey-locating senses, and efficient or fast modes of locomotion that allow the search to continue for long periods and allow the predator to catch up with fleeing prey.

The many specific foraging adaptations (and equally many specific antipredation adaptations in the intended victims) are ingenious and interesting. We review some of them later in the chapter. First, we examine recent attempts to discover whether or not general rules exist that apply to all foragers regardless of where they fit in the schemes outlined previously.

OPTIMAL FORAGING

Feeding is not a single behavior but a large collection of functionally related behaviors. Included are several stages of searching, locating, or encountering; pursuing, capturing, or otherwise obtaining; handling; and finally ingesting. How do animals accomplish the job of food input when faced with numerous choices and various environmental challenges? A general energy budget formula for feeding requires that net energy gain (i.e., energy from food minus energy spent searching, etc.) during a given period be greater than zero. This creates several biological problems that must be solved with appropriate behavior:

1. What to eat and how to recognize it
2. Where to search and how long to search there before searching elsewhere
3. What movement pattern to follow while foraging
4. How to deal with prey that resist
5. When to eat (and when to stop)

The fourth problem, what to do with prey that resist, involves ways of finding, catching, and handling prey. These topics are covered later in the chapter. The fifth problem, when to start and stop feeding, is largely a matter of internal, physiological control. It is discussed briefly later in this chapter and at more length in Chapter 21.

How does an animal deal with the first three problems listed? Are there general solutions for foraging behavior? Can one generalize about foraging whether it is a lion searching for and attacking a zebra or a squirrel searching for and cracking a nut?

These questions are all related to an important, recent, and rapidly growing body of theory referred to as *optimality theory*. The main development of this field has related to the topic of feeding (hence its introduction here) under the name of *optimal foraging theory* (OFT). In essence, OFT borrows from the subject of human economics and looks at feeding from a benefit/cost standpoint. OFT states that animals should feed or change their feeding behavior in a manner that maximizes the benefits and minimizes the costs. The ultimate biological measure of this is fitness, that is, reproductive output of an animal and its descendants. Reproduction, however, is usually difficult to assess and does not lend itself to short-term analysis. Thus energy in calories ingested minus calories spent during a given time period, or other forms of currency, have been used instead.

Specifically, OFT presumes that animals should forage to maximize the rate of caloric intake per time spent foraging, E/t. Time spent foraging can be divided into two parts: search time, time spent locating prey; and handling time, time spent catching, killing, manipulating, and ingesting food. Thus foragers are presumed to maximize $E/t_s + t_h$.

It is obvious that when food is available in limited supply, efficient foragers that maximized E/t would have a natural selection advantage over others that did not. But even if food is not limited, OFT workers argue, efficient foragers are still favored. They have more time available for other important behaviors. Additionally, since foraging also exposes the forager to predators, efficient foraging may minimize that vulnerable period.

The concept of optimal foraging is natural and reasonable, so much so that it seems almost trivial. Furthermore, the basic idea has been inherent or implied in the theory of natural selection since the time of Darwin. This kind of thinking, although not necessarily identified as optimality theory, has played a large role in ecological, evolutionary, and population dynamic theory for many years (e.g., Cole 1954). But OFT does differ in its emphasis on the behavior of individual animals rather than the characteristics of a species. Those individuals that can make optimal choices in their daily foraging should have a large selective advantage over those that cannot, and the ability would be expected to evolve. The development of optimal foraging theory is considered to have begun with two important papers, one by MacArthur and Pianka and one by Emlen, which were published back to back in the journal *The American Naturalist* in 1966.

These two papers have been followed by a large number of others. Schoener (1971) provided a review of the theory of feeding strategies through 1971. Two major reviews of optimal foraging were published subsequently by Pyke et al (1977) and Krebs (1978). A symposium in 1978 was devoted to the topic (Kamil and Sargent 1981) with the royalties from the book funding a subsequent conference and publication (Kamil et al 1987). Krebs and McCleery (1984) and Stephens and Krebs (1986) provide concise introductions and overviews of the subject. Interest and publications since these reviews have continued to proliferate.

All of the interest and activity, unfortunately, have also generated a plethora of jargon, controversies, misunderstandings, and confusing rules and principles (hypotheses) that have yet to be sorted out. All of this is normal and a healthy sign of a new, active, and evolving area in science. But it does create some difficulties in understanding and following the subject by persons who are not actively involved in it themselves and who often lack some of the auxiliary background, such as for this case in economics and mathematics. It also continues to raise the question of whether we are really dealing with a genuinely new and significant subject or only new terminology and ways of talking about old subjects—sort of the scientific equivalent of designer clothes. Stephens and Krebs (1986) address this point at various places in their book, including the following remark (1986:170): "In the 1970s 'the foraging ecology of the wren' became 'foraging strategies of the wren' and in the 1980's the same research might be labeled 'constraints on foraging in the wren.'"

OFT has been concerned largely with three categories of choice, corresponding to the first three of the questions raised earlier: (1) what to eat (optimal diet or food type), (2) where to find it, particularly when food is hidden and not distributed uniformly in the habitat (optimal patch choice and allocation of time in patches), and (3) optimal directions, patterns, and speed of movements (optimal search paths).

Considerations of optimal foraging in particular and optimal behavior of any kind in general can be broken into three basic components. Stephens and Krebs (1986) call these *assumptions* related to decisions, currencies, and constraints, respectively. They are:

1. The problem or choice being considered, such as what to eat or where to find it
2. The item or currency being measured and what is to be optimized, such as calories per hour, number of food particles being filtered per minute, or number of nuts opened per hour
3. Constraints on the choices and payoffs in the particular situation

An example of what is meant by a constraint is the limited memory ability of butterflies in their choice of flowers for obtaining nectar. As long as a given species of flower is available, individual butterflies of some species feed consistently at that single species of flower, a phenomenon originally noted by Darwin. Lewis (1986) studied the cabbage butterfly *(Pieris rapae)* and showed that the butterflies had to learn how to extract nectar from a flower. Successive feedings on a particular flower species took less time and became more efficient. When given a second species of flower, the butterflies had to learn how to feed from that one. However, learning to feed from a second one interfered with their ability to feed from the first species. This suggests that amount of memory limits learning to one flower species at a time. Switching flower species results in learning costs; it is better for butterflies to stay with a single species as long as it is available.

Figure 9-1 Specialized foraging behavior in the dung beetle *(Kheper aegyptiorum)*. In the sequence of behavior diagrammed here **A**, the beetle first cuts a ball of dung from a dropping, **B**, rolls it to a suitable location by pushing backward while walking on the front legs, **C**, buries it by digging a hole from under it, **D**, then lays an egg in it, covers it over, and leaves.
From Heinrich, B., and G.A. Bartholomew. 1979. Scientific American 241(5): 146-156.

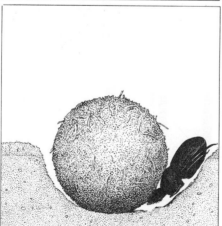
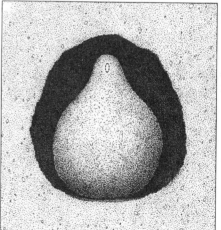

Much of the treatment of optimal foraging is highly mathematical and beyond the scope of this introductory book. Many of the findings, however, have provided remarkable confirmation of various theoretical predictions, and a few examples are given.

Optimal Food Types

The type of food an animal chooses may be governed by optimality principles. Animals can be classed by diet on a broad continuum from generalists, such as earthworms, which eat almost anything they can ingest, to others that are specialists. Dung beetles, for example, are relatively specialized in foraging on dung, with the aid of specialized behavior (Figure 9-1). Oldsquaw duck *(Chanqula hyemalis)* are specialized for feeding on invertebrates, primarily amphipods, after diving to depths of 46 m (150 feet) or more underwater (Peterson and Ellarson 1977). Another specialized predator is the Everglade kite, which eats one species of snail, the apple snail. There is clearly strict prey selection operating in these species, the predators usually selecting on the basis of specific cues. Many species, such as herbivorous insects or snakes, find and identify their prey on the basis of its odor. Some species show geographical variation in apparently innate prey preferences (Figure 9-2). Some species are so highly specialized that they feed only on particular parts of a single species of plant or animal.

Figure 9-2 Prey preferences of newborn garter snakes (*Thamonphis sirtalis*) from different geographical locations. Responses were measured as attack behavior shown toward water extracts from various prey species. Responses were inferred to be innate and appropriate for the particular locality. From Burghardt, G.M. 1970. Behaviour 36:246-257.

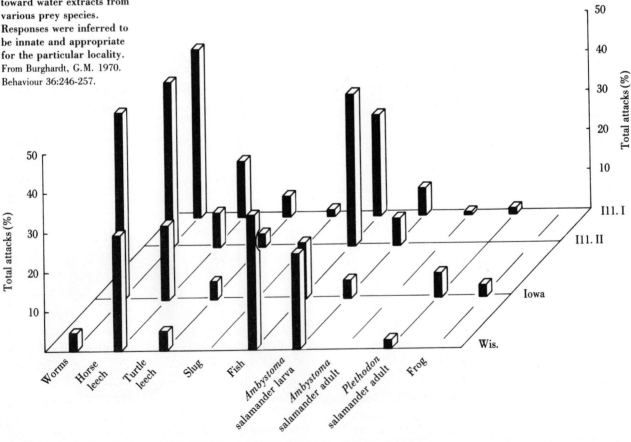

Generalists have more types and sources of food available to them, but they may be less efficient at using any particular source. To use an old cliche, a generalist is a "Jack of all trades but master of none." In terms of OFT, generalists usually have shorter search times for finding prey but longer handling times for dealing with a particular prey than would be the case for a specialist on that prey species. Laverty and Plowright (1988) demonstrated a clear example of this in the foraging behavior of a specialist bumblebee *(Bombus consobrinus)* compared with two generalist species of *Bombus.* Naive *B. consobrinus* quickly and efficiently found the nectar when first presented with their specialty plant *(Aconitum,* monkshood), whereas naive individuals of the other two species probed extensively in the wrong places on the flower and often gave up before becoming successful.

Within the range of a species' adaptations for feeding, animals may choose their particular food on an optimal basis. OFT predicts that animals should choose the most profitable food items among those available, be more selective when profitable items are more common, and ignore unprofitable food outside the optimal range no matter how common it is (Krebs 1978). In food-rich environments, a narrower range of foods should be taken; in food-poor environments the diet should be broader. These predictions have been demonstrated now in numerous cases. Pinon jays *(Gymnorhinus cyanocephalus),* for example, assess the quality of pinon seeds *(Pinus edulis)* and take only the good ones (Ligon and Martin 1974).

An excellent example of the subtleties in optimal choice of food items is provided by northwestern crows *(Corvus caurinus),* which feed on whelks—large, shelled marine mollusks (Zach 1978, 1979). To get into a whelk the crow obtains it from the water's edge, flies inland (where shells do not bounce back into the water), and drops the shell onto a rock (Figure 9-3, *A* to *C*). The larger the whelk the more likely the shell is to break, but the more energy is required to carry and drop it from a height. If a shell does not break, the crow is faced with the problem of whether to try again until it breaks or go get another (they repeatedly drop each until it finally breaks). In all aspects, including size and weight of whelk, choice of dropping site, height of drop for different-size whelks, and number of drops, the crow behavior conforms to predictions of OFT. In choice of size, for example, crows choose the larger whelks and ignore the smaller ones, although the smaller ones are more common and easier to carry (9-3, *D*). The smaller ones are more difficult to break and require more travel and more drops; hence the larger ones are more profitable.

Meire and Eroynck (1986) provide an example involving oystercatchers *(Haematopus ostralegus,* a shore bird) selecting mussels. This resembles the northwestern crow example except that the birds' choice of mussels is more complex. The optimal choice of mussels for oystercatchers depended not only on overall size (shell length) but also shell thickness, whether shells were overgrown by barnacles, and how much time was wasted on shells that were not opened.

Recognition of basic differences in adaptation among species and the constraints imposed by those adaptations is important to understanding optimal foraging. Search and handling times for a given food type vary among forager species with different adaptations. However, basic evolutionary adaptations only roughly tune the organism to the environment, and additional fine tuning may be up to the individual. Ability to alter foraging behavior varies from species to species even among close relatives. In *Peromyscus* mice, Drickamer (1972) showed that *P. leucopus* is more flexible than *P. maniculatus* and that the flexibility develops as the animals mature. Fur-

Figure 9-3 Whelk dropping behavior by northwestern crows. **A,** The crow flies from a perch to the water's edge where it searches for live whelks, then flies inland to make repeated drops until the shell breaks. After breaking a shell and eating the whelk, the crow either returns for another or goes back to its perch. **B,** Some crows drop the whelk from the top of the flight, which gives added height, but the crow cannot see where the shell drops. Others tip forward and lose height but can see better where they are dropping the shell. Several tradeoffs are involved in this behavior. **C,** Examples of common types of breaks in dropped whelk shells. **D,** Differences in the frequencies of whelk sizes available versus sizes actually chosen by the birds. The larger whelks are fewer, more difficult to find, heavier and require more energy to carry, but they break more easily and hence provide a greater net benefit.

A from Zach, R. 1979. Behaviour 68:106-117. B to D from Zach, R. 1978. Behaviour 67:134-148.

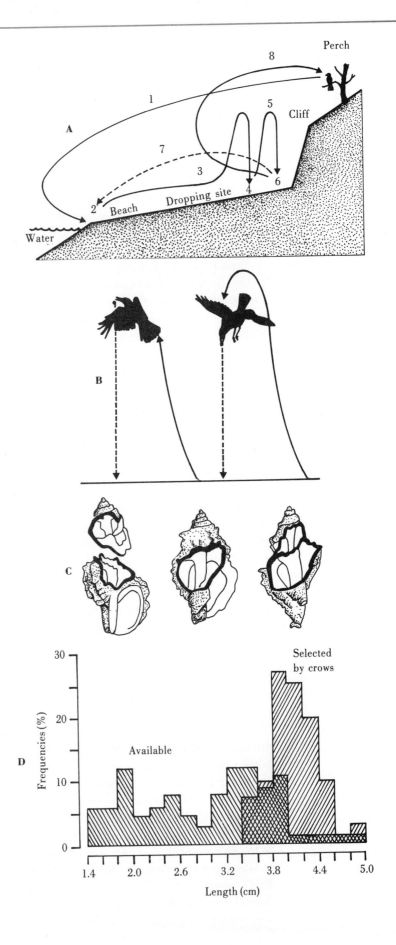

thermore, feeding flexibility correlates with habitat conditions, with *P. leucopus* living in a wide variety of habitat types. Fitzpatrick (1981) showed that differences in optimal foraging behavior in flycatchers (birds, *Tyrannidae*) vary with body size and are also related to the various other factors. Moermond (1979), during investigations of diet specialization in insects from an ecological viewpoint, concluded that herbivorous species that appear to be generalists over large geographical ranges may be specialists at the local level. This could result from geographical genetic differences or from individual flexibility in dealing with food species present.

Predators that eat a wider variety of prey frequently concentrate on particular species at particular times, although they may take many different species through the year. The predators often do not take animals in proportion to their availability; they may take animals of one species and seem to ignore or refuse animals of another species that are equally abundant. Such apparently idiosyncratic behavior merits further discussion in the sections that follow.

Prey Switching

Many predators concentrate on one species of prey for a period of time, then suddenly switch to another type of prey. Many insectivorous birds and fish, for example, change the species of insects they are eating and refuse those they ate previously, although these still may be abundant. Hyenas (Kruuk 1972) and African wild dogs (Schaller 1972) hunt different large prey, for example, wildebeest or gazelles, then switch to something else such as zebra. Furthermore, depending on which prey they are pursuing, they may start the hunt with different pack sizes and different prehunt rituals. After hunting wildebeest for several days and switching to zebra, the predators may leave the den and travel past or through large herds of wildebeest, ignoring them completely, on their way to hunt for zebra. In some cases, such as perhaps with insectivores, different prey may possess different nutritional characteristics, and the switch might be explained by dietary needs. In others, however, such as those involving large mammalian prey, the dietary differences seem to be insignificant. Furthermore, it is not always a matter of simply taking easier or more generally preferred prey when it becomes available; the predators may subsequently switch back to the first prey type. Anthropomorphically, the predators sometimes seem as if they were bored and want a change.

Switching has sometimes been accounted for by proximate explanations. Some of the switching has been interpreted in terms of different appetitive behaviors (Chapter 2); that is, there may be different specific appetites (or drives) for different prey, and after one appetite has been satisfied, its threshold is raised and another type of prey becomes preferable. To say that predators switch because they have different appetites, however, runs into the circular logic of many drive concepts when appetite is inferred from the switch. A better understanding of what occurs inside the animal is needed.

Pygmy owls in the laboratory exhibit different tactics for hunting and killing small birds than for mice (Scherzinger 1970). The owls hunt birds by hiding in dense foliage and ambushing them or by dashing through vegetation. They kill the birds by grasping them with the feet and talons. The owls hunt mice, on the other hand, by watching from high, open perches. They kill mice by biting the head and neck with the beak. Relevant to our discussion of different drives, there may be separate neural mechanisms for the two types of prey, and at different times one or the other may predominate.

Switching also has been studied in the drinking behavior of laboratory rats (Morrison 1974). When offered more than one solution to drink, the rats concentrate on one, switch to another, and switch back again. The observations are consistent with (but do not prove) the interpretation that palatability declines with consumption until it falls below the palatability of an alternate solution.

Switching is not universal. It has already been noted that some predators stick to the same type of prey throughout their lives. In other species and on different occasions, some predators are reluctant to switch to different prey, sometimes even showing obvious fear of strange potential prey. It is clear that we are far from understanding the real causes, proximate or ultimate, of prey switching phenomena.

Search Image and Associated Concepts

This discussion of the identification and choice of food items is closely related to the concept of *search image*, first proposed by von Uexküll (1934, also see Schiller 1957) and elaborated by Tinbergen (1960). According to this view, animals learn certain stimulus characteristics of food they encounter, which improves their chances of finding similar items in the future. It also involves a process of selective attention or perceptual change, by which an animal searches for a particular object. The idea is supported by the apparent nonrandom choice of prey items (i.e., not in statistical proportion to their availability), the fact that animals may not take a particular prey item although it is available but then suddenly switch to it, and the further observation that a prey is ignored or selected irrespective of its position in the habitat (e.g., part of a tree), its novelty (some new items may be taken readily), or its conspicuousness. In fact, it is with respect to the last point that search image may be most important; it may permit a predator to find items easily and efficiently that otherwise are difficult to detect. Search images seem to be less valuable with obvious prey that require less searching. Humans searching for relevant items among other meaningless material, such as for particular words or phrases on a printed page, report that the relevant items stand out, whereas the other material is blurred (Neisser 1966). Human hunters, birders, nature photographers, and wildlife researchers know that experience permits one to spot distant or hidden animals to a degree that can surprise even the observer. In fact, they are sometimes likely to mistake sticks or pieces of vegetation for camouflaged animals, instead of vice versa.

The issue of selective perception raises the topic of internal neural control and integration (Chapters 16 to 21). The search-image concept is attractive, but there are several problems with it, among which are confusions with other closely related ideas. Krebs (1973) discussed search image relative to different types of learning: how useful the term is depends on the forms of learning with which it is associated. Phrases synonymous or nearly synonymous with search image include learning to see, shifts of attention, sharpened peaks of generalization gradient, and others.

Holling (1959), considering predation from an ecological standpoint, introduced the concept of *functional responses* in predators (Figure 9-4). Functional response refers to a change in the number of prey taken per predator. When functional responses are combined with numerical responses (change in the number of predators present), one gets the total predatory response to a prey population. Some responses in the presence of alternate prey are consistent with the notion of search image; that is, with an increase of prey density there is little increase in predator attention to that prey until suddenly there is a rapid increase in the numbers of

Figure 9-4 Predation responses to increases in numbers of density of prey animals. **A,** Numerical responses as shown by two genera of shrews (*Sorex* and *Blarina*) and deer mice (*Peromyscus*) to changes in sawfly cocoon density. With increased numbers of cocoons there were increased numbers of *Sorex* and *Peromyscus*, up to a point, preying on them. *Blarina* showed little numerical response. **B,** Functional responses. These refer to the numbers of prey taken by individual predators. In this case, *Blarina* responded the most, up to a point, and *Sorex* the least, just the opposite of the numerical response.
Modified from Holling, C.S. 1959. Canadian Entomologist)1:293-320.

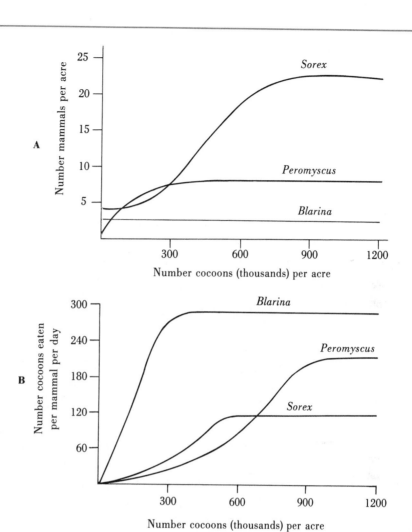

that prey taken per predator. With further increases in prey density the numbers taken per predator level off as the predators become satiated. The combined effects of acquiring the selection and then becoming satiated produce an S curve.

There are several variables that complicate or potentially detract from the concept of search image, and some persons (e.g., Guilford and Dawkins 1987) believe that the search-image hypothesis remains unproven. For example, similar numerical patterns may arise from other phenomena; one is the common phenomenon of *area-concentrated search*. Many species of prey animals live somewhat concentrated in clumps. As an evolved or learned response to this, many predators, encountering a prey item, change their pattern or speed of traveling and searching so that they concentrate their efforts in the same area (see section, "Search Paths"). Guilford and Dawkins (1987) refer to this as a *change in search rate*. The effect, which resembles that from a search image, is that the predator stays in the vicinity of a higher concentration of prey, encounters them at a higher probability, and takes more than would be expected from their average density over a larger area. A closely related concept is *profitability of hunting*. According to this idea, which is based on the principles of OFT, a predator should concentrate only on those areas that produce an optimal return of resources for the effort. When the return drops in a particular patch of habitat, the animal should go elsewhere, perhaps to other similar patches (see section, "Optimal Patch Choice"). This could also produce an effect similar to search image; that is, the predator is not searching the environment randomly but is concentrating on those areas that are most productive, thus increasing its probability of encountering some prey relative to others.

There are additional problems in interpreting search image. Social facilitation, whereby predators take what they see others taking, somewhat like a fad among humans, can lead to a nonrandom selection of prey items. There may be a training bias in which animals simply choose what they are familiar with from previous experience. The predators might be able to detect and find all items equally well but stick with the familiar, again producing results resembling searching based on a search image. (This would amount to a resistance to switching, however, and would not be expected to produce an S-shaped functional response curve.)

Fear of new items (*neophobia*) is another phenomenon that could resemble search image by restricting what an animal preys on. Many predators refuse to take some potential prey apparently because they literally are afraid of them. Eagles and many other large predators, unless extremely hungry, may show explicit signs of fear and distress, including vocalizations and attempts to escape or withdraw, when exposed to unfamiliar but otherwise suitable potential prey species. Curio (1976 and references contained therein) cites numerous examples of predators, including owls and mustelids, that have been observed to give alarm calls and respond with panic at the sight of novel prey. In such cases the predator gradually may have to become familiar with new prey by scavenging on parts of or whole carcasses of dead animals; even these may be approached with caution and hesitation. (Other predators, such as some of the small *Accipiter* hawks, on the other hand, may tackle novel prey items seemingly with abandon and carelessness, including larger and potentially dangerous prey such as various gulls and rooster pheasants.)

Finally, one must consider negative experiences that predators have with some types of prey. Bad-tasting prey may cause taste aversion, and injury from dangerous prey may cause aversive conditioning (Chapter 20), both of which may lead to the avoidance of prey that otherwise would be pursued. This may shift the predator's concentration to other forms of available prey and again may produce an end result that resembles search image. The opposite may happen if a hungry predator eats a harmless mimic of such an aversive prey, then switches to eating ones it has learned are sometimes acceptable.

The whole topic of search image and related concepts deserves much more study and clarification. Some persons continue to use the concept of search image, at least in restricted ways (e.g., Alcock 1973, Pietrewicz and Kamil 1979, Lawrence 1985 a, b). It appears that the search-image concept is not dead, but needs a thorough examination in the light of optimality theory.

Prey Selection: Are Only the Weak, Old, or Young Taken?

In the preceding sections the problem of prey recognition, that is, how predators find and recognize potential prey, is discussed. The associated concepts imply that predators are looking for common cues or animals that look (or sound or smell) alike. Now we discuss a different aspect of predation—how particular individual items of a given prey type are chosen. One proposal is that predators select odd or different prey. In the case of prey selection, as opposed to prey recognition, a predator has found so many prey that it is faced with the problem of choosing which to try to take.

Alternately, there may not be several prey present simultaneously from which to choose, but over a period of time the predator may encounter several potential prey of the same species. Potential prey may vary in their ease of capture or likelihood of fighting back and inflicting injury. In this case, it may be advantageous for the predator to somehow discriminate between those that are worth the effort and those that are not. Can predators make this distinction, and do they?

One of the stock answers in popular biology is that predators take only the sick and unfit, thereby preserving the quality of the prey and maintaining the balance of nature. Dawkins (1982) calls this the *BBC theorem*, since it appears so often in televised nature documentaries. It is unlikely that predators are acting in the interests of the prey or some abstract general notion about nature, even de facto. But their own interests in not wasting effort or avoiding injury would produce similar effects. There is some evidence that, at least in some species and under some circumstances, predators definitely select odd or different prey (e.g., Mueller 1977). On the other hand, there are numerous instances in which predators appear to be taking from the prey population at random across the various age and sex classes (Pearson 1966). Three-spined sticklebacks preferred odd-colored prey (*Daphnia*) depending on the density of prey (Ohguchi 1981). They chose odd prey at high densities of prey and common prey at low densities. Prey selection in many predators is influenced by several factors, such as ease of capture, in addition to or instead of oddity per se. One cannot generalize that predators always take the sick, injured, and odd, only that sometimes they do. One factor that may come into play is whether a forager is an ambusher or a hunter: hunters may often have to make a choice, whereas ambushers take what comes along.

Optimal Patch Choice

In the problems of choosing optimal patches and deciding how long to stay in them, first considered by MacArthur and Pianka (1966), the time spent searching includes the time spent in patches plus the time spent traveling between patches. The profitability of a patch often changes with consumption; that is, an animal depletes the food in a patch until there comes a point where it becomes more profitable to leave that patch and go to another. Furthermore, different patches initially vary in quality. Animals should search within a patch until it is no longer profitable to do so, then travel as quickly and directly as possible to the next patch. The OFT statement of what a forager should do has been called the *marginal-value theorem*: to maximize the rate of gain, leave a patch when the rate of gain becomes marginal, that is, when it becomes equal to (or less than) the long-term average for that type of habitat.

Charnov (1976) described the basic mathematical aspects and provided some simple illustrations (Figures 9-5 and 9-6). The foraging patterns of most species investigated so far have conformed to the predictions. Bumblebees, for example, travel about and spend time in the most profitable places (Figure 9-7). Different hypotheses for explaining how the animals determine profitability thresholds have been the subject of much recent research (e.g., Krebs et al 1974, Hodges 1981,

Figure 9-5 The marginal value theorem. The curve shows how energy intake rate declines as a patch is depleted. Foragers should leave a patch after a certain time (the giving up time, GUT) when the rate of energy intake falls below what a forager would expect if it went to a different patch. The GUT is determined in the graph as a line to the X axis, from the tangent point where a line from the average travel time to a new patch intercepts the energy intake curve. GUT is affected by the quality of the patch and by the travel time.
Modified from Charnov, E.L. 1976. Theoretical Population Biology 9:129-136.

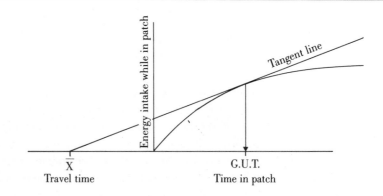

Figure 9-6 Optimal foraging among patches. **A,** Random distribution of two types of patches in a simplified illustrative model. A forager spends time in a given patch until the average rate of energy intake for that patch drops to or below the average rate that could be expected from the habitat as a whole; then the animal moves on to another patch. **B,** Time spent and energy intake in different environments should depend on the average quality of patches. In high-quality environments, foragers should leave patches earlier.
Modified from Charnov, E.L. 1976. Theoretical Population Biology 9:129-136.

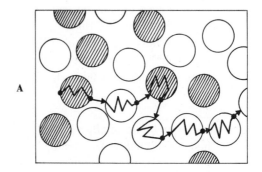

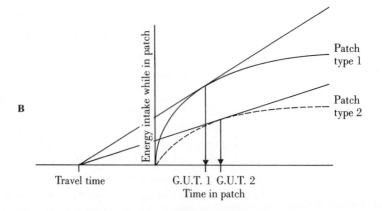

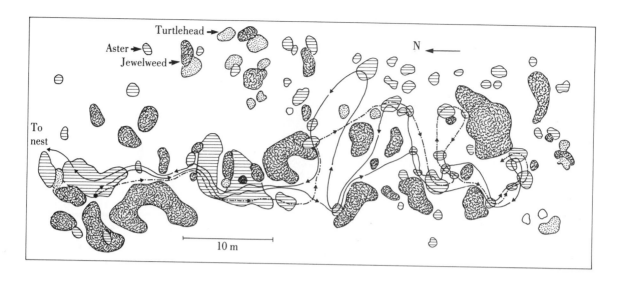

Figure 9-7   Diagram of actual foraging paths by an individual bumblebee worker (*Bombus fervidus*) in the same area on two different days (*solid* and *dashed lines*). The bee spent much time within each patch, then traveled directly and quickly to the next patch. (Many movements within patches are not detailed in this illustration.) This bee primarily visited aster, as did some other individuals, but other individuals specialized in other patch types, such as jewelweed. From Heinrich, B. 1979. Bumblebee economics, Harvard University Press, Cambridge, Mass.

Schmid-Hempel 1986, Kamil et al 1988). Arditi and Dacorogna (1988; also see review in Kacelnik and Bernstein 1988) extended the theory to situations where food resources vary in a more continuous fashion rather than just in discrete patches.

Aside from the time spent, cost depends on the foraging activities themselves. Depending on prey density, two different foraging methods may differ in their costs, as theoretically illustrated by Krebs (1978) (Figure 9-8). Osprey *(Pandion haliaetus)* search for prey using several different methods: watching while perched, while gliding or flying forward, and while hovering in a stationary position in the air. Hovering is the most costly method in terms of energy but overall leads to successful catches 50% more often than do dives from other kinds of flight (Grubb 1977). The difference varies, however, depending on weather and visibility conditions so that under some conditions it would be more or nearly as profitable to hunt via the less costly methods. Pyke (1981 a) has demonstrated expected differences in foraging methods by comparing the energetics and feeding of hummingbirds that hover and Australian honeyeaters (Meliphagidae), which perch. The costs of hovering increase faster as body size increases than do the costs of perching. Thus smaller birds would be expected to hover more than larger ones.

The cost of foraging may vary not only among methods but also with distance traveled, particularly if the animal returns to a central place after obtaining food (Orians and Pearson 1979). Kramer and Nowell (1980) showed that chipmunks *(Tamias striatus)* fill their pouches with different amounts of food with the optimal load size depending on the travel time from the central location. Choice of feeding site also appeared to depend on travel time, as expected.

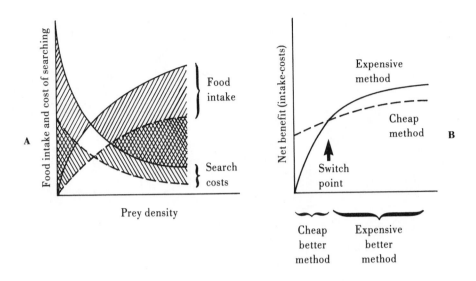

Figure 9-8 Methods of foraging depending on cost of method and density of prey. Theoretically it may be advantageous to use one method rather than another under different circumstances. **A,** The cross-hatched areas, similar to hourglasses on their sides, represent the net benefit, or intake minus cost. To the left of the break-even point the animal is spending more energy than it is taking in (negative benefit). When net benefit is positive, there may be situations in which a cheap method, such as walking, is better than an expensive method such as running. **B,** The cheap method may result in lower intake, but because the cost is much lower, the net benefit may be larger than for the expensive method. Depending on the shapes of the benefit curves, however, there may be a point where it is better to switch to the expensive method.
Modified from Krebs, J.R. 1978. Optimal foraging: decision rules for predators. In: Krebs, J.R., and N.B. Davies, editors. Behavioural ecology: an evolutionary approach. Blackwell Scientific Publications, London.

Search Paths

When in a patch, OFT predicts animals move in a pattern that maximizes the rates at which they encounter suitable food items. What that pattern is will be a function of the abundance and spacing pattern (evenly spread or clumped) of the food items.

Where food occurs in clumps, a commonly observed search pattern is area-restricted searching. In this pattern, foragers move in a quick direct manner out of areas where they are not encountering prey; by contrast, they search intensively in short circular paths in an area where prey encounter rates are high (Smith 1974). The difference between these two kinds of search paths can be quantified using a directionality index that varies in the value from 1.0 (long straight lines) to 0.0 (short endless circling) (Cody 1974). In switching from ranging (high-directionality movements) to intensive searching (low-directionality movements) foragers may change any or all of three aspects of their behavior. First, they may change from making only slight shifts in direction to making sharp turns. Second, they may switch from alternating left and right turns in succession to sequencing turns randomly or even stringing together a sequence of all left turns or all right turns. Finally, they may slow down their speed of movement. Any or all of these changes result in the forager remaining in an area where prey have recently been found. A number of animals, from birds to bumblebees, have been shown to exhibit such changes in search path (see Schmid-Hempel 1986 for examples).

Learning and Optimal Foraging

Optimal foraging theorists have not concerned themselves much with questions about the development of optimal foraging, whether it is innate, learned, or both. The case of shell dropping by northwestern crows, cited earlier, was interpreted by Zach (1978) as learned, whereas that in gulls is less flexible (but still presumably optimal) and perhaps innate (Tinbergen 1953). Partridge (1978) found that English blue tits and coal tits showed innate preferences for the habitats where they foraged most efficiently (oak versus pine trees, respectively). Even with these innate preferences, the birds foraged even more efficiently after a period of experience in that habitat. Through learning, animals can discover some unique ways to increase their food supplies. Examples include crows that have learned to drop nuts in front of automobile tires for cracking, birds taking insects from spider webs and the radiator grilles of cars, and woodchucks that learned to obtain aquatic vegetation by climbing trees that had fallen into a lake (Fraser 1979).

A more recent twist in the relationship between optimal foraging and learning concerns the role that information gathering for optimization may play in learning, rather than the other way around. Animals frequently have incomplete information for making proper choices about foraging (or other activities). How they deal with this problem may be related to a well-known (but poorly understood) phenomenon in animal learning known as *partial-reinforcement extinction effect*. If an animal is always reinforced for a particular response and then the reinforcement is stopped, the behavioral response also stops (or becomes extinct) rapidly. However, if the reinforcement is only partial and occasional, particularly if it is unpredictable, then the behavioral response is more persistent and disappears much more slowly after the reinforcement stops.

More generally, the advantages of optimal foraging may be an environmental selection factor favoring the evolution of learning abilities. Especially in variable environments, foragers with well-developed learning abilities may have an advantage over ones with less flexible tactics. Among the most remarkable memory feats of animals are those shown by animals that cache food: store it for later retrieval and use. Such caching behavior is discussed later in the chapter.

Stephens and Krebs (1986) explored the topic of optimal foraging and learning and reviewed the available references. The picture is just starting to emerge, and more information, discussion, and development can be expected in the near future.

Why Are Foragers Sometimes Not Optimal?

In spite of the promise that optimality theory offers, it has not been without problems. Some critics have argued that the theory amounts to circular reasoning, that only examples that fit the predictions have been set forth, and that there are too many ifs, ands, buts, and conditional statements. Examples of conditional statements can be found in many tests of optimality theory. Pyke (1981 b), for example, notes a significant difference between observed and predicted behavior and suggests the "difference could have been due to the birds' transient occupancy of the study area." In another paper Pyke (1981 c) comments, "These movement patterns are consistent with the expectations of optimal foraging theory only if the hummingbirds cannot or do not determine the directions of possible inflorescence and if they cannot assess independently the sizes and distances of possible inflorescence." Kushlan (1978) demonstrated nonoptimal, nonrigorous foraging in great egrets (*Casmerodius albus*) wherein the birds used less profitable foraging methods over more profitable ones. Jaeger et al (1981) described a number of factors involved in whether or not red-backed salamanders *(Plethodon cinereus)* foraged optimally. Some tests of optimal diets have been clearly negative (Emlen and Emlen 1975).

Many of the questions about the role of optimal foraging involve differing interpretations of data that are not yet sufficient to resolve the issue. A good example involves continuing discussion on why hummingbirds hover. Hovering while feeding always costs more energy than perching during feeding, but if hovering results in a higher net rate of energy gain (that is, if it pays off proportionally more than perching because of time saved), then hovering would be worth the added cost and should be selected (e.g., Pyke 1981). Miller (1985), however, proposed that hovering resulted instead because the structure of many flowers does not permit birds to perch while feeding and thus they were forced to hover to obtain the nectar, irrespective of optimality considerations. In essence the availability of perches would be a constraint.

Hainsworth (1986) suggested that the distance traveled to the feeding site also needed to be considered. He provided numbers and calculations as follows:

> The power for forward flight for a 3.0-g hummingbird is about 0.637 W, the power for hovering is about 0.75 W (Wolf and Hainsworth 1971), while the power for perching is about 0.157 W (Hainsworth and Wolf 1978). Let a hummingbird obtain 62.8 J of energy from a feeder on a visit and assume it takes 1.0 s at the feeder if it hovers but 1.10 s if it perches. If the 3.0-g hummingbird flies 20 m round-trip from perch to feeder and back at a flight speed of 2 m/s, the total time if it perches is 11.1 s, the total cost if it perches is 6.545 J, and the rate of net energy gain is 56.25/11.1 = 5.07 J/s. If the hummingbird hovers, the total time is 11.0 s, the total cost is 7.122 J, and the rate of net energy gain is 55.68/11.0 = 5.06 J/s. For this case perching is marginally more efficient. Even if the difference is not detected by the birds, they may adopt the least energy-demanding behavior and perch instead of hover.
>
> Now consider the case where distance is shorter. If a round-trip distance is 4 m, the total cost with perching is 1.447 J and the rate of net energy gain is 61.35/3.1 = 19.79 J/s. The total cost with hovering is 2.024 J, and the rate of net energy gain is 60.78/3 = 20.26 J/s. Here, hovering is the more effective behavior.

Miller (1986) replied to Hainsworth on several points: (1) questioning energy values obtained from laboratory conditions, (2) questioning whether hummingbirds in a natural, variable environment can detect the slight (e.g., 0.47 J or 2.3%) difference, (3) emphasizing other data showing that hummingbirds perch to feed if a perch is available, and (4) restating his point that constraints such as floral architecture still need to be considered, whether or not hummingbirds forage optimally otherwise. (Plants may have evolved to make hummingbirds forage in certain ways for the plants' benefits, for example, to ensure efficient pollination.)

The main point from this example for our purposes is that optimality theory is the subject of many continuing studies and discussions and involves a number of unresolved issues. Hainsworth's suggested experiment could validate or falsify the optimal foraging model for this particular situation. However, even after this case is settled, there are still many others like it waiting to be resolved.

Another point from the hummingbird example, raised by Miller (1986), but that needs to be considered more generally is just how well we can measure or estimate the actual costs and benefits of foraging behavior. Bennet (1986) stresses this problem in his conclusion to a paper on measuring behavioral energetics:

> We cannot trust our intuitive sense of how expensive a behavior might be. Some activities are surprisingly costly. Who, for example, would have anticipated that a

calling frog raises its metabolic rate 400 to 2400% above resting levels (Bucher et al 1982; Taigen and Wells 1985)? Conversely, other expenses may seem low.... It is simply not acceptable to assign metabolic costs by sheer guesswork ... Additionally, the source of metabolic support may be poorly anticipated. The avoidance of anaerobic metabolism by diving animals (Seymour, 1982) and its use by calling frogs (Pough and Gatten, 1984) are both surprising findings. The study of behavioral energetics must be an empirical exercise until we know considerably more than we presently do.

Many of the apparent failures of OFT, however, clearly are not the result of insufficient data or inaccurate measurements but involve situations where the initial, simple models are obviously wrong. It is profitable to look at some reasons for these failures of OFT models, after which we can reassess the value of the optimality approach.

1. Animals may not forage optimally in terms of maximizing E/t, because high-calorie foods may lack other necessary nutrients. An animal may have to eat a variety of foods that ensures a balanced set of nutrients but lowers the rate of energy intake. For example, Belovsky (1978) found moose gained energy at higher rates when eating terrestial plants but had to spend a certain amount of time eating water plants to meet minimum sodium requirements.

Note that moose are herbivores. Herbivores face more difficult challenges than carnivores in many ways. Much of the organic matter in vegetation is cellulose, which is indigestible for most animals. Worse, plants have evolved a tremendous array of secondary plant compounds, antiherbivore chemical defenses. Different species have their own specific chemicals, some of which are directly toxic and others (such as the tannins of oak leaves) that lower the food value of the rest of the plants' organic compounds. Plants may be imbalanced in amino acids or other specific nutrients. (For example, remember that humans can synthesize only 12 of the required 20 amino acids, and must obtain the other 8 in food. Meat eaters get them in an uncomplicated manner; vegetarians must be sure to eat a variety of plant foods, for example, mixing rice, low in lysine and isoleucine, with beans, low in tryptophan.) Generalist herbivores may be able to tolerate somewhat a variety of toxins but may not be able to take too much of any single kind; specialist herbivores may have evolved virtually complete tolerance for the compounds of their food plant but have a harder job searching. All of these considerations mean that an optimal diet in the fullest sense may not be the one that maximizes E/t for a herbivore.

2. Foragers may appear to be less than optimal to an observer because an observer has a false idea of what the animals' foraging behavior has actually been selected to do. Pyke (1979), for example, found that golden-winged sunbirds did not forage for nectar in flowers in a way that would maximize energy intake but behaved exactly as predicted if what they were doing was minimizing the cost of foraging.

Caraco et al (1980 a) found that yellow-eyed juncos were risk-sensitive foragers whose major objective was to minimize the risk of starvation. If given experimentally a choice of conditions of equal average returns, but one safe and steady and the other boom and bust, the juncos preferred the first. An example would be situation A that provides two food items every unit of time, versus situation B that provides four items 50% of the time but no items 50% of the time. Different species may differ in their ability to tolerate dry spells, making them more or less risk-sensitive (small warm-blooded animals such as shrews or small birds may be especially risk-sensitive). The concept of risk-sensitive foraging is reviewed by Stephens and Krebs (1986) and Real and Caraco (1986).

3. The word *risk* introduces another reason foragers may not be maximizing E/t: they are themselves vulnerable to predation. A number of studies show that the presence of predators alters the way foragers behave: sticklebacks attack *Daphnia* in lower-density areas when they have to watch for a kingfisher (Milinski and Heller 1978), and food selection by juvenile salmon is impaired in the presence of predators (Metcalfe et al 1988).

4. Foragers do not reproduce as a result of foraging alone: optimal foraging may be sacrificed so that a more generally optimal balance of different behaviors is achieved. Great tits combine foraging and territorial defense in the best possible way (Kacelnik 1979); backswimmer bugs (*Notonecta hoffmanni*) balance prey pursuit underwater with trips to the surface to obtain oxygen.

5. An important drawback in OFT models is that they consider only the individual forager and its food, without considering the presence of other competing foragers. But sticklebacks, for example, behave differently when foraging in the presence of other sticklebacks than when foraging alone (Milinski 1988). A particular influence may be an animal's social status. Dominant animals may be able to forage optimally, whereas subordinates may have to forage in a way that allows them to keep an eye out for predators (see examples in Chapter 13) or for harassing dominants (as with juncos, Caraco et al 1980 b). In such social competition situations, game theory (see Chapter 13) may be a better way to model behavior than is optimality theory.

6. A naive OFT model assumes that an animal has perfect information about its environment (knows the average foraging rate in a patch and the average travel time to new patches, for example). But foragers are not omniscient: to acquire information requires sampling, and to track a changing situation requires continual sampling. Some lack of optimality may be a necessary cost of acquiring the information a forager needs so as to do the best it can. Studies by Smith and Sweatman (1974) confirm that great tits track their environment continually and can make quick switches in foraging behavior to deal with changed circumstances.

7. Finally, we always have to remember that the information animals have is limited by their nervous system capabilities. Unlike OFT workers using sophisticated mathematics, natural selection is likely to have endowed animals with generally effective, but perhaps not quite optimal, rules of thumb. To take a nonforaging example, Dawkins and Brockmann (1980) initially believed that great-golden digger wasps (*Sphex ichneumoneus*) were making a serious behavioral mistake in fighting over jointly provisioned nests on the basis of how much prey each wasp itself had added rather than the total amount of prey present. However, a different model of their behavior based on a more realistic view of what the wasps could be expected to sense fit the observed results well.

In addition to these specific points about optimal foraging, the list of reasons that organisms are imperfectly adapted generally also applies here, as outlined in Dawkins (1982). For example, we cannot expect animals to behave optimally in an unnatural situation to which they have not evolved responses. Inglis and Ferguson (1986) found that starlings would continue to use normal but expensive foraging techniques, for example digging for hidden mealworms, even when visible, unhidden mealworms were provided. Abundant, concentrated, helpless prey are rare in nature, where foragers usually have to work hard.

Do these examples mean that OFT is a failure? Not at all; to think that would be to misunderstand what OFT is intended to do. Its proponents do not assume that animals are perfect or that they are failures if OFT models do not fit observed behavior. OFT models are really meant to do two things. First, they force students of foraging to examine closely the energy costs and benefits of different food items and of different foraging tactics. Second, when a model fails, one assumes the fault is in the model, not the animals, and one is led to look for the constraints of optimal foraging that caused the model to fail. The search for these constraints may well give important information of the seven kinds just considered above. Even if the first model fails, understanding advances. As Stenseth (1988) remarked, "…optimal foraging is emerging from a troubled adolescence into an uncertain future," but like all good scientific approaches, its ultimate downfall (if that occurs) is likely to be due to the information that was discovered because of its use.

CACHING BEHAVIOR

Many species of both invertebrates and vertebrates store or cache food items. Numerous descriptions and interpretations of storing behavior exist in the literature, for example, Anderson and Krebs (1978) and Balda (1980). Recently caching behavior has been subjected to contemporary interpretations.

Storing food outside their own bodies reduces the need for animals to carry around extra weight, establishes a store of resources for times of scarcity, hides resources from competitors, and may release the animal from foraging behavior so it has more time available for other activities. Male bowerbirds *(Amblyornis macgregoriae)*, for example, store fruit in nearby vegetation during the season when they are maintaining and displaying in bower areas (Pruett-Jones and Pruett-Jones 1985; see discussion in Chapter 11). The males do not use the stored food as part of the bower decorations, and they do not store food at other times of the year. Similarly, females do not store fruit, and immature males store little. The Pruett-Joneses (1985) suggest that the fruit storage permits more time for interacting with females and reduces vulnerability of bowers to marauding by other males.

Caches, however, pose risks of loss to thievery, spoilage, and forgetfulness. Caches require an ability on the part of the owner to relocate them and retrieve the stored food. Most studies of cache retrieval have shown that they are relocated on the basis of spatial memory rather than trial and error or some sensory cues emanating from the location (Bunch and Tomback 1986, Balda et al 1986, and McQuade et al 1986). The capacity of some animals, particularly birds, for remembering storage locations numbering in the hundreds or even thousands is phenomenal.

Morphological and behavioral specializations for caching behavior vary among even closely related species. Among corvids (the crow family), for example, Clark's nutcracker *(Nucifraga columbiana)* is a highly specialized seed storer with a specialized structure for carrying seeds and specific ways and places where they are stored, whereas the scrub jay *(Aphelocoma coerulescens)* lacks such morphological and behavioral specialization but shows a more general form of caching behavior (Vander Wall and Balda 1981). The relationship of food storing to learning is just beginning to be explored. This, as well as references to many additional examples, is discussed by Shettleworth (1984).

Optimal scattering of food storage sites may involve numerous small caches to balance losses to other competitors from having too much food per site against the cost of establishing and remembering too many sites. Stapanian and Smith (1978) studied these factors in fox squirrels storing black walnuts and found that observed values were consistent with predicted optimal values based on losses of experimentally buried nuts. They also noted correlations between cache and tree distributions, which suggested coevolutionary relationships between nut burying by the squirrels and nut production by the trees.

PREDATOR AND ANTIPREDATOR BEHAVIOR

Predation and the avoidance of predation strongly influence the behavior of almost all animals (Figures 9-9 and 9-10). Predator behavior includes the capture and overcoming of the prey by using a variety of techniques, depending on the species involved, such as pursuit, ambushing, trapping, and various forms of trickery. Characteristic prey behavior may include means of avoiding or escaping from predators, hiding and camouflage, and various ways of being undesirable, hence left alone. Speed and alertness are seen in both predator and prey and are believed to be direct evolutionary consequences of predation. These involve high levels of muscular coordination, keen senses, and sophisticated nervous systems.

Such sophisticated sensory, neural, and defensive systems amd complex behavioral tactics appear to result from the evolutionary conflict among different species. If there were no bats, it is unlikely that moths would have bat detectors. Gradual improvements in biological machinery and tactics are the very earmarks of evolution. Predation is responsible for a large proportion of these changes. Predators and their potential prey, in keeping up with or ahead of each other, are slowly but constantly improving in their abilities. Those that do not cannot survive and reproduce. The predator-prey evolutionary phenomenon has even been compared with the spiraling arms races of human military powers, where an advance on one side sets up strong pressure for an evolutionary counter-adaptation by the other (Dawkins and Krebs 1979).

The effects of predator-prey coevolution on animal behavior and morphology are

Figure 9-9 Leopard attacking a baboon. Photographed at the moment just before contact. The leopard is braking itself and the baboon responds defensively.
Photograph by John Dominis, Life Magazine © 1965 Time.

Figure 9-10   A sampling of predators with prey. **A,** Jumping spider with fly. **B,** Sharp-shinned hawk with starling. **C,** Bobcat with rabbit. **D,** Mink with mouse.
A courtesy Lyn Forster. See also Forster (1982). **B** to **D** courtesy Tom Brakefield.

so pervasive that it even has been theorized to be one of the main factors in, for example, the evolution of the vertebrates. Many evolutionary expansions and radiations of new forms have come after animals evolved new weaponry (such as jaws rather than round mouths) or new guidance and movement systems (such as balance organs, streamlined bodies, rayed fins that permit fine control of swimming).

At this point we can leave the more academic aspects of predation and appreciate the evolutionary outcome: an incredibly rich diversity of hunting and antipredator behaviors. Most readers may be familiar with many of these behaviors; our purpose is to bring many together in one place and remind you of the diversity that millions of years of coevolution has produced.

Hunting and Capture Methods Of Predators

Listed below are some of the hunting and capture methods used by predators. These may be used together in various combinations. Many predators possess a diverse repertoire of predatory behaviors and use different hunting methods, some of them situation specific, at different times.

1. **Groping and flushing**. Many predators, such as octopuses, various arthropods, and other animals using their feet or feelers, work their way through the environment groping and feeling for possible prey. One recalls the familiar picture of a raccoon feeling around under water for prey items. In some cases the prey is caught as quickly as it is encountered, and in other cases, it is flushed and then chased. Many bird and mammal predators may crash through bushes or tall grass, flushing victims that are then chased and caught.

2. **Stalking and ambushing**. Mantids and true chameleons are familiar examples of stalking and ambushing, as are many of the cats. Some predators use other tactics in conjunction with stalking, and various fish wiggle bait or items that attract the curiosity of unsuspecting prey (Figure 9-11). Some are aggressive mimics (i.e., "wolves" in "sheep's" clothing). Some mantids, for example, resemble harmless leaves or even flowers. In a complicated, three-way, ant-aphid predator relationship a predacious lacewing larva attaches aphid "wool" to its body so that it resembles the aphids (Figure 9-12). The larvae, masquerading as aphids, escape detection by the guarding ants and proceed to feed on the aphids. (Ant-aphid relationships are discussed further in Chapter 10).

Ambushers, such as rattlesnakes, often will not pursue potential prey even if provided the opportunity, unlike other species, such as cobras,

Figure 9-11 Anglerfish of the genus *Antennarium*, in luring posture, displays a bait that bears a remarkable resemblance to a small fish. When a prey animal is attracted close to the lure, the anglerfish jumps forward and snaps it up. From Pietsch, T.W., and D.B. Grobecker. 1978. Science 201.

Figure 9-12 Woolly aphids and predacious lacewing larvae—wolf in sheep's clothing. **A,** Typical habitat of woolly alder aphid (*Prociphilus tesselatus*). An aphid colony appears white on branch of alder bush in foreground. **B,** Close-up of part of an aphid colony with a larva of *Chrysopa slossonae* (*arrow*). **C,** *Chrysopa* larva in its normal, wax-covered (shielded) condition. **D,** Ant protecting aphids by biting an attacking finger. **E,** Ant imbibing a droplet of honeydew delivered by an aphid. **F,** Ant biting a shielded *Chrysopa* larva that was released in its vicinity. **G,** Ant biting a denuded larva that it has just detected. **H,** Denuded larva applying plucked wax to its rump with the head. From Eisner, T., et al. 1978. Science 199:790-794.

Figure 9-12 For legend see opposite page.

actively chase and catch their prey (Radcliffe et al 1986). Rattlesnakes and other ambushers, however, may move near prey animals not in active pursuit but to create a context from which to ambush (Hennessy and Owings 1988).

3. **Chase and pursuit**. This is done from a distance.

4. **Interception of flight path**. This is a variation on outright chase and pursuit where a predator anticipates a flight path and moves at an angle or by a different route so that it meets the prey at a particular point.

5. **Exhaustion of prey**. Some predators will not outrun or outfly their prey but simply stay with them and cause them to continue moving or otherwise harass them until the victim is finally exhausted. This technique is used not only in cases where the prey may be dangerous but also where it has other effective short-term defenses.

6. **Tool use to get prey**. The most obvious users of tools are woodpecker finches and chimpanzees, both use twigs or stems from plants to pry out certain prey items from crevices. Some vultures use rocks to break into eggs. One can consider other structures external to an animal's own body as tools when they aid prey capture. Classic in this category would be the webs of spiders. A few other species also use webs or sticky substances, and some, such as the ant lion, build pits or other kinds of traps.

7. **Communal hunting**. Many predators hunt socially and use their joint efforts to overpower, confuse, or exhaust prey, or to communally round up their victims. Familiar examples are wolves, several other species of canids, killer whales, brown pelicans, and communal spiders. Communal hunting has been shown to increase the numerical success ratio of prey capture, and it is particularly advantageous in dealing with large victims that could not be handled by individual predators alone. Curio (1976, modified from Schaller 1972) tabulated the increase in success rate of lions with increased numbers of animals hunting together (Table 9-1). Other advantages of group hunting, as compiled by Curio, include more economical consumption of carcasses, consumption with less interference from other competitors, improved ability to feed young with large carcasses (often by carrying part back), easier stalking of herds without stampeding, and sharing of food sources. The foraging advantages of social life are considered at greater length in Chapters 13 and 14.

Table 9-1 The relationship of hunting success to the number of lions stalking or running

No. of Animals Hunting	Thomson's Gazelle		Wildebeest and Zebra		Other Prey		Total	
	No. Hunts	Success (%)	No. Hunts	Success (%)	No. Hunts	Success (%)	No. Hunts	Success (%)
1	185	15	33	15	31	19	249	15
2	78	31	17	35	11	9	106	29
3	42	33	16	12.5	5	20	63	37
4 to 5	42	31	16	37	4	25	62	32
6+	15	4.1	21	43	7	0	43	33
TOTAL	362		103		58		523	

From Schaller, G.B. 1972. The Serengeti lion, University of Chicago Press, Chicago.

Table 9-2 Examples of hunting success of various predators*

Predator	Prey	No. Attempts	Successful (%)	Comments	Reference†
Woodruffia metabolica (Holotricha)	*Paramecium* sp.	Many	14	2.2%/encounter; attempt = dilate mouth to engulf prey	Salt (1967)
Busycon carica (Gastropoda)	*Venus mercenaria*	26	58		Carriker (1951)
Cuttlefish	Shrimp	?	90		Messenger (1968)
Largemouth bass	Fish	85	94	Prey fishes without evasive movements	Nyberg (1971)
Forster's tern (*Sterna forsteri*)	Fish, four species	1538	24		Salt and Willard (1971)
American kestrel	Rodents (insects?)	?	33	On familiar hunting ground	Sparrowe (1972)
Osprey	Fish	469	80 to 96	Both dives and snatches from surface	Lambert (1934)
Various raptors (*Falco columbarious, F. peregrinus, Accipiter nisus, Haliaetus albicilla*)	Birds	688 (60 to 260)	7.6 (4.5 to 10.8)	On migration	Rudebeck (1950, 1951)
Black bear	Salmon	1481	38.6	During 310 fishing sequences	Frame (1974)
Wolf	Moose	77	7.8	From all moose tested	Mech (1970)
Spotted hyena	Wildebeest calf	108	32	Similar for wildebeest adult, Thomson's gazelle, zebra	Kruuk (1972 b)
Puma	Deer, elk	45	82	Excluding aborted hunts	Hornocker (1970)
Cheetah	Thomson's gazelle	87	70	Only fast chases tallied	Schaller (1972)
Chimpanzee	Mammals, six species	95	40	Including primates	Teleki (1973)
	Olive baboon (*Papio anubis*)	18	36	Adolescent victims <2 years	Teleki (1973)

From Curio, E. 1976. The ethology of predation. Springer-Verlag, New York.
*Additional examples are provided by Schaller 1972: Appendix B.
†See Curio (1976) for references.

Table 9-3 Hunting success in relation to type of prey

Predator	Prey	No. of Attempts	Successful (%)	Comments	Reference*
Hierodula crassa (Mantidae)	Fly walking	898	63.0	70.2% strikes/contact	Holling (1966)
	Fly flying	112	13.4	33.1% strikes/contact	
Herring larva, 35 to 42 days	*Artemia nauplii*	81	100		Rosenthal (1969 b)
	Larger plankton, *A. metanauplii*	303	96.5		
Red fox	Rodents	58†	25 to 100	Depending on snow conditions	Palm (1970)
	Larger prey‡	9	0		
Wild dog	Thomson's gazelle <2 months	22	95	Mean = 70%, including other prey except zebra	Schaller (1972); see also Estes and Goddard (1967), H. and J. van Lawick-Goodall (1970)
	Thomson's gazelle >2 months	47	49		
Cheetah	Thomson's gazelle fawns	31	100		Schaller (1970),
	Thomson's gazelle adults	56	54		Schaller and Lowther (1969)

From Curio, E. 1976. The ethology of predation. Springer-Verlag, New York.
*See Curio (1976) for references.
†162 km tracked.
‡Roe deer *(Capreolus capreolus)*, European hare *(Lepus europaeus)*, red squirrel, birds.

The improved success of communal hunting raises the general question of success rates in different tactics of hunting. Success is difficult to quantify because it depends on many variables and subjective judgments on the part of the observer. Observing predation under natural conditions is difficult. One is not always sure even if an act of predation has been attempted, for assessment of whether or not it was successful. Nonetheless there are some crude estimates for different predators under different situations (Table 9-2). Table 9-3 compares success rates depending on the type of prey. As can be seen, success rates span the entire range, from 0% to 100%, with some under 19%, a few over 90%, and most somewhere in between. The range is so wide that it provides little useful information, particularly in view of all the variables and subjectivity that enter into the tallies. All that can be safely concluded is that the success rate for predators is rarely high and may be surprisingly low. Most of them miss their prey much of the time and must keep trying over and over. Many may fail too often and starve.

Handling and Killing Tactics

Once a predator has made physical contact with a prey and brought it down, the job often is not done; the victim has to be subdued and killed. There are four main techniques: (1) use of teeth, beak, or other mouthparts to bite or tear, (2) use of claws to puncture or tear into the animal, often in conjunction with squeezing, (3)

Figure 9-13 Corn snake killing a mouse. Constriction patterns vary among snakes. Some species coil tightly around the prey, whereas some use coils consisting of lateral twists and loops of the body. Different species use relatively stereotyped patterns of constriction. Photograph courtesy Tom Brakefield.

use of constriction, such as with body coils in snakes (Figure 9-13) (Greene and Burghardt 1978, Greenwald 1978) or otherwise smothering or suffocating the prey, which is perhaps just a variation of the preceding, and (4) use of poison or some form of stinging. The last category includes coelenterates, stingrays, venomous snakes of several types, spiders, cone snails, scorpions, shrews, and a few other animals with poisonous bites or stinging structures. A relatively rare method of immobilizing or killing prey involves the use of electrical shock, primarily in a few cartilaginous fish (e.g., electric and torpedo rays) and some electricity-producing bony fish (e.g., electric eels and electric catfish).

Antipredator Defense Tactics

The evolved diversity of predatory tactics is matched or perhaps even exceeded by the diversity of defense tactics and behaviors. As in the case of the predators' tactics, the following different techniques may be used in combination:

1. **Escape by fleeing and outrunning, outswimming, or outflying the predator.** This is one of the basic and more familiar tactics. It undoubtedly has led to much of the fleetness and agility seen in animals. In some species the intended victims literally may jet away, as in cephalopod mollusks and scallops, or they may twist away or jump or fly into a different habitat, such as diving into water or the air or climbing a tree.

2. **Advance warning systems.** Many prey species have exceptionally acute sensory systems and are able to detect the predator before it comes dangerously close. The bat detectors of noctuid moths are discussed in Chapter 1.

3. **Unpredictable movements.** These acts sometimes are called *protean displays*, as used by Curio (1976). (The term *protean*, from the Greek god Proteus who could change his appearance at will, has not always been used consistently in the literature; also see item 6 d.) Fleeing animals may zigzag, jump, turn, change direction of flight, and otherwise move in

an unpredictable fashion that is difficult for predators to follow success-fully. An example of this tactic is given in Chapter 1 as one technique by which moths evade bats.

4. **Hard armor.** Many prey species have shells or other forms of hard or tough outer covering or manage to block themselves behind a shield of some form.

5. **Spines.** Examples of spiny animals include porcupines, hedgehogs, por-cupine fish, the crown-of-thorns starfish, and numerous others.

6. **Employing offensive defenses.** There are many related tactics.

   a. **Fighting back.** Many prey species are extremely aggressive them-selves. Some insects and many rodents, for example, can inflict seri-ous bites and may turn and attack their attackers. Some salamanders are able to repel even shrews (Brodie 1978). Prey that are as large as or larger than their predators have equal or greater strength and can usually fight back.

   b. **Retaliation with shock or poisonous stings, bites, etc.**

   c. **Use of borrowed poison.** Some mollusks consume coelenterates and then use their poison as protection against their own predators. Mon-arch butterflies, in a more familiar example, incorporate poisonous glycosides into their bodies from their food plants, milkweeds (*Asce-pias* sp.).

   d. **Surprising the attacker.** Numerous species have different tricks for creating sudden visual displays such as suddenly exposing eyespots on the wings, or loud noises such as claps, vocalizations, and hisses. Some animals quickly change body posture, shape, and size or spread their wings or legs. (These startle responses also sometimes are called *protean displays*, since they involve a sudden change in appearance.)

   e. **Chemical defense.** The classic chemical user is the skunk, but many other animals, particularly some insects (Figure 9-14), fight back with noxious sprays and odors or are poisonous, as in the case of the mon-arch butterfly.

   f. **Aposematism or warning displays.** Many species, particularly those carrying chemical defenses, also possess conspicuous warning signals. These include the monarch butterfly's bright orange-and-black pattern, the skunk's black-and-white pattern, rattlesnakes' rattling noises, Arc-tiid moths (Chapter 1), and numerous others.

   g. **Mertensian mimicry.** The prey may reverse the tactic and resemble something dangerous to the predator—sort of a sheep in wolf's cloth-ing. Some snakes, for example, may shake their tails, hiss, flatten their necks, and engage in striking or more subtle movements that re-semble those of other, dangerous species. Burrowing owls make a hissing sound that resembles rattlesnake rattling (Rowe et al 1986). Some flies resemble more dangerous bees. The dangerous model may be so lethal that, if the real thing were encountered, the victim would not have a second chance. This form of mimicry receives its name from Mertens (1956), who participated in an ongoing controversy over mimicry in coral snakes and whether the most poisonous forms were

Figure 9-14 Bombardier beetle defense. Bombardier beetles, of the Carabidae subfamily Paussidae, spray a hot, quinone-containing secretion at potential predators. The secretion is formed by mixing component reactive chemicals from two-chambered glands. It is aimed and sprayed with the aid of flanges on the body surface, which direct the spray toward its target. **A,** Beetle discharging on chemical indicator paper in response to stimulation of left foreleg and **B,** right hindleg. **C,** Tip of abdomen in lateral view of beetle immobilized while discharging toward a foreleg. **D,** Same view of a beetle immobilized while discharging toward a hindleg. Movable lip, lowered in D from its position in C, is shown in center of photograph. From Eisner, T., and D.J. Aneshansley. 1982. Science 218:84.

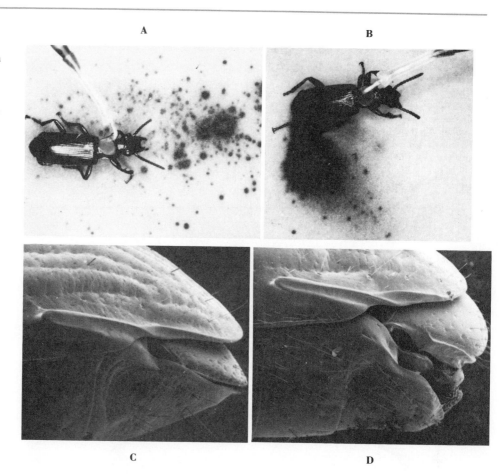

the models or the mimics (for a summary discussion of that problem, see Wickler 1968 and Greene and McDiarmid 1981). The terminology and classification of mimicry in general, incidentally, is unsettled. Balsbaugh (1988) discusses many different forms and categories in one family of beetles, the *Chrysomelidae*.

7. **Camouflage.**
   a. **Camouflage patterns and disruptive coloration.** In a few extreme cases the animals actively cover themselves (Figure 9-15).
   b. **Protective mimicry.** This is a camouflage involving resemblance to a specific object. Some species resemble flowers, leaves, dead objects, and even bird droppings. Mysterud and Dunker (1979) speculated that the ear tufts on owls (which are predators themselves) may serve to mimic the facial expressions of mammalian predators that they may encounter in their nesting environment (Figure 9-16). In other striking examples, juvenile lizards (*Eremias lugubris*) mimic the

Figure 9-15 Camouflage by decorator crabs. Decoration behavior and hooked setae, which help hold materials, are shown for *Oregonia gracilis*. The crabs manipulate pieces of vegetation and debris with their mouthparts and chelae and then attach them to various parts of the body. From Wicksten, M.K. 1980. Scientific American 242(2):146-154.

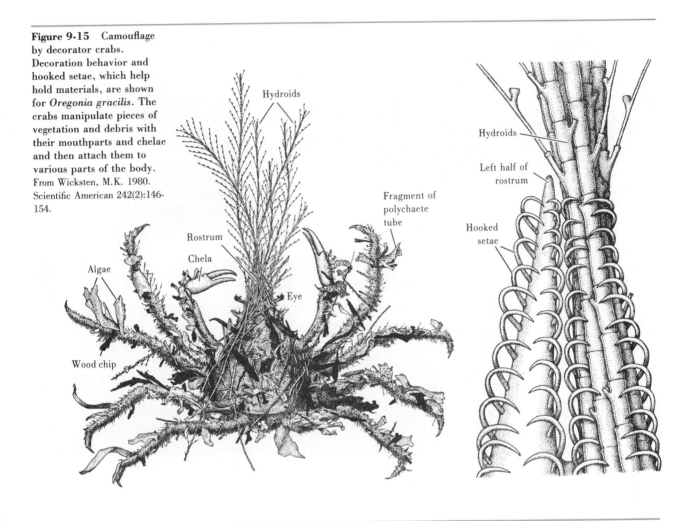

Figure 9-16 Owl facial mimicry of predators inhabiting their nesting areas. **A,** *Lynx lynx.* **B,** *Martes martes.* **C,** *Vulpes vulpes.* **D,** *Bubo bubo.* **E,** *Asio otus.* **F,** *Asio glammeus.* From Mysterud, I., and H. Dunker. 1979. Animal Behaviour 27:315-317.

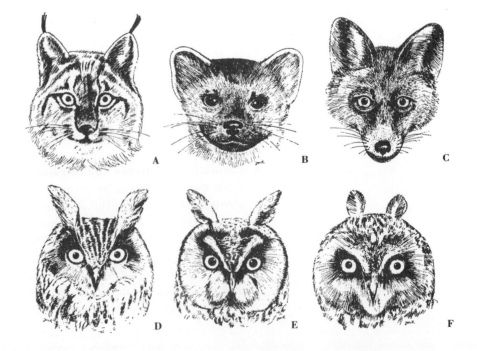

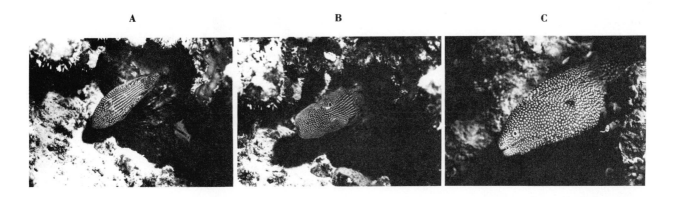

Figure 9-17   Plesiopid fish (*Calloplesiops altivelis*) mimicking a noxious moray eel
(*Gymnothorax meleagris*) when frightened by a predator. **A,** *Calloplesiops altivelis* (total length
approximately 15 cm). Normal posture. Note the ocellus at the posterior base of the dorsal fin.
B, Intimidation posture. The fish is now facing away from the camera. **C,** *Gymnothorax
meleagris* (total length approximately 1 m, head length approximately 15 cm.). Posture when
confronted by a diver.
From McCosker, J.E. 1977. Science 197:400-401.

**Figure 9-18   A, Front view
of zebra spider, *Salticus
scenicus* and B, posterior
view of its mimic, the
snowberry fly, *Rhagoletis
zephyria*.**
From Mather, M.H., and B.D.
Roitberg. 1987. Science
236:308-310.

pattern and posture of noxious beetles (Huey and Pianka 1977), a plesiopid reef fish when frightened resembles the head of a moray eel
(Figure 9-17), and a fly resembles a spider (Figure 9-18).

c. **Disorientation cues.** Many prey species have false heads or prominent markings located in the least vulnerable parts of the body, such
as at the tip of the tail, which may fool and misdirect attacks by the
predator. In one extreme case a butterfly has an inconspicuous body
with an entire false body marked on the tips of the wings; even the
shape of the wings points toward the false markings (Figure 9-19).

d. **Batesian and Mullerian mimicry.** Many potential prey animals are
edible but mimic inedible animals by carrying the same warning
colors and patterns. These are Batesian mimics. Some mimics among
insects are so realistic that they even have fooled entomologists and
have been misclassified in insect collections. Some flies and beetles,

Figure 9-19 False body markings on butterflies of the genus *Thecla*. The wing pattern displays a false head with striped pattern that draws attention away from the vulnerable true body and head.
Modified from Wickler, W. 1968. Mimicry in plants and animals. McGraw-Hill, New York.

False head and antennae

Wing stripes lead toward false head

Figure 9-20 Examples of Batesian mimicry among insects. Many harmless insect species resemble dangerous or noxious species, as indicated.
Modified from Wickler, W. 1968. Mimicry in plants and animals. McGraw-Hill, New York.

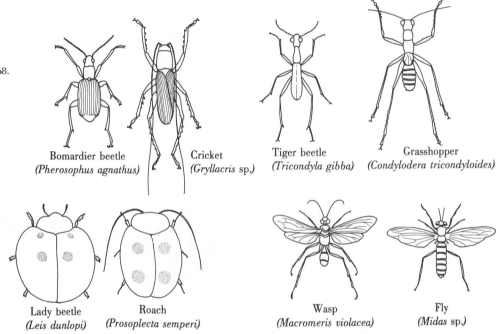

Bombardier beetle
(*Pherosophus agnathus*)

Cricket
(*Gryllacris* sp.)

Tiger beetle
(*Tricondyla gibba*)

Grasshopper
(*Condylodera tricondyloides*)

Lady beetle
(*Leis dunlopi*)

Roach
(*Prosoplecta semperi*)

Wasp
(*Macromeris violacea*)

Fly
(*Midas* sp.)

Figure 9-21 Generalization and discrimination of mimics by predators. Background patterns of different colors and shapes were used. Artificial prey—insect-shaped pastry either edible (soaked in water) or distasteful (soaked in quinine hydrochloride)—were placed on these backgrounds. Wild birds, mostly starlings, were permitted to take the food items in natural settings. Colors and patterns of backgrounds are shown along with mean number of items that were eaten. The noxious and perfect mimics were eaten least. Imperfect models were generalized and also partially avoided. The lack of complete avoidance, however, indicated that the birds were discriminating between perfect and imperfect mimics.
Modified from Morrell, G.M., and J.R.G. Turner. 1970. Behaviour 36:116-130.

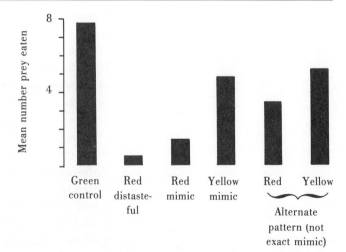

Mean number prey eaten

8

4

Green control

Red distaste-ful

Red mimic

Yellow mimic

Red

Yellow

Alternate pattern (not exact mimic)

for example, resemble various hymenopteran insects (Figure 9-20). Many of these tactics work as a result of learned taste aversions; the predator learns from killing or at least getting a bite from one animal to leave similar-appearing ones alone.

If the first animal encountered is an edible mimic, the predator does not acquire an aversion and may continue killing animals that look the same, perhaps including several more edible ones, until it encounters one or more poisonous forms. Thus for Batesian mimicry to be successful, predators should encounter a minimal proportion of the real thing; and in general, mimics cannot outnumber the inedible forms (or must appear later in a season than the inedible ones).

Several studies (e.g., reviewed in Morrell and Turner 1970) have shown that, at least among vertebrates, predators can learn to discriminate and generalize concerning mimics. These are two different aspects of recognition or abstraction. Discrimination is the ability to distinguish between items, and generalization is the perception of common characteristics (Figure 9-21). With further trial and error or increased hunger, predators may be able to discriminate better until only perfect mimics are avoided; that is, they can learn to distinguish only partial mimics (Figure 9-22).

Mullerian mimics are inedible themselves and share their appearance with other inedible species. This forms a sort of double protection by which two or more species carry similar warnings. Batesian and Mullerian mimicry technically differ from Mertensian mimicry in that they involve taste related edibility as opposed to other forms of prey defenses.

8. **Confusion effects.** A few species prevent attack by confusing the predator. Octopuses eject a cloud of black ink from which they escape while the predator cannot see. Geese in the presence of an attacking eagle may stay at the surface of the water and splash water wildly with their wings, which surprises and confuses the eagle.

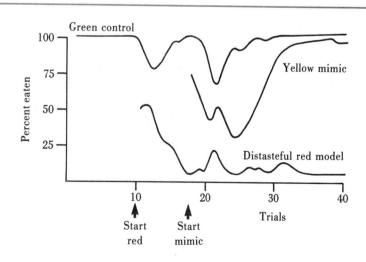

Figure 9-22 Sequence of bird choices of food items. The noxious red model and perfect red mimic were not eaten by the wild birds after a few brief trials in which the birds experienced the bad taste. The birds generalized and partially avoided yellow mimics but learned to discriminate and eventually accept the imperfect mimics. Control items were green.
Modified from Morrell, G.M., and J.R.G. Turner. 1970. Behaviour 36:116-130.

Figure 9-23 American toad hiding by burying itself under sand.
Photograph courtesy Tom Brakefield.

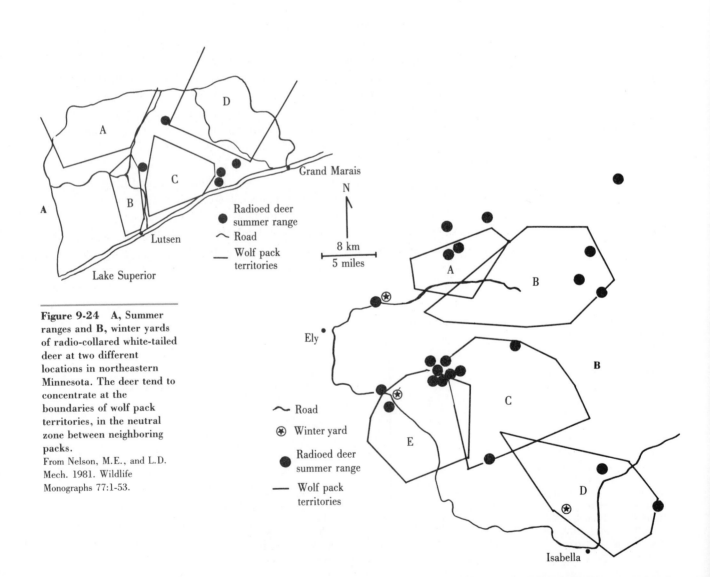

Figure 9-24 **A,** Summer ranges and **B,** winter yards of radio-collared white-tailed deer at two different locations in northeastern Minnesota. The deer tend to concentrate at the boundaries of wolf pack territories, in the neutral zone between neighboring packs.
From Nelson, M.E., and L.D. Mech. 1981. Wildlife Monographs 77:1-53.

9. **Hiding or seeking cover.** This is a well-known tactic that probably ranks with simple fleeing as a major means of protection. A large number of species try to get under or behind cover during an attack by a predator, and many stay in cover during vulnerable periods, such as while sleeping or when young are present. Many species bury themselves in the substrate (Figure 9-23). A few animals do not go into cover but actually go into the open when under attack. Prime examples are caribou during the winter in forested lake country. When surprised or frightened, as by a wolf pack, caribou run from the woods out onto the ice and snow of a frozen lake.

10. **Safety in numbers.** One of the most common defense measures is to group together with other individuals. If any are taken, there is a good chance that it will be another—the selfish herd principle (Chapter 13). But in many cases the simple presence of the group prevents further approach by the predator so that no individuals are lost. One of the best-known group-defense tactics is shown by musk-oxen, which form a tight circle with the young in the middle and their formidable, armored heads all facing out. Chapter 13 contains an extended discussion of the antipredation advantages of social life.

11. **Occupation of neutral zones between the territorial boundaries of predators.** This tactic is exemplified by deer living between the territorial boundaries of neighboring wolf packs (Figure 9-24). The wolves avoid these buffer zones between territories to avoid encounters, often fatal, with neighboring packs. The deer use these zones as refuges (Mech 1977, Nelson and Mech 1981).

PROXIMATE FACTORS OF PREDATION

Foraging is not always distinct from nonforaging behavior. A forager that is resting or engaged in some nonforaging activity may suddenly encounter an opportunity to catch something and take advantage of it. Or a forager that is roaming about, perhaps returning home from a visit to a territorial boundary, may begin to search for something to eat. Likewise, serious foraging may turn into a halfhearted effort or the individual may be interrupted or distracted, such as by its mate. Some female arthropods (e.g., mantids and spiders) mate and eat their partner at the same time.

Much of this chapter considers ultimate explanations of foraging behavior, but it is also possible to consider some of the proximate factors involved. The most obvious is hunger on the part of the forager. Hunger is difficult to define and quantify because it involves internal neurophysiological mechanisms. For the present discussion, however, hunger will be used in the familiar sense. Operationally, it generally is measured in terms of time since previous feeding, amount previously eaten, and an animal's normal or average body weight. This view of hunger is oversimplified (see Chapter 21), but it will do for now. Much of what applies to feeding in well-studied species such as the fly and rat (Chapter 21) undoubtedly applies also to other species. There are probably some phylogenetic effects on the control mechanisms of feeding, such as between insects and mammals, but a number of common phenomena also probably exist.

Thomas Eisner

Thomas Eisner
Bombardier Beetles

Bombardier beetles are the skunks of the insect world. When attacked, they emit a jet of defensive spray. The fluid is produced by two large glands that open at the tip of the beetle's abdomen. By revolving the abdominal tip, the beetles are able to aim the spray accurately at the enemy (see figure below). The spray contains potent chemical irritants called quinones. A single beetle can spray upward of twenty times before depleting its glands. Ejections are accompanied by audible pops. Most predators are instantly thwarted by the discharges.

The most remarkable property of the spray is that it is hot, as hot in fact as boiling water (100° C). This is because the quinones are generated explosively at the moment of ejection, by the mixing of two sets of chemicals ordinarily stored separately in the glands. Each gland consists of two compartments. In one compartment (the reservoir), precursors of quinones called hydroquinones are stored, together with hydrogen peroxide. In the other compartment (the reaction chamber), there are special enzymes. To activate the spray, the beetle simply squeezes fluid from the reservoir into the reaction chamber. This results in an instantaneous enzymatic liberation of oxygen from hydrogen peroxide and an oxidation of the hydroquinones to quinones by the freed oxygen. The liberated oxygen also provides the propellant, causing the reacting mixture to pop out. The heat that accompanies the formation of the spray can be felt by humans. Early explorers reporting on large bombardier beetles from the Amazonian jungle noted that these animals induce a strong burning of the fingers when being picked up.

Despite the effectiveness of the spray against such enemies as frogs, birds, and ants, some predators can cope with bombardier beetles. Orb-weaving spiders encase the beetles in silk when they fly into their webs, immobilizing their abdominal tip. Prevented from aiming their spray, the beetles are quickly killed and eaten. Similarly, certain horse-fly larvae that lie in ambush beneath the surface in mud, catch bombardier beetles with their mouth hooks and pull them partly into the soil. Shielded from the spray by the mud, the larvae have no difficulty subduing the beetles.

Perhaps one of the most common problems for some foragers, especially carnivorous predators, is the feast-or-famine nature of the natural situation. When food is at hand (or paw), there may be much more than the animal needs for its immediate nutrition. Subsequently the animal may be forced to go with little or no food for relatively long periods of time. Other species tend to have a more constant or stable food supply. But for predators near-starvation and actual starvation may be routine facts of existence. As a result, some predators may gorge themselves much more than other animals when given the opportunity. Spotted hyenas *(Crocuta crocuta)*, for example, may eat up to seven times as much (Kruuk 1972, cited in Curio 1976) during a big meal as they would normally eat on a subsistence diet. Some fish, snakes, and frogs will eat so much that they cannot get it all into the stomach at once, and the remainder has to wait in line, with tails or other parts of the food extending out of the mouth until digestion can make room for it.

At the other extreme, most large predators are able to go many days or weeks without eating. A hyena followed continuously by Kruuk (1972) went up to 5 days between meals, and Mech (1970) reported that wolves can go at least 17 days without eating. Eagles may go 2 weeks or more between meals in some cases. Some large snakes can live for a year without feeding. Predators, especially the young and inexperienced, may frequently go too long between meals and starve: whereas predators are a major source of mortality to their prey, a major cause of natural mortality among predators may be starvation.

The ability to gorge when food is available, however, is tempered by considerations other than physical capacity, particularly in smaller predators. Animals that are too heavily loaded down may become more vulnerable to predation themselves. Excessive gorging is observed mainly among the larger predators, which have few if any predators.

The complete sequence of a predatory act, from initial sensory contact with the prey to eating, can be broken into several component acts and movements. Some of these are associated with the level of hunger, and some are not. Furthermore, those that are associated with hunger may vary from species to species. For example, hunger may alter the thresholds at which predators chase or strike prey or the size of the animal they will tackle. Curio (1976) lists several traits that may be altered by hunger level in several different species. He modified and extended a comparison among a few representative species made by Holling (1966) (Figure 9-25). An ambushing predator such as a mantid may require a higher level of hunger before pursuing prey than predators that normally pursue their prey. Vertebrates appear to have a higher threshold for eating than for capturing, so that they are more likely to capture without eating than are arthropod predators. Furthermore, recovery after performance of different component acts, such as searching, lurking, chasing, catching, killing, and eating, occurs at different levels of hunger for each component (Leyhausen 1965). This suggests that an act of predation is not a single, unified behavior but rather a collection of components.

Some of these components may not be affected by hunger level at all, depending on the species. Speeds of stalking and pursuit, for example, are relatively constant regardless of hunger level. Similarly, the time required to eat a given amount of food may be fairly constant in many species, but not in others. Once a mantid strike is released, it is fairly mechanical, and the success rate of mantids does not depend on hunger level (Holling 1966). In many vertebrates, however, vigor and

Figure 9-25 Comparative, subjectively determined hunger thresholds at which predators engage in different behaviors. Note that mantids, which normally do not pursue their prey, will capture and eat at a lower threshold than is required for active pursuit, whereas the stickleback and cat, which normally do pursue prey, will pursue or stalk at lower thresholds than those at which they eat. Also, some predators may capture prey but not eat them if the predators are below the hunger level at which they normally eat.
Modified from Curio, E. 1976. The ethology of predation. Springer-Verlag, New York. Based in part on Holling, C.S. 1966. Memoirs of the Entomological Society of Canada 48:1-86.

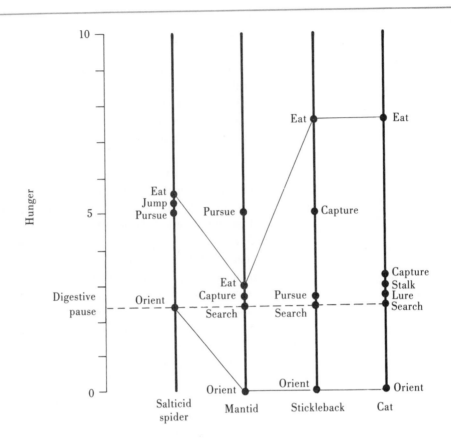

persistence of an act of predation tend to increase with hunger, and the success rate may go up.

Some foragers, particularly among spiders, some insects, birds, and mammals, store excess prey as previously discussed. Whether or not storing varies with hunger, however, appears quite variable and depends on the species. In some cases the handling of prey before storage involves different behavior from the handling of prey to be eaten immediately. This suggests that separate internal mechanisms are involved. Curio (1976) provides several examples among birds and mammals.

In addition to hunger, there are many other proximate factors that affect foraging. It may be strongly affected by both diurnal and seasonal rhythms: predators hunt and kill at some times but not at others, and they may change their tactics and prey preferences on a regular basis. Diurnal versus nocturnal patterns are easily identifiable, with predator and prey activity patterns obviously interrelated. Circannual rhythms, aside from those clearly associated with prey abundance, have received little attention but may be involved in variations in foraging.

Competing species of predators also may influence each other. Short-eared owls on the Galapagos Islands, for example, hunt night and day on islands where the Galapagos hawk *(Buteo galapagoenis)* is absent but only during the night where the hawk is present. Owls that do fly during the day where hawks are present are attacked immediately by the hawks (de Vries 1973).

Figure 9-26 Syrian woodpecker foraging behaviors: self-feeding versus feeding young. Frequencies differ depending on the circumstances of feeding.
Modified from Winkler, H. 1973. Oecologia 12:193-208.

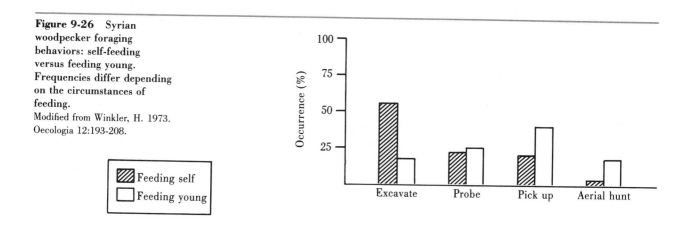

A major proximate factor in foraging is the presence of dependents. A mate or offspring that requires feeding will increase greatly the amount of foraging and also may alter the type of food and the manner in which it is handled. Norton-Griffiths (1969), after a thorough study of parental feeding in oyster-catchers, concluded that self-feeding and the feeding of young shared common internal mechanisms. This does not appear to be the case, however, in other species where the parent may or may not feed itself during the same period when it is feeding young and where the specific foraging behaviors may be quite different. Raptors, for example, pluck and decapitate their prey before giving it to a mate or young in a manner not seen when the birds are simply feeding themselves. Likewise, whitethroats (*Sylvia communis*) remove the heads of flies before feeding their offspring but not before feeding themselves (Sauer 1954). Winkler (1973) showed marked differences in the foraging methods used by Syrian woodpeckers that were feeding themselves and those feeding their young (Figure 9-26).

One somewhat intangible factor in predation, not closely tied to hunger, has been called *readiness to hunt*. In birds of prey, this sometimes is referred to as *yarak* or *sharp-set*. It has been proposed that prey species can recognize the appearance of raptors that are ready to hunt (e.g., Hamerstron 1957). But that seems to be disadvantageous for the predator and should be selected against; Grier (1971) was unable to find any evidence of hunting readiness in the postures and external appearance of hawks before they were offered food, although they quickly showed readiness after food became visible. Although the visibility of hunting readiness in the absence of potential prey may vary (e.g., obvious in African hunting dogs and absent in raptors), most people working with predators generally attest to its presence. In captivity, snakes are particularly prone to alternating periods of eating and not eating, to the point of near starvation in some cases. In a manner that does not always seem to correlate well with body weight or other measures of hunger, some predators will be ready, if not almost on a hair trigger, and others will completely ignore opportunities. In anthropomorphic terms, some predators at some times seem to anticipate and relish a good hunt. At other times or with other individuals in the same situation, predators may appear as if they could not care less about an opportunity. Lack of response may depend in part on the experience of past failures and successes.

Readiness to hunt generally involves a higher level of excitement and alertness, occasionally to the point that a predator performs predatory acts even in the absence of prey. These vacuum behaviors (Chapter 2) may include the full sequence of chasing, catching, killing, and eating—all with an imaginary prey. Furthermore, some components of predatory behavior will be displayed at high levels of excitement when not appropriate, such as "killing" a prey object given to an animal when the prey already is dead. Similarly, predators deprived of certain components of their behavior may engage in it excessively when later given the opportunity. Birds of prey, for example, typically pluck some of the feathers and hair from bird or mammal prey before eating it. If fed only meat rather than whole animals for a period of time, raptors sometimes ignore meat even if quite hungry when also given an object that can be plucked. Furthermore, they may pluck it much more than normal.

SUMMARY

Feeding is of vital and universal importance to animals. It also presents a host of problems, however, including what to eat, where to find it, how to find it, and how to go about obtaining it. For most species there is also the occasional to constant threat of becoming someone else's food. Animals solve these problems in an incredible diversity of ways. One way of viewing and organizing this behavior involves a relatively recent body of theory known as *optimal foraging theory*. It is based on economic analysis of costs and benefits of foraging and suggests that animals should find optimal solutions to the various problems listed above. Many animals do indeed appear to be doing just that, and there are numerous examples that match the predictions. There are many others, however, that do not seem to match well and some that are even contrary to predictions. As a result, optimal foraging theory has received close scrutiny, is undergoing revision to become more inclusive and realistic, and is currently an active area of research, discussion, and debate. There also are some alternate and associated concepts, such as the notion of *search image*.

Many animals cache, or store, food. This activity provides a number of potential benefits but also involves some hazards, such as loss to thieves, spoilage, or forgetfulness. Food storage requires that animals possess the ability to relocate and retrieve the food. This is usually accomplished by spatial memory, which many species, particularly among birds, have in large capacity, allowing remembrance of specific hiding places numbering into the hundreds or even thousands.

Predators possess a great variety of behaviors for locating, capturing, handling, and killing prey. Prey possess an equal or greater diversity of methods to avoid being eaten.

Foraging can be discussed from a proximate and an ultimate perspective. Foraging behavior may vary with hunger level, time of day or season, presence of hungry dependents, presence of competitors, or with the state of internal physiological mechanisms that affect a forager's readiness to hunt.

FOR FURTHER READING

Books

Curio, E. 1976. The ethology of predation. Springer-Verlag, New York.
  Reviews predation behavior from a variety of perspectives.
Feder, M.E., and G.V. Lauder. 1986. Predator-prey relationships. Perspectives and approaches
  from the study of lower vertebrates. University of Chicago Press, Chicago.
  *This book resulted as an outgrowth of a symposium but is not a typical symposium proceeding. It
  emphasizes speculation, ideas, and interdisciplinary interactions rather than volumes of data to
  encourage "collaboration, new concepts, and new hypotheses." It is a stimulating book covering a
  wide variety of topics related to foraging and predation.*
Kamil, A.C., J.R. Krebs, and H.R. Pulliam. 1987. Foraging behavior. Plenum Press, New York.
  *This book also is a conference publication but contains a variety of important papers on optimal
  foraging, diet choice, patch use, caching, foraging and learning, and other topics. (It is an
  expensive book, $115, which may discourage most persons from purchasing it for personal
  libraries, but your public or institutional library may carry it.)*
Stephens, D.W., and J.R. Krebs. 1986. Foraging theory. Princeton University Press, Princeton,
  N.J.
  *The currently acknowledged primary (and affordable) source for persons wanting a concise and
  lucid discussion of foraging theory.*

A Selection of Original Papers

Fraser, D.F., and F.A. Huntingford. 1986. Feeding and avoiding predation hazard: the behavioral
  response of the prey. Ethology 73:56-68.
Guilford, T., and M.S. Dawkins. 1987. Search images not proven: a reappraisal of recent
  evidence. Animal Behaviour 35:1838-1845.
Hainsworth, F.R. 1986. Why hummingbirds hover: a commentary. Auk 103:832-833.
Inglis, I.R., and N.J.K. Ferguson. 1986. Starlings search for food rather than eat freely-available,
  identical food. Animal Behaviour 34:614-617.
Laverty, T.M., and R.C. Plowright. 1988. Flower handling by bumblebees: a comparison of
  specialists and generalists. Animal Behaviour 36:733-740.
Lewis, A.C. 1986. Memory constraints and flower choice in *Pieris rapae*. Science 232:863-865.
Meylan, A. 1988. Spongivory in hawksbill turtles: a diet of glass. Science 239:393-395.
Miller, R.S. 1985. Why hummingbirds hover. Auk 102:722-726.
Miller, R.S. 1986. Response to F.R. Hainsworth. Auk 103:834.
Real, L., and T. Caraco. 1986. Risk and foraging in stochastic environments. Annual Review of
  Ecology and Systematics 17:371-390.

10

OTHER INTERSPECIFIC BEHAVIORS

Most of this book focuses on the behavior of individual animals or interactions among animals belonging to the same species. However, there is also a fascinating and vast array of behavioral interactions between animals of different species (and between animals and plants). The previous chapter discussed direct trophic interactions between animals of different species, that is, where one eats another. This chapter continues with a look at the other, indirect trophic and nontrophic, ways that animals (and animals and plants) interact.

Various interspecific interactions occur among virtually all taxonomic groups. The most complex, however, occur primarily in two groups: birds and highly social insects. In addition to being interesting per se, such relationships raise a number of intriguing questions. How do complex interspecific interactions with their associated stereotyped behaviors arise evolutionarily? Given that such interactions evolve at all, why are they not more common? Why are the most complex interactions seen predominantly in birds and insects? What, if any, relationships exist between the different types of interspecific interactions? This chapter surveys and describes the diversity of interactions, discusses associated evolutionary hypotheses where appropriate, and attempts to provide an overall synthesis.

INTERSPECIFIC INTERACTIONS AMONG ORGANISMS IN PERSPECTIVE

The majority of animals of different species living in an area have little to do with one another. Most of those living around us, for example, pay little or no attention to us. Similarly, aside from a handful of biologists, most humans hardly even notice other species. Some species are large or otherwise conspicuous, such as many birds, or they receive our attention because they are pests or used for sport, food, fur and other material resources, utilitarian purposes such as beasts of burden, or provide entertainment, relaxation, and companionship as pets. Even with all of that, the total number of species that most humans interact with is quite small and constitutes only a tiny fraction of the total present in any given area. The fact is that organisms of most species simply go on about their business and do not interact with most of the other organisms surrounding them.

At the same time, however, virtually all animals do interact with some other species. At the minimum there is the problem of obtaining food or keeping from becoming food. Beyond these trophic relationships, which were discussed in the previous chapter, there are many other types of interactions between species that affect all or most animals.

The diversity of such interspecific interactions is so great that it almost defies classification. Wheeler (1911, cited in Thompson 1982) expressed the problem well when referring just to parasitism, which he described as "an extremely protean phenomenon, one which escapes through the meshes of any net of scholastic definitions in which we may endeavor to confine it." Other types of interspecific relationships are equally difficult to define and classify. Nonetheless, it is still possible to identify and discuss some general categories. Depending on whether the interactions are beneficial, negative, or neutral to the various participants, the categories can be viewed as in Figure 10-1 and the box on page 292, which further

Figure 10-1 Categories of interspecific interactions, classified according to whether the relationship is beneficial (+), disadvantageous or harmful (−), or neutral (0) to each of the participants respectively.

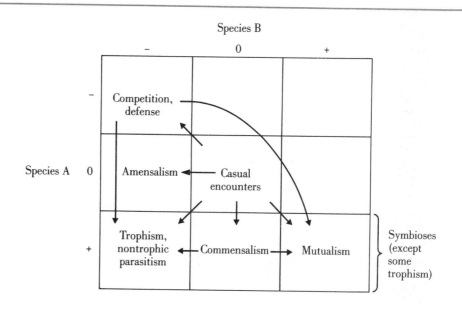

Classification of Interspecific Interactions

    I. Casual, chance encounters
    II. Amensalism
   III. Competition
        A. Exploitative
        B. Interference
   IV. Defense
    V. Commensalism
        A. Plesiobiosis
        B. Interspecific groups (mixed flocks, etc.)
        C. Trophic commensals
        D. Phoresy
   VI. Mutualism
        A. Interspecific groups
        B. Trophobiosis
        C. Nontrophic symbioses
        D. Parabiosis
  VII. Direct trophism
        A. Grazing
        B. Direct trophic parasitism
        C. Predation
 VIII. Indirect trophic and nontrophic parasitism
        A. Kleptoparasitism
        B. Xenobiosis
        C. Temporary social parasitism
        D. Dulosis
        E. Inquilinism

---

subdivides many of the categories. Briefly, the various categories can be characterized as follows.

Casual or chance encounters and incidental aggregations of individuals from different species are often of no significance to any of the animals involved. However, they provide important starting points for the evolution of other relationships and associated behaviors. Occasionally, usually accidentally, interactions are harmful to one of the participants while being irrelevant to the other. This is referred to as amensalism (*mensal* is derived from the Latin, *mensa*, which literally means "table;" hence amensal means "not at the table;" see commensalism). In many interspecific interactions there are disadvantages and costs to both sides; these include competition and defense. In commensalism (literally "together at the table"), one species benefits and the other is unaffected. Mutualism differs from commensalism in that both species benefit. There are several situations in which one participant benefits at the expense of the other. In addition to the direct trophic relationships discussed in the previous chapter, there are several forms of indirect trophic and nontrophic parasitism that involve one-sided, or mostly one-sided, benefits. These include various forms of social parasitism such as nest parasitism. The boundaries between trophic and nontrophic parasitism are not always clear cut. Nestlings of some species of nest-parasitic birds may kill the young of the host species, but they do not eat them. Some social parasites in the nests of many insect species also may kill and consume eggs, young, or other individuals of the host species.

The interactions that bring animals of different species into close relationships throughout much or all of their lives, particularly commensalism, mutualism, and the various forms of parasitism, are known as *symbioses* ("living together"). The term *symbiosis* has often been used when referring to mutualistic relationships, and the two have been considered synonymous in the minds of many persons; however, symbioses also include many interactions that are not mutualistic. Wilson (1971) reviewed social symbioses among insects, the taxonomic group in which such relationships are most numerous, complex, and highly evolved. He subsequently (1975) expanded the treatment to other, noninsect organisms including vertebrates.

In the remainder of this chapter, we expand on the descriptions and discussions of these categories of interspecific interactions, including various evolutionary considerations, and attempt to integrate the relationships among the different categories, to the extent that such relationships exist and are understood, into a single overall picture. Figure 10-1 and the box on page 292 provide a summary of the categories and hypothesized evolutionary relationships, indicated by the arrows in Figure 10-1, that one can refer to while reading the following sections.

CASUAL, CHANCE ENCOUNTERS, AMENSALISM, COMPETITION, AND DEFENSE

Animals of many species meet; their paths cross with little or no significance; and the animals may hardly notice each other. They might meet coincidentally at a migration point or a nonlimiting resource such as a watering hole. Or they may by chance simply end up at the same place at the same time.

Sometimes an encounter results in harm or a loss to one of the species, although not affecting the other. This is called *amensalism*. It does not involve an obvious conflict over a resource of mutual interest, although it might involve a resource of one of the participants.

Amensalism does not seem to be important evolutionarily, and indeed it does not seem to be very prominent in nature. When it does occur, it probably is mostly incidental or accidental, such as when large grazing animals trample eggs of ground-nesting birds, trample or break up vegetation (e.g., elephants tearing down trees) that provides food or cover for other animals, or when animals otherwise accidentally damage or destroy something of relevance to another species.

Perhaps one good example of nonaccidental amensalism, although it was not considered as such by the authors who described it (Nuechterlein and Storer 1985), involves interspecific aggression and killing by steamer-ducks (*Tachyeres patachonicus*) in Argentina. These are large, tough, and aggressive ducks with hardened, horny orange knobs on the wrist region of the wings that males use in territorial fighting. They also attack and may kill other species of waterfowl, species that do not seem to be competitors or otherwise pose any threats to the steamer-ducks. Furthermore, the steamer-ducks are more aggressive than appears necessary for simple territorial defense, and it seems unlikely that the males are confusing other species with their own. Nuechterlein and Storer found the behavior difficult to explain and could only speculate on possible advantages, if any. They considered, for example, "Possibly males victimize birds of other species in order to display their belligerency and fighting abilities to their females." Or, "Possibly observational learning is important, and holding a 'public beating' enhances the effectiveness of their territorial displays."

There may be some advantage of such behavior that has not been considered or perhaps there is no selective advantage and the behavior just exists in a nonadaptive sense (Gould and Lewontin 1979). Thus the steamer-duck is currently a candidate for use as an example of amensalism.

In many encounters, however, particularly over limited resources such as food, nesting sites, or other protective cover, animals of different species may interact directly. This is known as *competition*. Clashes over food are particularly frequent. For example, a badger and a cougar may happen across the same carcass and fight over its possession, using stereotyped behavior patterns. The interactions of large mammalian and avian predators and scavengers over prey, such as wintering bald and golden eagles in North America or several species of mammals and vultures in Africa, can be spectacular and are familiar to most persons. Other less spectacular but familiar fights over food include interactions among different bird species and between birds and squirrels at winter bird feeders. Less familiar but equally important battles over food and other resources occur commonly among different species of animals from many other taxonomic groups.

During competition over food, animals of many species not only display typical behavior toward those of other species, they also win in the encounter more often than others. At bird feeders, for example, some bird species such as house sparrows and grosbeaks commonly establish dominance over other species. Interactions among several different species of birds and mammals at a seed bait station are shown in Table 10-1. It seems likely that such competition has led or at least contributed to the evolution of associated behavior such as removal of food from

Table 10-1 Results of encounters among antelope squirrels of known rank, quail (single or in groups), and single individual cottontails, scrub jays, and wood rats*

| Opponents | | Supplantations by | |
A	B	A	B
Alpha, beta, or gamma squirrel	Gambel's quail (1 to 3 individuals)	19	12
Alpha, beta, or gamma squirrel	Gambel's quail flock	0	30
Alpha, beta, or gamma squirrel	Cottontail	5	11
Alpha, beta, or gamma squirrel	Scrub jay	13	2
Lower-ranking (delta or below) squirrel	Gambel's quail (1 to 3 individuals)	7	63
Lower-ranking (delta or below) squirrel	Gambel's quail flock	0	42
Lower-ranking (delta or below) squirrel	Cottontail	0	17
Lower-ranking (delta or below) squirrel	Scrub jay	4	10
Alpha squirrel	Wood rat	9	0
Lower-ranking (delta or below) squirrel	Wood rat	1	17
Gambel's quail flock	Cottontail	17	3
Gambel's quail (single or pair)	Cottontail	0	15

From Fisler, G.F. 1977. Animal Behaviour 25:240-244.
*Squirrel data are the combined results of encounters between single squirrels and each opponent (single or group).

the original site, taking food to secluded or hidden areas for consumption, and caching or covering excess food (Chapter 9). Many species bury extra food; leopards carry theirs up into tree branches; and some species urinate, defecate, or otherwise mark their food and prevent others from consuming it.

Aside from food, the resources most likely to be limited and generate behavioral competition between species are holes or cavities for nesting or protection. Many hole-nesting birds and small mammals fight for occupancy; and again, some such as the European starling are commonly more successful than others.

Competition may be of two types: *exploitation* or *interference* (Park 1954, Miller 1967, 1969). In exploitation competition the animals do not interact directly and physically but merely reduce each other's resources by exploiting them first. Exploitation competition is consequently of interest more from a purely ecological standpoint. Interference competition, on the other hand, involves more direct, behavioral interaction such as territorial behavior, fighting and other aggression, or release of toxic or noxious substances. When ring-necked pheasants are introduced to areas with native prairie chickens (*Tympanuchus cupido*), for example, cock pheasants harass the cock prairie chickens, and hen pheasants lay eggs in prairie chicken nests; both activities interfere with prairie chicken reproduction.

Defensive interactions occur in circumstances where one or both of the participants stand to lose, including amensalism, competition, most trophic circumstances, and nontrophic parasitism. Both sides actually lose when one or both are defensive, even if they both survive, because the defense eliminates or at least decreases the benefits that one participant would have gained otherwise, and it costs the other participant, which has to produce the defense. Depending on the situation, however, the loss is usually unequal or asymmetrical; that is, one loses more than the other. Both defenses and counter-defenses may be sophisticated. The toxic chemical defenses of plants, for example, may be circumvented by unique physiological pathways of animals that feed on them (Rosenthal et al 1978) or even by mechanical countermeasures. Some insects, for example, feed on latex-producing plants by cutting leaf veins and blocking the latex flow to the outer parts of leaves, which are subsequently eaten (Dussourd and Eisner 1987).

COMMENSALISM

Plesiobiotic relationships are those in which two species commonly live in close relationship with each other but where the benefit to one or the other is slight or not understood. Some species of social insects, for example, live close together, even in compound nests, but they keep their broods separate and otherwise live their own lives. Such relationships have not been well studied, and with further information many instances may turn out to belong to another category. Even if there is no benefit to either participant or if the relationship is one sided, plesiobiosis may preadapt the species for the evolution of more complicated relationships.

Interspecific groups (including flocks, herds, and schools) are those where groups of two or more species associate with each other. This category of commensalism, commonly referred to as *mixed groups*, is similar to plesiobiosis except that the associations tend to be more transient in time and place. Most of these associations tend to be loose and not consistent, but some can be quite stable. Mixed-species groupings are seen in marine and some freshwater fish, foraging and migrating birds, roosting bats, some mixed troops of primates, some dolphins, and open-plains ungulates. A few mixings extend even beyond class boundaries, such as between groupings of some monkeys and birds. There has been some attempt to classify the importance of different species based on their roles in a mixed group (Moynihan 1962): nuclear species are those primarily responsible for the attraction and cohesion of the mixed group; attendant species are the regular joiners; and accidental species are those that occasionally, but not commonly, become involved.

What benefits occur in mixed-species groups and to whom they accrue are not well understood. In many cases, it seems almost as though species' distinctions are not relevant in these situations and the individuals from various species are behaving as if they belonged to the same species. This notion may be unsettling to some persons, particularly those accustomed to good-of-the-species thinking. The benefits may go primarily to individuals of one species (i.e., commensal) or to all (i.e., mutual). Because benefits can be to one or all, the category has been listed under both commensalism and mutualism (Wilson 1975). The major hypothesized benefits for mixed-species groupings are similar to those of intraspecific groupings (Chapter 13): improved foraging efficiency and reduced predation.

It is not clear how mixed flocks can improve foraging efficiency, but circumstantial evidence supports the notion. Mixed foraging flocks in birds seem to occur during periods when conditions are harshest and food supplies are scarcest, as during winter or in marginal habitat. Less successful individuals, regardless of species, that otherwise might fail on their own may be able to take advantage of the experience and success of others. More eyes watching may make it more likely that food, which can be used by all, will be discovered. Among mixed herds of ungulates, individuals of different species may actually facilitate each other's feeding by feeding on different parts of the vegetation (McNaughton 1976). Animals of one species may remove those parts of plants that are obstacles or are indigestible to animals of other species. Ungulates also flush insects that are caught and eaten by birds, such as brown-headed cowbirds and cattle egrets.

Antipredation advantages amount to extensions of those proposed for grouping within a species, namely the selfish-herd principle, the confusion principle, and increased awareness of possible danger (Chapter 13). Moynihan (1962) suggested that antipredation is the main advantage of mixed-species flocking in the tropics. Numerous recent reports support the role of antipredator functions for grouping, whether the grouping is intraspecific or interspecific.

Munn (1984) intensively studied mixed flocks of birds in the lowland rain forests of South America where up to 70 species may forage together in a single flock. There are two basic flock types: canopy flocks and understory flocks. In both cases, there are a few core species that are stable flock members and that maintain a joint territory. Individual birds live their entire lives in the same flock. Other species join on a more temporary basis. Each flock includes a sentinel species that mostly watches and sounds alarms at the appearance of a predator, such as a hawk. The sentinels generally are faster than other flock species and quickly catch insects that are flushed and escape from the more methodical foragers. When sentinels are not present, such as when they are occupied with their own nesting, the other species forage less in exposed positions (Munn 1984).

Trophic commensals are those that live essentially off the garbage of another species. Remoras, or "shark suckers," for example, attach themselves to sharks and ride along to pick up scraps of food from feeding sharks. Whole armies of scavenging birds, mammals, and arthropods may trail along after predators to clean up the remains. Most nests, dens, and hives, whether of birds, mammals, or social insects, have an entourage of arthropod scavengers, including numerous isopods, nonparasitic mites, collembolans, beetles, flies, and others. These commensals may be accepted and ignored by their host, or they quickly get out of the way and avoid direct encounters. Some interact directly with their hosts, using the same communication signals and displays. Some silverfish (insect) commensals that associate with tropical army ants even follow along during raids. Occasionally the chain becomes more complex. Ant-butterflies follow army ants and feed on the droppings of birds, which also follow the army ants. The birds eat insects flushed by the ants. The butterflies follow the birds following the ants, but the butterflies apparently use cues, possibly the trail odors, from the ants (Ray and Andrews 1980).

Unlike some animals such as remoras or trophic parasites that attach themselves to other animals and ride along with the primary benefit being feeding or something other than just the ride, there are many animals that use other animals primarily as a source of transportation. This is referred to as *phoresy* (Latin, *phorus*, Greek, *pherein*, "to bear"). The riders simply use the other animal to get from one place to another, do not cause the carrier any harm, and ride at virtually no cost (in most cases the animals involved are small enough that weight is an insignificant factor). As reviewed briefly by D.S. Wilson (1980, including several references to previous reviews), phoresy is a common phenomenon that apparently has "evolved hundreds of times in as many taxa."

Phoresy is particularly prevalent with mites that hitchhike on animals to get from one food item to another. There are some flower mites, for example, that travel between flowers on the bills of hummingbirds (Colwell 1973). In another coevolved system involving mites, several species of mites that live in dung ride on dung beetles to get from one dung pat to another (Costa 1969, cited in Wilson 1980). When they arrive at a new dung pat, the mites get off the beetle. The beetles and mites live and reproduce separately in the dung, but their life cycles are synchronized. The mites take less time than the beetles to complete their reproductive cycle. However, they then undergo a resting stage, from which they emerge just as the next generation of beetles emerge from their pupa. The new mites climb aboard the newly emerged beetles and ride to the next dung pat.

MUTUALISM

Most forms of mutualism, where both species benefit, involve a benefit of food (hence the term *trophobiosis*) for one of the participants. To the other participant goes one or more benefits, which, depending on the situation, include protection (from predators, weather, etc.) and even dental and body hygiene. Body cleaning (Figure 10-2) is best known among several species of fish in which some clean others, but it also is seen in some reptiles and a number of large mammals over which certain birds forage and pick off parasites. Some species, such as cleaner fish of the genus *Labroides*, are obligate or full-time cleaners (Potts 1973), whereas many others are only facultative or occasional cleaners (Sulak 1975; Brockmann and Hailman 1976). Specific, stereotyped behaviors often are involved in the interaction between the cleaner and its host.

Humans occasionally are included in trophobiotic interactions. One example

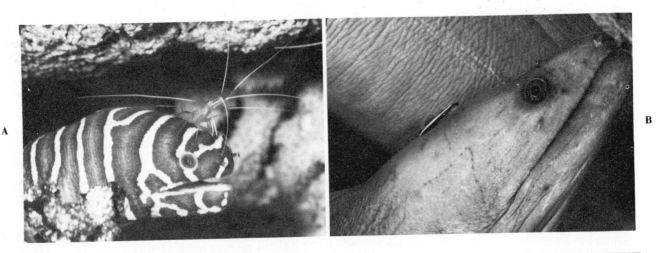

Figure 10-2   Cleaners removing parasites from the bodies of hosts. **A,** Cleaner shrimp cleaning a zebra moray. **B,** Neon goby cleaning a green moray. This cleaner is an obligate cleaner. Many species of fish, including some familiar ones, are facultative, or occasional, cleaners. Some fish species mimic cleaners and have become parasitic, nipping bites of flesh from the host instead of cleaning them.
From Animals, animals. © Zig Leszczynski.

involves native people and a bird, the greater honeyguide, in parts of Africa (Isack and Reyer 1989). The birds lead people to honeybee hives in trees, rock crevices, and termite mounds. The people break into the hives to get honey and leave the bee larvae and wax, which the birds then consume. Ancient rock paintings suggest that this interaction may have been going on for at least 20,000 years. Even the scientific name of the bird, *Indicator indicator*, reflects the relationship.

The humans and birds have developed, through evolution or learning (Chapter 20), elaborate interspecific communication signals. The people, such as the Borans of Kenya, use loud whistle sounds, made by blowing through their hands or various instruments, to attract the birds. The birds in turn fly close to the people and emit a special call to indicate their presence. Then the birds use a combination of flight and perching patterns plus different calls to indicate the direction and distance to the hive. They also signal when they have arrived and are in the immediate vicinity of the hive (which is not always obvious).

Earlier records and information about the honeyguide-human interaction were anecdotal, and persons considered the story to be a myth. However, a careful 3-year field study by Isack and Reyer (1989) using a regression analysis of the various behaviors confirmed the validity of the relationship. Furthermore, by observing from camouflaged positions, the researchers were able to gain more insight into the phenomenon. For example, birds scouted out areas and found hives by visiting them early in the morning when it was cool and the bees were still docile. The birds would fly into hive entrances and peer into them briefly. If the people could not (or would not, as in one of the tests during the study) find the hive after they had arrived in the immediate vicinity, the bird would give up at that site, start with a new sequence of behavior, and lead the people to a different hive.

When guided by the birds, the Boran people were able to find hives on the average in only 3.2 hours whereas they required an average of 8.9 hours to find hives when unguided. The benefit to the birds was twofold: (1) they had much greater access to hives (96% of all hives were accessible to them only with human help), and (2) the use of smoke by people to clear out the bees greatly reduced the chances of the birds getting stung by the bees.

In some of the more advanced forms of mutualism a few species literally tend to and husband other species. Some species of hermit crabs, for example, attach sea anemones to their shells and, when they change shells, take the anemones along with them. Another well-known mutualism is found between sea anemones, which provide protection, and anemone fish (also called *clown fish*), which provide food. The relationship is induced by several specific chemicals released by the

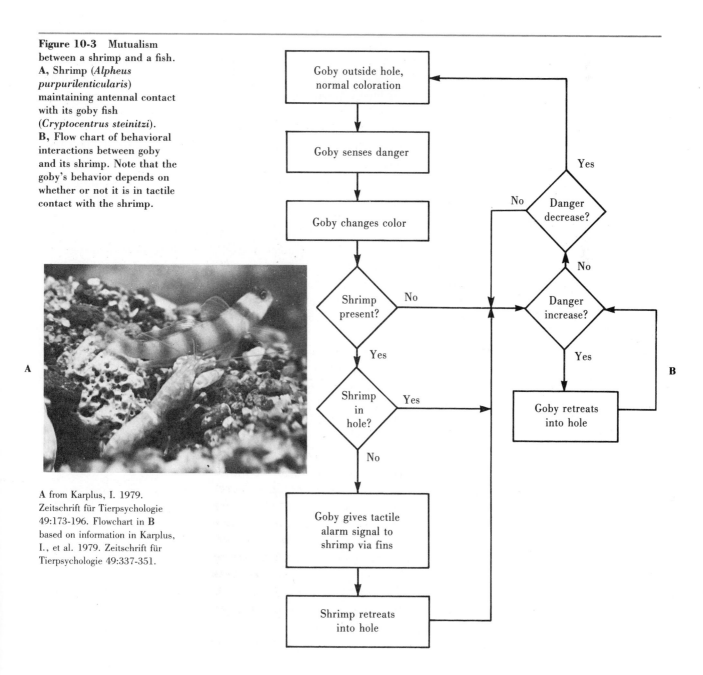

Figure 10-3 Mutualism between a shrimp and a fish. **A,** Shrimp (*Alpheus purpurilenticularis*) maintaining antennal contact with its goby fish (*Cryptocentrus steinitzi*). **B,** Flow chart of behavioral interactions between goby and its shrimp. Note that the goby's behavior depends on whether or not it is in tactile contact with the shrimp.

A from Karplus, I. 1979. Zeitschrift für Tierpsychologie 49:173-196. Flowchart in **B** based on information in Karplus, I., et al. 1979. Zeitschrift für Tierpsychologie 49:337-351.

anemones (Murata et al 1986) and the chemical nature of mucus on the fish, which does not stimulate the firing of the anemones' nematocysts.

Several species of ants and some stingless bees milk aphids and a few other types of insects for honeydew, a sugar-rich fluid that is excreted by many insects that feed on plant juices. The plant feeders extract only part of the nutrients and excrete the rest, something of an affluence in the presence of abundant resources.

Some ants and bees simply follow the plant feeders and consume the honeydew, whereas many other species have evolved a mutualistic relationship, including signals from ants, which stroke or tap aphids with their antennae when it is time to feed.

In the most advanced cases the aphids have lost all of their normal defense mechanisms, and protection is afforded by their ant caretakers. Some aphids have morphological adaptations that aid the ants feeding on the honeydew. In addition, the life cycles of the two species are synchronized, and the ants are well coordinated in their care of the aphids. They take them underground into their own nests in the winter, transport them to new plants, and move developing aphids to appropriate parts of plants during the nymphal development. Such advanced levels of mutualism are not found among other species, particularly among vertebrates, except for the relationships among humans and their domestic animals.

Not all mutualistic relationships involve feeding. In one case involving a type of goby fish and a shrimp, for example, the shrimp dig holes that they both live in (Figure 10-3, A). The goby provides warning of danger. The relationship includes evolved tactile signals (Table 10-2) (Figure 10-3, B). The shrimp maintains contact with the goby through its antennae.

Parabiosis involves species that live so closely together that they behave and function as if they belonged to the same species. They share in mutual defense, foraging, actual feeding, and all activities except the rearing of offspring. Good examples are known at the present only for a few species of ants in Central and South America; however, more cases may be discovered.

Table 10-2 Interindividual two-act sequences in interactions between gobies and *A. rapax*

Gody Initial Acts	*A. rapax* Following Acts						
	Flee	In Burrow	Manipulate Objects	No Change	Plough	Sit	Withdraw
Dorsal erect	0	4	1	3	3	0	1
Flee	9	13	0	0	0	0	0
Guard	14	934	109	19	939	52	691
Move away	0	20	4	22	21	1	22
Nip sand	5	16	2	14	12	2	13
Pectorals wave	0	9	2	3	8	0	5
Sit away	3	8	0	0	7	1	4
Tail beat	3	0	1	1	0	0	1
Tail flick	119	17	3	134	22	82	65
Tail wave	0	22	3	25	15	1	38
Withdraw	1	15	2	3	10	2	6

From Prestons, J.L. 1978. Animal Behaviour 26:791-802.

INDIRECT TROPHIC AND NONTROPHIC PARASITISM

Contrary to many people's understanding, there are many more ways that one species takes advantage of another as a parasite than just living in or on an animal and directly eating parts of it. The word parasite can be traced back to the Greek *parasitos*, which in turn was derived from *para* and *sitos* ("beside-food") and which originally meant "eating beside or at the table of" someone else. The term then acquired the familiar meaning of eating the host itself. The meaning has subsequently been broadened to include various indirect trophic and nontrophic types of parasitism. A few parasites defy classification by fitting more than one category.

Kleptoparasitism

Some types of trophic parasitism are closer to the original meaning of the term than to the more familiar contemporary concept of parasitism. That is, there are many species that do not consume parts of their hosts directly but rather steal the host's food supply. This type of robbery is kleptoparasitism. Included in this category are hyenas, which steal food from African wild dogs and other large predators, many birds that rob food from other species and occasionally from certain mammals (reviewed by Brockmann and Barnard 1979), and some fish and a number of arthropods that take food from other species. In some cases the parasites will eat both the hosts and the hosts' food.

Food stealing probably occurs most frequently with highly social insects. Many species of insects and mites parasitize various ants, bees, and termites, and sometimes steal from each other or enter nests and consume eggs or larvae. Ants of many if not most species and many beetles and other insects rob from another if given the opportunity, but some have become obligatory parasites; that is, they depend on this way of life and it is the only way they live. Some kleptoparasites of ants wait along odor trails and ambush the ants as they return with food supplies. One subgenus of ants that nest next to the nests of larger ants and parasitize their broods are known collectively as *thief ants*. There are some species of termites that live in and consume the walls of other termite nests. (Technically, because they are not stealing food, they are not really kleptoparasites in the usual sense.) As Wilson (1975) put it, "Some termites have termites in their houses!"

Xenobiosis

Most of the trophic parasites are treated with hostility if detected and captured by the hosts. In xenobiosis, however, the robbers may become tolerated guests and even exchange communication signals with the hosts, although the relationship is still parasitic and the benefits are in one direction only. The best examples come again from a number of different ant species.

Temporary Social Parasitism

From this point one encounters more advanced forms of parasitism—forms that generally involve more than mere feeding. Temporary social parasitism involves living occasionally as a parasite but otherwise being independent. This form of parasitism has been identified among a number of social insects and in nest (egg and brood) parasitism among birds.

Among insects, temporary parasitism occurs when a queen of one species invades the colony of another species, removes the resident queen, and then appropriates the resident workers for her own uses. The new parasitic queen lays her eggs, and the former workers help care for them. The new workers that result increase in number and eventually replace the old ones of the parasitized species. The former workers are not replaced by their own species because their queen is gone. When all the former workers are gone, the colony functions on its own.

Temporary social parasitism is found in a number of ants, some social wasps, and a few bumblebees. There are several ways, depending on the species, by which the parasitic queens gain entrance to the parasitized colony and kill the resident queen. Some force their way in; some more or less sneak in and associate with the workers until they have acquired the odors of the colony; and some use the normal communication signals of the parasitized species. In some cases (e.g., in the mound-building ant, *Formica exsecta*, which occasionally parasitizes *F. fusca*) the queen may lie down and tuck in her legs, as in the pupal posture, when approached by a worker of the other species. The worker then picks up the queen and carries her down into the nest. Different species also use a variety of ways to kill the resident queen. These usually involve strangling or biting of the neck with the mandibles after using different techniques to get the opponent in a vulnerable position.

Some temporary social parasites go to extremes as hyperparasites (i.e., as parasites of parasites). Some ants of the subgenus *Dendrolasius* parasitize the nests of free-living *Chthonolasius*, which had been acquired by parasitism in the first place by *Chthonolasius* taking over nests of the subgenus *Lasius*.

The evolution of social parasitism in insects has been traced through comparative studies, particularly of the social wasps (references in Wilson 1975). It is thought to begin with occasional (or facultative) parasites within the same species, whereby queens attack other colonies of their own species. This eventually may lead to occasional attacks (in subsequent generations) on colonies belonging to other species. Further evolution may then lead to obligate relationships where one species becomes totally dependent on the other species.

Temporary social parasitism of a roughly similar nature is seen also among a few birds. Nest parasitism, in which the eggs of one species are laid in the nests of another species, is practiced by about 80 different species in 7 families or subfamilies (reviewed by Lack 1968, Meyerriecks 1972, and cited in Wilson 1975). Best-known nest parasites are European cuckoos and American cowbirds, for which the behavior is obligatory. However, there also are many others, including the black-headed duck (*Heteronetta atricapilla*) and a few species that are only occasional nest parasites. The parasites' eggs are incubated by the host, and the young are subsequently brooded and fed or otherwise cared for.

The complexity of the avian nest parasite relationship depends on the species. In some it is fairly simple, with the female merely laying eggs in other birds' nests and leaving; the hosts do not discriminate the presence of an odd egg or chick but simply accept them and raise them along with their own.

In other cases, however, there are various degrees of coevolution, whereby changes occur in both species. In most of the obligate nest parasites, the females of the parasitic species possess a number of physiological and behavioral traits that improve their ability to parasitize. They may survey the neighborhood for suitable nests, be able to sneak into nests while the owners are away temporarily, or otherwise intimidate the owner; they may destroy nests that are too far advanced so that the owner has to start over; and they can usually lay eggs quickly.

The parasite's eggs themselves may be thicker shelled, can withstand dropping, and usually develop slightly faster than the host's, so that the parasite's eggs are the first to hatch. Then, depending on the species, the parasitic chicks may maneuver the host's eggs or chicks on its back and dump them out of the nest (e.g., European cuckoos) or kill them with a hooked beak (e.g., the honeyguide, *Indicator indicator*).

Among many species the hosts recognize odd eggs and young and either destroy or abandon them. This apparently has led to a high level of mimicry in several species of parasites. The eggs of the European cuckoos resemble the eggs of the hosts, with different females producing eggs that mimic those of particular host species. Females parasitizing a particular host species and laying the appropriate type of egg belong to a given gens (pl. gentes). This has posed an interesting genetic problem, because females that produce eggs and lay them in the nests of different species all belong to the same species and even the same populations. Furthermore, a single male cuckoo may fertilize several females of the different gentes. The genes controlling which gens a female belongs to may be sex linked, as with baldness and hemophilia in human males. (In birds the female has the unmatched pair of sex chromosomes, which is the opposite of the case in mammals.)

Mimicry in some cases extends to the begging signals that the chicks give to the adults. The young of the combasson and widow birds of Africa have similar markings on the inside of their gaping mouths, which are recognized by the adults. The species are closely related, however, and the markings were probably present in both, a preadaptation for the parasites and not a matter of convergence. (The egg mimicry of the cuckoos clearly is a case of convergence.) In some cases the parasitic species' offspring may learn the song of the host species and use that to ensure that male and female from the same gentes mate, as with the Viduine finches.

The evolutionary sequence leading to nest parasitism in birds, as revealed by comparative studies among closely related species, is thought to begin with birds using abandoned nests of other birds or appropriating the nests of others—but still incubating and rearing their own eggs and young. Next they may abandon their eggs in some cases, a few of which get incubated by other birds, but continue to take care of their own in other cases. In several species that lay large numbers of eggs, particularly among Galliformes (e.g., pheasants) and some waterfowl, hens may have dump nests, where they lay large numbers of eggs early in the season. Several hens and even different species may dump in the same nest. Egg dumping or other forms of abandoning eggs in other's nests may lead to facultative parasitism, whereby the species is occasionally parasitic but not dependent on that mode. Such a pattern could evolve, through many generations, into greater and greater dependence on parasitism until the species reaches the point of complete dependence, that is, obligatory nest parasitism.

In a well-studied complex situation investigated by Smith (1968), there appears to be a mixture of parasitism and mutualism between several bird species and the giant cowbird (*Scaphidur oryzivora*) of South and Central America. This cowbird has gentes, as in the European cuckoos: three produce eggs that mimic the eggs of different genera of oropendolas (birds), one mimics the eggs of a cacique (another bird), and one is not a mimic but a dumper that produces a generalized sort of cowbird egg. Females of the mimic gentes are shy and elusive, whereas those of the dumper gens are more aggressive and may barge right into the nests of the hosts.

The different egg-producing types and associated behaviors of this cowbird species are a form of polymorphism. But the hosts also are polymorphic; some discriminate strange eggs, and others do not. The cause of this apparent polymorphism was found to be a problem with a local botfly (*Philornis* sp.), which infects bird nests and kills many of the nestlings. Nestling cowbirds protect themselves (and, additionally, any hosts) by snapping at adult botflies and by preening nestmates, which removes botfly eggs and maggots.

There are yet more participants in this tangled story: large colonies of social wasps and stingless bees also live in the region. The wasps and bees somehow repel the botflies. The result (and presumed cause of the polymorphisms) is that some oropendolas and caciques nest near wasp or bee colonies, receive botfly protection from them, and discriminate against the cowbirds—treating them as parasites. Other oropendolas and caciques, however, do not nest near the wasps and bees but accept the presence of the cowbirds. They live with the cowbirds in a mutualistic relationship: the hosts provide incubation, brooding, and food, whereas the cowbird nestlings provide botfly protection (Smith 1968).

Dulosis and Inquilinism

The most advanced forms of parasitism, exemplified by several species of social insects, particularly among the ants, are dulosis and inquilinism. The difference between the two is that in dulosis, ants of one species capture the pupae of another species and carry them back to the nest to become workers for the first species, and in inquilinism one species totally depends on and lives permanently with another species.

Dulosis has also been referred to as *slave-making* or *domestication*, terms borrowed from human use. Of the two, *slave-making* has been used more frequently but may be less appropriate because in human usage slaves are from the same species, whereas domestication involves other species. Even domestication may apply only where long-term genetic, evolutionary change has occurred; in other cases perhaps *taming* would be a better term (as in dogs being domesticated by humans whereas wolves are tamed only when brought into human captivity). Because of the problems with these everyday terms, we recommend using the less familiar *dulosis*. At least one case of true intraspecific slavery, incidentally, has been reported in ants (Figure 10-4).

Dulosis in ants was first reported by Huber (1810). Since then the subject has received much attention. Darwin, in *The Origin of Species*, proposed that dulosis began from ants of one species raiding another species for pupae that were used for food, but a few of the pupae survived and became workers. Wilson (1974), in an alternate explanation, suggested that the initial evolutionary steps involved a combination of territoriality and brood tolerance; that is, territorial ants fought with and killed other adult ants that nested too close, thus eliminating the adults, but they tolerated the presence of the pupae, which then emerged, and the workers began working for the winning colony.

Species that initiate dulosis usually conduct their raids with the aid of odor trails. Workers that discover a nearby colony lay a trail back to their own nest, which is then followed by others on the raid. The raiding ants have well-developed mandibles that are used for killing any resisting opponents. In some cases, such as *Formica subintegra*, the workers further disarm their opponents with propaganda substances (Regnier and Wilson 1971). The chemicals are produced in large Du-

Figure 10-4 True slaves (of the same species) in ants (*Mymecocystus mimicus*). Individuals of opposing colonies are displaying in tournament fights. Behavior shown here includes **A**, stilt-walking and head-on; **B**, lateral; and **C**, lateral with antennal drumming. The winning colony, after tournaments involving hundred of individuals, raids and enslaves the losing colony.
From Hölldobler, B. 1976. Science 192:912.

A

B

C

four's glands (Chapter 15). They cause alarm and dispersal in the defending workers but do not create adverse effects in the workers emitting the substances.

Dulosis and the raiding of other nests may in a few instances result in the raided queen also being captured and enslaved, with continued production of foraging slave workers. This may be followed by an evolutionary reduction in the repertoire of behavior such as foraging by the raiding workers and the eventual evolution of a species that permanently depends on the other species. This extreme state of complete dependence is inquilinism. Wilson (1971, 1975) has proposed that the above route is one of three major evolutionary pathways that may lead to inquilinism. The others are plesiobiosis to xenobiosis to inquilinism and temporary parasitism to inquilinism.

Once a species has become inquilic, it rapidly may degenerate (on an evolutionary scale) to the point where all but the basic reproductive tasks are delegated by it to the host species. In the extreme case (e.g., the little ant *Teleutomyrmex schneideri*) the queens have concave undersides that fit the shape of their hosts; their legs are adapted for grasping and holding on; and they ride around on the backs of the host queens. The little ant queens receive nourishment from the host workers, and they somehow physiologically induce the host queen to produce only workers rather than any reproductively capable offspring.

The evolution of social parasitism in insects appears to be related to temperate zones. Richards (1927) and Hamilton (1972) proposed that it involves closely related species living in neighboring latitudes. Their ranges subsequently may merge. Differences in the timing of their life cycles may result in one emerging sooner than the other, with the first taking advantage of the other. Wilson (1971) proposed that cooling, whether seasonally, nightly, or latitudinally, and associated immobility may dull the responses of host species, making them more susceptible to the raiding species.

INDIRECT TROPHIC AND NONTROPHIC RELATIONSHIPS BETWEEN ANIMALS AND PLANTS

Nature seems to have little respect for our arbitrary taxonomic boundaries, and most of the kinds of relationships described above that animals have with other animals also extend into the other kingdoms of organisms, particularly into the plants and fungi. Evolutionary developments involving behavior do not depend on the consciousness of the participants and hence can occur with organisms such as plants that do not even have a nervous system. The examples are exceedingly diverse, numerous, and exist in both directions: animals that gain at the expense of plants and vice versa, including animal parasites of plants and plant and fungi parasites of animals.

Direct trophic interactions are certainly the most familiar, including both herbivores that eat plants and carnivorous plants that eat animals, but many other interesting interactions exist as well. There are several species of animals, including leaf cutter ants, some termites, and humans, that cultivate plants or fungi for various uses (primarily food). There are numerous other commensal and mutualistic relationships.

Daniel Janzen, an active and prominent researcher of animal-plant relationships since the 1960s, provided a classic description of a mutualism between ants and acacia plants (Janzen 1966, 1967). Ants of the species *Pseudomyrmex ferruginea* and the bull's-horn acacia *(Acacia cornigera)* have an obligatory interdependence. The acacia provides the ants with food and shelter in large thorns that the ants hollow out. The ants in return protect the acacia from other, herbivorous insects that can inflict more damage. Neither species survives well under natural conditions without the presence of the other. Janzen (1966), for example, showed that *A. cornigera* plants where ants were experimentally kept off suffered greatly reduced growth because of damage from other insects. Control plants with ants, or experimental plants returned to ants, showed normal growth and development. The relationship is continuous throughout the year with the *P. ferruginea* ants being active 24 hours a day, unlike most other ants. The bull's-horn acacia has green leaves throughout the year, which provides a continuous supply of food for the ants.

There are other ant-acacia relationships involving different species, but they are facultative, not obligatory like the one described previously; that is, both the ants and acacias can exist independently of each other. Unlike the acacias, which are guarded by ants, many acacia species without ants have toxic compounds in their leaves.

An apparent plant-animal mutualism between an endemic tree species *(Calvaria major)* and the now extinct dodo bird *(Raphus cucullatus)* on the island of Mauritius may have led to the near extinction of the tree (Temple 1977). *Calvaria* has large, tough seeds that appear to require passage through the digestive tract of a large bird before they germinate. A few of the mature trees continue to survive, but there may have been no natural germination since the extinction of the dodo over 300 years ago, in 1681. Temple was able to artificially germinate some *Calvaria* seeds by force feeding them to domestic turkeys and recovering those that were subsequently regurgitated or passed in the feces.

Some of the most complex animal-plant relationships involve pollination, with orchids being the best known and most highly specialized. As early as 1793 Sprengel (cited in Darwin 1889) described the actions of parts of orchid flowers and discovered that insects are necessary in their pollination. Since the time of Darwin and his interest in flower and insect relationships, there has been a voluminous amount of research and literature on the subject. Faegri and Pijl (1966), for example, cite a single handbook on pollination, published from 1895 to 1905, that covered 2972 pages. For more recent, additional references, see Pijl and Dodson (1966).

There are estimated to be well over 20,000 species of orchids, comprising about 7% of all species of flowering plants. As one might expect, there is a wide diversity of pollination methods among them. Most are specialized for one type of pollinator, ranging from birds to bees or other kinds of insects, including moths, butterflies, dipterans, and beetles. Many are adapted for pollination by a single species of animal. The orchids attract the pollinators with nectar, then attach the pollen to their bodies in a mass, the pollinarium, that is subsequently carried to another flower. An example for a simple bee-pollinated orchid is shown in Figure 10-5.

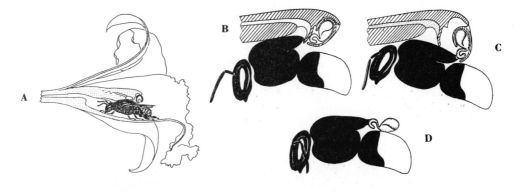

Figure 10-5   Mechanism for attachment of pollinarium in a simple bee-pollinated orchid flower. **A** and **B,** Position of bee in relation to the column after the bee has completely entered the flower. **C,** As the bee backs out, the dorsal thorax pushes against the rostellar flap and sticky liquid from the stigma cements the base of the pollinarium to the bee. **D,** Position of the pollinarium as the bee flies to another flower.
Drawing by G.P. Frymire. Figure and legend from van der Pijl, L., and C.H. Dodson. 1966. Orchid flowers. Their pollination and evolution. University of Miami Press, Coral Gables, Fla.

The variety of orchid flower structures and associated pollinator interactions is mind boggling. Many of the orchids force the animals into restricted passageways as a means of attaching the pollen mass. Some of the potential pollinators, such as carpenter bees, which have strong mandibles, however, bypass the flower mechanism and simply cut or tear their way into the base of the flower, obtaining the nectar without picking up the pollen. But as an evolved defense against such breaking and entering, some Asian orchids have additional nectaries outside the flower that attract particular ants that, in turn guard the base of the flower and force the bees into the normal route for obtaining nectar and picking up the pollen.

A final well-known example of animal-plant interaction involves deceptive mimicry on the part of the plant. Butterflies of the genus *Heliconius* feed on neotropical vines belonging to the genus *Passiflora*. The larvae also eat other *Heliconius* eggs. As a result, female butterflies carefully avoid laying eggs at sites on *Passiflora* plants where other *Heliconius* eggs already exist. A number of *Passiflora* species in turn have evolved structures that resemble butterfly eggs, which prevent butterflies from laying eggs there. Whether this should be considered "communication" between the plants and the butterflies is a matter open to debate (see Chapter 15). However, it leads us to a general phenomenon in interspecific interactions: the use of a species' communication signals by another interacting species.

TAPPING THE COMMUNICATION LINES OF OTHER SPECIES

As suggested at a number of points in this chapter, highly advanced interspecific interactions often involve the use of one species' signals by another. This may not only occur between closely related species but also may involve convergence of distant species, such as between the cuckoos and their hosts. In insects the ability may extend to other orders of insects or even to other groups of arthropods. Some bolas spiders, for example, capture only one type of prey: male moths that they attract by emitting the moth's sex pheromones (chemical attractants), as if the spiders were female moths (Stowe et al 1987).

One of the best-documented cases involves a beetle *(Atemeles pubicollis)*, which is a social parasite in colonies of ants of the genera *Formica* and *Myrmica* (Figures 10-6 and 10-7). This fascinating relationship was investigated thoroughly by Holldobler (1970, 1971). The beetles locate the ants by the airborne odors of the colonies. They prefer to winter with *Myrmica*, which maintain larvae throughout

Figure 10-6 Drawing of a larval *Atemeles* beetle being fed by its *Formica* ant host. From Hölldobler, B. 1967. Zeitschrift für Vergleichende Physiologie 56:1-21.

Figure 10-7 *Atemeles* beetle using ant communication signals to gain access to ant colonies. As shown in this sequence, **A,** the beetle antennates with the ant; **B to D,** secretes appeasement substances and then adoption chemicals; **E,** curls up into a pupal posture, and is carried by the ant to its brood chamber. **F,** Also shown are the positions of the various glands.
From Hölldobler, B. 1970. Zeitschrift für Vergleichende Physiologie 66(2):215-250.

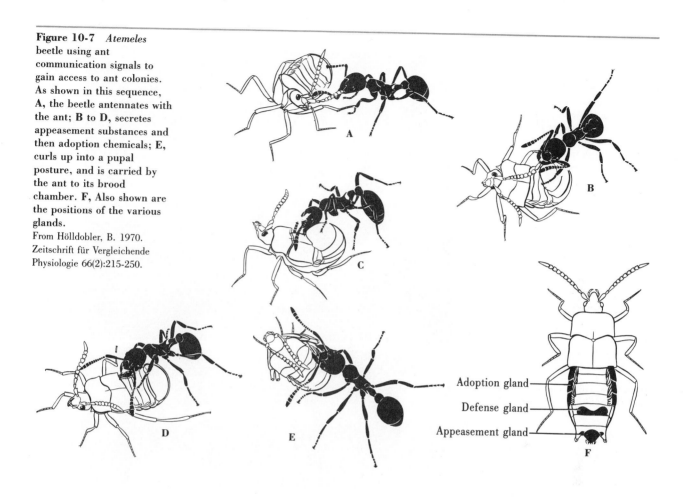

the fall, winter, and spring; then they switch to *Formica*, which have larger colonies and more larvae during the summer. The beetles mingle with ants by using the ants' communication signals.

When the beetles first encounter a worker ant, they present to the worker an appeasement gland, located at the tip of the abdomen. This calms the worker, and the beetle then presents its adoption gland, which the ant licks. Subsequently the beetle is picked up and carried by the ant into the colony. If the appeasement and adoption signals fail and the worker ant becomes aggressive, the beetle has a system for bailing out of the tight situation: defensive glands that repel the ants. Once inside the ant colony, the beetles communicate with the ants by tactile signals and are fed by the ants in the same fashion that the ants normally feed each other. The larvae of the beetles have glands that cause the ants to pick them up and place them with the ants' own larvae, which the beetle larvae then proceed to eat.

In addition to the *Atemeles* genus, there are numerous other beetles and other arthropods that parasitize social insects by tapping the chemical and tactile messages of the hosts. The social insects are particularly susceptible to such intrusion, according to Wilson, because of (1) "the relative impersonality of insect societies," and (2) "the narrow sensory Umwelt of [the] hosts." (For a description of Umwelt, see Chapters 1 and 16.) Wilson (1975) quotes a statement by Wheeler, who spent much of his life studying social parasitism, that compares the situation of code-

Figure 10-8 Tachinid fly (*Euphasiopteryx ochracea*) females approaching and attacking a cricket (*Gryllus integer*) using the cricket's song. In this series of photographs the cricket is dead but is mounted above a speaker through which the songs are being played. From Cade, W. 1975. Science 190:1312.

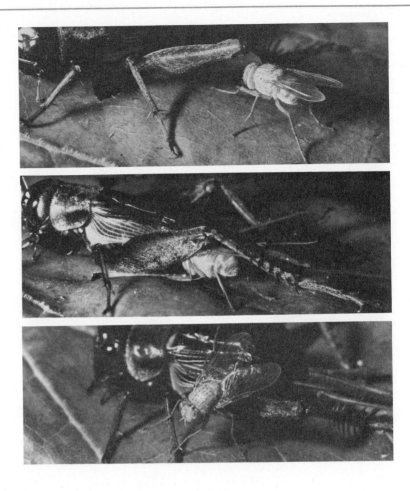

breaking and the subsequent effects with what would happen if humans were to "delight in keeping porcupines, alligators, lobsters, etc., in our homes, insist on their sitting down to table with us and feed them" to the neglect of our own children.

Exploiting the lines of communication of other species is practiced not only by parasites but also by predators (keeping in mind that the distinction between parasites and predators is not always clear). Some predators locate their prey by the prey's intraspecific vocalizations and visual displays. Examples include tachinid flies (Figure 10-8) and frog-eating bats (Tuttle and Ryan, 1981).

Fireflies belonging to the beetle family Lampyridae court members of the other sex by communicating with flashes of light. The males, which generally outnumber the females manyfold, fly about advertising themselves with species-specific flashes. Important characteristics of the flash are duration, frequency, and pattern of flashes, as well as repetition of sequences (see Figure 15-6). The patterns of males are now known for over 130 species around the world. Different species using different flash patterns often live sympatrically. Females, which wait on the ground or perch on vegetation, answer males passing overhead with relatively simple flashes that depend partly on precise timing for species identification. When a male detects an appropriate female response, it hovers near the female. The two continue to flash to each other; then the male lands, walks to the female, and attempts to copulate.

In most species the behavior is basically as just described, without further complication or elaboration. The flash patterns are sufficiently distinct and stereotyped that species can be separated by the taxonomist. In some species, however, individual fireflies vary their signals (Carlson and Copeland 1978). A few species show variation to such an extent that they have caused taxonomic confusion. Some of the variation consists of males varying their signals under different contexts, such as for searching versus actual courting, or variation used to more precisely locate the females.

Other variability, however, has been discovered to consist of aggressive mimicry on the part of the females, the femmes fatales, in a few predatory species of the genus *Photuris*. These females mimic the responses of females belonging to other species, attract and catch a proportion of their males, and eat them (Figure 10-9). Individual females may mimic the responses of three or more other species (Figure 10-10).

Figure 10-9 Femme fatale. A predatory female firefly (*Photuris versicolor*) is eating a male firefly of another species. When males advertise themselves by flashing, the predatory females attract them by mimicking their own females' response flashes.
Photograph courtesy James Lloyd.

Figure 10-10 Luminescent signals of fireflies. Response used by predator female is shown below the female answer she mimics. Vertical bars at right indicate observed individual repertoires; *n* is the number of females exhibiting the repertoire. Capture rates (percent) are adjacent to prey species. The flash rate of the *Photuris congener* female is variable and not well understood.
From Lloyd, J.E. 1975. Science 187:452.

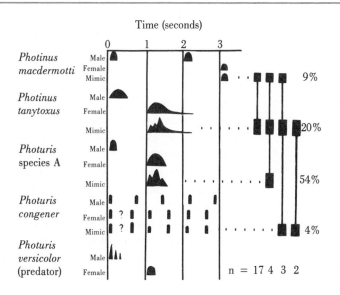

The femmes fatales generally alternate between reproductive and predatory behavior. They first use their own species-specific patterns to attract correct males, with which they mate. After copulation, they switch to calling other species' males, which are eaten. Later they may revert to correct courting responses. The behavior has been studied extensively under both field and laboratory conditions (Lloyd, 1975, 1981). Techniques have ranged from the use of simple stopwatches and penlights for simulating the flashes to complex photomultiplier and electronic recording systems. During field experiments where the researcher was artificially interacting with various femmes fatales females, subjects occasionally would interrupt the experiment to answer passing wild males (Lloyd 1975).

To further complicate the femme fatale story, males of some of the predatory species also mimic the other species by using their male patterns (Lloyd 1980). These males normally use their own species-specific flashing, but, appropriate to the location and seasonal timing of other species that may be present, they occasionally mimic the other males. More than 15 hypotheses have been offered to explain the evolution of such mimicry on the part of males. After sorting through the pros and cons of the various explanations, Lloyd (1980, 1981) leaned toward an explanation that basically consists of males tricking females that are attempting to trick other males. The males may mimic other males until they are called in, then switch to their own species-specific patterns and induce the female to change from predatory to reproductive behavior. Other explanations also may be partially valid. For example, there may be an element of "kamikaze-copulation."

Humans also use species-specific communication signals to attract other species, such as with artificial game calls (e.g., duck calls or elk bugling) used during hunting or when insect pheromones are used to attract insects during control efforts.

COEVOLUTION AND A SYNTHESIS: EVOLUTIONARY RELATIONSHIPS AMONG THE INTERSPECIFIC RELATIONSHIPS

How do the various types of interspecific interactions arise, and are different categories related in any way to each other? Most or all of these interactions undoubtedly begin with casual, incidental encounters, as shown by the arrows in Figure10-1 (page 291). Furthermore, as in any evolutionary development, the interaction also requires a sufficiently high frequency of encounters and the genetic opportunities on the parts of the participants to permit it to arise and become established. That is, the components of the interaction must already exist within the normal variation of behavior or arise by mutation and then become fixed either by chance or by selective advantage. Beyond that, the relationships and coevolutionary routes between different forms of interspecific interactions, if and when such routes exist in the first place, are not always so obvious.

In an attempt to bring some semblance of theoretical order to the incredible array of observed interspecific relationships, John N. Thompson (1982) investigated the ecological factors that may favor such interactions, irrespective of the taxonomic boundaries. He set forth his findings as a framework for further discussion and research rather than definitive conclusions. Some of these were presented earlier for trophic interactions in Chapter 9.

Thompson suggests that coevolution between interacting species is least likely for competition because, unlike other interactions, "selection does not act on any competing species specifically to increase the likelihood that an individual will encounter a competitor." Rather, selection among competitors would be expected to cause them to diverge, which then creates the "White Knight's dilemma" (from Lewis Carroll's *Through the Looking Glass*). The dilemma is that one has to infer the presence of competition from failure to use a resource, like the White Knight, who had a beehive that bees did not come to and a mousetrap that mice would not come to, because, according to the White Knight, "I suppose the mice keep the bees out—or the bees keep the mice out, I don't know which." Thus it is often difficult to infer competition in the first place and even more difficult to infer coevolution that may have resulted from competition. There have, however, been recent experimental demonstrations that competition occurs. The artificial manipulation of densities, both reductions and increases, of suspected competitors among bird species results in changes in reproductive fitness and distributions of the participants (Leonard and Picman 1986, Gustafsson 1987). Also, recall the competitive interactions that affect habitat use among mice (Chapter 7).

Thompson proceeds to reason that coevolution among competitors, if and when it does occur, would be most likely with increasing chances of encounters, namely in environments with fewer rather than more species and when species are less mobile and less able to avoid competitors. Furthermore, the environment should be relatively stable and benign over long periods of time, or fluctuations in resources may keep potential competitors below the levels where competition would occur on a regular enough basis for significant selection to occur. The conditions favorable for coevolution among competitors would not guarantee coevolution but merely increase the probabilities of encounters under which it could take place. Some resources, such as the presence of other mutualistic organisms, however, might keep species competing in close enough association over sufficient time periods that coevolution would be much more likely; hence competition over mutualists would be a good place to look (Thompson 1982).

Thompson suggests that commensalism and mutualism are derived primarily from antagonistic encounters that are inevitable and unavoidable. (If the interaction can be avoided or prevented, then one would expect an escalation of defense or avoidance mechanisms and behavior.) He states (1982:61): "If it is unlikely that individuals can avoid a specific antagonistic interaction, then selection will favor individuals that have traits causing the interaction to have at least less of a negative effect on them. This selection regime sets the stage for the evolution of the interaction toward commensalism or mutualism." As an example, under warm, humid conditions pollination of plants by animals might be favored over wind pollination (Raven 1977)—if the animals do not eat all of the pollen first; hence plants that provided nectar that the animals would eat instead of the pollen would have a clear selective advantage. Similarly, there would be an advantage to fruits that are eaten instead of the seeds, seeds that pass through the digestive tract undigested, or situations in which large numbers of seeds are produced and carried about by animals, with only a portion being consumed and the remainder surviving to possible germination.

Mutualism being derived from antagonistic relationships may also extend to cases involving three or more species. The cowbird-oropendula-botfly situation would be an example. A similar case involves burying beetles, mites, and *Calliphora* flies (Springett 1968). The beetles and flies are competitors over dead mice, on which they both lay their eggs and the larvae develop. However, the flies are the superior competitors, and when flies are present the beetle larvae do not survive. The mites the beetles carry, however, eat fly eggs and permit the beetles to survive.

Other conditions that favor or predispose the development of mutualism, according to Thompson's (1982) synthesis, include the following: (1) environments that favor intermediate levels of survival (conditions with low survival rates do not favor encounters and when they do occur may pose high risks of associating with the potential partner, and conditions with high survival do not gain from additional advantage); (2) environments causing physical stress, such as low levels of nutrients, in which the partner may offer a large advantage by providing resources; and (3) situations involving social species that have complex interactions already in their repertoire. Concerning the last point, for example, a bibliography by Bequaert (1922) before the recent interest in social insects already contained over 1100 citations on relationships between ants and plants.

Thompson (1982) reviews other hypotheses and considerations from a number of other workers attempting to explain interspecific interactions. Other persons more recently have discussed these interactions from contemporary viewpoints. Davies et al (1989), for example, considered interspecific brood parasitism in birds and ants as evolutionary arms races, in which each of the two opposing sides must keep up with or get ahead of the other side if it is to survive the interaction. Davies et al noted a number of remarkable similarities in the brood parasitisms of cuckoos and ants, including parasite manipulations and deceptions of hosts, and parasites taking advantage of general behavioral rules of the hosts such as "feed all brood in your nest" and "work for the colony you mature in." A few of their conclusions based on this line of thinking include a number of reasons why hosts often seem defenseless (e.g., because the costs have become too high, because counter adaptations are not worth the investment if the enemy is rare, or because the host is lagging in the arms race). Davies et al also suggested why social parasitism is not more common: essentially because the benefits often barely exceed the costs, such as searching for suitable hosts, competing with other parasites for hosts, and the costs of a continuing arms race in which the hosts fight back. In other words, few parasites get a free ride.

One aspect that has not received much theoretical attention, aside from that reviewed by E.O. Wilson mentioned previously, is why the most advanced forms of interspecific relationships (excluding humans) seem to predominantly involve birds and the highly social insects. Is that just an artifact of those two groups receiving more attention in general, or are there certain factors that predispose those two groups to more complicated interactions? We suspect there are several important factors, including that both groups are numerous, hence affording a high frequency of both reliable encounters and evolutionary opportunities; they are highly social, with a high degree of parental care and complex communication signals that can be tapped; they both have nests that provide reliable locations for interactions to occur; the nests contain reproductive objects (such as eggs, larvae, pupae) for which parasites can substitute their equivalent items; and there also are numerous other species in the suitable size range that provide ample evolutionary opportunities on their side (i.e., there is a high probability that the necessary behavioral variation for the interaction to arise exists somewhere among the many candidate parasites). Elephants and whales, by contrast, are highly social and have a high degree of parental care and communication, but they do not remain at fixed sites and they are relatively unique in terms of size so there simply are not many opportunities for other species to socially parasitize them.

All of this remains a rich area for further theoretical development and research.

SUMMARY

Interspecific interactions and behaviors form a large category ranging from relatively neutral situations involving casual encounters to various forms of direct competition, predation, and parasitism, in which one or both sides suffer a loss from the interaction. Defensive behaviors, seen in a variety of interactions, similarly cost both sides in an encounter. In many cases, organisms of different species live together in close association throughout much of their lives (symbioses). Advantages of such living arrangements may accrue to both participants (mutualism), be one sided with advantages to one and basically not affecting the other (commensalism), or be harmful to one of the participants (parasitism). These and other different categories and a number of subdivisions are summarized as follows (+ = benefit or advantageous; − = loss or disadvantageous; 0 = no significant positive or negative effect on the participant; first for one species, "species A," and the second for the other, "species B"):

    casual encounters (0, 0)—the encounter is without proximate significance but may be important ultimately

    amensalism (0, −)—a harmful encounter for one of the participants, usually accidental

    competition (−, −)—generally involves limited resources (particularly food, cavities for nesting, or other protective cover)

        exploitative—participants use the same resource but without direct contacts or interactions with each other

        interference—participants directly interact (frequently fight) with each other, often with stereotyped behaviors

    defense (−, −)—behavior that costs both sides, one in producing the defense and the other in reduced or lost benefits or in producing countermeasures

    commensalism (+, 0)—benefit to one without harm to the other

        plesiobiosis—participants live in close contact but benefits not well understood

        interspecific groups—such as flocks, herds, schools, etc., consisting of more than one species, often involving feeding

        trophic commensals—the advantage involves food

        phoresy—one species uses another for transportation

    mutualism (+, +)—both species benefit

        interspecific groups—as for the similar category in commensalism

        trophobiosis—involves food, with both sides benefiting

        nontrophic symbioses—the benefits are other than food, (e.g., shelter, warning of danger)

        parabiosis—species that share all activities except reproduction

    trophism (+, −)—feeding (grazing, parasitism, predation); see previous chapter

    nontrophic parasitism (+, −)—one species takes advantage of another in ways (mostly) not involving feeding on the host

        kleptoparasitism—robbery of the host's food or other resources

        xenobiosis—the host tolerates and may even encourage the parasite

        temporary social parasitism—occasional social parasitism, such as nest or brood parasitism

        dulosis—(also, perhaps inappropriately, called *slave-making* or *domestication*)—one species appropriates the workers of another species

inquilinism—an extreme form of dulosis in which the parasite totally and permanently depends on the host

The last two categories are found primarily in various species of ants.

Complex interspecific interactions occur not only among different animal species but also between animals and plants and fungi. Examples are the numerous complex interactions between orchids and insects.

Most types of complex, evolved interspecific relationships probably originate from chance encounters. Their course of evolution then depends on a host of factors related to the ecology, frequency of encounters, and genetics—evolutionary predispositions and opportunities provided by the phylogenetic histories and behavioral variation of the species involved. The most complex forms of interspecific interactions seem to occur predominantly in two groups—birds and highly social insects—perhaps for a number of speculated reasons. A number of the specific interactions have also received theoretical consideration. Three hypotheses have been proposed for the evolution of inquilinism in ants, for example: (1) via intermediate levels of raiding of other species, followed by dulosis; (2) by simply living in close association with other species, then going through intermediate levels of evolution; and (3) by an evolutionary route involving trophic parasitism. Most of the theoretical relationships between different types of interspecific interactions are summarized by arrows in Figure 10-1.

FOR FURTHER READING

General

Davies, N.B., A.F.G. Bourke, and M. de L. Brooke. 1989. Cuckoos and parasitic ants: interspecific brood parasitism as an evolutionary arms race. Trends in Ecology and Evolution 4:274-278.
Comparative analysis of similarities in social parasitism of these two unrelated groups, using an arms-race analogy.

Thompson, J.N. 1982. Interaction and coevolution. John Wiley & Sons, New York.
A concise but thorough review and analysis of the possible ecological and evolutionary factors involved in the development of interspecific interactions among organisms, regardless of taxonomic boundaries. This book helps synthesize the current knowledge on the subject and sets the stage for further discussion and research.

van der Pijl, L., and C.H. Dodson. 1966. Orchid flowers. Their pollination and evolution. University of Miami Press, Coral Gables, Fla.
A moderately detailed, readable, and excellent description of orchids and their diverse structures and interactions with various pollinating animals. Includes chapters on mimicry and evolution of orchids.

Wilson, E.O. 1971. The insect societies. Belknap/Harvard University Press, Cambridge, Mass.
A major review of insect social behavior that includes coverage of many interspecific relationships.

Wilson, E.O. 1975. Sociobiology, the new synthesis. Belknap/Harvard University Press, Cambridge, Mass.
Another major publication by Wilson, this book extends the coverage of social behavior to groups other than insects, again including many interspecific situations.

Recommended Original Papers

Dussourd, D.E., and T. Eisner. 1987. Vein-cutting behavior: insect counterploy to the latex defense of plants. Science 237:898-901.

Gustafsson, L. 1987. Interspecific competition lowers fitness in collared flycatchers *Ficedula albicollis*: an experimental demonstration. Ecology 68:291-296.

Isack, H.A., and H.U. Reyer. 1989. Honeyguides and honey gatherers: interspecific communication in a symbiotic relationship. Science 243:1343-1346.
A fascinating and highly recommended paper.

Leonard, M.L., and J. Picman. 1986. Why are nesting marsh wrens and yellow-headed blackbirds spatially segregated? Auk 103:135-140.

Munn, C.A. 1984. Birds of different feather also flock together. Natural History 93(11):34-42.

Murata, M., K. Miyagawa-Kohshima, K. Nakanishi, and Y. Naya. 1986. Characterization of compounds that induce symbiosis between sea anemone and anemone fish. Science 234:585-587.

Smith, N.G. 1968. The advantage of being parasitized. Nature 219:690-694.
    A classic paper on complex, tangled interspecific interactions.

Stowe, M.K., J.H. Tumlinson, and R.R. Heath. 1987. Chemical mimicry: bolas spiders emit components of moth prey species sex pheromones. Science 236:964-967.

Williams, K.S., and L.E. Gilbert. 1981. Insects as selective agents on plant vegetative morphology: egg mimicry reduces egg laying by butterflies. Science 212:467-469.

11

REPRODUCTIVE BEHAVIOR
SEXUAL REPRODUCTION AND
SEXUAL SELECTION

For biologists, the ultimate measure of an organism's success is the number of surviving offspring it produces. Thus an understanding of the behaviors involved in reproduction is crucial for biologists with an evolutionary perspective. Apart from their evolutionary significance, however, reproductive behaviors also include some of the most complex and spectacular actions of animals.

For most animals, reproduction is sexual, involving gamete (egg and sperm) production followed by fertilization. The widespread existence of sexual reproduction is something of an evolutionary puzzle, but given its existence, sex leads to sex differences in types of reproductive effort, in morphology, and in behavior between large-gamete producers (females) and small-gamete producers (males). In this chapter, we show how the concept of sexual selection, first presented by Charles Darwin and developed a century later by Robert Trivers and others, explains these sex differences in behavior.

Sexual selection involves competition to gain access to mates, often but not always among males, and competition to choose the best possible mates, often but not always among females. There are many forms of mate competition and mate choice—we examine some of the ways in which competition for mates occurs and some of the criteria used in mate choice. New and traditional explanations for the courtship behaviors of animals are considered. Finally we look at the "exceptions that prove the rule," instances of competition among females and choice by males that are understandable in the context of sexual selection theory.

318

REPRODUCTIVE BEHAVIOR OF THE RED-WINGED BLACKBIRD: A CASE IN POINT

One of the most familiar of North American birds is the red-winged blackbird, *Agelaius phoeniceus*. Its reproductive behavior is described by Searcy and Yasukawa (1983). In the spring, male red-wings leave large overwintering flocks and attempt to establish territories in marshes or wet grasslands. Male red-wings are jet black in color, with brilliant red and yellow shoulder patches, or epaulets (Figure 11-1, *A*). Territory establishment involves use of the epaulets in shoulder-spread displays directed at intruders or males in neighboring territories, supplemented by loud vocalizations and, if necessary, chases and physical combat. The most familiar vocalization is the conc-a-ree song, but males have one to seven additional song types in their repertoires. Competition for territories is fierce, especially for prized marsh territories with high densities of cattail *(Typha)* vegetation. Many males cannot obtain such territories and must either accept inferior territories or become nonterritorial floaters. Absence of a male from his territory for as little as 15 minutes may result in a takeover by a floater or a territorial neighbor.

Experiments show that larger males tend to win the chases and combats between males; other experiments show that songs and shoulder-spread displays help repel intruders without actual fighting. Peek (1972) and Smith (1976, 1979) muted male red-wings by cutting the hypoglossal nerve that controls the birds' sound-producing structure, the syrinx, or by puncturing the intraclavicular air sac. (Cutting the nerve mutes birds permanently; puncturing the sac mutes birds for 2 to 3 weeks. Control birds received equally stressing but nonmuting sham operations.) The muted birds suffered more intrusions and tended to lose part or all of their territories. Yasukawa (1981) performed speaker-occupation experiments, in which territorial males were removed and replaced with loudspeakers that produced either red-wing songs or control silence. Intrusion rates were significantly lower when songs were present, and were lowest when a varied repertoire of songs, rather than a single song type, was played.

Figure 11-1   A, Male and B, female red-winged blackbirds, showing extreme sexual dimorphism.
Photographs courtesy Ed Bry, North Dakota Game and Fish Department.

A

B

Peek and Smith also covered the epaulets of some territorial males with black polish (controls received clear substances instead). Intruders appeared more often, responded less to threats, and usually ousted the resident when confronted with an epaulet-covered bird instead of a control resident.

Female red-wings arrive in breeding areas after male territories are established. Females are less brightly colored than males, being a dull, streaked brown (Figure 11-1, B). They are also smaller, 20% shorter than males, and 50% lighter (35 g compared with 65 to 80 g). Females settle into male territories, build nests there, and mate mostly with the resident male. There are exceptions: Bray et al (1975) vasectomized territorial residents, but a high percentage of their females nevertheless laid fertilized eggs. As many as 20% of female matings may involve extra-pair copulations (EPCs) with neighboring residents or floaters. Territory choices are important for females, because the young will be raised there, and the territory must contain lots of food (mainly emerging dragonflies) and cover to provide protection from predators (as many as 50% of clutches are completely lost to predators).

Red-winged blackbirds are polygynous: a male with a superior territory may have more than one female mate nesting there. On the other hand, many males do not breed, either because they cannot establish territories at all, or because their poor territories attract no females. The range of harem sizes in territorial males is large, from 0 to 15 females per territory. The number of offspring for males is related in a linear fashion to harem size, so a male with 15 females has 15 times as many offspring as a male with one female. Studies by Lennington (1980) and others showed that successful males were ones whose territories were large and had high cattail densities, where food was abundant and rates of nest predation were low. That females tend to choose territories rather than males per se is shown by the fact that certain locations tend to be preferred by females year after year, despite the fact that those areas are possessed by different resident males in different years.

This is not to say that the males that possess large harems are a random subset of the male population. Breeding males tend to be older, experienced birds that have large song repertoires and sing at high rates. These traits help males establish large territories in high-quality areas. Females may be able to locate good territories quickly by paying attention to the displays of males, since high-quality territories tend to be occupied by high-quality males.

With several females nesting in some territories, one might expect some aggression between females. Such competition does occur, but it is less intense than that between territorial males. Females have a chatter song, but it is given at lower rates than male songs. Females are not known to exclude others from settling in the same male territory, but early females (primary females) can delay breeding by latecomers through aggressive behavior. This results in some variation among females in number of offspring produced, but the range is much less than the fifteenfold variation among males.

The burden of parental care in red-winged blackbirds falls mainly on the females. They do all of the incubating of eggs and the majority of the feeding of the nestlings and fledglings. Males do provide two types of parental care: they guard the nest against predators, and in some populations they feed the young enough to affect nestling survival rates. In such populations the experienced males with large territories that have large harems tend to feed the young at higher rates than do less successful males on poorer territories. Thus the potential parental assistance of males may be another factor in the preference of females for certain males and territories.

The patterns just described for red-winged blackbirds are fairly typical for sexually reproducing animals (although not necessarily for birds as a group). Males are often larger, more conspicuously colored, more aggressive, and devote more effort to obtaining mates and less to parental care than females do. With this example in mind, we can begin to explore reproductive behavior in animals more generally. We begin by asking a very basic question: why does sexual reproduction exist at all?

WHY SEX?

Sex is a complicated business. At the cellular level, chromosomes must be condensed, duplicated, matched up with partners, broken into pieces that are swapped between partners, reassembled, separated from their partners, and split into two separated chromatids before unwinding into dispersed form (refer to the discussion of meiosis and recombination in any good general biology book, see Chapter 4). At the organismal level, sex involves the development of specialized primary and secondary sexual characters. Sexual behavior involves the expenditure of large amounts of time and energy, and its conspicuousness often increases the risk of predation (not to mention the danger from sexually transmitted parasites and pathogens). Worst of all, from an evolutionary standpoint, sexual reproduction is a particularly inefficient method of passing on one's particular alleles. The paradox of sexual reproduction was stated forcefully by Williams (1975), who wrote, "…

Figure 11-2 The spread of asexual mutants in a previously sexual population. The asexual females (dark figures) and sexual females (open figures) are assumed to produce 10 offspring each, of which 2 survive to breed. The proportion of asexual females in each generation is shown at the right.
From Trivers, R.L., 1985. Social evolution. Benjamin/Cummings, Menlo Park, Calif.

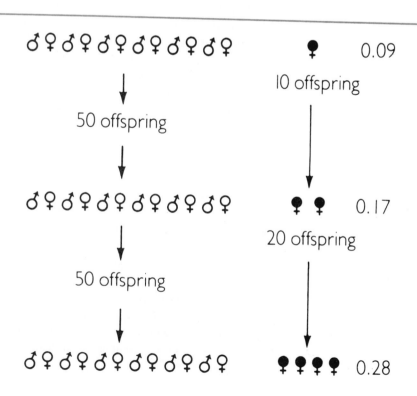

the prevalence of sexual reproduction in higher plants and animals is inconsistent with current evolutionary theory."

Consider a mutant asexual (parthenogenetic) female in a sexual species as explained by Trivers (1985). This mutant produces only daughters; the surviving ones themselves produce only daughters. By contrast, sexual females produce offspring that consist of equal numbers of sons and daughters. The sons do not produce any new offspring, but merely contribute genes to offspring produced by other females' daughters. Therefore, as Figure 11-2 shows, the reproductive output of asexual forms doubles that of sexual forms every generation. It is as if there are two merchants with equal amounts of money to spend in buying goods for sale, but one of them always pays twice as much as the other—we would not expect these two to prosper equally. (Sexual females get to purchase only half as many genes in the next generation per amount of egg-producing material as asexual females do.) This twofold genetic cost of sexual reproduction has been called the *cost of sex* and the *cost of meiosis*, but it is perhaps best thought of as the *cost of making males*.

To explain the continued existence of sexual reproduction, evolutionary biologists need to find a benefit of making males that is greater than the twofold cost. Until the 1960s, biologists thought they knew the answer (or rather, answers). Sex was advantageous to evolving species for two long-term reasons, first explained by the geneticist Muller (1932).

Sex is in essence a mechanism for creating new genetic combinations. The process of crossing over in meiosis creates new combinations of genes on particular chromosomes. When eggs and sperm combine, new combinations of chromosomes come together in the zygote. Thus each offspring in a sexual species presents a new genetic combination, different not only from both parents but also from all other members of the species. According to Muller, the first advantage of this gene shuffling was to provide new combinations that would allow evolution in a species to keep pace with changing environments. Old genotypes, suited to old conditions, could be quickly replaced by new genotypes providing better combinations for new conditions. Asexual species have to wait for the unpredictable, infrequent appearance of a series of favorable new mutations and probably become extinct while waiting. Sexual species can quickly assemble new, fit gene combinations.

Muller's second advantage for sex involves getting rid of some genetic variations. A new unfavorable mutation in an asexual female is passed on to all her offspring, which must suffer the consequences. In a sexual female, only some offspring inherit the bad allele. When they die, the species is cleansed of the bad genes, whereas other offspring that do not possess the bad mutation perpetuate the female's lineage. This mechanism has been called *Muller's ratchet* (Maynard Smith 1978).

In the 1960s, biologists realized that Muller's explanations were not really sufficient. They were essentially group-selection explanations, in which sex was advantageous to species over the long course of evolution. (See Chapter 14 for a discussion of group versus individual selection.) The trouble is that, although evolution of a species proceeds over the long term, natural selection on individual organisms proceeds every generation. Muller's explanations require sexual females to pay a 50% cost in fitness every generation, so that at some point in the indefinite future the species will be able to evolve rapidly or get rid of new bad mutations. The short-term advantages of asexual reproduction should lead to the loss of sex in the short term. Clearly, biologists required new, short-term, individual-selection explanations for sex.

Williams tackled the problem in his 1975 book, *Sex and evolution*. He pointed out that short-term advantages for sex must exist in at least some species, since many species undergo alternating cycles of asexual and sexual reproduction (examples include aphids and rotifers). Because these species are capable of suspending sex for a time, they are presumably able to evolve the capacity to do without it at all. That they have not done so suggests that it must frequently be advantageous to reproduce sexually. Williams used a striking analogy, that of a lottery, to explore what the advantages might be.

Imagine a lottery in which $1000 is paid out for every correct guess. Tickets cost $1, and you have $100 to spend. If you know the winning number, clearly your best strategy is to submit that as your guess 100 times and win 100 times. Williams says that is the asexual strategy: if environmental conditions are stable and predictable, you should be asexual and produce offspring all of whom are identical to you. The fact that you have survived to reproductive age indicates that you have a good genotype for the existing conditions.

Now imagine a slightly different situation: you still have $100 to spend but do not know what the winning number is. In this case, your best strategy is to submit 100 different guesses, so as to increase your chance of having at least one winning entry. This, said Williams, is the sexual strategy. If future conditions are unpredictable, your success now does not necessarily predict success of the same genotype in the future. A female should produce genetically diverse offspring, to increase the chance that at least one will have a genotype well suited to whatever the new conditions turn out to be.

After Williams, the problem became one of finding ways in which environments could vary so much every generation that sex should be continually favored as an individual strategy. Sex, after all, produces a genotype in one generation, but subsequently breaks it up with meiosis in the next generation. If environments were sometimes variable and sometimes stable, the pattern mentioned above of alternating periods of asexual reproduction with periods of sex would be best. Yet many animals, including most vertebrates—all birds and mammals—are exclusively sexual. This is puzzling, especially since birds and mammals, with their well-regulated homeostasis, seem to be able to control their conditions better than other organisms—their worlds seem less variable, not more variable, than those of lower organisms. Williams concluded that sexual reproduction in the higher vertebrates was not the best arrangement, but was simply present as an evolutionary holdover. Sex had been favored in early vertebrate ancestors and had simply been retained in their modern descendants, perhaps because the required developmental changes to do away with it are too extensive to evolve easily.

Others continued the attempt to find individual selection advantages, asking how environments could differ in such a consistently chaotic way so as to make exclusively sexual reproduction favored even in such organisms as the higher vertebrates. Bell (1982) suggested that a good way to attack the problem was to look for slight differences in ecological abilities of sexually produced individuals. If slight differences existed in habitat patches used by members of a species, the genetically variable offspring of a sexual female could occupy a number of these patches and could flourish without competing with one another. The progeny of an asexual female, however, being genetically identical, would be well suited to only one or a few patches, and would all compete with one another. Thus sibling competition plays a key role in Bell's hypothesis. Borrowing an image from the last paragraph of

Darwin's *The Origin of Species*, Bell calls his idea the *tangled bank* hypothesis (you are urged to look up the original passage and evaluate the appropriateness of that label).

The currently most-favored individual selection hypothesis was proposed by Jaenike (1978) and Hamilton (1980). They emphasize that the environment of an animal includes other species, as well as physical factors such as the weather. Natural enemies such as parasites and predators differ from the physical environment in an important way: not only are they hostile, but they also evolve through natural selection. Thus success of one host or prey genotype at one time may not predict continuing success of that genotype at later times—the parasite or predator continues to evolve new genotypes better suited to overcoming existing host or prey defenses. Jaenike and Hamilton suggest that sexual reproduction might be favored continually to allow a female to collect for her offspring any newly appearing genetic defenses against the continually evolving biological enemies of her species.

A recent explanation for sex points out that, in a diploid asexual species, a new favorable recessive gene requires two independent mutations at the same site before the new gene can be expressed. A change from genotype AA to genotype Aa does not change the phenotype; an additional mutation from Aa to aa is required. In a sexual species, however, an individual of genotype Aa can have some aa type of offspring, by mating with another Aa individual (Kirkpatrick and Jenkins 1989).

These recent hypotheses are still too new for a final evaluation of their success in explaining the existence of sex. They do, however, exemplify science at its best, discovering problems to contemplate and new creative explanations to test. As Trivers (1985) says in his excellent chapter on the evolution of sex, "Twenty years ago biologists hardly knew there was a problem where sex was concerned. Since then a whole new world has opened up ... Although we have made considerable progress, much work remains to be done ... We may still be in for some surprises concerning the real meaning of sex."

WHY MALES AND FEMALES?

The preceding discussion of the twofold cost of sex applies only to species in which one sex contributes only its genes to the offspring. The condition in which one sex produces large, nonmotile eggs (whose cytoplasm provides all the materials for the offspring's early growth and development), whereas the other sex produces small, motile sperm (that contribute only DNA to the offspring), is called *anisogamy*. Not all sexual species are divided into egg producers (females) and sperm producers (males). In some algae and protozoans, all sexually reproducing individuals produce gametes of equal and intermediate size. When these combine, each gamete contributes an equal amount of cytoplasmic material to the new individual's growth. Such equivalent gametes are called *isogametes*; the condition in which all individuals produce such isogametes is known as *isogamy*.

Isogamous species gain the benefits of sex (whatever they are) without incurring the 50% cost of producing males. It seems puzzling, then, that all true plants and animals (as well as some other organisms) are anisogamous. Therefore it is important to consider the reasons for the evolution of anisogamy. This is especially so because current theory maintains that it is the existence of anisogamy that leads to the differences in morphology and behavior between males and females that we saw in the red-winged blackbirds (Trivers 1972).

Figure 11-3 How disruptive selection could have led to anisogamy. In an originally variable population, selection is presumed to have favored producers of large or small gametes but to have disfavored producers of intermediate-sized gametes. From Thornhill, R., and J. Alcock. 1983. The evolution of insect mating systems. Harvard University Press, Cambridge, Mass. After Daly and Wilson. 1978.

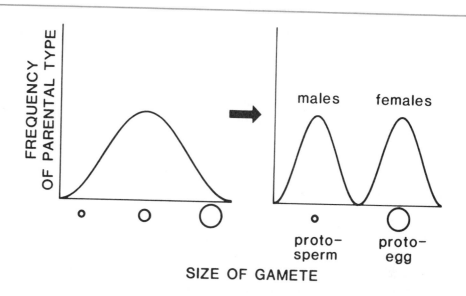

Parker, Baker, and Smith (1972) proposed that the evolution of anisogamy came about through a process of disruptive selection, the phenomenon in which intermediate types in a species are less fit than the extreme types. They considered that in an original isogamous species that simply released gametes into the environment (sea water), two alternate but mutually incompatible gamete production strategies would be favored over the production of isogametes. One strategy would be to produce larger than average gametes. Although fewer such gametes could be made by an individual, the survival rates of such gametes and the subsequent zygotes derived from them would be higher and favor their production. On the other hand, an alternative tactic of producing large numbers of small, active gametes would increase the number of fertilizations that an individual's gametes would participate in, compared with intermediate-sized gametes. Thus isogamete producers would not have the survival advantages of large gametes (proto-eggs) or the fertilization advantages of small gametes (proto-sperm): isogamete producers would have lower fitness than proto-females or proto-males. Eventually an anisogamous population would evolve, in which only females and males, with their egg and sperm anisogametes, were present (Figure 11-3). Fisher (1930) showed that natural selection usually favors equal numbers of males and females in anisogamous species.

Although other models for the evolution of anisogamy have been proposed, most ethologists accept the basic correctness of Parker, Baker, and Smith's model. It has received some support from comparative studies of sexual reproduction in algae (Knowlton 1974). Whatever the reasons for anisogamy's initial evolution, it turned out to be a crucial trait, responsible for major differences in the behavior of the two sexes. Such differences were first emphasized in the writings of Darwin.

SEXUAL SELECTION

In *The Origin of Species*, Darwin (1859) introduced his mechanism of natural selection to explain how the complex morphology and behavior of animals adapted

Figure 11-4 Fighting between males in several mammalian species: **A**, rabbits; **B**, Masai giraffes; **C**, bighorn sheep.
A from Mykytowycz, R., and E.R. Hesterman. 1975. Behaviour 52:104-123. Photograph courtesy R. Mykytowycz and E.R. Hesterman. **B** by Leonard Lee Rue III. **C** by Len Rue Jr.

A

B

them so well to their particular environments. Animals have the traits they do because their ancestors that possessed those traits survived (and subsequently reproduced) at greater rates than other animals that possessed different traits. Although natural selection provided a convincing explanation for most characteristics, other traits seemed to pose a problem. In many species, individuals of the male sex possess traits that seem likely to lower, not increase, the chance of surviving. In most birds and mammals, for example, males are larger than females, making them more vulnerable in times of food shortage. Males of many species, such as the peacock, are brightly colored and possess plumage or other ornaments that seem to make them more conspicuous to predators. Males of some species, such as red deer stags, engage in ferocious combat using exclusively male weapons such as horns or antlers (Figures 11-4 and 11-5). These combats are costly in terms of energy expended. Additionally, they not infrequently result in injury or even death and make the participants vulnerable to predators.

To explain such male weapons, ornaments, and conspicuous behaviors, Darwin postulated an additional evolutionary mechanism, which he called *sexual selection*. Although he introduced the concept in *The Origin of Species*, he developed the idea at great length in his later book, *The Descent of Man and Selection in Relation to Sex* (1871).*

Darwin defined sexual selection as "the advantage which certain individuals have over others of the same sex and species solely in respect to reproduction." Darwin

*Modern readers are often puzzled to discover that less than half of the *The Descent of Man* is about human evolution, and more than half is about the sexual behavior of animals. The explanation is that Darwin felt that human racial differences in appearance were the result of sexual selection, from different races having different ideas of what was physically attractive. Thus he felt the need to explain his concept of sexual selection fully before applying it to people. Modern physical anthropologists are more likely to invoke natural selection than sexual selection as the reason for the evolution of the external differences between human groups, although the species-wide body differences between men and women are commonly attributed to past sexual selection. See the fascinating paper by Guthrie (1970) for speculations on the origin of such physical differences between men and women.

C

Figure 11-5 Examples of head structures evolved under mate competition sexual selection. **A**, moose skull (weight 17 kg, most of which is antlers). **B**, Male Hercules beetle from South America. **C**, Lucanid beetles (*Chiasognathus granti*); male on left, female on right. Photographs by James W. Grier.

saw this happening in two ways. First and more obvious, males competed with each other to gain access to females. This process of competition, Darwin felt, led to the evolution of the greater pugnacity of males, as well as their greater size and possession of "weapons of offence and defence." Large size, effective weapons, and pugnacity would be important to success in fights with rivals. On the other hand, the colorful plumage or ornaments of males, and their elaborate premating displays, evolved through a process of female choice of mates, with females competing to select the most attractive males as mates. Apart from such choice behavior, females would be more strongly affected by natural than by sexual selection. They would therefore be more cryptic in appearance and cautious in behavior and would devote more time to parental care and less to mate seeking, than do males. Thus, Darwin postulated, male ornaments and weapons would evolve despite their negative effects on survival; their positive effects on mating success would more than counterbalance the increased mortality rates.

Darwin's concepts of male competition and female choice met with varying degrees of acceptance. His male competition mechanism was immediately recognized as a major force in animal behavior, and its importance continues to be acknowledged. On the other hand, female choice met with little acceptance. The codiscoverer of natural selection, Alfred Russel Wallace, argued against it during Darwin's lifetime. When the modern "Neodarwinian" evolutionary synthesis of mendelian genetics and darwinian selection was constructed in the 1930s, the author of the most important evolutionary text, Julian Huxley, dismissed female choice (or *epigamic selection*, as he called it) as unimportant (*Evolution: The Modern Synthesis*, 1942). One major objection to female choice was that it seemed to be anthropomorphic, attributing aesthetic feelings and standards of beauty to female animals. We now realize that a female need not "think a male is beautiful" for her nervous system to be stimulated by his displays, leading to a positive response. It is perfectly conceivable that natural selection could have led to the evolution in female animals of nervous systems that cause them to behave unconsciously in a way that superficially looks like behavior involving aesthetic choices.

Perhaps a more serious shortcoming of Darwin's presentation of the concept of sexual selection was that, although the observed sex differences in morphology and behavior are real and widespread, Darwin was not able to provide a convincing theoretical explanation for why they should exist. Why should evolution have favored pugnacity in males and mate discrimination in females? The best answer Darwin could provide was that male animals usually had "stronger passions." This only begs the question; why should males have stronger sexual passions than females (if indeed they do)?

Not until a century after publication of *The Descent of Man and Selection in Relation to Sex* did an ethologist provide a convincing theoretical explanation for the two patterns that Darwin had so extensively documented. Appropriately, the breakthrough came in a paper that was part of a book celebrating the centennial of *The Descent of Man*. The paper was titled "Parental investment and sexual selection" (1972), written by Trivers.

Parental Investment and Sexual Selection

For Trivers, anisogamy—which defines the sexes—also explains sex differences in morphology and behavior. In anisogamous species, in which all of the material contributions to the zygotes come from the female's eggs, the number of offspring produced is limited by the number of eggs females can make. Because sperm lack nutritional materials, males can make many more sperm than females can eggs. In fact, in some species a single male may make enough sperm to be able in theory to fertilize all the eggs of all the females of the entire species. (In *Homo sapiens*, a woman makes only a few hundred eggs in a lifetime, whereas a man may make 100 million sperm each day.) To use the language of economics, the nutrition-filled eggs of females are a resource in short supply compared with the abundant sperm of males. In such a resource-limited situation, we should expect males to compete to combine their sperm with the scarce eggs and to evolve the means to come out ahead of other males in this competition for fertilization.

There is another way to see why competition to get mates is expected to evolve in males rather than females. That is to ask, for an individual of each sex, how can the reproductive fitness of that individual best be increased? For males, each additional female copulated with represents a quantum increase in number of offspring produced. For females, additional copulations do not (under most circumstances) bring such quantum jumps in number of offspring; females need the resources that can be turned into eggs or that can increase the survival of eggs or young; they do not need additional sperm. Competition between females will usually be over food, oviposition sites, or territories, not over males. So, in anisogamous species, males evolve to compete for matings, females usually will not.

The nutrients in eggs are the most common form of what Trivers (1972) called *parental investment* (PI) in offspring. Trivers defined PI broadly, as "any investment by the parent in an individual offspring that increases the offspring's chance of surviving (and hence reproductive success) at the cost of the parent's ability to invest in other offspring." Putting more yolk proteins into current eggs limits the number of additional eggs a female can make. Many other forms of parental investment exist. Retaining eggs in the body is PI; providing embryos with food through a placenta is PI; building nests or other protected shelters for eggs or young is PI; staying with and defending eggs or young against predators or parasites is PI; feeding young is PI; devoting time and energy to the instruction of the young or their incorporation into a social group is PI. All of these activities take time, energy, and materials that consequently limit the total number of young that can be produced.

Because PI can take forms other than yolk in eggs, PI can also be made by males. Although males cannot make eggs directly, get pregnant, or lactate, they can establish territories, build nests, incubate eggs, and defend and feed young. They can even contribute to the making of eggs by feeding females before ovulation. So anisogamy necessarily leads to male competition and female choice only in species in which there is no courtship feeding before egg-laying and no PI after egg-laying. In other species, both males and females can make PI, and the degree of male competition depends on the relative extent of PI in a species by males and females. It is sometimes, although rarely, even the case that PI by males is greater than by females (some examples are given later in the chapter). In such cases, females may evolve pugnacity and males discrimination.

Because pugnacity or discrimination can evolve in either sex, the old terms *male competition* and *female choice* should be discarded as general labels for Darwin's two forms of sexual selection. *Intrasexual selection* (for competition) and *intersexual selection* (for choice) are widely used alternatives. These terms, however, may also be misleading and inappropriate because both forms can involve competition among the members of a single sex. We prefer the simple and unbiased terms *mate competition sexual selection* and *mate choice sexual selection*.

Trivers' parental investment as a limiting resource concept leads intuitively to the conclusion that individuals of the non-PI sex will evolve to be overtly competitive, but why should the PI sex be choosy? There should be an excess in members of the non-PI sex available over the numbers actually needed. The available members of the non-PI sex will probably vary in a number of ways. If the members of the PI sex can benefit by mating with non-PI individuals possessing certain traits rather than others, choosiness or discrimination should be favored and will evolve. Examples of such beneficial traits include more food to offer before copulating, a better territory in which to breed, or perhaps superior genes to add to the haploid set already in the eggs. Later we look at the kinds of traits that are chosen in mates by members of the PI sex; at this point we only point out that choice is expected to evolve, provided of course that it does not cost too much in terms of delay in reproduction or exposure to natural enemies.

Trivers' PI model for sex differences in reproductive behavior led to an explosion of interest in sexual selection. One study showed that the percentage of papers about sexual selection increased from only 4% in the early 1970s to over 15% in the 1980s (Burk 1986). We now turn to some of the examples of animal sexual behavior that have emerged from this renewed interest. We begin by looking at how individuals (usually males) compete for mates; then we look at the kinds of characteristics that other individuals (usually females) choose when accepting mates.

Mate Competition, Including Mate Guarding and Cuckoldry

Competition for mates can take an almost bewildering variety of forms, especially male-male competition for females (female-female competition for males is discussed in a later section). One way to organize and discuss this diversity is to classify the forms into two broad categories, precopulation competition and postcopulation competition. Mate competition often includes mate guarding, as introduced in Chapter 5. When a competing individual manages to provide its gametes to another individual's mate, such as a different male fertilizing the eggs of a female that already has a mate, it is referred to by some as *cuckoldry*, a term derived from nest parasitism in cuckoos (although that phenomenon is interspecific, see Chapter 10).

Precopulation Competition

In some species, males simply scramble to be first to locate receptive females and fight over them when two locate a female at the same time. An example of such a species is the thirteen-lined ground squirrel *Spermophilus tridecemlineatus*. As described by Schwagmeyer and Woontner (1985), these ground squirrels are highly dispersed and have a short breeding season (10 to 12 days). Furthermore, an individual female is sexually receptive for only a few hours of one day. Males range widely in search of females, and frequently several locate one simultaneously. When that happens, chases result, with one male ultimately driving away the others and copulating with the female. (After he leaves, however, the female may be discovered by a latecomer male—most females mate more than once during their several-hour period of sexual activity). In these squirrels the most successful males are not necessarily the most dominant. They may instead be simply the ones that cover the most ground (Schwagmeyer 1988).

In many crustaceans, females can only be mated immediately after a molt, when they are still soft-bodied. Males search for females and on contact are able to sense whether or not females are near to the next molt. (Odor signals given off by the females seem to be the cue used by the males.) In the amphipod *Gammarus pulex*, if a male accepts a female, he carries her around for as long as a few days, until she molts. After molting the pair copulate, with the male leaving afterwards. Other males attack the paired male throughout the precopulation phase; large males are more successful in fighting off attackers or in taking over females as attackers (Ward 1988). Thus unlike thirteen-lined ground squirrels, these amphipods defend their access to females and remain with them for extended periods, not just when they are actually receptive.

If for some reason, such as antipredation protection (see Chapter 13), females occur in groups, a male or males may attach themselves to the group and defend it against other males, thereby gaining exclusive mating access to the females as a harem. This seems to be the basis for inclusion of adult males in most mammalian social groups, from lions (Schaller 1972) to horses (Berger 1986) to deer (Clutton-Brock et al 1982) or monkeys (Hrdy 1977); however, such groupings are also found in other animals, some as different from mammals as cockroaches (Gautier 1974).

If the estrus periods of females in such groups are relatively synchronized, males may attempt to control the group only during the actual mating season. Such a pattern is seen in the Scottish red deer studied by Clutton-Brock et al (1982) (Figure 11-6). The fall rut lasts about a month, and dominant stags attempt to keep together as large a group of hinds as possible for as long as they can during that period. In addition to herding the ever restless hinds and mating with them, the stags must fight off intruding stags. This defense includes not only the familiar antler-crashing charges but also deep-chested roaring (at an average rate of 2.7/minute) and parallel-walking intimidation displays. Even the strongest stags pay a tremendous cost—stags lose 20% of their weight and become rutted out, just when harsh winter conditions are approaching. Also, the risks of injury are appreciable: 23% of harem-holders are injured each year and 6% are permanently injured.

If female receptivity is less synchronized, a male or males may stay with the female group throughout the year, as in lions or monkeys. In some cases a single-male group is observed, as in langur monkeys; in others a multimale group exists, as in lions or baboons (Kummer 1968). The factors responsible for this difference are not completely known but among them are length of the breeding season and intensity of competition. In primates, Ridley (1986) has shown that single-male groups are associated with extended breeding seasons (greater than 2 months) whereas multimale groups are associated with short breeding seasons (less than 2 months). In lions, harems (prides) are held by coalitions of males (see Chapter 14): single males are unable to hold prides at all, and larger coalitions retain hold for longer periods than smaller ones (Packer and Pusey 1982).

Figure 11-6 A red deer stag with his harem. From Animals, animals © Robert Maier.

Females are in a sense defendable if numbers of them can be encountered at a certain location, even if they are encountered one at a time instead of all at once. If males can locate a place or resource where females will turn up, it is worth fighting to defend that place even if females are not present at the moment. The range of such female-predicting resources is shown by a paper by Alcock et al (1978). They noted that males of many bee and wasp species can be observed defending territories against other males in four general types of rendezvous sites: (1) female foraging areas, such as clumps of flowers, (2) emergence sites, where newly emerging virgin females first appear, (3) nesting sites, where reproductive females can be found coming to and fro, and (4) flyways near conspicuous landmarks that females use for orientation on their flights from nest to foraging sites and back.

Such resource defense occurs in many other groups of animals. Male dragonflies and damselflies defend territories along the banks of ponds and streams to which females come for oviposition (Waage 1984). Male tephritid fruit flies such as the apple maggot (*Rhagoletis pomonella*) defend fruits to which egg-laying females are attracted (Prokopy 1980). Many other examples among insects are given in the book by Thornhill and Alcock (1983). The previously mentioned male red-winged blackbirds defending their territories are doing this.

Perhaps the best-known example of such male-male competition for female-required sites in vertebrates occurs in elephant seals. Dominant male elephant seals establish themselves on the beaches where females come to give birth and nurse the newborn calves. After birth, females mate and conceive the calves of the following year. Giving birth on isolated island beaches probably ensures safety of the cow and calf from both marine and terrestrial carnivores. Successful dominant males are able to defend large harems. In a northern elephant seal *(Mirounga angustirostris)* population studied by LeBoeuf and Peterson (1969), the four most dominant males, representing only 6% of 71 males in the area, obtained 88% of 120 observed copulations. Similarly, in a southern elephant seal *(M. leonina)* population studied by McCann (1981) the top-ranking bull obtained nearly 40% of 331 observed copulations.* With such spectacular reproductive success as a prize, it is not surprising that competition is intense. Males challenge one another by roaring or making postural threats; if one does not back down a fight ensues, with the bulls facing off and hurling their massive bodies together, digging into each other's heavily padded chests with downward stabs of their canine teeth. Enormous amounts of energy are expended, and risks of injury are substantial; some injuries lead to death (Cox 1981).

Larger elephant seal bulls tend to become dominant, and this no doubt has been the selection pressure leading to the enormous sexual size dimorphism in these species: bulls are 450 to 650 cm long and weigh up to 7000 kg, whereas cows are

---

*It is obvious that the intensity of sexual selection is high in elephant seals, but the question of how best to quantify that intensity has been the subject of much theoretical interest. As in elephant seals, variation in reproductive success may differ between the sexes: males either father many offspring (if a harem-holder) or none (if not a harem-holder), whereas females vary less in number of offspring (most obtain mates, and differences in female fecundity are small compared with differences in male mating success). This was first pointed out by Bateman (1948) and was an important stimulant to Trivers (1972) in his development of parental investment theory. Thus Wade and Arnold (1980) recommended using the variance of reproductive success as a measure of the intensity of sexual selection. Sutherland (1985) criticized this approach, favoring instead the use of the operational sex ratio, first devised by Emlen (1976). This is the ratio of time spent by males seeking mates to time spent by females seeking mates. For good discussions of how to measure intensity of sexual selection, see Sutherland (1985) and Wade (1987); for a discussion of whether one should measure intensity of sexual selection, see the amusing but lucid paper by Grafen (1987).

Figure 11-7 A fur seal
bull (in center) and harem of
cows; note the sexual
dimorphism.
Photograph by Leonard Lee Rue
III.

only 300 to 350 cm long and weigh up to 900 kg (Walker et al 1975). Striking
sexual dimorphism has been shown to correlate with high levels of mate competition
in several groups of mammals (Alexander et al 1979) (Figure 11-7), and increased
mating success by larger males is common for many animal groups (see Thornhill
and Alcock 1983 for examples in insects). We must be careful not to assume that
males larger than females is a universal rule. Natural selection may favor large size
in females, for example if large size increases fecundity. There are a number of
species of birds and mammals in which females are larger than males. In most
other animals, such as insects, it is usual for females to be larger than males.
Which sex will be larger depends on whether the sexual selection advantages of
large size in males are greater than the natural selection advantages of large size
in females. It is worth remembering that the largest animal that ever lived is a
female: blue whales are the largest animals of all time, and in whales females are
larger than males.

Not all areas defended by sexually active males contain resources that females
need. Sometimes males defend symbolic display territories in aggregations called
leks—examples are found in grouse, coral reef fish, fruit bats, and fruit flies.
Although they do not contain resources, competition to acquire and defend such
lek territories is still intense, and large and experienced males often come out
ahead. The evolution of leks is discussed in Chapter 12.

In our discussion of mate competition so far, we have emphasized actual fighting,
or at least threatening, between males. Precopulation mate competition can also
occur by competitive mate searching (as in thirteen-lined ground squirrels) or
competitive mate signaling. Successful reproduction begins with members of the
opposite sexes getting together; usually one sex signals and the other shows a
locomotory response. In some animals, females signal and males respond—silk-

worm moths are well-known examples. In most animals, males signal and females respond, as in crickets, frogs, and birds. Because males are usually the high-stakes sex and females the discriminating play-it-safe sex, it is expected that the more dangerous role will be adopted by males. Usually this is the conspicuous calling role, not the less conspicuous responding role. (Female moths may signal because pheromones are hard to eavesdrop on (see Chapter 15) whereas bats make nocturnal flight by responders relatively dangerous, see Greenfield 1981.)

Competitive mate searching has been discussed by Parker (1978). Rules for profitable searching are similar to those for foraging animals, as discussed in Chapter 9. Key considerations include the temporal and spatial distributions of receptive females.

As an example of competitive mate searching, consider the flights of mate-seeking males of the common firefly (actually a beetle in the family Lampyridae), *Photinus pyralis*, as studied by one of us (Burk) with help from several undergraduate students. As with most North American species of fireflies, *P. pyralis* males fly for short periods each night (approximately 50 minutes in this species). Males flash at a species-specific rhythm and look for specifically timed response flashes from a nonflying female perched on a blade of grass. *P. pyralis* males begin flying about 13 minutes before sundown when there is a light intensity of about 750 lux. At the beginning of the flight period they fly slowly in somewhat circular paths, averaging only 20 cm above ground. By the end of their activity period, during which they will have flown 500 meters and flashed 500 times, it is completely dark. At that time they are flying four times as fast, going in relatively straight paths, and averaging 1.2 m above ground (Figures 11-8 to 11-11). These changes are most easily explained as adaptations to the differing ambient light intensities. When it is still light enough to see the ground, *pyralis* males search for females in promising-looking spots. After it is too dark to see the ground, males cover as much ground as possible while flashing to spread their light to as many potential mates as they can. Such efficient change in tactics has no doubt evolved under intense sexual selection pressure: Lloyd (1981) has estimated that on a given night there are probably 50 male fireflies searching for every receptive female present (females mate only once whereas every male probably searches every night).

Figure 11-8
Representative *Photinus pyralis* flight paths. Note that some are short and circular, others are long and linear. Each dot represents a flash.

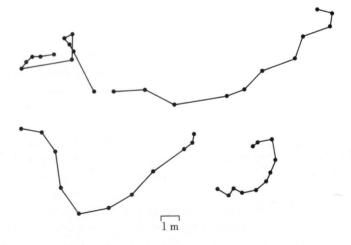

1 m

Figure 11-9 Increase in directionality (straightness) of *P. pyralis* flights as the evening proceeds.

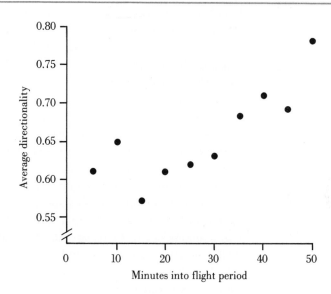

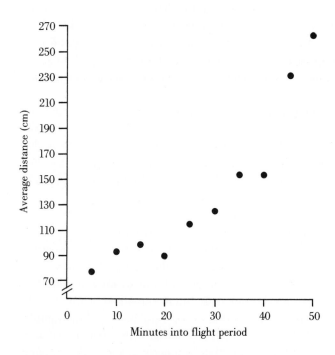

Figure 11-10 As the evening proceeds, *P. pyralis* males fly faster. Thus the distance between flashes increases.

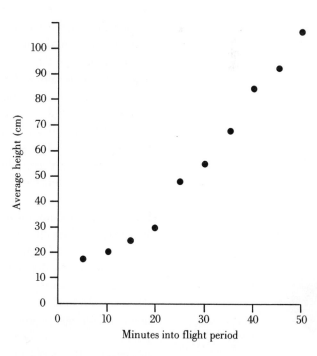

Figure 11-11 *P. pyralis* males fly higher as the evening proceeds.

Mate-attracting signals figure prominently in mate competition. Courtship signals sometimes also attract rivals, and fights may result (reviewed by Burk 1988 and Thornhill and Alcock 1983). Signals considered to be mate attractants, such as the songs of birds, croaks of frogs, or chirps of crickets, may also act as keep-away threats to rivals (Borgia 1979, Cade 1979, Robertson 1986).

There may also be competition in signal production between males. The songs of calling males can be incredibly loud (*Hyla versicolor* frogs call at 109 dB, measured 50 cm from the caller; Wells and Taigen 1986). This seems to show that males often attempt to out-shout rather than out-fight each other. Females are often preferentially responsive to loud males; this has been shown, for example, for crickets (Forrest 1980) and frogs (Halliday 1987).

More subtly, signaling males may compete by timing signal production carefully. In many species of calling frogs and katydids, males producing rhythmic songs often synchronize their croaking or chirping (Greenfield and Shaw 1983). Such precise synchronization may help attract a larger number of females or, perhaps more likely, maintain a species-specific signal code. But frequently the songs are not precisely in phase—distinct leaders begin to produce signals slightly earlier than followers do. In *Ephippiger* sp. katydids studied by Busnel (1967) and *Pterophylla camellifolia* katydids studied by Shaw (1968), the identity of specific individuals as leaders or followers was consistent, and in *Ephippiger* the leaders were aggressively dominant. A similar situation exists in Pacific tree frogs, *Hyla regilla*, except that in this species calls occur in bouts followed by periods of silence. Again, certain males began bouts, and they also continued to call longer than others so that they ended bouts as well (Whitney and Krebs 1975). It seems likely that females find it easier to locate a specific signaler when only he is calling and that dominant individuals are able to carve out some competitor-free signal time by leading or by starting and ending bouts. Presumably they enforce this system through actual aggression, as needed.

Another form of signaling competition involves interference with the signals of others. This may destroy species-specific signal rhythms, thus preventing a female from responding. The interfering male would benefit by keeping that female in the pool of potentially available receptive females that might later be attracted to the erstwhile interferer. One example of this is found in *Photinus* fireflies (Lloyd 1981).

Postcopulation competition

In 1981, Hanken and Sherman published the results of a paternity analysis of 38 litters of pups in Belding's ground squirrel, *Spermophilus beldingi*. By comparing blood proteins in the pups and various possible sires, they were able to resolve paternity in 27 litters of the 38 cases. They came up with the striking finding that 21 of the 27 were multiply sired; in 78% of cases, litters contained a mix of full- and half-siblings, resulting from the mother mating with several males. This example points out that even when a male has achieved a copulation with a female, he has not ensured that he has fertilized all her eggs. Mate competition goes on even after copulation.

The most common name for such competition derives from an important paper by Parker (1970): *sperm competition*. Parker was studying *Scatophaga stercoraria* dungflies. Females lay eggs in fresh cowpats; males also arrive at such cowpats and attempt to mate with the females. After males mount females, matings continue for 30 to 35 minutes; during this time other males attempt to oust the paired male and take over the female. Using irradiated males that produced inviable sperm (thereby producing infertile eggs), Parker was able to show that the eggs of a female that had been mated by two males were predominantly fertilized by the second male (about 80% versus 20% for the first male). This last-in, first-out pattern of sperm usage has been called *sperm precedence* and is quite common, although not universal.

Parker (1970) pointed out that sperm competition and sperm precedence are probably particularly important in insects, because female insects can store sperm in organs called *spermatheca* and use that sperm to fertilize eggs far into the future. Copulation by one male does not exclude successful copulation also by other males.

The most amazing example of sperm competition comes from Waage's (1979) studies of the damselfly *Calopteryx maculata*. As mentioned previously, male dragonflies and damselflies defend suitable oviposition sites, and females must mate with the resident male before being allowed to oviposit. Thus females mate many times, usually once per egg-laying visit. Waage compared amounts of sperm in females that had mated once and twice, surprisingly finding that they had equal amounts stored in their spermathecae. Waage then compared amounts of sperm in females before, during, and after second matings. Although females that had mated once or twice had plentiful sperm, females interrupted early on in a second copulation had no sperm in their spermatheca. Waage then killed and dissected pairs still in copula, and solved the puzzle of the missing sperm. As shown in Figure 11-12, the intromittent organ of the male has a head with lateral horns extending. During the first part of copulation, marked by conspicuous undulatory movements of the male, the horns are extended into the female's spermatheca. Sperm there

Figure 11-12 Male and female genitalia representing five damselfly genera. The distal segment of each male penis (center) is shown to scale with the corresponding female genitalia (right), and enlarged to show details (left). bc = bursa copulatrix; ovid = oviduct; st = spermatheca; vag = vagina. Chitinous structures are shaded. The valve between the vagina and bursa copulatrix is not illustrated.
From Waage, J.K. 1984. In: R.L. Smith, editor. Sperm competition and the evolution of animal mating systems. Academic Press, New York.

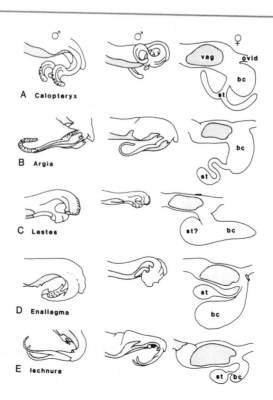

are removed into scooped-out flanges of the penis, and are probably held in place by backward-pointing hairs. Thus the penis scoops out and packs away previously present sperm before refilling the female's spermatheca. Not surprisingly, 100% sperm precedence is observed in *C. maculata*. Since 1979, Waage has examined a number of other damselfly species and has discovered that such sperm removal organs are widespread in the damselfly suborder of the Odonata (Waage 1986).

Sperm removal has been observed in vertebrates too. Davies has conducted a fascinating study of reproduction in a common English bird, the dunnock. In dunnocks, a male precedes copulation with an extended (2-minute) session in which he pecks the female's cloaca (Davies 1983). As this proceeds, the cloaca becomes pink and distended and makes strong pumping movements. Eventually a mass of sperm from previous matings is ejected, after which the male mounts and a brief (less than 1 second) copulation occurs.

It is obviously not in the interests of males that latecomers should displace their sperm. Thus it is not surprising that adaptations to reduce sperm competition have occurred. The most widespread is probably postcopulatory guarding, which has been most extensively studied in dragonflies and damselflies.

Guarding males stay with female damselflies until they have completed oviposition of their current batch of eggs. In some species the male retains a grip on the female's neck as she oviposits; this is called *tandem* or *contact guarding*. In *Calopteryx maculata* and some other damselflies, males merely perch nearby—noncontact guarding (Alcock 1979).

The relative merits of the two types seem to be as follows: contact-guarding males make takeover virtually impossible, but noncontact guarders are free to leave the female if further opportunities for mating arise. Which a male should do depends on how many male competitors are around (strengthening the tendency to contact-guard) and how many additional females are likely to enter the male's territory while the first is egg laying (more females making noncontact guarding a better tactic).

Perhaps the ultimate in contact guarding is to remain in copula for long periods; many insects do this. The record, held by walkingsticks, is 79 days (Sivinski 1978). In some species, copulations last longer when more rivals are present—McLain (1980) showed this for the southern green stink bug, *Nezara viridula*.

An alternative to guarding is a "chastity belt." In snakes and many insects a male plugs the female's genital opening with gelatinous secretions that prevent further mating, at least until the plug has dissolved, by which time the female may no longer be receptive (Parker 1984). Acanthocephalan worm males seal up a female's vagina with a gluelike cap, after filling the vagina with sperm. If a male of these internal parasites encounters another male inside a host's gut, it omits the sperm and seals up the rival's genital region with the cap alone (Abele and Gilchrist 1977). This effectively removes that rival from the mating population for weeks.

Male *Heliconius erato* butterflies transfer an antiaphrodisiac chemical to females at mating. When approached by subsequent males, females release this repellent odor and the males desist from courtship (Gilbert 1976). Why should latecomer male *H. erato* give up so easily? Presumably the release of the odor by a female means she will actively resist mating attempts, making it better for a male to immediately give up and go elsewhere. Such an antiaphrodisiac system presumably can work only if it is in the female's interests, as well as those of the first-mating male. This makes an important point. Sperm competition and antisperm competition mechanisms in males coevolve with female adaptations. In some cases, it may be best for females to mate only once: they save time for other things, are probably less exposed to predators, and reduce the hassle and possible injury of additional persistent male mating attempts (Parker occasionally saw female dung flies drown in dung pats because of the struggles of males to hold and take them over). In such cases, females should cooperate with antisperm competition mechanisms of males. On the other hand, it may pay a female to mate multiple times if that ensures complete fertilization of her eggs, gives her access to additional resources, or gives her a chance to diversify her offspring (discussed later in this chapter). In those cases, antisperm competition mechanisms face hostile coevolution in females. Walker (1980) has discussed the variety of ways in which females can keep their options open; Knowlton and Greenwell (1984) have stressed the importance of keeping the female's point of view in mind.

The most dramatic instance of postcopulatory conflict between males and females involves the phenomenon of infanticide by takeover males. This was first brought to the attention of ethologists by Schaller (1972) who observed it in lion prides. Infanticide has been reported since then in a variety of other mammals, including rodents, horses, and primates (review by Hrdy 1979). An example that best demonstrates the essential points of infanticide is Hrdy's study of Hanuman langur monkeys, *Presbytis entellus* (1977 a).

Hanuman langurs live, as many primates do, in groups composed of a number of related females and their offspring. Associating with this group of 25 or so monkeys is (usually) a single adult male, which mates with all receptive females but must constantly defend his position against marauding all-male bands attempting to oust him. In Hrdy's study, average tenure of a male with a group was 27.5 months.

When a band of males ousts a resident, fights between band members eventually result in a single male ousting the others and becoming the new resident. What then happens is chilling, as in a case reported by Hrdy (1977 b).

On August 12, 1972, Hrdy watched a group of females as they fed in some trees near a school. The females were obviously nervous about the presence of Mug, a newly resident male. Whenever Mug climbed into a female's tree, she left. At about 4 PM Mug suddenly charged at a female named Itch, grabbing at her infant Scratch, clinging to her belly. Itch fought back, assisted by three other females. Mug was driven off, but Scratch was left covered with bloody cuts. Three weeks later Mug attacked Scratch again, when he accidentally fell out of a jacaranda tree 50 feet away from where Mug was sitting and watching. Again Mug was driven away from the infant by Itch and another female. By September 10, 1972, Scratch was dead, killed by severe bites to the head and abdomen by Mug.

This was not a unique event. Figure 11-13 shows the patterns of male tenure in two of Hrdy's troops from 1971 to 1975 (Hrdy 1977 a): most changes in resident male were followed by infanticide. Why?

As with many mammals, female langurs delay ovulation while they are lactating. Thus if a female has a nonweaned infant, it may be many months before the new male can impregnate her and start his offspring. Worse, even after conception, there will be 6 to 7 months of gestation and 13 to 20 months of lactation before that youngster is capable of independent survival. Total time from conception to independence is a minimum of 19 months and a maximum of 27 months (Hrdy 1977 b). A male can "expect" a tenure of only 27 months. If he waits for a newborn

Figure 11-13 Summary of histories of two langur monkey troops, showing probable instances of infanticide following male takeovers. From Hrdy, S.B. 1977. American Scientist 65:40-49.

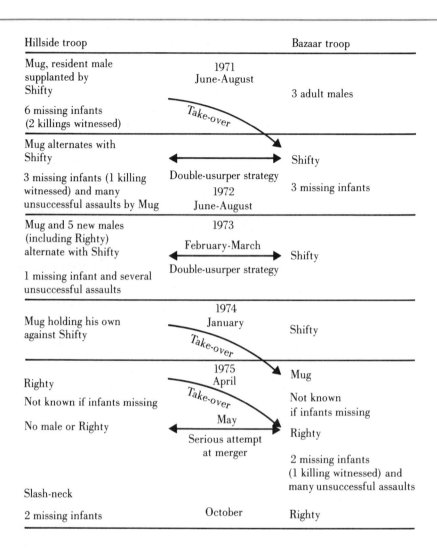

infant to be weaned before he impregnates its mother, his own infant will only just be arriving as he is being ousted. If infanticide begins in a population, a new resident male is in a bind; he must kill nonweaned infants, causing their mothers to resume ovulation, or his infants will not be old enough to escape death from his successor. So noninfanticidal males would pass on fewer genes than infanticidal ones. Infanticide makes grisly sense in terms of mate competition.

Can females do anything? Clearly it is not in their interests that infants be killed. There are a variety of counter-tactics by females. As the example showed, females try to fend off male attacks on infants. Some females leave the troop until infants are weaned, sometimes keeping company with the ousted male—the infant's father. However, the balance of circumstances favors the infanticidal male; preventing infanticide requires constant vigilance by several females, but successful infanticide requires only a few seconds of carelessness and a quick bite with the canine teeth.

Pregnant females often show pseudoestrus, copulating repeatedly with the new male. This may effectively confuse paternity of the newborn later on. An infant appearing 5 months after a takeover could be from the previous resident, or it could be a premature baby sired by the new resident; probably the new male cannot afford to take the risk of infanticide, because it just might be his offspring.

The one thing that females cannot evolve is a tendency to refuse to copulate with infanticidal males—to do so would only limit their total lifetime number of offspring, compared with nonboycotting females. Whereas humans could work out and enforce such a strategy, "selfish genes" (Chapter 14) in nonrational animals will not allow a self-sacrificing attempt to stamp out infanticide.

If infanticide makes such grisly sense, why is it not more common? There are probably several reasons. First, in many species females are bigger than males and may be able to physically prevent it. Second, note that the advantage of infanticide depends on a high frequency of takeovers and a relatively long lactational delay of ovulation. Where neither of those conditions holds, males gain less by being infanticidal, but they still incur costs, such as risks of injury from protecting females.

Infanticide by males may help explain the puzzling Bruce effect in mice. Bruce (1960) showed that urine from a strange male caused pregnant mice to reabsorb their embryos. In nature, urine from a strange male might follow a takeover; reabsorption may represent a reduction of losses by a female. If the infants would be killed, better to stop things immediately, recycle the materials of the current embryos, and get on with the next litter sired by the new male. A similar explanation may be given for spontaneous abortion by wild mares after mating with a strange stallion (Berger 1983); although the resources represented by the aborted young are lost, at least no additional resources are put into what is probably a doomed infant.

Alternative mating tactics

It may appear that all members of the competitive sex in a species compete for mates in the same way. This is not the case. Since the mid 1970s it has become apparent that in a wide variety of animals, a number of alternative mating tactics appear. The following two examples demonstrate such tactics.

Large male bullfrogs, *Rana catesbiana*, defend and call from territories along the edge of the shoreline or mats of vegetation in ponds (Howard 1978). These large males wrestle and fight with other males; 78 of 87 fights between different-sized males were won by the larger of the two contestants. Male size is related to age; in fights between frogs of different ages, the older male won 32 of 33 fights. Such older, larger territorial males have the greatest mating success—there was a significant positive correlation between male size and mating success. Also, larger males mated with larger, more fecund females, and embryos from eggs laid in larger males' territories survived better than those from territories of smaller males, indicating that larger males had higher-quality sites. Large males do, however, incur some costs. Predation by snapping turtles is greatest on these males.

What can intermediate- and small-sized bullfrogs do? Intermediate-sized and intermediate-aged males are opportunistic; they call as territorial males do but do not defend their position. If challenged they do not fight but instead move away to call from a different place. They have intermediate mating success.

The youngest and smallest males are parasitic (or satellite) males; parasitic males selectively adopt positions near large territorial males but do not call or move around. When females arrive in response to the calls of the territorial males, the parasitic males attempt to intercept them. They rarely succeed; only 2 of 73 matings observed by Howard (1978) were by parasitic males. Older territorial males spend a great deal of time chasing the parasitic males but rarely catch them.

Thus in bullfrogs, young small males act as parasites, but if they survive and grow they may eventually become opportunists and ultimately territory holders, achieving an increase in mating success as they move up.

Panorpa scorpionflies feed on dead arthropods; male *Panorpa* defend such carcasses and emit pheromones to attract females. When a female arrives, the carcass is offered to her as a nuptial gift; she feeds while he mates with her. Dead arthropods are in short supply, however, and males compete for them; most end up in the possession of large males (Thornhill 1981).

A male without a dead insect, but which has been feeding regularly, can produce an alternative nuptial gift. He secretes saliva that is deposited onto a substrate and advertised by pheromone emission. Females prefer dead insects, so this is a less suitable option.

A third option exists for males that neither have a dead insect nor can secrete a salivary mound; these males attempt to force copulations on apparently unwilling females, making use of a clamplike structure on the male's abdomen. However, these attempts at forced copulations rarely succeed.*

Thus in *Panorpa*, as in bullfrogs, three alternative tactics with varying likelihoods of success exist. Unlike bullfrogs, *Panorpa* scorpionflies cannot grow after their adult molt, so they may be more permanently restricted to a particular tactic.

Both of the examples given are instances of what have been called *conditional strategies;* it is assumed that males have the potential to behave in any of the available ways, with conditions dictating which is adopted (Dawkins 1980). A single strategy along the lines of "if very big, be territorial; if intermediate, be opportunistic; if small, be parasitic" is imagined to be present in the behavioral repertoire of an individual.† In most cases of conditional tactics, one is clearly preferable; the others represent making the best of a bad job (Dawkins 1980). Van den Berghe (1988), however, has found an alternative tactic, piracy, in a Mediterranean fish, the peacock wrasse, that is actually better than the usual winner, the territorial tactic. In peacock wrasse (*Symphodus tinca*), territorial males coexist with males adopting less successful satellite and interceptor tactics (satellites are like bullfrog parasitic males; interceptors attempt to spawn with females away from usual territory sites). However, a fourth tactic exists—piracy. Pirates, the largest males, take over the nests of territorial males during the last third of the nests' spawning period and during that time mate with females. Their mating success is equal to that of the nest's rightful owner. They then abandon the nest. Pirates avoid the costs of nest construction and of postmating parental care; in fact, their offspring are guarded by the rightful owner, who cannot abandon the nest because two-thirds of the eggs in it were fertilized by him.

An alternative possibility to conditional tactics would be a behavioral polymor-

---

*Thornhill (1980) titled his paper "Rape in scorpionflies," which elicited a response from Gowaty (1982). She argued that the phenomenon of rape in humans is different from the alternative tactic of *Panorpa* scorpionflies, and argued against use of commonly used but easily misunderstood terms like *rape*. The alternative is a less vivid jargon phrase like *extra-pair forced copulation*. What do you think?

†In a number of species, individuals are born with the capacity to change sex during their lifetime. In some species a *protandrous* (first male, then female) pattern is followed; in others a *protogynous* (first female, then male) pattern is followed. The pattern that is found in a species depends on the relative advantage in that species of being large as a male (so as to be dominant) or a female (so as to have greater fecundity). For a full discussion of sex change in animals and the behavioral consequences, see the July-August 1987 issue of *Bioscience*.

phism in which different tactics were performed by genetically different males. The population would contain a mixture of males, each pursuing a single pure strategy. Unlike best-of-a-bad-job tactics, such a genetic polymorphism could persist only if, on average, the reproductive success of all tactics was equal. This would be most likely to occur if one tactic was somewhat favored when rare but disfavored when more common; such frequency-dependent selection would maintain the genetic types in some equilibrium proportions.

The best candidate for such an equilibrium mixture of genetically alternative tactics is in the coho salmon studied by Gross (1985). In these fish, females mature and return to spawning streams at age 3. Two different life histories exist in males (Figure 11-14). Some males mature at age 3, growing to large size and developing exaggerated hooked jaws for use in fights; these are known as *hooknose* males. By contrast, other males mature at age 2, when they are as small as 30% of the size of hooknose males; these males are known as *jacks*. Hooknose males fight to establish dominance, with the winners gaining access to nests that females have excavated and are spawning in. Jacks hide in refuges (rocks, debris, or shallow areas) but dash out to deposit sperm over eggs in nests as opportunities arise. Gross found that jack fertilization was only two thirds as successful as that of hooknoses; but note that genes for the jack strategy would pass themselves on every 2 years, whereas genes for the hooknose strategy would pass themselves on only every 3 years. Overall, the strategies are close to being equally successful ways of passing on genes. Breeding studies cited by Gross suggest that age of maturity in coho salmon males is heritable.

Modes of mate competition are summarized in the box on page 344.

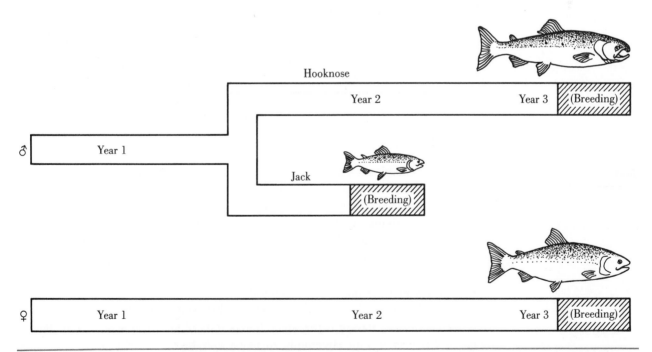

Figure 11-14 Alternative life histories in coho salmon. Females mature and return to breed at age 3 years. Hooknose males also mature and breed at age 3; jack males mature and breed at age 2. All three types die shortly after breeding.
From Gross, M.R. 1985. Nature 313:47-48.

Common Modes of Sexual Selection

I. Mate competition
  A. Precopula
    1. Scramble
    2. Signal
    3. Defense
  B. Postcopula
    1. Sperm replacement
    2. Guarding and plugs
    3. Infanticide
  C. Alternative tactics
    1. Opportunist
    2. Satellite
    3. Force Copulator
II. Mate choice
  A. Passive attraction
  B. Active discrimination
    1. Correct species
    2. Fertilization assurance
    3. Compatible genotypes
    4. Material benefits
      a. Resources of direct benefit to chooser
      b. Parental care by chosen
    5. Good genes
      a. Runaway (sexy son)
        1. Fisherian
        2. Arbitrary
      b. Indicator traits
        1. Beneficial to sons (macho sons)
        2. Beneficial to offspring of both sexes (overall vigor, including parasite resistance)

Mate Choice

Although general acceptance of Darwin's mate choice sexual selection mechanism was slow, the last several decades have seen an impressive increase in the attention given by ethologists to this process. The resultant studies show a number of ways in which discriminating females might have greater reproductive success than females that mate randomly. Included in the box is a classification that is intended to clarify various proposals about how mate choices might be made.

Choosing a mate of the correct species

A consequence of the intense competition between males is a relative lack of refinement on the part of males about the identity of potential mates. Some flowers fool male insects into attempting to mate with them, thereby achieving cross-pollination, and male beetles have been photographed attempting to copulate with beer bottles (Figure 11-15). Presumably mate competition sexual selection has resulted in low stimulus thresholds for the release of male courtship and copulation;

Figure 11-15 Male of an Australian buprestid beetle trying valiantly but unsuccessfully to copulate with a brown beer bottle, which shares some of the key stimuli of fecund females. Photograph by D.T. Gwynne. From Thornhill, R., and J. Alcock. 1983. The evolution of insect mating systems. Harvard University Press, Cambridge, Mass.

no potential opportunity is to be missed, and the costs of making mistakes may be low—only time and quickly replaceable sperm is lost (but an exception would be if poor discrimination makes a male more likely to succumb to predators, as discussed by Burk 1982).

For females, the cost of making bad choices may be high. A female that allows a male of another species to mate with her may find that an entire clutch of eggs, fertilized by sperm from that male, is inviable. Instead of a little time lost, she suffers a quantum decrease in her lifetime fitness, because her eggs cannot be replaced as cheaply or quickly as a male's sperm. Thus an important first criterion for mate choice is to ensure that a male is of the correct species. This undoubtedly has been the selective force leading to the species specificity of long-range or initially produced sexual signals. Such species-specific codes are present in the songs of birds, frogs, and crickets, in the visual displays of crabs, lizards, and fireflies, and in the pheromones of moths, as well as in other displays in many other animals (see many examples in the book by Sebeok 1977). One example involves the calling or attraction songs of male *Teleogryllus* crickets. The only calling songs that are similar between species are from species occurring in different areas, where there is no possibility of confusion. In areas such as Japan and Australia, where more than one species occur, the songs are easily distinguishable. Once species-specific pairing has occurred through female response to calling songs, no possibility of confusion will arise. Thus selection has not caused diversification of *Teleogryllus* courtship songs, which are produced only after a female has contacted a singing male. These songs are extremely similar in different species of *Teleogryllus* (Burk 1983).

Ensuring fertilization

A female must choose a mate that can adequately fertilize her entire batch of eggs. This explains some observed patterns of mate choice. Burk and Webb (1983) found that female Caribbean fruit flies, *Anastrepha suspensa*, normally showed a 2 to 1 preference for large over small males as mates. Sivinski (1984) found that these females preferred a small virgin male over a large male that had just mated, because it takes 2 hours after mating for a male to be able to produce another full ejaculate. Similar results have been found in *Drosophila* flies and in fish (references in Sivinski 1984). In frogs, males and females must match up their cloacas, so that as the female lays eggs the male can cover them with sperm (Figure 11-16). In several

Figure 11-16 Dwarf American toad. Note how the cloacas of the smaller male and larger female must be aligned well for fertilization to occur as eggs are laid. Photograph by John H. Gerard.

species, size-assortative mating occurs, probably at least in part to assure a female that the male can line himself up properly and fertilize all the eggs (Davies and Halliday 1977).

Choosing a compatible genotype

After she has found a male of the right species with an adequate supply of sperm, some authors have suggested that a female might choose a male whose genotype is particularly compatible with hers, thus producing genetically well-endowed offspring (this is a separate mechanism from the general good-genes mechanism described later). For example, one explanation for the well-known rare male preference of female *Drosophila* uses this approach. As reviewed by Ehrman and Probber (1978), females of several species of *Drosophila*—and other animals—show a disproportionate tendency to mate with male types that happen to be rare in a population. (Rare males are apparently distinguished from common males by differences in odor. These can be created artificially by rearing flies in spice-scented media. For more on rare-male-choosing behavior see Dal Molin 1979 and Spiess and Kruckeberg 1980.) Because a female is on average likely to be a common type, her offspring by a rare type of male would have a greater than usual proportion of heterozygous gene loci and might gain from heterozygous or hybrid vigor.

Another compatible genes possibility has been put forward by Bateson (1982). Given a choice of mates, female Japanese quail prefer cousins over closer relatives or nonrelatives. This might make sense if some optimal degree of inbreeding was favored. If different populations of a species lived in slightly different habitats, natural selection might lead to their having different genotypes, each adapted to the local situation. Thus hybrid offspring would end up ill-adapted to any situation. On the other hand, mating with too close a relative would bring about the well-known risk of genetically inferior offspring possessing rare harmful recessive alleles in homozygous combination. Mating with cousins might be a good compromise.

Both the increased heterozygosity and optimal inbreeding models have been advanced as explanations for the local dialects observed in many species of birds (Krebs and Kroodsma 1980). Like the other examples in this category, these ideas remain inadequately tested.

Choosing material benefits (PI)

It was a presumption of Trivers' PI model that reproductive success of members of the investing sex was not limited by number of mates but was limited by shortage of resources to invest (food, nest sites, etc.). If a female could get resources from males in return for mating, and if she could get more resources from some males than from others, mate choices based on offerings of material benefits should evolve. A number of examples of mate choice based on material offerings have been observed recently in many different animal groups.

Perhaps the best-studied examples are found in *Hylobittacus apicalis* hangingflies and katydids, studied respectively by Thornhill (1980) and Gwynne (1981). Like their relatives the *Panorpa* scorpionflies, hangingfly males offer arthropods to females before mating (Figure 11-17). However, hangingfly males catch live insects rather than scavenge dead ones (they also steal them from other males). Thornhill found that smaller prey were discarded by males after the males fed on them; they were too small to offer to females. Larger prey were offered instead. Females copulated with prey-offering males but exercised discrimination by varying copu-

Figure 11-17 Mating pair
of black-tipped hangingflies
(*Hylobittacus apicalis*). The
female (right) and male (left)
are both holding a nuptial
gift (fly) offered by the male.
From Thornhill, R. 1980.
Scientific American 242(6):162-
172.

Figure 11-18 Mating
success (number of sperm
transferred to the female) in
Hylobittacus apicalis
depends on copulation
duration, which in turn
depends on the size of the
nuptial gift offered to the
female.
From Thornhill, R. 1980.
Scientific American 242(6):162-
172.

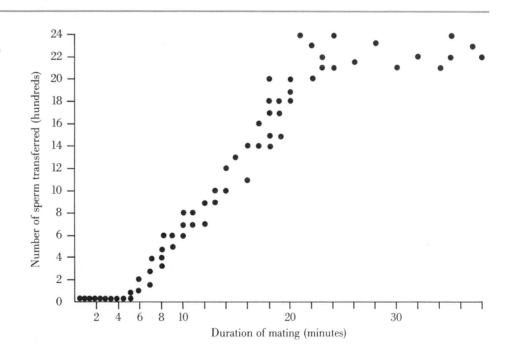

lation times. If the prey was too small, the female stopped feeding and mating after about 5 minutes; if the prey was large, she continued to mate for an average of 23 minutes. Experiments by Thornhill showed that the number of sperm transferred was correlated with duration of copulation (Figure 11-18). No sperm were transferred during the first 5 minutes; there was a direct relationship between copula duration and number of sperm transferred between 6 and 23 minutes; and no additional sperm were transferred after 24 minutes. Thus females obtain full sperm loads from males with big prey, no sperm from males with small prey, and partial sperm loads from males with medium-sized prey. Interestingly, males break off copulations that go on for more than 23 minutes; if there is any meat left in their big insect, they could still use it to transfer sperm to a different female. The more sperm a female gets, the more eggs she lays before going back to mate with another male, so actual fertilization numbers correlate with amount of sperm transferred.

Gwynne studied cricket relatives in the family Tettigoniidae, known in the United States as katydids and elsewhere in the world as bush crickets. The famous Mormon cricket, *Anabrus simplex*, is well known as the plague of the early Mormon settlers of Salt Lake City (Gwynne 1981). Like many other katydids and a few true crickets, Mormon cricket males transfer sperm to females in a two-part package known as a *spermatophore*. A bulb like ampulla contains the actual semen, and a large spermatophylax is the posterior part of the spermatophore (Figure 11-19). The spermatophore, a product of secretions from male accessory glands, is transferred to the female during mating; after copulation ceases, sperm empty out of the ampulla into the female's spermatheca, and she plucks off and eats the spermatophylax. When she finishes that, she pulls out and also eats the depleted ampulla. The nutritional donation is substantial; males put about 25% of their precopulation weight into the spermatophore. In another katydid, *Requena verticalis* (a Western Australian species), Gwynne has quantified the effects of such offerings on female fitness (Gwynne 1988): females that fed on multiple spermatophores had more and larger (better viability) eggs. In a common North American katydid, *Conocephalus nigropleurum*, Gwynne (1982) found that larger males produced larger spermatophores, and females preferentially approached and mated with large calling males over small calling males.

Courtship feeding, as in the above two examples, is found in a large variety of other animals and takes a variety of forms: spermatophores in butterflies, salivary

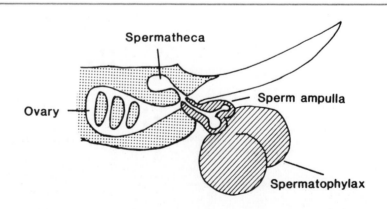

Figure 11-19 Schematic drawing of the posterior part of a female katydid, *Requena verticalis* (stippled), showing the male's spermatophore (striped) as positioned after mating. After copulation the female eats the spermatophylax while sperm empty from the sperm ampulla into her spermatheca. She then eats the ampulla too. From Gwynne, D.T. 1988. Evolution 42:545-555.

secretions in flies, glandular secretion in cockroaches, and fish in seagulls and terns (for insect examples see Thornhill and Alcock 1983; for courtship feeding in birds see Wiggens and Morris 1986). In common terns (*Sterna hirundo*), it has been shown that the courtship feeding rates of males correlate significantly with the size of clutches and of eggs produced by females and with hatching and fledging success of chicks (Nisbet 1973). Therefore females should (it has not yet actually been shown that they do) mate preferentially with males that courtship-feed at higher rates.

Material benefits can also take the form of paternal care, and females can pick mates on the basis of their paternal qualities. In terns, male courtship feeding also correlates significantly with subsequent feeding rate of the chicks (Wiggins and Morris 1986). We saw in red-winged blackbirds that the males with the best territories also devoted the most care to the young. Mate choices may be related to paternal abilities in other species too. In the three-spined stickleback fish, females mate preferentially with males that are most aggressive against other male stickle-backs. These same males are the most aggressive defenders of the nest site against egg-eating predators (Huntingford 1976).

Perhaps the most common manifestation of female mate choices based on differences in material benefits is seen in species where females enter resource-containing territories defended by males. We saw in red-winged blackbirds that females seem to be picking specific locations as much as particular males; the same has been shown for dragonflies (Jacobs 1955). Of course, it is not usually the case that females will have to choose between picking a good territory or a good male; the result of mate competition sexual selection will usually be that the dominant males are in possession of the best resources, as already mentioned for red-winged blackbirds and bullfrogs. We explore resources and mate selection further in the discussion of polygyny and polyandry (Chapter 12).

Choosing good genes

None of the preceding explanations for discrimination by females between possible mates seems unreasonable, although additional empirical confirmation of those possibilities would be helpful. For this last category, however, considerable controversy has been generated, from Darwin's day to the present.

You will remember that Darwin postulated mate choice sexual selection to account for what might be called the *spectacular ornaments* of the males of some species. Examples include the elaborate tails of peacocks or male widowbirds, the decorated constructions of male bowerbirds, and perhaps the elaborate songs of wrens and nightingales. The problem is—and has been for the past 130 years—what does a female gain by choosing such mates? It is obvious why a female should mate with a conspecific male or with a male offering a generous nuptial gift of food but not so obvious why she should mate with a male that is more brightly colored or has a longer tail.

The first successful model of such choice was produced by Fisher in 1930. Fisher's model went as follows. Imagine that some male trait was correlated with overall fitness; for example, a male might have a longer tail because his genotype made him better at getting food. If a gene arose that made a female more likely to mate with males with longer tails, the offspring of females possessing such a gene would tend also to get the good genes that led the male to be able to grow a tail: female tail-choice genes would spread. Any genes that caused males to grow longer tails would also spread, because choosy females would pick out males that had such genes. Thus genes for choosing long tails and genes for growing long tails would become involved in a correlated process of positive feedback. Tails would lengthen even beyond the point where they would begin to have negative consequences, for example making their possessors more vulnerable to predators. Tail length would continue to increase until such negative effects became so strong that they would counteract any additional mating advantages of having even longer tails. Females could not escape from this runaway process, because once other females chose mates on the basis of long tails, a female that mated with short-tailed males would produce unsexy short-tailed sons and would have no grandprogeny from them. To have sexy sons, females would have to mate with the longest-tailed males available.

Fisher's model presumed that originally the trait being chosen was an indicator of what might be called *true vigor*, but more recently Kirkpatrick and Lande have shown that that need not be true (Kirkpatrick 1982; Lande 1981). A variety of processes might result in a genetic coupling of choice genes and trait genes, getting the runaway process started. In such a case the traits chosen by females could be said to be completely arbitrary, not related to the viability of their possessors (including the choosing female's offspring), but retained simply because of the mate choice advantages.

A number of authors (e.g., Borgia 1987) have argued that females choosing on the basis of traits that are related to true vigor or viability should do better than females that choose on the basis of an arbitrary trait. This viewpoint has been called the *indicator trait* model, because the expression of the trait indicates that the male being chosen has genes that make it more vigorous or viable in some way. We present three alternative kinds of indicator traits, a classification somewhat different from others that have been recently suggested. (For a full discussion of models of female choice, see the papers in Bradbury and Andersson 1987.)

First, a female might choose males with particular traits because they indicate possession of certain good genes that would be beneficial to offspring of both sexes. Examples might be mating with older males, since living to old age might indicate possession of genes for getting food, escaping predators, etc., that would benefit sons and daughters (Manning 1985). Another possibility, proposed by Hamilton and Zuk (1982), would be a choice of males that have particularly bright or good-condition plumage or pelage, or ones able to produce exhaustive courtship displays for extended periods, since those might indicate superior genes for disease and parasite resistance. A third possibility would be a preference for large males, since both sons and daughters might benefit from being big.

Second, a male's traits might indicate possession of genes that would be beneficial for sons to have but that would not significantly affect daughters. These would be similar to Fisherian sexy son traits but would be nonarbitrary. Borgia (1987), for example, has argued that females should mate with the most dominant males because they could thereby acquire for their sons genes that would predispose them to success in mate competition sexual selection. Burk (1988) called this the *macho son model*. Such traits would be expected to be condition-dependent; they would be such that they are difficult to fake, being developed only by genuinely high-quality males.

Finally, Trivers (1985) and Seger and Trivers (1986) have argued that females might choose among males on the basis of traits that are relatively unimportant to their sons but important to their daughters. Most of their examples, such as large male size or ability to grow energetically expensive structures such as antlers, also seem to fit a macho son model; but they suggest that such traits might evolve more easily than that model suggests because any costs incurred by sons in developing the traits would be overcome by gains in females, which inherit the physiological vigor that allows such traits without actually producing the traits—they can put that physiological capacity toward other activities, such as making more or better eggs or young.

Halliday (1978) raised one difficulty for all good-genes models, whether of the runaway or indicator trait type. This is the problem of the depletion of genetic variance. Population geneticists have argued for some time that traits under strong selection should show little existent genetic variance because past selection will have eliminated that variance and fixed the gene pool for the superior allele. Put another way, the more important the trait, the lower its current heritability should be (see Chapter 4 to review the concept of heritability). In the context of mate choice, Halliday suggested, the problem is as follows: originally, when choice evolved, females that chose males with certain genes were favored. Soon, because of that history of selection, all males would be genetically alike, having the chosen genes. At that point it no longer makes any difference whether females choose or not. Because all males would, for example, have the same tail-length genes, any differences between males would be the result of environmental rather than genetic differences and would not be passed on to a female's progeny.

Although this objection created a lot of concern in the following years (for example, see Thornhill 1980), it was somewhat reminiscent of discussions among population geneticists before the development of electrophoresis in the 1960s. They assumed that most individuals would be genetically fixed for a common wild type at most loci; but electrophoresis showed that in most species 30% to 40% of genes in a species vary. Many ethologists suspected that, as with other genes, variation in sexually selected genes would prove to be present. In fact a number of variation-preserving factors have been identified, which would make continued mate choice genetically profitable. A particularly good review of such factors is Cade (1984). His list includes the following:

1. New mutants
2. Polygenic determination of traits
3. Conflicting natural and sexual selection
4. Variation in selection pressures (i.e., variation in the environment)
5. Genetic linkage between favored and disfavored genes
6. Continuing coevolution with natural enemies

After this consideration of how (and why) good-gene female choice might evolve, let us turn to some actual studies, ones that deal particularly with questions of (1) do females choose, (2) are those choices heritable, (3) are the traits chosen heritable, (4) what is the nature of the traits chosen, and (5) are the offspring that result from choice more successful than the offspring from randomly mating females?

Examples of choice
Studies showing that females mate preferentially with certain males rather than others, and that such patterns are the result of female responses rather than merely male-male competition, are accumulating rapidly (see Thornhill and Alcock 1983 for examples in insects, Searcy and Andersson 1986 for examples of choice based on male songs). In some cases, females respond to some males because they

Figure 11-20 Male (above) and female (below) long-tailed widowbird, *Euplectes progne*.
From Majerus, M.E.N. 1986. Trends in Ecology and Evolution 1:1, 3-7.

broadcast sexual signals more intensely or for longer periods (the response of some female crickets mentioned previously is an example); Parker (1983) has called this *passive choice*. There are also examples of females responding to certain males that are qualitatively rather than quantitatively different, which Parker calls *active choice*. We do not find this distinction particularly important from an evolutionary perspective because good-genes choices could be of either type. (That is, a male might sing louder or better because of superior genes.)

One of the clearest demonstrations of female choice based on male ornaments is an experimental study of pairing in long-tailed widowbirds, *Euplectes progne*, carried out in Kenya by Andersson (1982). As in red-winged blackbirds, female widowbirds are a mottled brown color and have short tails, whereas males are jet black and have red epaulets. Unlike male red-winged blackbirds, however, male widowbirds have exceptionally long tail feathers, the longest can measure up to 50 cm long (Figure 11-20). Males defend territories and females express choice by settling in male territories, where they subsequently raise their two to three young without help from the male. Successful males have a number of active nests (females with eggs or young) in their territories. Andersson experimented on nine groups of four birds, with males in each group carefully matched with respect to quality of their territories. In each group one male had his tail shortened to 14 cm, with another male having those feathers attached to his tail so as to lengthen it by an average of 25 cm. The third and fourth males acted as controls, one having no operation performed on its tail and the other having its tail cut and reattached. There were no significant differences between males before the experiment, but after the manipulations the birds with elongated tails had significantly more active nests in their territories than the other males. This pattern was the result of female choice and not male competition, because males with shortened tails did hold their territories successfully after the shortening and did not become less active than other males in courtship. Thus females seemed to make choices based on length of a male's tail.

Inheritance of choice and of chosen traits

Majerus and associates (Majerus 1986) showed a genetic basis for differences in female choice in the ladybird beetle (ladybug) *Adalia bipunctata*. These ladybugs are polymorphic, with black and light-colored forms (Figure 11-21). Some females, of both dark and light forms, mated preferentially with dark males. That differences in mating preference among females had a genetic basis was shown by artificial selection. Majerus and colleagues were able to increase the level of the dark-male mating preference from 20% to 50% in only three generations of selective breeding.

Figure 11-21 Melanic (right) and nonmelanic (left) forms of the two-spot ladybird beetle, *Adalia bipunctata*. A melanic male is shown mating with a nonmelanic female (center). From Majerus, M.E.N. 1986. Trends in Ecology and Evolution 1:1, 3-7.

Analysis of the results of matings from females chosen after ten generations of selection led Majerus and colleagues to suggest that different alleles at a single locus were responsible for differences in mating preference in this species.

Hedrick (1986) showed that female *Gryllus integer* field crickets choose among males on the basis of their calls: preferred males produce long-duration trills, whereas nonpreferred males produce trills that are more frequently interrupted by periods of silence. Hedrick (1988) was then able to show that call duration was a significantly heritable trait: the father-son heritability estimate for the call-duration trait was 0.75, measured in two different ways; brother-brother correlation measured the same two ways was 0.69 and 0.76 (heritability is discussed in Chapter 4). Thus Hedrick showed that variation existed for a trait under female choice selection and that the mate-attractive trait could be passed on to the choosing female's progeny.

Nature of chosen traits

Andersson showed preferences for long-tailed males, Hedrick for long-calling males, Majerus for dark males. Other studies have focused on the types of traits chosen to evaluate some of the proposed theoretical grounds for good-genes mate choice. As mentioned previously, Hamilton and Zuk (1982) suggested that brightly colored or vigorously courting males might be chosen by females looking for males with genes predisposing them to success in fighting off parasites. This leads to two hypotheses, one interspecific and one intraspecific. Interspecifically, bright plumage should be correlated with high rates of parasitism—bright plumage is only a reliable badge of immunity if the male has indeed been challenged by parasites. But intraspecifically, for parasitized species, parasite load should be higher in drab or less vigorous males than in bright or highly vigorous males. Hamilton and Zuk (1982) found support for the interspecific hypothesis: a survey of North American passerine birds showed that attack by blood parasites such as malaria protozoans was highly correlated in birds with brightness of male plumage and complexity of male songs. Read (1987) tested the interspecific hypothesis for European passerines and an expanded set of North American passerines and confirmed the correlation in parasite prevalence and brightness across species. (See a recent reassessment of these results in Read and Harvey 1989, Zuk 1989.)

The intraspecific hypothesis was only weakly supported for satin bowerbirds (*Ptilonorhynchus violaceus*) studied by Borgia (1986 a). Males display to females from special thatch constructions called bowers (discussed later), and although bower-building males are less parasitized than nonbuilders, there is no significant correlation between mating success and parasite load among bower-holders. Two studies strongly confirm the intraspecific hypothesis. Kennedy et al (1987) found that female guppies *(Poecilia reticulata)* mated preferentially with less parasitized males because they were able to court more vigorously than heavily parasitized males. Zuk (1988) found that less parasitized male crickets (*Gryllus veletis* and *G. pennsylvanicus*) enjoyed greater mating success, since they had enough body resources at hand to produce multiple spermatophores and retain females for multiple matings, but heavily parasitized males had less spermatophore-producing stamina. Collections in the field showed that males collected in the company of females were less parasitized than males collected alone.

These results seem to support an argument for choice based on overall quality, but other studies seem more compatible with a macho son model. As in most other bowerbird species, male satin bowerbirds construct bowers of sticks, from which they display and in which they mate with any females attracted (Borgia 1986 b). The bowers are often decorated with shells, fruit, feathers, or other conspicuous objects; even human artifacts such as coins, bottle tops, and pieces of glass are often used. After mating, females leave and raise their young elsewhere without male assistance, therefore, some sort of good-genes criterion seems to be operating.

Borgia found that mating success in male bowerbirds was correlated with age, quality of bower (neatness of construction and number of decorations), and complexity of songs produced (complex satin bowerbird songs consist of a broad-band mechanical sound followed by mimicry of the songs of other species [Loffredo and Borgia 1986]). These factors in turn were correlated with dominance. Male bowerbirds repeatedly destroy each other's bowers and steal decorations. Dominant males are able to destroy other birds' bowers more often and suffer bower destruction less often than subordinate birds, and they accumulate larger numbers of decorations on their well-kept bowers. Borgia found that when he experimentally placed blue feathers on bowers in his study area, an initial random distribution of feathers was quickly replaced by the accumulation of all the feathers on the bowers of only a few males. Thus, although age and song complexity play some role, females seem to choose among males on the basis of sufficient dominance to maintain a neat and attractive bower. Borgia (1979) has argued that many ornaments like the peacock's tail will turn out to be—like bower decorations—provokers of male aggression or structures used in threat displays, so that choice of the brightest is often equivalent to choice of the most dominant: mate competition and mate choice will coincide with resulting macho sons and vigorous daughters.

Benefits of choice

The key factor for determining whether good-genes choices are arbitrary or indicative of good traits is whether the offspring produced by choosy females are superior to those of nonchoosy females in any way other than sexiness to females. If choice is arbitrary, choice offspring will not grow faster or survive better but if choice is based on an indicator trait, they might. Tests of this idea so far have not been convincing. Partridge (1980) presented some *Drosophila melanogaster* females with a number of males in a cage and other females were assigned mates at random. She found that the offspring of females with a number of males to choose from were competitively superior as developing larvae over offspring from randomly mated males. However, Schaeffer et al (1984) failed to replicate these findings in studies of *D. melanogaster* and *D. pseudoobscura*, and Kingett et al (1981) pointed out that any nonrandom mating in Partridge's cages might have been the result of male competition rather than male choice, with dominant males excluding subordinates from mating. Boake (1986), studying *Tribolium castaneum* flour beetles, showed that some males produced pheromones that were more attractive to females than others, but found there was no correlation between a male's attractiveness and the fitness of his offspring, as measured by the developmental rate of his offspring (Boake showed that developmental rate was important to fitness and that genetic variation for developmental rate existed in her population of beetles).

Simmons (1987), however, did produce a study showing that a positive result in offspring quality was more likely to be the result of female mate choice. He presented *Gryllus bimaculatus* cricket females with either an assigned small male, an assigned large male, or a choice of males. This design partly overcomes the objections to

Partridge's, because large size is associated with dominance in crickets (Simmons 1986). (However, in the choice group male competition between equally large males could still bias outcomes.) Simmons found that females from the choice group laid more eggs than females from either of the assigned groups and, more significant, the offspring of the choice females developed to maturity faster than the offspring of females that had mates assigned, whether they were large assigned males or small assigned males.

The study of mate choice is an extremely active field of research, with many theories postulated and many questions unanswered. Frequent referral to such journals as *Animal Behaviour* and *Behavioral Ecology and Sociobiology* over the next few years should provide exciting new insights and answers.

THE FUNCTIONS OF COURTSHIP

The explosion of interest in sexual selection over the past 20 years has greatly altered ethologists' views on the purpose of precopulatory, or courtship, behaviors. It is worthwhile to briefly contrast what might be labeled the *classic ethological* and *sexual selection* views of courtship.

The following functions, chosen from the textbook by Immelmann (1980), can be taken as representative of previous views. For Immelmann, four major classes of function exist for courtship behaviors:

1. Sex and species recognition: courtship allows individuals to ensure that they are dealing with a member of the opposite sex but the same species.
2. Mate attraction and mating orientation: long-distance sexual signals bring widely separated males and females together, and shorter-range courtship behaviors bring about precise alignment of male and female, ensuring that a sperm reaches the egg.
3. Synchronization of mating and parental behavior: the courtship behavior brings male and female physiology to similar motivational states, so that effective cooperation is brought about. Thus if a female spawns, the male will too; if an incubating or brooding female leaves her nest to feed, the male will take her place to assure the survival of the eggs or young.
4. Overcoming of aggression: the breeding season is often a time when males are extremely aggressive toward other individuals; courtship behaviors help prevent attacks by aggressive males on females during the initial stages of pair formation, protecting the female until the male is habituated to her presence. Immelmann describes courtship feeding as serving this function; food-begging by a female releases a male's paternal rather than aggressive behavior, giving her safety from attack while pair formation proceeds.

More recent viewpoints would not consider this list wrong but rather incomplete. Mate identification, attraction, and orientation are important, but courtship would also allow comparisons among conspecific males. Points 3 and 4 in the above list, are proximate explanations that could be supplemented with ultimate ones. Courtship is primarily a period in which prospective mates are assessed:

1. Is the mate of the correct species and sex?
2. Can the mate deliver an adequate number of sperm (or eggs; see below on male choice of mates)?
3. Is the mate's genotype compatible?

4. Will the mate deliver material benefits?
   a. Are there (how many?) nuptial gifts?
   b. How good is the mate's territory?
   c. How good does it look as a parent? (Courtship feeding rate may predict parental feeding rate; aggression against rivals may predict aggressive defense of young against predators.)
   d. Is it committed to helping parentally? (Prolonged courtship in species with short breeding seasons may tie up a male for so long that there are no unpaired females remaining after he has mated with this one; thus after mating, it may be less profitable to desert a female than to stay and help her raise the brood [Knowlton 1979, Emlen and Oring 1977]).

5. Does the mate have good genes? Is it dominant, old, brightly colored, or vigorous?

Again we stress that neither of these views of courtship is of itself entirely right or wrong. To some extent they demonstrate the distinction between proximate and ultimate explanations of behavior. Also, they are not mutually exclusive: both are probably needed for a complete understanding of courtship in all its aspects.

SEX ROLE REVERSALS

Earlier in the chapter we discussed the terms *mate competition* and *mate choice* for the two types of sexual selection. Yet, rather conventionally, our examples so far have involved mate competition between males and mate choice by females. Trivers' PI theory, however, speaks not of females and males but of limiting and nonlimiting sexes. If and when males make more PI than females in a species, Trivers' theory would predict a reversal of usual sex roles. Males would become the discriminating and females the assertive sex. A number of cases of such sex role reversal have emerged; here we present two.

We mentioned Mormon crickets earlier as an example of a species with expensive nuptial gifts from males to females. In high-density populations, where food is scarce, male spermatophores become a major source of food for females attempting to make eggs; a female's reproduction can become limited by her ability to mate with spermatophore-offering males. Gwynne (1981) has shown that in this situation sex role reversal occurs. When a male has gathered enough resources to produce a spermatophore, he climbs into a bush and begins to sing. Hungry females are attracted, and when more than one arrives, fights between females are quick to occur. When a female reaches the male, she mounts him; but of 45 such mounts observed by Gwynne, 29 were disrupted by the male pulling away after only 1 to 2 minutes. Gwynne found that accepted females weighed significantly more than rejected females (3.77 g versus 3.20 g): males were exercising mate choice. Gwynne found that there were strong positive relationships between female size and female ovary weight or fecundity. Thus male choices appear to be adaptive: PI by males is given to females that will produce more progeny for the male.

Sex role reversal is seen in other organisms, including vertebrates (Trivers 1985 lists a number of cases). One of the best-studied instances is in the spotted sandpipers *(Actitis macularia)* studied by Oring and associates (Oring 1986). Female sandpipers arrive on the breeding grounds 3½ days earlier than males and fight to establish territories. Some females, especially young ones, are driven away from breeding areas and are prevented from mating by this competition. When males arrive, females compete for them using agonistic (contesting) displays and physical combat. As might be expected, individuals of the competitive sex are larger: females average 50 g in weight, males only 41 g (Oring and Lank 1986). Successful females establish a pair-bond with a male but, after laying a clutch of eggs for him, desert and fly away. While this male does all the incubating and brood care, the female attempts to establish another territory and mate with another male. Females may pair with as many as four males but help only their last mate with incubation and chick protection. The size of territory held by a female is highly correlated with her probability of attracting a male, and the number of offspring of females is primarily a function of the number of mates they attract. So far there is no clear evidence of male mate choice in spotted sandpipers.

Female competition for mates can be expressed in many of the same ways as male competition. Emlen et al (1989) have found even infanticide by females of other females' chicks in the sex-role-reversed bird they studied, the wattled jacana (*Jacana jacana*). As with langur monkey and lion males, infanticide by jacana females occurs when a challenger ousts a previous territory holder and takes over her mates.

In sex-role-reversal species, female choice may continue to exist. For example, in the moorhen, *Gallinula chloropus*, dominant large females prefer short, fat males as mates (Petrie 1983); such males are in better condition to withstand the demands of incubation, during which males lose up to 10% of their body weight.

Such instances of sex role reversal are rare but of great interest. They are the exceptions that prove the rule, cases that show the generality of Trivers' PI theory of sex differences.

SUMMARY

The evolutionary advantage of sexual reproduction is not completely understood, but it may be the advantage that it gives a species in the race to keep up with its biological antagonists. Whatever the advantage of sex, once it was present, a disruptive selection process probably led to sexes: differentiation of gamete producers into two types, one producing eggs and one producing sperm (the condition known as anisogamy). According to Trivers' parental investment theory, this evolution of sexes inevitably led to gender differences in sexual behavior, with the more-investing sex (usually females) evolving mate discrimination. Trivers' explanation provides a theoretical basis for the two types of sexual selection (differential success among members of a sex in mating behavior) pointed out by Darwin: mate competition sexual selection and mate choice sexual selection.

Competition in the noninvesting sex manifests itself in searching, signaling, and fighting directly over females or indirectly over resources or other sites where females gather. Competition may occur before, during, or after actual mating (as in the case of sperm competition). Within a species, alternative forms of competition may exist; some are balanced genetic alternatives, although most represent differences between preferred tactics used by dominant males versus best-of-a-bad-situation fallbacks practiced by subordinate males.

Criteria for mate discrimination by the investing sex include the correct species, material offerings, and choice of good genes. Material offerings may include food, breeding territories, parental assistance, and others. In some species, sufficiently large offerings are made by males that these become a source of competition among females. Trivers' PI theory can account well for such sex role reversals. Good genes mate choice is less well understood, and may represent choice for arbitrarily evolved characters (sexy sons), macho sons, or for traits useful in offspring regardless of sex. One exciting possibility of the last kind is the Hamilton-Zuk idea that choice of vigorous and/or colorful mates may help choosers obtain genes for parasite resistance.

Courtship behaviors of animals may be interpreted according to the perspectives of classical ethology or sexual selection. Although such differing interpretations may appear to be in conflict, in large part they reflect the difference between proximate and ultimate explanations of behavior and are not mutually exclusive.

Although in most species mate competition is more conspicuous in males and mate choice in females, sexual differences in behavior are ultimately the result of differences in parental investment. Thus instances of sex role reversal involving competitive females and discriminating males are seen in species where males show more parental investment than females.

FOR FURTHER READING

Borgia, G. 1986. Sexual selection in bowerbirds. Scientific American 154(6):92-100.
Not only an account of Borgia's work, but also a defense of the good-genes school of mate choice.

Bradbury, J.W., and M.B. Andersson, editors. 1987. Sexual selection: testing the alternatives. John Wiley and Sons, Chinchester.
The definitive, up-to-date set of papers on sexual selection, theory and evidence.

Diamond, J. 1985. Everything *else* you've always wanted to know about sex ... Discover 6(4): 70-82.
Entertaining account of modern sexual selection ideas.

Diamond, J. 1988. Survival of the sexiest. Discover 8(5):74-81.
Popular account of mate choice sexual selection.

Gwynne, D.T. 1983. Coy conquistadors of the sagebrush. Natural History 92(10):70-75.
Popular account of Gwynne's studies of Mormon crickets.

Halliday, T. 1982. Sexual strategy. University of Chicago Press, Chicago.
Beautifully illustrated popular presentation of sexual selection.

Moranto, G., and S. Brownlee. 1984. Why sex? Discover 5(2):24-28.
Popular presentation of theories on the individual selection advantage of sex.

Ryan, M.J. 1985. The Tungara frog. University of Chicago Press, Chicago.
Comprehensive study of sexual selection in a single species of frog.

Searcy, W.A., and K. Yasukawa. 1983. Sexual selection in red-winged blackbirds. American Scientist 71:166-174.
Good introduction to this much-studied bird.

Thornhill, R., and J. Alcock. 1983. The evolution of insect mating systems. Harvard University Press, Cambridge, Mass.
Not only reviews all aspects of sexual selection in insects, but also discusses the key concepts of sexual selection and how to test them.

Trivers, R.L. 1972. Parental investment and sexual selection. In: B. Campbell, editor. Sexual selection and the descent of man. Aldine, Chicago.
The paper that started the modern explosion of interest in sexual selection.

Trivers, R.L. 1985. Social evolution. Benjamin/Cummings, Menlo Park, Calif.
Excellent overview of sociobiology by one of its greatest practitioners. Chapter 13 is the best brief presentation anywhere on the topic of the evolution of sex.

Zuk, M. 1984. A charming resistance to parasites. Natural History 93(4):28-34.
Popular introduction to Hamilton and Zuk's parasite-plumage explanation for mate choice.

12

REPRODUCTIVE BEHAVIOR MATING SYSTEMS AND PARENTAL BEHAVIOR

Sexual selection theory provides an explanation for differences in sexual behavior between male and female animals. It also leads to an understanding of the many ways in which members of the nonlimiting sex compete for reproductive success and the criteria used in choice by members of the limiting sex as they attempt to maximize their reproductive success. In actual practice, however, members of each sex must interact with members of the other sex when reproducing. These interactions may involve conflicts when the ways in which members of each sex are trying to achieve maximum reproductive success limit the options available to members of the other sex. The way in which male and female sexual behaviors interact in a species is known as its mating system; a number of mating systems exist. The type (or types) present in a species depends on the species' environment, especially the distribution of resources necessary to members of the limiting sex and the species' population density. In the first part of this chapter, we survey mating systems and their relationship to a species' environment.

Closely related to the concept of mating systems is the question of how extensive parental care is in a species and which of the sexes provides it. These phenomena also turn out to be greatly influenced by a species' environment.

The chapter ends with a discussion flowing from the topic of parental care: parent-offspring conflict. The interests of parent animals do not always coincide with those of individual offspring. We look at three examples of such conflicts and how they are resolved in nature.

MATING SYSTEMS

**Male versus Female
Interests**

It is apparent from the discussion in Chapter 11 that the interests of a male and a female in a sexual interaction need not coincide. For instance, a female confronted with a subordinate male may not choose to mate with him, but he may still attempt to mate either by displaying in such a way as to suggest that he is better than he really is or by attempting a forced copulation. The manner in which male and female interests play themselves out in a species is known as the mating system. A number of different mating systems exist. But before we explore their diversity, let us look at three different resolutions of male-female sexual conflicts of interest.

In the Ceratopogonid midge fly, *Probezzia concinna* (Figure 12-1), females appear to dominate the interaction (Downes 1978). As in many flies, sexually mature male midges gather in mating swarms, males flying in station over species-characteristic swarm-marker landmarks. Female midges enter the swarms to mate, but mating takes an unusual form in *P. concinna*. As the smaller male attaches his genitalia to the larger female, she pierces his head with her mouthparts and dissolves and sucks out the contents. After mating, the male genitalia break off and remain in the female's genital tract for some time. This is not surprising, since a male that has made the supreme PI of giving up his life to provide substances for his offspring would certainly be selected to prevent sperm displacement by a subsequent male. The male genitalia must come off for egg laying to occur, and it is hard to see how subsequent remating by these cannibalistic females can be prevented. Since remating provides another easy meal, it certainly should occur from a female's point of view. At present not enough is known about the details of this striking instance of the conflict between the sexes.

In yellow-bellied marmots the males dominate the interaction. As described by Downhower and Armitage (1971), these large, burrowing rodents are active during the summers in Alpine meadows in the Rocky Mountains. Social groups consist of a territorial male and one to seven females that constitute the male's harem. (Some

Figure 12-1 A, A female ceratopogonid midge feeding on the male during mating. B, The detached male genitalia attached to the female's abdomen after mating.
From Downes, J.A. 1978. Memoirs of the Entomological Society of Canada 104:1-62.

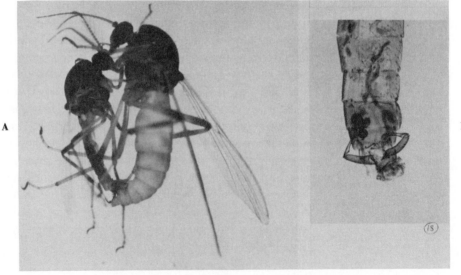

A

B

Figure 12-2 Reproductive success versus harem size in yellow-bellied marmots. The lines show expected reproductive success per individual in harems of various sizes, for males and females. The dots show the actual relationship in seven harems studied over a period of time.
From Downhower, J.F., and K.B. Armitage. 1971. American Naturalist 105:355-370.

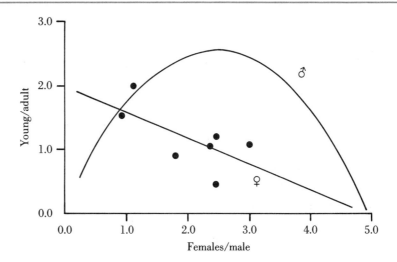

females, excluded from territories by other females, attempt to reproduce as single parents but do poorly.) Downhower and Armitage found that reproductive success, measured in number of litters produced or number of yearlings raised, was greatest for females in harems of one (i.e., monogamous pairs). Reproductive success per female declines significantly as harem size increases (Figure 12-2). However, since a male's reproductive success is equal to the total of all the females in his harem, male reproductive success is highest for harems of two to three females. Most harems were observed to have two or three females; apparently the male optimum is achieved. Females attempt to oust each other from the territory, but the territories are large enough that subordinate females can sometimes avoid dominants by staying at the opposite corner. Also, it is possible that the harem-holding male intervenes to suppress attacks by resident females on newcomers.

Most commonly, however, we might expect actual mating patterns to reflect some compromise outcome. A good example of this is seen in toads, *Bufo bufo*, studied by Davies and Halliday (1977). The optimum mate for a female toad is a male slightly larger than herself, since this results in the best matching of male and female cloacas and the highest percentage of fertilization of eggs. The optimum mate for a male is a female slightly larger than himself, since larger females lay enough additional eggs to offset a slightly lower success at fertilizing eggs. The observed pairings reveal males and females mating with others almost exactly the same size, a compromise between male and female optima (Figure 12-3).

Mating System Types

Keeping in mind the possible conflicts of interest between males and females, we can begin to explore the variety of mating systems seen in animals. Although there is almost infinite variation, a fairly satisfactory classification system involves four major mating systems. Our classification is partly based on an important paper by Emlen and Oring (1977) but also incorporates the results of more recent work. The four major mating systems are as follows:

1. Polygyny—in which successful males have a number of mates, but females usually have only one (at least for a given breeding season)

Figure 12-3 Optimum *(lines)* and observed *(dots)* size relationships between mating partners in *Bufo bufo* toads.
From Davies, N.B., and T.R. Halliday. 1977. Nature 269:56-58.

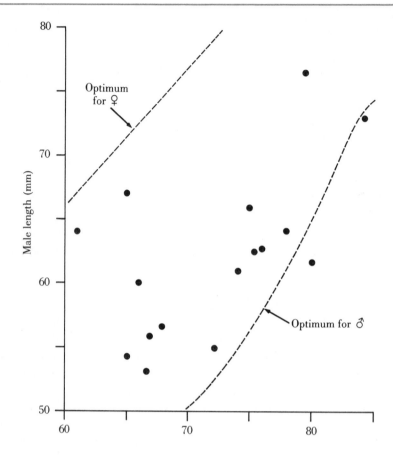

2. Monogamy—in which both males and females usually mate with only a single member of the opposite sex, at least during a single season (the pattern in which a pair is maintained for one season, but each member pairs up with a new partner in subsequent seasons, is known as *serial monogamy*)
3. Polyandry—in which successful females mate with several males, but males usually have only one mate
4. Polygynandry—in which both males and females normally mate with several members of the other sex

Polygyny

Assuming for the moment that males are the non-PI sex and females the PI sex, polygyny might be considered the ideal mating system from a male's point of view. There are also circumstances under which it is best for females. The evolution of polygyny is discussed well in the previously mentioned paper by Emlen and Oring (1977). They contend that the evolution of polygyny depends on what they call the *environmental potential for polygamy*. This environmental potential depends on the extent to which multiple mates or resources critical to them are economically defendable. Spatial clumping of resources and temporal asynchrony of available mates make polygyny possible. Emlen and Oring recognize the following three forms

Figure 12-4 The polygyny threshold model. The fitness of a female is assumed to depend on the quality of the territory she resides in. If differences between the quality of already occupied versus unoccupied territories exceed B (the polygyny threshold), a newly arriving female should opt for polygyny in the better territory over monogamy in the worse territory.

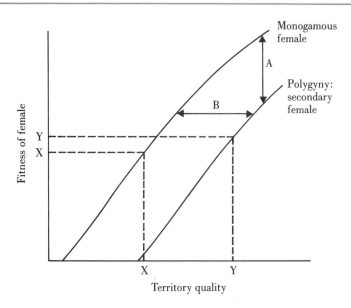

of polygyny: resource defense polygyny, female (harem) defense polygyny, and male dominance polygyny.

Resource defense polygyny occurs when the resources that are essential to females are clumped and therefore defendable by males. Territory defense by red-winged blackbirds constitutes an example. Good territories are localizable, and a male that can defend one can mate with many females attracted to it.

Why should a female be part of such a polygynous group? This question motivated Verner and Willson (1966) and later Orians (1969) to develop one of the most stimulating models in ethology over the past quarter century. Called the *polygyny threshold model* (Figure 12-4), the model poses the dilemma faced by a female deciding on which male territory to settle. Should she settle where an earlier female has settled (if the other female lets her) and become the second mate of a polygynous male? Or should she settle with an unpaired male, becoming his monogamous partner? Although the second seems intuitively better, it might be that the unpaired male has a poor territory and the paired male a very good territory. If the difference in territory quality exceeds some threshold amount—the polygyny threshold—the female is better off to settle in the territory of the paired male than that of the unpaired male. The superiority of a good territory with respect to resources may more than overcome such factors as a lack of parental assistance from a polygynous male.

The polygyny threshold model leads to predictions that field workers can test. Garson et al (1981) made the following four such predictions: (1) secondary females should rear as many offspring as monogamous females mating at the same time, (2) quality of territory should correlate positively with harem size, (3) males on whose territories females settle first should obtain the most mates, and (4) the earliest-settling females, since they have the widest choice, should be most successful. (See the discussion in Oring 1982.)

One study that upholds all four predictions is Pleszczynska's (1978) study of polygyny in lark buntings, a bird of the North American prairies. She found that, in 35 comparisons among contemporaneous primary and secondary females, 23 secondary females fledged as many or more offspring compared with monogamous females. She found that amount of cover was the best predictor of territory quality and was actually able to increase nesting success experimentally by adding plastic leaves to increase the shade in some territories. Using this criterion, she was able to correctly predict for 52 of 58 territories whether the male would end up polygynous or monogamous. She also found that the earliest females to settle fledged more young than latecomers.

Not all applications of the polygyny threshold model have been so supportive, but in many cases, tests of it have led to increased understanding of the mating system of a particular species (Oring 1982).

Female (harem) defense polygyny, the second form of polygyny, occurs when females are aggregated and therefore become defendable. Such aggregation may occur for a variety of nonsexual reasons, as discussed in Chapter 13. If females are aggregated and a male can maintain exclusive access to them, he may obtain substantial numbers of matings. Examples of female defense polygyny in red deer and elephant seals were discussed previously.

As with resource defense polygyny, saying that males monopolize females in female defense polygyny should not be taken to mean that females cannot express choice. A female in a red deer harem may be able to escape and join the harem of some other male (Clutton-Brock et al 1982). Even a female elephant seal can influence who mates with her, as shown by Cox and LeBoeuf (1977). If a female is mounted by a dominant beachmaster, she allows him to mate without a struggle, but if she is mounted by a subordinate while the dominant is temporarily absent from her vicinity, she struggles and vocalizes loudly, attracting the attention of the dominant bull. He then comes and drives away the subordinate. Thus, by inciting aggression against subordinates, the female elephant seal ensures that only high-quality males sire her offspring.

Male dominance polygyny is the third form of polygyny. This is quite different from the other two in that there are no monopolizable resources or female groups. Rather, females choose males freely, and polygyny results from a consensus among different females regarding which males are most desirable. Perhaps the purest examples of good-genes female choice occur in instances of this type of polygyny. Particularly striking are mating systems involving phenomena called *leks*.

The word *lek* appears to be derived from the Swedish word *leka*, to play (Oring 1982), as a result of the energetic displays of males involved in this form of polygyny. As used by modern ethologists, a lek is a communal display aggregation of males, each defending a small area in which it displays, to which females are attracted solely for mating. Leks were originally described for birds, especially open-country species of grouse such as the European black grouse and the North American sage grouse and prairie chicken. Recently, however, leks have been described for various species of amphibians, fish, mammals, and insects (see reviews by Wittenberger 1981, Bradbury 1985, and Loiselle and Barlow 1978). Bradbury, the most influential current student of leks, describes lek mating systems as requiring the following four criteria (Bradbury 1985): (1) there is no male parental care, (2) males occur on territories that are spatially clustered, (3) these territories contain no resources that influence female attraction or mate choice, and (4) females have freedom to choose mates within the cluster of male territories.

Current discussion of leks focuses on the following two questions: (1) why do

Figure 12-5 A male sage grouse *(Centrocercus urophasianus)* displaying on its lek.

they form at all? and (2) why do they form where they do? Before considering those points, we should describe lek behavior. To do so, we choose perhaps the best-studied lekking species, the sage grouse, *Centrocercus urophasianus* (Wiley 1978, Gibson and Bradbury 1986).

Male sage grouse are highly sexually dimorphic. Males weigh 2.5 kg, have a bright yellow expandable throat pouch surrounded by a collar of white feathers, and possess a fan of pointed, brightly tipped tail feathers (Figure 12-5); females weigh only 1.2 kg and are cryptically colored. From February to May, males gather before sunrise each day at traditional locations. There each spends 3 to 4 hours defending his small (5 to 10 m diameter) territory against other males and producing sexual displays. These include, in particular, strut vocalizations, in which a male inflates the throat sac, then compresses it by contracting chest muscles, resulting in a loud pop as the air is released (Wiley 1978). Strut vocalizations are performed at a rate of one every 10 seconds during the active period (Krebs and Harvey 1988). Female grouse visit the groups of 20 to 100 males in April, usually arriving in small groups. Females tend to aggregate near the center of the lek and wander among male territories for some time. Eventually each mates—usually only once—then leaves to nest away from the lek, sometimes as much as 4 km away. Females do all the incubating and rearing of the six to eight young produced. (See Figure 12-6 for lekking behavior in two other grouse species.)

How females choose is somewhat controversial. Wiley (1978) reported that older dominant males occupied central territories that were the preferred mating sites of females. Gibson and Bradbury (1986), however, reported no significant correlation between male location and mating success but did note that male mating success

Figure 12-6 Bird leks as seen in two closely related grouse. **A to C,** Prairie chicken *(Tympanuchus cupido).* **D to F,** Sharptailed grouse *(Pediocetes phasianellus).* In each series of three pictures the first shows several males on the lek; the second is a close-up of a displaying male; the third shows two males fighting at the boundary of their territories.
Photographs courtesy Ed Bry, North Dakota Game and Fish Department.

A

D

was correlated with the rate at which males could perform the demanding strut vocalizations (Hartzler 1972, as reported by Gibson and Bradbury 1986, observed a similar correlation). Whatever the exact criteria, female sage grouse mate preferentially with dominant, vigorous males.

Why do leks form? Why should males display in immediate proximity to their intense rivals? Bradbury and Gibson (1983) list the following seven possible factors, four focusing on the males and three on females:

1. Males may clump to reduce predation while displaying. It is well known that displaying males are subject to predation (Burk 1982), but so far

there is little evidence that lekking males are more or less safe than other
displaying males. This hypothesis is compatible with the fact that, in
grouse, species that live in open country lek but forest-living species do
not (Wiley 1974).

2. Males clump to increase signal range or amount of time signals are being
   emitted. As Bradbury's (1981) analysis shows, however, increase in signal
   range does not increase proportionately with number of males in a lek.
   Although a large lek, discernible over a large distance, may attract more
   total females than a small lek, it is unlikely on purely physical-acoustic

grounds to attract as many females per male. That ratio is, of course, the crucial one from a darwinian viewpoint.

3. Males may require specific display habitats that are limiting and patchily distributed. Various studies point out that many apparently suitable sites seem to exist but are not used (sites do not seem to be highly limiting). Also, some leks carry on in traditional locations even when those locations have been greatly modified (for example, by placing an airport runway nearby).

4. Males may clump near hotspots through which large numbers of females are likely to pass. This not only addresses why a lek should form, but also where, so we discuss it later.

5. Females prefer clumped males because this reduces predation on them. (Possible reasons are discussed in the next chapter.) The same comments made for (1) apply here.

6. Females prefer males that clump in less desirable habitat patches, thus leaving more desirable patches free for exclusive use by the breeding female. This was proposed by Wrangham (1980) but has been criticized by Bradbury and Gibson (1983) as not being supported by any good examples.

7. Females prefer larger clumps of males because this facilitates mate choice. Alexander (1975) argued that, as with other female-chosen sexual traits, a tendency to clump might evolve in males through advantages to females that mated only with a male that was part of a group. Benefits to such females might include a better overall mate obtained in less time, with consequent savings in energy and reduced risk of predation.

Although it may simply reflect the current widespread interest in mate choice sexual selection, (7) is probably the reason most widely quoted by ethologists for the evolution of lekking. It is not incompatible with some of the other possibilities (4 and 6), especially when we consider the next question: why do leks form where they do?

Several reasons for the specific siting of leks were hinted at previously. Lack (1968) suggested that leks formed at particularly safe sites, to reduce predation; Wrangham (1980) suggested that leks form in less desirable habitat patches. However, little specific evidence has been produced in favor of these reasons. Recent discussions of lek-siting began with a model proposed by Bradbury (1981). Working from a model of ecological resource use developed by Fretwell (1972), Bradbury envisioned leks evolving through a female choice process. If females preferred males in clumps, it would be reasonable to suppose that a female encountering clumps of different sizes would visit the largest clump within her home range. This would lead to a few large, widely spaced leks with few or no males displaying between leks. If female home ranges overlap, the stable distance between leks would be approximately one female home-range diameter. The advantage of the model was that it led to specific predictions that ethologists could test.

Unfortunately for this female preference model (Bradbury and Gibson 1983), most of the evidence collected went against these predictions. Thus in 1983 Bradbury and Gibson proposed an alternative, which they called the *hotspot model*. In this model, certain locations would experience high numbers of visiting females. Those might be locations containing resources attractive to females or transit points between a number of such places. Males establishing display territories at such hotspots would expose themselves to the most females; any factors that prevent a single male from ousting other males and monopolizing such a spot (such as high male density or female preference for clumped males) would lead to the evolution of leks in these hotspots.

Even more recently, Bradbury and Gibson's hotspot model has been challenged by Beehler and Foster (1988), who proposed the *hotshot model*. They stress male competitive interactions in the dynamics of lek formation. As they see it, leks form rather randomly with respect to ecological parameters but nonrandomly with respect to social ones. Dominant males, with conspicuous attractive displays, lure females to their territories. Other, less dominant males gather around these hotshots; eventually a cluster of males—a lek—forms. Beehler and Foster cite a number of features of male-male and male-female interactions in leks that support their model, such as disruption of courtships by dominant males and shifts of positions by lekking males to get closer to other males that are courting females successfully.

It seems likely that lek formation occurs for several reasons. Consider for example, lek formation in an insect, the Caribbean fruit fly. In this species, although isolated displaying males are common, females usually mate with males in leks, supporting a female-preference model of some kind (Hendrichs 1986). Leks form each afternoon, apparently with one male establishing a one-leaf territory in a fruit tree and releasing a sexual attractant odor. This odor attracts not only females but also other males, which cluster around the original male, subsequently jostling with each other for territories; this lek-formation process agrees with the hotshot model. Yet the location of leks is not random—they form in fruit trees that are hosts for the fruit-boring activities of females, and they form on the lower periphery of the tree where newly arriving females pass as they colonize the tree, in compliance with the hotspot model (Burk 1983). Similar multiple explanations may be applicable for other species, too.

Monogamy

Humans are a basically monogamous species, and we are fascinated by long-term monogamous bonds in other animals, such as geese and swans (Lorenz 1979). This should not obscure the fact that monogamy is, overall, a relatively rare phenomenon in animals. To understand why, it may help to state clearly the grounds on which monogamy must be based. For males, monogamy evolves when circumstances prevent them from achieving polygamy or when increases in survival of one set of young from male parental behavior outweigh decreases in total number of young sired from being monogamous instead of polygamous. For females, monogamy evolves if there are no resource-accrual or other advantages of remating or if females are never better off as a second mate of one male than as the sole mate of another. Following Wittenberger (1981), we can convert these into the following four specific explanations for the evolution of monogamy:

1. Monogamy evolves because male help is essential to female success. For species breeding in harsh conditions, it may be impossible for a female to raise any young without assistance from a male. For example, a penguin female needs a mate to incubate the eggs and warm the chicks when she goes out to the sea to feed, and a herring gull female needs a mate to guard her eggs or chicks from predators or other gulls that are cannibals. In such species a male that deserts a female after mating to search for additional matings obtains no offspring from the first mating. Similarly, any

additional matings provide no additional offspring; any number times zero
still gives zero as the total. Lack (1968) considered this to be the reason
for monogamy in most birds. Unlike most animal groups in which polygyny
is most common, more than 90% of bird species are monogamous.
However, Wittenberger has argued that complete reproductive failure by a
deserted female is unlikely in all but a relatively small number of species
facing exceptionally difficult circumstances. For most birds, Wittenberger
argues in favor of the next explanation.

2. Monogamy evolves because the polygyny threshold is never reached in a
   species. Remember, polygyny may be favorable from a female's point of
   view if there is wide variation in male territory quality. A female may be
   better off as the second mate of a male with a good territory (even without
   his parental help) than as the sole mate of a male with a poor territory
   (even with his parental help). Conversely, if marginal habitats are not
   sufficiently worse than prime habitats, females should always choose to
   settle in the territories of bachelor males. Wittenberger argues that this is
   the case for most birds—breeding habitat is not severely limited, and
   polygynous females would do sufficiently worse than monogamous females
   because of the lack of male help that monogamy evolves as the exclusive
   mating system. This explanation may also apply to some of the few
   mammalian species that are monogamous (forest-living antelopes and such
   dog species as coyotes, foxes, and jackals).

3. Monogamy evolves because female aggression prevents males from
   achieving polygyny. We saw earlier that for marmots, polygyny was best
   for male harem-holders, but monogamy was preferable for females. In
   some such cases, monogamy may be the usual mating system as the
   outcome of aggressive behavior by dominant monogamous females. Second
   females may be driven away by resident females in territorial species (such
   behavior may be more common than currently realized because it may be
   somewhat difficult to observe). In more social species a single breeding
   female may allow subordinate females (often the breeding female's
   relatives) to remain in the group but suppresses their reproductive
   physiology by asserting dominance over them. This pattern is well known
   in some primates such as marmosets, in many social canids such as
   wolves and African wild dogs, and in some cooperatively breeding birds
   (see references in Wittenberger 1981).

4. Monogamy evolves through male defense of individual females. Although
   polygyny could be considered to be the ideal mating system for males of
   many species, circumstances for some species make "a female in hand
   worth two in the bush" for males. In such species a male remains with a
   single female and guards her against other males for the duration of her
   breeding cycle. In these species, males show no parental care, although
   their mere presence may occasionally deter predators or parasites. A
   number of factors increase the likelihood of such mate-guarding monogamy
   (Alcock 1989). They include male-biased sex ratios, high overall male
   density, a high degree of synchrony among females in receptivity to males,
   and female-required resources that are not easily monopolized by males.
   Wittenberger (1981) uses this explanation for monogamy in some
   crustaceans and insects, as well as some vertebrates such as explosive-

breeder toads and frogs and waterfowl. Females, it should be noted, may derive benefits from being guarded, if this cuts down on harassment from other males and allows them to get on with offspring production and care.

Before leaving the subject of monogamy it is appropriate to briefly discuss the breakup of pairs or divorce (Diamond 1987). Even excluding high rates of mate loss because of mortality (50% to 73% in migratory American songbirds) and serial monogamy (in which pairings typically last for only one breeding attempt), the breakup of established pairs is relatively common. In well-studied bird species, it averages 21%—not high by Hollywood standards but appreciable nonetheless. Several studies of monogamous-pair breakup suggest it is adaptive. It usually follows reproductive failure in the previous year; although in oystercatchers studied by Harris et al (1987), one bad year after previous successful years was less likely to result in breakup. Furthermore, at least for the kittiwake gulls studied by Coulson (1966), long-term reproductive success increased for birds that had broken from previous partners and subsequently formed new pairs with different partners (for the oystercatchers, males that repaired did no better than in their first pairing, however, females increased their reproductive success by 61%—suggestive but not significant because of small sample size).

Polyandry

We are using the term *polyandry* in a sense that is somewhat more restrictive than is often the case. For us it does not include all cases of multiple partners for females but only those in which it is usual for females but not males to have multiple mates. Given all we have said about the advantages of multiple matings to males, it is probably not surprising that polyandry is quite rare. It is found in a few birds and fish (seahorses and their kin) but is absent in most animal groups.

The spotted sandpipers discussed previously are perhaps the best-studied polyandrous species. Their sex-role-reversed behavior, typical of polyandrous species, has been sufficiently discussed already and serves as a representative of other polyandrous species. Here we focus on the evolution of this rare mating system.

Actually, as pointed out by Oring (1986), polyandry takes two forms. In classical polyandry, exemplified by the spotted sandpiper, individual males set up breeding sites, and the female divides her attention among the males. In cooperative polyandry, a group of males shares a single breeding effort with one female. The most famous example of cooperative polyandry is probably that of the Tasmanian native hen *(Thibonyx mortierii)*, a large flightless gallinule studied by Ridpath. Groups in this species commonly consist of a pair of brothers breeding with a single female, with all the birds sharing in territorial defense, nest building, incubation, and care of the chicks. The reason for distinguishing the two types of polyandry, as Oring (1986) sees it, is that they have probably arisen in different ways, and their evolution has to be understood with different primary questions in mind.

For classical polyandry the key question is, why do males tolerate being deserted by females? There seem to be two answers. (1) In some cases, males may be better than females at providing one-parent care, perhaps because they have not had to strip body reserves to lay eggs and are in better condition than females. If so, desertion of the nest in search of additional breeding opportunities might be more likely by females than by males. A successful outcome for the first brood is more likely if the male is left to rear it. There is little evidence bearing on this hypothesis. (2) Deserting females might be more likely than deserting males to obtain second

Lewis Oring

Lewis Oring

In Defense of Long-Term Studies

I have been studying spotted sandpipers since 1969; and since 1971, when I noticed that some females were polyandrous (i.e., they had more than one mate), I have been studying their social system (Oring et al [in press], Behavioral Ecology and Sociobiology). Friends ask, "Haven't you learned enough about that bird?" My wife says, "I'll believe you're finished when I see it!" Come next spring, I will be going back to Pelican Island, in the center of Leech Lake, Minnesota, for my twenty-second year studying spotted sandpipers. Why study one species so long?

When it comes to long-lived species, long-term studies are crucial to understanding variation in mating systems and reproductive success. Studies such as Fred Cooke's on snow geese and Tim Clutton-Brock's on red deer are the studies that truly advance an understanding of social biology. In fact, if I had my way, we would replicate such fine studies across each species' ecological range.

Not everyone appreciates the value of long-term studies. Recently I was shocked when a finishing PhD student at one of North America's premier universities told me he could not see any reason for studying animals for their lifetime—why not just multiply annual reproductive data times the life span? He is a product of an epidemic that grew with sociobiology and was fostered by ever tightening research budgets. Find a hypothesis to test and write a paper with a sexy title—preferably with a colon. That is how you get a job.

Now, this is not to say that I believe every study should run for 20 years. In general, in species where individuals are short-lived, annual data are useful in estimating lifetime patterns. However, the success of this approach depends on the uniformity of mating patterns across the life span. If the pattern is one of relatively little variation, the primary need for lifetime data is to determine longevity. With the exception of revealing the time of breeding onset, long-term studies might yield little in terms of age-specific variation in mating patterns. However, if mate acquisition varies with age or experience, within an age class, or with resource availability and if longevity is highly variable, then lifetime

mates. This would occur if the sex ratio of breeding individuals is heavily male-biased and perhaps also because the readiness of females to lay more eggs is more variable than the readiness of males to incubate. Because the female is producing the eggs, there is more certainty on her part about the completion of egg laying. Conversely, there is uncertainty on the male's part; there may remain from his viewpoint the possibility of further eggs to inseminate, whereas, in fact, there will be no more eggs. Using a somewhat anthropomorphical analogy, she can desert him when he is still expecting to have another egg of hers to fertilize.

For cooperative polyandry the key question is, why do two or more males share reproductive opportunities? Oring provides the following list of three ecological and one mitigating factor as partial explanations: (1) If conditions are extremely harsh, it might take not two parents but three to bring a brood to completion; or at least trios may do much better than duos. (2) The number of available breeders may be so large relative to restricted breeding habitats that the only option open to some data are essential for describing the mating system and determining lifetime reproductive success.

Long-term studies are also necessary when rare events (such as catastrophes) regulate population processes or to determine if unexpected observations are the norm, rather than a rare anomaly. For example, in the western gull in southern California, a female-biased sex ratio results in female-female pairing (Hunt et al 1980, Auk 97:473-9). Only the availability of long-term data allowed confirmation that this was a frequent event and showed why and how this behavior was locally maintained.

Over the past decade, the area of greatest expansion in mating system research involved the use of molecular biological techniques to determine whether adults attending eggs were indeed the parents. Here, too, the field has suffered from the use of shortcuts. Molecular techniques should be applied when behavioral observations reveal problems requiring these sorts of data. Parental exclusion/inclusion studies must be conducted in concert with long-term studies involving detailed behavioral observations. Only then can we differentiate between fundamental alternatives, for example, whether there has been an extra-pair fertilization while a firm pair-bond was intact or whether the pair-bond changed during the midst of the fertile period.

Now, more than 100 years after Darwin categorized mating systems as monogamous, polygynous, polyandrous, and promiscuous, the study of mating systems thrives. Knowledge of within-species variation in mating patterns, including alternate reproductive strategies, has mushroomed. Unfortunately, we still understand relatively little about how temporal and spatial variation of mates and other critical resources leads both to the evolution of particular mating patterns and to their expression. There is a critical need for carefully controlled experiments exploring factors regulating mating patterns and for the application of molecular biological techniques to mating system research. However, for these studies to lead to an understanding of the ecological and evolutionary bases of mating patterns, they must be done in concert with long-term, detailed behavioral observations.

For additional readings see Clutton-Brock (1988) and Newton (1989).

males may be to join groups. (3) If food is sparse and female home ranges large, males may not be able to maintain exclusive control over females, which may in turn have reasons for mating with several males (see the dunnock example in the section on polygynandry). (4) All of these factors seem to be more likely to come into play if the cooperative males are relatives (discussed under kin selection and cooperation in Chapter 14). At present there is insufficient evidence to evaluate properly the relative importance of these various factors in the evolution of this rare and interesting mating system.

Polygynandry

By now it should be clear that obtaining multiple mates will often be adaptive for male animals. On the other hand, it has usually been the assumption of ethologists that multiple matings are not adaptive for females. They can get all the sperm they need from a single mating or mate and would be wasting time and energy and risking predation if they pursued additional matings or mates. Where the phenomenon of multiple matings by females has been mentioned, it has usually been as the flip side of resource-defense polygyny: females remate only because they have to, when access to an important resource such as food or oviposition site is given to a female by a resource-holding male in return for a copulation. The undesirable (for anthropomorphic reasons) term *prostitution polyandry*, has been used for this pattern (Wolf 1975). We have defined polyandry more restrictively—if this pattern is considered polyandry, a better term for the pattern just described would be *resource access polyandry*.

In fact the belief that it is rarely advantageous for female animals to seek multiple mates may be based on little more than vestigial Victorian prejudices held by mostly male ethologists (Hrdy 1981, Burk 1986). There are a number of possible reasons in favor of multiple matings by females. Halliday and Arnold (1987) list no fewer than the following 10 reasons:

1. Multiple mating allows females to collect a number of nuptial gift nutritional offerings in those species where males provide them (or gives females multiple access to other important resources).
2. Multiple mating may secure the cooperation of several males in tending the brood (or at the least secure their neutrality and prevent such phenomena as infanticide).
3. Multiple mating may reduce the costs of sexual harassment.
4. Multiple mating may reduce the socially disruptive effects of sexual competition among males in social groups.
5. Multiple mating may be a hedge against infertility of the first mate.
6. Multiple mating may be required to obtain enough sperm to fertilize an entire clutch and is required for long-lived females when all the sperm from prior matings have been used.
7. Multiple mating may guard against deleterious effects caused by genetic deterioration of sperm stored for long periods after earlier matings.
8. Multiple mating may provide an advantage by producing a more genetically diverse set of offspring.
9. Multiple mating may promote competition among spermatozoa so that only the most vigorous ones achieve fertilization.
10. Multiple mating may simply be the nonadaptive expression in females of a trait that is strongly selected for in males and therefore becomes a nonsex-limited feature of a species' behavior (a view strongly asserted by Halliday and Arnold).

Multiple mating by females may be the norm not the exception. If so, the most common mating system would not be polygyny (as we have defined it) but instead a mating system in which both males and females have multiple mates. The old term *promiscuous mating* might be used, but we dislike its connotation of random pairing; multiple-mating females would continue to be discriminating, and would be most unlikely to mate randomly with respect to various mate choice criteria. Clearly an alternative term would be useful. Such a term is *polygynandry*.

As traditionally used, polygynandry referred to bird species in which there were several males and several females with long-term bonds in a breeding group. (A similar situation exists in lion prides, as discussed in Chapter 14.) There is no reason this term cannot be appropriately applied also to species in which both males and females mate a number of times but in which no pair bonds are present.

Examples include species like damselflies and apple maggot flies, which combine resource-defense polygyny in males and resource-access polyandry in females. The presence or absence of pair bonds is not considered a fundamental factor in lumping examples together in other mating system types such as polygyny.

What is fast becoming the best-known example of polygynandry comes from a study of the mating system of an English songbird, the dunnock or hedge sparrow (*Prunella modularis*). We have already mentioned the odd cloaca-pecking prelude to copulation in this species as a vertebrate example of sperm competition. As revealed by the work of Davies and his colleagues (Davies 1983, 1985, Davies and Lundberg 1984), dunnocks not only exemplify polygynandry but are also a fascinating species with which to conclude our discussion of sexual behavior. Almost everything mentioned in this and the previous chapter occurs in this single species.

Dunnocks are small birds (wing length approximately 7 cm) that feed mainly on small insects and seeds on the ground in dense vegetation. Davies and colleagues studied them from 1980 to 1983 in the botanical garden of Cambridge University, a 16-hectare area with a mosaic of different habitats, some densely vegetated and others more open. In the winter the birds have overlapping home ranges and sometimes congregate at rich feeding patches. In the spring, both males and females establish independent and overlapping territories that are defended against other dunnocks of the same sex. Males establish as large a territory as they can. Females seem to establish a territory that is large enough to provide a good number of feeding sites—the better the food supply of an area, the smaller the female territories are. Males associate with the females whose territories overlap theirs—the actual mating arrangement depends on the size of female territories and the way in which they overlap male territories. Monogamy, polygyny, polyandry, and polygynandry are all present in this single population (Table 12-1). The most common arrangements are monogamous pairs and polyandrous trios (two males, one female). The former occurs when a female territory is about the same size and overlaps well with a male territory. The latter occurs when a female territory extensively overlaps two male territories: both males associate with the female and begin to intrude on each other's ground

Table 12-1 Dunnock mating systems: mating combinations at the start of the breeding season for 3 years

Mating Combination	Frequency			
	1981	1982	1983	Total
Unpaired ♂	1	1	3	5
Polyandry 3♂♀	—	1	1	2
2♂♀	4	12	8	24
Monogamy ♂♀	13	11	10	34
Polygyny ♂2♀	2	—	—	2
Polygynandry 3♂2♀	—	1	—	1
2♂2♀	1	1	6	8
2♂3♀	—	1	1	2
2♂4♀	—	—	1	1
Breeding population				
No ♂	26	46	48	
No ♀	23	31	38	

From Davies, N.B. and A. Lundberg. 1984. Journal of Animal Ecology 53:895-913.

Figure 12-7 Examples of different mating combinations in dunnocks. Lines show male song territories, dots show female home ranges. Solid versus dashed lines and open versus filled dots reflect different individuals.
From Davies, N.B. and A. Lundberg. 1984. Journal of Animal Ecology 53:895-913.

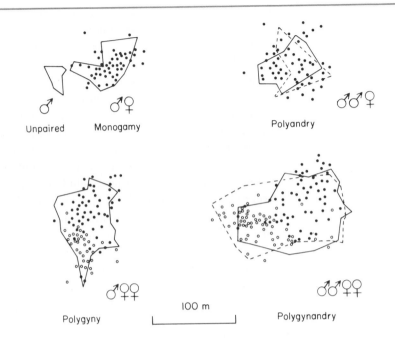

(Figure 12-7). Eventually the two territories fuse, but one of the males (the alpha male) is always clearly dominant over the other (the beta male). Some males with large territories exclusively overlap two female territories and achieve polygyny. Polygynandry results when several males fuse territories that overlap with the territories of several females (female territories do not fuse but remain exclusive to single females). A variety of polygynandrous groups was observed, from three males and two females to two males and four females. The balance of different arrangements is determined by food distribution and abundance. High food availability keeps female territories small and leads to more monogamy and polygyny; low food availability makes female territories larger and leads to more polyandry and polygynandry. Davies and Lundberg (1984) supplied extra food to some areas and observed the expected shifts toward polygyny and monogamy.

Female dunnocks perform all the nest-building and incubation duties. When females approach the onset of egg laying, males begin to follow them tenaciously and to solicit copulations. Polygynous and monogamous males, naturally, have reasonably assured access to their females (although such females also solicit copulations from other males in neighboring territories). In polyandrous and polygynandrous groups the alpha male tries simultaneously to stay with the female and drive away the persistent, ever-harassing beta males. Alpha males succeed often enough to obtain about 60% of the copulations, but alpha males lose the females often enough and for long enough that beta males get about 40% of the copulations. Success of betas is higher in polygynandrous groups than in polyandrous groups; this is because the alpha males cannot simultaneously be with both or all of the females. Beta males' success is also strongly affected by the degree of synchrony in the nesting activities of females. Where females are synchronous, beta males are quite successful (achieving some copulations in nine out of nine cases), where females are asynchronous, beta males are much less successful (achieving copulations in only two of seven cases).

Success by beta males is not only a function of their activities. Females in polyandrous and polygynandrous groups actually try to give alpha males the slip and actively seek out copulations with beta males. The result of all this is a blur of females, alpha males, and beta males flashing in and out of the undergrowth until females begin incubating, at which point things tend to settle down again.

Why do females seek out matings with males other than the alpha? There are at least three apparent reasons. The most important is that males will help to feed the chicks of a female only if they have copulated with the chicks' mother. Thus a female that has mated with beta males obtains their parental help in addition to that of the alpha male. In 28 of 29 cases, alpha males of trios and polygynandrous groups fed females (the exception was a polygynandrous male with two females with chicks at the same time—he fed only one family, whereas the beta male fed the other). In 18 cases of 29 the beta males copulated, and they fed the chicks in 16 of these (one of the exceptions was the polygynandrous case just mentioned where the beta male fed one family but not the other). In the 11 cases where beta males did not copulate, they did not feed the chicks. This help from beta males makes a difference. Females with this help lay more eggs (apparently they are able to gauge how much help they can expect to receive from the number of males they copulate with), and their chicks are fed at higher rates and end up heavier than those of females that receive help from only one male.

Two other benefits of multiple matings are gained by females. First, there is

Table 12-2 Summary of costs and benefits to male and female dunnocks of different mating combinations		Costs	Benefits
	Presence of more than one male in mating combination (polyandry or polygynandry)		
	To female	Increased harassment in egg-laying period leads to (1) decreased feeding rate (2) unhatched eggs. Interference of breeding attempt if some males fail to copulate.	Increased parental care if more than one male copulating leads to larger optimal clutch size and better chick survival.
	To male	Increased competition for mating leads to (1) longer and more intense mate guarding (2) unhatched eggs. Interference of breeding attempt if other male(s) fail to copulate. Loss of paternity if other male(s) do copulate.	Increased chick survival if several males help to rear brood.
	Presence of more than one female in mating combination (polygyny and polygynandry)		
	To female	Share help of male(s) in chick rearing with other female(s). Aggression versus other female(s) and possible interference from them in breeding attempt	None.
	To male	Increased mate guarding.	Increased production of chicks.

From Davies, N.B. 1985. Animal Behaviour 33:628-648.

strong suggestive evidence that beta males peck and destroy eggs of clutches from females they have not copulated with. Female dunnocks lay several clutches per summer, and infanticide would cause a female to start over with a new clutch that the beta male has a new chance to help sire. So solicitation of copulations with beta males may prevent infanticide. Also, females that are continually chased and harassed during the prelaying period are much more likely to produce eggs that ultimately do not hatch, presumably because of the stress. Mating with beta males reduces harassment.

The end result is a strong conflict of interest between females and males. Males do best as polygynous males, females as members of polyandrous trios if they can get parental help from beta males (Table 12-2). Beta males and females contrive to copulate, whereas alpha males do their utmost to prevent that. The actual ability of males to monopolize females in monogamous or polygynous relationships depends ultimately on the food supply (Davies and Lundberg 1984, Davies 1985).

The dunnock studies exemplify the marvelous complexity of sexual interactions in animals. One cannot help wondering what generations of Cambridge clerics would have thought if they had realized what was going on in the bushes all around them as they took their afternoon strolls in the botanical garden.

PARENTAL CARE

For many species of animals, especially marine invertebrates, reproductive behavior ends with the release of the gametes. For other animal species, however, including most insects and vertebrates, sex is followed by at least some parental solicitude toward the offspring. The extent of this solicitude varies widely, from merely putting the fertilized eggs in favorable locations, to staying with them until they hatch or even longer, or feeding, protecting, and instructing the offspring for extended periods of time. The peak in parental care is undoubtedly found in the eusocial insects, birds, and mammals, where parental care involves extensive brooding, feeding, construction of nests or lairs, vigorous defense from predators, and facilitation of the offspring's entry into the species' social structure. Many examples of parental activities have already been given; others are presented here and in Chapters 13 and 14 (Figure 12-8). As a highly parental species ourselves, we find parental care by other species fascinating.

Parental care is not universal, however, and it is worth considering why it is found in some species and not in others. What are the ecological correlates of parental care? That is the first of three topics we take up in this section. We then consider variations in the identity of care givers: why do some species show only maternal care, others only paternal care, and others biparental care? Finally we ask: why does conflict sometimes break out between parents and offspring? Such conflict, from minor squabbles over weaning to actual cases of parental infanticide, poses a challenge to ethologists who consider offspring production to be the ultimate measure of an animal's success.

The Ecology of Parental Care

Even within specific animal groups there may be great differences in the amount of parental care provided. A female housefly may deposit a thousand eggs rather haphazardly in roughly suitable areas, whereas a female tsetse fly rears one offspring

Figure 12-8 Examples of parental care among species. **A,** Egg brooding by male water bug *(Abedus herberti)* with newly hatched nymph nearby. **B,** Adult urchin *(Strongylocentrotus franciscanus)* with juveniles under its spine canopy. Juveniles of a closely related species in the same genus are not commonly found under the adults. **C,** Lycosid spider carrying young by specialized abdominal hairs. **D,** Oral birth from gastric brooding by the frog *Rheobatrachus silus.*
A from Smith, R.L. 1979. Science 205:1029. B from Tegner, M.J., and P.K. Dayton. 1977 Science 196:325. Figure courtesy Mia Tegner and Paul Dayton. C from Rovner, J.S., et al. 1973. Science 182:1153. Photograph courtesy Jerome Rovner. D from Tyler, M.J., and D.B. Carter. 1981. Animal Behaviour 29:280-282. Photograph courtesy M.J. Tyler.

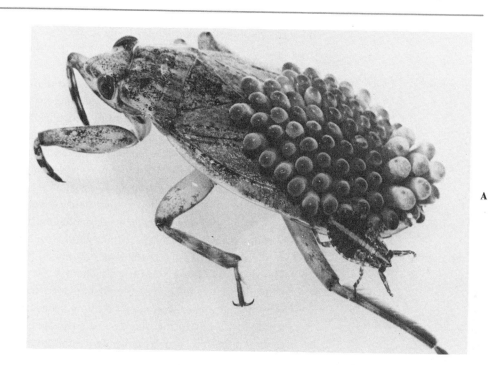

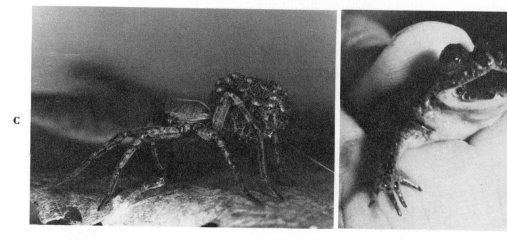

at a time inside a womblike organ, giving birth only when the larva is fully grown and ready to pupate immediately. What factors favor the evolution of parental care? In his book *Sociobiology*, E.O. Wilson (1975) lists the following four factors favoring such evolution:

1. Stable, saturated habitats. Following standard ecological theory, Wilson argues that stable, favorable habitats increase the intensity of intraspecific competition. Population density of a species builds up until the habitat is saturated with individuals. For a young animal to establish a home range or to be able to breed in such a situation it needs to be large, socially skilled, and efficient at finding and exploiting resources. Extended parental feeding, protection, and assistance produce offspring with a better chance to succeed in such a high-density world. Although this idea seems reasonable and has existed in the ecological literature for some time, it still has little strong support from actual studies. However, it may well apply to some cases of extensive parental solicitude, as perhaps in the Florida scrub jays we discuss in Chapter 14.

2. Unusually stressful physical environments. Some habitats are so physically extreme that unassisted eggs or young seem to have little chance of survival. Earlier in the chapter we suggested that penguin eggs and chicks in Antarctica probably require two committed parents to survive. In another example, parental care has evolved in a few species of frogs—eggs are incubated in skin pouches, the mouth, even in the stomach of one species (Figure 12-8, *D*). In almost all of the cases the frogs are terrestrial rather than aquatic breeders that have evolved unusual care to deal with what to an amphibian egg is an extremely harsh terrestrial environment.

3. Specialized diets. Some prey require such skill to capture or process that a long period of practice or apprenticeship is required before an animal can completely support itself—parental help in the learning period is essential. A classic example is found in the oystercatcher, *Haematopus ostralegus*, a wading bird that frequently specializes on shellfish such as mussels or cockles. Feeding on these mollusks requires extremely skilled methods, such as hammering precisely on the weakest point of the shell or stabbing quickly inside a mollusk that has its shell slightly open while feeding (Norton-Griffiths 1969). Young oystercatchers learn these techniques by associating with and watching their parents, and they do not become good enough foragers to raise their own chicks until they are 4 years old.

4. Predator pressure. This is probably the most frequently operating reason for the evolution of parental care. Many studies have shown that protection of eggs and young by one or both parents greatly reduces predation rates. To pick one example, Tallamy (1984) studied a lace bug, *Gargaphia solani*, in which some females abandoned egg masses after depositing them. Other females remained with the eggs and resulting hatchlings through their entire juvenile period. Deserting females laid more than twice as many eggs as guarding females (parental care costs a parent something), but maternal defense increased juvenile survival sevenfold. In Chapter 13, we show that predator pressure is important in the evolution of other social interactions, as well as those between parent and offspring.

Who Provides Parental Care?

Given that ecological circumstances favor provision of parental care, a new question arises: why is the care sometimes provided by both parents and in other cases by only one? If by one, why is it sometimes by the male and in other (more numerous) cases by the female?

There are a number of ways to tackle these questions, perhaps the best is to pick a group of animals that displays all the possible combinations (no care, maternal care, paternal care, biparental care). This would eliminate the possibility of confusing the effects of current ecology and past evolutionary history. Such confusion might result, for example, if one were to ask why biparental care is the usual rule in birds and maternal care is the rule in mammals. These patterns may reflect a conservative retention of ancestral behavior rather than recent adaptations to existing circumstances. Among vertebrates, fish and amphibians demonstrate the full range of possibilities to a greater extent than other groups, so a number of workers have used them as focal groups, with fish receiving more extensive study.

Parental care is known from 87 of the 422 families of bony fishes (Osteichthyes) (Gross and Sargent 1985). In over 95% of the care-giving species, parental care takes the form of guarding the eggs and fry from predatory fishes (often conspecifics, as in the case of bluegills mentioned in Chapter 13). Maynard Smith (1978) and Ridley (1978) each pointed out a distinct pattern to the identity of the parent giving care, a pattern that was confirmed and extended by Gross and Shine (1981). The latter also noted the same pattern for amphibians with parental care. As shown in Table 12-3, for fish with external fertilization (both males and females release their gametes into the water, where fertilization occurs), there is no parental care in 100 families, care by males in 61 families, and care by females in 24 families. In other words, for externally fertilizing fish with parental care, the male provides the care in 72% of the cases. On the other hand, female care is found in 14 families of fish with internal fertilization (the male deposits the sperm in the female's body, where fertilization occurs—this is unusual in fish and amphibians but is the norm in reptiles, birds, mammals, and insects). Male care is found in only 2 internally fertilizing families, with no care in 5 families. Thus it is the female parent performing the care in 88% of cases. The same pattern holds for amphibians. For those with external fertilization there is male care in 14 families, female care in 8, and no care in 10 (male care in 64% of cases of care). With internal fertilization there is

Table 12-3 Distribution of male and female parental care with respect to mode of fertilization in fishes and amphibians. The table shows number of families; a single family may appear in more than one category, but is listed under "no parental care" unless care is completely unknown in the family.

Fertilization	♂ Parental Care		♀ Parental Care		No Parental Care	
	Internal	External	Internal	External	Internal	External
Fish	2	61	14	24	5	100
Amphibians	2	14	11	8	0	10
TOTAL	4	75	25	32	5	110

From Gross, M.R., and R. Shine. 1981. Evolution 35:775-793.

male care in 2 families, female care in 11, no care in none (female care in 85% of cases).

Three explanations for these patterns have been proposed, termed the *paternity certainty hypothesis*, the *order of gamete release hypothesis*, and the *association hypothesis*, respectively. These hypotheses have been carefully evaluated by Gross and Shine (1981) and Gross and Sargent (1985).

The paternity certainty hypothesis was proposed by Ridley (1978). The idea is that a parent should care only for offspring that are really its own. For a male, fertilization occurs "before his very eyes" in externally fertilizing species, but he can never be sure just what is going on inside the female's reproductive tract in the case of internal fertilization (remember our discussion of sperm competition). On the other hand, a female should almost always be assured that the eggs are her own. The differences in certainty of paternity for males should favor care by males with external fertilization, by females with internal fertilization.

Gross and Shine (1981) point out two weaknesses of this hypothesis, one empirical and one theoretical. First, studies do not show that sperm competition and cuckoldry (Chapter 11) are less likely in externally fertilizing species; they seem to occur in both cases (see examples in our section on alternative male tactics in Chapter 11). Second, Maynard Smith (1978) pointed out that, if certainty of paternity is low for one mating, it would also be low for any additional matings a male obtained by not caring for his first clutch. If there is so much promiscuous mating, it might actually lower the cost to a male of staying home and being parental—his gains from additional matings may be less than his gains from caring.

The order of gamete release hypothesis was advanced in 1976 by Dawkins and Carlisle (the latter was at the time an undergraduate animal behavior student). If (as we saw in lace bugs) parental care lowers the total number of offspring one can have, we might expect each sexual partner to attempt to desert after mating, leaving the restrictive care to the other. Thus which sex shows care is a matter of which one is left "holding the bag" after gamete release. With internal fertilization the female is left with the eggs and sperm, and the male can desert. With external fertilization the female spawns first, and can leave before the male deposits the sperm.

The major problem with this idea is that it does not account for the pattern observed in species with external fertilization and simultaneous release of eggs by the female and sperm by the male. This is actually the usual situation (50% to 70% of known cases, Gross and Shine 1981). In the order of gamete release hypothesis, we would expect half of such cases to result by chance in male care and half in female care. However, the actual data show male care in 36 of 46 cases (78%, a highly statistically significant pattern). Thus males are just as likely to show care in externally fertilizing species with simultaneous gamete release as in those with an egg-first order.

The association hypothesis was proposed by Williams (1975) and is the one upheld by Gross and Shine (1981) and Gross and Sargent (1985). In this idea, members of one sex find themselves in closer proximity to the embryos after fertilization than do members of the other sex. This predisposes them to become the care-giving sex if ecological circumstances favor the evolution of care. In species with internal fertilization the female is automatically more closely associated with the embryos, thus the correlation in fish, as well as the maternal care predominance in insects and mammals. What circumstance would cause males to be in continuing proximity to eggs they had already fertilized? The answer seems to be territoriality. As we have seen, in many species males defend areas to which females are attracted for mating. If females oviposit in a male's territory, as they often do in fishes, a male persisting in that place will be close to his offspring. It may be relatively easy in such situations for the male to chase off predators while still attempting to attract additional females. Polygynous territoriality poses two benefits to males, allowing them to gain something from parental care while losing little in lost mating opportunities.

What about biparental care? It is rare in insects, somewhat more common in fish (22% of species with care) and mammals, and the usual situation in birds. If we again concentrate on groups where no single way of behaving dominates, we find that biparental care usually evolves from maternal care in mammals, but from paternal care in fish (Gross and Sargent 1985). What does a male mammal gain by becoming a second caregiver that offsets his loss of additional matings, or a female fish gain by becoming a second caregiver that offsets her lost opportunities to obtain more resources that she can turn into eggs?

In the case of fish, where parental care usually means protection, a female would have to add sufficient extra protection beyond that already provided by the male that the increased survivorship of the young would offset her decrease in fecundity. Although it sometimes may be the case that two parents provide more protection than one, the fact is that only 22% of fish families with parental care have biparental care. Fecundity must usually outweigh offspring protection for female fish.

In mammals, males seem to evolve as second caregivers where they can carry some of the offspring (as in primates such as marmosets) or feed them (as in wolves or African hunting dogs, which regurgitate meat to the pups). It is unusual to see a paternal mammal that only guards the young. This perhaps emphasizes the greater food demands of endothermic (warm-blooded, high-metabolism) offspring.

The almost universal biparental care of birds may reflect the high demands of chicks for food. Two birds may well double the amount of food they provide, although it is unlikely that they double the amount of protection (Gross and Shine 1981). Perhaps the absence of biparental care in mammals, especially herbivorous mammals, reflects the difficulty that most male mammals would have in providing appropriate food. Females, which lactate, have solved that problem, leading to the question posed by Daly (1979): "why don't male mammals lactate?" Carnivores like wolves do provide food in the form of regurgitated meat brought back from a kill in their stomachs.

PARENT-OFFSPRING CONFLICT

Most human parents can attest to the fact that parents and their offspring sometimes find themselves in conflict over the way in which the parents provide resources and other care. In 1974, Robert Trivers produced an important paper that examined this phenomenon of parent-offspring conflict in animals.

Trivers' discussion proceeded from a genetic asymmetry: a parent is equally related to each of its offspring, but an offspring is twice as related to itself (it obviously has 100% of its own genes) as to its siblings (who share only 50% of their genes with it). Therefore at times we should expect to see conflicts between offspring and parents, since all offspring are expected to want somewhat more than

Figure 12-9 Weaning conflict. A sequence of drawings from photographs indicating a weaner's unsuccessful attempt to steal milk resulting in it being chased out of the harem. In the last illustration the weaner is seen resting outside the harem with a group of other weaned pups. From Reiter, J., et al. 1978. Behavioral Ecology and Sociobiology 3:337-367.

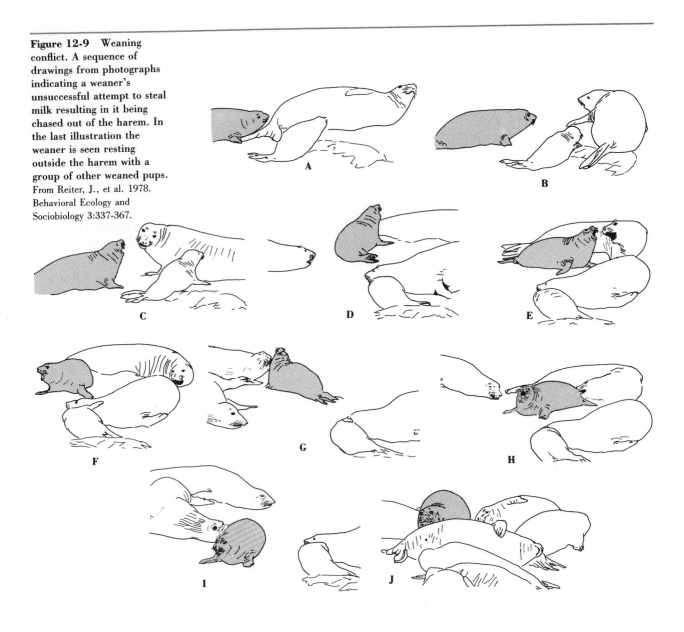

their fair share of parental investment, whereas parents are expected to want to distribute it widely and evenly. It need not even be the case that several offspring are simultaneously present; a parent may retain resources for potential future offspring even though current offspring may want the parent to give the resources now rather than hold them for future young. Trivers' model applies particularly well to a phenomenon known in mammals, weaning conflict, in which a lactating mother tries to wean her offspring of milk, while they attempt to prolong the period of suckling (Figure 12-9). Studies have shown that the period of lactation is enormously demanding physiologically on the female, much more demanding even than the period of pregnancy. A female cannot begin setting aside body reserves for the next litter until she has weaned the current one (Clutton-Brock et al 1989).

Figure 12-10 Parent-offspring conflict with changes in ratio of cost to parent/benefit to offspring, as proposed by Trivers (1974). The subjective index is merely to provide scale to help illustrate the ratio. Shapes of the curves shown here also are arbitrary. Modified from Trivers, R.L. 1974. American Zoologist 14:249-264.

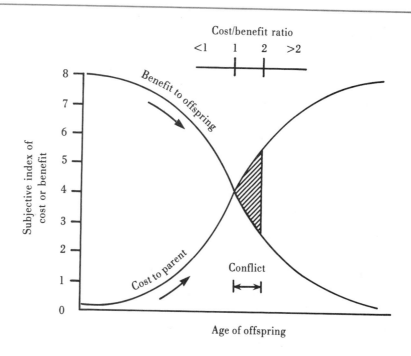

The model shown in Figure 12-10 illustrates the parent-offspring conflict. Up to a certain point the current offspring's benefits from additional care (such as lactation) are so large that the parent gains much more by helping this offspring than by terminating care. As the infant grows, it reaches an age where it can do almost as well independently as by receiving more help. At that point the parent may gain more from investing in new offspring than by continuing to help this one. Yet because of the genetic asymmetry, it is advantageous for the current offspring to become independent only when the cost-to-parent/benefit-to-current-young ratio is 2 to 1, instead of the 1 to 1 point favored by a parent. During the time represented by the shaded area on the figure, the parent is expected to attempt weaning, and the offspring is expected to resist weaning.

An excellent example of how the behavior of mothers and infants changes during this period is found in studies of rhesus monkeys by Robert Hinde (1977). As seen in Figure 12-11 the percentage of time an infant is in contact with its mother declines as the infant ages. Concomitantly, there is an increase in the number of times a mother rejects the infant's approaches (Figure 12-11, *A*). As seen in Figure 12-11, *B*, for the first 15 weeks or so of the infant's life, mother-offspring contact is usually broken by the infant, wandering off to explore. After about 15 weeks, contact is usually broken off by the mother wandering away, attempting to wean the infant and promote its independence.

Where there is such parent-offspring conflict, is there an expected winner? Not on theoretical grounds, although Richard Alexander (1974) argued that parents would win. His argument was based on the fact that all successful individuals will be both offspring and, later, parents themselves. He argued that an offspring that successfully manipulated its parents when young would lose out when it was a parent; its offspring would inherit its ability to manipulate, and beat it. So, over

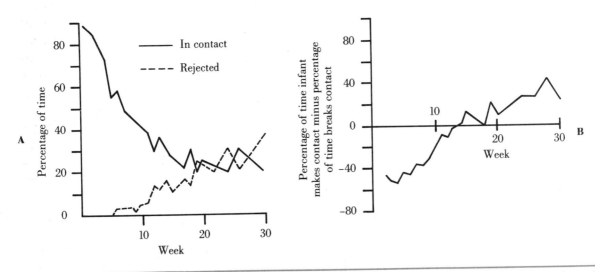

Figure 12-11   Changes in mother-infant relationship over time in captive rhesus monkeys. **A,** Percentage of time mother and infant are in contact declines over time. (*solid line*). Frequency of maternal rejection of the infant's approaches increases over time. (*dashed line*). **B,** The individual responsible for breaking contact changes from the infant (early) to the mother (late). From Trivers, R. 1985. Social evolution. Benjamin/Cummings, Menlo Park, Calif. After Hinde, R.A. 1977. Philosophical Transactions of the Royal Society of London, B 196:29-50.

the long run, the most successful reproducers would be parents who did not get manipulated. Parents should always come out ahead evolutionarily. However, we have already seen that there is a possibility for varying, conditional behavioral tactics, and Alexander's parents always win tactic would not be as good as a conditional tactic that programs individuals to be successfully manipulative offspring when young and successfully manipulative parents when old.

On pragmatic rather than theoretical ground, however, one might expect parents to win. Being larger and more powerful, they are in a better position to withhold food than an infant is to take it. On the other hand, infants may be deceptive and manipulative. Only a baby bird for example, knows exactly how hungry it is. It may beg vigorously, as if it were in desperate need, while in fact it may actually be in less danger of starving than a nest mate. Selection on offspring to deceive may be balanced by selection on parents to be able to assess real need on the part of offspring. Either or neither participant may achieve outright success. Unlike a coevolution in predator-prey arms races (Chapter 9), however, the competition is somewhat muted because they share genes.

Parents and offspring may be in conflict over issues other than the amount of food offered or the duration of care. In social insects, there may be differences over whether or not an offspring becomes a worker or a reproducer. There may be differences over dispersal, with parents prodding offspring into dangerous emigration and offspring preferring the safer option of staying in the natal area. When there is control over sex of offspring, as in hymenopteran insects (wasps, bees, and ants), parents and offspring may differ about the sex an individual offspring should be. In some extreme cases, parents and offspring may even differ about whether an offspring should be allowed to live or die. We close our examination of parent-offspring conflict with a consideration of these two cases. First we examine conflict between social insect queens and their worker offspring over the sex ratio of the new reproductive siblings of workers. We then examine the striking case of parental infanticide in birds.

Sex Ratio Conflict in Social Hymenoptera

Species in the insect order Hymenoptera have an unusual sex-determination mechanism, called haplodiploidy. Males are derived from unfertilized eggs and therefore have only one set of chromosomes (haploid); females are derived from fertilized eggs and have two sets of chromosomes (diploid). The ancestral hymenopterans (and the large majority of modern species) are solitary-living parasites of other insects, in whose eggs or larvae they lay eggs. A parasite may have to search widely for a mate and for hosts, and haplodiploidy probably evolved as a way of allowing a female wasp to take advantage of a host even if she had not yet managed to find a male. Whatever its original function, the presence of haplodiploidy is full of significance for hymenopteran behavior—it may explain why all the social insects except the termites are members of this single order (extended discussion in Chapter 14).

One thing that haplodiploidy does is create unusual patterns of relationship between members of a family. These are discussed extensively in Chapter 14; at this point we will merely tell you that the situation is unlike that in normal diploid species, where parents and offspring are half related to one another, as are all full siblings to one another. In haplodiploids it is usual for queens to be half related to their sons and daughters, but females are three quarters related to their sisters and only one quarter related to their brothers, as pointed out by Trivers and Hare (1976). This sets up a conflict between the queen and her daughter workers over what ratio of new queens and drones (male reproductives) the queen should produce. The queen gets the best genetic return on her egg investment when a sex ratio of 1 to 1 is produced in the new reproductives. However, sister queens are three times as valuable genetically to workers as brother drones, so the workers should favor a ratio of 1 drone to 3 queens.

Who wins? One might suppose the queen is in the determining position, because she can control whether eggs are fertilized or not and therefore whether diploid females or haploid males are produced. However, Trivers and Hare argued that in fact the workers are in the better position to achieve their end. Regardless of the sex ratio of eggs produced, workers perform all the brood care. They can dote on future queens and neglect future drones to produce a final 1 to 3 ratio. Trivers and Hare supported their conclusion with data from a number of ant species, showing that reproductive sex ratios tended to be 1 to 3. As a convincing special case, they showed that 1 to 1 ratios existed in slave-making ant species. In these species, queens' offspring are reared by captured workers from the colonies of other ant species; as a result, there are no conflicting genetic interests on the part of worker offspring.

Trivers' and Hare's results were criticized by Richard Alexander and Paul Sherman (1978) on both theoretical and empirical grounds. Theoretically, they pointed out that in cases of inbreeding between brother drones and sister queens, a female-biased sex ratio would also be favorable to queens. Because her sons would compete only with one another for matings with sisters (local mate competition), a queen in such cases should produce fewer sons and more daughters. Thus there is an alternative explanation for female-biased sex ratios in hymenopterans. Empirically,

Alexander and Sherman questioned the validity of some of the data from previous studies used by Trivers and Hare to evaluate their ideas.

Although the report by Alexander and Sherman was a valuable contribution, studies performed since its appearance have been done on species that do not show local mate competition and on which careful data collection was performed. A number of these (e.g., Ward 1983) confirm that, as Trivers and Hare suggested, workers achieve a 1 to 3 ratio.

Other recent studies, however, also show that the queen sometimes wins. Bob Metcalf (1980) found a 1 to 1 ratio of newly produced reproductives in the wasp *Polistes metricus*. Queens produce males early in the season, when few or no workers have yet emerged and therefore cannot affect the production of males. Later the queen lets her worker daughters help her produce new queen daughters. This example points out how ecological considerations and a species' natural history affect the way that reality imposes itself on theory.

Actually, the best solution of all for a female hymenopteran would be to help her mother produce new sister queens (related to her by 75%) and to produce new males herself. Her sons, produced by unfertilized eggs she might lay, would be twice as related to her (50%) as the brothers her mother would make (25%). Therefore another potential conflict exists between hymenopteran queens and workers, this time over who gets to make the next generation of males. A recent review of such worker reproduction (Strassmann 1988) shows that usually the queen wins this conflict. In a majority of species of ants, workers do not lay viable haploid eggs, but there is a substantial minority of cases where they do: Bourke (1988) lists 45 examples. In 27 of 40 cases where the details of worker egg laying were known, such laying occurs only after the queen has died. These cases do not represent a loss of control by the queen, but rather a last-gasp reproductive burst by a queenless colony that is on its way to dying out. In 13 of 40 cases, however, workers are apparently making sons "under the queen's nose," and a substantial number of the new drones may come from that source.

Infanticide in Birds

In a number of bird species, especially raptors (hawks and owls), the parents routinely lay more eggs than the number of offspring that usually survive (Stinson 1979). Unlike the situation in most birds, incubation of the eggs begins with the laying of the first egg, so the chicks hatch asynchronously and are subsequently different in size, with the eldest being the largest. Especially in species that lay two eggs, the eldest chick is given the majority of food by the parents. The result for some species in an average year is that the younger chick gradually starves. It may then be fed by the parents to the survivor. In fact, it is not uncommon for the older chick to peck and harass the younger, indirectly or directly speeding up its demise. The parents seem, at the least, to tolerate this siblicidal aggression and in a sense may be said to encourage it by producing an asynchronously hatching brood and by differentially feeding the older chick. Although parents probably do not actually kill the younger chick, their selective neglect can fairly be called *infanticidal*.

Two major hypotheses have been advanced to explain this phenomenon. One, the insurance hypothesis, suggests that production of a second (or in larger broods an additional) chick is parental insurance against the failure of the first chick (or chicks) to survive. Should the first egg fail to hatch, the first chick be defective, or the first chick be lost to disease or predation, the parents have a second (or replacement) chick already in place, to which they can switch their parental attention. In some species, death of the younger chick almost inevitably happens soon after hatching (e.g., lesser spotted eagles and black eagles, Stinson 1979); the insurance hypothesis seems to apply to these species.

In other species, such as the tawny owl, survival of the younger chick or chicks depends on the amount of food available. In years of low food availability the later-hatching chicks starve, but parents have not wasted much food or effort on these failures, and they can salvage part of that effort by feeding the carcass to the surviving young. On the other hand, in exceptionally good years the parents have enough food to satisfy the needs of the younger and the older offspring. In better condition, the younger chick or chicks can survive harassment from their siblings and survive to fledge successfully. In these species, producing more offspring than can normally survive represents an opportunistic strategy by the parents, keeping their options open in case it happens to be an unusually good year.

In each case, however, the parents and the younger chicks are in a situation of conflict. In most cases the advantage to the parents of a potential sacrifice of one chick is to produce the largest total number of well-fed chicks, whereas the possibility of the victim chick starving and being cannibalized would obviously not be to that chick's advantage. However, if food is in particularly short supply, so that it will be difficult to raise even one chick, natural selection could favor a suicidal self-restraint on the part of the smaller chick. Better that some of its genes survive in the body of its sibling than that none of them survive when both chicks die (O'Connor 1978).

A fascinating study of the dynamics of brood reduction through siblicidal aggression and parental neglect has been carried out by Douglas Mock and colleagues, studying two nonraptor species, great egrets and great blue herons, in Texas. Both species are fish-eaters with a monogamous mating system and both sexes participating in all aspects of parental care. In both species the usual brood size is three or four chicks. Yet the species differ in one respect: older great egret chicks frequently attack and kill younger siblings, whereas great blue herons rarely do. In natural nests studied by Mock (1984), 8 of 12 nestling deaths in egrets were caused by siblicide, whereas only 1 of 11 nestling heron deaths was the result of siblicide. Mock, however, was able to show that siblicidal aggression was potentially present in both species, but was especially stimulated by the circumstances of the great egret nest. Egret parents regurgitate fish on the nest floor, provoking chicks to fight for it; heron parents put their bills into a chick's bill and transfer fish directly, making squabbles for food pointless. Mock did cross-fostering experiments, putting heron chicks in egret nests and vice versa. As predicted, herons now attacked each other: six of six deaths of herons in egret nests resulted from siblicide. On the other hand, egrets in heron nests continued to fight, although at somewhat lower rates, and six of eight deaths were caused by siblicide.

Mock and Parker (1986) discovered an additional twist to the egret infanticide story, one that is highly relevant in this section on parent-offspring conflict. Mock and Parker assumed that the opportunistic explanation was the correct one to account for tolerance of siblicide by egret parents. A prediction from that explanation would be that, as brood size is reduced in sequence from four to three to two to one, likelihood of survival of the remaining chicks would increase because the available food would be shared among fewer mouths. (Thus the increased survivorship of

Figure 12-12 Survivorship of great egret chicks in broods of different sizes. From Mock, D.W., and G.A. Parker. 1986. Evolution 40:459-470.

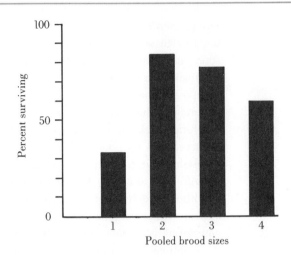

individuals in smaller broods would compensate parents for the reduction in brood size brought about by siblicide.)

Mock and Parker discovered that this was true but only up to a point (Figure 12-12). Chicks in groups of three survived better than those in groups of four, and chicks in groups of two survived better than those in groups of three. But chicks in groups of one (singletons) survived only about one third as well as chicks in groups of two. The reason for this surprising result was soon discovered: parents often abandoned nests with only one chick, to make a renesting attempt elsewhere. The abandoned chick naturally died. Such nest abandonment, frequent with one-chick nests, was otherwise rare (it occurred, for example, in only 6% of nests with two or more chicks). It appears that parents tolerate siblicide as long as it leaves them two healthy chicks surviving, but when only one chick is left the chances are better with having more than one chick in a renesting attempt, so the adults commit infanticide by abandonment. Not surprisingly, rates of sibling aggression decline precipitously when a brood declines to only two. However, sometimes (about 30% of the time) singletons survived. The determining factor was timing in the season: renesting attempts succeed only if there is enough time left in the breeding season to complete the new attempt. Thus parents abandon singletons early in the season but make the best of their bad one-chick situation later (early singletons had a 26% survival rate, late singletons a 55% survival rate).

The egret example demonstrates the complexities and subtleties involved in the interactions among members of a social group, even a nuclear family. This is a theme we explore in more depth in the next two chapters, which examine the ethology of sociality.

SUMMARY

The combination of male and female tactics in a species results in the species' mating system or systems. Four general mating systems exist: polygyny, monogamy, polyandry, and polygynandry. Ecological factors determine what the pattern of resources required by the investing sex will be; this in turn determines whether or not members of the noninvesting sex will be able to monopolize such resources and impose their preferred mating system on members of the other sex. As ecological situations vary, mating systems may vary even within a species, as shown in dramatic fashion by an English bird, the dunnock.

Sexual behavior may be followed by parental care, although extensive care is present in a minority of animal species. Parental care seems to evolve in response to high density of competitors, high rates of predation, extremely harsh physical environments, high demand for food by the young, and the need of young to be nurtured while difficult skills are mastered. Where care is provided, it may be provided by only one sex or by both; if by one sex it may be provided by the male or the female. The form of parental care that evolves depends on the relative benefits of care in terms of survival of the young, compared with the costs of lost mating opportunities (for males) or lost opportunities to gather resources and produce a larger number of eggs (for females).

Parents and offspring may find themselves in conflict, because there is an asymmetry in genetic relationship between parents and offspring versus an individual offspring and its siblings. Either partner may come out ahead in this conflict, or an intermediate compromise may evolve. Extreme cases of conflict include weaning conflicts in mammals, queen-worker conflict over the reproductive sex ratio and over who produces the males in the social hymenopterans, and conflict over siblicide and infanticide in birds. Social dynamics can be complex in animals, a point expanded on in the next chapter.

FOR FURTHER READING   Alexander, R.D., and P.W. Sherman. 1977. Local mate competition and parental investment in social insects. Science 196:494-500.
Along with Trivers and Hare (1976) and Trivers (1985), an excellent example of a scientific debate.
Davies, N.B. 1985. Cooperation and conflict among dunnocks, *Prunella modularis*, in a variable mating system. Animal Behaviour 33:628-648.
Classic, lucidly presented paper on variability in mating systems due to ecology.
Emlen, S.T., and L.W. Oring. 1977. Ecology, sexual selection and the evolution of mating systems. Science 197:215-223.
The classic paper on the ecology of mating systems.
Mock, D.W. 1985. Knockouts in the nest. Natural History 94(5):54-61.
Popular presentation of Mock's study of siblicide in egrets and herons.
Mock, D.W., and G.A. Parker. 1986. Advantages and disadvantages of egret and heron brood reduction. Evolution 40:459-470.
Combines theory and data to explore siblicide and infanticide in egrets and herons.
Trivers, R.L., 1974. Parent-offspring conflict. American Zoologist 14:249-264.
Another Trivers paper that opened up a whole field of study.
Trivers, R.L. and H. Hare. 1976. Haplodiploidy and the evolution of the social insects. Science 191:249-263.
Raises the issue of parent-offspring sex ratio conflict in the hymenopterans.

13

BEHAVIOR IN SOCIAL GROUPS
COSTS, BENEFITS, AND COMPETITION

Humans are the most social of all the vertebrates. As such, we tend to assume that animals living in large groups are more advanced than those living solitary lives. Recent research tends to rebut that assumption. Whether or not a species lives a solitary or social life depends on its ecology—the resources it needs to survive and reproduce and especially the way such resources are distributed in space and time. Social living has inevitable and occasional costs, as well as common and occasional benefits. In this chapter, we explore the costs and benefits of life in social groups, focusing on a well-studied species, the cliff swallow. We develop a classification for the range of behaviors, competitive and cooperative, shown by group-living animals. Competitive phenomena such as dominance hierarchies and territorial defense are discussed from several perspectives, traditional and recent. Chapter 14 continues our discussion of behavior in social groups, focusing on cooperative and altruistic behaviors.

COSTS AND BENEFITS OF COLONIALITY IN CLIFF SWALLOWS

Cliff swallows *(Hirundo pyrrhonota)* (Figure 13-1) do everything socially. These acrobatically flying insectivorous birds build nests of dried mud on cliffs or—increasingly—under the eaves of bridges, buildings, or highway culverts. The nests are usually found in colonies, with colony size ranging from 1 to 3500 active nests, the average being 355 nests. Cliff swallows breed throughout much of western North America from May to August; during the rest of the year they are either in their overwintering range from southern Brazil to Argentina and Chile or in migratory transit. Since 1982, cliff swallows have been studied in southwestern Nebraska by Charles Brown, now of Yale University (joined during the study by Mary Bomberger Brown). This continuing study, involving among other things the individual leg-banding of 25,000 cliff swallows, has emerged as a classic examination of the costs and benefits of living in association with large numbers of conspecifics.

Other studies of the benefits of sociality suggest two major advantages: increased protection from predators and increased success at hunting. For cliff swallows, antipredator advantages seem to be minor. The most important predators are bull snakes, which enter colonies and move from nest to nest, eating the occupants. Performing experiments that involved reeling in plastic snakes attached to fishing lines, Brown and Brown (1987) found that snakes were detected farther away by large colonies (with many vigilant eyes) than by small colonies. When cliff swallows detect a predator, they mob it, swirling in circles above it and vocalizing continuously. (Unlike the related solitary-living barn swallow, they do not swoop down close to the predator; Brown and Hoogland 1986.) Unfortunately, mobbing seems to have little effect on bull snakes, which plow relentlessly on into the colony. Such mobbing may, however, deter other predator species, and in large colonies the increased likelihood of seeing a snake coming may give adult birds a decreased

Figure 13-1 Pair of cliff swallows gathering mud for nest building.
Photograph courtesy Ed Bry, North Dakota Game and Fish Department.

chance of being trapped inside the nest by a snake (eggs and chicks are lost, regardless).

Cliff swallows may be vulnerable to predators when visiting mud puddles to collect nest material (see Figure 13-1). Although there may be several suitable puddles, cliff swallows in a particular colony all seem to use the same one, with the result that there is often a group of swallows present. The larger the mud-gathering group, the more cliff swallows there are that are looking up for predators at any given moment. As the vigilance is shared, however, individual birds do not have to look up as much and have more time to spend gathering mud.

The major benefit of colonial living in cliff swallows seems to be in prey location (discussed in Chapter 3). Cliff swallows feed mostly on small flies that are concentrated in swarms. Although such swarms can contain huge numbers of insects (enough to support 500 swallows), swarms usually stay together in one place for only 20 to 30 minutes. Where a swarm will be at any given time is unpredictable. Cliff swallows that live in large colonies, however, do not always have to locate such swarms by chance. If they do not currently know where such a swarm is, they can monitor incoming colony mates. Successful foragers return to the colony (for example, to feed nestlings) with a visibly full load of insects. When such a forager leaves to go back to its insect swarm, other birds can follow it to the feeding site. Brown (1986) looked at the likelihood of following another forager, and of being followed, for birds that had been successful or unsuccessful on previous trips. As shown in Table 3-3, successful foragers followed another bird out of the colony only 17% of the time but were followed 44% of the time. Unsuccessful foragers followed 75% of the time but were followed only 10% of the time. In other words, successful foragers were 4 times as likely to be followed, and unsuccessful foragers were 4 times as likely to follow.

At present there is no firm evidence that this information sharing is anything but a passive exploitation phenomenon. However, Stoddard (1988) believes that when foraging is especially poor, successful foragers give a rarely heard bugs call to alert the rest of the colony; Brown (personal communication) also suspects that cliff swallows sometimes actively communicate with others about food availability. Such communication would seem to be advantageous in the long run because finding swarms initially is a matter of chance, most insect swarms contain much more food than one bird can exploit, and an individual swallow will sometimes be a follower and sometimes be followed.

Information sharing has dramatic effects on the amount of food delivered to nestlings. Cliff swallows in large colonies are so successful as a result of more information being available that they are able to feed nestlings 15 times an hour, as opposed to 10 times per hour in small colonies. The size of the load delivered is substantially larger: 0.7 g per trip in large colonies versus 0.2 g per trip in small colonies. Brown collected loads by putting a pipe cleaner around a nestling's throat so it could not swallow, then removing a food load after a parent delivered it to the nestling's mouth. (He did this only a few times per nestling before removing the pipe cleaner, so no harm was done to the chick.)

It is not known to what extent these differences in foraging success affect adult birds, as opposed to chicks. They may be important, since adults in small colonies lose about 10% of their weight over the course of the breeding season whereas adults in large colonies do not (Brown 1989), and adult weight probably affects survival rate during migration.

Students of animal social behavior consider social life to entail two almost unavoidable costs: increased competition with conspecifics for resources and increased rates of parasite or disease transmission. Because of the special distribution of their food, cliff swallows probably do not increase food competition by breeding in colonies. They may suffer increased interference in mud-nest building as a result of coloniality, but that has not been shown thus far.

The major cost of coloniality for cliff swallows (and it is a big one) is increased levels of parasitism by blood-sucking ectoparasitic bugs and fleas. Brown and Brown (1986) showed that the larger the colony, the more heavily infested with bird fleas *(Ceratophyllus celsus)* individual swallows were. However, the Browns have not detected any serious consequences from flea infestation. The situation with regard to another blood-sucker, the swallow bug *(Oeciacus vicarius)* (a member of the same insect family as the human bedbug), is much more serious. Again, the larger the colony the greater the bug infestation level per individual cliff swallow. With swallow bugs, the greater the level of infestation the smaller the chick (Figure 13-2). In some colonies, individual nests may contain as many as 2500 bugs, and chick mortality was virtually 100% for chicks with 5 or more bugs on them (Brown and Brown 1986). The effects of swallow bugs are seen most dramatically by comparing chicks from large unmanipulated colonies with those from large colonies that were fumigated with insecticide by the Browns (Figure 13-3). (Fumigation had no effect on nestling size in small colonies.)

The effects of swallow bugs on adult condition are not known yet. One probable cost of bugs to adults relates to nest building. In small colonies, returning breeders often reuse old existing nests. In large colonies, old nests probably already contain large bug infestations (bugs remain in the nests over winters in a dormant state). Starting a new nest means that birds at least start off parasite free, but breeding is delayed and an energy cost is paid in building the new nest.

Figure 13-2 Effect of swallow bug infestation on cliff swallow nestling weight at 10 days of age. Means ±1 standard error are shown. Numbers above each bar show the number of nestlings sampled in the colony; numbers below the bar, the number of nests sampled. r_s = −0.39, p < .001. From Brown, C.R., and M.B. Brown. 1986. Ecology 67:1206-1218.

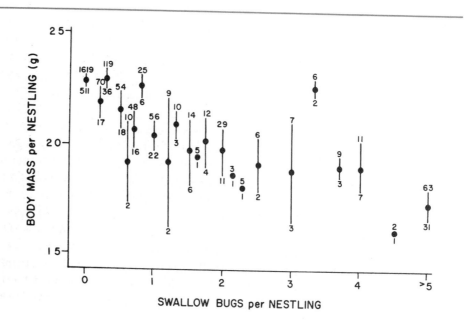

Figure 13-3   Typical 10-day-old cliff swallow nestlings from a nonfumigated nest (left) and a fumigated nest (right), showing the effect of swallow bug parasitism.
From Brown, C.R., and M.B. Brown. 1986. Ecology 67:1206-1218.

To summarize, birds in large colonies feed their chicks several times as much food as do birds in small colonies, but chicks in large colonies do not end up any bigger on average than chicks in small colonies. Most of that extra food merely goes to feed swallow bugs. In fact, from the Browns' studies it is possible to say that living in a small colony is better than living alone, but it is not obvious that life in a big colony is better than life in a small colony.

Colonial living opens up new reproductive options for cliff swallows; whether these are considered to be costs or benefits depends on perspective. Many female cliff swallows are intraspecific brood parasites. Interspecific brood parasites, like the cuckoos and cowbirds discussed in Chapter 10, lay eggs in the nests of other species. Intraspecific brood parasites, by contrast, put their eggs in the nests of other pairs of their own species. Brown (1984) discovered early on in his study that brood parasitism was occurring. As with other birds, female cliff swallows can only lay one egg a day; yet when he did daily egg counts in study nests, Brown commonly found two eggs added over a 24-hour period. Intensive direct observation showed that females were darting into neighboring nests and laying single eggs while the residents were away.

Subsequently the Browns (1988) discovered a second form of brood parasitism, one that is, so far as currently known, unique among birds. After laying has ceased, females continue to place their eggs in other birds' nests, but at those times parasitism actually involves transfer of an egg in the parasite's bill, from its nest to the host's. Thus brood parasitism not only occurs during the laying phase but continues right on up to the time broods hatch.

Brood parasitism is correlated—at least to some extent—with colony size. Both egg-laying and egg-transfer parasitism are practically unknown in colonies with 10 or fewer nests and quite common in larger colonies, although the extent does not increase with colony size beyond the initial difference at a nest size of 10 (Brown and Brown 1988, 1989).

Overall, rates of brood parasitism were high. Brood parasitism was assessed in a variety of ways, including observations of two eggs appearing within 24 hours in one nest, direct observation of laying or transfer, and inference of parasitism from analysis of blood proteins of nestlings and putative parents. The best estimate combining all these methods gives a parasitism rate of 43% (Brown and Brown 1989). Nearly half of cliff swallow nests contain eggs that are not the offspring of at least one of the resident pair. (Although they cannot be the female's they could be the male's, if he had an extra-pair copulation with the parasite.)

What are the benefits and costs of brood parasitism? For parasites, birds that lay eggs in another's nest still produce a normal clutch size in their own nest, so they increase their total egg production, on average by one or two eggs. For egg-transferring parasites, no additional eggs are laid, but the parasite may gain by not "putting all its eggs in one basket." Clutches are often lost in their entirety, to bull snakes or to swallow bugs. By spreading eggs around a female swallow may achieve some success even if her own nest is wiped out. Supporting this hypothesis is the finding that rates of brood parasitism are higher in some colonies than others, ones where nest failure rates are unusually high. In such colonies, there would be a selective advantage to spreading the risks around.

However, parasites do pay a cost that lowers their overall, net benefit. To parasitize someone else, a bird has to leave the nest. When away, another parasite can parasitize that bird. For 24 nests of known parasites, parasitism rates were 54%, and 46% of those parasitized nests were parasitized more than once.

Although parasitism is a partial blessing for parasites, it is an unmitigated disaster for hosts. Cliff swallow females normally stop laying when their usual clutch size is present in a nest. If one egg is from a parasite, that means the host lays one less egg, an average cost of 25%, since the usual clutch size is four. Cliff swallows also seem to be completely incapable of discriminating between their own eggs and those of a parasite, so they waste time and energy in misdirected parental care.

Not only do strangers' eggs show up in cliff swallows' nests, but sometimes eggs get tossed out by intruders as well (Brown and Brown 1988). The extent of such egg tossing was strongly correlated with colony size, and in large colonies affected 15% of nests. In some interspecific brood-parasitic birds, parasite females toss out a host egg before laying their own, thus maintaining the existing clutch size and not raising the hosts' suspicions. Of 10 cliff swallows directly observed entering others' nests and tossing eggs, however, only one was a female. The other nine were males from nearby nests. The Browns (1988) propose the following explanation. Male swallows are known to attempt extra-pair forced copulations with neighboring females (this is common in cliff swallows and has also been described in detail for another highly colonial swallow, the bank swallow [*Riparia riparia*] [Beecher and Beecher 1979]). Female swallows continue to mate as long as they are still laying eggs; by tossing an egg a neighboring male induces a female to lay a replacement, keeping her potentially available for extra-pair copulations for at least one more day.

The final number of young produced by a nest that suffered egg tossing was

GUEST ESSAY

Charles Brown

Charles Brown
Optimal Colony Size in Cliff Swallows

One of the most striking features of coloniality in cliff swallows is the great range in colony sizes found in a single population. In Nebraska, where Mary Bomberger Brown and I have studied cliff swallows for the last decade, we have seen colonies ranging from a mere two pairs nesting together up to massive assemblages of 3700 pairs all on the same bridge. Many of the ecological costs and benefits of living in colonies of various sizes are described in this chapter, yet we still do not have a clear understanding of why these birds choose to nest in colonies of so many different sizes. Since expected reproductive success may vary widely among different colony sizes, we must be able to explain the different preferences of individuals regarding where to nest if we hope to achieve any sort of insight into why coloniality has evolved in cliff swallows and other species.

Investigating the birds' preferences for various colonies requires knowing something about the colony size where their expectation of success is highest, that is, the optimal colony size. An optimal colony size can be defined as the one with the greatest net positive difference between the benefits and costs of coloniality. For example, imagine that individual foraging efficiency increases rapidly with colony size, which our work on information centers would suggest. In large colonies, however, this benefit levels off because individuals can only spend so much time feeding. As colony size increases beyond that at which birds are feeding at maximum efficiency, there is no further foraging gain from living in increasingly large colonies (Figure 13-A). Large colonies are hospitable for ectoparasites, and as colony size increases the transmission of these insects between nests is enhanced. In extremely large colonies the ectoparasites are transmitted between nests at so great a rate that the cost of ectoparasitism increases faster than the foraging benefit. For cliff swallows that experience these sorts of cost-benefit curves (Figure 13-A), the optimal colony size—the one where their expectation is greatest—is neither a small nor a large colony but one of medium size.

So why, then, do not all cliff swallows live in the one optimal colony size? This reasoning might lead one to predict that the birds should sort themselves such that all colonies were the same size where everybody in the population had the same expected reproductive success. However, it is possible that not all individuals have the same expectation of success in a given colony size. For example, imagine that individual birds differ in their ability to find food, perhaps because of age, experience, or even innate genetic ability to forage successfully without help from others. A cliff swallow that is adept at foraging may have little need to use his neighbors to find food, and therefore that individual would gain little foraging advantage from living in larger colonies. Its benefit curve would rise slowly and probably level off at relatively small colony sizes (Figure 13-B). The costs this bird would suffer from ectoparasite infestations in large colonies would still be substantial. As a result, the good forager's optimal colony size would be smaller than that for a bird that relied more on information-center-related foraging benefits. The net gain might be roughly the same for each type of individual, but these gains come at different colony sizes (see Figures 13-A and 13-B).

This example considers only the benefit of enhanced foraging in colonies and the cost of increased ectoparasitism. The same sort of analysis could be applied

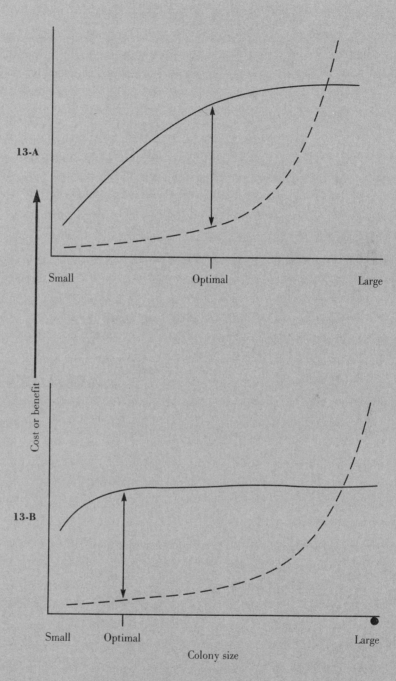

to the whole suite of cost and benefits that result from coloniality. With many different individuals in a population with their own cost and benefit curves, depending on their individual attributes, one could conceivably get an entire range of colony sizes that are optimal for different individuals. This range could account for the variety in colony sizes we observe in the field. Of course, testing this hypothesis requires information on how cost and benefit curves vary for different individuals within the population. Unfortunately, such information is still lacking.

essentially the same as the number of young produced from nests that did not (3.0 versus 2.9, Brown and Brown 1988), so the victimized female bird loses only whatever physiological strains result from her having to lay one additional replacement egg. For her mate, however, egg tossing has a greater cost if the egg tossing neighbor becomes the forced-copulating sire of the replacement chick.

For cliff swallows the net results of coloniality must be beneficial, since coloniality persists. The cliff swallow example does show that social living has costs and makes it plausible that some species remain solitary because those costs outweigh the benefits for them. Furthermore, the cliff swallow example shows that behavioral interactions between members of social groups are a complex mixture of cooperation, competition, tolerance and exploitation. Those conclusions stand as themes for the remainder of this chapter. In the next sections, we consider the costs and benefits of social living more generally than in our cliff swallow example. A list of these is given in the box on this page.

Costs and Benefits of Sociality

I. Costs of sociality
   A. Widespread costs
      1. Increased competition for resources and mates
      2. Increased rates of parasitism and disease
   B. Occasional costs
      1. Increased conspicuousness to predators
      2. Risk of inbreeding
      3. Risk of misdirected parental care
      4. Risk to offspring from conspecifics

II. Benefits of sociality
   A. Widespread benefits
      1. Foraging benefits
         a. Information access
         b. Increase in capture rates
         c. Capture of larger prey
         d. Defence of prey
      2. Antipredation benefits
         a. Increased vigilance
         b. Selfish-herd effect
         c. Dilution effect
         d. Confusion effect
         e. Mobbing and group defence
         f. Alarm calling
   B. Occasional benefits
      1. Protection against hostile conspecifics
      2. Thermal advantages
      3. Locomotory advantages
      4. Reproductive stimulation (Fraser Darling effect)
      5. Increased reproductive behavior options

COSTS OF SOCIALITY

Widespread Costs

"There is no automatic or universal benefit from group living. Indeed, the opposite is true: there are automatic and universal detriments ..." wrote Richard Alexander in an important paper on the evolution of social behavior (Alexander 1974). Living in a social group means living in close proximity with one's greatest ecological competitors, one's conspecifics. To a much greater extent than members of other species, members of one's own species need the same foods, nest sites, hiding places, display perches, etc. Conspecifics of the same sex need the same gamete-producing resources or the same limited parentally investing mates. Thus living within a group almost automatically increases competition for the necessities of survival and reproduction. This was the first of Alexander's (1974) two automatic and universal detriments.

The extent of competition for food is dramatically shown by Mexican free-tailed bats *(Tadarida brasiliensis)*. Twenty million of the bats live in the world's largest bat aggregation, Bracken Cave near Austin, Texas. Keeping alive this number of bats requires the consumption of 250,000 pounds (125 tons) of flying insects each night. Faced with 20 million competitors, a Mexican free-tailed bat may have to fly 100 miles away from Bracken Cave in search of insects (Tuttle 1988). Similar competition for food is seen in an African social spider, *Agelena consociata*, one of only 60 out of 30,000 species of spiders that live in groups. Reichert (1985) found that the amount of food consumed by a spider per day was negatively correlated with the size of a feeding group (Figure 13-4). More surprising from a darwinian

Figure 13-4 Prey consumption by *Agelena consociata* spiders for individuals housed in cages with groups of various sizes. Bars represent mean total weight of prey consumed per day; lines at top of bars indicate standard errors. From Reichert, S.E. 1985. Florida Entomologist 68:105-116.

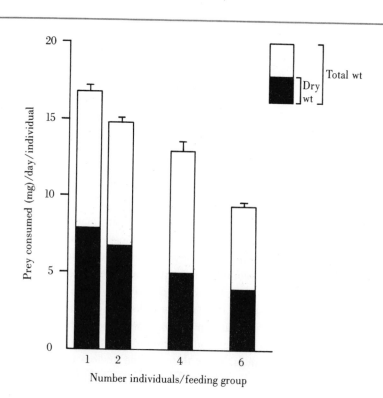

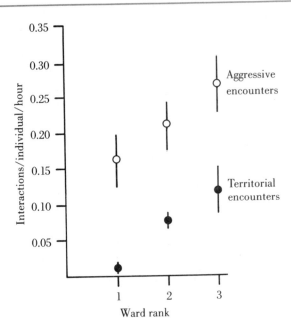

Figure 13-5 Effect of colony size on aggression in black-tailed prairie dogs. Wards were ranked from 1 (small) to 4 (large). Statistical test = Kendal rank correlation. From Hoogland, J.L. 1979. Behaviour 69:1-35.

perspective, the number of eggs produced per individual also went down with increased feeding group size; less food means fewer eggs produced. (We defer until the next section the revelation of what *A. consociata* gains from group living.)

Increased competition for ecological necessities leads to aggressive behavior, both overt fights and aggressive displays. Such behavior is energetically demanding, increases the risk of serious injury, distracts the participants' attention from predators, and takes time away from other necessary activities. All of these costs must be higher for animals in social groups. Increased aggression in social groups was found in white-tailed prairie dogs *(Cynomys leucurus)* and black-tailed prairie dogs *(C. ludovicianus)* by Hoogland (1979 a). Larger colonies had several times as many aggressive interactions per individual per hour than in smaller colonies (Figure 13-5). We note in Chapter 11 the intensity of male-male competition for mates in large aggregations such as the leks of sage grouse and the breeding beaches of elephant seals.

The second automatic and universal cost of group living is increased rates of parasitism and disease. With individuals packed close together there are many opportunities for parasites and pathogens to be transmitted through air and water, direct contact between group mates, or indirect contact via nests or substrates such as soil, sand, or cliff faces. Thus infestations can quickly spread to become epidemics. We have seen how parasitism from swallow bugs is the major cost of coloniality for cliff swallows. Hoogland has observed similar phenomena in prairie dogs. Major ectoparasites of these colonial ground squirrels are rodent fleas; Hoogland (1979 a) found that the number of fleas per burrow increased significantly with increases in colony size for both white-tailed and black-tailed prairie dogs. Although fleas are only a minor irritant to cliff swallows, they are a major danger to prairie dogs, since they are the vectors for bubonic plague. Ubico et al (1988) found that five different flea species (out of ten found in prairie dog burrows) carried the plague bacterium (*Yersinia pestis*). Plague epidemics have been reported as frequent events in all four prairie dog species; colonies can be entirely wiped out by annual mortality rates that are as high as 99% and that produce losses that are far greater than can be replaced by annual reproduction.

Increased parasitism and disease rates do seem to be a widespread cost of group living; in addition to presenting his own observations with prairie dogs, Hoogland (1979 a) also reviews a number of studies with similar findings in other mammalian social groups.

Occasional Costs

Although costs of resource and mate competition and costs of increased disease and parasite transmission may be virtually automatic for group-living animals, there may also be other costs present for particular species. For example, a group of animals, both because of the larger area occupied and because of the combined stimuli from the behavioral displays of so many individuals, may be a more conspicuous target for predators. Hoogland (1979 a) found this to be true for prairie dogs. Taking himself as a model predator, Hoogland found that the number of conspicuous vocalizations (mostly territorial calls), colony-revealing burrow mounds, and actual individuals exposed were all correlated positively and significantly with colony size (Figure 13-6). Andersson and Wicklund (1978) examined the potentially higher predation on clumped individuals experimentally. Their subject species was the fieldfare *(Turdus pilaris)*, a close relative of the American robin and European blackbird. Fieldfares are found nesting solitarily and in colonies. Andersson and Wicklund put out artificial nests, in aggregations and alone, and noted how many of them survived for varying periods of time. As Figure 13-7 shows, experimental colonial nests suffered much higher rates of predation than experimental solitary nests.

Animals in groups may be more likely to attract the attention of predators, however, animals in groups frequently have subsequent antipredator advantages,

Figure 13-6   Effect of colony size on vocal conspicuousness to predators in black-tailed prairie dogs. From Hoogland, J.L. 1979. Behaviour 69:1-35.

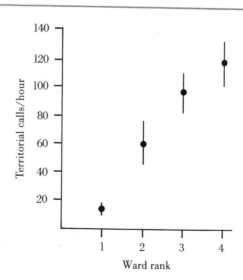

Figure 13-7 The pattern of predation at solitary and colonial artificial fieldfare nests.
From Andersson, M., and C.G. Wiklund. 1978. Animal Behaviour 26:1207-1212.

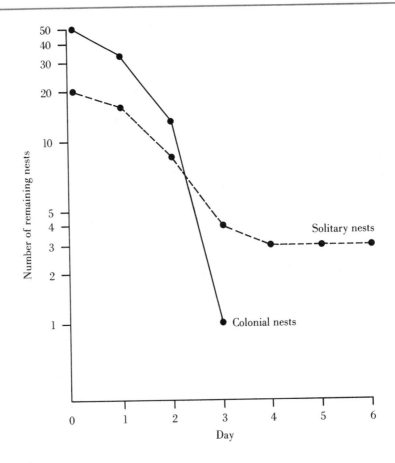

which often overcome the conspicuousness effect and actually make animals in groups safer than animals living alone. This turns out to be true for both prairie dogs and fieldfares.

A second occasional hazard of group living is the possibility of inbreeding, matings between close relatives, with well-known deleterious genetic consequences. For most group-living animals, however, mechanisms seem to be in place that reduce this hazard. In most species, members of at least one sex usually emigrate out of the birth group to take up adult residence elsewhere (in birds, females usually leave; in mammals, males usually leave, although there are exceptions such as African hunting dogs, chimpanzees, and humans; Greenwood 1980). More specific mechanisms also exist, which accounts for the lack of incestuous matings where relatives do coexist. An example is in chimpanzees, where males in a troop mate with almost all females except their mother and sisters (Goodall 1986). In black-tailed prairie dogs, Hoogland (1982) found that inbreeding was rare, as a result of no fewer than four different mechanisms: (1) young males almost always leave the birth group and females remain in it; (2) adult males usually leave their breeding group before daughters reach breeding age; (3) yearling females are much less likely to come into estrus if their father is still present in their group (only 2 of 28 in such cases, as opposed to 13 of 50 females whose father was absent); (4) females in estrus usually avoid sexual advances from male relatives and seek out matings with nonrelated males.

Perhaps a more serious occasional cost of group living is that of providing misdirected parental care, that is, lavishing parental care on youngsters that are not one's own. This outcome may come about in a number of ways. The simplest is through accidental mismatching of parents and young in a large colony. The most dramatic example of this is found in Mexican free-tailed bat aggregations, consisting of millions of individuals. McCracken (1984) decided to examine previous reports that females of this species would indiscriminately suckle any youngster at random. Using analysis of blood proteins from blood samples taken from adult female bats and youngsters they were suckling, McCracken found that most mothers were in fact suckling their own young, but an appreciable minority, at least 17%, were not. McCracken suggests that these mispairings are the result of the incredibly difficult task faced by nursing female bats. They leave their young together in creches containing high densities of pups, then fly out of the cave to hunt down insects all night. Returning from the hunt, they have to fly back into the cave and pick out their own pups in the creche for their twice-a-day feedings. McCracken (1984) suggests that "it can be argued that females actually do well to find and nurse their own young as frequently as 83% of the time!"

In most species, such mispairings are prevented by the mutual exchange of signals leading to the ultimate imprinting of parents and offspring onto each other. Comparative studies of this process in birds show that the development of individual-specific signal production and of parent-offspring recognition is timed to coincide with the period when mixups might be expected to occur. In precocial birds (species in which the chicks are able to leave the nest soon after hatching) such as waterfowl and gulls, eggs are not recognized but 1- to 2-day-old chicks are (although in kittiwake gulls, where chicks are nest-bound by the precarious placing of nests on cliff ledges, imprinting is a much later affair, only being complete when chicks are 5 weeks old and getting ready to fledge; Cullen 1957). In altricial birds (species in which the chicks remain in the nest for extended periods), recognition develops about the time of fledging, flying from the nest; in bank swallows this is at around 15 to 17 days of age (Beecher et al 1981). Studies by Beecher et al of parent bank swallows shows they accept transfers of strange chicks up to that time, which also corresponds to the time of development in chicks of individually unique signature calls.

However, misdirected parental care may be the result of factors other than accidental shuffling of chicks or pups. A male may find himself tending offspring that are the result of a copulation by his mate with another male (Cuckoldry, Chapter 11). This may be the result of forced extra-pair copulations, as in bank swallows and cliff swallows or may be the result of the female actively seeking out extra-pair copulations (reasons for the existence of extra-pair copulations, forced or sought, are given in Chapter 11). In any case a male may be devoting parental care to a litter of which only some or none are his offspring. The extent of this phenomenon is still unknown. Most of the examples of multiply-sired litters cited in a paper by Hanken and Sherman (1981) involve rather promiscuous species; in the 1-male polygynous black-tailed prairie dog, Foltz and Hoogland (1981) found only 2 litters out of 70 that were clearly not the product of a resident territorial male.

A female (and probably in most cases her mate as well) may find herself looking after someone else's progeny if intraspecific brood parasitism is present in a population. As many as half of cliff swallow nests may contain eggs laid or carried there by a nonresident female. Such exploitation of others' parental efforts by brood-parasitic female conspecifics, although a recently discovered phenomenon, is emerging as a widespread one. Examples are now known from several species of waterfowl, starlings, bee-eaters, moorhens, and European barn swallows, as well as cliff swallows (references given in Brown and Brown 1989).

A fourth and final occasional cost of coloniality, one that is being discovered in an increasing number of species and that is quite serious when present, is the risk of fatal attacks on offspring by conspecifics. This can occur for a variety of reasons. It may be a manifestation of intrasexual selection (e.g., infanticide of nonweaned infants by new harem-holding males in langur monkeys, egg tossing by male cliff swallows, or egg pecking by male dunnocks). It may be the result of simple cannibalistic predation, where conspecific young are an accessible, easily dispatched prey item, as in herring gulls (Parsons 1971; incidentally, Polis 1981 reviews cannibalism generally, which he reports has been described in more than 900 published papers covering 1300 animal species). Infanticide may also be a manifestation of reproductive competition between females, especially when dominant females kill the young of subordinates—a famous example is African wild dogs (van Lawick and van Lawick-Goodall 1970).

Infanticide turns out to be the major source of infant mortality in black-tailed prairie dogs (Hoogland 1985). About 51% of litters were partially (13%) or completely (38%) lost in this way. Four kinds of infanticide were observed by Hoogland. The most important was by lactating females of the same colony as the victim (usually relatives of the victim): 40 of 133 litters were affected. These females usually cannibalized the victims, and as a result were better fed and in better condition than *nonmarauders*, as Hoogland (1985) referred to them. Marauders were more likely to successfully wean a litter, had larger litters at weaning, and had heavier young at weaning. They may also have improved their offspring's ultimate reproductive success by clearing the area of future competitors. The second class of infanticide involved nonmaternal females, which apparently had abandoned their young, resulting in infanticide of the abandoned pups (sometimes even by the abandoning mother and her relatives); this accounted for the loss of 15 litters. A third class of infanticide, involving 14 litters, was performed by newly immigrant males; although this does not speed up estrus in the victims' mothers (unlike the case of langur monkeys or lions), it does seem to increase the survival rate of the male's offspring the next year, perhaps because they experience reduced competition. The final class of prairie dog infanticide, affecting 2 litters, was by newly immigrant females; this can probably also be viewed as a case of clearing the field of future competitors of the marauder's offspring.

Thus group living has several widespread and even more occasional costs, some of which have drastic effects on the reproductive fitness of group members. We may have painted such a bleak picture of social living that you wonder why any animals live in groups. There are many compensating advantages for group-living animals, and we now turn to those.

BENEFITS OF SOCIALITY

Widespread Benefits   Alexander (1974) suggested that "an exhaustive list of the selective backgrounds of group living may contain no more than three general items ..." These were (1) extreme localization of some resource, such as safe sleeping caves for hamadryas baboons (Kummer 1968), (2) foraging advantages when seeking foods of certain types or found in certain distribution patterns, and (3) lowered susceptibility to predation. The first of these can be said to represent a passive grouping by necessity rather than an active grouping by behavioral choice. Individuals in the first case do not gain by being in the presence of others, whereas in the latter two they presumably do. Because of these distinctions we do not discuss passive grouping any further. Recent students of animal sociality tend to concur with Alexander (1974) that foraging and antipredation advantages still constitute the "big two" benefits of active sociality but also point out some other minor advantages that come into play for particular species.

Foraging advantages

When food is locally abundant but unpredictably located in space and time, an animal in a group may obtain information about the location of current feeding sites from other members of the group. In cliff swallows, this currently is thought to be an exploitive rather than a communication process; but in some other groups (ones composed of relatives) there is active transmission of information from individuals that have it to others that do not.

One example of such information sharing has been described by Erick Greene (1987). He studied a group of ospreys *(Pandion haliaetus)* in a colony on an island in Cow Bay estuary, Nova Scotia. These fish-eating eagles prey mainly on four species of fish. Three occur in schools, which are unpredictably distributed in a fashion similar to that of the insect swarms attacked by cliff swallows. The fourth species, the winter flounder, is a nonschooling species with a random distribution. When an osprey has caught a fish and is returning to the colony with it, the fish is held in the bird's feet and is conspicuous to other birds in the colony. This visually apparent prey information is sometimes supplemented by the returning osprey, which calls persistently and performs an undulatory flight display. It does so only when returning with useful information, that is, it has caught a fish of a schooling species. The response is dramatic: other colony members leave quickly, depart in the direction from which the forager came, and subsequently catch their own fish faster than when they depart naively, without gaining such information (Figure 13-8). These generalizations hold true only when foragers return with a schooling prey, not a winter flounder.

Undoubtedly the most spectacular example of information-sharing group foraging is found in honeybees, with their renowned waggle-dance communication system, described in Chapter 15. The nectar and pollen food of honeybees is an excellent example of a widely scattered, briefly abundant but rapidly depleted resource. Studies of honeybee foraging, such as those by Visscher and Seeley (1982), demonstrate that bees can efficiently locate and exploit food as far as 10 km away from the hive, in any direction of the compass. Over the course of a year an average hive "rears 150,000 bees and consumes 20 kg of pollen and 60 kg of honey ... To collect this food, which comes as tiny, widely scattered packets inside flowers, a colony must dispatch its workers on several million foraging trips, with these foragers flying 20 million kilometers overall" (Seeley 1985).

The major problem faced by cliff swallows, ospreys, and honeybees is finding their food; for other social foragers, group living may be more of a help in catching and keeping it. It has been shown for a number of species that the percentage of hunts that are successful goes up as the number of hunters does (see review by

Figure 13-8 Departures by ospreys leaving a colony. **A,** Orientation of departing ospreys depends on whether a returning bird arrived in the colony with a prey from a schooling fish species in the last 10 minutes. **B,** So does the number of ospreys departing. Asterisks indicate statistically significant orientation. For ospreys departing after another returned, 0 degrees (top of circle) indicates flight in the direction whence the returnee came.
From Greene, E. 1987. Nature 329:239-241.

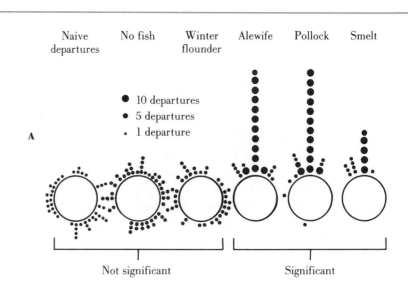

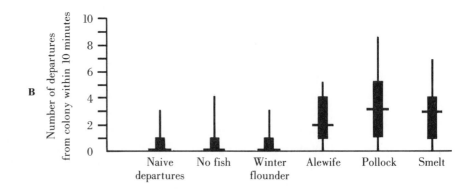

Packer and Ruttan 1988). Of course, for an individual forager to come out ahead, prey capture rates in groups must climb fast enough to compensate for the fact that food will be shared. For hunting in pairs to be favored, pairs must be twice as successful, etc. In spotted hyenas, single hunters trying to catch wildebeest calves succeeded on only 15% of 74 tries, mainly frustrated by the defensive efforts of the mother wildebeest. Pairs of hyenas succeeded in 74% of 34 attacks, one hyena catching the calf while the mother was distracted by the other (Kruuk 1972). Gulls often forage in groups for fish; Gotmark et al (1986) studied this by having black-headed gulls forage for fish in a 5 × 5 m swimming pool. They found that individual foraging success rose as group size went from one to three to six. The reason was that fish fleeing from one gull often swam toward another gull, increasing overall capture rates. This is an example of what has been called the *beater effect*, in which prey are flushed by one individual and caught by another (Wittenberger 1981). (This also explains some interspecific following such as why cattle egrets hunting for insects follow large herbivores.)

A recent detailed study on cooperative hunting in which individual success rates were higher for group foragers was conducted by Bednarz (1988) of Harris' hawks

(*Parabuteo unicinctus*). Their major prey are jackrabbits, hunted by hawk family units, consisting of a male and female plus up to four nonbreeding auxiliaries. Number of kills per individual and energy intake per individual are larger for larger groups. The following three group-hunting tactics were used: (1) surprise pounce by several hawks at once on an unsuspecting prey, (2) flush and ambush of prey in hiding, and (3) relay attack, in which hawks pursued rabbits for several minutes with different hawks alternating as lead pursuer.

Jackrabbits, at 2.1 kg, are three times as large as male Harris' hawks (680 g) and twice as large as female hawks (960 g). This leads to another advantage for group foragers. Prey individuals that are too large to be caught and killed by an individual forager may be taken by a group. Thus wolf packs can specialize on adult moose that no single wolf could handle (Mech 1970), lion prides can catch adult buffalos that single lions do not even bother to try for (Schaller 1972), and army ants can catch katydids, cockroaches, and even nestling birds (Wilson 1971).

Once one has captured prey, one also has to hang on to it long enough to eat it. Other predators and scavengers may attempt to drive away the original owners, and group hunting may permit greater success at retaining kills. Such a trend seems to exist in the African plains for both hyenas and lions, which routinely steal kills from smaller groups of their own and other species (Kruuk 1972); black-backed jackals lose 30% of their kills to spotted hyenas (Gittleman 1989). In fact the need to defend scavenged or killed prey may explain why lion prides are the size they are.

Early considerations of group hunting by lions (which is unusual for a cat species) seemed to show that lions hunting as pairs would be more successful than solo lions but that individuals would do worse in groups larger than two (Schaller 1972, Caraco and Wolf 1975; Figure 13-9). Reanalysis of those results and addition of data from other studies led Packer (1986) to question whether lions were ever more successful at making kills when in groups (Packer 1986). In any case the usual number of lionesses in a hunting group is four to eight, bigger than optimal size even if that optimal size is two. Packer (1986) suggests that scavenging is more

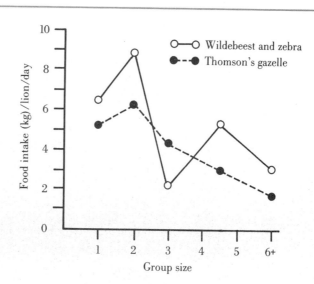

Figure 13-9   Rates of daily food intake for lions in different size groups resulting from hunting either Thomson's gazelle or wildebeest/zebra.
From Packer, C. 1986. In: D.I. Rubenstein and R.W. Wrangham, editors. Ecological aspects of social evolution. Princeton University Press, Princeton, N.J.

Figure 13-10 Amount of silk building per individual in *Agelena consociata* spider colonies of different sizes. From Reichert, S.E. 1985. Florida Entomologist 68:105-116.

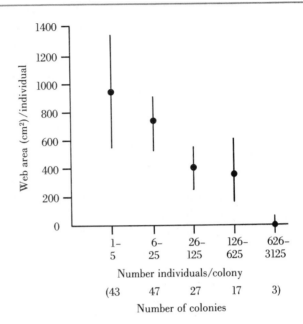

profitable than hunting for lions, and larger groups are present than would be needed to kill prey to defend discovered carcasses from hyenas and other lions.

In addition to these advantages of group foraging, there may be others that apply in rare cases. For example, the African social spider mentioned earlier *(Agelena consociata)* catches less food per individual in groups than when solitary. Groups share a communal capture web, and the costs of web building are greatly reduced for colonial individuals (Figure 13-10). During the African rainy season, nests are destroyed frequently by heavy rains, and the energy savings of reconstructing part of a damaged large communal web as opposed to completely rebuilding a small solitary web can make the difference between death and survival for a spider.

Antipredation advantages

The old saying that "there's safety in numbers," appears to be true; animals in groups often experience reduced risks of predation compared with solitary animals. This may more than compensate for any increased conspicuousness to predators that the grouping creates. There are a number of ways in which grouping provides antipredation benefits, several of which may occur simultaneously. Indeed, it is often easier to show a reduced risk of predation to group members than to isolate a single cause for that reduction.

Perhaps the most obvious advantage of being in a group is that, with many eyes vigilant for predators, they are likely to be spotted at a greater distance and are less likely to make a successful attack, if indeed they attack at all after being spotted. This vigilance, or "many-eyes effect," has been observed in a number of species—we have already remarked on its presence in cliff swallows. Its effectiveness was demonstrated by experiments carried out by Kenward (1978). An experienced falconer, Kenward released a trained goshawk *(Accipiter gentilis)* near pigeons in flocks of various sizes and recorded the subsequent events. As Figure

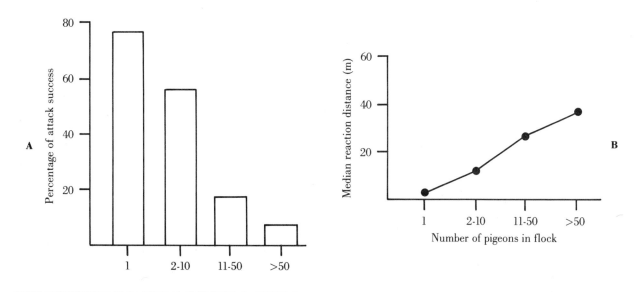

Figure 13-11   A, Goshawks *(Accipiter gentilis)* have lower success rates hunting pigeons in larger flocks. **B,** This is because the larger the flock, the greater the distance from which it reacts to the goshawk.
From Kenward, R.E. 1978. Journal of Animal Ecology 47:449-460.

13-11 shows, pigeons in larger flocks spotted and reacted to the goshawk at much greater distances than solitary pigeons or ones in small flocks. Consequently the success rates of the goshawk declined from almost 80% for solitary pigeons to less than 10% against flocks of more than 50. Not only is survival increased by living in a flock, individual flock members may each spend less time being vigilant—time that may then be spent on other activities.

A second antipredation benefit of grouping, although known previously, was brought to widespread attention by Hamilton in 1971 in a paper titled "Geometry for the selfish herd." Hamilton posed the hypothetical situation of a group of frogs randomly spaced around the edges of a lily pond, preyed on by a water snake that pops up above the water and grabs the frog closest to it. Hamilton showed that the safest arrangement for the frogs was a cluster, with each frog trying to be on the inside rather than an exposed edge of the cluster. Aggregation for the purpose of putting others between you and a predator has been called the *you-first* or *selfish-herd* effect.

For a selfish-herd effect to be demonstrated, individuals in groups should suffer less predation than solitary individuals and, for individuals in groups, central individuals should be safer than peripheral ones. Andersson and Wicklund (1978) demonstrated this for their artificial fieldfare nests. These experiments were an extension of the ones described previously, which showed that colonies of artificial nests were predated more heavily than solitary nests. In these new experiments, Andersson and Wicklund put artificial nests near natural fieldfare nests, with some near solitary nests, some near nests on the periphery of colonies, and some near nests inside colonies. As Figure 13-12 shows, the number of nests surviving predation increased in the order expected according to the the selfish-herd hypothesis.

Figure 13-12 The pattern of predation at artificial fieldfare nests near solitary nests, and on the inside and outside of fieldfare colonies. From Andersson, M., and C.G. Wiklund. 1978. Animal Behavior 26:1207-1212.

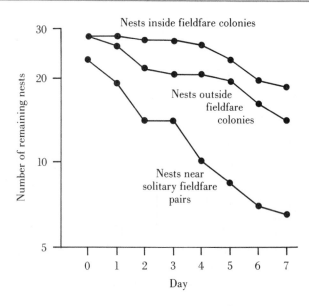

The vigilance and selfish-herd effects often interact. For example, one might expect that individuals at the periphery of a group are at greater risk and therefore spend more time being vigilant. Such a pattern has been found in, among other species, starlings (Jennings and Evans 1980; Figure 13-13) and black-tailed prairie dogs (Hoogland 1979 b).

A classic instance of the selfish-herd effect involves nesting in bluegill sunfish *(Lepomis macrochirus)* (Dominey 1981, Gross and MacMillan 1981). Eggs in bluegill nests are eaten by snails and by other fish, including conspecific cannibals. Snails are found at twice the density in solitary nests as in peripheral nests of colonies, and they are present in turn at twice the density in peripheral nests as in central nests (Table 13-1).

The bluegill example demonstrates another antipredation benefit of grouping, one that helps produce the selfish-herd periphery versus center effect. When a predatory fish approaches a bluegill nest, the resident male attacks it to drive it away. When a central nest is approached, other males from nearby nests may join in, creating a more effective multiple deterrent. Such group responses are not observed when predators approach peripheral or solitary nests (see Table 13-1).

Although in the bluegill example (and in cliff swallows) there is no coordination of such group defense, it is seen in other species, constituting a third antipredation advantage of grouping. Group mobbing of predators, often coordinated by inciting vocalizations, is shown by many bird species. Musk oxen are famous for their defensive circles presented against wolves (Figure 13-14). The collective colony defense of social insects is unforgettable to anyone who has blundered head first into a paper wasp nest or accidentally kicked a fire ant mound. Collective defense is currently front-page news, as African honeybees, with their ferocious defensive behavior, invade the United States from South and Central America.

Sometimes groups of prey coordinate their behavior in a way that may be as much a coordinated escape as a coordinated defense. For example, when a falcon

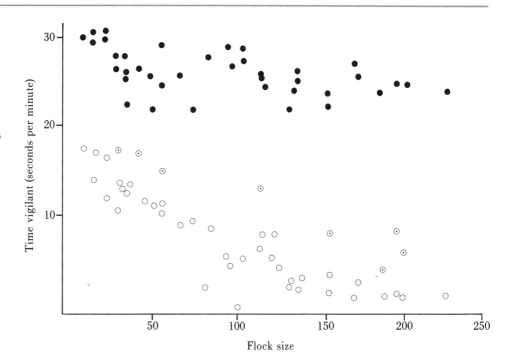

Figure 13-13 Vigilance time in relation to flock size and position in starlings. Birds in each flock were categorized in three positions: central (○), midway (◉), or peripheral (●). Each point represents the mean time that birds in that position had their heads up and were watching for danger.
From Jennings, T., and J.M. Evans. 1980. Animal Behaviour 28:634-635.

Table 13-1 Effect of nest location on predation and costs of antipredator defence in bluegill sunfish colonies

	Nest Position		
	Solitary	**Peripheral**	**Central**
Mean number of predator chases/nest[a]	10.4	8.7	1.5
Percentage of group chases	—	8.2%	50%
Mean number of fish predators attacking[b]	5.1	5.9	1.9
Mean number of snails/nest	29.7	13.7	6.9

[a]11 nests were observed for 110 minutes at each location.
[b]7 nests were observed for 5 minutes each at each location.
From Gross, M.R. and A.M. MacMillan. 1981. Behavioral Ecology and Sociobiology 8:163-174.

Figure 13-14 Musk oxen defense ring against predators.
Photograph by L.L. Rue III.

Figure 13-15 A, Flock of European starlings, undisturbed. B, Reaction of starlings to peregrine falcon. From Tinbergen, N. 1951. The study of instinct. Oxford University Press, Oxford, England.

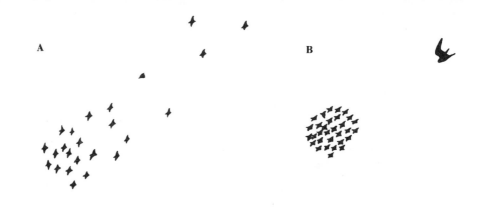

Figure 13-16 Movements of a coordinated group of animals in response to attack by a predator (at dark arrow). Similar responses are shown in many groups of animals under attack, including schools of fish, flocks of birds, and herds of ungulate mammals. The spreading of the group as the predator approaches has been described variously as flash expansion or fountain effect. The fleeing animals then regroup and continue their escape as a tight school, flock, or herd. Modified from Partridge, B.L. 1982. Scientific American 246(5):114-123.

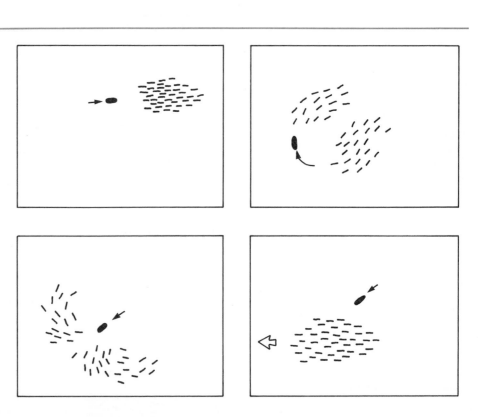

approaches a flock of starlings, the starlings bunch together into a much tighter flock (Figure 13-15). This could be a simple selfish-herd response, but it has also been proposed that starlings clustered in this way pose a risk of injury to a falcon that would swoop down through such a cluster, perhaps breaking a wing.

In other cases, individuals in groups are clearly attempting to confuse approaching predators, a fourth antipredator effect of group living. It is difficult for predators to pick a single target and keep up with it when a group of prey are escaping in a chaotic, flash-expansion, or fountain-effect pattern (Figure 13-16). This confusion effect can be demonstrated with some balls and a friend. Throwing your friend a single ball presents no challenge; he or she can probably catch two thrown at the same time. If you throw three or more at the same time, note the decreased likelihood that your friend will catch even one. This confusion effect of group escape has been found in cephalopod mollusks and fish attacking fish schools (Neill and Cullen 1974), and in a predatory fish, the jack *(Caranx ignobilis)* attacking anchovies *(Stolephorus purpureus)* (Major 1978). In the latter case, because of the chaotic escape behavior of anchovy schools, individual jacks had little luck catching schooled anchovies. The chaotic escape backfired when jacks hunted in schools. Anchovies veering away from one jack frequently veered right toward the mouth of another.

Bluegill sunfish are highly synchronous breeders, with spawning episodes in a colony occurring over a period of only 1 to 2 days and eggs hatching later with similar synchrony. This may be an example of a fifth antipredation advantage, known as the *dilution effect*. The principle at work is that the risk of predation for an individual may go down as the number of individuals in a group goes up. This may occur for a variety of reasons. If predators are territorial, a large group will overlap only one or a few predators instead of many. If vulnerable young are produced all at once, predators (even in feeding themselves to satiation) may eat a smaller proportion than would be the case if there were a year-round supply of vulnerable young in smaller numbers at any given time. Thus the dilution effects can be spatial (clustering of individuals in small areas) or temporal (clustering of large numbers of individuals in time) or both.

There are numerous examples of the temporal dilution effect. Estes (1976) found that all wildebeest calves are usually born during a 2-week period at the beginning of the rainy season in East Africa: calves born at the peak are less frequently captured by predators such as lions and hyenas than calves born before or after the peak. Some insects emerge in spectacular numbers. These include mayflies (millions of individuals emerging over only a few days of a year) and periodic cicadas (millions of individuals emerging over a period of only a few weeks once every 13 or 17 years, depending on the species and location). Some bamboo plants flower and set seeds only once in 100 years, perhaps the most extreme case of the temporal swamping of predators if that is the cause (there may be a different explanation).

There are also good examples of spatial dilution effects. Monarch butterflies in large, dense overwintering clusters (see Chapter 8) are safer than ones in smaller clusters (Figure 13-17), and tungara frogs preyed on by frog-eating bats (see Chapter 15) are safer in large choruses than in small ones.

You may be somewhat confused regarding the difference between the selfish-herd and spatial dilution effects. Although it probably is not possible to draw an absolutely clear-cut distinction between the two, we find the following explanation helpful. A selfish-herd effect is manifested when animals are safer in groups per se than when alone, and when individuals are safer in central than peripheral locations in groups. On the other hand, if individuals in larger groups are safer than those in smaller groups, it is best to think of that as a dilution effect. The former is essentially a positional advantage, the latter a purer safety-in-larger-numbers phenomenon.

A sixth antipredation advantage is one that builds on the many-eyes effect; this is the production of warning or alarm calls, so that when one individual spots a predator, all members of the group are immediately informed. Not all vigilance effects are supplemented by alarm calls. Davis (1975), for example, observed that

Figure 13-17 Relation of colony size to predation at five overwintering sites of the monarch butterfly in Mexico.
From Calvert, W.H., L.E. Hedrick, and L.P. Brower. 1979. Science 204:847-851.

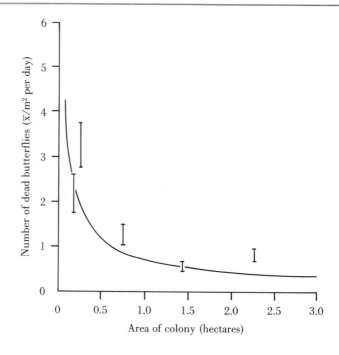

pigeons merely kept close watch on other flock members, and when one suddenly began to take off in response to spotting a predator, others quickly sprang into flight also. Davis demonstrated this effect experimentally by laying out an electrified grid and baiting it with birdseed. Pigeon flocks soon learned to feed at that location. Davis would switch on the current when a single bird's feet were touching the grid. The startled pigeon would jump, not intending to fly off. When that pigeon jumped without warning, the rest of the flock immediately exploded into escape flight.

Why alarm calls should evolve is a problem of much discussion (see Chapter 14). In some cases, alarm callers seem to increase their own personal safety while also increasing the safety of other flock members. In other cases, by attracting the predator's attention to the caller, the caller seems to help others but increase its own risks. In either case, other flock or herd members usually gain by being warned.

Perhaps the most remarkable recent example of the adaptiveness of alarm calling involves vervet monkeys, *Cercopithecus aethiops* (Seyfarth et al 1980, Seyfarth and Cheney 1982). Vervet monkeys give acoustically distinct alarm calls to at least three and possibly five different predators. These alarm call types are: (definitely) eagle, leopard, and python, and (possibly) baboon, and human. Appropriate responses to these predators differ, and typical responses by troop members to alarm calls, whether real or tape recordings played back by experimenters, correspond to those that would be appropriate. Martial eagles pick monkeys off the ground or branches of trees; response to an eagle alarm is to run into heavy brush. Leopards lie in ambush in such brush; response to a leopard alarm is to run up a tree. Pythons are dangerous only if they catch a monkey unaware: the vervet's response to a python alarm is to look around to try to locate the snake. Young vervets seem able to classify potential predators into appropriate categories but only gradually learn to call only when actual predators are sighted (Table 13-2; infants, for example,

Table 13-2 The number of times different species elicited leopard, eagle, snake, and baboon alarms from vervet monkeys in each age-sex class

	Leopard Alarms	Eagle Alarms	Snake Alarms	Baboon Alarms
Adult ♂♂ N = 34	**Leopard** 19; Lion 3; Cheetah 2; Jackal 1	**Martial Eagle** 5; African Hawk Eagle 1; Eagle Owl 1	Black Mamba 1; Tortoise 1	
Adult ♀♀ N = 88	**Leopard** 23; Lion 2; Cheetah 3; Jackal 1; Martial Eagle 2; Baboon 1	**Martial Eagle** 27; Black-chested Snake Eagle 10; African Hawk Eagle 5; Eagle Owl 7; Tawny Eagle 1; Vulture 7; Stork 1	**Python** 2; Black Mamba 2	Baboon 1
Juveniles N = 88	**Leopard** 7; Lion 2; Cheetah 2; Hyena 3; Bateleur 1; Cobra 1; Baboon 4	**Martial Eagle** 14; Black-chested Snake Eagle 10; African Hawk Eagle 10; Eagle Owl 4; Tawny Eagle 3; Goshawk 2; Bateleur 5; Vulture 8; Secretary Bird 1; Stork 1; Spoonbill 1; Baboon 1	Black Mamba 1; Mouse 1	Baboon 5
Infants N = 38	Warthog 1	**Martial Eagle** 1; Black-chested Snake Eagle 3; African Hawk Eagle 3; Eagle Owl 1; Tawny Eagle 2; Goshawk 2; Bateleur 2; Vulture 3; Stork 1; Heron 1; Goose 1; Pigeon 1; Lilac-breasted Roller 1; Ground Hornbill 1; Falling leaf 1	Black Mamba 6; Green Mamba 7; Shaking Vine 1	Baboon 1

N = number of alarms by individuals in each age-sex class. Major Predators are indicated in boldface type. From Seyfarth, R.M. and D.L. Cheney. 1982. Zeitschrift für Tierpsychologie 54:37-56.

gave an eagle alarm to a falling leaf and a snake alarm to a shaking vine). Similarly, adults almost always make the specific appropriate response to alarm calls, whereas juveniles are more likely to get mixed up.

Occasional Benefits

Although there is still a consensus among ethologists that foraging and antipredator advantages are the most important benefits of grouping, a number of other occasionally important advantages have also been pointed out. We mention five.

The first of these has recently been advocated by Wrangham and Rubenstein (1986) as being so generally important as to join the two previously discussed, and make it a "big three." This benefit is protection against hostile conspecifics. Wrangham and Rubenstein believe that polygynous groups may form so that the grouped females may use the resident male as a "hired gun" to protect them from harassment from other males (including such extreme forms of sexual harassment as infanticide). Where there is a multimale group, the females may be gaining even greater protection. Putative examples are the polygynous social groups of horses, lions, gorillas, and chimpanzees. Also, as Wrangham and Rubenstein point out, neither the marauders nor the protectors need be males; groups of related females (matrilines) may stay together to protect each other or to compete with other matrilines. Examples might include lionesses, ground squirrels, and monkeys. We have already pointed out that defense of prey against hostile conspecifics may be an important factor in group formation of carnivores such as lions and hyenas, where the basic social units are matrilines.

Figure 13-18 Colony
metabolic rate in honeybees
as a function of ambient
temperature. Note that as
temperature falls below 20°
C, cluster formation allows
heat to be conserved even
with low metabolism.
From Seeley, T.D. 1985.
Honeybee ecology. Princeton
University Press, Princeton, N.J.
After Southwick, E.E. 1982.
Comparative Biochemistry and
Physiology 71:277-281.

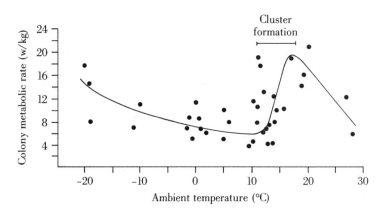

A second occasional benefit of grouping is seen in those species that gain thermoregulatory or moisture-conserving advantages from clustering (examples are reviewed in Wittenberger 1981). Especially for small warm-blooded animals, limiting heat loss is a major survival challenge. Since heat loss is proportional to surface area-to-volume ratio, animals may lose proportionally less heat by clustering together, reducing that ratio by exposing less body surface to the ambient environment. Examples of clustering to prevent heat loss on cold nights are known from songbirds and hibernating bats. Perhaps the most remarkable social thermoregulators, though, are the social insects. The environment of a termite mound or a beehive may be almost as controlled and constant as the interior of a warm-blooded bird or mammal. This may be a passive result of well-engineered nest construction or may involve constant behavioral adjustment. The core of a honeybee hive is maintained at around 35° C even during cold winter periods. In the summer, bees cool overheating hives by everting drops of water from their crops and fanning their wings to increase evaporative cooling. In cooler periods, honeybees cluster together and increase their metabolic rates to retain heat and generate increased heat (Figure 13-18). For more on bee thermoregulation, see Seeley (1985) and Southwick and Heldmaier (1987).

Another possible advantage of grouping lies in energetic gains during locomotion. The rather regular geometric arrangements of fish schools and bird flocks (such as the V formations of flying geese) have led to the hypothesis that individuals may gain from being in the slipstream of others moving through a viscous medium. Analyses suggest air is not sufficiently viscous for that to be a major reason for flying in V's, at least for an animal the size of a bird flying at a bird's speed, but Weihs (1973) did suggest there would be several energy-saving hydrodynamic advantages to a schooling fish that positioned itself well. Specifically, schools should consist of crystal like diamond formations (Figure 13-19), with a fish positioned midway between two fish in the row ahead of it, about one half to one body length diagonally away from its nearest neighbor. However, the most complete test of Weihs' hypothesis failed to provide strong support. Partridge and Pitcher (1979) found that fish schools did not maintain rigid structures and that fish were not usually correctly positioned to gain hydrodynamic advantages. Partridge (1982) has argued for an antipredator function for the usual spatial relationships of schooling

Figure 13-19 Diagram of hypothetical fish school. The dotted line indicates Weihs' predicted diamond lattice. Swimming in those positions would theoretically gain a fivefold energy savings resulting from several hydrodynamic benefits. Fish C should be halfway between fish A and B. Semicircles with arrows indicate vortices in the water.
From Partridge, B.L., and T.J. Pitcher. 1979. Nature 279:418-419.

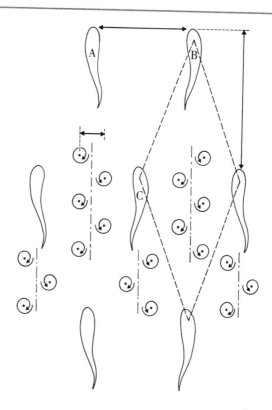

fish. However, schooling fish are understandably difficult to study in natural situations, and we list hydrodynamic advantages as at least a theoretically possible benefit of grouping.

A fourth occasional advantage of grouping derives from the greater social stimulation an individual receives by being near many displaying conspecifics. The result of such multiple stimulation may be that males and females in colonies come into full reproductive condition earlier in the year than do solitary individuals. Since an early start to breeding may give parents increased chances of producing additional broods or may allow parents to feed and nurture young for a longer period before migration or the onset of winter, this socially stimulated early start may increase reproductive success. This rapid onset of reproduction in larger groups is traditionally known as the *Fraser Darling effect*, after Frank Fraser Darling (1938) who documented it for colonial seabirds. The Fraser Darling effect can also be a proximate mechanism for achieving colony-wide breeding synchrony, increasing the likelihood of a beneficial temporal dilution effect. Advantages of early breeding and of temporally synchronized breeding are well documented in birds; a thorough review of supporting evidence is provided by Wittenberger (1981).

A final potential advantage of grouping lies in the increased number of reproductive options that are available to group-inhabiting individuals. A dominant male may have a chance at extra-pair copulations; subordinates may act as satellites or seek forced copulations (see Chapter 11). Females may seek out extra-pair copulations (for advantages of doing so, see Chapter 12). They also may be able to become intraspecific brood parasites, putting their eggs in other birds' nests. Examples of many of these phenomena, along with an assessment of their advantages, are found in the discussion of the costs and benefits of coloniality in cliff swallows.

Of course, this last category includes a class of behaviors that are beneficial to their performers but are harmful to their victims. They could be classified as both benefits and costs of grouping. This again makes the point, raised at the end of the parental behavior section of Chapter 12, that the interactions between members of social groups are varied and complex. We turn to an examination of such social interactions.

A CLASSIFICATION OF SOCIAL INTERACTIONS

The range of behaviors produced by animals in social groups is so large that to list them all is probably impossible. A more sensible way to arrange a discussion of behavioral interactions in social groups is to take advantage of a classification scheme proposed by Hamilton in 1964. Hamilton recognized four basic kinds of social interactions, based on the consequences of the interaction for the direct fitness of the two interacting individuals (see box on this page). Interactions in which the originator of the interaction benefits whereas the recipient loses are called *selfish* or *competitive behavior*. Where the originator and recipient both gain, the interaction is called *cooperative*. If the originator loses but the recipient gains, *altruism* has been observed. (Note that in ethology altruism is defined only according to the consequences of an action; conscious motivation to be self-sacrificing is definitely not implied, as it might be for a social psychologist speaking of human altruism.) Finally, if an animal does something that is harmful to itself but harms another more, that could be called *spiteful*.

As discussed, competitive interactions are expected to be virtually universally present in social groups. In the remainder of this chapter, we discuss various forms of social competition and several ways to analyze them. Spiteful behavior, although theoretically possible, could evolve only in special situations, is not considered to be a widespread and important type of social interaction (Hamilton 1970, Knowlton and Parker 1979) and is not discussed further in this book. Cooperative behaviors are not evolutionarily surprising within a species, since we have seen that cooperation can be extensive even between members of different species (Chapter 10). Altruistic behaviors, on the other hand, seem paradoxical at first glance in darwinian evolutionary reasoning in which individuals attempt to maximize their own chances of surviving and reproducing. Studies of the evolution of cooperation and the problem of altruism comprise the core of the exciting and controversial subdiscipline in

Outcomes of Social Interactions

		Originator of an act	
		Gains	Loses
Recipient	Gains	Cooperation	Altruism
	Loses	Selfishness	Spite

ethology known as sociobiology. Therefore we devote the next chapter to the exploration of cooperation and altruism. We turn now to competition.

AGONISTIC BEHAVIOR

Competition in social groups may involve actual fights or merely the production of threat displays; both can be considered examples of aggressive behavior. Such assertions of social dominance may be mirrored by specific behaviors that indicate submission or appeasement by the threatened animals. Such submissive behaviors may in fact be more apparent in social groups than are aggressive behaviors (Rowell 1973). The term *agonistic behavior* includes all of the behavior seen in competitive interactions, whether submissive or aggressive, involving actual fights or only ritualized displays.

RANK ORDERS AND DOMINANCE HIERARCHIES

In the 1910s and 1920s the Norwegian zoologist Schjelderup-Ebbe studied the events that occur when a number of unfamiliar hens are placed together to form a group (Schjelderup-Ebbe 1935). At first there is a great deal of overt fighting over resources such as food. As time passes a number of changes occur. First, the amount of fighting declines markedly. Secondly individuals seem to adopt and persist in particular roles in encounters over resources. For any given pair of birds, one asserts its dominance over the other at resources, usually with no more than a quick peck or the threat of one, and the other hen defers to the first and moves away. Each hen seems to come to know all the others and to know whether she can assert herself or must defer. Thus a peck order is gradually established. One bird, the alpha (α) can peck all the others and is pecked by none. The beta (β) can peck all but the alpha and so on down the list. The lowest hen can peck no one but must defer to all others. Hens seem to retain their rank in such orders for fairly long periods of time and, if one is removed, remember their group mate and their dominance relative to the isolated individual for several weeks.

The phenomenon described by Schjelderup-Ebbe seemed to provide an answer to the problem of how a social group could maintain itself despite competition among its members. Lorenz (1935) and especially an influential group of zoologists at the University of Chicago led by Allee (Allee et al 1949) incorporated and greatly extended Schjelderup-Ebbe's work in the 1930s and 1940s. (The Allee group came to be called the *Chicago school of animal sociologists* or, irreverently, the *great AEPPS* [pronounced "apes"], from the authors of their most important work: *Principles of Animal Ecology*, by Allee, Emerson, Park, Park, and Schmidt, 1949.)

Allee and others observed similar orderly rankings of access to resources in a variety of species, including birds, mammals, and even some invertebrates. They examined the effects of hormones on aggressive behavior and studied the effects of physical characteristics (and social experiences) on the positions of individuals in such peck orders. Since not all animals peck, they introduced the terms *rank order* and *dominance hierarchy* as more generally applicable synonyms.

Largely as a result of the influence of Lorenz and the Allee group, the importance of dominance hierarchies as the organizational backbone of animal societies became firmly established in the minds of ethologists by the 1950s and 1960s. However, the last 20 years have seen a number of lively controversies as many of the generalizations (or overgeneralizations) and conclusions of the animal sociologists have been challenged by revisionists (see review by Bernstein 1981). The result has been the clarification of many of the issues discussed. We take a number of these in turn. (1) How is dominance measured and how are dominance hierarchies demonstrated? (2) What are the defining characteristics of true dominance hierarchies? (3) What are the mechanisms whereby dominance hierarchies form? (4) What attributes determine an individual's place in a hierarchy? (5) What are the functions of dominance hierarchies? Arising from the last is a question that has been fruitful in the development of a key tool for analyzing animal social behavior: (6) Why is animal aggression so restrained, often involving displays rather than overt fighting?

Dominance and Its Measurement

Merely defining dominance has been controversial (Bernstein 1981). The definition that seems appropriate to us is the following: Dominance is the assertion of priority of access to an important resource by one individual over another. It is important to note that dominance necessarily refers to an asymmetry in the relationship between two individuals and cannot be considered an inherent attribute of an isolated individual. Evidence of winning and losing during encounters are conventionally that the individual that moves away from the site of the encounter has lost; the one that remains has won. Once a consistent pattern of outcomes is seen in encounters between two individuals, one can refer to the usual winner as *dominant* and the usual loser as *subordinate*, but note that these refer only to that dyadic relationship. Obviously an individual can be dominant in one relationship and subordinate in another.

If all social interactions involved agonistic behavior, either fighting or display, determining dominance would be easy. In most groups at most times, interactions may be more subtle. Watching a group of monkeys, we may see one animal drinking, then stop and wander away in a slow and rather relaxed manner. A minute later another may sidle up and have a drink. Unless we had watched that group for some time and knew the history of the relationship between those two individuals, we would not know whether what occurred was the result of a dominance-subordinacy relationship or just two independent actions. Determining dominance relationships often means looking carefully at asymmetries of all kinds between the behavior of two individuals, with the hope that one can calibrate those asymmetries to actual agonistic outcomes.

A diverse set of behaviors has been used to detect dominance patterns: agonistic outcomes, initiation of encounters, supplanting of others from resources, biases in grooming, differences in position within a group, etc. Problems arise when different measures correlate poorly or not at all with agonistic outcomes or each other. In many cases the solution adopted has been to adapt a composite measure of some kind (see the excellent discussion by Richards 1974). An important point was made by Rowell (1973), who pointed out that acts of submission or withdrawal by subordinates were often more observable than acts of assertion by dominants—she argued that *subordinacy hierarchies* might in fact be a better term than *dominance hierarchies*.

Once enough data have been acquired on the pairwise relationships of individuals, the observer may tabulate those results to construct a rank order for the entire

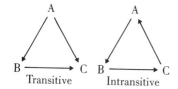

Figure 13-20 The configuration of individual dominance relationships in transitive and intransitive triads. Arrows point from dominant to subordinate. From Chase, I.D. 1982. Science 216:439-440.

group. The resulting table is shuffled to put all individuals in order from the one with the most wins at the top to the one with the fewest at the bottom. The table usually exhibits some reversals. A reversal is a win by the individual that usually lost—not all relationships are all-or-none. The individuals are arranged in the order that minimizes ambiguity caused by nonlinear or triangular relationships. Such relationships occur when animal A is dominant over B, and animal B over C, but animal C is dominant over A. Triadic (three-individual) patterns that are linear are transitive; nonlinear or triangular ones are intransitive (Chase 1982; Figure 13-20).

When the best order has been determined, one can refer to a particular individual's location in the order. If the numerical rank is used (first, second, twenty-third, etc.), that is referred to as the *ordinal rank*. If one knows the win-loss record or value of the dominance index used, one may be better able to predict the outcome of future encounters between any two individuals than if one knew only their ordinal ranks. It may be more helpful to know that individual A wins 90% of his fights and individual B 40% than to know that A is ranked second and B twelfth. Thus a cardinal dominance rank based on such an index may be desirable (see discussion and methods in Boyd and Silk 1983).

It should be pointed out that construction of a rank order in this way may be useful to the observer but not a process that the animals in the group use. There is disagreement over whether or not animals are actually aware of their own or other group members' ranks. Some, such as Bernstein (1981), feel that pairwise or dyadic relationships are the only level of awareness present for most animals. Others, such as Seyfarth (1981), argue that, at least for some intelligent animals such as monkeys and apes, complex interaction patterns show that group members are in some sense aware of each other's ranks.

Attributes of Dominance Hierarchies

By the beginning of the 1970s any group for which a rank order could be constructed was considered to possess a dominance hierarchy, yet there are problems with this line of thought. For example, one might question whether the same social structure is implied for a group of individuals from a species such as a monkey where solitary individuals are never present, as opposed to a laboratory-constructed group of four individuals from a species that in nature would never show such an aggregation. In each case a rank order of dominance could be constructed. One also could question whether the same social structure is present in hens, where individuals behave differently in encounters depending on who the opponent is, as in crickets, where each individual fights in his individual manner regardless of the opponent.

Our proposal is that the term *dominance hierarchy* be restricted to what could be termed real social structures whereas *rank order* be used for any tabulation of the outcomes of a series of competitive interactions between dyads. This would correspond with Wilson's (1975) definition of a dominance hierarchy as "the physical domination of some members of a group by other members, in *relatively orderly and long-lasting patterns*" (italics added). Our proposal is that to merit the dominance hierarchy designation, the patterns revealed in a rank order should be both linear and stable.

Linear means that rank order has few or no intransitive relationships among its constituent triads. (This is not to say reversals cannot be present; remember that a transitive relationship may include reversals if the dominant does not always dominate). Linearity as an attribute of dominance hierarchies has been recognized and even quantified since the work of Landau in the 1950s (Landau 1951 a, 1951 b, 1968). Landau developed an index he called the *hierarchy index*, h, to quantify a hierarchy's orderliness or linearity. Landau's h is given by the following formula (Landau 1968):

$$h = \frac{12 \sum (V_i - V_t)^2}{N(N^2 - 1)}$$

where V is the number of individuals dominated, V_i being the value for the ith member of a group, and V_t is the number dominated by each member of a perfectly equal group (in the event of ties, each individual is assigned a score of ½). N is the group size. Landau's h has a value of 0 for a perfectly equal group, 1.0 for a perfectly linear group; Landau considered a group to be significantly linear if $h \geq 0.9$.

Landau's h has been used to study mechanisms of hierarchy formation and to document increasing linearity over time (Landau 1968, Bekoff 1977). Later workers have looked at various factors that influence h and its usefulness (Appleby 1983, Nelissen 1986).

Stability, the persistence of a particular rank order over time, has received less attention than linearity, perhaps because it requires a much longer period of observations to assess. There are several ways to assess stability. One could use correlational statistics, as Keiper and Sambraus (1986) did in comparing stability of hierarchies in horses: one would plot the rank of each individual at time A against that same individual's rank at time B. One could also use a statistical test of concordance, asking if two rank orders, calculated at different times, agree with each other. Burk (1979) attempted to develop a stability index to complement Landau's linearity index. Burk's index was labeled b, for bump index, after Oxford boat races, in which boats begin in linear order and each tries to catch, or bump, the one ahead. (A successful crew does this each day for the four days of the races, thus getting four bumps and ascending four places in the rankings.) Burk's b is merely the index of the number of observed dominance reversals from time A to time B, divided by the maximum possible number of such reversals:

$$b = N_o/(n - 1) + (n - 2) + (n - 3) + \dots + (n - n)$$

where N_o is the observed number of dominance reversals and n is the size of the group. The b ranges in value from 1.0 for a completely upturned rank order to 0.0 for a perfectly stable rank order. Burk (1979) considered b = 0.2 to indicate a significantly stable order. He used the index to demonstrate that laboratory-assembled groups of male crickets produced rank orders that were highly linear every day but that were unstable from day to day (and that therefore, by our criteria, should not be called *dominance hierarchies*).

Mechanisms of Hierarchy Formation

The consensus of current ethologists is that dominance hierarchies do not suddenly crystallize but gradually emerge as each pair of individuals determines its dominance relationship and then transitive or intransitive relationships among triads develop. As Chase (1982 a) puts it, hierarchy formation is like the assembly of a jigsaw puzzle, where the pieces are the triadic relationships represented in the entire rank order. Development of the jigsaw puzzle model by Chase (1982, 1986, Chase and

Figure 13-21 The four
possible sequences in the
formation of the first two
dominance relationships in
triads. 1 = first relationship
established, 2 = second
relationship established.
From Chase, I.D. 1982. Science
216: 439-440.

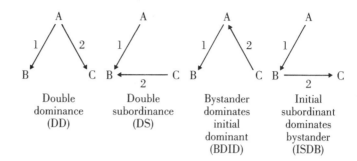

Double
dominance
(DD)

Double
subordinance
(DS)

Bystander
dominates
initial
dominant
(BDID)

Initial
subordinant
dominates
bystander
(ISDB)

Rohwer 1987) has recently focused the attention of ethologists on the process of rank order formation, not just the final outcome.

Chase (1982 a) pointed out that there are only four possible sequences in the development of a dominance triad (Figure 13-21). By convention, the dominant in the first relationship to be formed is labeled A and the subordinate is labeled B. The completion of an ABC triad can involve A subsequently dominating C (double dominance, DD), B subsequently submitting to C (double subordinacy, DS), C subsequently dominating A (bystander dominates initial dominant, BDID), or B subsequently dominating C (initial subordinate dominates bystander, ISDB).

Examination of the four possible sequences leads to some surprising results. DD and DS sequences inevitably lead to transitive, linear triads, regardless of the nature of the third dominance relationship. On the other hand, BDID and ISDB sequences lead to transitive, linear triads half the time and intransitive, triangular triads half the time, depending on the third relationship established. For example, in sequence BDID, a transitive triad is established if C goes on to dominate B, but an intransitive triad is the result if B goes on to dominate C.

Just as dyadic dominance relationships constitute the "jigsaw pieces" of a triad, so triads constitute the building blocks of the next level of organization, a tetrad. For example, a tetrad of individuals ABCD includes four triadic groups: ABC, ABD, ACD, and BCD. Each of those triads in turn came about through a DD, DS, BDID, or ISDB sequence completed by a third dominance relationship.

What Chase's jigsaw puzzle approach provides is essentially a null hypothesis, an expected chance background against which actual hierarchy formation can be viewed, which in turn allows one to hypothesize about the social mechanisms involved in the establishment of dominance relationships. For example, one would expect by chance that 25% of all sequences observed should be DD, 25% should be DS, 25% should be BDID, and 25% should be ISDB. Furthermore, as Appleby (1983) pointed out, the chance probability of getting a transitive linear order in a group of three is 75% (all of the DD and DS plus 50% of the BDID and ISDB sequences). (For groups of four the chance probability of perfect linearity goes down to 38%, for groups of five to 12%, and so on; Appleby 1983.)

Chase (1982 b) observed hierarchy formation in groups of 3 or 4 chickens, comparing the above chance predictions with real results. He found a highly non-random pattern in the hens: in groups of 3 hens, 17 of 23 triads were of the DD type, 4 were of the DS type, and there were only 1 each of the BDID and ISDB types. About 91%, not 50%, of the sequences were of the types assuring linearity.

Similar results occurred for the groups of 4 hens. DD was present in 33 of 55 triads, DS in 15, BDID in 3, and ISDB in 4; 87%, instead of the expected 50% of sequences, were of the 2 types guaranteeing linear rank orders. In fact, in his 14 different groups of 4, all 14 turned out to be perfectly linear (only 38% or 5 of 14 would have been expected).

Recently, Chase has extended his approach to cover the actual behavior used in pairwise dominance relationship establishment (Chase 1985) and to hierarchy formation in groups larger than four (Chase and Rohwer 1987). The jigsaw puzzle methodology promises to be a fruitful framework for future studies, in spite of Slater's (1986) criticism that Chase did not pay careful attention to the individual characteristics of the interacting individuals and the psychological factors at work in the interactions. Chase's approach, which is basically descriptive, shows that highly nonrandom patterns are at work, and stimulates ethologists to discover the reasons for that result. We next consider some of those factors.

Factors Affecting Position in a Hierarchy

At first glance one might assume that an individual's position in a rank order is purely a function of its ability to win fights, what Parker (1974) called its *resource holding potential* (RHP). Many studies show that larger and heavier individuals do tend to win a majority of fights. Quite early on in the study of dominance hierarchies, however, Landau (1951 a) suggested that differences in RHP alone would not suffice to produce linear hierarchies in large groups. Essentially the problem is that as group size increases, the average difference between individuals in physical traits declines. In large groups the physical differences between two similar individuals will be so small that they may not be strongly determinant of outcomes in encounters between the two. In fact, given the limits on an individual's ability to accurately assess another's size and compare it with its own, any differences may not be discernible. Thus a fair number of reversals and intransitive relationships should be expected between similar individuals, unless fight outcomes are rigidly determined by even small differences in RHP. Later analyses, especially by Chase (1974), confirmed the need for social factors to supplement physical ones in producing the commonly observed highly linear rank orders of animal groups.

A number of such social or psychological factors have been identified. The effect of previous experience in encounters is one. A strong tendency for a winner to keep winning and a loser to keep losing was observed in mice as early as 1942 by Ginsburg and Allee. This confidence effect is now known in a number of birds, mammals, fish, and insects (reviewed in Chase 1985). A striking example is found in field crickets (Alexander 1961, Burk 1979). Burk, for example, studying the species *Teleogryllus oceanicus*, found that every 1 of 49 crickets studied had a higher winning percentage in fights after a win than in fights after a loss. In one group of 4, the top-ranking cricket won 93% of 153 fights after a win, but only 39% of 31 fights after a loss; the bottom-ranking cricket won 33% of 27 fights after wins, but only 19% of 83 fights after losses. In *T. oceanicus*, the "remembrance of things past" extends at least 5 fights back: an average cricket that has just won 5 fights in a row is almost 90% likely to win the next one, whereas a cricket that has lost 5 in a row is more than 80% likely to keep losing (Figure 13-22).

The confidence effect might explain why DD and DS sequences are so common and BDID and ISDB sequences so rare. A social factor, called the *bystander effect*, might also explain these patterns. If a third party observes a dominance interaction

Figure 13-22 The effect of strings of wins and losses on male cricket *(Teleogryllus oceanicus)* fighting success. The dots show the total percentages of fights won, based on all observed encounters of 26 male crickets in 6 laboratory groups. The minimum length of run means that all runs of at least this many wins or losses are included in this percentage.
From Burk, T.E. 1979. An analysis of social behavior in crickets. Ph.D. thesis, Oxford University.

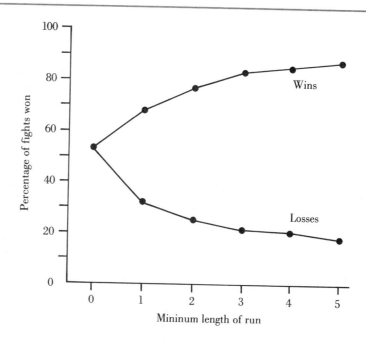

between two individuals, it may be more likely to attack the subordinate and defer to the dominant, thus falling into a linear order. Such bystander effects have not been specifically documented, mainly because studies have not been designed in such a way as to reveal them, but some studies strongly suggest their presence, for example, studies of rhesus monkeys by Barchas and Mendoza (1984).

A third psychosocial factor that would help stabilize dominance hierarchies is individual recognition by group members of each other, which would tend to inhibit reversals and reinforce established relationships. The hens studied by Schjelderup-Ebbe and the Chicago group obviously knew each other, and it came to be assumed that this was the case for all animals exhibiting rank orders. Guhl (1962) went so far as to proclaim: "The existence of a peck order is evidence that birds recognize one another, otherwise pecking would be promiscuous and unidirectional pecking would not occur."

That conclusion is not justified. Barnard and Burk (1979) showed that linear hierarchies could be produced in a variety of ways not involving true recognition of individuals per se. Two in particular are the confidence effect and assessment of an opponent's RHP based on physical characteristics and threat displays. Individuals may defer not to "individual A" but rather to "that large, threatening individual in my group." Dominant individuals in cockroaches (Breed 1983) and mice (Apps et al 1988) smell different than subordinates, for example, and a fellow group member might be responding not to "Joe" but to "anyone who smells dominant." Careful experiments, involving switches of known and unknown dominants and subordinates, are required to distinguish such assessment of the class of the opponent from true individual recognition (see Beaugrand and Zayan 1985).

In fact, once individuals begin to assess each other's future behavior on the basis of their current displays, the situation is ripe for the evolution of bluff and counter bluff. If all it takes to become a dominant individual is to smell like one, selection may favor release of dominance-indicating pheromones even by individuals of low RHP. In competitive social interactions the ability to recognize true abilities may come to involve noting more and more about other group members, ultimately knowing them as distinct individuals (Barnard and Burk 1979). Thus individual recognition may evolve in the context of social competition and dominance hierarchies without being an inevitable component of them.

A fourth psychosocial factor involved in dominance position is aid from others in the group. Such alliances are widespread. Often they involve groups of close relatives (in mammals, members of the same matriline) assisting each other in competition with nonrelatives or more distant relatives. In fact, in many primate societies the major determinant of a female's dominance rank is the rank of her mother or matriline (Walters and Seyfarth 1986).

Allies need not be relatives, at least in primates. Subordinate male olive baboons may form alliances to attack dominants (Packer 1977); similar coalition building occurs among male chimpanzees on their way up the dominance ladder (de Waal 1982). Female baboons may form friendship alliances with particular males that assist them and their offspring in competitive interactions (Smuts 1985). In fact, as we discover more about the complex network of competitive interactions in primate social groups, the less it seems like ethology and the more it seems like social psychology or politics, as reflected by the titles of such recent books as *Chimpanzee Politics* (de Waal 1982) or *Sex and Friendship in Baboons* (Smuts 1985).

Advantages of Dominance and Consolations of Subordinacy

The past few years have seen a dramatic shift in the consensus of ethologists as to the functions of dominance and dominance hierarchies. Schjelderup-Ebbe, Lorenz, and the Chicago school viewed them as vital organizational phenomena that made social life possible and led to a number of results that were beneficial for the group and its species. This is shown by the following quote from Lorenz's book *On Aggression* (1964):

> Let us ... consider in what ways intraspecific aggression assists the preservation of an animal species. The environment is divided among the members of the species in such a way that, within the potentialities offered, everyone can exist. The best father, the best mother are chosen for the benefit of the progeny. The children are protected. The community is so organized that a few males, the "senate," acquire the authority essential for making and carrying out decisions for the good of the community.

There are two (at least) problems with this masculine utopian vision. First, current understanding of natural selection suggests that evolution favors traits good for individual organisms (or their genes) as opposed to those that are for the "good of the community" or the "preservation of the species." This is a major topic developed at length in the next chapter. Second, the statement seems clearly presented from the point of view of the dominant: he or she gets food, he or she reproduces, he or she "acquires authority." The picture may not look so harmonious from the subordinate's point of view. The statement is reminiscent of the old line that fascism is good because under it the trains run on time.

A modern view would probably suggest that in most cases dominance hierarchies and rank orders have no function per se. They are an emergent pattern that an observer can construct, one that reflects the real underlying phenomenon of mate and resource competition between the individuals of a group. Except in a few cases probably restricted to primates and perhaps carnivores, an animal is not aware of ranks but only of which other individuals it can dominate or must defer to.

Therefore let us look briefly at the benefits of dominance from the standpoint of the dominant's increased access to mates or resources. Then we can look at why subordinates tolerate being dominated, why they stay in a group and act submissively rather than leaving the group.

Benefits of being dominant   That it is beneficial for dominants to gain priority of access to resources or mates seems obvious and should not require documentation. Abundant documentation does exist; a recent compilation can be found in Huntingford and Turner (1987).

Dominant individuals in groups commonly show greater reproductive success than subordinates. For example, in a herd of bison *(Bison bison)* studied by Lott (1979), the top 13 of 26 males obtained 30 of 39 matings. For southern elephant seals *(Mirounga leonina)* studied by McCann, the top-ranked bull (of 63) obtained 38% of 331 observed copulations, the top 5 bulls together obtained 88%. The effect holds for females too. In red deer *(Cervus elephas)*, because of the polygynous mating system, a dominant male has more reproductive success than subordinates or than females; dominant females produce a significantly son-biased sex ratio, thereby gaining more grandprogeny than do subordinate females (Clutton-Brock et al 1982).

Correlations between dominance status and various measures of reproductive success have been found in many studies of primates (Silk 1986), although there have also been many failures to obtain such correlations. Harcourt (1989), for example, noted that of 14 studies that showed a difference between dominant and subordinate female primates in an age of onset of breeding or fecundity, dominants had higher reproductive success in 13. Part of the reason for occasional lack of correlation or even negative correlations between rank and reproduction may lie in the difficulty of measuring fitness. What one would really like to measure is an individual's total lifetime production of mature, successful, high-quality progeny. Since that is difficult to do, simpler measures such as total number of copulations (for males) or births (for females) are often substituted. These may be misleading. For example, Hausfater (1975) found that in baboons *(Papio cynocephalus)* a female in estrus was mated equally often by dominant and subordinate males, but most offspring were probably sired by dominant males, since they usually mated with the female on the day she was most likely to ovulate. Similarly, Gouzoules et al (1982) found that female dominance rank was negatively related to infant survivorship but positively related to birth rate.

Dominants also commonly obtain more food than subordinates. This may come about by dominants occupying better feeding ranges, as in female red deer (Clutton-Brock et al 1982), or taking food from subordinates, as happens in lions and oystercatchers (Ens and Goss-Custard 1984). It may come about most frequently because dominants feed in safer locations and therefore can devote a greater proportion of time to feeding and less to vigilance than subordinates. Two good examples of this have been found in bird flocks. Murton et al (1971) found that subordinate wood pigeons *(Columba palumbus)* were at the exposed head of foraging groups, dominants in the safer center. As a result, dominants made many more pecks for food than subordinates per unit time, maintained their weight better, and were much less likely to starve (Table 13-3). Ekman (1987) found that dominant willow tits *(Parus montanus)* excluded subordinates from the upper areas of the pine trees

	Location	Mean Weight (grams)	Feeding Rate (pecks/min.)	Adrenal Condition (mean nuclear volume/ cortical cell (μm³)
Table 13-3 Effect of flocking and flock position on feeding and physiological condition in wood pigeons	Middle of flock	484	89	61
	Front of flock	410	74	85
	Solitary birds	360	53	110

From Murton, R.K., A.J. Isaacson, and N.J. Westwood. 1971. Journal of Zoology, London 165:53-84.

in which these birds forage. Those areas have denser foliage and are safer; as a result subordinates must spend a significantly greater proportion of time scanning for predators (mainly pygmy owls and sparrowhawks). Even so, subordinates experience higher predation rates.

Occasionally, dominants gain in other ways. Wood pigeon subordinates, for example, are under a great deal of stress from their vulnerability to predators and harassment by dominants. This is reflected in their enlarged adrenal gland cells (see Table 13-3) and high rates of mortality from the resulting stress syndrome.

However, it is worth pointing out that being a dominant may also have costs. We point out in Chapter 11 that dominant male bullfrogs are more heavily preyed on by snapping turtles (Howard 1978); dominant female vervet monkeys also experience higher predation rates (Cheney et al 1981). In great tits and pied flycatchers, dominant birds have higher metabolic rates and must take in more food than subordinates (Roskaft et al 1986).

Consolations of being subordinate   The benefits of dominance hierarchies are one-sided. Why do subordinates stay and pay the costs of being subordinate? Answers come when we consider their range of three options. First, they could emigrate and attempt to join another group. Emigration is dangerous and newcomers to groups usually start at the bottom of the hierarchy, so little would be gained. Second, they could leave and live as solitary individuals. As bad as being a subordinate in a group is, being a solitary individual is usually worse. For example, remember in our selfish-herd example of blue-gill nest aggregations that solitary nests fared worse than peripheral ones (see Table 13-1). Table 13-3 shows that solitary wood pigeons feed less, are smaller, and are under more stress than subordinates in flocks. Ekman's study of willow tits showed that solo birds spent about 30% more time scanning for predators than flock subordinates did. Perhaps the third option, remaining in the group and deferring to dominants, does not seem too bad. There are some gains in safety and food, and a few mating opportunities may arise. In addition, prospects may improve; many subordinates are young animals, "hopeful dominants" (West Eberhard 1975), that may rise in rank and achieve the benefits of dominance if they survive long enough at lower ranks. Also, since many animal groups consist of close relatives, the selfish-herd protection that dominants gain from their presence or the food they find that dominants take from them may indirectly help pass on subordinates' genes (see the section on kin selection in the next chapter).

This last point helps answer another question that might be raised. Since subordinates do get some matings and do take some food, why do dominants not drive them out? First, as just stated, dominants may use subordinates as protective cover or food finders and thus gain some direct benefits. Second, the subordinates may be relatives and the dominants may gain some genetic advantage by not killing subordinates or driving them out to be lost to starvation or predators. Third, it may not be worth the time, energy expense, and slight risk of injury to drive them out, since in most cases a quick dominance display will get the dominant animal first access to resources anyway.

EVOLUTION OF RESTRAINED AGGRESSION AND USE OF GAME THEORY MODELS

One function attributed to dominance hierarchies by Schjelderup-Ebbe, Lorenz, and the Chicago school was the reduction of actual fighting. As individuals came to "know their place," disputes could be settled quickly with threat and submission displays, and expensive and dangerous overt fighting would be avoided. This leads us into an issue related to the previous section, the issue of why animal aggression is so relatively restrained and ritualized (if indeed it actually is). This has been a subject of much debate in its own right, and that discussion also led to development of a highly useful tool for the analysis of animal social behavior.

The restraint shown by fighting animals was an article of faith among the early ethologists. The passage from Lorenz (1964) on dominance hierarchies quoted previously continues:

> Though occasionally, in territorial or rival fights, by some mishap a horn may penetrate an eye or a tooth an artery, we have never found that the aim of aggression was the extermination of fellow members of the species concerned.

Similarly, the following passage appears in an article titled "The Fighting Behavior of Animals," by I. Eibl-Eibesfeldt (1961):

> A complete investigation of fighting behavior must take account, however, of another general observation. Fights between individuals of the same species almost never end in death and rarely result in serious injury to either combatant. Such fights, in fact, are often highly ritualized and more nearly resemble a tournament than a mortal struggle. If this were not the case—if the loser were killed or seriously injured— fighting would have grave disadvantages for the species.

The views of the early ethologists have been challenged on two grounds. First, as pointed out by Geist (1971) and emphasized by Wilson (1975), injury and death in aggressive contests is certainly not a merely human phenomenon but is actually widespread in animal species. It is worthwhile to quote Wilson (1975) on this point:

> The evidence of murder and cannibalism in mammals and other vertebrates has now accumulated to the point that we must completely reverse the conclusion advanced by Konrad Lorenz in his book *On Aggression* ... murder is far more common and hence "normal" in many vertebrate species than in man. I have been impressed by how often such behavior becomes apparent only when the observation time devoted to a species passes the thousand-hour mark. But only one murder per thousand hours per observer is still a great deal of violence by human standards. In fact, if some imaginary Martian zoologist visiting Earth were to observe man as simply one more species over a very long period of time, he might conclude that we are among the more pacific mammals as measured by serious assaults or murders per individual per unit time, even when our episodic wars are averaged in.

A recent list of recorded instances of injury and death in animal fights is provided by Huntingford and Turner (1987), and it is quite a long list. For example, 5% to 10% of male musk oxen die in the rut each year (Wilkinson and Shank 1976), whereas in red deer 23% of mature stags show signs of serious injury (blinding, leg or antler breakage), and 6% are permanently disabled (Clutton-Brock et al 1982).

A second objection to the conclusion of the early ethologists is theoretical rather than empirical. Although they may have underestimated the amount of overt fighting, they were certainly right that most encounters involve only display and few encounters end with serious injuries. However, the "for the good of the group or species" explanation for restraint is not compatible with the current belief that behavior must primarily benefit the behaving individual by propagating its genes (see extended discussion in Chapter 14). If most fights are restrained display contests, there must be an individual selection basis for the restraint shown.

In an attempt to discover an individual advantage for restraint by interacting competitors, Maynard Smith and Price (1973) decided to use the analytical technique of game theory modeling, borrowed from economics. (Actually, the technique had been used previously by biologists, notably Fisher [1930] and Lewontin [1970].)

Game theory is a method that is appropriate when the consequences of behavior for one animal depend on what others are doing. It is particularly appropriate for the analysis of social interactions. (The optimality approach discussed in Chapter 9, by contrast, is best suited to games played by individuals against the environment rather than against conspecifics.) The basic approach is to specify two or more tactics or ways of behaving, and to set values for the payoffs for the tactics, as specified in a table or pay-off matrix. The ultimate objective is to find a tactic that cannot be bettered if it becomes widely adopted. Such a tactic is said to be an *evolutionarily stable strategy** or ESS.

To model the problem of restrained fighting, Maynard Smith and Price constructed two tactics called *hawk* and *mouse*; but since this was in the Vietnam War era when the news was full of "hawks" and "doves", the mouse soon became a dove. What follows is a version of the hawk-dove game presented in a review paper by Maynard Smith in 1976. (Also note that this terminology does not involve different species but rather different behavioral types of individuals within the same species.)

The hawk tactic represents an animal that does not use threat displays but rather engages in all-out fighting until its opponent is killed, injured, or flees, or until the hawk itself is injured. A dove, on the other hand, never escalates its behavior beyond mere displaying. It continues in a contest until the opponent leaves or attacks; if attacked, a dove flees immediately to avoid injury.

Now that the tactics have been devised, payoffs need to be established. Maynard Smith set the value of winning the contest and getting the contested resource at +50 arbitrary units (this could represent calories or fitness chips ultimately cashed in for offspring). The payoff for losing is 0 points. The cost of being seriously injured in a fight is set at −100 (i.e., getting hurt costs you more than getting this resource helps you, which seems reasonable). Finally, when two doves meet, a long display contest ensues, so we need to recognize some cost for prolonged wasting of time and energy, say −10 points.

---

*Perhaps a better name would be evolutionarily stable *tactic* since the ploys used in game theory are more analogous to the military concept of tactic than the concept of strategy—look them up. In either case the terms are anthropomorphic and objectionable to many biologists (Louw 1979). However, *strategy* is the term used in the early discussions on the subject (e.g., Maynard Smith 1972), and its use has since become firmly entrenched.

Table 13-4 Payoff matrix
for the hawk-dove game

Attacker	Opponent	
	Hawk	**Dove**
Hawk	½ (50) + ½ (−100) = −25	+ 50
Dove	0	½ (50 − 10) + ½ (−10) = + 15

After Maynard Smith, J. 1976. American Scientist 64:41-45.

With these payoffs we can construct a payoff matrix (Table 13-4). This gives the average outcome for each participant in each kind of encounter. When a hawk meets a dove it wins every time, getting a payoff of +50 while the dove gets 0. When two doves meet there is a prolonged encounter. The eventual winner gets +40 (+50 for winnng −10 points for time wasted); the loser gets −10 (time wasted but no resource gained). An average dove will win half of such contests and lose half of them, so its average payoff is the average of +40 and −10, or +15. When two hawks meet they fight until one wins and the other is seriously injured. The winner's score is +50, the loser's −100. Since an average hawk will win half the time and lose half the time, the average payoff for a hawk against another hawk is the average of +50 and −100, or −25.

Is either hawk or dove an ESS? Perhaps surprisingly, the answer is no. In a population composed entirely of doves a mutant or immigrant hawk would win every fight and have lots of baby hawks: the proportion of hawks would quickly increase, but so would the proportion of doves in a population initially composed only of hawks. In such a population a dove would always get a score of 0; but hawks would be injuring each other and scoring −25. Nothing is better than less-than-nothing, and we would expect doves to out-reproduce hawks. In other words, no pure tactic is an ESS in the hawk-dove game.

There is an ESS, however, that is an unbeatable mixture of hawks and doves. With these payoffs, the ESS mix is $7/12$ hawks and $5/12$ doves (see box on page 434). There are two ways of achieving this mix: (1) There could be a population where each individual always uses only one tactic. Thus a mix of $7/12$ hawks and $5/12$ doves would be stable; having more or fewer of either type would result in disparate success of the two tactics until the mix returned to $7/12$ and $5/12$; (2) There could be a population where every individual sometimes acts as a hawk ($7/12$ of the time) and sometimes acts as a dove ($5/12$ of the time). In such a population, no mutant or immigrant that played hawk and dove in a different proportion would score as well as the existing individuals, each of which is $7/12$ hawk and $5/12$ dove.

In other words, the ESS could be a mixture of pure tactics or a population uniform for a mixed tactic.

Let's back off and see what we have learned from our hawk-dove, game theory model. First, all-out aggression is not the expected evolutionary outcome from the standpoint of individual contests. What should evolve, given the payoffs specified, is a mix of attacks and displays, which is what animals actually show.

Second, and perhaps of even more significance, we find that evolution in such a contest situation will not produce the best conceivable outcome. If you plug the

Finding the ESS in the Hawk-Dove Game

1. Let h be the proportion of hawks in the population. The proportion of doves will be $(1 - h)$.

2. At equilibrium, the average payoff for a hawk ($\overline{H}$) will equal the average payoff for a dove ($\overline{D}$).

3. The average payoff for a hawk is:

$$\overline{H} = -25(h) + 50(1 - h)$$

(-25 per trial against hawks multiplied by the h proportion of fights against them, plus 50 per trial against doves times $(1 - h)$ proportion of fights against them. The -25 and $+50$ come from our payoff maxtrix, see Table 13-4.)

4. The average payoff for a dove is:

$$\overline{D} = 0(h) + 15(1 - h)$$

5. We can solve for h by setting $\overline{H} = \overline{D}$ because of point 2:

$$0(h) + 15(1 - h) = -25(h) + 50(1 - h),$$
$$\text{or } 15 - 15h = -25h + 50 - 50h,$$
$$\text{or } 60h = 35,$$
$$h = {}^{35}\!/_{60}, \text{ or } {}^{7}\!/_{12}$$

6. The ESS is a mixture of $^{7}/_{12}$ hawks and $^{5}/_{12}$ doves

From Krebs, J.R., and N.B. Davies. 1987. An introduction to behavioural ecology, ed 2. Sinauer Associates, Sunderland, Mass.

$^{7}/_{12}$ and $^{5}/_{12}$ values for h and $(1 - h)$ into the formulas in the box on this page, you find that at equilibrium, the average payoff is $+6.25$. In a population of doves, the average payoff (from Table 13-4) would be $+15$. The problem with a conspiracy of doves is that it is vulnerable to invasion by cheating hawks. Only in a species that could arrive at agreed rules and enforce sanctions on cheaters—that is, only in humans—could the optimum be achieved. For nonrational animals the evolved outcome is the ESS, the tactic or mix of tactics that is immune to invasion.

The hawk-dove game is, of course, overly simplistic. Other, subtler tactics can be devised and tested. Among the ones tried by Maynard Smith and Price were bully, which displays if its opponent attacks but attacks if its opponent displays; retaliator, which displays unless attacked, in which case it returns the attack; and prober-retaliator, which behaves like retaliator but also occasionally makes a probing attack in case its opponent is a dove that would retreat if attacked. The outcome of any particular game depends on which tactics are used and on the values of the payoffs assigned; for example, if the cost of injury is less than the value of a contested resource, playing hawk becomes an ESS in the hawk-dove game. For the results of a game theory model to yield insight into the behavior of real animals, the payoff values of tactics relative to each other must be realistic. The tactics specified, although oversimplistic compared with the real behavior of animals, must correspond to what might evolve. As Krebs and Davies (1987) put it, a contestant with a machine gun will win every time, but a gun-toting tactic is obviously not something we could reasonably expect to evolve for nonhuman animals.

An interesting game theory model examined by Maynard Smith (1976) tested the hawk and dove tactics against a third tactic called *bourgeois*. An individual using that tactic fights hard if it is the owner of a resource (such as a territory) but retreats if it is an intruder and is challenged. This corresponds to a common observation in territorial encounters, where residents are much more likely to win contests than are intruders. Bourgeois is an ESS in this game: a population of hawks is invadable by dove and bourgeois, a population of doves is invadable by hawk and bourgeois, but a population of bourgeois animals cannot be successfully invaded by either hawks or doves.

We have introduced the hawk-dove-bourgeois game for two reasons. First, like the hawk-dove game, it throws light on a phenomenon often seen in real animal contests, the *resident-wins effect*. Second, it is an example of what is called a *conditional tactic*. Conditional tactics include two or more options, linked by if-then statements. Bourgeois says, "If resident, attack; if intruder, retreat." This is quite different from the mixed tactics described previously, in two ways. A mixed tactic, such as "be hawk $7/12$ of the time, be dove $5/12$ of the time," does not specify a change in behavior according to the situation. Such an animal should act as if it flipped a slightly loaded coin before each contest to determine whether it should play hawk or dove that time. The user of a conditional tactic does not act according to some random probability device but assesses the situation and acts accordingly.

The second difference between mixed and conditional tactics lies in the relationship between the options in terms of payoffs. In a mixed tactic the outcomes of the alternatives must be equal in payoff, on average, for the mixture to be stable. In a conditional tactic, one option (being a resident) is clearly better than the other (being an intruder), but it is the total average payoff for an individual using that tactic that determines whether or not it is an ESS. For further discussion of pure, mixed, and conditional tactics, see Dawkins (1980), Cade (1980), Austad (1984), and Waltz and Wolf (1984).

Many of the alternative male mating tactics mentioned in Chapter 11 can be viewed in the light of game theory. Most are conditional tactics. A male bullfrog seems to be playing a single conditional tactic that goes, "if small, be a parasite; if medium-sized, be a nonterritorial calling male; if large, be a territorial male." Only a few of the alternative tactics seem to involve mixtures of pure tactics, or a population uniform for a mixed tactic. An example that seems to fit the former is the hooknose-jack dichotomy in coho salmon (Gross 1985) discussed in Chapter 11. As expected in a mixed-tactic ESS, the two pure tactics seem to give roughly equal payoffs and to be at least partly the result of genetic differences.

Since its introduction about 15 years ago, the ESS game theory approach has become widely used to study interactions where the participants have conflicts of interest. Good general discussions of the approach can be found in the book by Maynard Smith (1982) and the paper by Parker (1984). We conclude our discussion of ESSs with a final example, one that may exemplify a population uniform for a mixed-tactic ESS and one that shows that the ESS approach is applicable in contexts other than fighting.

Female great golden digger wasps *(Sphex ichneumoneus)* sting and paralyze katydids, which they store in underground burrows. After three or four have been collected, a female lays an egg on them, then seals off the burrow entrance and goes away to start over; the egg hatches into a larva that consumes the katydids, pupates, and emerges from the burrow as an adult the next summer. Brockmann found that female wasps come by the burrows in two ways—they sometimes dig one themselves and sometimes enter one that has already been dug. As a result of this second habit, some females end up sharing a burrow with another female. (The two usually do not encounter one another, since each spends most of its time out hunting. When they do meet, they fight fiercely; Dawkins and Brockmann 1980.)

Brockmann et al (1979) tested a mixed ESS model for the nesting behavior of the wasps, using extensive data on the outcomes of nesting collected over several years by Brockmann. Their model involved two tactics, founding and joining. However, when predictions of the model were tested against the actual data, they were decisively rejected. This caused Brockmann et al to reconsider the behavior of the wasps, on which they realized that a female wasp entering an existing burrow did not know if it was occupied by another female or had been abandoned. (Apparently suitable abandoned burrows are common; reasons for abandonment by original owners are not always known, but may include disturbance by ants, parasitic flies, or humans.) Thus the true alternatives were not founding versus joining, but actually digging a burrow versus entering an existing one. The latter is good if the burrow is abandoned: lots of time and energy that would have been spent digging is saved. It is not so good if the burrow is being used by the founder; the entering wasp may get kicked out and lose any katydids she put in before being ousted. Using this mixed-tactic digging versus entering model, Brockmann et al (1979) found that 59% of decisions were to dig, leading to reproductive success of 0.96 egg laid per 100 hours. The other 41% of decisions were to enter, leading to a not statistically different success rate of 0.84 egg per 100 hours. All females seemed to both dig and enter, and there seemed to be no conditions that correlated with which behavior a female chose for a particular nest. The best explanation seemed to be a mixed-tactic ESS, with each wasp using the mixture "dig 59% of the time when starting a burrow; enter 41% of the time."

ASSESSMENT AND ASYMMETRICAL CONTESTS

While Maynard Smith was developing the game theory approach for studying animal contests, Parker was pursuing a parallel track. Maynard Smith's model assumed that contests were symmetrical; that is, the fighting abilities and backgrounds of the two contestants were the same, the only differences being in tactics used. Parker (1974) considered the more usual case in which contests would be asymmetrical. In particular, he argued that where individuals differed in ability to win fights, that is, resource holding potential (RHP), individuals should evolve the ability to assess each other's RHP and behave accordingly. If greater than another in RHP, an individual should be aggressive; if lesser in RHP, an individual should retreat. Parker interpreted ritualized display as a period in which two contestants assessed each other's RHP (and in which each perhaps tried to outbluff the other by trying to look as big and ferocious as possible). If contests are settled on the basis of assessed RHP differences, we would again see mostly display contests rather than fights. Actual fights would occur when individuals are closely matched in RHP or when they are unable to assess it accurately.

In 1976, Maynard Smith and Parker got together and combined their approaches in a paper titled "The logic of asymmetric contests." The conclusion of their joint effort was that, if asymmetries existed, then settling contests on the basis of those asymmetries before escalated fighting occurred would be an ESS. They recognized three classes of asymmetry:

1. Asymmetries in RHP. In this, probably the most common case, the individual that assesses its RHP as being lower should immediately withdraw. We discuss a number of examples in this and the preceding chapters in which larger contestants for mates or resources were successful over smaller rivals. Many examples are given in Thornhill and Alcock (1983) and Huntingford and Turner (1987).

2. Asymmetries in resource value. The resource may have more value to one contestant than to the other. In that case, one contestant should fight harder than the other. If so, Maynard Smith and Parker showed, the ESS for the contestant that assesses the value of the resource to be lower is to withdraw. There are a number of reasons why a resource may be of more value to one contestant than another. A starving individual may need a food item much more than a well-fed individual. An older individual with less future life ahead of it has less to lose by fighting hard for a resource than a younger individual that has to weigh current gains against the future losses if injured or killed. A territory resident may fight harder than an intruder; the resident is familiar with the territory and can immediately profit from its resources at high rates. An intruder, even if it wins, will not be able to exploit its gains at a high rate until it learns its way around the territory. A mate-guarding male may know more about the reproductive value of a female than a rival attempting a takeover does, and may be willing to fight harder if it knows the female is of high value. An excellent example along these lines comes from Austad's study of fights in the bowl-and-doily spider (*Frontinella pyramitela*) (Austad 1983). Males live only 3 days as sexually mature adults and range widely in search of virgin females. Unlike the situation for many species, as discussed in Chapter 11, there is first-male sperm priority (with two matings, 95% of eggs are fertilized by the first male). When a male encounters a female there is a preinsemination stage during which the male determines if the female is sexually mature. The male then copulates with the female by passing sperm from his palps, which in spiders are used as intromittent organs. A virgin female is worth 40 eggs to a male; by 7 minutes of mating he will have put in enough sperm to fertilize 90% of her eggs, by 21 minutes enough for 99%. The value of a female to a male differs between mating males and latecomer males. A latecomer does not know if a female is virgin, partly inseminated, or fully inseminated, so it acts as if it were approaching an average female with an egg value of 10 eggs (the average number available from the total encounters with all types of females). As the "resident" interacts, he acquires information about this female—if she is a virgin, she is worth 40 eggs to him. As he mates, she becomes worth less to him in the future because most of her eggs are already accounted for. After about 5 minutes she is worth less than the average female he might encounter if he left her. Residents fight more or less tenaciously depending on their experience with the female, whereas intruders always fight as if it were an average female. As a result, one would predict, and Austad (1983) did find, that residents persist and win most fights up to about 5 minutes after mating begins but withdraw and lose fights longer

Figure 13-23 Percentage of fights won by the resident male in bowl-and-doily spider fights over females. Only the resident "knows" the value of the female in terms of eggs yet to be fertilized. When that value is high (up to 7 minutes of the insemination phase of copulations), the resident fights hard and usually wins. When that value is low (after 7 minutes of insemination), the resident gives up the female in most fights. P.I. = preinsemination.
From Austad, S.N. 1983. Animal Behaviour 31:59-73.

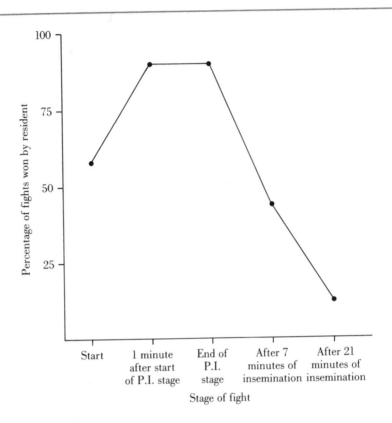

than 5 minutes (Figure 13-23). (Incidentally, there is a large-male advantage too. These findings are specifically for males of the same size.) Austad's study clearly reveals the importance of getting (or failing to get) accurate information in the assessment process.

3. Uncorrelated asymmetries. The third type of asymmetry that should be used to settle contests short of overt fighting, according to the analysis of Maynard Smith and Parker, is somewhat surprising. If the cost of fighting is relatively high or the value of the contested resource relatively low, and if the two contestants do not differ in RHP or in the assessed value of the resource, it would be an ESS for the contestants to settle the outcome on the basis of any asymmetry that can be identified. Such an asymmetry unrelated to RHP or resource value is called an *uncorrelated asymmetry*; settling contests on the basis of such asymmetries is the animal equivalent of flipping a coin to see who wins.

Not many examples of the use of uncorrelated asymmetries to determine outcomes of contests have been forthcoming, which is perhaps not surprising, since in most cases contestants will differ in RHP or value of resource. The ones that have been identified so far all seem to be based on the asymmetry of previous residence versus recent intrusion. There are several reasons why a resident in a territory might win most fights against invaders—some have already been mentioned, and we discuss them further in the next section. Those reasons would not predict that residents should win 100% of the time. Where such an absolute resident-wins outcome is seen, it seems likely that the contests are really using something like Maynard

Smith's (1976) bourgeois tactic, and letting the resident versus nonresident uncorrelated asymmetry settle the contest. An example may make this seem more plausible.

Davies (1978) studied territorial contests in a European woodland butterfly, the speckled wood butterfly *(Pararge aegeria)*. Male speckled wood butterflies defend sunspots, lighted areas on the forest floor. From these they sally forth to court any passing females. Males unable to get sunspot territories at any given time (about 40%) fly up to the forest canopy and search widely for females there. Most courtships involve sunspot rather than canopy males, which makes sunspots sound like a valuable resource, but sunspots are temporary features; since the sun's angle overhead changes, sunspots move, merge with one another, and disappear. So residents lose their territories and intruders often move into newly appearing ones; 90% of the intruders seen in territories by Davies were later spotted as owners of their own spot. So at any given time, a sunspot may not be worth running much of a risk of injury for.

In keeping with that assessment, few actual fights were observed. Intruders seemed to enter a sunspot mainly to see if it was occupied; if so, they immediately left. (Intrusions to occupied spots occurred only half as often as entering vacant spots.) Residents won 210 out of 210 such encounters. Davies performed experiments in which he caught a resident, allowing an intruder to come in and occupy the spot. Davies then released the original resident (Figure 13-24). In 10 out of 10 trials, the new resident always retained the territory and the original resident gave up without a fight and flew away. The only extended fights (spiraling dogfights lasting over 10 times as long as usual interactions) occurred when the resident-intruder asymmetry was absent and both butterflies "thought" they were residents. This happens naturally when sunspots merge; Davies managed to provoke it five times by carefully slipping an intruder into one corner of a sunspot without the

Figure 13-24 The experimental procedure and outcome of experiments showing that the resident always wins in territorial encounters between speckled wood butterfly males. From Davies, N.B. 1978. Animal Behaviour 26:138-147. (drawing by Tim Halliday).

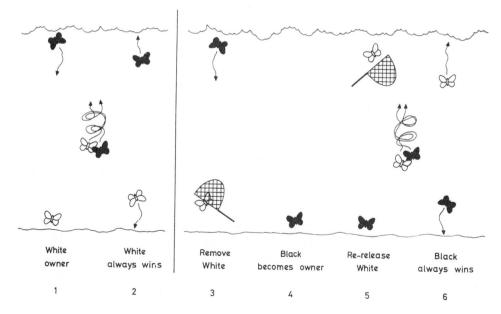

White owner	White always wins	Remove White	Black becomes owner	Re-release White	Black always wins
1	2	3	4	5	6

resident noticing. When the two males noticed each other, each was residing in the spot.

From an original attempt to explain the restraint in animal contests, we have moved to a discussion of ESSs and the dynamics of how individuals settle contests by the use of asymmetries so as to gain resources while minimizing the cost of contesting. Work since that of Maynard Smith and Price (1973) and Parker (1974) has greatly increased our understanding of how animals fight and has led to much interest in the evolution of agonistic communication. Thus from the original good-for-the-group view of restrained aggression much insight has been gained, exemplifying a point made by Darwin, who once said that false theories are good rather than bad for science, since they stimulate much productive work to come up with better alternatives.

The speckled wood butterflies were asserting dominance not as members of a stable group but in the context of territorial residence and defense. Territories and territorial behavior are the subject of the next and last section of this chapter.

TERRITORIALITY

A *territory* is a defended area (Noble 1939). Despite the simplicity of that definition, some confusion has existed about territories and territorial behavior. We begin our discussion by defining some terms. The area within which a nondispersing animal carries out its normal range of activities is a home range. In some species the entire home range area is defended by an individual, pair, or group against conspecifics; in these species the home range and territory are the same. In other species, only a central core area within the home range is defended; the home range is larger than the territory. In yet other species the home range is not defended at all; in these species home ranges exist but territories do not.

Territories are maintained by territorial behavior, or territoriality. Territoriality has four aspects (Huntingford and Turner 1987). First, the territorial animal's aggressive behavior is restricted to a particular place. This, for example, differentiates it from maintenance of a personal space around an individual wherever it is, as can be seen in bird flocks perched on telephone lines. Second, the defended area is used exclusively by the defender(s). Third, use of agonistic behavior is involved. This may not be frequent and may consist mainly of displays rather than fights. As with dominance hierarchies, avoidance of residents may be more apparent than expulsion of intruders. Thus some, such as Davies (1978), have used a non-random, overdispersed, or spaced-out spatial pattern of individuals in a species, instead of site-restricted defense, as the criterion for recognizing territoriality. However, since our use of agonistic behavior includes avoidance and submission, as well as defense and dominance, this third aspect is still justified. Fourth, dominance turns to submission when residents cross into another's territory. An individual may be confident and aggressive on one side of a boundary but cautious and passive on the other side.

To some ethologists it might seem strange that we discuss territoriality in a chapter devoted to behavior in social groups. To them, especially those influenced by the early ethological tradition, dominance and territoriality would be quite distinct manifestations of agonistic behavior, applying to social and more solitary animals, respectively. Yet we believe there are at least three good reasons for

Figure 13-25 Examples of
territories in a variety of
species and environmental
conditions. **A,** Boundaries of
communal group ranges of
the busy-crested jay in an
area of rain forest and
coffee plantations in
Nicaragua. Area shown is
approximately 700 by 900 m.
B, Areas defended by four
territorial fish, dwarf
cichlids, in a 470 L
aquarium. Markings at edge
were used to identify
locations. **C,** Hexagonal
territories, the boundaries of
which are visible in the sand
substrate, in a tightly
packed nesting group of
Tilapia fish.

A from Hardy, J.W. 1976.
Wilson Bulletin 88:96-120. B
from Black, C.H., and R.H.
Wiley. 1977. Zeitschrift für
Tierpsychologie 45:288-297. C
from Barlow, G.W. 1974. Animal
Behaviour 22:876-878.
Photograph courtesy George W.
Barlow.

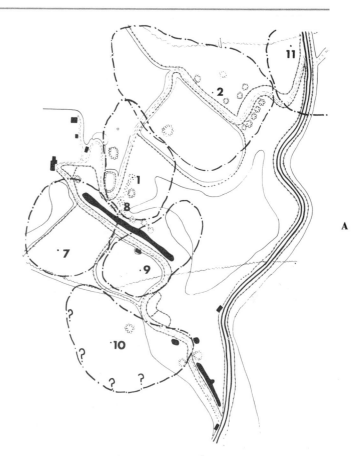

A

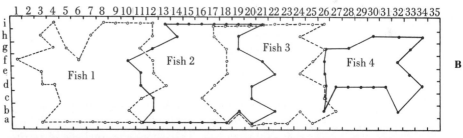

B

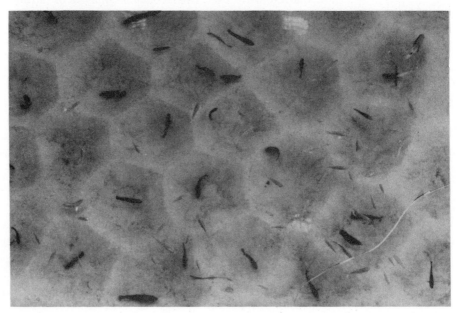

C

discussing territoriality at this point. First, the same phenomena of threat, submission, assessment, attack, and retreat are seen in both. Second, they are not mutually exclusive. Within an aggregation, such as a lek, there may be individual territories. A territory may be defended by a social group against other social groups (within all of which group dominance interactions are seen). Third, the social structure in a population may grade from one to the other and back again, according to circumstances. In several dragonflies, for example, males defend territories when population densities are low but show nonsite-restricted dominance interactions when densities are high (Pajunen 1966, Ueda 1979). (Explanations for such phenomena are considered next.) Examples of territories that illustrate these points are shown in Figure 13-25.

Economics of Territoriality

The importance of territories in the social lives of some animals has been recognized in an informal way for hundreds of years, as indicated by the Chinese saying, "There can only be one tiger to a hill." The English clergyman-naturalist Gilbert White (1788-89) observed territoriality in crickets. He loved to hear their chirping, and tried to establish large numbers of males in the chinks of a stone wall near his vicarage. The attempt failed, however, because "... the first that got possession of the chinks would seize upon any that were obtruded upon them with a vast row of serrated fangs." Scientific focus on territoriality began with the publication of *Territory in Bird Life* by H. Eliot Howard in 1920. For the next four decades the main question pondered by students of territoriality, such as Hinde (1956), was, "What are territories for?" The search for answers to this question revealed a number of types of territories, for mating, raising young, feeding, or for a combination of purposes. Yet the identification of a function for territories in particular cases really did not advance understanding regarding why some species are territorial and others are not. Progress on that front began with a paper by Brown in 1964.

Brown applied the cost-benefit approach to territoriality. He argued that the presence or absence of territoriality would depend on whether the gains from exclusive use of some localized site would outweigh the costs of defense, of maintaining that exclusive use. This economic analysis of territoriality has been fruitful.

The general benefits and costs of territoriality are listed by Huntingford and Turner (1987). Benefits include:

1. Decreased risk of predation because spaced-out individuals or nests are harder for predators to find.
2. Decreased risk of parasitism and disease.
3. Foraging advantages:
   a. Food lasts longer if fewer animals are exploiting it.
   b. Renewable foods are depleted less and can be harvested at higher rates.
   c. Food supplies may fluctuate less in a territory.
4. Increased number of mates (see Chapter 11).
5. Offspring rearing advantages:
   a. Young raised in superior sites.
   b. Decreased risk of predation or cannibalism.
   c. Decreased risk of intraspecific brood parasitism.

Against those benefits must be balanced the costs of acquiring and defending the territory:

1. Acquiring a territory may involve ousting a previous owner.

2. Maintaining a territory is energetically expensive, due to the continuing cost of displays.
3. Maintaining a territory takes time away from other activities.
4. Defending a territory may increase the risk of being injured.
5. Territories used for mating only may not have enough food, leading to loss of condition.

From this cost-benefit approach, one can predict that territoriality will exist when resources are what Brown (1964) called *economically defendable*. Factors affecting this will be the spatial distribution of resources, their abundance and renewability, and the density of rivals to be excluded (see discussion in Davies and Houston 1984). To exemplify these considerations, we present two studies that involve this cost-benefit approach.

Gill and Wolf (1975) studied territoriality in a nectar-feeding bird, the golden-winged sunbird *(Nectarinia reichenowi)* in the mountains of Kenya. Sunbirds defend territories with about 1600 flowers of the plant *Leonotis nepetifolia*; territory size is highly variable, from 6.7 to 2300 m², depending on flower density. By measuring the nectar content of flowers, the metabolic costs of various activities, and the time spent in various activities, Gill and Wolf could assess territoriality's costs and benefits. Territory defense cost sunbirds 3000 calories per hour and took up about 0.28 hours per day (Krebs and Davies 1987), giving a total cost of 728 calories per day. When a bird defends a territory, its flowers can replenish their nectar without intruders depleting them; territoriality results in the owner getting more nectar from a flower per visit. This in turn means the resident needs to forage for much less time to satisfy its daily needs. For example, if undefended flowers have 2 μl of nectar, a bird needs to forage for 4 hours; if defended flowers have 3 μl of nectar, only 2.7 hours of foraging are necessary. Foraging costs 1000 calories per hour whereas resting quietly costs only 400. Thus territoriality saves 600 calories an hour times 1.3 hours saved, or 780 calories. The calories saved outweigh the costs of defense and territoriality pays. However, if intruder pressure is low and nectar availability high (so that most flowers will not be at risk of nectar depletion and will be full even if not defended), the cost may be greater than the gains from defense. Indeed, under those conditions Gill and Wolf did see days when sunbirds did not defend territories.

Our second example also shows economically derived variation in territorial defense. Davies studied winter feeding territoriality in pied wagtails *(Motacilla alba)* along the Thames River near Oxford, England (Davies and Houston 1981). In the cool, short winter days, these small insect-eating birds need to eat an insect the size of a midge or mosquito every 3 or 4 seconds to stay alive—natural selection can surely be expected to have made these birds efficient foragers.

Riverbank territories are defended by dominant male wagtails, which feed on insects that died over the river and are washed up on the bank. Such insects constitute a renewing food source, like the nectar in the previous example. Subordinate birds feed in flocks, usually at flooded pools in riverside meadows. Non-territorial birds also attempt to intrude undetected in riverside territories. The advantage to territory owners seems to be a long-term one. Food abundance varies greatly, and on some days territory owners actually fed at lower rates than flock birds, but owners have a more assured food supply, are less likely to experience prolonged periods of food shortage, and are less likely to starve over winter.

Nevertheless, as food abundance varies, territory owners show four different patterns of behavior. At very low food densities they leave their territories to join the feeding flocks, returning at intervals, however, to reassert their claim. At low food densities they always stay in their territories and evict intruders as quickly as they spot them. At very high food densities they abandon any attempts to oust intruders—the cost of defense presumably is not worth the gains.

The most interesting behavior of owners occurs when food abundance is high and intrusion rates fairly high. Under those conditions residents allow a second bird, a female or juvenile, to settle as a satellite territory owner. This has marked costs for the owner. When the owner is by itself a given patch of riverbank has 40 minutes between visits to renew its supply of insects; when the satellite is also present, only 20 minutes elapse between visits, so the owner gets fewer insects. The owner also benefits from the satellite's presence because the satellite helps defend the territory against other intruders. As much as 50% of defense may be done by satellites; the average proportion is 38%. Behavior by residents toward satellites changes markedly. On one day they may be tolerated, on the next driven out, and on the next tolerated again. Using measured prey availability data, Davies and Houston calculated how many extra or fewer items per minute an owner would get with a satellite present on different days. When that gain or loss was considered against the number of chases (attempts by the resident to oust the satellite), a strong relationship was observed.

Thus, not only does the cost-benefit analysis help one understand territoriality, it even helps explain what might otherwise seem to be puzzling variation and inconsistencies in territoriality, as in sunbirds and wagtails.

Who Wins Territorial Contests

Territoriality has been observed in large numbers of species in all the major classes of vertebrates, as well as in many insects (Baker 1983) and a few other invertebrates. As a result, many studies have examined the factors that affect the outcome of territorial contests. Two major determinants stand out. Winners tend in many studies

Table 13-5 Resident wins effect in male *Teleogryllus oceanicus* crickets

Male	Percentage of Fights Won		
	Away From Burrows	As Burrow Resident	As Intruder
A-1	65.2	94.2	41.5
A-2	48.1	81.9	12.8
A-3	24.1	84.8	11.4
A-4	62.9	72.1	9.7
B-1	25.2	80.5	14.3
B-2	34.1	87.5	7.1
B-3	51.7	91.2	10.1
B-4	30.4	63.6	1.0
B-5	90.1	97.0	41.9
TOTAL	50.0	86.4	13.6
(N of Fights)	(513)	(1025)	(1025)

From Burk, T. 1979. An analysis of social behaviour in crickets. Ph.D. thesis, Oxford University.

Table 13-6 Effect of size and residency in territorial encounters in *Anastrepha Suspensa* flies

Size of Resident	Size of Intruder	Number of Fights Where:		
		Resident Won	Intruder Won	Percentage Won by Resident
Large	Small	30	5	86
Large	Large	28	13	68
Small	Small	25	7	78
Small	Large	11	21	34

From Burk, T. 1984. Florida Entomologist 67:542-547.

to be larger than losers, or residents larger than intruders (a number of examples can be found in Huntingford and Turner 1987).

The second major effect is the resident-wins effect. The effect can be striking. For example, Burk (1979) found in the field cricket *Teleogryllus oceanicus* that residents won over 86% of fights at defended burrows. All of the crickets studied did worse in fights away from burrows and much worse as intruders at burrows (Table 13-5).

All three of the asymmetries pointed out by Maynard Smith and Parker (1976) may be involved in resident-wins outcomes. Residents may be residents because superior RHP allowed them to win previous encounters. Thus they might be a bigger than average set of animals, likely to win the majority of current encounters. Second, because of their familiarity with their territories, residents may be better able to profit from the territory than an intruder would, so there may be asymmetries in value of the resource to residents and intruders. Finally, residence may be used as an uncorrelated asymmetry, as in speckled wood butterflies and perhaps crickets.

It is not uncommon for the size and residence effects to interact in a particular species. Table 13-6 gives an example from one of our studies. In Caribbean fruit flies *(Anastrepha suspensa)*, the larger males tend to win territorial fights; but where the resident and intruder are the same size, the resident usually wins.

The Role Of Keep-Out Displays

In keeping with the general importance of agonistic displays as opposed to overt fighting, it should come as no surprise that many studies have shown the importance of keep-out displays in deterring territorial intruders. In Chapter 11, we mentioned that both the red epaulets and the conc-a-ree song are important keep-out signals of territorial male red-winged blackbirds (Smith 1972, 1979, Roskaft and Rohwer 1987). Similar functions have been discovered for cricket songs, fish visual displays, and mammalian pheromones, as well as many other signals. We close this chapter with mention of an experimental demonstration of the importance of bird song as a keep-out signal.

Classic studies by Krebs (1971) had shown that high-quality oak forest territories were in short supply for English great tits *(Parus major)*; when territorial pairs are removed by trapping or shooting, intruders have entered and established themselves as replacement pairs within a few hours. Intruders must have constantly been nearby to have exploited an opening so quickly, yet few actual fights were seen. Were the territorial songs of the owners acting as effective deterrents? Krebs and associates tested that idea by again performing removals. This time, however, some removed pairs were replaced by loudspeakers playing great tit songs at normal rates. (Other pairs were replaced by nothing or by control sounds.) Song-containing territories stayed empty for several days, unlike the immediately occupied controls (Krebs 1977). In a follow-up series of experiments, Krebs et al (1978) played a single great tit song type in some emptied territories, a repertoire of four song types in others (total amount of song was the same in the two treatments). Control territories were reoccupied by the end of 1 day, single-song territories by 2 days, repertoire territories by 3 days. It may be that repertoires indicate the presence of older, experienced birds that are more formidable residents. Alternatively, Krebs et al (1978) suggest that by singing many songs, residents may be trying to suggest the area is super-saturated with territorial residents, an area definitely to be avoided by potential intruders. Krebs et al (1978) call this possibility the *Beau Geste effect*, after the French Foreign Legionnaire hero of the Wren novel, who propped up dead legionnaires around the fort and ran from one to another firing rifles, thus convincing attacking Bedouins that the fort was adequately manned.

For Darwin, intraspecific competition was the major driving force in evolution. Thus we should not be surprised to observe it in animal social groups. Cooperative and altruistic behaviors are more surprising in darwinian animals. We explore those social interactions in the next chapter.

SUMMARY

As demonstrated by cliff swallows, living in a group has both costs and benefits. Major costs include increased competition for resources and increased rates of parasitism and disease. Groups may also attract predators, increase risks of inbreeding, and expose parents to intraspecific brood parasitism and young to cannibalism. Major benefits come from foraging and antipredation effects. Group foragers may share information, make kills more often, take larger prey, and be able to defend food sources better than solitary foragers. Animals in groups may be safer because of increased vigilance, selfish-herd protection, dilution effects, cooperative defense, confusion effects, or alarm calling. Animals in groups may also gain protection against hostile conspecifics, have more reproductive options, stimulate each other via displays, and experience thermal and locomotory advantages.

The interactions of social animals can be classified as competitive, spiteful, cooperative, or altruistic, based on the effects of the interaction for the two participants. Spiteful interactions are expected to evolve rarely; cooperation and altruism are the subject of the next chapter. Competition involves attacks, withdrawal, threat, and submission; collectively these constitute agonistic behavior.

Rank orders and dominance hierarchies reflect the outcome of competitive interactions in groups. Unlike the former, the latter must be stable over time and linear, with few triangular or intransitive patterns. The jigsaw puzzle method allows one to study the establishment of dominance hierarchies. Dominance relationships depend not only on differences between individuals in fighting ability (resource holding power, RHP) but also on a number of social and psychological factors, such as confidence effects, bystander effects, individual recognition, and assessment of opponents. Dominant individuals usually gain feeding, survival, and reproductive advantages over subordinates, but subordinates may stay in groups because dominants are their relatives, because living solitarily may be even worse, or because they may in time become dominants.

Early ethologists stressed that animal contests were restrained, involving little fighting and much displaying. Application of the game theory methods and the evolutionarily stable strategy (ESS) concept show that a mix of fighting and display is the expected result of natural selection acting on competing individuals. Contests may be settled on the basis of assessed asymmetries between the opponents in RHP, resource value, or of an uncorrelated nature.

The distinction between social-group dominance and territoriality is not an absolute one, and we discuss territoriality in this chapter in relation to other competitive social interactions. Such defended areas should be found when the benefits of exclusive access to particular defendable resources (food, breeding sites, mates, etc.) outweigh the costs involved in maintaining exclusive access. As ecological circumstances change, territoriality may appear or be abandoned even within a particular group or individual. Territorial contests are often won by individuals that are larger and/or are established residents. As with dominance interactions in social groups, displays are as important as actual fighting in territorial contests.

FOR FURTHER READING

Alexander, R.F. 1974. The evolution of social behavior. Annual Review of Ecology and Systematics 5:325-383.
   A classic paper on the costs and benefits of group living.
Brown, C.R. 1986. Ectoparasitism as a cost of coloniality in cliff swallows *(Hirundo pyrrhonota).* Ecology 67(5):1206-1218.
   Details the tremendous impact of swallow bugs.
Brown, C.R., and M.B. Brown. 1986. Cliff swallow colonies as information centers. Science 234:83-85.
   Classic example of the information exchange benefit of grouping.
Brown, C.R., and M.B. Brown. 1989. Behavioural dynamics of intraspecific brood parasitism in colonial cliff swallows. Animal Behaviour 37:777-796.
   Details the complex story of the extensive brood parasitism in cliff swallow colonies.
Chase, I.D. 1982. Behavioral sequences during dominance hierarchy formation in chickens. Science 216:439-440.
   Concise introduction to the jigsaw puzzle approach.
Clutton-Brock, T. 1982. The red deer of Rhum. Natural History 91(11):42-47.
   Introduction to this well-studied population, in which dominance status affects both male and female reproductive success.
Davies, N.B., and A.I. Houston. 1984. Territory economics. In: J.R. Krebs and N.B. Davies, editors. Behavioural ecology: An evolutionary approach, ed 2, Sinauer, Sunderland, Mass.
   Good review with clear examples of the cost-benefit study of territoriality.
Hoogland, J.L. 1979. The effect of colony size on individual alertness of prairie dogs (Sciuridae: *Cynomys* spp.) Animal Behaviour 27:394-407.
   Demonstration of the vigilance and selfish-herd effects in prairie dogs.
Hoogland, J.L. 1985. Infanticide in prairie dogs: lactating females kill offspring of close kin. Science 230:1037-1040.
   Account of this dramatic cost of group living.
Huntingford, R., and A. Turner. 1987. Animal conflict. Chapman and Hall, London.
   Highly recommended up-to-date and well-rounded study of agonistic behavior.
Krebs, J.R., R. Ashcroft, and M. Webber. 1978. Song repertoires and territory defense in the great tit. Nature 211:539-542.
   Account of the loudspeaker replacement experiments on the role of song in territory defense.
Lorenz, K. 1964. On aggression. Bantam Books, New York.
   Although modern ethologists do not agree with many of the conclusions, this controversial book set the stage for recent studies of agonistic behavior.
Maynard Smith, J. 1982. Evolution and the theory of games. Cambridge University Press, Cambridge, England.
   Summary of Maynard Smith's use of game theory and evolutionarily stable strategies.

Maynard Smith, J., and G.A. Parker. 1976. The logic of asymmetric contests. Animal Behaviour 24:159-175.

Important paper on the grounds by which animal contests are settled.

McCracken, G.F., and M.K. Gustin. 1987. Batmom's daily nightmare. Natural History 96(10): 66-73.

Popular account of the dilemma faced by a nursing Mexican free-tailed bat trying to find her youngster among the millions in the caves occupied by this species.

Simon, C. 1979. Debut of the seventeen-year-old cicada. Natural History 88(5):38-45.

Popular story of perhaps the most famous and amazing example of the temporal dilution effect.

Smuts, B. 1987. What are friends for? Natural History 96(2):36-45.

Popular account of the complex network of amicable and competitive relationships in a baboon troop.

Weatherhead, P.J. 1985. The birds' communal connection. Natural History 94(2):34-41.

Popular account of the costs and benefits of communal roosting in birds.

Wrangham, R.W., and D.I. Rubenstein. 1986. Social evolution in birds and mammals. In: D.I. Rubenstein and R.W. Wrangham, editors. Ecological aspects of social evolution. Princeton University Press, Princeton, N.J.

Modern overview of the major costs and benefits of group living.

14

BEHAVIOR IN SOCIAL GROUPS
COOPERATION AND ALTRUISM

From the darwinian perspective, animal behavior should always promote the reproductive success of the individual performing the behavior. Thus the competitive social interactions described in the previous chapter are not surprising. Less expected, however, are altruistic behaviors, in which the behaving individual's survival or reproduction may be reduced while another individual benefits. Even cooperative behavior, in which both the actor and another benefit, is somewhat surprising, since reproductive fitness is always relative, a matter of how well one does compared with others. Providing explanations for the evolution of cooperative and altruistic behaviors has been called the central challenge of the relatively new branch of ethology called sociobiology, which is the evolutionary study of animal social behavior. In this chapter six different explanations for cooperation and altruism are discussed from a general or theoretical point of view. Then a number of specific cases of social cooperation and altruism are examined to see whether the proposed explanations seem to account for them. Finally, attempts to apply the same explanations to the evolution of human social behavior are briefly discussed.

AN INTRODUCTORY PHENOMENON: ALARM CALLS IN BELDING'S GROUND SQUIRRELS

Imagine a bright, cool July morning at Tioga Pass Meadow, 3000 m up in the Sierra Nevada mountains of California where a population of Belding's ground squirrels (*Spermophilus beldingi*; Figure 14-1) have been studied for many years by Paul W. Sherman and associates (Sherman 1977, 1985). Consider two members of a colony of these rodents, a 2-year-old male and a 2-year-old female. A goshawk suddenly appears from behind a nearby crest. The male and female squirrels seem to spot the hawk simultaneously. While bolting for the nearest burrow each gives a single-note whistle vocalization (Figure 14-2). Suddenly the whole colony is in alarm, all the squirrels dashing chaotically for cover. Confused by the melee, the goshawk flies away.

Later that day, when the squirrels are again out and feeding, a coyote comes ambling up the slope toward the colony. The male, although spotting it first and watching it carefully, makes no sound. As the coyote gets a little nearer, the female sees it. She gives a distinctly different trilled vocalization (Figure 14-3) than the whistle she gave on sighting the goshawk. The coyote, hearing it, suddenly veers at full speed toward her. She runs frantically to and down the nearest burrow, just ahead of the coyote. After some half-hearted poking and digging around the burrow entrance, the coyote gives up and trots off.

The whistle response to the goshawk, given by both squirrels, seems to have been individually beneficial. In the ensuing dash for cover, callers gain from a

Figure 14-1 Belding's ground squirrel at Tioga Pass, California.
From Sherman, P.W. 1977. Science 197:1246-1253.

Figure 14-2 Sonagram of Belding's ground squirrel antipredator single-note whistle.
From Leger, D.W., S.D. Berney-Kay, and P.W. Sherman. 1984. Animal Behaviour 32:753-764.

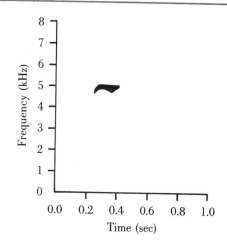

Figure 14-3 Sonagram of Belding's ground squirrel antipredator trill.
From Sherman, P.W. 1977. Science 197:1246-1253.

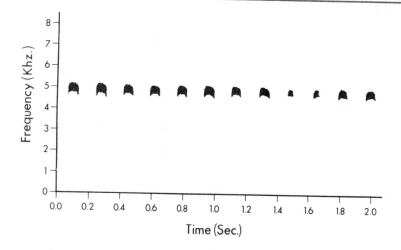

confusion effect and are less likely to be caught (Sherman 1985). The trill given in response to the coyote by the female only seems to have actually increased the caller's likelihood of being caught, while warning others of the coyote's presence (Table 14-1).

Unlike the whistle, the trill was an altruistic behavior. Individuals in many species give such dangerous alarm signals. A variety of other altruistic behaviors are known, including helping to rear others' offspring, helping others gain matings, sharing resources with others, defending others from attackers, and forgoing one's own reproduction to work on others' behalf. How can such behaviors evolve in a world described by Darwin as an intraspecific struggle for existence? No question has been so stimulating of thought and observation in the field of animal behavior in the past three decades, and a number of explanations have been proposed. We discuss six.

Table 14-1 Alarm calling and survival in Belding's ground squirrels at Tioga Pass, Calif. All data are from observations made during attacks by hawks ($n = 58$) and predatory mammals ($n = 198$) that occurred naturally during 1974 to 1982.

Category	No. of ground squirrels		
	Captured	Escaped	Percent captured
Aerial Predators (single-note call)			
Callers	1	41	2%
Noncallers	11	28	28%
TOTAL	12	69	15%
Terrestrial Predators (trill)			
Callers	12	141	8%
Noncallers	6	143	4%
TOTAL	18	284	6%

From Sherman, P.W. 1985. Behavioral Ecology and Sociobiology 17:313-323.

EXPLANATIONS FOR COOPERATIVE AND ALTRUISTIC BEHAVIOR

Mistaken Identity

We saw in the last chapter that in large breeding colonies of the Mexican free-tailed bat, such as Bracken Cave in Texas, as many as 17% of mother bats may be suckling babies that are in fact not their own (McCracken 1984). Since lactation is physiologically costly to a female bat, one can argue that these bats with "adopted" young are being altruistic. Selfishly, the females would gain by not suckling them, whereas the adopted young clearly benefit enormously from being suckled. Yet this example is best seen not as a threat to the darwinian edifice but rather as a simple case of mistaken identity. The lactating females all have babies of their own somewhere, and it is likely that they are simply mistakenly feeding one of the millions of other babies in the cave, thinking it is in fact their own. Some instances of seemingly altruistic behavior may simply prove that animals are not perfect at recognizing other members of their social group.

Manipulation

Mistakes are one thing; exploitation is another. In the last chapter, we saw that up to half of the nests in a large cliff swallow colony contain eggs put there by a female other than the female of that pair. Again we see a pair of cliff swallows brooding, protecting, and feeding chicks that are the offspring of others. This costly burden of parental care, to the benefit of some other pair's youngster, could also be said to fit our definition of altruism. Yet, as with the Mexican free-tailed bats, the brood-parasitized cliff swallows seem to have no altruistic purpose but seem to be working on behalf of what they "think" to be their own offspring. Such manipulations of the behavior of others, as by intraspecific brood parasites like cliff swallows, interspecific brood parasites like cuckoos, cowbirds, or widowbirds, or slave-maker ant species (see Chapter 10), do not seem to pose a theoretical challenge. They simply represent the selfish success of some competitors at the expense of others.

Group Selection

Until about 25 years ago, many ethologists would not have felt that altruistic behaviors posed a challenge for evolutionary biologists. For people such as Konrad

Lorenz, altruistic and cooperative behaviors (and restrained aggression and dominance hierarchies) were present because they were beneficial to the groups or species to which the altruists or cooperators belonged. Groups that contained altruistically behaving individuals would tend to stay in existence because of the actions of the altruists; for example, overall rates of predation on group members would be lower if some members gave alarm calls. By contrast, groups composed entirely of "selfish" individuals would die out faster; for example, lack of alarm calls would reduce the group's size through high rates of predation, leading eventually to group extinction.

Differential survival rates of groups with altruists over groups without would therefore lead to the spread and ultimate fixation of genes for altruistic behavior in a species gene pool. Such a process of evolution by differential survival of differing groups can be called *group selection*. Although a group selection model for the evolution of many social behavior phenomena was implicit in much of the writing of people like Lorenz and W.C. Allee, the group selection model was made explicit by its forceful presentation in a significant book, *Animal Dispersion in Relation to Social Behavior*, by V.C. Wynne-Edwards (1962).

Wynne-Edwards was particularly concerned with the phenomenon of population regulation. He felt that when a species increased in numbers, altruistic self-restraint in breeding would allow some groups to survive, whereas other groups consisting entirely of selfishly reproducing individuals would overpopulate their resource base and die out. Territoriality, dominance-subordinacy interactions, helping others to breed while not breeding oneself, and a variety of other phenomena were interpreted as group selection-favored behaviors that prevented overpopulation and group extinction.

Wynne-Edwards' book was significant not because its conclusions were widely accepted but rather because it made explicit the group selection thinking shared by so many others, and because it stimulated rebuttals and searches for alternative explanations for breeding restraint and other forms of altruism.

The most important rebuttal was *Adaptation and Natural Selection* by G.C. Williams (1966). In addition to providing individual-selection alternative explanations for the phenomena described by Wynne-Edwards, Williams argued for the general ineffectiveness of group selection as compared with individual selection. Think carefully about how group selection would work. In a group of altruists, a selfish individual would flourish. It would be warned of predators without giving any risky warnings itself; it would reproduce when others were abstaining, etc. If the differences in behavior between altruists and selfish individuals resulted from differences in alleles for social behavior, the "selfish alleles" would be reproduced at much higher rates than the "altruistic alleles." Eventually the selfish alleles would eliminate the altruistic ones in the group's gene pool. In the terms of the previous chapter, reproductive selfishness is an ESS compared with reproductive altruism in a group. For behaviors to evolve by group selection, groups "contaminated" by selfish genes must die out faster than such groups can send out selfish emigrants to invade groups composed only of altruistic genes.

Group selection—at least of this classical type—requires extremely high rates of differential group extinction and extremely low rates of between-group emigration to lead to fixation of altruistic alleles. (There is also a problem regarding how groups composed entirely of altruists would arise in the first place; but this could happen in small groups through such chance evolutionary effects as genetic drift or founder effects, as described in any general biology or genetics textbook.) Simulation and mathematical models suggest that the required combination of extinction and emigration rates is rare; almost always, individual selection outpaces group selection and determines the behavior of individuals in social groups.

However, Williams (1966) went further, arguing that selection acts at the level of the gene rather than the individual. That is, some behaviorally related genes influence their owners in ways that cause copies of those genes to be reproduced at higher rates than is the case for alternative alleles causing bodies to behave in different ways. For Williams, individuals last for only one generation (so do genotypes—which are broken up by meiosis in sexually reproducing organisms). Alleles, however, persist over many generations in a species' gene pool. Natural selection, according to this view, is the competition between alternative alleles to predominate in a gene pool. Bodies, paraphrasing Samuel Butler's famous remark about chickens and eggs, are just the genes' way of making more genes. This striking way of thinking about the evolution of behavior, known as *the selfish gene approach*, is best presented in the book of that title by R. Dawkins (1976).

For most cases, as it turns out, it really does not matter whether we view behavior as being for the survival and reproduction of the behaving individual or for the spread of the behavior genes. After all, the usual best way for a gene to propagate itself is to promote the survival and reproduction of the body it is in; but in some

Figure 14-4 Intrademic group selection process. A stands for altruists, S for selfish individuals; $\overline{A}$ is the proportion of altruistic alleles at a social behavior gene on the gene pool; $\overline{S}$ is the proportion of an alternative selfish allele at the same locus.
From Wilson, D.S. 1975. Proceedings of the National Academy of Science, U.S.A. 72:143-146.

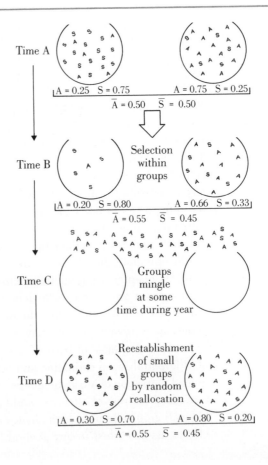

cases, as we will see in this chapter, altruistic behavior becomes more understandable from the gene selection view than from the individual selection view.

D.S. Wilson recently proposed a different kind of group selection model, which he calls the *intrademic group selection model* (Wilson 1975, 1983). Wilson's model is shown in Figure 14-4. Classical group selection involves discrete, isolated groups. Intrademic group selection involves alternation between one part of the life history spent in smaller groups and another part of the life history where greater intermingling of individuals occurs. With (1) such a life history pattern and (2) some groups containing a greater proportion of altruists than others, group selection could lead to the spread of altruism alleles. In the smaller groups, selfish individuals would out-reproduce altruists, as expected. However, groups with high proportions of altruists would increase in size faster or decline in size more slowly (because of the benefits of the altruism) than groups with few altruists. Thus, when individuals from all the groups gather, more will come from groups with altruists than from groups without altruists. A paradoxical result occurs: altruists do worse than selfish individuals in every group, but the proportion of altruists in the entire population increases. (Study Figure 14-4 carefully to convince yourself of that.)

The general importance of Wilson's model is not yet known; although it has been used to account for some observed phenomena (Colwell 1981), it has not been applied to the altruistic behaviors that are of greatest interest to sociobiologists. This may merely reflect the stigma attached to the term *group selection* in the post-Williams era. Some, such as Maynard Smith (1976), consider Wilson's model to be a case of altruism toward relatives (kin selection—discussed later), since perhaps the most common way for groups to differ in proportions of altruists would be for groups to be composed of relatives (some families would contain altruism genes, others not).

The overall consensus of sociobiologists has been that classical group selection is not important and intrademic group selection's general importance is not obvious. They have mostly turned to other explanations for cooperation and altruism.

Subtle or Long-term Selfishness

At several points in the last few chapters we stress the increase in understanding that comes from long-continued studies of a species. In Chapter 11, it is emphasized that the fitness consequences of various reproductive tactics needed to be calculated not for a single breeding season but over an entire reproductive lifetime. In Chapter 13, reference is made to E.O. Wilson's (1975) "thousand-hour rule," which states that only long-term studies reveal the existence and importance of relatively rare behaviors. In the same chapter's discussion of subordinates' behavior, it was suggested that young individuals accept subordinacy in the short term to eventually achieve dominance later. Sociobiologists have made a similar point with respect to cooperative or altruistic behavior. It may well be that instances of such behavior induce dominants to allow subordinates to remain in a group, help a subordinate form alliances that may be helpful later, or allow a subordinate male to establish a relationship with females that later pays reproductive dividends. In other words, cooperation or altruism may be part of a calculated-risk strategy, in which an animal incurs survival or reproduction losses in the short term but reaps greater survival or reproduction advantages down the line. Such instances of cooperation or altruism might better be considered a component of subtle or long-term selfishness. In our examples, there are several instances of this idea.

Kin Selection and Kin Recognition

Darwin was aware of the challenge to his theory posed by altruistic behavior. In particular he was troubled by the case of sterile worker castes in social insects. Individuals in such castes do all the foraging, defense, nest building and maintenance, and brood care for the colony, but only the reproductive queens and drones pass on their genes. How could genes for becoming sterile or for showing worker behavior be passed on in a species' gene pool if workers do not reproduce? Darwin, in *The Origin of Species* (1859), said that this case "at first appeared to me insuperable, and actually fatal to my whole theory."

Darwin not only identified a key problem, he also arrived at the gist of the solution.

> This difficulty, though appearing insuperable, is lessened, or, as I believe, disappears, when it is remembered that selection may be applied to the family, as well as to the individual.... Thus I believe it has been with social insects: a slight modification of structure, or instinct, correlated with the sterile condition of certain members of the community, has been advantageous to the community: consequently the fertile males and females of the same community flourished, and transmitted to their fertile offspring a tendency to produce sterile members having the same modification.

For Darwin, the key was that the altruism benefited the workers' reproductive relatives, which possessed the genetic tendency to produce (under certain conditions) sterile worker offspring.

The exact ways in which such altruism toward relatives would evolve were not worked out for more than a century after Darwin's first attempts, although to some extent biologists were aware of the issue. In a possibly apocryphal story, the eminent geneticist Haldane is alleged to have thumped down his pint of ale in an English pub and announced, "I would gladly lay down my life for two brothers or eight cousins." In 1964, however, Hamilton worked out in two important papers the details of the altruism-toward-relatives phenomenon.

Consider the fate of a recently arisen rare allele that in some way predisposes an individual animal to perform in an altruistic way, for example, to warn others by giving a risky alarm call when a predator is seen. If the altruist is surrounded by nonrelatives and is eaten as a result of his alarm calling, an allele for alarm calling is lost, whereas alternative alleles for not alarm calling survive in the bodies of the animals the predator did not eat. So genes for such randomly directed altruism would decrease in frequency in the gene pool of the species.

However, consider the fate of such an allele in an animal living with its parents and two siblings. As before, imagine that the animal gives an alarm call and is eaten as a result. In this case, however, the production of the alarm call may actually have resulted in a net gain in survival of the allele for altruism. The altruist must have inherited the allele from one of the parents, each of which has a 50% chance to be the one to possess it. Similarly, if the altruist had that allele, each of the two siblings also has a 50% chance to have inherited that allele from a parent. Thus four individuals each with a 50% chance of having the altruism allele were saved, or on average two copies of the allele were saved when one was lost, a net of $+1$. Contrast what would happen to an alternative allele for nonalarm calling: the selfish possessor of such an allele would silently hide while the predator ate the parents and siblings. Thus one copy of the nonaltruism allele would be saved, but on average two would be lost, a net of -1. Genes for altruism toward relatives can actually increase in gene pools at the expense of genes for selfishness, because of the effects of the altruism gene on other bodies that possess the gene.

That was the insight of Hamilton in 1964. He showed that, to understand the evolution of genes for altruism, one had to look at the consequences of the gene both for the behaving individual and for the relatives of that individual. The fitness consequences of a behavior include not only the direct consequences for the behaving individual's survival and reproduction, but also the effects of that individual's actions on the survival and reproduction of its relatives. In a sense the altruist can claim as part of its fitness the number of extra offspring each of its relatives had because of its altruistic behavior, devalued by how closely related each was. Thus, Hamilton argued, the correct measure of fitness* for an altruist was not merely its personal fitness, but rather its inclusive fitness. As pointed out by Grafen (1982), it is easy to misdefine inclusive fitness (ten out of ten animal behavior textbooks did so), so we will be as careful and explicit as we can. Inclusive fitness is (1) an individual's reproductive success, plus (2) the extra reproductive success its relatives had because of its behavior, devalued in each case by the coefficient of relatedness of the relative to the individual, minus (3) the extra offspring the individual had (if any) because of the help it received from its relatives.

The reason relatives' extra progeny have to be devalued is that, the more distant the relative, the less likely it is to have inherited from a common ancestor the same allele for altruism. Each parent and full sibling has one half chance of having the same allele as each offspring. On the other hand, grandparents, grandchildren, aunts and uncles, and half-siblings only have one fourth chance of having the same allele. First cousins have only a one eighth chance of sharing the same allele by descent. The chance that another individual shares an allele with one because of

*The term *fitness* is confusingly used in a number of ways by biologists. For a valuable discussion of these uses, see Chapter 10 in Dawkins (1982).

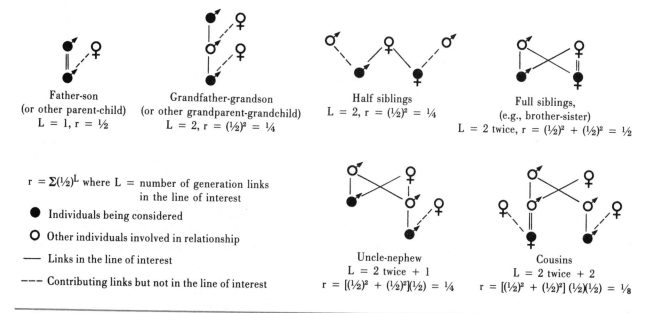

Figure 14-5   Coefficient of relatedness. Study these examples to understand how the coefficient is calculated.

common ancestry is called the *coefficient of relatedness*, r*. It was r that Haldane was intuiting when he made his statement in the pub (if he really made it). How to calculate r is shown in Figure 14-5.

To keep the explanation simple, we have so far used hypothetical examples in which altruists lost their lives to save the lives of recipients. Of course, altruistic interactions more commonly involve lesser costs to altruists and benefits to recipients. Hamilton developed a general formulation of the necessary conditions for the evolution of altruism, which has come to be called *Hamilton's rule*:

$$B/C > 1/r,$$

where B is the benefit to the recipient, C is the cost to the altruist actor, and r is the coefficient of relatedness. (B/C is sometimes expressed as K, in which case Hamilton's rule becomes $K > 1/r$.) Hamilton's rule states, for example, that altruism toward full siblings evolves only if the average gain by the siblings is greater than twice the cost to the altruist ($1/r = 2$ in this case). Note that whether or not an altruistic behavior evolves is influenced by three factors, B, C, and r. Thus even close relatives may not be helped in some cases (B/C is too small, perhaps because B is too small), whereas fairly distant relatives may be helped in other cases (B/C is big, perhaps because C is small). The general point to be made is that it is necessary to understand the ecological circumstances that affect B and C and to be able to calculate r, in any given instance.

Two names have been given to the altruism-toward-relatives phenomenon analyzed by Hamilton. Maynard Smith (1964), focusing on the consequences in natural selection of genes' effects on kin, called it *kin selection*. Alexander (1974), focusing on the behavior—preferential treatment of relatives—borrowed the colloquial term for such favoritism in humans, *nepotism*. For our purposes these can be considered synonymous.

One consequence of adopting Hamilton's viewpoint is the realization that parental care of offspring is perhaps just a particularly common type of nepotism or kin selection. Such a fresh perspective often leads to new and interesting questions. In this case the question is, why is parent-offspring nepotism so much more common than other forms, such as sibling-sibling? The answer cannot lie in r, since r = 0.5 for parent versus offspring and for sibling versus sibling (at least for full siblings). The answer is more likely to lie in the B/C ratio (Dawkins 1976). Children can be thought of as superbeneficiaries; they may gain much more from a given act (high B) than an adult would. A gram of mosquitoes is a lifesaver for a baby cliff swallow but no big deal to an adult. Second, such benefits may not cost parents too much (15 minute's foraging for a cliff swallow). On the other hand, the costs of helping may be high for young pairs of siblings, whereas the benefits of helping may well be less for the same pairs when they are grown up. Thus $B/C > 2$ may be more commonly true for parental than sibling altruism.

With the development of the ideas of inclusive fitness and nepotism, the classification of social interactions becomes more complex. Figure 14-6 shows how inclusive fitness ideas can be incorporated into the classification scheme presented in the box on page 420.

Although Hamilton's inclusive fitness and kin selection ideas were developed in

---

*Coefficients of relatedness are easily misunderstood. They are often considered as "percentage of the total genome shared by two individuals," a mistake that leads to several fallacies. For an excellent discussion, see *Twelve Misunderstandings of Kin Selection* by Dawkins (1979).

Figure 14-6 Hypothesized connection, via changes in gene frequencies, among various social relationships and kin selection. Shared genes are represented by degree of shading. Numbers of offspring indicate inclusive fitness. Bananas represent resources; sticks represent acts that reduce the fitness of another individual.
Modified from Wilson, E.O. 1975. Sociobiology, the new synthesis. Belknap/Harvard University Press, Cambridge, Mass.

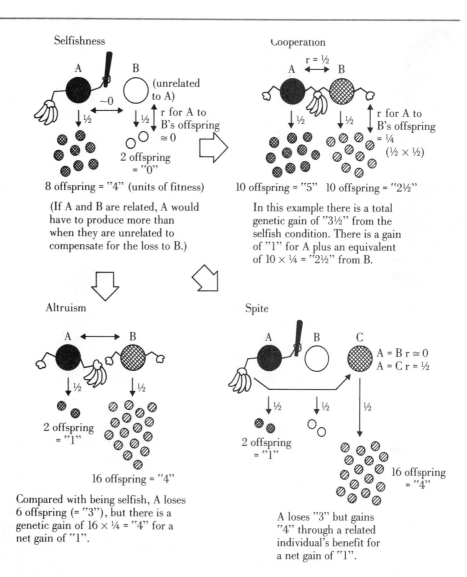

Selfishness

A   B (unrelated to A)

~0

$\frac{1}{2}$ $\frac{1}{2}$ r for A to B's offspring $\cong 0$

2 offspring = "0"

8 offspring = "4" (units of fitness)

(If A and B are related, A would have to produce more than when they are unrelated to compensate for the loss to B.)

Cooperation

r = $\frac{1}{2}$

A   B

$\frac{1}{2}$ $\frac{1}{2}$ r for A to B's offspring = $\frac{1}{4}$ ($\frac{1}{2} \times \frac{1}{2}$)

10 offspring = "5"   10 offspring = "2½"

In this example there is a total genetic gain of "3½" from the selfish condition. There is a gain of "1" for A plus an equivalent of $10 \times \frac{1}{4}$ = "2½" from B.

Altruism

A   B

$\frac{1}{2}$ $\frac{1}{2}$

2 offspring = "1"

16 offspring = "4"

Compared with being selfish, A loses 6 offspring (= "3"), but there is a genetic gain of $16 \times \frac{1}{4}$ = "4" for a net gain of "1".

Spite

A   B   C

A = B r $\cong$ 0
A = C r = $\frac{1}{2}$

$\frac{1}{2}$ $\frac{1}{2}$ $\frac{1}{2}$

2 offspring = "1"

16 offspring = "4"

A loses "3" but gains "4" through a related individual's benefit for a net gain of "1".

the 1960s, they began to have a really large impact in the 1970s with their application in a number of particular studies and their propagation by books such as the ones by Wilson (1975) and Dawkins (1976). Inevitably, as they became popular, these ideas came in for some criticism. One of the silliest criticisms was by Sahlins (1976), in a book called *The Use and Abuse of Biology*:

> ... It needs to be remarked that the epistemological problems presented by a lack of linguistic support for calculating r, coefficients of relationship, amount to a very serious defect in the theory of kin selection. Fractions are of very rare occurrence in the world's languages ... they are generally lacking among the so-called primitive peoples ... I refrain from comment on the even greater problem of how animals are supposed to figure out how that r [ego, first cousins] = 1/8. The failure of sociobiologists to address this problem introduces a considerable mysticism in their theory.

What Sahlins is saying is that even some human cultures do not possess the concept of fractions, much less do animals have it. Without knowing about fractions, neither people nor animals could evolve nepotism, he argues. Presumably he also thinks that animals cannot reproduce without a course in sex education and that only spiders that understand geometry and structural engineering can build webs. Of course, the fact is that natural selection acts on genetically varying individuals, so that those that for whatever reason behave in adaptive ways survive and pass on the genes that made them (quite unconsciously) behave in those ways. Just as people can build chess-playing computers that look as if they were thinking, so natural selection builds animals that act as if they were doing mathematics. Therefore, just as the computer does not have to think like a grand master to compete with one, a bee does not have to be as clever as a cultural anthropologist to favor its relatives.

Sahlins' comments do bring up an interesting topic, which is how do animals come to recognize their relatives and discriminate them from nonrelatives? Although it breaks somewhat the continuity of our discussion of the different explanations for cooperation and altruism, it is appropriate at this point to digress and discuss the actual mechanisms of kin recognition in animals.

Four mechanisms of kin recognition have been proposed, although the fourth remains a purely theoretical possibility (Holmes and Sherman 1983). The first mechanism is recognition by place or spatial distribution. That is, someone is your relative if you find him or her in a certain place. This seems to be a common rule used by parent birds to govern their treatment of eggs or chicks, "If it is in my nest, it is my egg." Earlier, for example, we mentioned the Beechers' studies of the development of parent-offspring individual recognition in bank swallows. Such individual recognition, based on unique signature calls, develops only after the chicks reach the age of 2 weeks. Before that parents accept and behave nepotistically toward any bank swallow egg or chick that appears in their nest. (Such behavior, of course, is what makes intraspecific and even interspecific brood parasitism successful.)

A second mechanism of kin recognition is known as the *association mechanism*. Here kin are recognized as individuals one associated with at some specific time, usually early in life. The true individual recognition of parents and offspring using signature calls shown by bank swallows would exemplify this. An early, conclusive demonstration of this mechanism was by Porter et al (1978), who studied spiny mice *(Acomys cahirinus)*. Young mice placed in an arena prefer to huddle together with siblings familiar to them rather than nonsibling strangers. The crucial experiment is to separate sibling mice at birth and establish litters that are composed of unrelated pups. When mice were placed in an arena after such manipulated rearing, Porter et al (1978) found that pups preferred to huddle with their nonrelated littermates rather than with true siblings separated from them at birth and reared in other litters. There is some support for an unconscious association mechanism in humans, from studies of marriage patterns in Israeli Kibbutzim (Shepher 1971). Young children of a kibbutz are raised communally, sharing meals, baths, dormitory sleeping arrangements, etc. Shepher's studies show that there are subsequently almost no marriages between nonrelated children brought up in a situation of such family-like intimacy. It may well be that unconsciously these individuals come to treat each other as siblings and therefore unacceptable marriage partners.

A third mechanism can help explain how an individual might recognize as a relative another it has not shared a nest with—even one it has never seen before. This is called *phenotype matching*. In this mechanism an individual compares the characteristics of an unfamiliar individual with those of some standard referent. If the phenotypes match, the stranger is treated nepotistically. If not, the stranger is treated as a nonrelative. Such a mechanism seems to be in operation in sweat bees *(Lasioglossum zephyrum)* (Greenberg 1979, Buckle and Greenberg 1981). The small colonies of this species are guarded by a female bee that blocks the entrance, admitting colony members (usually sisters) and rejecting strangers. Greenberg (1979) challenged guard bees with bees from other nests that differed as a result of careful inbred matings. He found that the more closely related an intruder was to the guard, the more likely it was to be admitted. Perhaps the guard compares the intruder with the nest mates it knows. To evaluate that possibility, Buckle and Greenberg (1981) tested guards reared in special ways. Some guards were brought up in colonies composed only of their sisters. Others were brought up in colonies composed partly of their sisters and partly of nonrelatives transferred in before adulthood from another colony (and therefore accepted as kin by the first two mechanisms discussed). Later, guards from these pure or mixed broods were challenged. Guards from pure broods rejected nonrelated intruders. But guards from mixed broods admitted not only bees genetically similar to themselves, but also strangers from the same original colony as their adopted broodmates. Bees must be following a rule like, "This bee resembles one of my nest mates and therefore is a relative."

The phenotype matching shown by the sweat bees involves a separate referent individual (broodmates in the example given)—we can call this form *other-matching*. It is also possible that an individual could compare a stranger's characteristics not with those of a third party but directly with its own—*self-matching*. In studies of kin recognition in Belding's ground squirrels, Holmes and Sherman (1983) found evidence for such self-matching.

Male Belding's ground squirrels cannot recognize their relatives, but females can (reasons for this sex difference are explained in a later section). Holmes and Sherman also performed cross-fostering experiments with newborn squirrels, creating mixed broods partly composed of siblings and partly composed of nonrelatives. For females, four classes of individuals were established: sisters reared together, sisters reared apart, nonsisters reared together, and nonsisters reared apart. When the squirrels were 8 months old, pairs were placed together in an arena and the amount of aggression between individuals was recorded. In nature, related female squirrels show much less aggression to and much more cooperation with each other than with nonrelatives. The results are shown in Figure 14-7, *A*. The association mechanism is obviously at work, since nonsisters reared together fight little (no more than sisters reared together). But sisters reared apart also fought significantly less than nonsisters reared apart. This implies that, despite never having seen a real sister, a female ground squirrel can somehow identify one. The only really obvious way to do so is to compare the stranger's traits (looks, smell, etc.) with one's own, and act appropriately depending on the degree of similarity.

A fourth proposed mechanism was presented by Dawkins (1976), who called it the *green beard effect*. (Hamilton [1964] had spoken of a similar concept, *recognition alleles*.) The whole purpose of identifying relatives, from the standpoint of an altruism allele, is to be altruistic to other possessors of that allele. Relatives, as we have seen, can be treated as probable possessors (in varying degrees). But is there any other way of spotting possessors of the altruism allele? Dawkins said this could be done if the gene had pleiotropic effects, leading not only to altruism toward

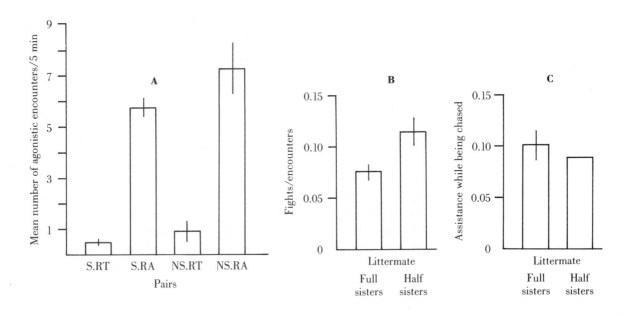

Figure 14-7   Kin recognition in Belding's ground squirrels. **A, Laboratory experiments: mean number (±1 S.E.) of agonistic encounters between pairs of yearling Belding's ground squirrels in arena tests. Nonsiblings reared together (N.S.Rt) are no more aggressive than siblings reared together (S.Rt). B and C, Field observations: aggression and cooperation among yearling females. Full sisters are less aggressive to one another (B), and assist each other more (C).** From Krebs, J.R., and N.B. Davies 1987. An introduction to behavioural ecology, ed 2. Sinauer, Sunderland, Mass. After Holmes, W.G., and P.W. Sherman. 1983. American Scientist 71:46-55.

other possessors of the allele but also to some obvious identifying trait—in Dawkins' thought experiment, a green beard. Thus an altruist with a green beard should behave nepotistically not just to relatives but rather more specifically to all individuals with green beards. The green beard effect would be difficult to separate from phenotype matching—careful experiments would be required. However, it is not obvious that it could be an ESS anyway. That way of behaving seems to be readily invaded by a mutant who grows a green beard but is not itself altruistic. Such a mutant would be helped but never pay any cost of helping. The green beard effect therefore remains at this point only an interesting thought exercise (see discussion in Holmes and Sherman 1983).

One effect of Hamilton's kin selection ideas has been a renewal of interest in the developmental mechanisms of kin recognition. An additional result of such studies has been the demonstration, which contradicts Sahlins (1976), that animals can discriminate relatives with different coefficients of relationship. For example, although Belding's ground squirrel females are tolerant of all sisters, they are more tolerant of full sisters (r = 0.5) than of half sisters (r = 0.25), even when the relevant individuals are part of the same litter (Holmes and Sherman 1983, Figure 14-7, *B*). Similarly, Getz and Smith (1983), working with honeybees *(Apis mellifera)* and using intruder introduction techniques similar to those of Buckle and Greenberg, found that unfamiliar half sisters were much more likely to be bitten than unfamiliar full sisters.

Reciprocal Altruism

A sixth explanation for apparently altruistic behavior was put forward in 1971 by Robert Trivers, who called it *reciprocal altruism*. Trivers argued that an animal might exhibit altruistic, self-sacrificing behavior for another individual in the expectation that the assisted individual would reciprocate in the future. Thus the original altruism is a sort of unconscious loan that is eventually repaid, so that over the long run there is no net disadvantage.

Such a mechanism seems to be quite vulnerable to cheating by individuals who, having been helped, refuse to reciprocate. Trivers (1971) was aware of this complication, and considered that reciprocal altruism would evolve only in species where individuals could recognize one another and in which some sanctioning discrimination could be applied to individuals who failed to reciprocate. Much of Trivers' paper in fact was devoted to reciprocal altruism in human behavior and related phenomena such as cheating, moralistic aggression, gratitude, sympathy, and guilt.

Trivers' idea of reciprocal altruism entered the collection of ideas discussed by sociobiologists and was invoked in several cases (for example, Packer's [1977] study of male-male alliances in olive baboons). However, interest in reciprocation increased after Axelrod and Hamilton (1981) applied a game theory analysis to the problem.

Axelrod, a political scientist, and Hamilton, an evolutionary biologist, began by presenting a classical social science model posing the problem of cooperation versus selfishness, known as *the prisoner's dilemma* (Figure 14-8). Imagine two suspects arrested by the police and held in separate jail cells with no possibility of communication. Each is told that there is a weak case against him, which is enough to put him away for a short time. However, each is encouraged to inform against the other, in which case the informer is set free and the other is sent to prison for a long time. However, if both "squeal," each is sent to prison for a intermediate-length term. The traditional solution in the prisoner's dilemma game sees both

Figure 14-8 The prisoner's dilemma game. The payoff to player A is shown with illustrative numerical values. The game is defined by T > R > P > S and R > (S + T)/2. Axelrod, R., and W.D. Hamilton. 1981. Science 211:1390-1396.

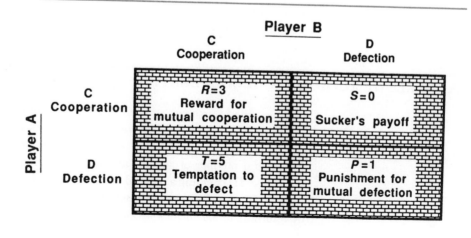

prisoners informing. For prisoner A, if B defects, A is better off defecting than holding out. But even if B holds out, A is still better off defecting than holding out. Exactly equivalent reasoning holds for player B. So, even though both holding out would be better than both defecting, both defecting is the ESS and is what we should expect to happen.

The prisoner's dilemma game seems to show that cooperation or reciprocation can never compete with selfishness. On the other hand, Axelrod and Hamilton (1981) pointed out that the classical prisoner's dilemma game is a one-shot affair. What would happen in a similar situation if individuals interact repeatedly? For example, imagine two individuals who engage in barter. One leaves his goods at a prearranged place in a forest. The other comes and picks up what the first has left. Should he now leave goods in return, or selfishly take the proffered goods without making the exchange? Does reciprocation pay if there is a possibility of a continuing series of interactions?

To find out, Axelrod invited game theorists in sociology, political science, economics, mathematics, and biology to design strategies for a computer tournament of the reiterated prisoner's dilemma game. The winning strategy, the one with the highest average score in the computer tournament, turned out to be the simplest, called *tit-for-tat*. This simple strategy called for the first party to cooperate in the first interaction, then do in subsequent interactions whatever the other party did in the previous interaction. If the second party defected, tit-for-tat defected the second time; if the second party cooperated, tit-for-tat cooperated. Thus a strategy based on reciprocity of cooperation or defection seemed evolutionarily stable. Axelrod and Hamilton (1981) attributed tit-for-tat's success to two features. First, it was retaliatory, which discouraged continued defection by others; second, it was forgiving, so that it did not lose out on benefits of future cooperation by carrying a grudge.

The Axelrod-Hamilton work, like all game theory studies, does not aim to describe nature in detail. It merely presents logical likelihoods, and helps one focus on the important parameters of any particular situation. Tit-for-tat's success shows that reciprocal altruism could evolve without requiring animals to use anything more sophisticated than simple behavioral rules of thumb. Later analyses, such as by Boyd and Lorberbaum (1987) and Axelrod and Dion (1988), have shown how the outcome of the repeated prisoner's dilemma game is affected by factors such as the number of players, range of choices, variation in payoffs, approaching ends to interactions between individuals, and population variables such as demographic composition and population density. These analyses make the situations considered more realistic but do not weaken the initial conclusion that reciprocation is a valid possible explanation for apparently altruistic behavior.

EXAMPLES OF ALTRUISM AND COOPERATION

Having armed ourselves with an arsenal containing six possible explanations for altruistic and cooperative behaviors, we now look at real examples from studies of selected species to see which explanations seem to apply in particular cases. Although a number of species are mentioned in passing, we mainly discuss five examples: (1) Belding's ground squirrels, (2) Florida scrub jays, (3) lions, (4) vampire bats, and (5) eusocial insects, especially bees, wasps, and ants. We begin by returning to the species we started the chapter with, Belding's ground squirrels.

Belding's Ground Squirrels

At the beginning of the chapter, the mountain setting of Belding's ground squirrel social behavior was described, as well as the two kinds of alarm calls the squirrels produce. To understand the results of Sherman's studies of their behavior (Sherman 1977, 1980, 1981, 1985), we need to also consider the following information.

Belding's ground squirrels are active in their subalpine meadows from May to September, hibernating the rest of the year. Although neighboring squirrels interact frequently, their burrows are not grouped into well-defined aggregations. Females are reproductively mature at age 1 and rear only one litter of four to six young per season; females live 4 to 6 years. Males are not reproductively mature until they are age 2 and usually live only 2 to 3 years. Females tend to remain in the general area where they were born, although some females do take up residence farther away. Males move away permanently from their birth area during their first winter and usually end up living about ten times as far from the home burrow and relatives as females do. Males do not establish permanent areas or relationships with females; they stay in an area through the next annual breeding season, then leave that area to spend the next winter and subsequent breeding season elsewhere. As a result, not only are males not usually near their parents or siblings, they are also usually nowhere near any offspring either. During the short breeding season, individual females are receptive for only 4 to 6 hours, during which time they run among males and incite competition among them. Females frequently mate with several males (and subsequently produce litters containing a mixture of full and half siblings); old, heavy, dominant males achieve the greatest mating success.

Cooperative behavior in Belding's takes several forms. As we have seen, whistle calls to aerial predators do not seem to be altruistic, but trill calls to terrestrial

Figure 14-9 Expected and observed frequencies of alarm calling to aerial (upper) and terrestrial (lower) predators by various age and sex classes of Belding's ground squirrels. Expected values were determined by assuming that animals called randomly, that is, in proportion to the frequency with which they were present in censuses of all above-ground animals taken just before the appearances of predators. The number of individuals that gave calls (n) is shown along with the statistical results comparing expected and observed calling frequencies.
From Sherman, P.W. 1985. Behavioral Ecology and Sociobiology 17:313-323.

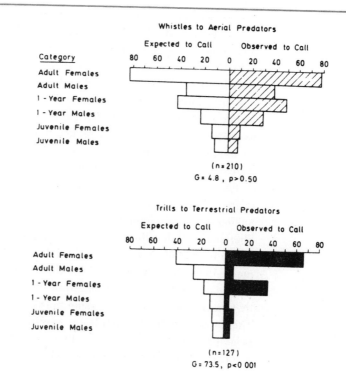

Whistles to Aerial Predators

Expected to Call   Observed to Call

Category
Adult Females
Adult Males
1 - Year Females
1 - Year Males
Juvenile Females
Juvenile Males

(n = 210)
G = 4.8 , p > 0.50

Trills to Terrestrial Predators

Expected to Call   Observed to Call

Adult Females
Adult Males
1 - Year Females
1 - Year Males
Juvenile Females
Juvenile Males

(n = 127)
G = 73.5, p < 0.001

Figure 14-10 The effects of kinship and residency on the frequency of alarm calling to aerial (*left*) and terrestrial (*right*) predators by female Belding's ground squirrels. The total number of times that calls were given (n) by females in each paired category is shown, along with the results of G-tests comparing the distribution of calling in each category with that expected if calls were given randomly. Reproductive females were either pregnant, lactating, or living with weaned young. From Sherman, P.W. 1985. Behavioral Ecology and Sociobiology 17:313-323.

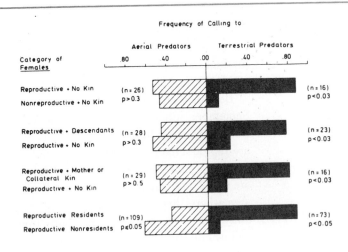

predators are dangerous and do seem to be altruistic. To gain insight into the reason for the production of trills, Sherman analyzed the likelihood of their production by individual squirrels of different age-sex classes. As Figure 14-9 shows, males were infrequent trill callers; also underrepresented were juveniles of both sexes. On the other hand, greatly overrepresented as altruistic trill callers were yearling and older females. Sherman (1985) was able to break down the yearling-and-older-females category into finer divisions, as shown by Figure 14-10. This shows that alarm calls are given by reproductive females in an area of prior extended residence, but not by females that are nonreproductive and have no living relatives or by females that have just moved into a new area where none of their relatives are present. These results agree perfectly with a kin-selected nepotism explanation for alarm calling. Those individuals that do not overlap spatially or temporally with relatives (males, nonreproductive and nonresident females) do not call, whereas ones with relatives to be helped do call. And none of these patterns holds for the individually beneficial whistle calls (see Figures 14-9 and 14-10).

Alarm calling is not the only manifestation of nepotism in Belding's. Reproductive females defend their territories ferociously against conspecifics (for a good reason, as we will see later). Closely related females (mother-daughter pairs, siblings) fight each other less often, permit each other into defended areas more often, and codefend areas and cooperate in chasing intruders more often than is the case with distantly related individuals (Figure 14-11). Note particularly that there are limits to ground squirrel nepotism (Sherman 1980). Aunts and nieces, cousins, and grandmothers and granddaughters do not treat each other differently than nonrelatives. Sherman explains this in terms of ground squirrel demography. Given the usual length of ground squirrels' lives and the usual spatial distribution of ground squirrel populations, it is unusual for female Belding's ground squirrels ever to encounter certain classes of relatives. The squirrels have not evolved nepotism toward relatives they never encounter. (The same holds true for males, which show no nepotism.) This is an excellent example for making the point that ecology (in this case demography) is as important as coefficients of relatedness in determining animal social behavior. For example, aunts and nieces (r = 0.25) and grandmothers and grand-

Figure 14-11 A, Female-female fights during the gestation period, categorized by genetic relationship between combatants. Fights were over nest burrows and defense perimeters around them. An encounter occurred when females came within 0.5 m of each other. Between-individual means (*bars*) and standard deviations (*thin lines*) are given, along with days (*d*) and hours (*h*) of observation. Note that close relatives fought less frequently than did distant kin or nonkin. **B,** Chases of variously related trespassers by resident females during the lactation period. Note that close relatives were chased less often than distant relatives or nonrelatives.

From Sherman P.W. 1980. The limits of ground squirrel nepotism. In: G.W. Barlow and J. Silverberg, editors. Sociobiology: beyond nature/nurture? Westview Press, Boulder, Colo.

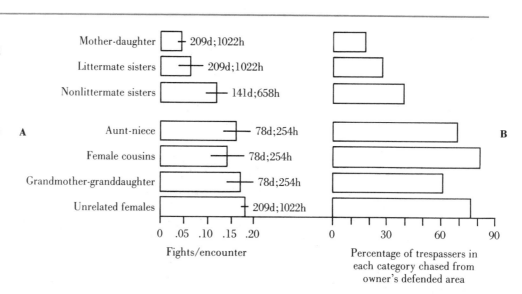

daughters (r = 0.25) are not nepotistic, whereas half sisters (r = 0.25) are; the reason is the differing likelihood of the different pairs encountering each other.

One final act of nepotism in Belding's ground squirrels is somewhat negative: closely related females, unlike other squirrels, do not eat each other's pups. Infanticide is common in this species, leading to the death of at least 8% of all pups (Sherman 1981). Nearly all cases of infanticide are committed by two types of squirrels: (1) yearling males, which on the average face a short life and have greater reproductive success if large; growing males need protein, and some of them get it by eating nonrelated infants and (2) females newly resident in an area; these females did not eat their victims and were probably clearing an area of competition for their future offspring. It is from such infanticidal individuals that territorial females are attempting to protect their offspring, leading to the chasing and fighting of nonclose relatives described previously.

Comparative information from studies of other ground squirrel species can be used to check on the conclusions drawn from Belding's ground squirrels. For example, long-term pair bonds and residency are shown by male black-tailed prairie dogs; they are near their own offspring (or the females that produce them). Predictably, Hoogland (1983) found that unlike selfish noncalling male Belding's, male prairie dogs give alarm calls.

The conclusion from this example is that kin selection is a necessary and sufficient explanation for the observed patterns of altruism and cooperation (and for observed instances of their absence).

Florida Scrub Jays and Pied Kingfishers

Florida scrub jays *(Aphelocoma coerulescens coerulescens)* are beautiful blue and gray birds that live in a specific and increasingly scarce habitat, oak scrublands along the ridge crests of central Florida. A population on the Archbold Biological Station has been studied for 20 years by G.E. Woolfenden, J.W. Fitzpatrick, and colleagues (Woolfenden and Fitzpatrick 1984). Florida scrub jays are one of 200 to 300 species of birds that are communal breeders (Brown 1987). That means that breeding birds are regularly assisted by nonbreeding helpers at the nest,* which engage in such apparently altruistic behaviors as feeding the young of the breeding pair, protecting those young, and defending the breeding pair's territory. During the past two decades, detailed studies of such communal breeders have shed much light on the evolution of apparently altruistic behaviors (J.L. Brown 1987).

In Florida scrub jays, only one pair breeds in an individual territory. Some pairs get no assistance. About half the pairs at any given time have from one to six helpers (for pairs with at least one helper, the average number of helpers is 1.78). Helpers vary in sex and age, but almost all helpers (90%) are offspring of at least one of the breeding pair. Most (64%) are helping their own mother and father. Yearling and 2-year-old helpers are equally often males and females, but older helpers (up to 7 years old) are almost all males (26 of 34 in Woolfenden and Fitzpatrick 1984). To some extent, this increasing male bias reflects the fact that males do not have as strong a tendency to disperse as do females. Older male helpers are more active in helping than are younger or female helpers (Stallcup and Woolfenden 1978).

Helpers really do help. They engage in antipredator vigilance, alarm calling, and mobbing. They also feed nestlings, in some nests providing as much as 30% of the food. The net result is that eggs in nests with helpers are 44% more likely to produce fledglings than eggs in nests without helpers (47% versus 33% probability of fledging); breeders with helpers are 10% more likely to survive from one year to the next (85% versus 77% survival); and pairs with helpers produce 51% more fledglings than pairs without helpers (2.39 per nest versus 1.58 per nest). The latter effect is more marked for first-time breeders than for experienced pairs (Table 14-2).

So far, helping at the nest by scrub jays sounds like kin-selected nepotism, but there are aspects of the situation that we have not yet mentioned, such as other options available for potential helpers. Although the breeding habitat is usually saturated with jays, an occasional breeding opportunity opens up when one or both

*Similar communal breeding is found in other groups than birds; carnivores, such as black-backed jackals (Moehlman 1979), provide one set of examples. Since the term *helpers at the nest* is established, ethologists talk about helpers at the nest in species that do not have nests.

Table 14-2 Fledging success as related to breeding experience of Florida scrub jay pairs		Pairs without Helpers			Pairs with Helpers*		
		No. pairs	Fledgl./ pair	Prop. successful	No. pairs	Fledgl./ pair	Prop. successful
	Both novices	17	1.24 ± 1.39	.47	5	2.20 ± 1.30	.80
	Mixed[b]	35	1.34 ± 1.21	.63	30	2.47 ± 1.94**	.77
	Both experienced	64	1.80 ± 1.25[c]	.73	118	2.38 ± 1.37**	.81
	TOTAL	116	1.58 ± 1.25	.66	153	2.39 ± 1.50***	.80*

From Woolfenden, G.E., and J.W. Fitzgerald. 1984. The Florida scrubjay. Princeton University Press, Princeton, N.J.
[a]Pairs with helpers significantly more successful than pairs without: *, $p < .05$; **, $p < .01$; ***, $p < .001$.
[b]No difference between novice ♂ and ♀ mates; therefore samples pooled.
[c]Experienced pairs show significantly higher fledgling production than pooled samples containing novices ($p < .05$) except in the case of pairs with helpers.

Table 14-3 A scrub jay helper would do better by setting up its own breeding territory if it could find one, instead of helping its parents at home	Option		Result
	1. Stay at home and help	Young produced by experienced parents with no help	1.80
		Young produced by pair with helpers	2.38
		Extra young due to presence of helpers	0.58
		Average number of helpers	1.78
		Genetic equivalents to a helper (r to nestlings = 0.43)	0.14
	2. Go off and rear own young	Young reared by first time breeders	1.24
		Genetic equivalents (r to own offspring r = 0.5)	0.62

After Woolfenden, G.E., and J.W. Fitzgerald. 1984. The Florida scrub jay. Princeton University Press, Princeton, N.J.

of a breeding pair dies. When this happens, a nonbreeding male or female from a nearby territory can move into the vacant spot and become a breeder. Table 14-3 shows the reproductive outcome of such first-time breeders, compared with their inclusive fitness reproductive success if they stayed and helped a relative. (Note that $r = 0.5$ for a breeder and its offspring; $r = 0.43$ on average for a helper and the nestlings it is helping.) The table shows clearly that, if a helper gets a chance to become a breeder instead, it should do so, as the genetic payoff is four times as great. Woolfenden and Fitzpatrick do find birds acting in this manner. When vacancies arise, helpers usually leave home to set up on their own. When a disease killed almost half the adults in 1979 and 1980, most young birds became breeders in the vacant territories instead of choosing to stay home and help.

For female scrub jays, moving to a recently vacant territory is virtually the only way to become a breeder. They are constantly seeking out such openings and also seem to give up helping as a fall-back option by the time they are 3 years old. Why do they stay and help that long? They are safer in their own home territory than as wanderers, and they do gain some inclusive fitness by helping. It may be, however, that a certain amount of help is the price they have to pay the breeding pair to be allowed to stay home where it is safe and to consume some of the territory's resources.

Males, however, have an additional way of becoming breeders, called *territorial budding*. Helpers assist their parents in aggression at territorial boundaries. Pairs with helpers can increase their territory's size at the expense of nonhelped pairs, and pairs with many helpers gain land at the expense of pairs with fewer helpers. When a territory becomes large enough, the oldest and most dominant male helper begins to restrict his activities to one part of the territory. Soon he begins singing to attract a mate and sets himself up in that area as a new territorial breeder. By helping, he puts himself in a position to eventually inherit a territory of his own, budded off from his parents'. And since his chance of doing so is a function of the number of other, subordinate helpers there are, there is an ultimate incentive to work hard as a helper, since it increases the number of younger brothers and sisters he has.

About half of surviving male helpers eventually inherit part or all of a parental

Figure 14-12 The acquisition of three breeding territories (*B, C, D*) by three male Florida scrub jay helpers. Shaded areas represent land inherited by two sons (*b, c*) and a grandson (*d*) of the breeding pair occupying territory A. Dashed lines denote incipient territorial boundaries. Small arrows identify regions of forthcoming territorial expansion. (The small rectangle represents the laboratory building of the Archbold Biological Station.) From Woolfenden, G.E. and J.W. Fitzpatrick. 1978. Bioscience 28:104-108.

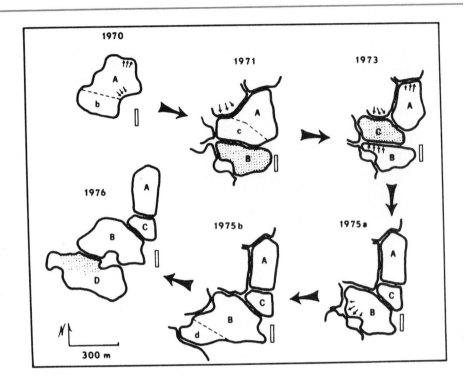

territory; this is the usual way for males to become breeders (Woolfenden and Fitzpatrick 1984). The pattern of territorial expansion and budding in one family is shown in Figure 14-12.

A detailed look at helping in Florida scrub jays leads one to question whether either male or female helpers are really being self-sacrificing; females stay and help for only a year or two, and staying increases their survival rate. Males may stay and help longer, but their help eventually pays off in terms of getting their own budded territory. Helping seems to be more a form of rent or a tactic for a long-term climb up the ladder to personal reproduction than an act of altruism. For both males and females it seems a means to an end, not an end in itself. It is true that individuals spread their genes both by their own reproduction ("direct fitness," Brown 1987) and through increased success of others ("indirect fitness," Brown 1987), but subtle selfishness seems more the driving force for the evolution of helping than kin selection.

In fact a trait will evolve because of the total effect it has on inclusive fitness; it is usually impossible if not pointless to parcel out the direct fitness versus indirect fitness components. In some species, however, it is possible to discriminate the two effects. A striking example is found in another communally breeding bird, the pied kingfisher of East Africa.

Pied kingfishers *(Ceryle rudis)* have been studied by H. Reyer at Lake Victoria and Lake Naivasha in Kenya (Reyer 1984). Conditions are different for these plunge-diving, fish-eating birds at those two lakes, and their behavior differs accordingly. At Lake Victoria breeding pairs defend an area around their nest sites, but there are common feeding grounds for the entire population. Two kinds of helpers are sometimes present to assist breeding pairs. Primary helpers are with the breeding pair from the beginning of the breeding season, whereas secondary helpers, although present in the area sooner, are tolerated by breeders only about 3 to 7 days after hatching of the pair's eggs. Pairs have only one primary helper, but there can be up to four secondary helpers (sometimes in addition to a primary helper). Primary helpers are more helpful; they spend more time guarding the nest against predators and deliver better fish at higher rates to nestlings than do secondary helpers. Primary helpers turn out to be the 1- to 3-year-old sons of at least one member of the breeding pair. They are helping to raise full and half siblings (average helper-nestling r = 0.32). Secondary helpers are males that are not close relatives of the breeding pair (average helper-nestling r = 0.05; it is above zero because there is a certain amount of within-colony inbreeding in these populations, making average individuals somewhat related).

The presence of helpers increases the number of fledglings produced. Pairs without helpers rear 1.8 fledglings on average. Pairs with a single primary helper only rear 3.6; pairs with secondary helpers only (1.45 of them on average) rear 3.7; pairs with both types (1 primary and 1.2 secondary on average) rear 4.6. Looking at the primary only and secondary only cases, a primary helper means 1.8 extra young on average, whereas each secondary helper means 1.3 extra young.

The presence of two or more helpers also increases the survival of the breeding female; with this much help she begs fish from the secondary helpers, which feed her in addition to the nestlings. Male breeders, however, are not fed by helpers; a small increase in survival rate when breeders are present is due to male breeders not expending as much energy and being exposed to less danger feeding the nestlings when helpers are also doing so.

What are the consequences of helping for the helpers? For primary helpers there is an inclusive fitness gain. This comes about for two reasons, the first being the extra chicks reared and the second being the increased survival rates of his parents because of his help. This comes, however, at two costs. The first is decreased survivorship. Nonfeeding birds have a 70% annual survival rate, secondary helpers a 73% rate; but primary helpers have only a 47% survival rate, resulting from their extra effort. Second, only 33% of primary helpers gain mates in the next year.

Secondary helpers, on the other hand, gain little in inclusive fitness from the increased success of the pair they help (since r = .05 to the nestlings). Their help does not, however, lower their survival rate and 89% become breeders themselves the next year. Commonly this happens when the male of the pair dies, in which case a secondary helper is positioned to become the female's new mate. (This rarely can happen for primary helpers, since the widow is often their mother; in only two cases where the widow was a stepmother did primary helpers mate with her.) What about the options of not helping at all? There are two of these. One, being a breeder, turns out to be the best option of all, as was the case for Florida scrub jays. Delaying, by neither breeding nor helping, turns out to be the worst option of all. A delayer does not breed, gain indirect reproductive success as a primary helper, or put itself in a good long-term situation as a secondary helper.

Ryer's (1984) conclusions are as follows: (1) The best option for a young bird is breeding, followed by primary helping, secondary helping, and, last, delaying. (2) The primary helping strategy is better than the secondary helping strategy if a young bird has parents surviving, because of the inclusive fitness gains from their increased survival and reproduction resulting from his help. (3) Primary helping is a kin-selected behavior; secondary helping is a behavior evolving as a form of individually selected subtle selfishness.

Lions

The social behavior of lions *(Panthera leo)* is remarkable in many respects. Lions are the only highly social cat species. We have already mentioned the infanticidal intrasexual competition among males. The mating system is a polygynandrous one. Most significant for this chapter, lions are among the most cooperative of social vertebrates.

A typical lion group, as studied since 1966 in the Serengeti Park and Ngorongoro Crater in Tanzania (Packer et al 1988), consists of one to eighteen adult females or lionesses (average number = 6), their cubs and subadult (1½- to 4-year-old) offspring, and one to seven adult male lions (most commonly two or three, Figure 14-13). A lion group really consists of two separate groups, the group of lionesses and a group of males.

The group of lionesses is known as a *pride*. Without exception these lionesses are relatives: mothers and daughters, grandmothers and granddaughters, sisters, aunts and nieces, cousins (Table 14-4; Packer et al 1988). Most lionesses for their entire lives remain in the pride in which they were born; even the few females that leave the pride set up ranges in areas near or adjacent to the birth pride's area, where they may form a new pride (if several leave together or if one leaves and rears daughters to adulthood). Females have never been seen to associate with unrelated lionesses (Hanby and Bygott 1987).

The result of this philopatry (attachment to the familiar area) is that the pride consists of a network of females whose average coefficient of relatedness to other

Figure 14-13   **A typical lion pride in the Serengeti Park, Tanzania. The group is composed of two brothers, five lioness relatives, two 3-year-old males, an 18-month-old juvenile, and two 5-month-old cubs.**
Drawing by Sarah Landry. From Wilson, E.O. 1975. Sociobiology: the new synthesis. Belknap Press, Cambridge, Mass.

Table 14-4 Kinship in male coalitions and female prides	Always Composed of Close Relatives	Ever Include Nonrelatives
Male coalitions	8	9
Female prides	19	0

From Packer, C. 1986. The ecology of sociality in lions. In: D.I. Rubenstein, and R.W. Wrangham, editors. Ecological aspects of social evolution. Princeton University Press, Princeton, N.J.

pride members is high (average r = 0.15, somewhat higher than for first cousins; Bertram 1976). Prides occupy stable and well-defined areas that are defended against other prides. Some prides retain an identity even as successive generations of lionesses live and die.

Cooperation among lionesses of a pride is extensive. No dominance hierarchy exists among the lionesses. They hunt together, defend their area together, and tend to enter estrus and give birth synchronously (Bertram 1976, Packer and Pusey 1983). When synchronous births occur, the cubs are pooled and reared, defended, and even suckled communally by all the mothers. Since all of these behaviors involve close relatives, an explanation of kin-selected nepotism seems obviously correct.

The group of male lions associating with a pride at a given time is called a *coalition* (Bygott et al 1979). Coalitions are formed when groups of similar-age subadult males leave a birth pride together. Coalitions therefore, are often composed of full or half brothers, cousins, etc. Bertram (1976) calculated the average relatedness between males in a coalition as r = 0.22, nearly the level of half brothers.

Males in coalitions cooperate almost as extensively as females in a pride. After emigrating from the birth pride, they hunt and scavenge together for several years as they reach their full, vigorous maturity (and grow the typical bushy mane). Then they cooperate in an attempt to oust other coalitions that are associating with prides. If they succeed, they defend their females collectively, and all are able to mate with the females. No clear dominance hierarchy exists among the males of a coalition (Packer et al 1988). When a female comes into estrus, males race to get to her first, but the losers respect the consortship achieved by the winner; fights are rare, occurring only when two males get there first at the same time. Since estrus lasts for an average of 4 days and several females are often in estrus at the same time, all the coalition males mate frequently.

So far, cooperation among coalition males again sounds like kin-selected nepotism. And indeed, that is how it was originally, exclusively presented (Bertram 1976; Bygott et al 1979). But the story turns out to be somewhat more complicated. Packer and Pusey (1982) pointed out that some coalitions include unrelated males; the most recent data (Table 14-4) show that about half of coalitions contain at least one nonrelated male. (If one asks what percentage of males in coalitions are related, it is higher than Table 14-4 suggests since, for example, a trio might consist of two brothers and a nonrelative. In that example, one out of one coalitions contain nonrelatives, but 67% of males are interacting with a relative. Looking at the data that way suggests that 74% of coalition members are with relatives, overall; Bertram 1983.)

All the behavior shown by coalition males to one another is exactly the same to

Table 14-5 Reproductive success of male coalitions in lions

Coalition Size	Proportion That Stay Together	Proportion That Gain Residence	Average Observed RS per Resident Male (n)	Expected RS Per Male
1	—	1/16	6.500 (2)	0.406
2	all	6/14	2.136 (11)	0.915
3	all	6/12	4.249 (12)	2.125
4	all	3/4	3.625 (2)	2.719
5-7	3/8	3/3	9.206 (4)	3.452

From Packer, C., L. Herbst, A.E. Pusey, J.D. Bygott, J.P. Hanby, S.J. Cairns, and M.B. Mulder. 1988. Reproductive success in lions. In: T.H. Clutton-Brock, editor. Reproductive success. University of Chicago Press, Chicago.

relatives and nonrelatives. So, in addition to nepotism, cooperation and reciprocation seem to be shown by coalition males. Why are nonrelatives admitted into coalitions?

The answer seems to be that success in ousting another coalition and gaining a pride of females—which is the only way to achieve reproduction for male lions— is a function of coalition size. Large coalitions may even be able to control two prides simultaneously. As shown by Table 14-5, single males virtually never reproduce, whereas many duos and trios do, and virtually all larger coalitions do. So, from the point of view of a subadult that was the only male to leave his birth pride in a given year, the alternative is to join another or others or fail to reproduce at all. If a sufficiently large coalition of relatives (three or more) exists, nonrelatives are not accepted. A pair of relatives or a singleton, however, will join forces with a nonrelative.

Not fully understood is why nonrelatives are found in many small coalitions (two or three), but are never found in large coalitions (four or more). Packer et al (1988) presented two possibilities. One is based on the fact that the building of coalitions among nonrelatives takes time. Gains from larger coalitions may not offset the penalty of delay in getting a pride. A more interesting possibility involves inclusive fitness. A nonrelative allied with two brothers gets one third of the matings, but a nonrelative allied with three brothers gets only one fourth: a point of diminishing returns may be reached. But a relative associating with three others gets not only direct fitness from his own fourth of the matings but also indirect fitness from the three fourths of the matings his relatives get because of his help.

So lionesses cooperate, and males cooperate. In an established group, both are nepotistic toward the cubs (the males more so than the females; males let cubs feed at kills whereas females will not do so until they are full). The lionesses and males are not related, so conflicts of interest arise between those two groups. Males do little hunting; they get their food by waiting for females to kill or find a carcass, then driving the females away until the males are full. More seriously, newly arrived coalitions kill all the nonweaned cubs sired by the previous coalition, as mentioned in Chapter 11. Although genetically beneficial to the new males, this is obviously a disaster for the pride females. Infanticide provides an important additional reason for sociality and cooperation in both sexes. Larger coalitions retain prides longer, so that their cubs grow to subadulthood and are not killed by infanticidal replacement males; prides with up to ten females seem to be able to defend cubs better and suffer lower rates of cub loss from infanticide after takeovers. (Larger prides suffer higher rates of infanticide, perhaps because the potential prize of so many lionesses attracts large numbers of coalitions bent on taking them over; Packer et al 1988.)

So the kings and queens of beasts exemplify the complexity of life in social groups—killing coexists with selfless assistance of relatives and allies. The cooperation shown reflects both nepotism and reciprocation. A mix of nepotism and reciprocation also exists in the next example.

Vampire Bats

Vampire bat *(Desmondus rotundus)* females aggregate in stable groups in day roosts in hollow trees. (They do their famous blood-sucking feeding at night.) As with lions, young males emigrate from their birth group and females join their maternal group. As a result, the groups (composed of 8 to 12 adults with any dependent offspring) consist mainly of relatives. However, unrelated females also join groups, at an average of one female every other year. Gerald Wilkinson, who studied vampires in Costa Rica, found the average r among females in a group was 0.11 (Wilkinson 1984).

Somewhat surprisingly, female vampires are quite altruistic. Their form of altruism is perhaps even more surprising—they donate blood to each other. More accurately, they regurgitate to each other blood obtained from their blood meals. The bats that are fed are bats that have been unable to obtain a meal on previous nights. These unsuccessful bats are most often the young ones. Wilkinson (1984) found that females less than 2 years old failed to feed 33% of the time, older females only 7% of the time. But even females as young as 8 months old would regurgitate to others after a successful night.

Wilkinson analyzed his data to see if bats were more likely to feed their relatives (as predicted if nepotism is involved) or more likely to feed other bats with which they associated more often (as predicted if reciprocation is involved). It turned out that relatedness and association were highly and equally correlated with blood regurgitation. Of course, these factors are confounded in the field; a female's associates are usually her relatives. To tease out the two effects, Wilkinson performed an experiment.

He created a study group including three females from one population and three females, an infant, and a male from another. Except for one grandmother-grandoffspring pair, none of the bats were related, so any effects must be due to reciprocation. Wilkinson removed one bat each night while the others were fed for 2 hours. He then reintroduced the removed bat and observed regurgitations.

Two trends were important. First, 12 of 13 feedings were to individuals from the same, not the different, original population. This indicated that an element of individual history between two individuals is important, suggesting reciprocation. Supporting that suggestion, Wilkinson found that bats that had previously been recipients subsequently were more active blood donors.

The second major trend was that the bats that were fed gained more from being fed than the donors lost by feeding them. As Figure 14-14 shows, a female that has just fed is over 50 hours from starving if she got no more food. Donating blood equivalent to 5% of her body weight to a needy associate does not put her much closer to starving ("time lost" in the figure). For a bat that missed feeding one day, however, starvation is approaching soon, and a 5% gain in weight from a donated meal can push her many hours back from the brink ("time gained" in the figure). In fact, Wilkinson estimated that average donors were 40 hours from starvation,

Figure 14-14 Blood sharing in vampire bats. The curve indicates how starvation approaches as time passes and weight is lost since a feeding. Note that a recently fed donor bat is not harmed (time lost) as much as a near-starvation recipient is helped (time gained) by regurgitation of a given amount of blood. From Wilkinson, G.S. 1990. Scientific American 262(2): 76-82.

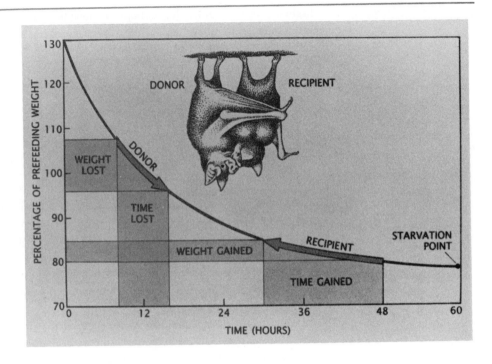

average recipients only 13 hours away—they might not have made it until the next night. So reciprocation pays—you may need help one night, and you are more likely to get it from someone you saved from starvation on a previous night; and, after all, it does not cost too much to earn that undying gratitude.

We now turn to the champion animal altruists, the eusocial insects.

Eusocial Insects

Only a saint comes close to an average ant, termite, or social bee or wasp in cooperative and altruistic behavior. The eusocial or highly social insects, including all the ants and termites and many species of wasps and bees, possess all three of a set of defining traits:

1. Overlap of generations—at least one generation of adult offspring remains in the nest with the mother or both parents.
2. Cooperative brood care—young are cared for (fed, defended, sheltered, etc.) by individuals other than the parents.
3. Existence of nonreproductive castes—unlike the first two traits, which are present in (for example) vertebrate species with helpers, existence of such sterile individuals is a virtually unique and defining characteristic of the eusocial insects.

The eusocial insects manifest cooperation and altruism in an incredible number of ways. For example, they cooperate to build elaborate nests that allow them to maintain constant conditions of temperature and humidity. Group foraging allows sharing of effort and information and the consequent efficient harvesting of widely scattered food resources (Figure 14-15). The communication involved in social

Figure 14-15   A living bridge formed of the bodies and legs of the weaver ant. From Hölldobler, B., and E.O. Wilson. 1977. Science 195:900-902.

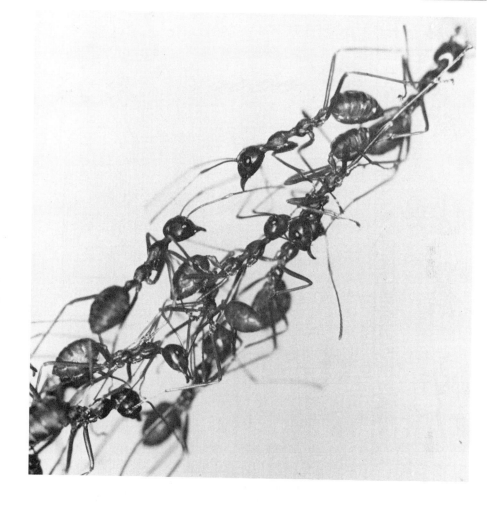

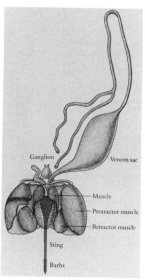

Figure 14-16   Altruistic suicide in honeybees. When a honeybee stings, the barbs of the sting become lodged in the victim's flesh; as a result, the entire sting apparatus (the apparatus includes muscles, odors that attract other guard bees, and the venom sac) tears out, fatally wounding the bee. Even after the attacking bee has departed, the muscles continue to work, driving the barbs farther in and pumping venom into the wound. From Gould, J.L., and C. Gould. 1989. The honey bee. W.H. Freeman, San Francisco.

insect foraging is among the most complex known, as discussed in Chapter 15. The nest or hive is aggressively, cooperatively, and sometimes even suicidally defended. When a honeybee worker stings, for example, she leaves sting, poison gland, and a set of still-contracting muscles behind as she pulls away. Thus, at the cost of her life, she continues to pump poison into the body of a threat to her nest (Figure 14-16).

The most amazing phenomena in the social insects are the evolution of behavioral and morphological castes specialized for various functions and the evolution of sterility. We have already seen how the evolution of sterility in eusocial insect workers posed a challenge to Darwin that ultimately led to Hamilton's development of the inclusive fitness and kin selection concepts. The evolution of a division of labor in social insect colonies is almost as difficult to envision.

In some species, such as the honeybee, the workers are not different morphologically (polymorphic), but there are behavioral specializations (polyethism). As Figure 14-17 shows, young honeybee workers spend their first few days cleaning cells in the hive; later they build honeycomb, tend the hive's brood, and put wax caps on cells where the grubs are ready to pupate. Then comes a period of time

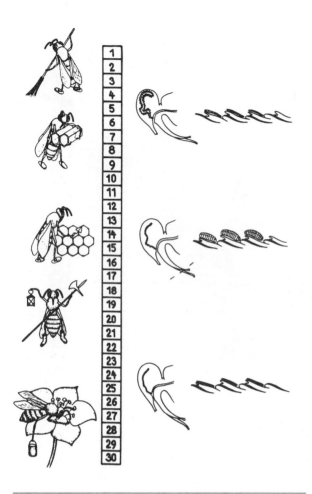

Figure 14-17 The schedule of a worker bee, arranged on the basis of her age in days. Her activity at a given time is correlated with the development of the nurse glands (in the head) and the wax glands (ventral in the abdomen).
From Lindauer, M. 1961. Communication among social bees. Atheneum, New York.

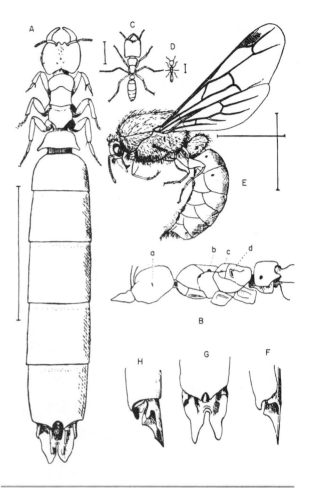

Figure 14-18   The castes of the driver ant *Dorylus helveolus* drawn at the same magnification: **A,** Queen from above. **B,** Queen from the side; vestige of eye (*a*); vestiges of wings (*b, c*); propodeal spiracle (*d*). **C,** Major worker. **D,** Minor worker. **E,** Male.
From Wilson, E.O. 1971. The insect societies. Belknap Press, Cambridge, Mass. After Emery, C. 1895. Zoologische Jahrbucher 8:685-778.

acting as guards near the hive entrance, followed at the end of their lives with the most demanding and dangerous job, foraging. Patrolling around the hive is done throughout life. (Although often busy, workers in fact spend the largest amount of their time resting.)

In the ants and termites, behavioral division of labor is supplemented or replaced by the evolution of morphologically different castes: majors or soldiers for defense, media or foragers for food gathering, minors or nurses for brood care (Figure 14-18). In some species, specialized castes have even developed as living doors or food-storage vats. An ant or termite colony may have up to millions of workers of four or five castes all working together to assist the queen (in bees, wasps, and ants) or queen and king (termites) in their reproduction (Figure 14-19). Social insect queens may develop into bloated egg factories cranking out millions of eggs, one every few seconds, over lifetimes lasting up to several years.

Figure 14-19 The interior of a typical nest of the higher termite *Amitermes hastatus* of South Africa. The primary queen and primary male sit side by side in the middle cell. To the lower left can be seen a secondary queen. In the chamber at the top are reproductive nymphs, characterized by their partially developed wings. Workers attend the queens and are especially attracted by their heads, to which they offer regurgitated food at frequent intervals. Other workers care for the numerous eggs. A soldier and presoldier (nymphal soldier stage) are seen in the lower right chamber, and worker larvae in various stages of development are found scattered through most of the chambers. Drawing by Sarah Landry. From Wilson, 1971 a). Wilson, E.O. 1975. Sociobiology: the new synthesis. Belknap Press, Cambridge, Mass.

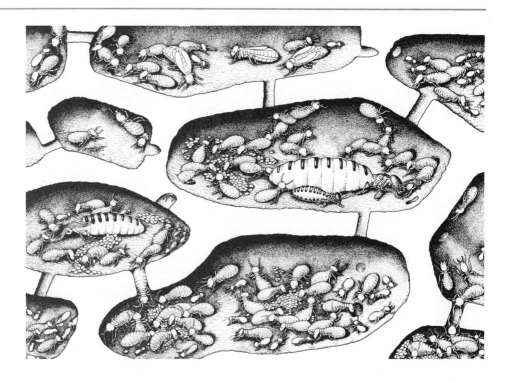

Nothing remotely like this degree of cooperation and social organization has evolved in vertebrates or the other invertebrates, leading to a focus on why it evolved in the eusocial insects.

If one looks at the taxonomic distribution of eusociality in the insects, an odd fact emerges. Although there are over 30 insect orders, eusocial species exist in only two, the Isoptera (termites) and Hymenoptera (bees, wasps, and ants). Furthermore, a detailed examination of which groups in those orders are eusocial leads to another striking result. Apparently eusociality evolved from a less social state only once in the cockroachlike ancestors of the termites, but has evolved 11 different times in various groups of Hymenoptera (Wilson 1975; that figure has been questioned, however, by Evans 1977). Something unusual seems to prevail in the hymenopterans that predisposes them to become eusocial.

In his 1964 paper on kin selection and inclusive fitness, Hamilton suggested that the "something" was an especially high tendency for kin-selected altruism because of the unusual sex-determination system of hymenopterans, known as *haplodiploidy*. Under haplodiploidy, males develop from unfertilized eggs and therefore have only one chromosome set; they are haploid. Females develop from fertilized eggs; they are diploid. Haplodiploidy may have originally been favored in Hymenoptera because early hymenopterans (and the vast majority of species even today) are small parasitic wasps that lay eggs in other insects. Parasites often exist in low densities and have unusual difficulty finding mates. Being haplodiploid allows a female wasp that has located a host to go ahead and make use of it (laying unfertilized eggs to make males) even if she has not found a mate yet. (It also allows her to control the sex ratio of her offspring if she has mated; see Charnov 1982.)

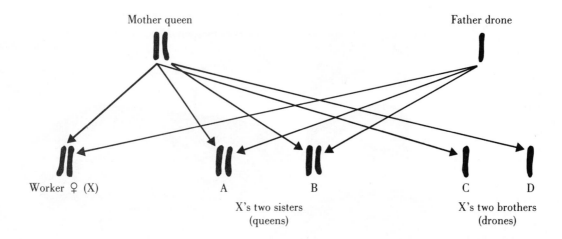

I. Relatedness of X to A: (½ of genes from mother × probability of 1 of sharing allele)
        + (½ of genes from father × probability of 1 of sharing allele) = 1.0

Relatedness of X to B: (½ from mother × probability of 0) + (½ from father × probability of 1.0) = 0.5

Average relatedness to sisters: (1.0 + 0.5)/2 = 0.75

II. Relatedness of X to C: (½ of genes from mother × probability of 1 of sharing allele)
        + (½ of genes from father × probability of 0 of sharing allele) = 0.5

Relatedness of X to D: (½ of genes from mother × probability of 0)
        + (½ from father × probability of 0) = 0.0

Average relatedness to brothers: (0.5 = 0.0)/2 = 0.25

Figure 14-20 Diagram showing degrees of relatedness with haplodiploidy. The chromosome bars represent different alleles of a gene for altruism. A worker female of a specific genotype is compared with various relatives to determine her chance of sharing an allele.

Haplodiploidy sets up unusual degrees of relatedness, as shown in Figure 14-20 and Table 14-6. Males have no sons (they do have grandsons); each daughter of a male gets all of his genes, and all sisters with the same father get an identical one half of their genome from that common father. Since they also share half of the genes that they receive from their common diploid mother, the total relatedness of females to their sisters is 0.75, not the 0.5 that is seen in diploid-diploid species such as humans. On the other hand, the relatedness of females to their brothers is only 0.25—less than the 0.5 of diploid species. This is, of course, because any genes obtained by females from the father cannot be present in the fatherless brothers.

Why might this affect tendencies to be altruistic? Compare a female's relatedness to her offspring (0.5) with her relatedness to her sisters (0.75). From the selfish-gene perspective, a sister is a greater carrier of a female hymenopteran's genes than an offspring. A gene for altruism toward sisters will be propagated faster than

Individual	Mother	Father	Relatedness (r) of that individual to its:					
			Daughter	Son	Sister	Brother	Niece	Nephew
Female	0.5	0.5	0.5	0.5	0.75	0.25	0.38	0.38
Male	1.0	—	1.0	—	0.5	0.5	0.25	0.25

Table 14-6 Degrees of relatedness under haplodiploidy (with outbreeding and single mating by females)

a gene for altruism toward offspring. This will be true if the female's help increases the number of new queens her mother makes by at least two thirds the number of fertile offspring the worker could have produced herself.

Because sisters are the best carriers of their genes and only their mother can produce sister queens, haplodiploidy may predispose female hymenopterans to stay home and help their mother rather than go off and try to reproduce on their own.

Since 1964, there has been a lively debate among entomologists and sociobiologists regarding the significance of haplodiploidy in the evolution of eusociality in insects (see Krebs and May 1976 and Evans 1977 for contrasting views). The debate has not been about whether haplodiploidy is important, but rather about whether it is the sole or major factor involved.

Points in favor of haplodiploidy's significance

Preponderance of hymenopteran eusociality   As already pointed out, elaborate societies evolved at least 11 different times in the order Hymenoptera, but only once in all other insects (in the cockroachlike common ancestor of the modern termites). The strong correlation between eusociality and hymenopterans implicates the unusual sex determination mechanism of hymenopterans as the causal factor.

Selfishness of male hymenopterans   As a glance at Table 14-6 shows, any biasing effect of haplodiploidy on social behavior should have different effects on females and males. Females are more closely related to sisters than to offspring (0.75 versus 0.50), but males are more closely related to offspring than to siblings (they have no sons, but are related by r = 1.0 to daughters and r = 0.5 to both brothers and sisters). One would predict, then, that the effects of haplodiploidy would predispose female hymenopterans to be altruistic helpers but males to be selfish individual reproducers. The behavior of male ants, wasps, and bees (drones) bears this out. They contribute little or nothing to the labor of the colony but do beg food from the workers, which tend them and tolerate them at least until the mating season is over, at which time they may be driven out by the workers (Wilson 1971). In fact, drone in everyday usage connotes a lazy, parasitic individual. As would be expected, however, drones are highly energetic and aggressive pursuers of mating opportunities (see review of drone behavior in bees by Currie 1987). In the diploid termites, both males and females are altruistic workers.

Sex ratio conflict in Hymenoptera   We mention in Chapter 12 that queens and workers in Hymenoptera are in conflict over the sex ratio of new reproductives produced, with queens preferring a 1:1 male-to-female ratio but workers preferring a 1:3 ratio. In some species the workers seem to win in this conflict, in some species the queen seems to win, and in others there seems to be some intermediate outcome. We are now in a position to understand this conflict. Table 14-6 shows that females are equally related to sons and daughters ($r = 0.5$). Sex ratio theory (Fisher 1930, Charnov 1982) concludes that offspring should be produced in ratios equivalent to the relative value of each sex in propagating the parent's genes. A queen therefore should produce a 1:1 sex ratio. However, a worker is three fourths related to a sister and only one fourth related to a brother under haplodiploidy, so a worker should prefer the colony to rear a 1:3 ratio. Each brother cancels out a sister and reduces the advantage of helping instead of reproducing, from a worker's point of view. Thus, since conflict is a prediction derived from haplodiploidy's presence, the observation of conflict tends to confirm haplodiploidy's importance.

Also note from Table 14-6 that workers are one fourth related to brothers but one half related to sons. So the ideal arrangement for a worker hymenopteran is to help the queen produce new queens (the worker's sisters) but for the worker to make the males (her sons) by laying unfertilized eggs. Although somewhat rare, worker production of males is known in some species, as discussed in Chapter 12. Again, conflict over who makes the males is a prediction, sometimes confirmed, derived from the supposition that haplodiploidy is a key determinant of hymenopteran social behavior.

Sterile soldiers in aphids   Proponents of haplodiploidy argue that extremely high coefficients of relatedness lead to the evolution of altruism. Therefore any other cases where there is an association of high r's and altruism would support the analogous situation of haplodiploidy. A case of high degree of relatedness was discovered in aphids, plant-sucking insects of the order Homoptera.

In 1972, Aoki found that there were two kinds of first instar (developmental

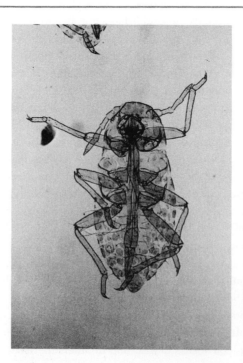 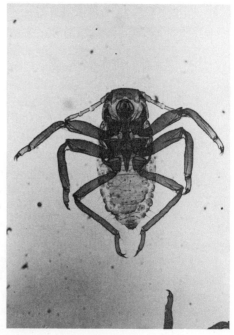

Figure 14-21 Soldier (*right*) and normal first instar larva (*left*) of the aphid *Colophina clematis*. Photograph by S. Aoki.

stage) larvae in the wooly aphid *(Colophina clematis)*. Normal larvae develop into normal second instar larvae and eventually into adults. But the second type never metamorphosed, instead eventually dying in the first instar. This type had hardened exoskeletons and modified mouthparts and vigorously attacked any insects threatening the aphid group. Aoki had discovered a sterile soldier caste in aphids (Ito 1989) (Figure 14-21).

This is relevant to haplodiploidy because these aphids are among a number of aphid species that are parthenogenetic; that is, the larvae are produced asexually by the adult female. As a result, r = 1.0 for larvae to the female and to each other. In fact, these aphids are as related to each other as your heart is to your brain. Just as we expect different organs to cooperate to maximize the whole body's reproduction, so might we expect evolution of sterile soldiers in these aphids if it increases the total number of surviving offspring produced by lowering predation rates on the soldier's sisters (note that all of these individuals are females).*

Since Aoki's original discovery, sterile soldiers have been discovered in nine other species (Ito 1989). Although all ten species are in the family Pemphigidae, they are found in three different taxonomic tribes, which suggests that soldier aphids, like worker hymenopterans, may have evolved independently more than once.

There is, then, evidence supporting the importance of haplodiploidy in eusocial insect evolution. We now turn to evidence that shows there is no inevitable connection between the two.

Evidence against haplodiploidy's exclusive significance

Nonhaplodiploid eusocial insects   Haplodiploidy cannot be required for eusocial insects, because the diploid termites are quite as advanced socially as any hymenopteran. (For a general overview of termite eusociality, see Howse 1970.) In fact the amazing parallels between termites ("white ants") and ants in caste evolution, nest construction, variety of foraging techniques, etc., constitute one of the most striking instances of convergent evolution, the evolution of detailed similarities in morphology, physiology, or behavior in unrelated species resulting from natural selection adapting them to similar environments.

There have been some attempts to find genetic analogies to haplodiploidy in the termites (Myles and Nutting 1988), but most sociobiologists have concluded that ecological rather than unusual genetic factors are most important. Two of these, predation pressure and the importance of nest construction, may also apply to hymenopterans and are discussed. One that may be uniquely important in termites relates to their dependence on gut-living symbiotic protozoans to digest the cellulose in the wood termites eat. When termites hatch from eggs, they have to be "inoculated" with protozoans by eating other termites' feces, which forces a prolonged parent-offspring association, one of the three requirements for eusociality.

Noneusocial haplodiploids   Haplodiploidy does not inevitably lead to eusociality. The majority of hymenopterans are solitary parasitic wasps, as mentioned previously.

---

*One should not be puzzled that genetically identical soldiers and normal first-instar aphids can be phenotypically so different. Remember that phenotypic outcome depends on how genetic programs are played out in developmental environments. Presumably chemical inputs from the adult female channel some aphid larvae along one potential genetic track rather than another. The situation is similar in hymenopterans and termites: reproductives are not genetically different from workers but were simply reared differently. In honeybees, for example, fertilized eggs reared in normal brood cells and given a normal diet become workers; those reared in larger cells and fed a special diet ("royal jelly") develop into new queens (Wilson 1971).

There are also noneusocial haplodiploid mites, whiteflies, scale insects, thrips, and beetles. Thus, even if haplodiploidy predisposes its possessors toward altruism, additional (perhaps ecological) factors must determine whether or not that predisposition is acted on by evolution.

Polyandry by hymenopteran queens   The coefficients of relatedness in Table 14-6 assume that a queen mated only once. What happens if a queen is polyandrous—mates with more than one male? If, for example, a queen mates with two males and their sperm mix together in the queen's sperm storage organ (spermatheca), a worker offspring of that queen would find that half her sisters are full sisters ($r = 0.75$) and half are half sisters ($r = 0.25$). Thus her average relatedness to sisters is 0.5, just as her average relatedness to any offspring of her own would be. Note that this does not take into account the devaluing effect of rearing brothers instead of sons. If the queen mates more than twice, r goes below 0.5 on average for a worker and her sisters.

Do hymenopteran queens mate more than once? The answer needs to be discovered for many species, but for some the answer is clearly yes. Honeybee queens mate on average with as many as 17 males (Laidlaw and Page 1984), and the sperm (about 6 million per male) mix fairly completely in the process in which about 5.5 million in total are stored for later use (Laidlaw and Page 1984). (Sperm mixing is an issue because if sperm from different males were stored in one-male clumps, all the females in a nest at a given instant would likely be full sisters.)

Thus, at least for honeybees, explanations other than haplodiploidy are required. These could be ecological, or they could be phylogenetic. It could be that when eusociality first evolved in honeybees' ancestors monogamy existed, but as colonies became larger queens laid more eggs and needed more sperm from multiple males. By this time, however, full worker sterility mechanisms had also evolved so workers could not opt out of helping. How general a problem polyandry is will be determined by the findings of future studies.

Cooperative nest founding   So far we have discussed the evolution of eusociality using a model in which daughters tended to stay and help their mother instead of leaving to breed themselves. But many eusocial insect colonies, especially in the small, primitively eusocial colonies of some bees and wasps, form in a different way. Same-aged females associate and cooperate with each other to construct a nest and produce a first brood of offspring, which then become daughter workers. Usually in this cooperative nest-founding process, only one of the foundresses becomes a breeding queen, with the others becoming an initial crew of helpers.

Haplodiploidy might still be involved in this form of eusociality, since a helper sister in haplodiploid species, helping to rear nieces, would be related by $r = 0.38$ to those nieces. In a diploid species a female is related to her nieces by $r = 0.25$. But why help at all if relatedness to offspring is $r = 0.5$—higher than relatedness to nieces even with monogamous haplodiploidy?

A further difficulty arises when one considers that the associating foundresses are not inevitably sisters. M.J. West Eberhard (1978 a) reports that in 15 species of wasps that had been studied intensively, group members were close relatives. In six other species, however, groups of foundresses sometimes included nonrelatives. What is going on?

An answer was provided by Lin and Michener (1972), developed further by West

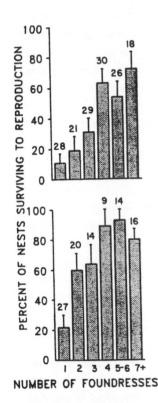

Figure 14-22   Colony survival in *Polistes annularis* for different foundress association sizes. Results from two different years are presented. Error bars denote standard errors. Sample sizes (colonies) are given over the bars.
From Queller, D.C. and J.E. Strassmann. 1988. Reproductive success and group nesting in the paper wasp, *Polistes annularis*. In: T.H. Clutton-Brock, editor. Reproductive success. University of Chicago Press, Chicago.

Eberhard (1978 b), Gamboa (1978), and Queller and Strassmann (1988). The answer involves the need to establish a nest and defend it against natural enemies before any reproduction can be achieved. Lin and Michener (1972) argued that multiple foundresses might be needed to get a nest built and to protect its food stores and larvae from predators (especially other social insect species such as ants). That nests are more likely to survive if there are multiple foundresses has been shown for several species, for example for the wasp *Polistes annularis* by Queller and Strassmann (1988) (Figure 14-22). Gamboa (1978) also pointed out that nests with multiple foundresses are less likely to be usurped by intraspecific competitors.

Putting these considerations together, West Eberhard (1978 b) proposed a *mutualistic loser* hypothesis. Foundresses, usually relatives but perhaps joined by nonrelatives (as with lion coalitions), join together to get a nest started. Each has a chance of subsequently dominating the other females, becoming the egg-laying queen and forcing them to become workers helping her rear her first brood of future worker daughters. Competition is delayed until near the time eggs can be laid, at which point dominance struggles erupt. One female wins; the others become mutualistic losers, probably too worn out to try again elsewhere so late in the season and in their lives. Of course, a consolation for some mutualistic losers is the kin selection benefit from having helped their relatives succeed in breeding, a consolation that is greater for haplodiploid species than for diploid species.

Perhaps, if a conclusion from our discussion of the eusocial insects is required, it might be like this: as with Belding's ground squirrels, Florida scrub jays, and lions, in eusocial insects genetic and ecological factors both need to be taken into account to understand the evolution of altruism. What haplodiploidy probably does is alter the cost-to-benefit ratio, with genetic factors making it somewhat more likely in the Hymenoptera for ecological circumstances to favor altruism.

A HUMAN SOCIOBIOLOGY?

In the last four chapters, we have seen how an individual or gene-selection focus has led in the last 25 years to a deeper understanding of animal reproductive and social behaviors. In 1975, E.O. Wilson (Figure 14-23) summarized the work of Hamilton, Trivers, Williams, Maynard Smith, and others in a massive synthesis with the title *Sociobiology: the New Synthesis*. In 26 chapters, Wilson laid out the theory and empirical evidence from nonhuman animals. In Chapter 27 ("Man: from sociobiology to sociology") Wilson considered the prospects for a human sociobiology. Arguing that previous efforts by anthropologists, sociologists, and psychologists to explain human behavior had partially failed because of the lack of a central organizing theory, he suggested that sociobiology was the missing theory that would bring coherence to the human social sciences.

Wilson recognized the plasticity of human behavior but argued that human behavioral diversity was constrained by certain basic patterns or predispositions that are part of our evolutionary heritage. Wilson suggested that past natural selection might be responsible for the existence in modern humans of such phenomena as reciprocation-based economic systems, differences between sexes in behavior and social roles, nepotism, widespread patterns of ethical behavior, indoctrinability, xenophobia, and tribal aggression (warfare).

Not surprisingly, reactions to Wilson's vision of a future sociology based on

Figure 14-23 Edward O.
Wilson, author of
*Sociobiology, the New
Synthesis* (1975), which
created a surge of interest in
the biology of social
behavior.
Photograph by Lilian Kemp.

sociobiology were mixed. Some social scientists and biologists welcomed it as a
source of new hypotheses to be tested. Critics of three kinds reacted vigorously in
opposition. Some philosophers, social scientists, and biologists felt that the attempt
was premature (too little empirical knowledge on which to construct theories) or
futile (since contemporary humans even including the remaining hunter-gatherer
groups no longer live in the ecological setting under which we evolved, and thus
we can never judge the adaptiveness or otherwise of different human social behaviors). The second and third set of critics reacted more vocally. Some social
scientists (e.g., Sahlins 1976) castigated Wilson for lacking awareness of the social
science literature and for a naive imperialistic attitude toward other disciplines. A
group of radical scientists and others castigated Wilson for consciously or unconsciously providing a rationalization for racism, sexism, or exploitive capitalism, for
propagating a new and dangerous form of social Darwinism (Allen et al 1976).

For several years the controversy raged in the public arena and in scholarly
publications. Wilson (1976) accused his critics of academic vigilantism, of at-

tempting to suppress scientific research for left-wing ideological reasons. More recently both sides have retreated somewhat and the debate has been more mannerly. Wilson acknowledged that some of his writing was incautiously phrased, whereas his critics have conceded that in some cases they reacted more to what they were afraid he might have been saying than to what he actually said.

We do not intend to discuss human sociobiology at length in this text, but the topic deserves some comment and further references. Attempts to study humans sociobiologically seem to have developed along two lines. Wilson (1975, 1978) and Lumsden and Wilson (1983) represent a developmental approach, looking for human behaviors that are present universally across cultures (and therefore are perhaps genetically based—would you agree with that conclusion?). They have also looked mathematically at how biological and cultural evolution might affect each other. A second approach, associated with R.D. Alexander (1979), evaluates human behavior, using a functional approach. This approach does not focus on questions of instincts versus learning but rather asks if the behaviors seen in particular cultures seem to be ones that maximize reproductive fitness in those cultures.

For books presenting the sociobiological approaches, we recommend Wilson (1978), Alexander (1979), and M. Daly and M. Wilson (1983). Some specific studies to be looked at might include Dickemann (1985), which reviews a number of studies; Thornhill and Thornhill (1983), which examines rape from a sociobiological perspective; and Daly and Wilson (1988), which presents sociobiological studies of family violence. For a critical perspective, we highly recommend the book by Kitcher (1985). Although hostile to human sociobiology, Kitcher's detailed analyses of a number of examples are rigorous yet fair, and help one to understand the level of science required for a really convincing human sociobiology to develop.

SUMMARY

Understanding the evolution of cooperative and altruistic behaviors is a central challenge for sociobiology, the darwinian study of animal social behavior. Some cases of apparent altruism are in fact simple mistakes, for example when Mexican free-tailed bats mix up their own pups with others'. Other cases represent successful exploitation of some individuals by others, as in intra- or interspecific brood parasitism. Four explanations have been proposed for more genuine acts of self-sacrifice. Group selection explanations, favored until 25 years ago, are now considered not to be evolutionarily stable strategies against individual or gene-selected alternatives. Kin selection, or nepotism, involves altruism toward relatives; it may evolve because relatives are likely (probability equal to r, the coefficient of relatedness) to possess the same alleles for nepotism as the altruist. Relatives can recognize each other by location, past association, or matching of phenotypes (a fourth possibility, of "green beard" recognition alleles, has not been conclusively supported). Reciprocation may evolve in species where association between known individuals lasts for some time. A final explanation for altruism is that short-term sacrifices may in fact bring longer-term advantages, a phenomenon sometimes called *subtle selfishness*.

These explanations can be applied to a number of specific examples. Cooperation (and its absence in some cases) in Belding's ground squirrels involves nepotism toward relatives that for demographic reasons are likely to be in close contact with each other. Helping at the nest in both Florida scrub jays and pied kingfishers involves some nepotism and a large measure of subtle selfishness, since in both species some helpers increase their chance of reproducing individually as a result of their help. In lions, nepotism explains most of the cooperation between males in coalitions and all of the cooperation between females in prides. However, cooperative reciprocation also exists in those coalitions that include nonrelatives that have banded together so as to have some chance of gaining a pride. Reciprocation coexists with nepotism in the case of vampire bats, too, when successful feeders sometimes regurgitate blood meals to less successful long-term associates that are close to starvation.

The most extreme animal altruists are the eusocial insects. Because eusociality evolved 11 times in Hymenoptera and only once elsewhere, speculation about the evolution of eusociality has focused on the effect of the unusual male-haploid, female-diploid mechanism of sex determination in that order *(haplodiploidy)*. In addition to the over representation of hymenopterans in eusocial insects, the importance of haplodiploidy is suggested by the lack of altruism by male hymenopterans, by queen-worker conflict over sex ratio and male production, and by the analogous phenomenon of sterile soldier evolution in parthenogenetic aphids. Support for factors other than or in addition to haplodiploidy comes from the fact that the eusocial termites are not haplodiploid; that many haplodiploid insect species are not eusocial; that with polyandry coefficients of relatedness are lowered and call haplodiploidy's role into question; and that among cooperative nest founders even nonrelatives may sometimes cooperate. Ecological factors such as high rates of predation on newly established nests may be as important as genetic considerations in explaining eusocial insect evolution.

Attempts to develop a sociobiology of human behavior have been and remain controversial. Although a number of hypotheses to be tested have arisen, it remains to be seen whether a rigorous, scientific human sociobiology can be developed.

FURTHER READING

Alexander, R.D. 1979. Darwinism and human affairs. University of Washington Press, Seattle.
Examination of human sociobiology from a functional viewpoint.
Axelrod, R., and W.D. Hamilton. 1981. The evolution of cooperation. Science 211:1390-1396.
Exploration of the evolution of cooperation using game theory approaches, with tactics like tit-for-tat.
Bertram, B.C.R. The social system of lions. Scientific American 232(5):54-65.
Introduction to the social behavior of the lions of the Serengeti.
Dawkins, R. 1979. Twelve misunderstandings of kin-selection. Zeitschrift für Tierpsychologie 51:184-200.
Important paper that clarifies many difficult ideas associated with kin selection and coefficients of relatedness.
Fitzgerald, J.W., and G.E. Woolfenden. 1984. The helpful shall inherit the scrub. Natural History 93(5):55-63.
Popular account of helpers in Florida scrub jays.
Grafen, A. 1982. How not to measure inclusive fitness. Nature 298:425-426.
Explains what inclusive fitness is and how most textbooks define it incorrectly.
Grafen, A. 1984. Natural selection, kin selection and group selection. In: J.R. Krebs and N.B. Davies, editors. Behavioural ecology: an evolutionary approach. Blackwell Scientific Publications, Oxford, England.
Clear account of various alternative explanations for the evolution of altruism.
Hamilton, W.D. 1964. The genetical evolution of social behavior. I and II. Journal of Theoretical Biology 7:1-15.
Landmark papers that introduced inclusive fitness and what came to be called kin selection to ethologists.

Holmes, W.G., and P.W. Sherman. 1983. Kin recognition in animals. American Scientist 71:46-55.

Good account of mechanisms of kin recognition with relevant examples, especially Belding's ground squirrels.

Kitcher, P. 1985. Vaulting ambition. MIT Press, Cambridge, Mass.

Rigorous yet fair, highly critical examination of human sociobiology. Required reading for all sociobiologists.

Krebs, J.R., and N.B. Davies 1987. An introduction to behavioural ecology, ed 2. Sinauer, Sunderland, Mass.

Approachable introduction to the topics covered in our Chapters 9 to 15; the chapter on cooperation in the social insects conveys well the complex hypotheses that have been proposed as explanations for their altruism.

Packer, C., L. Herbst, A.E. Pusey, J.D. Bygott, J.P. Hanby, S.J. Cairns, and M.B. Mulder. 1988. Reproductive success in lions. In: T.H. Clutton-Brock, editor. Reproductive success. University of Chicago Press, Chicago.

Up-to-date summary of the more than two decades long study of the Serengeti lions.

Pennisi, E. 1986. Not just another pretty face. Discover 7(3):68-78.

Popular account of research on the naked mole rat—superficially ugly but perhaps the only eusocial mammalian species.

Pusey, A., and C. Packer. 1983. Once and future kings. Natural History 92(8):54-63.

Popular account of coalitions in male lions.

Sherman, P.W. 1977. Nepotism and the evolution of alarm calls. Science 197:1246-1253.

A must-read paper for three reasons: its portrayal of the time commitment made by some field workers, its clear presentation of a kin-selected phenomenon, and most of all its example of the logic of hypothesis testing in animal behavior.

Sherman, P.W. 1980. The limits of ground squirrel nepotism. In: G.W. Barlow and J. Silverberg, editors. Sociobiology: beyond nature/nurture? Westview Press, Boulder, Colo.

Shows how kinship and demography interact in the evolution of social behavior.

Trivers, R.L. 1971. The evolution of reciprocal altruism. Quarterly Review of Biology 46:35-57.

Yet another Trivers classic.

Wilkinson, G.S. 1990. Food sharing in vampire bats. Scientific American 262(2):76-82.

Good account of blood donation by vampire bats.

Williams, G.C. 1966. Adaptation and natural selection. Princeton University Press, Princeton, N.J.

The definitive critique of group selection and original proponent of the selfish-gene approach.

Wilson, E.O. 1971. The insect societies. Belknap Press, Cambridge, Mass.

Still the authoritative work on the eusocial insects.

Wilson, E.O. 1975. Sociobiology: the new synthesis. Belknap Press, Cambridge, Mass.

The influential and controversial work that gave its name to the evolutionary study of social behavior.

Wilson, E.O. 1978. On human nature. Harvard University Press, Cambridge, Mass.

Completes Wilson's sociobiology trilogy with an explicit application of his ideas to humans.

Wilson, E.O., and B. Holldobler. 1990. The ants. Harvard University Press, Cambridge, Mass.

Called the most thorough and synthetic review of any group of organisms. A beautiful book at home either on the coffee table or on the lab bench.

Winston, M.L. 1987. The biology of the honey bee. Harvard University Press, Cambridge, Mass.

One of several recent good introductions to this most studied and fascinating of eusocial insects.

Woolfenden, G.E., and J.W. Fitzgerald. 1984. The Florida scrub jay. Princeton University Press, Princeton, N.J.

Book-length presentation of the long-term studies of the ecology and helping phenomenon in this species.

15

ANIMAL COMMUNICATION

Animal communication has been one of the most confusing, frustrating, important, and exciting topics in the field of behavior. It has been confusing because of numerous difficulties related to its definition, measurement, and theoretical development. Recent conceptual progress with the subject of communication, however, appears to be unifying the fragments of our understanding and making the topic more tractable. The subject of communication is important because it involves the means by which animals (and in some cases plants and animals) interface and interact with each other. It is the means, in essence, through which an organism with a nervous system reads and responds to another organism. This has provided an incredible range of opportunity for the operation of natural selection and as a result has produced a wide and often bewildering array of behaviors. Because no animal—even the most solitary—is totally isolated from other organisms, all animals interact with other organisms in at least some minimal way and hence the subject of communication applies to every animal alive now or in the past.

In this chapter, we describe and define communication as well as we are able (it defies a strict and universally accepted definition); we provide several examples; and we trace the theoretical development of the subject. After a period of several years that might best be described as groping for meaning, persons working with animal communication appear to be finally constructing a conceptual framework that places more meaningful and satisfying boundaries around the topic.

COMMUNICATION IN HUMANS

Humans have a vast capacity for communication through our complex spoken and written language. Is this type of communication unique to our species, or do we merely possess a highly advanced form of a facility that can be found in other, perhaps all other, species as well?

The details of human language are beyond the scope of this book, and we are not concerned here with human language per se. Similarly, we do not delve deeply into the learned versus innate aspects of language and language development, although these are considered in several other chapters. It is biologically reasonable to believe that human language evolved from simpler forms of communication and shares many of the same principles, properties, and constraints of other forms of communication. Also, it is worth noting that human communication includes many components, both verbal and nonverbal, that appear to be unlearned, understood universally among our species, and are most likely analogous and homologous to communication in other species. Examples include some patterns of intonation (such as rising pitch when curious, high pitch when submissive), blushing, smiling, sticking out the tongue, and approaching aggressively with a raised clenched fist.

Before discussing communication in other animals, in its diversity of manifestations and theoretical implications, we begin with a classic and popular case of animal communication: the repertoire of signals shown by honeybees, with particular emphasis on their dance language. Honeybee communication is fascinating in its own right. However, it also serves as an excellent example to illustrate many of the properties and principles of animal communication in general.

COMMUNICATION IN OTHER SPECIES, A CASE IN POINT: HONEYBEES

That honeybees have a dance language is known even to the general public, although most persons do not know the details. Equally famous is the man, Karl von Frisch, who spent a lifetime studying honeybee behavior. These studies were partly responsible for his receiving the Nobel Prize, along with Konrad Lorenz and Niko Tinbergen, in 1973. Von Frisch investigated the bees for many years on his own and then was joined by several others, including Martin Lindauer, Harold Esch, and Warwick E. Kerr, who have become nearly as well known themselves. The stories of the research and findings are published in several places and at different levels of detail and technicality (von Frisch 1967, Esch 1967, Lindauer 1971, and references contained therein).

In brief, foraging honeybees *(Apis mellifera)* perform a dance that symbolically codes information on the distance and direction of food from the hive. This dance recruits other workers which then are able to travel to the distant food site. There are two basic dances, the round dance, in which the dancing bee circles to the left and then to the right to indicate nearby food sources (Figure 15-1), and the waggle dance, for distant food sources. The waggle dance consists of a figure-eight looping dance with a straight run performed in the middle of the loop. The waggle, in which the abdomen is waggled rapidly back and forth, occurs during the straight run. While the forager performs this dance, other recruit bees gather alongside and behind it and follow it through the dance, keeping their antennae in contact with it.

Figure 15-1   Round and waggle dances of the honeybee.
From Frisch, K. von. 1967. The dance language and orientation of bees. Harvard University Press, Cambridge, Mass.

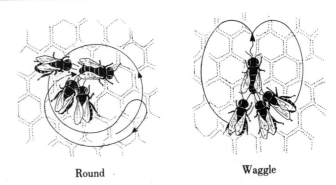

Round   Waggle

In addition to waggling the abdomen, the dancing bee emits sounds, thought to be produced by the wings, that serve to incite the following bees to forage (Wenner 1964, Esch 1967). The waggles and sounds have been studied with many ingenious techniques, including the placing of small magnets on the bees' abdomens and recording induced electromagnetic currents (Esch 1967) and by painting white spots on the abdomens and following the dance photographically with series of flashes (Wenner 1964). Sounds have been studied with the aid of small microphones and speakers, natural, modified, and artificial playbacks and by using both airborne and substrate-borne sounds. It appears that the bees are not receptive to airborne sounds but rather sense the vibrations through their feet or by direct contact with the antennae. During silent dances (i.e., lacking the sounds) other bees follow the dancer but subsequently do not leave the hive and forage. During the dance the recruits also sense the odor of the specific food source, which permits them to locate it exactly once they have used dance information to arrive at the general vicinity.

Distance to the food source from the hive is indicated in the dance by the duration of the straight run, which is correlated with the number of waggles per straight run or the characteristics of the sound that is emitted during the run. Direction is coded inside a dark hive by the angle by which the straight run departs from straight up. This angle represents the horizontal angle outside the hive that the food is from the azimuth of the sun (a vertical line drawn to the horizon, Figure 15-2). In other words, cues from gravity, a sort of negative geotaxis, on the vertical surface inside the hive are substituted for cues from the sun in a horizontal plane outside the hive.

This general pattern of dance communication is used by all bees of the genus *Apis*, but different species and races vary in the distance coding, that is, when they switch from the round to the waggle dance and how many waggles they make per straight run (Figure 15-3). In addition, there are two other variations on the theme. In one the Italian honeybee *(A. mellifera ligustica)* incorporates a different dance, the sickle dance, between the round dance and the waggle dance. In the other variation the dwarf honeybee *(A. florea)* dances on a horizontal surface and in view of the sun or open sky. Lindauer tried to force dwarf bees to dance on a vertical surface by tipping their hive and eliminating horizontal surfaces, but they would not dance on a vertical surface and apparently were unable to transpose from light to gravity.

Figure 15-2 Horizontal orientation of the waggle dance to the sun outside the hive and vertical orientation with respect to gravity inside the hive. The angle from the vertical or from the sun indicates the direction of the resource.
Modified from Frisch, K. von. 1967. The dance language and orientation of bees. Belknap/ Harvard University Press, Cambridge, Mass.

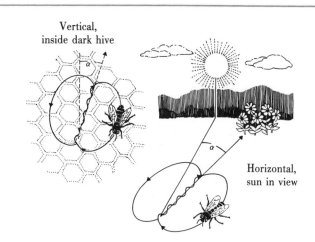

Figure 15-3 Different species and subspecies of honeybees vary in their use of the dance language, as indicated.
Modified from Frisch, K. von. 1962. Scientific American 207(2):78-87.

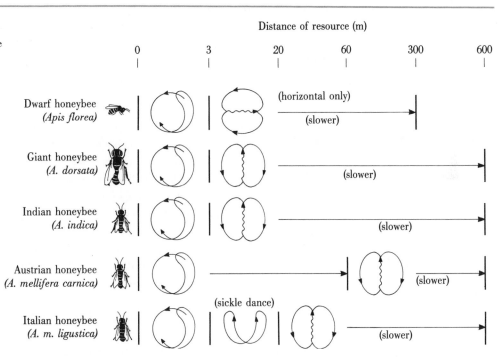

The evolution of honeybee dance language has been elucidated by comparative study of many other genera of bees, as well as ants and other flying insects. The family Apidae, to which honeybees belong, consists of three tribes: the Apinae or honeybees, Meliponinae or stingless bees, and Bombinae or bumblebees. Bombinae are not known to use a dance language at all. Among various species of stingless bees, however, there are a variety of recruitment techniques that suggest steps by which the advanced honeybee language evolved. Foragers of some species of stingless bees return to the hive and excitedly dash around, bumping into other bees, passing on bits of food, and spreading the odor. They then lead the recruits all the way back to the food source. In some species of stingless bees the returning foragers stop every few meters and leave strong scent marks, which are secreted by the mandibular gland, on rocks or vegetation. They subsequently use this scent trail in going back from the hive to the food site. This behavior is similar to the use of scent trails by ants, which are in the same order (Hymenoptera). In other species of stingless bees the foragers do not take the recruits all the way to the food source but only lead them part way, after which the recruits continue in the same direction to the food while the initial forager returns to the hive. Many of the stingless bee species use sound to indicate distance to the food source.

The initial excitement or waggle movements may have been derived from a general tendency of flying insects, including moths, to rock back and forth after landing. The amount of rocking varies with the length of time the insect has been flying. The function, if any, of rocking outside bee language is not known; it may serve some physiological function related to respiration or muscle physiology.

The role of light versus gravity has been investigated by shining bright lights or using mirrors to reflect the sun into the hive. The result is that when light is available, the honeybees use that rather than transposing to gravity. For example, imagine that a food source is directly in line with the sun. Inside the dark hive the dancers perform the straight run of the waggle dance straight up. If, however, a mirror is placed under the hive and the sunlight is directed up from below, the dancers point the straight run down, toward the light (Esch 1967). When the mirror is removed, the dancers reorient the dance up again. If a normally vertical hive is tipped so the bees have only horizontal surfaces, they will not dance in the dark, but they will dance horizontally if given a view of the sun or a bright light.

Focusing on the recruitment dance language of bees, one must not overlook the other communication capabilities of honeybees. In addition to their round and waggle dances, they have a less well-known dance, the vibration dance, and they possess a large number of pheromones (chemical messages) and also use sound in several other contexts. During the vibration dance a worker vibrates another bee by grasping it with its legs and rapidly vibrating its abdomen up and down. This dance is performed before foraging and is thought to be related to short- and long-term food availability in a manner somewhat analogous to the preforaging or prehunt activities of African wild dogs or human pep rallies (Schneider et al 1986 a,b).

Honeybees in the hive make at least 10 distinct, different sounds, some of which are merely byproducts of their activities and some of which are signals. Some of the sounds are types of pipping or beeping, which, depending on the particular sound, may either alert or soothe the bees in the hive.

During the replacement of a queen in a hive, after the original has been lost, the candidate queens emit sounds known as *tooting* and *quacking*. In the absence of a queen, several eggs or developing larvae are moved to larger queen cells and fed royal jelly, causing them to develop into new queens. The first one to emerge proceeds to go around and locate the cells of others; attempts to uncap the cells and kill her potential competitors by stinging. The workers, however, may block her from the other queens, at which time she begins tooting. The toots become quite loud and can be heard by a human as far as 2 or 3 m from the hive. Meanwhile, as the other developing queens attempt to emerge, the workers keep pushing them back and resealing them in their cells. They begin to quack. The messages apparently announce to the workers that there is one free queen and one or more others attempting to get out. The workers release the quackers one at a time to fight with the free one. Eventually only one remains. She then flies from the hive for the nuptial flight with the drones. After mating, the queen returns to the hive and begins egg production. Included in the pheromones are at least two inhibitors, known as queen substance, that are produced by the queen and that prevent any new queens from being developed as long as she is present.

During the 1960s a honeybee language controversy developed. The importance of bee and food odor to the lives and behavior of honeybees was well known; in fact, von Frisch's early work had concentrated on odor, and it was only several years later that he stumbled across the dance. There was no doubt that the dance contained information on the distance and direction of food (or other important resources such as a new hive site). A human observer can easily decipher the distance and direction (this makes a good laboratory exercise if one has access to an observation hive). What came to be at issue, however, was whether the bees themselves actually use the information contained in the dance. The question was raised by Wenner and his associates beginning in 1967 (Wenner 1971). They believed, from good evidence, that honeybees were recruiting others and that the others were able to locate the food based on cues from the odor alone. The criticisms, charges, countercharges, and debate continued for nearly a decade, until the matter finally was settled by Gould (1975).

The history of the controversy and its final resolution are interesting in their own rights, reminiscent of the bat echolocation story (Chapter 1) and the pursuit of the information-center hypothesis described in Chapter 3. In this case a critical, experimental test settled the issue (as opposed to the continuing correlational approaches on the information-center hypothesis).

It might be possible to test whether other bees use the information by providing artificial dancing bees, but attempts to recruit bees artificially with model dancing bees largely failed until recently (Michelsen et al 1989). Esch (1967), for example, constructed an animated dummy bee, driven by a motor and tape loop. The model successfully stimulated other bees to follow the dance, but they would not leave the hive to search at the place indicated by the dance. Esch concluded that there was more to it than just the waggling dance. Gould's solution (1975) was ingenious, involved, and complex. The gist of it was to use the fact that the bees prefer a light cue rather than gravity if given the choice. The light has to be at a particular threshold brightness or the bees retain their orientation to gravity. Furthermore, the bee's sensitivity to brightness is determined not by any or all of its eyes but by the three simple eyes, or ocelli, located between the compound eyes. If these are covered, bees are still capable of seeing for flying and foraging, but they are six times less sensitive to light in the hive. The experiments consisted of covering the ocelli of some bees and not others and using light of proper brightness in the hive so that some bees (with covered ocelli) oriented toward gravity and others (with ocelli functional) oriented toward the light.

Bees were provided food at various locations, the angle of the light in the hive was controlled, and possible feeding locations for recruits were provided at numerous positions in a circle around the hive. All procedures were rigorously and exhaustively controlled and monitored with automatic recording devices that excluded foreign objects such as ants from interfering and that permitted accounting for alien bees, flies, etc.

The results were clear and unambiguous: recruits went to the locations that were predicted, and they could be tricked or misled by combinations of covered ocelli and artificial lights. For example, foragers with covered ocelli would dance with respect to gravity in the presence of an artificial light. Recruits without covered ocelli interpreted the dance as oriented to the light. These recruits then traveled to the stations indicated by orientation to the light rather than to the station where the forager had actually fed. These results established without doubt that the recruits can and, at least part of the time, do use the symbolic information coded in the dance.

Why, then, were there discrepancies between the results of the two groups of researchers, von Frisch's group on one side and Wenner's on the other? Having confirmed that the bees can use the dance language, Gould experimented further, incorporating the techniques of both groups. In addition to a number of minor differences in method, there was one major difference: Wenner trained bees with concentrated sucrose solutions and the same scents that were used subsequently in experimental tests. The other workers used more dilute solutions and different scents, if any scent at all, than scents that would be used subsequently. By experimenting with combinations of conditions, Gould (1975) discovered that both groups of researchers were correct. He concluded, "Depending upon conditions, honey bee recruits use either the dance language *and* odor information, or odors alone" (emphasis Gould's). He stated further, "By their very different training technique, von Frisch and Wenner may have been sampling two stages of the same process: exploitation of an abundant food source. Von Frisch's experiments could be seen as examining the early phase, while Wenner's would be exploring a later phase."

Several lessons can be drawn from the controversy over honeybee language, lessons that have been encountered previously (e.g., with bats and echolocation). First, different techniques of research may produce different results during investigations of the same phenomenon. Second, one often (in fact, usually) does not have the full story. Third, the intricacies of behavior, such as food recruitment in social bees, may be much more complex and subtle than supposed. Bee communication also provides a basis for discussing communication by other animals.

WHAT IS COMMUNICATION?

Background

Communication (Latin, *communicare*, "to share") in the common, everyday human meaning of the term refers in its essence to a sharing of information between two or more individuals through the use of coded signals. In spoken or written human language the signals are in the form of words. *Information* means a reference, correspondence, or correlation to something else. The words *dog* and *frog*, for example, refer (correspond or correlate) to two different kinds of animals.

This basic or informational view of communication was the starting point for considering animal communication (e.g., Marler 1959, Smith 1977). Animal signals were thought to function primarily as carriers of information between cooperating senders and receivers. According to this view, natural selection should improve accuracy and reduce ambiguity of the signals (e.g., Wilson 1975).

Signals are coded in that they are arbitrary and do not inherently mean anything on their own, but they acquire meaning and are governed by rules that are understood by the participating individuals. The rules in the case of the words *dog* and *frog* are according to the English language and depend on the order of the letters. If you reverse the letter order you get different meanings, in these examples, "god" and "gorf."

Assuming that you know the rules of English (or you would not be reading this book), the meanings of the words *dog* and *frog* are immediately clear because they are so familiar. They might not seem to be arbitrary, but to anyone who does not know English, *dog* and *frog* are meaningless. To someone who knows only English, words in other languages are also meaningless, for example, *chien* and *grenoville* (French) or *aso* and *palaka* (Tagalog, Philippine).

The rules of human language and other forms of communication can be numerous, complex, and often arbitrary. Some words (such as adjectives) may modify the meaning of other words (e.g., nouns). In English, this depends on word order, such as *house dog* versus *dog house*, whereas in a different language the rules can be different (in French: "chien domestique" versus "la miche du chien").

Sometimes a given signal can have more than one meaning. Unfortunately, this has happened with the words we use to discuss the subject of communication, which has led to misunderstanding and confusion. One of the problem words has been the central term *information*, which has at least two different meanings in the context of animal communication. We will return to the special problems involving the word *information* (also see Chapter 4), but first we consider a host of other aspects of animal communication.

When ethologists first began considering animal communication, several controversies developed over just what communication is or is not in other species. Points of contention and confusion involved the extents to which signals must be intentional (witting, conscious, etc.), modulated, adaptive (and to which participants), verbal, and so on before they qualify as communication. Should communication be defined so that it is only intraspecific, or can animals from different species communicate with each other? Should communication have certain specified functions? Is communication in other animals qualitatively different from that of humans, or is the difference only a matter of degree (i.e., quantitative)? Do the participants have to be behaving (or even alive—what about the message in the appearance of a dead butterfly or a dead skunk, for example)? What qualifies as a signal or message in the first place? Whittaker and Feeney (1971) classified the continuum of chemical interactions that are possible between organisms (see box on page 523). Which of these should qualify as communication? Do the receivers have to show an outward response, and how does one detect communication if there is no outward response? Many messages, including human communication, can be received and recognized with no apparent change in outward appearance. Completely valid messages might say "leave me alone," "do not move," or "go on about your other business." There may be tremendously long time lags between the time the message is received and the time the response is noted.

As an extreme example, consider the appearances of flowers that attract pollinating insects or plants that repel insects (Figure 15-4). The plants obviously are not doing it consciously because they do not even possess a nervous system; they are not modulating the signal; and they do not belong to the same species or even the same kingdom as the receivers. Should this be considered communication?

In addition to the previously described problems, it became obvious that animal (and plant) signals possess several attributes that do not fit well if one simply considers the mechanical transfer of information from one organism to another. For

Figure 15-4 Coevolution of plant "messages" and butterflies. *Heliconius* butterfly larvae feed on the leaves of *Passiflora* vines. The larvae also are cannibalistic toward each other, which has led to female butterflies avoiding leaves with eggs already present for oviposition sites. Some *Passiflora* vine species have evolved leaves with fake egg spots, which results in butterflies avoiding these leaves. Should such "signals" of the plant be considered "communication?" Modified from Gilbert, L.E. 1982. Scientific American 247(2):110-121.

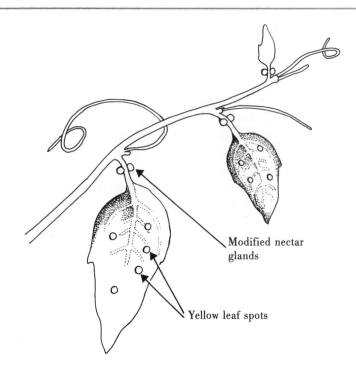

Modified nectar glands

Yellow leaf spots

example, signals are not always directed at a particular receiver or in many cases toward any receivers of which the sender may be aware. There often is much redundancy and repetition of displays and calls, as anyone knows who has tried to sleep amidst the constant racket of songbirds in the spring or singing insects in the summer. Furthermore, some messages are considered to be deceptive: organisms may withhold or exaggerate information; they may send false information, such as on the strength or intentions of the sender; or they may mimic something that they are not. Although not involving much force or energy in themselves, signals can mobilize much force or energy from the receiver. With the right signal one can move whole herds of elephants. A signal can make the difference between pushing an elephant and getting it to move itself. One can save vast amounts of energy or accomplish tasks that otherwise would be impossible by stimulating another animal to behave appropriately.

In view of these seeming noninformational aspects, Dawkins and Krebs (1978) proposed an alternate theory of animal communication, that its primary function was not to convey information but to manipulate other organisms, as occurs in advertising. However, even manipulative communication, including situations with false or deceptive information, requires signals with information and meaning. Thus Krebs and Davies subsequently (1984) modified their stance to incorporate all the attributes of animal signals.

Wiley (1983) moved the whole subject forward conceptually by further balancing the roles of information versus manipulation. He discussed these concepts and placed them in a broad context of current views of social behavior.

It was noted earlier (page 497) that the term *information* has been a special problem. *Information* in the familiar sense as discussed previously refers to the meaning or content of the signal. This is the semantic meaning. However, *information* also has another less common or familiar meaning that is statistical. The statistical meaning refers to the amount of uncertainty an outside observer has when trying to predict how one animal will respond to the actions of another animal. If the second animal's behavior becomes more predictable after certain behavior by the first animal, then information is said to have been transmitted between the two. The uncertainty is measured from a large number of observations using mathematical techniques of probability theory. (The actual techniques are described and discussed in more detail on pages 506 to 512.) It took a while for biologists to recognize the problem(s) that stemmed from these two different meanings. Some confusion still lingers.

When ethologists first considered the subject of animal communication, they naturally carried into it their familiar everyday understanding of communication and information in the semantic sense. A focus on communication per se, incidentally, developed some time later and was built on an earlier accumulation of knowledge about particular cases of animal communication, such as bee dance language and bird song. Von Frisch's early pioneering work on honeybees, for example, was concerned with the bees themselves and their behavior, not with the general topic of communication. The focus for bee language did not shift from the bees to the topic of communication until later (e.g., Haldane and Spurway 1954, Wilson 1962).

Thus the initial biological interests in and concepts of communication were based on the semantic information aspects. In an attempt to quantify information, however, the early workers borrowed statistical techniques from formal, mathematical information theory (Wiener 1948, Shannon and Weaver 1949), which also was based on human communication. This inadvertently tangled together the two meanings of *information*—semantic and statistical. In the case of human communication, on which both meanings were based, it was not really a problem. The problem arose when communication was considered for other species.

All of this was further complicated by a subtle form of anthropomorphism—an assumption that a transfer of "information" in the semantic sense is possible between nonhuman animals in the first place. The transfer of semantic, meaningful information between other animals is what we might be interested in and want to understand. We would like to know what is really going on among communicating participants in other species. However, although semantic information transfer might or might not really occur in other species, we do not know and probably never can know for certain. Perhaps there is no meaning in the human sense for other animals. Perhaps communication in other species is purely a matter of sequences of action and reaction resulting from innate evolved traits, conditioned learning, or a combination of the two. This raises the whole problem of anthropomorphism in general and knowing what goes on inside another animal's head, as discussed elsewhere in this book (Chapters 1 and 2). In other words, we as humans, whether we like it or not, are inescapably forced to be outside observers of other animals, their minds, and all that goes along with that—including possible consciousness and understanding of meaning, etc.

Therefore at best we can only observe and quantify the actions and reactions of other animals to each other and draw correlations or infer functions between those behaviors and their environmental and social contexts. If one wishes to call such correlations or functions *biological meaning*, one is free to do so. However, in view

James E. Lloyd

James E. Lloyd

A Firefly's Flash, an Experimentalist's Delight

Many of the earth's 1900 known firefly species use bright flashes of light for sexual communication. Unlike insects that proclaim their sexual status and intentions with obscure (to us) twitches of bristles, pulses of sound, or medleys of perfume (pheromones), luminescent fireflies emit signals we humans are especially sensitive to. This means that as poets or naturalists we easily tune in to the entreaties and "conversations" of fireflies. As intruders, with a penlight we can "talk" with them.

By flashing a light in imitation of the male pattern (below), one can elicit a flash response from a female of that species; and by answering a flying male's pattern with the correct flash-response, one can attract him. If (as often happens in the middle of the night) a researcher gets a bright idea or theory about insect signals, rival tactics in mate competition, or risks involved in signaling, experimentation can begin immediately, often with inexpensive equipment. I think that light-signaling fireflies can be to the study of insect communication what drosophila fruit flies are to the study of insect genetics.

When you find yourself on a summer lawn or meadow within the range of *Photinus pyralis* (illustration on page 501), from the northern edge of Florida to New York, west to Kansas, experiment: Within 30 minutes after sunset you will see J-shaped, dipping streaks of yellow light over open ground (below, number 8). These will be repeated each 5 to 7 seconds as the males fly here and there, signaling then waiting, for a female's answering flash from low vege-

tation. Get inside their signal system by flashing a half-second flash about 2 seconds after the male's flash . . . put the lighting-tip of a penlight against the ground a few feet in front of him. He will fly closer, flash again, hover, wait. Count 2 seconds and answer again, pushing the penlight further into the ground as he gets closer, to make its flash dimmer. If you do it right, after a few flash-response episodes (n = 6 − 10), the male will be standing on the penlight, waving his antennae ("sniffing"). A next and experimental step is to test the importance of the 2-second delay in the female response. Answer males in the same penlight-fashion, except use a half-second delay; then try a 3-second delay.

At the next twilight, beginning a few minutes before your "Big dipper fireflies" start, walk around their activity space flashing half-second flashes in imitation of the male flashes you saw the night before. Pause after each flash to wait for an answer at the 2-second window—the critical part of the species' code. Especially, aim your light at clumps of grass and bushes, and under boughs and shrubs that overhand the edge of the open space (If you get a green answer, inspect the emitter closely and discreetly, because *Photuris* firefly females mimic the answers of *Photinus* females, attract males and eat them; if it is a *Photuris* female that answers, it is worth an evening's watch to see if she catches a male). When you find a Big Dipper female, note where she is, so with correctly-timed flashes, you can lead a male to her, and she can begin answering him. You may contribute to the perpetuation of the Big Dipper at that site, by promoting an earlier production of 50 to 100 fertile firefly eggs. (Time is risk.)

Nearly always there are many more sexually active males than females, and competition is keen. Suppose a male sees another male approaching a female, might he try to cut in? Experiment! Hint: use one penlight as a male and another as a female, in the view of a mate-seeking male. Does he charge in? Does he give deceptive female like answers to the (bogus) rival male's flashes? Why might he approach with caution? How might a male distinguish a predator's flash from the flash of a potential mate? Might this be the historical (evolutionary) significance of the Big Dipper's 2-second delay? Have predators gotten inside the system already? Someday someone with a penlight may find out!

of the semantic quicksand of using loaded words like *meaning*, it might be safest to simply recognize them as correlations between the behaviors of different animals, refer to them as *functions*, and let it go at that.

Following the lead of Wiley (1983), Markl (1985) further clarified the roles of manipulation and information; and he carefully distinguished the two separate meanings of *information*. All of this cleared the way, or at least clarified the issues, for an improved concept of animal communication.

The Current View of Animal Communication: The Basic Model

To understand the current biological view of animal communication, it is important to remember the distinction between the two meanings of *information* (semantic versus statistical), remain alert to the subtle anthropomorphic problem of trying to interpret what a signal means to another animal, and remember that we can only be observers of the interacting behaviors of other animals as discussed in the section above.

Perhaps the best general definition of communication currently available is that given by Burghardt (1970 b): "Communication ... occurs when one organism emits a stimulus that, when responded to by another organism, confers some advantage (or the statistical probability of it) to the signaler or its group." This definition is not as exact as many people might like. For example, it does not specify what should constitute a stimulus or a response, and it does not suggest how the advantage should be measured, particularly so that it is not simply inferred in a circular manner. This definition is probably the most inclusive and acceptable that we are likely to get.

The basic characteristics of the current biological concept of communication include a signal (coded information or message), a sender, and a receiver. Other attributes include the following:

1. The sender and receiver may belong to different species, although they usually belong to the same species.
2. The process is normally adaptive in some way to the sender. It may or may not be advantageous to the receiver.
3. The sender and receiver must possess the appropriate structures to respectively send and receive the message. This does not imply that the sending or receiving is conscious, voluntary, or even neural. Even in humans, where voluntary communication is least disputed, there may be facial or other expressions that are more or less involuntary, are universally understood—regardless of oral language—and may speak louder than accompanying words.
4. There is a distinction between a signal and its function (or meaning). The same signal can function in many different ways depending on the context and the receiver(s).
5. The process of receiving signals potentially changes the behavior of the receiver. Only by noting some such change can we be certain that communication has occurred.

As an example, consider the appearance of a brightly colored male monarch butterfly, a species that is poisonous and tastes bitter if eaten. The scales on the wings reflect light in different patterns at different wavelengths (colors). Thus there are several messages; the butterflies are equipped to send the messages; and there may be several separate functions. One message in the range of light that is visible to vertebrates functions as a warning to potential predators; it "says" that the butterfly tastes bad. There are different functions in the ultraviolet wavelengths sensed by other butterflies concerning species and sexual identification.

The butterfly is unlikely to be aware of the signals, but both types of receivers, vertebrates and other butterflies, can see their respective messages and act accordingly. The messages clearly are adaptive to the butterfly, the primary recipient of benefits in this case. The message to the potential predator increases the butterfly's chances of survival. The message to other butterflies increases the sender's chances of reproduction.

On the receiver's end, the message saves the predator from a bad meal and possible sickness. If the receiver is another butterfly, the meaning and value of the message depend on the species and sex of the receiver. The signals may enhance the chances of reproduction for a female monarch or reduce the chances of unnecessary interactions and wastes of time and energy for males or for butterflies of a different species.

In this example the monarch's messages are adaptive for all concerned, but the adaptive value varies: the predator warning is much more adaptive for the sender than for the receiver. For the sender his whole life is at stake; for the predator it is but one meal or perhaps just a bite of potential food versus a minor case of sickness. In other words the messages evolved in the monarch primarily as a result of reduced predation and enhanced reproduction and not to save the predator from getting ill or to save other species from wasting time attempting to reproduce. If the latter functions exist, they are adaptive to and evolved at the receiver's end. In other words, both sending and receiving species have coevolved.

There may also be other players in the scene. In the case of monarch butterflies, other harmless and edible species may mimic the monarch as illegitimate senders; other predators that have evolved physiological defenses against the monarch's bad taste and poisons may use the conspicuous signals to locate a meal; and biologists may be standing by, with camera and notepaper, as neutral observers.

Figure 15-5 Diagram of a general model of communication. The intended target, or primary pathway, is shown in bold type.

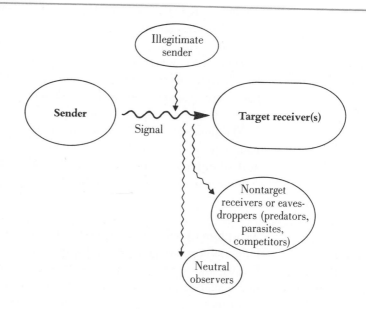

This current view of animal communication is illustrated in Figure 15-5. The remainder of the chapter inspects the details of various signals, behavioral interactions in animal communication, and a host of evolutionary ramifications. Present understanding of animal communication is incomplete and we still face several problems, including difficulty in quantifying and measuring communication. The next section briefly considers some methods to do that, problems within them, and implications of these problems for our knowledge of communication.

MEASURING SIGNALS

The quantitative study of communication has involved the following four basic approaches:

1. Infer functions by correlating assumed or hypothesized signals with various observed contexts under natural conditions.
2. Use experimental manipulation of signals to demonstrate presence or absence of hypothesized effects and hence inferred function.
3. Use Markov analyses and statistical tests for measuring conditional probabilities of behavior, that is, to test for significant changes in the behavior of one individual (the hypothesized receiver) in response to other behavior (the potential signal) of another (the sender).
4. Attempt to measure or provide an index to the amount of information, in the statistical sense, via statistical information theory.

We briefly describe each of these approaches.

Inferring Functions by Correlating with Contexts

Signals and communication often are inferred by observing the situations in which they occur. Early studies did (and pilot studies do) this subjectively by simply watching animals and thinking about what was (is) observed. If a male bird gives certain vocalizations only during the breeding season when inside the territory, and females do not make such sounds, it is inferred that these calls or songs function to attract females and repel other males. Subsequent studies usually quantify the behaviors (signals) and contexts more precisely. Brown (1964), for example, considered the vocalizations of Steller's jays under a variety of aggressive and non-aggressive situations (Table 15-1). Von Frisch basically used this approach in associating the waggle dances of honeybees with food gathering by other honeybees. He inferred (and was later shown to be correct by Gould's experiments) that "scout" bees were "communicating direction and distance information" to "recruited" foragers. (Inferences shown in quotes.)

The problems with this approach are that it is correlational at best—that is, it cannot prove cause and effect—and in many cases similar signals may be used in several different contexts. The "wah" call of Steller's jays, for example, occurs in a number of situations (see Table 15-1).

Experimental Manipulation to Demonstrate Presence or Absence of Effects

As with most science in general, correlations are useful in suggesting hypotheses for possible cause and effect relationships, but correlations cannot go much further. Ideally, the testing of hypotheses involves experimental manipulations. Many experimental manipulations have been conducted with animal communication. One of the most ingenious and convincing experiments was Gould's test of whether other

Table 15-1 Inferred signals: comparison of behavioral acts of Steller's jays with the contexts in which they occur*

Context	Number of Observations						
	Rattle	Musical	Growl	Wah	Too-leet	Shook	Total
By supplanter	25	29	4	8	5	16	87
By supplantee	2	—	—	—	—	—	2
At individual but no supplanting	12	19	5	5	1	22	64
Aggressive sidling	—	2	6	2	7	32	49
In fight	—	—	—	1	—	2	3
After fight	1	—	—	1	—	—	2
Mobbing	1	—	—	31	—	—	32
Just alighted	—	—	—	26	—	2	28
On picnic table	—	—	—	28	—	—	28
Appeasement	—	—	—	1	—	—	1
Answering at a distance	—	—	—	—	—	2	2
Courtship	8	2	—	—	—	—	10
TOTAL	49	52	15	103	13	76	308
Context unspecified	78	88	2	131	17	77	393

From Brown, J.L. 1964. University of California Publications in Zoology. 60:223-328.
*This example lists the contexts in which Steller's jays gave various calls. Several of the vocalizations occur in several contexts with much overlap and variability. Thus only general inferences are possible. The rattle and musical calls are given mostly by one individual supplanting another or in courtship; the wah call is used mostly to maintain social contact, and the shook call is used by aggressive individuals toward others.

bees obtain information from the waggle dance, as discussed on pages 494 to 495.

The context, sender, receiver, or the signal itself may be modified artificially. This approach generally is used to test hypotheses formed from the observations under more natural conditions. What is modified and how depend on the modality, the species, the situation, and the signal in question. The following list includes a variety of examples of this general approach. All of these require imagination, ingenuity, and, when working with modalities such as chemical, auditory, or electrical fields, the technical ability to measure or modify the signals. Often the sensory systems of the animals themselves are used to measure the response to a stimulus; one simply taps into the nerves coming from the animal's sense organ. The results of using an insect's antenna to monitor responses to various chemicals, for example, are known as *antennagrams*.

- Isolate chemical signal extracts from the animals or their deposits to determine molecular composition or to present the chemicals in artificial, altered, or other controlled contexts.
- Create synthetic molecules for chemical signals. The molecules can be modified at different points, then presented to the animals to determine potency and which aspects of the molecule are most important.
- Use speakers for sound signals along with actual, artificial, or variously modified recordings. Emlen (1972), for example, modified the vocalizations of indigo buntings by cutting tape recordings and splicing the segments into new patterns. More recent computer technology permits signals to be modified electronically. By presenting these modified songs to birds, one can isolate the critical aspects of the song.

- Use physical models for visual signals. Such models frequently have been remarkably simple, often consisting of little more than a piece of wire and yarn, a painted stick, or pieces of cardboard. The common use of simple models in early ethological research and the ready responses to them by many animals led to many of the early concepts of releasers (simple cues as the important triggers of behavior) and innate releasing mechanisms (discussed and illustrated in Chapter 2).
- Modify the appearance of the animal itself. In common flickers, for example, the male has a facial "mustache" marking that females do not have. Noble (1936) painted mustaches on females and covered the mustaches on males. This reversed the responses from other flickers so that males were treated as females and vice versa. The mustache can be inferred to be a signal that identifies sex. Similarly, in numerous other instances, prominent markings of animals have been added, eliminated, exaggerated, or otherwise modified.
- Block the signal in either the sender or receiver. Covering or otherwise eliminating markings was just mentioned. At the receiving end, vision can be blocked with lenses, eye covers, or opaque screens. For chemical signals the openings or surfaces of glands can be shellacked or otherwise blocked; the receiver's sense organs can be blocked; or appropriate nerves in the receiver can be cut. Receivers can be deafened to block auditory signals. With appropriate combinations of techniques, some modalities can be blocked selectively. Animals can be placed, for example, in soundproof chambers with windows for vision. One problem with this technique is that the experimental manipulation may alter or otherwise disturb the animal's behavior generally.

Statistical Tests of Conditional Changes

An important feature of our concept of communication is that the behavior of a receiver changes in response to signals from a sender. If so, it should be possible to detect such changes.

The behavior of the receiver should be different after receiving signals than when no signals are present. Mathematically this can be expressed as:

$$P(x_2|x_1) \neq P(x_2)$$

where

Individual	A	B
Behavior	x_1	x_2
Probability of act occurring	$P(x_1)$	$P(x_2)$

In words the equation means that the probability of an act, x_2, occurring in individual B given that act x_1 has occurred in individual A is different from the probability of x_2 occurring in the absence of x_1. A would be the sender, B the receiver, x_1 the message, and x_2 the response.

For a simple hypothetical example, consider three different behaviors. Their frequencies of occurrence during a given time period are given in Table 15-2. Behaviors one and three appear to have been altered by the signal. Behavior one was decreased, and behavior three was enhanced, whereas the frequency of behavior two was unchanged.

Table 15-2 Alteration of behavior frequency in one individual as a result of a signal from another individual

	Observed Frequency of Behavioral Acts		
	Behavior One	**Behavior Two**	**Behavior Three**
In absence of signal	300	500	2
After signal	100	500	200

Table 15-3   Comparison of frequencies of acts in one animal following acts by another animal: staged encounters between captive mantis shrimps*

Initial Act (X, given by individual A)	Following Act (Y, by individual B)									Total	p(i)
	Approach	**Meral Spread**	**Lunge**	**Strike**	**Chase**	**Grasp**	**Coil**	**Avoid**	**Does Nothing**		
Meet	0(0.16)	3(1.9)	0(0.28)	5(1.8)	0(1.6)	0(0.08)	4(2.7)	2(3.5)	0(2.0)	14	.04
Approach	0(0.46)	**15(5.7)**	0(0.81)	5(5.2)	1(4.7)	1(0.23)	10(7.9)	8(10)	1(5.9)	41	.12
Meral spread	1(0.66)	10(8.1)	2(1.2)	3(7.5)	**0(6.8)**	0(0.33)	12(11)	**28(15)**	**3(8.5)**	59	.17
Lunge	1(1.36)	2(4.4)	2(0.63)	0(4.1)	0(3.7)	0(0.18)	**17(6.1)**	9(7.9)	1(4.6)	32	.09
Strike	1(0.73)	4(9.0)	0(1.3)	14(8.2)	**1(7.5)**	1(0.37)	13(12)	**28(16)**	**3(9.3)**	65	.18
Chase	0	0	0	5	0	0	0	0	0	5	.01
Grasp	0	1	0	2	0	0	1	2	0	6	.02
Coil	0(0.32)	6(3.9)	2(0.55)	6(3.5)	0(3.2)	0(0.16)	6(5.4)	8(6.9)	0(4.0)	28	.08
Uncoil	0(0.24)	8(2.9)	1(0.41)	3(2.7)	1(2.4)	0(0.12)	0(4.0)	1(5.2)	7(3.0)	21	.06
Avoid	1(0.95)	**0(12)**	0(1.7)	**2(11)**	**38(9.7)**	0(0.47)	**5(16)**	**2(21)**	**36(12)**	84	.24
TOTAL	4	49	7	45	41	2	68	88	51	355	1.01†
p(j)	.01	.14	.02	.13	.12	.01	.19	.25	.14	1.01†	

From Dingle, H.A. 1969. Animal Behaviour 17:561-575.
*Observed frequencies are shown to the left with expected frequencies in parentheses. This set of data is more complex than the hypothetical illustration in Table 15-2 in that it contains more behaviors, and the potential signals have not been identified beforehand. Thus all acts are considered for both participants, and expected frequencies are based on overall frequency of occurrence (marginal totals as a proportion of grand total). Acts in which following behavior was significantly altered (tested by chi-square) are in italics, and their significant data are in bold; these acts could be considered communication signals. Whether they increased or decreased can be determined by comparing the observed with the expected values.
†1.01 rather than 1.00 results from rounding error.

Because real behavior is normally more complex and variable, however, changes or differences usually require statistical techniques to detect. An actual example and classic study of interactions analyzed with this approach, aggressive communication in mantis shrimps *(Gonodactylus bredini),* is shown in Table 15-3 (Dingle 1969). This example illustrates the statistical tests and specific behaviors that may serve as signals as shown by altered responses by the receiver. A number of similar studies have been conducted by other persons, e.g., Steinberg and Conant (1974), working with grasshoppers, and Burk (1979), working with crickets.

**Information-Theory
Measures**

Using cybernetic (Greek, *steering* or *control*) information theory to measure or index the amount of information, in the statistical sense, in communication is perhaps the least familiar approach for most persons encountering the subject of communication for the first time. This approach has been used with some degree of success and deserves more attention but has many limitations. Also, there have been and continue to be many excellent studies of animal communication that do not use the information-theory approach, for example, Hyatt and Salmon (1978), Marler et al (1986 a, b), and Gyger and Marler (1988). The information-theory approach is closely related to the testing of conditional changes as described previously, but it attempts to measure or estimate how much information, in terms of reduced uncertainty, is transmitted.

The first reference to the application of this technique for animal communication that we have been able to find is Haldane and Spurway (1954, not available in many libraries—including ours—but cited in Wilson 1962). Haldane and Spurway attempted to measure or, more properly, estimate the amount of information (in the semantic sense) contained in the honeybee dance language. Wilson (1962) subsequently attempted to perform a comparable analysis for the amount of information (also in the semantic sense) in the trail odors of fire ants and additionally to refine the application for honeybees. In retrospect (now that we have more clearly distinguished between the two meanings of *information*), it probably was not appropriate to mix the two uses and, perhaps because of that, their results are not convincing. The significance of their work, however, is that they tried a new approach and started others thinking and working with it, which led to further and more appropriate (in the statistical sense) uses of information theory. Examples of some of the first statistical-sense uses include communication in the interactions of rhesus monkeys (Altmann 1965) and hermit crabs (Hazlett and Bossert 1965).

The cybernetic-statistical approach depends on breaking information into discrete packages, the smallest of which is the two-state (0-1, on-off, yes-no, up-down, left-right, etc.) binary digit or bit. Although this needs to be viewed in the statistical sense, not the semantic sense (because the system is being viewed from the outside, not the inside), we nonetheless start discussing it in terms of semantic information to set the stage for explaining it in the statistical sense.

To begin with, imagine that we are inside a communication or signaling system and have knowledge of the actual content, that is, semantic information. Perhaps the most familiar example nowadays is the computer. To represent a standard alphanumeric character in the computer, we need to allow for 92 possibilities: 26 standard letters of the alphabet twice (for upper and lower case), plus 10 numbers (0 to 9), plus punctuation marks and other characters as seen on a standard computer keyboard. That requires more than 6 bits or 6 two-state combinations ($2 \times 2 \times 2 \times 2 \times 2 \times 2 = 64$) but fewer than seven ($2^7 = 128$). To be more precise, it requires $\log_2 92 = 6.52$ bits. (Many hand calculators do not have a $\log_2$ function. To convert from $\log_{10}$, $\log_2 X = (\log_{10} X / \log_{10} 2 = \log_{10} X / 0.3)$; that is, simply obtain the logarithm, base 10, of the number and divide by 0.3.)

Because bits come in discrete packages, 92 items need a minimum of 7 (which leaves room for $128 - 92 = 36$ more characters, such as ö, ü, etc.). To allow room for even more characters and for historical reasons, 8 bits (the familiar *byte*) are used for 256 characters, as seen in the standard computer ASCII table.

Now consider another system, a standard deck of playing cards with 52 cards in it. How many bits would it take to represent all of the cards? The answer is more than 5 (= 32 combinations) but fewer than 6 (= 64 combinations) or $\log_2$ 52 = 5.70. That would be the number of bits as an engineer would see it. The actual way that playing cards are set up, however, is slightly less efficient: there are 4 suits (2 bits), face-versus-nonface cards (1 bit), which face card (jack, queen, or king, $\log_2$ 3 = 1.58 bits), and, if not a face card, 10 possibilities ($\log_2$ 10 = 3.32) for a total of 7.90 bits (the additional 2.20 bits comes from breaking the system into additional categories, suits, etc., rather than a simple set of 52 items). Because bits are discrete, one would need a minimum of 6 bits for 52 simple items or 9 bits for the deck of cards (2 for face cards and 4 for nonface cards plus the 2 for suit and 1 for face-versus-nonface cards).

We can apply all of this to something more biologically relevant: movement outward from a given point such as a nest, hive, or other home base. How many bits will it take for information on which direction to travel? Right away we encounter a problem: direction is continuous, not discrete. Thus we will have to arbitrarily break it into discrete segments, say the 8 principal compass positions: N, NE, E, SE, S, SW, W, NW. That requires a minimum of 3 bits: 1 bit for the first 4 of the 8 possibilities versus the second 4, then (within the 4 chosen) 1 bit for the first 2 choices versus the second 2, then (within the 2 to which it has been narrowed) 1 bit for the first versus the second choice. (Note that the way we normally represent these 8 directions is, as in the case of cards, somewhat inefficient from a pure information standpoint and requires 4 bits of information. The main or cardinal points of the compass, N, E, S, and W, really should be represented in this scheme as NN, EE, etc. In other words, there are two character positions and each has four possible characters, hence, 2 times 2 bits.) If, however, direction is broken into 360 degrees as in the standard (human) compass, to get to the correct degree will require $\log_2$ 360 = 8.49 bits. That is a big difference from the 3 bits if only 8 segments are considered.

A similar problem occurs if we attempt to consider distance, which is also continuous rather than discrete. Attempts to estimate information content for such factors as direction and distance are faced with an open, ambiguous situation that could render outcomes useless or close to useless. Furthermore, different organisms faced with systems that appear similar, such as bees, ants, and birds moving outward from a starting point, may be operating on entirely different bases—playing with different decks of cards, as it were.

Bees communicate direction and distance in a manner in which it seems the receiving bee would be on her own and could easily go off in any direction. Ants, on the other hand, are following a much more distinct trail. Birds, such as vultures or gulls leaving a roost, may simply follow each other. Following a trail does not necessarily require direction or even distance information but only the ability to detect the trail and an end point. It is potentially as simple as a train following a track until it comes to the end and requires only 2 yes-no bits of information (on trail or not and at end point or not). The bee system clearly requires more information than simple trail following, namely "keep the position of the sun fixed over certain ommatidia of the eye as translated from the scout's waggle dance and travel approximate distance (also obtained from waggle dance) until encounter odor of food carried by scout." How that relates to the meaning of *direction* as we humans conceive of direction can be debated.

To summarize, there are two significant problems facing any attempt to measure semantic content. First, to measure content within the system we must be inside the system, that is, understand exactly what details are involved. We can do that with computers and playing cards—because they are our systems—but we cannot do it with bees, ants, birds, or other nonhuman species and not even with all human systems. Secondly, much of the real world of animal communication exists in a continuous rather than a discrete state. That occurs even in the case of animal signals themselves (pages 514 to 516). Any attempt to break a continuous situation into a discrete one involves arbitrariness, which, until new techniques are developed, renders quantification by the present system of information theory somewhat questionable.

We have described two (of several) limitations on using information theory for animal communication, and we have introduced the concept of bits. By recognizing and appreciating these limitations, we can now consider some ways in which information theory can properly be applied to animal communication (although there are still some further limitations lurking in the shadows). It is time to move to the other, statistical meaning of *information*. The term *information* is so misleading in this sense that it probably is a misnomer and what we are really looking at perhaps should better be considered as diversity—diversity of behavior, as follows.

Because we are stuck outside the system as observers, we can only describe what we can see the animals doing outwardly, that is, their behavioral actions and reactions. We can count these things; and within a given context, such as courtship or feeding, we can usually quantify the diversity of such behavior—somewhat like counting the elements in a computer or deck of cards and trying to infer the possible combinations.

Information in this sense can be viewed as a reduction of uncertainty. There is much uncertainty, for example, about which cards are contained in an opponent's hand during a card game. As the game proceeds and cards are laid out on the table, however, you become less uncertain about what remains in the hand. (You remain somewhat uncertain until all the cards are out.) Animal behavior is somewhat like a game of cards: the repertoire contains several possibilities, and an observer is uncertain which behavior an animal will engage in next. When information is transmitted in the system, however, the observed behavior becomes less uncertain, that is, more predictable.

In the process of measuring information as reduction of uncertainty, we are not able to say what specific information is being transmitted (even though we might very much want to know that); at best we can attempt to estimate how much, if any, information has been transmitted. This is done by assessing changes in diversity (of behavior) in response to (or, in mathematical terms, conditional on) the behavior given by a potential signal sender. The usual symbol used for this diversity (or information) is H (after R.V.L. Hartley, a pioneering mathematician in the subject of information theory). H is the number of bits it would take to describe a given repertoire of behavior. Because H is based on samples of behavior and thus is only an estimate, it is usually shown as $\hat{H}$ ("H hat").

Thus, to begin with, we could estimate the number of bits required by our simple formula: $H = \log_2$ (number of behaviors in repertoire). However, because behaviors do not occur with equal frequencies, that is, animals are more likely to behave in some ways than in other ways whether communicating or not, the estimate has to be tempered with a consideration of each behavior's probability of occurrence. This complicates the formula somewhat and changes it to:

$$\hat{H}(A) = -\sum_{i=1}^{n} p(i) \log_2 p(i) \text{ bits}$$

where A is an animal's repertoire composed of n behaviors and p(i) is the probability of the ith behavior, i.e., p(i) = frequency of i/total frequency of all of the behaviors. The maximum value of $H(A) = \log_2 n$. $\hat{H}(A)$ is a measure of the uncertainty of A's behavior.

Communication, however, involves two animals—a sender (A) and a receiver (B)—so we can calculate both $\hat{H}(A)$ and $\hat{H}(B)$. To illustrate this we can use the same set of data that was used for the statistical test (Table 15-3). A's behavior is shown in the leftmost column as "meet" ... "avoid" with the total frequencies and p's shown in the two last columns on the right-hand side. There are 10 behaviors, so $\hat{H}(A)$ max = $\log_2 10 = 3.33$.

$$\begin{aligned}\hat{H}(A) &= -(.04 \log_2 .04 + .12 \log_2 .12 \ldots + .24 \log_2 .24) \\ &= -([.04][-4.66] + [.12][-3.07] \ldots + [.24][-2.07]) \\ &= -([-.19] + [-.37] \ldots + [-.50]) = 2.96 \text{ bits}\end{aligned}$$

Similarly, $\hat{H}(B) = -(.01 \log_2 .01 \ldots + .14 \log_2 .14) = 2.75$ bits.

The next step is to determine the conditional uncertainty, that is, whether the amount of uncertainty in B's behavior is reduced (or becomes more predictable) after A's behavior. That is, some of A's behavior may influence B's and can be considered as candidate signals. The conditional uncertainty in B's behavior, or B's behavior given A's behavior, is determined from the cell frequencies in a matrix such as Table 15-3 with the following formula:

$$\hat{H}(B|A) = -\sum_{i,j} p(ij) \log_2 p(j|i) \text{ bits}$$

where p(ij) is the joint probability of the two behaviors (e.g., for meral spread by B after approach by A it is 15/355 = 0.04) and p(j|i) is the probability of the jth behavior by B given that the ith behavior of A has occurred (e.g., for the meral spread example it is 15/41 = 0.37). Note that for cell frequencies of 0, p(ij) $\log_2$ p(i|j) = 0. Thus for our example of mantis shrimps:

$$\begin{aligned}\hat{H}(B|A) &= -(0 + (3/355) \log_2 (3/14) \ldots + (36/355) \log_2 (36/84)) \\ &= -(0 + (.008)(-2.230) \ldots + (.101)(-1.227) \\ &= -(0 + (-.019) \ldots + (-.124) = 1.97 \text{ bits}\end{aligned}$$

If the behaviors of A and B are statistically independent, H(B|A) would be equal to $\hat{H}(B)$. ($\hat{H}[B|A]$ is never larger than H[B]). If B's behavior is completely determined by A's behavior, B's behavior would be completely predictable by knowing A's behavior; that is, there would be no uncertainty or $\hat{H}(B|A) = 0$.

The difference between H(B) and H(B|A) is a measure of the "information transmitted from A to B," symbolized by large $\hat{T}(A$ to $B)$ or as usually presented, $\hat{T}(A;B)$. That is, $\hat{T}(A;B) = H(B) - H(B|A)$ bits. For the mantis shrimp example, $\hat{T}(A;B) = 2.75 - 1.97 = 0.78$ bits.

There is still some subjectivity in the measure because the behavioral repertoire itself usually is identified arbitrarily. One observer may include a different number, n, of behaviors than another observer, and that would influence the magnitude of the calculation. Therefore the reduction in uncertainty resulting from transmitted information can be viewed as a proportion or percentage of the total uncertainty.

This is called the *normalized transmission* (Steinberg and Conant 1974, Steinberg 1978) and is symbolized by small $\hat{t}(A;B)$:

$$\hat{t}(A;B) = \frac{\hat{T}(A;B)}{\hat{H}(B)} = \frac{\hat{H}(B) - \hat{H}(B|A)}{\hat{H}(B)}$$

Because the value of t is a proportion, it is without units and ranges from 0 to 1.00 or, if expressed as a percentage, from 0% to 100%. For the mantis shrimps, $\hat{t}(A;B) = 0.78/2.75 = 0.28$ or "the transmitted information in A's behavior reduced the uncertainty of what B would do by 28%." The term *transmission* is somewhat abstract or even misleading in this statistical usage. Markl (1985) suggests that "it might be better to translate T as 'tightness' (of correlation) rather than as 'transmission.'"

Although measuring information in this way may seem too abstract, the results can be informative. For example, Burk (1979) found that $\hat{t}$ was typically higher for aggressive communication in species where males are capable of injuring each other than in species with only ritualized agonistic behavior.

For further explanation, discussion, and examples of the use of statistical information theory in animal communication, see Dingle (1969, 1972), Steinberg and Conant (1974), Steinberg (1977), and Markl (1985). Losey (1978) and Fagen (1978) discuss additional statistical aspects such as confidence limits of H.

Measurement of Communication in Perspective

The various approaches described have been more or less successful at providing insight into animal communication depending on the situation. Each method has strong and weak points.

Much of our understanding of animal communication is based on qualitative rather than quantitative observation of behavior. Quantification of signals in interactions between animals obviously requires much time and effort. Calculations for statistical tests of significance and estimates of H are tedious, subject to both human and rounding error, and can become unwieldy for large matrices involving many behaviors and more than two participants. Computers are helpful even for relatively small sets of interaction data, and they are virtually required for larger data sets.

Determining which behaviors to include in an analysis and how to split or lump them can inject a fair amount of subjectivity and bias into even the best attempts to quantify communication. Baylis (1976), for example, in analyzing courtship sequences of cichlid fishes, suggested that it is necessary not only to consider sequences of behavior between two individuals but also to include sequences within individuals where more than one act may be shown by the first individual before the second individual responds. That immediately doubles the size of the interaction matrix and further complicates both the calculations and the interpretations.

Analyses of interactions between animals may include some behaviors that would not seem to qualify as communication. If one animal physically strikes another and the second strikes back, should the striking itself be considered as an act of communication? The second animal's behavior is influenced by the first animal's behavior, but not all observers would consider the striking action as a communication signal. Nonetheless, such interactions would be included in a measure of H or a statistical test of significance, as in Table 15-3 and the previous sections.

The results of many studies using these four basic approaches form the basis for the remainder of this chapter. However, attempts to quantify communication have provided only partial solutions, and many questions remain unanswered. Further quantification and perhaps new approaches would be helpful to increase our understanding of the subject. We return to this topic at the end of the chapter. In the meantime, we proceed into the next sections using results from the potpourri of methods described previously.

GENERAL PROPERTIES OF ANIMAL SIGNALS

The nature of an animal signal depends on the physical source of energy and the manners in which it is produced or modulated. Animal communication may involve molecules, mechanical waves (sound), actual contact (i.e., touch or tactile), light, or electrical fields. The content of a signal is carried by a system through variation or changes of its state; that is, it is modulated. The simplest form of modulation is a simple binary, digital, or two-state condition such as on-off, yes-no, go-go on, presence-absence. A light, for example, can blink on and off. Examples of digital, or discrete, signals are shown in Figures 15-6 and 15-7. This simple means of coding information is only one of many ways by which signals may vary. The following is a list of the major ways by which signals, regardless of modality, are formed, modulated, and received by various animals:

Figure 15-6 Discrete signals: firefly flash patterns in nine *Photinus* species. Some predatory fireflies, incidentally, mimic the signals of other species to attract and capture them (see Chapter 10).
Modified from Lloyd, J.E. 1966. Miscellaneous Publications, Museum of Zoology, University of Michigan 130:1-95.

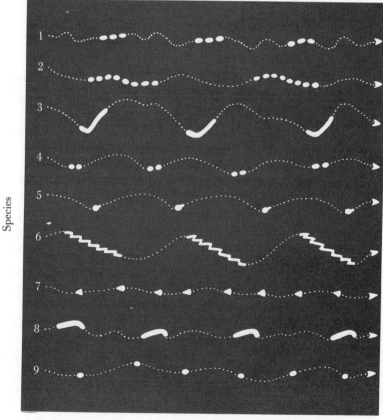

Time

Figure 15-7 Discrete signals (for the most part): head-bobbing patterns in the courtship displays of seven *Sceloporus* lizard species. Although mostly discrete, the bobbing also is somewhat continuous as height of bobs and length of time in different positions occurs at intermediate ranges. Modified from Hunsaker, D. 1962. Evolution 16:62-74.

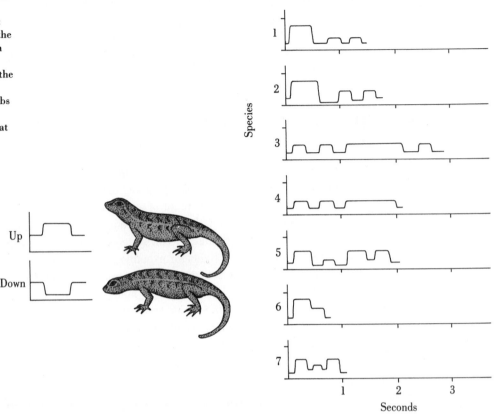

1. **Qualitative differences.** These include different molecular configurations for chemicals, different colors for light, and different pitches for sound. These characteristics are affected by ecological factors and must account for background noise that could reduce signal effectiveness. Different fireflies, for example, emit yellow or green light, depending on whether the species is active at dusk or dark, respectively (Lall et al 1980). Hunter and Krebs (1979) describe how bird songs vary, depending on whether they are living in forests or more open woodlands.

2. **Quantitative or intensity differences.** These include concentrations of chemicals, brightness of light, and loudness of sounds. The intensity of a message can occur in discrete, digital jumps as just discussed or by gradual, graded, or analog modulation (Figure 15-8). Some messages consist of combinations of digital and analog modulation (Figure 15-9 and see further discussion on page 528).

3. **Directional information.** Depending on the message, the situation, and the sensory capabilities of the receiver, the receiver may be able to determine the location or source of the signal.

4. **Patterns.** Patterns can be spatial, as in familiar complex arrangements of forms and colors (e.g., the appearance of the butterfly wing), or a complex combination of chemicals or sound pitches. The pattern can also be temporal with time sequences. Examples of temporal patterns include

Figure 15-8 Continuous signals: crest raising in Steller's jay. The crest can be raised or lowered over the entire range of movement. Meaning of the signal at different angles is indicated in the diagram. Modified from Brown, J.L. 1964. University California Publications in Zoology 60:223-328.

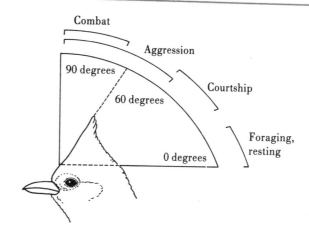

Figure 15-9 Signals formed by composite discrete and continuous messages. A, Zebra messages where threat (ears back) versus greeting (ears up) is indicated by ear position, and magnitude of the signal is graded by how widely the mouth is held open. B, Ant messages where odor trails guide following behavior, and head movements recruit to food (lateral movements) or nest site (backward-forward movements).

A modified from Trumler, E. 1959. Zeitschrift für Tierpsychologie 16:478-488, and Wilson, E.O. 1975. Sociobiology, the new synthesis. Belknap/Harvard University Press, Cambridge, Mass. B from Hölldobler, B. 1971. Zeitschrift für Vergleichende Physiologie 75:123-142.

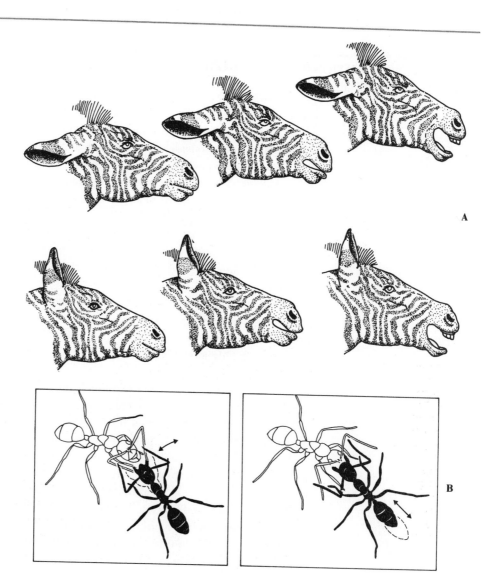

familiar sequences of sound as in bird and cricket songs or in human language. Visual examples include firefly flash patterns (see Figure 15-6) or the head-bobbing sequences performed by lizards (see Figure 15-7). Pattern recognition by the receiver requires both advanced sensory capability to detect variation or changes and a well-developed nervous system to store and interpret the information (see Chapters 16 and 17).

Messages may be modified not only by the sender but also by the environmental context. The same message can have different functions under different circumstances. The molecular queen substance in bees, for example, can lead to the development of new queen larvae if no queen is present, keep workers from rearing new queens if a queen is present, keep workers sterile, or serve as a sex attractant. The songs of male birds may function one way to a female and quite differently to another male. The same songs may have different functions at different seasons. The same vocalizations that repel male red-winged blackbirds as part of territorial behavior and aggression during the nesting season, for example, serve to attract males together outside the breeding season (Brenowitz 1981).

Next we consider the diverse forms of communication seen among animals.

SOME GREAT COMMUNICATORS: COMPLEXITY OF COMMUNICATION IN VARIOUS ANIMAL GROUPS

From a purely descriptive standpoint, it is clear that the complexity of communication varies tremendously throughout the animal kingdom. We survey this range of complexity, and consider some of its evolutionary aspects.

Humans clearly possess the most highly advanced communication ability of any living species. The differences between humans and nonhumans may be largely a matter of degree, and there must have been some evolutionary development and continuity. The gap may be due to rapid evolution of early *Homo* brains and language, with the extinction and loss of intermediate forms. Several different species of early man, living during the same time period on earth, may have possessed spoken language, but only one, the direct ancestors of modern humans, survived.

Among other mammals, highly complex signals are seen in whale vocalizations (for description and discussion, see Payne 1983), dolphins, and fairly complex interactions are found in some nonhuman primates. Beyond that, mammalian communication is only moderately complex and not particularly remarkable. Most modalities, including visual, acoustic, olfactory, and tactile, are used but generally not at levels of complexity that are much different from those seen among other vertebrates or the arthropods. Mammalian visual signals do not reach the height of complexity seen in cephalopods (discussed later), nor are the visual or acoustic signals of mammals as rich and complex as those given by some birds.

One feature of communication in some mammals, however, is noteworthy: the use of different vocalizations in reference to different types of predators. The use of different warning calls for different kinds of threats has now been documented in at least two unrelated species, California ground squirrels (*Spermophilus beecheyi*) and vervet monkeys (*Cercopithecus aethiops*).

California ground squirrels have two basic alarm vocalizations, chatters and whistles (Leger and Owings 1978, Leger et al 1979, and Owings and Leger 1980;

compare with the Belding's ground squirrels in Chapter 14). The chatters are used predominantly in the presence of ground predators and the whistles with large birds of prey. Furthermore, with variations of acoustic and temporal characteristics of the chatter call the ground squirrels can distinguish among snakes, badgers, and a general category of cat-canid predators. Interestingly, the chatter call is also used (and distinguished) under a different, social context in agonistic interactions between ground squirrels.

A similar situation has been reported for vervet monkeys (Seyfarth and Cheney 1980). As mentioned in Chapter 13, the species uses four different calls in the presence of their four primary predators, leopards, eagles, pythons, and baboons. Playback experiments, in which recordings of the different alarm calls are played back when no predators are present, evoke the appropriate escape responses. For example, leopard alarms cause the monkeys to run for and climb into trees, eagle alarms cause them to look up or run under bushes, and snake alarms cause them to look down. The monkeys clearly distinguished among the different meanings of the calls.

The other vertebrates that are recognized as having highly complex signals are birds, particularly songbirds of the order Passeriformes. Bird vocalizations have received much attention in behavioral studies, partly because of interest in their own right and partly because of their importance in developmental studies. Much of the interest in bird song involves the ontogeny of bird vocalizations, a subject that has played a major role in understanding of learning. Bird song therefore is discussed in detail again in Chapter 19. Bird vocalizations fall into two general categories: calls and songs. Calls usually are short and simple. They are used for most of the signal functions of birds. Songs, on the other hand, generally are more elaborate and complex. They are given by males, except in a few rare instances, and mostly during the breeding season, although they also may be given sometimes during the fall (by temperate songbirds) when there may be a slight resurgence of gonadal hormones (Chapter 18).

It is common to wonder why birds sing. On a proximate level it is because of hormonal influences on the appropriate parts of the brain and the presence of relevant external stimuli. From the ultimate standpoint it is believed that the primary function in the evolution of song is the identification of the singer. The complexity of song probably has been influenced also by the strong forces of sexual selection (discussed in Chapter 11). Song is thought to serve as territorial identification to other males and sexual attraction to females. (The territorial function of great tit songs, for example, was discussed in Chapter 13).

Within the broad framework of identification, however, and depending on the species and the complexity of the song, songs may carry potential identity information of several different types. (This does not mean that receiver birds use all the information or that it is all necessarily functional.)

1. **Species recognition.** The basic song pattern is thought generally to be used by the birds within a species to recognize their own kind. It differs among species but is relatively constant within the species. The basic pattern tends to be more complex in areas with more species.

2. **Local dialects.** These are similar to dialects in human speech and may be either functional or merely consequences of slight geographical separation between populations and the drift in expression that occurs over generations of copying the song through learning.

3. **Individual recognition.** Some characteristics of the song vary within the species but are constant for any given male. Many birds appear able to recognize individuals by their vocalizations, as is familiar in human speech.

4. **Motivational variation.** There also may be much variability among songs given by any one individual. Characteristics that vary in this manner include loudness, length, or frequency of repetition. These variations are believed to relay information on the internal state of the bird.

Further discussion of bird song, aside from a brief mention of mimicry, can be found in a discussion of song development in Chapter 19. A wealth of detailed information exists for bird song with numerous studies of particular species and groups of birds. Interested persons should consult an ornithology text, books on bird song, or bird and behavior journals.

Aside from birds, mammals, and social insects, another group that has an amazingly complex and fascinating communication ability is cephalopod mollusks. The visual communication capabilities of cephalopods are thought to rival or exceed those of many vertebrates, with upward of 35 different communication patterns. Communication in this group of mollusks, which is only partially understood, has been reviewed by Moynihan and Rodaniche (1977). Cephalopods are relatively large (the largest invertebrate, at 450 kg, is a cephalopod, the giant squid) and have flexible bodies, good vision, well-developed brains, and the ability to change their appearance rapidly through the use of chromatophores and other cells at the surface of the body. They are able to spread and withdraw, raise, lower and curl, and even change the surface texture of various body parts, change colors, darken and lighten, and form a variety of patterns on the body surface. A few of the varied appearances are illustrated in Figure 15-10. Cephalopods are able to change appearance even more than their vertebrate counterparts, fishes (compare with Figure 15-11). Fishes have much less flexible bodies and fewer appendages with which to work.

Many of the cephalopods, particularly the squids, are quite gregarious; virtually all use complex patterns of courtship; and most are social to some degree. In

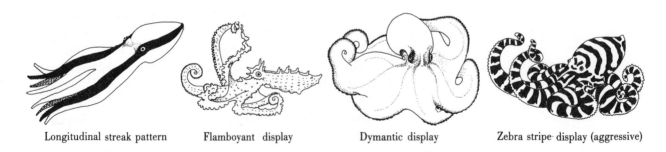

| Longitudinal streak pattern | Flamboyant display | Dymantic display | Zebra stripe display (aggressive) |

Figure 15-10 Coloration and body shape displays in octopus. Individual cephalopods are capable of expressing pigmented patterns from light to dark in a wide array. Pigment changes can be accompanied by changes in body shape. Apparently homologous patterns and displays are expressed among various species of both octopuses and squids.
Modified from Moynihan, M.H., and A.F. Rodaniche. 1977. In: Sebeok, T.A., editor. How animals communicate. Indiana University Press, Bloomington, Ind.

addition to their social characteristics, cephalopods are both predators and favorite prey. Thus communication signals may be combined in various ways with considerations of camouflage and mimicry. Some of the differences in signals appear to depend on context. In the squid *(Sepioteuthis sepioidea)* there are several alarm patterns. Two of the most common, transverse bars and longitudinal streaks, may occur depending on whether the animals are in open water near the surface or toward the bottom near cover. When a group is in an intermediate situation, some individuals show the bars and some the streaks under threatening conditions, after which the whole group rapidly swims off together. The signal flexibility in cephalopods is so great, according to Moynihan and Rodaniche (1977), that an individual can send more than one signal at a time: "A squid in the midst of a group, for instance, can transmit at least three or four different 'messages' absolutely simultaneously, to completely different individuals and in different directions by assuming different color patterns on different parts of the body."

In addition to the visual patterns employed by cephalopods, some—particularly those living in deep water or those that possess shells *(Nautilus)*—may use touch and olfaction for communication. Some of the deeper-water cephalopods also appear to have developed light-producing organs for communication. The only sense that does not appear to be used for communication by cephalopods is hearing. These animals lack hearing organs, do not possess air bladders that are sensitive to sound as in some fishes, and most do not possess hard parts that could be used for sensing, transmitting, or producing sounds. Thus their lack of use of hearing may simply be an evolutionary constraint; that is, their bodies were not predisposed toward hearing. Moynihan and Rodaniche (1977) also suggest that sound communication by cephalopods might be risky and thus disadvantageous. Some sound propagates long distances in water. With the animals being highly sought prey, sounds might unnecessarily expose them to predators.

PHYLOGENETIC COMPARISONS OF COMMUNICATION

One question, which is partly stimulated by viewing humans at a communication peak, is whether there is a clear trend from simple to advanced in the phylogeny of communication. One cannot answer for sure, but it appears to be "no."

The communication of most invertebrates has received little attention. Thus much information is lacking and of that which exists there have been few if any reviews for organisms other than arthropods and mollusks. If and when there is communication in the simpler animals, chemical and tactile signals appear to dominate, as would be expected from the limited sensory capabilities of these organisms. Depending on how one views the nature of the social interactions in colonial coelenterates, the degree of social behavior is also limited in the lower invertebrates. One would not expect much communication in the simpler animals in the first place. The simple interactions that do occur among these animals can be and probably are accommodated by the simplest of signals. Such signals may be highly species specific, but a given species would not have to possess much of a repertoire.

Above the simpler invertebrate forms, the wide diversity of more advanced organisms cannot be placed together on a single ladder or path of related, increasing advancement. The most advanced communication (in terms of repertoire size, diversity, complexity, and specificity) occurs in species with the most sensory and

Figure 15-11 Visual, poster coloration in a cichlid fish *(Aequidens paraguayensis)*. From Timms, A.M., and M.H.A. Keenleyside. 1975. Zeitschrift für Tierpsychologie 39:8-23.

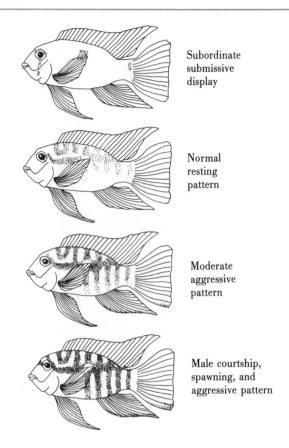

Subordinate submissive display

Normal resting pattern

Moderate aggressive pattern

Male courtship, spawning, and aggressive pattern

social development and where the surrounding biological community is most complex. The roles of physical and biological environmental factors in the evolution of communication are discussed further later in this chapter.

Numerous examples of convergence in communication exist. Cichlid fish (Figure 15-11) use visual signals, changes in body patterning and color, that are similar to those seen in cephalopods, but fishes and cephalopods are not at all closely related. Cephalopods are much different from their other molluscan cousins, and cichlid communication is much different from that found in some other fish even in the same order. The communication of fish in complex marine reef communities may rival and resemble the communication of insects and birds in complex tropical rain forests far more than it resembles the simpler communication of fish in an arctic lake.

Collias (1960) classified bird and mammal calls in five groups: food calls, warning calls, sexual calls, parent-young calls, and group-movement calls. He noted the apparent convergence among species within these categories. Kiley (1972) discussed convergence in 14 context categories among 50 different ungulate species for which information was available.

Eugene Morton (1977, Hopson 1980) suggests that there are even some basic underlying structures in bird and mammal (including human) vocalizations that correlate with emotion or mood. In particular, he has identified three common categories: the growl, bark, and whine (Figure 15-12). The growl is a harsh, low-

Figure 15-12 Generalized "growl," "bark," and "whine" vocalizations of birds and mammals. The frequency and time are variable, depending on species.

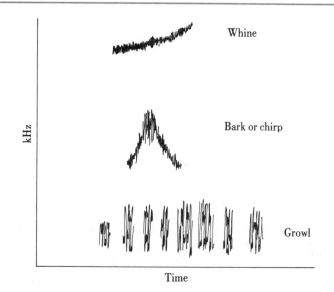

frequency vocalization associated with anger. The higher-pitched, often rising-in-pitch whine is given by immature individuals and others when fearful, friendly, or appeasing. The high- versus low-pitched sounds may be an example of antithesis (this concept is explained in detail later in this chapter). The low-pitched growls may be the vocal equivalent of puffing up and exaggerating one's size, as lower-pitched sounds are otherwise associated with bigger individuals. Phenomena such as shouting may also fall into this category (Zahavi 1979). In between the high- and low-pitched sounds are the barks, which are chevron shaped, rising and then falling in pitch, which indicate interest or curiosity. The characteristic chevron pattern shows up even in the chirp of a Carolina wren; when a recording of the chirp is slowed down, it sounds remarkably like the bark of a small dog. In another example, the trill alarm of Belding's ground squirrels has a typical chevron shape. The same patterns also underlie the emotional intonations of human speech. For example, "I love you" and "please" are normally said in the rising and higher-pitched whine pattern. "Wow" produces the chevron pattern, and "I'll kill you," if meant, comes out in a typical low-growl pattern. In fact, it is difficult to say such phrases convincingly without using the proper intonations.

The comparative study of communication, in spite of an already vast literature on the subject, is still in its infancy. Most information is concentrated and biased toward only a few taxonomic groups. Many issues, definitions, and approaches still must be clarified before meaningful comparisons can be made. Tantalizing questions and suggestions abound, but we still are not quite sure what to do with them. Brown pelicans, for example, are highly social and live in colonies near other colonial species, surrounded by potential predators and competitors. Yet, according to one of the most exhaustive studies of brown pelicans (Schreiber 1977), the pelicans form pairs and reproduce with a repertoire of only five signals. Other birds (hence close phylogenetic relatives) that also are colonial and live alongside the pelicans (under similar environmental constraints) have much larger and more complex repertoires of signals. Why are there such marked differences?

Wilson (1972) simply counted the number of signal categories or displays known in honeybees and other social insects and found them to range from 10 to 20 per species. Moynihan (1970) compared 30 species of vertebrates and found the number of displays to range from 10 to 37 per species. There appears to be more variability among species within a class, such as within fish or mammals, than between classes. The average number of displays was slightly greater in mammals than in birds and in birds compared with fish, but the mammals with the fewest (deer mouse [*Peromyscus maniculatus*] and a monkey [*Aotus trivirgatus*], each with 16 displays) had smaller repertoires than the fish with the most (*Badis badis*, which had 26 displays). Of the mammals studied, the rhesus monkey has the largest repertoire with 37 displays.

A major problem in comparing animals on a basis even as simple as just counting displays is that there are no universal, standard ways of categorizing behavior (Chapter 3), and different observers may get quite different results even with the same species. Different techniques can produce not only relatively minor, quantitative differences in results but also major qualitative differences, as was seen in the controversy over honeybee communication. Another complication arises when a signal has different meanings to different individuals or in different contexts: is that a single display or a number of displays? Such problems make it difficult to compare communication between different species, and there remains much work to be done in this field.

In spite of the problems in comparative studies of communication, however, there has been much progress and considerable theoretical development. The following sections consider the evolutionary aspect of communication with the sensory modalities and influence of the physical environment. Other influences, particularly those resulting from the presence of other organisms, are then discussed.

SENSORY CHANNELS AND CONSTRAINTS ON SIGNALS FROM THE PHYSICAL ENVIRONMENT

The physical environment provides numerous options and raw materials that may be used to make signals: molecules, light waves of various frequencies, electrical fields, and mechanical waves, vibrations, or movements of objects including air, water, and solid objects. All that is required is that organisms have the capacity to "tune into," that is, sense, these sources. (Senses per se are discussed in Chapter 16.) The physical environment, obstacles that it contains, and the physical properties of the various media such as light, sound, and molecules not only provide opportunities for signals but also impose limitations on them. These opportunities, limitations, and the resulting signals in various media are surveyed in the following sections.

Chemical Communication

Molecular messages generally fall into two main classes: releasers, which affect another's behavior (such as alarm and trail substances), and primers, which affect another's physiology (such as queen substance in bees and odors in the urine of male mice that cause abortion of fetuses in females impregnated by other males—the Bruce effect mentioned in Chapter 11).

Chemical messages, depending on how far one wishes to extend the definition

Classification of Chemical Interactions Between Organisms*

Intraspecific Chemicals
Pheromones

Releasers (affect behavior directly)
Primers (affect physiology such as growth and development and may affect behavior indirectly)
Autoinhibitors (prevent others from settling or living too close)
Autotoxins and wastes (may be toxic, inhibitory, or without selective effect)

Interspecific Chemicals (Allelochemicals)
Allomones (favor emitter against receiver)

Repellents
Escape substances (as with ink clouds used by cephalopods)
Suppressants
Venoms
Inductants (like primers among pheromones)
Counteractants (neutralize chemicals from other)
Attractants (baits or lures)
Depressants and wastes (selectively neutral to emitter but disadvantageous to receiver)
Kairomones (favor receiver, with or without advantage to emitter)
Attractants (to food, to nest sites, for pollination, etc.)
Inductants
Alarm or *warning* signals
Stimulants

Modified from Whittaker, R.H., and P.P. Feeney. 1971. Science 171:757-770.
*Whether or not these are considered as communication depends on one's viewpoint. If considered as messages or "information-carrying" molecules, they are known technically as semiochemicals. Names of categories most commonly accepted as communication are in italics.

of communication (see box on this page), generally are called *pheromones*. They may well have been the first signals used by primitive organisms. They may be ancestors or descendant substances of internal hormones—at least they seem closely related in modes of production and action and differ mostly in whether they carry the information internally or from animal to animal. The same chemical may be used internally and externally by different organisms. Chemicals probably are the most universal of communication signals. Even animals that rely heavily on other modes also commonly use molecular messages.

The advantages of chemical communication are that it transmits through darkness and around obstacles; it has the greatest potential range (at least in air—often carrying for several kilometers); and it may be stable and last a long time, such as with many of the pheromones used in territorial boundary marking by deer, rabbits, and wolves (Figure 15-13). Chemical communication can transmit into the future and does not depend on the continued physical presence of the sender. Another advantage is that pheromones are relatively private in that the receiver needs specific receptors to receive them; natural enemies find it hard to eavesdrop.

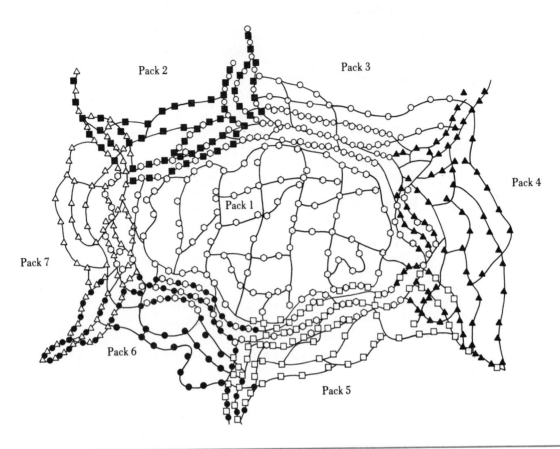

Figure 15-13   Travel routes *(lines)* and raised-leg-urination scent mark positions *(dots)* of a wolf pack and six neighboring pack territories. The different packs are represented by different symbols (combination of shape and whether open or dark). Territory boundaries are clearly visible in this diagram.
Modified from Peters, R.P., and L.D. Mech. 1975. American Scientist 63:628-637.

The disadvantages of molecules, depending on the circumstances, are that they are slow to transmit (the larger the molecule, the more slowly it diffuses) and they may either fade too rapidly and be lost or not fade quickly enough and interfere with subsequent messages.

Other considerations (not necessarily advantages or disadvantages) for chemical messages are that different molecules (hence signals) require different glands for their production and release. Black-tailed deer, for example, have six different pheromone-producing glands (Figure 15-14). Many social insects are literally walking batteries of glands (Figure 15-15). Graded aspects of the messages depend on the concentrations of the chemicals. The concentrations may depend not only on how much was produced or released but also on the fading with time (Figure 15-16), distance, and wind conditions (Figure 15-17). With time the molecules may diffuse into the environment, or the molecular configuration may change. The time that a signal lasts, the distance it travels, and the ability to localize or pinpoint it depend in part on what is called the Q/K ratio, where Q is the emission rate and K is the threshold concentration at which animals respond. Sexual attractants have high Q/K ratios, whereas alarm and trail substances have low ratios.

Figure 15-14 Pheromone-producing glands in black-tailed deer. Glands include metatarsal, anal, and interdigital. Scents from the metatarsal are rubbed onto other regions of the body, such as the forehead, which in turn is rubbed on twigs. Odors also are left on the ground and diffuse into the air.
Modified from Muller-Schwarze, D. 1971. Animal Behaviour 19:141-152.

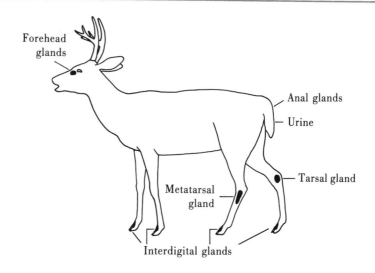

Figure 15-15 Pheromone-producing glands in insects: honeybee and ant. Many of the glands are believed homologous among species, and some are unique to different species. Dufour's gland is important in trail marking.
Modified from Wilson, E.O. 1965. Science 149:1064-1071.

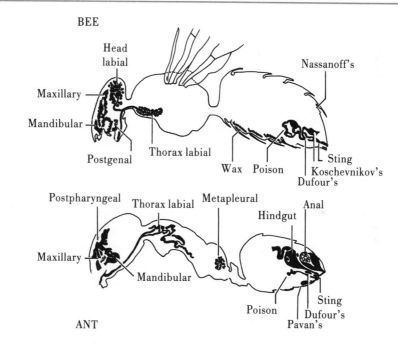

Figure 15-16 Signal fading with time as indicated by male hamster attention (sniffing time) toward defensive ("flank marking") pheromone artificially deposited in corners of their cages. Males ceased responding to male pheromones that were between 40 and 50 days old but continued to respond to female vaginal odor marks 100 days old.
Modified from Johnston, R.E., and T. Schmidt. 1979. Behavioral and Neural Biology 26:64-75.

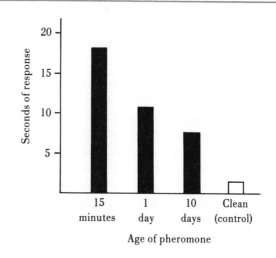

Figure 15-17 Active space, where the Q/K ratio is sufficient to produce a response, of a pheromone such as the gypsy moth sex attractant. Although the odor covers a larger area, it is sufficiently dense to produce a response only within a particular region. A breeze may carry the active space far downwind, but turbulence from strong wind may rapidly reduce the concentration and reduce the active space. Modified from Wilson, E.O., and W.H. Bossert. 1963. Recent Progress in Hormone Research 19:673-716.

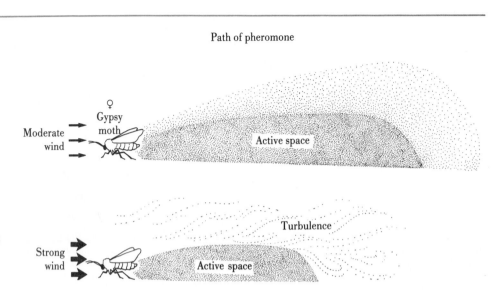

Pheromone characteristics also depend on their molecular composition. The molecular alphabet consists largely of three letters, carbon, oxygen, and hydrogen, the most common compounds of virtually all biochemicals. The constraints on the size and amount of molecular specificity are:

1. Amount of information to be coded
2. Stability, in terms of length of time and gradient
3. Metabolic cost of synthesis, storage, and transport (including volatility)

Insect pheromones, for example, usually involve 5 to 20 carbon atoms. The smaller ones include alarm substances that have molecular weights of 100 to 200. They should diffuse or disperse rapidly (to spread the message), and they do not have to be specific or carry much information—just a simple message such as "Danger!" Therefore they can (because of low information content) and should (for fast diffusion) be small molecules. Insect sex attractants, on the other hand, need to carry more information and be highly specific, but speed of diffusion is less of a factor. Accordingly, those molecules are larger, having molecular weights of 200 to 300.

Auditory Communication

The advantages of using sound for communication are many. It can go around obstacles, such as in a forest, or be used in the dark; it is much faster than chemical communication. It is much more flexible (is easily modulated and a variety of sounds can be produced by a single organ such as the larynx), and it allows much complexity, permitting lengthy and complex messages. Sound waves can carry a great deal of information.

Sound also has disadvantages. Sound is subject to interference or attenuation, and it distorts with distance. Thus only the simplest of messages can be sent over long distances in the air. (Low-frequency, complex signals are claimed to travel for miles in the ocean.) Interference may arise not only from echoes and previous messages but also from other sounds and vibrations (background noise) in the environment. Major disadvantages of auditory communication are that it requires considerable energy to produce signals, they must be produced repeatedly, and they are often easy for natural enemies to overhear.

The flexibility and complexity permitted by sound are familiar in human speech and in the vocalizations of cetaceans (whales, etc.) and birds. Auditory communication via patterned sound waves also is used by other mammals, reptiles, amphibians, many fish, many species of insects, and a number of marine arthropods.

The ability to produce and interpret large amounts of information carried in sound requires advanced sound-producing and sound-sensing organs and a well-developed nervous system. Most organisms, aside from primates, cetaceans, and birds, apparently do not have the neural capabilities or else have not invested (in the evolutionary sense) much of their nervous system in the analysis of complex auditory discrimination. (Bats and owls of course have done so for prey capture rather than or in addition to communication.)

Among pairs of animals of several species of birds and some mammals, the male and female sing in duet, that is, together. Crane duetting is illustrated in Chapter 4. The duets in some species are meshed so perfectly that the sound seems to come from a single individual.

Sounds may be transmitted by the air or water or via the substrate. Many spiders, for example, communicate by vibrations of the web or other substrate. Mole rats *(Saplax ehrenbergi)* use seismic signaling to communicate; they bang their heads against their underground substrate to produce signaling vibrations (Rado et al 1987). Water strider bugs communicate by setting up vibration waves on the water surface (Wilcox 1979).

Tactile Communication	This form of communication generally is found in situations of close bodily contact between individuals, as might be expected. Tactile communication often occurs in conjunction with olfactory, visual, or auditory signals. Three important general functions that involve tactile signals are:

1. **Reciprocal feeding.** This involves passing food from one individual to another, usually from mouth to mouth. It occurs among many social insects, some birds, and a few mammals. The request for such feeding may be made by contact signals. For example, some insects tap their mouthparts together, and in many species of birds, such as gulls, chicks peck at the beaks of their parents.
2. **Grooming and initiation of grooming.**
3. **Initiation of physical transport.** Some species, particularly among social insects, may pull on another's mandible or bite at the neck of another to initiate carrying of one by the other.

The advantages and disadvantages of tactile communication are difficult to compare with those of the other sensory modalities. The major constraint obviously is that direct contact is required. Tactile communication might be viewed as a vibration sense, something like an auditory sense at zero distance. Assuming that direct contact is involved, tactile communication would require less specialized structures for sending and receiving messages than does hearing. Almost any external part of the body could be used for touching or bumping another animal, and most animals have numerous touch receptors on the surfaces of their bodies. All that is required is the generation or evolution of a message-carrying function for these contacts; the signals must acquire meaning or, in other words, be interpretable.

Visual Communication

The advantages of visual communication include: (1) it is transmitted instantaneously; (2) it may carry a large amount of information, assuming the receiver's eyes and brain are capable of processing it all; (3) it is highly directional, permitting the source to be located; and (4) some aspects, such as body coloration, are permanent, involving only an initial expenditure of energy and needing to be produced only once (although the subsequent uses of some structures entail much movement and energy expenditure).

Visual displays often involve movements of extensions of the body, such as the head, ears, legs and wings, and tail. Elephants make much use of their trunks, ears, and heads for communication. Tail movements, sometimes in association with rump attention, serve for communication in many species. Kiley-Worthington (1976), in an extensive review of mammalian tail movements, reported that tail (and head) elevations were associated with muscle tonus related to movement. In general, upright postures are correlated with preparation to move, alertness, and warning. Lowered tails are correlated with relaxation, fear, nonaggression, and submission. Lateral, wagging movements are generally associated with locomotion, intervals between other behaviors (Hailman and Dzelzkalns 1974), and frustration and inhibition. There are numerous species-specific differences, and both ritualization and exaggeration are evident in many cases. In particular cases, there have been a variety of hypothesized functions such as deer tail-flagging for warning (Alvarez et al 1976), follow-contact signals, or as means of displaying dominance.

The chief disadvantages of visual communication are: (1) it cannot be used over too great a distance, depending on the visual acuity of the receiver; (2) it is blocked easily by obstacles so that the sender and receiver must be in direct line of sight; and (3) it requires light or the production of light.

Vision is perhaps the sensory channel most easily dispensed with for communication (although it still may be vital for flying, predation or avoiding predators, and living in trees). Almost no species relies wholly on vision for communication.

Electrical Communication

A number of electrical field-producing and field-sensing fishes, such as in the African family Mormyridae, communicate with each other via electrical signals in a manner analogous to the use of sound by other animals (e.g., Hopkins 1974, Kramer 1978). Hagedorn and Heiligenberg (1985) compared the signaling modulations of electric organ discharge (EOD) rates in two different genera of gymnotid fish and found the signals to be remarkably similar. Examples of modulated EODs are shown in Figure 15-18, and diagrams of comparable discharge patterns are given in Figure 15-19. The fish in both genera established dominance hierarchies with EOD signals and used EOD "chirps" in courtship.

Communication by Multiple Channels

Complete messages may depend on combinations of component signals. Simple signals, that is, with a single meaning, may be composed of several components. A defensive threat, for example, may simultaneously involve sound, exaggerated movements, and perhaps even offensive odors. Another way in which combinations of signals are used is to modify the meaning of one signal by information carried in other signals. The lowered and outstretched neck of a goose, for example, may mean a threat if accompanied by a hiss or a welcome if accompanied by a cackle (Figure 15-20). Other good examples can be seen in zebras and ants (Figure 15-

Figure 15-18 Modulations
in instantaneous frequency
characterize chirps in the
electric fish *Apteronotus
leptorhynchus*. **A,** A chirp of
short duration; **B,** and **C,**
increasingly longer chirps;
these are characterized by a
steeper and higher rise in
instantaneous frequency and
by an interruption of the
fundamental electric organ
discharge cycle. A weak
signal of higher frequency
appears for the period of the
interruption.
From Hagedorn, M., and W.
Heiligenberg. 1985. Animal
Behaviour 33:254-265.

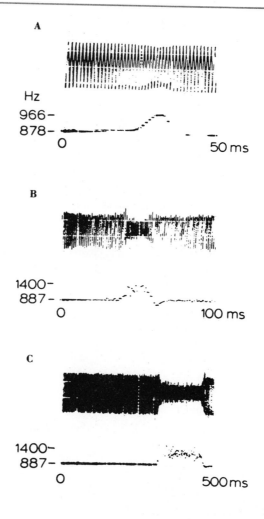

Figure 15-19 A schematic
representation of the signals
used in courtship in the
electric fishes *Eigenmannia
virescens* and *Apteronotus
leptorhynchus*. The abscissa
displays time, the ordinate
frequency, with individual
scales for each row of
signals.
From Hagedorn, M., and W.
Heiligenberg. 1985. Animal
Behaviour 33:254-265.

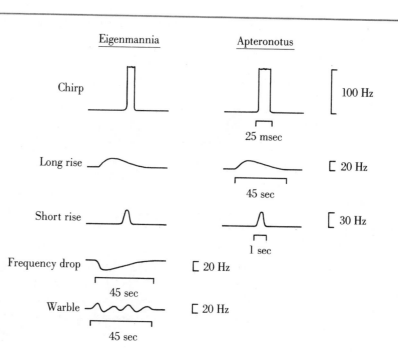

Figure 15-20 Greylag goose postures in a variety of situations. Modified from Tinbergen, N. 1965. Animal Behavior. Time-Life Books, New York.

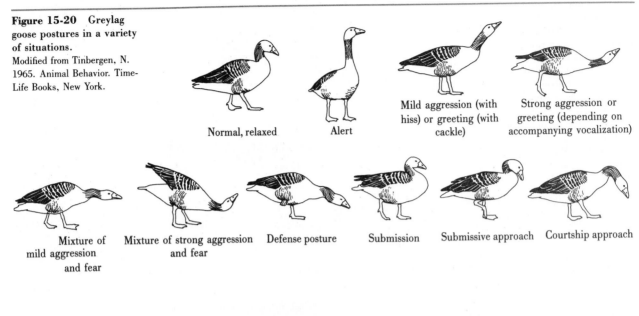

Normal, relaxed   Alert   Mild aggression (with hiss) or greeting (with cackle)   Strong aggression or greeting (depending on accompanying vocalization)

Mixture of mild aggression and fear   Mixture of strong aggression and fear   Defense posture   Submission   Submissive approach   Courtship approach

Figure 15-21 Metacommunication. Play-intention movements of lions. The lowering of the anterior half of the body indicates that subsequent actions are meant in a playful rather than a serious manner. Similar invitation to play is seen in canids.
From Schaller, G.B. 1972. The Serengeti lion. The University of Chicago Press, Chicago.

Figure 15-22 Postural indications of status in wolves. The middle wolf in the background is a female in a normal nonrank-displaying posture. The three in the foreground are, left to right and forward, top- to lowest-ranking males, respectively. Rank is indicated by position of tail, head, ear, leg, and fur plus associated movements. From Zimen, E. 1976. Zeitschrift für Tierpsychologie 40:300-341.

9). Note that in these examples one of the channels is discrete, and the other is graded. In the zebra's message, for example, the position of the ear is either forward or back, hence discrete, whereas the mouth can be open to a variable, graded extent.

Some of these complex, multiple-channel messages where the meaning is modified by other signals or context are referred to as *metacommunication*. Metacommunication is a two-stage communication wherein the first signal modifies or alters the meaning of subsequent messages. In essence, it creates a context. Two of the main categories of communication that have been dubbed metacommunication are:

1. Invitation to play. The first message is that what follows is only "in jest" or "for the fun of it." This is seen in cats, dogs, humans, and others (Figure 15-21).

2. An indication of social status. The message sender indicates by its overall posture, carriage, and actions where it is on the social ladder. Examples for wolves are shown in Figure 15-22.

The total communication repertoire for any given species may involve more than one modality. And within a particular taxonomic group, such as a family, order, or class, different species may vary considerably in the proportion of their total communication repertoire that is transmitted via one modality or another. As might be expected, the modalities used by any one species depend on environmental and life-history considerations. The proportions of their signals that fall into different sensory modalities can be placed on a series of continua (Figure 15-23).

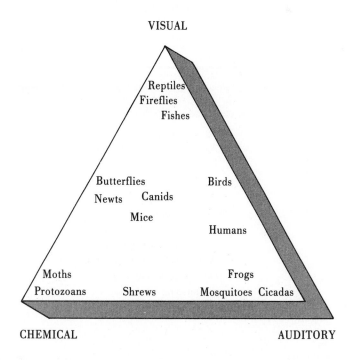

Figure 15-23   **Role of different sensory channels, on a proportional basis, in the total communication repertoire of various types of animals.**
Modified from Wilson, E.O. 1975. Sociobiology, the new synthesis. Belknap/Harvard University Press, Cambridge, Mass.

THE EVOLUTION OF ANIMAL COMMUNICATION

General Considerations

How do animals begin using signals in the first place, how does communication evolve, and what forces it into particular patterns? Otte (1974) describes the general picture for the origin of signals:

> The usual preconditions of evolving a new signal are (a) the existence of a characteristic, X, in one class of organisms, A, relevant to the activities of another class, B; (b) the existence of a receptor system in B that permits them to perceive X; and (c) a survival or reproductive advantage to A resulting from B perceiving X. Characteristic X becomes a signal as soon as selection begins to favor its display because of its effect on B. Perhaps most signals are postulated to have arisen from characteristics originally having other functions, but we cannot rule out the existence of signals that arise de novo. The sudden appearance of a new color trait to which other organisms respond is an example. The basic requirement is that the new attribute, however rudimentary, have positive selective value at the outset in the context of information transmission.

Otte (1974) proceeds to give examples using defense signals in predator-prey relationships and in grasshoppers and crickets.

On the basis of contexts in which messages are given, several general, common functions of signals have been categorized. The list is more or less arbitrary (see box on this page). It is taken largely from information in Brown (1975) but has been modified to account for recent information. Smith (1977) and others, incidentally, have objected to the functional classification of messages and have provided an alternate list. For a review of Smith's treatment of communication, see Hailman (1978 c).

Commonly Observed Contexts or Inferred Functions of Animal Communication*

Survival (group)
  Assembly and recruitment
  Leadership and following
  Contact signals
  Alarm and distress
Social spacing
  Individual and/or caste recognition—status symbols
  Aggressive threat and inhibition
  Defensive threat
  Submission

Reproduction
  Courtship and pair formation
  Pair maintenance
  Stimulation (e.g., of ovulation)
  Precopulation
  Nest relief and exchange
  Synchronized hatching
  Parent-young
Miscellaneous and general
  Incitement to hunt and forage
  Grooming
  Greeting
  Initiation of physical transport
Metacommunication
  Invitation to play
  Status signaling (also listed above under social spacing)

---

*There may be several specific signals in any category for any species.

Figure 15-24 Similarity of alarm calls given by different species of birds. The long drawn-out horizontal call is given under conditions of likely danger. The gradual starting and stopping of the call create a ventriloquial effect, which makes it difficult to locate the source of the call. The sharp vertical calls, on the other hand, are given during mobbing or scolding circumstances where the bird may even be attacking the predator and its position is already known.
Modified from Marler, P. 1959. In: P.R. Bell, editor. Darwin's biological work. John Wiley & Sons, New York.

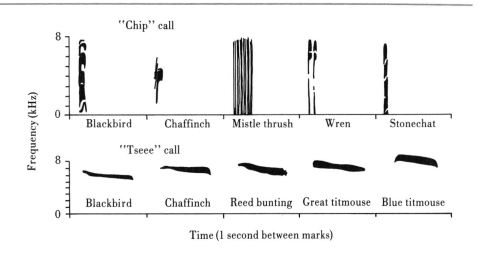

The difference between alarm and distress calls is the difference between alerting to possible trouble and signaling that one is in trouble. Distress calls may be calls for help, or one might argue from inclusive fitness that they could signal, "I'm in trouble; save yourself!" These calls are similar in many species of birds (Figure 15-24), thus are generalized, and may be understood by other species. Some convergence may exist even among widely different organisms. The distress screams of chickens, woodpeckers, rabbits, some ungulates, and humans, just to name a few examples, are remarkably similar to a casual listener. The alarm calls are thought to be of a sound quality that makes them difficult to locate, although there have been conflicting observations and opinions.

Aggressive and defensive threats differ in their intent and often in their appearance. Aggressive threats may use intention movements and incomplete motions to attack. Defensive threats, on the other hand, frequently use more symbolic signals that are less apparent to an outside and uninformed observer. In cats, for example, the defensive threat involves a sideways, arched-back posture, raising of the hair, backing movement, a grimacing facial expression, and hissing—the classic Halloween-cat posture. Although initially uninformed, most animals, such as dogs, that ignore the message and press forward usually learn quickly what it means. Within the species the messages are more likely to be understood innately. Defensive threat postures commonly involve an exaggeration of body size, which can be seen not only in the cat but also in, for example, many owls and some insects that spread their wings when confronted. Frilled lizards expand a collar of skin around the neck, raise their bodies, and hiss.

Reproduction signals, such as in courtship and pair formation, may be specific (as discussed in Chapter 11) and often are easy to locate; they may be highly directional. Most of the communication of other animals that people are commonly familiar with or likely to encounter is directly or indirectly related to reproduction.

In the miscellaneous category, there are several less familiar signals. Incitement to hunt or forage occurs with bees and can be seen in the ritualized prehunting behavior of African wild dogs. Some insects, such as ants, may request or initiate physical transport (e.g., to new nests) by tapping their antennae. The one to be carried folds up in the pupal position and is then picked up and carried by the other. Functionally similar, or analogous, messages also can be observed in other species, particularly young mammals, including humans. Social grooming is quite common in mammals and some other groups; it obviously is derived from cleansing behavior. This type of grooming appears important in social bonding and is performed even when the individuals are clean and possess no parasites.

Evolutionary Origins of Displays

It is chiefly in displays that one finds the behavioral, evolutionary characteristics of ritualization and exaggeration. Most displays are thought to have their evolutionary precursors in the basic locomotion, maintenance, and feeding movements of animals. An example of a display in mallard ducks that is clearly derived from preening is shown in Figure 15-25. Hailman (1977) includes intention movements from general locomotion along with those from several other categories of movement, plus autonomic responses as major sources for the evolution of visual displays (see box on page 535).

Conflict behavior (when an animal is stimulated to both flee and fight at the same time) is also thought to be an important source. The animal may hesitate briefly, perhaps in an intention movement, or engage in a redirected or displacement activity (Chapter 2) in the presence of its opponent. If the result is selectively advantageous to the actor, even by only a slight amount, the behavior may become incorporated into the species' repertoire. During many subsequent generations the signal may change and become more prominent and effective. Some displays may become so altered that they no longer resemble the original, nonsignal behavior.

Figure 15-25 Speculum-flashing or preen-behind-wing display in the mallard duck. The speculum is the brightly colored patch with contrasting borders on the secondary feathers of the wing. During the display males orient laterally toward the female, raise the secondaries on that side in a preening-type motion, and run the bill behind them, occasionally making a rattling noise. The display clearly seems to have been derived from basic preening movements.
From McKinney, F. 1975. The behavior of ducks. In: E.S.E. Hafez, editor. The behavior of domestic animals, ed 3. Bailliere Tindall & Castle, London.

Intention Movements and Other Effector Output as Sources for Visual Displays, as Reported in the Literature

Intention Movements of Skeletal Action Patterns	Autonomic Responses
Maintenance and basic activities	Pilomotor actions of fur and feathers
Locomotion	Sleeking
Rolling and wallowing	Fluffing and ruffling
Scratching	Respiratory responses
Bill wiping	Yawning
Preening and grooming	Vasoresponses
Orientation of sense organs	Flushing
Foraging and feeding	Blanching
Reproductive behavior	Ocular responses
Mounting and copulation	Pupillary action
Penile erection and lordosis	
Nest building	
Parental care	
Agonistic behavior	
Fighting and combat	
Fleeing, flight, and protective responses	
Antipredator behavior	

From Hailman, J.P. 1977. Optical signals. Indiana University Press, Bloomington, Ind.

Signals in other contexts may develop from other original movements or borrow from signals that already are established. Huxley (1914), who first introduced the idea of ritualization, described, for example, European great-crested grebe courtship display, which involves mutual headshaking. The headshaking is believed to have derived from an appeasement display whereby two birds move their beaks away from each other, which in turn initially may have arisen from turning to flee. The mutual headshaking is then incorporated into a different display, the "penguin dance," which includes diving and the presenting of water plants by the two birds to each other. The part of the display involving the water plants is thought to have arisen from displacement behavior where the birds pick up vegetation during a conflict between hostility and sexual attraction.

Antithesis

A common property of many animal signals is known as *antithesis*. This means that messages with certain meanings are frequently expressed by postures and behavior patterns that appear opposite. A classic example, first noted by Darwin (1859), can be seen in the communication of dogs (Figure 15-26, *B*). Another good example is provided by gulls (Figure 15-26, *A*). Although this principle seems intuitively clear, opposite meanings and signals are not always so obvious on closer inspection. What is really the opposite of fear: anger, calmness, joy, confidence, or something else? These may be real emotional states with physiological bases, such as sympathetic or parasympathetic components of the nervous system. The subtleties,

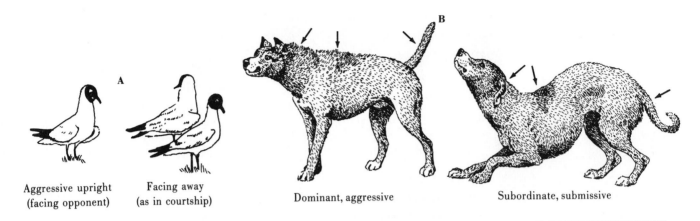

Aggressive upright
(facing opponent)

Facing away
(as in courtship)

Dominant, aggressive

Subordinate, submissive

Figure 15-26   Classic examples of antithesis in communication: **A**, Gulls, **B**, Dogs. Essentially opposite postures are used in the expression of "opposite" conditions.
A modified from Darwin, C. 1873. Expression of the emotions in man and animals. D. Appleton, New York.
B modified from Tinbergen, N. 1959. Behaviour 15:1-70.

however, are less clear and may easily become entangled in semantic or anthropomorphical problems.

The Nature of Signals Depending on Who's Listening: Selective Forces of the Biological Environment

After it was pointed out (Wiley 1983) and realized (e.g., Krebs and Dawkins 1984) that communication was not a matter of either information transfer or manipulation but could be both, a broader view and more coherent concept emerged. This view also permitted a better evolutionary interpretation. Rather than focusing narrowly on particular evolutionary constraints, such as the influence of physical factors like distance and attenuation from obstacles on sounds, it became necessary to view the full set of simultaneous and often opposing factors on the evolution of signals. Mate-choice sexual selection or distance between sender and receiver could select for loud calls, for example, whereas the presence of predators and competitors could select against such calls. Many factors and selective pressures can be present and opearating simultaneously.

Markl (1985), following Wiley's (1983) lead, categorized the selective advantages and disadvantages to sender and receiver in different dyadic (two-party) and triadic (three-party) relationships depending on whether or not the relationships are mutualistic and cooperative (Tables 15-4 and 15-5). Dyadic relationships are generally focused on target senders-receivers, whereas triadic relationships bring in outside participants that have a greater potential of exploiting the system, whether outside receivers or senders. In this expanded view of communication, incidentally, the information side of the old information-versus-manipulation dichotomy fits into the mutualistic or cooperative category in which selective advantages accrue to both sender and receiver.

In the intended or targeted dyadic communication, there is usually a reciprocating relationship in which the two participants trade roles as sender and receiver. That is, first one individual is sender and the other is receiver. Then they alternate and the receiver becomes sender and vice versa. This tradeoff creates a potential balance

Table 15-4
Communication dyads: bilaterally tuned relationships between sender and receiver. The profit balance, or total benefits minus total costs of interaction, for each of the participants is given as: gain (+) or loss (−) of fitness equivalents because of the communicative interaction of dyad members.

Type	Sender	Receiver	Examples of relationships
1	+	+	Intraspecific: mutualistic cooperation (e.g., male-female, parent-offspring, members of social group). Interspecific: symbiosis, aposematic signaling. Perfect, if gains are equal for sender and receiver; imperfect, if one gains more than the other (e.g., in agonistic communication).
2	+	−	Exploitation of receiver by sender. Intraspecific: ritualized combat (victory by threat) or social dominance hierarchy; cuckolded male investing alloparental care; immature males gaining experience by courting female who has to carry interference costs. Interspecific: interspecific interference; dominance by threat.
3	−	+	Exploitation of sender by receiver. Intraspecific: interloper attracted by signals of male competitor. Interspecific: detector of palatability of mimetic prey which concentrates on hunting for the mimic.
4	−	−	Wasted efforts of sender and receiver. Intraspecific: disturbed and unsuccessful courtship; both opponents of fight—after undecided threat duel—severely damaged, none of them winning resource; fighting or courting pairs attracting predator. Interspecific: inexperienced predator suffering from killing dangerous aposematic prey.

From Markl, H. 1985. Fortschritte der Zoologie 31:163-194.

Table 15-5
Communication triads: exploitation of communication dyad (I) by third participant (II) who can tune in as receiver II or as sender II. Profit balance of interactants: gain (+) or loss (−) of fitness equivalents because of communicative interaction between member of dyad and exploiter.

Type	Dyad Sender I	Receiver I	Participant II as Sender II (Mimic)	Receiver II	Examples of relationship
1	−			+	Sender I exploitation by predator or parasite which detect prey/host (= sender I) by their signals/cues which are selected to influence receiver I; food parasite/competitor attracted by food calls.
2	+	model		+	Generalized cooperation, intraspecific in courtship chorus, lek displays (attraction of males by male displays); interspecific in mixed flocking or cross-specific alarm calls.
3		−	+		Mimicry, detrimental for receiver, e.g., Batesian mimicry (receiver loses opportunity, search effort) or aggressive mimicry.
4	model	+	+		Mimicry, advantageous for receiver, e.g., Müllerian mimicry or cichlid egg mimicry.

From Markl, H. 1985. Fortschritte der Zoologie 31:163-194.

between the abilities of each side to manipulate the other, gives rise to a need to establish credibility, and increases the likelihood of honest exchange and correct information.

There still is plenty of room for one side to take advantage of the other in dyadic interactions, however, and triadic situations are wide open for exploitation, which is unavoidable in some cases. With exploitation comes the possibility of deceit, lying, and misinformation in the communication process. Furthermore, the characteristics of signals may vary depending on whether a relationship is mutual or exploitive and whether or not there are eavesdroppers or illegitimate senders. Thus the surrounding biological environment—other living organisms of the same and other species—can profoundly affect the evolution of communication.

The presence of deceit and misinformation, whether deliberate and conscious or not, in animal communication is fascinating and now well documented (e.g., Trivers 1985). Mimicry and camouflage, etc., in predator-prey interactions are familiar and have already been discussed in Chapter 9. However, deceit may also be common in intraspecific relationships. The lying and misinformation at times can be quite subtle and at other times may be surprisingly brash.

Some birds mimic other sounds, including the songs and calls of others birds, as well as other sounds from the environment. Vocal mimicry has been hypothesized to function in many ways, including the reduction of interspecific competition by acquiring and taking advantage of other species' alarm and territorial calls. Hindmarsh (1984, 1986) considered a number of alternate hypotheses and proposed that vocal mimicry in birds has no specific function per se but rather is a result of song learning by males under the pressure of sexual selection and simply permits them to incorporate a ready source of new sounds in the construction of sexually selected complex songs.

Intraspecific deceit occurs most commonly in relation to reproduction and in intraspecific competition over resources. Males, for example, may appear as females or immatures, enter the territories of other males, and sneak copulations (see Chapter 11). Animals may pretend to be bigger, better, or of a different social status than they actually are, and in various and diverse ways cheat and bluff (see Chapter 13).

One fascinating manner in which deceit crops up is through the use of alarm calls, "crying wolf." Munn (1986) reported two species of insectivorous birds that lead mixed flocks, serve as sentinels against hawks, and occasionally give false alarm calls (Figure 15-27). These birds feed on insects flushed by other birds. Sometimes when the others have an easy source of insects the sentinels distract them with a false alarm call, then proceed to go after the insects themselves. Moller

Figure 15-27   A to D
Tracings of spectrograms of true alarm calls, E to H, false alarm calls, I to L, and other (nonalarm) calls of the flycatching birds *Thamnomanes schistogynus* and *Lanio versicolor*. True alarms are given in response to real danger; false alarms during a chase after arthropods; and other calls in entirely different contexts. The nonalarm rattle (I) was used as a control in the experiment. The true and false alarms of *L. versicolor* are characterized by descending notes with simple, strong harmonic structure; both characteristics are not found in other calls. Note that for *L. versicolor*, the timespan is halved, whereas the frequency range is doubled compared with the results for *T. schistogynus*.
From Munn, C.A. 1986. Nature 319:143-145.

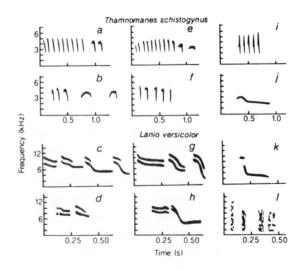

(1988) demonstrated a similar situation with great tits during winter feeding. False alarms would disperse or distract competitors at a food supply, whether the competitors were other great tits or other species such as sparrows.

Apparent deceit has also been reported in primates. Richard Byrne and Andrew Whiten (see Lewin 1987) reported a young male baboon that had been watching an adult female dig plant roots. He looked around, screamed as if being attacked, attracted his mother, which chased the other female off, then proceeded to eat the plant roots. Another case reported by Shirley Strum (also in Lewin 1987) involved a male baboon that had just killed an antelope, which it normally would not share with others. A female baboon, however, approached and started grooming the male. After his attention had been distracted from the antelope, the female snatched the food and ran.

The success of many types of deceit such as bluffing (in threat displays, sneaking of copulations, and some other forms of mimicry) is thought to be frequency-dependent. That is, the mimic must be much less common than the real thing, the model, or it quickly loses its effectiveness. One exception involves mimicry of deadly objects such as some snakes. Otte (1984:398) proposed that in such cases " ... the greater the potential harm to predators, the greater the protection afforded to mimics, and the model need not be numerically more abundant than the mimic."

The ramifications of deceit and misinformation for the evolution of communication are many. Andersson (1980), for example, suggests that a species repertoire of threat displays can become more diverse and complex through a sequence in which an action associated with actual aggression (such as baring the teeth before biting) takes on a display function, becomes ritualized and exaggerated, then loses value as an indication of the signaler's true likelihood of fighting because of bluff. That display is then replaced by a new signal that goes through the same sequence. This process of signal acquisition, ritualization, devaluation, and replacement proceeds until a large repertoire of threat displays accumulates and until further proliferation becomes limited by the availability of suitable behavior patterns.

An important influence of exploitive outsiders is that the communication between the "intended" target participants of a mutualistic dyadic interaction should become more secret, restricted, and less vulnerable to outside evesdropping and exploitation. Thus it will become more efficient, confined to narrower channels with less widely understood coding (i.e., it should be encrypted—as in the scrambling of satellite television transmission), and it should become quieter or otherwise less conspicuous. In addition to being less open to exploitation by natural competitors and predators, incidentally, these types of signals also become much more likely to be overlooked, and they are difficult for biologists to study. This may have led to increased attention to the loud and conspicuous signals and a bias in our understanding of animal communication. Perhaps more attention should be directed to the more subtle, cooperative communication signals of animals. It should be noted, however, that we have not been completely blind to such communication; honeybee recruitment communication is a prime example.

Costs of Communication

Relatively few studies have been made on the energetic costs of communication. For some modalities, such as visual, the energetic costs are practically unknown and difficult to assess. Perhaps the modality that has received the most attention is acoustic communication. Sound is an accessible medium for researchers because it is easily recorded for subsequent measurement and documentation. Furthermore, many animals produce calls in confinement where the associated energetic costs can be measured through oxygen consumption. In addition to physiological, energetic costs, vocalizations are easily detected by nontarget receivers and thus expose the signaler to costs of competition and predation or parasitism. Vocalizations can be costly, and they are energetically inefficient.

Recent estimates of the energetic efficiencies for calling insects and frogs range from less than 1% to 2% or 3% at most (MacNally and Young 1981, Ryan 1985, Kavanagh 1987). This inefficiency of vocalization apparently results from the transitory nature of sounds and the need for repetition and constant production, a substantial loss to friction in the sound-producing mechanisms, and often an unavoidable mismatch between the best sounds for a situation and the anatomical resources available to the animal (e.g., loud and relatively low-frequency calls from small animals).

In spite of the inefficiency, animals can manage to produce some incredibly intense sounds. Small frogs, crickets, katydids, and cicadas, for example, produce calls of over 90 to nearly 120 dB at close range (1 to 50 cm from the caller) (references in Burk 1988). Sustained sounds at that level at close range are capable of damaging human hearing.

Loud output through an inefficient mechanism requires a tremendous input of energy. Various studies (reviewed and referenced in Burk 1988) indicate that calling frogs and insects increase their metabolism during calling from 4 to 20 times over their resting rates. Some male frogs expend as much as 86% of their assimilated energy on calling (MacNally 1981) and can lose over 1% of their weight per day (Halliday 1987). Some insects may lose 2% to 3% of their body weight during 1 to 2 hours of calling (Dodson et al 1983). In some insects calling ranks close to flying in terms of energy demand (Prestwich and Walker 1981).

Costs at the time of sound production are not the only physiological costs. There can be significant costs to developing, maintaining, and carrying around the equipment. The trunk muscles used for vocalizations by male frogs, for example, may comprise as much as 15% of the total body weight (Taigen et al 1985). The comparable muscles in females of the same species amount to only 3% of the body weight.

Nonphysiological costs also may be considerable. Mortality rates of callers from predators attracted to the calling have been estimated at 1% or more per hour for some frogs (Ryan 1985) and over 90% per season for some insects (Cade 1979, Burk 1982).

What makes calling worth these incredible costs? Burk (1988) and Ryan (1988) suggest that it is runaway sexual selection. Mate-choosing females of many species (e.g., in frogs, see Halliday 1987) have been shown to respond the best to males that call the loudest, at the highest rates, and for the longest duration. This apparently has created a spiraling coevolutionary "arms race" that proceeds until it reaches the limits (in terms of physiological and nonphysiological costs) of the system.

The highest-intensity, most conspicuous sounds are given by unmated males. At the close approach of a female the males typically stop the loud calling and, depending on the species, frequently switch to less conspicuous kinds of courtship displays.

APE LANGUAGE STUDIES: DO ONLY HUMANS POSSESS LANGUAGE ABILITIES?

A lively debate has developed over the past several years about whether human language, whatever it is, is or is not unique. The controversy has centered on recent claims that great apes, particularly chimpanzees and gorillas, are capable of rudimentary forms of humanlike communication except that they lack vocal ability. On one side are those who say that the apes can communicate like humans. They include Beatrice and Allen Gardner, who initially worked with a chimpanzee named Washoe and who made the original claims. On the other side are persons such as Herbert Terrace and colleagues, who worked with a chimpanzee named Nim, who maintain that apes are not capable of humanlike communication. In between are others, including several working with chimpanzees, who are less dogmatic. The controversy has spread to include symbolic capabilities of other animals as well, such as horses, dogs, pigs, goats, parrots, pigeons, and even people.

To the immediate participants (i.e., the Gardners and Terrace) the debate appears to be quite serious and even has become emotional. Some outside, third-party observers, however, have been watching and reporting almost with an attitude of amusement and levity. Articles have included catchy titles such as "The great ape debate" (Benderly 1980) and "Does man alone have language? Apes reply in riddles, and a horse says neigh" (Wade 1980). Several articles have been preceded by short poems or literary quotes; the style of writing has been almost tongue in cheek in many cases; and endings have often been amusing or philosophical, such as, "If only Koko could write." Wade (1980) began one paragraph, "A chimpanzee who asked not to be quoted by name told *Science* ... 'Those who live in the academic jungle shouldn't ape the law of the jungle.'"

In line with the tone already set by others, the present status of the debate is introduced with the following story. A cat lover once claimed that she could speak and understand cat language. Several of her friends did not believe the story and decided to put her to the test. So they brought her a cat, waited for it to meow, then asked her what it had said. She replied she did not know because it was Persian.

In brief, there have been several ape-language projects and two major approaches. The initial findings and to a lesser degree the subsequent controversy have received wide publicity, so many of the names involved are likely to be familiar.

The Gardners, working with Washoe and then other chimpanzees, attempted to teach the apes to use the American Sign Language (ASL) of the deaf and compared their results with language acquisition in children (Gardner and Gardner 1969 to 1978). This signing approach (or variations of it) was subsequently used also by Roger Fouts at the University of Oklahoma (Fouts 1972, Fouts and Mellgren 1976), by Penney Patterson (1978), who worked with a gorilla named Koko, and by Terrace and colleagues (e.g., Terrace et al 1979), who used Nim (Figure 15-28). Nim, incidentally, is short for Neam Chimpsky, after a famous linguist, Noam Chomsky. The apes were raised, to various extents, in a homelike setting where they interacted with humans in a variety of ways, including attempts to use ASL. Logistical problems with Nim, however, included shortage of time, volunteers who had difficulty coping with the chimpanzee, scratches, and torn clothes. This led to Nim being taught by a large number of persons, which may have complicated the results. On the other hand, Terrace used video recordings of Nim's progress, which permitted subsequent reanalysis.

Figure 15-28 Symbolic communication in apes. Nim answering Joyce in response to Joyce's question, "Who?" From Animals, animals. © H.S. Terrace.

The second major approach uses external objects, or lexigrams, rather than signs. David Premack used a variety of plastic chips that were associated with English words (Premack 1972). Duane Rumbaugh and Sue Savage-Rumbaugh used a computer keyboard with a variety of symbols (Rumbaugh 1977, Savage-Rumbaugh et al 1980). Rumbaugh and associates first worked with a chimpanzee known as Lana and then two subsequent subjects, Sherman and Austin; the work was conducted at the Yerkes Regional Primate Center in Atlanta, Georgia, and the symbolic computer language was dubbed Yerkish.

Those using the signing approach, except Terrace's group, have been adamant that they have solid evidence that the apes understand the symbols and can form rudimentary sentences. Terrace and colleagues have been adamant that they cannot and that "In sum, evidence that apes create sentences can, in each case, be explained by reference to simpler nonlinguistic processes" (Terrace et al 1979). Those using the lexigraphic approach have generally taken somewhat of a middle ground. Savage-Rumbaugh et al (1980), for example, concluded:

> Symbols have merely served to replace or accompany non-verbal gestures the chimpanzee would otherwise employ ... Thus, it appears that chimpanzees, even with intensive linguistic training, have remained at the level of communication they are endowed with naturally—the ability to indicate, in a general fashion, that they desire another to perform an action upon or for them when there exists a single unambiguous referent.

Even the Rumbaughs have been entangled in the controversy, including a heated exchange with Thomas A. Sebeok in 1980 at a conference under the auspices of the New York Academy of Science. (The conference was described as "a celebration of deception in all its varieties.")

Figure 15-29 Symbolic communication between two pigeons. Pigeons were trained to spontaneously communicate with each other using color-coded keys on a sustained basis and without continuing human intervention. **A,** Jack *(left)* asks Jill *(right)* for a color name by depressing the WHAT COLOR? key. **B,** Jill looks through the curtain at the hidden (from Jack) color. **C,** Jill selects the symbolic name for the color while Jack watches. **D,** Jack rewards Jill with food by depressing the THANK YOU key. **E,** Jack selects the correct color as Jill moves toward her reward. **F,** Jack is rewarded with food.
From Epstein, R., et al. 1980. Science 207:543-545.

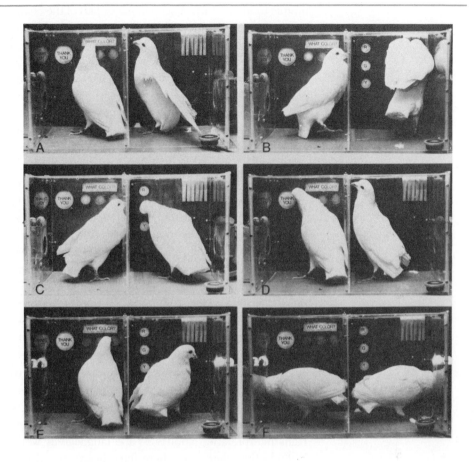

Other researchers have used other species. Pepperberg (1981), for example, worked with an African grey parrot and obtained results similar to the chimpanzee work. Epstein et al, including B.F. Skinner (1980), taught two pigeons, Jack and Jill, to communicate with each other using symbols (Figure 15-29). Epstein et al (1980) concluded:

> We have thus demonstrated that pigeons can learn to engage in a sustained and natural conversation without human intervention, and that one pigeon can transmit information to another entirely through the use of symbols. It has not escaped our notice that an alternative account of this exchange may be given in terms of the prevailing contingencies of reinforcement ... The performances were established through standard fading, shaping, chaining, and discrimination procedures (see Chapter 20). A similar account may be given of the Rumbaugh procedure, as well as of comparable human language.

Thompson and Church (1980), who maintained that Lana's behavior (in particular) could be explained by paired associative learning and conditional discrimination learning (Chapter 20), went one step further and simulated the entire process and results in a computer.

Charges and countercharges have been hurled back and forth between those who maintain that apes (and other animals) have language and those who believe they do not. The highlights of the arguments, as reviewed in part by Marx (1980), follow. Much of the debate is philosophical, concerning whether one views humans as unique from the start. A second major part of the controversy is semantic. Apparently no one is able to define what language is, or at least different workers do not accept the others' definitions. This is essentially the same problem previously discussed in defining communication. Stokoe (cited in Benderly 1980), remarked on the issue of chimpanzee signing, "When language is reduced to that level, then to me it is not language.... (The problem is that) everyone knows what languages are, but no one knows what 'language' is."

The third major source of disagreement concerns methodology. Terrace, for example, believes that what appears to be sentence construction by signing apes may simply be a result of the apes cuing to subtle, unconscious responses on the part of the researchers, a phenomenon known as the *Clever Hans effect*.

Clever Hans was a horse that lived around the turn of the century and supposedly was able to perform arithmetic calculations. When given a mathematical problem, Hans would tap the correct answer with his hoof. The Clever Hans phenomenon was explained by a psychologist, Oskar Pfungst. It was discovered that, rather than calculating the answer himself, Hans was simply tapping until he perceived subtle cues from his owner that he had tapped enough. The owner's cues were unintentional, and the owner was not aware that he was giving them. By 1937 at least 70 other cases of "thinking" trained animals were known including dogs, cats, a goat, and even a "mind-reading" pig. A French horse named Clever Bertrand seemed even more psychic than Hans; he was blind. (Bertrand probably used auditory cues.) The Clever Hans interpretation was applied to claims during the 1950s that dolphins could be taught to communicate with humans. Terrace and several others, including experts on zoo animals, animal trainers, skilled magicians, and several zoologists working with animal communication in general, believe that the Clever Hans interpretation applies to ape language as well.

From the other side, there have been several complaints that Terrace used too many and incompetent volunteers in the training of Nim and that other aspects of Nim's environment were not comparable with those of the other sign-using researchers. Fouts complained further that Terrace, a student of Skinner, was simply using Skinner's techniques of operant conditioning (Chapter 20), which would be expected to produce a passive, imitative subject. The Fouts maintain that their chimpanzees were different, but they will not permit Terrace access to their data and threatened to sue Terrace for copyright infringement for using films of their work that were produced for television. Rumbaugh and Savage-Rumbaugh (reviewed in Wade 1980) have said that Sebeok does not understand their methods and that the criticisms simply reveal what they consider to be their critic's incompetence.

A major problem underlying the signing technique is that ASL is a whole separate language, having originated in France and developed linguistically on its own. ASL signers in the United States who do not know French can easily communicate in sign language with French signers who cannot speak English. ASL has its own structure and properties, which cross the boundaries of other spoken and written languages. ASL has its own grammar, which is not well understood (by linguists) and which clearly is not the same as in oral languages. All that is known for sure is that ASL works as well for those who use it routinely as any other human language. People who understand ASL as their original language, primarily the deaf or children of deaf persons, can communicate as completely and subtly in ASL as other persons can with spoken language. People who have not acquired ASL naturally (i.e., by growing up with it) rarely master the language. This applies, unfortunately, to the ape researchers, and as noted by Benderly, what they are using is not ASL but a form of pidgin sign English: "This rather impoverished form is really not a language, but a manual code for English."

One aspect of the problem is that no method has been devised for knowing what another animal knows, the basic problem described in Chapter 1 and earlier in this chapter. Benderly (1980) gives a clear example of how this applies to language:

> An American diplomat newly arrived in South America once watched as his 10-year-old son Tommy, who knew no Spanish, played soccer with some neighborhood boys. Tommy spotted a hole in the opposing line. "Aqui! Aqui!" he shouted, and a teammate kicked the ball to him. Later, while complimenting Tommy on picking up the language so quickly, the father asked what "aqui" meant. "I don't know, Dad," Tommy said, "but it sure gets me the ball in a hurry."... Is this all that the apes have learned to do, or have they truly entered into the cognitive world of human users of language? In Tommy's case we can ask; in Washoe's we cannot.

The most recent exciting developments in the ape language field come from studies by Savage-Rumbaugh and colleagues, working with Kanzi, a pygmy chimpanzee or bonobo *(Pan paniscus)*. This species, not as familiar as the common chimpanzee *(Pan troglodytes)*, is rare both in the wild (in Zaire) and in captivity, but recently has attracted much attention from primatologists, since in many aspects of its social behavior it resembles humans more than common chimps do (Savage-Rumbaugh 1988).

The work with Kanzi, born in captivity in Atlanta in 1980, was designed with the ape language controversy well in mind. All work was videotaped, and careful single- and double-blind procedures were used to eliminate the possibility of Clever Hans effects.

Kanzi's performance has been remarkable. With the Rumbaughs' standard use of a geometric symbol system, Kanzi has mastered the use of 150 symbols. In a new twist in ape language studies, Kanzi can touch a symbol on a keyboard attached to a speech synthesizer, so that the appropriate English word is produced. Kanzi's observers not only use a keyboard to communicate with Kanzi but also use verbal English. Unlike previous studies that focused on production ("Can an ape make a sentence?"), the work with Kanzi has focused equally on comprehension ("Can an ape understand a sentence?"). As Savage-Rumbaugh points out, comprehension often develops faster than production in human infants. Kanzi is not caged but is with human companions throughout the day, can visit his mother whenever he wants, and stays with her at night.

As of 1988, a speech comprehension test showed that Kanzi responded appropriately 100% of the time to 109 of 194 spoken test words, and 75% of the time to another 40. He did nearly as well when words were provided to him by the speech synthesizer. He responded appropriately to 105 of 107 action-object utterances ("Kanzi, go get me a spoon"). He appears to possess some syntactic comprehension ability, since he responds differently to a word depending on its function in a sentence. Savage-Rumbaugh summarizes Kanzi's progress by saying "Kanzi appears to be the first ape who is truly acquiring language ..." (Savage-Rumbaugh 1988).

If that conclusion holds up, it raises the question of why Kanzi has developed more language than the other apes studied. Is it that he is a different species? Is it that he was exposed to a language environment at an earlier age? Answers to questions like these may emerge with further studies; at the moment all that can safely be said is that new excitement has been infused into an already highly controversial field.

There are many points of contention among those involved in the ape-language controversy, but we have covered enough. Do apes and other nonhuman animals possess humanlike communication capabilities, even if only in limited degree? In light of the quoted remarks by Epstein et al (1980), one might even ask whether humans possess language.

Regardless of which way one views this issue, however, there is still a substantial difference between humans and all others species, even if the difference is only quantitative. Other species use communication largely or entirely during direct interactions; and in most cases the signals carry information about the animals themselves. The number of such signals in a repertoire varies from species to species, but depending on how signals are identified and counted, the maximum number per species in the wild apparently does not exceed 40 or so (e.g., Moynihan 1970). Human communication is at least related to that of other animals in that humans also possess a certain repertoire of nonverbal, species-typical signals. The underlying structure of emotional intonation in human speech may share characteristics with vocalizations in other vertebrate species, as suggested by Morton.

Human communication has much more. In our language, humans make extensive distinctions and references to other things, places, events, and abstract concepts that may be external and far removed from ourselves in time, space, and even reality. Human language is such an important part of existence that it undoubtedly has shaped much of what the species is and may well be the most important, unique hallmark that distinguishes humans from all other animals.

Deese (1978) proposed that even consciousness is directly related to elaborate human use of language: "One of the chief linguistic functions of consciousness is to enable us to monitor our own speech. We listen to ourselves and we can tell, most of the time, when we have made a mistake, and we correct it."

Honeybees with their dances and ants with their odor trails, as well as many species with alarm signals, refer to other things (e.g., food and enemies) external to themselves. Even here the signals are limited, and the external items referred to are not conceptually distant. The difficulty in teaching apes, which involves much time, effort, and research money, in addition to the whole controversy over the results, simply indicates that even our closest (living) relatives are severely limited compared with humans. Even if the difference is only quantitative, it is so great as to almost qualify as qualitative. Whether or not it really is qualitative may be little more than an academic issue. Philosophically, one might ask whether it is a profitable research strategy in the first place to try to teach other animals a communication system, such as human language, to which they are not adapted. Instead perhaps humans should learn more about the communication systems that animals actually use. This brings us to some thoughts on future possibilities for studying animal communication.

FUTURE DIRECTIONS AND OPPORTUNITIES FOR UNDERSTANDING COMMUNICATION

Although the conceptual framework for understanding animal communication has moved forward by great strides and we have facts and information for many species and contexts, several aspects of the subject could profit greatly from further work and theoretical development. Rather than focusing so much on human language, although that is certainly an important, interesting topic, and trying to compare other animals with humans, perhaps we should do the reverse: investigate human communication in the light of knowledge from other species.

Studies of other species clearly need broader phylogenetic coverage, beyond groups such as honeybees, other hymenopteran and orthopteran insects, passerines and a few other birds, great apes and a few other primates. In all cases, more attention needs to be paid to the quieter, less conspicuous cooperative communication between individuals. An accumulation of such information across a wide phylogenetic base would permit better comparative studies and the development of new principles and theories.

Underlying all of the above is a need for improved methods of quantification and measurement of communication. Fresh thought should be given to creative, new approaches to permit better comparisons across species and situations than are possible with the current hodgepodge of techniques. The use of cybernetic information theory and H, for example, has numerous limitations and is vulnerable to much subjectivity. A substantial increase in quantitative information should permit useful revision, testing, and furthering of the theoretical aspects.

One topic in communication that has barely been tapped involves the cost of communication. This really has been measured only for acoustic signals. We need to find ways of assessing costs for other types of signaling to investigate the generalities derived from acoustic signals. The cost picture needs to be expanded to include subtle, inconspicuous communication and the noisier, conspicuous types.

SUMMARY

The subject of animal communication is complex, fascinating, and in many aspects controversial. The subject has, however, matured considerably from a conceptual standpoint in recent years. The varied and complex communication of honeybees, including round, waggle, and vibration dances, as well as numerous sounds and molecular signals, is used as a case in point for several of the problems and principles in communication. Some of these include problems of defining and measuring communication, the basic sender-signal-receiver principles of communication, advantages and limitations of using different sensory channels, and the selective forces of different (and often opposing) physical and biological factors in the environment. Several of these factors are described and discussed in the chapter.

Although honeybees are used as a prime and initial example of communication in animals, numerous other species and groups are introduced and discussed in the chapter as well. Although some persons in the past questioned whether communication even exists in species other than humans, we have come to realize that the world is awash in animal communication. One has only to sit and listen to birds, crickets, etc. and then consider all of the nonvocal signals and all that we cannot readily detect. We are usually tuned just to our own perceptual world, our Umwelt. However, we are now probing and slowly coming to understand the communication of other species. We have far to go. The chapter concludes with a consideration of ape language and a discussion of future directions and opportunities for the subject of animal communication as a whole.

FOR FURTHER READING   Andersson, M. 1980. Why are there so many threat displays? Journal of Theoretical Biology 86:773-781.
This theoretical paper discusses how threat displays may arise, then lose their efficiency through use in bluffs and have to be supplemented with additional displays.

Dingle, H. 1972. Aggressive behavior in stomatopods and the use of information theory in the analysis of animal communication. In: H.E. Winn and B.L. Olla, editors. Behavior of marine animals. Vol. 1: invertebrates. Plenum Press, New York.
One of the early uses of information theory in animal communication, applied to stomatopod crustaceans. Dingle discusses various aspects of the calculations involved in information analysis.

Frisch, K. von. 1967. The dance language and orientation of bees. Belknap/Harvard University Press, Cambridge, Mass.
A thorough, detailed, and major review of research on the dance language of honeybees. In addition to containing a wealth of information, this book provides valuable insight into the research process and the pursuit of a fascinating topic over many years.

Gould, J.L. 1975. Honey bee recruitment: the dance-language controversy. Science 189:685-693.
This paper reports the critical experimental test that resolved the dance-language controversy and demonstrated that the bees do convey information to each other via the dance under some but not all circumstances.

Gyger, M., and P. Marler. 1988. Food calling in the domestic fowl, *Gallus gallus*: the role of external referents and deception. Animal Behaviour 36:358-365.
A consideration of the possible use of deception (and alternate explanations) by roosters calling hens to food items. This paper does not use information-theory analysis.

Hagedorn, M., and W. Heiligenberg. 1985. Court and spark: electric signals in the courtship and mating of gymnotoid fish. Animal Behaviour 33:254-265.
An original research journal article that describes the use of electric signals in the courtship behavior of two different species (and genera) of weakly electric fish.

Hindmarsh, A.M. 1986. The functional significance of vocal mimicry in song. Behaviour 99: 87-100.
Based on an analysis of mimicry in starlings, Hindmarsh proposes that vocal mimicry of nonspecies sounds results from mistakes in song learning and has no evolutionary, functional significance.

Krebs, J.R., and R. Dawkins. 1984. Animal signals: mind-reading and manipulation. In: J.R. Krebs and N.B. Davies, editors, Behavioural ecology: an evolutionary approach. Sinauer Association, Sunderland, Mass.
This chapter presents the view of animal communication as manipulation, with comparisons of information concepts.

Markl, H. 1985. Manipulation, modulation, information, cognition: some of the riddles of communication. In: B. Hölldobler and M. Lindauer, editors. Experimental behavioral ecology and sociobiology. Fortschritte der Zoologie 31:163-194. G. Fischer Verlag, New York.
Reviews the background of information, manipulation, and other views of communication and proposes that the functions may differ under different circumstances.

Munn, C.A. 1986. Birds that 'cry wolf.' Nature 319:143-145.
An investigation of the possible deceptive use of alarm calls by birds that distract other individuals and enhance their own ability to capture food.

Otte, D. 1974. Effects and functions in the evolution of signaling systems. Annual Review of Ecology and Systematics 5:385-417.
Reviews the origins, functions, and selective factors of animal signals.

Ryan, M.J. 1988. Energy, calling, and selection. American Zoology 28:885-898.
This paper discusses how several factors, including sexual selection, the physics of signals, species recognition, and receptor physiology, constrain mating signals.

Savage-Rumbaugh, S. 1988. A new look at ape language: comprehension of vocal speech and syntax. Comparative perspectives in modern psychology: Nebraska symposium on motivation, 35:201-255.
A recent, thorough review of the subject of ape language, with consideration of syntax comprehension.

Schneider, S.S., J.A. Stamps, and N.E. Gary. 1986. The vibration dance of the honey bee. I. Communication regulating foraging on two time scales. II. The effects of foraging success on daily patterns of vibration activity. Animal Behaviour 34:377-391.
The vibration dance is different, less known, and more poorly understood than the waggle dance. This paper reports the results of investigations into the daily and seasonal occurrence of the vibration dance and its association with food availability and foraging activity, perhaps as a form of incitement behavior (as in a pep rally).

Sebeok, T.A., editor. 1977. How animals communicate. Indiana University Press, Bloomington, Ind.
*A somewhat dated but nonetheless major, if not classic, contribution to our understanding of
animal communication. This large book consists of chapters by many authors covering a wide
range of animal groups and topics in the subject.*

Steinberg, J.B., and R.C. Conant. 1974. An informational analysis of the inter-male behaviour of
the grasshopper *Chortophaga viridifasciata*. Animal Behaviour 22:617-627.
*This paper represents one of the early uses of information theory in communication analysis and
includes much discussion and explanation of the techniques.*

Steinberg, J.B. 1977. Information theory as an ethological tool. In: B.A. Hazlett, editor.
Quantitative methods in the study of animal behavior. Academic Press, New York.
Describes and discusses the calculations involved in information theory statistics, with examples.

Wenner, A.M. 1971. The bee language controversy. Educational Programs Improvement
Corporation, Boulder, Colo.
*This book presents Wenner's side of the honeybee dance language controversy, disputing the use
of the dance by honeybees as a source of information on distant locations. See J.L. Gould's
article for the resolution of the controversy.*

Wilson, E.O. 1962. Chemical communication among workers of the fire ant *Solenopsis saevissima*
(Fr. Smith). 1. The organization of mass-foraging. 2. An information analysis of the odour trail.
3. The experimental induction of social responses. Animal Behaviour 10:134-164.
*One of the earliest attempts to use information analysis in animal communication, with a
thorough description and discussion of the associated behavior.*

PART THREE

INTERNAL CONTROL AND CHANGES IN BEHAVIOR

16

INTRODUCTION TO THE NEURAL, SENSORY, AND OUTPUT ASPECTS OF BEHAVIOR

The previous section of the book focused on the outward, observable aspects of behavior, with much emphasis on ultimate factors. We now turn to the more proximate causes and the internal control and integration of behavior. These subjects are first broken into several components, after which we attempt to synthesize them into as coherent a picture as current knowledge permits. This chapter considers the basic mechanics of nervous systems, including the senses and output, as a foundation for subsequent chapters. Along with the basic anatomy and physiology, this chapter also considers how the phylogenetic backgrounds and environmental surroundings of animals contribute to the differences in their senses and other parts of their nervous systems. Chapter 17 explains the fundamental principles of neural processing, that is, how nervous systems tune into the appropriate inputs from outside of animals, combine that information with the internal states and characteristics of the animals, and select and carry out the appropriate outputs.

A nervous system, however, is only one of two major components in the overall integration of behavior. Chapter 18 discusses the second major internal control system, the endocrine system. Although these two systems are closely associated and interact intimately with each other, and some persons consider them to be a single system (the neuroendocrine system), there are sufficient differences to justify treating them separately. The nervous system is involved primarily in rapidly occurring events with response times that may be only a few milliseconds, whereas the endocrine system primarily coordinates the slower responses over seconds, hours, and days, sometimes involving periods of weeks, months, or even years.

Also, there are some major differences in the structures and mechanisms of the two systems. In chapters 19 and 20, we consider changes of behavior in the individual, ontogenetic or developmental changes (including maturation), and learning per se. That is followed by a synthesis of all of these topics by using the control of feeding behavior as a case in point in Chapter 21.

THE ANATOMY AND PHYSIOLOGY OF BEHAVIOR

The internal mechanisms of animal behavior have provided an ongoing, exciting, and challenging frontier. Until recent decades, an animal and its behavior had to be viewed as something of a "black box," with people able to do little more than observe from outside the animal, correlate inputs and outputs, and speculate about what was going on inside (discussed from a historical perspective in Chapter 2). However, much research effort, combined with an increasing arsenal of modern technology, has produced significant advances. We now have a considerable body of knowledge on the subject and can even make some intelligent guesses in areas that we do not yet understand completely. There are still many unknowns and challenges, but that, after all, is what keeps science fresh and moving.

The complete picture of behavioral integration involves dynamic interactions between the animal and its environment through an interplay of the nervous and endocrine systems and the nervous system's connections to muscles and various glands. The dynamics include maturational changes that occur as animals grow and develop plus other changes that can be classified as various types of learning. To break this big picture into manageable pieces, we begin with the nervous system and its components, starting at or near the point where most general, introductory biology texts leave off. If you would like additional information for introduction or review, we recommend a basic biology text (see list at end of Preface).

NERVOUS SYSTEMS

Figures 16-1 and 16-2 present a conceptual overview of the nervous systems. Figure 16-1 breaks a generalized nervous system into its three major functional components: input, processing, and output, each with its own subcomponents, associated terms, and principles. This figure and its parts and labels serves as a basis for the following discussion and as a summary at the end. Parts that are not familiar initially should become meaningful. Thus keep this figure as a reference point through the chapter.

Of the different types of cells that make up the nervous system, those that primarily process the information and are of greatest interest for behavior are the neurons. Neurons of different animals and even within any one animal vary tremendously in size, measuring from a fraction of a millimeter to several meters in length. Individual motor neurons innervating the tail region of a blue whale, for example, extend all the way from the spinal cord to the most distant muscle. Axon diameters range from about 0.001 to 0.2 mm, and the somas or cell bodies vary in diameter from around 2 to over 500 micrometers (μm).

Such numbers are out of the range of everyday experience and, thus, almost meaningless to most people. To put such dimensions in more familiar terms, imagine the axon of a human motor neuron, roughly 8 μm or slightly larger than a red blood

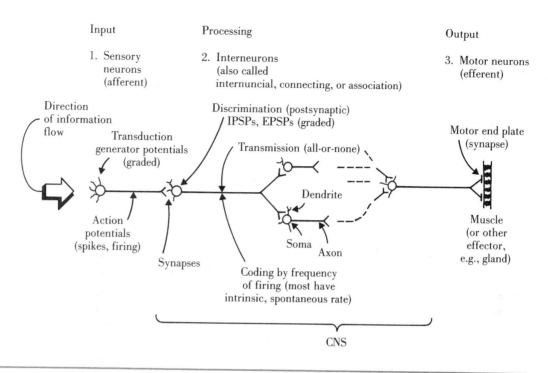

Figure 16-1   Conceptual overview of the three major functional components of the nervous system, with associated terms.

Figure 16-2   Basic arrangement of invertebrate nervous systems contrasted with that of vertebrates. The central nervous system of invertebrates (among those that have a centralized system) is mostly ventral, whereas that of vertebrates is dorsal.

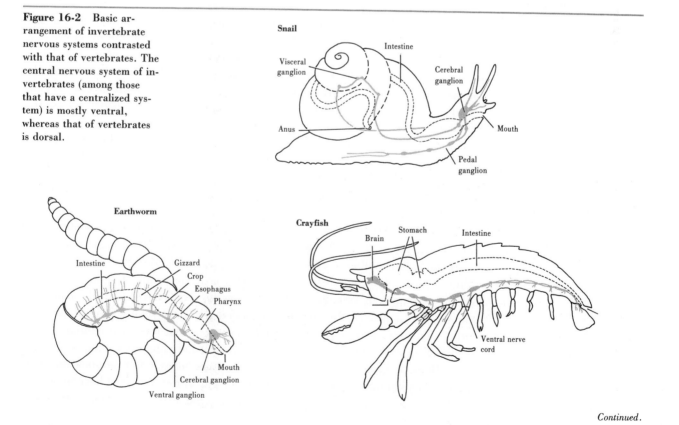

Continued.

Figure 16-2, cont'd Basic arrangement of invertebrate nervous systems contrasted with that of vertebrates. The central nervous system of invertebrates (among those that have a centralized system) is mostly ventral, whereas that of vertebrates is dorsal.

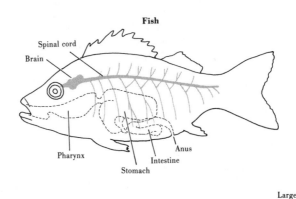

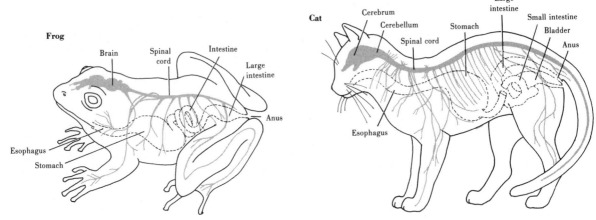

cell, being the size of a common 19 mm (¾-inch) garden hose. If that axon belonged to a motor neuron running from the spinal cord to a finger muscle in an average-sized woman (1.7 m or 5½ feet tall with an arm 60 cm or 2 feet long), at the scale of the garden hose the axon would be 1.7 km or a little over a mile in length. By the same scale, the cell body would be 11.9 cm (4.7 inches) in diameter or slightly larger than a softball. The woman's brain to this scale would be about 1.7×10^{13} ml (596,949,824 ft³), roughly the volume of 10 Houston Astrodomes.

Perhaps even more amazing is the thickness of the axon's membrane. A 19 mm garden hose has a wall 2.4 mm (³⁄₃₂ inch) thick, equivalent to the thickness of 33 pages of this book. To the same garden hose scale, the axon's outer membrane (actual thickness about 8 nm or 0.008 μm) would be only one fourth the thickness of a single page.

Figure 16-2 illustrates the general differences in the nervous systems among animals on the two major branches of the phylogenetic, evolutionary tree: protostomes (including mollusks, annelids, and arthropods) and deuterostomes (including echinoderms and vertebrates). In brief the protostomes have a ventral, basically paired, solid nervous system, whereas the deuterostomes have a dorsal, single, and hollow nervous system. Protostomes in general also have relatively fewer and proportionally bigger neurons, with a fixed number so that the same neurons can be identified, named, and numbered among different individuals of a species (Figure 16-3). Because insects and other protostomes have many fewer neurons and fewer synapses per neuron on the average compared with vertebrates, yet have to perform all of the behaviors required for living, they tend to use the same neurons in

Figure 16-3 Identified neurons in the supraesophageals ganglion of a larval tobacco hornworm moth *(Manduca sexta)*. The large cells visible (in boxed areas) are medial 1a neurosecretory cells. Smaller, more numerous neurons, which would require staining to make them readily visible, also can be individually identified and named.
Photograph courtesy Gerald G. Holt, USDA Metabolism and Radiation Research Laboratory, Fargo, North Dakota.

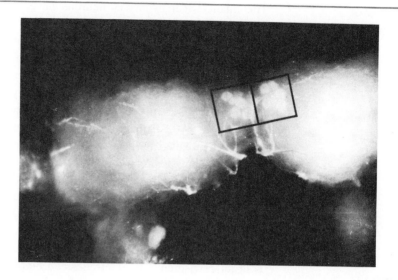

different and efficient ways, a principle known as *neural parsimony* (Roeder 1948).

Before continuing with the basic principles of how nervous systems operate, we briefly consider some of the methods used to obtain this information.

NEUROPHYSIOLOGICAL METHODS

Knowledge in neurophysiology has proliferated over the last 50 years, but the roots of the understanding of neurons go back a surprisingly long way, to at least Malpighi in the seventeenth century. Although there were many early misunderstandings and misinterpretations, some progress was being made through the entire nineteenth and early twentieth centuries.

Visual Techniques

The development of magnifying lenses and microscopes with modest magnification was a great breakthrough for studying biological structures. Present-day light microscopes are capable of around 1000 × magnification. The next major step was the development of the electron microscope, which permits magnification up to 250,000 ×. The limits are not really an issue of magnification but rather resolution, the smallest dimensions that can be distinguished. The shortest visible wavelengths of light permit one to resolve objects to about 0.2 μm, whereas the electron microscope takes resolution down to a theoretical limit of about 4 or 5 A°, or 0.0004 to 0.0005 μm. (Recall that the motor neuron axon diameter in the example earlier was 8 μm and its outer membrane was 0.008 μm.)

Neurons are not normally pigmented. Thus to be seen, in addition to being magnified, some of them also must be preserved (fixed), stained, and occasionally sectioned. There are numerous techniques for staining neurons, including procedures that stain only parts of neurons such as the cell body. Some neurons, particularly of invertebrates, are large enough that they can be separated out, seen, and worked with even while they remain alive and unstained.

Recording of Bioelectrical Events

The bioelectrical events of neurons have been recorded using several different techniques. Voltage differences and current flow have been measured by inserting one recording electrode into the cell and keeping another outside. Electrodes are the points or terminals of an electrical source or recording device. Metallic wire is mostly unsuited for making small electrodes; they usually are made instead by carefully drawing out heated microcapillary glass tubing and filling them with electrically conducting salt solutions. The size of the tip of a modern microelectrode is from 0.1 to 0.5 μm. Some of the first recordings from the insides of neurons were accomplished by inserting an electrode through the cut end of a giant squid axon. This giant axon (300 to 800 μm) was used initially because of its large size. More recent techniques involve penetrating the neuron's membrane directly. These methods obviously demand much practice, patience, and skill. The minute size of the electrical events, in millivolts or 1/1000 volt, also requires specialized amplifiers that do not interfere with the events.

If the simple presence of a neuron's impulses, rather than their actual magnitude, is of interest, then both electrodes can be placed on or near the surface of the cell or its extensions. When the two electrodes are placed on the same neuron at separate points, they measure the relative difference between the two points instead of the difference between the inside and outside of the neuron. It is also possible to place one electrode on the cell and the other in the external medium away from the neuron. If several axons are in contact with the electrode, the impulses of all are recorded together, and the recorded event is called a *compound action potential*. Probably the simplest technique of all is to use a large (relatively) wire electrode that is bent into a hook. An entire nerve or part of a nerve can be lifted from the surrounding tissue with such an electrode, and the activity of all of the neuronal impulses that pass through it can be simultaneously recorded.

The amplified impulses can be displayed visually with an oscilloscope on its cathode ray screen, marked on paper with pens, or detected acoustically via clicks on a speaker. The sequence can be tape-recorded for later replay, sent to electronic counters, or forwarded directly to a computer for analysis.

Relatively recent developments in neurophysiological techniques have shown that neuronal activity can also be recorded optically (Tobias 1952, Ross et al 1974, Grinvald 1985). These techniques may permit the activities of smaller neurons and larger numbers of neurons to be recorded simultaneously.

Stimulation, Ablation, and Lesioning

Much information and insight into the properties and pathways of neurons have been gained by stimulating one part of a nervous system and recording subsequent activity elsewhere. For example, sensory endings can be stimulated in a variety of manners to discover the characteristics to which the neurons are sensitive. The resulting impulses, called *evoked potentials*, can be recorded directly from sensory neurons or traced into other neurons further along in the system. Likewise, parts of the brain, spinal cord, or ganglia can be stimulated, depending on the organism, to discover which muscles contract. Stimulation is accomplished by electrodes that supply electrical currents to the neurons. Instead of stimulating various regions, various parts of the system can be damaged or interrupted by cutting (lesioning) or otherwise destroying (ablating) segments to determine the effects. Effects also can be followed by injecting or implanting chemicals such as radioactive-labeled precursors, hormones, and poisons.

Figure 16-4   Stereotaxic
instrument and surgical
setup to place electrodes in
specific regions of the rat
brain.
Photograph by James W. Grier.

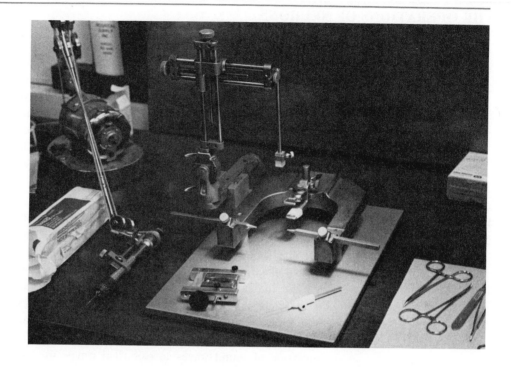

The placement of electrodes, lesions, or introduced substances may be relatively easy in small nervous systems or on surface preparations. Neurons or regions of neurons that are deep in the brain are usually located with the aid of precision tools, stereotaxic instruments (Figure 16-4), which permit exact location in three-dimensional space. Stereotaxic maps and atlases of the brain are available for many organisms. After the effects of stimulation or implants have been observed in live organisms, the organisms are killed, and the brain or other parts of the nervous systems are removed, fixed, and cut into thin sections for viewing under a microscope to determine the exact locations of the electrodes or implants.

**Experimental
Manipulation and
Analysis**

Electrical and chemical processes and events by which neurons operate have been investigated through a variety of experimental techniques such as submerging iso-lated neurons in solutions of different chemical compositions. In addition to a large number of important ions, including Na^+, K^+, Cl^-, and HCO_3^-, there are many biochemicals involved in neural functioning, particularly at synapses.

One technique that has been important in conjunction with such work is voltage clamping. Normal events in the neuron, such as changes in voltage, occur too rapidly to allow the measurement of all components of the event. The technique of voltage clamping holds voltages at experimentally controlled levels, permitting the determination of electrical and chemical processes occurring in the neuron at that voltage. Voltage clamping is accomplished by using the usual electrodes to record voltage plus additional electrodes to artificially feed current into or out of the cell, as determined through a feedback apparatus.

We now return to the functioning of neurons and nervous systems.

THE OPERATIONS OF NEURONS

Neurons have two primary functions: (1) to process and (2) to transmit information. Processing is also referred to as *discrimination*, a kind of simple yes-no decision making that depends on summing up inputs that are either for or against firing the neuron. The decision, to fire or not to fire, is not conscious on the part of the neuron. It occurs as a result of properties of the neuron in conjunction with its input. The whole system operates by the combined actions of millions of unique, individual neurons interacting with other neurons in a generally orderly fashion, including priorities and hierarchies that usually resolve conflicts.

Discrimination by Neurons

The two main types of input that form the basis for neuronal decision making are (1) transduction of sensory input from outside the nervous system and (2) input from other neurons, which usually occurs at a synapse. In both cases, these inputs produce electrical potentials in the neuron that are local and graded. They occur primarily in dendrites and at the cell body. They are called *graded* because, unlike the all-or-none property of action potentials, the size of the potential is proportional to the input (i.e., the greater the input the greater the change in potential).

Transduction involves a change from one energy form to another, such as the conversion of sound waves to electrical waves in a microphone. In the nervous system, it refers to the conversion of various inputs (light, sound, odors, touch, etc.) into nerve impulses or action potentials by sensory cells. The graded potentials from sensory input are named *generator potentials*. Potentials caused by synaptic input from previous neurons are aptly termed *postsynaptic potentials (PSPs)*. Post-

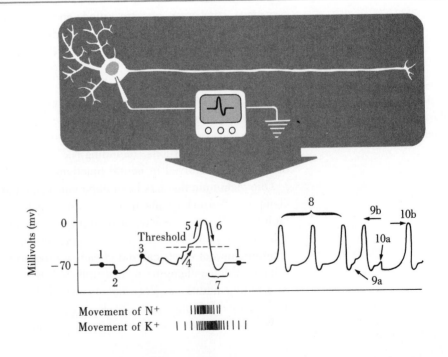

Figure 16-5 Terminology and events associated with changes in neuronal membrane potentials. 1, Resting potential; 2, more negative—hyperpolarization—as in IPSP; 3, less negative—depolarization—as in EPSP; 4, summation; 5, action potential—inward rush of Na^+; 6, return to negative state—outward rush of K^+; 7, overshoot—refractory period; 8, intrinsic rate of firing—spontaneous depolarization; 9 a, stimulus of EPSP; 9 b, increase in rate of firing; 10 a, inhibition — hyperpolarization by IPSP; 10 b, delay in firing.

synaptic potentials, depending on the characteristics of the particular synapse, are either excitatory (EPSP) or inhibitory (IPSP). EPSPs lead to a slight amount of depolarization, that is, less polarization. IPSPs cause hyperpolarization or increased polarization.

Graded potentials are not only local in the cell, they are also transitory; that is, the effect of the potential or charge diminishes or decays exponentially with distance and time (Figure 16-5). The greater the change in potential, the greater the area affected and the longer it lasts. The spread of the potential involves the combined electrical-chemical properties of the neuronal membrane. It is passive, and it is termed *electrotonic*.

The effects of graded potentials are also additive. A given graded potential, for example, a PSP, lasts for awhile and then disappears. If, however, another PSP comes along before the first is gone, the effect of the second adds to the remaining effect of the first. Additional PSPs can come either from a sequence of impulses in the same synapse or from different locations on the neuron at neighboring synapses. The effects are referred to respectively as *temporal* or *spatial summation*, which add together for the total summation (Figures 16-5 and 16-6). Most neurons receive input from a large number of synapses (up to an estimated 80,000 in a type of vertebrate neuron known as the *basket cell*) and are thus subject to a constant barrage of EPSPs and IPSPs, all of which are thus summed over time and local space.

If enough EPSPs add together to depolarize the neuron beyond the threshold potential, then it fires; that is, the action potential and its host of membrane changes and ion flow occurs (see Figure 16-5).

Transmission of Decisions

The change in the membrane that accompanies an action potential at a particular point affects the surrounding membrane. This sets off a similar change in the neighboring regions, and by this means the spike does not diminish, as in graded potentials, but is maintained or propagated in an active chain reaction. It travels throughout the neuron, normally from the cell body to the end of the axon. The process of propagation is somewhat analogous to the burning of a firecracker fuse.

Imagine a neuron with spikes being propagated down it in rapid succession (Figure 16-7). The number of spikes per unit of time is the spike frequency. This rate or frequency of firing is the means by which neurons normally code information. With increased stimulation one usually finds an increased rate of firing. The amplitude of a spike is not important for coding information; each spike is all or none, that is, either present or absent.

Neurons do not generally code information by simply firing or not firing but rather by changing their rate of firing. Under natural conditions in the intact animal, depolarization may occur as a result of generator potentials or postsynaptic potentials, or happen spontaneously; that is, many neurons do not really have a resting potential but rather a resting or spontaneous rate of firing. A common role of generator and synaptic potentials is not to cause an action potential per se but to alter the neuron's spontaneous or endogenous firing rate. If a cell is depolarizing intrinsically, EPSPs speed up the rate of depolarizing via the normal summation process and cause the cell to fire sooner and increase the frequency of spikes. Similarly, IPSPs can reduce the frequency. See Figure 16-5 (9 a, 9 b, 10 a, and 10 b).

Figure 16-6 Local nature and summation of graded potentials. An individual postsynaptic potential diminishes with time and distance from its initial occurrence. Strength of the potential (depolarization) is indicated on the surface of the neuron by amount of shading. Each series of neurons represents a sequence in time. **A,** Single EPSP, fading with time. **B,** Two EPSPs at the same site. **C,** Two EPSPs arriving at different locations. **D,** A combination of temporal and spatial summation.

Single EPSP

Single EPSP from incoming spike

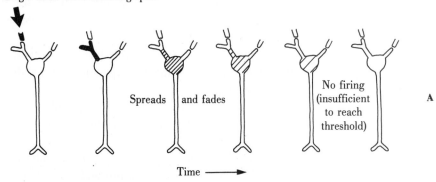

Temporal summation

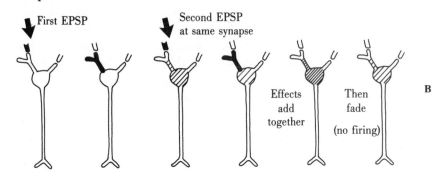

Spatial summation

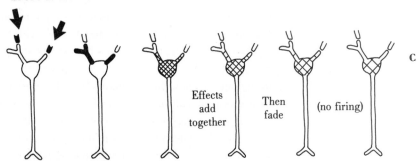

Spatial and temporal summation

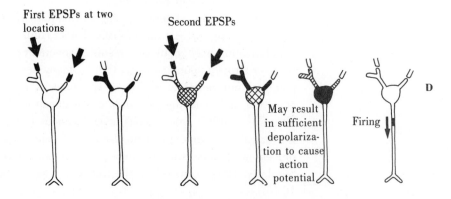

Figure 16-7 Repeated action potentials in a neuron. After one spike passes a particular point on the axon's surface and the membrane has recovered, another spike can follow soon thereafter. Note that the impulses are recorded in sequence from left to right on the screen. Within a given neuron, several spikes may be traveling down the axon simultaneously. The frequency of spikes over time is the means by which neurons code information.

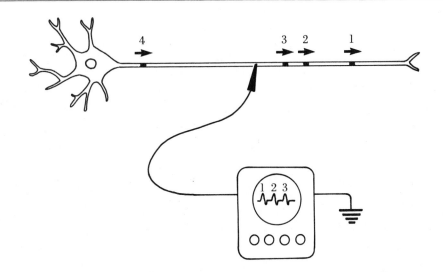

By looking at spike frequency, one can "break the code," tap the nervous system, and begin looking for eventual correlations with behavior. It should be stressed, however, that this is only a start. A complete picture requires information on the whole system. Attempting to gain much information from a single neuron is like trying to understand the operations and behavior of a city by tapping only one phone line.

The speed at which an action potential travels down the axon varies from a few centimeters per second to around 100 m per second (roughly 200 miles per hour). In a competitive and dangerous world where speed counts the advantage of rapid transmission is obvious. The speed of transmission is affected by several factors, such as the size or diameter of the axon (the larger, the faster), the temperature (the warmer, the faster), and whether or not myelin is present. Myelin is a fatty insulated wrapping around the axon which permits the action potential to leap electrically (saltatory conduction) from point to point along the axon without being slowed down by the normal biological membrane process that otherwise would occur in the intervening spaces. Myelination permits a combination of small diameter and rapid transmission of action potentials in the axons. A myelinated frog axon 12 μm in diameter conducts at the same velocity (25 m per second) as an unmyelinated squid axon nearly 30 times the diameter. Although large diameters increase speed, they take up more space, which creates some disadvantages that are discussed in the next chapter.

When the action potential reaches the end of the axon, the information is transmitted to the next neuron across the synapse by various chemical transmitters. The processes of release, diffusion, and reception of the transmitters takes about 0.5 msec. That is fast, but it is much slower than the propagation of action potentials within a neuron. Each synapse slows down the total message, and a response involving many synapses is considerably slower than one involving only a few.

In addition to the more familiar chemical synapses, there are also some less common electrical or gap junctions between parts (including neighboring dendrites and somas) of some neurons. They involve membranes that are in contact with each other, permitting the local graded potentials to pass from one neuron to another and creating complex local circuits. Depending on the characteristics of the particular gap junction, the current can flow in either direction or in just one direction. The passage of current remains passive and thus is not easily blocked. It occurs without the delays seen in chemical synapses.

A controversy developed during the first half of the twentieth century over whether transmission from one neuron to another was electrical or chemical, the so-called spark versus soup schools. It turned out that both were correct for different connections.

Many chemical substances have been identified as neurotransmitters. Two of the most familiar are acetylcholine and norepinephrine. Once a transmitter has done its job of transmitting the message across the synapse, it is about as useful as old news and can create problems if it remains and accumulates. There are several means by which transmitters are inactivated and recycled: enzymatic destruction, active uptake into the axon terminal or neighboring nonneuronal glial cells, or diffusion into the intercellular spaces. Acetylcholine is metabolized by cholinesterase; the choline is actively reabsorbed into the terminal; and new acetylcholine molecules are manufactured. Norepinephrine apparently is reabsorbed intact and repackaged in new vesicles to be reused.

Synapses and their complex molecular and membrane interactions are highly susceptible to a wide variety of exogenous chemicals, that is, chemicals introduced into the animal from outside by oral, injected, or absorbed means. These include a number of naturally occurring animal and plant toxins and artificially synthesized compounds such as: (1) some components of venoms; (2) hallucinogens; (3) some anesthetics and depressants used in veterinary, zoo, and farm animal management; (4) a number of lethal chemicals used by humans for pest control and; (5) psychoactive chemicals like cocaine (which blocks the uptake of norepinephrine in certain synapses and also fosters the release of dopamine), caffeine, nicotine, and Valium (diazepam), to mention a few.

The means and pathways by which numerous neurons connect and work with each other to process and integrate complex information requires further elaboration. That is the entire topic of Chapter 17. In the remainder of this chapter, we consider how the information gets into and out of the nervous system; that is, what are the inputs and outputs, how do they work, and what are their roles in behavior?

SENSES AND BEHAVIOR

The concept of the Umwelt, an animal's own sensory and perceptual world, was introduced in Chapter 1, and the importance of different senses for various modes of communication was stressed in Chapter 15. Senses provide an animal's access to and first filter of environmental input. They help determine not only what an animal tunes into but also what it does not. Because an animal's Umwelt is so important to its behavior and because humans are so immersed in our own (primarily visual and auditory) Umwelt and we do not normally or easily consider those of other animals, we discuss the Umwelt further.

Technically, the Umwelt as a sensory and perceptual world is only a crude approximation of von Uexküll's (1909) original concept. Originally the Umwelt meant much more, including also the behavioral or effector (muscular and glandular) output of the animal. For insight into the original concepts proposed by von Uexküll, see the translation by Schiller (1957), including her introductory note. Tinbergen (1951) used an alternate term, *Merkwelt*, to refer to the perceptual world. Umwelt, however, has gained widespread acceptance primarily in association with the perceptual world, so we continue to use it in that context.

To expand on the Umwelt for a moment, imagine a young couple, for example, two college students, on a walk in the woods with a dog on a warm spring afternoon. Within several meters of them in that same woods are many other species of animals. Try to imagine how the world "appears" to each of them.

First consider the humans. They may not seem to be sensing much; they are mostly paying attention to each other and largely are unaware of their surroundings. Much of this stimulus filtering that occurs in humans and other animals is central; that is, it involves neural integration at higher levels within the central nervous system. Central filtering of input is extremely important and adaptive because even with sense organs being limited to certain types and ranges of input, the total amount of sensory information available is nonetheless potentially overwhelming. Central stimulus filtering, however, is another topic. Because it involves higher levels of neural integration, it is discussed in Chapter 17. Here we consider only peripheral stimulus filtering, that is, what the senses are capable of receiving if the individual is paying attention.

Thus, if and when they pay attention to them, the couple can sense a large number of phenomena in the environment with which most persons are familiar. There are the sights and colors of leaves, branches, and tree trunks, blue sky and clouds overhead, sounds of breezes, singing birds, and their own footsteps. There are a number of odors that they detect faintly, such as her perfume, his body odor (which she may detect but he himself may not), and the smells of the damp woods. You can fill in the rest of the picture.

What about the other animals? The dog is lower to the ground, for one thing, and has a different perspective simply for that reason. In addition, the dog sees somewhat more poorly and lacks the visual resolution that humans have. The dog receives a rich combination of smells much more strongly than the people and also hears higher pitches, such as those of nearby insect sounds. Scurrying under the leaves is a shrew that sees even more poorly than the dog but probably has keener senses yet for smell, perhaps hearing, and touch. The shrew's world is dominated by chemical senses, touch, and vibrations received from around its body, particularly through vibrissae (sensory hairs) around the facial area of the head. It is visually aware of light and dark but not much more.

The shrew is hunting earthworms, which in turn are tuned primarily to odors and a world full of vibrations in and on the soil. Vision is practically nonexistent for the earthworm; it is aware only of the presence or absence of light. As the people walk by and their shadows fall across the earthworm, the worm's world suddenly becomes darker and is filled with a tremendous number of vibrations. The earthworm is aware of light in a similar manner that humans are aware of temperature (heat). Further down the path is another animal, a copperhead snake, that senses objects through vision with typical vertebrate eyes, through radiated heat from the objects by using infrared-sensitive pit organs on the head, and through odors that it picks up with flicks of the tongue.

As night falls and the woods become dark, the two people have more difficulty finding their way about. The light on which they depend has rapidly disappeared.

The change to night will be less noticeable to the shrew, the dog, and many of the other animals around them. At the bottom of a nearby muddy stream is a small catfish that senses little change in light or temperature but is surrounded by a world of currents and water pressure changes, aquatic sounds (some coming in from the stream bed and bank), and a rich world of chemical cues, which may be detected at many places over the fish's body.

The senses of other animals, just like our own, clearly constrain the types of environmental input to which an animal has access, and as a result the stimuli to which it can and will respond behaviorally. Before we can truly understand the behavior of other animals, we must realize that we may be unaware of many of the stimuli to which they are sensitive and responding and vice versa. Many of the sensory aspects of our lives that are so familiar to us that we take them for granted are completely irrelevant to the behavior of other animals. How can we deal with this problem?

There are various ways of assessing the sensory worlds of other animals, as introduced in Chapter 1. We discuss these methods, then present a survey of what is known about senses in general. That should provide a broader perspective and better insights into the behavior of other animals. In a way, this is the opposite of anthropomorphism (Chapter 1); rather than trying to understand them from a human viewpoint, we can try to imagine ourselves in their worlds.

Methods of Analyzing the Senses of Other Animals

Because we cannot relate directly to the Umwelts of other animals, we are forced to use indirect means and infer what information is going into their systems. The methods used to accomplish this fall broadly into the following four major categories:

1. **Evoked potentials: neural activity recorded from sensory neurons.** Using microelectrodes inserted directly into axons of sensory neurons or external electrodes placed nearby, changes in impulse frequency can be measured and associated with different stimuli impinging on the sense organ.

2. **Blocking, interruption, or removal of sense organs or neural connections.** This is the method Spallanzani used (e.g., plugging the bat's ears, covering the eyes, etc.) when he inferred that bats were using hearing to navigate. By interfering with the normal sensory processes, one can observe subsequent changes in the animal's behavior.

3. **Learned responses in conjunction with sensory stimuli.** This is perhaps the most commonly used technique in the study of sensory discrimination. Learning (Chapters 2 and 20) may involve conditioned reflexes (e.g., Pavlovian or classical conditioning) or be instrumental. If food is preceded by a signal, for example, dogs become conditioned to salivate on hearing the signal. The signal can be changed, for example in intensity or pitch, until the dog no longer shows a response.

   Von Frisch was a pioneer in the use of this technique to investigate animal senses, including hearing in fishes and visual sensitivity in bees (e.g., von Frisch 1914). Von Frisch showed, for example, that bees could discriminate blue color but that they are color blind to red.

   Tavolga and Wodinsky (1963, reprinted in Tavolga 1976) used instrumental learning (Chapter 20) to determine the precise intensity thresholds of marine teleost fish to single-tone sounds. They placed fish in a tank with two chambers separated by a shallow barrier. This tank was similar to avoidance conditioning shuttle-boxes used with rats. Sound of a particular tone was presented, followed 10 seconds later by a mild electric shock. Sounds were presented in a carefully controlled and monitored manner through underwater speakers. Most of the different species of fish learned to swim across the shallow barrier to avoid the shock 90% of the time after 5 to 7 days of 25 trials per day. After the avoidance behavior in the fish was established, the loudness of the sound at a particular frequency was successively decreased from test to test until the fish ceased to avoid it, at which time loudness in subsequent tests was gradually restored. This technique is known in sensory psychophysics (physical aspects of psychology) as the *staircase method* (Cornsweet 1962). The whole process was repeated at other frequencies of sound to provide a profile of the different species' hearing abilities.

4. **Natural responses in conjunction with experimental stimuli.** Rather than use artificial, conditioned responses, many workers have investigated senses via the animal's own responses, for example, the turning of flying moths in response to bat cries (Chapter 1). Perhaps one of the best-known responses, used in tests of visual discrimination, involves movements of the eyes, head, or whole body of an animal toward moving objects. This frequently is accomplished by placing the subject inside a rotating cylinder with patterns placed on the inside of the cylinder. When the cylinder moves, the animal will turn to follow the pattern—if it can see it.

One of the most striking uses of a natural response was demonstrated by Moericke (1950), who investigated color vision in peach aphids. These aphids stab at a substrate if, and only if, the material is placed on a green

Figure 16-8
Demonstration of negative afterimage. Color the center part of the star red and the border blue with felt-tip pens. You can also color the star green and the border yellow. Gaze directly at the dot in the center for a minute or more under bright light and then look at a white piece of paper or a white wall. A negative afterimage should be seen.

background. He then exposed them for a length of time to a background of purple (the color complement of green). Subsequently they would stab at a neutral gray background—a behavior they otherwise would not display. This demonstrated a negative-color afterimage in the aphids. To experience a similar effect yourself, see Figure 16-8.

Another striking and more frequently used response for investigating senses uses animals' natural intraspecific responses, such as the movements of males toward females and territorial interactions between males. A female's odor, for example, can be simulated and then modified to find the thresholds of the male's ability to detect it.

Senses, Their Basic Features, and Evolutionary Considerations

Although we usually think in terms of the so-called five senses (taste, touch, hearing, vision, and smell), there are more than just five. Furthermore, among the various species of animals there are many variations on any particular kind of sense. A systematic list of the various senses is provided in the box on this page.

Natural selection has clearly favored an appropriate match between animal senses and different ways of living under various environmental conditions. Which is matched to which, however, can sometimes be an elusive question, like asking which came first, the chicken or the egg. Both behavior and senses are subject to variation and evolution. Thus changes in one can lead to changes in the other or they may both change together, more or less simultaneously. Among a group of animals with sensitive olfaction but poor vision, for example, those that are active at night might survive and reproduce better than those that are active during the day. Among those active at night, selection might favor those with the best olfactory abilities, whereas for those active during the day, certain types of vision would be selectively advantageous. Some environments definitely favor or hinder the evolution of some senses. Vision would be hindered where light is completely or mostly

---

Classification of Animal Senses, Based on Form of Energy Involved

   I.   Mechanoreception
      A. Touch/tactile—pressure
      B. Proprioceptive (internal, usually subconscious)—bending, stretching
      C. Inertial/equilibrium—statocysts or other weight or movement detectors, including fluid (or air bubble) movements against hair cells
      D. Vibrational/hearing—wide variety of structures, in some cases including amplification
  II. Photoreception—vision and simple light detection
 III. Thermoreception—heat and infrared light
 IV. Electroreception—electric senses
  V. Magnetoreception—magnetic field senses
 VI. Chemoreception—chemical or molecule detection
      A. Taste
      B. Smell/olfaction

precluded such as in caves and muddy or deep water. Electrical senses, on the other hand, would be favored in conductive aquatic environments but would not work in terrestrial settings.

These arguments can be applied to all senses or to particular aspects of any given sense, such as whether high- or low-frequency hearing is better under particular circumstances. This issue also is discussed in Chapter 15 in relation to the use of various senses in different forms of communication. Most general biology textbooks (Preface) provide good introductions to sensory systems, including human vision, hearing, balance, touch, and internal proprioceptive senses. The structure and basic aspects of these basic mammalian (and, to some extent, other vertebrate) senses are not covered here (except as they pertain to neural processing in Chapter 17); we merely survey the highlights of various senses and cover some points that receive little or no attention elsewhere.

A Brief Survey of Animal Senses

Hearing and other vibrational senses have been studied extensively and are familiar in vertebrates. In all vertebrate classes the sense of hearing depends on a portion of the inner ear, an outgrowth known as the lagena, or, in its most highly developed form, the cochlea, which is derived from and closely related to the system involving balance. It includes fluid that transmits vibrations and hair cells, which detect the fluid movement and transduce the energy into nerve impulses. The inner ear, which contains receptors that are important for both balance and hearing, appears related to the lateral line system in fish in which clusters of hair cells are located in tubes or sunken channels in the skin of the fish. These hair cells sense movements and low-frequency vibrations in the water at the surface of the fish. In the most primitive fish the hair cells are distributed over the surface of the body rather than in specialized lateral lines.

Bats are not the only mammals that can hear high, ultrasonic pitches. At least 23 species other than bats are known to be capable of hearing above the 20 kHz upper limit of humans (Hess 1973 b). Several, including chimpanzees, hear up into the range around 30 kHz, and many small mammals, including mice and shrews, hear up to the range of 90 to 120 kHz. Porpoises and seals may produce and hear underwater sounds up to around 180 kHz, although seals apparently have an upper limit for airborne sounds around 22 kHz (Mohl 1968). Sound travels nearly five times faster in water than in air, and because of this and other differences, mammalian hearing may be different for water and air, and upper-frequency limits in the two media may not be strictly comparable.

Hearing in invertebrates is usually claimed only for arthropods, particularly among insects where several different mechanisms of pressure-sensitive chambers and vibration-sensitive hairs, antennae, and other body extensions are found (Figure 16-9). Hearing is also likely in several crustaceans and spiders, some of which produce sounds believed to be used in communication.

From the point of advanced, specialized hearing structures, sensing of vibrations appears to merge with a more general sense of touch and vibration among both vertebrates and invertebrates. The more general sense is mediated usually through body mechanoreceptors but may include specialized detectors on the limbs, between body parts, and on various hairs, antennae, and vibrissae. Many organisms, including lower invertebrates, are sensitive to low-frequency vibrations in or at the surface of the ground or water. Such vibrations in the substrate are used, for example,

Figure 16-9 Hearing organs of insects. Insects use a variety of structures for hearing, including pressure chambers with a tympanic membrane, movement receivers such as hairs on antennae that extend from the body and vibrate with sound waves, or combinations of pressure chambers and projecting structures. **A,** Cricket with its ears on its legs just below the knee. In the cricket, incidentally, a tracheal tube connects between the ears on the two sides of the body, permitting sound to travel from one side to the other. **B,** Pressure-sensitive mouthpart palps for hearing in hawkmoths (Sphingidae). The moth's right palp *(left arrow)* is intact, whereas the other *(right arrow)* is broken open to show air chamber. Eyes are to sides, and tongue is in center. A heavy layer of facial scales (hairlike) had to be removed to view the palps. **C,** Vibration-sensitive antennae on the face of fruit fly. **D,** Projecting rocking club structure *(light object at point of arrow)* connected at its base to the tympanum of the ear on the thorax of a water boatman. This structure is best seen on fresh specimens; it commonly is damaged during drying, as in collections. For further information on insect hearing, see Michelsen (1979). Photographs by James W. Grier.

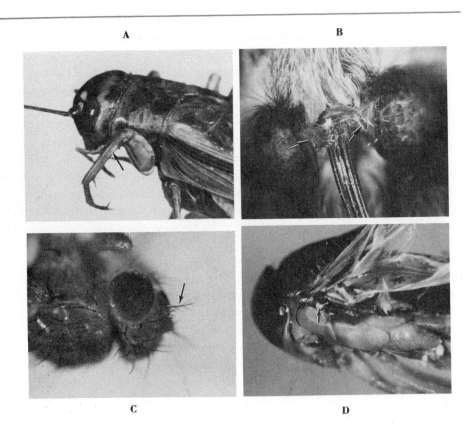

by water striders, many spiders, coelenterates and mollusks, and probably most underground dwellers, including earthworms, snakes, and burrow-dwelling mammals.

Vision also depends on vibrational energy and might seem as though it should be considered under other vibrational senses. The vibrations have a much shorter wavelength, however, and the transductional process is significantly different. Vision is based on light, a narrow band in the electromagnetic spectrum. Light is one of only two general types of electromagnetic energy that reach the surface of the earth from outer space. The other form involves radio waves. All other wavelengths are filtered out by the earth's atmosphere. Hence it probably is not a coincidence that biological organisms tuned into light wavelengths during the course of evolution.

The wavelengths of light are so small, less than a 1/1000 mm, that rather than interacting with and vibrating whole structures (as with sound), light interacts with matter at the molecular level. Biological light sensors thus must contain light-absorbing, light-sensitive pigments or photochemicals.

The sensitivity of different organisms to light and the types of structures they possess for sensing it vary considerably. Most animals are sensitive to light in one manner or another. The protozoans possess photochemicals or light-sensitive organelles and have a diffuse light sense. *Euglena,* a flagellate protozoan, has 40 to 50 orange-red granules in the stigma or eyespot near the flagellum. When exposed to light, it swells and affects the direction in which the flagellae beat. Coelenterates,

Figure 16-10 Examples of eyes among different animals. **A,** Scanning electron micrograph of a scallop eye peering out from among the folds of the mantle. **B,** Weevil. **C,** Asiatic land crab. **D,** Frog. Close-up view of surface of arthropod compound eyes: **E,** Square ommatidia of crayfish eye; **F,** Hexagonal ommatidia of noctuid moth. A courtesy of P.P.C. Graziadei. B, E, and F, courtesy North Dakota State University Electron Microscope Laboratory. C from Animals, animals. D courtesy John H. Gerrard.

Figure 16-11 A graphic concept of how a flower might appear through the compound eye of an insect. This is only a crude, speculative representation, however, as humans do not yet know how the perception might be processed and "appear" in the insect's CNS. **A,** As seen through a vertebrate eye. **B,** As seen through an insect compound eye.

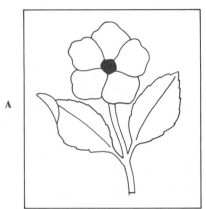
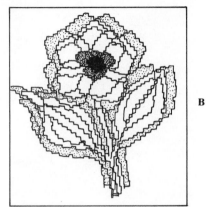

As seen through vertebrate eye   As seen through insect compound eye

annelids, and most of the other invertebrates have at least light-sensitive cells located somewhere on or in their bodies. Mollusks, arthropods, and vertebrates possess a variety of eye structures (Figures 16-10), some quite advanced and with image-focusing lenses. The view through different eye structures may be strikingly different (e.g., Figure 16-11). However, we do not know how the final perception is processed internally in many species, and the simple view through the eyes may be misleading. Visual processing in mammals is discussed in detail in Chapter 17.

Whether or not other species can distinguish color has been a question of perpetual interest to the lay public and many biologists alike. Von Frisch (1914) was one of the first to resolve the question in a nonhuman species. Since that time, training methods, responses inside rotating cylinders, and even some evoked potential recordings have been applied to a wide variety of animals to determine whether they have color vision.

Color vision has been demonstrated in bees and many other insects, cephalopod mollusks, many fish, many amphibians, diurnal but not nocturnal reptiles and birds, and some mammals, including primates, squirrels (but not many other rodents), many carnivores, and a few others that have been tested such as pigs and horses. Color vision, however, varies, and it is not simply a question of whether or not an animal possesses it. Furthermore, it appears to vary even among individuals of the same species. Several primates, including some individual humans, have little or no ability to discriminate red. Cats, although demonstrating some ability to discriminate colors, show much better brightness and pattern discrimination.

Eyes are capable of responding to a wide range of light intensity, depending in part on different visual pigments and types of sensory cells. The basic cell types are rods, more sensitive and used primarily at night, and cones, less sensitive and used during the day and for color vision. Most invertebrates that rely heavily on vision apparently have only one type and thus are active either by day or night, but not both. Vertebrates, however, depending on the species, usually have populations of both cell types and in some instances may be active both day and night. Most vertebrates have both rods and cones, but the proportions vary; diurnal species have cones predominantly, whereas nocturnal species have a much higher proportion of rods. Persons interested in the chemistry of vision should consult a physiology text or the review by O'Brien (1982).

One of the most interesting techniques for detecting light at extremely low levels involves "recycling" the light. After the light enters the eye and has passed through the sensory cells, a reflective surface at the back of the eye sends it back through the sensory cells for a second pass. This is the reason for the shine that can be observed in the eyes of many nocturnal animals, including deer, cats, and some moths. Diurnal species, on the other hand, may be faced with the opposite problem—too much light in many circumstances. Accordingly, the back surfaces of the eyes of chiefly diurnal species have become black, which absorbs the light.

Heat and infrared sensing are similar to vision in that they are widespread among animals and vary from a generally dispersed, diffuse sense in many animals to an advanced, highly developed sense in others. Heat and light sensing differ, however, in two important aspects. First, advanced senses are quite common for light but not for heat. Although most animals probably have some temperature sense, it is almost universally of the diffuse sort, like that of humans. Species or groups that have the advanced capabilities and highly specialized organs include a number of arthropods that are parasitic on warm-blooded vertebrates, a few types of snakes (the pit vipers, boas, and pythons), and a few other species. Night moths have infrared receptors by which the males apparently locate the females at close range, after first using olfactory modalities. The Australian brush turkey (Megapodiidae, *Leopoa ocellata*) is a bird that has an acute temperature sense and maintains a nest

Figure 16-12 Infrared heat sensors in snakes. **A,** Rattlesnake, showing the infrared detecting pits in front of the eyes and beneath the nostrils. **B,** Infrared "shadows" cast in the pits of a pit viper, depending on how the head is oriented toward the heat source.
A by James W. Grier. B modified from Gamow, R.I., and J.F. Harris. 1973. Scientific American 288(5):94-100.

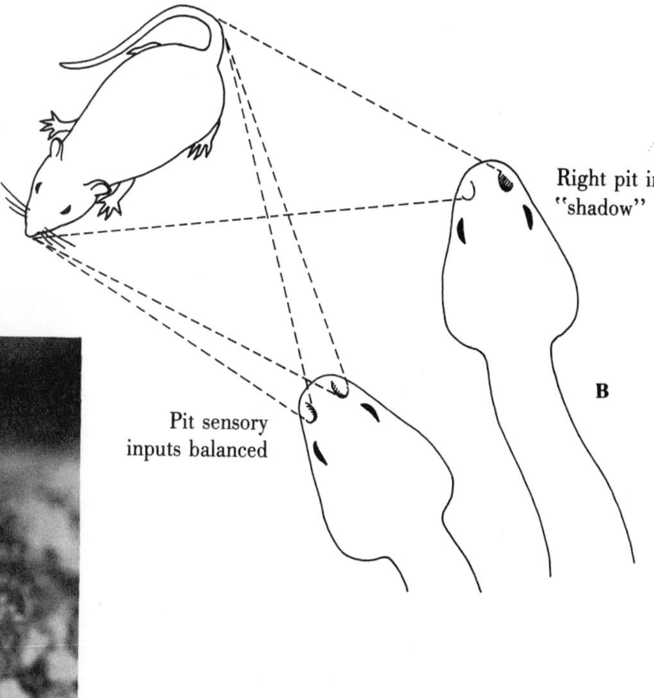

Right pit in "shadow"

Pit sensory inputs balanced

temperature at 33° C through the use of decomposing leaf litter—not its own body heat. The nest activities are performed by the male while the female goes on to lay more eggs in the nests of other males. The bird frequently tests the temperature inside the nest mound by taking samples of the material in its mouth. If the mound has become too warm, the bird kicks material away from it to let it cool; or if it is too cool, it kicks more material onto the nest.

The second major difference between heat and light sensing involves the transduction process. Heat detection, rather than using photochemical molecular changes, uses the thermal properties of molecular motion in some manner that is not well understood. Some sensors, such as in the night moths, snakes, and many of the arthropods, use radiated heat that is sensed from a distance. The means by which snake pits localize a heat source are shown in Figure 16-12. Direction is determined through the differences in maximum and minimum areas (shadows) of stimulation. Distance is sensed from the difference in the angle toward an object from two different positions, as in binocular vision. For further information on infrared sensing in snakes, see Gamow and Harris (1973).

One of the most fascinating senses is the ability to sense electrical fields. Most organisms are unable to do so, except perhaps for the vague uncomfortable feeling that is detected indirectly by other sensors in a strong field. Many fish, a mammal—the platypus—and some amphibians can detect the electrical fields produced by muscular activity in their aquatic prey (Bullock and Heiligenberg 1986, Griffiths 1988). A number of specialized electrical fish, including many that commonly are sold in pet shops, are able to produce their own electrical fields and then sense changes that occur in the fields caused by nearby objects. A few species—electrical skates, rays, and eels—can generate enough electricity to use in defense and for stunning prey. Most species, however, appear to use it merely for sensing their environment and for communication and interaction with others of the same species

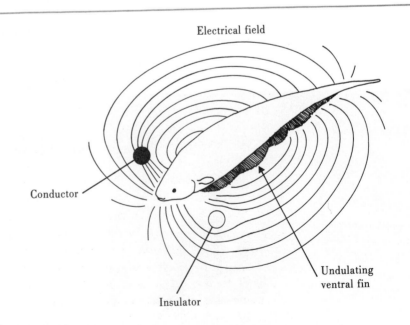

Figure 16-13 Electrical senses. Electrical field surrounding a fish, showing perturbations caused by conductive or insulative objects. A rigid body and undulating fins minimize the fish's disturbance of its own electrical field. The fish shown here is of the genus *Sternopygus* (order, Gymnotiformes). Several fishes, including sharks and skates, possess electrical senses.
Modified from Lissman, H.W. 1963. Scientific American 208(3):50-59, and Matsubara, J.A. 1981. Science 211:722-725.

Electrical field

Conductor

Insulator

Undulating ventral fin

(Moller and Serrier 1986, Hopkins 1988). That these fish can detect differences in electrical fields and conductivity of objects was originally demonstrated by two-choice tests in which the animals were trained to choose, for example, a conductor over a nonconductor. Figure 16-13 illustrates the effects in the electrical fields.

The chemical senses, including taste and olfaction, are almost universal among animals. For many species almost their entire Umwelt is based on chemical senses. Various chemicals are vital to life itself, and possibly every single animal has at least some chemical sense. Even the simplest, one-celled protozoans must select the molecular food substances they take into their bodies and the microclimates of the chemical environments in which they exist. The sensing of various molecules is used in the detection of and orientation to food and the avoidance of noxious environments and substances. Chemicals commonly are used in communication, for example, as sex attractants, territory markers, and as warnings or repellents.

As a biochemical instrument for the analysis and recognition of molecules, even the human nose is in many cases much better than the most sensitive and sophisticated of man-made instruments (although many chemicals are dangerous to health or may damage the nose). Yet our chemical sense appears to be as crude and insensitive compared with those of some species as their sense of vision or hearing is compared with ours. Our own poor chemical senses have hampered our views and research on the sensory capabilities of other species, and until recent years humans commonly assumed that other animals used chemical senses much as we do ourselves.

Many other animals differ significantly from humans not only in their sensitivities to chemicals in general but also in terms of the specialization of their sensory organs, location of the organs, and the evolved specificity to which particular chemicals or classes of chemicals can be sensed. Many aquatic organisms and some arthropods have receptors all over the body surface or on legs, antennae, and other structures away from the mouth and "nose."

The mechanism of odor and taste transduction is quite different from all other modalities discussed so far. It depends on molecular configuration and characteristics such as charge. Because of the chemical nature of the sense, the sensory organs must have a moist surface for solubility. Also, the operation of chemical senses involves molecular movement and the probability of the particular sensory cells encountering molecules of the proper characteristics for stimulation. The greater the evolved match between the sensor and the molecule and the larger the surface area, or number of sensors, the greater the chance that the molecules will be detected. The chances also increase with greater numbers of molecules, that is, the concentration. For a recent review of the neurobiology of the chemical senses, see Finger and Silver 1987.

One of the best-studied chemical sensors is that of the silkworm moth *Bombyx mori* (Schneider 1974). Bombykol is a sex-attractant pheromone that attracts males to females (Figure 16-14). Males move to and attempt to copulate with any object emitting bombykol, regardless of sight, sound, or touch, and they ignore females from which they cannot detect bombykol. Bombykol is an unsaturated fatty alcohol, trans-10-cis-12-hexadecadien-1-ol. Furthermore, only the trans-cis isomer is effective. Other geometrical isomers of the same compound stimulate males only weakly. Unlike many olfactory systems where a variety of odors are discriminated, the bombykol system responds only to a single kind of molecule. It is a finely tuned system indeed. The sensory cells are located on the male antennae (and only on

Figure 16-14 Male
silkworm moth *(Bombyx
mori)*, responding to
olfactory but not visual
stimuli from a female. The
female is releasing the sex-
attractant pheromone
bombykol.
Modified from Schneider, D.
1974. Scientific American
231(1):28-35.

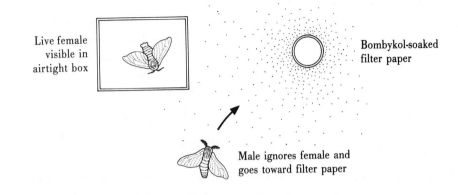

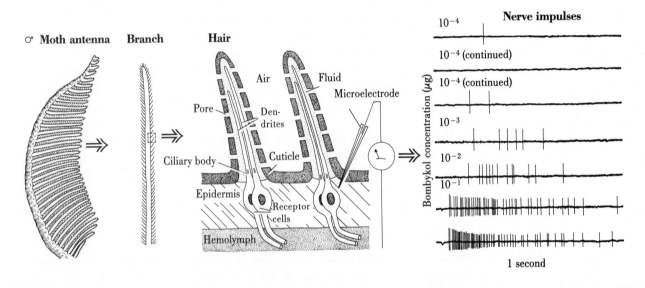

Figure 16-15 Increased magnification of the chemical sensory organs on the antenna of the
silkworm moth *(Bombyx mori)* and neural output in response to varying concentrations of the
sexual attractant bombykol.
Modified from Schneider, D. 1974. Scientific American 231(1):28-35.

the male) and respond proportionally to increasing concentrations of the pheromone,
as shown in Figure 16-15.

Many forms of energy, including most of the electromagnetic spectrum above
and below the region of light and heat, appear not to be used by any Earth-based
biological organisms. Several species commonly used in psychological laboratories,
including rats, mice, cats, rhesus monkeys, and pigeons, have been shown to
respond behaviorally to low levels of ionizing radiation (e.g., Haley and Snider
1964). The mechanisms involved are not well understood and probably are indirect,
related to nausea and taste aversion-like effects (discussed in Chapter 20) or some
unknown effects related to sleep arousal in the brain.

Radio waves, although reaching the surface of Earth through our atmosphere, appear not to have been used naturally by any living organisms on Earth. The only, but substantial, use has involved their artificial production and reception by humans. The physics of radio waves, their wavelengths, energy needed for production and modulation, and complex circuitry needed for processing do not seem to make radio waves accessible to biological systems.

Use of Emitted Energy for Sensing

The source of energy for the sensory systems of many species of animals is the animal itself. The individual emits the energy by which it then senses the environment or a component of the environment. Examples include bats, which emit cries and then listen for the echo, and electrical fish, which generate an electrical field, then detect changes resulting from objects within the field. Other examples of animals that use sonar are many of the marine mammals and the oilbird (Figure 16-16).

The use of sound underwater presents some intriguing problems and opportunities. Sound does not reflect well until it encounters a medium that differs in density from the one in which it is traveling. Much biological tissue is largely composed of water and hence would transmit rather than reflect sound underwater; such tissue would be largely or partially "invisible" to sonar. The sonar echoes from many fish probably reflect off their internal air bladders and perhaps bones rather than from their body surface. This would give sonar-using animals such as dolphins something like "x-ray vision." They might even be able to "see" (hear the echoes of) air bubbles in the digestive tracts of their young.

Figure 16-16 Oilbird *(Steatornis caripensis)* using echolocation to avoid an obstacle placed in its path. The bird was captured with this flash photograph just as it came to a momentary stop and was hovering in midair inside a dark cave. From Konishi, M., and E.I. Knudsen. 1979. Science 204:426.

Figure 16-17 Use of emitted light by flashlight fish (*Photoblepharon palpebratus*). It has been postulated that the light functions to obtain prey, deter predators, communicate intraspecifically, and aid general seeing. Fish has luminescent organ lid open.
From Alex Kerstitch/Tom Stack and Associates.

Some aquatic insects use water surface waves. Whirligig beetles *(Gyrinus* sp.*)* whirl rapidly and detect changes in the returning wavelets. The sensory cells that are used by the beetles are in fine hairs at the second joint from the base of the antennae (reviewed in Griffin 1959). They may be able to detect waves as small as 4×10^{-7} mm, less than a millionth of a millimeter. The beetles are unable to orient or avoid obstacles when these particular hairs are removed. The beetles whirl intermittently, apparently waiting between periods of whirling for the return of waves echoing off obstacles, each other, and other objects.

A few deep-sea fish and perhaps some other marine organisms produce their own light by which they see (Figure 16-17). They may have exceptionally sensitive and specialized eyes.

A number of organisms produce light but apparently do not use it for their own sensory purposes. The function, if any, of the luminescence in many marine organisms is not well understood, although there are several hypotheses. In other cases, such as fireflies and some deep-sea marine animals, the luminescence clearly serves a communication function (Chapter 15).

Using Combinations of Sensory Input

Few species rely solely on a single sensory mode for any given behavior or other sensory function. Even as simple a situation as body orientation, knowing which end is up, usually involves input from more than one sensory source. Body orientation in most vertebrates uses a balance organ, or semicircular canals and associated structures, but only in part. Proprioceptive input from the weight of different body parts, the stretch of various muscles, and vision also play important parts. "Up" for fish, for example, usually involves the brightest part of their environment because of daylight entering the water from the surface. If, as in an aquarium, illumination is brightest at the side, many fish tip to orient their dorsal surfaces to the brightest light level. The tipping is rarely complete, however, but rather at an angle that vectors, or compromises, the input from vision and that from the semicircular canals (Figure 16-18). Likewise, body orientation in humans depends in part on the perceived horizon. If the horizon is gradually shifted so that the change is not detected, as in so-called gravity houses at amusement parks, our

Figure 16-18 Fish vertical orientation in respect to the combination of input from light (normally coming only from above) and gravity pulling from below. **A,** Fish in normal upright position with light from above. **B,** Fish tipped in a vectored response to light from the side and gravity from below.

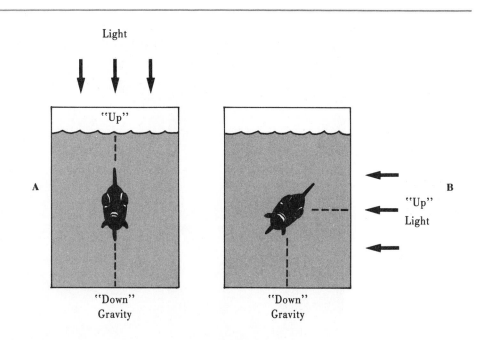

sense of orientation may be altered and, for example, water may appear to run uphill.

With combinations of input there is a certain amount of redundancy, and on occasion some of the input may be eliminated without noticeable impairment. However, if all input is eliminated or if the system places high priority on an important sense that is lost, serious problems may develop. Confusion with balance, or dizziness, can leave one with the sensation that one is stationary but the rest of the world is spinning even when it can be seen that it is not. In extreme cases, all sense of orientation is lost and vertigo develops, an uncomfortable and alarming state. This may occur in flight when clouds, snow, or similar obstructions obscure the horizon and turning motions interfere with the inertial senses. Jet pilots on complex turns and maneuvers sometimes have to learn to live with vertigo and trust only their instruments for long periods of time. Vertigo develops sometimes with damaged or diseased nerves from the inner ear, and it also may be triggered by extreme nausea or emotional stress.

Sensory input from different modalities may be weighted in importance. There generally is a hierarchy, as was seen in honeybee orientation (Chapter 7) and bird compasses (Chapter 8). If high-priority inputs are available, even if they are incorrect, lower-priority inputs may be ignored. Furthermore, the hierarchies of input may switch within the same individual under different circumstances. A sense that is used heavily in one behavior, such as in detection and defense against a predator, may be quite different from the sense used in a different behavioral situation, such as searching for food.

Attributes of Senses

For any particular kind of sense, there are numerous attributes of the sense that may be involved. To help understand this point, look at or imagine another person

Figure 16-19 Flower appearances under visible *(left frames)* and ultraviolet *(right frames)* light. These views are as they would appear to a vertebrate eye if one could see into the ultraviolet. Insects can see ultraviolet, but their eye structure does not permit the detailed vision that humans possess. For a suggestion of how the flowers might look to an insect, see Figure 16-11. The dark centers of the flowers in ultraviolet are thought to serve as targets or guides for foraging insects.
From Silberglied, R.E. 1979. Annual Review of Ecology Systematics 10:373-398.

not too far away from you who is talking. It is clear to you where that person is, and you could easily point to him or her. Furthermore, you can distinguish much detail in the person's appearance. If you close your eyes and just use your ears, can you still locate and point to him or her (including when the person moves to different locations)? Now, keeping your eyes closed and if the person makes no sounds, hold up a hand and try to determine where the person is by body temperature (you most likely cannot do it). Now return to vision: open one eye but keep the other closed and try to bring your two forefingers together, or try to put your finger on something a few centimeters or inches away. How accurately can you do it compared with using both eyes? If you use your ears, can you determine which of two sounds is of higher pitch? Can you specifically identify the pitch of any particular sound (such as musical notes C, A, etc.)?

The basic attributes of any particular sense are sensitivity, ability to localize the source, and pattern discrimination. Sensitivity involves whether the sense is present or absent. Some insects, birds (Chen et al 1984), and perhaps a few other animals, for example, can sense ultraviolet light (Figure 16-19) and the polarity of light whereas we cannot. On the other hand, humans can see red color but many other species cannot. Many species can sense the Earth's magnetic field; it is not yet certain whether humans can. If something can be sensed, then the next consideration is the magnitude of intensity and resolution of differences that can be discriminated and whether the ability to detect differences is absolute or only relative. The difference between intensity versus resolution is the difference between loudness and pitch in sound or between brightness and color in vision.

The difference between absolute and relative sensing occurs with most senses; heat detection provides a good example. Temperature sense in humans is both relative and imprecise. Thus it is possible for different parts of the body to sense the same temperature differently. Bathwater, for example, commonly feels hotter to the cold feet than to other, warmer parts of the body. Likewise, it is disconcerting to have one's hands adapted to different temperatures, such as by holding one in warm water and one in cold, then picking up an object that is of an intermediate temperature; it will feel cold to one hand and hot to the other. Many other species, however, have been shown to possess temperature senses that are both absolute and extremely precise. Mosquitoes have been reported (Herter 1962) to be sensitive to as small a temperature difference as $0.002°$ C and fish to as small a difference as $0.02°$ C (reviewed by Hess 1973 b). A few rodents, bees, and fish have been trained to choose a particular temperature that does not depend on a relative difference with a previous or background temperature. Honeybees maintain the temperature inside their hives, and various species of African termites control the interior temperatures of their mounds, as described in Chapters 6 and 13. Some lice have an absolute temperature sense that is so precise that they have been used in primitive societies to indicate a person's body temperature and health. As long as a person had the normal complement of lice, he or she was healthy, but if a fever developed or the body temperature went too low, the lice would leave that person and move to others (cited in Milne and Milne 1964).

Ability to localize the source of a stimulus varies greatly. We use vision, which is accurate, and hearing to a degree; but our ability to localize sources of heat is quite poor in comparison. Other animals, however, may have a highly accurate sense of heat localization but do not use vision or hearing.

Pattern discrimination may involve both spatial aspects, such as visual acuity

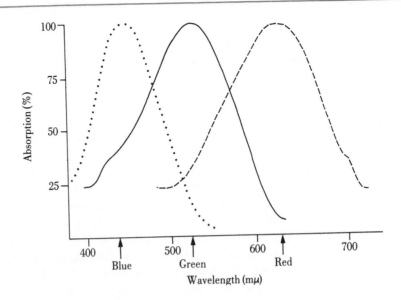

Figure 16-20 Three-point cone sensitivity in goldfish, a species believed capable of color vision. Color sensation is believed to result from combinations of input from these three sensors. Yellow, which is between green and red, for example, would be sensed by moderate stimulation of green cones plus moderate stimulation of red cones.

Modified from MacNichol, E.F. Jr. 1964. Scientific American 211(6):48-56.

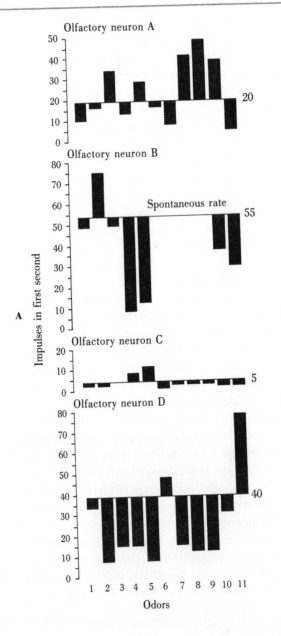

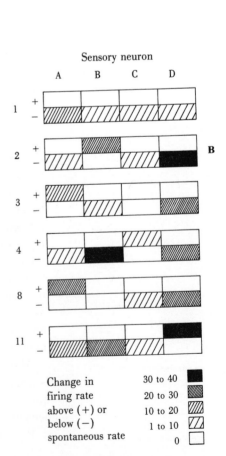

Figure 16-21 Identification of chemical cues by combinations of sensory input. **A,** Firing rates of four olfactory receptors in the tobacco hornworm larva *(Manduca sexta)* in the presence of 11 different odors. **B,** Examples of simultaneous sensory input from the four sensors in response to different odors. Combinations of input from just these four permit the distinction of many unique odors based on their total sensory "complexion."

Modified from Dethier, V.G. 1971. American Scientist 59:706-715.

and complexity of the pattern that can be distinguished, and temporal aspects, such as sound patterns (e.g., in human music), patterns of light flashes (as in fireflies), or movement patterns (as in lizard head bobbing) over time.

In a few senses, such as hearing in mammals, resolution depends on input from a large number of cells that respond differently (because of their position along a vibration gradient in the case of hair cells in the inner ear). Most senses, however, use only a few basic cell types that, by working together, permit the fine discrimination of differences in stimuli. This principle is somewhat analogous to the construction of many different words from only a few basic letters. In color vision, for example, there are not particular retinal cells for each color that can be distinguished. Instead there are only two to four categories of color-sensitive cells, depending on the species. Humans and most other species with color vision have three types of cones for color vision, but butterflies and at least one species of fish (the Japanese dace, *Tribolodon hakonensis*) have been shown to possess four (Harosi and Hashimoto 1983). Figure 16-20 illustrates the three-pigment visual sensitivities for the goldfish. Similarly, in olfaction there may be only a few kinds of sensory cells that work together to distinguish among a wide variety of different molecules, as shown in Figure 16-21.

MUSCULAR AND GLANDULAR OUTPUTS

The final output of neural processing in all animals involves the contraction of muscles, release of glandular secretions, or other effector output such as cilia or flagella. The nervous system can stimulate secretions from both endocrine (ductless) glands such as the adrenal gland (i.e., when animals are fighting) and exocrine (with ducts) glands (such as salivary, sweat, odor, and some digestive glands). Secretions result from a variety of mechanisms in response to neural impulses such as membrane or vascular changes in the glands and in some cases with assistance from skeletal or smooth muscles.

Figure 16-22 Motor endplates. The connections between a motor neuron and the muscle fibers it inervates.

Photograph by James W. Grier.

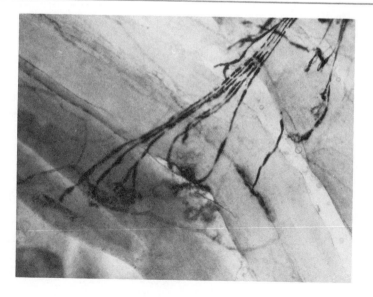

Although glandular secretions occur and may be important in particular situations, most neural output occurs through muscular contractions. Muscles and muscular contraction are fascinating subjects. Our present interest for behavior, however, is with the connections between the nervous and muscular systems. A particular movement, hence behavior or part of a behavior, depends on which muscles contract and the spatial-temporal patterning of the contractions. This is largely under the control of the nervous system, as finally mediated via the motor neurons. These stimulate muscular contraction through the motor end-plate, a synapse between the axon of the motor neuron and the surface of the muscle fiber (Figure 16-22). An action potential, similar to that seen in neurons, is initiated in the muscle fiber. This action potential triggers a sequence of chemical reactions that culminate in the observed contraction (Figure 16-23). A given fiber (in familiar twitch muscles) either contracts or does not contract (all or none). Individual muscles are composed of large numbers of fibers, and the overall graded strength of contraction in a muscle depends on what proportion of its fibers are contracting at any one time.

The nervous system exerts exquisite, specific control over the pattern of muscular contraction: which particular muscles contract, to what extent, when, and in what sequence. Furthermore, the higher control centers can direct the patterns to different sets of muscles, which can be easily demonstrated in ourselves by a person's handwriting style. The writing style of most persons is unique. Writing normally uses a large number of muscles in the hand and forearm, with letters being formed by an exact sequence of contractions. If a person writes in large letters on a wall

Figure 16-23 Time sequence of action potential and contraction in a muscle. The first event is an action potential, similar to that which occurs in a neuron. The contraction of the fiber then develops subsequently and more slowly.

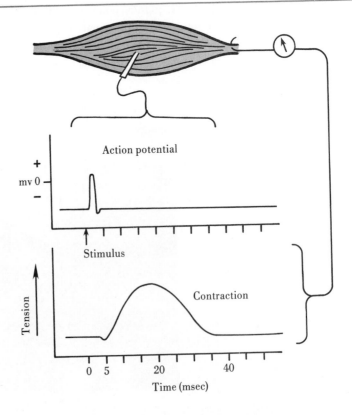

Figure 16-24 Writing styles of two different persons writing with either the hand (A and C) or arm and shoulder (B and D). Styles may differ widely among different persons, but for any one person there is less variability, depending on which set of muscles is used.

or blackboard, however, completely different muscles, largely in the upper arm and shoulder, are involved. Yet the writing style normally shows little or no change (Figure 16-24). This is because the nervous system is directing the writing, and it is sending the instructions to different muscles. Compare your own hand and arm writing with that of someone else.

Although the preceding paragraphs describe the general picture of muscular contraction, nature has found much room for variation. Neural parsimony in arthropods, for example, also occurs at the muscular output. It is achieved in a variety of ways, but basically by sending different information down the same axon rather than using different axons for different information. In both vertebrates and invertebrates a given motor axon may innervate several muscle fibers, but axons typically innervate proportionally more fibers in invertebrates. In many cases, all the fibers in a whole muscle may be innervated by only one or a few axons' impulses, and recruitment of muscle fibers is accomplished by different patterns or frequencies of firing. If the pattern rather than just the frequencies of impulses is important, the nerve-muscle synapses are said to be pattern sensitive. In the crayfish claw opener muscle, which is innervated by a single axon, a specific train of impulses results in a particular strength of contraction. If the impulses are grouped in pairs, the muscle contracts more strongly. In vertebrates, on the other hand, recruitment is mostly accomplished by adding more motor neurons, which in turn fire more fibers.

The all-or-none principle in muscles is strictly true only for twitch muscles—fibers that contract quickly with rapidly propagated action potentials. There are more slowly contracting fibers and fibers in which only local regions of the fiber contract, thus showing graded contractions. This includes a number of arthropod muscles, some vertebrate smooth muscle, and even the slow fibers of vertebrate striated muscle. Muscle fibers vary tremendously in their properties and as a result are difficult to classify. Categories of fast or slow, red or white, voluntary or autonomic, smooth or striated, etc., are only partially useful. Muscle fibers also vary in size (up to 1 mm diameter in a barnacle, *Balanus nubilis*) and length of time they can remain contracted. The shells of some mollusks are held closed by "catch" muscles that can remain contracted for days.

Some muscles are capable of both speed and precise fine control, as in human fingers typing, whereas others are involved more in slower, power movements such as in lifting weights. Precision control muscles generally have a higher ratio of neuron-to-fiber innervation. In many muscles, different muscle fiber types (e.g., slow and fast) are intermixed.

Another means of varying recruitment is via polyneuronal innervation. Different axons from different neurons may impinge on a given muscle fiber. Some are excitatory and some are inhibitory, and a given axon may have endings that deliver fast (phasic) short bursts of impulses and other endings that deliver slower (tonic) trains of impulses. These in turn affect the speed of the ensuing contraction and thus affect the development of strength in the muscle. The summed effects of polyneuronal innervation are reminiscent of discrimination normally associated with the nervous system. Some observers have quipped that crabs can think in their legs.

The frequency of rapid wing movements of some insects seems to exceed the capacity of neuronal action to fire them. Indeed, some are not caused to contract at each beat by neural impulses. The flight muscles of some of the more primitive insects connect directly to the wings and are contracted in standard fashion. In more advanced forms, however, the flight mechanism is complex and varies among species (Pringle 1957). In essence the muscles are connected indirectly—not to the wing but to the elastic thorax. This snaps back and forth (the click mechanism), which moves the wings, via the lever principle, and also stretches the flight muscle on the rebound. This stretch, rather than a neuronal action potential, causes the next contraction, and the movement continues. The role of the nervous system is only to trigger the initial contraction, periodically maintain the process, and apparently modify the frequency to achieve controlled movements. A similar system is involved in the sound-producing tymbal organ of cicadas.

This overview provides a background for the next steps in understanding how the nervous system controls behavior. Of the three main components, input, processing, and output, the influences of input and processing are perhaps the most important causes of differences in behavior among different animals. We thus devote the remainder of this part of the book to sensory input and neural processing, with emphasis on processing.

SUMMARY

The nervous system can be conceptually broken into its three major components: input (sensory), processing, and output (primarily muscular contractions but also glandular secretions).

The basic anatomy of nervous systems is of two major types: (1) the ventral, solid, basically paired nerve cord central nervous systems with determined numbers of neurons found in the protostome invertebrates and (2) the dorsal, hollow, single spinal cord system with indeterminate numbers of neurons found in vertebrates (deuterostomes). Insects (and other invertebrates) show neural parsimony: the ability to efficiently use many fewer neurons than vertebrates to accomplish the tasks of the nervous system.

The many techniques used to investigate nervous systems include: visual techniques (light and electron microscopes, with a variety of associated techniques such as staining); recording of bioelectric events such as action potentials with microelectrodes; stimulation, ablation, and lesioning of neural tissue; and a variety of experimental techniques such as voltage clamping.

Individual neurons have two primary functions: (1) to process or discriminate incoming input, either transduced sensory input or from previous neurons, and (2) to transmit decisions, via action potentials (also called *impulses* or *spikes*). Input from previous neurons, because it has come across a synapse, is referred to as *postsynaptic potentials (PSPs)*. These may be either excitatory (EPSPs) via local depolarization in the postsynaptic neuron or inhibitory (IPSPs) via hyperpolarization. EPSPs and IPSPs add together in summation. There are two types of summation: temporal, which involves subsequent inputs over time at the same synapse, and spatial, which involves inputs from different locations, (i.e., different synapses).

Coding of neural information is through the rate or frequency of action potentials. Many neurons have a spontaneous rate of firing, which may be increased or decreased by input. The speed (which is separate from frequency) at which action potentials travel down the axon is increased in axons that are larger in diameter, warmer, or myelinated. When an impulse arrives at the end of a neuron, it continues onto the next neuron through electrical gap junctions or the more common and familiar chemical synapses, through the release of one of a number of types of neurotransmitters.

The Umwelt, or perceptual world, which depends partly on sensory input, varies greatly among different species and even among different individuals of the same species. The Umwelts of other species may be quite foreign to our own and difficult for most of us to envision. However, those Umwelts influence the entire behavioral picture of other animals and thus must be considered. The senses form the first, peripheral stages of stimulus filtering. Additional central stimulus filtering occurs through central nervous system processing (discussed in the next chapter).

Senses can be assessed through a variety of techniques, including: evoked potentials, blocking, interruption or removal of organs or connections, and learned or specific natural responses. Natural selection strongly favors the use of particular senses in different behaviors under different environmental conditions. The senses include: hearing and other vibrational senses, vision, heat and infrared sensing, chemical senses, and the use of emitted-energy sources. Most animals use multiple senses in combination. Senses, regardless of type, have a number of attributes such as sensitivity, ability to localize the source, and pattern discrimination.

The nervous system can stimulate glandular secretions through a number of mechanisms. Muscular contractions are stimulated by action potentials at specialized synapses, motor endplates. The ensuing action potential in the muscle fiber triggers the physical contraction.

FOR FURTHER READING

Bullock, T.H., R. Orkland, and A. Grinnell. 1977. Introduction to nervous systems. W.H. Freeman, San Francisco.
A standard, but now somewhat dated, introductory reference for nervous systems. (See also Kandel and Schwartz 1985.)

Camhi, J.M. 1984. Neuroethology. Nerve cells and the natural behavior of animals. Sinauer Association Sunderland, Mass.
From among several books on neuroethology currently available, this is considered one of the most useful. It is concise, clear, and covers the subject all the way from elementary principles of the individual neuron to numerous examples involving a wide range of animals.

Ewert, J.P. 1985. Concepts in vertebrate neuroethology. Animal Behaviour 33:1-29.
A review of neuroethology, available in a journal article, based on vertebrate examples (primarily toads) by one of the leaders in the field. This article is based on a Niko Tinbergen lecture.

Hopkins, C.D. 1988. Neuroethology of electric communication. Annual Review of Neuroscience 11:497-535.
An updated review of electrical senses and communication, including both neural and behavioral (and evolutionary) aspects, in weakly electrical tropical freshwater fishes. This article touches on several different issues covered in this chapter.

Kandel, E.R. 1976. Cellular basis of behavior. W.H. Freeman, San Francisco.
Another standard introductory reference, along with Bullock et al, for the relationship between neurons, nervous systems, and behavior. For a more recent text, see the next listing.

Kandel, E.R., and J.H. Schwartz, editors. 1985. Principles of neural science. Elsevier, New York.
A (massive) introduction to the subject. Although this book is thorough and, as a result, large, it is well written and edited and serves as an excellent source for further reading, whether by persons with little previous knowledge about neurons and nervous systems or by experienced persons who want to review, update their knowledge, or branch into less familiar topics.

Okanoya, K., and R.J. Dooling. 1988. Hearing in the swamp sparrow, *Melospiza georgiana*, and the song sparrow, *Melospiza melodia*. Animal Behaviour 36:726-732.
A recent example of using learned responses to measure and interpret sensory capabilities.

17

NEURAL PROCESSING PATHWAYS THROUGH THE NERVOUS SYSTEM

The centipede was happy quite
Until a toad in fun
Said, "Pray, which leg goes after which?"
That worked her mind to such a pitch
She lay distracted in a ditch
Considering how to run.

Nineteenth century,
Anonymous

Neurons do not function in isolation, acting individually, but as vast networks of many neurons working together. They accomplish this by sifting, sorting, and passing on information based on a number of principles and properties of the component neurons and of the system as a whole. This chapter develops and explains those principles through a number of examples. Although this book is on animal behavior in general rather than human behavior in particular, examples using human cases—including visual and language pathways—are particularly useful for demonstrating general principles that are thought to operate in all or most neural processing.

THE SIGNIFICANCE OF CENTRAL PROCESSING

We have now come to the most challenging and perhaps most intriguing part of the picture: how the nervous system actually processes information to produce the multitude of functional, observable behaviors in animals. It is one of the most complex of all biological processes, yet it goes on quietly behind the scenes inside the animal.

It frequently is claimed that the human brain, for example, contains around 10 billion neurons. However, based on a density of around $100,000/mm^2$ of surface area (given in Hubel and Wiesel 1979), there may be at least 10 billion neurons in the cerebral cortex alone. The cortex is only part of the brain. Recent guesses regarding the number of neurons in the whole brain range as high as 1 trillion. We conservatively use the traditional figure of 10 billion for discussion purposes. For other parts of the nervous system it is estimated (e.g., Nauta and Feirtag 1979) that in humans there are about 2 million to 3 million sensory neurons and roughly an equivalent number of motor neurons. Although 4 to 6 million sensory and motor neurons may seem like a lot, in comparison with the estimated 10 billion total, that is less than 0.01 of 1%; 99.99% of the neurons are interneurons. The number of synapses on any one neuron may be as high as 80,000. Other species of animals may not have as many neurons as humans, but the principle is the same. Most of an animal's neurons and the bulk of the nervous system are devoted to the processing of neural information, not input and output.

Details of the structure and arrangement of most nervous systems are beyond the scope of this text, and specific parts of the nervous systems in different animals are mentioned only where necessary. Persons interested in more detail can refer to texts covering particular taxonomic groups, such as vertebrates or various invertebrates, for the basic anatomy of animal nervous systems. The anatomy of the human nervous system, with emphasis on the brain, is covered to varying degrees in most general biology texts, human anatomy and physiology texts, and most psychology texts.

In this chapter, we use a series of examples to illustrate different principles of central nervous system processing. These principles are pulled together and summarized at the end. The material presented is merely a small, introductory sampling from an area of neurophysiology that has come to be known as *neuroethology* or part of a larger aspect called *physiological ethology*; it is the study of the neural basis of functional behavior (as opposed to just studying the nervous system by itself). Neuroethology is currently an extremely active area of advanced research. We begin with a relatively simple example by returning to the bat and moth interaction where we left off in Chapter 1.

CENTRAL STIMULUS FILTERING: MOTH HEARING AS A CASE IN POINT

The moth ear has proved to have the following advantages for neural and behavioral research: (1) moths are relatively common and easy to obtain; (2) the ears are relatively easy to get at; (3) each ear has a total of only two acoustic sensory neurons (A_1 and A_2 cells) plus one proprioceptive sense cell (the B cell), which greatly simplifies the interpretation of results; and (4) the responses of these sensory cells are typical, hence representative, of sensory cells in many other organisms and

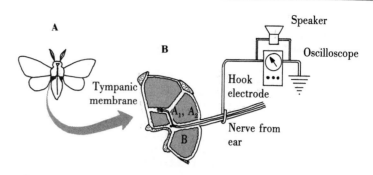

Figure 17-1 Setup for recording impulses from a noctuid moth ear. Based on photographs and discussion in Roeder, K.D. 1967. Nerve cells and insect behavior. Harvard University Press, Cambridge, Mass.

Recording from one ear

more complex sensory structures. The following is based on photographs and extensive discussion in Roeder (1967), which remains a classic book on the subject.

A moth's ear is sensitive to sounds in the frequency range of 3 to 150 kHz, but it is most sensitive to the range of 50 to 70 kHz. Within this range, however, the moth is tone-deaf; it apparently cannot distinguish pitch but reacts in a similar manner to all sounds. Recall that bat calls range in frequency from 20 to 100 kHz, in the same range as the moth's hearing. (Human hearing is in the range 0.04 to 20 kHz.)

With the use of an extracellular electrode on the auditory nerve and a recorder (Figure 17-1), one can infer what a bat sounds like to a moth. Note that we are recording a compound action potential from three axons: A_1, A_2, and B.

A recording from one moth ear under quiet conditions is diagrammed in Figure 17-2, *A*. The periodic spikes are from B cell.

In the presence of a pure continuous tone of varying intensity, responses are as shown in Figure 17-2, *B*. The following characteristics of the response can be seen:

1. There is a brief delay between the start of the sound and the onset of the action potentials at the site where they are recorded.
2. The spike frequency, that is, the number of action potentials per unit of time, slows as a result of adaptation of the neuron.
3. Increases in sound intensity produce an increase in the initial spike frequency.
4. At low sound intensities only one cell, A_1, responds; and then at higher sound intensities A_2 joins in. In some instances, A_1 and A_2 occur simultaneously and produce a double peak.
5. The spike frequency of B stays the same regardless of sound intensity.

With short bursts of sound (Figure 17-2, *C*), rather than with continuous tones, the adaptation does not occur. There is still a brief delay between the start of the sound and the start of action potentials, however, and another characteristic becomes noticeable: the spikes continue briefly after the sound has stopped. The brief continued firing is called *after-discharge*. The spike frequency of the B cell remains unchanged.

With increased intensity of short bursts of sound (Figure 17-2, *D*), three effects can be observed: increased spike frequency, a shorter delay between the start of sound and start of spikes, and a longer after-discharge. The B cell continues to fire at a relatively steady, unchanged pace.

Figure 17-2 Compound action potentials from sensory neurons A_1 and A_2 and the B cell of a moth ear under a variety of conditions to illustrate principles of neuronal coding of information. Refer to text and study this illustration in conjunction with that information.
Based on photographs and discussion in Roeder, K.D. 1967. Nerve cells and insect behavior. Harvard University Press, Cambridge, Mass.

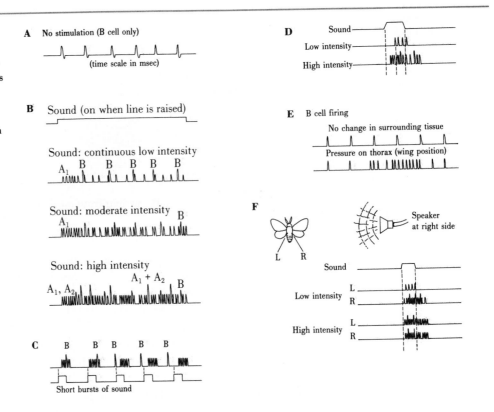

The moth has no independent means of assessing the sound other than the series of spikes coming from those few neurons. A low spike frequency could arise from several different situations, such as a distant, low-intensity signal in the sensitive range of detection, a loud signal that is outside or on the edge of the moth ear's sensitivity range, or sensory adaptations and the decline in a cell's sensitivity. There seems to be no way for the moth to distinguish between these and other alternatives, but there is more to the picture.

The results of recordings made simultaneously from both ears are shown in Figure 17-2, F. Unless both ears are saturated, the neural response starts sooner, has higher spike frequency, and lasts longer on the side from which the sound is coming. It is not far from one side of a moth to the other, and the sound will hit both sides at practically the same time; thus the delay in arrival time is too small to be useful. The ears are pointing in different directions, however, and receive a given sound at different intensities. This permits a means of determining at least a directional component in the sound, via the combination of input from right and left sides. In addition to spike frequency in particular neurons, the nervous system has access to combinations of inputs, in this case from A_1 left, A_2 left, A_1 right, and A_2 right. These findings, incidentally, are what prompted the search for the directional behavior, which is described in Chapter 1.

A complete moth still has not been described. Recordings were made from both ears, but the wings were removed to get at the neurons, and the moth was fixed in a holding device. Ideally we would like to record from the neurons of a free-flying moth being chased by a real bat, but that would be difficult and most likely be too uncontrolled for interpretable results. Some additional information came from the artificial application of pressure to the thoracic membranes near the ear, as might occur during movement by the wing muscles. This pressure caused a change in the B-cell spike frequency (Figure 17-2, E); thus it was suggested that such internal (proprioceptive) information might also be of use to the moth, giving it data on the position of its wings.

The recording setup was further improved by leaving the wings on the moth in different positions and presenting the sound from all around the moth. There were several technical problems, such as the prevention of echoes and the need for many standardized recordings in short periods of time. The whole system needed to be mechanized.

The solutions were ingenious. In brief, the researchers used a mechanical stage to hold the speaker in different positions and then varied the sound (automatically via negative feedback) to keep a constant nerve response. This permitted the sensitivity to be measured indirectly by the sound intensity required to trigger the response (e.g., if it needed a louder sound, the ear was less sensitive under the given conditions).

From these results the sensitivity was recorded as on a sphere surrounding the moth. It was discovered that the wings block or baffle the sound, depending on their positions (Table 17-1). In general the ears of the moth were most sensitive to sounds coming from the same side as the ear when the wings were up (thus giving side-to-side directional information) but only to sounds from below, regardless of which ear was tested, when the wings were down (thus giving information relative to dorsal-ventral direction of the sound source).

From this it can be seen that, from a total of only six sensory cells, two acoustic and one proprioceptive on each side of the body, the moth potentially can extract sufficient information to localize the source of the sound. Furthermore, with moving wings a "flicker" would result in the sound pattern from all locations except from directly behind the moth—the best position for a fleeing moth to keep the bat.

Table 17-1 Summary of sensitivities of the noctuid moth's ears to sounds coming from different directions, depending on the moth's wing positions.		Wings Up		Wings Down	
	Direction of Bat from Moth	Right Ear Sensitivity	Left Ear Sensitivity	Right Ear Sensitivity	Left Ear Sensitivity
	above to right	high	low	low	low
	below to right	high	low	high	high
	above to left	low	high	low	low
	below to left	low	high	high	high
	directly behind	high	high	high	high
	directly ahead	moderate	moderate	moderate	moderate

Summarized from hearing-sensitivity maps of noctuid moths in Roeder, K.D. 1967. Nerve cells and insect behavior. Harvard University Press, Cambridge, Mass.

Asher E. Treat

Kenneth Roeder

Modified from Huber, F., and Markl, editors. 1983. Neuroethology and behavioral physiology. Springer-Verlag, Berlin.

Asher E. Treat

Moth Hearing and Bat Sounds: The History of a Collaboration

On a summer night in the early 1950's a hiker on the Appalachian Trail where it skirts the Tyringham Cobble in western Massachusetts might have been puzzled by the sight of a grown man tossing wet wash cloths into the paths of fluttering moths as they approached a light in the yard of a country house. Enchanted by the intricate and beautiful structure of noctuid tympanic organs, I was trying to test the often suggested, but as I supposed, unproved idea that such organs were sensitive to high-frequency sounds. With help and encouragement from Donald Griffin and others I had been using crude methods and equipment to probe the reactions of moths both in stationary flight and in free flight under attack by the native local bats. The wash cloths, I thought, might show me whether a bat-sized but silent object would evoke the dodging, ducking, diving actions that were so often seen when a real bat was attacking.

A paper describing some of my experience had been accepted for publication when, early in 1954, I was present at a meeting of entomologists where a copy of Roeder's new book, *Insect Physiology*, was being circulated. When the book reached me I hastily searched it for some reference to lepidopteran tympanic organs. There was nothing in the text, but in the bibliography appeared the mysterious entry: "F. Schaller and C. Timm, 1950 . . . Sound reactions, moths." As soon as possible I looked up the reference and learned to my dismay and chagrin that my supposed "discovery" of ultrasonic sensitivity had been thoroughly and elegantly anticipated at the University of Mainz in 1948! I withdrew the accepted paper and a few months later submitted a revised version confirming and extending some of the German results by kymographic recordings of responses in stationary flight. As published, this paper included a brief paragraph on the only real discovery that I could claim: that of the parasitic mites that frequently deafen one but not both of a noctuid's ears. Roeder's book had saved me from the embarrassment of having made a false claim to the first proof of ultrasonic sensitivity in a moth.

My interest in the tympanic organs survived this near disaster and I continued the work during the summer of 1954. On a visit to my attic laboratory in Tyringham, Dr. John Pappenheimer suggested that an electrophysiological study of the tympanic organs would be rewarding. Unequipped with either knowledge or apparatus for such studies, and with Roeder's book in mind, I decided to write to Professor Roeder and ask his help. In July of 1954 I did so. The response was prompt and cordial: Come to Tufts next week.

In the smelly basement of Barnum Hall I found Ken besieged by students, colleagues, and visitors including the photographer Roman Vishniac who was filming a piece on insect flight for Life magazine. Despite the distractions Ken took time to listen to my story, but he asked his then post-doctoral student Edward Hodgson to guide me through the complexities of apparatus and procedure in our first attempt at recording from the tympanic nerve. The effort was a dismal failure. At my ill-advised suggestion we had placed the gel electrode directly on the sensillum, immobilizing it of course, and thus rendering it totally unresponsive. Ken looked in on our unhappy doings, and although then knowing little about moth ears he suggested that with an electrode more centrally placed we might have had better luck. I was skeptical and discouraged, as is clear in the portrait of the two of us that Vishniac took at this juncture.

Neither of us knew the innervation of the tympanic organ, and anyway my time had run out, my day in court was over and ahead of me was the long drive back to Tyringham with my wife and youngster to share my gloom.

The next winter I spent some hours dissecting preserved noctuids and learning to recognize and expose the nerve trunk that carries the acoustic fibers, and in the following summer, 1955, with some trepidation I wrote again to Ken asking for another chance. Again he was all encouragement, and on July 26, with Lila Schraeder (a former student) and several freshly caught moths, I set out once more for Medford. When we reached the laboratory Ken was ready for us and prepared to give his own full attention to the experiment. He had set up a loud speaker to transduce audibly any nerve impulses that we might evoke, but I was not hopeful and was expecting another failure. Our first moth was a male of *Apamea amputatrix* (Fitch), the handsome adult of the yellow-headed cutworm. We divided it sagitally, implanted the indifferent lead, lifted the nerve on a silver hook electrode and sounded the Galton whistle. There was an immediate sharp burst of spikes on the oscilloscope screen and a loud ripping noise from the speaker. The emotional effect on all of us was electrifying. We were astonished at the sensitivity of the preparation. It reacted to high-pitched sounds from any source, the jungling of keys or coins in a trousers pocket. At one train of spikes on the taped record Ken's voice announces, "That was Asher blowing his nose."

Both of us were excited and eager to explore the possibilities that all this suggested and we quickly made plans for another session a month later. On this occasion we used a more sophisticated source of ultrasound which, however, was less portable than the Galton whistle. In one experiment, still using the whistle and with Ken taping and narrating, the nerve continued to respond to the whistle blasts while I carried the little instrument farther and farther down the corridor and even into the lavatory at its opposites end. Before the day was over, we had clear evidence of the two acoustic fibers with their different thresholds, and of a third fiber, previously unknown and still problematic, which we later traced to a cell associated with the *Bügel* of Eggers (1925) and which we therefore named the B cell.

Ken's enthusiasm was infectious. In his boyhood he had been an eager collector of moths and he was delighted to return to these insects as subjects for his physiological and behavioral studies. His interest kindled a growing friendship and sparked a period of intermittent collaboration and correspondence that continued as long as he lived.

In the course of one of our early sessions at Tufts a bat was released in the laboratory and tympanic response to its echolocating cries were noted. Experience with bat-moth interactions in Tyringham led me to suggest to Ken that we try some recordings in the field. He was not immediately taken with the idea. "We've shown that the moth can hear the bat," he said. "What's to be gained by doing the same thing outdoors?" Eventually, however, he began to see possibilities in the notion, and in July, 1958, for the first of several such endeavours, he loaded all the necessary equipment into his car and drove out to Tyringham. In some of these early field sessions we were joined by Donald Griffin, Fred Webster, and others interested in the behavior of bats. In later years, of course, most of Ken's elegant field studies were done in his own yard in Concord, but on his many subsequent visits to Tyringham, always with his wife Sonja, he often spoke nostalgically about our early adventures in the Berkshires.

Figure 17-3 Output from moth interneurons during the process of stimulus filtering and identification of bat calls. Different neurons respond to similar input (from the sensory neurons) in different ways. The combination of these different outputs permits the system to abstract and extract information.

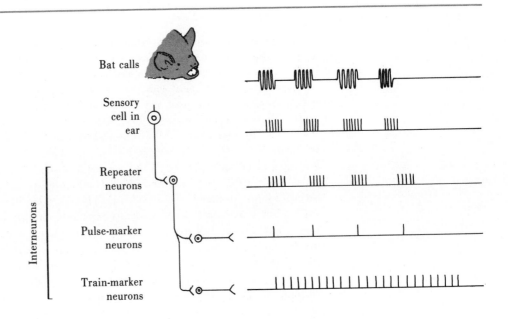

Further research (see Chapter 1 and Roeder 1967) confirmed that the information on sound intensities and location actually was useful to the moth. It was also determined that moths can detect bats at a distance of about 43 to 130 m, whereas the bat cannot detect the moth until it is within about 2.5 to 4 m.

This still is not the whole story. Bat chirps have not yet been distinguished by the moth's nervous system from all the other sounds in the environment, such as cricket chirps. Furthermore, once the input has been filtered and identified as bat calls, there remains the problem of translating this into the appropriate, directed, muscular-controlled flight response.

From studies that penetrated the moth's nervous system further, via evoked potentials, it was discovered that different interneurons filter the input in different ways (Figure 17-3). There are repeater neurons that basically relay the initial A-cell input, relatively unchanged, to the nervous system, whereas the other neurons respond differentially to different parameters of the input. Pulse-marker neurons respond only once per ultrasonic sound pulse and only if primed by intervening periods of silence. Train-marker neurons maintain a continuous output, at their own spike frequency, during a series of sound pulses. These and other neurons are believed to sequentially and selectively filter through the characteristics of the initial input, somewhat like using a combination lock, until the right combination is reached that identifies a bat call. The moth example has thus shown how the intensity and identity of sensory input is coded by spike frequency and how the system begins to sift and filter the input to extract relevant information from it.

The process of hearing on the bat's side also has been investigated (e.g., Fuzessery and Pollak 1984, Kobler et al 1987), but it is much more complex. We continue first with other simple systems and then consider the more complex vertebrate pathways, focusing mostly on vision rather than hearing. Although these might seem like pieces belonging to different puzzles, they nonetheless fit together in the end to give a good picture of how nervous systems operate.

PROCESSING SPEED AND MORE ON STIMULUS FILTERING: COCKROACHES

Our next example involves another simple picture of neural integration that illustrates different aspects of processing (and it reinforces some of the points derived from the moth ear). It involves the startle response of cockroaches, which permits the rapid detection and evasion of potential predators. The account presented here is a brief summary of material reviewed by Roeder (1967; also see Camhi 1984).

Cockroaches run at the slightest disturbance and are fast, as anyone who has ever tried to squash one can attest. Unlike the fairly specific bat response of moths, the evasive response of cockroaches is more generalized. The response is characterized by speed rather than specificity or information content; that is, cockroaches do not take time to identify the source or location of the stimulus.

The sense organs that trigger the response are *cerci*, hairlike projections on the tip of the cockroach's abdomen. The cerci are sensitive to slight air movements, including low-frequency sounds. When the cerci are stimulated, the cockroach jumps and runs, scurrying off to a dark hiding place.

Total response time from a puff of air to the start of a jump was measured in cockroaches by an ingenious setup as shown in Figure 17-4. The cockroach was affixed to a support with a drop of hot wax; then the support was connected to a phonograph pickup, which in turn was connected to an amplifier and oscilloscope. Another pickup was connected to a flag that was placed near the cockroach's cerci to detect the stimulus. The cockroach was given a little ball to hold in its feet. When a puff of air was given to the cerci and the cockroach jumped, kicking away the little ball, the pickup would record the rebound of the cockroach.

The research was not without problems. Cockroaches often would not accept the setup quietly or readily but instead walked, cleaned themselves, or otherwise

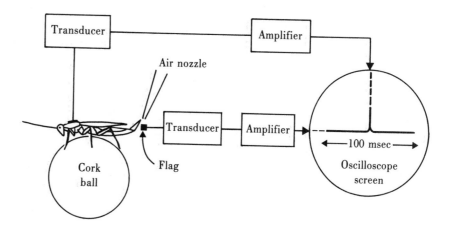

Figure 17-4   Diagram of apparatus to measure the time required for cockroaches to respond to puffs of air directed at their cerci. Air from the nozzle simultaneously strikes the cerci and flag on the electronic transducer to indicate time of the stimulus. As the cockroach attempts to jump from the ball it is holding, the rebound of its body stimulates the electronic pickup to which it is attached. The difference in stimulus and rebound times indicates the length of time needed for the cockroach's nervous system to respond. For a test of your own response times, see page 598.

From Roeder, K.D. 1967. Nerve cells and insect behavior. Harvard University Press, Cambridge, Mass.

Measuring Human Response Times

Some human behavioral response times can be measured easily in a variety of ways. If one has access to a digital timer accurate to hundredths of a second (such as a sporting event timer), an oscilloscope, or a physiological strip-chart recorder, one person can hit the switch or button that starts the timer then a second person can turn it off as quickly as possible.

A method that requires a minimum of equipment uses the physics of falling objects. It is one of the oldest techniques used to measure response times (Carrard, cited in Luce 1986, studied subjects catching falling canes in the late nineteenth century). It is also the basis of the old tavern bet that one cannot catch a dollar bill dropped between one's fingers before it has fallen out of reach of the closing fingers. (That trick works incidentally, i.e., the dollar cannot be caught, if the bill is held so the middle of it is between the open fingers. If the bill is held so the fingers are closer to the bottom edge, there is more of the bill to fall through the fingers; it takes longer; and most persons can catch it.) The time required to stop a falling object once one sees it start can be measured using any stiff piece of paper or other object, such as a thin plastic ruler. For this exercise to work so that the response time can be timed precisely, the fingers should be held at the bottom edge or level with a specified mark on the object to permit the distance the object falls to be measured.

The basic technique is for one person to hold the object at the top, with the open fingers of the subject level with the bottom or indicated mark. As soon as the subject sees the other person let go of the object, the subject stops it by closing his or her fingers. The distance from the starting point to the stopping point on the object (measured from the same part of the finger) can be converted to time from the equation or measurements shown below. This basic response can be timed several times to determine the minimum or average response time for the subject. The average minimum time required for a simple finger-closing response to a visual stimulus for an average young adult to middle-aged person after a few practice trials is usually around 130 msec.

$$\text{Distance} = (\tfrac{1}{2}) \text{ (acceleration due to gravity) (time}^2)$$

or

$$\text{Time} = \sqrt{\frac{(2)\text{ (distance [in centimeters]) (seconds}^2)}{980.6 \text{ cm}}}$$

Distance (Centimeters)	Time (Milliseconds)	Distance (Centimeters)	Time (Milliseconds)	Distance (Centimeters)	Time (Milliseconds)
1	45	11	150	21	207
2	64	12	156	22	212
3	78	13	163	23	217
4	90	14	169	24	221
5	101	15	175	25	226
6	111	16	181	26	230
7	119	17	186	27	235
8	128	18	192	28	239
9	135	19	197	29	243
10	143	20	202	30	247

vibrated the pickup, or even kept kicking the little balls away before the time of the desired stimulus. Good recordings were, however, finally obtained. The measurements of 23 successful cockroach reactions averaged 54 msec over a range of 28 to 90 msec. (Human response times are usually much slower, around 130 msec at the fastest, for example, to respond to and stop a falling object—see box, page 598.)

Next a number of subsequent spike recordings and experiments were performed at various locations inside the cockroach. The parts of the nervous system and the minimum measured times for the messages to traverse different individual segments are shown in Figure 17-5. These times, plus other characteristics of the spikes, revealed several important characteristics of the neurons and the whole system.

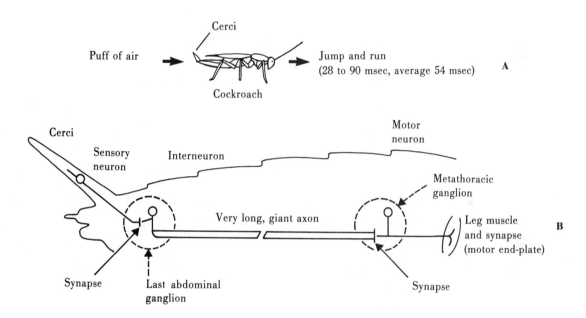

Times (in milliseconds) measured at different points:

Response time of cercal sensory neuron	0.5
Conduction time (short axon)	1.5
Synaptic delay	1.1
Conduction in long, giant axon	2.8
Synaptic delay	4.0
Conduction in motor axon	1.5
Neuromuscular synaptic delay	4.0
Development of muscular contraction	4.0
TOTAL (close to observed but lower)	19.4 msec.

C

Figure 17-5 Cockroach startle response times, as measured in Figure 17-4, and times for neural events measured with electrodes placed at different locations within the cockroach nervous system. **A,** Sequence of events. **B,** Diagrammatic section of cockroach body. **C,** Times (in milliseconds) measured at different points.

Based on information in Roeder, K.D. 1967. Nerve cells and insect behavior. Harvard University Press, Cambridge, Mass.

First, the discrimination or filtering of the input is accomplished by passage through neurons with different properties, as was seen in the moth. Furthermore, spatial and temporal summation are involved. The giant fibers do not fire unless they receive impulses from several of the cercal sensory axons more or less simultaneously, that is, via spatial summation; motor neurons do not fire until they receive two or more successive impulses from the giant fibers, that is, via temporal summation. The practical implications of this finding are that slight air currents do not trigger the escape response but those from more substantial input, such as a predator or an approaching folded newspaper, do.

Each of the neurons in line has a unique response to the preceding neurons. Motor neurons, for example, maintain a significant afterdischarge of spikes that continues long after the giant fibers cease firing. This afterdischarge is believed to help maintain the response, that is, to help keep the cockroach running after the senses serve to trigger the response. The sustained running carries the cockroach out of the range of danger. Although some discrimination of the input is occurring in this example and although more sensory neurons are involved (150 cercal neurons versus the 6 moth ear neurons), the discrimination does not appear to be nearly as complex as in the moth. Translated roughly to English, the message in the cockroach is "Danger!"; the message in the moth's system is more like "A distant cruising bat approaching from the upper left!"

The cercal and giant fiber axons could fire 200 to 300 impulses per second for extended periods, but the synapses fail after just a few seconds at this frequency. The synapses were much more disrupted by various anesthetics and drugs than by the action potentials. For the distance involved, the synaptic transmission was much slower than for axonal transmission. Thus of the two main functions of neurons, discrimination at the synapses is much more time-consuming and subject to disruption than is the transmission of the impulse.

Finally, although the speed is least variable in the axons, the speed varies, depending on characteristics of the neuron, particularly its diameter. The impulse travels the length of a giant fiber in 2.8 msec, which is estimated to be nearly 10 times faster than in the other normal-sized axons. This speed is interpreted as beneficial: it may make a sufficient difference (even if only 1% to 2%) in the overall response time to mean the difference between life and death. Over long periods of evolutionary time the faster cockroaches, with the larger axons, evidently survived and reproduced in greater numbers than the slower ones.

There is a cost. The large diameters take up space that could be used by a larger number of small axons. Recall that part of the information processing may occur through different combinations of signals. With more units the increase in number of possible messages is multiplicative, not additive. Fewer axons mean many fewer possible combinations. Nature rarely goes to one extreme or the other, however, but usually strikes a compromise, even in cockroaches. Thus the ventral nerve contains not only giant fibers but also a number of smaller axons (Figure 17-6), which also must contribute to the overall processing, discrimination, and probably much of the variability among responses.

Compared with the moth, the cockroach takes sensory input from a larger number (150) of similar sensory neurons and boils it down to a relatively simple, rapidly transmitted message. The result is a response time of around 54 msec. The moth, on the other hand, uses fewer (6) but dissimilar sensory neurons and apparently extracts more information by more internal processing, but it takes more time. The

Figure 17-6 Cross section of cockroach ventral nerve cord showing both giant axons and axons of smaller, normal size *(smaller specks)*. This view is as if one were to cut across a telephone cable carrying a large number of individual wires. The different wires would represent the axons of individual neurons.
Based on photographs in Roeder, K.D. 1967. Nerve cells and insect behavior. Harvard University Press, Cambridge, Mass.

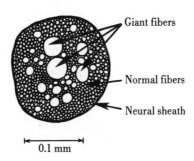

turning responses of moths to bat calls (see references in Roeder 1967) range from 75 to 252 msec, averaging approximately 140 msec.

Response time obviously is important in the evolution of behavior. Further evidence for the validity of these inferences would be comforting. One would predict, for example, that predators and prey should be closely matched in speed. In their long-term interactions each side would be constantly changing in the process of staying ahead of or even with the other side. This is sometimes viewed by analogy to an "arms race" (e.g., Chapter 9). Because of fast prey, faster predators survive and reproduce more than slower predators. This tends to push up the average speed of the predators. In turn, this increases the pressure on the prey, causing an upward shift in the speed of prey. Because increased speed has costs (such as a loss of ability to discriminate), the pace of the evolution would not be expected to be excessive, nor would either side be far ahead of the other. Large differences between the abilities of the participants should cost more than the advantages and thus not be predicted.

Unfortunately, the match between moths and bats is hard to evaluate. Moths detect bats well in advance of the bat's detecting the moth, and there are measurements of the moth's response time, but bats fly faster than moths, and the bat's response times and ability to connect with the moth are not known. All things considered, the two should be closely matched. For cockroaches, one has only their side of the picture. Information does exist, however, for another predator-prey interaction: that between mantids and flies (also reviewed in Roeder 1967). Mantid strikes from ambush take 50 to 70 msec to complete, and laboratory-raised mantids miss about 10% to 15% of their strikes. Measurements of fly response times show that it takes 45 to 65 msec for them to start escaping. The match in timing is remarkably close.

In addition to the filtering-discrimination and speed aspects, there are several other important principles in neural processing. We continue with a much different animal and different behavior to present the next principles.

DIVERGENCE, CONVERGENCE, AND PARALLEL PATHWAYS AS ILLUSTRATED BY CARDIOVASCULAR OUTPUT AND GILL WITHDRAWAL IN *APLYSIA*

In some invertebrates such as the marine snail *Aplysia* the small and consistent number of neurons has permitted the complete circuit with every neuron involved to be traced. Two examples are the control of cardiovascular output (Figure 17-7)

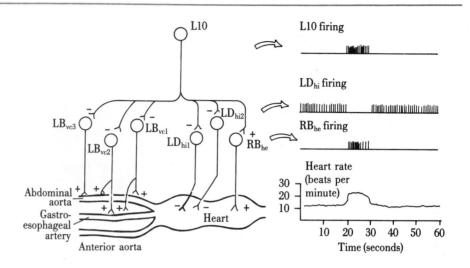

Figure 17-7 Control of cardiovascular output in *Aplysia*: effects from a single higher-order identified neuron, L10. Schematic diagram showing the relationship and connections of L10 to motor neurons and the heart region, and the effects of activity in L10 on two other neurons and heart rate. The labels (e.g., L10, LB$_{vc3}$) identify individual neurons.
Modified from Kandel, E.R. 1979. Scientific American 241(3):66-76.

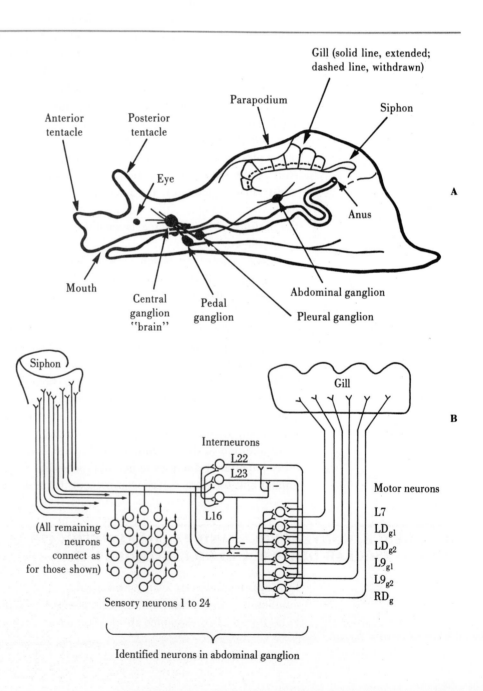

Figure 17-8 Neural control of the gill withdrawal reflex in *Aplysia*. **A,** A snail showing the position of the gill when extended in normal position and after withdrawal following stimulation of the siphon. **B,** Schematic diagram of neurons involved in the withdrawal reflex.
A modified from Kandel, E.R. 1979. Behavioral biology of *Aplysia*. W.H. Freeman, San Francisco. B modified from Kandel, E.R. 1979 b. Scientific American 241(3):66-76.

Figure 17-9 Principles of synaptic pathways. Passing of information among neurons is indicated.

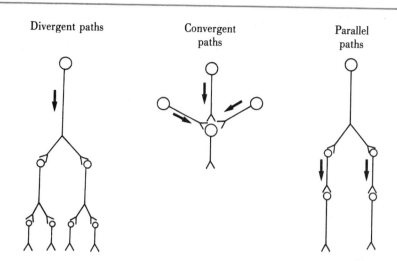

Divergent paths   Convergent paths   Parallel paths

and the gill withdrawal response (Figure 17-8) in *Aplysia*. Cardiovascular output is controlled by an interneuron, L10. The snails have their gills located on their posterior dorsal surface. If the siphon is touched, a reflex action causes the gill to be pulled in.

Both cases illustrate that even the simplest actions involve numerous neurons, divergence of information from one neuron to several others, convergence of inputs from several sources onto single neurons, parallel pathways, and combinations of inhibitory and excitatory synapses that are summed together. Simplified diagrams of the general pathways are shown in Figure 17-9. Not only are the specific neural connections known, but progress has been made in understanding, at least for these systems, how the pathways may be modified (i.e., involving learning) for the gill withdrawal reflex. The subject of behavioral modification in the individual, is covered in Chapters 19 and 21.

CIRCUITS IN OTHER INVERTEBRATE PATHWAYS

There are now many other examples of neural pathways that have been described for various invertebrates, including simpler organisms such as jellyfish, other pathways in *Aplysia*, and several arthropods. Some insects, such as honeybees (e.g., Menzel and Mercer 1987), have been investigated extensively. Perhaps the most completely known organism of all is a small nematode worm, *Caenorhabditis elegams*, which has a total of 302 neurons of 118 specific types and 8000 synapses connecting them, all of which have been identified and traced (see Lewin 1983 for an overview). The genetics, development, natural history, and simple behavioral repertoire of the animal are also well known—in some cases with known genetic mutations traced all the way through the neural circuitry to the observed behavior. These numerous other examples for the most part simply confirm the principles we have already shown in the examples above. Further references are listed at the end of this chapter.

For the next principles of neural processing we wish to discuss, we turn to vertebrate systems.

VERTEBRATE PATHWAYS: REFLEXES AND BEYOND

Vertebrate nervous systems are far more complex and involve many more neurons than the invertebrate examples. Furthermore, all of the neurons and all of the connections involved in the circuits have not been traced as completely, so that much of our knowledge is inferential. Nonetheless these systems have received much study, are surprisingly well understood, and provide many insights into neural processing. Vertebrate systems provide the full spectrum from simplest to most complex pathways. We start with the simplest of all, the simple reflex, then use that as a basis to build a more complete picture.

The simple reflex path consists of two types of neurons: sensory neurons connected directly to motor neurons. It is somewhat analogous to a doorbell with a push-button switch for the stimulus and a buzzer for the response. There are only two types but usually large numbers of neurons of each type. Although simple, such reflexes are relatively uncommon. They are found primarily in the terrestrial vertebrates (plus a few invertebrates such as coelenterates). In the terrestrial vertebrates, they are mostly postural reflexes and probably evolved secondarily. Without the supporting medium of water, sudden slips of a relatively heavy animal from a tree, the edge of a cliff, or the like can be fatal. The individuals that survive these slips either are small, with low mass and slight momentum so that crashing is not serious in the first place, or can react quickly and catch themselves before falling. The fewer the synapses, the faster the information is transmitted; some muscle stretch receptors appear to have evolved direct synapses on motor neurons in the spinal cord (Figure 17-10). The receptors are stimulated during a sudden stretch, as when a slip occurs,

Figure 17-10 Simple, two-neuron stretch reflex. An unexpected stretch, such as from an accidental slip, stimulates a proprioceptive stretch receptor in the muscle, which synapses directly to a motor neuron in the spinal cord, which in turn causes a reflex contraction of the muscle. The sequence of events can be traced by following the numbers.

a limb moves, and the muscles are pulled unexpectedly. This is the basis, incidentally, for the knee-jerk reflex.

Even the simplest two-neuron system shows some of the basic principles of neural pathways and integration, particularly the filtering and modification of the signal. Not just any stimulus or any degree of stimulation will result in a muscular contraction. The stimulus has to involve a mechanical stretch of a certain minimum threshold. Greater stimulation may lead to greater contraction by enlisting greater numbers of muscle fibers.

In addition, the pathways of neurons are not isolated even in the simple reflex. The sensory neurons connect not only to motor neurons but also to interneurons through collateral branches. These interneurons may in turn contact the motor neurons or other neurons. Some of the axons cross to motor neurons on the other side of the spinal cord, and some travel up to the brain itself.

Most of the reflexes that are found in vertebrates involve not just two neurons (sensory and motor) in a given path but also contain one or more intervening interneurons. There may be much crossing from side to side in the spinal cord, and a lot of information passes to and from the brain. A cross section of the spinal cord (see Figure 17-10) shows both gray and white matter. The gray matter consists of motor (in the ventral horn) and interneuron somas plus processes (dendrites and axons) traveling relatively short distances within the local neighborhood. For most of the distant travel, such as to and from the brain, the processes move into the outer white regions. The white appearance results from the white fatty material in the myelin. Thus this white matter represents myelinated fibers carrying impulses rapidly over relatively long distances.

To put this into an example, imagine that you are walking through the kitchen and happen to bump an elbow on a hot stove (Figure 17-11, A). The first response, a reflex, is to jerk your arm away. The jerk, which was mediated and integrated through neurons in the spinal cord, occurred a few milliseconds before you were consciously aware of it. The pathways from the temperature-sensing neurons to the motor neurons are shorter and faster than those going up to the brain. The signals also went to the brain, so you may have sensed the pain, become aware of the jerk, and perhaps responded with an outburst of profanity.

Now imagine two different scenarios. In the first you are carrying a pan that is full of near-boiling grease when you bump the stove. This time you do not jerk your arm away. Rather you move it slowly and tolerate the burn a little longer. An explanation would be that there is increased information coming down the spinal cord from the brain, caused by the awareness of the pan of hot grease and the consequences of spilling it, so that impulses from the higher, conscious levels in the cerebrum are inhibiting the action of the motor neurons that would otherwise lead to the jerk. The presence of these IPSPs cancels out (in the summation process and involving several interneurons) the incoming EPSPs from the temperature sensors in the elbow. The pathways back up to the brain and pain centers are not inhibited; you become aware of the burn; and the same swearing responses blurt out. Diagrams of these various routes are illustrated in Figure 17-11, B.

In the next scenario (Figure 17-11, C), which may be with or without the pan of hot grease, you are not alone but have a new friend from college standing by. This time, as the higher centers become aware of the burn and evaluate the responses, other centers inhibit the vocal swearing response. You may swear to yourself, but you also quickly consider that you do not wish to embarrass your

Figure 17-11
Diagrammatic cross section of the spinal cord, sectioned at the point where a spinal nerve is connected. Scenarios of reflex pathways illustrate basic paths of neural information to, from, and within the central nervous system. In this illustration, darkened neurons indicate activated pathways; open neurons indicate inhibited or nonactivated pathways. A, Accidental bumping of a hot stove. B, As in A but with a pan of hot grease that should not be spilled. C, As in B but in the presence of company.

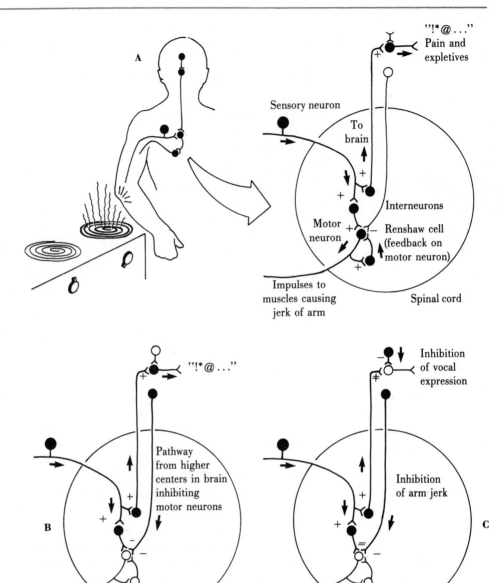

friend and you must keep calm, with much of that processing going on subconsciously. The centers that would lead to the stimulation of motor units in the vocal apparatus are not stimulated or the swearing may have already begun, but other centers cut it short or modify the words to something less objectionable.

The preceding example involves arm movements, which do not seriously interfere with balance. Now consider, however, that instead of bumping your arm on the stove, you step barefooted on a sharp tack or start to step over the edge of a drop-off that you did not realize was there. This leads to a reflex that causes the foot and leg to be jerked rapidly back. The remainder of the body's movement and

Figure 17-12 Collateral pathways and crossing of sensory information to other motor neurons in the mammalian spinal cord, diagrammatic.

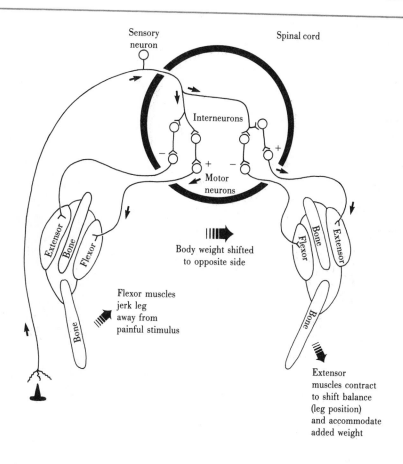

weight is already going forward, on the premise that the jerked leg was going into its next position. If nothing else intervenes and the remaining movement continues, balance will be lost with the changing center of gravity, and you will fall down. Falls do not always occur. Instead they may be caught by the simple reflex. The stretch receptors signal that stretch is occurring faster than it is supposed to, and the appropriate motor units are excited to fire, causing contraction of various muscles that quickly reverse the motion of the body and restore the center of gravity to a new balance. Also, collateral branches cross the spinal cord (Figure 17-12) and synapse with appropriate motor units on the opposite side of the spinal cord to cause the appropriate corrective action.

The response times of some human actions are relatively easy to measure compared with those of moths and cockroaches. A readily accessible, simple behavioral response by which one can easily demonstrate, measure, and explore the consequences of different and increasingly complex processing effects on response times involves the time it takes for a person to stop a falling object (see box, page 598). The basic pathway involves sensing the event, neural impulses traveling to the brain where they are processed, instructions sent down the spinal cord and out the arm to the muscles of the hand, and the muscles contracting. A visual stimulus such as seeing an object start to fall is familiar to most people and provides an easy pathway to start with. After determining the response times using a visual

Figure 17-13 Example of a pathway of information from the brain down the spinal cord to motor neurons. Other pathways are used for different movements. Similar but different pathways are involved in carrying sensory information.
From Eccles, J.C. 1977. The understanding of the brain, ed 2, McGraw-Hill, New York.

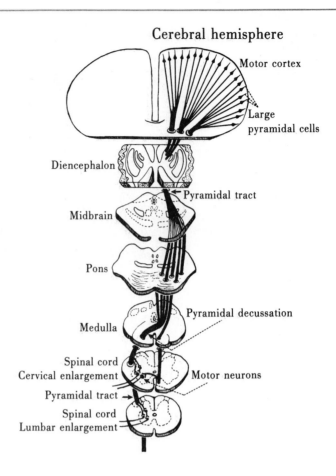

stimulus, one can try an auditory stimulus, with the subject's eyes closed and listening for a vocal signal or the sound of a timer switch. To illustrate the effects of interference and inhibition by other processing, or the additional time required for more complex processing, one can add conditional signals to indicate "yes" or "no" instructions for responding, or conduct the exercise while the subject is mentally performing mathematical calculations, reciting the alphabet or poetry, or discussing a topic. The response times now, including time to evaluate instructions, should be much longer and there may be more mistakes. If the falling-object technique is used, it may become difficult or impossible to catch the object at all.

Response times by themselves are of only limited usefulness for determining specifically what is going on inside the nervous system (Luce 1986). As used here they illustrate both the minimum response times of humans (as compared with cockroaches, for example) and the general effects of more complex processing or inhibition by competing pathways.

Much work has been done in recent years, and many of the specific pathways up and down the vertebrate spinal cord are now well mapped and understood (Figure 17-13).

Reflexes are believed to differ from more complex forms of behavior mostly in degree, that is, quantitatively. Reflexes are at one end of a continuum of coordinated movement. At the other end are longer sequences of more complex and sophisticated behaviors such as predator-prey interactions, long-distance migration, courtship, and other social interactions. These higher levels of behavior involve additional numbers and sets of interneurons in the immediate neural pathway, plus central command units that initiate some movements or parts of movements, feedback pathways based on additional sensory inputs that help coordinate and refine the movement, and processes that resolve conflicts between competing pathways.

Based on information from simpler cases, such as sensory input and reflexes, it seems likely that conflicts in general are handled by one of the major processes of neural integration: inhibition. Different stimuli are clamoring and competing to get through the nervous system but only one or a few actually get through. When a stimulus does make it through, in essence by shouting the loudest, it not only stimulates or excites certain pathways, but it inhibits other pathways via IPSPs. This results in a neural filtering process whereby everything does not happen at once. Because some pathways are inhibited, they may be stopped and not proceed at all or they may simply be held up and have to wait their turn, which adds to the total processing time.

Some of the centers of neurons that control more complex behavior are not located in the brain but reside mostly or completely within the spinal cord. If frogs receive a slight irritation on the back, for example, they will direct the hind leg to that point and scratch it. A frog completely deprived of the brain and all the higher faculties will still correctly orient the leg and scratch at an irritation on the back. Scratching reflexes in other vertebrates may be similarly coordinated within the spinal cord. Many of the basic locomotion patterns such as swimming, walking, running, and flying also are controlled within the spinal cord. An example of behavioral control outside the brain in invertebrates involves copulatory movements by male mantids. The senses and internal programming for these movements are located below the head of the male. His brain contains neurons that normally inhibit copulation. When the male's head is removed, copulatory movements are disinhibited. While the female proceeds to eat the male's head, the remainder of his decapitated body moves into the correct position and copulates with the female. When females do not eat the male, the inhibiting neurons in the male's head apparently are disinhibited by other means. In addition to these examples of frog scratching, locomotion movements, and mantid copulation, there are also several other behaviors that are controlled outside of the brain. However, most complex patterns of behavior in animals are controlled by the brain.

MAPPING THE PROCESSING PATHWAYS IN VERTEBRATE BRAINS

Various types of information have permitted us to map particular pathways and various functions to particular places in the brains of different animals. It has been known for over 100 years that certain parts of the human brain, for example, are associated with particular neural functions. Localized injuries and damage to the human brain were known to consistently affect a person's sensory or motor abilities, depending on the specific location. Electrodes placed at the surface of the brain in work done on other species, particularly cats and various primates, demonstrated similar topological properties in their brains. In particular, if mammals are exposed to a variety of stimuli, such as sound, flashes of light in front of the eyes, or touch

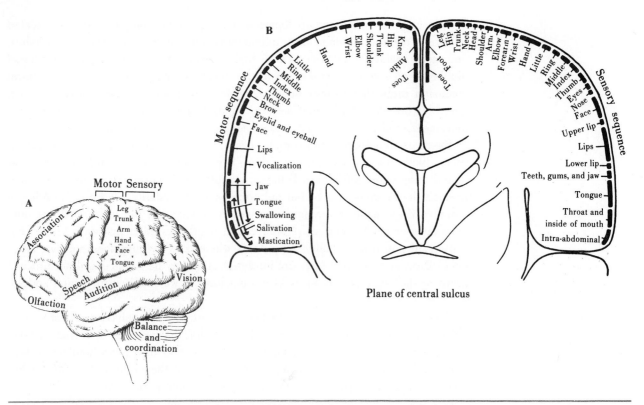

Figure 17-14   Graphic map of sensory and motor regions of the cerebral cortex in humans. A, External view. **B,** Cross-sectional view.

A from Hickman, C.P., Jr., L.S. Roberts, and F.M. Hickman. 1982. Biology of animals, ed 3, The C.V. Mosby Co., St. Louis. B from Rasmussen, T., and W. Penfield. 1947. Federation of American Societies for Experimental Biology, Federation Proceedings 6:452.

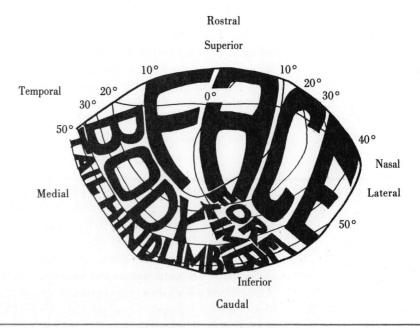

Figure 17-15   Somatic sensory map of the optic tectum region of the iguana (*Iguana iguana*) **brain.**

From Gaither, N.S., and B.E. Stein. 1979. Science 205:595.

at different points on the body surface, effects can be detected at recording electrodes in certain positions. If electrical impulses are given at different locations on the surface of the brain through stimulating electrodes, muscular contractions can be elicited. A stimulus at one location, for example, will cause a leg to jerk, whereas at another position, a paw may be caused to twitch. During recent years, in the course of brain surgery, medical doctors have had the opportunity or need to record from or stimulate various regions of the human brain.

Results from all of these sources have demonstrated fairly consistent locations of particular functions and have permitted the mapping of several important regions of the brain. Figure 17-14 shows the sensory and motor maps for the surface of the human cerebral cortex. Note that the amount of area occupied for different regions of the body is not proportional to their actual surface area; some, such as the facial and hand regions, receive proportionately much greater representation than other areas, such as the back and legs. Not only are there differences from top to bottom and anterior to posterior but also between the right and left sides, an indication of asymmetry between the two sides of the cerebral cortex. Similar mappings have been accomplished in a great many other species, such as iguanas (Figure 17-15) and barn owls (Knudsen and Konishi 1978).

The recently developed technique of positron-emission tomography (PET) permits the mapping of activity in different brain regions of normal, healthy living subjects. A special radioactive tracer is injected intravenously and is taken up by those brain areas that are most active metabolically. The emissions are detected by scanners and converted to metabolic maps by a computer. Examples are shown in Figure 17-16. The latest developments of the technique even allow some computerized three-dimensional representation of brain activity (Hibbard et al 1987).

These maps of vertebrate brains are important because they demonstrate that at least some functions of the brain are compartmentalized; that is, certain brain areas

Figure 17-16 Positron-emission maps of the human brain under different visual experiences. These pictures result from radiographic emissions from 2-^{14}C deoxy-D-glucose being metabolized by active regions of the brain and recorded photographically from outside and above the head. Drawings on left represent cross sections of brain illustrating the primary (PVC) and associative (AVC) visual cortices. In the remaining columns, lighter areas represent regions of highest metabolic activity. From Phelps, M.E., et al. 1981. Science 211:1446.

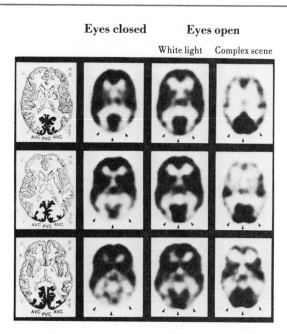

Eyes closed   Eyes open

White light   Complex scene

are particularly involved with controlling some functions and not others. They also give one some points of contact with neural processing. Thus one now has information at the level of the senses, contractions of the muscles, and regions of sensory and motor processing at the surface of the cerebral cortex. However, we must go deeper and consider additional principles of neural integration.

ABSTRACTION AND PATTERN RECOGNITION: VERTEBRATE VISION

Although one can discuss simple tracing and mapping of pathways through the nervous system, remember that the pathways are not just open roads or relays; rather different interneurons are filtering and modifying the messages. This leads to a principle, illustrated in a simple way first in the moth ear, known as *abstraction*. The verb *abstract* literally means to draw from. Different neurons in the pathway abstract different bits of information from the incoming impulses, and together they may derive a generalized picture. The noctuid moth with only six sensory neurons in its ears, counting the stretch receptors, is able to identify a bat call from among the bewildering hodgepodge of other environmental sounds. This is because, in addition to the sensory neurons, there are different interneurons. Some relay the sensory information with little change; some fire only when the pulses fall within the range given by bats; and some mark the length of trains of such pulses. Taken together they mean *bat*. This unlocks the pathway for a particular behavioral response much like the proper combination of numbers in a combination padlock. One can get an even better picture of this sequence by looking at the processing of visual information in vertebrates.

To understand the basic problem, consider what happens when an object, such as the letter A, is viewed. The image projected onto the retina is not simply passed onto the brain intact and unaltered, like a television picture to its screen. For a system such as that to work and permit abstraction on a simple basis, the image would have to be more constrained. If either the eye or the object is moved, for example, closer or farther away so that its apparent size changes or if the object is rotated or shifted up and down or sideways, then the image will fall on different sensory cells. Somehow the system must abstract the qualities that make the object look like the letter A over a range (within limits) of sizes, rotation, and other variations in appearance. Also consider other letters such as N, which turns into Z when rotated, or M, which turns into W. For other species and situations, abstraction might take the form of recognizing, for example, *bug*, or distinguishing edible bugs from inedible or harmful ones, or visually identifying mates and predators. All might occur at different distances, at different positions in the visual field, against differing backgrounds, and so on. Abstraction thus involves much more than simple passing of the picture from one point to another.

The visual pathway in vertebrates begins at the retina of the eye (Figure 17-17, A). The retina is actually an extension of the brain, being derived embryologically from the neural tube. Accordingly, it is more complex than most other sensory structures. At the same time it is removed from the rest of the brain and is thus simpler and easier to get at; it has provided relatively easy access for understanding the vertebrate brain.

The retina, which is about as thin and fragile as a spider web, is composed of five general types of neurons, arranged in layers (Figure 17-17, B). The cells are

Figure 17-17 Initial retinal processing of visual information in the vertebrate retina. **A,** Diagram of the eye showing pathway of light and position of retina. **B,** Close-up of retinal structure showing variety of cells and the pathway of neural impulses away from the sense cells and toward the brain via the optic nerve.

Modified from Polyak, S. 1941. The retina. University of Chicago Press, Chicago.

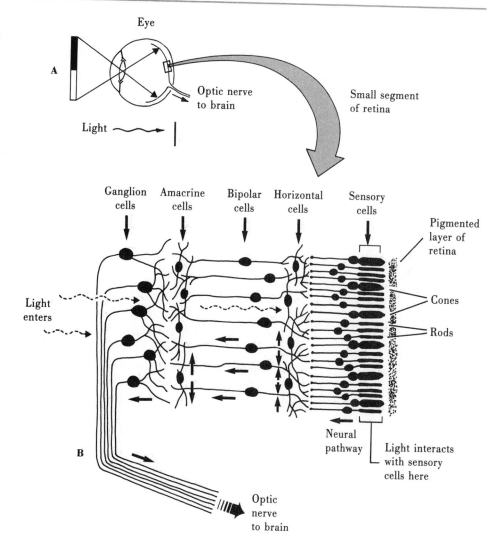

(1) the sensory cells (rods and cones of various types), (2) the horizontal cells, which carry information horizontally or across the retina between neighboring sensory cells, (3) the bipolar cells, which further integrate information and carry it to the next layer, (4) amacrine cells, which also integrate horizontally and may help detect directional movement, that is, fire when sensory information crosses the visual field in one direction but not the opposite, and (5) the ganglion cells, which synthesize qualities or abstractions from the previous cells and then forward them to the brain via the optic nerve; that is, the optic nerve is composed of axons from the ganglion cells. There are no sense cells at the point of the retina where the optic nerve leaves the eye, which creates a blind spot (Figure 17-18).

Rods and cones, as well as a few other receptor cells such as the hair cells involved in vertebrate hearing, do not operate quite as simply as the general pattern of most neurons. They are short and release transmitter chemicals without the need for action potentials and long axons. Furthermore, when stimulated, they do not

Figure 17-18 Demonstration of the blind spot and visual processing in the sensory field of the human retina. Close the right eye and focus the gaze of the left eye on the hunter riding the horse. Slowly move the book page back and forth at different distances from the eye. At one point the fox will become invisible because the image is falling on the point in the retina where the optic nerve leaves the eye and there are no sensory cells. If, however, one switches eyes, closes the left eye, and focuses the right eye on the fox, the horse and rider cannot be made to disappear completely. This is because the horse and rider occupy a larger area, extending beyond the blind spot, and the visual processing system fills in some of the missing information from retinal areas surrounding the blind spot. This visual processing is described in following figures and text.
From Lorenz, K. 1952. King Solomon's ring. Thomas Y. Crowell, New York.

respond by becoming depolarized and releasing more transmitter but by the opposite—becoming hyperpolarized and releasing less transmitter than when resting. Again one is reminded that nature is not obliged to keep things simple, and there is much variety in the nervous system. Beyond the sensory cells, the other cells in the visual sequence operate in the more familiar pattern with action potentials that code information by rates of firing.

The extent to which different species take advantage of the initial processing in the retina depends on the species and is related to their environment and life-style. Frogs show a high degree of early processing, with several general categories of information being synthesized by the ganglion cells themselves before the input is sent down the optic nerve (Figure 17-19). The overlapping area of sensory cells covered by a particular ganglion cell is called the *excitatory receptive field* (ERF). Particular ganglion cells and their optic nerve outputs are classed according to the type of information (i.e., visual quality) about what is being detected. In addition to those illustrated in Figure 17-19, there are brightening detectors for any increase in brightness, particularly in blue wavelengths.

The outputs of these classes of ganglion cell fibers go to different regions of the brain (Figure 17-20). The first four go to different regions of the midbrain and the optic tectum, and the fifth goes via the dorsolateral geniculate region of the diencephalon and from there to the cerebrum. All of this processing is further integrated, including perhaps some minor modification from previous, learned experience, and the appropriate motor responses are chosen. Most of this is believed to be accomplished in the midbrain of the frog. Then, depending on whether the abstraction comes up with *bug*, *solid object in the way*, or *approaching large object (possible predator)*, the frog will, respectively, aim and steer its tongue at the object, avoid the object while moving, or jump away and attempt to escape. Note that much of the visual abstraction occurs directly in the retina, and the synthesized information is relayed largely into the midbrain and onto the appropriate motor output. It is not simply relayed, however, but does receive some processing as emphasized by Ingle and Crews (1985) and Roth (1986). It all works quite well; frogs have been around for many millions of years; and they continue to populate Earth wherever suitable habitat remains.

Figure 17-19 Output of different ganglion cells in the frog retina in response to different visual events. Based on Maturana, H.R., et al. 1960. Journal of General Physiology 43:129-175.

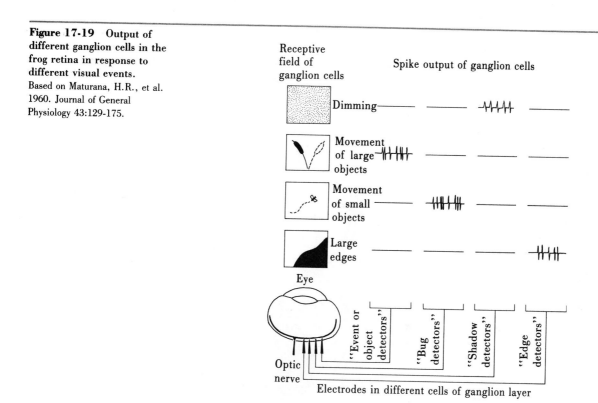

Figure 17-20 Basic visual pathways from the retina to the brain: frogs versus cats. Modified from Tinbergen, N. 1965. Animal behavior. Time-Life Books, New York.

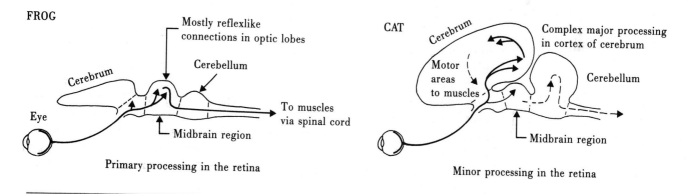

Other vertebrates, such as mammals, have similar visual systems, with some modifications. The proportions of retinal ganglionic responses vary. Mammals tend to have more dimming detectors (called *off sensors*) and brightening detectors *(on detectors)*, with fewer ganglion cells responding to more complex features. In rabbits and some squirrels, for example, only about 34% of the optic nerve fibers carry complex responses, such as to movement in the ERFs. In cats, only about 8% of optic nerve fibers are for complex classes and the remaining 92% for on and off detectors. (It was earlier thought that mammal eyes were entirely dominated by on or off detectors, but the more complex types have been documented and quantified.)

Figure 17-21 Visual pathway from the retina to the brain in primates (as seen from above the head). The right visual field from both eyes goes to the left side of the brain and vice versa for the right side. Note that input travels through the brain stem and lateral geniculate body, and some information goes to the superior colliculus, although the majority goes to visual areas of the cortex. Modified from Polyak, S. 1957. In: H. Kluver, editor. The vertebrate visual system. University of Chicago Press, Chicago.

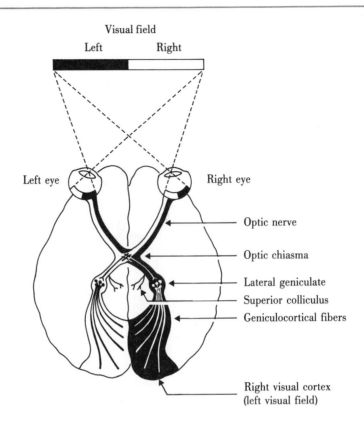

In other words, cats (and to variable extents other mammals, including primates) rely less on retinal processing of the information and shift the job more into the brain itself, where presumably there is much greater capacity for abstraction.

In addition, where the information is processed in the brain has shifted in the course of evolution leading to the mammals. Most of the visual information from the mammalian eye goes to the dorsolateral geniculate region of the diencephalon and from there onto the visual cortex of the cerebrum (see Figure 17-20), where extensive abstraction occurs. There is much less of the simple relay that takes place as in the frog midbrain.

So how and where does all this activity take place? Among carnivores and primates with binocular vision, information from half of each retina goes to each side of the brain. Among other mammals and other vertebrates that have been investigated, all the information from each eye simply crosses entirely to the opposite side of the brain. The information goes first from the optic nerve to the dorsolateral geniculate region, then to the visual cortex (Figure 17-21). The two basic categories of visual information, from the ganglion cells, are on-center, off-surround and off-center, on-surround. (Figure 17-22 shows patterns of action potentials depending on the patterns of light falling in the ERF.) Similar responses are shown both at the ganglion cell level and at the dorsolateral geniculate.

In the visual cortex the brain tissue is organized in columns and arranged in complex patterns, where there are simple cells, complex cells, and hypercomplex cells. All of these steps sequentially abstract additional qualities out of the information. This is explained with diagrams (Figure 17-23).

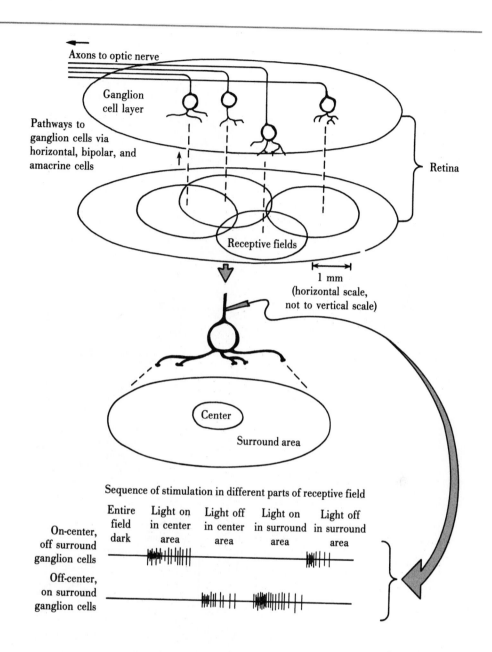

Figure 17-22   On-off centers of visual processing in the mammalian retina showing ganglion cell firing under different stimulation of the receptive fields. Based on Hubel, D.H. 1963. Scientific American 290:54-62.

It is interesting to note, incidentally, that the human cortex is only about 2 mm thick, although it is quite irregular and has a surface area of about 200,000 mm². Just as amazing as what takes place in the weblike retina is what has resulted from that little 2 mm sheet of tissue: society, architecture and great buildings, art, science, and much more.

From the visual cortex, where the information (such as for the letter *A*) has been extracted, impulses go to other regions of the cortex and brain, get combined with other information, and lead eventually to muscular output and observable behavior.

This description, although useful for illustrating several important principles, is still a bit oversimplified. The excitatory receptive fields (ERFs), also referred to as the *classical receptive fields* (CRFs), do not completely describe the extent or reach of the area covered by ganglion cells. The ERFs (CRFs) are also affected by events

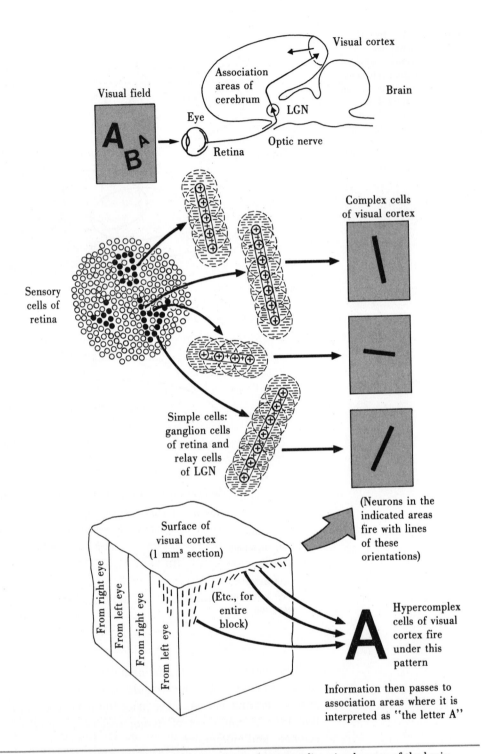

Figure 17-23 Higher-order visual processing in the mammalian visual cortex of the brain. Study these diagrams to understand the details. The broad outline of what is occurring is explained in the text. Images are inverted by the lens onto the retina. LGN is the lateral geniculate nucleus. Different parts of the illustration refer to other parts as indicated by the arrows.

Based on Hubel, D.H., and T.N. Wiesel. 1979. Scientific American 241(3):150-162, and Bullock, T.H., et al. 1977. Introduction to nervous systems. W.H. Freeman, San Francisco.

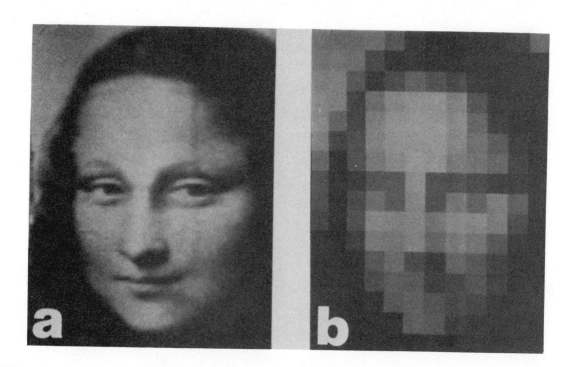

Figure 17-24 Pictures of the Mona Lisa. Visual processing of the right picture emphasizes the blocks and inhibits or interferes with interpretation. Viewing the picture from a distance or by squinting, however, will make the Mona Lisa more visible.
From Morrone, M.C. et al., 1983. Nature 305:226-228.

occurring beyond them, in what is called the *total receptive field* (TRF), as reviewed by Allman et al (1985). For another example of the true complexity of visual processing, the visual pathway is not just one-way from the retina to the brain; there are also reciprocal feedbacks from the higher levels back down to the lateral geniculate nucleus that modify and often enhance the incoming stimulation (Harth et al 1987). The details of these and other advanced aspects of visual processing are beyond the scope of this book. Interested persons can refer to the references plus Marrocco (1986), Dowling (1987), and Livingstone and Hubel (1988).

Before leaving the topic of vision, we use two quick exercises to further illustrate the delays and other effects created by inhibition and more complex processing. In the first, look at the two pictures in Figure 17-24. Under normal vision, the second picture does not appear to possess much recognizable detail and without having the first picture to compare it with, one might not recognize it at all. However, if you squint at the picture or view it from a greater distance, it becomes clearly recognizable. The cause of this phenomenon is not well understood. It may be that processing of edge information at early stages or higher levels of processing that identify blocks inhibit other processing of the total picture. Viewing from a greater distance or through squinting may stop the interfering inhibition. Alternately, the reduced amount of information from a distance or through squinting may cause increased feedback enhancement as discussed previously.

For the second example, see Figure 17-25, which is based on a similar, colored illustration in Tzeng and Wang (1983). Name the color of the bar or word as quickly as possible while someone else times the response (a watch that displays seconds is sufficient). Look at part A first and identify whether the bars are black, gray, or white. Then look at part B and repeat the exercise with the letters. Part B is much more difficult and takes longer because recognition of the words inhibits and interferes with recognition of black-gray-white when the two contradict each other. In human neural processing in general, language-related pathways (see discussion later in this chapter) often take precedence over and inhibit other pathways, including auditory pathways (Whalen and Liberman 1987).

Other senses have their own pathways. Olfaction, for example, comes in at the anterior end of the cerebrum through the olfactory nerves, and audition arrives from cranial nerve VIII. Figure 17-26 illustrates the auditory pathway.

Figure 17-25   Inhibition from contradictory neural processing where language-related pathways take precedence over and inhibit nonlanguage pathways. **A,** Verbally identify, as rapidly as possible, whether each of the bars is black, gray, or white. **B,** Do the same for the letters of the words.
Modified from Tzeng and Wang. 1983. American Scientist 71:238-243.

Figure 17-26 Auditory pathway in the human brain.
Modified from Picton, T.W., et al. 1973. Electroencephalography and Clinical Neurophysiology. 36:179-190.

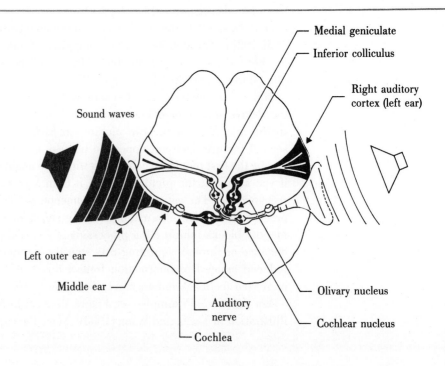

MOTOR CENTERS AND PATHWAYS

Just as the sensory side of the pathways does not involve simple passing of information but includes modification and abstraction at various steps, the motor side also involves sequences, but somewhat in reverse. At the motor side, as far as can be inferred, the information often starts or is primarily implemented (after a stimulus from elsewhere) as something of an abstract general command, often in command neurons. This is analogous to executive decision making. Then it is broken down in an administrative fashion (i.e., by delegation of specific tasks) until it reaches the lowest levels of spinal motor neurons and finally the muscles themselves. One can use stimulating electrodes inserted into the nervous tissue and, depending on level or pathway, cause simple muscle twitches or release entire patterns of behavior.

Holst and St. Paul (1960, 1963), for example, were able to insert stimulating electrodes in chicken brains and elicit whole, functional movements. When the electrodes were closer to particular centers, they could be stimulated with less current, and the response occurred sooner, that is, with a shorter latency period, than when the electrodes were farther from a particular area. With the electrode in a particular region, the outcome depended on the strength of the current.

At lower centers for simple behaviors, such as turning the head, standing up, or walking, two centers could be stimulated simultaneously, and the movements would add together or cancel out, depending on their mechanical compatibility. A chicken, for example, could be stimulated to stand up and turn its head to the right at the same time. If the centers to turn the head to the right and turn the head to the left were stimulated simultaneously, the chicken would do nothing.

Higher up in the neural center hierarchy, centers for complex behaviors were found. These include sleep, which if stimulated led to a sequence of the chicken looking around, yawning, sitting down, fluffing the plumage, retracting the neck, and closing the eyes. The complex sequence depended in part on what the chicken was doing when the current was turned on, the remainder of the sensory environment, and its internal physiological condition. If the chicken was already sitting down, for example, the sequence was shorter. If it was eating, the chicken would stop eating and perhaps walk before starting to sleep. Unlike the simple behaviors, such as turning the head, the complex centers did not add together but rather inhibited each other so that only one would be active at a time. If one center was stimulated before another, the first was expressed and the second was inhibited. If two were activated at the same time, one would suppress the other, which either would not be expressed at all or would wait until after the first was completed. Sometimes they would alternate. A chicken stimulated to simultaneously reconnoiter and feed would alternately make pecking movements and raise its head and look around. In a few instances, opposing behaviors such as aggressive pecking and fleeing could be stimulated equally so that a different behavior was expressed. In the case of a hen stimulated to be aggressive and flee at the same time, she would run back and forth with her wings up and scream.

From this emerges a picture whereby there is a hierarchy of administrative levels on the motor side. Higher command centers do not direct the muscle-by-muscle movement but somehow direct lower centers, which direct still lower centers until the level of specific muscle movement is reached. The lower centers are used in common by several of the higher centers. The running motions of a chicken, for example, can be incorporated into higher functional categories of running toward food, running toward a mate, or running away from an enemy. This interpretation

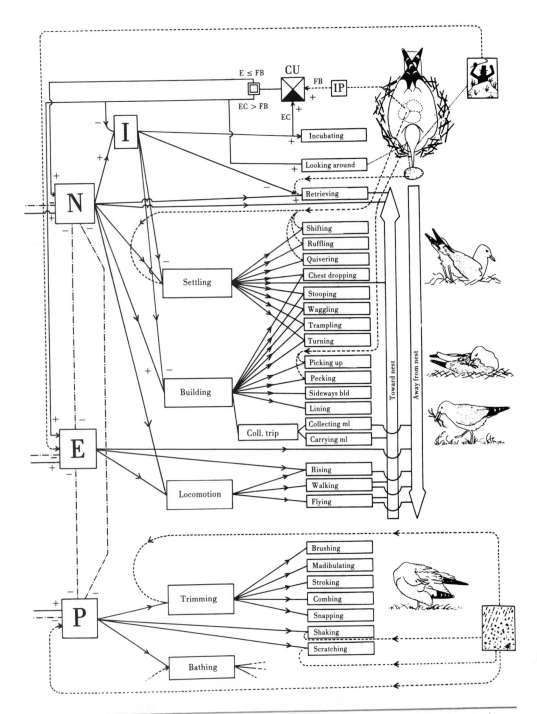

Figure 17-27 Hypothesized stimulatory and inhibitory relationships for interruptions in the incubation behavior of the herring gull. *N*, Incubation system; *E*, escape system; *I*, inhibition input unit. For other symbols and detailed explanation, see Baerends (1976). Compare this contemporary interpretation of functional organization of behavior with the classic Tinbergen releaser model (Figure 2-15).
From Baerends, G.P. 1976. *Animal Behaviour* 24:725-738.

is partly compatible with Tinbergen's synthesis of hierarchies of centers (Chapter 2). It differs, however, in that it is not just a matter of releasing inhibited drives; the whole picture is more complex and may include stimulation or excitation as well. Research into command processes and central pattern generation is continuing (e.g., Schoner and Kelso 1988).

Contemporary, updated models of the functional organization of behavior (Figure 17-27) are approaching physiological mechanisms more and more closely. Inhibition is not just a matter of something needing to be released in the classic view of IRMs but rather is a process that different pathways, particularly at the higher levels, perform on each other. The neural message that gets through is the one that in a sense manages to shout the loudest and inhibit the others.

This still is not the whole story. Some of the aspects that are not understood yet are: (1) how learning fits into this scheme (Chapters 19 to 21), (2) how voluntary processes arise, and (3) how everything is integrated at the highest levels. The voluntary control of behavior in animals is one of the least understood aspects of neural processing and is one of the main targets of present research and interest. The intrinsic, spontaneous firing of some neurons, such as the command cells, may stimulate some voluntary actions. Some spontaneous activity may not be able to get through the inhibition of other traffic except under certain circumstances or during lulls. Some activities, that appear voluntary may simply be a result of delayed responses or be stimulated by subtle or internal (e.g., hormonal) effects (Chapter 18), or unrecognized stimuli and hence not really be voluntary. Voluntary behavior is of interest not only because it is so poorly understood but also because of its relevance to initiative and creativity, free will, and consciousness in humans and perhaps other species.

One aspect that has become fairly clear is that there is no single highest center, the so-called hidden observer or man at the top running the control panel. Rather there are several highest centers. They may have connections and communicate, in the sense of inhibiting each other, but all parts otherwise are thought to operate independently of each other. The apparent overall control of behavior may be either just an illusion or simply a property of the whole.

PROCESSING WITHIN PATHWAYS AS ILLUSTRATED BY HUMAN LANGUAGE

As processing proceeds within any given pathway, from one hierarchical level to the next, the path involves different regions or centers performing different functions. This was suggested, for example, by the results of artificially stimulating different behaviors in chickens. Another example and one of the best illustrations of this principle involves human language. Language pathways have been traced completely from input to output, and it has been shown that different regions of the brain are responsible for different properties in the final output of language.

Much of the understanding of language pathways has resulted from injuries and disease that affected language in different ways. Through the years there have been numerous accidents, diseases, birth defects, and other disorders that have led to various damage to the brains of humans. These problems, observations on their effects, and attempts to correct or otherwise deal with them have led to a wealth of information and insight into the probable workings of the brain. Four examples of such disorders are Broca's aphasia, Wernicke's aphasia, prosopagnosia, and

Gilles de la Tourette's syndrome. Epilepsy and attempts to minimize its effects by cutting the corpus callosum, a band of tissue that connects the right and left hemispheres of the cerebrum, also have aided understanding of the human brain by providing the so-called split-brain experiments. Following the results of the split-brain work, researchers have found means of simulating the experiments in normal subjects by splitting and screening the sensory input rather than by surgically dividing parts of the brain. Minor everyday problems, such as slips of the tongue during normal speech, have also provided some insights into language processing (Motley 1986).

Broca's aphasia, named after Paul Broca, who studied the problem beginning in the 1860s, is a speech disorder that results from damage to part of the cerebral cortex, shown in Figure 17-28. The speech of a person with such damage is broken and difficult to follow. Words are not spoken smoothly, and there is a tendency for nouns and verbs to predominate and modifiers are lost (e.g., *dog runs* versus *the black dog runs fast*.) The patient comprehends reasonably normally, but speech output is impaired.

Problems associated with Wernicke's area (see Figure 17-28), named after Karl Wernicke, who studied patients in the 1870s, are different from those in persons with Broca's aphasia. In Wernicke's aphasia the speech is smooth and sounds normal superficially—but it generally comes out as nonsense. Even if one pays close attention, it is difficult or impossible to understand what a person with Wernicke's aphasia is trying to say. Wernicke's area is thus believed to be the region where language is interpreted, that is, takes on meaning.

A tract of nerve fibers, called the *arcuate fasciculus*, connects Broca's and

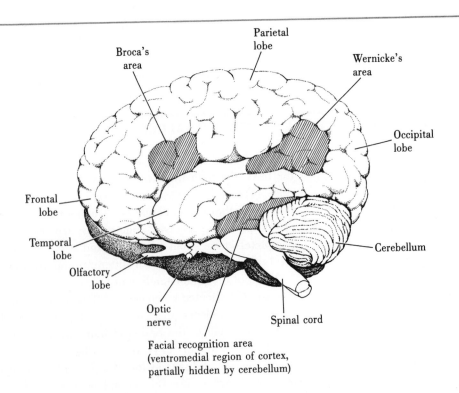

Figure 17-28 Language and facial recognition areas of the human brain. The sizes of these areas differ between the right and left sides of the brain, occupying more space on the left side in many cultures.

Wernicke's areas of the cortex. Thus it is inferred that the language pathway is connected in the middle at this point, but what about the input and output? Visual and auditory input were discussed previously. Language information can enter visually when one reads something written or through the auditory system when one hears something spoken. In the case of blind persons, other pathways, such as using touch for Braille, may be developed. The components are abstracted, as discussed previously, then forwarded to association areas and onto Wernicke's area for comprehension. In the case of vision the abstracted information goes from the primary visual area to the angular gyrus for association and then to Wernicke's area (Figure 17-29). From Wernicke's area, if something is to go to output, perhaps

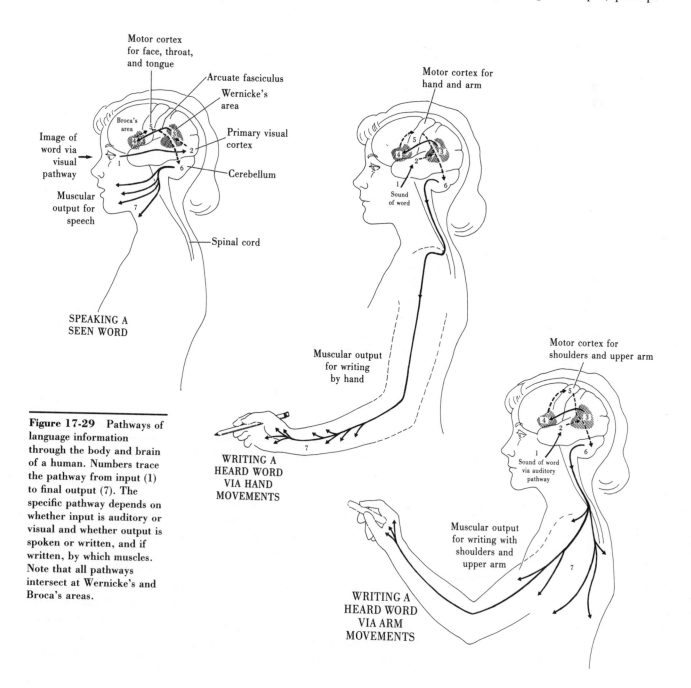

Figure 17-29 Pathways of language information through the body and brain of a human. Numbers trace the pathway from input (1) to final output (7). The specific pathway depends on whether input is auditory or visual and whether output is spoken or written, and if written, by which muscles. Note that all pathways intersect at Wernicke's and Broca's areas.

after further processing and other input from other parts of the cortex, it is given meaning and then sent to Broca's area for administrative cleaning and polishing. From there one can infer that it goes to the motor cortex.

Which part of the motor cortex the information is sent to depends on the specific form of output. If it is to be spoken, the chain of information goes to motor areas that direct word formation by the muscles in the larynx, throat, face, and jaw. If it is to be written by hand, the information goes to centers controlling the forearm and hand. If the information is to be written larger, such as on a blackboard, then the output goes to centers controlling muscles of the upper arm, shoulder, chest, and back, as illustrated earlier (Chapter 16, Figure 16-24).

Therefore one can see that Wernicke's and Broca's areas are at a common intersection between different paths of input and output. Language information can come in by one or more channels and go out by several channels.

However, language is not just cold, logical chains of information. It may take on subjective and emotional flavors, which are contributed from other regions of the brain. Injuries or problems in another part of the cortex lead to a syndrome known as Gilles de la Tourette's syndrome. With this condition the speaker is a compulsive swearer and continuously speaks in a manner that is socially unacceptable. Such individuals are unable to inhibit swearing except with utmost effort. Thus it appears that there are regions in the cortex associated with language of an emotional content and with normal (to various degrees) social inhibition.

Damage to another region of the cortex (see Figure 17-28) leads to a condition known as *prosopagnosia*. In this case the unfortunate persons are unable to identify faces. If they are shown pictures of familiar faces or see the persons, they are unable to say who it is. The problem from this region of the cortex, however, is completely visual. If the person hears a familiar person speak, he or she can readily identify the voice. Thus there appears to be a region of the cortex devoted specifically to facial recognition and another area for speech recognition. It is likely that other species have similar specialized regions of the brain appropriate for individual recognition. Parents and offspring of penguins, for example, are able to identify each other vocally even in the midst of nesting colonies that may have millions of individuals. A facial recognition region of the brain has recently been demonstrated in sheep (Kendrick and Baldwin 1987). The basic process involves abstraction and identifying or recognizing a specific example from a class of items.

RIGHT VERSUS LEFT SIDES OF THE CEREBRUM: BRAIN LATERALIZATION

Many aspects of neural processing in some species of animals, particularly in mammals, involve differences between the two sides of the cerebrum. Broca was the first to detect these differences in humans. Injury to the left side of the brain produced much more serious aphasia (inability to speak) than injury to the same parts of the right side. Understanding of this phenomenon increased greatly with research in other primates and humans from split-brain work conducted by Sperry and his associates (Sperry 1970, 1974, Nebes and Sperry 1971).

There is a prominent band of neural tissue, containing about 200 million nerve fibers, that connects the right and left sides of the cerebrum, called the *corpus callosum*. Although it may be carrying as many as 4 billion or more impulses per second, no one was really sure of its function. Some persons thought it served just to hold the two sides of the brain together. A few unfortunate persons with severe

Figure 17-30 Examples of results from split-brain testing. **A,** Normal person reads and reports *Hat band*. **B,** Normal person reads and reports *Nut* and also reaches behind a screen and correctly selects a nut by touch from among several objects not visible to the subject. **C,** In a split-brain patient, only information going directly to the left hemisphere can be verbally reported because the cut corpus callosum does not permit passage of information from the right hemisphere. **D,** The patient is not able to report seeing any words, but the left hand is still directed by the right hemisphere, through neural pathways that cross to the other side below the corpus callosum, to select the correct object. Modified from Sperry, R.W. 1970. Research Publications, Association for Research in Nervous and Mental Disease, vol. 48, and Sidtis, J.J., et al. 1981. Science 212:344-346.

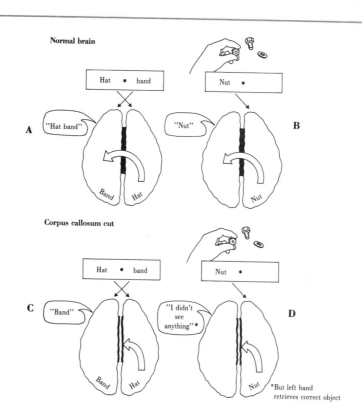

and constant epilepsy, who could not be cured with medicine, were able to recover substantially by having the corpus callosum cut. The epilepsy involved neural electrical storms that traveled back and forth between the two sides and apparently built up intensity in the process. Superficially the patients appeared normal after the operation of cutting the corpus callosum. Further testing, however, revealed surprising effects (Sperry 1970, Sidtis et al 1981).

Patients were asked to look at a dot in the center of a viewing screen while words were briefly flashed to either the left or right of the dot (Figure 17-30). Because of the visual pathway described earlier, words on the left go to the right hemisphere and vice versa. Information in the right hemisphere is then normally transferred via the corpus callosum to the left side for synthesis with information already there. In the case of split-brain patients, however, that transfer was blocked. The results are shown in Figure 17-30. Although the patients could not report information in the right hemisphere, it was clearly there and could be acted on.

After the identification of problems with patients who had the corpus callosum cut, techniques were developed for simulating the results in normal persons. The procedures involve presenting photographic images so quickly that there is not enough time for the information to travel across the corpus callosum. From this research it was discovered that the two sides of the cerebrum function somewhat as two separate computers with different functions. The left side (at least in most persons raised in Western culture; see Sibatani 1980) is involved with series or sequential and analytical processing, which includes language and mathematical types of analysis. The right side, on the other hand, is more involved with visuospatial processing, which enables one to look at things more holistically and synthetically. This side of the brain involves more of the abstract, artistic, and subjective aspects of thinking in humans. The corpus callosum is the normal communication channel that exchanges information back and forth and, in the complete process, permits the integration of all parts.

Many of the properties of human awareness and consciousness are associated with the left hemisphere and may be tied to language. Deese (1978) proposed that consciousness developed as a process associated with language to monitor, detect, and correct routine errors in speaking. Persons with split brains or those allowed input only to one side seem unaware of input they cannot process in language, although the brain on the other side can still understand and perform. This has been shown by the ability to pick up an object that could not be identified verbally. The hand performed properly, but the person could not state what had been done. Connections to lower, emotional regions of the brain also still function normally. A woman shown a nude photograph to the right but not to the left hemisphere became embarrassed but did not know why.

The topic of language processing and consciousness in humans calls to mind earlier discussions in this book about the differences between language in humans and communication in other animals (Chapter 15) and the problems of self-awareness and anthropomorphism (Chapter 1). Do other animals constantly live, even when awake, in a mental world that would seem like ours when we are asleep or anesthetized? It is an intriguing question that is still beyond our grasp. For further discussion of this fascinating subject, see Eccles (1970, 1977) and Crick (1979).

Interest, research, and publications on brain lateralization in humans and many other species have continued (e.g., Geschwind and Galaburda 1984, Witelson 1985). It is likely that some of the differences in organization of the brain are developmental and depend on which inputs are encountered first during compartmentalization processes. The subject has also stimulated much popular interest (in some cases carrying the implications beyond what is warranted by the scientific results; McKean 1985).

CONTRIBUTING COMPONENTS AND OTHER PATHWAYS

There are three other major topics that are relevant to the preceding discussion of neural processing and pathways in vertebrate brains: (1) the role of the cerebellum, (2) the reticular activating system, and (3) the limbic system.

Based initially on clinical information from injuries and other problems in humans, then as confirmed and further researched in a number of other species, it has been shown that the cerebellum is the major point of administrative handling of muscle control; that is, commands come down from higher centers in the cerebrum (of mammals) or elsewhere (e.g., midbrain in amphibians), and the cerebellum works out the details. The cerebellum is second only to the cerebrum of humans in size, number of neurons, and complexity of neurons. The Purkinje and basket cells, neurons with an incredible amount of dendritic branching (Figure 17-31), are found in the cerebellum. There are numerous known feedbacks between the cerebrum and cerebellum so that administrative functions of the cerebellum seem not only quite reasonable but highly likely. For a good introduction to processing of motor control by the cerebellum, see Eccles (1977).

Figure 17-31   Purkinje cell.
Modified from Bullock, T.H., et al., 1977. Introduction to nervous systems. W.H. Freeman, San Francisco.

There may, however, be significant differences among the different vertebrates. There is a rough correlation between proportional size of cerebellum and amount of activity and agility. Birds and some tree-climbing mammals, for example, have large cerebellums. Within particular groups, such as some of the amphibians, some of the more active and agile species have smaller cerebellums than those of their sluggish relatives. There is much yet to be learned about the cerebellum and most other parts of the vertebrate brain.

The reticular activating system (RAS) is a loose net of neural connections largely in the brain stem but extending into the cortex. It appears to function importantly in alertness and attention. It does not connect specifically to any of the other pathways but rather with essentially all of them. It is somehow involved with monitoring, sorting out, and mediating all the pathways and maintaining some order in all the traffic. The RAS may be the closest thing to the "hidden observer." When the RAS is activated, the animal is alert and awake. Sleep, on the other hand, involves a reduction of activity in the RAS. The RAS seems to serve as something of a central switchboard operator or relay system.

The limbic system is a collection of regions, or nuclei, in the brain that borders on and overlaps with (depending on authority) the brain stem and the outer layers of the cortex. The limbic system appears to be involved with emotion and motivation, including anger, rage, and affection. The brain is formed of general layers that are complex and intricately connected and show an array of homologies among different groups of vertebrates. The first region of the forebrain, the telencephalon, can be subdivided into three basic sections: the first or inner, concerned basically with the primitive sense of smell; the second, known as the striatum; and the third, outer or mantle layer, which is the pallium or cortex. These may be subdivided further, for example, into archistriatum, paleostriatum, and neostriatum and archipallium, paleopallium, and neopallium or neocortex. During the course of vertebrate evolution the brain has gradually acquired new layers and parts, building on top of the old, and there has been some transfer of function from one region to another. Because of the remodeling, the limbic system does not correspond entirely to any single part of the forebrain.

The limbic system is "lower" in the brain and more primitive than our language centers. The presence of these different functions in different parts of the brain emphasizes that humans have inherited a highly evolved organ with advanced abilities built on ancient characteristics and behaviors, with such faculties as human language being the most advanced of all. Even our most advanced neural processing still retains connections with and input from other, older parts of the system.

The highest, most complex levels of human cognition have long received interest and discussion. Most of the issues, go beyond the level of this book. Persons interested in pursuing the subject further on their own can take courses in psychology or cognitive science or refer to recent texts in those areas. A few recent general articles on the subject are Sternberg (1986), Kihlstrom (1987 a, also see 1987 b and associated comments from others), Posner et al (1988), and Kosslyn (1988).

BRAINS VERSUS COMPUTERS

Biological nervous tissue is much different than the materials and structure of human-made computers. In living systems, chemistry plays a much larger role in the discrimination and transmission processes, that is, nerve cell potentials depend on living membranes, a host of biochemical and physical chemical processes, and even the conduction of action potentials is an electrochemical process, rather than largely electrical or electronic as in computers. Thus only a limited analogy can be made between living nervous systems and the computers designed by humans.

Nonetheless, there are many general principles of information processing, such as branching or divergence of information, parallel pathways, merging or convergence, large numbers of units that permit numerous combinations to form different patterns of meanings, and a number of other features such as hard wiring versus soft wiring and memory storage (see Chapter 20). Because of these shared general properties, biological nervous systems and computers are increasingly being used as models of each other. Studies of advanced computer processing have yielded and likely will continue to yield some insight into biological neural processing.

Artificial intelligence has long been a topic of interest and research, but there have been significant difficulties, slow progress, and it has remained largely in the realm of science fiction. Continued attention and effort devoted to it, an accumulation of understanding, and greatly improved computer capacities, however, are leading to rapidly expanding progress and even greater interest and research.

The subject goes under the banner of several names: neural networks, parallel processing or parallelism, parallel distributed processing, connectionism, AI (artificial intelligence), and robotics. By starting with simple parallel networks, then gradually increasing the complexity of pathways and number of network components involved, and exploring the outcomes, numerous possibilities and principles are being discovered and a whole new set of terminology is emerging. Persons who wish to pursue this subject may start with Rumelhart and McClelland (1986) and Cauldill (1988), both of which are listed at the end of the chapter, then refer to (in Literature Cited section at the end of the book): Ullman (1986), Hopfield and Tank (1986), Tank and Hopfield (1987), Goldman-Rakic (1987), and Waldrop (1988 a,b).

NEURAL PROCESSING PRINCIPLES: A RECAP

This series of examples ranging from moth ears to human language pathways, although only a small sampling from the literature on neural processing, has nonetheless illustrated most of the major principles on which the nervous system is currently considered to operate. The principles, mostly from this chapter but also including some from Chapter 16, can be summarized as follows:

1. Input into the nervous system is initially filtered by the selective reception of sensory cells that are responsive to only certain, specific forms of information. This is peripheral stimulus filtering. How specific any particular cell is depends on a number of conditions. For example, some sensory cells such as the hair cells involved in mammalian hearing, because of their location and the mechanics of the inner ear, are tuned specifically to only particular frequencies of sound. Other cells, such as the A_1 and A_2 cells in the moth ear or the sense cells in the cerci of cockroaches, are stimulated by a broader range of vibrations. Even in the latter cases, however, the sensory cells still respond only to a restricted range of environmental input. The sensory cells convert the external inputs into changes of membrane potential. The general conversion process is called *transduction*.

2. The primary method of coding information in the nervous system is by the frequency of action potentials, also referred to as the *rate of firing* or *spike frequency*. This rate of firing changes with the magnitude of input, which may either excite and speed up or inhibit and slow down the firing, via the process of spatial and temporal summation. Summation is the basis of discrimination or decision making at the level of the individual neuron.

3. Processing of information, including central stimulus filtering, even that gathered by only a small number of sensory neurons (such as in moth ears), involves a large number of interneurons. This is not merely a description or statement of fact. It is what makes complex processing possible, through the following:

   a. Different neurons respond to similar input in different ways. Thus neurons do not merely pass input unchanged along to the next point. Because of this a given stream of input can be split up, with divergent pathways, and sent to different neurons with different response and output properties that extract and filter different characteristics from any given input. In the case of moth hearing, for example, there are repeater neurons, pulse-marker neurons, and train-marker neurons. In another example, vertebrate vision, there are complex cells that respond to diagonal patterns of different orientation.

   b. There are parallel pathways within the nervous system that permit different packets of processing, including large numbers of similar processing events, to occur simultaneously.

   c. The results of different parcels of processing can be brought back together, via convergent pathways, to combine a variety of similar information from different sources (such as different parts of the visual field) or of different kinds (such as diagonal lines of different orientation to form a letter, e.g., *A*, or different features of a sound that mean *bat* to a moth). The combination of divergence, parallel pathways, and convergence permits the process of abstraction, or the generalization and extraction of "meaning" in a set of input.

   d. Different meanings depend on different combinations or patterns of neural inputs. The larger the number of neurons and pathways present, the greater the number of combinations that are possible.

   e. Motor output depends on a hierarchical top-to-bottom process that is somewhat the reverse of abstraction. It depends on command neurons or central pattern generators that produce general instructions that are then passed down to more specific units in an administrative fashion, including feedbacks from and adjustments of the resulting movement and behavior as it develops. Motor neurons and muscle fiber contractions or glandular secretions are the last points in the path, much as sensory cells were the first points on the input side.

   As an example, language may be commanded, that is, formed and given meaning, at a higher level such as Wernicke's area of the cortex and then be polished up and prepared for final output at a lower level such as Broca's area of the cortex. Higher command centers may command that an animal run toward food, toward a mate, or away from an enemy; then lower centers take care of the actual job and ensure that the movements are smooth, coordinated, and corrected on the basis of changed, incoming sensory input.

By analogy, these various functions can be viewed somewhat as command, administrative, and clerical. In a sense, the reticular activating system (RAS) of the vertebrate brain is the telephone operator or central switchboard that answers incoming calls and directs them to the appropriate destinations. (Furthermore, in this analogy, when that switchboard closes down, the animal goes to sleep.)

f. Different parts of the nervous system are organized and localized for different specific functions, such as for particular sensory or motor jobs. Many of these have been specifically mapped in vertebrate brains. In addition to the major processing and association areas and centers of particular activities, many of the specific pathways from one part of the nervous system to another have also been mapped.

g. Speed, a highly advantageous property in neural processing, can be obtained with larger-diameter axons. However, larger axons take up more space and cut down the number that can fit in a given area of the nervous system, hence reducing the number of possible combinations. Other means of increasing speed without increasing axon diameter include increased temperature and myelin.

4. A major feature of neural processing involves inhibition of different pathways by each other. With divergence and convergence, neurons in the different pathways send output to and receive input from the other paths. Depending on priorities and on which pathways can generate the greatest amount of processing activity, some pathways inhibit others through IPSPs. That prevents everything from happening at once and results in only particular pathways being expressed at any particular time. (In addition to inhibition, various hormones help influence the priorities of various behavioral pathways. These are discussed in the next chapter.)

SUMMARY

After information enters the nervous system of animals, it is filtered, processed, and combined with other information and internal characteristics already present through large numbers of interneurons that form the bulk (generally over 99%) of the nervous system. The large numbers of neurons act in different ways on the sensory information being processed, not merely by passing it simply from one point to another but to abstract general features and meaning, which then may lead to the appropriate behavioral responses. Output proceeds from central command neurons or motor units, in a fashion that is somewhat the reverse of abstraction, to the lowest level of motor neurons and muscle fibers. A series of examples from moth ears, cockroach startle responses, and snail reflexes, to vertebrate reflexes and human language pathways illustrate the major principles that are involved in central processing. Those principles are listed in the recap section that immediately precedes this chapter summary.

A topic of much current interest and investigation involves artificial neural networks and the reciprocal modeling of biological nervous systems and human-made computers for each other. Although that topic is largely beyond the scope of this book, it promises future advances in understanding how nervous systems operate and bears close watching.

FOR FURTHER READING   **References on Neural Pathways and Processing**

Bullock, T.H., R. Orkland, and A. Grinnell. 1977. Introduction to nervous systems. W.H. Freeman, San Francisco.
A standard introductory reference for nervous systems.

Camhi, J.M. 1984. Neuroethology. Nerve cells and the natural behavior of animals. Sinauer Association, Sunderland, Mass.
From among several books on neuroethology currently available, this is considered one of the most useful. It is concise, clear, and covers the subject all the way from elementary principles of the individual neuron to numerous examples involving a wide range of animals.

Ewert, J.P. 1985. Concepts in vertebrate neuroethology. Animal Behaviour 33:1-29.
A review of neuroethology, available in a journal article, based on vertebrate examples (primarily toads) by one of the leaders in the field. This article is based on a Niko Tinbergen lecture.

Kandel, E.R. 1976. Cellular basis of behavior. W.H. Freeman, San Francisco.
Another standard introductory reference, along with Bullock et al, for the relationship between neurons, nervous systems, and behavior.

Kandel, E.R., and J.H. Schwartz. 1985. Principles of neural science. Elsevier, New York.
A major, thorough introductory text on the nervous system. In addition to covering basic topics in neurons and senses, as described in the previous chapter, this text also covers higher levels of processing, integration, and neural pathways.

Kobler, J.B., S.F. Isbey, and J.H. Casseday. 1987. Auditory pathways to the frontal cortex of the mustache bat, Pteronotus parnellii. Science 236:824-826 (plus cover illustration).
Although Chapter 17 considers only the hearing pathway of moths, this article pursues the neural connections on the bat's side of the bat-moth interaction, for persons who might be interested.

Livingstone, M., and D. Hubel. 1988. Segregation of form, color, movement, and depth: anatomy, physiology, and perception. Science 240:740-749.
An excellent article that further considers visual processing in primates.

Nolen, T.G., and R.R. Hoy. 1984. Initiation of behavior by single neurons: the role of behavioral context. Science 226:992-994.
An accessible, original paper that provides another concise and excellent, contemporary example, somewhat parallel to that of moths and bats, of neuroethology in an insect (flying crickets) responding to ultrasonic sounds.

Roeder, K.D. 1967. Nerve cells and insect behavior, revised edition. Harvard University Press, Cambridge, Mass.
The classic introduction to the neural basis of behavior. It formed the basis for parts of this chapter and is highly recommended reading for all interested persons.

Rose, G.J., R. Zelick, and A.S. Rand. 1988. Auditory processing of temporal information in a neotropical frog is independent of signal intensity. Ethology 77:330-336.
Much as a moth can discriminate bat calls against a noisy background, frogs can discriminate subtle details of their own species' calls over a wide (10,000-fold) range of intensity and against a loud background of similar calls from other species.

References on Neural Network Topics Including Artificial Networks

Anderson, J.A., and E. Rosenfeld, editors. 1988. Neurocomputing: foundations of research. MIT Press, Cambridge, Mass.
A collection of readings from the professional (and technical) literature that ranges from early speculations around 1890 to current papers.

Rumelhart, D.E., and J.L. McClelland. 1986. Parallel distributed processing. Explorations in the microstructure of cognition. Two volumes. MIT Press, Cambridge, Mass.
The first major resource on the subject of neural networks and connectionism. Content ranges from introductory to advanced levels.

Sijnowski, T.J., C. Koch, and P.S. Churchland. 1988. Computational neuroscience. Science 241:1299-1306.
A discussion and general overview, with examples, of using computer models to simulate brain processes.

Wasserman, P.D. 1989. Neural computing: theory and practice. Van Nostrand and Reinhold, New York.
This is perhaps the best reference to introduce the subject and provide avenues for interested persons to delve deeper. Includes a historical review and brings one as up-to-date as possible in this rapidly expanding field.

In addition to the references above, there are several periodical publications dealing with neural networks. These include *Advances in neural information processing systems* published by the Neural Information Processing Society (NIPS), and publications by the IEEE Annual Meeting on Neural Networks and the International Neural Network Society. There are also occasional articles in the magazines Byte and AI Expert.

18

THE INTERNAL CHEMISTRY OF BEHAVIOR

The previous two chapters discussed the basic operations of the nervous system, with particular attention to neurons, sensory inputs, muscular (and other) output, and neural pathways. The nervous system can respond quickly to changing conditions within milliseconds, second and minutes, up to a period of hours. Maturation and learning, as described in the next two chapters, may greatly extend the time frame and flexibility of the nervous system for some categories of responses. However, there are other categories of events and processes in life that require slower or longer term coordination and integration of behavior. These events and processes generally involve a large number and diverse variety of other chemicals beyond chemical synapses and their molecular transmitters. These chemicals work in conjunction with and through the nervous system to help coordinate and integrate behavior. They coordinate an animal's internal physiology and mediate between the animal and its external environment, including both the physical environment and other organisms. These chemicals operate on a time scale ranging from seconds and minutes to hours, days, weeks, months, and even years—for example involving sexual maturation in some species such as ourselves. This chapter focuses on the nature of these chemicals and how they affect behavior. Examples range from relatively simple behavior such as thirst and drinking to complex reproductive behaviors.

THE SIGNIFICANCE OF LONG-TERM COORDINATION AND INTEGRATION OF BEHAVIOR

Individual animals that can synchronize their lives over the long-term with the activities of other organisms, including social behavior within their own species, or with seasonal changes and other aspects of the external environment clearly have better chances of surviving and reproducing than those individuals that do not. The long term synchrony among different animals often is quite striking and can be shown to result from coordinating processes among the animals. That is, the synchrony breaks down when the animals are separated from each other.

An interesting example involves molting and associated behavior in collembolans, small insects commonly known as springtails. Leinaas (1983) discovered that molting group-living collembolans (of the genus *Hypogastrura*) synchronize all of their molts, regardless of age, apparently via some unknown chemical means.

Leinaas kept collembolans under constant environmental conditions in the laboratory. He held them in small chambers with sand and plaster bottoms and pieces of algae-covered bark for food and cover. He created colonies of mixed ages (determined by size of the individuals) that ranged from approximately 20 to 50 individuals. The molting process from premolt to molt required 3 to 5 days with all members of each group molting within a couple of days of each other, as shown in Figure 18-1, *A*. Approximately a week separated the completion of one molt period from the start of the next. If a synchronized colony was split up (Figure 18-1, *B*), the new daughter colonies would drift apart and molt at different times from each other but remain synchronized within the colony. On the other hand, if different colonies were combined, they would become synchronized with each other (Figure 18-1, *C*).

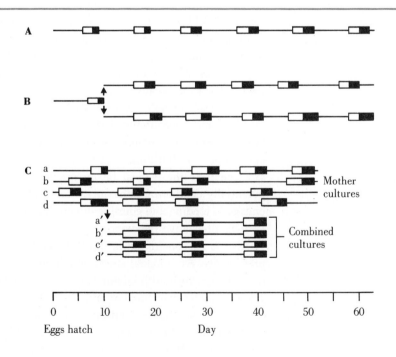

Figure 18-1 Synchrony of molt in group-living collembolan insects. **A,** Sequence of molts in an undisturbed culture. Open rectangles, premolt periods; solid rectangles, periods of molting from first to last individual in the group; horizontal lines, intermolt periods. **B,** Drifting apart of two groups from the same mother culture. **C,** Formation of synchrony in a culture formed by combining individuals from mother cultures that were not synchronized with each other.
From Leinaas, H.P. 1983. Science 219:193-195.

The animals remained synchronized if kept in the dark, thus precluding visual cues, and different subgroups in a container under separate pieces of bark molted together, suggesting that tactile cues were not involved. Auditory communication was not eliminated but does not seem likely. The most reasonable explanation for synchrony within groups involved pheromones.

Leinaas speculated that the adaptive significance of group molting is to coordinate long-distance (for collembolans) movements to new locations. The animals are inactive for a period before molting, and synchronized molting would greatly facilitate group travel.

In many other insects, molting at a particular stage of their life history is synchronized among different individuals in response to the same external, environmental cues. Examples include synchronized emergence from subimago to full adult in cicadas or from pupa to adult in midge flies. Whether coordinated directly among individuals or indirectly through environmental cues, such coordination may offer significant selective advantages as described in Chapter 13. In addition to social coordination, there are usually obvious advantages to coordinating with seasonal and other physical environmental changes, food supply, etc. Such coordination is important for virtually all living animals.

There is far more than coordination per se. All categories of behavior such as feeding, seasonal movements, and reproduction, each involving its own coordination, must be integrated with each other. If one stops to think about what is required, it is amazing that coordination and integration occurs at all. Yet it not only exists, it is practically universal in the lives of animals.

Consider for example, the life of a typical migratory bird living in the northern hemisphere such as the cliff swallow described in Chapter 13. Hatching occurs during a relatively short period of time that depends on the optimal integration of at least three major factors; social, suitable weather, and food supply. The young bird must then grow and develop rapidly if it is to achieve flight capability in time for moving on with others of the species. Migration must be coordinated with the seasons, and birds must have sufficient energy reserves—which requires additional feeding and fat storage at the proper time. Migration itself involves specific behavior, traveling long distances in the proper direction. After living for several weeks in its wintering area the bird must coordinate and integrate its behavior and physiology with the season, weather, and other members of its species to migrate back to the nesting area. Sexual maturity and subsequent reproductive behavior, including nest building and coordination with a mate, must also be integrated. Throughout the nesting period the behaviors change from one appropriate category to the next: courtship, nest building, egg laying and incubation, and rearing of offspring. At any one time there must be an integration of priorities among overlapping and potentially conflicting behaviors. Most individuals, for example, do not attempt to engage in feeding and courting or copulating simultaneously but must decide among various alternative behaviors.

Also to be considered is the fact that life does not always proceed smoothly or predictably. An unusual storm may interrupt migration or some stage of reproduction. An unexpected predator may destroy the eggs, nestlings, or one of the mates. The successful coordination and integration of behavior must include the ability to adjust to these stresses and uncertainties in life.

How is all of this accomplished internally? What are the proximate, physiological mechanisms? Most of them involve specific internally secreted chemicals that interface closely with the nervous system to cause appropriate behavior at the correct times. We present a number of different examples later in this chapter. First we discuss several general aspects and principles.

TERMINOLOGY

The classic view of internal chemical coordination involves the concept of another system, the endocrine system, being more-or-less distinct from the nervous system. However, the distinction between the two systems is a bit arbitrary and oversimplified. The concept of two systems and much of the accompanying terminology came from earlier times when we knew far less about what was really going on. When we look more closely at the components, the difference between nervous and endocrine blurs and the dichotomy is only partially useful. Furthermore, many of the invertebrate organisms, such as arthropods, have relatively few ductless endocrine glands outside of the nervous system, although they still show the full range of chemically mediated phenomena as in vertebrates. Even the definition of *endocrine* has caused some debate (e.g., Anctil 1984 and response by Krieger 1984).

Thus, at least for our mostly behavioral interests (as opposed to other more strictly physiological aspects such as the control of glucose or other substances in the blood), it is perhaps better to just distinguish between the neural and chemical aspects of control and integration. Even this, however, encompasses a continuum from synaptic transmitters between neurons to hormones, with several other chemical phenomena in between. In a sense (pun intended), this continuum can be extended conceptually to include intraspecific chemical coordination among members of a species via pheromones and various interspecific interactions.

At the moment there is no good name for this hodgepodge group of chemicals as a whole. They are not all hormones, transmitters, neurosecretions, or neurochemicals. The term *semiochemicals* has been used for information-carrying molecules, but that term has so far been used mostly in the context of pheromones, (i.e., intraspecific communication between animals). *Ethochemicals* might be considered, but that term may imply more than some neurophysiologists would desire. We therefore refer to the different types of chemicals by their familiar names, (e.g., *hormones, transmitters, pheromones*), or lump them collectively as *behavioral chemicals*.

HISTORY AND METHODOLOGY

The study of the chemical aspects of behavior is relatively recent compared with other topics. This is a new and extremely active field of research. Much of the reason for relatively slow progress earlier and the explosion of information during recent years is the development of a battery of new techniques and technology that permits small amounts of complex molecules to be isolated and analyzed. Many of these chemicals are produced by and secreted from neuronlike cells in the nervous system, known as *neurosecretory cells* (Figure 18-2). These cells have been less accessible than other internally secreting organs, the endocrine glands such as the adrenal gland or gonads.

The idea of chemical transmission of nervous impulses at synapses is itself a

Figure 18-2 Neurosecretory cells in the brain of a larval tobacco hornworm moth. Photograph courtesy Gerald G. Holt.

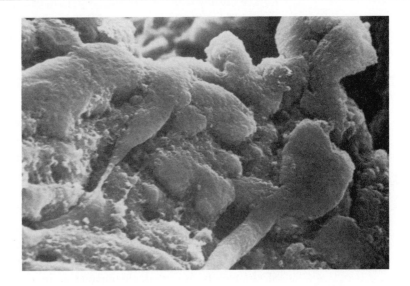

relatively recent proposal (Loewi 1921); and the 1920s can perhaps be viewed as the starting point for our understanding of behavioral chemicals. Further progress was slow in coming. Ernst Scharrer (1928) reported glandlike neurons in the hypothalamus of a fish. Over the next 60 years he and his wife, Berta Scharrer, conducted a series of pioneering studies and publications on secretory tissues in the nervous system (B. Scharrer 1987). Ernst focused on vertebrates and Berta on invertebrates. She was the first to report secretory activity in the nervous system of an invertebrate (ganglia of the marine snail *Aplysia*; B. Scharrer 1935). A large number of other workers also became involved. However, neural and endocrine aspects were being considered more or less separately, with focus on either neural processing via transmitters per se or endocrine effects on other tissues of the body outside the nervous system. It was not until the 1950s and early 1960s that the real interdependent relationship and presence of mutual signals between the nervous and endocrine systems came to be realized and accepted (E. Scharrer 1952, E. Scharrer and B. Scharrer 1963). That set the stage for a whole new generation of investigation and understanding of neurosecretion and the physiology of behavior.

Along with the conceptual advances there was a rapid proliferation of new technology, including: (1) improved light and electron microscopes with which to better see the tissues, cells, and parts of cells—including the vesicles containing the chemicals, (2) several types of sophisticated chromatography and other physical techniques (such as nuclear magnetic resonance) for the separation and identification of complex molecules, (3) a number of methods for tagging and measuring specific molecules and their receptor sites by fluorescent or radioactive markers, and (4) use of immunological techniques of creating antibodies that bind to particular substances. A basic method used in endocrine work from early on has been the bioassay. Solutions thought to contain the hormones or other chemicals of interest are injected into the animal or applied to cells in tissue culture to see if they produce the expected effects. All of this has provided a wealth of information from which we give a brief overview and sampling in the remainder of this chapter.

THE PRIMARY INTERNAL CHEMICALS THAT AFFECT BEHAVIOR

Several major kinds of biochemicals affect behavior in a number of different ways. The number of such known chemicals is now so large, amounting to well over a hundred, that a comprehensive list is not meaningful for introductory purposes. A few examples and their molecular structures are shown in Figure 18-3. Persons wanting larger listings can refer to Barchas (1978), Krieger (1983), or a current physiology or endocrinology text.

Pheromones, which are transmitted between animals of the same species, have already been discussed in relation to communication (Chapter 15) and senses (Chapter 16); thus they are not considered further here. Rather we focus on the naturally occurring chemicals within the animal's body. There are five major types (although the first four are closely related and overlap with each other): behaviorally active amino acids, amines, catecholamines, peptides, and steroids. There are also a few other miscellaneous types of molecules such as insect juvenile hormone (page 647), which is a terpene and not a member of any of the above five categories. Some others, such as the prostaglandins, have also been implicated as having behavioral effects, but for now we concentrate on the five categories that have clearly been identified as having roles in behavior.

Structures, Origins, and Properties of the Behavioral Molecules

The basic structures of the behavioral molecules are highlighted in Figure 18-3. Amines are molecules with $-NH_2$ groups. Amino acids have a carboxyl, $-COOH$ structure in addition to the $-NH_2$. These are the familiar amino acids, some of which (e.g., glycine, glutamic acid) are behaviorally or neurally active. Amines are derived from amino acids by decarboxylation, that is, by removing the carboxyl group. Monoamines have a simple $-NH_2$ group, and catecholamines are a particular type of monoamine that has a catechol (1,2 dihydroxybenzene) group in the structure. There are three catecholamines: dopamine, norepinephrine (noradrenalin), and epinephrine (adrenalin), with each one being derived from the previous one, respectively. Dopamine is derived from DOPA (*di*hydro*xy*phenyl*al*anine, which in turn is derived from the amino acid phenylalanine) (Figure 18-4).

The catecholamines have important effects both in the nervous system (as transmitters and in other ways) and in many other tissues of the body. Acetylcholine is another important transmitter that is derived from choline—which in turn comes from a parent amino acid, serine. Many of the behavioral molecules are peptides, that is, polymers or chains of amino acids. Neurons and pathways within the nervous system are often labeled by their associated types of molecules by adding the suffix *ergic*, e.g., peptidergic, cholinergic, adrenergic, etc.

Various amino acid-derived molecules are produced both in endocrine tissue outside the nervous system, such as the adrenal gland and gut, and in neurosecretory cells. Neurosecretory cells are like neurons in that they are inside the nervous system, receive input from other neurons via synapses, and display action potentials. However, their large axons do not synapse with other neurons but rather release their chemicals into the blood at special sites known as *neurohemal* organs. The amino acid-derived chemicals, in addition to being produced both in and outside the nervous system, also affect particular cells at sites both within and outside the nervous system.

One of the important discoveries related to the amino acid-derived class of

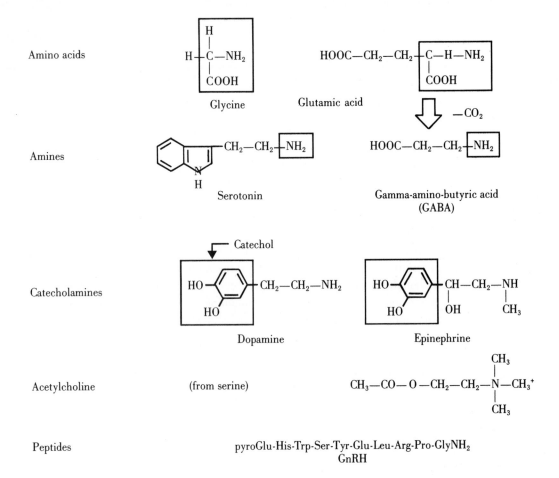

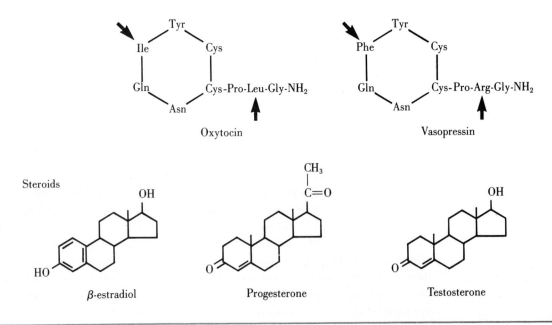

Figure 18-3 Molecular structure of a sampling of amino acid-derived behavioral chemicals and steroid hormones. The distinguishing segments of the different types of molecules are indicated by boxes. LH, FSH, prolactin, and other proteinaceous hormones are large, long-chained sequences of around 140 to 200 amino acids and cannot be easily illustrated because of their sizes.

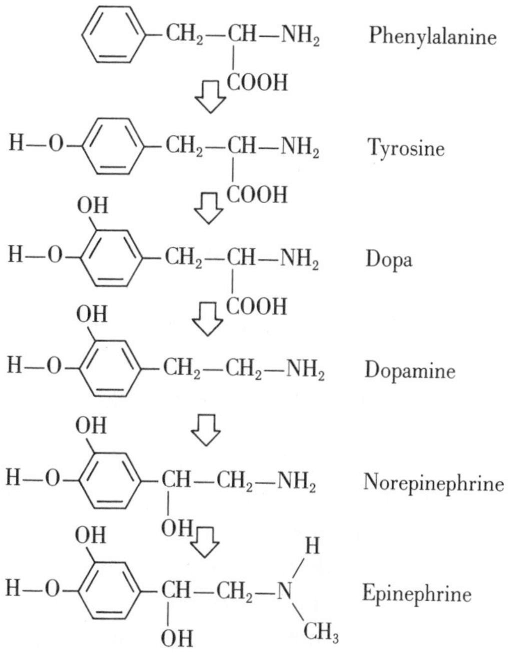

Figure 18-4 Relationships among the catecholamines and their derivation from phenylalanine. The difference in each molecule from the previous one can be seen by comparing the structures.

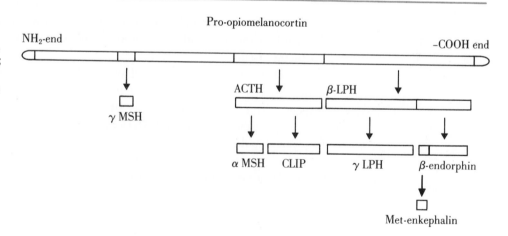

Figure 18-5 Pro-opiomelanocortin precursor polypeptide and a number of active peptides derived from it. MSH, melanocyte-stimulating hormones; ACTH, adrenocorticotropic hormone; LPH, lipotropic hormone; CLIP, corticotropin-like intermediate peptide.
Generalized from several sources (Kreiger 1983, Martin 1985, and references therein).

molecules is that they commonly originate from larger precursor peptides, which are then cleaved into a number of different, smaller, active molecules. Perhaps the classic model of this phenomenon is pro-opiomelanocortin (POMC) from the vertebrate hypothalamus and pituitary and the adrenal medulla. It breaks down into several different active fragments, including ACTH, β-endorphin, and met-enkephalin (Figure 18-5). The latter two help reduce pain, which is discussed later in the chapter. Another example is a large peptide resulting from a single gene in the marine snail *Aplysia*. This peptide is cleaved by enzymes into a hormone related to egg laying (ELH) and at least five other peptides, two more (in addition to ELH) of which produce effects on the snail's neurons and muscles related to egg-laying behavior (Scheller et al 1983). (These are discussed in more detail later in the chapter.)

According to Scheller et al (1983), a large precursor polyprotein that breaks down into a number of separate, distinct peptides offers several advantages over the alternative of having them produced separately, which usually occurs with other proteins in the body. First, synthesis and release of the various proteins are better coordinated among them all. Second, with the possibility of splitting the precursor different ways into different peptides at different times, there is greater flexibility in the number of combinations of peptides that can be released. Third, the production and release of a variety of chemicals after the point of RNA processing allows more temporal flexibility and control in the cells. Finally, the presence of inert or nonfunctional peptides interspersed among active sequences provides another source of variability in addition to mutations and recombinations of genetic material per se.

The steroid molecules form a distinct major class of molecules of their own. They have a four-ring nucleus, as shown in Figure 18-3. They are all derived from cholesterol (for details and relationships, see a physiology or endocrinology text). Steroids that serve as hormones (not all do) are produced in and released from endocrine tissue outside of the nervous system. However, they may profoundly affect neurons at specific sites within the nervous system (as well as producing effects elsewhere in the body). They are lipid soluble and can pass through cell membranes, hence moving easily from the blood into various tissues, including the brain. A major principle of neuroendocrine interaction is that the steroids are commonly produced in response to inputs from the nervous system, that is, by stimulation from particular neurosecretions. Thus there is an important interaction, frequently involving feedbacks, between the nervous system and steroid-hormone-producing tissues.

One of the major features of behavioral chemicals of all kinds is that relatively minor differences in chemical structure may have major differences in biological response. This can be seen readily by closely inspecting differences among some of the chemicals in Figure 18-3. Oxytocin and vasopressin (or antidiuretic hormone), for example, differ only in two of their amino acids, yet they produce different biological effects: oxytocin affects uterine contractions and the process of milk release from the mammary glands of female mammals, whereas vasopressin affects kidney function and the circulatory system. Likewise, for steroids, testosterone is similar to the estrogens chemically but produces differences that are male versus female biologically.

Another major feature of behavioral chemicals, aside from the close-acting synaptic transmitters, is that they are broadcast throughout the body, but only certain other cells respond. The specific message is carried in the molecule rather than the pathway. Neural impulses are basically all the same (as action potentials), but they go to specific points. This mode of transmission is known as *anatomical addressing*. Chemical messages, on the other hand, go to many points, but the message is specific and only certain target cells respond. This mode of transmission is *chemical addressing*. The difference between these modes of transmission is like sending a message over a telephone line where only a particular phone receives it versus a message like a fire alarm that is broadcast widely but only a few, such as firemen and perhaps news people and a few curious onlookers, respond.

How do these information-carrying molecules code their messages, and how are they received and transduced into biological effects such as observable behavior? These mechanisms are becoming fairly well understood and are briefly described next.

Modes of Action of the Behavioral Molecules at the Cellular Level

There are two fundamentally different ways in which behavioral chemicals exert their effects. The amino acid-derived types affect their target cells at specific molecular receptor sites at the cell's surface. There may be short-term membrane effects such as PSPs. Through a complex sequence of molecular interactions including a secondary internal messenger such as cyclic AMP (cAMP) or cyclic GMP (cGMP) there can be long-term effects on protein synthesis (Figure 18-6, *A*). The second messenger initiates a cascade of subsequent steps in which each can involve many other molecules, thus causing a multiplying (or magnification) effect that might be better described as an avalanche rather than a cascade. For an example of the molecular details at the receptor site, see Levitzki (1988).

Many of the effects mediated by second messengers when they occur in neurons can modify the functional properties of the neuron. Thus some of these amino acid derivatives are considered neuroregulators or neuromodulators. They may arrive at cells via the blood, local diffusion, or at the synapse—often in conjunction with the transmitter chemicals. The release of more than one chemical, that is, transmitters plus neuroregulators, from a given axon is easy to understand because they may be jointly derived from the same larger precursor.

The steroid hormones, on the other hand, enter the cells, such as specific neurons in the brain, and combine with receptor molecules inside the cytoplasm. The hormone-receptor complex moves to an acceptor site on the DNA in the nucleus and affects subsequent expression of the genes via the messenger RNA and protein

Figure 18-6 Modes of action at the cellular level for amino acid-related versus steroid messenger molecules.

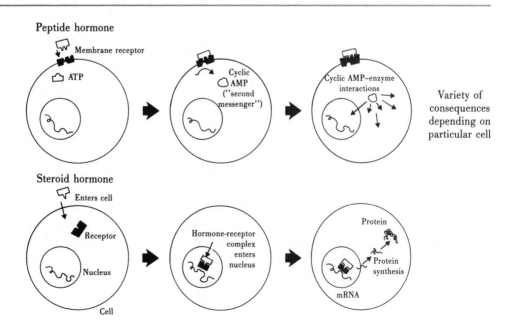

synthesis (Figure 18-6, *B*). This in turn may alter the functional properties of those particular neurons.

In both cases, it is easy to see how, by interacting with the machinery of gene expression and protein synthesis, hormones exert effects that span seconds, hours, days, and months and may even make permanent changes. In some cases, hormones can affect the organization and functioning of the brain itself. A particularly important example involves sexual differentiation of the brains of mammals, including humans (pages 661 to 662). Next we look at several specific examples involving behavioral chemicals.

CASE STUDIES OF BEHAVIORAL CHEMISTRY

Thirst, Sodium Appetite, and the Control of Drinking Behavior

Maintaining the proper amount of water inside the body is critical to the existence of all animals. Accordingly, all species have one or more means for controlling water intake and outflow, depending on the environment and the internal and external conditions at any particular time. Organisms living in fresh water must contend with a constant inflow of water because of osmotic pressure, and they normally must expel water. Terrestrial and many marine animals, on the other hand, have the opposite problem; they are usually losing water and must replenish it. Most terrestrial animals get their water by drinking, which involves specific behaviors of searching for and consuming it. (Some terrestrial animals get by on metabolic water, that is, obtained from their food, and they rarely or never have to drink.)

Probably every human has been thirsty at one time or another, and it is reasonable to assume that many other animals experience a similar sensation. At least from what we can observe, many other animals engage in drinking behavior under similar environmental and physiological conditions, including the experimental introduction of substances that cause humans to experience thirst. The experimental introduction of some substances provokes drinking behavior even in animals that are otherwise well hydrated. One of these substances is polyethylene glycol (PEG), a colloidal solution that can be injected subcutaneously into an animal to reduce blood volume by osmosis (the molecules are too large to cross capillary membranes hence draw water out of the capillaries).

A separate but somewhat related physiological and behavioral concern involves the body's sodium concentration and the animal's appetite for salt. Both water and salt are important for maintaining the correct fluid osmolality and blood volume.

An important chemical thought to be related to thirst and sodium appetite is angiotensin II (Figure 18-7). This is an 8-amino-acid peptide (or octapeptide). It is derived by enzymatic action from a 10-amino-acid peptide, angiotensin I, which in turn is cleaved from an even larger molecule (a glycoprotein) that goes under several names, (e.g., *angiotensinogen*). Angiotensinogen is produced in the liver and evidently elsewhere (including perhaps the brain). Angiotensin II produces a host of effects related to water balance throughout the body, including the stimulation of sodium appetite and perhaps thirst in the brain. Some of the other effects include vasoconstriction of arterioles (which increases blood pressure) and the release of aldosterone from the adrenal cortex, which causes the kidney to reabsorb more sodium and water. Angiotensin II may produce many of its effects through cGMP, although the precise mechanisms are not yet certain. After (or while) exerting their

Figure 18-7 Sequence of interactions in the endocrine motivation via thirst of mammalian drinking behavior.

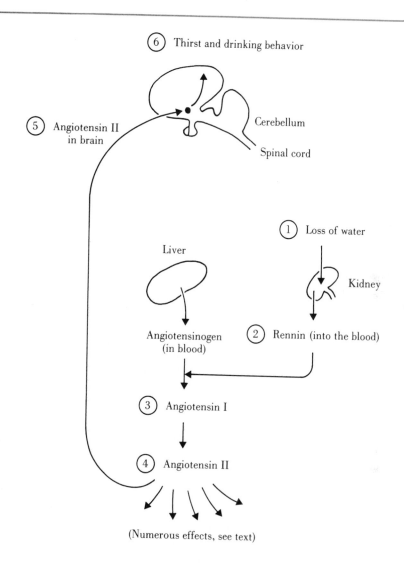

(6) Thirst and drinking behavior

(5) Angiotensin II in brain

Cerebellum

Spinal cord

(1) Loss of water

Liver

Kidney

Angiotensinogen (in blood)

(2) Rennin (into the blood)

(3) Angiotensin I

(4) Angiotensin II

(Numerous effects, see text)

effects, Angiotensin II molecules are broken down into smaller, inactive fragments.

How does angiotensin II produce sodium appetite and—if it does—thirst, and what is the chain of events leading up to it? The pathway for sodium appetite is not clear; it appears to involve aldosterone and, in some unknown way, oxytocin (Stricker and Verbalis, 1988). Thirst, on the other hand, appears to occur partly through direct neural connections to the brain from stretch receptors in blood vessels in response to blood volume (Stricker and Verbalis, 1988). However, the sensation of thirst may also result from neurons possessing angiotensin II receptors in a region of the brain near the hypothalamus. There has been some argument about how or if angiotensin II gets across the blood-brain barrier, specialized capillaries that restrict most large molecules from entering the brain. Alternatives are: (1) it does cross the barrier, (2) it acts at regions that lack the barrier, or (3) there is separate production and metabolism of it within the brain. There are several lines of evidence for the last idea. Whatever the precise pathway, the targeted neurons send impulses to a region of the brain (location not known) where the sensation of thirst is perceived and that motivates appropriate water searching and drinking behavior.

This sequence is triggered by a reduction of body water, which reduces cellular and extracellular fluid, blood volume, and blood pressure along with changes in the concentrations of sodium and potassium. Intra- and extracellular dehydration involve partially independent and additive mechanisms. These effects in the kidney (and perhaps elsewhere, such as in the brain) lead to the release into the blood of renin, a specific enzyme that cleaves angiotensin I from angiotensinogen. Thus physical and osmotic changes from a reduction of the body's water or sodium levels initiate a sequence of chemically mediated steps that lead to (1) various physiological events that conserve water, and (2) an identifiable motivation, via the sensation of thirst or appetite for salt, to behave in such a way as to replenish the body's supply of water, sodium, or both. The main points of this process are summarized in Figure 18-7.

Growth, Development, and Molting in Arthropods

As arthropod animals grow and develop they must periodically shed or molt their exoskeleton to increase their body size. The process of molting (ecdysis) is accompanied by a set of appropriate but complex behaviors. The molting individual must move such that (1) the old skin is anchored to the substrate then loosened and split, (2) the body is moved forward and out of the skin via peristaltic waves and other actions that include further expansion and splitting of old skin and extrication of appendages, (3) final expansion of new cuticle and immobility during hardening, and (4) a new period of activity which may include eating the old, shed exoskeleton. Carlson and Evans (1977) recorded electrical activity from 22 muscles and simultaneously videotaped the behavior for frame-by-frame analysis of cricket molting. They discovered that the entire sequence consisted of large numbers of neural controls and motor programs with their own patterns, cycles, and bouts of activities. In addition to the basic, main programs there are backup programs (in case certain parts such as antennae get stuck), optional programs, and alternate programs (to deal, for example, with molting from vertical versus horizontal substrates). Tamm and Cobb (1978) showed that the stage of molt affected aggressive behavior and the social dominance status of lobsters.

Molting is coordinated not only by the nervous system but also with neurohemal structures, nonneural endocrine tissue, and various messenger molecules (Evans 1984). There is considerable variation, however, among the different classes of arthropods, such as insects versus the crustaceans.

In insects (Figure 18-8, *A*), appropriate sensory inputs to the nervous system

Figure 18-8 Comparison of molting neurochemistry in insects versus crustaceans. Major neuroendocrine organs and messenger chemicals in **A**, insects and **B**, crayfish.

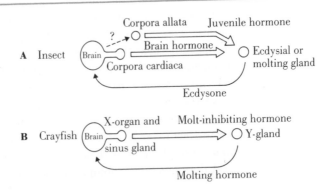

stimulate neurosecretory cells located in the brain to release a hormone called *brain hormone* (BH), or *ecdysiotropin*. The neurosecretory axons end in paired structures called the *corpora cardiaca*, a neurohemal organ analogous to the neurohypophysis of vertebrates. The brain hormone circulates via the hemolymph ("blood") in the hemocoel (arthropods have an open circulatory system without vessels) where it reaches the ecdysial or molting gland (also called the *prothoracic gland* in some insects). That gland then produces a steroid hormone, ecdysone, which initiates a host of molt-related physiological changes and presumably the associated behavior (the precise mechanisms related to behavior are less well understood than for vertebrate situations).

Insects have a second major hormone involved in the molting process, known as *juvenile hormone* (JH). It is produced in the paired corpora allata, which are analogous to the adenohypophysis of vertebrates and are under control of the brain in a manner that is not yet completely understood; it may be via direct nervous system control, circulating chemicals, or both. The corpora allata can be both stimulated and inhibited by the brain. Juvenile hormone, when present with ecdysone, permits growth and molting to occur, but it prevents maturation (hence the animal remains juvenile). At the time of maturation, juvenile hormone is reduced and the animal changes to the mature stage (with metamorphosis in some species).

Molt in crustaceans (Figure 18-8, *B*) differs from molt in insects in two important ways: (1) rather than the first hormone being stimulatory and released periodically like insect BH, the hormone is inhibitory and is released most of the time—being periodically reduced by the brain; and (2) there is no counterpart of juvenile hormone. The crustacean neurosecretory cells are located in the x-organ in the eyestalk and release their product, *molt-inhibiting hormone*, at another point, the sinus gland, also in the eyestalk. Molt-inhibiting hormone inhibits another gland (the *Y-gland*, a molting gland in the head region) from producing its hormone, *molting hormone*—the counterpart to insect ecdysone. When the brain, under appropriate sensory stimulation, inhibits the molt-inhibiting hormone (a double negative), the molting gland is released to produce its molting hormone. That hormone then produces various molt-related physiological effects and associated behavior.

Coordination of Reproduction

Egg-laying behavior in *Aplysia*

The general pattern of neurosecretion and chemical control of behavior in mollusks is only partially understood. In some cases, such as reproductive behavior in *Aplysia*, many of the details are well understood whereas some aspects remain unknown. The following has been distilled from Scheller et al (1983).

Egg-laying behavior in *Aplysia* involves a highly stereotyped sequence of movements, legitimately viewed as fixed action patterns (FAPs; see Chapter 2). The animal stops moving and feeding, increases its respiratory movements, extrudes a string of eggs, takes the egg string in its mouth and covers it with mucus, then waves the head back and forth, laying the egg mass in an irregular pile that sticks to the substrate. This sequence appears to be under the control of a series of behavioral chemicals of which at least a part of the sequence is now known and is remarkably analogous to induced ovulation in mammals. Induced (or reflex) ovulation is adaptive as one method of ensuring fertilization as an alternative to synchronizing the male and female through external cues.

The sequence begins with copulation, which stimulates the atrial gland, an

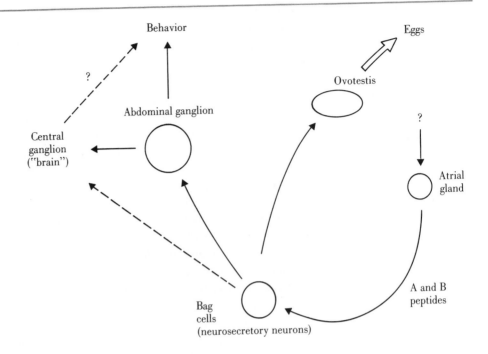

Figure 18-9 Sequence of molecular interactions in *Aplysia* egg laying. For additional perspective, see Figure 17-8.

endocrine gland in the snail's reproductive tract, to release two peptides, A and B. These two peptides stimulate neurosecretory cells known as *bag cells* near the abdominal ganglion (see Figure 17-8, *A*). The bag cells respond by releasing several neuropeptides, the products of the cleaving of a large precursor polyprotein resulting from a single gene, as discussed previously (page 641). These include the egg-laying hormone (ELH), four different small peptides called *bag-cell factors* (BCFs), and other unknown peptides. The circulating ELH causes contraction of smooth muscles in the ovotestis, which in turn causes the eggs to be expelled. ELH in combination with the other peptides, or various other of the peptides acting alone, also stimulate or inhibit specific neurons located in the abdominal ganglion. The firing patterns of these neurons are thought to produce the observed egg-laying movements. A summary of these steps is shown in Figure 18-9.

There are, however, several questions still unanswered. Unlike the controls of cardiovascular output and gill withdrawal reflex (Chapter 17), the full set of neurons involved, the connections between them, and the specific effects of the different peptides on particular neurons are not yet known. In other words, except for the influences of a few of the peptides on the firing of a few of the neurons, we do not know which peptides elicit which behaviors. Scheller et al (1983) speculated that some of the same peptides in different combinations could be involved in behaviors other than egg laying. Feeding, for example, also involves side-to-side head movements, and the same peptide could be responsible for such movements in both feeding and egg laying, with the difference in final outcome, depending on which other peptides are released along with the one (or more) causing head waving. Other questions are: what stimulates the atrial gland to produce A and B peptides; and what role do other parts of the CNS, such as the central ganglion (brain), play in the full set of behavioral movements? Are any steroids involved? (So far no steroids have been implicated.) In spite of all the unknowns, however, there is no doubt that a host of behavioral chemicals are extremely important in the egg-laying behavior. The nervous system per se is clearly not acting alone.

Reproduction in arthropods

Arthropods show great diversity in the chemistry of reproduction. Just among insects there is tremendous variation in the underlying anatomy and physiology. Thus it is difficult to give a generalized introductory picture for whole groups such as insects. Insect reproduction relies on a basic system, involving the central nervous system, corpora cardiaca, and corpora allata, as illustrated for molting, but in many different ways, depending on the type of insect.

Although there is an extensive body of knowledge on some parts of the endocrinology of insect reproduction, including much information on neurosecretion, the total picture that includes the neural and behavioral aspects remains poorly known. This problem partly stems from the variation among different species. Juvenile hormone, for example, has been shown to be important in the calling behavior and pheromone production and release of armyworm moths *(Pseudaletia unipuncta)* but not in two other species of moths (Cusson and McNeil 1989). The neural and behavioral aspects of insect reproduction deserve increased attention. Persons wanting further details on the physiological aspects of insect reproduction should refer to Raabe (1982, 1986).

Reproduction in vertebrates

Reproductive behavior in ring doves.   One of the first clear examples of the role of hormones in animal behavior was provided by the work of Lehrman on ring doves (e.g., Lehrman 1964). When a male and female ring dove are placed together, the male first begins a bow-and-coo display. If they are provided with nesting material and other conditions are suitable (i.e., they are not frightened or stressed, have sufficient food, the day length is appropriate, etc.), the two doves go through a sequence of behavioral stages whereby they pair and copulate, build a nest, lay eggs, incubate, hatch and raise the young, and then abandon the young and start all over again. It takes about 7 days from the start of the bow-and-coo courtship behavior of the male until the female lays the first egg and about 6 weeks for the full cycle.

There are many ways in which this sequence could be interpreted. Perhaps, in the anthropomorphic view, the birds want to have some babies, understand what needs to be done to accomplish that, and set about doing the job, step by step. However, we have already mostly rejected that view (Chapter 1). Alternately, there may be a simple chain of complex events that set up a series of stimuli such that, in the presence of the proper stimulus, the next step is triggered. This interpretation also appears incorrect. Birds given stimuli, such as eggs in a nest or chicks that are begging food, will not necessarily behave properly unless they are in the correct hormonal condition.

Each stage of the process involves a particular set of hormones, stimuli, and behaviors. Considering just the stage leading up to incubation, for example, the stimuli available to the birds and their internal hormonal conditions can be altered experimentally, as was done by Lehrman. A summary of his experiments is as follows (for references and details see Crews and Silver 1985):

1. If given a normal mate and nesting material, a female will begin incubating eggs (her own or others that are provided artificially) in 5 to 7 days, but not before. If given eggs in a nest before day 5, she will completely ignore them or take the nest material to build her own nest.
2. If given a normal mate but no nesting material to work with, it will be about 8 days before the female will incubate eggs that are provided to her.
3. If given a castrated male for a mate, the female will not incubate eggs at any time. A castrated male given hormone injections, however, serves as a normal male and stimulates a female to lay eggs and incubate.
4. If a lone female dove is given a nest and eggs, she will not incubate at any time. She ignores them.
5. If a lone female dove is given hormone injections, the results depend on the hormone:
   a. With progesterone she will begin incubating within 3 hours.
   b. With estradiol she will incubate after 1 to 3 days.
   c. Testosterone injections produce no effect; she will not incubate.

Incubation is only one of several behaviors involved in the whole reproductive sequence. In the case of feeding behavior, whereby the dove produces crop milk, a nutrient-rich substance produced by the lining of the esophagus that the mother regurgitates to feed the young, prolactin is the required hormone. The whole set of interacting factors—behaviors, hormones, and internal, physiological conditions—is shown in Table 18-1.

If the stimuli are not appropriate for the birds' particular hormonal state, the birds generally either completely ignore the stimuli or become aggressive, destroying

Table 18-1 Summary of events during reproduction in the ring dove

| | Day | Behavior | | External Stimulus | Major Hormones | Major Physiological Effects |
		Male	Female			
1 week	0 (start)	Bow and coo		Mate	Testosterone in ♂	
	1	Nest call	Nest call		LH and FSH (pituitary)	
	2	Nest build (bring materials)	Nest build (build)	Nest material	Estrogen and progesterone (gonads)	Ovary and oviduct development
		Copulation	Accepts copulation			
	7		Egg laying			
2 weeks	~9	Incubation	Incubation	(Eggs in nest)	Prolactin	
	~23 (hatch)					Crop development
2½ to 4 weeks		Feeds young	Feeds young	Young begging		
	40	(Goes back to start)				

the eggs or killing the chicks. These outcomes should not occur either if the adults "knew" what they were doing or simply were responding to particular stimuli. Thus the proper hormonal conditions in conjunction with certain stimuli are required before particular behaviors, such as nest building, incubation, or chick feeding, will be expressed.

Other persons have continued to investigate the connections between hormones and reproductive behavior in ring doves (e.g., Silver 1977, 1978, Silver and Barbiere 1977) and in many other birds and other types of vertebrates.

The same general processes seen for ring doves can be observed in most vertebrates, although there is some variation in patterns. In some cases, such as in many mammals, there may also be modifications resulting from learning. The next well-understood example involves reptiles, based on extensive research by David Crews, a student of Lehrman.

Reproductive behavior in anole lizards.   The green anole *(Anolis carolinensis)* is a small iguanid lizard common throughout the southeastern and south central regions of the United States. It is the familiar chameleon (not true chameleons, which belong to a different family) often sold in pet shops. Their normal breeding period runs from around March or April to July or August. They are reproductively inactive the rest of the year, except for a period of male territorial behavior beginning around late January and February. They are dormant from around late September or October into January and February. The males are the first to emerge from winter dormancy. They engage in a period of territorial displays, which includes the projection of a prominent flap of skin, the dewlap, along the throat. The males' testes begin development to the point of producing spermatids but stop before producing fully developed spermatozoa until after the females appear, at which time testicular development resumes. The females emerge in March and April, at which time the males spend less time in aggressive behavior with each other and start courting the females, also using the dewlap (Figures 18-10 and 18-11, *A*).

Figure 18-10 Displaying male anole lizard (left) with dewlap skin under chin extended. Female is to the right side of the photograph. Photograph courtesy David Crews.

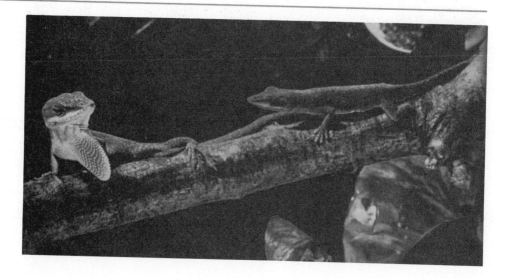

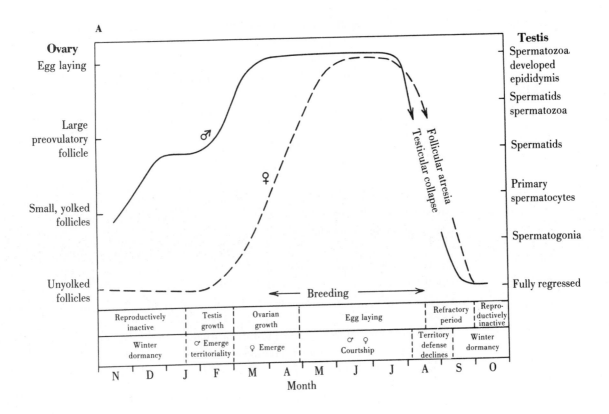

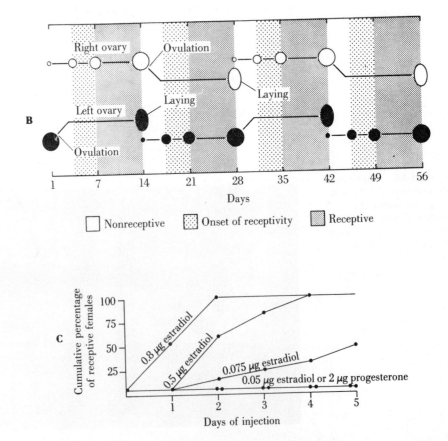

Figure 18-11 Effects of male behavior and female hormones on anole lizards. A, Sequence of behavioral and physiological events in annual reproductive cycle of *Anolis carolinensis*. B, Normal sequence of female behavior patterns and ovarian development during the breeding season. C, Behavioral receptivity of females to males after the females' ovaries were removed and they received injections of various female hormones of different concentrations. A and B from Crews, D. 1975. Science 189:1060. C Modified from Crews, D. 1975. Science 189:1060.

The ovaries of female anoles begin developing after males commence courtship behavior; then the right and left ovaries alternately produce eggs. The females also cycle between being receptive and nonreceptive to the males (Figure 18-11, *B*).

Crews (1975) demonstrated experimentally that ovarian development of the females depends on male courtship behavior. The ovaries of females exposed only to courting males developed early and continued to show follicular growth. However, if females were exposed only to aggressive males, their ovaries failed to develop at all. Females exposed to conditions that were reversed from normal, that is, first to courting, then to aggressive males, showed early ovarian development followed by early regression. Crews also showed, in a manner somewhat similar to Lehrman's work with doves, that injections of the appropriate hormone at a sufficient dosage led to receptive behavior by the females. Progesterone did not stimulate receptive behavior, whereas a sufficient dose of estradiol did (Figure 18-11, *C*). For a review of further details see Crews and Silver (1985).

General patterns of vertebrate reproductive chemistry and behavior.   The basic picture of neural-endocrine control of vertebrate reproduction that has emerged from studies such as those cited previously and numerous others is shown in Figure 18-12. Hormones from several glands, including the pituitary and the gonads, affect reproductive behavior. Adrenal hormones released during emergencies or stress may interfere with reproduction. In addition to interactions with hormones from glands other than the gonads, the neurons in various parts of the brain are affected by endogenous rhythms (e.g., circadian) and—via the senses—external stimuli.

The external stimuli vary from species to species and include environmental conditions of the physical habitat (e.g., effects of rainfall on desert species, photoperiod, temperature, water pH), vegetation (e.g., odors from plants, chemicals

Figure 18-12 Basic pattern of neuroendocrine integration in vertebrates. There are, however, several departures from this basic pattern among various species.

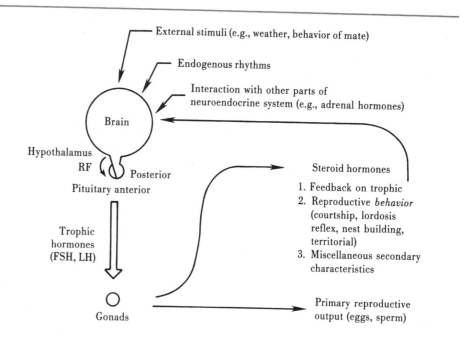

in plant food), and the behavior of other animals, such as mates, offspring, and social peers or superiors.

All of these stimuli are routed through the brain presumably in familiar neural fashion, involving particular pathways, regions of the brain, and so forth. Some of the neural pathways lead to neurosecretory cells in the hypothalamic region at the ventral side of the diencephalon (i.e., ventral to the thalamus). Some neurosecretory output, such as oxytocin, extends into the distal parts of the neurohypophysis where it is released directly into the general circulation. Other output is released more locally (from the median eminence) into the hypothalamic portal blood vessels and carried a short distance to the adenohypophysis (anterior pituitary). The chemicals released over this short path are called *releasing factors* or *releasing hormones* (RHs). At the adenohypophysis they stimulate or inhibit additional hormones, of which some act directly and others, generally called *tropic* (from *feeding*) *hormones*, stimulate yet other hormones. Follicle-stimulating hormone (FSH) and luteinizing hormone (LH) have both direct and tropic effects (and hence are called *gonadotropic hormones*). The releasing factor from the hypothalamus that stimulates their release is gonadotropic-releasing hormone (GnRH). FSH and LH travel via the bloodstream throughout the body and are picked up by cell membrane receptors at appropriate tissues, particularly in the gonads.

The gonadotropic hormones then cause glandular development in the gonads, depending on which sex is involved. This results in primary reproductive output, eggs or sperm, and a different set of hormones, the sex steroids such as estradiol and testosterone. The steroids travel throughout the body via the circulatory system and produce several secondary effects, including the familiar morphological differences between males and females in dimorphic species. They also feed back on the nervous system in two important ways. They modify neural output in particular regions of the brain, which affects behavior directly, and they interact with the pathways that lead to the hypothalamic neurosecretory cells. The modified behavior of the animal may serve as a stimulus for the behavior and reproductive development of another individual that may be a mate or potential mate. For several additional examples of sexually dimorphic behaviors and their underlying neuroendocrine mechanisms, see Kelley (1988).

The target neurons responsible for particular behaviors, in response to particular hormones, are located in particular regions of the brain. These neurons possess the proper cellular receptors to interact with the specific hormones. Many of these receptors are located in the brain stem and limbic systems. Some of the receptor areas have been located with the aid of radioactive-labeled hormones. Tritium-labeled estradiol has been injected into the bodies of female rats, for example. It was given time to be picked up by the brain; then the rats were killed and the labeled areas of the brain identified (McEwen 1976). In another example tritium-labeled testosterone demonstrated hormone-concentration regions in auditory and vocal centers in male frog brains (Figure 18-13).

Various reproductive behaviors and the extents to which they are expressed at different times have been studied in most groups of vertebrates but particularly in the birds and mammals. An example for the ringed plover is shown in Figure 18-14, based on an early study (Laven 1940). More recent studies of several species of birds have firmly established the role of hormones in enabling these behaviors.

Progress in measuring and understanding the presence of various hormones in birds was stimulated by the development of new analytical techniques for low levels

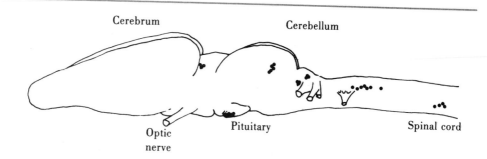

Figure 18-13 Uptake of androgen by auditory and vocal centers in the brain of clawed frogs *(Xenopus laevis)*. Dots indicate locations of uptake as revealed by autoradiographic studies.
From Kelley, D.B. 1980. Science 207:553-555.

Cerebrum

Cerebellum

Optic nerve

Pituitary

Spinal cord

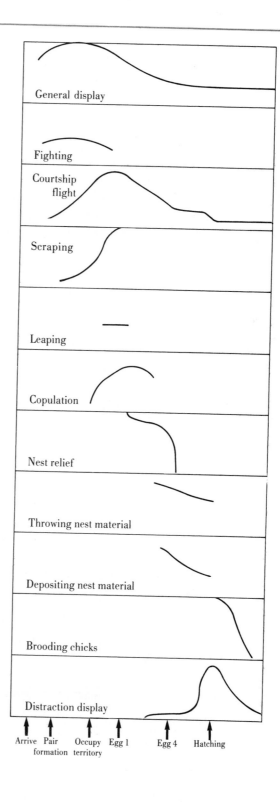

Figure 18-14 Several behavioral components are involved in the reproductive cycle of the ringed plover. Frequency of each behavior is shown by the height of the line.
Modified from Laven, H. 1940. Journal of Ornithology 88:183-288.

General display

Fighting

Courtship flight

Scraping

Leaping

Copulation

Nest relief

Throwing nest material

Depositing nest material

Brooding chicks

Distraction display

Arrive Pair formation Occupy territory Egg 1 Egg 4 Hatching of hormones in the blood. The most important methods involved radioimmunoassays and competitive-protein binding assays. Such techniques were first used in studies of birds in captivity (e.g., Follet, Scanes, and Cunningham 1972), then wild birds that were captured and sacrificed (Temple 1974), and finally on free-living birds that were repeatedly captured for blood samples and returned to the wild (Wingfield and Farner 1976). One of the prominent and most productive leaders in this field has been John C. Wingfield. He and several of his colleagues and students have extensively investigated and documented the roles of several hormones in various reproductive behaviors for one or both sexes of a number of different species including the starling (Ball and Wingfield 1987), white-crowned sparrow (Wingfield and Farner 1978), house sparrow (Hegner and Wingfield 1986 a,b, 1987 a,b), song sparrow (Runfeldt and Wingfield 1985), and in comparative studies of several species (Sherwood et al 1988). In one experiment, treated free-living song sparrows were given estradiol implants and controls were given empty (placebo) implants. The treated females and the untreated males mated to them continued their reproductive behavior for up to 3 months longer than the controls. The males of the treated females continued to have elevated levels of testosterone and maintain their aggressive territorial behavior toward other males or experimental playbacks of tape-recorded song after other males stopped responding territorially. These and related studies have also contributed to our understanding of the role of hormones

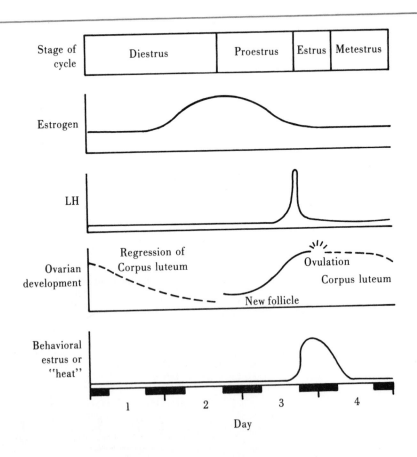

Figure 18-15 Normal estrous cycle and behavior of the female white rat. Black bars indicate periods of nighttime darkness. Modified from McEwen. B.S. 1976. Scientific American 48-58.

in learning in birds (e.g., song learning in adulthood, Nottebohm et al 1987), as discussed in Chapters 19 and 20.

The general pattern of estrus in mammals (e.g., Figure 18-15, for white rats) is well known; the hormonal and ovarian changes are accompanied by particular behaviors. The females of several species of mammals display a rigid downward-arched posture known as *lordosis* during estrus (or heat), at which time they are receptive to males. At other times females may not only be unreceptive but even aggressive toward males and actively reject any attempts to copulate. In at least one lizard species *(Holbrookia propinqua)*, females display stereotyped behaviors, bright colors, and aggressive behavior toward males during stages of the ovarian cycle where copulation is not appropriate. These changes of color and behavior are induced by combinations of particular hormones (Cooper and Crews 1987).

Details of the interaction between reproductive behavior and hormones vary considerably, depending on the extent and nature, if any, of parental care, whether eggs are laid externally or developing embryos are carried internally (e.g., most mammals), and the timing and patterns of ovulation and embryonic development (Crews 1987). In mammals, for example, there are different patterns of ovulation in relation to courtship and copulation. In some mammals with induced ovulation, including rabbits and a few species from some other orders, ovulation is triggered a few hours after copulation. In most other species, however, ovulation occurs at periodic intervals or in response to particular cues from the physical environment. Some birds such as jungle fowl (ancestors of the domestic chicken) and some waterfowl, unlike the doves, develop and lay eggs whether a male is present or not. For further information on reproduction and hormones in vertebrates, see van Tienhoven (1968), Blum (1968), Crews (1987), Norris and Jones (1987), or an endocrinology text (e.g., Martin 1985, Hadley 1988).

Some of the hormonal effects occur early in the life of the animal and produce permanent changes in the organizational anatomy and physiology of the nervous system. This is particularly important in the sexual development of some animals such as the mammals, some birds and reptiles, and many insects among the invertebrates. Inherited genetic predispositions direct early hormones, which in turn direct the eventual expressed sex of the individual. In many other animals (both vertebrates and invertebrates), however, sexual expression is much more labile, as with the reef fishes or in several of the reptiles. The sex of many turtles, crocodilians, and some other reptiles, for example, depends on the temperature at which their eggs are incubated (Raynaud and Pieau 1985).

Departures from the common patterns: snakes in the grass, unisexual lizards, and other vertebrate reproductive diversity.   Although variation on the general theme of vertebrate reproduction has already been mentioned in the previous sections, the diversity is considerable, raises a number of questions about our general understanding, and deserves further discussion. Our original picture of vertebrate endocrine-behavioral interactions was based primarily on studies of a few species of birds and mammals, with occasional studies of other species, such as the green anole, that seemed to confirm the pattern. However, later studies of a wider variety of species have stretched the basic pattern almost to the breaking point.

Red-sided garter snakes *(Thamnophis sirtalis parietalis)* in Manitoba, Canada, for example, show a unique separation of sexual behavior—including mating—from the gonadal and endocrinological events (Crews and Garstka 1982, Crews

Figure 18-16 Mating ball of red-sided garter snakes. A large number of males have gathered around a female emerging from hibernation and are attempting to copulate.
Photograph courtesy David Crews.

1983 a, b). These snakes mate immediately on emergence from their winter hibernacula or dens. Emerging females are surrounded by large numbers of males, forming mating balls (Figure 18-16). What is particularly remarkable, however, is that mating occurs before either the testes of the males or the ovaries of the females have developed (Figure 18-17). Furthermore, the mating behavior of the males does not appear to depend on hormones in the familiar fashion. Castration of males does not prevent mating, and injection of testosterone does not induce mating in otherwise inactive males. Male behavior instead is triggered by warming after a period of several weeks of cold. Adult males that have been warmed after the cooling period engage in sexual behavior regardless of the presence of testosterone or a large number of other hormones that were tested (Garstka et al 1982).

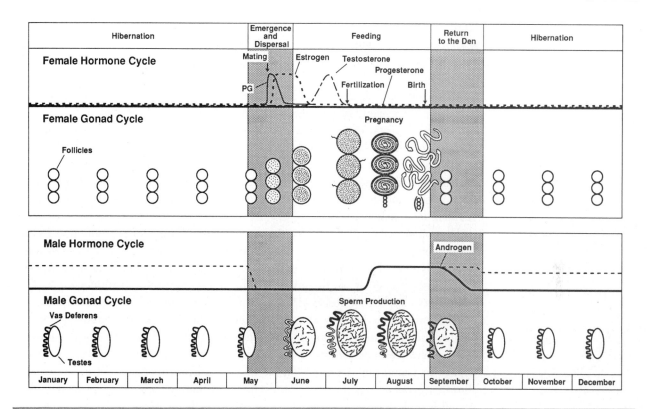

Figure 18-17   Annual reproductive cycles of red-sided garter snakes. Note that mating occurs before the testes and ovaries develop. Also compare with Figure 18-11.
Figure courtesy David Crews.

Mating before the gonads of either sex have developed seems like a useless exercise. However, it obviously works as there are lots of these snakes, sometimes several thousand in a relatively small area. How it works is that during copulation the males inseminate with sperm that was produced the year before and stored over the winter in their vas deferens. After insemination, the females store the sperm in their reproductive tracts until later in the summer when their ovaries have developed and the ova are ready for fertilization.

There may still be some other unknown hormonal factor in the blood of these male snakes that is involved with mating behavior. Blood serum taken from a sexually active male and infused into a noncourting male leads to courtship behavior in a few days (Crews 1983). Also, patterns of male garter snake behavior may be organized in the brain early in life, somewhat similar to the situation in mammals (discussed later). If either castrated males or females are given slow-release testosterone implants shortly after birth, they court adult sexually attractive females in 4 weeks, behavior that does not otherwise occur in newborn or juvenile male or female snakes at any time. Furthermore, if the hormone is removed and the animals are hibernated, only those castrated males that received androgen court females on emergence. This indicates that the early infusion of androgen organizes the brain so that the change in temperature in the spring activates mating behavior.

Curiously, female red-sided garter snakes show unexpectedly low levels of progesterone during gestation and significant cycles of testosterone, with a peak during vitellogenesis. These observations are not well understood and require further research (see results and discussion in Whittier et al 1987). Female red-sided garter snakes also do not show the sustained elevated levels of corticosterone that are typically seen in wild vertebrates after capture and related stress. Whittier et al (1987) noted that this species is easily bred in captivity and another "stress-related correlate, aggression, is completely absent in red-sided garter snakes."

In other aspects, except for a lack of territoriality or aggressiveness on the part of either males or females, the snakes' behavior is more-or-less familiar. That is, females secrete a pheromone that attracts males; the successful male deposits a cloacal plug in the female during copulation that both physically and behaviorally (by a pheromone) repels other males; and male behavior seems to stimulate female gonadal development.

The important point raised by the garter-snake pattern of reproduction is that sexual behavior and reproductive chemistry may be at least somewhat disassociated from each other. The separation of behavioral sex (or proception, Beach 1976) from gonadal sex is also shown by unisexual* lizards of the genus *Cnemidophorus*. This genus includes both duosexual* and unisexual species. The unisexual species consist entirely of females that reproduce parthenogenetically, that is, without fertilization by males. Although males are not present, however, the females alternate between malelike and femalelike behavior and, when behaving as males, engage in pseudocopulation. That is, they mount on the backs of other females showing receptive female behavior in typical fashion for other closely related duosexual lizards (Figure 18-18). These lizards have been investigated extensively in a series of studies by David Crews and his colleagues (Crews 1982 b, 1983 a, Moore et al 1985). The unisexual lizards display receptive femalelike behavior primarily during the vitellogenic stage of ovarian development and malelike behavior

---

*To avoid ambiguity with the terms *homosexual/heterosexual/bisexual*, the following are used here (Crews 1982 b): *duosexual*, functional male and female with fertilization of the ova by sperm (also termed *gonochoristic*); *unisexual*, parthenogenetic reproduction by females with unfertilized eggs; *asexual*, reproduction without meiosis.

A   B

Figure 18-18   Copulation and pseudocopulation in **A**, duosexual and **B**, unisexual *Cnemidophorus* lizards.
Photographs courtesy David Crews.

primarily following ovulation. Further, femalelike behavior can be stimulated by administering estrogen, and malelike behavior can be stimulated with progesterone or testosterone (Crews 1982 b).

Unisexual lizards maintained alone, without the presence of other male-acting females, reproduce successfully. However, their productivity is markedly lower than when exposed to male-acting females. Thus it appears that pseudocopulation stimulates or facilitates reproduction, and the behavior of a "mate" seems to affect the internal physiology in unisexual species just as it does in duosexual species. Crews (1982 b) has even speculated that behavioral sex may have evolutionarily preceded gonadal sex by offering selective advantages of social synchrony of reproduction rather than the other way around, that is, rather than the behavior evolving later to synchronize the sexes.

Atypical reproductive patterns do not occur only among reptiles. In a few birds, such as some of the shorebirds and in a few other species, the males are primarily responsible for incubation of eggs and parental care of the young. The females display intense intrasexual selection for males, may establish and fight for territories, and frequently are polyandrous (Chapter 12). These sex-role reversals, however, have not been shown to involve a simple reversal of the normal hormonal conditions. The respective male and female steroids are found essentially at the expected normal male and female levels (Fivizzani and Oring 1986, Fivizzani et al 1986, Oring et al 1988). Hormonal effects related to the malelike behavior of the females, rather than being caused by malelike absolute levels of testosterone, may result instead from increased sensitivity of receptors in the females to the normally low levels of testosterone that are present. Mated female spotted sandpipers (*Actitis macularia*), for example, show significantly higher levels of testosterone than unmated females, but they are still much lower than in male spotted sandpipers (Fivizzani and Oring 1986). Prolactin has been shown to be higher in male spotted sandpipers during incubation than in females, in somewhat of a reversed condition from other species; however, prolactin is involved in incubation and parental care in both sexes and is normally stimulated by the presence of eggs or young, and the differences may be a consequence rather than a cause of the reversed roles.

Sexual differentiation of the mammalian nervous system.   In mammals the female and male nervous systems differ in the production of particular behaviors and the patterns of neurosecretion. LH production in females, for example, is cyclic with prominent peaks, such as just before ovulation (see Figure 18-15, page 656), whereas LH is produced at more constant levels in males. The sexual differentiation of the mammalian brain, in the few species that have been studied, occurs somewhere around the time of birth. In humans it is just before birth. In rats it occurs a few days after birth, which has made the phenomenon easily accessible for research outside the mother rat's body.

The sexually undeveloped rat brain is female. Development of the male brain results in response to a sudden spurt of testosterone secreted by the young male's testes around the time of birth. If a male is castrated just before the event, its behavior remains feminine for the rest of its life. It will not become aggressive toward other males and will show, for example, the female lordosis response if given estradiol. A male that is castrated but then artificially given testosterone at the right time behaves as a male later if supplemented with the proper adult male hormones. A female, on the other hand, if given testosterone at the critical period, develops a masculinized brain. The results of these experiments are illustrated in

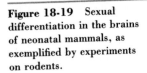

Figure 18-19 Sexual differentiation in the brains of neonatal mammals, as exemplified by experiments on rodents.

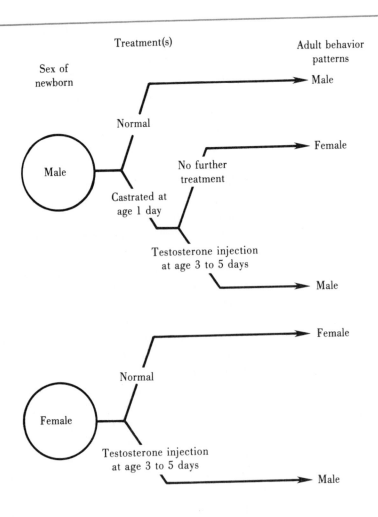

Figure 18-19. Female fetuses located next to males in the uterus sometimes become partially masculinized (Meisel and Ward 1981, Saal et al 1983).

The masculinization of the brain, incidentally, is slightly more complex than just described, and testosterone is only the intermediate chemical in the process. Paradoxically, it is estradiol, a "female" sex hormone that actually causes the masculine development. This comes about by testosterone entering the neurons in the appropriate regions of the brain and being converted intracellularly into estradiol. Why does the female's own estradiol, in young females, not cause the same effect? First, the levels of estradiol in young females are low, and second, the liver of females produces an estrogen-binding protein, alpha-feto-protein, which binds up any endogenous estradiol and prevents it from causing male differentiation in the female brain (McEwen 1976). The enzyme is produced only for a short period of time, which occurs during this critical period of the brain's development.

Stress in Mammals and Birds

Stress is a subject that has received much attention and investigation, particularly in mammals, from a physiological standpoint, and partly from psychological interests in the stresses of contemporary life in western human culture. Stress generally refers to the perturbations or displacements of a system from its normal position.

Stress to some degree is a natural part of life and involves numerous uncertainties and vagaries in the environmental conditions to which all organisms are exposed. A late spring snowstorm in temperate latitudes, for example, can disrupt the normal feeding, reproducing, and other activities of animals that get caught in the storm.

Relatively immediate, transient stresses or emergencies in mammals, where the phenomena have been studied most thoroughly, are handled by a combination of the sympathetic side of the autonomic nervous system and two important hormones from the medulla of the adrenal gland: epinephrine (adrenalin) and norepinephrine (noradrenalin). The sympathetic responses generally heighten the senses (e.g., dilate the pupil of the eye), increase respiration and heart rate, inhibit digestive activity, mobilize energy resources within the body (e.g., release glucose from the liver), and redirect blood flow away from less critical areas (e.g., digestive system to critical muscles). Such responses can be rapid (milliseconds) because they involve the nervous system. Some of the nerves of the sympathetic system connect to the adrenal medulla where they stimulate the release of epinephrine and norepineph-rine. These hormones supplement and extend the various neural effects.

Epinephrine is involved in the release of glucose from liver cells and fatty acids from fat cells, increases heart rate and stroke volume, and affects smooth muscle cells in the circulatory system involved in the redistribution of blood flow. Nor-epinephrine, which is released independently of epinephrine, is involved in in-creasing the blood pressure. Along with the other neural responses (such as in-creased heart rate), the nervous system of animals in emergency situations also usually invokes specific flight-or-fight behavior patterns, that is, escape or aggres-sion, depending on the circumstances.

Long-term stresses (after-effects of short-term emergencies in some cases) involve a different route and different hormones. Senses and central neural integration lead to neurosecretion of releasing hormones from the hypothalamus that in turn stimulate the adenohypophysis to release ACTH (adrenocorticotropin). ACTH circulates through the blood and stimulates the cortex of the adrenal gland to produce and secrete a variety of steroid hormones, corticosteroids, that affect glucose metabo-lism, salt and water balance, and the immune system. Importantly for our interests, these hormones also directly or indirectly affect behavior and, reciprocally, can be affected by behavior.

Along with other, nonsocial sources of stress such as inclement weather, insuf-ficient food, and aversive factors such as predators, an animal's social environment may create stress. Social stress can include isolation and too few social interactions, too many social interactions such as from crowded conditions, negative interactions, and low social status. These social factors have long been associated with the production of corticosteroids in mammals and birds (Selye 1956, Christian 1960, Christian et al 1965, and numerous others [e.g., references in Wingfield et al 1982 and Wingfield and Silverin 1986]). (The anatomy of the mammalian and avian adrenal glands is different but the hormones are homologous and their molecular structures are similar.)

Regardless of the source of stress, whether social or from other factors, there often is also an inhibition of reproduction and reproductive behavior, including reduced levels of LH and (particularly in males) reduced testosterone. It has been suggested that corticosteroids may directly inhibit either the gonads or the hypo-thalamic-pituitary-gonadal pathway.

However, as is the case with reproduction, there is much variation, depending on species and even among individuals. Factors that are stressful for one, such as temperature extremes, may have little effect in another. There may be variation in stress response, depending on the season. Wingfield et al (1982), for example, discovered that white-crowned sparrows responded to various stresses differently during the winter than during the breeding season and, in some cases, differently than observed in other species of birds—particularly domestic species.

The connection between corticosteroids and reproduction is not direct but only correlational in at least one species. Wingfield and Silverin (1986) experimentally implanted corticosterone, one of the corticosteroids, in one group of free-living male song sparrows while giving a control group only empty implants. The males were then challenged with simulated territorial intrusions by placing caged male song sparrows into their territories and playing tape-recorded sparrow songs. The researchers also periodically captured the males to weigh them, assess their fat deposits (visible through the skin), and sample their blood for a variety of hormones. The birds with corticosterone implants had higher levels of corticosterone—as expected from the treatment—and showed greatly reduced territorial behavior. They also had increased levels of body fat. However, their levels of LH and testosterone were not significantly different and their overall body weight was not changed. Wingfield and Silverin concluded that corticosterone was affecting territorial behavior independently of testosterone. That would be advantageous if it meant that a stressed bird could quickly resume reproduction with its hypothalamic-pituitary-gonadal system ready at full operating level after the period of stress passed, rather than having to restart a depressed and possibly refractory reproductive system which could take weeks or even require waiting for the next reproductive season. The curious observation that the corticosterone-treated birds fed more and increased their fat deposits without changing overall weight required some explanation. Wingfield and Silverin speculated that the increased feeding and weight from fat deposition were counteracted by a weight loss from glyconeogenesis and breakdown of muscle protein caused by corticosterone.

Although there is an association between corticosteroids and behavior, the full pathway and mechanism of action (i.e., directly through particular neurons and brain centers or indirectly through other hormones or something else) was still not known at the time this book was written. However, it is an area of active research and much information may be forthcoming in the near future.

MOTIVATION, PLEASURE, PAIN, AND EXOGENOUS MOOD-ALTERING CHEMICALS

As probably everyone who has passed the age of puberty or owned a dog or cat in heat knows, sexual behavior is a strong and often difficult desire to suppress. At the end of Chapter 1 we mentioned that most animals, including ourselves, do not engage in sex for the deliberate or conscious purpose of producing offspring. To the contrary, humans often take extraordinary measures through birth control and abortion to avoid producing offspring. Why, then, do animals engage in sex? At least in our own species in which we can describe the sensations, people do it because it feels good; that is, it produces pleasurable sensations. Much the same can be said about many of the other things we do, including eating and a number of compulsive behaviors. What is the basis of those feelings, and do they exist in other animals as well?

Konrad Lorenz and other ethologists were interested in drive and action-specific energy or potential (Chapter 2). Many psychologists, meanwhile, were attempting to understand motivation. The animal and its nervous system remained a black box, and no one could quite put a finger on what was driving or motivating any particular behavior. The problems of dealing with internal, subjective feelings and anthropomorphism, as discussed previously (Chapters 1 and 2), caused many persons working with behavior to shy away from the area of drive and motivation. (Some ethologists and psychologists were not interested in those topics in the first place.)

We still are far from a complete picture of motivation and the proximate causes of behavior. However, much progress has been made, several pieces of the puzzle seem to be falling into place, and we may be getting closer to the real basis of feelings, emotions, and other aspects of motivation. For many behaviors, it is now clear that internal chemistry plays an important role. This chapter has already provided several examples, such as the role of angiotensin II in thirst and drinking (of water, not alcohol—which is a separate issue), progesterone in incubation behavior in doves, and estrogen in contributing to the receptivity of female anole lizards.

Not only do chemicals affect specific categories of behavior, such as drinking, migration, and reproduction, but chemicals influence and integrate behavior in general ways. An animal's overall moods and levels of motivation, for example, may be affected significantly by specific chemicals. Of great interest during recent years have been findings related to the phenomena of pleasure and pain. The subject has been of particular relevance to humans because of problems related to drug use in human culture, especially in present-day Western society. Both legal and illegal drugs have assumed an increasing importance in human daily living and welfare. Ironically, research into those problems has led to whole new insights into the basic chemical machinery of behavior.

Research has progressed along several lines. One of the first insights into internal mechanisms came in the 1950s when Olds and Milner (1954) discovered the so-called pleasure centers in the brain. Rats with electrodes inserted into these regions would press bars that resulted in self-stimulation of those areas with electrical impulses. No other reward was required; the rats pressed the bars simply to stimulate those regions of the brain.

Along another line of inquiry, it was asked why the brain should be so receptive to certain narcotic drugs such as opium and morphine. Why, for example, would evolution have produced receptors in the nervous system with the molecular specificity for these particular chemicals derived from plants? It was hypothesized that it was not for external chemicals but rather for naturally occurring molecules of a similar nature that were already within the system. Searches soon turned up several such molecules. They have been named *endorphins* (for endogenous morphine) and *enkephalins* (for in the head). There are several different molecules in this class, varying in size and shape.

These chemicals are internal, naturally occurring morphines or opiates. They only resemble the external opiates, however, and there are some important differences. For one, they break down more rapidly in the body, and the effects from particular molecules do not remain as long. For another, they are natural parts of the system and in a sense belong and fit better in the whole integration process. Because of this, they may be required for normal operation of the nervous system.

Figure 18-20 Hypothesized pain pathway in the central nervous system of vertebrates. P substance carries pain information from sensory neurons to neurons in the pain pathway of the spinal cord. Release of P substance may be blocked, however, by presynaptic enkephalins released from neurons in a pathway from higher centers such as the pleasure center. Externally introduced narcotics may simulate the enkephalins, and naloxone will occupy the enkephalin receptors but not block the release of substance P with natural or external analgesic routes. This enkephalin-P substance pain pathway, however, is only part of the total picture of pain.

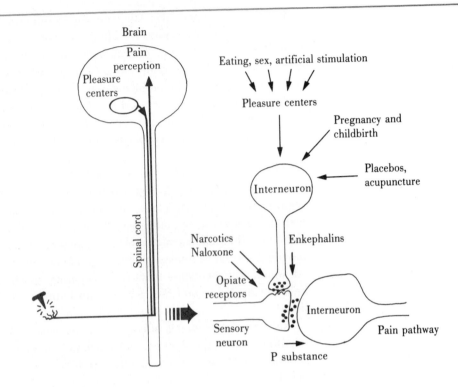

In this sense, every individual vertebrate is drug dependent. (A similar opiate system has also been found in snails; Kavaliers et al 1983.)

Under normal conditions these naturally occurring chemicals are thought to function in two main areas: motivation and control of pain. They may be important in motivation by providing the mechanism, through the pleasure centers, for stimulating an animal to do something. Certain behaviors are rewarding or pleasure producing. These may include such fundamental behaviors as eating and sexual activity.

In the control of pain, some of the chemicals apparently serve as natural analgesics that reduce the perception of pain by inhibiting the sensory pathways of pain up to the brain. This mechanism, as currently interpreted, is illustrated in Figure 18-20. Pain has an important function and is highly advantageous. It warns that something is wrong and corrective behavior is required. When pain is dulled, one is only covering over the real problem, which may remain or even get worse.

Why should evolution have produced a natural pain suppressant? It has been proposed that in some instances, such as childbirth in mammals, the pain cannot be avoided and the signals should simply be suppressed (pregnancy and childbirth are important events associated with the release of natural opiates). In other situations, there may be critical emergencies where survival itself is at stake and the animal might die if it were to stop and pay attention to the pain; in this instance it would be more advantageous to suppress the pain and wait until later to deal with the problem. In other words, the short-lived natural, internal opiates might get the animal through in a tight pinch.

The opiate receptors are located, among other places, on the axons of sensory neurons in the spinal cord. These receptors also bind to two other types of chemicals: exogenous narcotics and an intriguing chemical called *naloxone*.

Naloxone has proved extremely useful as a key to unraveling part of the story about the mechanism of pain and pleasure. Naloxone, although it fits the receptors, does not produce the narcotic effects; it simply plugs the sites and prevents other chemicals from occupying the receptors and producing their effects. Thus in the presence of naloxone, the pain pathways are not inhibited and the pain is felt.

The implications of these findings although still somewhat speculative, are many and may be significant. Not only may several drug addictions be related, but many other compulsive behaviors may have a related underlying mechanism. For example, compulsive overeating or compulsive sexual behavior may be indirectly related to heroin addiction. The difference is that the opiates are being turned on internally via the pleasure centers rather than being received from outside through a needle under the skin. The behavioral problems can be nearly as compulsive and difficult to correct in either case. This is speculative, however, and we have an incomplete understanding of the whole mechanism. Furthermore, the opiate pathways do not account for all pain suppression or analgesia. Some types of analgesia appear to involve pituitary hormones, with or without opiates, and some involve neither hormones nor opiates. Thus at least four categories and presumably mechanisms of natural analgesia exist (Watkins and Mayer 1982). (For further information on analgesia and details concerning the endorphins, see Terman et al 1984 and Casy and Parfitt 1986.)

The complete picture, particularly in human behavior, appears to go far beyond a simple interpretation of pain and pleasure. Some people are addicts, and some are not. With the normal operation of pleasure and pain, some persons become compulsive and some do not. Why are there differences? In the case of alcoholism, for example, there are numerous factors that involve a complex of family and social interaction and developments during the life of the individual. The neural mechanisms of alcoholism, incidentally, although being intensively studied, are still poorly understood and may be much different than in other types of addiction (Barnes 1988).

The opiate pleasure/pain-suppressant system may be involved in many, perhaps even most, "high-producing" phenomena, including many of the pleasures of daily living. This system may come into play in producing pleasure after long-term participation in otherwise uncomfortable activities such as jogging. Additionally, the endorphins are thought to be responsible for placebo effects (where persons feel less pain because they think they have gotten a pain-killer when they actually did not) and phenomena such as acupuncture, where pain is suppressed by placing pins in particular positions in the skin. Naloxone, incidentally, blocks both the placebo and acupuncture effects.

It has been proposed that some drug addiction may result from exogenous narcotics replacing and suppressing the natural systems. The normal operation of the nervous system apparently depends on at least a minimum level of these types of chemicals. When the external sources are denied, the natural sources may have been suppressed and are not present, or are present at insufficient levels; the nervous system no longer operates properly; and the individual suffers withdrawal and even death in extreme cases.

It should be emphasized that not all mood-altering chemicals operate through the opiate system just described. Many exert their effects at other points in the complex set of neural and endocrine interactions. Some resemble synaptic transmitters and stimulate postsynaptic receptors, as if there were increased firing of presynaptic neurons, and the circuits are hyperactivated. Mescaline, for example,

David Crews

David Crews

Behavior—Neuroendocrine Adaptations

A concept of the ecology of mechanisms is developing in behavioral endocrinology. Such a perspective helps account for the many significant differences observed in the neuroendocrine controlling mechanisms of common laboratory and domesticated animal species, as well as wild species. Because the brain receives and integrates stimuli from both internal and external environments, the neural mechanisms regulating reproduction and sexual behavior ultimately are as much an adaptational response to the environment as any phenotypical character. This is clear from the study of unconventional animal models. This work has shed light on the evolutionary rules that have shaped the neuroendocrine mechanisms controlling reproductive processes and revealed something of their origins. By using a comparative and multidisciplinary approach, we find that although different species use homologous neural structures to mediate fundamental units of species-typical mating behavior, the neuroendocrine mechanisms controlling the expression of sex-typical behavioral and physiological processes are not universal.

A key to understanding species variation in neuroendocrine controlling mechanisms lies in appreciating the constraints that have given rise to the diversity in reproductive patterns. Reproduction is constrained by the physical and social environment, rate-limiting steps in certain physiological systems, and the organism's evolutionary history. These constraints have influenced various components of the reproductive process, namely the production of gametes, the secretion of sex steroid hormones by the gonads, and the timing of mating behavior, and hence the neuroendocrine mechanisms controlling each of these processes. This can be illustrated by considering naturally-occurring variation in seasonal reproduction.

Environmental constraints. Most animals are seasonal breeders, reproducing during those periods most propitious for the survival of both parent and young. Environmental constraints on reproduction are especially severe when the environment is (1) harsh but predictable, allowing only a brief favorable period for reproduction or (2) harsh and completely unpredictable. Under these conditions, animals must respond rapidly and directly to physical changes to mate, to reproduce, and for the young to grow sufficiently to survive the upcoming harsh conditions. In both kinds of environments the breeding season typically is extremely brief and the species exhibit explosive or opportunistic reproduction. Evidence indicates that specific physical or behavioral cues, rather than increasing concentrations of sex steroid hormones, trigger sexual behavior in such species.

Physiological constraints. Physiological processes may also constrain the evolution of reproductive mechanisms. I consider just two aspects, namely the time necessary to grow a gamete and the temperature dependence of gonadal activity. In most of the male mammals and birds studied to date, spermatogenesis apparently cannot be compressed into less than 6 weeks. Some mice and ground squirrels can produce mature sperm in as little as 31 days, but they appear to be the exceptions to this rule. A similar time constraint applies to the female, although some small rodents are capable of generating ova in as little as 3 weeks. In ectothermic vertebrates, it is common for gamete maturation to take many months or even years.

The developmental constraint of a 6-week period for gametogenesis does not pose problems for species living in environments in which favorable conditions are predictable and/or prolonged. Hence, it is common to find that gonadal activity and mating behavior are temporally associated in vertebrate species living in temperate and tropical regions. Animals exhibiting such an associated reproductive pattern have been the most frequently studied vertebrates.

The second aspect is the temperature dependence of gametogenesis and steroidogenesis. In extreme environments, many nonmigrating mammals hibernate. At the temperatures characteristic of deep torpor in mammals (e.g., $\leq 5°$ C), spermatogenesis and androgen secretion does not occur. Thus, in hibernating endotherms living at high altitudes and/or latitudes, such as turkish hamsters and golden-mantled ground squirrels, gonadal activity occurs during periodic arousals during hibernation. However, ectotherms in extreme environments do not exhibit periodic arousals as in the case in mammals. For example, red-sided garter snakes are active, albeit slow, throughout hibernation. It is well established that ectotherms have evolved species-specific thermal preferenda for metabolic activity with both upper and lower thresholds. In temperate ectotherms, gametogenesis is inhibited and the gonads do not respond to gonadotropin stimulation at the low temperatures characteristically experienced during the winter. However, it is clear that in extreme environments with very low or very high temperatures, ectotherms reproduce successfully (e.g., the body temperature of the tuatara rarely exceeds 12 ° C).

In many species living in harsh but predictable or harsh but completely unpredictable environments, mating must occur as early in the spring as possible so that young can be produced by late summer. In some species, implantation is delayed or there is an embryonic diapause. In other species, particulary ectothermic vertebrates, it is common for gamete growth to be temporally dissociated from breeding activities. In some species this dissociated reproductive pattern is extreme as in garter snakes in which gonadal recrudescence occurs only after all mating activity has ceased. In the male the sperm used during mating were produced the previous summer and stored, whereas in the female, mating induces ovarian growth.

Phylogenetic constraints. The evolutionary history, or phylogeny, of the species acts as a third category of constraints predisposing future trajectories and not others. Simply put, what has come before determines to a large extent what will follow. If closely related species share a similar reproductive pattern despite living in different environments, then it is likely that the underlying neuroendocrine mechanisms are fundamental to successful reproduction in this group.

Research focusing on neuroendocrine adaptations have shown that (1) gonadal growth and sexual behavior are not necessarily functionally associated, (2) sexual behavior need not depend on increased levels of gonadal steroid hormones, (3) the inititation, maintenance, and termination of sexual behavior are independent and controlled by different cues, and (4) different proximate cues can be used by males and females of the same species to regulate similar reproductive (behavioral) events.

resembles norepinephrine and dopamine. LSD and psilocybin resemble serotonin. LSD is particularly potent, although its exact mode of action is not known. The most widely used stimulant drugs, methylxanthines (e.g., caffeine and theophylline, which are found in coffee and tea), compete with adenosine in the brain and reduce adenosine's normal inhibiting effects on neurotransmitters, which leads to increased neurotransmission and neuron firing. Several drugs, such as cocaine and a number of pharmaceutical antidepressants, operate by blocking the normal metabolism and uptake of certain transmitters at the synapses, which amplifies the effects of those transmitters. Many of these substances, incidentally, are derived from plants where the effects may be either coincidental or, more likely in most cases, have an evolutionary origin as general defenses against herbivores.

There are several other examples of internal chemicals that clearly or apparently are involved in the process of motivating behavior. Some of them may also be associated with pleasure and pain, as indicated previously. One more case we discuss involves the control of feeding behavior. That, however, also involves considerations of development and learning in many species. Thus we stop at this point, look at development and learning in the next two chapters, and return to the subject of feeding behavior in Chapter 21.

CONCLUSIONS AND EVOLUTIONARY CONSIDERATIONS

It is clear that many of the chemicals in an animal's body act on the nervous system to affect behavior. Behavior, through the senses and nervous system, reciprocally influence much of the body's chemistry. The overall functional actions and interacting pathways of behavioral hormones, other neurochemicals, and neurons are best understood in the three groups of the most complex animals: vertebrates, arthropods, and mollusks—in the order listed (i.e., vertebrates have received the most investigation and are best understood). The general pattern that emerges from these diverse groups of animals is shown in Figure 18-21. In short, the brain processes and integrates information, then secretes neurochemicals that either affect subsequent neural processing or travel by the bloodstream (-hemal or -humor) to stimulate other secretory tissue, the products of which may feed back on and affect the nervous system. This frequently involves a sequence of several interacting glands or neural sites (Figure 18-22).

The importance of a close interface of the nervous system and its short-term responses with distant endocrine glands and longer-term responses in animals with the most complex behaviors can be seen by the independent evolution of distinct neurohemal organs such as the hypothalamus-pituitary complex of vertebrates and brain-corpora cardiaca-corpora allata complex of insects (see Figure 18-23) and other analogous structures in crustaceans (Figure 18-8). Animals with less complex behaviors also have correspondingly less complex structures, although neurosecretion and specific endocrine glands (e.g., the neurosecretory bag cells and atrial gland of *Aplysia*) are still present. However, our understanding remains fragmentary. Many of the techniques such as radioimmunoassay are relatively new. They also are expensive, require much time and effort to understand and apply, and may require a relatively large sample size of animals at some point. Much research remains to be done among a wider variety of species under natural, free-living conditions in those groups already receiving the most attention (mammals, birds,

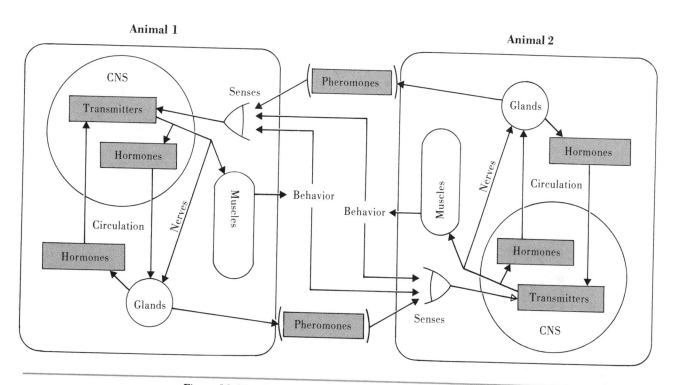

Figure 18-21 General relationships between the intra- and interanimal components of behavior and chemicals. The three major categories of behaviorally important chemicals are transmitters, hormones to and from the nervous system, and pheromones between animals. The behavior of an animal, resulting from patterned muscular output, may affect, via the senses, chemical interactions both in the behaving animal and in other animals.

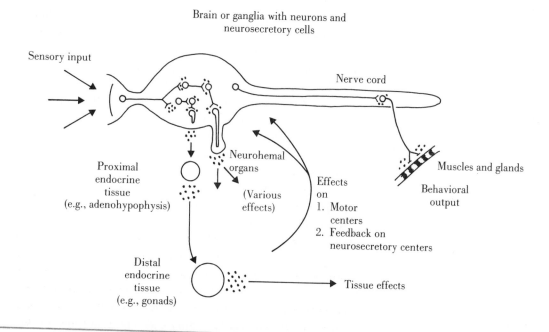

Figure 18-22 Generalized relationships among the central nervous system, neurohemal organs, other endocrine tissue, behavioral chemicals, and behavior.

Figure 18-23   Analogous neurohemal structures **A**, in vertebrates, **B**, compared with insects.
From Orchard, I. 1984. In: A.B. Borkovec and T.J. Kelly. Insect neurochemistry and physiology, Plenum Press, New York.

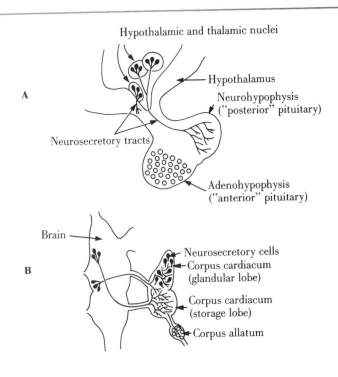

insects) and in other groups such as cephalopod mollusks (which have complex behavior) that have not received so much investigation along the line of behavioral chemistry.

The patterns of evolution regarding the structure of molecules used to carry the internal messages are somewhat variable and only partially understood. Within the vertebrates there are quite consistent patterns and homologies among molecular structures and functions, to the point where some of the hormones are identical or nearly so and virtually interchangeable among the different taxonomic classes from the mammals and birds to the repitles, amphibians, and even extending to the fishes in some cases, although fish were not discussed in this chapter. This is not to say, however, that all specific hormones are the same for all vertebrates. There can be considerable variability in some cases. There are at least five forms of vertebrate gonadotropic releasing hormones (GnRH) from the hypothalamus (mammalian, chicken I, chicken II, pigeon, and salmon). Chickens have at least two forms, as indicated by the names in the list, and starlings have been discovered to have more than one GnRH (Sherwood et al 1988). In other words, there can be variability in the molecular structure of a hormone even within a given species.

Steroids have been discovered as useful hormone molecules in two major groups, arthropods and vertebrates. The role of steroids in other groups is less clear; they have not been shown to be important in *Aplysia*, for example. The presence of amino/peptide/protein types of hormones is widespread, as might be expected from the chemical nature of these substances, and there is a large list of hormones, transmitters, and other behavioral molecules that are of one of these types. Molecules in this group frequently or perhaps generally are manufactured in the body from larger precursor molecules that are split up into a variety of products. Juvenile hormone in insects appears to be fairly unique as a terpene.

Apparently what matters most in a message-carrying molecule is that there is a match between the molecule and its respective receptor. As long as the coding-decoding works, the particular shape or material involved in this key-and-lock system is relatively unimportant. Although some types of molecules seem to be better suited than others, they are not exclusive.

There are numerous unanswered questions in the biochemistry of behavior; we have only a small sample of knowledge from the great diversity of species that exist; unexpected surprises keep arising with new techniques, research, and explorations into the subject; and this is an active field of investigation. Our knowledge is likely to continue to expand rapidly in this area in the future.

SUMMARY

Behavior is coordinated not only by the nervous system but also by a large variety of chemicals. Behaviorally important chemicals within the body fall into two general classes: amino acid-derived types (including peptides, proteins, and others) and steroid hormones. The former generally operate through cell membrane receptors and an intermediate messenger chemical, such as cAMP or cGMP, acting within the cell. Steroids enter their target cells directly to exert their subsequent influence. There are also other types of hormones such as juvenile hormone, a terpene.

Chemical transmitters carry information from neuron to neuron within the nervous system. In addition, neurosecretory cells within the nervous system, after receiving input directly from other neurons, and often through specialized neurohemal organs such as the pituitary gland (in vertebrates) or corpora cardiaca (in insects), release hormones into the circulation for other parts of the body. Many types of chemicals alter the properties of neurons and hence the subsequent behavior. Through this mechanism, external and internal stimuli exert an influence on several facets of an animal's life, including growth and development, migration, reproduction, social interactions, and even adjustments to stress.

Similarly, hormones released by other parts of the body, such as the gonads, are received at target neurons within the central nervous system and affect behavior. The interactions both ways, brain to body and vice versa, as well as interactions with other animals, relevant environmental stimuli, and the present internal state of the animal, are numerous and complex. Examples include the control of thirst and water intake behavior, molting in arthropods, and reproduction. The expression of reproductive behavior in many animals, for example, depends on hormones from glands that were stimulated by other hormones from the brain, which in turn were stimulated by sensing the appropriate behavior of a mate and other environmental cues that vary with the species. Examples can be seen in doves, lizards, and others. Analogous hormonal interactions can be found in virtually all multicellular animals.

Chemicals exert their effects not only on immediate behavior, but some hormones impose long-term effects by organizing the brain during critical periods in early development. Some chemicals, such as the endorphins, are involved normally in pleasure (which may serve the proximate function of motivation), and pain (which serves to warn of problems). Some exogenous, externally introduced chemicals, such as the opiates, resemble, simulate, and may interfere with some of these natural neurochemicals. Other mood-altering, psychoactive chemicals exert their effects in other ways.

FOR FURTHER READING

Crews, D. 1983. Alternative reproductive tactics in reptiles. BioScience 33:562-566.
A concise discussion of the differences in reproductive behavior and endocrinology among green anoles, red-sided garter snakes, and whiptail lizards.

Crews, D., editor. 1987. Psychobiology of reproductive behavior: an evolutionary perspective. Prentice-Hall, Englewood Cliffs, N.J.
A collection of reviews on reproductive behavior and its internal neuroendocrine control in various vertebrates from fish to mammals.

Crews, D., and W.R. Garstka. 1982. The ecological physiology of a garter snake. Scientific American 247(5):158-168.
A clearly written, detailed account of the reproductive biology of the red-sided garter snake.

Evans, H.W. 1984. Insect biology. Addison-Wesley, Reading, Mass.
This book includes a good review of the roles of hormones in insect reproduction and development.

Fivizzani, A.J., and L.W. Oring. 1986. Plasma steroid hormones in relation to behavioral sex role reversal in the spotted sandpiper, *Actitis macularia*. Biology of Reproduction 35:1195-1201.
Original, clearly written journal paper on the relationships between steroid hormones and sexual behavior in the spotted sandpiper.

Kelley, D.B. 1988. Sexually dimorphic behaviors. Annual Review of Neuroscience 11:225-251.
A review providing numerous examples, with consideration of the underlying neuroendocrine basis, of sexually dimorphic behaviors in vertebrates.

Krieger, D.T. 1983. Brain peptides: what, where, and why? Science 222:975-985.
A good overview of neuroactive peptides, with emphasis on vertebrates. (Although the article is excellent overall, a few points are overgeneralized or misleading; see the comments and response by Anctil 1984 and Krieger 1984.)

Laufer, H., and R.G.H. Downer, editors. 1988. Endocrinology of selected invertebrate types. Invertebrate endocrinology, vol. 2. Liss, New York.
An up-to-date account of endocrine aspects of invertebrate biology, with much emphasis on neural and behavioral effects, including the regulation of egg laying in Aplysia.

Lehrman, D.S. 1964. The reproductive behavior of the ring dove. Scientific American 211:48-54.
A classic publication on the interdependence of behavior and hormones.

Martin, C.R. 1985. Endocrine physiology. Oxford University Press, New York.
A thorough, clearly written, and up-to-date reference on endocrinology.

Moore, M.C., J.M. Whittier, A.J. Billy, and D. Crews. 1985. Malelike behaviour in an all-female lizard: relationship to ovarian cycle. Animal Behaviour 33:284-289.
An original journal paper on the unisexual lizards.

Oring, L.W., A.J. Fivizzani, M.E. El Halawani, and A. Goldsmith. 1986. Seasonal changes in prolactin and luteinizing hormone in the polyandrous spotted sandpiper, *Actitis macularia*. General and Comparative Endocrinology 62:394-403.
An original journal paper on two more hormones in spotted sandpipers to accompany the one by Fivizzani and Oring.

Raabe, M. 1986. Insect reproduction: regulation of successive steps. Advances in Insect Physiology 19:29-154.
A thorough, up-to-date review with emphasis on the variety of patterns, for persons wanting examples and details of insect reproductive endocrinology and behavior.

Runfeldt, S., and J.C. Wingfield. 1985. Experimentally prolonged sexual activity in female sparrows delays termination of reproductive activity in their untreated mates. Animal Behaviour 33:403-410.
An interesting recent demonstration of the endocrine and behavioral effects on birds and their mates caused by experimentally altering the hormone levels of the females.

Scharrer, B. 1987. Neurosecretion: beginnings and new directions in neuropeptide research. Annual Review of Neuroscience 10:1-17.
A stimulating and superbly written perspective on the history and contemporary status of the field of neurosecretion, written by one of the pioneers and architects of the field. This article conveys a considerable amount of information and understanding in addition to the overview perspective.

Scheller, R.H., J.F. Jackson, L.B. McAllister, B.S. Rothman, E. Mayeri, and R. Axel. 1983. A single gene encodes multiple neuropeptides mediating a stereotyped behavior. Cell 32:7-22.
An original research paper that describes the behavior and chemical relationships to the extent that they are known, and provides a thorough analysis of the genetics and molecular structure of the peptides involved. Although highly technical in parts, the paper is well written and highly readable.

Wingfield, J.C., and D.S. Farner. 1978. The endocrinology of a natural breeding population of the white-crowned sparrow *(Zonotrichia leucophrys pugetensis)*. Physiological Zoology 51:188-205.
An original paper and excellent example of a detailed study using radioimmunoassays and a competitive protein-binding assay (for corticosterone) to correlate several hormones with their associated behaviors.

Wingfield, J.C., and B. Silverin. 1986. Effects of corticosterone on territorial behavior of free-living male song sparrows *Melospiza melodia*. Hormones and Behavior 20:405-417.
An experimental study of the effects of a stress hormone, corticosterone, on reproductive behavior and other hormones.

19

ONTOGENY OF BEHAVIOR

During growth and development, animals (except protozoans) go from single cells through intermediate stages to the point where they are functional organisms consisting of a large number of differentiated cells. When and how does behavior enter this picture? Even after birth or hatching, animals rarely display all of their behavioral repertoire at the start. Different behaviors become manifest at different times in the individual's life, more or less in orderly sequence. Behaviors shown by young animals may be replaced by or overlap with adult behaviors. An enigmatic category of behavior shown in some young mammals and birds is play. Many behaviors in many kinds of animals can be modified by learning processes. What determines all of these various changes, and how do they occur? To address these questions one must deal with the intertwined subjects of ontogeny (the development of an individual), learning, and memory. These topics are considered in this and the next chapter. The primary focus of this chapter is ontogeny.

THE APPEARANCE, DEVELOPMENT, REPLACEMENT, AND OTHER CHANGES OF BEHAVIOR DURING AN INDIVIDUAL'S LIFE

Most animals do not start life with their full set of behaviors. Rather, new behaviors appear at different times and earlier behaviors may change or disappear, as shown in Figure 19-1. The changes may differ among closely related species (for example, compare the trout and salmon in Figure 19-1), and there may be many differences among individuals of the same species.

In some cases a behavior may occur only briefly during a particular stage of an animal's life. Worker honeybees, for example, perform different activities depending on their ages. After first emerging and until 1 to 3 weeks of age, new workers perform hive duties such as building and cleaning comb cells, storing food, and caring for the developing brood. Older workers eventually become foragers outside the hive. In between the stages of being hive bees or foragers, many spend a brief period of 1 to 6 days as guard bees, which wait at the hive entrance, greet incoming foragers, and fly out to inspect and repel any intruders. The ontogeny and characteristics of guard behavior were studied by Moore et al (1987). Guard bees assume specific postures (e.g., antennae forward, wings held out, and forelegs raised) and behave in ways not shown by hive bees and foragers. Guard behavior lasts for such a short period and is so stereotyped that Moore et al (1987) suggested that experience is not involved.

The sudden appearance of apparently fully formed behavior is frequent among many species, including vertebrates, and in some cases has been shown to be resistant to experience and environmental variability. For example, less than 7% of the variance in head-bobbing displays of at least one species of lizard (Sceloporus undulatus) results from experience (Roggenbuck and Jenssen 1986). In many other behaviors, however, experience and learning may play important roles in behavioral changes.

The roles of genes and the environment in forming behavior were discussed in a general way in Chapter 4. How does development take place during the life of an individual animal, and how does it relate to behavioral changes that can occur throughout life, including in adults?

Unfortunately, these questions have no easy answers. Nature, in the course of evolutionary branching, has not stuck with one simple plan for development. Different organisms develop in many different ways. They use different schemes of embryological or asexual development processes, acquire particular characteristics at different stages and in different sequences, and develop fairly directly into adult form in some species and only indirectly—often through quite different larval stages—in other species.

Many people now view learning as a more diverse phenomenon than was formerly suspected. The internal, molecular workings of learning and memory have proved extremely intractable and have provided some of the greatest challenges to biology and psychology.

The ability of biological systems to store large amounts of information in the nervous system is impressive. Consider the role of memory in the daily lives of humans. Regardless of amount of formal education, everyone knows an incredible number of places, facts, and skills. Before the advent of writing, many persons remembered so much of their genealogy that it would take many hours or days to

Figure 19-1 Sequential appearance of behaviors during the lives of individuals of different species. Open bars, common occurrence in most individuals during the indicated ages; dark broken bars, ranges resulting from variability among individuals and preliminary and subsequent traces of the behavior.

Modified from Dill, P.A. 1977. Animal Behaviour 25:116-121; Fernald, R.D., and N.R. Hirata. 1979. Zeitschrift für Tierpsychologie 50:180-187; Candland, D.K. 1971. In: Moltz, H. The ontogeny of vertebrate behavior. Academic Press, New York; Altmann, J., and K. Sudarshan. 1975. Animal Behaviour 23:898-920; Archibald, G.W. 1976. Ph.D. thesis. Cornell University, Ithaca, N.Y.

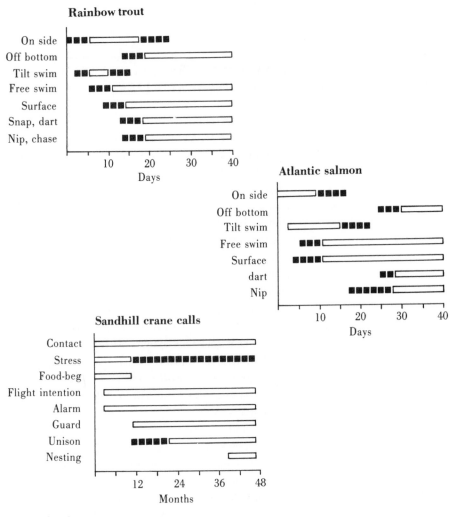

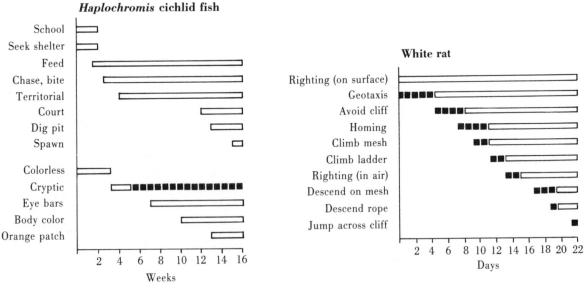

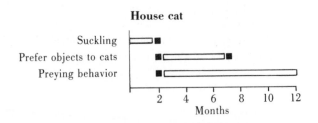

quote it all. Most persons have had the experience of recognizing faces, voices, or musical phrases, perhaps only from the first few notes, after many years since the last exposure to them.

Are the mechanisms of memory and learning the same in all animals? The way an ant remembers how to run through a maze may be vastly different from the way a rat remembers (Schneirla 1959). Either the ant or the rat may tackle spatial problems quite differently from how they deal with learning and remembering in feeding or social functions.

Because of the complexity of the subjects, the amount of available literature, and the usual limitations of a textbook, the material in this and Chapter 20 represents a limited sampling and overview of these subjects. Ontogeny and a few learning topics are covered in this chapter. The remainder of the learning and memory material is discussed in Chapter 20.

EMBRYOLOGICAL AND NEONATAL ROOTS OF BEHAVIOR

The rates and stages of development at hatching or birth vary greatly among animal species. There may be stages variously defined as larval, imago, adult, embryonic, fetal, neonatal, postnatal, immature, juvenile, and subadult. Some young are altricial at birth or hatching; that is, they are helpless on their own and require parental care and attention before they can leave the site of birth or hatching, such as a nest or den. In most species, including the vast majority of invertebrates and vertebrates, the young may be precocial (Figure 19-2). They can move about, feed, and have at least a limited capacity to deal with the hazards of life on their own. Some precocial young appear to be like miniature adults. The terms *altricial* and *precocial* only define points on a continuum. Different species may occur almost anywhere between one extreme and the other.

Figure 19-2 Precocial young. White-tailed deer fawn—20 minutes old. Photograph by Leonard Lee Rue III.

Even the point of birth is somewhat arbitrary; marsupials are born essentially as fetuses and must crawl into the mother's pouch to complete their early development. Because of this tremendous diversity, it is difficult to generalize. Behavioral developments that take place before birth or hatching in some species may occur afterward in other species.

A major difference between vertebrate and invertebrate embryos concerns the contraction of muscles before hatching. Embryonic muscles of vertebrates show considerable twitching, jerking, and other movements. Among amniotes (reptiles, birds, and mammals), the amniotic membrane also contracts and moves. Invertebrate embryos, however, are virtually motionless while packed in the egg and, among those that go through a pupal stage, again in the pupae. Their earliest movements are seen in the larval stages or larval instars. The beginnings of behavior thus usually take place within the embryo or soon thereafter in vertebrates and after hatching in invertebrates. Most of the information until the last few decades came from vertebrates.

Many of the earliest notions about behavioral development, along with development in general, were based on the idea of *preformation*. Under this concept the organism is fully formed and complete even in the sperm or egg, and development amounts primarily to growth. This idea now is of little but historical interest, having been replaced by the basic concept of *epigenesis* (Needham 1959). Epigenesis states that various characteristics, including behavior, are not present initially but form and become visible during development.

Along epigenetic lines, however, there have been several different opinions concerning the early development of behavior. Opposing viewpoints have roughly followed the old nature versus nurture dichotomy. Most of the early workers observed embryos and contemplated what they saw without attempting to do any experiments. Different persons focused on different species and then speculated and attempted to generalize to all vertebrates or all animals.

The nature side of the argument, that early behavior is essentially predetermined, can be traced to 1885 when Preyer published a large volume on behavioral embryology. He and later workers observed that early movements begin before the embryo is capable of responding to external stimuli and that there are rhythmic activities that do not correspond to outside stimulation. One viewpoint that developed, known as *autogenous motility*, stated that the early motions were intrinsic to the muscle tissue or developing motor neurons. The movements were not believed to be either activated or modified by sensory input or feedback.

A view based partly on species with an earlier development of reflex pathways became known as the *reflexogenous concept*. Under this viewpoint neuromuscular pathways are believed to be predetermined, but movements are not shown until pathways develop; movements reflect the pathways; and movement is sensitive to external stimulation. Subdivisions of this and some other views depended on the orientation of the researcher. Neuroanatomists tended to work conceptually from the inside out, viewing the structure as of primary interest and the behavior as only secondary, whereas behaviorists worked from the outside in, considering the structure as of secondary importance. An early neuroanatomist, Coghill (e.g., 1929), proposed that movement begins as a total, undifferentiated pattern, out of which discrete or independent parts develop. Other workers (e.g., Swenson 1929, Windle 1944), still in the reflexogenous camp, suggested it was the other way around—that parts of the behavior develop first, then add together to form functional patterns.

Evidence that accumulated during the 1930s and 1940s made the autogenous and reflexogenous views difficult to generalize. One way to deal with the diverse information was to propose more abstract generalizations, such as systemogenous behavior. This viewpoint (e.g., Carmichael 1963, Anokhin 1964) focused more on the behavioral system and stated in essence that the movements needed for survival under the particular environmental situation would be developed before birth. Emphasis was not on the particulars of how development occurred; it was believed that the outcome could vary from species to species, depending on the environmental factors the species encountered at birth. The various outcomes for the different species were still seen to be predetermined.

Meanwhile, on the nurture side of the argument, there developed a probabilistic viewpoint about behavior in the embryo. The main proponents of this school were Holt (1931) and Kuo (1939, 1967). Building on early information (during the period from 1910 to the 1930s) about nerve growth, such as dendrites growing toward axons from which they received the most stimulation, the probabilistic view stated that early behavioral development was not predetermined and certain. Rather, it was viewed as being only probable, depending on a host of stimulating factors. These included mechanical agitation such as from heartbeats and amniotic contractions, internal and external sensory input, hormones, and musculoskeletal effects of use in conjunction with previous developmental history. The early behavioral substrate was viewed as a blank slate (the tabula rasa associated with Hobbes and Locke; see discussion in Bolles 1979), which became channeled by external and internal events until it developed into an observable outcome. Various outcomes were possible, although some, namely those generally observed, were considered most probable.

Other major contrasts between this view and that of the predeterminists were whether structure only influenced function (predeterminist) or if effects could go both ways (probabilist) and whether development proceeded in discontinuous jumps (predeterminist) or more smoothly and continuously (probabilist). Gottlieb (1970) provides further details and references. The relationships between these and the other schools of thought are diagrammed in Figure 19-3.

Investigation into these issues lessened for a few years, but with improved techniques for working with embryological and neuromuscular tissues, the pace picked up again, and there is now a wealth of recent information. A sampling of these findings follows.

Basic movement patterns in a wide variety of species appear to develop normally, even when they are deprived of the normal sensory input. Examples among arthropods include locust flight (Kutsch 1974) and lobster locomotion (Davis 1973). Chick embryos with cut sensory nerves appeared to display fairly normal activity in their leg movements during most of the subsequent embryonic development (Hamburger et al 1966). These observations support the autogenic viewpoints established earlier on the basis of rhythmic movements and movements that occurred before the establishment of complete nerve-muscle connections.

In a classic demonstration of the presence of developed behavior before any opportunity for sensory feedback or rehearsal, Bentley and Hoy (1970) investigated the species-specific song production of crickets (Teleogryllus spp.). The songs are produced by stridulatory movements of the forewings by adult males. Immature crickets go through 9 to 12 instars, with the wings becoming functional only at the final, adult stage. By lesioning an appropriate part of the brain in late (but not yet

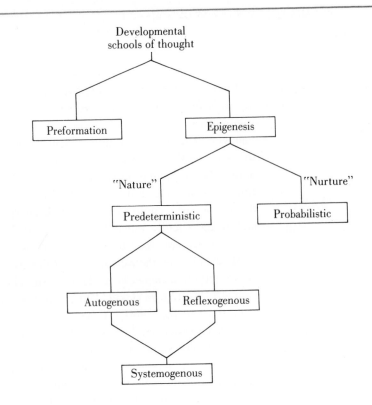

Figure 19-3 Relationships among various schools of thought concerning the ontogeny of behavior. Of the ones shown here and depending on the species involved, the autogenous, reflexogenous, and systemogenous schools have received the most experimental support.

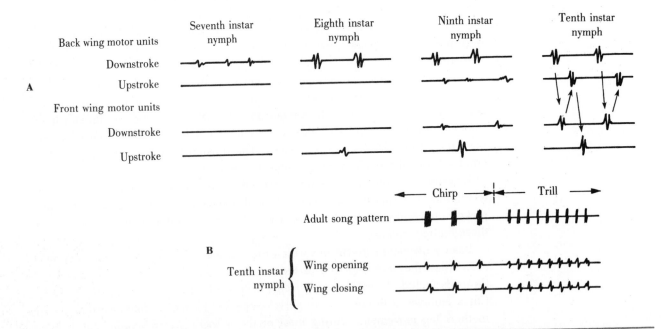

Figure 19-4 Development of nerve networks for flight and singing in crickets is completed during the last few larval stages. **A,** Motor neuron firing patterns in the tethered nymphs placed in a wind tunnel. Although the wings are not developed, the conditions stimulate the motor neurons to fire in typical fashion, as if the wings were functional. At first the pattern is only partial. Then the pattern develops for other muscles. At the final instar, before the adult stage and before the wings emerge fully, the full motor pattern is present. Upstroke and downstroke neurons alternate, and the back wing leads the front wing, as indicated by arrows. **B,** Muscle impulses during calling song: top trace is adult song pattern; bottom is from a tenth instar nymph with a brain lesion that disinhibits the output patterns. Tenth instar nymphs normally do not attempt to sing. Also see Figure 4-3.

Modified from Bentley, D., and R.R. Hoy. 1974. Scientific American 231(2):34-44.

adult) instars, Bentley and Hoy discovered that the larval wings, almost too small to be recognized as wings, were set in motion. Electrodes inserted into the wing muscles showed contraction patterns characteristic of those that produced the song in adults (Figure 19-4). The neural program needed to produce the species-specific song was entirely present in the nervous system but normally was inhibited from expression by another part of the nervous system. The inhibiting region apparently was removed by the brain lesion, which permitted the premature release of the behavior.

Tracing movements back to their earliest origins is difficult for three main reasons. First, there may be several movements occurring simultaneously or nearly simultaneously so that it is difficult to know which ones are related and which are only passive. Second, embryos and their muscular movements are so small that the movements may be almost imperceptible and, even if visible, difficult to photograph or otherwise measure. Third, the size, proportions, and postures of the growing embryo change considerably so that it can be difficult to compare measurements from one time to another.

One solution to these problems is to record electromyograms (EMGs) from individual muscles. This is similar to recording action potentials from neurons. It permits the contractions to be recorded electronically and directly. One can investigate the developing patterns of contractions and compare them with patterns in subsequent, fully developed behaviors. An operational, functional behavior requires coordination at several levels: intramuscular, intermuscular, intrajoint, interjoint, and interlimb. Accordingly, patterns of contraction can be studied at all these levels.

Anne Bekoff studied the development of intra- and interleg motor output patterns of variously aged chick embryos, hatching chicks, and walking chicks in a series of ongoing studies first reported in 1976 (Bekoff 1976), with continuing follow-up publications (e.g., Bekoff et al 1987 and references cited therein). She implanted fine (approximately 50 μm diameter) wire electrodes with long, flexible leads into various leg muscles so normal movements were relatively unrestricted. Her recording setup and a sample of results are illustrated in Figure 19-5. The results appear to demonstrate that patterns of contraction similar to those shown in coordinated hatching movements are already present by at least the ninth day of development, when reflex arcs are not yet present and functional. (Chicks hatch after 21 days of development.) That would suggest that the development of the behavior is primarily autogenous.

However, later studies comparing hatching and walking movements—which both use the same basic intraleg motor circuits—of normal chicks versus those with sensory pathways from the legs cut suggest that sensory inputs indeed play an important role in the development of the leg movements (Bekoff et al 1987). Furthermore, sensory input from the position of the neck, which is bent while the chick is inside the egg and hatching but not after hatching, during walking, influences the motor output. During hatching the chick uses episodic bursts of synchronized kicking movements. During walking the movements are more cyclic and sustained within each leg and alternate between the two legs. The difference in motor output depends on whether the neck is bent and on sensory inputs from the legs themselves. Thus there are elements of both autogenous and reflexogenous development.

Some coordination develops long before it seems to be needed; early coordination

Figure 19-5 Schematic diagram of apparatus used by Bekoff to investigate muscular output of developing chicken embryos. Recording setup and results: similarity of movement patterns between **A**, 9 and **B**, 20 days of development. Intensified movements of these types are used during hatching at age 21 days. Based on photographs and illustrations in Bekoff, A. 1978. In: G.M. Burghardt, and M. Bekoff, editors. The development of behavior. Garland STPM Press, New York.

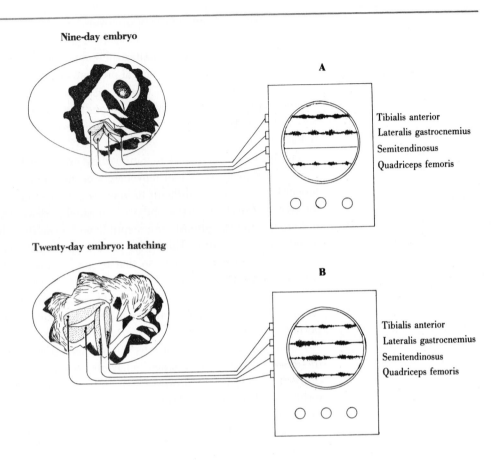

Nine-day embryo

A

Tibialis anterior
Lateralis gastrocnemius
Semitendinosus
Quadriceps femoris

Twenty-day embryo: hatching

B

Tibialis anterior
Lateralis gastrocnemius
Semitendinosus
Quadriceps femoris

may develop without the need for sensory feedback; and coordinated movements may be assembled gradually before they become obvious to more superficial observations. Central pattern generators exist among a wide variety of vertebrates and invertebrates, as discussed in Chapter 17, and some of their outputs may be expressed early and used in some of the first movements of the animal. Sooner or later, however, it generally becomes necessary to modulate the motor outputs with sensory inputs.

Another approach to analyzing the early development of behavior and locating the cells that are initially responsible for directing the behavior is through a technique called *fate mapping*—following and correlating mutant or abnormal lines of cells from early embryonic stages into the adult. The location of the responsible site is known as the *behavioral focus*. Hotta and Benzer (e.g., 1970, 1972) and subsequent workers have constructed fate maps for a number of mutant alleles affecting behavior in *Drosophila melanogaster*. These include abnormal electroretinograms, artificially induced leg movements, flight, and more complex courtship and mating behaviors. In males, for example, courtship orientation and wing vibrating can be mapped to locations in the head, whereas copulation movements map to the thorax. More detail is expected when additional mutants and markers become available.

All these lines of evidence strengthen the inference that many patterns and components of behavior develop early in life and are under significant central nervous system control. This central control, it seems reasonable to assume, is in turn directed by relatively unmodifiable genetic instruction. Thus there is strong support for some form of predetermined development, either autogenous or system-ogenous, in at least some behaviors in some organisms. This seems reasonable to expect. Consider feeding, for example. If an animal did not start with some innate behavior, such as sucking in mammals, it could easily starve to death before learning what to do. For many species, the innate behavior is all there is. Even for animals in which learning is important in the development of their behavior, a stereotyped starting point often forms the basis for subsequent learning. Furthermore, learning—which permits behavioral flexibility and modifiability—is adaptive primarily in circumstances where the environment is relatively variable and unpredictable. The environment of a developing embryo and during early life under the care of a parent (for those species with parental care), however, varies little from individual to individual and hence is predictable. Thus, even in species that possess a significant amount of learning capacity, it is reasonable to expect learning to be less important during the embryonic and earliest stages of life than in later stages.

MATURATION AND BEHAVIORAL CHANGE

Many of the changes that are observed in an animal's behavior can be traced to maturation of the underlying structures. *Maturation* refers to normal changes during growth and development that cannot be correlated with environmental variability and experience per se, but they are still changes and may give the appearance of learning. Other changes, such as refinement and perfection of the behavior, may involve learning or only represent additional maturation.

Practice, as will be seen shortly, is a phenomenon that leads to an improvement in behavior in many cases. It commonly is thought to involve increasing completion, organization, or reinforcement of neural pathways responsible for the specific movements, perhaps through changes in dendritic synapses. Practice may also produce improvement through the exercise effects of strengthened muscular and skeletal tissue. Practice is not well understood, and it is difficult to place in relationship to maturation and learning. A topic that may be closely related, play, is discussed in more detail shortly.

The effects of hormones on behavior, as described in Chapter 18, can be viewed as maturational changes in many instances. Such would be the case with the neonatal or fetal surges of androgen that masculinize the mammalian brain. Other hormonal effects that cause behavioral changes from one growth stage to another include going from one larval stage to another and from larvae to pupae, as well as changes seen in puberty and premolting and molting behaviors. Hormones and, in a sense, maturation also play a large role in starting, developing, and stopping reproductive and migratory behaviors.

Clear examples of maturational changes are fairly numerous. One involves color preferences in amphibians. Frightened frogs (*Rana* sp.), for example, jump toward water, showing a strong preference for blue and an avoidance of green. Tadpoles of the same genus, however, move toward green, which under natural conditions would be underwater vegetation and protective cover. The preferences have been demonstrated behaviorally and correlated with evoked potentials recorded from electrodes placed at different points in the optic pathway, from the retina to the

Figure 19-6 Changes in the color preference of frogs' (*Rana temporaria*) phototactic responses as the animals matured from tadpoles to frogs. These curves are similar for both behavioral choice and physiological, evoked potential responses; that is, frogs responded by jumping toward different choices in the frequencies as indicated, and electrodes inserted in the visual pathway showed neural firing rates in similar proportions. Modified from Muntz, W.R.A. 1962. Journal of Neurophysiology 25:699-720, and Muntz, W.R.A. 1963. Journal of Experimental Biology 40:371-379.

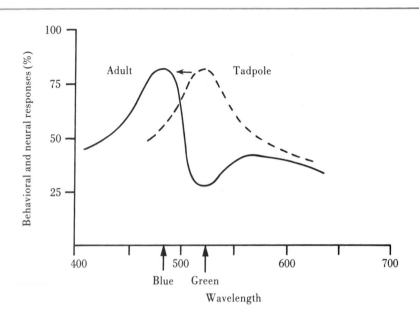

diencephalon (Muntz 1962 a,b, 1963 a). The change in this response with age (Figure 19-6) corresponds to a change in the sensory cells and visual pigments in the retina of the frog's eye. The first cells to develop absorb green wavelengths of light. Later two additional types of retinal cells develop: one believed to act synergistically with the first set, plus another type, all of which create the new blue preference. It appears that the development of the frog's color preference behavior is strongly influenced by the maturation of the sensory cells. These conclusions are strengthened by similar findings in other amphibians, some of which have similar preferences and sensory cells and some of which have different preferences and correspondingly different visual pigments (Muntz 1963 b).

Cuttlefish *(Sepia* sp.*)*, a cephalopod mollusk, prey on small shrimp, among other things. Adult cuttlefish learn from punishment, such as from electrical shock, not to attack. Young cuttlefish do not seem to learn to stop attacking. They continue to attack in spite of electrical shock or when the shrimp are protected behind glass so that the young cuttlefish fail to catch them (Wells 1958). Young cuttlefish continue to attack until physically exhausted.

Young individuals do display some changes in their attack behavior. The full behavior has four main components: when first presented with a shrimp, the cuttlefish shows a latency before giving any response; then the eye closest to the shrimp turns toward it; next the cuttlefish turns its body and both eyes toward the shrimp; finally it attacks by throwing out the tentacles to grab the shrimp. All but the first component vary little, taking about 10 seconds regardless of age or experience. The first, latency stage, however, shows a marked change with experience. A naive cuttlefish requires about 2 minutes before showing any response to its first shrimp. The second shrimp elicits a response in 40 to 50 seconds. The latency rapidly declines further until, after about five trials, it is 10 seconds or less. Although superficially the improvement in response time might appear to be learning, the change is quite consistent, apparently unmodifiable, and probably amounts to a practice effect, a change in attention or other central stimulus filtering (Chapter

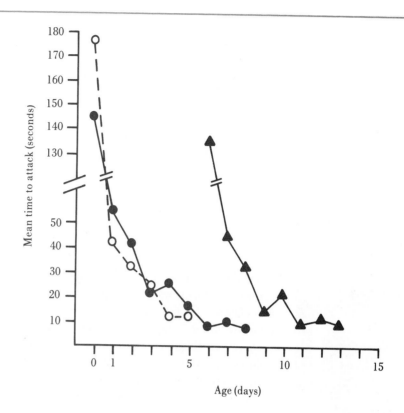

Figure 19-7 Latency of attack of cuttlefish on shrimp. The latencies decline rapidly with repeated trials, but the pattern of decline does not depend on success of capture, starvation, or age within a range of 1 to 5 days.
Modified from Wells, M.J. 1958. Behaviour 13:96.

● Rewarded for attacking

○ Not rewarded for attacking

▲ Starved for 5 days and then rewarded for attacking (1 trial per day)

17), an inhibition of initial fear, or perhaps some form of maturation specific to the attack behavior. The change in response time occurs whether the cuttlefish is rewarded or not and whether it is initially well fed or starved (Wells 1958) (Figure 19-7).

What about the change from young cuttlefish that do not "learn" to stop attacking to adults that do? This also appears to result from maturation. The region of the adult cuttlefish brain known to be involved in learning, the vertical lobe, does not develop until later in the cuttlefish's life.

Another classic example involves chickens, ground-dwelling birds that obtain their food by pecking at it. The chicks are not fed by the parents, although hens may peck toward food, indicating its location, or occasionally even hold items in their beaks for the chicks. For the most part, however, chicks must feed themselves from the start, and they do quite well if artificially raised without an adult.

At first the chicks are not accurate and miss many of their pecks at food. With time, however, there is a considerable improvement. Cruze (1935) held chicks in the dark to reduce their movement and fed them powdered food by hand for various periods of time. He then tested their accuracy in pecking at grain. After the initial test he allowed them to feed naturally in light and then retested them 12 hours later. Accuracy improved in all cases after practice, suggesting a practice effect. It also improved among older chicks that had never practiced before (Table 19-1). This suggests an improvement from maturation, whether within the sensory, neural, or muscular systems. The improvement may have resulted from increased strength and ability of the leg and neck muscles.

Table 19-1 Number of missed food pecks by chicks of different ages before and after a 12-hour practice period*

	Age When Removed from Dark (Hours)				
	24	48	72	96	120
First attempt	6.04	4.32	3.00	1.88	1.00
After 12 hours of practice and 12 hours of rest†	—	1.96	1.76	0.76	0.16

Modified from Cruze, W.W. 1935. Journal of Comparative Psychology 19:371-409.

*Values represent the mean number of misses out of 25 pecks averaged for 25 chicks.

†Actual age at second test is same as birds in next group at their first test. Birds removed from the dark at age 48 hours, for example, were tested first upon removal from the dark, allowed to eat naturally for 12 hours, held without food for 12 hours, and then retested.

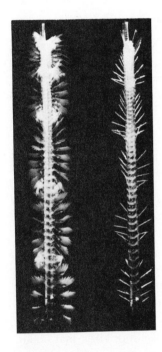

Figure 19-8 Stroboscopic photographs of wing flapping of a 13-day intact chick *(right)* and a wingless chick with plastic prostheses *(left)*. The chicks were dropped gently a short distance to stimulate the wing flapping. This study indicated that sensory feedback from the intact wing was not necessary for the wing-flapping pattern.
From Provine, R.R. 1979. Behavioral Neural Biology 27:233. Photographs courtesy Robert R. Provine.

Further evidence for the maturation effect plus an indication that reward or sensory feedback was not required was obtained in an ingenious experiment by Hess (1956). He fitted chicks with rubber hoods and prismatic lenses that deflected their vision by 7 degrees. Control chicks wore plastic lenses that did not deflect their vision. Results were recorded in soft Plasticine modeling clay with a nail for a target; each peck left a dent in the Plasticine. The chicks were first tested and then allowed to feed (while wearing the glasses) for 3 days, then retested. Both groups showed tighter clustering, hence, precision in their pecks, but those in the treatment group continued to miss the target—they simply missed it in a more consistent manner. The treatment chicks had managed to get some food while on their own, but barely enough, and many were losing weight. They had not learned to compensate for the deflected vision, and the tighter clustering of pecks was interpreted to result from some type of maturational response. Roughly similar pecking responses and results have been shown in gulls (Hailman 1969, Margolis et al 1987).

Another behavior in which the role of maturation has been studied involves wing-flapping exercises by nestling birds. The young of most birds stand and pump or flap their wings before they are able to fly. As early as 1873 Spalding showed that the wing flapping might not be necessary. He put young swallows in cages so small that the young were unable to open their wings. When released at an age when they should be able to fly, the caged swallows flew as well as other young that had been able to stretch their wings and exercise. Grohmann (1939) obtained similar results by raising an experimental group of pigeons in cardboard tubes that prevented flapping wing movements.

Provine (1979) experimentally investigated wing flapping in chicks of domestic chickens. He amputated the wings of one group and fitted them with limb prostheses made from soda straws. Chicks were dropped a short distance under carefully standardized conditions and photographed stroboscopically (Figure 19-8). The mean numbers of wing flaps by chicks with wings intact and chicks with the plastic wing prostheses were not only statistically similar, they were virtually identical. Proprioceptive effects from the muscles, located in the breast, and possible effects from neonatal movements before amputation could not be ruled out. It was clear, however, that postnatal sensory feedback that would be associated normally with an intact wing was not necessary for the flapping patterns.

A final example of maturational effects on behavior and the differences between species is the development of prey-handling ability in predators. Mueller (1974) investigated prey recognition and predatory behavior in young American kestrels *(Falco sparverius)*, a small falcon. These birds typically prey on mice, among other things, and kill them in a stereotyped falcon fashion by biting the neck soon after capture. Acquisition of components of the behavior might be accomplished by familiarity and learning while under the care of their parents, by simply pursuing any small moving object and being rewarded or reinforced to concentrate on those they have success in catching and eating, or by having the patterns of behavior develop by maturation of the nervous system. Young kestrels in the wild seem to feed mostly on insects at first. If given live mice prematurely, they show either fear or no response. This information suggests that the behavior requires maturation.

Mueller hand-reared and hand-fed young kestrels to prevent experience with whole animals or live prey. Then at ages 10 weeks to 1 year the kestrels were tested on four types of items: tissue-paper balls and tissue model mice, stuffed mouse skins, dead mice, and live mice. In each case, except the live mice, which were moving, the objects were presented either stationary or moving. The falcons showed few responses to the paper objects and then only in a playful manner. The stuffed mouse likewise received little attention. The dead mice were taken by a third of the kestrels, which attacked and bit them in typical fashion, as if they were live. With live mice, however, eight of nine birds attacked the first time with responses that were "intense, rapid, sustained and well-oriented" with capture and killing being expert and indistinguishable from adult behavior. All birds achieved consistent, expert performance in fewer than six trials. Mueller inferred that experience was unimportant in the recognition, capture, and killing of mice by American kestrels.

Raber (1950), in studying the ontogeny of predatory behavior of owls *(Asio otus)*, obtained similar findings, except there appeared to be a critical period within which live prey had to be taken. If live prey were not taken during that period, the birds would not attack live prey at a later time if given the opportunity. Thorpe (1948) demonstrated prey-killing behavior in another owl *(Athene noctua)* as a vacuum activity (Chapter 2); the owls would display the entire predatory sequence, including killing the imaginary prey, in the absence of any prey stimulus.

The ontogeny of predatory behavior in mammals is less clearly a case of maturation and may involve more learning, depending on the species. Eibl-Eibesfeldt (1956, 1963) and Leyhausen (1956, 1965) have suggested that experience and learning are necessary for orienting bites to the necks of prey by mustelids (weasel family) and cats. Gossow (1970) and Eaton (1970), however, believed instead that maturation may be more involved in at least some of these mammals.

Eibl-Eibesfeldt (1951) studied the ontogeny of nut opening in squirrels. Young squirrels inexperienced at opening nuts have inherent initial movements for gnawing and prying. However, they become much more efficient with experience. Different routes of success at opening nuts led to a diversity of behaviors among different individual squirrels. That suggested that learning was involved rather than just the maturation of a stereotyped behavior.

It is clear that the roles of maturation and learning in the development of behavior vary widely. Variation can be seen even among animals in the same class, such as mammals. Furthermore, maturation and learning are not mutually exclusive categories but are related. Maturation of the most extreme form, with no learning involved, can occur and could be viewed as a subtly disguised form of innate behavior (instinct, genetic, nature, etc.). It simply is innate behavior that takes some time to develop. On the other hand, there may be some learning of novel behaviors that have little or no specific maturational or innate substrate. As discussed in Chapter 4, most behavior probably lies somewhere between the two extremes.

EARLY EXPERIENCE AND THE DEVELOPMENT OF BEHAVIOR

**Deprivation,
Environmental
Enrichment, and
General Effects**

Being deprived of various stimuli or opportunities for learning seems to pose few problems in the ontogeny of some behaviors in some species, as indicated in parts of the preceding section. Deprivation may have some effects, however, as in the case of the owls observed by Raber (1950). In some instances the effects of deprivation are striking. This section deals with a few examples where deprivation caused a significant impact on subsequent behavior. The greatest effects of deprivation have been observed in mammals, from which all of the following examples come. Examples of deprivation effects in birds are discussed later under the separate section devoted to the development of song in birds.

Kittens have been the subjects of extensive research on the effects of deprivation of vision and visually guided behavior (e.g., Wiesel and Hubel 1963). When deprived of normal visual experiences for the first 2 to 3 months of life, kittens are functionally blind thereafter. In various experiments, they have been deprived of sight by being kept in darkness, by wearing diffusing hoods, and by having the eyelids carefully sutured closed, an operation that is easily reversed. In the eye-suturing experiments (e.g., Wiesel and Hubel 1965), either one or both eyes were sutured; sutures were subsequently removed and eyes were opened; or eyes that had been open were subsequently sutured closed. The results were clear in all cases: the visual pathway in the brain deteriorated for all eyes deprived during early life. Effects included reduced size of neurons in the lateral geniculate body and reduction of connections in the visual cortex of the brain on the appropriate side. Kittens with single eyes sutured showed a loss of binocular vision, with the loss occurring on the affected side of the brain. Opening of the eyes after these effects had occurred led to only slight, insignificant recovery. Similar deprivation of vision, for comparable periods of time, in adult cats did not cause similar losses. Visual experience is important for developing visual pathways not only in vertebrates but also in invertebrates. Mimura (1986) demonstrated visual deficiencies in visually deprived flies similar to those seen in cats.

Lack of use during the developmental period may affect not only the sensory pathways but the entire sensorimotor response in a general way. This has been studied in several species with a variety of techniques. Perhaps one of the most ingenious and best-known studies involves self-produced responses in kittens in which an active kitten and a passive kitten were exposed to the same visual environment and visual movements (Held and Hein 1963). Through the use of a merry-go-round apparatus (Figure 19-9), one kitten generated the movements and the other rode in a gondola. Mechanical linkages duplicated all of the active kitten's body movements at the gondola for the passive kitten. The procedure was repeated with 10 pairs of kittens, each pair coming from a different litter. Kittens were placed in the apparatus for 3 hours per day for several weeks. The remainder of the time

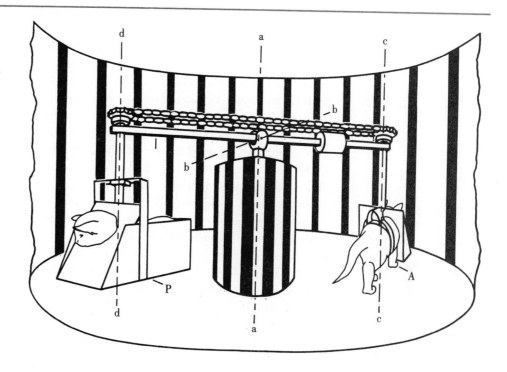

Figure 19-9 Apparatus for equating motion and consequent visual feedback for actively moving (A) and passively moving (P) kittens. From Held, R., and A. Hein. 1963. Journal of Comparative and Physiological Psychology 56:872.

they were kept in the dark with their mothers and litter mates. At the end of the period they were tested in two principal ways: (1) with a visually guided paw placement test in which they were carried toward a horizontal surface (toward which a normal kitten reaches its paws before contact) and (2) with a visual cliff test in which kittens were placed on a narrow platform over a piece of glass with a patterned surface immediately under the glass on one side and a similar surface 30 inches below on the other side. At the end of the experiment all of the active kittens responded to the paw-reach and visual cliff tests normally, whereas none of the passive kittens showed the reaching response or discriminated between the shallow and deep sides of the cliff test. Held and Hein concluded that "self-produced movement with its concurrent visual feed-back is necessary for the development of visually-guided behavior." They eliminated anatomical or physiological deterioration of the systems by showing normal blinking and pupillary responses in the passive kittens and by obtaining subsequent recovery after the passive kittens were permitted to move normally in a lighted environment.

In experiments involving puppies, Melzack and Scott (1957) showed that environmental deprivation during early life can lead to abnormal behavioral responses associated with pain. They used several litters of Scottish terriers, with each litter divided into one of two groups, a control group where the pups were raised as pets in private homes and the treatment group where pups were raised in diffusely lighted isolation cages, deprived of normal sensory and social experience. This treatment lasted from the time of weaning until about 8 months of age. The dogs were then tested in a variety of ways 3 to 5 weeks after being released from isolation, and some were retested again 2 years later.

The normally raised dogs showed normal avoidance responses to pain induced by electrical shock, pinpricks, and heat, whereas all of the deprived dogs showed significantly different, nonadaptive responses. The deprived dogs showed local reflex responses to the noxious stimuli, indicating that the senses functioned at least to some extent, but the dogs showed no organized avoidance, attempts to get away, or any signs of emotional distress. The deprived dogs also were abnormal otherwise in their responses to objects in the environment. For example, they would walk repeatedly into water pipes along the wall of their testing rooms.

Probably the best-known, most widely publicized studies of deprivation involve rhesus monkeys deprived of a normal, live, soft mother and normal peer relationships (Harlow 1959, Chamove et al 1973). Several aspects of growth, development, and individual and social behavior were investigated experimentally by comparing young monkeys raised under normal (although captive) conditions with those raised in various combinations of soft and wire surrogate mothers, no peers, different categories of peer relationships, and peers but no mothers. The experimental monkeys developed normally from a physical and physiological standpoint but not behaviorally. Young deprived during the first 6 to 12 months of life showed a host of problems and inabilities in emotional and social behaviors. The problems lasted throughout the lives of the affected individuals.

In some species, such as rats and humans, the lack of normal social interaction, particularly the contacts and touching between infants and their mothers, may also affect physical growth and development. This may result from increased levels of β-endorphin interfering with growth hormone during maternal deprivation (Barnes 1988).

Deprivation is a relative concept. Although some of the cases cited previously might be viewed as rather severe, total deprivation lies at one end of a continuum with environmental enrichment at the other end. The normal situation is somewhere in between. Normal kittens, for example, would be reared with their mother and littermates in surroundings that stay relatively constant day after day. Eventually, however, the kittens are able to move about and explore. They then are exposed to a variety of objects and environments by their own movements or by being carried about by their mother. Even with such exploration and movement the kittens are exposed to new things within limits. It is possible to artificially enrich their experiences by introducing a wider variety of novel objects, sights, and sounds, carrying them in vehicles, and changing the general surroundings. Would such experiences have the opposite effect of deprivation? What could be learned about the normal development of behavior?

The implications of enriched versus deprived rearing are obviously of profound importance and interest in human development, and there is a large body of human-related literature and studies. The findings have been fairly consistent regardless of whether the mammal studied was a human, another primate, carnivore, or rodent.

A few notes of caution are in order. The subject is fraught with semantic and methodological differences of approach and opinion. Terms related to deprivation include *isolation* and *restriction*. Enrichment, on the other hand, has included environments referred to as *free, unrestricted, enriched,* or *complex*. It is rarely clear just what *normal* means, particularly when dealing with the domesticated, laboratory-reared animals commonly used in research. Meyers (1971) attempted to obtain a degree of uniformity in recognizing the polarity by classifying environments simply as restricted or complex. Methodology has varied widely, with various mazes commonly being used to measure behavioral performance. Furthermore, as usual, theories about the underlying mechanisms abound. Thus, although some findings are fairly consistent, closer inspection of the field reveals a morass of unsettled issues. The consistent, if not somewhat superficial, findings are considered, as well as the important anatomical and physiological differences between deprived and enriched forms of rearing.

It appears to be a fairly safe conclusion that an enriched environment between the time of birth and puberty leads to enhanced problem-solving ability in rats (and many other species of mammals). On the other hand, a rat's performance can be degraded by a deprived environment. Furthermore, these effects appear to be genuinely related to the performance in question (e.g., ability in a maze) and are not artifacts from differences in sensory ability, emotionality, or simple exploratory behavior. Further discussion and references are provided by Meyers (1971).

The most concrete effects of deprivation versus enrichment can be seen in the nervous tissues of the rats. Most information along these lines stems from the pioneering work of Rosenzweig et al (e.g., 1972) at the University of California at Berkeley. They reared rats generally in one of three environments: (1) enriched with 10 to 12 animals in a group with various objects, frequent handling, and experience in mazes, (2) social control with 3 animals living together in a standard laboratory cage, and (3) isolated control where animals were housed individually in cages with three solid sides.

Enriched-environment rats showed significant differences in the cortex of the brain compared with the isolated controls. Depending on which neural characteristic was considered, the social-control rats were generally somewhere intermediate. Enriched environments during rearing led to heavier and thicker cortical tissue, larger neurons, increased glial tissue, and increased dendritic branching. Information on acetylcholinesterase (AChE) levels implies that there may be increased synaptic activity in the neural tissue of enriched-environment animals compared with the deprived individuals. (AChE is the primary enzyme that deactivates acetylocholine after synaptic transmission.)

Deliberate, experimentally caused abnormal behavior has been studied for many years. Perhaps the name best associated with work in the area of experimental neuroses and abnormal behavioral problems in animals is that of the Russian physiologist-psychologist Ivan Pavlov. Pavlov began research early in the twentieth century; he was initially interested in the physiology of digestion. This led him into the subjects of conditioning and learning, for which he is best known and which were discussed earlier in this book. He went on to study abnormal behaviors and was able, for example, to produce nervous breakdowns and brainwashing in dogs. These were achieved by such means as strong stimuli, extending the length of aversive stimuli, frequent presentation of aversive stimuli, switching positive and negative stimuli to cause uncertainty and confusion (some effects referred to as *collision*), and a number of related techniques. This early work has been followed by much research into similar conditioning effects, hormonal effects, neural pathways, corticovisceral or psychosomatic interrelationships, and how behavioral problems can be either caused or prevented. Unfortunately or fortunately, come aspects of this work have been applied to humans in some instances (e.g., Sargant 1957).

The effects of early experience and social surroundings, particularly involving parents and peers, have received much research. The work has tended to concentrate on primates and canids. The early work of Harlow and colleagues (Harlow and Zimmerman 1959) has become classic. The implications of this research for normal biological mechanisms are incorporated in portions of this book. An excellent review that focuses on the abnormal behavioral aspects is provided by Sackett (1968), who concluded that behavioral impairments could result from improper or inadequate interactions with either mother or peers. Peer relationships seemed to exert the greatest effects. Deficiencies were mostly emotional, motivational, and social rather than intellectual. Furthermore, there was a critical or vulnerable period consisting of the first 6 months of life. Although the effects can be permanent and extend into adult life, they are not necessarily irreversible. Suomi and Harlow (1972), for example, have treated some of the problems through the use of "therapist" monkeys.

It is thus clear that early experience in the life of the individual may have important effects on its subsequent behavior. These effects of experience are different from what would be considered as either simple maturation or learning. Learning, as commonly defined, usually refers to more specific changes in behavior. These developmental effects are general.

Figure 19-10 Examples of play as shown by **A,** lion cubs, **B,** domestic dog pups, and **C,** juvenile rhesus monkeys.
A and B from Animals, animals. A © Margot Conte. B © Paula Wright. C from Fagen, R.M. 1981. Animal play behavior. Oxford University Press, New York. Photograph by John Bishop.

A closely related topic involves the ontogeny of emotional behavior (Candland 1971). The conclusions from work on the ontogeny of emotional behavior are quite similar to what was considered earlier: experiences during early life may have significant effects on various emotions later. These effects may be all too familiar in human experience, and they have been amply demonstrated in a variety of animals under objective study. Again, although there may be differences in emotions attributable to other factors, such as inheritance, there are clear effects in many cases that result from early experience. As with general problem-solving ability, most persons would not associate these effects with what would traditionally be called *learning*. (Traditional categories of learning are discussed in the next chapter.)

Play Behavior in Mammals and Birds

Although play behavior is familiar to everyone, it remains a puzzling topic: no one is really sure just what it is or why animals do it. It is common in mammals (Figure 19-10) and some birds. Some human play shows striking similarity to that of other mammals, particularly other primates (Figure 19-11). Play is normally observed in young animals and may be associated with development; hence it is placed in this chapter.

As observed in its basic form among animals, depending on the species, play consists of frolicsome leaps, running and quick turns, rolling about, climbing, and other exaggerated or otherwise unconventional movements, plus knocking, kicking,

Figure 19-11   Similarity of human play to that of other primates. **A and B,** Play-wrestling by lowland gorillas in an upright position. **C,** Play-wrestling by Sherpa boys, Nepal.
A from Animals, animals. © M. Austerman. B and C Fagen, R.M. 1981. Animal play behavior. Oxford University Press, New York. Photographs by John Bishop.

throwing, dropping, and wrestling with other objects—including other animals. Play may become quite elaborate, particularly in social situations. There are several types of play and a number of general characteristics that can be described for play. Fagen (1981) provides an excellent review of the subject. He sifted through the natural history literature and documented many cases and categories of play.

Play often involves other individuals, including those from different species. Play frequently seems to contain an element of "thrill" or surprise seeking, as seen in the following examples, which involve Lowe's guenon (*Cercopithecus campbelli*) monkeys at play, as summarized by Fagen (1981) from several sources (see Fagen for references):

> Social play includes arboreal and terrestrial chasing and wrestling. A monkey may vault over another's back. One monkey may present itself as if to solicit mounting, then suddenly flee at the instant its partner attempts to mount. In an arboreal game two to five monkeys scramble up to a swaying tree branch and wrestle there. One monkey may abruptly fall off the swinging bough, which then springs upwards, knocking other monkeys off balance. Free falls are frequent in play. One monkey may drop down by stages through small trees or underlying bushes. Subadults appear to "seek thrills"..., breaking or tearing open the nests of stinging weaver ants *(Oecophylla longinodis)*, jumping around frantically while the ants attack them, then returning immediately to the nest to repeat the process. Juvenile Lowe's guenons play interspecifically and reciprocally, chasing and being chased by wild squirrels *(Heliosciurus gambianus)* or fleeing from a charging (but tame) mongoose *(Crossarchus obscurus)*, then returning to "tag" the mongoose and flee again. The monkeys also pursue other mammals and birds, apparently for the sole purpose of scaring them.... The habit of teasing hornbills *(Tockus fasciatus)* seems best developed. The hornbills roost in a large tree. Once they are settled, the juveniles leap toward them and shake the branches until the hornbills take flight. The monkeys then wait in the tree until the birds return. When the birds have again settled, the monkeys repeat their mock attack.

Fagen (1981) interprets the basic problem in past scientific attempts to study play as being a lack of adequate theory. An underlying theory or understanding, if only rudimentary and even if wrong, helps guide the activities that are called *science*. If the scientific process is working properly, it will eventually identify and eliminate or correct a theory that is wrong.

The problem of formulating a theory about (or even a definition of) play stems from two major underlying causes. First, play seems to have no immediate or obvious function. Because of traditional biological beliefs that attributes arise only under selective pressure, it is easy to think that play should have some advantages or function or else it should not have evolved. If so, however, the advantages have been difficult to determine and associate with the actions of play.

There may be alternate explanations for behavior rather than play, but the causes are so subtle that they are easily overlooked or, when considered, cannot be easily separated from each other. The following example was cited as play in Fagen (1981). It involves interactions between ravens and wolves (Mech 1966):

> As the pack traveled across a harbor, a few wolves lingered to rest, and four or five accompanying ravens began to pester them. The birds would dive at a wolf's head or tail, and the wolf would duck and then leap at them. Sometimes the raven chased the wolves, flying just above their heads, and once, a raven waddled to a resting wolf, pecked its tail, and jumped aside as the wolf snapped at it. When the wolf retaliated

by stalking the raven, the bird allowed it within a foot before arising. Then it landed a few feet beyond the wolf and repeated the prank.

At first glance this certainly seems like a valid case of play. However, there are at least two other possible interpretations: (1) it may be a form of avian mobbing like crows chasing owls or small songbirds chasing cats in the back yard, which serves to draw attention to them and reduce their hunting effectiveness or even drive them from the immediate neighborhood; or (2) it may be related to foraging behavior and the scavenging of meat by ravens at carcasses of animals killed by wolves. Similarly the apparent play chasing between some birds of prey and other birds, such as hawks and crows, may be simply inept and inexperienced but hungry young raptors chasing inappropriately large prey while the intended prey in turn chases the raptor in mobbing behavior (Verbeek 1985). Thus some behaviors may have other functions or explanations, but they are not obvious and inadvertently get labeled as play.

The other major problem facing play research and understanding has been the difficulty of dealing with play experimentally. Trying to experiment with play, for example, by preventing an animal from doing it, may inadvertently affect many important aspects of the animal's life, biology, and development. Some of the aspects affected actually may be correlated with play, and some may be totally irrelevant but are inescapably caught up in the experiment. This has proved a nasty and persistent problem and has produced disagreement over opposing hypotheses.

General attributes of play

Play seems to be restricted among various taxonomic groups.   Few if any instances of play have been observed among invertebrates or poikilothermic vertebrates. Thus it appears to be confined primarily to birds and mammals and among these is most conspicuous in mammals. Ortega and Bekoff (1987) surveyed the presence of three types of play behavior (locomotor, object, and social) among birds. Play is seen most frequently among altricial species requiring much parental care, species with the greatest forebrain development, predators, highly social species, and species with the greatest abilities of vision and manipulation.

Play lacks apparent, external goals.   This point has already been mentioned, and it is perhaps one of the most obvious characteristics of play behavior. One or more objects, including other animals, may be involved (and may even be modified or damaged) in an animal's play, but no recognizable function is accomplished with the object. The animal simply plays with the object—it is hard to avoid the trap of circular self-definition.

Not only does the play itself appear to lack a goal, but it also generally is engaged in only in the absence of other goal-directed situations.   Other more obviously functional behaviors may inhibit or interrupt play. An animal may break off a bout of play at the appearance of danger, food, a chance to mate, or other opportunities or needs but not the other way around.

Although play may be demanding of energy and time and thus involve work in a physical sense, it generally seems to lack seriousness or purpose in terms of human connotations (i.e., something that needs to be done). It does not generally occur during times of other behavior, such as actual predation, escaping from predators, eating, or reproduction. In addition, older members of a group that are engaged in less playful, more goal-directed activities may attempt to discourage those individuals that are playing.

The movements of play generally are borrowed from other behaviors. There may be unique motor patterns used as play-soliciting signals preceding or during social play (see Chapter 15). Beyond these, there are few if any movements seen in play that are not also seen in other contexts. The movements include those displayed in predatory behavior, social fighting, fleeing, reproduction, and eating. Loizos (1966), in a qualitative assessment of the use of various motions in play, suggested that these borrowed motions may be reordered in sequence with many possible permutations, exaggerated, repeated, fragmented and shown in incomplete sequences, or displayed in combinations of these (e.g., exaggerated and repeated or fragmented and repeated).

Bekoff (1976) stresses that these characteristics have not yet received adequate quantitative analysis. From the few data available (e.g., for black bear cubs, *Ursus americanus*, and New England coyote-canids, *Canis latrans*), however, there is little or no evidence that movements are exaggerated in play, and in some acts the opposite is true. Bekoff suggests that Markov models, information-theoretic analysis, and measures of conditional uncertainty provide possible means for quantitatively assessing the sequences of movements and for comparing the amount of order or randomness in play versus nonplay behaviors using similar movements. The limited information that is available (e.g., on vervet monkeys, *Cercopithecus aethiops*, black-tailed deer, infant canids, and young rats) supports the notion that play movements may be reordered in sequence but has not permitted a measure of whether motions in play are more random than in nonplay.

Play generally is seen more frequently in young animals than in adults. Adults of some species, however, may play occasionally, particularly when the other needs, such as feeding, have been met.

Play is most likely to be shown, at least in the few mammals studied, under relaxed and familiar conditions. The mammals studied include lions (Schenkel 1966), polecats (Poole 1966), and humans (Hutt 1966). Young mammals may require familiar surroundings, the presence of the mother, or a familiar object ("security blanket") before they engage in play. In the absence of the mother, for example, young lions will hide and show no play or exploratory behavior.

Play is potentially dangerous and costly. The risks and costs of play have not been well quantified and have been the subject of ongoing debate (e.g., Martin 1982, 1984 versus Lee 1984). The topic is just beginning to be subjected to optimality considerations (Chapter 9). From the meager published information that Fagen was able to summarize, however, it is clear that play can be costly. Accidents may be common during play and may lead to temporary injury, permanent disability, and even death. In a study of confined ibex *(Capra ibex)*, for example, at least 5 of 14 kids sustained injuries during play that produced visible limps (Byers 1977). Most injuries related to play in all species, including humans, result from falls.

Animals not only expose themselves to accidents while playing but also may become more vulnerable to predation. Juvenile vervet monkeys *(Cercopithecus ae-*

thiops), for example, were caught by baboons *(Papio cynocephalus)* most often when playing in groups away from the adults (Hausfater 1976). This may be a factor in the lack of play by some species, such as lions, unless they are in the presence of adults or in familiar surroundings. Playing animals also may run risks from their own species when play escalates into serious, aggressive fighting or if it attracts punitive intervention by other individuals.

In a less direct sense, play also may be costly in terms of time and energy. Again, good quantitative data are few, but available studies (reviewed in Fagen 1981), mostly from primates and other confined animals, show that young animals commonly spend 1% to 10% and sometimes 50% or more of their time in play. This may be physically demanding, is often exhausting, and consumes considerable metabolic energy. Time and energy spent on play are not available for other activities.

Evolutionary theory suggests that play has some selective advantage that outweighs the costs and potential risks. Otherwise it seems logical that play should be selected against.

Play appears to involve feelings of pleasure.   This is a subjective and difficult point. The interpretation is derived from the human perspective and hence is anthropomorphic. Also, some interpretations are redundant and circular: "Something is pleasurable because the animals feel pleasure." Such aspects of behavior have received considerable academic discussion. Nonetheless, the notion that other animals may experience pleasure seems reasonable. Physiological evidence exists for pleasure centers in the brain (Chapter 18). The presence of pleasure sensations in other species, whether or not associated with play, can be argued theoretically from two evolutionary standpoints:

1. If humans experience pleasure and humans evolved, then it is possible, if not likely, that our relatives also possess the sensation. That is, it is reasonable to assume a continuity of evolution.
2. Ultimate factors require proximate mechanisms. The physiological process that is subjectively called *pleasure* could be an important proximate mechanism; that is, pleasure simply may be the name that is given to the neurological process that motivates animals to eat, engage in sexual and grooming behavior, and perhaps play, as discussed in the previous chapter (pages 664 to 670).

Play often contains elements of surprise and apparently the seeking of the unexpected or thrills.   Perhaps the most familiar instances of anticipation and surprise can be seen in the ambushing and pouncing play of kittens. They crouch and hide behind objects, with the tips of the tails twitching, then suddenly jump or strike out with their paws. Dogs often race circles around each other or their master, coming as close as possible, then run past as if daring to be tagged or caught. Tree squirrels and many other species often run up and tag another animal and then race away. The young of many ungulate species seem to delight in chasing birds that are on the ground. They chase after them until they fly, wait for them to land elsewhere, then chase and flush them again. Fagen (1981) cites a great many more instances of "thrill seeking." This aspect of play alternatively could represent (in anthropomorphic terms) an escape from boredom. Boredom is an intriguing phenomenon; however, it is not well understood even in human behavior.

Play is probably a category of behavior that includes many specific types and appears to overlap in some cases with exploration.   There are at least two kinds of exploration: specific and diverse (Hunt 1966). Futhermore, they may change over time from one category to the other. In specific exploration the animal's senses and attention are directed toward the object, and, depending in part on the complexity of the object and number of senses it stimulates, the animal's attention decreases in time with repeated exposure to the object. This type of exploration might be viewed as learning rather than play.

Diverse exploration, however, seems to fall more into the class of what most persons would call play. In cases where the properties of the object are known, senses are directed more away from the object, and attention is given to doing things with the object. The behavior seems to shift from "What does it do?" to "What can I do with it?" This has led some persons (e.g., Hutt 1966) to distinguish between exploration as acquisition of information and play where learning is incidental. Exploration and play are difficult to distinguish in human infants but diverge and become more distinct as the individual grows older. Exploration also appears to differ among species and may be correlated with predatory and nonpredatory modes of life (Hutt 1966). Rodents, for example, more readily explore new environments than objects in familiar environments, whereas the opposite is true of predators, including humans.

Diverse exploration may be associated strongly with creativity in humans. Dissanayake (1974), as well as earlier thinkers such as Herbert Spencer and Friedrich Schiller, proposed that art originated from play. At least subjectively, there appear to be strong correlations among humor, play, creativity, and the development of thinking ability (Adams 1974). The converse also appears to apply; persons who are more serious or socially inhibited (often described as "good" and "obedient" children) may be more reluctant to play and are less creative (Hutt 1966).

Why do animals play?

Although one cannot be certain of the emotions of animals of other species, it is fairly clear that many animals play because it is "fun" (i.e., pleasurable). This is at best simply the proximate cause of the behavior.

Of most interest are the ultimate, evolutionary causes (assuming that they exist) for behavior. There are candidate hypotheses. All but the most recent one—proposed by Gordon M. Burghardt and discussed at the end of this section—have been discussed by several persons, for example, Smith (1977) and Caro (1981), and have been reviewed extensively by Fagen (1981) and Burghardt (1984).

The various theories of play are: expenditure of surplus energy, pleasure (only), arousal/stimulation, practice, exercise, social functions, and the ectotherm-endotherm transition.

Surplus energy theory.   Schiller (the eighteenth century poet), Spencer during the nineteenth century, and others proposed that play results from the use of excess energy in animals. Most attempts to deal with this theory, however, met with little success. Müller-Schwarze (1968) found no significant increase in play actions of previously play-deprived black-tailed deer except for an increase in running speed of one subject. Chepko (1971) obtained inconclusive or insignificant differences between play-deprived and nonplay-deprived young Toggenberg goats *(Capra hircus)*. Again, it is a difficult hypothesis to test.

On logical grounds alone, excess energy, that is, energy not required for immediate or short-term needs, seems more likely to be physiologically metabolized and stored as fat rather than being wasted. Furthermore, as considered by Bekoff (1976), "The attributing of play to an excess of energy does not lead to any further clarification of the characteristics of play." Further consideration of this theory, however, led Burghardt (1984, 1988) to propose a new theory for play behavior that we return to shortly.

Pleasure theory.   Play appears to give pleasure to the participants. The pleasure may even be shared by onlookers. Aside from the proximate aspects, pleasure might be selectively advantageous in itself. The pleasure theory, which can be traced in one form or another back at least to Pycraft (1912), suggests that play occurs only for the immediate pleasure or fun of it and that there is no other adaptive significance. However, this would be difficult if not impossible to test. One may be able to discard the idea that play exists for pleasure only by elimination, that is, by adequately demonstrating that an alternate explanation exists and that there is a reasonable function or other ultimate causal factor.

Arousal/stimulation theory.   Based on earlier psychological notions of drives, Ellis (1973) suggests that play helps elevate an animal's level of arousal. In this view the behavior is involved in generalized stimulus seeking related to exploratory behavior that may expose it to other necessary aspects of life. Again, that is a vague concept and would be difficult to experiment with. Bekoff (1976) references and comments on semantic problems with the term *arousal*. One possibility might be to measure levels of arousal in the reticular activating system or from surface brain waves as during measures of sleep and then correlate them with play behavior. Until such measurements are obtained and carefully interpreted, there is not much more that can be said about a possible arousal function for play.

Practice theory.   Practice seems to be one of the most intuitive explanations of play, based on our present understanding of learning, but evidence does not support the theory. According to the practice theory, through practice and repeated performance in a nonserious setting, an animal learns or neurally perfects movements and behavior needed in more serious contexts. This concept was suggested by Groos (1898) and has been maintained by subsequent authors (e.g., Aldis 1975). One problem, again, is that it is difficult to deprive animals of play to see if this interferes with the performance of other behavior.

In the few cases where naive or socially deprived animals have been studied, there is little evidence that subsequent behavior such as prey killing or copulation suffers. Vincent and Bekoff (1978) attempted to deal with the problem by correlating performance in play with performance of prey killing in infant coyotes. They found, however, no significant correlations involving frequency of participation in play, participation in agonistic interactions, or frequency of most motor patterns. Only the frequency of pouncing in play was correlated with prey-killing ability.

In other cases the array and performance of motor patterns in play versus other behaviors are confusing, but they do not seem to support the practice theory. In the aggressive play of polecats (Poole 1966), for example, the patterns of play develop with age, but as they appear, various components of the patterns emerge in stereotyped and complete adult form with no subsequent modification. Thus many borrowed behaviors that showed up in play appear in already complete form, without need for practice. On the other hand, some motor patterns that are important in the serious context are either absent, for example, in the play of young spotted hyenas, *Crocuta crocuta* (Kruuk 1972), or are unimportant in the play behavior, such as stalking in lions, *Panthera leo* (Schaller 1962).

Thus there has been little evidence to support the practice theory of play. Nonetheless, the lack of evidence (as with other theories) does not by itself discount the theory. Sample sizes in studies have been small, which makes it difficult to reject a false null hypothesis, and only a few species have been carefully studied. Even if practice is not involved in the frequency of expression of certain movements, it still might affect the efficiency or finer aspects of the movement.

Before proceeding to the next theory, it might help to stress the difference between two of the theories already mentioned: the arousal theory and the practice theory. Both concentrate on use of the nervous system in play. They differ in that arousal involves more general, nonspecific stimulation of the nervous system, whereas practice focuses on specific neural patterns affecting the performance of particular movements. The next theory concerns general, nonspecific stimulation of the muscular system.

Exercise theory.   According to the exercise theory, the use of muscles and associated (e.g., cardiovascular) systems in play stimulates the development and maintains the physiological condition of these systems in the absence of risky, serious contexts. Then in the presence of serious situations, which are encountered more frequently when an animal grows up and loses the care and protection of its parent(s), the animal is in better physical condition and able to cope with the demands of serious situations. This theory differs from the practice theory in that it emphasizes physical endurance systems rather than the neural coordination aspects. Persons suggesting general neural or sensory stimulation as a function of play (i.e., the arousal theory) also generally have pointed out the possibility of muscular and cardiovascular effects (references in Fagen 1976).

Although also skimpy on direct experimental data, the exercise theory seems like a good candidate for at least partially explaining play. The selective advantages of strength and endurance are obvious. The muscular and cardiovascular systems appear to require development and maintenance and suffer from disuse as much as or more than the nervous system.

Social functions theory.   Play with other animals, primarily belonging to the same species, represents play in its most elaborate, sophisticated form. The animals may play in a rough-and-tumble manner but normally not so roughly as to seriously hurt each other. Social play often shows a degree of cooperation and even what might be called *fairness*. Larger and tougher animals may play by restraining or handicapping themselves (Figure 19-12). Cats generally keep their claws retracted during play; bears cuff and wrestle using their paws in a manner that does not expose their sharp claws to scratching and tearing of the other; and horned animals either do not use their horns during play or else use them only gently. When using their mouths and teeth, playing animals usually only nip or mouth their play partners rather than truly biting in a way that would break the skin or cause damage.

Social play usually is accompanied by a set of relatively unambiguous communication signals, discussed in Chapter 15 under the topic of metacommunication.

Figure 19-12 Self-handi-capping among animals of unequal size and strength. When animals that are not matched in physical ability play, the larger and stronger often restrains itself and does not use its full strength in the interaction.
From Fagen, R.M. 1981. Animal play behavior. Oxford University Press, New York. Photograph by John Bishop.

Figure 19-13 Mammalian play-face. The open-mouth grin is shown by many species during play. Shown here are mother and infant chimpanzees playing.
From Van Lawick-Goodall, J., and H. Van Lawick. 1965. National Geographic 128:802-831.

The relaxed, open-mouth ("smile" in primates) play-face is almost universal among mammals (Figure 19-13). A southeast Asian primate, the douc langur *(Pygathrix nemaeus)*, has a bright yellow face, white whiskers, blue eyelids, and brown and chestnut on the remainder of its head. During play, and only during play (or playful preliminaries before copulation), this langur closes the eyes and displays the blue eyelids toward others.

Other visual signals include distinct postures, such as the familiar play-bow (probably derived from stretching) shown by dogs and lions and a particular position of the tail shown by bovids and equids. Some primates bend over and look between their legs. The signals may take the form of movements such as rolling around on the back or gamboling play-gaits seen in many species (perhaps derived from solo play).

Play signals are not confined to vision; various species may use signals in any of the sensory modalities. There are a number of auditory play signals plus many that use the tactile senses, with various body contacts, poking, tagging, and jostling. Olfaction has received the least attention by humans (who do not use olfaction to the degree that many other animals do), but there is some evidence that other animals use play-smells. Wilson (1973) applied ether extracts from the bodies of playing short-tailed voles *(Microtus agrestis)* to the bodies of nonplaying voles. Nonplaying voles with the extracts then elicited play from other voles, whereas nonplaying voles with ether alone did not.

The essence of the play signals, as judged by the reactions of conspecifics, is the general message that "this is done in fun or jest and not to be taken seriously; I am only kidding or joking." However, most species have a number of different signals. The need for several is not clear. Such redundancy may ensure that the chances of misunderstanding are minimized, or there may be much more to the message than simple play versus nonplay. The various signals, or gradations of signals, may carry information on interest and motivation in play, strength, or social status of the players or other messages (perhaps even including deception) that have not yet been considered.

As a set of behaviors, most play seems to be stable; that is, it appears to be a solid ESS. Cheating, bullying, refusing to play, eruptions of play-fights into genuine, serious fights, and other nonplay or antiplay actions exist and can be observed, but they do not occur frequently and on the whole do not seem to be overly disruptive.

If one were to consider the selection of play purely from the standpoint of the individual, one would expect many conflicts of interest between individuals of different sizes, strengths, sexes, and—particularly—different ages. If play were to function primarily for physical or neurological training, as has been argued, then animals of different ages should have greatly differing needs for and interests in play. Some of this is seen, but it is not the whole story, as stated by Fagen:

> Older animals, especially certain individuals, appear to become increasingly conscious of status, increasingly unwilling to lose play fights, and increasingly unwilling to accept a subordinate position in play, whereas younger playmates appear tense or anxious when older individuals solicit play from them.... These changes may suggest that as physical ability improves through play, play must be more like true fighting in order to further improve ability, behavior of immatures becomes a better predictor of adult behavior (including competitive ability), animals begin to assess each other, to misinform each other, and to cheat in play fights, the cost of play increases and the benefits of play decrease, trust breaks down, and play decreases in frequency. In this sense, play may be said to contain the seeds of its own destruction. The facts are not quite this simple, however.

Although examples of the ontogenetic transition from play to more serious living can be cited, there still are many counter-examples and enough variation, including older animals playing well with much younger individuals, to cloud the picture and indicate that more thought and research are needed.

One might expect the breakdown in play to involve similar-aged individuals competing for status, whereas younger animals would not pose a threat; hence play would not be disrupted. Sociobiological theory, including kinship, inclusive fitness, and game theory, are being considered for possible resolutions for some of these problems. (It is perhaps appropriate that game theory be applied to play behavior.) It has been shown, for example, that playmates of mixed ages are generally siblings or that third-party intervention in play that is not going well is by a relative of the loser. Intervention that disrupts play, on the other hand, usually is by individuals that are not closely related to the players.

Social play has been hypothesized to facilitate the acquisition of social and communication skills or otherwise provide information about other individuals with which an animal is playing (Gomendio 1988). Alternatively, social play may help develop the physical condition for future encounters (Caro 1988). Males of many species, for example, engage in play more than females, as if in preparation for future intrasexual combat. All of these hypotheses (practice, exercise, etc.) however, are similar to those in a nonsocial context, merely applied to a social setting.

The ectotherm-endotherm transition theory.   The most recent and significantly in-novative theory about why mammals and some birds but not other animals play was proposed by Burghardt (1984, 1988) after he had considered various implications of the surplus energy (resource) theory. His approach was to "look at play from the reptile up, as it were, rather than, as is usual, from the human down" (Burghardt 1988:140).

Most of the modern reptiles and their earliest ancestors that also gave rise to the separate phylogenetic lines leading to mammals and birds are (and were) ectothermic and lack(ed) parental care. Mammals and birds, however, independently acquired endothermy and—most likely in conjunction with the endothermy— parental care. The ectothermic-endothermic contrast is correlated with a constellation of characteristics (listed in Burghardt 1984, 1988), including the fact that neonates (new born or newly hatched young) of ectotherms are, with few exceptions, precocial and independent. They must obtain their own food, shelter, other resources, protect themselves from predators, and possess perceptual and motor systems that are more adultlike right from the start. Furthermore, because they are ectothermic they have lower metabolic rates, more sporadic and shorter periods of activity, and a low energy life-style and physiology that needs less exercise. They generally are relatively unsuited to the conditions involved in play.

The young of mammals and birds, on the other hand, are altricial or relatively altricial compared with the young of reptiles. This leaves them with a period of life in which adultlike behavior is not only unnecessary but even inappropriate. As a result they have different behavior patterns that are, in addition, based on a higher degree of learning than occurs in reptiles (although reptiles are also capable of learning to a limited degree, Burghardt 1977 b). As mammals and birds become adult and independent, they acquire more complex behavior (and also the need for fully capable muscular and cardiovascular systems). In fact the adult behavior of mammals and birds is more complex than the adult behavior of reptiles. Before becoming adult there is a much greater difference in the behavior pattern of young and adults than occurs in the reptiles as shown in Figure 19-14.

During the transition from ectotherms without parental care to endotherms with parental care, the ancestors to mammals and birds would have lost the selective ad-

Figure 19-14   A topographic landscape comparing the complexity of behavior in young with adult reptiles and mammals, suggesting an evolutionary continuity and reorganization of some behaviors during the ontogeny of behavior.
From Burghardt, G.M. 1988. In: E.M. Blass, editor. Handbook of behavioral neurobiology, vol. 9. Plenum, New York.

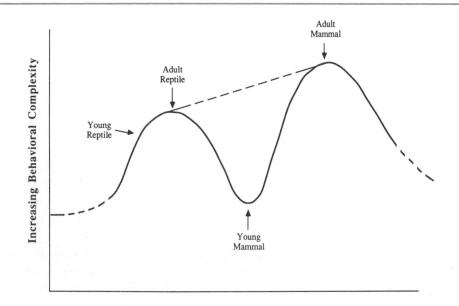

Gordon M. Burghardt

Gordon M. Burghardt
Play and the Comparative Method

Play and playfulness in animals are topics that attract much attention from the public at large and students in animal behavior classes. The study of animal play, as well as human play, has had a distinctly checkered history. Play seems to be "not serious" although scientists take what they study very seriously indeed. In addition, the lack of a clear role for play in the life of an animal, compared with feeding, locomotion, mating, and avoiding enemies also made it seem of secondary importance. Yet 20 years ago, when several students and I raised and studied two orphaned black bear cubs (Burghardt, in press), the sophistication, variety, and physical energy involved in their play had an enormous, though delayed, impact on my thinking about the relation between evolution and development. The richness of social play in bear cubs particularly intrigued me since adult bears are rather solitary outside mother-cub associations. Thus play as simple practice or social learning of adult behavior seemed unlikely.

Research with bears was a respite from my primary focus on behavioral development in reptiles. In my years of watching neonate reptiles of all kinds in both field and lab, I saw plenty of exploratory behavior and complex neonatal social interactions but nothing resembling the physically vigorous repetitive behaviors we label play in mammals and some birds. Occasional references to play in insects, fish, and reptiles are best evaluated as uncritical anecdotes or misinterpreted behavior based on lack of accurate natural history information.

It is essential to analytically separate the causes from the functions of any behavioral phenomenon, including play. So much emphasis has been placed on the enigma of why animals play that the proximate mechanisms were ignored. Herbert Spencer's surplus energy theory, derided by functionalists, seemed to me to provide an important insight: play is a by-product of physical well-being and opportunity (Burghardt, 1982). This led me to compare the life history and physiological characteristics, at the class level, of reptiles and mammals.

vantage of retaining specific adultlike behaviors during the immature stages, but they would have still possessed the behavior patterns. That sets the stage, according to the theory proposed by Burghardt, for the development of play. Not only would there be less selection for keeping the behaviors adultlike, there would be a new selective environment involving parental care and dependency, which is more like the process of domestication (discussed in Chapter 22). This could lead to deterioration of the behavior patterns, changes of response thresholds, broadening of effective stimuli, and, along with increased metabolic capacity resulting from endothermy, produce a marked reorganization of the behavior patterns in young animals.

This theory does not require play to be functional, although it does not exclude the possibility of functions such as exercise. The ectoderm-endoderm transition theory also makes possible several predictions. Burghardt (1988) lists and discusses 19, some of which are related to metabolic considerations and others related to developmental aspects. Smaller species of animals should be expected to play less

Many of these seemed to be related to the reptile-mammal dichotomy in play. For instance, take metabolic rate, which is about 10 times lower in reptiles, and relative precocity at birth, which is much lower in mammals. Although it is impossible to prove that the reptile-mammal play dichotomy is explained directly by such differences, the fact that metabolic rate and degree of precocity vary within mammals allows for comparative tests. That is, if the metabolic difference between mammals and reptiles is responsible for the difference in play, then mammals with low metabolic rates should play less, and less vigorously, than mammals with high metabolic rates. Similarly, mammals differ in how dependent they are on their parents at birth. Thus those with longer or greater dependence should play more than more precocial mammals. If energetic constraints are a factor, animals that live in a medium in which locomotion is less energetically costly, such as water, should play more and, if living in both terrestrial and aquatic environments, should play more in water. Aquatic mammals are the most playful and do play more in water than on land.

Although the comparative data on play are far from complete, qualitative tests support the view that life history and physiological factors underlying reptile-mammal differences also are associated with differences among endothermic vertebrates.

There will be exceptions to any global integrative view of a behavioral phenomena be it feeding, aggression, mating systems, or play. It is the exceptions that often prove most useful in clarifying, amending, or discarding a theory, but a comparative perspective and theory is necessary to both force us to examine diversity and to gather critical data to evaluate theories. Lacking this, problems arise, such as when studies of learning are limited to domestic rats, pigeons, and college students. Ultimately a comparative perspective is needed even for considering differences within a species. Consider the problems and controversy that arise, for example, when universal health care guidelines are based on research done only with one sex, nationality, race, or socioeconomic class.

than larger ones, for example, because of the increased energetic costs as the volume-to-surface-area ratio decreases. As an example of a prediction along developmental lines, relatively precocial species should play less than those that are more altricial.

Before leaving the various play theories, it is important to emphasize that there is no necessity for a single explanation for all play. There are many different species, and they face a diversity of environmental challenges. Factors underlying the development of what is called play in some species might be quite different from those causing similar-appearing behavior in other species. Gomendio (1988) notes different developmental trends in types of play and suggests that play may be a heterogeneous category of behavior even within a single species. Although the exercise and ecto-endotherm transition theories appear to be the best general explanations at the moment, they need not be universal, and they may appear best only because competing ideas lack data and theoretical development.

Development and the Acquisition of Specific Changes in Behavior

Up to this point we have been discussing mostly general effects of development in behavior, such as general sensory abilities, emotional and social reactions, and speed of attack in predators. The distinction between general and specific is not always clear and obviously forms a continuum with much overlap. Some changes in behavior clearly involve quite specific outcomes, such as the type of animal with which an individual will mate or the specific sounds or songs that a bird or other animal makes. The specific aspects often show many similarities with the general aspects discussed earlier. A few examples are described next. An important component of the picture, the critical period, first needs to be discussed.

Critical or sensitive periods

Recall, from Chapter 18 and earlier discussions, that certain events in an animal's life often have to occur at a particular time during life for effects in the animal's behavior to be realized. Examples include the sexual organization of male brains in mammals, resulting from a surge of testosterone during a brief period around the time of birth, development of visual and sensorimotor pathways in young but not older kittens, and the development of normal emotional, social, and predatory aspects of behavior. The period of time during which the events impose their significant impacts on the behavior of the animal is called the *critical* or *sensitive period* of the animal's life. The phenomenon of critical period shows up repeatedly in respect to both general and specific behavioral changes and should be kept in mind in the following sections. Bateson (1979) provides further discussion of critical periods per se.

Imprinting

Perhaps the most familiar form of behavioral development in many animals is imprinting, the process of a young animal forming an association or identification with another animal, object, or class of items. Imprinting received much attention

Figure 19-15 Apparatus used in the study of laboratory imprinting consists primarily of a circular runway around which a decoy duck can be moved. In this drawing, a duckling follows the decoy. The controls of the apparatus are in the foreground.
From Hess, E.H. 1959. Science 130:134.

beginning with Lorenz (1935) and has been the subject of much research and review (e.g., Immelmann 1972, Hess 1973 a, Hess and Petrovich 1977). It has been recognized or at least implied for centuries even in nursery rhymes such as "Mary had a little lamb" (also see comments in Chapter 2).

In the best-known examples of imprinting, involving the following of a parental object (filial imprinting), a young animal forms an impression of its parent soon after hatching or birth and follows that object for a period of time thereafter. Under natural circumstances the object is the actual parent, and the following is adaptive. Under artificial conditions, however, if the true parent is not present, the young animal may imprint on almost any other moving object nearby, including humans (e.g., Figures 2-3 and 19-15). Two important early findings of imprinting studies were that (1) many species, particularly precocial ones, show a marked critical period during which exposure to the imprinting object must occur (Figure 19-16) and (2) the strength of the imprinting depends on the effort expended by the young animal in following the object (Figure 19-17); that is, the imprinting process is not merely passive; the young animal itself must actively participate.

In addition to filial imprinting, many other categories of imprinting have now been identified. These include imprinting that leads to eventual choice of sexual partner, or sexual imprinting, imprinting to types of food that will be eaten, and,

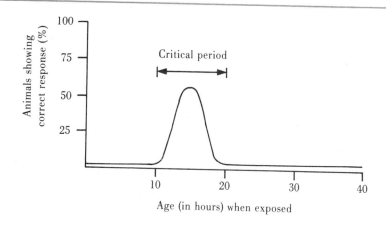

Figure 19-16   Critical age for laboratory imprinting in mallards expressed as the percent of animals making perfect scores.
Modified from Hess, E.H. 1973. Imprinting. Van Nostrand Reinhold, New York.

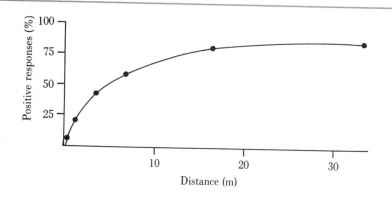

Figure 19-17   Strength of laboratory imprinting as a function of distance traveled by ducklings, with exposure time held constant.
Modified from Hess, E.H. 1973. Imprinting. Van Nostrand Reinhold, New York.

it has been proposed, imprinting to specific habitat or nesting substrate that will be occupied later in life. Imprinting to the odors of a home stream, for example, is described for salmon in Chapter 7.

As might be expected, there is much variability in the roles that imprinting plays in the lives of different species. Some species, such as golden eagles (Durden 1972), show sexual imprinting onto other species but slowly revert to their own species if given the opportunity. Some species show sexual imprinting that is not reversible, and some show no sexual imprinting whatsoever—they appear to choose the correct class of mating partners and the correct sex of the correct species, if such is present, or they simply will not engage in reproductive behavior. Variation in the types and extent of imprinting occurs markedly even among relatively closely related groups such as waterfowl.

Although studied most extensively among birds, imprinting is also well known among other vertebrates, particularly mammals. In the most specific sense, it is responsible for individual recognition, particularly between mothers and offspring, and is based on individual sights, sounds, or odors encountered soon after birth or hatching (the critical period). The mother-offspring bonding is reciprocal in some cases and involves a critical period during which the mother learns which offspring to accept or reject thereafter. Mammals also show some forms of sexual imprinting. Fillion and Blass (1986), for example, raised infant male rats with mothers that had artificial lemon-scented nipple and vaginal odors. The males were then not exposed again to those scents until adult, at which time they were given either normal sexually receptive females or sexually receptive females with lemon-scented vaginal odors. The males readily ejaculated with the scented females but not with normal females.

Some invertebrates also show imprinting or imprinting-like phenomena in relation to some behaviors. Examples include prey choice by larval dragonflies (Blois and Cloarec 1985), recognition of conspecific adults by ants (Moli and Mori 1984), and identification of conspecific cocoons by ants (Carlin et al 1987, and references therein).

An understanding of imprinting has proved quite useful in applied cases of animal behavior (Chapter 22). It forms the basis for cooperative (as opposed to forced) artificial insemination of many species of wildlife (e.g., Grier 1973), and it has been used judiciously where sexual imprinting problems are not involved or expected in cross-fostering offspring to other species for reintroduction into the wild. Persons wanting further information on imprinting should refer to Hess (1973) or Hess and Petrovich (1977).

Development of bird vocalizations

The means by which birds acquire the ability to produce specific patterns of sound has attracted much attention and has elucidated the ontogeny of behavior perhaps better than any other single topic. Bird vocalizations are varied and complex, providing extensive detail and means of identification, and the birds themselves are relatively easy to raise and manipulate in captivity. Through the use of magnetic tape recordings and contemporary electronic technology, it is relatively easy to accurately reproduce sounds, create artificial analogs, and experimentally manipulate most aspects of the sound. As a result, there currently exists an immense and continually growing body of literature on the subject. Pioneering work was done by Marler (1952), Thorpe (1958), Konishi (1965), Emlen (1972), and many others.

Interest, research, and discussion on the ontogeny of bird song has continued unabated up to the present. For an example of recent references see Marler and Peters (1988 a,b), Irwin (1988), and Pepperberg and Neapolitan (1988). Because of space limitations, only a cursory summary of the findings are presented here. Persons wishing further details should consult any of the many available reviews on the subject (e.g., Marler and Mundinger 1971, plus references in Marler and Peters 1981).

There is much variation in the vocalizations of different species, of different individuals of a given species, and within an individual. Some birds have only a few simple vocalizations, and others may have a rich and changing repertoire. In general the simpler and nonreproductive sounds are referred to as *calls*. The term *song* usually is reserved for the more complex reproductive vocalizations and, in the minds of most ornithologists, just for species in one order, Passeriformes, commonly known as *perching birds* or *songbirds*.

Among other, nonpasserine orders of birds, it appears that vocalizations, including those used in reproduction, are largely impervious to environmental effects. These birds, when mature and in reproductive condition, generally give species-typical vocalizations regardless of whether raised in isolation, in the presence of other species, or under other types of experimental manipulation such as deafening. Examples of such birds include chickens, and doves and cranes as discussed in Chapter 5.

The Passeriformes, however, present quite a different picture. Some, such as the brown-headed cowbird, have a relatively typical song (although cowbirds also incorporate flexibility, as is discussed shortly). At the other extreme are birds that have a more variable natural repertoire and in some cases can accurately mimic the sounds of other organisms, including humans. Mynah birds and a few of the related starlings, for example, can mimic to the extent that a listener can identify the gender and often even the individual human that the bird is mimicking.

The group of birds that has been most instructive regarding the development of behavior, however, includes those that are intermediate between the two extremes— birds that modify their songs during development but only within limits. There are many birds in this group, but various species of sparrows have received the most study. The results of different studies, using different techniques, different species, and by different persons in different locations, have provided different pieces to what seems to be emerging as one puzzle or one picture. Because of this, we have synthesized a general or hypothetical case rather than citing different examples for different parts of the story. Persons wanting specific examples from which this synthesis was built should refer to the reviews cited earlier. Details vary depending on which species one is considering, but the general pattern appears to fit most species.

A generalized summary of various research is shown in Table 19-2. From this has been derived the interpretation illustrated in Figure 19-18. The complete, normal sequence in the acquisition and ontogeny of song in this group of birds is as follows. While still juveniles or young adults (depending on the species), the young male hears the song of its own species from other birds around and close to it. During a critical period some form of memory (sometimes referred to as a *template*) of the sound pattern is established within the bird's nervous system. The bird does not sing or attempt to repeat the sound itself during this time; it simply has captured a memory of the sound. Later, often after a period of several months, when it begins

Table 19-2 Generalized results of experiments on song learning among many passerine birds*

Condition and Treatment(s)			Fledgling → Critical Period	→ Sub. → Rehearsed →Primary song Song Song	Results
Normal wild birds	Male		Normal		Normal song
	Female		Normal		No song
Female hearing own species			Injected with testosterone		Sings like male
Male isolated from own species	Kept in quiet surroundings			⊢ ← Isolated after critical period	Normal song
			Isolated before critical period → ⏐		Develops own song
	Presented with recordings of song	Normal song from own species	Before → ⏐		Develops own song
			⏐ ← During → ⏐		Normal song
				⊢ ← After critical period	Develops own song
		Backward song from own species	(During critical period)		Sings backward
		Song from different species	(During critical period)		Develops own song—not normal and not other species
Surgically deafened	Before critical period		Deafen → ⏐		No song
	After critical; before full primary			Deafen	Song deteriorates
	After full primary			Deafen	Normal, lasting song

*See text for explanation.

Figure 19-18 Generalized pattern of events in song development in several species of songbirds.

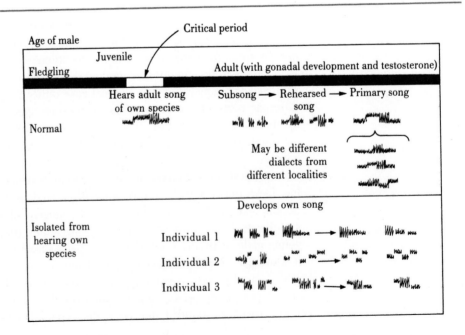

to mature sexually and under the influence of testosterone on the nervous system, the bird begins to vocalize. Initially the song is not complete but consists only of bits and pieces, the *subsong*. These components eventually (over days or weeks) blend and begin to resemble the song that the bird heard as a juvenile. This song is then rehearsed and perfected until it becomes the full primary song.

Apparently during the rehearsal period the bird is listening to its own song, comparing it with the pattern in its memory, and correcting it until the two match. After this point the song becomes fixed and will not subsequently change. Before becoming fixed, the song pattern is vulnerable. If prevented from hearing their own rehearsals, via deafening, birds slowly lose the correct pattern that was developing, and the song deteriorates. If they never heard a song in the first place, they develop and fix their own concoction. If they hear another song, artificial, abnormal, or from another species, whether or not they memorize and subsequently copy it (which process is involved or whether both processes are involved is not known) depends on how close the pattern is to the natural song for that species. In other words, there are limits, or a range, regarding the patterns that will work. Generally they have to be from the normal song or closely resemble it. There often is much variation in the natural song, however, and local dialects in song patterns may develop. Marler and Tamura (1964), for example, demonstrated dialects in three populations of white-crowned sparrows within a 100-mile distance near San Francisco.

The ranges of variation that different species permit vary considerably. Many species show some variation but do not accept the songs of other species or radically different artificial patterns. Some not only accept but naturally copy the songs of other species. At the other extreme, some have a narrow range and quite stereotyped pattern.

The singing or responses to singing by these birds is clearly under hormonal influence. Normally males sing and females do not, but the females respond to the male songs in various ways. Females given testosterone, however, go through the sequence of song development as if they were males and eventually sing. Female responsiveness to male song follows an analogous pattern to that of male singing; that is, the specific songs to which females respond when they become sexually mature depend on the patterns they heard when they were young (e.g., Baker et al 1981, Searcy and Marler 1981). Other research, involving male canaries (Nottebohm 1981, 1984, 1986, 1989), has shown that changes in the acquisition of song (including new song in subsequent years—canaries are more flexible than many of their relatives) is accompanied by measurable changes, including the development of new neurons in specific regions of the brain. Additional findings include the importance of gonadal steroid hormones in the development of bird song (e.g., Heid et al 1985, Weichel et al 1986, Nottebohm et al 1987). Also for some species the nature of the object (tutor) from which the birds learn their song may be important (e.g., Clayton 1988). Many species require more than hearing just tape-recordings of their species' song; the young may need the visual presence of live conspecific tutors before they learn to sing, even if they were not raised by conspecifics and even if the conspecific tutor does not sing the correct song.

There is still more to the development of song in birds. Even in the case of those with the most stereotyped songs, experience and environmental influences may affect the singing. The classic case of stereotyped song is that of the cowbird (*Molothrus ater*). The species is a brood parasite. Females lay their eggs in the nests of well over 100 other species, which then raise the young. Young cowbirds thus are reared in the presence of other species and do not have early experiences hearing their own species. Thus it has always been assumed that their songs are under strict genetic control. Research with isolated, hand-reared individuals seemed to confirm these beliefs. Isolated males give typical songs, and isolated females respond to them.

It has been discovered (West et al 1981) that there nonetheless is variation among cowbird songs and that their potency or effect on other male and female cowbirds varies considerably. The more potent songs stimulate more female sexual responses and also cause a much greater aggressive response from other males. Normally the potency varies with the individual male's position in the dominance hierarchy of the cowbird group it joins after leaving its foster parents. Through fighting and other social interactions, the male learns its place in the local social setting and thereby is affected in how potent a song to sing; that is, variation in song is influenced by social experiences. In artificial manipulations the most dominant males were moved from their own groups of cowbirds to others. When they gave their most potent songs in the new groups, they were killed by males already established in those groups. Meanwhile the groups from which they came shifted in their absence, and new males became most dominant and began singing the most potent songs. Cowbird song has become the focus of ongoing interest in the ontogeny of and effects of experience on song development under conditions that are not superficially typical of song learning in other species. If interested in further details, see West and King (1985), King et al (1986), Rothstein and Fleischer (1987), and Rothstein et al (1988).

Birds are not the only organisms in which unexpected cases of learning have been discovered. Environmental influences, critical periods, and related phenomena sometimes show up in species of animals where not anticipated and sometimes do not appear where they might be predicted, at least in those taxonomic groups of animals with the most complex nervous systems. This brings us to a general consideration of learning in the next chapter.

SUMMARY

Some behaviors occur only early in life, to be replaced subsequently by other behaviors. There is much variability among species in the sequencing of behavior, presence and absence of different behaviors, rates at which they develop, and the influence of environmental variability on behavioral variability. Interpretation and generalization of observations about the development of behavior are complicated by major developmental differences among animals.

Maturation, which involves normal growth and development of underlying anatomical parts and pathways, plays an important role in the initial appearance and changes of many behaviors. Some of the changes that might otherwise be interpreted solely as learning often can be accounted for largely as maturation of underlying systems. Maturation and learning, however, are not mutually exclusive. The effects of environmental variability and experience on development of maturing behaviors vary greatly.

The development of normal behavior has been studied in part through the opposite techniques of environmental deprivation and enrichment. In many cases, these appear to impose little effect on the behavior in question. In others, however, particularly in the ontogeny of many mammals, the effects are striking.

Play behavior is fascinating and familiar, but the function or functions, if any, are not well understood. Several explanations have been proposed for the evolution and presence of play behavior. They include the following play theories: surplus energy, pleasure, arousal, practice, exercise, social and a newly proposed explanation related to the transition from ectothermic reptiles without parental care to mammals and birds (along two separate line of evolution) that are endothermic and have parental care. According to the ectotherm-endotherm transition theory, play would amount to a reorganization of behavior in young animals from previous independent, adultlike behaviors to a different set of behaviors that develop under conditions of dependency.

Effects of sensory and environmental input, whether normal or resulting from either deprivation or enrichment, frequently occur only during critical periods in the individual's life; that is, particular inputs or deprivations produce an effect if presented at one time but not at another. Critical periods have been shown most clearly during imprinting phenomena and during the process of song learning among several species of birds. Critical periods also occur, however, in many other, less obvious instances of behavioral development in some species.

Plasticity of behavior and the effects of experience and environmental variability differ greatly from species to species, sometimes even among closely related species. Furthermore, what is modifiable in one species may not be in another species, but the latter may have different components of its behavior that are modifiable; that is, different species vary in which aspects of behavior are plastic.

FOR FURTHER READING

Bateson, P. 1979. How do sensitive periods arise and what are they for? Animal Behaviour 27:470-486.
A thorough review and discussion of sensitive periods and various hypotheses concerning them.

Burghardt, G.M. 1988. Precosity, play, and the ectotherm-endotherm transition. In: E.M. Blass, editor. Handbook of behavioral neurobiology, vol. 9:107-148. Plenum, New York.
A description and discussion, including predictions, for a new theory of play in mammals and birds.

Burghardt, G.M., and M. Bekoff, editors. 1978. The development of behavior: comparative and evolutionary aspects. Garland Publishing, New York.
A collection of papers devoted to the ontogeny of behavior and ultimate considerations.

Carlin, N.F., R. Halpern, B. Hölldobler, and P. Schwartz. 1987. Early learning and the recognition of conspecific cocoons by carpenter ants *(Camponotus sp.)*. Ethology 75:306-316.
An original research paper providing experimental results, comparison with other species, and a discussion relating the results to the general topic of slave-making and advanced forms of social parasitism (see Chapter 8).

Caro, T.M. 1988. Adaptive significance of play: are we getting closer? Trends in Ecology and Evolution. 3(2):50-54.
A brief status report of current information and hypotheses.

Fagen, R.M. 1981. Animal play behavior. Oxford University Press, New York.
A thorough and extensive review of play behavior. A major treatise on the subject.

Fillion, T.J., and E.M. Blass. 1986. Infantile experience with suckling odors determines adult sexual behavior in male rats. Science 231:729-731.
An interesting report on the effects of odors on eventual sexual preferences.

Gomendio, M. 1988. The development of different types of play in gazelles: implications for the nature and functions of play. Animal Behaviour 36:825-836.
An original research paper with both descriptive data and theoretical discussions. Well written and readable.

Hailman, J.P. 1969. How an instinct is learned. Scientific American 221:98-106.
A classic article on maturation and learning.

Harlow, H.F. 1959. Love in infant monkeys. Scientific American 200:68-74.

Another classic article, this one on the effects of maternal deprivation.

Irwin, R.E. 1988. The evolutionary importance of behavioural development: the ontogeny and phylogeny of bird song. Animal Behaviour 36:814-824.

A review of the literature and theoretical discussion of bird song, its development in the individual, and comparative, evolutionary aspects.

Marler, P., and P. Mundinger. 1971. Vocal learning in birds. In: H. Moltz, editor. The ontogeny of vertebrate behavior. Academic Press, New York.

A major review of the ontogeny of song in passerine birds.

Marler, P., and S. Peters. 1988 a. Sensitive periods for song acquisition from tape recordings and live tutors in the swamp sparrow, *Melospiza georgiana*. Ethology 77:76-84.

A recent original research paper in a continuing program of research by one of the pioneers of the field (Peter Marler) and a colleague. This paper provides results for the swamp sparrow and contrasts them with previous information from another member of the same genus, the song sparrow.

Marler, P., and S. Peters. 1988 b. The role of song phonology and syntax in vocal learning preferences in the song sparrow, *Melospiza melodia*.

Another investigation into song learning of the song sparrow, probing deeply into a number of details of the phenomenon. An excellent example of the current state of the art.

Martin, P. 1984. The time and energy costs of play behaviour in the cat. Zeitschrift für Tierpsychologie 64:298-312.

This paper reports the results of observations on a familiar playful animal, the domestic cat. It provides not only an example of relevant information for the subject of this chapter but also an example of techniques, ranging all the way from simple, direct observation to measurements requiring more sophisticated equipment (e.g., an oxygen analyzer).

Mimura, K. 1986. Development of visual pattern discrimination in the fly depends on light experience. Science 232:83-85.

An interesting report that parallels the findings on the effects of light deprivation in vertebrates.

Moore, A.J., M.D. Breed, and M.J. Moor. 1987. The guard honey bee: ontogeny and behavioural variability of workers performing a specialized task. Animal Behaviour 35:1159-1167.

An original research paper describing changes in behavior in worker honeybees, with emphasis on guard behavior.

Nottebohm, F., M.E. Nottebohm, L.A. Crane, and J.C. Wingfield. 1987. Seasonal changes in gonadal hormone levels of adult male canaries and their relation to song. Behavioral and Neural Biology 47:197-211.

A research paper that reports on the relationship between hormones (particularly testosterone) and song learning in adult male canaries, a species that shows in essence continuing ontogenetic development in its behavior.

Roggenbuck, M.E., and T.A. Jenssen. 1986. The ontogeny of display behaviour in *Sceloporus undulatus* (Sauria: Iguanidae). Ethology 71:153-165.

A descriptive report of behavior that appears in essentially fully developed form and that is determined through variance to be "almost totally innate." An example of a behavior that develops with little role played by learning or environmental modification.

Rothstein, S.I., D.A. Yokel, and R.C. Fleischer. 1988. The agonistic and sexual functions of vocalizations of male brown-headed cowbirds, *Molothrus ater*. Animal Behaviour 36:73-86.

A detailed analysis and discussion of three separate vocalizations of these cowbirds.

Smith, P.K., editor. 1986. Play in animals and humans. Basil Blackwell, New York.

A collection of 14 chapters by various authors active in the field of play behavior. Another major reference (in addition to Fagen 1981) on the subject of play.

West, M.J., A.P. King, and D.H. Eastzer. 1981. The cowbird: reflections on development from an unlikely source. American Scientist 69:56-66.

An excellent overview and introduction to the effects of experience on singing by cowbirds.

West, M.J., and A.P. King. 1985. Social guidance of vocal learning by female cowbirds: validating its functional significance. Zeitschrift für Tierpsychologie 70:225-235.

Further work on the effects of experience on vocal learning of males in the presence or absence of female cowbirds. Females do not sing but nonetheless clearly affect the outcome of singing by males.

20

LEARNING

Learning has long been one of the most fascinating and at the same time challenging and frustrating topics in behavior. The major problem has been our lack of understanding of the internal mechanisms of learning and memory. Memory, a necessary component of all learning, has been particularly elusive. Other components and processes of learning have similarly remained unknown. As a result, learning has been a black box about which we have only been able to make inferences (guesses) based on observed changes in behavior as they relate to particular behavioral conditions and inputs. Different workers in the field of learning have frequently disagreed in their interpretations. This chapter provides a description of some of the categories of learning, differences in learning observed among different groups of animals, and an overview of the relationship between the outputs versus inputs of learning.

The previous chapter ended with a discussion of critical periods, imprinting, and the development of bird song. These topics are important to the contemporary view of learning. In fact the ontogeny of behavior merges inseparably with the larger subject of learning so that this chapter amounts to a continuation of the previous one. In addition to imprinting, of which bird-song development in part is an example, there are many other categories of learning that have been identified.

WHAT IS LEARNING?

A universally accepted definition of learning does not exist. For our purposes the following is used: Learning is a specific change or modification of behavior involving the nervous system as a result of experience with an external event or series of events in an individual's life. By focusing on external events, one can exclude fatigue and maturational or purely developmental processes. By specifying the nervous system, we exclude other changes that might affect behavior, such as a broken leg. Within the framework of what normally is considered learning, however, there are at least 10 possibly unique types of learning. Whether or not they are really unique or are all merely variations on one or a few general underlying processes is a matter of ongoing debate. At least at the cellular, neuronal level there are many ways for neural processing to be modified.

LEARNING CATEGORIES AND TERMINOLOGY

Under various circumstances and in different species the outward expression of learning varies considerably, and this has led to the classification of different categories of learning. Just as there is no generally agreed on definition of learning, there is no one system of classification, and the categories vary widely among different authorities. Thus the following categories are not fixed, and there may be much overlap from one category to another. They simply include types of learning commonly recognized and discussed in the fields of biology and psychology. These categories of learning have also been collectively lumped into larger categories of nonassociative learning (habituation and sensitization), associative learning (classical and operant conditioning plus taste aversion), and complex learning (the remaining categories) (Mackintosh 1983). For expanded discussions on these categories see Domjan and Burkard (1982), Mackintosh (1983), Roper (1983), or any of numerous other available reviews and books.

Habituation

Habituation involves the gradual fading of an unlearned response to a stimulus that proves to be safe or irrelevant. The initial response usually is one associated with danger, such as fleeing, crouching and becoming immobile, or some form of startle response. The stimulus is generally something new or unfamiliar to the animal. After repeated occurrence without significant meaning to the animal, the stimulus loses its novelty and is ignored. This probably is the most primitive and universal form of learning. It can be seen in virtually all species and is an important means of dealing with an otherwise overwhelming amount of environmental sensory input, most of which is irrelevant to the animal. Even protozoan species have demonstrated a clear ability to habituate that is not simply a matter of temporary fatigue.

Habituation is the type of learning involved in the early hawk-goose study of fear responses, as discussed in Chapter 2. In that case, birds habituated to the appearance of other birds that were not hawks or to hawk shapes that did not attack. One of the earliest and more interesting demonstrations of habituation (Peckham and Peckham 1894) involved orb-weaving spiders. If a vibrating tuning fork is held near the web of these spiders, they will drop from the web on a thread or web material. With subsequent presentation of the stimulus, the spiders drop shorter and shorter distances until they cease to drop at all. Behavioral responses to pain are relatively immune to habituation, as indeed one might expect them to be; pain normally alerts the animal to a problem that requires attention.

Sensitization

Sensitization is, in its outward appearance, approximately the opposite of habituation. That is, with sensitization the animal shows an increased response to repeated stimuli. This generally involves highly relevant stimuli, such as encounters with predators, unlike the irrelevant stimuli involved in habituation. Internally it is thought to involve different underlying neural processes.

A good example of sensitization is birds in a flock that experience an attack by a hawk that captures and carries off one of the members, struggling and screaming. Thereafter the birds show increased alertness and escape responses at the appearance of a hawk. Another example of sensitization occurs in the marine snail, *Aplysia*, as described later in the chapter under mechanisms of learning.

Classical Conditioning

Classical conditioning focuses on changes in the stimuli that elicit behavior and is based on natural or normal stimulus-response systems of species. The basic stimulus and response are referred to as the *unconditioned stimulus* (US), such as food, and the *unconditioned response* (UR), such as salivation. When the US is properly paired with a novel or different stimulus, such as the sound of a bell, the new stimulus, which is referred to as the *conditioned stimulus* (CS), may elicit the behavior, now referred to as the *conditioned response* (CR).

The bell-food-salivation example is easily recognized as being associated with the Russian physiologist and psychologist Ivan Pavlov. This form of conditioning is often called *Pavlovian conditioning* because of his well-known work on the subject. In the original Russian writing on classical conditioning, the terminology should have been translated as *conditional* rather than conditioned. The term *conditional* better reflects the meaning that the conditional stimulus depends on (is conditional on) pairing with the initial (unconditional) stimulus to elicit the response. The first translations used the term *conditioned*, however, the term *conditional* is slowly working its way into the vocabulary.

Although usually studied in the laboratory with artificial CSs such as ringing bells or flashing lights, classical conditioning clearly is a natural phenomenon with real biological value. Under natural conditions animals may learn to associate the presence of food or other items with other normally occurring stimuli. One of the major characteristics of classical conditioning seen in connection with this point is known as *sign tracking* or *autoshaping*; that is, learned behavior will become shaped by or track the associated stimuli or sign of importance to the animal. Predators, for example, learn to cue into and follow the auditory, olfactory, or visual signals from potential prey.

In classical conditioning the specific nature of the pairing between CS and US is important. The strength of the conditioning depends on several factors, including (1) the consistency with which the US follows the CS, (2) amount of time between the US and the CS (the sooner the better, usually within a range of seconds or minutes), and (3) the particular relevance or strength of the US to the animal at the time it is presented. The strength of food as the US, for example, is more important to a hungry animal than to one that is satiated.

Several other important terms are used in connection with conditioning. The normal sequence, called *forward conditioning*, is for the CS to be presented before the US. If the reverse occurs (which generally does not lead to any learning), it is called *backward conditioning*. If the CS and US are presented in random sequence with variable intervals in between so that there is no clear association between the two, the situation has been referred to as *pseudoconditioning* or, more recently, as a form of *sensitization*. (This phenomenon and coincidences between events may contribute to human superstition.) If changes in behavior are observed with backward or pseudoconditioning, they generally are considered as artifacts and not signs of bona fide learning.

When a clear association between the CS and the US ceases to exist so that the US occurs without being preceded by the CS, the animal's response to the CS may decline and disappear completely. If so, the conditioning is said to have been extinguished, or extinction has occurred. This situation, in which the CS no longer occurs with the US, is different from the phenomenon of *forgetting*, which is a decrease in the conditioning resulting from an intervening lapse of time. Analogous phenomena, which may or may not share an underlying neural substrate, can be seen in losses of habituation. A loss of habituation as a result of a lapse of time without the stimulus is referred to as *recovery*, which is similar to forgetting. A change of habituation as a result of a stimulus becoming relevant or because of an interaction with a new stimulus is called *disinhibition*. An example of classical conditioning involves earthworms with weak vibration as the CS and light as the US (Ratner and Miller 1959) (Figure 20-1).

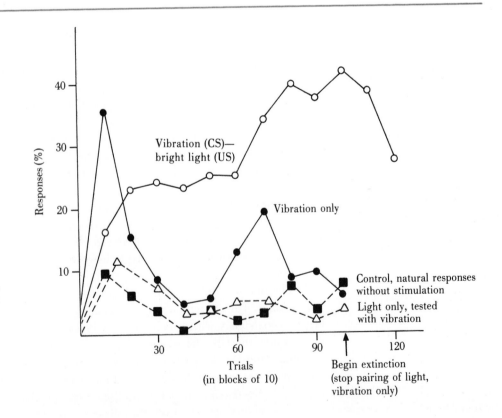

Figure 20-1 Learning versus related phenomena in earthworms *(Lumbriculus variegatus)*. Bright light was used as the US, a mild vibration as the CS. The classical conditioning group received the two stimuli paired in the standard fashion. The sensitization group received vibration only. The pseudo-conditioning group received 10 trials of light only, followed by 5 of vibration only, alternating throughout the series of trials. The control group received no bright light. Responses were measured by testing the worms with weak vibrations. Modified from Ratner, S.C., and K.R. Miller. 1959. Journal of Comparative and Physiological Psychology 52:102-105.

Operant Conditioning

Operant or instrumental conditioning is a second major form of conditioned learning. In this type of conditioning, actions, called *operants*, are instrumental in producing certain consequences in the environment of the animal. The consequence is called a *reinforcing stimulus* or *reinforcer*. This type of conditioning is involved in maze learning and what many (e.g., Thorpe 1963) call *trial-and-error learning*.

This type of conditioning has been the subject of a vast amount of research by numerous workers using many kinds of methodologies. There are subtle differences between instrumental and operant conditioning, and there is an array of terminology. Persons interested in the details should consult a contemporary text on the psychology of learning (e.g., Domjan and Burkhard 1982).

One early example of operant conditioning was demonstrated in planarians in the Netherlands by van Oye (1920). He showed that planarians could gradually be trained and perform better than untrained controls in tests where they had to travel up the side of their container, across the underside of the water surface, and down a wire to get to food suspended in the middle of the water (Figure 20-2).

A fascinating controversy and considerable confusion over planarian operant conditioning arose over later experiments with conditioning. At first the arguments were the usual in-house disagreements and bickering that seem to accompany studies in any area of research. The real fireworks started, however, when McConnell et al (1959) first reported that the specific learning could be passed on to regenerated planarians. If planarians are cut in half (or even smaller pieces), each part regenerates the missing portion. McConnell et al reported that if trained planarians were cut up, the regenerated individuals would learn faster than naive regenerates. This initiated a flurry of worm-running experiments that produced more and more surprising and unbelievable results and a journal, *Worm runner's digest*, that were not taken seriously by many biologists.

Figure 20-2 Conditioning of planarians to reach food. **A,** Naive animals do not reach food item suspended below the water's surface. **B,** Animals reach food suspended immediately at the surface by traveling along the undersurface. With gradual lowering of the food the animals learn to travel down the wire suspending the food. **C,** Animals readily travel to suspended food by using the wire. Modified from Corning, W.C., and S. Kelly. 1973. In: W.C. Corning, et al, editors. Invertebrate learning. Vol. l. Plenum Press, New York.

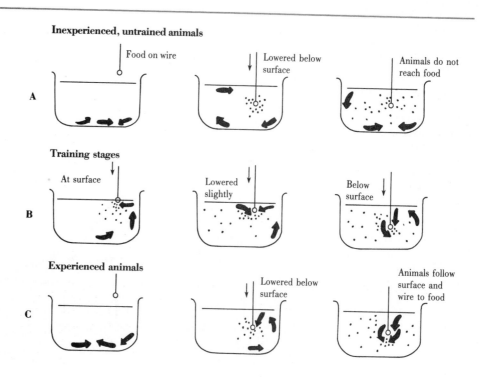

The most surprising experiments came with subsequent cannibalism studies (for a review, see Fjerdingspad 1971). Trained planarians were chopped up and fed to untrained planarians, and the previously untrained animals were then able to learn much faster. All of this suggested a transfer of learning that had some molecular basis, perhaps even specific memory molecules. This would be astonishing if true, but several persons found it incredible, and the research came under close scrutiny and criticism. In particular, some of the early work had not properly controlled for pseudoconditioning, fatigue of the animals, and other factors.

As a result of the problems, many of the learning experiments were redone more carefully. There were some negative or ambiguous results, but the majority seemed

Figure 20-3   Double discrimination learning of planarians. The US was shock, and the CSs were either light or vibration. **A,** Responses of animals given light paired with shock or vibration without shock, followed by extinction (neither paired with shock), then the reverse—vibration paired with shock or light neutral, followed by extinction trials. **B,** Responses of different subjects given pairings in reverse of those in A; vibration was paired with shock first and then light (with a period for extinction after each). Evidence of conditioned learning is provided by (1) acquisition of response only for appropriate stimulus, (2) more rapid acquisition of response in stage 3 than in stage 1, (3) extinction after shock was stopped, and (4) more rapid extinction in stage 4 than in stage 2. Modified from Block, R.A., and J.V. McConnel. 1967. Nature 215:1465-1466.

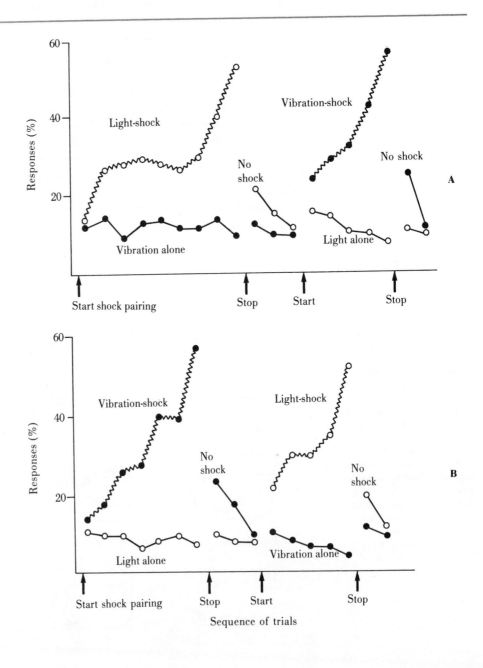

to clearly support an interpretation of bona fide learning. An example of one of the most carefully controlled and clearest studies is shown in Figure 20-3 (Block and McConnell 1967).

Further studies of possible molecular transfer of learning brought even more surprises, seemingly showing specific learning effects. McConnell and Shelby (1970), for example, trained planarians to go to one arm or another of a T maze; then they chopped them up and fed them to naive animals in four groups. Group I was fed animals trained to go to one arm; group II was fed animals trained to go to the other arm; group III was fed animals from both of the trained groups; and group IV was a control group, fed untrained planarians. Groups I and II, fed on animals trained for one arm or the other, learned to go to the proper arm much faster than the controls, whereas group III took longer than the controls. They showed more head-waving at the junction of the maze, as if undecided or confused. Other workers, however, were not always (or in some cases ever) able to confirm these results.

The possibility of molecular transfer, focusing primarily on RNA but also with some attention to protein, was further studied using techniques such as blocking RNA synthesis and extending the studies to other animals like crabs and rats. (Planarians have a simple digestive system that does not destroy macromolecules as in more complex animals. Thus studies using crabs and rats have had to rely on injections of material rather than simple, cannibalistic feeding.) The results of these supposed memory transfer experiments still have not been interpreted or understood to everyone's satisfaction.

Although the issue of memory transfer still is not settled, disagreement over whether or not planarians can learn has largely subsided. Most persons accept that conditioning exists in at least some species under some conditions. Planarians served to draw attention to the importance of considering simpler, invertebrate animals in the study of learning. Corning and Kelly (1973) conclude, "The flatworm may not achieve the position in psychology that the fruit fly gained in genetics and the bacterium achieved in molecular biology, but it certainly has stimulated interest in researching animals other than the rat."

In spite of some severe criticism of the research (Bitterman 1975) a good example of operant conditioning seems to have been well demonstrated in octopuses, particularly one species *(Octopus vulgaris)*. Earlier work involved discrimination learning whereby octopuses, which normally rush out to grab food items, could be conditioned to avoid food items under certain circumstances. Octopuses would refuse food that was paired with visual cues followed by shock (Figure 20-4). Subsequent work involved finer details of visual and tactile discrimination (Figure 20-5).

Several examples of operant conditioning have been demonstrated in spiders. It was shown as early as 1884 (Dahl) that spiders would avoid tackling bees, certain beetles, or otherwise edible insects coated with turpentine. Bays (1962) and Walcott (1969) trained different species of spiders to associate dead flies that had been dipped in either sugar or quinine solutions with sounds of different pitch. After the association had been learned, unflavored glass beads were substituted for the flies, and the spiders either discarded or bit the beads, depending on the pitch of the sound that was presented.

Operant conditioning also has been shown in several other arthropods. The use of conditioning in bees, such as for studies of color discrimination, is classic and

Figure 20-4 Conditioned learning in octopus. **A,** A crab and geometric figure (white square in this case) are presented to an octopus. **B,** If it takes the crab to eat but then is shocked by electrodes on the white square, **C,** it subsequently will not take crabs when presented with the white square. **D,** If the crab and square are moved toward the octopus, it not only will not attack but will retreat. Crabs presented alone will be attacked. By using similar and related techniques, octopuses can be trained to take crabs and avoid fish, vice versa, and discriminate among many different objects both by vision and by touch (see Figure 20-5 and 20-13).
From Boycott, B.B. 1965. Scientific American 212(3): 42-50.

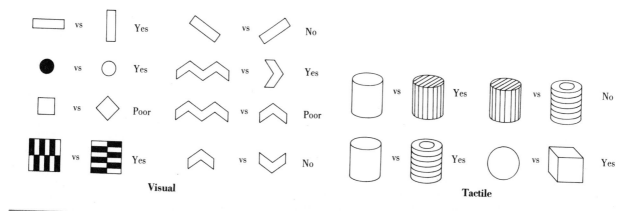

Figure 20-5 Examples of some of the visual and tactile stimuli used to test octopus learning and discrimination abilities. Octopuses could be taught to discriminate between smooth and grooved objects and, with some effort, between round and cube-shaped objects. They would not distinguish grooved objects, however, on the basis of vertical or horizontal grooves. Limits also were found for their ability to distinguish on a visual basis. Octopuses were tested also on a variety of natural objects, such as smooth and rough-shelled mollusks, which they could easily distinguish.
Modified from Sanders, G.D. 1975. In: Corning, W.C., et al, editors. Invertebrate learning, vol. 3. Plenum Press, New York.

well known. The specifics of learning in bees may be less familiar but are quite revealing. As revealed by von Frisch (1967), bees learn the color of flowers on which they land only during the 2 seconds before they land. In a different situation involving learning the hive location, honeybees relearn the position of their hive each morning but only with their first trip out. Bees will return to hives that were moved at night before the first trip, as is known to beekeepers around the world. If the hive is moved after the first trip, even with the bees inside, the bees subsequently become lost. Work on the details of honeybee learning has continued (e.g., J.L. Gould 1985, 1988).

In an interesting twist of learning in insects, grain beetles *(Tenebrio molitor)* retain learning through metamorphosis (e.g., Alloway 1972). This is intriguing because most of the larval neurons, at least peripheral ones, are replaced during metamorphosis.

Instrumental conditioning differs, at least conceptually, from classical conditioning in that instead of focusing on the connection (contingency or contiguity) between two different stimuli (CS and US), the emphasis is on the contingency between the instrumental response or behavior of the animal and the reinforcer or consequence of that behavior. Instrumental conditioning is somewhat similar to classical conditioning in that the contingency or pairing of interest can shape the animal's behavior. As in classical conditioning, there are many important factors such as frequency of pairing and consistency of the association. The instrumental response and the reinforcer can be paired consistently or only part of the time. If only part of the time or with only a percentage of the responses, it can be on a fixed (e.g., every tenth time) or variable schedule. Furthermore, the time interval between behavior and reinforcer may be constant or variable. Interestingly, conditioning under variable schedules is more resistant to extinction than under fixed or continuous and predictable schedules.

From comparative work with different species, it has been discovered that not all are equally capable of learning or becoming conditioned to the same stimuli or at the same rates. Some species seem to learn better under some circumstances, and others are better in other situations. To biologists this fits well with concept of *evolutionary adaptiveness*, which varies among species, and the idea of *biological constraints*. In psychological parlance, these evolved predispositions have gone under names such as *preparedness*, *relevance*, or *belongingness*.

The adaptiveness of conditioned forms of learning, in animals capable of it, is obvious. In classical conditioning, it would be to an animal's advantage to associate or anticipate noxious, dangerous stimuli that should be avoided. Conversely, positive stimuli may prove advantageous as predictors of useful events or resources. An animal that learns to associate a certain sound, for example, with potential food may become much more efficient and successful in obtaining food. Similarly, any animal that can assume some degree of control over its surroundings or at least in how it relates to its surroundings, in a framework of instrumental conditioning, would have a distinct selective advantage over animals that could not.

Operant techniques have also become important in behavioral ecology. Optimal foraging studies using captive animals, for example, rely heavily on animals that have learned to forage under research conditions.

Taste Aversion

With taste aversion an animal associates sickness or severe discomfort in the digestive system with an item that has been ingested. The animal subsequently avoids further ingestion of similar items, even to the point of starvation if that is the only food available. Such learning is highly advantageous because it prevents poisoning. Taste aversion learning appears to be widespread and general, at least among vertebrates. Taste aversion learning appears superficially as another form of classical or operant conditioning (or both). It differs markedly, however, in the opinion of some learning theorists and perhaps is best viewed as a separate category of learning.

Although widespread, the phenomenon of taste aversion is also quite variable among species and perhaps even individuals. Its expression depends in part on the normal food-sensing modalities of the species (e.g., whether by olfaction or vision). It also may come into play at different stages of predation, depending on the species. Some predators, for example, continue to kill but not eat objects with noxious associations, whereas others avoid the objects altogether. Different animals also may vary in the strength of the aversion, depending on a host of factors such as species, previous experience with the food (before poisoning), and perhaps other factors such as age, sex, and level of hunger.

Many species of organisms have acquired, through the course of evolution, the ability to produce taste aversions in other animals that feed on them. Well-known examples include monarch butterflies and many other invertebrates plus some amphibians, particularly toads, and a few reptiles such as gila monster lizards (Figure 20-6), which have poison glands in their skin. The natural sequence for the acquisition of taste aversion learning is for an animal to ingest an object or substance that produces sickness, usually including nausea and vomiting. Through the coupling of vision or olfaction or perhaps other sensory modalities with the taste of the sickness-producing agent, the animal becomes averted (i.e., learns to avoid future encounters).

Figure 20-6 Gila monster—a lizard with poisonous skin (and also a poisonous bite).
Photograph courtesy T. Brakefield.

Table 20-1 Differences between food aversion and other forms of associative learning

Characteristic	Food Aversion Learning	Other Associative Learning
Time and repetition	One-trial learning with CS-US delays up to 24 hours	Requires repetition and delays between CS-US of no more than a few seconds
Relevance	Poisons more readily associated with foods than with lights and sounds	Shock more readily associated with lights and sounds than with tastes
Specificity to training environment	Food aversion generalizes readily to a new environment	Shock avoidance generalizes less readily to a new environment
Maturation of learning abilities	Weanling rats equal to adults	Weanling rats worse than adults
Effect of active versus passive exposure to CS	Less learning if taste is force-fed than if actively ingested	Passive presentation of CS is effective; active participation not considered important
Learning under anesthesia	Yes	No
Synaptic pharmacology	Cholinergic and anticholinergic drugs have little or no effect on learning; aversions may depend on histaminergic synapses	Cholinergic and anticholinergic drugs have large effect; role of histamine unknown

From Kalat, J.W. 1977. In: N.W. Milgram, L. Krames, and T.M. Alloway, editors. Food aversion learning. Plenum Press, New York.

The phenomenon can be elicited not only by natural noxious agents but also by a variety of artificial means, including hypodermic injection of sickness-producing chemicals and, apparently, radiation-induced sickness. The organism receiving such treatment becomes averted to food that may have been ingested before the induced sickness.

Research on taste aversion in a variety of species has revealed several important differences between this form of learning and other types of avoidance conditioning.

Latencies between the stimulus (noxious food) and response (sickness with vomiting, etc.) can be quite long, up to 2 hours or more. The aversion usually is formed with a single encounter, and it is highly resistant to extinction. When taste-averting stimuli and other forms of avoidance-producing stimuli are presented together, the distinction between the two forms of learning becomes particularly noticeable.

There are also some similarities between taste aversion learning and other avoidance learning. In both cases the learning is greater with bigger differences between the natures of the CS and US, greater novelty of CS, increased intensity of US, more initial fear of CS, and the occurrence of repeated pairings. Learning is weaker in both cases if the animal has previously been exposed to CS without pairing with US. Finally, there is much variability among species in both cases. The similarities, however, are far outnumbered by the differences (Table 20-1). For further comparisons and references, see Kalat (1977).

Latent or Exploratory Learning

Latent or exploratory learning involves an animal using experience gained at one time in the modification of behavior at a much later time. Information and memories that may not be immediately useful are nonetheless retained and may become useful subsequently. This is particularly important for spatial information. In a laboratory setting, for example, rats that have been allowed to spend time in and explore a maze without food or other reward learn to negotiate the maze for reward at a later time much faster than rats that are introduced to the maze for the first time. The advantages of such learning, in a sense the prior acquisition of knowledge, would be clear in the wild. It gives an animal enhanced access to resources and the additional advantage of residing in a known home range (see discussion in Chapter 7). This may serve as the basis for much of the exploratory behavior observed in animals, including humans and other primates.

Place or Spatial Learning

Place or spatial learning concerns an animal learning its surroundings and familiar place or familiar path (Chapter 7). It involves an ability to become oriented in space and if necessary to reorient to new locations. The value of this type of learning is obvious, so much so that it may be taken for granted. It may be one of the most common, if overlooked, forms of learning and may involve some of the most primitive and ancient parts of nervous systems. Many persons would classify this ability as instrumental, trial-and-error learning, as in learning to negotiate mazes, but it is not clear whether different forms of learning are involved or not.

Many examples of spatial learning occur in the arthropods. For many species it involves a home base of some type—a cavity or den or a resting and hiding place on a web, hive, etc. In addition, it may involve the position of a food item. Bees are well known for their ability to return to nectar and pollen sources (Chapter 7). A classic study by Tinbergen (Chapters 2 and 7) involved the ability of a wasp to return to nest sites. Numerous other examples involving insects, such as ants, cockroaches, and others, can be included. Among Crustacea, many lobsters, crayfish, and crabs have been shown able to return to a consistent, familiar hole or hiding place. Several studies with spiders (e.g., Bartels 1929, Peters 1932) showed that spiders learn and remember the locations of prey on their webs; even if it is artificially removed without disturbing the web, the spiders return to the exact place where the prey was located and search repeatedly for the missing item. Subsequent

Figure 20-7 Disorientation of spiders (*Agelena* sp.) in response to learned orientation to light. If the direction of a light source is reversed after a spider leaves its retreat and arrives at its food, it travels in the opposite direction in an attempt to return to its retreat.
Modified from Bartels, M. 1929. Zeitschrift für Vergleichende Physiologie 10:527-593.

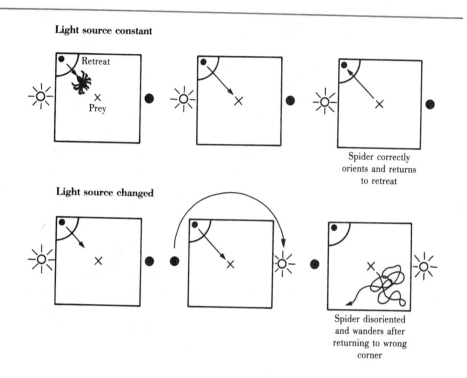

searches may become less specific, with the spider ranging slowly farther and farther from the site from which the prey had been taken.

The orientation cues that are involved in spatial learning in spiders frequently are tactile or visual. In one genus of spider, *Agelena*, the animals become disoriented if the source of light is changed while they are away from their retreat (e.g., Bartels 1929) (Figure 20-7). Further studies (Gorner 1958) demonstrated much variability in the disorientation (e.g., percentage of spiders becoming disoriented under different circumstances), and, perhaps more important, it makes a difference when the lights are changed. If the lights are switched before or soon after the spider leaves its retreat, there is little or no disorientation. The disorientation becomes worse and worse, however, the longer the spider has been out when the lights are changed. Disorientation is greatest if the lights are changed after the spider has arrived at its prey item.

Maze learning in ants has been studied extensively by several people. Studies by Schneirla and associates spanned a period of 40 years (e.g., Schneirla 1929, Weiss and Schneirla 1967, with numerous publications in between). The general method has been to use various mazes, runways, and gates through which ants had to travel to get from their nest on one side to food on the other. Visual cues have been controlled by artificial lighting, and chemical cues are manipulated by interchanging or replacing liners in the passageways.

One can barely do justice to this much research in a few short sentences, and there are many considerations and qualifications (such as a trait in ants called *centrifugal swing*) (Schneirla 1929) that cannot be covered here. In brief, however, ants were found to use several cues—visual, chemical, and kinesthetic (movement and sequences of position)—in various combinations. They could get by without some cues, depending on which stage of learning was involved. Chemical cues, for example, and somewhat surprisingly, visual ones (as normally happens when ants are underground) are not necessary. Evidence of kinesthetic cues was demonstrated by lengthening segments of well-learned mazes; ants would attempt to turn at the point where previous turns had been located.

Two important generalizations resulted from the ant studies. First, learning progresses through relatively distinct stages. The first one or two stages involve essentially trial and error, with the elimination of entrance into blind passages. The final stage involves integration of the behavior into a more coherent, smooth whole by which the ant efficiently negotiates the passages.

Second and perhaps most important for a comparative view of learning, maze learning in the ants is situation specific; that is, each part of a maze is learned separately, and new mazes must be learned from the start. Even learning a maze on the way to food does not improve learning an identical maze on the return trip. There appears to be little ability to generalize or transfer learning in one situation to a new situation, even when the situations are otherwise identical. Situation specificity in the learning of arthropods has also been shown in several other groups and species; recall, for example, the details of color learning by bees approaching and landing at feeding sites and the possibly similar phenomenon of spiders becoming visually disoriented, depending on when the lights are switched.

Cultural or Observational Learning

Cultural or observational learning has been documented in many of the higher vertebrates, particularly birds and mammals. In this form of learning, one animal learns to do something by watching or otherwise sensing what another animal is doing, including not only overt gross actions but in many cases even subtle mannerisms. Through culture, information can pass quickly from animal to animal and effectively bypass the long periods required for evolution and inherited acquisition of new behavioral traits. As an analog for the information-carrying agents in inherited traits (genes), Dawkins (1976) proposed the term *meme* for the information elements, at least in human ideas, of cultural transmission. Culturally learned behaviors are most likely to be learned from parents or peers, and several social factors may be involved.

Interest in the analogies between cultural and genetic transfer of information and the interaction between the two modes is increasing and may provide direction for much research in the near future. For expanded treatment of the topic Bonner (1980) gives a readable popular introduction; Cavalli-Sforza and Feldman (1981) present a more in-depth, technical treatment; Hutchinson (1981) considers the implications for humans.

Culture, at least in rudimentary aspects and as a form of learning, has been found to be quite widespread among mammals and some birds. Examples of cultural transmission include the learning of potato and grain washing by Japanese monkeys, milk-bottle lid removal by European tits, and predator mobbing by passerine birds. In the predator-mobbing example, birds learn to mob objects that they see others mobbing (e.g., Curio et al 1978 a,b) (Figures 20-8 and 20-9). As in other forms of learning, there seems to be a predisposition for mobbing objects of a certain type (such as owls), but via learning and through artificial manipulation, the birds can be taught to mob other things such as nonpredatory birds and even plastic bleach bottles.

The cultural transmission of techniques for obtaining food has received much

Figure 20-8 Experimental setup to investigate acquisition of mobbing by European blackbirds. The presentation box was moved into view by a pulley operated by the experimenter and then rotated to expose the stimuli to the subjects. An owl, which elicits a strong mobbing response, was shown to the mobbing bird (teacher) while a novel object, either a honeyeater bird, a nonpredatory and unfamiliar species to blackbirds, or a plastic bleach bottle was presented to the observing learner blackbird.
Modified from Vieth, W., et al. 1980. Animal Behaviour 28:1217-1229.

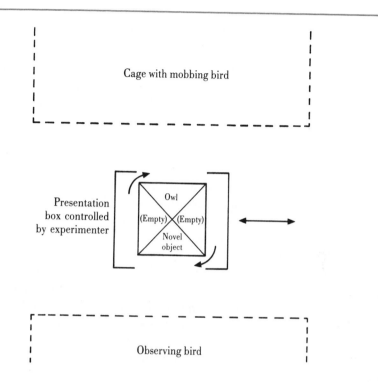

Figure 20-9 Results of European blackbird mobbing studies. Sequence of trials: (1) empty box presented to control for effect of box movement and appearance, (2) presentation of novel object, without teacher seeing the owl, (3) presentation of owl and novel object to teacher and learner, respectively, and simultaneously, and (4) test of learner's response to novel object alone. Mobbing responses were standardized to that initially shown to an empty presentation box. Decreased responses to repeated presentations of empty box or novel subject alone represent habituation. Note that the birds learned to mob even the empty plastic bottle, but the response was not as strong as toward a novel species of harmless bird.
Modified from Curio, E., et al. 1978. Zeitschrift für Tierpsychologie 48:184-202.

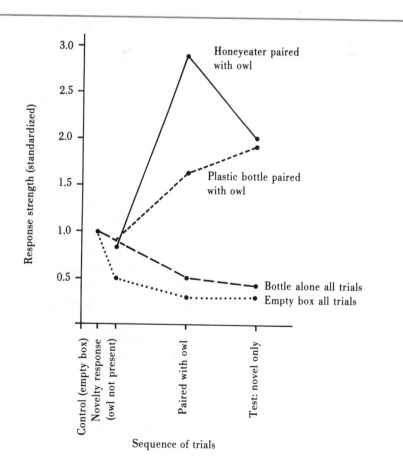

attention, particularly in birds (e.g., Palameta and Lefebvre 1985, Sasvari 1985, and Lefebvre 1986). At least in pigeons, however, the birds must actively participate in obtaining their own food in the process of learning by observing others (Giraldeau and Lefebvre 1987). Pigeons that shared food (scrounging) from the efforts of others did not learn the techniques and performed as if they had not observed. When prevented from further scrounging, they subsequently learned from observation.

Imprinting

Imprinting is recognized as one of the major forms of learning. Imprinting involves learning the properties of a stimulus object toward which the animal subsequently directs its otherwise normal feeding, social, or other behavior, depending on the species and situation. The learning period generally occurs early in the individual's life, often during only a narrow critical period. This form of learning is discussed in Chapter 19.

Insight Learning

Insight learning is perhaps the most advanced form of learning, at least from the *Homo sapiens'* chauvinistic viewpoint. In this form of learning the individual derives new behavior or solutions to problems by insight or thinking about them. Previous specific experience is not involved, although general and closely related experiences may help. Rather, the modified behavior and actions are new to the situation. This probably is the least widespread and least understood form of learning. It may involve processes of mental modeling and simulation; that is, the individual thinks about something and tries it out mentally before putting it into overt form. Scientific hypotheses are formed by insight learning.

Insight learning and mental modeling, however, are by no means confined to scientists or persons with formal education. A familiar but less-recognized form of such thinking occurs when one thinks about future personal interactions and relationships with the world and other humans. Examples are "If I bend this lever, I wonder if the machine will work?" and "If I wear such and such clothes to the party next weekend, what will so and so think of me?" Most persons are good at rehearsing likely conversations and events in advance. It may be that complex social relationships played a major role in the evolution of insight capabilities.

Mental simulations and hypothesizing form a type of trial-and-error situation that can be repeated, modified, and worked out in the comfort of one's own mind, without the risks of the real world. Thus it seems much more efficient and less hazardous than real trial and error; the advantages seem immense. If the advantages really are great, why are not more animals capable of greater insight learning than they appear to be? If as in computer simulation the ability requires considerable processing and memory capacity, it may be that humans are one of the few living species that possess sufficient mental ability. We do not, however, have a monopoly on insight learning—just the major share of it. Novel solutions to problems such as reaching places with difficult access can be seen in other primates, other mammals, and a few birds, including members of the crow family (Corvidae).

Learning-Set Formation

Learning-set formation is a form of learning that may be closely related to insight learning, perhaps differing mostly in degree or just in name. It has been studied fairly extensively in other animals, particularly primates and a variety of other mammals. In learning sets, animals generalize from previous learning and solve closely related learning tasks more rapidly. In essence the animals "learn to learn" or learn how to improve their learning. The learning set involves a set of principles or a strategy. For example, if food is found under an object, the animal will continue to search under similar objects, but if not, it switches to a different object. This particular strategy is called a *win-stay, lose-shift strategy*. Many persons consider this type of learning as an extension of instrumental conditioning.

A related, perhaps converse phenomenon is a type of learning known as *learned helplessness*. If an animal is faced with an intractable problem for which there is no solution, it may simply give up and in essence learn that it cannot learn. As an example with dogs (Seligman 1975), animals were given shock treatment from which some could not escape and some could. Those that could not escape eventually stopped trying and then would not escape even when given the opportunity.

Overlaps and Combinations of Types of Learning

Overlaps and combinations of types of learning occur in many situations, and it is not always clear in which category to place particular instances or whether the categories are really distinct in the first place. Classical and instrumental forms of conditioning share many properties in common, and many persons believe they may just be different aspects of the same process. Other instances of overlap between categories were indicated earlier.

In one example involving instrumental learning, cultural transmission, and what appears (at least superficially) to be some elements of insight learning, mother cats aid (teach?) their kittens in the process of learning how to capture prey (Caro 1979, 1980). When kittens are between 4 and 12 weeks old, mother cats go through a sequence in which they first bring dead prey to the kittens and eat it in their presence; next they bring dead prey but do not eat it. Then they bring crippled but live prey for the kittens to play with and eat. Finally they bring unharmed prey and release it in front of the kittens. If the prey escapes, the mother recaptures it, brings it back, and releases it again. The kittens follow the female into the field and begin capturing their own prey, after which the mother no longer takes part in interactions over prey with the kittens. It should be noted, however, that when no kittens are present, female cats frequently bring uneaten and often still living prey to their human associates, or perhaps they bring it to their home as a vacuum activity. Thus it is not certain how much, if any, insight learning is really involved.

Problem solving or insight learning in house mice *(Mus musculus)* depends on genetic and cultural factors (Mainardi et al 1986). Success among individual mice in problem solving ranged from only 3.9% for naive offspring of unsuccessful parents to 32.3% for others that were offspring of spontaneously successful parents and reared in the presence of their parents solving the problem.

THE EVOLUTION AND PHYLOGENETIC RELATIONSHIPS OF LEARNING

An attempt to survey the various phyla of animals for the role and extent of learning in different species is difficult. First, there has been a disproportionate amount of study involving a few species. Some, such as the rat and human, have been studied extensively, whereas several other phyla, most orders of animals, and the vast majority of species have gone virtually untouched.

The second major problem is that there are so many differing opinions about what constitutes learning and how to properly measure it. Bitterman, for example, in a scathing attack on studies of invertebrate learning, remarked (1975:144), "We are seeing ... a diminishing respect for that tradition among comparative psychologists, who find it fashionable now to denigrate their past (about which they know less and less) and to stand in awe before the loose anthropomorphic and teleological models of the ethologists." Beyond the rhetoric there are some valid criticisms. The main points that Bitterman (1975) stresses are: (1) greater attention to and familiarity with the findings of the intensively studied species, such as the rat, are needed because many of the pitfalls have been recognized, and (2) studies must become more efficient and objective to remove observer bias. Lahue and Corning (1975) suggest a third major problem—that even as differentiated as learning classifications have become, we still may not have a fine enough resolution of various categories. They believe that when several criteria for a given form of learning are considered, it will not be the same among different groups of animals. Using habituation as an example, Thompson and Spencer (1966) have classified 9 (or 10 if the eighth is split further) criteria for identifying the presence of habituation. In their phyletic survey, Lahue and Corning show that different criteria are satisfied in different species and that available information also shows marked differences in underlying mechanisms. In view of this divergence in habituation, they remark, "Comparative statements based only upon behavior [are] of little value—it is similar to announcing that two species are the same because both can fly. As Jensen (1967) points out, man could be characterized ... as a featherless biped that talks and would accordingly be in the same grouping as a plucked trained parrot" (pages 167-168). Nonetheless, comparative studies of both homologous and analogous characteristics remain important, as discussed in Chapter 5; this includes the study of learning. Corning et al (1973-75) provide a good comparative survey. Further discussion is available in Marler and Terrace (1984).

The existence of learning, aside from habituation and sensitization, in the simplest forms of animals has been ambiguous and controversial. Cnidarians, for example, are relatively simple animals, but they are only deceptively simple from a learning standpoint. The general view of cnidarian behavior since the turn of the century has been that cnidarians, as well as many other simpler organisms, behave in a straightforward, simple reflex manner. Parker (1919), for example, stated that the sea anemone is "a delicately adjusted mechanism whose activities (are) made up of a combination of simple responses to immediate stimulation." As late as 1952, Pantin stated that cnidarians, more than any other animal, brought us close to "a complete analysis of the structural units on which behavior is based." Some, such as Jennings (1905, 1906), knew better and argued to the contrary of that position.

In addition to responses to external stimuli, cnidarians show endogenous, often cyclic, patterns of behavior and much variability from individual to individual and from species to species and group to group. Compounding this difficulty is a host of miscellaneous problems. The movements and responses of cnidarians may be slow—sometimes too slow to be observed by eye—and may tax both the patience of the observer and seemingly the capabilities of the animal itself.

Such complications have resulted in many problems and alternate explanations for cnidarian learning studies. Early experiments (e.g., Nagel 1894, Parker 1896) suggested that certain anemones would learn to discriminate between filter paper soaked in crab meat juice and crab meat. The anemones would move the pieces of filter paper to the mouth, swallow them, and then eject them. Eventually the tentacles would respond more slowly or simply not carry paper to the mouth; then the tentacles would reject the paper altogether. The anemones, however, would continue to accept crab meat. Only the tentacles that were tested would show the effect; other tentacles would still accept filter paper (until they too were repeatedly stimulated). The slowing of response by specific tentacles also is shown normally toward pieces of food. Later studies (reviewed in Rushforth 1973) attempting to confirm that the anemones (or their tentacles) would learn to distinguish between flavor-soaked paper and real crab meat, however, produced highly variable and conflicting results. Some workers believed that the findings could be explained by sensory adaptation, habituation, or accumulation of substances on the tentacles or in the surrounding water.

An apparently overlooked aspect of the anemone behavior, incidentally, whether it involves learning or not, is how does a cnidarian recognize an indigestible item and expel it in the first place? With the relative sparseness of the nervous system, relative lack of specialized neurons, and lack of a well-defined central nervous system, how do anemones discriminate whether something is food or not? If they monitor the progress of digestion, how do they keep track of the length of time something has been present in the gastrovascular cavity without being digested?

Other attempts to show conditioning or more advanced forms of learning have been plagued similarly by complicating factors, conflicting results, and differing interpretations. Perhaps the clearest demonstration of more complex learning in cnidarians involves swimming reactions in a sea anemone of the genus *Stomphia*. When exposed to contact or chemicals from particular surface regions of two species of starfish, *Hippasteria spinosa* or *Dermasterias imbicata*, or the nudibranch, the anemone *Stomphia* elongates its body column (Robson 1961) (Figure 20-10). The response develops over a period of seconds; then, after landing in a new location—if the animal moves at all—the contractions slow down and quit after a few minutes. The anemone becomes inactive and eventually reattaches to the substrate. The nudibranch is a predator of these anemones, and the reaction seems to be one of escape. Aside from occasionally bumping into the anemones, however, the starfish are not predators of anemones and pose no threat to them; the reaction of the anemone to the starfish is a bit puzzling. Furthermore, the chemicals from the bodies of the nudibranch and starfish are quite different and stimulate the anemones at different regions. Starfish stimulate only the anemone's tentacles, whereas the

Figure 20-10 Sequence of swimming and resettling by the anemone *(Stomphia coccinea)* after contact with the starfish *Dermasterias.* Modified from Rushforth, N.B. 1973. In: W.C. Corning, et al, editors. Invertebrate learning, vol. 1. Plenum Press, New York, and Robson, E.A. 1961. Journal of Experimental Biology 38:343-363.

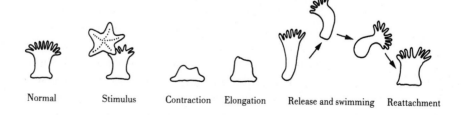

Normal   Stimulus   Contraction   Elongation   Release and swimming   Reattachment

nudibranch stimulates only the body column of the anemone. Another starfish, *Henricia*, does not stimulate *Stomphia* to swim.

It is not clear why the anemones respond as they do to stimuli from the starfish *Hippasteria* or *Dermasterias*. Nonetheless, this response has been used as a possible conditioning response. Ross (1965) used contact with the starfish *Dermasterias* as the unconditioned stimulus. He tried three conditioned stimuli, that is, stimuli presented before the US: (1) a pipe cleaner dipped in food extract (from a clam) and touched to the tentacles, (2) contact with the nonstimulating starfish *Henricia*, and (3) gentle pressure applied to the base of the anemone. There was no evidence of classical conditioning or of any interaction between US and CS for the pipe cleaner and *Henricia*. Pressure to the base led to a reduction in the subsequent swimming response, an effect that Ross termed *conditioned inhibition*. However, pressure at the base may have had some other mechanical or neural effect. As a group cnidarians may represent organisms with the earliest types of neurons (perhaps protoneurons) and nervous systems. Studies of modifiability of these systems still are confusing.

Mollusks learn about as well as might be expected from their phylogenetic position. Cephalopods, however, seem to be outstanding exceptions. Octopuses seem more advanced in some respects even than arthropods, and although there are no direct evolutionary relationships, they act much like many vertebrates. These two groups, cephalopods and vertebrates, provide an outstanding example of evolutionary convergence in respect to vision and the outward expression (but not the underlying detailed structure) of learning. The similarities may be related to speed of movement (other mollusks are famous for being slow) and ability (via the advanced eye) to sense events at a distance rather than just at the surface of the body. Cephalopods also have appendages, the tentacles, capable of manipulating things in the environment, like many of the vertebrates.

In other mollusks (reviewed by McConnell and Jacobson 1973 and Willows 1973), there have been few learning studies except in the gastropods, including *Aplysia*. There have been many claims of learning in snails, including one rather farfetched one about a pet snail that learned to recognize the voice of its owner (Dall 1881). Close scrutiny of the various evidence reveals convincing support only for habituation and sensitization types of learning.

Arthropods comprise most of the species (perhaps 80% to 90%) and possibly most of the individuals of animal life found on Earth. In the face of such numbers and diversity, relatively few species have been studied for learning. However, the learning of some species has been and continues to be intensively studied. These include spiders, a few familiar crustaceans, and hymenoptera (ants, wasps, and bees). Among hymenoptera, the honeybee has been the "white rat" of arthropod studies.

A major aspect of learning shown by invertebrates is the compartmentalization of learning ability. Vertebrates, including humans, also show this but not to the same extent. As reviewed by Krasne (1973), for example, crayfish and lobsters are able to avoid adversaries by flapping their abdomens and swimming away. Furthermore, they can sense and avoid obstacles while doing so. However, they seem unable or "unwilling" to learn to use these same patterns of movement to swim to food that they otherwise cannot reach. The characterization of insects as mechanical and robotlike is not completely accurate because they do have capacity for some learning, but relative to the flexibility seen in cephalopods and vertebrates, the robot characterization for insects may not be misleading. There appears to be no convincing evidence of learning-set, cultural, or insight types of learning in arthropods.

The diversity among organisms and the particular types of learning found in different species of arthropods have raised an important concept concerning the biological relevance of particular learning capacity. This or a similar notion will surface again when the biological constraints of learning among vertebrates are considered. The relevance of different kinds of learning to different natural histories has received increasing emphasis during recent years. Krasne (1973), for example, tabulates differences in morphology, senses, and ways of living for 22 groups of crustaceans alone. He says:

> (We) would anticipate that an animal such as the lobster, living in a fixed burrow from which it emerges at night to forage, might well learn something about the topography of its home range and the locus of its burrow. We might therefore anticipate that it would be capable of learning its way through mazes in the laboratory. On the other hand, we would not be surprised at the absence of such an ability in a small planktonic creature such as a copepod, which is forever wafted here and there by water currents.

Studies of learning in many of the invertebrates, particularly protists, sponges, cnidarians, and echinoderms, have been difficult and relatively unrewarding to the investigator, in spite of studies going back to Jennings (1907) and even earlier. Willows and Corning (1975) summarized the situation as it relates to echinoderms: "Definitive conclusions concerning echinoderm learning must be forestalled until better preparations and perhaps more patient investigators appear."

Among vertebrates that have been studied, there are no clear correlations between learning characteristics and phylogenetic relationship. It appears that all vertebrates, from fish and birds to many species of mammals, are capable of rapid avoidance learning in feeding situations with noxious stimuli (of many sorts). In classical conditioning with light and shock, the time fish and pigs require to learn is about the same or slightly less than that needed by rhesus monkeys. Rats surpass monkeys in learning simple mazes, and they are about equal to humans in complex mazes. Fish, chickens, and several mammals have been shown to learn discrimination tasks at roughly the same rate. Rats learn visual discrimination so much faster than chimpanzees that Hebb (1958:454) remarked that a "large brain like a large government may not be able to do simple things in a simple way." There are many differences between species and even between some strains or races of the same species so that in one situation one learns more quickly and in another situation the other learns faster. There are several learning tasks that other vertebrates can do better than humans. Humans, however, are still considered to have a greater array of complex learning abilities than any other animal.

What factors correlate differences in learning among different vertebrates? Warren (1973) listed life history (e.g., predators versus herbivores), sensory dominance (different species are tuned to different sensory modalities—rhesus monkeys to visual cues, which they learn more rapidly than auditory, whereas cats learn better in response to sounds), and response availability. As an example of the last aspect, Warren describes differences between two different kinds of fish. In a light-shock avoidance training test, goldfish learn to escape much more rapidly than Siamese fighting fish (Otis and Cerf 1963). Goldfish are bottom-feeding scavengers, and their normal response is to flee in the presence of aversive stimuli. Siamese fighting fish, on the other hand, are predators and are typically aggressive; they normally stay where they are or fight back in response to threats. Thus in the training methods used, goldfish would be expected to flee much more quickly than Siamese fighting fish; the normal responses in the species' repertoires are quite different.

Warren attempted to compare information on learning sets but concluded that there have been too many confounding and complicating variables (such as species differences in visual capabilities and in response to visual versus nonvisual cues) to "yield uncontaminated measures." As a result Warren (1973) stated, "One can therefore never safely conclude that any quantitative difference in learning set performance, however large, is a valid indication of a difference in learning capacity, rather than a reflection of species differences in adaptation to the arbitrary demands of the test situation." It is much like the problem of trying to eliminate cultural biases from human IQ tests, only more so. Warren concludes (1973:500-501):

> The classical approach to the comparative psychology of learning was based on oversimplified and obsolete ideas concerning natural selection and phylogeny. Animal species are subjected to selection for survival and success in reproduction, not for the degree to which they manifest progressively more humanlike capacities for learning and problem solving in the Skinner box or WGTA (Wisconsin General Testing Apparatus).

The basic point of this discussion is that each surviving species has become adapted for living by whatever means responds most adequately to selection pressures. Specific learning capacities must have been subjected to the same kind and degree of selection as any other trait of the organism, and therefore specific learning capacities can be expected to vary with taxonomic status as do morphological traits.

LEARNING PROCESSES VERSUS LEARNING PHENOMENA

So what sense can be made of this information distributed so unequally across the spectrum of animal species? First, it is important to review the questions and working assumptions that implicitly or explicitly have stimulated discussions, to the point in some cases of outright controversy. One general question has concerned whether particular behavior is innate (genetic, instinct, nature) or learned (nurture). This is based on the biologically naive and improper assumption that it must be either-or. All behavior is unavoidably the joint product of both genetic and environmental factors, as discussed in Chapter 4. What really is of interest and the proper way of phrasing the question is: "To what extent can the observed variability in particular behavior be attributed to genetic versus environmental variability among several different individuals?" or, for a given individual, "To what extent can the behavior of that animal be modified as a result of experience or surrounding, environmental events during its lifetime?" Two separate questions are being dealt with—one concerning variability in populations and the other concerning development in an individual.

Maturation and modifiability of normal developmental processes complicate the picture. Examples of developmental changes include modified tube-webs of male salticid spiders initially forced to construct webs in abnormal spaces. One might also view the ontogeny of bird song in this light. What is encountered is a biological continuum from what was classically considered as development into what was classically considered as learning. Across species and groups of species there is a continuum in the extent to which behavior such as bird vocalization is modifiable by experience. Nature simply is not obliged to come in neatly divisible packages with unambiguous boundaries as one might like; continuums are a fact of life.

Aside from the nature-nurture problem, the other major set of questions and assumptions about learning has concerned whether the characteristics of learning are universal and whether differences between different animals are only a matter of degree. That is, when modifiability of behavior as a result of experience does occur, is it the same whether it occurs in rats, humans, pigeons, snails, or worms? Can the characteristics of learning that are discovered about rats be applied also to humans (and pigeons, snails, etc.) or vice versa? There is a constant problem of analogy versus homology. Although not rejecting the idea that learning might be universal, many persons are not yet willing to accept the proposition; they instead maintain an agnostic position on the topic.

As has been seen, there are many categories of the outward expressions or phenomena of learning. Among different groups of animals, learning of a particular type, such as habituation, may have some common attributes but also much variation. Do similar appearances of learning mean that the underlying, neural processes are also the same? Could one simply be seeing convergence, as Bitterman (1975) phrased it, "whether the resemblance between vertebrate and invertebrate learning is any more profound than that between the hand of an ape and the claw of a lobster?"

Consider the groups that possess the greatest capacities for learning: cephalopods, arthropods, and vertebrates, groups that are only distantly related to each other and none of which is in a direct line of descent from either of the others. What were the common ancestors like, and did these major groups all inherit their capacity for learning from a common ancestor? Were learning abilities derived independently? These questions cannot be answered for sure. The common ancestors are long gone, and all there is to work with today are a few fossils (which tell little or nothing about learning) and a few modern-day remnants of ancient groups.

Some insight can be gathered from one group, the birds. Perhaps one of the most enlightening conclusions from studies of birds is the amount of variability in the extent and ways that song can be modified even within a single order of birds (Passeriformes). Birds are certainly a diverse lot, but from the perspective of variation in all other forms of animal life, birds constitute an extraordinarily uniform group. It is therefore remarkable that there is so much variation among species in the degree to which their vocalizations may be modified. This implies that learning (development or however one wishes to view bird song) is not necessarily conservative.

When all of this evidence is taken together—similarities in greatly (and anciently) diverged groups, differences among closely related groups, situation-specific learning, compartmentalization, and the growing list of categories of learning— it seems a good guess that learning has evolved on numerous occasions and because of different specific advantages. Similarities in phenomena probably only represent convergent evolution. There is a high probability that underlying neural processes are different.

It is important to emphasize that a guess is merely a guess, not an answer. In spite of the diversity seen in basic neurons, there are many shared characteristics, including ionic and other biochemical properties such as some transmitters, among virtually all organisms. Many of these common attributes probably resulted from processes occurring in groups of neurons, hence above the level of the basic neuron. Many of the higher forms of learning appear to be associated with advanced senses for distant events, perhaps abilities to manipulate objects, and other factors such as predation. Nonetheless, perhaps the roots of higher learning processes were present in the earliest ancestors of animals and what is seen today is related, that is, homologous. Perhaps there are constraints on nervous systems so that when modifiability does occur, it occurs only in certain basic ways, and the processes end up being the same (or similar to some extent) whether independently evolved or not. If this is so, then the whole issue becomes merely academic.

All that can be said at the present is that we really do not know what the underlying processes are for most organisms, and we still are struggling with and sorting out the characteristics of the phenomena. All three of the major behavioral disciplines—psychology, ethology, and neurophysiology—have contributed and continue to contribute significantly to the contemporary understanding of learning. Psychology developed careful designs and controls of learning studies and showed how to measure and differentiate subtle characteristics of learning phenomena. Ethology led to a broad evolutionary perspective, looking at a wide variety of organisms, considering problems of convergence, analogy, and homology, and stressing the roles of natural, ecological, and evolutionary forces in the differences among species. Neurophysiology is providing access to difficult tissues at the microscopic level and is opening the door to deal with high levels of complexity. For a major review of the natural history of learning that includes both psychological and ethological viewpoints and covers a wide range of invertebrate and vertebrate taxa, see Marler and Terrace (1984).

In the meantime, we continue to deal with learning mostly as phenomena not as processes. The detailed inner working of the brain as a whole remains a black box. Most progress in understanding the neural substrate of learning amounts to unwrapping only the outermost layers of that black box.

At the level of treating learning as a phenomenon, one of the most satisfying contemporary interpretations of learning was proposed by Mayr (1974). It encompasses all of the variability, similarities, biological constraints, developmental aspects, critical periods, and other matters discussed earlier. It is a simple analogy drawn from understanding of computer programs. Without specifying the underlying mechanisms, Mayr interprets learning as follows:

> A genetic program which does not allow appreciable modifications during the process of translation into the phenotype I call a *closed program*—closed because nothing can be inserted in it through experience. Such closed programs are widespread among the so-called lower animals. A genetic program which allows for additional input during the life span of its owner I call an *open program*. Even this improvement in terminological precision does not remove all our difficulties. A particular instinctive behavior act is, of course, never controlled directly by the genotype but rather by a behavior program in the nervous system which resulted from the translation of the original genetic program. It is particularly important to make this distinction for the open program. The new information acquired through experience is inserted into the translated program in the nervous system rather than into the genetic program because, as we know, there is no inheritance of acquired characters.
>
> An open program is by no means a tabula rasa; certain types of information are more easily incorporated than others.... Whether natural selection will favor the

evolution of an open or a closed program for a given behavior depends on the circumstances. For instance, the shorter the life span of an individual, the smaller the opportunity to learn from experience.... The situation is radically different in species with a more or less extended period of parental care. Here, the fixed responses of the newborn can be quite few in number, being limited primarily to adequate responses to the parents.

The longer the period of parental care, the more time will be available for learning, hence, the greater the opportunity to replace the closed genetic program by an open program. The great selective advantage of a capacity for learning is, of course, that it permits storing far more experiences, far more detailed information about the environment, than can be transmitted in the DNA of the fertilized zygote.

Under what circumstances is a closed genetic program favored and under what others an open one? ... the genetic program for formal signaling must be essentially closed; to state it more generally, selection should favor the evolution of a closed program when there is a reliable relationship between a stimulus and only one correct response.... On the other hand, noncommunicative behavior leading to an exploitation of natural resources should be flexible, permitting an opportunistic adjustment to rapid changes in the environment...Such flexibility would be impossible if such behavior were too rigidly determined genetically.

The longer the life span of an individual, the greater will be the selective premium on replacing or supplementing closed genetic programs by open ones.... The direction of many evolutionary pathways, thus, is clear. It often leads to personally acquired information to an even greater extent.... There are two prerequisites for this to happen ... (a) greater storage capacity than is needed for the carefully selected information of a closed genetic program... (that is,) a larger central nervous system ... (and) prolonged parental care. When the young of a species grow up under the guidance of their parents, they have a long period of opportunity to learn from them—to fill their open programs with useful information on enemies, food, shelter, and other important components of their immediate environment.

All animals (including the simplest and most sedentary, such as the sponges) must have at least some opening in their behavioral programs. At the minimum, for example, all animals respond to food stimuli and mobile animals must at least use sensory input to avoid bumping into obstacles. Beyond that, it appears that further opening of behavioral programs to outside input depends on specific requirements and details of the programs, specific life-styles, reliability of stimuli or cues, and general attributes such as storage capacity and parental care. (Many organisms have parental care, but many do not also have the necessary storage capacity for holding much outside information.)

Mayr's interpretation of learning may not be perfect in all detail and may need further development for better resolution, such as to accommodate human language. At a simpler level (than human language), the interpretation has not been developed to the level of resolution necessary to handle all the details even of bird communication. Furthermore, the closed-open program model has not been extended yet to such aspects as insight learning and consciousness. Mayr's closed-open program model provides a useful, if only heuristic, framework from which to view learning.

The likely independent evolution of learning and memory in different taxa of animals could mean that the processes of learning are different among these different groups. The means by which spiders habituate to irrelevant vibrations of their webs, for example, may be quite different from the internal processes by which young birds habituate to the patterns and shapes of other birds flying overhead. Whether the mechanisms are quite different, roughly similar, or the same, we do not know at present. The only way to know is to pursue and discover the mechanisms of learning in each of the different groups. The mechanisms are now becoming known in some of the organisms with simpler nervous systems. The best known are marine slugs of the genus *Aplysia*.

MECHANISMS OF LEARNING AND MEMORY

Habituation and Sensitization in *Aplysia* and Related Studies

The gill withdrawal reflex of *Aplysia* described in Chapter 17 (see Figure 17-8) has been the subject of intense study by Kandel and his associates (e.g., Kandel 1979 a,b, 1987). These studies have elucidated the mechanisms of basic forms of learning in this mollusk and suggest avenues of approach for other organisms. The gill withdrawal reflex shows the learning effects of both habituation and sensitization. Habituation results from repeated simple touching of the siphon; the gill withdrawal reaction decreases and eventually ceases. If a noxious stimulus is applied to the head at the same time the siphon is touched, however, sensitization occurs and the gill withdrawal increases. Habituation effects have been shown to occur over periods ranging from only a few minutes to several days and weeks (Figure 20-11). Sensitization can reverse any previous habituation.

Habituation, at least over the short term, has been shown to involve a reduction

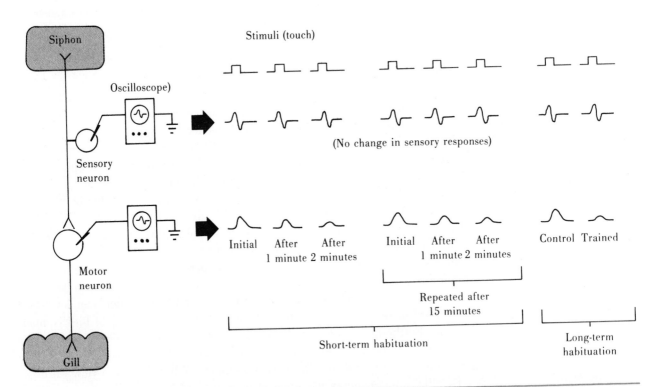

Figure 20-11 Habituation in a motor neuron of the marine snail *Aplysia*. Repeated stimulation of a sensory neuron leads to repeated firings of that neuron but a decline in the postsynaptic response of a motor neuron with which it synapses. Long-term (days and weeks) habituation also can be seen in recordings from the motor neuron.
Modified from Kandel, E.R. 1979. Scientific American 241(3):66-76.

Figure 20-12 Sensitization in the gill-withdrawal reflex of *Aplysia*. Sensitization is a form of learning and memory in which the response to a stimulus is enhanced. A stimulus to the head activates neurons that excite facilitating interneurons, which end on the synaptic terminals of the sensory neurons. These neurons are plastic, that is, capable of changing the effectiveness of their synapse. The transmitter of the facilitating interneurons, thought to be serotonin, modulates the release of sensory-neuron transmitter to the excitatory interneurons and motor neurons.
Modified from Kandel, E.R. 1979. Scientific American 241(3):66-76.

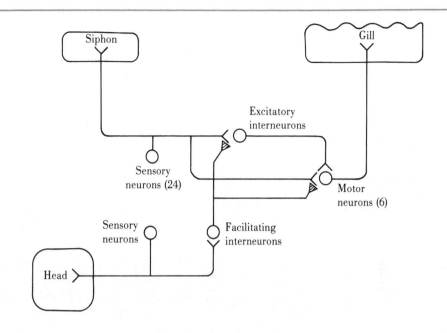

of synaptic transmitter release from the sensory neurons. The sensory neurons still fire (see Figure 20-11), but the action potential results in less transmitter being released. This is thought to be caused by less calcium inflow during action potentials, with calcium being partly responsible for synaptic vesicles binding to the release sites in the presynaptic neurons. As a result of less transmitter being released into the synapse, there are insufficient EPSPs to fire the motor neurons (e.g., L7), and the response diminishes or habituates.

Sensitization, on the other hand, involves a different mechanism that is mediated by facilitating interneurons. These interneurons connect between the sensory neurons of the head, the motor neurons, and the excitatory interneurons. A simplified diagram is shown in Figure 20-12. The facilitating interneuron makes presynaptic connections to the siphon sensory neuron axons. The facilitating interneuron, by release of its transmitter (serotonin), increases the amount of transmitter (acetylcholine) that is released in turn by the siphon sensory neuron. This increases the EPSPs in the motor neuron and additional excitatory interneurons. The increased excitation leads to increased muscular contraction and an increase in the gill withdrawal, hence sensitization.

Depending on the nature of the stimuli, one can produce inhibition, disinhibition, and sensitization at the level of the synapses. Thus one can account for changes in behavior over time or learning, in a relatively simple neural pathway from input to output. Similar modifications of synaptic transmissions have been shown in other invertebrates and even in a few isolated vertebrate preparations. The possibility that such mechanisms form the basis for higher orders of learning and memory is promising and intriguing. Kandel and co-workers, plus others in other laboratories, are actively pursuing these leads to higher and more complex parts of the nervous system in the *Aplysia* and related mollusks.

Another mollusk being used as a major subject for study is the marine snail, *Hermissenda crassicornis*. This snail is being studied by Daniel L. Alkon and his

GUEST ESSAY

William W. Beatty

William W. Beatty

Learning and Memory from Mollusks to Mammals

Studies by Kandel and his colleagues represent the most complete account of the neurophysiological and molecular mechanisms of any form of learning in any organism.

Even if their findings apply only to habituation and sensitization in *Aplysia*, their work merits and has received much acclaim. However, it is natural to wonder to what extent mechanisms that subserve behavioral changes in *Aplysia* may underlie both simple and more complex forms of learning and memory in other organisms. The answer to this question is not and may never be fully known. Nonetheless the problem is worth considering.

Two observations favor the view that mechanisms identified in *Aplysia* are at least somewhat general: (1) the basic operations of neurons as studied by electrophysiological, pharmacological, or biochemical techniques seem to be highly similar across species and (2) with few exceptions temporal relationships between CSs and USs that foster rapid and stable learning are similar across stimuli and species.

Kandel has assumed that changes in the strength of the gill withdrawal response arise from facilitation or inhibition of the activity at existing synapses rather than from the formation of new or the loss of old neural connections.

associates (e.g., Alkon 1983, 1984, 1988). These studies have focused on associative learning. Squire (1987) provides a review of many of the details for both *Aplysia* and *Hermissenda*. In addition to the scientific details there is also a fascinating human element in this research, including intense competition and conflicts among some of the researchers. Persons interested in the historical and personality aspects may enjoy reading the book, *Explorers of the black box*, by Susan Allport (1986). For additional references see Carew et al 1983, Scholz and Byrne 1987, Dale et al 1988, Schacher et al 1988, Marchus et al 1988.

Memory and the Elusive Engram

The ability to learn, regardless of the type of learning involved, depends on memory, that is, the ability of the nervous system to store information or change the manner in which the system processes or responds to particular input. The capacity of memory storage in the nervous system is impressive. Examples of human memory were noted at the start of Chapter 19. Large feats of memory have been shown also for other mammals and some birds. Some food-storing birds, for example, may remember the locations of thousands of stored seeds over periods of several months (Sherry et al 1981, Shettleworth 1983). How is all of this accomplished? What is the biological mechanism of memory? Is there a specific location in the brain where memory is located? These questions have provided some of biology's most baffling mysteries and greatest continuing challenges.

The physically coded representation that originally was assumed to exist for each specific memory in the brain has been called an *engram,* a German term originally

Although this assumption is in accordance with his data, it is well established that neurons and synapses proliferate (and are lost) during the course of normal development. In a few cases (e.g., learning of songs by canaries), reasonably good correlations between changes in the number of neurons and behavior have been established. This suggests that learning and memory involve formation of new connections, as well as modification of existing circuits.

In fact, changes in the number of neurons or their connections may be important for learning in *Aplysia* as well. Carew (1990) and his colleagues have shown that the capacity for habituation appears earlier in the ontogeny of this animal than does the capacity for sensitization. In a rough way, this parallels the developmental course of learning in most organisms in which simpler forms of learning appear earlier in development than more complicated forms.

Finally, it must be recognized that the information processing capacities of large sets of neuron chains arranged in a parallel fashion (which exist in the mammalian nervous system) may be quite different than those of simpler circuits, even though the large parallel sets are composed of many simple circuits.

These considerations suggest that the study of learning and memory in organisms like *Aplysia*, which have relatively simple nervous systems, will continue to have important implications for understanding the neurobiological mechanisms of learning in general.

For additional reading see Carew (1990).

proposed by Richard Semon in 1904 (Semon 1904, Schacter 1982) and given widespread attention by Karl Lashley (1950). In view of recent information, discussed shortly, the concept of the engram may not be particularly useful or actually represent an identifiable physical entity as in the case of the term *gene*. Partly for the aura of mystery that surrounds the term, the engram serves as a convenient starting point for this discussion.

Much of the problem of studying memory has involved the location of it within the nervous system. In 1825, Gall (reviewed by Squire 1987) proposed that specific memories were located in specific places in the brain. Evidence such as from work on rats by Lashly (1950) and others, however, suggested instead that memory was diffusely distributed throughout the nervous system. In attempts to find a location for memory, different parts of the brain were removed. The extent of loss caused by such removal seemed to depend on the amount of neural tissue removed, not its location, as if the memory exists everywhere simultaneously. This phenomenon has been shown in animals as diverse as mammals and cockroaches (Chen et al 1970). In the case of the cockroaches, intact cockroaches, headless insects, and isolated segments of the insects were classically conditioned to avoid electrical shock. Initial learning occurred in all cases but was not very good unless the head and brain were present. Once learning had occurred, all segments of the nervous system showed evidence of and retained the learning. Recent work on this problem using double-labeled molecular markers (2-fluorodeoxyglucose) suggests that there may be 5 million to 100 million neurons involved in memory of visual events in cat brains, with the cells widely distributed through the brain (John et al 1986).

Other work, starting with Wilder Penfield's studies of human epileptic patients during neurosurgery (e.g., Penfield and Jasper 1954), implied that local areas of the brain were indeed important. As Penfield and others stimulated various parts of the cerebral cortex, patients reported that they experienced various visual, auditory, and other perceptions that were interpreted as memories of past experiences. Subsequent work (Halgren et al 1978, Gloor et al 1982) with stimulating electrodes in various parts of the brain has focused on the role in memory of the hippocampus and amygdala, components of the limbic system that connects between the cerebral hemispheres and the thalamic region. It now appears that several regions and complex pathways in the brain, from the senses to and from the cerebrum and cerebellum—with the hippocampus and amygdala in between—are involved in memory.

Basically similar concepts have also been proposed for some complex invertebrate memory systems. A model for learning and memory in the octopus has been proposed (e.g., Boycott 1965) whereby abstractions lead to memory cells in specific regions of the brain and in which new information is compared with previous experience. Then, depending on the outcome, appropriate motor pathways are stimulated and inappropriate ones are inhibited (Figure 20-13). Similar models have also been pursued in other organisms such as bees (e.g., Menzel and Erber 1978).

The best current view is that a given memory is located neither diffusely and redundantly in several different places nor in a single specific location, but rather it involves a complex circuit and process that extends throughout various regions of the central nervous system. Interception of the circuit at various points—such as in the cortex or the hippocampus—can either stimulate or interfere with the memory, depending on the nature of the interception (stimulating electrodes or ablation, for example). Particular pathways may vary among different species, even among different individuals of the same species, and among different kinds of memories and learning. This is an area of intense current research with many of the details being worked out and contested (e.g., Thompson 1986, Bloedel 1987, Thompson 1987).

Regardless of where memory is located, it is clear that something in the nervous system must change for memory to occur. What is the nature of the change? Evidence involves both (1) relationships between inputs and outputs and (2) physiological and morphological changes in neurons.

Based on a wealth of relatively consistent information, the basic model for the operation of memory, as a phenomenon if not a process, is as follows. Initially there is a transient sensory trace, which may last less than ½ second or so, followed by short-term memory (STM) or working memory. Depending on the circumstances and species, the information may reside in the STM for a few minutes to hours and then either disappear or become fixed in long-term memory (LTM), or what might be referred to as *library memory*. The process of transferral from STM to LTM is not understood but is thought to involve some kind of rehearsal or repetition—perhaps some kind of reverberation whereby the information cycles several times through the same pathways and then becomes consolidated into LTM. Recall of the information depends on some type of retrieval process, perhaps a matching process by which the impulses of new input or internally generated impulses (a mental image) are compared with the stored template of information or impulse patterns. Proper recall requires a full complement and sequence of internal events. When recall does not occur, it may be because of one or more problems: fading of the memory trace (whatever and wherever it is), failure of consolidation, or inability

Figure 20-13 Hypothesized memory and learning system in the octopus, based on numerous brain-lesioning experiments and differential interruption of memory.
A, Diagram of nervous system, including major parts of the brain.
B, Components of the memory and learning system with indications (in parentheses) of brain parts where found. According to this scheme, information from the eyes is combined with input from other senses in memory cells. These memories are then compared with future visual input, and, depending on the match, either attack or retreat responses are stimulated whereas opposite behaviors are simultaneously inhibited (arrows and negative signs in center of illustration).
Modified from Young, J.Z. 1965. Proceedings of the Royal Society of London (Biology) 163:285-320, and Boycott, B.B. 1965. Scientific American 212(3): 42-50.

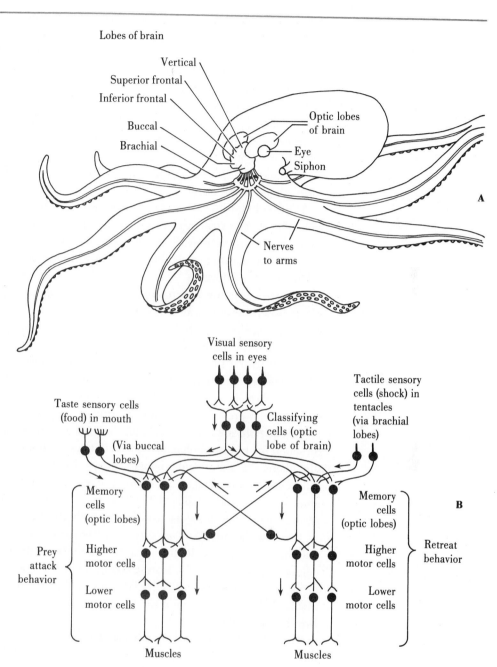

to retrieve the stored information. Improvement of human memory generally attempts to deal with the last problem (e.g., Ericsson and Chase 1982).

The evidence for two different memories, STM and LTM, is widespread. Many events that create amnesia, including electroconvulsive shock, concussions to the head, and interruptions or interference by other significant events or intense sensory input, for example, cause a loss of the most recently acquired information but not previously stored memories. This suggests that those events interfered with a transient memory, that is, wiped out the trace before it became stored more permanently.

In general, memories appear to become more and more resistant to loss with the passage of time (although recall can also become more difficult with time if memories become misplaced, as it were, among a large number of other memories). The general characteristics of the STM versus LTM phenomenon are found over a wide diversity of organisms, including mammals, fish, and insects, and even in isolated parts of nervous systems such as segments of the spinal cord (e.g., Chamberlain et al 1963). The commonly accepted model of STM-LTM is not without some problems, however. When different forms of learning trials were run using pigeons with stimuli that would reduce STM, STM was interfered with but LTM still took place (Maki 1979). Hence the relationship between STM and LTM may be more correlational than causal.

At the cellular level there have been two schools of thought concerning memory: molecular versus structural mechanisms. The molecular line of thinking received much impetus from the early, controversial planarian studies described earlier plus findings that learning seemed to correlate with increased levels of RNA and protein synthesis. According to this view, memory is somehow encoded into "memory molecules," whereby information is stored in a manner perhaps analogous to that which occurs in DNA. Memory recall would then require some sort of mechanism for decoding this information and returning it to a neural basis, which seems to create logistical problems for the model.

The structural line of thinking proposes that there are structural changes in the neural pathways with alterations resulting from use or disuse, which would facilitate future replays over those pathways that become established. Although this view has not been proved and we are far from a complete understanding or being able to demonstrate a specific case, the bulk of recent evidence seems to lean in the direction of structural changes. In particular, it appears with learning and memory that there are significant changes in dendritic spines in the synapses between different neurons (Figure 20-14). For examples with more information and discussion, see Coss and Globus (1978) and Greenough (1975). Also note that the focus is on synaptic sites, as discussed previously for *Aplysia*.

Much of the attention on synapses focuses on several biochemical events, including the role of calcium and, in some cases, glutamate (as a transmitter) receptors. These can lead to lasting changes in the properties of the synapse, referred to as *long-term potentiation* (LTP). A number of synaptic transmitters may be involved. For further details see Lynch and Baudry (1984), Black et al (1987), and Brown et al (1988).

The findings of increased RNA and protein synthesis are compatible with the model of structural changes because structures themselves are composed of molecules. Although the molecules would not encode the memory directly, their use in the building of new structures would still require protein synthesis. In addition to structural molecules of the synapses themselves, there may also be extracellular proteins, called *ependymins,* that are outside of the synapse but that contribute to changes in synapses (Shashoua 1985).

Most of the attention on changes within the brain have focused on changes within existing neurons, particularly in dendritic spines and the nature of the synapses. It was formerly thought that whole neurons themselves could not be added or replaced once the system was originally formed. It is clear now, however, that entirely new neurons can be added throughout life in at least some organisms such as birds. These changes along with other dendritic and synaptic changes, often in

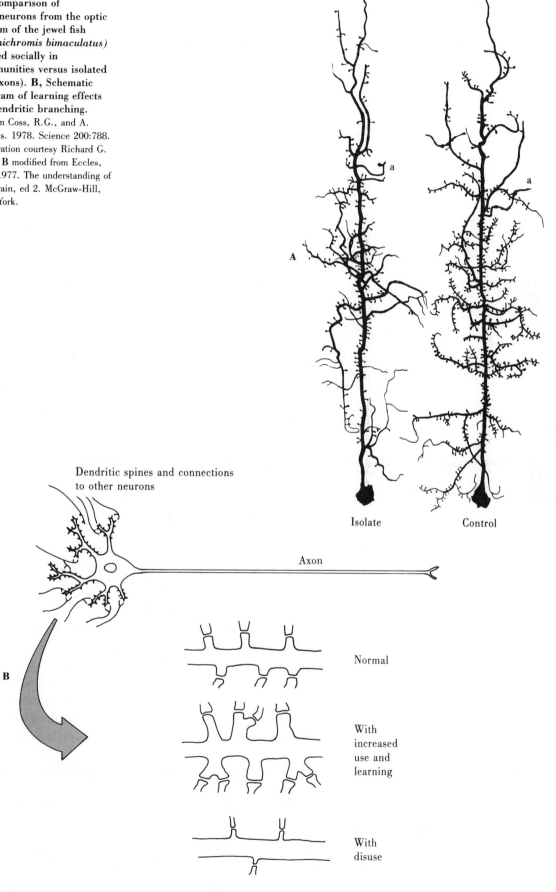

Figure 20-14 Learning and dendritic branching. **A,** Comparison of interneurons from the optic tectum of the jewel fish *(Hemichromis bimaculatus)* reared socially in communities versus isolated (*a*, axons). **B,** Schematic diagram of learning effects on dendritic branching. A from Coss, R.G., and A. Globus. 1978. Science 200:788. Illustration courtesy Richard G. Coss. **B** modified from Eccles, J.C. 1977. The understanding of the brain, ed 2. McGraw-Hill, New York.

A

Isolate

Control

Dendritic spines and connections to other neurons

Axon

B

Normal

With increased use and learning

With disuse

conjunction with hormones (Chapter 18) appear to play a significant role in learning, such as the learning of new songs by adult male canaries. The pioneering work on this subject has been by Fernando Nottebohm and his colleagues (e.g., Nottebohm 1984, 1989, Patton and Nottebohm 1984, Nottebohm et al 1986). A popular description of this work, including a background account on Nottebohm and his long-term interest in bird song, is given by Montgomery (1990).

Mammals are still considered to be unable to add new neurons after original establishment of the nervous system (e.g., Rakic 1985), but it is an open question (e.g., Nottebohm 1985). Whether or not mammals can replace whole neurons, there is abundant evidence for the changes in dendrites, synapses, and some role for hormones (e.g., Kurz et al 1986, Aoki and Siekevitz 1988). In addition to the implications for the basic topics of learning and memory, this information also provides promise and hope for correcting human neurological problems of various sorts (Rozenzweig 1984, Nottebohm 1985).

In conclusion, investigations into the black box of learning and memory have been and are producing many insights and surprizes. We may discover new properties of these complex networks not yet envisioned (John et al 1986). It is at this point where analogies or even directly related findings from computer models and artificial intelligence can provide help (Thompson 1986, Kolata 1987). All that can be said for certain is that we still do not understand exactly what is occurring in the process of learning and memory, but this is a rapidly expanding field that is generating almost feverish excitement and activity. We seem to be closing rapidly on a much better understanding of the subject.

SUMMARY

The subject of learning merges with that of ontogeny and the development of behavior. Learning is defined as a specific change or modification of behavior involving the nervous system as a result of experience with an event or series of events in an individual's life. The current list of learning and learninglike phenomena includes habituation, sensitization, classical conditioning, instrumental or operant conditioning, pseudoconditioning, discriminant learning, taste aversion learning, latent or exploratory learning, cultural or observational learning, imprinting, insight learning, and learning-set learning. Several of these categories overlap. Many can be subdivided into smaller categories or they can be lumped into larger groupings such as nonassociative, associative, and complex types of learning.

From a phylogenetic survey of learning it is clear that learning does not exist on a single, progressive scale simply leading up to a peak in humans. Instead learning appears as a set of branching and independently evolved phenomena that depend on life history characteristics, sensory capabilities, particularly ability to sense distant events, and the ability to manipulate objects via extensions of the body.

In an attempt to synthesize the current information and understanding of learning, different forms of learning can be considered as phenomena (i.e., the observable characteristics) as opposed to processes (i.e., underlying neural mechanisms). Such a comparison has involved two central controversies: the nature-nurture problem and whether learning processes are universal and species differ only in degree or if the underlying processes themselves vary. The nature-nurture or instinct-learning question has two separate aspects: variability of behavior in populations and development of behavior in individuals. When phrased in this manner, it becomes clear that there are continuums, such as in song learning among different species of birds, and nature is not neatly divisible into instinct versus learning categories.

Concerning the problem of universality of learning among different species, it seems likely that many advanced forms of learning that appear similar among animals may have been independently derived and simply represent convergent evolution.

Mayr has characterized learning as the extent of openness in the translation of genetic information to neural systems in producing information-control systems that are somewhat analogous to computer programs. Behavioral programs that do not permit much input are said to be closed, whereas those that permit input are open.

The actual mechanisms of learning and memory are finally starting to yield their secrets to a broad front of investigations that range from laboratory studies of simple animals like marine snails to clinical studies of humans. Memory is considered to involve complex circuits among various parts of the brain such as the senses, cerebrum, cerebellum, hippocampus, and amygdala of mammals. At the cellular level memory appears to involve changes in synapses and dendrites. This is an active field of intense, ongoing studies.

FOR FURTHER READING   Alkon, D.L. 1988. Memory traces in the brain. Cambridge University Press, New York.
A review of recent studies.
Allport, S. 1986. Explorers of the black box. W.W. Norton, New York.
A lively and readable account of the "search for the cellular basis of memory," including the personalities involved in the research.
Brown, T.H., P.F. Chapman, E.W. Kairiss, and C.L. Keenan. 1988. Long-term synaptic potentiation. Science 242:724-728.
An excellent introductory, review article on LTP.
Bitterman, M.E. 1975. The comparative analysis of learning. Science 188:699-709.
A critical and penetrating analysis of learning techniques and findings for various taxa of animals.
Bonner, J.T. 1980. The evolution of culture in animals. Princeton University Press, Princeton, N.J.
A review and overview of observational learning among animals.
Boycott, B.B. 1965. Learning in the octopus. Scientific American 212:42-50.
One of the classic readings on learning.
Corning, W.C., J.A. Dyal, and A.O.D. Willows, editors. 1973, 1975. Invertebrate learning. Plenum, New York.
A major phylogenetic survey of learning among nonvertebrate animals.
Domjan, M., and B. Burkhard. 1982. The principles of learning and behavior. Brooks/Cole, Monterey, Calif.
A general text and overview of learning.
Giraldeau, L., and L. Lefebvre. 1987. Scrounging prevents cultural transmission of food-finding behaviour in pigeons. Animal Behaviour 35:387-394.
An original research paper on cultural learning in pigeons and some of the factors that promote it or prevent it from occurring.
Gould, J.L. 1988. A mirror-image "ambiguity" in honey bee pattern matching. Animal Behaviour 36:487-492.
A journal paper and example demonstrating research and results of work on some of the particular aspects of learning in honeybees.
Hess, E.H. 1973. Imprinting. Van Nostrand Reinhold, New York.
Classic review and standard reference on the phenomenon of imprinting (primarily in birds).
Hess, E.H., and S.B. Petrovich, editors. 1977. Imprinting. Dowden, Hutchinson, & Ross, Stroudsburg, Pa.
A collection of readings on the subject of imprinting.

Mackintosh, N.J. 1983. General principles of learning. In: Halliday, T.R., and P.J.B. Slater, editors. Animal behaviour. Vol. 3, pages 149-177.
A discussion of some of the categories and properties of learning.

Mainardi, D., M. Mainardi, and A. Pasquali. 1986. Genetic and experiential features in a case of cultural transmission in the house mouse *(Mus musculus)*. Ethology 72:191-198.
Original research results of an experimental study of the genetic and culturally learned factors contributing to problem-solving learning in house mice.

Marcus, E.A., T.G. Nolen, C.H. Rankin, and T.J. Carew. 1988. Behavioral dissociation of dishabituation, sensitization, and inhibition in *Aplysia*. Science 241:210-213.
Further research, details, and insights (including the questioning of some previous conclusions) concerning molecular and neuronal mechanisms of nonassociative learning in Aplysia.

Marler, P., and H.S. Terrace, editors. 1984. The biology of learning. Springer-Verlag, Berlin.
A review of ecological, evolutionary, and other factors associated with learning over a wide range of invertebrate and vertebrate taxa, including humans. Includes a number of fundamental and theoretical considerations.

Mayr, E. 1974. Behavior programs and evolutionary strategies. American Scientist 62:650-658.
An extensive discussion of behavior and learning from a broad phylogenetic-life history and evolutionary perspective.

Montgomery, G. 1990. A brain reborn. Discover 11:48-53.
A popular account of Fernando Nottebohm and his research.

Nottebohm, F. 1989. From bird song to neurogenesis. Scientific American 260:74-79.
Nottebohm's own summary account of his work, findings, and the implications.

Paton, J.A., and F.N. Nottebohm. 1981. Neurons generated in the adult brain are recruited into functional circuits. Science 225:1046-1048.
An original research report on the subject.

Rakic, P. 1985. Limits of neurogenesis in primates. Science 227:1054-1056.
Report on research using radio-labeled thymidine that appears to confirm that new neurons are not added to the brains of postpubertal primates (using rhesus monkeys).

Roper, T.J. 1983. Learning as a biological phenomenon. In: Halliday, T.R., and P.J.B. Slater, editors. Animal behaviour. Vol. 3, pages 178-212.
Gives an ethological perspective to the psychological categories of learning.

Rosenzweig, M.R., E.L. Bennett, and M.C. Diamond. 1972. Brain changes in response to experience. Scientific American 226:22-29.
A classic reading on the effects of experience on neural tissue.

Sasvari, L. 1985. Different observational learning capacity in juvenile and adult individuals of congeneric bird species. Zeitschrift für Tierpsychologie 69:293-304.
An experimental investigation of differences and similarities among young and adult birds from five species of tits (Parus) and thrushes (Turdus). Increased learning capacity appeared to be a function of maturational processes and increased proportional brain weights.

Squire, L.R. 1987. Memory and brain. Oxford University Press, New York.
A review of recent studies, with some historical background. (Also see Allport.)

Thompson, R.F. 1986. The neurobiology of learning and memory. Science 233:941-947.
A review and essay on the contributions of computer modeling and artificial intelligence computational networks toward our understanding of biological memory mechanisms.

21

INTERNAL INTEGRATION OF COMPLEX BEHAVIOR FEEDING AS A CASE IN POINT

Chapter 20 considered learning and memory as phenomena and, to the extent that we are rapidly coming to understand them, as processes. However even at the most, learning is only one component in the whole picture of internal integration of behavior. The other components discussed in the previous chapters (16 to 19) include the structure and physiology of the nervous system including senses, processing, and output, neurochemical and endocrine aspects, plus maturation and other ontogenetic factors. How do all of these work together to control complex behavior on a proximate basis?

One category of behavior that is complex and has received much attention along these lines is feeding. (The ecological and other ultimate aspects of feeding are considered in Chapter 9.) Although feeding might seem to be fairly simple, it really is complex. In many animals, it is more complex than reproduction and for at least some animals it might be the most complex behavior of all. This chapter focuses on some of the possible internal integration of feeding.

Portions of this chapter were contributed by William W. Beatty.

ATTEMPTS TO UNDERSTAND THE INTERNAL CONTROL OF FEEDING

Feeding is an activity of enormous importance to all animals, as discussed in Chapter 9. There is much diversity in the kinds of foods eaten, the ways in which they are obtained, and the temporal pattern of feeding. Some filter feeders feed more or less continuously, extracting nutrients from a dilute medium that typically contains only 4 to 5 mg of organic matter per liter. Other animals feed intermittently, alternating periods of feeding with rest and other behaviors. What, if anything, governs this behavior?

We now are dealing in part, incidentally, with the related topics of drive and motivation. These topics have a somewhat embroiled, controversial history, as outlined in Chapter 2. A recent view of motivation relative to pleasure centers is mentioned briefly in Chapter 18. Persons wanting further information on motivation per se should consult discussions elsewhere (e.g., Bolles 1975).

The evidence that a particular level of body weight, hence feeding, is maintained in particular animals is quite convincing. The adult rat, for example, has a remarkable ability to control its food intake to maintain reasonably constant energy reserves in the face of changing environmental conditions. Enforced exercise (usually accomplished by forcing the rat to run on a treadmill) greatly increases energy expenditure. The rat compensates for this drain on fuel reserves by a proportionate increase in caloric intake up to a certain point. If more than about 5 hours of running is demanded per day, body weight is not maintained; instead the animal controls its weight at a somewhat lower than normal level. Rather precise caloric adjustments in intake also occur in response to variations in environmental temperature, which affects the energy expended to keep warm. Body heat is a significant factor for a small mammal such as the rat. Changes in temperature cause appropriate changes in feeding; rats eat more in cold environments, less in hot.

Variations in the caloric density of the diet also initiate appropriate alterations in feeding behavior. Precise caloric control of intake occurs when solid diets are diluted by as much as 25% with nonnutritive bulk (Figure 21-1). Increases in consumption occur up to 75% dilution. On liquid diets diluted with water, caloric control is even better.

The rat's control system, as in many other animals, including humans, has one significant flaw. It is vulnerable to the sensory quality of the diet. Fed only the nutritive but (evidently) bland chow used in most laboratories, the rat controls its weight nicely. However, if given an assortment of sweet and fatty foods, rats overeat and become obese (Sclafani and Springer 1976). The effective ingredients to cause dietary obesity in the rat are familiar. They are available in abundance and variety at any grocery store. Conversely, the rat controls its body weight at a lower than normal level if forced to subsist on a foul-tasting diet (such as a standard laboratory diet mixed with bitter quinine). The animals totally refuse such diets for a time and then, with considerable reluctance, consume them in quantities that are adequate to maintain life, albeit at a lower than normal level of body weight (Peck 1978).

The fact that body weight is maintained at a relatively constant level has lead many researchers to postulate that there is some sort of internal set-point mechanism that achieves this end. Because fat is the major energy reserve in the body, it has been proposed that somewhere in the body there is a lipostat, a mechanism that measures the amount of fat, or something else in the fat cells themselves that

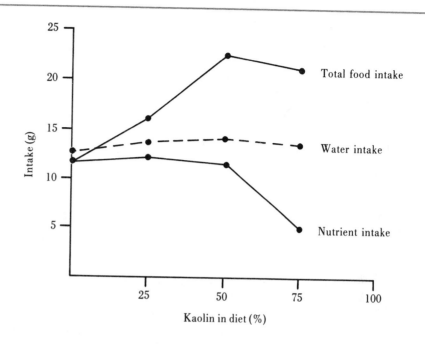

Figure 21-1 Effect of dilution of the diet with nonnutritive kaolin in normal male white rats. The animals maintained relatively constant nutrient intake and body weight for dilutions up to 50% by increasing the total food intake.
Modified from Kennedy, G.C. 1950. Proceedings of the Royal Society of London 137:535-548.

initiates appropriate behavioral and physiological responses to correct deviations from the body fat set point (see overview by Kolata 1985).

As has been seen, there is fairly good evidence that food intake is altered to compensate for changes in body weight above or below the set point. In adult mammals, weight and fat are well correlated. Direct support for the lipostatic theory is sparse, although there are a few reports indicating that removal of the body fat organs initiates increases in food intake and increases in the amount of fat in remaining fat organs (Mrosovsky and Powley 1977).

Although little progress has been made in understanding the signals that the hypothetical lipostat detects, it has become increasingly clear that bodily energy reserves are vigorously defended. In addition to alterations in feeding that have been described, changes in metabolism are involved in conserving energy stores. If food is scarce, basal metabolism declines. Conversely, if the organism overeats, the body initially stores the excess as fat. If overeating persists, weight does not increase further. Apparently the body increases its metabolic rate and burns up the extra calories (Figure 21-2).

Bolles (1980), however, challenged the view that regulatory mechanisms exist. In this view of feeding, any controls, particularly satiety, evolved because they protected the organisms from hazards in the environment. One of the big hazards is the occasional presence of too much food, which could lead to bursting. Likewise, other mechanisms such as taste aversion protect the individual from ingesting poisonous substances. Mechanisms involving learning with respect to feeding can provide the animal with a means for evaluating whether or not something is worth the effort, based on previous experience.

Bolles proposed that, instead of being regulated, an animal's food intake and body weight simply represent an equilibrium or balance of evolved species-specific

Figure 21-2 Weight gain in a human subject after a prolonged period of greatly increased caloric intake. Despite the fact that the subject nearly doubled his caloric intake, he never gained more than 14% above his initial weight. After prolonged over-feeding even a 40% increase in intake above his initial consumption did not sustain his modest weight gain.
Modified from Sims, E.A.H. 1976. Clinical Endocrinology and Metabolism 5:377-395.

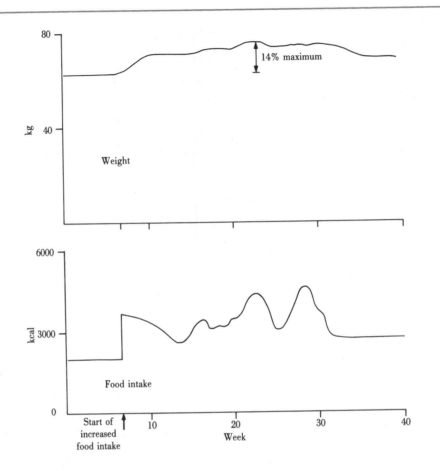

Figure 21-3 A hypothetical feeding system proposed by Bolles for the control of body weight. Items with negative symbols reduce feeding with an increase in that component; others increase the amount of feeding.
From Toates, F.M., and T.R. Halliday. 1980. Analysis of motivational process. Academic Press, London.

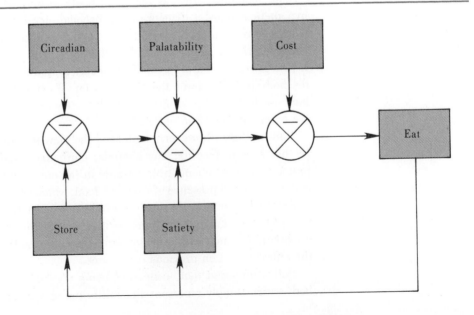

characteristics in a particular environmental context. The amount of food eaten, in this view, is the summation of several factors (see also Figure 21-3):

Eating = A(Palatability) + B(Satiety) + C(Storage) + D(Cost) + E(Circadian)

Thus as palatability, for example, is modified, so is food intake. With increased palatability the animal eats more and gains weight to a new equilibrium. On poorer-tasting diets the animal loses weight. As the cost of obtaining and processing the meals goes up, food intake goes down. In an illustration provided by Bolles, consuming a zebra is shown to be a major and dangerous undertaking for a lion. The job is not over once the zebra is captured and killed; zebras come wrapped in leather. If lions have to catch their own zebras, they remain relatively sleek. If provided with easier meals, however, lions become fat and heavy.

In this sense, feeding might be viewed somewhat analogously to human economy where the amount of money in the bank is a balance between expenditures and income. Most persons desire a positive balance with a certain level of safety and savings (analogous to fat storage) for the future. Some individuals go on saving almost without limit, whereas others increase their expenditures or stop working so hard once there are adequate savings. Part of the economy equation, like palatability, depends on how easy or acceptable it is to obtain the money by a given method. There is no set amount of money in the bank toward which everyone carefully regulates. The amount in the account is simply a balance between inflow and outflow and varies from one individual account to another and, for any one account, from time to time. This viewpoint already has been discussed from the ultimate standpoint in Chapter 9, under the topic of optimal foraging. What is of concern now is the proximate means by which animals achieve some balance in their feeding behavior.

For a simple analogy consider a bucket that holds 10 L. If it is filled to overflowing and any excess spills out, it still holds only 10 L. If, however, a valve is placed in a water line leading into the bucket, the water flow can be turned off when the bucket fills up and before it overflows. Now the bucket holds 10 L but for a different reason from that in the first case: a mechanism is present for controlling the input.

To refine the picture further, imagine that the valve is used to turn water on when the level drops to 5 L and off when it reaches 8 L. It can also be used to keep the level more precise, between 7.5 and 7.7 L, or right at 7.6 L. One hand can be used to turn the valve on and off, or different hands can be used, one for on and one for off. The amount of water, the 7.6 L, might be arbitrary and by chance, or perhaps it was determined by an order from the city water department.

In this analogy, what does *regulation* mean: the fact that the bucket contains only so much water, the presence of a mechanism to control input, the control of amount within any limits or only within precise limits, or that some factor (e.g., the city water department) specifies the particular level? The problem emphasizes the ambiguity of a common word such as *regulation* and exposes some of the semantic problems faced. It also breaks the topic of interest into a set of more specific questions that can be applied to feeding behavior: are there control mechanisms; if so, how precise are they; how do they compare among different species; and what, if anything, determines the particular level at which weight and feeding are expressed?

It is on the last aspect that Bolles' view differs from the commonly held notion. Bolles says that the expressed level is simply a balance of several factors, with different factors possibly involving different mechanisms. One can consider the separate mechanisms, but it is a waste of time to look for a single level-determining mechanism, commonly referred to as a *set point*.

The opposing view states that the level is determined (although it is not known how) by some master factor to which the contributing mechanisms respond. In this view, one can investigate both the contributing mechanisms (analogous to the valves and hands that turn them) and the master, set-point mechanism (analogous to the city water department). This view does not require the set point to be at the same level in all individuals—only that each animal has a set point or master mechanism.

In the remainder of this chapter, we will consider the component control mechanisms, which are known to exist, and remain alert for the presence of a master control, which may or may not exist. Feeding in different but well-studied organisms is described and compared.

CONTROL OF FEEDING IN LEECHES

Although feeding is generally complex, a good point to begin is in animals in which the behavior and its internal mechanisms are least complex. One relatively well-studied case involves the medicinal leech, *Hirudo medicinalis*. This is an annelid worm, distantly related to earthworms. As with most groups of animals, there are many different kinds of leeches with a corresponding diversity of ecological requirements, types of food, and behaviors. Not all leeches, for example, are blood suckers on warm-bodied hosts. The medicinal leech is, however, and we now know a fair amount about it. A good overview and references are provided by Lent and Dickinson (1988), from which the remainder of this section is distilled.

This species has been viewed by the general public both negatively and positively. It evokes a lot of squeamish aversion and dislike from people who may encounter it or fear encountering it in the water. It also has played a significant role in the history of medicine from ancient times up to the present where it is now used to remove excess blood in tissue after some types of surgery. Its saliva and genetic material also serve as sources for a number of important pharmacological agents including some used in the control of tumors.

The medicinal leech's feeding behavior in the wild is relatively simple. When hungry they attach to an exposed substrate near the water surface where they wait for ripples from suitable warm-blooded hosts. After detecting suitable ripples they swim with an undulating motion to the potential host, attach and crawl about. On encountering a warm area, they bite using three serrated, rasping jaws. If that site does not produce blood, the leech explores further and tries another location. When successful it injects a lubricating, clot-inhibiting saliva, and begins sucking blood with rhythmic contractions of muscles. The blood is pumped into a crop region of the digestive tract. After feeding for several minutes and increasing their weight by seven to nine times, they detach, move to deeper water, and hide under or between objects where they may remain for several weeks or months (up to a year in the laboratory) until they again become hungry. When satiated, these leeches not only are not attracted to warm bodies but will actively avoid them, for example if artificially placed on or touched with a warm surface.

What controls all of the component behaviors, including the shifts between hungry and satiated patterns? Distension of the body and the related proprioceptive feedback from stretch receptors plays a major role; but it does not operate alone in a simple mechanical manner. In the case of the leech, we are dealing with a relatively simple nervous system: ventral paired nerve cord (typical of protostomes, see Chapter 16) with 32 ganglia, each of which have about 400 neurons. Leeches possess neurons with both electrical and chemical synapses (Chapter 16). The chemical synapses use several familiar transmitters including acetylcholine, GABA, serotonin, and some others. Serotonin is particularly important in leeches; it is produced in large quantities in the largest neurons, Retzius cells, as well as in some of the other neurons.

Serotonin is a key factor in leech feeding behavior. The presence of serotonin, whether released naturally as a transmitter or applied artificially by an investigator produces several feeding related effects: it stimulates mucus secretion from the skin, softens body wall muscles and increases their ability to distend, generates the undulatory swimming patterns, evokes biting, and stimulates salivation. The effects can be evoked by the presence of serotonin even after interfering with the normal neural routes. The behaviors and other effects do not occur and cannot be stimulated in the absence of serotonin—such as when the neurons are experimentally depleted of their serotonin by 5, 7-dihydroxytryptamine (5,7-DHT). With replacement of serotonin after treatment by 5,7-DHT, the feeding responses resume.

This and further research have provided a nearly complete picture of the internal control of feeding in the medicinal leech. Only a small number of neurons and neural pathways are involved. Appropriate stimuli (ripples in the water and warm surfaces) cause a few specific neurons to release serotonin. The serotonin produces a host of effects, depending on the location. Inhibition from distension of the body then stops the release and further production of serotonin. This inhibition overrides competing stimulation (e.g., from warmth on the mouth region). After distention subsides the neurons start releasing (or producing and releasing if they were depleted) serotonin. The picture for more complex feeding in more complex animals, however, is not so simple, as can be seen in the next examples.

CONTROL OF FEEDING OF AN INSECT: THE BLOWFLY

Blowflies are relatively large flies, and as is common among insects, there are several species. Largely through the pioneering efforts of Vincent Dethier and his students, a vast amount of information is available on the feeding behavior of one species of blowfly, the black blowfly *(Phormia regina)* (Dethier 1976).

A typical blowfly *(Phormia)* weighs 25 mg when full grown and has a life span of 42 to 60 days under optimal conditions. Although the fly usually eats complex meals that contain carbohydrates, fats, and proteins, it can maintain itself entirely on a carbohydrate diet to supply energy. Females cannot reproduce, however, unless they eat a protein meal at the appropriate time in the reproductive cycle.

The fly's major method of getting around in the world, flying, consumes enormous amounts of energy. Energy to power flight is derived almost exclusively from the aerobic oxidation of glycogen (a polymer of glucose) stored in muscle and from oxidation of the fly's blood sugar, trehalose (a disaccharide of glucose). Blood trehalose probably is the most important fuel. Flies flown to exhaustion resume flying almost immediately after consuming a carbohydrate meal or after injection of sugar into the blood.

**Feeding Behavior and
Patterns**

A completely satiated blowfly is relatively inactive. As deprivation increases, the fly's activity also increases (Figure 21-4) until the fly finds a suitable meal or dies of starvation (about 2½ days). Although it may seem paradoxical that the fly should expend more and more of its dwindling energy supplies, a positive relationship between activity and deprivation is generally observed in animals. Presumably the increased activity increases the probability that the fly will discover a suitable meal to replenish its energy deficit before the increased drain on energy leads to death. The mechanism, although risky, seems to work fairly well, judging by the world's fly population.

During daylight, when blowflies are generally active, the fly alternates between

Figure 21-4 Activity levels of three individual flies over a 5-day period in the absence of food, under constant light conditions. Although faced with decreasing energy supplies, they became more and more active until they died of starvation. In other words, they did not become inactive and conserve energy.
From Green, G.W. 1964. Journal of Insect Physiology 10:711-726.

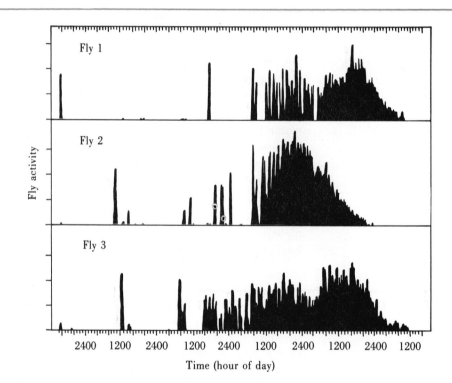

Figure 21-5 Typical pattern of feeding and blood sugar in a fly.

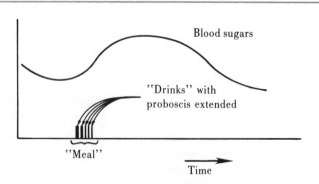

periods of random flight and rest. While airborne, the fly's behavior is governed by visual stimuli, wind currents, and olfactory stimuli. Flies and other insects orient in flight toward the direction of the wind. This tends to place the fly in an optimal position to detect food odors, although it still is a matter of debate whether wind currents or odors are more important in guiding flight. Both mechanisms may contribute.

Once on the ground the fly starts to walk. Its behavior is then controlled by olfactory stimuli (if they are present). Unacceptable odors lead to an abrupt cessation of walking, followed by flight, turning away, or frantic grooming. Acceptable odors lead to a variety of behaviors, including stopping, extending the proboscis (the fly's mouth), grooming, and approach. If the surface does not contain a volatile (and hence odorous) chemical, the fly walks randomly until it encounters a liquid with its legs (Dethier 1976).

Suppose a hungry fly encounters a source rich in carbohydrates. In the wild this might be a nectar-filled flower, an open wound on a herbivore, or a bit of decaying material. (Little is known about the feeding habits of wild blowflies.) In the laboratory, more artificial but better controlled food sources, such as solutions of known sucrose or glucose concentrations, ordinarily are used. If the fly lands in the sugar-water solution, it immediately extends its proboscis and begins to feed (actually drink). Feeding continues without pause for perhaps 60 seconds if the sugar solution is fairly concentrated. During the next minute the fly may feed for 20 seconds or so. Then it pauses for a few minutes, drinks again for about 10 seconds, pauses for a few minutes, and drinks again for 10 seconds or so (Figure 21-5). Feeding then ceases for 30 minutes to 2 hours, even if the fly remains standing in the sugar-water solution.

Control of Feeding

Initiation of feeding

Feeding (and drinking of water as well) is initiated by sensory stimulation. As already mentioned, olfactory stimuli sensed by receptors in the antennae determine the fly's approach to potential sources of food and water at a distance. After the fly locates a meal, olfactory inputs are relatively unimportant.

Ingestion is controlled by three sets of chemoreceptors located on hairs (Figure 21-6). One set is located on the tarsi of the fly's legs, the second on the lobes of the labellae at the distal end of the proboscis, and the third set is within the lobes of the labellae in the interpseudotracheal papillae.

Each hair contains five receptors, one of which is a mechanoreceptor. The other four cells are chemoreceptors, the dendrites of which extend down the shaft of the hair. The tip of the hair has a small opening or pore permitting liquids to make contact with the chemoreceptor dendrites (Figure 21-7).

Electrophysiological recordings from individual hairs suggest that each of the four chemoreceptors responds to a different class of chemicals. One receptor responds to sugars, primarily simple pentoses and hexoses. A second responds to water. The third responds to salts, especially sodium chloride, and the last chemoreceptor remains something of a mystery. It has been called the *anion cell*. It seems to respond to sodium salt anions of long-chain fatty acids, but its exact chemical specificity remains unclear.

What role do these chemoreceptors play in the control of the fly's feeding behavior? Receptors in the tarsi make first contact with liquids that might satisfy the

Figure 21-6 Sensory receptors on the labellae (lips) of flies. **A,** Scanning electron microscope photograph of the surface of the labellum showing the fringe of chemosensory hairs of a housefly *(Musca domestica)* proboscis. **B,** Diagram of side of labellum showing the numerous long chemosensory hairs of a blowfly (see Figure 21-8 for basic anatomy).
A Courtesy North Dakota State University Electron Microscope Laboratory. B from Wilezek, M. 1967. Journal of Morphology 122:175-201.

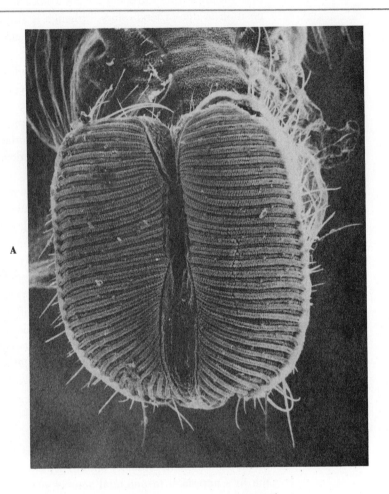

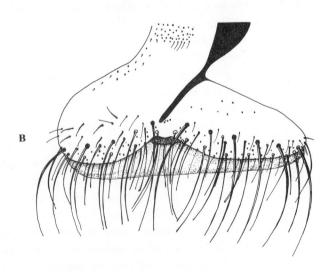

Figure 21-7 Close-up of a
chemosensory hair (minus
the long middle section) and
associated sensory neurons.
From Dethier, V.G. 1976. The
hungry fly. Harvard University
Press, Cambridge, Mass.

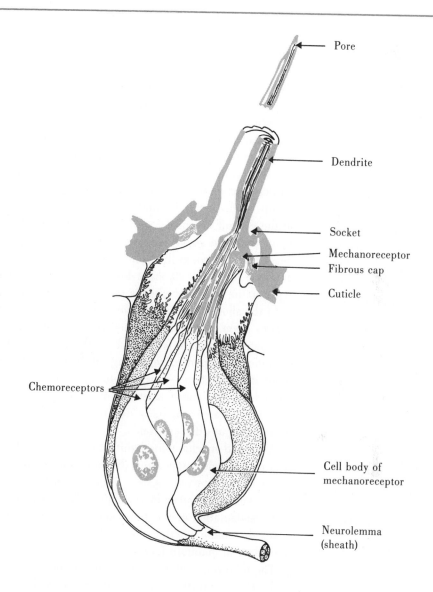

Pore

Dendrite

Socket

Mechanoreceptor

Fibrous cap

Cuticle

Chemoreceptors

Cell body of
mechanoreceptor

Neurolemma
(sheath)

fly's hunger. If the solutions stimulate the sugar receptor, the fly extends its proboscis. Varying degrees of extension are possible, depending on the strength of
stimulation of tarsal sugar receptors (and also the strength of inhibitory factors).
Suppose the fly extends its proboscis fully. This brings the hairs on the surface of
the labellar lobes into contact with the sugar-water solution. These hairs also contain
sugar receptors. When the sugar receptors of the labellar hairs are stimulated, the
labellar lobes are opened, and the fly begins to suck. Opening of the labellar lobes
exposes the receptors in the papillae, which perform a final evaluation of the solution
before it is ingested.

Thus the initiation of feeding is controlled by four sets of chemical receptors.
Olfactory receptors make the initial determination. Then taste receptors on the legs,
labellae, and papillae are exposed in sequence. The potential meal is screened for
suitability, and as might be expected, the papillae are the most demanding. Certain
substances that are acceptable to the tarsal and labellar receptors are rejected by

the papillae. Both stimulation and rejection are extremely sensitive. Sucking can be initiated, for example, by stimulating a single labellar hair (i.e., one sugar receptor).

Cessation of feeding and satiety

The salt and anion receptors obviously are important in protecting the fly from ingesting unpalatable food, but they are not involved in controlling feeding in the presence of palatable substances such as sugar-water solution. What causes the cessation of feeding, and what maintains the state of satiety?

Electrophysiological studies indicate that chemoreceptors emit a burst of activity on first contacting a solution and quickly reduce their firing to a lower, relatively steady level. Both the initial burst and the steady state evidently convey information to the central nervous system, but the maintenance of ingestion requires sensory input from the sugar receptors above a certain minimum level. Sugar receptors cease firing within milliseconds after chemical stimulation is removed; sucking may continue for 10 to 20 msec longer; then it also ceases. Thus continuous sensory input to the sugar receptor is required to maintain ingestion, but it is not enough.

In the short run, that is, over a period of a few seconds or minutes, two mechanisms seem to determine whether or not input from the sugar receptors will be adequate to maintain feeding. First the receptors themselves exhibit adaptation, a reduction in response, beginning within a second or less. The receptors also exhibit rapid disadaptation—a return toward the initial high level of response. Under normal circumstances when the fly's behavior is under control of many different receptors located on many different hairs, different receptors are adapting and disadapting at different times.

A second mechanism, habituation, a longer-lasting depression of the sensory response, is also important. Habituation is presumed to arise at synapses deep inside the fly's nervous system. Again, in natural feeding there are dynamic changes in the state of central synapses supplied by peripheral receptors. Together adaptation and disadaptation of receptors and habituation and dishabituation of central synapses determine whether the total amount of sensory input exceeds the threshold necessary to maintain feeding.

The following factors also are important:

1. **Deprivation.** This generally reduces the acceptance threshold for palatable substances (i.e., the minimum concentration that will be ingested). Greater deprivation also seems to increase the range of acceptable substances, probably by overriding inhibitory mechanisms.
2. **Concentration of the sugar.** The amount ingested is an increasing function of concentration, at least up to the point where the viscosity of the substances limits the rate of sucking.
3. **Palatability of the substance.** This relates to sensory perception and is not the same as the nutritional value of the food.

Perhaps the most convincing demonstration of the last point comes from studies in which flies are given a choice between a highly stimulating but completely nonnutritive substance such as sugar fructose and a nutritive but relatively non-stimulating substance such as mannose or sorbitol. Flies invariably choose the nonnutritive but stimulating substance and starve to death. Under natural conditions the substances that stimulate the receptors normally occur together with, and hence serve as reliable indicators of, the nutritional substances.

Adaptation and habituation interact with the fly's state of deprivation and the nature of the carbohydrate to determine the fly's feeding behavior during the first few minutes after it encounters a suitable meal. But adaptation and habituation are neuronal processes that typically have a time course of seconds or minutes. What accounts for the fact that the fly ceases to feed for a period of hours although food is readily available?

To answer this question a brief description of the digestive system of the blowfly is necessary. The digestive system is divided into three parts: the foregut (consisting of the esophagus and crop), the midgut, and the hindgut (Figure 21-8). Fluid is sucked into the esophagus. Waves of peristalsis drive the fluid initially into the midgut (some goes into the crop as well). When the midgut is full, valves close and shunt fluid into the crop. As the midgut empties, fluid is regurgitated from the crop into the esophagus and back into the midgut.

Any number of mechanisms involving various portions of the gut or the blood might be invoked to explain the long-term inhibition of feeding. Blood sugar level or some other humoral factor that changes as a result of nutrient absorption might be involved, but experimental data rule them out.

To summarize a great deal of painstaking research quickly, injections of glucose throughout the gut and into the blood do not affect feeding. This is rather remarkable because such injections supply energy and will restore active flight in a fly flown to exhaustion. Similarly, other possible humoral factors seem to be unimportant.

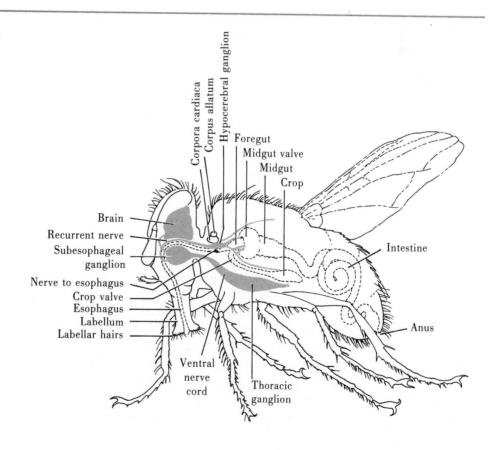

Figure 21-8 Internal anatomy of *Phormia* showing relationships of digestive system, associated nerves, and central nervous system. Modified from Dethier, V.G. 1976. The hungry fly. Harvard University Press, Cambridge, Mass.

Figure 21-9 Parabiotic blowflies. These flies have been surgically joined at their thoraxes to permit interchange of blood. They are allowed to exercise on rolled screen wheels suspended from rods and onto which they hold and walk.
Modified from Dethier, V.G. 1976. The hungry fly. Harvard University Press. Cambridge, Mass., and Belzer, W.R. 1970. Ph.D. thesis. University of Pennsylvania, Philadelphia.

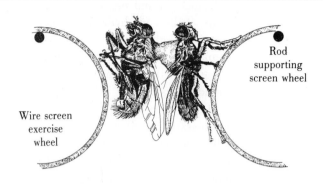

Rod supporting screen wheel

Wire screen exercise wheel

Flies whose "bloodstreams" are connected (parabiosed, Figure 21-9) feed as individuals. Consider, for example, that two members of a parabiotic pair of flies are starved, and one is fed to repletion. The fed partner fails to feed when offered a suitable carbohydrate meal. The starved partner exhibits normal acceptance thresholds and feeds normally, despite being exposed to any potential humoral satiety factors.

These and other experiments suggest that satiety in the blowfly is not related to the nutritional consequences of the ingested meal. Why then does feeding remain suppressed long after sensory adaptation and habituation have dissipated?

The answer is that there are two sets of inhibitory inputs, both arising from abdominal receptors and both of which can override excitation from the sugar receptors. One inhibitor is a set of stretch receptors located in the region between the crop and the midgut. These receptors monitor the passage of fluid from the foregut to the midgut. Signals from the receptors inhibit feeding via the recurrent nerve (see Figure 21-8). The existence of such a mechanism is supported strongly by the results of severing the recurrent nerve. Flies so prepared overeat and in some instances may ingest so much that they burst. The duration of the first drink is perfectly normal; that is, the fly stops drinking after a minute or so, presumably because of sensory adaptation and habituation. Unlike the normal fly, once adaptation and habituation have dissipated, it resumes feeding to excess.

A second system of feeding inhibition involves stretch receptors located in the body wall. These receptors are distended as the crop fills, although they are not actually located in the crop. These receptors send inhibitory inputs via the ventral nerve cord. If this nerve is cut, overeating also occurs, but the pattern is different from that after recurrent nerve severing. Flies with the ventral nerve cord severed are reported to drink one long meal that terminates when the fly bursts.

Thus the basic control of feeding (at least feeding on carbohydrates) in the blowfly turns out to be appealingly simple. Sensory stimulation initiates and maintains feeding until adaptation and habituation combine to terminate a drink. Disadaptation and dishabituation result in recovery of feeding, which is short lived. Eventually stretch receptors monitoring distension in the crop and at the junction of the foregut and midgut provide longer-term inhibitory control. It is even possible to draw a flow chart for this set of interactions (Figure 21-10), as if one were to write a computer program to operate the fly's feeding behavior.

As elegant as this picture is, there are some disquieting questions. Neither the

Figure 21-10 Flow chart of the control of feeding in the blowfly.

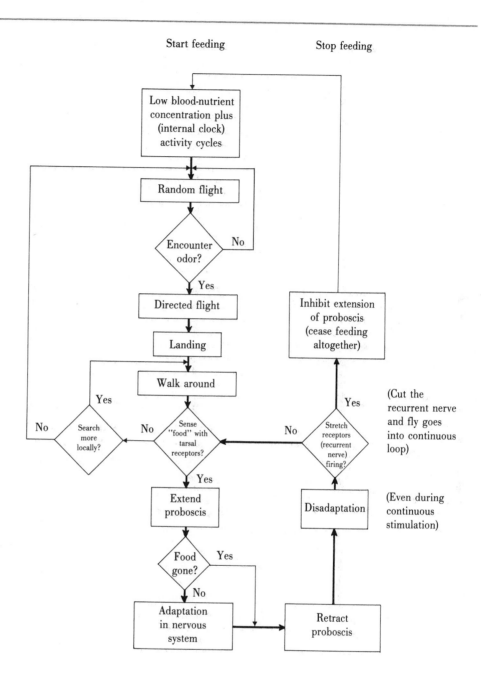

excitatory nor the inhibitory mechanisms involve monitoring of any factor directly related to available energy or potentially available energy. The sugar receptor responds nicely to substances that the fly cannot metabolize—so nicely that the fly may starve to death in the presence of other adequate but "unattractive" energy sources. Similarly, the major long-term inhibitory mechanisms also do not monitor energy. Perhaps this is not a real problem. Natural stimuli that activate the sugar receptor usually will be nutritious also or occur with nutritious substances. The fly's feeding system obviously works.

There is one other disconcerting fact in this otherwise simple story. Deprivation exerts powerful effects on feeding. This occurs partially because the empty crop does not release fluids to stimulate foregut stretch receptors inhibiting feeding. Then there are the effects on acceptance thresholds for both palatable and unpalatable substances, mentioned earlier. Eating and activity rates continue to increase after the crop is completely empty. In spite of the negative results from nutrient injections and the parabiotic experiments, this suggests that somewhere nutrient reserves may be monitored. Where and how (and in fact whether they are) are not known.

In contrast to the wealth of truly elegant work on peripheral mechanisms of blowfly feeding, almost nothing is known about what goes on deeper in the central nervous system. How, for example, do inputs from the recurrent nerve and the ventral nerve cord inhibit feeding? These and other questions await empirical answers.

The general picture of feeding in blowflies appears to also apply in at least some of the other insects that have been studied (e.g., locusts, Simpson et al 1988). What about other, unrelated animals such as vertebrates? Feeding in rats has received considerable attention. Next we consider that case.

CONTROL OF FEEDING IN A MAMMAL: THE RAT

The control of feeding in mammals is a great deal more complex than in the blowfly and appears to involve much more learning and memory. Although the control of food intake and body weight has been studied in many mammals, the most complete information exists on the rat, particularly laboratory strains derived from *Rattus norvegicus*. Fortunately enough data exist from other species to suggest that the major features of control of food intake and body weight are fairly similar among various mammals, including humans.

From a biological perspective rats are enormously successful creatures. No doubt a portion of the rat's biological success results from the fact that they are omnivores; like humans, they eat a variety of plant and animal food. Nonetheless, rat feeding behavior is quite discriminating. The following are several aspects of the rat's food selection habits worthy of note:

1. Rats are quite suspicious of unfamiliar foods. When first exposed to a strange food, rats initially eat little of it. This behavior is termed *neophobia* (literally, fear of the new) and may in part explain why rats are notoriously difficult to poison.
2. Rats have well-established and perhaps some unlearned preferences with regard to foods. Although omnivores, they prefer foods that taste sweet to humans. They also like mildly salty foods. On the other hand, they avoid sour and bitter-tasting substances, eating them only as a last resort. This aspect of rat feeding behavior also may have adaptive value, since sweet-tasting substances generally are rich in substances that can provide energy, whereas bitter substances are often poisonous.
3. The rat's nutritional requirements are similar to those of humans. As a consequence rats long have been used by nutritionists to assess commercially prepared foods.
4. Rats (and mice) display obesity syndromes caused by a single recessive gene, fa/fa for fatty (ob/ob for obese in the mouse). Genetically obese

animals seriously overeat, produce elevated levels of insulin, and gain weight, nearly all of which is fat.

Feeding Behavior and Patterns

The fact that the rat can adjust its intake rather precisely to compensate for changes in caloric density of the diet and variations in energy expenditure might lead one to suppose that feeding is initiated when some sensor detects that energy reserves have fallen below a certain threshold and continues until reserves are replenished. Studies of patterns of meal taking in freely fed rats, however, do not support this expectation. There appears to be no relationship between the size of a particular meal and the length of time since the preceding meal. The experiments were performed under conditions of constant diet formulas and reasonably constant energy expenditure by the animal, so the state of energy reserves should be related directly to the length of time since the last meal (Le Magnen 1971). The duration of feeding during a given meal, however, seems to depend more on the taste properties of the meal than on the depletion of the animal's energy reserves. It is similar to humans stuffing themselves on a big Thanksgiving meal but still having "room" for dessert, even if it hurts.

After the rat has eaten a large meal it is likely to wait longer before taking another meal than if it has eaten a small meal. Thus apparently rats (and presumably many other mammals) achieve their long-term control of food intake by varying the interval after a meal, depending on how much they ate, rather than by varying the amount to be eaten, depending on how long it has been since the previous meal. In other words, if a rat eats a lot, it waits longer until eating again. If it eats only a small amount, it tries to eat again sooner.

This may seem confusing, but the outcome makes sense if one considers that the laboratory rat's ancestors lived in a world where food resources may have been scarce, depending on environmental conditions and competition. In such a world, it would not be adaptive for the animal to pass up a good meal that it happened on, even if it had fed recently. On the other hand, having gorged itself, the rat has no reason to go out seeking food again soon.

The foregoing interpretation of rat meal patterns is speculative. Taste quality complicates the picture. At present there are few detailed data on meal patterns in rats under natural conditions or good comparative studies of feeding patterns and states of nutrition in many other mammals. Furthermore, there appear to be no studies that compare the effects of free-running feeding patterns, as discussed previously, with effects of imposed regular feeding schedules, such as with regular daily meals in many human cultures or in the food provided to domestic animals on a regular basis.

It should be clear, however, that feeding patterns in rats are not as simple as in flies. Some of the differences may be only superficial. Perhaps because flies have smaller bodies and are flying machines, they carry less fuel and run out sooner and hence have more predictable feeding patterns. Or because flies and rats are so unrelated, there may be major differences in underlying mechanisms. What is known about mechanisms controlling feeding in the rat?

To begin with, the factors cannot be easily divided into those that initiate and those that stop feeding, as is the case with the blowfly. Instead there are numerous interacting factors, and it is not known yet just how they achieve the final control of feeding. The important factors can be divided into two groups: peripheral (sensory, digestive tract, liver, en route from the viscera, hormonal, and genetic) and central nervous system (hypothalamic centers, and pathways). There is space for only a brief review of a few of the highlights from extensive research. Persons interested in more detail should see Carlson (1987) and Le Magnen (1986).

Peripheral Controls

Much as with the fly, olfactory, taste, and digestive system sensory factors are also present in the rat. Taste, olfaction, and other factors include some that have not been well examined. We really do not have any idea of what importance (if any) the sight of food is to the rat. Not much more is known about the role of olfaction in controlling feeding. Most knowledge is based on studies aimed at determining the importance of taste.

Taste stimuli exert powerful control over the rat's behavior. In the most revealing arrangement (Snowdon 1969), food was delivered to the stomach only as long as the rat depressed a lever. Rats feeding in this way maintained original body weight, but they did not grow as normal rats do, and there were several other differences. In a maze-learning situation, for example, oral rewards promoted much more rapid learning than intragastric rewards (Miller and Kessen 1952). Taste appears to be responsible mainly for the rewarding effects of food or, in less stodgy terms, the joy of eating.

Theories of hunger historically (e.g., as proposed by the famous American physiologist Walter Cannon) placed great emphasis on stomach sensations. Stomach contractions were assumed to stimulate receptors in the stomach, which in turn gave rise to the sensation of hunger (or "hunger pangs"). Most persons would swear that they "feel" hunger in their stomachs, and many studies have confirmed that gastric motility is well correlated with subjective sensations of hunger in humans.

However, many lines of evidence make it clear that neither the stomach nor sensory inputs from it are essential to hunger or satiety. Surgical removal of the stomach or a section of its nerve supply, for example, does not greatly disturb control of feeding or appetite. Human patients whose stomachs have been removed because of disease or injury continue to experience hunger. Although they must eat smaller, more frequent meals, their feeding behavior seems otherwise normal. This does not mean, however, that the stomach is unimportant under more normal circumstances.

The possible role of stretch receptors in satiety was considered early. Both humans and rats were induced to swallow balloons that could be filled with air or water, thereby distending the stomach. As might be expected when the balloons were pumped full, the subjects lost interest in eating. It was quickly noted that the experience of having one's stomach filled with air was aversive to humans (and also seemed aversive to rats), so the experiments were hardly definitive. However, rats tolerate a fair amount of gastric distension that accompanies eating diets diluted with nonnutritive bulk, so most workers concluded that gastric distension was not important, except under extreme conditions.

For many years it has been known that the introduction of hypertonic fluids into the stomach or the duodenum through tubes suppressed feeding in hungry rats (Schwartzbaum and Ward 1958). The effects are seen even when nonnutritive substances (such as NaCl) are used, so the inhibition of feeding most likely occurs either because of the activation of osmoreceptors or because of distension produced by entry of water into the gut across the osmotic gradient. It is not clear which of these mechanisms is involved, but the evidence tends to favor distension (e.g., Collins and Davis 1978). The duodenum is more sensitive to these inhibitory effects than is the stomach (Snowdon 1975).

There also is evidence that chemoreceptor mechanisms important to satiety are localized in the gut. Gastric loads of nutritive substances are generally more effective in inhibiting feeding than nonnutritive loads (Snowdon 1975). Some research indicates that the stomach may contain specific nutrient detectors (Deutsch et al 1978).

In sharp contrast to the picture for the blowfly has been the discovery that hormones from the intestine may be partially responsible for satiety in mammals. Three substances have been implicated: enterogastrone (Schally et al 1967), cholecystokinin (CCK) (e.g., Antin et al 1975, Kraly et al 1978), and bombesin (Gibbs et al 1978). Other hormones are also involved in indirect ways.

Major control over the absorptive and fasting phases of metabolism is exerted by hormones produced by the pancreas, the adrenal medulla, the adrenal cortex, and the pituitary gland. Branches of the autonomic nervous system in turn regulate the release of these hormone products and exert direct controls over metabolism.

The major hormonal control over the absorptive phase of metabolism is exerted by insulin, a protein hormone produced by islet cells in the pancreas. Insulin is secreted in response to a rise in blood levels of either (1) glucose or (2) amino acids. Insulin promotes the storage of glucose and other nutrients as fat.

Several other important hormones that affect energy and food metabolism, hence (directly or indirectly) feeding behavior, include glucagon, epinephrine, growth hormone (somatotrophic hormone), and certain hormones of the adrenal cortex (called glucocorticoids, of which corticosterone is most important in the rat). In general, these hormones promote the release of energy from bodily stores.

The hormones of the gonads also influence feeding, growth, and body weight regulation. Testosterone stimulates feeding and body weight gain and reduces the proportion of body fat, apparently by facilitating growth of muscle. Estradiol reduces food intake, body weight, and body fat. Both testosterone and estradiol can cause these changes when injected directly into the brain in small quantities, so they may regulate feeding by influencing brain control mechanisms. Alternatively, the effects of the gonadal hormones may result largely from influences on peripheral metabolism.

Investigations of other peripheral controls have included a large number of experimental manipulations such as self-administered feeding via intravenous catheters, cutting and exteriorizing the esophagus so that the mouth is separated from the gut (sham feeding), infusion of a variety of nutrient and nonnutrient solutions into the hepatic portal vein (between the intestine and liver), cutting the vagus nerve, which supplies the viscera, and perfusion or blood-mixing experiments as in the parabiosed flies. The results of these various studies indicate that the mouth, stomach, duodenum, liver, vagus nerve, and hormonal and chemical factors all play important roles in the control of feeding. When any one or a few of these at a time are eliminated, the rat still controls its food intake but in all cases imperfectly.

The Hypothalamus and Central Mechanisms

Unlike the paucity of information concerning central nervous system mechanisms in controlling fly feeding, there is much information for possible central mechanisms in the rat. Unfortunately, much of it is conflicting or ambiguous. The historical record provides a revealing journey through the perils and problems of trying to understand neural mechanisms in complex brains such as those possessed by mammals.

Early workers in the physiology of behavior believed, or perhaps more accurately hoped, that the brain would be compartmentalized simply and discretely into functional units that agreed with ways of categorizing behavior. This view of brain function can be traced historically to the ideas of the phrenologists, Gall and Spurzheim, in the early 1800s and is in fact observed in the central projections of the major sensory systems (Chapter 17). This view, however, does not apply to the control of feeding, as revealed by extensive research.

For many years it appeared that, in its most basic outline, the control of feeding in mammals such as the rat might be quite similar to the control of feeding in the blowfly. Recall that in the blowfly feeding is excited by the detection of sugars by chemoreceptors. Feeding ceases when the chemoreceptors adapt and when an inhibitory state of satiety is produced by foregut and body wall receptors that monitor distension and connect to the brain via the recurrent nerve and ventral nerve cord. The dual center model of feeding in mammals, first proposed by Stellar (1954), basically is a similar mechanism. It postulates the existence of an excitatory feeding center in the lateral hypothalamus (LH) and an inhibitory (satiety) center in the ventromedial hypothalamus (VMH).

If given unrestricted access to reasonably palatable food in the home cage, a rat with VMH lesions begins to overeat almost immediately. This overeating (which may result in a doubling or even tripling of daily food intake) persists for 3 to 12 weeks after surgery. During this dynamic stage body weight increases rapidly and may reach twice the preoperative level of that of an intact control. Nearly all of the excess weight is fat; lean body mass is not increased and may in some instances be reduced. Thereafter the rat enters the static stage, and food intake returns to nearly normal levels. Body weight is maintained indefinitely at the elevated level (the static obese plateau).

If a static obese rat with VMH lesions is starved and thus caused to lose weight, another dynamic stage of overeating and rapid weight gain occurs as soon as food is again made freely available. This experiment can be repeated indefinitely with identical results; each time the VHM rat overeats voraciously until it regains its static obese body weight level, indicating that the effect of the lesion is permanent (Mook and Blass 1968).

The fact that overeating persists only until a certain higher than normal weight level is reached suggests that the VMH lesion has elevated the level at which body weight (or body fat) is being controlled. Several other lines of evidence support this conclusion (Figure 21-11) (e.g., Hoebel and Teitelbaum 1966). In addition, the level of feeding in lesioned rats is affected by taste quality and energy demands such as from exercise or cold.

Even if palatable food is available, however, overeating and obesity do not develop after VMH lesions if the animal has to exert much energy to obtain the food. There are a variety of ways of manipulating effort; all of them point to the same conclusion.

Information about the other implicated region of the hypothalamus, the LH just lateral to the VMH, began appearing in the early 1950s. Anand and Brobeck (1951) reported that bilateral lesions in the LH resulted in aphagia (no eating) and death. Within a few years other workers confirmed these observations but also showed that death from LH lesions was not the rule; indeed recovery of feeding typically

Figure 21-11   Body weight and food intake of a rat that sustained a bilateral VMH lesion at day 0. Note the initial period of overeating (hyperphagia) and rapid weight gain followed by return toward normal intake as the rat's weight reached a static plateau about 100 days after surgery. After subsequent force-feeding, which drove the rat's weight above its plateau level, its food intake was reduced until it lost weight. After a period of forced starvation, food intake increased until the lost weight was regained. Modified from Hoebel, B.G., and P. Teitelbaum. 1966. Journal of Comparative and Physiological Psychology 61:189-193.

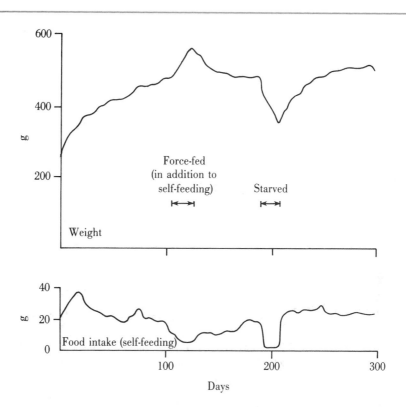

occurred provided the animal was artificially fed during the immediate postoperative period (Teitelbaum and Epstein 1962). However, feeding and drinking in LH-lesioned rats are in no sense normal, despite considerable recovery. The effects of LH lesions have been shown in a number of mammals in addition to rats. Cats, for example, normally attack and eat mice. LH-lesioned cats have little interest in mice, and in fact when handed a mouse directly, at least one experimental cat went to sleep with the mouse in its jaws (Wolgin et al 1976).

Despite recovery of the ability to survive on dry food and water, rats with LH lesions control body weight at chronically lower than normal levels, apparently indefinitely. Lower than normal weight control occurs regardless of whether the diet is highly palatable. The duration of aphagia after LH lesions is a function of the rat's relative weight at the time the lesion is made (Figure 21-12). When LH lesions were made in animals that had been eating as much as they wanted, they remained aphagic until they lost enough weight to reach about 80% of normal. In another group that had been preoperatively starved so that their weights were 60% to 65% of normal, lesioned rats resumed eating soon after surgery. They were in fact slightly overeating until they gained weight up to the 80% level. In a third group that was force-fed to make them heavier than normal, the duration of aphagia was prolonged. These animals also resumed eating when the lower than normal weight had been attained.

Despite a large number of subsequent studies on rats with VMH and LH lesions, however, there are numerous problems in interpretation. Different research laboratories have gotten conflicting outcomes on some aspects, and a clear picture has

Figure 21-12 Body weights of rats that received small (4 seconds) or larger (7 seconds) LH lesions. Before the lesions the rats were either fed as much as they would eat (ad libitum) or starved to reduce body weight at the time of the lesion below the level at which it would ultimately be regulated. Modified from Powley, T.L., and R.E. Keesey. 1970. Journal of Comparative and Physiological Psychology 70:25-36.

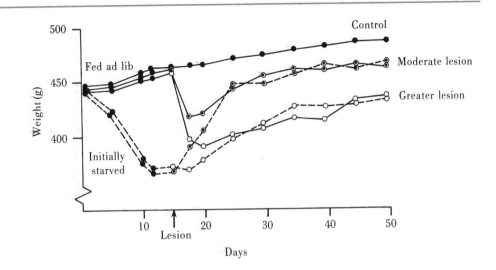

not yet emerged. One major problem with lesions in the hypothalamus, for example, is that in addition to the centers of neurons located there, both sensory and motor pathways from other parts of the brain run through this region. Lesions thus also damage those pathways, and that may be producing some of the observed effects (e.g., Zeigler and Karten 1974, Grossman 1975).

One line of evidence that originally seemed to support the concept of an LH feeding center was the demonstration that electrical stimulation of the lateral hypothalamus could elicit feeding and food-rewarded behaviors. Although some laboratories produced maps that appeared to have identified a center, the exact location of the center varied considerably from one laboratory to another. The reason for this soon became apparent. Electrical stimulation of a particular hypothalamic site could give rise to a variety of behaviors, depending on which goal objects were present in the testing situation (Valenstein et al 1970).

Experiments revealed that switching from one electrically elicited behavior to another was not limited to oral activities such as eating and gnawing. Animals could be switched from eating to copulating, behaviors that obviously involve different motor sequences.

So what is electrical stimulation of the hypothalamus doing if it is not activating discrete centers for this or that bit of behavior? The most reasonable explanation is that it activates pathways that are essential for all active, motivated behaviors. Given that these pathways are activated, stimuli (goal objects) in the immediate environment direct the appearance of specific behaviors. The broad spectrum of behavioral deficits produced by LH lesions (and other treatments) is consistent with the view that the LH area and the pathways that course through it are important to a wide range of active behaviors rather than part of a neuronal circuit that controls only feeding.

What of the cells in the LH area? Are they completely unimportant to the control of feeding as the preceding discussion seems to indicate? Various works suggest that these cellular groups may still play an important role. That role may be more directly related to feeding than to other motivated behaviors. Rolls (1978) has identified LH and neighboring cells of monkeys that seem to respond selectively to hunger and food-related stimuli. Whether or not similar cells exist in the rat and other species is not yet known.

The notion of two neural centers is difficult to defend. Although the neurons in the LH and VMH seem to have a role in controlling feeding, the brain mechanisms that control feeding clearly involve much more.

A different view of the VMH as stated by Powley (1977) emphasizes changes in hypothalamic control of visceral functions. Specifically, Powley suggests that the behavioral changes accompanying VMH lesions arise from an exaggeration of the normal patterns of digestive reflexes triggered by the sensations associated with food in mouth. Because of a general overresponse of visceral systems to the taste (and other stimuli) associated with food, the VMH animal is in effect locked into a positive feedback loop, and overeating is the result.

Another aspect of the problem is that, given minimal access to food, rats with VMH lesions are hyperinsulinemic (Frohman and Bernardis 1968). Insulin drives glucose and amino acids into cells and fosters their conversion to glycogen and fat. Insulin injections in normal animals induce overeating and obesity. Thus another reasonable hypothesis is that the overeating and obesity that accompany VMH lesions are caused by the hyperinsulinemia.

Experiments designed to evaluate this hypothesis indicate that hyperinsulinemia is a factor in VMH obesity, but it is not the whole story. Rats prevented from developing hyperinsulinemia (by destroying the insulin-producing cells in the pancreas and injecting fixed amounts of a long-acting form of insulin) still develop overeating and obesity when given VMH lesions (Vilberg and Beatty 1975). The magnitude of both effects is less in these animals than in animals that do become hyperinsulinemic.

Thus changes in insulin secretion contribute to the VMH syndrome, but other factors evidently are also important. Most likely the other responses are other activities of the viscera controlled by the VMH via efferent branches of the vagus nerve.

This does not, however, explain the static phase of the VMH syndrome. This is a troublesome problem for many theories because it is hard to explain why the dynamic period of overeating stops and especially why it stops at a precise level of obesity. Powley's explanation is that there are limits to how long the visceral systems can operate in an exaggerated manner; that is, overeating ends because the visceral systems become fatigued. Why does not the fatigue not dissipate and the overeating resume? The hypothesis awaits rigorous examination. The set point, if it exists, remains elusive.

OTHER ORGANISMS (INCLUDING HUMANS) AND CONCLUDING COMMENTS ON THE INTERNAL CONTROL OF FEEDING

In spite of conflicting views and many unanswered questions, the emerging view of the internal control of feeding in mammals, as exemplified by the laboratory rat, places great emphasis on a complex system of humoral, neural, and metabolic factors that together control feeding. More than likely the system is redundant in the sense that some of the controls could be eliminated without grossly altering the working system. What, in biological terms, is going on with all this complexity? Is it truly excessive, truly advantageous, or some of both?

Compared with the leech and the fly, the rat's internal control mechanisms for feeding appear excessive. The picture for the leech seems fairly complete. That is not the case for the fly, yet is it clear that control of feeding in flies is vastly simpler than in rats. The fly system appears to rely on two peripheral routes and minimal redundancy to stop feeding; one involves stretch receptors in the region between the crop and midgut that operate through the recurrent nerve, and the other (perhaps a backup) apparently uses stretch receptors in the body wall that connect to the ventral nerve cord. Feeding in flies is also stopped or, more properly, interrupted by temporary adaptation and habituation. This does not lead to long-term satiation.

The mammalian system (believed to be typified by the rat) involves peripheral mechanisms but apparently places far less emphasis on them than does the fly. The involved nerves (e.g., vagus in the mammal as opposed to the recurrent in the fly) are not homologous (from the same ancestral origin). Until more is known about the vagus nerve, one might argue that they are hardly even analogous (have the same function).

The mammal appears to rely much more on chemical—metabolic and hormonal—mechanisms in the control of feeding than does the fly. The amount of redundancy in the mammalian system seems to be disconcertingly excessive. The complex control of mammalian feeding also makes the system vulnerable to problems and complications, such as obesity and overeating in humans.

Obesity is not a single condition with a single cause but rather there are many causes and types of obesity and overeating. Some appear to be related to each other and to other common symptoms such as depression. Three that may be related in humans are disorders that involve depression and carbohydrate appetite: seasonal affective disorder (SAD), carbohydrate-craving obesity (CCO), and premenstrual syndrome (PMS). Studies also implicate two other internal chemical factors in the control of eating in mammals: the hormone melatonin from the pineal gland and the neurotransmitter serotonin as in the leech. Melatonin levels increase in the dark and decrease in the presence of bright light such as sunlight. It affects human moods and subjective feelings of well being and activity levels. Serotonin, a neurotransmitter derived from dietary tryptophan (an amino acid) influences appetite and as a consequence feeding behavior. SAD, CCO, and PMS all involve cyclic patterns (annual, daily, and monthly respectively), apparently related to light levels. The interrelationships between melatonin, serotonin, and all of the other factors are not yet understood. For further discussion and references see Wurtman and Wurtman (1989).

The control of food intake obviously is important, and at least some backup equipment can be useful or even critical. Given the number of biochemicals involved in food and metabolism, there clearly is some advantage to fine tuning the system to different substances, although how an animal achieves a balanced diet could be a whole new problem. That there are finely controlled differences in diet and feeding behavior based on specific factors has been shown in the starling (del Rio and Stevens 1989). Starlings lack the digestive enzyme for sucrose (sucrase) and get sick if they ingest sucrose. Consequently they eat food with other types of carbohydrates such as fruits with glucose and fructose. They avoid food with sucrose, unlike some other birds such as hummingbirds that select high sucrose nectars.

The difference between leeches or flies and rats or starlings may partially be one more case of neural parsimony in insects (Roeder 1967) (Chapter 16). Vertebrates may also have acquired some excess neural and endocrine baggage through the eons. Most of the complexity in the internal integration of feeding behavior in these more complex animals is probably related to a diversity of diet and feeding conditions in present day species or their distant ancestors.

SO WHAT?

Scientists have identified general neural pathways for many behaviors from cockroach startle responses to human language; we have learned much about relatively isolated neurons, groups of neurons, their diversity, and physiology; and we have a surprisingly complete picture now for the internal control of behavior and even of the mechanisms of learning for some of the simpler systems, such as in *Aplysia* and nematodes. Based on this, we can make tantalizing guesses about some of what may be going on in higher systems. The persistent problems in attempting to completely understand even something as seemingly simple as starting and stopping eating in mammals stand as a humbling reminder that we are not there yet. Scientific curiosity, however, seems to remain as persistent as the problems. Research is continuing, and further progress may be expected. Stay tuned.

SUMMARY

We are approaching a fairly complete understanding of the internal control of some behavior in some organisms, such as feeding in leeches and blowflies, but we are still far from understanding it in groups such as mammals with more complex nervous systems and apparently a greater number of components involved in the internal control, including a large number of hormones and a greater role of learning.

Feeding in medicinal leeches is stimulated by simple cues such as ripples in the water and the touch of warm surfaces on their lips. These and related physiological aspects are mediated primarily through a single neurotransmitter hormone, serotonin, from only a few specific neurons. Distension of the body inhibits further release and production of serotonin, and consequently feeding. The control of feeding in blowflies is more complex, also involving stretch receptors and apparently hormonal factors, but it is simpler than in vertebrates such as the rat. The internal control of feeding in mammals includes many more peripheral and central factors, including several hormones and neurotransmitters (including serotonin—as in the leech). Much of the complexity in the controls of vertebrate feeding undoubtedly arises from the increased complexity of the diet, ecological considerations, and specific nutritional requirements and constraints (such as the lack of a particular digestive enzyme, sucrase, in starlings).

Progress is rapidly being made in our understanding of the internal integration of feeding behavior and for several other categories of behavior, for which feeding may serve as a model. The total picture of internal integration includes several components: neural structure, processes, and pathways, numerous behavioral chemicals, maturation and other ontogenetic aspects, and in many cases learning. In spite of the complexity, problems, and lack of understanding for the operations of complex nervous systems, however, rapid progress is being made and a more complete picture is emerging.

FOR FURTHER READING

del Rio, C.M., and B.R. Stevens. 1989. Physiological constraint on feeding behavior: intestinal membrane disaccharides of the starling. Science 243:794-796.
A description of the impact of a specific factor on feeding behavior in a vertebrate.

Dethier, V.G. 1976. The hungry fly. Harvard University Press, Cambridge, Mass.
A review and synthesis, in highly readable form, of findings on the control of feeding behavior in blowflies.

Garcia, J., W.G. Hankins, and K.W. Rusiniak. 1974. Behavioral regulation of the milieu interne in man and rat. Science 185:824-831.
A classic article on vertebrate food preferences and related feeding behavior.

Le Magnen, J. 1986. Hunger. Cambridge University Press, Cambridge, U.K.
An up-to-date review of information on the internal control of feeding in the rat (primarily). Includes the historical aspects of research in this subject.

Lent, C.M., and M.H. Dickinson. 1988. The neurobiology of feeding in leeches. Scientific American 258:98-103.
An excellent summary of the internal control of feeding in the medicinal leech.

Simpson, S.J., M.S.J. Simmonds, A.R. Wheatley, and E.A. Bernays. 1988. The control of meal termination in the locust. Animal Behaviour 36:1216-1227.
An original research article that makes a good companion to the work done on blowflies (see Dethier 1976).

Wurtman, R.J., and J.J. Wurtman. 1989. Carbohydrates and depression. Scientific American 259:68-75.
A description of the internal, physiological correlates of three common cyclic human disorders involving feeding behavior: seasonal affective disorder (SAD), carbohydrate-craving obesity (CCO), and premenstrual syndrome (PMS).

PART FOUR

ADDITIONAL TOPICS

22

BEHAVIORAL RELATIONSHIPS BETWEEN HUMANS AND OTHER ANIMALS

Most of this book is devoted to the behavior of animals without considering human interactions, except those that involve the historical and methodological aspects of research (Chapters 2 and 3 plus other mentions throughout the book). Chapter 10 deals with interspecific relationships among animals but mostly excludes those involving humans. In this chapter, we focus specifically on the myriad of interactions between other animals and humans, from ancient times up to the present day, and involving not just a few interested biologists but the whole human species. Modern society is currently debating the roles of humans in the lives of other animals in relation to ecological concerns, animal welfare, animal rights, and the ethics of what humans can or should do with animals; this chapter discusses those issues. At the end of the chapter we provide some general comments to give insight and direction for the practical problems of managing animals, whether in captivity or in the natural environment; these comments are based on behavioral considerations and related nonbehavioral aspects (such as health of the animal).

INTERACTIONS BETWEEN HUMANS AND OTHER ANIMALS: A SURVEY AND HISTORICAL OVERVIEW

The relationships between people and other animals and the opinions different people have about other species and how we humans should interact with them are extremely diverse. In response to other animals, humans have hugged and hated them, cuddled and clubbed them. Along another dimension, the views of people about other animals range over a continuum from naive, uncritical anthropomorphism (Chapter 1) to seeing animals purely as objects to be used or treated in any way one wants. Many people are concerned about animal welfare, but it is not always clear what welfare means or how to measure it. Beyond welfare per se, some people believe that other animals should have complete rights and protection equal to (or in some cases even greater than) human rights. Other persons, however, believe that only humans have rights.

This diversity of attitudes and actions toward other animals not only leads to a lack of consensus but also has created much confusion and heated debate. Before considering the present-day behavioral relationships among humans and other species and the issues related to those interactions, it is important to first understand the historical background of our current state of affairs.

Animal Threats to Humans

In the natural environment, where every animal looks out for the welfare of itself or relevant others such as relatives, mutualists, etc. (Chapters 10 to 15), humans are just one more species. The natural world is not a particularly benign place, and there are various other species of animals that can kill or otherwise harm us. This includes some large predators such as a few sharks, some of the big cats, bears, and a few crocodiles, as well as many animals capable of significant defenses, including venomous snakes, arthropods, and cnidarians, plus large animals that can harm us by virtue of their size, teeth, or claws. In addition, many animals pose indirect threats to human welfare in diverse other ways by being competitors for resources, vectors for disease, parasites, or simply nuisances.

The adaptive response by humans, acquired mostly through cultural learning from parents or peers, is to develop negative attitudes, including fear, toward those species causing problems. A few responses such as a fear of snakes may be partially innate (Morris and Morris 1965). Up until the last century or so, humans were in the minority of animals, and negative attitudes toward some animals had survival value. During the past century or so, however, with the explosive growth of human populations, technology, development of the Earth's land and water resources, geographical expansion, and cultural developments, the tables have turned, and the human species now has the upper hand. The populations (and habitats) of many of the more dangerous animals have been greatly reduced or even eliminated. There are still places where one must be alert to dangerous predators and the presence of venomous species; and we still have some pests, parasites, and disease vectors (e.g., deer ticks carrying Lyme disease). Now, however, most humans live in locations and have more sheltered lifestyles in which they are not exposed to dangerous animals. Nonetheless, even habits and attitudes that are entirely learned develop an inertia that dissipates slowly, and many of the negative attitudes toward some animals persist.

In a few cases, threats from other animals remain, and some negative attitudes may still be adaptive. This creates a source of conflict between people with vested interests, such as livestock owners facing losses from wolves or other large predators, and other people without the vested interests who view the animals differently or who think that society's need to preserve biodiversity and vanishing wildlife should take priority over individual economic interests. In other words, in a world of rapidly diminishing wilderness and natural quality, the value to society of what remains may exceed the threats and costs. This issue has created an arena of heated debate and politics.

Hunting, Trapping, and Fishing

Capturing and killing other animals for food or other resources is another area of human life that, like the problem of animal threats, has undergone major changes in attitudes and behavior. Our earliest ancestors were hunters and gatherers who depended on the resources provided by other animals, not for sport or entertainment but for basic livelihood. Our ancestors were omnivorous and included animals in their diet. That affected human biology and sociology both in our genetics (even in our morphology—we still have meat-tearing canine teeth, for example) and in our learned culture.

The skills, techniques, and attitudes toward hunting, fishing, and trapping were passed on from one generation to the next. Children grew up accompanying their parents during the capture, killing, butchering, and preparation of the meat and byproducts. This began to change with the development of agriculture, division of labor in society, growth of cities, and a reduction in the proportion of the human population that was directly involved with killing and processing animals. A large proportion of the Earth's human population continues to include meat in their diet today, but few are directly involved in the bloody job of killing and butchering. Also, much of the meat, including fish, now comes from animals specifically raised for food purposes rather than from wild populations.

Some of the human desire (and need?) to kill, or at least pursue, some species remains. There are still some native peoples throughout the world for which these activities continue as an important source of livelihood. Now, however, most other people hunt, trap, and fish only on a supplemental basis or as a source of sport or relaxation. Some people maintain that the pursuit of other animals is an inherited behavior that should be permitted to continue, especially if the animals are scientifically managed and not in danger of extinction. Other people disagree vehemently, believing that hunting, trapping, and fishing are outmoded cultural relics that have no place in modern life. Many people are unaware of, or uninterested in, the arguments; they simply enjoy and intend to continue hunting, trapping, and fishing.

Regardless of one's personal opinion on pursuing and killing other animals, it is a fact of life that many people around the world currently do it. However, because of expanding human populations, diminishing populations of some other species, differences of opinions, reduced opportunities to find and pursue other animals, changed environments under which children grow up, and changing cultural attitudes, this realm of human behavior has become the subject of much debate, politics, and change.

The hunting, trapping, and fishing uses of animals have generated a whole field of activity and research related to the protection, conservation, and management of game species. Much of the development was in response to overharvest of the animals, particularly during the latter half of the nineteenth century and first half of the twentieth century. However, some of this practice goes back for several centuries with the employment of specialized persons in the protection and management of game species for royalty and wealthy persons. From this developed game wardens and their modern counterpart, conservation officers, legislation and regulations, and an entire profession of wildlife management and research. During the last few decades these interests also joined forces with more traditional biologists, particularly in the areas of ecology and behavior, and expanded their interests—including conservation and protection—to include nongame species (animals that traditionally or legally were not pursued for sport).

Livestock

The story for livestock animals raised for purposes of food and other resources parallels that of hunting, trapping, and fishing. As human culture became more sedentary, with greater numbers of people living in cities, domesticated animals became the primary source of meat. Domestic animals did not require pursuit to the same extent as wild animals, but before they could be consumed they still had to be killed and taken apart.

Although not all humans were directly involved with livestock production, many were, particularly at the family level. In other words, children grew up in an environment in which animals were raised, killed, and butchered. They learned to do these activities themselves, and they acquired various attitudes regarding such activities. This included views on how animals could and should be handled and cared for, and how humans relate to animals. Because of the large number of people involved, these attitudes remain deeply seated in modern society.

Livestock production is a major industry throughout the world; it has a place in formal education in the public schools in rural areas and at many universities; it involves the participation and vested interests of millions of people; and it influences how many people think about and relate to animals. Livestock production and meat-eating, like hunting, trapping, and fishing, are facts of human life and are likely to exist for a long time to come.

However, both the methods of producing livestock and society's attitudes have been changing. Production has become more and more mechanized; larger numbers of animals are involved; the average amount of physical space devoted to each animal has been reduced; and the persons involved in the production often have decreasing personal interaction with the animals. In spite of the millions of people involved in livestock production, there are now many billions more people who are not directly involved, including many vegetarians who have no desire to use meat or animal products. Many people have reduced or eliminated their consumption of meat for health or ecological reasons. Most people who are not directly involved in livestock production are far removed from it in terms of generations, environments, and lifestyles. The majority who eat meat get it from the market, grocery store, or restaurant. They have little or no cultural continuity with livestock production, little or no understanding and appreciation of it, and frequently have acquired or developed other attitudes about how humans should behave toward other animals (often modeled after attitudes toward other humans). Increasing numbers of people are starting to express their differing opinions. That has already caused some changes in the relationship between humans and livestock animals.

Zoos, Circuses, Other Entertainment, Education, and Research

The history of the use of animals in menageries and zoos is long and fascinating. This aspect is also somewhat intertwined with using animals in circuses or otherwise for entertainment, as well as the use of animals in education and research. Zuckerman (1980) provides a review of the general history of zoological gardens and case studies of 24 modern zoos. Other useful discussions of the roles of zoos include books by Hediger (1964, 1968). The term *menagerie* is derived from the French *menager*, which refers to the general management and care of a household. In the earliest use of menagerie, it meant the enclosure attached to a home where the family livestock was kept. Louis XIV, in the seventeenth century, added wild animals to the menagerie attached to his palace at Versailles and established a new custom. The word *zoo* dates from 1847, when it was derived as an abbreviation for the Zoological Gardens of London.

Although our terminology is relatively recent, the practice of keeping wild animals in captivity goes back to before recorded history. By the time of ancient Egypt certain wild animals were considered sacred; specimens were kept in temples and parks; and animals were mummified and kept in special mausoleums. The Egyptians and later the Romans held animals in high esteem and used them in triumphal processions. With this came the training of camels, elephants, and even lions and such species as the oryx to pull chariots. Collections were set up in Mesopotamia, India, and China. As early as 1100 BC there was a collection in China known as the Garden of Intelligence. The taming and training of animals were developed to a high degree, to the point of teaching baboons to play musical instruments and pick different characters of the alphabet on cue (Zuckerman 1980). More and more animals were collected and paraded before the public.

The Romans, however, then proceeded over a few hundred years to transform public interest in wild animals into, as Zuckerman describes it, "a depraved and brutal cult." Professional schools were established for training gladiators, and the public slaughter of wild animals and men became national sport. For centuries people crowded into stadiums, amphitheaters, and coliseums to watch the butchery—shows often went on for months on end. Trajan is recorded as holding games for 4 months, during which 10,000 men and 11,000 wild animals were killed. Nero held one event with a company of horsemen against 400 bears and 300 lions. In some cases the coliseums were flooded, and gladiators in boats fought hippopotamuses, seals, and crocodiles.

All of this obviously required the keeping of large numbers of animals in captivity. Augustus, during a 15-year period, kept 3500 wild animals. From 79 to 81 AD Titus was said to have kept and "used up" 5000 animals.

A temporary stop was put to all of this in 325 by Constantine, the first Roman emperor to become a Christian. But the gladiatorial and animal combats were resumed in the sixth century by Justinian, an otherwise scholarly and serious Roman emperor, after he took a mistress, then wife, named Theodora. She was the daughter of a bear keeper in Constantinople and, according to the reports, was rather wild herself. When gladiatorial and animal fighting resumed as a sport, the only major difference from the earlier periods was that it became illegal for priests and bishops to watch. These animal contests were held at least into the twelfth century.

Relics of this violent past in today's culture can still be seen in bullfights, cockfights, dogfighting, circus lion-tamers and ringmasters with their whips, and the general machismo of people who wrestle or otherwise associate with ferocious beasts.

The next phase in the evolution of zoos, although they were not called that then, began about when organized gladiatorial games came to an end. It became fashionable for kings and monarchs to have, give, and be accompanied by collections of wild animals. The tradition of the royal collection in England can be traced back to at least 1235 AD. This collection remained until the Zoological Gardens were opened in 1828.

Elsewhere there were royal zoos throughout Europe, including Paris, Prague, and Vienna, and there are records of somewhat similar zoos even in Mexico before the arrival of Columbus. Vestiges of the use of animals in performances continued and grew in popularity with the growth of circuses, many of which traveled.

Next came the holding of animals for science and education. Although captive wild animals were kept for educational purposes in the Chinese Garden of Intelligence, mentioned earlier, and the zoo of Ptolemy II in Alexandria around 300 BC, the use of animals for other than recreation or pomp and power for the owner did not really begin until the Versailles menagerie in 1665. One of the early educational uses of collections of wild exotic animals was for obtaining fresh specimens (in the absence of refrigeration and preservatives) for anatomical studies. However, such studies did not really become important from a zoo standpoint until about the middle of the eighteenth century when zoos achieved a new prominence as resources for studies of taxonomy, anatomy, and physiology. The Zoological Society of London was granted a royal charter in 1829 as a scientific society, with its associated zoo—this zoo to be scientific from the start—for "the advancement of Zoology and Animal Physiology and the introduction of new and curious subjects of the Animal Kingdom." Such uses of zoos then spread around the world.

After World War I came a period of decay for zoos. The subjects of previous research at zoos, such as comparative anatomy and taxonomy, were considered to be less glamorous, and there was a constant clash between the use of animals for public display and for research. Zoological and biological research for the most part was being undertaken in different settings. There also were serious economic pressures, and social attention was directed to many other concerns. Zoos, for the most part, slid from the limelight and suffered from lack of attention and funding.

During World War II many of the zoos in the Old World were destroyed, which in many cases was a blessing in disguise. With the destruction of these old, inadequate, run down, and poorly equipped zoos, the stage was set for a fresh start and new designs.

After World War II the plight of zoos began to reverse, and the latest phase, one of commerce and conservation, began. Public interest started to increase for new zoos and displays. This increased demand and pressure for exotic animals opened a new world of animal collecting and commerce, which began to endanger the populations of some species. The rarer the animal, the more it was desired. This was accompanied by many other environmental problems, including severe loss of habitat for some species. At the same time there was a rapidly growing public awareness and shift in attitudes in favor of animal welfare, both in the wild and in captivity. Combined with new knowledge about the animals, particularly in the field of animal behavior, and the building of new facilities, zoos stepped into a largely new position in two important respects: (1) for the conservation of dwindling populations of wild species—in some instances as the only place they existed until they could be reintroduced into the wild—and (2) for the study of behavior.

Figure 22-1   Modern zoological facilities designed for multiple educational, scientific, and recreational purposes: a beaver exhibit at the Minnesota Zoological Garden. **A,** Beaver cutting aspen trees (which are placed in the exhibit daily by keepers). **B,** A beaver working on a dam with flowing water in the exhibit. **C,** Above and below water viewing of the beaver pond. **D,** Class studying the exhibit. In addition to these viewing facilities, the exhibit includes closed-circuit television from inside the beaver den for public and scientific viewing and has included blinds for extended scientific observation of the animals. The exhibit also is inhabited by other normal members of the pond community, including fish, turtles, and waterfowl.
Courtesy Tom Cajacob, Minnesota Zoological Garden.

The state of the art in philosophy, design, and operation of zoos today has been elevated to a high level of sophistication (Figure 22-1). The diverse functions of public display and recreation plus education and scientific study, including the opportunity for much behavioral research, have been brought back together in modern zoos and are smoothly integrated at a level not formerly known. Behavior became important both for the design of facilities and management and as a subject for further scientific study. In many cases, modern zoos provide some of the best circumstances available for studying animal behavior.

The training of animals is another activity that partly developed from zoos and circuses (Hediger 1964, 1968). The use of trained animals remains familiar in circuses and similar performances, including animals as actors in semidocumentary or entertaining motion pictures and television, in work animals, and even in the home.

In more utilitarian and academic uses, trained animals have figured importantly in the field of psychology, where thousands of rats and individuals of other species are trained for study every day. Much knowledge about bird orientation and navigation has been obtained from trained pigeons. Training has been useful in experiments designed to infer sensory and other capabilities of animals.

In military usage, there are many instances of trained animals, other than serving simply as beasts of burden or for riding. The use of carrier pigeons for messages is a classic illustration. Porpoises have been used to retrieve torpedoes and bombs from the ocean floor, gulls trained to follow enemy submarines (cited in Skinner 1960), pigeons used as guidance mechanisms for missiles (Skinner 1960), and pigs used to root out land mines. During World War II the United States partially developed, then abandoned, a project to have bats carry miniature incendiary bombs into enemy buildings (Feist 1982).

In livestock and zoo applications, animals can be trained to drink or eat from certain implements, follow a certain daily routine, and come or do other activities on command. There are probably few people who work with animals who do not use training to some extent.

Training is based on a deliberate or unconscious understanding of the various principles of learning, as discussed in Chapter 20. When training animals, one must consider the natural learning tendencies and abilities of the species. As discussed, most primates, for example, cannot be trained to new toilet habits without extensive effort. Many breeds of domestic animals have been selected for enhanced trainability for certain characteristics, such as dogs that retrieve objects, herd sheep, or lead blind people.

Other Personal Uses of Animals, Including as Pets

From the earliest times, people kept some animals for purposes other than killing, eating, and other resources. Even among those animals that were eventually to be killed and eaten, some served different uses first, as hunting assistants or partners (for both livelihood and sport), personal entertainment and relaxation, or simply companionship as pets. These nonconsumptive uses of animals included wild animals taken into captivity and tamed or otherwise trained, such as falcons, cheetahs, and various canids, and a few species that evolved into permanent, domestic relationships with people.

The nonconsumptive personal uses of animals have continued right up to the present time, complete with a host of accompanying attitudes and behaviors toward the animals. Domestic pets embody the most common, familiar, and widely accepted relationship between people and other species on a worldwide basis. As more and more of the human species shifted to city living and new life-styles, they carried their domestic pets, particularly dogs and cats, along with them.

The association between pets and humans may be important for both participants. Some people need every friend they can get, and the adage that "man's best friend is a dog" may not be far from the truth. Lorenz (1952) provides a delightful essay on the behavioral relationship between man and dog. Levinson (1968, also see Holden 1981 and Culliton 1987) discusses the relationship further and indicates that pets may be significant to the mental hygiene of some people, particularly children, the elderly, and people who are isolated or suffer emotional problems. Petting animals may substitute for social grooming, much of which has been lost in contemporary human culture.

The keeping of nondomestic species of animals in captivity, although it continues to exist and even expand, is not as widely practiced or accepted as the keeping of domestic animals. Many people are against the keeping of nondomestic animals for a variety of reasons: the status of natural populations, the accidental introduction of exotic species, unfamiliarity with the animals' behavior and anthropomorphic concerns about the animals' freedom, or, just the opposite, because of familiarity with the animals and awareness of their specialized needs.

From a behavioral standpoint, the keeping of nondomestic animals in captivity depends greatly on the species. Many species are highly specialized and require equally specialized housing, care, and food. They are difficult to keep in captivity; and only trained, experienced persons should attempt it. In addition, there are many species that can be dangerous if not kept or handled properly; they may pose a threat to themselves, their owner, neighbors and innocent bystanders, or the environment if they escape. On the other hand, there are some highly social domestic animals, such as dogs and horses, that may suffer more from unmet behavioral needs and be more difficult to care for than some nondomestic animals such as many species of fish and a few species of turtles and harmless snakes. Some nondomestic animals are quite adaptable and do well in captivity.

Some nondomestic species of animals are now being bred commonly and routinely in captivity. These provide new sources of animals that can take pressure off reduced wild populations. Furthermore, the housing, care, and feeding of such species is usually well understood and proper information can be obtained.

Nonetheless, the keeping of nondomestic animals in captivity, whether for personal or public reasons, remains a somewhat specialized endeavor. People frequently are ignorant of the animals' requirements; and the animals often suffer and die. It often is a difficult, sensitive, and controversial issue that will continue to receive discussion and debate, as in the cases of hunting, trapping, fishing, livestock production, education, research, and other captive uses of animals.

Many people, both amateur and professional, study or photograph animals. These activities often are conducted in association with teaching, writing, or other professions. A few well-known names are Konrad Lorenz, Niko Tinbergen, George Schaller, Jacques Cousteau, Jane Goodall, Diane Fossey, Frederick Kent Truslow, Alan and Joan Root, and several whose photographs appear in this book. The Roots filmed *The African Elephant, Year of the Wildebeest,* and *Mysterious Castles of Clay* (about termites). Today there are hundreds, if not thousands, of professional animal researchers and photographers. A considerable proportion of the people who work with animals professionally, incidentally, started as young children who kept animals as pets (Dewsbury 1985).

In virtually all cases an intimate understanding of the behavior of their subjects is essential to success. Examples of techniques in the study and photography of behavior are given in Chapter 3 and the Appendix.

Animals and Contemporary Western and Urban Society

Each of the previous sections describes the history of the particular relationship between people and other species up to the present time. This section reiterates three important historical trends that are common to all or most of the above: (1) the cultural attitudes and behaviors of people toward other species of animals have changed dramatically and continue to change through time, with perhaps the most rapid changes occurring during the past century; (2) during this same time period, the life-styles of the majority of the human population also have changed in a way that increases the isolation of most people from the natural world; and (3) relationships between people and other species of animals are receiving increased scrutiny, debate, and in many cases regulation and restriction.

The previous discussion brings us to the present time and sets the stage for exploring a number of related topics, including both the biology and politics of animal-human relationships. These are discussed next. We consider animals in captivity separately from those living under natural conditions.

ANIMALS IN CAPTIVITY

Domestication and Taming

What are the roles and general effects of domestication on behavior? Good discussions are provided by Fox (1968), Hale (1969), Ratner and Boice (1975), Clutton-Brock (1981), and Price (1984). The contemporary biological definition of *domestication* is "that process by which a population of animals becomes adapted to (humans) and to the captive environment by some combination of genetic changes occurring over generations and environmentally induced developmental events reoccurring during each generation" (Price 1984:3). This definition emphasizes (1) long-term evolutionary adaptation and (2) environmental influences resulting from a close relationship with humans over many generations. Breeding in captivity is a key ingredient. The organisms have been removed, at least partially, from the forces of natural selection and subjected instead to artificial (human-related) selective pressures. These pressures do not have to be deliberate or conscious; that is, the humans do not have to be aware of their effects.

Some of the environmental differences experienced by domestic animals compared with their wild counterparts result from care provided by their human benefactors. These differences include: improved nutrition; reduced competition for resources such as food, mates, and places to live; reduced predation by natural predators; reduced parasitism and disease or assistance in controlling them when they do occur; increased contact with humans; reduced variability in intraspecific social interactions; and differences in the physical environment, including climate, shelter, and space (Price 1984).

There is a distinction between domestication and taming. *Tameness* refers to the calmness and ease of approach and handling of individual animals or certain strains or species. Tameness is also sometimes referred to as *socialization* (to human beings).

The opposite of being tame is referred to commonly as being *wild*. Thus wild is used rather loosely in both contexts—nondomestic (referring to recent evolutionary background) and not calm in the presence of humans.

For examples of these points, consider the hunting relationships between humans and birds of prey or between humans and cheetahs. People have been forming an association, as in the sport of falconry, with these animals for thousands of years. Until just recently, however, these species did not breed in captivity. Humans simply kept replenishing their needs for individuals from wild populations, and it is unlikely that the association had much significant impact on the selection and evolution of the wild populations. Individual raptors or cheetahs, in the process of being handled and trained, may become quite tame; that is, they are tame but not domestic.

Among many domestic species, however, individuals that are not handled and raised in close association with humans may become "wild," that is, fearful, excited, and not calm in the presence of humans. Some of the least tame animals are from domestic species that grow up without being properly handled by people. On the other hand, many animals of nondomestic species are either naturally tame or can become tame with appropriate rearing and handling. Many individual animals are both domestic and tame. Domestic animals that return to living on their own in the wild are referred to as *feral*.

Domestic animals are not necessarily different from their wild counterparts in all aspects, only in specific traits that have been subject to artificial selection. Domestic canaries *(Serinus canaria)*, for example, differ from wild canaries, depending on whether the breed has been subject to selection for appearance, such as British border canaries, or for song. The songs of border canaries are much more similar to those of wild canaries than are those of the song canaries (Güttinger 1985). Domesticated animals also are not necessarily degenerate, in spite of biologists sometimes referring to white rats and domestic chickens as zoological monstrosities. Domestic animals have been selected under particular, human-associated circumstances, and within this setting they should be viewed as adapted and not as inferior (Boice 1973).

Adaptiveness depends in part on the perspective of the observer, environmental context, and recent evolutionary history. For example, broody behavior in a wild chicken, whereby the hen stops laying eggs and begins incubation, leads to survival and reproduction. In a chicken at an egg farm, on the other hand, broodiness leads to the stew pot.

Many of the common characteristics for which domestic animals, depending on the species, have been selected involve behavior directly or indirectly. Hale (1969) and Price (1984) discuss several preadaptations of species that favor domestication, including social groups that extend beyond the family, hierarchical social structure (as opposed to territorial), males affiliated with female groups, promiscuous mating patterns, imprinting of social bonds, precocial young, natural acceptance of humans and variable environments, and several other characteristics. Based on listings provided by several persons, particularly Fox (1968), Clutton-Brock (1981), and Price (1984), characteristics that are commonly selected for during the process of domestication include the following:

1. Morphological traits such as body size, outward appearance and pelage or plumage characters, and skull and other skeletal features.
2. Tameness or docility and ease of handling. Natural tameness may have been a factor in some species' entering the domesticated state in the first

place. It has been claimed that the difference between the wild and savage Norway rat and the docile domestic rat resulted from the mutation of a single gene (Keeler and King 1942). Regardless of the source of docility, animals that are easy to handle are generally favored whereas unmanageable individuals are usually culled out.

3. Increased tolerance of crowding with other members of their own species.

4. Ability to live under human conditions of housing and nutrition, along with regional and climatic conditions where different groups of people live. Hill and downs sheep, for example, have different grazing and fleece characteristics, which permit them to live in different climates along with the humans in those regions.

5. Characteristics of economic or utilitarian importance. These include high reproductive rates, production of useful items (such as eggs, milk, or wool), and rapid growth and efficient use of food.

6. Accelerated maturation, prolongation of infant stages (neoteny), or reduction of some adult characteristics (paedomorphosis). It has been suggested that this results in animals that are less specialized and more flexible behaviorally, which may facilitate care and enhance attractiveness to humans. Along with a reduction of secondary sexual characteristics (e.g., reduction in horns, thickness of skin, territorial aggressiveness), it may become easier to crowd the animals, handle them, and process their products. Some breeds of dogs are described as "perpetual puppies." Castration is used occasionally to keep male animals immature in some aspects.

7. Acceleration of evolutionary changes by selective inbreeding and hybridization in addition to rapid artificial selection. Comparing the pace of evolutionary changes wrought by humans with those imposed by nature might be debatable, but humans have clearly demonstrated the ability to rapidly select characteristics deemed most desirable.

The effects of domestication on behavior depend on the particular behavior in question. Hence egg laying may be increased, broodiness decreased, flight distances may be decreased, the threshold of canid barking responses may be altered, and activity involved in searching for food may be decreased, but other behaviors, such as the movements involved in copulation or maintenance, may be unchanged. Even within a species, such as chickens or dogs, some strains may be selected for fighting ability for sporting purposes, whereas in other strains fighting and aggressiveness are selected against.

Problems of Abnormal Behavior in Some Captive Animals

Animals in captivity frequently behave in ways not observed under natural conditions. This raises the issue of abnormal behavior.

Abnormal is difficult to define and it is a relative term; it depends on what one means by normal and under what circumstances. A careful observer who is familiar with the behavior of any particular species notices variations in the ways animals behave and occasionally sees something atypical, even among wild animals under natural conditions. Yet, how different does something have to be, and in response to what, to be considered abnormal?

For example, when placed in a particular condition of captivity, an animal might behave in a way that it would never act otherwise. However, all other members of the species might respond similarly in the same circumstance. Thus the response would be abnormal not because the individual is abnormal but because the situation is.

For our purposes we subjectively define *abnormal behavior* as significant behavior that is clearly uncommon or otherwise different compared with that normally observed in the species in its natural environment. We restrict the definition to significant behavior to exclude obvious trivial differences. A rat in a cage, for example, will drink from a water bottle, which it would not encounter normally in the wild; rats usually drink water from pools, streams, or dripping from objects. However, drinking from water bottles rather than natural sources is not significant to our present concerns.

The concept of abnormal behavior frequently includes a notion that, in addition to being uncommon or different, it also is maladaptive (inappropriate, aberrant, undesirable, or disorderly). Our definition does not exclude that but also does not require it. For further discussion of the concept of *abnormal,* see Fox (1968).

Abnormal behavior occurs in wild animals in their natural environment but it is, by definition, not common. That which is seen in the wild usually results from extremes in the natural variation of normal behavior, or from problems resulting from disease or injury that happened to be observed and recorded before the animal was lost from the population. There has been little attempt to synthesize descriptions of abnormal behavior in wild animals. Aside from occasional reviews of specific abnormalities, the information is mostly scattered as anecdotal short notes in the backs of journals.

A perusal of bird journals during one year, for example, revealed a number of interesting behaviors in wild birds never previously reported. For example, a raven *(Corvus corax)* was observed hanging upside down from tree branches by its bill and feet (Elliot 1977). Brown (1977) observed a male brown-headed cowbird *(Molothrus ater)* approaching and attempting to court two male purple martins *(Progne subis)*. And there were two unusual cases of abnormal nest building by eastern phoebes *(Sayornis phoebe)* encountered during a study involving 277 nests (Weks 1977). One nest was built in a continuous mat along the side of a bridge beam and measured 2.2 m (over 6 feet) long, and the second nest was composed of five separate cups: a normal cup in the center with two cups on each side descending in stairstep fashion.

Abnormal behavior is far more common and of more concern here, however, in relation to animals held in captivity. The problem of wild animals in zoos is classic. The problems, as reviewed by Meyer-Holzapfel (1968), Hediger (1964, 1968), and Morris (1964), are caused by a host of factors, including attempts to escape, fear-causing stimuli, lack of appropriate normal stimuli and boredom, lack of exercise, improper nutrition, inadequate housing or facilities, and intraspecific social problems resulting from forced groupings or separations of animals.

The abnormal behaviors that result from these various causes include abnormal escape or hiding reactions; feeding problems, including refusal to eat; stereotyped movements such as pacing and weaving or swaying; abnormal displacement or redirected behavior (Chapter 2); self-mutilation; a host of reproductive and social problems, including prolonged infantile behavior, especially food begging, or a reversion to juvenile behavior; and various forms of aggression, including toward people.

Aggressiveness toward keepers is a problem and can develop suddenly even in formerly calm animals. Many keepers have been killed as a result. Male animals usually become most aggressive during the rut or courtship period, whereas females may become most dangerous when tending offspring. These should be viewed in many instances as normal and adaptive behaviors, except that the aggression is directed toward humans. For reasons not understood, many animals form a strong antagonism toward specific people and not toward others—even when there has been no mistreatment involved. Serious aggression toward humans often develops, apparently as misdirected intraspecific aggression, by animals that were hand-raised and may be imprinted on humans. This is a common problem with hand-raised birds and ungulates. It is unwise to allow unfamiliar and unprepared persons to enter the confines of such animals; also it is generally a mistake to release such hand-raised animals back into the wild.

Imprinting may also create other problems related to an animal's misidentification of the species to which it belongs. Perhaps one of the more amusing examples, except to the people concerned, involved a hand-raised great horned owl *(Bubo virginianus)* that had been released in a rural cemetery and disrupted a subsequent funeral by flying down to the ground and joining the people.

Meyer-Holzapfel (1968) describes several examples of abnormal sexual behavior. A chimpanzee was reported to form a sexual relationship with a cat, and in a similar instance an immature female gorilla showed pseudomale behavior toward a dog. In one case a male brown hyena frequently attempted to copulate with its metal water bowl. Misdirected sexual imprinting by birds frequently results in sexual behavior being displayed toward other species, including humans.

The last category of abnormal behavior to be mentioned in association with zoo animals is what might best be described as apathy. Social animals kept in isolation or after separation from a mate or partner often become indifferent to their surroundings, have a poor appetite, and show various forms of disturbed or mildly abnormal behavior. This is particularly evident in monogamous birds and great apes (chimpanzees, orangutans, and gorillas). In the latter cases, even the facial

Figure 22-2 Unsuitable conditions at a wayside exhibit of animals. Photograph by James W. Grier.

Figure 22-3   Example of proper housing facilities at a modern zoo: a lynx exhibit at the Minnesota Zoological Garden. **A,** Artist's perspective of the naturalistic exhibit. **B,** Lynx in this exhibit.

Courtesy Tom Cajacob, Minnesota Zoological Garden.

expressions have been reported to be interpretable by humans as indicative of depression and mourning.

Some of the problems still remain at older zoo facilities and at smaller, usually private, wayside exhibits of animals (Figure 22-2). Fortunately, most of these problems are disappearing with enlightened practices of zoo management, construction of new facilities that are specifically tailored to the animals' needs (Figure 22-3), and improved understanding of animal behavior.

Abnormal behavior can be important also in captive research animals. In many cases such behavior is unwanted, unplanned, and unexpected and may interfere with the main goals and procedures of the research project. This happens particularly when the research conducted on animals is not behavioral and the researchers may pay little or no attention to the behavioral implications of their care and experimental treatments. On the other hand, abnormal behavior in animal research is frequently the actual subject of a project. Such behavior often is studied as a model of abnormal behavior in humans (who themselves might be considered captive in the artificial environments of modern life). Abnormal behavior that is deliberately caused and studied experimentally for insight into human behavior forms a massive body of literature in the field of abnormal psychology.

Domesticated animals frequently show abnormal behavior. Abnormal behaviors, causes, and solutions for domestic animals are discussed by several persons, including Fox (1968), Kiley-Worthington (1977), Craig (1981), and others, particularly those working with specific domestic groups such as horses (e.g., Waring 1983).

The situation with domestic animals differs significantly from the situations discussed previously not only by the presence of domestication per se but also in the numbers and variety of people involved. Many people understand their animals' needs and behavior quite well, but many others do not. Many of the abnormal behaviors seen in domestic animals result from ignorance, traditional ways of treating animals that cause problems, and a tendency to anthropomorphize excessively. In spite of selection for animals that tolerate captive conditions, the problems often are similar to those of nondomestic animals in captivity: escape-related behaviors, pacing and weaving, and a number of problems related to overcrowding—including cannibalism of young.

One characteristic that is perhaps remarkable in some domestic species is the ability to tolerate extremely limited, if not cramped, individual confinement. White rats, hamsters, mice, domestic rabbits, and poultry are routinely subjected to housing where there is barely room to turn around or lie down. Yet many strains have been selected for such a life and do quite well under these conditions with few signs of boredom, pacing, or other behaviors that might be considered abnormal.

Problems of fear and attempted escape can cause bizarre reactions and death under social conditions in domestic species that herd or flock. For example, cattle, sheep, turkeys, and some strains of white leghorn chickens may easily panic and stampede. The behavior may occur in some flocks or herds and not in others even from the same brood or litter. In chickens the problem is known as *chicken hysteria syndrome*, as described graphically by Ferguson (1968):

Whenever someone entered the pen a brief period of tense quietness would ensue to be quickly interrupted when a hen would emit a loud squawk and then dash to a new location. This would be followed immediately by blind running and milling around with loud squawking, by the entire flock. Hens that had been concealed within

Figure 22-4 Chicken hysteria. Many birds are crowded under the small feeder, and others are in the open nest boxes, some with heads showing above the edge. Most are quite tense, and at a signal from one of the hens they will dash frantically to a new location. From Sanger, V.L., and A.H. Hamdy. 1962. American Veterinary Medical Association 140:455-459. Courtesy Ohio Agricultural Research and Developmental Center.

community nests left these to participate in the panic behavior, at the end of which the majority of birds would once again be crowded silently under feeders or in nests and corners (Figure 22-4).

Many behavioral problems seen in domestic animals apparently are caused by modified environmental cues and related hormonal effects. In addition to effects of conflict and stress on the sympathetic nervous system, these problems may produce several reproductive system anomalies via interactions with epinephrine and adrenocortical steroids from the adrenal gland and ACTH from the pituitary. Males exhibit a number of reproductive deficiencies such as incomplete copulatory movements, excessive sexual activity, and sexual interest in inappropriate objects. Females are susceptible to a number of problems in estrous and maternal behavior.

Domestic animals are also subject to a wide variety of abnormal behaviors resulting from undernutrition and malnourishment. These may lead, in mild cases, to hyperactivity, increased searching behavior, or increased fighting and aggressiveness. In more severe cases, there may be developmental problems and actual interference with neural function. A good introductory review of nutritional problems is provided by Worden (1968).

Domestic pets are exposed to all of the problems previously mentioned plus more. People and pets involve by far the largest number of animal-human relationships and the lowest proportion of persons who really understand animal behavior. Ignorance, misunderstandings, anthropomorphism, and "old wives' tales" are rampant when it comes to pets. Deliberate mistreatment of pets is relatively uncommon, but inadvertent problems are quite numerous. People living in close association with social and highly domesticated species, such as the dog, may create subtle, psychological problems in their pets just as they cause similar problems among themselves.

Pets may suffer developmental and social problems from improper early handling and experience, inconsistency in training and treatment (particularly when given directions by different people), isolation and lack of social contact and attention, insufficient or excessive stimulation and handling, and improper housing and feeding (often of a human nature rather than what the animals need).

All of this results in a vast array of problems, including sudden changes in behavior of the pet, reproductive problems of many kinds, including false pregnancies, abnormal care-seeking (et-epimeletic behavior), autism and stereotypy, excessive fear, shyness and withdrawal, depression, hysteria, climactic fits, vices, many neuroses and psychopathic behaviors, plus psychosomatic conditions. The latter include anorexia nervosa (failure to eat, from a nervous condition), gastrointestinal ulcer, heart problems, vomiting, improper defecation or urination, various muscular spasms, other ingestional and digestional problems, respiratory problems, and even stroke and sudden death (reviewed in Schmidt 1968).

Many of the problems depend on the species or breed (particularly affecting those with much inbreeding) and may involve causes that are not readily apparent to the owner. Cats, for example, may suddenly stop using their litter box with the introduction of new cats or humans into the house or neighborhood. Male cats may begin spraying urine in the house even if neutered.

HUMANS AND NONDOMESTIC (WILD) ANIMALS IN NATURE

Managing Pests and Nuisance Animals

The common human approach to problems caused by pest or nuisance animals, usually animals that eat or damage things we want for ourselves or that are creating ecological problems, is to try to kill them or move them to a different location. Reducing or eliminating a population by killing or moving works in some cases but not with many pest species. Killing animals or trapping and transporting them away from the area may be effective with large birds and mammals, but it is notoriously ineffective for smaller mammals and many species of insects. Trying to eliminate some species is about as practical as trying to dip all the water out of the ocean with a bucket: every time one removes any, more move in to take its place.

An old, institutionalized folly along these lines, now largely abandoned, was the concept of bounties, whereby persons were paid to kill certain animals. Payment was usually made on presentation of evidence of the kill, such as the legs, ears, skin, or whole carcass. Several studies (reviewed in Palm et al 1970), however, showed bounty systems to be costly, wasteful, ineffective, and often fraudulent.

In addition to not being effective in many cases, lethal techniques of control create other problems. Many people object to the methods as being cruel and inhumane. Reasonable people may disagree considerably on the value of a species; some people regard it as desirable, whereas others consider it highly undesirable. Furthermore, there often are negative side effects, particularly with chemical pesticides that may kill or affect other species, including humans.

Thus the notion that humans can control certain species, in the sense of eliminating them, usually is a myth. Instead we may have to view pest control as making the best of an undesirable situation and search for the most acceptable compromise. This is referred to as *integrated pest management*. It involves using many different approaches to the problem and finding the best combination of different techniques at different times and places to minimize the impact of a pest population on our own interests. A knowledge of the behavior of the species generally is extremely useful in this endeavor.

Examples of the application of knowledge about animal behavior to pest management are becoming more numerous every day. A classic case involved the reduction of screwworm populations in the southwestern portions of the United States. The screwworm is the larva of a fly *(Cochliomyia hominivorax)*. Adults lay their eggs on cattle, and the maggots burrow beneath the skin, causing a reduction in health and productivity and damaging the skin for later use as leather. Behavioral research showed that the females of the species mate only once. So large numbers of males were raised in the laboratory, sterilized with radiation, then spread through the affected regions by being dropped in cardboard boxes from airplanes. These and other sterile male techniques, including the search for genetic strains that lead to subsequent sterility, have now been explored for a number of pest species. Another general technique in insect pest management involves the use of pheromones to attract animals so they can be killed or to permit the population to be sampled.

Game, Nongame, and Endangered Species Management

Although we might wish to reduce the populations of some species, there are others that we wish to increase in the wild. Again, many of the best solutions involve applications of knowledge about the behavior of the species.

Aldo Leopold generally is credited as the father of both wildlife management using scientific techniques and holistic environmental ethics. He published a now classic text, *Game Management* (1933), and started an academic department in wildlife ecology at the University of Wisconsin. Another of his books, *Sand County Almanac* (1949), is considered a classic of environmental philosophy.

Since that beginning, wildlife management has been approached through two main channels: ecology and, more recently and to a lesser extent, wildlife behavior. Initially attention was directed toward defining which species were most suitable for sport killing (the game species) and finding techniques to maintain a sustained yield. Because of the difficulties of working with such species on a large scale, the numerous environmental and biological factors involved in the populations, and the difficulty of accurately and precisely measuring populations, the real effects of wildlife management efforts are often obscure. Nonetheless, many of them were logical and seem to have worked. Some species, such as white-tailed deer, may be more numerous now than at any previous time, although some of the population growth may have been merely coincidental with efforts to manage the species. Regardless of its effects, wildlife management has become an important discipline: it has generated a large body of literature pertaining to wild animals, there are now agencies at most levels of government, concern for nongame species has emerged to accompany the interest in game species, and many specialty areas such as for endangered species have developed.

Endangered species management is a relatively recent development within the wildlife management profession. A conference on the management of endangered birds was held at Madison, Wisconsin, in 1977 to bring workers in the field together (Temple 1977). Subsequently there have been a number of meetings to discuss management of endangered birds and other groups. This field has tended to give more consideration to behavioral aspects of the problems and their solutions than do most other wildlife management endeavors. Endangered species are those believed to be in danger of imminent extinction. However, we frequently do not know the real size or vulnerability of populations, and there are also considerations related to economics and politics. Working with endangered species often involves much effort, cost, and logistical problems; the work generally is labor-intensive for the number of animals involved. This type of work with birds has been dubbed *clinical ornithology*.

HUMAN ETHICS, ANIMAL WELFARE, AND ANIMAL RIGHTS

The Situation

The relationships between humans and animals are currently governed by a hodgepodge of diverse viewpoints, personal and moral attitudes, regulations, and laws that vary from place to place and situation to situation, often without consistency and sometimes without apparent rational explanation. These issues are currently receiving intense scrutiny, discussion, and debate with a resulting proliferation of books and other published articles (see section at end of chapter).

The general topic of human ethical treatment of other species involves two separate but somewhat related issues: animal welfare versus animal rights. Animal welfare pertains to animal pain and suffering. Animal rights, on the other hand, involves a belief that animals have rights similar to humans and deserve equal (or greater) protection under the law. Many persons feel that the distinction is significant (Schmidt 1990), whereas others either do not make the distinction or believe that they are closely related and that the welfare issue is only a part of the larger rights issue (Fox 1980, Rollin 1981). It is believed that a majority of the general public (as well as researchers and educators) is sympathetic to and concerned about animal welfare, whereas a much smaller percentage of people is sympathetic to the animal rights issue (Schmidt 1990).

Part of the ethical issue of how we relate to and treat other species of animals involves anthropomorphism (Chapter 1), as well as a person's general understanding and beliefs about biology and evolution (Burghardt and Herzog 1980, 1989, Burghardt 1985). Humans may also carry some evolved predispositions for or against certain types of animals, depending on their threat to humans, size, humanoid features, juvenile appearance and behavior, phylogenetic relatedness, perceived intelligence, habits and communication, rarity, variability, similarity of emotional responses, and abilities of the animals to socially bond with humans (Burghardt and Herzog 1989).

Much of the interest in how we treat other animals has been generated by concern for the possible physical pain and suffering of animals, that is, animal welfare issues. Although the often heated contemporary public discussion initially and primarily focused on the fur industry, sport hunting, and the testing of commercial products (such as cosmetics) on animals, concern has spread to include animals used in food production, education, medical and scientific research, and occasionally zoos. The issues related to fur, hunting, and food and cosmetic testing are beyond the scope of this book. However, concern for animals used in research and education is directly related to the study of animal behavior and must be addressed.

Attitudes among persons toward the use of animals in research vary considerably. Although in a survey of students at one university overall opposition to the research use of animals was low (6.2%), there were significant differences among respondents, depending on their gender and academic major (Gallup and Beckstead 1988). How humans (and laws and regulations) treat species that are used in research also varies inconsistently with how the animals are labeled (Herzog 1988). For example, the treatment of common house mice *(Mus musculus)* in a single building on a university campus may depend on whether they are loose and considered pests, being raised as food for other animals, being used in conjunction with research on unprotected species, or are laboratory mice that are subjects of research themselves—in which case they are protected and their care and treatment are strictly regulated. Views, laws, and regulations also vary markedly, depending on geographical location.

The historical perspective at the start of this chapter provides some insight: (1) because of cultural differences, attitudes and resulting regulations or standards of conduct can be expected to vary—they will be different concerning the care of livestock animals, for example, depending on whether most of the population involved is rural or urban; and (2) because of changing attitudes and patterns of behavior toward animals, time is required for a prevailing consensus and associated regulations to develop, but in many cases there has not yet been sufficient time for that to occur. As a result, the whole issue of the ethical treatment of animals and animal rights is in a state of flux. Some major questions have been posed. There also has been much productive thinking and discussion. In our opinion, there are some reasonable perspectives and solutions.

The Questions

In any particular situation involving humans and other species, such as hunting, managing wild populations, or keeping animals in captivity for whatever purpose, there are several significant questions:

1. What should people be allowed or not allowed to do with the animals?
2. Who should decide what people can or cannot do; that is, who decides what is not acceptable?
3. How should we weigh human rights, benefits, or needs—including health, economics, and vested interests—as opposed to various considerations on behalf of the animals (animal rights)?
4. How should the conflicts of interests and attitudes between individual people and society as a whole be resolved concerning animal issues?
5. Given all the diversity and numerous dimensions involved, how does one determine the interests of society as a whole in the first place?

These are all relative issues with few absolutes and essentially no higher, international, or universal authority. Nature provides no guidelines and commonly seems rather inhumane. Some predators, for example, start eating their prey before killing it; and parasites, intraspecific aggression, famine, weather, and natural disasters can produce some rather gruesome outcomes. Even the Bible does not provide much guidance for the ethical treatment of other species. In Genesis 1:28, God gives humans dominion over the other animals but subsequently states or implies that we are to be responsible and treat the animals with care and respect (e.g., Genesis 9:8-17, for additional Biblical references see Rollin 1981). Furthermore, the Bible represents only the Judeo-Christian tradition and is not accepted by all people. It will probably be a long time, if ever, before all people can agree on how to treat other species of animals; people seem to have enough difficulty agreeing on how to treat each other.

In spite of the problems, there seem to be some logical and reasonable ways to approach these questions, including behavioral considerations. We begin with some undeniable relevant facts.

**Perspectives and
Solutions**

Regardless of where a person stands on various issues of animal rights and their ethical treatment, there are several realities that need to be recognized. First, as indicated previously, there is much diversity of opinion and relatively little consensus; differences in attitudes are not going to disappear quickly, if at all. Second, it is unreasonable to expect the total elimination of human uses of other species of animals; those uses span a continuum of acceptability so that as soon as one use is determined to be unacceptable, there will always be another slightly more acceptable one next in line for consideration. For example, many people are against shooting mammals, but they are not against catching fish. Someone might be against using mice in research but not against trapping them as pests. Third, some management of nondomestic animals in natural environments, including killing or removal from an area, is unavoidable. The alternative of letting nature take its course is often much less desirable than continued human intervention. Fourth, the most agreed-on aspects of animal treatment by persons in contemporary society are to avoid pain and suffering and promote health to the maximum extent possible.

In response to much of the concern being expressed about animal welfare and treatment, three major actions have occurred in a number of countries. (1) Many governmental agencies from federal to local have enacted legislation or regulations that restrict what can be done with various species of animals, including both captive and wild. (2) Most educational and research institutions and some agencies have implemented animal care committees that must approve certain types of work with some kinds of animals, particularly the warm-blooded vertebrates, before the proposed activities can proceed. (3) Most societies, institutions, and other organized groups involved with living animals, including organizations involved with hunting, trapping, fishing, and livestock production, now have well-thought out guidelines on how to work with animals (e.g., Animal Behavior Society 1986).

In addition to the above facts about contemporary human society, there are at least two points that can be made from the biological-behavioral standpoint. First, many species of animals if given proper conditions survive and reproduce well in captivity—so well that surplus numbers of the animals can become a problem. The kinds of animals kept in captivity that reproduce well range from domestic species such as cats that may spend their entire lives inside their owners' homes, to laboratory research animals, zoo animals, and many nondomestic animals in private collections, including aquarium fish. Thus captivity per se cannot be cited as a problem. Persons who object to keeping animals in captivity must base their complaints on more specific grounds.

Second, the nervous systems (including senses, processing capabilities and capacities, the resulting Umwelts, etc.) of different animals are different from each other and often much different from that of humans. There are also some general similarities. Both the differences and similarities need to be taken into account when deciding issues of animal rights.

One similarity involves the perception of pain in animals, at least among vertebrates. Pain is highly adaptive—it alerts the animal to problems—and there is considerable neural and biochemical evidence for its existence in many if not all vertebrates. Pain, however, is another phenomenon that occurs over a continuum from minor problems like itches, light bumps, pricks and light stabs, minor cuts, and slight temperature discomfort to major, long-term, excruciating pain that results from severe injuries, environmental problems (high or low temperatures, intense radiation, etc.), and some serious diseases. It certainly is reasonable to avoid causing significant pain and suffering for other animals whenever possible and unless there is clear and widely acceptable justification for it.

A general similarity in the behavior of all animals is that they have proximate behavioral needs and mechanisms, by virtue of their possessing nervous and endocrine systems. These cause (drive, or motivate, if you wish) all live animals to act in certain ways, such as seeking food, mates under some circumstances, perhaps cover in which to hide, etc. The specific behavioral needs depend on the particular species.

Having considered the general similarities, what are the differences in animal nervous systems? The major one is a large difference between the capabilities of abstraction (Chapter 17) of humans and other species. All species, including moths, frogs, and even sponges and protozoans, abstract information to some extent. None of the other species closely approach our abilities, for example, to process language as in reading and understanding written symbols. Those apparently capable of considerable abstraction, particularly the apes, elephants, and cetaceans (whales and dolphins), may approach our abilities, but it is unlikely that even they are capable of abstract thinking to the degree that we are. As a consequence, we must be careful to avoid uncritical anthropomorphism when considering animal rights and ethical treatment of other species.

There is a fine but important distinction between the points of the two previous paragraphs and an uncritical anthropomorphic point of view. It is this: every animal has behavioral needs but they are not always the same as ours, particularly our derived, abstract needs such as freedom. A domestic cat might fight being in a particular cage even if it is of sufficient size and even for short periods of time, but this is most likely because of fear of unfamiliar surroundings, something about the cage that we do not understand, or because of some unmet behavioral need such as to search for food, rather than a desire for freedom. The same cat might be quite content to rest quietly in the cage if a brown paper bag or box is provided, along with occasional social interaction and opportunities to explore or otherwise be stimulated, such as by having objects and space for playing. It is unlikely that the cat understands an abstract concept of freedom any more than it would understand trigonometry or one of Shakespeare's plays.

A particular person or group of people might not like to see cats in cages for aesthetic reasons, but that is a separate, additional issue. Aesthetics reside in the attitudes of the people (and their abilities to abstract personal likes and dislikes) and not in anything inherent in cats or their rights. Some other person or group of people might think that cats look nice in cages.

In addition to the difference in capacity for abstraction, there are several other differences among animal nervous systems, including senses. As discussed in Chapter 16, many animals are tuned to phenomena of which we are not even aware. Conversely, some factors that may seem important to us can be totally irrelevant to the animal being considered.

So how should people who need or want to work with animals proceed; and how might one think about the problems of animal rights and ethical treatment of animals, whether the issue is as mundane as choosing a new pet or class project, or a major consideration like bringing a rare, large animal into a zoo or determining how to manage a critically endangered species in the wild? Discussions on this subject have been both extensive and intensive. We particularly recommend M. Dawkins (1980), Dodds and Orlans (1982), and Donnelley and Nolan (1990). The latter two publications are the outcomes of two large meetings focused on the ethical uses of animals: a conference with 92 participants and a 2-year Hasting's Center project with 23 participants, respectively. The Hasting's Center project, incidentally, was composed of persons from the "troubled middle"—those who were not from either of the ideological extremes of animal liberationism or human welfarism but who were in the middle and wanted to ethically serve the conflicting interests of both people and other animals.

The following recommendations on how to ethically consider using animals are largely distilled and modified from the above sources, particularly Dodds and Orlans (1982).

1. The common sense and accountability principle—people who do anything with living animals need to conduct their activities responsibly by using discretion and common sense, and by being prepared to justify or account for their activities to other persons.

   Many of the current problems and controversies over the ethical treatment of animals have been created by irresponsible activities, lack of discretion and common sense, and either by assuming or hoping that no one else would find out or simply by not stopping to consider what would happen if the activities come to the attention of other persons, including the general public. If one chooses to work with animals, it helps to consciously think about what one is doing with the animals, the potential consequences, and the possibility of an unintended audience.

   Good tests for this point are to imagine being called on to discuss the activities in a public forum or conducting the activities in the presence of uninformed onlookers (allowing for one to explain in plain language what is taking place while the spectators watch). If the activity involves doing something with animals in their natural environment, imagine having a television crew or newspaper reporter tag along to record and publicize the activity. If the activity passes those tests, then it has a better chance of being acceptable.

   Some research labs, such as at the World Center for Birds of Prey in Boise, Idaho, and many zoos, such as the Dallas Zoo, actually have windows opening into their research areas or back working areas so the general public can watch their activities. Many places conduct tours of their facilities and activities. In addition to increasing accountability, this also provides an excellent opportunity for educating others about both the animals and the research or whatever other activity is taking place.

2. Consider the 3 Rs (replacement, reduction, refinement) for reducing problems with animal uses whenever possible. These principles, following Russell and Burch (cited in Rollin 1981), involve replacing the use of animals with other methods (such as computer simulation or plastic models), reducing the number of animals used, or refining the procedures to minimize pain, suffering, and stress. These approaches are not always feasible, but whenever they are they should be used.

3. The greater the potential harm to an animal resulting from a given activity, the greater must be the justification and accountability for the activity. The American Veterinary Medical Association has classified adverse states in animals into a series of categories: fear, anxiety, discomfort, distress, and pain. The magnitude of these along with other

considerations such as taxonomic classification of the species involved, purpose of an experiment or other use, and qualifications of the personnel can be used to assess whether a proposed activity is justified (Donnelley and Nolan 1990).

4. Established legal requirements and professional or other group standards of conduct need to be followed. Any person who acts entirely on his own these days can create significant problems not only for himself but also for many others with similar interests in the same kind of animals. In addition to conforming to established, published standards, persons working with animals should discuss proposed activities with peers, colleagues, and persons outside the discipline and who may have diverse viewpoints.

5. Animals should be kept in captivity or managed in the wild by techniques and under conditions appropriate to the species based on the best knowledge and expertise available for that species. It is the responsibility of persons working with the animals to seek out such information.

6. We need to continually strive for increased knowledge about the behavior and behavioral needs of other species. In addition to helping satisfy our own scientific curiosity, such information can improve the ways we interact with and treat those animals. In the meantime, it is perhaps better to provide more space and other resources than an animal needs rather than less.

7. All reasonable effort should be extended to avoid or minimize maladaptive abnormal behavior of animals as a result of our interactions with them, whether the animals are in captivity or natural environments.

8. Pain and suffering of other animals, out of humane respect, should be avoided or minimized to whatever extent is possible. Whenever animals are to be killed, it should be done as painlessly and quickly as possible using reasonable and recommended euthanasia or natural methods (such as feeding predatory species).

9. Persons working with animals should seek professional advice, such as from veterinarians, for health, behavior, or husbandry problems whenever they arise.

10. Persons and institutions working with animals should continually strive to improve and upgrade facilities and animal care, which may require the pursuit of increased funding, space, and other resources.

11. People on all sides of particular animal-treatment and ethical issues need to recognize and respect the diversity and backgrounds of other people's viewpoints. Presumably all sides share a mutual interest in the animals themselves; they differ in respect to how the animals should be treated. Polarity and heated debates only serve to reduce communication and rarely achieve rational solutions.

For persons interested in applied behavior, that is, using an understanding of animal behavior in the practical management of animals, whether in captivity or natural environments, we next expand on some of the previous points, particularly number 5. In all cases, there is one cardinal dictum pertaining to the successful application of animal behavior: become as familiar as possible with all aspects of the animal.

THE PRACTICAL MANAGEMENT OF CAPTIVE ANIMALS, INCLUDING BEHAVIORAL CONSIDERATIONS

Regardless of the purpose of captivity or the species involved, there are several concerns that must be addressed. Following is a discussion of these concerns, and together they may be viewed as something of a checklist to be considered before any activity with animals is initiated.

Legal Ramifications and Requirements

Legal requirements for the capture or other means of obtaining particular species, their transport, possession, and care must be considered. There may be permits or licenses and various forms of inspection that are needed. Minimum standards, such as in cage size and construction, frequently are involved (Pakes et al 1985, Tuffery 1987). In addition, there may be local or regional zoning laws that prohibit certain species from being held in certain neighborhoods or locations. Many institutions and agencies now have animal care committees that must be consulted and approval obtained before conducting any proposed work with live animals.

Housing Facilities

Except for a few species of domestic animals (some people object even there), few animals can or should be merely placed in a simple cage, room, or form of confinement. Not only size but shape, construction materials, location, and associated furniture and hiding places, if necessary, must be considered. Bigger is not always better, nor must the surroundings necessarily always be complex. House cats, for example, for reasons not completely understood, seem quite content to live in cages if provided with cardboard boxes to sit in, an appropriate, spacious view before them, a few toys with which to play, and, depending on the degree to which the individual has been socialized previously, someone or something to give them occasional attention. Many burrowing or secretive animals, such as most snakes, do quite well in captivity if provided a place to hide, food, and a proper range of temperature.

Specialized Handling and Moving Techniques and Equipment

Not all animals can be easily handled or picked up. Many are dangerous, too active or excitable, or simply too large and heavy. Such animals may require special tools, ropes, cages, anesthetics, lifting equipment, extendable manipulators, and sometimes protective clothing, guards, or shields. Most often these animals also require expertise, training, and familiarity with the animals' unique and often subtle behavioral characteristics. Persons who work with venomous snakes, for example, must be familiar not only with the general behaviors of snakes (including their unpredictability and skills as escape artists) but also know the particular striking directions and ranges, defensive versus offensive and feeding behaviors, intention movements, and many other behaviors of the specific types of snakes they are working with. Vipers, cobras, and mambas, for example, strike and otherwise behave differently from each other; and different species within any group may act differently than others in the same group. There is often individual behavioral variations. All aspects must be taken into account when handling or moving animals.

Nutrition and Feeding

A few species have generalized diets and can be fed broad carnivorous or herbivorous diets. Many others, however, have specialized types of food or places, times, or ways of eating that require special attention. Eastern hognose snakes (*Heterodon platyrhinos*) for example, in spite of being fairly tame and suitable for captivity, require a specialized diet of toads or, for most individuals, a long and difficult period of training to shift them to another form of food.

Behavioral Welfare (Psychological Concerns)

As discussed, virtually every species has behavioral requirements that need to be met. Often what is a concern for one species is irrelevant to another. Armadillos and badgers, for example, need a substrate in which they can dig; most rodents and many other mammals need things on which to chew; birds of prey need to pluck feathers or fur; and many animals require hiding places.

Prevention and Correction of Disease and Injury

There are general precautions and concerns related to animal health that pertain to almost all species. In addition, most species have their own peculiarities that must be known if they are to be kept successfully. There may be, for example, certain diseases or parasites to which they are particularly susceptible or environmental conditions that predispose them to problems.

Special Problems

Many species have unique needs or weaknesses of which anyone working with them should be aware. Extended periods of inactivity on certain substrates, for example, may lead to special foot or leg problems. Social incompatibilities may be expected under certain restricted circumstances, such as at particular seasons. Some territorial species, for example, show territoriality during the breeding season but then herd or flock together at other times of the year.

References and Specialists

To follow the cardinal dictum of getting to know the animal, there are two important steps: (1) spend time observing the animal itself and (2) consult knowledgeable sources. The best sources of information are the professional, specialized literature and specialists in the field, including serious amateurs. Information from the book racks at pet shops is variable; some pet shop books are incorrect or, at best, worthless, whereas others are quite good. The most useful publications are to be found on the library shelves of large institutions, including major or specialized universities, modern zoos and museums, and a few large city libraries. One of the best sources is the *International Zoo Year Book*, which began publication in 1960 and annually devotes a major section to the special captive management requirements of a particular group of animals, such as reptiles or ungulates. Examples of other good general, introductory references are Crandall (1964) and the Universities Federation for Animal Welfare (UFAW) handbook (1972). Other excellent handbooks are published, for example, by the Chemical Rubber Company in the United States. Young (1973) and Fowler (1978) are excellent sources for information and techniques on capture, restraint, and handling techniques for a large variety of species.

In addition to the published sources, there usually are interested specialists who work with particular species or groups of animals and who may be contacted by letter or telephone, if they have available time. A person interested in starting to work with a species or group of animals for the first time should spend a period of apprenticeship working with an experienced person.

Some of the more common specialty areas, aside from domestic livestock, common pets, and the more general zoo animals, are bees, tropical fish, marine fish, marine invertebrates, various reptiles, pigeons, gallinaceous game birds, waterfowl, furbearers, ungulates, canids, felids, primates, and birds of prey.

The management of livestock and research animals in captivity differs from that of wild animals in two important respects. First, the animal species are generally domestic and hence better adapted for the captive conditions. They have fewer specialized needs and are easier to work with in the first place. Second, there are many more people involved with these species, making it much easier to learn about the special requirements and particular behaviors of the animals.

For livestock animals the number of qualified people and good literature is vast, although there still is comparatively little deliberate attention paid to the behavior of the animals. Many universities have entire departments and curricula devoted to animal science, that is, livestock production, handling, and management. Several professional journals are available in these fields, and most libraries have entire sections of books devoted to horses, cattle, and poultry, for example. In many rural agricultural communities, there are clubs, such as 4-H, where one can learn about the proper care and management of these animals.

Work with livestock often is a family tradition, and in some communities there are probably more genuine experts than nonexperienced people in working with horses, cattle, sheep, swine, and poultry. There are many people who have spent a major part of their lives, for example, with horses and who understand their behavior and all of their subtle movements just as thoroughly or even better than some specialists working with an obscure, rare primate deep in a tropical rain forest. Waring (1983) provides a thorough review and discussion on horse behavior. Good starting references on the behavior and management of livestock animals include Hafez (1975) and Hart (1985).

With the common animals used in zoological and psychological research, such as mice, rats, hamsters, rabbits, cats, dogs, and pigeons, there are far fewer people working with the animals than in agriculture and livestock. But the numbers are still high, and one can find good, qualified persons and much literature at most universities and various other research institutions, including health and hospital laboratories. Excellent books on the subject include the UFAW handbook cited earlier.

What can be done to improve the application of knowledge about animal behavior to pets? The overall picture involves large masses of the general public with many long established misunderstandings about animals and their behavior. The best avenue for behavioral information about pets is through veterinarians. The need for veterinarians to have formal training in behavior, along with education in anatomy, physiology, and surgery, is perhaps summarized best by Taylor (1962): "If the veterinarian is to understand the whole mechanism of domestic animals (and that is one of the things expected of him) he cannot afford to leave out the functions of the cerebrum."

Familiarity with animal behavior helps with veterinarians' own handling and care of animals and also helps them to advise other persons. Small animal veterinarians often know the best methods for raising and training dogs and cats. This includes an awareness of sensitive periods, weaning, housebreaking, and specific training that may be associated with different breeds. Behavior is important not only for signs of behavioral problems per se, such as from improper rearing, training, or social interactions, but the symptoms of many diseases, parasites, and injuries are also behavioral. Through knowledge of animal behavior, it often is possible to save many pets that otherwise would be disposed of. Because many of the problems arise from the behavior of the pet owner, it is as important for the veterinarian to understand human psychology as it is to understand animal behavior.

In large animal practice, much of the same can be said; also, the understanding of behavior may facilitate capture, handling, and transport of the animals. Many large mammals, for example, often refuse to cooperate because of problems with strange odors, of which the handler (being human and relying more on vision) is not aware. Cattle that otherwise refuse to enter a loading chute or truck may readily move in if some of their familiar bedding, another animal with which they are familiar, or another familiar-smelling object is placed ahead of them. In another example, instead of leading them forward, horses can be maneuvered and transported in trailers more easily by backing them into the trailer and keeping them rear-facing while traveling (Cregier 1981).

Many of the problems in large animal practice, particularly involving reproduction (e.g., artificial insemination), arise from improper surroundings, including the social situation, or previous trauma. New stock often must be introduced to a herd or flock for replacement or for genetic considerations. However, disruption of the established social hierarchy can create serious, even fatal, effects. The problem is likely with the introduction of older males. New males usually are introduced best as youngsters or where no other males exist—the others perhaps having been removed. The elimination or reduction of these problems may require an understanding of the particular situation and the ability to analyze it.

In spite of occasional problems and misunderstandings, the present level of pet care in the Western world generally is quite good. In fact the average pet in some countries probably receives better care and nutrition than the average person on this planet.

As has been echoed as a common theme throughout this book, the more behavior and general biology one understands, the better one can understand and manage the behavior of one's pets. It is useful to know not only what should be done for the proper management of pets but also what should not be done and which things create problems in the behavior of the species.

BEHAVIOR AND THE PRACTICAL MANAGEMENT OF NONDOMESTIC ANIMALS IN NATURE

Behavioral considerations are crucial in working with nondomestic animals, whether in the wild or in captivity, and increasing attention is being focused on applying behavioral understanding to such efforts. Jeffrey W. Lang, for example, who has spent most of his research career with crocodilians and is a leading world authority on the group in general and on crocodilian behavior in particular, specifically addresses the implications of crocodilian behavior for the management of the various species (Lang 1988). Rather than being sluggish, stupid and relatively innanimate and uninteresting animals, crocodilians have a rich repertoire of behavior including many reproductive and other social, thermoregulatory, and foraging behaviors that must be considered for proper management of the various species. The same can be said of virtually all nondomestic animals.

Changes in the field of wildlife management during recent decades, in addition to an increasing awareness of the importance of animal behavior, include increased public awareness of and interest in nongame species of animals and an increased realization that much of the effort in wildlife management must be directed toward human activities and people management rather than just managing the animals themselves. One must treat the basic, human causes and not just the symptoms in the wildlife.

Thus the biological side of the problem has slipped slightly into the background as more and more effort is directed to the human-related aspects. Within the biological context, however, the behavior of the animals is seen as becoming more and more important in dealing with the species. There currently is no single reference devoted solely to the principles of applied behavior in wildlife management, but there is a growing body of behavioral information that could be applied to many species, and a general synthesis may be forthcoming. The main key to successful management of nondomestic animals in a natural environment (or what remains of it) is a combined understanding of animal behavior, ecology, and interactions with people for that particular species.

A good example involves the Uganda kob (Buechner 1974) (Figure 22-5). The management is based on several years of intensive study of a large number of animals, many of which were individually marked. The techniques allowed a careful

Figure 22-5 Male Uganda kob *(Adenota kob)* on its territory in a courtship arena (lek) with several displaying males. Photograph by Leonard Lee Rue III.

analysis of both the structured population dynamics and the animals' behavior. The hunting harvest is based on an understanding of the social behavior.

The territorial grounds, or leks, consist of a cluster of 30 to 40 small territories of males. Total size of the lek is only around 200 m in diameter. The lek is where the breeding action takes place, with probably fewer than 50 males breeding most of the 500 to 600 adult females. The system works well, with nearly 100% of the females being bred. Spread between the leks are larger, spaced-out, single-male territories. These single territories are considered important for maintaining the spacing and social stability of the herd. Although the males holding the single territories are able to successfully copulate only rarely, these single territories are preferred by females for their daily activities. There they are less harassed by young males, and they may be somewhat protected from predation. Most of the males are to be found in an all-male, bachelor herd. Turnover on the lek is rapid, and the active males that hold lek territories lose them and move to the bachelor herd. There they rest and feed before returning once again to the lek. In addition to the social, age, and sex structure of the population, the species' needs for food and protection from natural predators have been considered in the management of hunting in this system.

Removing animals from the bachelor herd would cause the most random losses from the gene pool and have the least impact on the genetics of the population. Hence hunting was permitted only for subadult males in the bachelor herd, with numbers to be spread over all the subunits. The herd was monitored from 1963 to 1969, during which time the cropping consisted largely of the subadult males taken out of the bachelor herd, a small number of trophy adults taken out of the bachelor herd, and a small percentage of females. In addition to the legal take, there was a fair amount of poaching, much of which was fawns that could be approached within spearing distance, and there were natural losses to predators, disease, and injury. As well as could be determined, the population remained stationary during the 6-year period. The regulated, managed exploitation by humans did not appear to impose any serious adverse effects.

Two examples using birds have been chosen to illustrate applied behavior in endangered species management. The first involves problems between two species. Puerto Rican parrots (Amazona vittata) have become endangered, apparently as a result of increased nest predation and competition. Intensive study by Snyder and Taapken (1977) showed that the most serious competitor was another hole-nesting bird, the pearly-eyed thrasher (Margarops fuscatus). These thrashers, which have shown rapidly increased populations during the past several decades, enter the parrots' nests and destroy the eggs and small chicks. Parrot nests were guarded directly by observers who frightened or shot the thrashers, but this required much manpower; some thrashers entered unnoticed; and there was an endless supply of thrashers—26 being shot at one site. Another technique involved removing the parrot eggs for artificial incubation and replacing them with plaster eggs—which the thrashers proceeded to fill with dents.

The solution came, however, after 3 years of experiments. It was found that the parrots and thrashers have slightly different preferences for nesting cavities. The thrashers prefer shallower holes, around 90 cm deep, whereas the parrots will accept deeper holes. The solution further used the thrashers' own social behavior. Preferred thrasher nest boxes were erected near parrot nest sites. This provided a suitable location for a pair of thrashers, which then did not bother the parrots. The intraspecific territorial behavior of the thrashers kept other thrashers out of the vicinity and away from the parrots.

In the next example, knowledge of numerous aspects of the behavior of falcons has been used in the management, conservation, and restoration of peregrine falcons to North America. The peregrine falcon *(Falco peregrinus)*, a favorite species for the sport of falconry for centuries and more recently for birding, suffered serious population declines and appeared to be headed toward extinction during the twentieth century (Hickey 1969). It was literally rescued, however, by a determined and dedicated group of falconers and avian conservationists through political and legal management (via endangered species legislation) and an active and aggressive program of captive breeding and reintroduction. In less than 20 years the species had been taken from an almost hopeless position in the wild, with much doubt that they could even be reproduced in captivity, to a point where captive breeding techniques became routine, several thousand birds had been returned to the wild, falcons were once again breeding on their own in the wild, and the species has been reestablished.

Knowledge and research on the falcons' behavior entered the captive breeding and reintroduction work at several points. To begin with, birds in captivity for potential breeding stock had to be maintained in good physical and behavioral condition. This was accomplished relatively easily through centuries of knowledge on the care and handling of the birds for the sport of falconry. The next aspects that had to be considered and in some cases researched included courtship and pairing behavior of the birds, environmental stimuli such as photoperiod, and factors related to the incubation and hatching of eggs.

Figure 22-6 Group of young peregrine falcon chicks being raised together before being placed with foster parents.
Courtesy of the Peregrine Fund, Inc.

Figure 22-7 An artificial hacking tower used for releasing young peregrine falcons to the wild.
Courtesy of the Peregrine Fund, Inc.

Once young falcons had been obtained, they needed to be reared in a manner that would promote their eventual survival and reproductive competence in the wild and prevent them from becoming imprinted or too tame toward humans. That was accomplished by raising them together in groups (Figure 22-6) then placing them with foster parents in captivity.

Reintroduction of the young to the wild was based on an understanding of philopatry, the tendency of many species of birds to return when sexually mature to the area from which they themselves fledged (see Chapter 7). A procedure called *hacking* was used in which the young birds are placed in natural or artificial eyries, in many cases involving city buildings, bridges, or specifically designed towers (Figure 22-7), before their being able to fly but from which they will eventually be able to fly freely. They were provided with food (without seeing the human feeders) until they eventually could fly and feed themselves. In some cases, human intervention also was necessary to simulate the role of falcon parents in protecting the young from natural predators such as the common and numerous great horned owls. Some captive-produced young have also been reintroduced to the wild by fostering to wild parents (Figure 22-8).

All of this was highly successful; reintroduced falcons survived and matured in the wild, then returned to start reproducing on their own; and the species is now doing well again in nature. For further information see Cade and Temple (1977), Barclay and Cade (1983), Cade et al (1988), and references therein. Without the input of behavioral knowledge, none of this would have been possible. Simply putting a male and female falcon together in a cage would not result in producing young, and simply tossing young birds to the wild would not have worked. Instead it required much attention and effort directed toward specific behavioral details. Similar captive breeding and reintroduction programs have now also been accomplished with a number of other species of birds of prey, including bald eagles and the Mauritius kestrel.

Figure 22-8 Wild female peregrine falcon with recently fostered young.
Courtesy Craig Himmelwright and Brian Walton.

SUMMARY

People and other animals have interacted in numerous ways since earliest times. Many of these interactions involved threats by some animals to human health, safety, resources, and vested interests. On the other hand, people have long pursued and killed other animals by hunting, trapping, and fishing. This led in part to development of the field of wildlife management. A few species of animals were domesticated as livestock to provide food or other resources and for a number of other uses, including as pets. Another long and intertwined relationship has involved the keeping of collections of animals in captivity. This practice led to developments such as zoos, circuses, other uses of animals in entertainment, and uses in education and research. Many of these interactions have included animal training in many different ways. Additional uses of animals have involved such activities as animal photography.

Each of the various interactions has generated particular attitudes and behaviors by the people toward the animals. These have been passed on from generation to generation or within particular segments of culture, often involving major portions of the human species. Within the past century, people's attitudes and behaviors toward other animals have changed considerably and increasingly are being modeled after views of how humans relate to each other. There currently exists an extremely wide diversity of opinion about how people and animals ought to interact. This raises a number of biological and sociological issues. Biologically, there are questions of the effects of domestication on animal biology, the problems of abnormal behavior caused by human interactions, and several ecological, conservation matters. Sociologically, there are the controversial and hotly debated matters of ethical treatment of animals, animal welfare, and animal rights.

The issues of animal welfare and animal rights also must be considered in the context of the biology of animal behavior and with respect to social concerns. The perception of pain is widespread among animals, at least among vertebrates, and is adaptive—it alerts the animal to problems. Along other lines, every animal has behavioral needs, but they are not always the same as ours. We need to address and learn more about the behavioral needs of other animals. Several points are recommended for consideration by anyone working with animals (see pages 804 to 805).

Also of concern is the practical management of animals, including behavioral considerations. Management techniques differ for captive versus noncaptive animals, as shown in examples of the management of the Uganda kob, the Puerto Rican parrot, and the peregrine falcon.

FOR FURTHER READING   **References on Animal Welfare, Animal Rights, and Ethics**

Animal Behavior Society, Ethical and Animal Care Committees. 1986. Guidelines for the use of animals in research. Animal Behaviour 34:315-318.
A discussion of several considerations to assist ethical judgments (and acceptability of manuscripts for publication) involving animals used in research.

Burghardt, G.M., and H.A. Herzog, Jr. 1980. Beyond conspecifics: Is Brer Rabbit our brother? BioScience 30:763-768.
A provocative essay on "ethological ethics" for guiding how we treat other animals.

Burghardt, G.M., and H.A. Herzog, Jr. 1989. Animals, evolution, and ethics. In: R.J. Hoage, editor. 1989. Perceptions of animals in American culture. Smithsonian Institution Press, Washington, D.C.
Considers the ethical dilemmas and numerous factors, including our own biology, that influence our uses of other species.

Dawkins, M. 1980. Animal suffering: the science of animal welfare. Chapman and Hall, New York.
A concise discussion about the debate and biological considerations of the animal rights and human ethics issues.

Dodds, W.J., and F.B. Orlans, editors. 1982. Scientific perspectives on animal welfare. Academic Press, New York.
Outcome of a conference to begin coordinating "a national effort by scientists to take the initiative for responsible use of animals in research." This book presents an overview of the subject and discussions and recommendations pertaining to the responsibilities of investigators, institutions, funding agencies, publishers, and public policy.

Donnelley, S., and K. Nolan. 1990. Animals, science, and ethics. Hastings Center Report, Special Supplement, May/June.
Discussion and recommendations based on a 2-year project involving 23 selected participants to study the ethical issues involved in using animals in science and education. This 32-page report can be obtained for $4/copy (less by quantity) from: Publications Department, The Hastings Center, 255 Elm Road Briarcliff Manor, NY 10510.

Fox, M.W. 1980. Returning to Eden: animal rights and human responsibility. Viking Press, New York.
A discussion of the issues by an animal rights advocate.

Gallup, G.G., Jr., and J.W. Beckstead. 1988. Attitudes toward animal research. American Psychologist 43:474-476.
This short paper (commentary) reports the results of a survey among college students of both sexes, different years in college, and varied majors. Overall opposition to animal research is low (only 6.2% felt it should be stopped) but varies significantly depending on gender and major. Also, students were generally not well informed of existing regulations and professional codes pertaining to animal research.

Herzog, H.A., Jr. 1988. The moral status of mice. American Psychologist 43:473-474.
A lucid, stimulating commentary on how our treatment of animals, including legal regulations, depends on how we label the animals (e.g., pest, pet, research animals, or food).

Rollin, B.E. 1981. Animal rights and human morality. Prometheus Books, Buffalo, N.Y.
An extensive discussion of philosophical aspects, moral theory, legal considerations, and examples of animal uses in commercial and academic research, education, and for pets. A stimulating and thought-provoking book that should be read by all persons interested in the subject because Rollin raises and discusses issues that should be considered. It is unlikely, however, that everyone will be receptive to his conclusions. In spite of his arguing against including plants and inanimate objects as objects of moral concern, it seems that his philosophical arguments of having interests and ends in themselves would also apply to plants and inanimate but historically, geologically, or ecologically significant objects and places. In addition, Rollin mounts an unnecessarily harsh attack on the use of animals in psychological research, an attack that may alienate much of the audience with which he wants to communicate.

Schmidt, R.H. 1990. Why do we debate animal rights? Wildlife Society Bulletin 18:459-461.
Discusses the differences and implications of considering animal welfare versus animal rights.

References on Other Topics Discussed in this Chapter

Cade, T.J. 1988. The breeding of peregrines and other falcons in captivity: an historical summary. In: Cade, T.J., J.H. Enderson, C.G. Thelander, and C.M. White, editors. Peregrine falcon populations, their management and recovery. The Peregrine Fund, Boise, Idaho.
Breaks the history down by recent decades and provides much insight into the progression of this most successful story.

Culliton, B.J. 1987. Take two pets and call me in the morning. Science 237:1560-1561.
Presents an overview, including some skepticism and balance, on the presentations at a National Institute of Health conference to evaluate the possible benefits of using pets as psychological therapy.

Fox, M.W., editor. 1968. Abnormal behavior in animals. W.B. Saunders, Philadelphia.
An older, but still perhaps the best (and almost only), book on the subject of abnormal behavior, particularly in captive animals.

Güttinger, H.R. 1985. Consequences of domestication on the song structures in the canary. Behaviour 94:254-278.
An original research paper investigating differences (and similarities), involving both genetic and learning aspects, of songs of wild canaries versus those selected primarily for appearance or for singing ability.

Hafez, E.S.E., editor. 1975. Behavior of domestic animals. Williams & Wilkins, Baltimore.
A standard, basic reference and introduction to the behavior of several species of domesticated agricultural and pet animals.

Hart, B.L. 1985. The behaviour of domestic animals. W.H. Freeman, New York.
A broad review of domestic animal ethology and behavioral physiology, including experimental aspects.

International zoo year book. Zoological Society of London, London.
An annual, periodical publication with much information on the care of particular groups of animals plus much other information. Available in most institutional libraries.

Lang, J.W. 1988. Crocodilian behaviour: implications for management. In: G.J.W. Webb, S.C. Manolis, and P.J. Whitehead. Wildlife management: crocodiles and alligators. Surrey Beatty and Sons, Australia.
An excellent example of applying behavioral knowledge. A description of crocodilian behavior followed by a discussion of implications for species management in both nature and captivity.

Lorenz, K.Z. 1952. King Solomon's ring. Thomas Y. Crowell, New York.
A collection of essays, anecdotes, and comments about Lorenz's own personal relationships and interactions with several other species (most of which were in a more or less captive situation).

Pakes, S.P. (chairman), et al. (16 other persons), Committee on Care and Use of Laboratory Animals of the Institute of Laboratory Animal Resources Commission on Life Sciences of the National Research Council. 1985. Guide for the care and use of laboratory animals. U.S. Department of Health and Human Services. Public Health Service, National Institutes of Health, NIH Publication no. 85-23.
A comprehensive and detailed guide covering institutional policies, housing and other husbandry aspects, veterinary care, physical plant, and a number of special considerations. Includes a large number of other references. Anyone working with animals in a laboratory should have access to this publication.

Palm, C.E., et al. 1970. Vertebrate pests: problems and control. National Academy of Sciences, Washington, D.C.
Includes many behavioral aspects in the discussion of problems with vertebrate pest control.

Sluckin, W. 1979. Fear in animals and man. Van Nostrand Reinhold, New York.
A consideration of fear and its behavioral causes and consequences.

Temple, S.A., editor. 1977. Endangered birds, management techniques for preserving threatened species. University of Wisconsin Press, Madison, Wis.
Discussions, including behavioral aspects, for the management and recovery of many different species. Presents an excellent perspective on both the amount of effort required and species-specific differences.

Tuffery, A.A. 1987. Laboratory animals: an introduction for new experimenters. John Wiley, Somerset, N.J.
An up-to-date resource, incorporating behavioral and ethical considerations, for anyone working with laboratory animals (not just newcomers).

Voith, V.L., and P.L. Borchelt, editors. 1982. Symposium on animal behavior. The veterinary clinics of North America/small animal practice., vol. 12, no. 4. W.B. Saunders, Philadelphia.
A discussion of behavior problems of cats and dogs as commonly encountered by veterinarians. Includes chapters on identifying and treating the problems. Also includes much interesting additional information on normal and social behavior, including communication, in cats and dogs.

Waring, G.H. 1983. Horse behavior. Noyes Publications, Park Ridge, N.J.
A thorough review of horse behavior.

Zuckerman, L., editor. 1980. Great zoos of the world. Westview Press, Boulder, Colo.
A reference for persons in the zoo side of applied animal behavior.

APPENDIX
SUGGESTIONS FOR OBSERVING LIVE ANIMALS

Animals of other species surround humans almost everywhere, even in large cities, and nearly everyone sees them whether intentionally looking or not. Meaningful behavioral observation of animals may not occur, however, without experience or suggestions on how to start. This appendix provides suggestions for watching animals to help engage the observer in seeing things that otherwise might be overlooked. The appendix should be read and used in conjunction with Chapters 3 and 22. Also refer to the list of recommended additional reading for Chapter 3. The videotape methodology teaching program and accompanying literature prepared by Hage and Mellen (1984) are exceptionally useful.

The observation of animals is limited only by imagination and the availability of the animals. Suggestions in this appendix represent only a few possibilities. A number of other field and laboratory workbooks (e.g., Price and Stokes 1975, and Biological Sciences Curriculum Study 1975) provide numerous additional exercises. The published, professional literature also provides a wealth of things that can be done; one can repeat or modify original studies that have been reported in behavioral and related journals, such as those on ornithology, mammalogy, ichthyology, and entomology. Most journals report studies on at least some easily accessible organisms, and some studies use a minimum of equipment. Repeated studies provide insight into the research process, focus attention on the subject of interest, and may even provide some unexpected surprises (such as different results from those reported).

Above all, the imagination should not be overlooked. With a little creativity, all sorts of interesting and meaningful behavioral observations on live animals can be conducted. Finally, observation does not have to be considered work; it is possible and recommended that watching animals be enjoyed. Perhaps one wishes to watch animals only for pure enjoyment and with little or no intent of doing a specific project or obtaining "meaningful" outcomes. Speaking for ourselves as persons who have spent considerable time just watching animals, in addition to doing specific projects at other times, we highly recommend watching animals for no other lofty purpose than simply watching animals. Other suggestions on specific projects aside from repeating published studies reported in the journals, using exercises published elsewhere, or devising one's own are given at the end of this appendix. They are simple and should be viewed only as starters or as bases for course projects.

Before live animals are used in behavior projects, experiments, and other laboratory or field experiences, however, a number of preliminary concerns need to be addressed. It is also important to consider such aspects as planning, methods, and equipment.

PRELIMINARY CONSIDERATIONS BEFORE OBSERVING ANIMALS

Legal, Ethical, Safety, and Humane Considerations

This currently is a complex world of ethics, rules, regulations, and laws pertaining to live animals, both wild and captive. Furthermore, many animals can be outright dangerous, physically or ecologically, and pose safety problems to persons involved in the project, to innocent bystanders in the vicinity, or to the local environment (such as when exotic species escape). For the welfare and protection of all concerned—instructors, students, property owners, neighbors, and the animals themselves—both instructors and students must be alert to and familiar with the possible legal, ethical, and safety ramifications of whatever is being done with the animals. Instructors should carefully supervise projects and clearly identify who is responsible in case of errors, problems, accidents, or violations. All persons working on independent exercises outside formal organized laboratory work or fieldwork are responsible for conducting everything properly, including having permits or licenses when necessary and obtaining permission to enter and work on private property. If in doubt, seek advice. Respect the rights of landowners and also be considerate of the animals themselves.

Two general areas in particular need attention: (1) direct physical manipulations of or interactions with animals (capture, handling, marking, restraint, and adding or removing anything to or from the an animal's body such as adding electrodes or taking blood, anything which may create pain or discomfort or which may otherwise affect the animal's health or welfare), whether the animals are in natural environments or captivity, and (2) working with animals in captivity per se. See expanded discussions in Chapter 3 and particularly the material in Chapter 22—which should be considered as required reading before proceeding.

Capture, handling, marking, etc.

Not everyone can or should capture, handle, mark, or otherwise directly and physically work with animals wherever and whenever they wish. Both expertise and permits may be required before one may engage in such activities. The individual animal's health, welfare, and behavior may be involved; some species populations may be threatened or endangered and unable to tolerate additional stresses; and in some areas, such as public parks, the marking or other manipulations of animals may destroy their aesthetic appeal to other persons. Accordingly and depending on the species and situation, a person interested in capturing, marking, and observing animals may have to gain experience first by working with other persons. Use discretion, and obtain federal, state, and local permits, including permission of local landowners.

Working with captive animals

The use of captive animals, whether in zoological parks, specialized research holdings, or private facilities, for behavioral observation has become routine and has led to much of the knowledge about animal behavior. However, keeping animals in captivity is now heavily regulated for many species and, in addition to the usual problems of observational interference, the conditions of captivity itself may create additional undesirable artifacts. (Depending on the nature of the observations, simplified environments and controls may be intentional, such as for the sake of controlled experiments.)

It is unlikely that few, if any, species other than humans have abstract concepts of freedom that pertain to captivity. Almost all animals, however, require some means of reasonable exercise, minimum housing and health considerations, some environmental diversity, and an absence of fear- and stress-inducing factors. Some species show poorly understood quirks in their captive psychology. House cats, for example, act restless, nervous, and upset in a variety of ways in some caged settings. With a change in the orientation of the cage so that only the view is different or if given certain simple furniture, such as a cardboard box to sit in or a brown paper bag to crawl into, the same cats become quite contented and seem to accept their captivity much more readily.

Highly social species require either others of their own species or surrogate interactions with humans, which may require many hours of a researcher's or an assistant's time. Various species of mammals and birds, with both highly advanced nervous systems and social behavior, seem to require the most attention. In highly unnatural surroundings, such as may be imposed in captivity, animal behavior can be affected adversely. The two most frequent symptoms are the extremes of (1) excessive activity or (2) inactivity and a simplified repertoire of behavior.

The holding and care of animals in captivity generally are controlled by various laws and regulations, which are designed primarily to prevent abuses of the animals and ensure proper cleanliness, health care, feeding, and humane treatment. In addition, professional groups have formulated policies and codes for the housing, care, and humane consideration of animals (e.g., Animal Behavior Society Animal Care Committee 1986). All of this must be rigorously considered and adhered to or, in addition to potential immediate problems for the animals and yourself, you may jeopardize the proper activities of other persons wanting to work with animals, as a consequence of adverse publicity and possibly uninformed, reactive legislation.

Choice of Species: the August Krogh Principle

When choosing the species with which to work and the circumstances of the exercise, it is important to recognize differences between species and situations. Some species and situations are easier to work with or more appropriate for a given problem (the so-called August Krogh principle) (Krebs 1975). Given the limited time during a course, however, some persons may not discover the problems until it is too late to change. Thus one should choose reliable species and circumstances to the best possible extent and avoid unusual, exotic, and unknown situations.

Time and Scheduling Problems

A combination of normal procrastination and lack of appreciation of the time involved in working with live animals, even on simple exercises, can create a real time crunch and unfinished work at the end of a course. Exercises should be organized and structured throughout the course; if not, one should be aware of the likely time problem at the onset and perhaps set intermediate deadlines.

Use of Literature

Depending on the nature of the project, references may need to be consulted. Particularly if done in the context of a course assignment, this should involve original journals, abstracts, and perhaps specialized books. An acceptable literature review should include more than books of pet shop quality and more than one can get simply from the card or computerized book catalog of a library.

Original papers in the scientific journals are useful in many ways—if one knows how to use them. Reading technical scientific papers is different from most other types of reading such as newspapers, nonreference books, and leisure reading. Unlike with those types of reading, one rarely sits down and reads a scientific paper straight through from beginning to end. Instead, there are different sections which provide different kinds of information and one reads only those of interest or jumps back and forth between sections, skipping some sections initially and referring back to them as necessary.

The title is the leading label for the paper and often is the only part read. The abstract (or summary) provides a concise overview of the paper's contents. It is usually the starting point (after the title) and it is often all one needs. Commonly, however, one then goes on to the introduction, skips to the results, then the discussion. Persons wanting an overview of the topic (but not being interested in the specific results of a particular paper) may read only the introduction and discussion. If someone is already familiar with the topic and not interested in another person's interpretations, but only wants to know the outcome of a particular experiment or project, they may read only the results section.

The methods and literature cited sections generally are just for reference while reading other parts of the paper. However, methods sections often are important (sometimes the only part a person wants) for details or new techniques that the reader might want to use. Also, the literature cited section is often useful to help search out other relevant articles. Thus, how a scientific paper is read depends on the kinds of information being searched for. Good journal articles also provide useful models for subsequent analysis of data and writing up the results.

Finding the articles that one might want or need in the first place also requires some skill. There is a battery of techniques: (1) literature or bibliography lists from other sources such as review articles, book chapters, literature cited sections of papers, other persons (including course instructors), colleagues, or published bibliographies; (2) periodic perusals of the tables of contents of current issues of relevant journals; (3) annual (or longer, e.g., ten-year) indexes of journals; (4) printed and computerized abstract and index service, such as *Biological Abstracts*; (5) the *Citation Index* to track subsequent publications that cite important earlier papers on a topic; and (6) serendipity—while browsing or looking for other items, one often stumbles across papers of interest that were not discovered previously. In several of these techniques, it helps not only to identify key topics and terms (e.g., in subject or key word searches) but also to learn the names of key authors; they may publish or be cited frequently and their names can be used to track certain topics.

None of these approaches, even using the computer, will uncover all of the relevant articles (computer searches only find about two thirds of what exists for any particular topic). Thus, the more techniques used, the more complete the list. Computer searches are by far the fastest, and one of the best starting points, but they also cost the most money. The other techniques cost less financially but require much more time, often more than one has available. If one has the financial resources or is short on time, then computers can be used for broad, general searches. However, they often are used more narrowly. Computer searches can be used to find a specific hard-to-locate item or small set of items on a restricted topic, by a certain author, or within a narrow time period.

The computer can be used to uncover a few significant or starting articles on a subject, then one can branch out more widely, using the other techniques. Serendipity is haphazard and not reliable by itself. If you are new to using the library and primary literature such as journals, including abstracts and computer assistance, consult your instructor or librarian.

LEARNING TO OBSERVE AND MEASURE ANIMAL BEHAVIOR

A person who is doing or may be doing research in behavior eventually will be faced with the problem of acquiring new techniques. New methods are learned, as has been stressed earlier, and may require some time and effort. In addition, there are some hints.

1. Be willing to try something new. Accept the uncomfortable feeling of being ignorant and a neophyte in an unfamiliar area; do not let pride get in the way; and simply watch a colleague or teacher and ask for help.
2. Learn to accept, if not relish, problems and chaos rather than avoid or fear them. Without problems there generally is little progress.
3. Be alert to methodology. Constantly ask oneself how someone else accomplished whatever he or she did. When reading the literature, do not just look at and accept the results and conclusions sections—read the methods sections both for specifics and for valuable ideas and insights.
4. Use imagination. The importance of ingenuity has been stressed repeatedly throughout this book. The problem with most people is not that they do not have an imagination but rather that they have not practiced using it.

METHODS OF OBSERVING, MEASURING, ANALYZING, AND REPORTING ANIMAL BEHAVIOR

For general references, see Brown and Downhower (1988), Lehner (1979), and Martin and Bateson (1986).

Planning the Work

Although the process of planning a project is not dealt with here, it is mentioned to note its importance. Behavioral observations are most meaningful and productive when given advance thought and planning, not just allowed to happen. Planning should include why the work is being done—goals and objectives—along with considerations of methods, necessary equipment and supplies, costs, and scheduling.

One aspect of scheduling that is frequently overlooked involves the actual time of observation or measurement: when and how much and whether it involves a sampling process (see Chapter 3). There are many ways to divide time—both the total time frame of the study and the time to be spent observing. This has led to several general categories, as reviewed by Altmann (1974). There are some guiding principles, and some techniques offer certain advantages over others. Much of how one proceeds with specific details depends on what is being measured behaviorally, the goals of the study, and numerous logistical considerations. The important point is to remove the subjective preferences of the observer and the possible bias that unconscious and ad libitum choices may create (see Chapter 3). The goal is to be as objective as possible while at the same time keeping statistics in perspective; statistics and sampling are simply tools and means to an end, not an end in themselves.

Observing without Interfering: Using Blinds

As described in Chapter 3, interference by the observer is an everpresent concern during studies and observations. Although not always simply frightened away or into attempted escape, an animal may nonetheless be distracted by an observer. Signs that an animal is distracted often are subtle and may depend on the species. Common indications include an animal being unusually still, even stiff, lack of normal response to surrounding natural stimuli, gazing directly toward the observer, stereotyped camouflage postures such as the stiff upright posture shown by owls (Figure 3-7), and intention movements (Chapter 7) or actual flight from the area. How disturbed an animal becomes often depends on distance from the observer, concealment, habituation, or a combination of those factors. Blinds provide a common form of concealment.

Blinds sometimes are quite elaborate, or they may be extremely simple and inexpensive. They may be built of local materials—reeds, branches, grasses, rocks, dirt, or snow—or of items as readily accessible as cardboard boxes, scrap construction materials, or blankets. Old large-appliance boxes and shipping crates work well for temporary blinds. For more permanent blinds there are a variety of custom-built structures, scaffolds, or tentlike devices that can be built or purchased (Figure A-1). Buildings or other structures that are already present can also be used. Figure A-2 was obtained from a house window. Figure A-3 was taken from a tree blind built from a single bedspread formed into a tentlike structure 10 m from the eagles' nest (Figure A-4). Vehicles of a type with which the animals are familiar provide a common and mobile form of blind; the use of landrovers for observations in Africa has become almost legendary.

The primary requirement for blinds is simply that they be functional, including attention to the comfort of the observer. Position is important in many respects, including lack of disturbance to animals, good viewing for the observer, access for entering and leaving, and such things as angles and lighting for observation or photography. A good blind also is constructed with provisions for protection of equipment from wind and rain. Blinds generally need to be constructed and in position well in advance of when they are needed to permit time for the animals to become accustomed to them.

Other means of direct observation of animals without their awareness of being watched include a variety of photographic and electronic video cameras, which are frequently camouflaged or otherwise hidden. Motion picture cameras can be housed

Figure A-1 Examples of designs for blinds or hides for the observation of animals.
From Pettingill, O.S., Jr. 1970. Ornithology in laboratory and field, ed. 4, Burgess Publishing, Minneapolis, Minn.

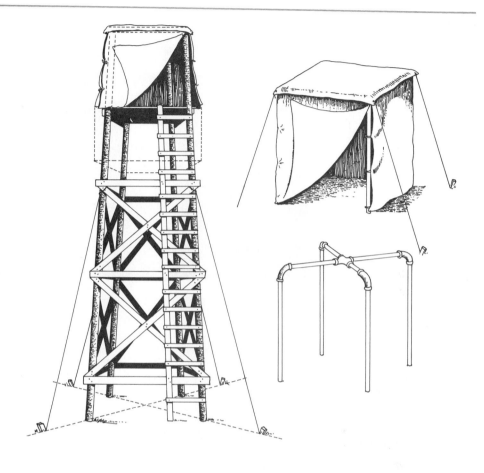

Figure A-2 Wild white-tailed deer fighting. Photograph was taken from window of a house used as a blind and from which the deer were not disturbed.
Photograph by James W. Grier.

Figure A-3   Photograph of
adult bald eagle on nest
feeding nestling. Photograph
was taken from a tree blind
located approximately 10 m
from nest.
Photograph by James W. Grier.

Figure A-4   Diagram of
tree blind used to
photograph eagles and
constructed of readily
available materials
(bedspread and scrap
lumber). A permanent tube
made from a painted empty
can for housing camera
lenses projects from the
blind. Note the single row of
large nails in the tree used
for climbing to the blind.
The photograph in Figure
A-3 was obtained from such
a blind.

in weatherproof boxes and operated at time-lapse speeds for extended observation during the absence of the researcher.

Data Recording and Analysis

Except for the initial stages of informal observation or pilot work, there are at least two bad methods of recording data: relying on memory and using general descriptive notebooks or logbooks. Memory is necessary at the time of an event and for temporary storage of new and unusual observations, but for recording routine data, the weakest ink is better than the strongest memory.

Logbooks also are useful for noting thoughts, occasional interpretations, and extraordinary events. However, general descriptive notebooks are too cumbersome and inefficient for the storage of routine data, in spite of their widespread use for such work. Trying to dig out the facts for later analysis can be a time-consuming, frustrating, and defeating job. If one is to record data manually, well-planned data forms or tables are far superior. Their construction in the first place focuses attention on which data are to be obtained versus other data that may be unnecessary; data are not as likely to be overlooked during the observing process; and specific items are much easier to find later because they have specific locations. Furthermore, it frequently is possible to include boxes or fields on the form for later computer key entry, perhaps by someone else who is not familiar with the project. Good data forms should be provided with a comments or miscellaneous section for noting the kinds of items for which most people use general notebooks.

Many, if not most, behavioral events happen too quickly or too many occur simultaneously for manual writing of data. At the other extreme, events may be so infrequent or occur so slowly that it would be a waste of time to sit by and keep track manually. In the first case, if the problem is not too serious, it may be possible to use a tape recorder and orally describe what is occurring without having to take one's eyes off the subject. Because of the importance of time to behavior, tape recordings, particularly if the stop, pause, fast-forward, or rewind functions are used, generally need to incorporate time information. A mechanical or spoken time signal can be added to the background or on a separate channel (Wiens et al 1970). Alternately, the recording can be played back in full sequence at the same speed it was recorded (usually a time-consuming process). Tape recorders, as with all mechanical and electronic devices, are subject to failure and various technical problems with subsequent loss of data. Manual recording can be expedited through the use of symbols or various types of shorthand invented for the occasion. It is important to provide a key somewhere with or near the data in case someone else has to decipher it or if one may not need or want to go over it until a later time. Things that seem as though they will never be forgotten have an uncanny way of disappearing when they are actually needed; cleverly coded sets of data can turn into a useless mess.

The next levels of sophistication in data recording involve multiple-key paper-chart recorders or direct computer entry. Persons working with the complex multiple interactions among groups of primates have developed this technology to a high level. For examples of sophisticated, computer-compatible recording techniques, see Fernald and Heinecke (1974), Butler and Rowe (1976), and Stephenson and Roberts (1977). The use of photographic, sound, and video tape for storage, replay, and general recording of data was described previously.

For many events, particularly the rapid, rare, slow, or routine spread over a long period, visual observation and associated recording may be unnecessary or inefficient. One may be able to turn the whole job over to automatic systems with the aid of mechanical or electronic counters or other digital and analog devices that measure and keep track of the events.

As in sampling, the methodology is only a means to an end, not an end in itself. The most important aspect of data recording is to decide which data are needed before they are gathered; then merely record in a manner that will make the data easiest to extract and analyze. Because of the financial and time costs involved, typical unexpected problems with gadgets and gimmicks, and the constant risk of technical breakdowns, the simpler the apparatus for a particular job, the better. Again, ingenuity can be a researcher's biggest asset.

Analyzing and Presenting the Results

Techniques for statistics and data analysis are outside the scope of this text. Consult with the instructor (if the project is part of a course assignment), a statistician, a general statistics text (e.g., Zar 1984), one of the behavioral methodology texts (e.g., Martin and Bateson 1986), or inspect the methods and results sections of published papers from similar behavioral research.

How the material is written for final presentation depends on the audience. If for publication or in a publication-style-format, see the instructions to contributors in the intended journal and also recently published articles as examples. For agency or institutional reports, consult the supervisors and any written instructions. In the case of class reports, consult the instructor.

In all cases, writing normally requires much time and work. As Mark Twain once quipped, writing is 5% inspiration and 95% perspiration. Expect first drafts to be rough, awful, and in need of revisions (perhaps several revisions). Rewriting is a fact of life for virtually all persons, including seasoned, experienced writers. Even when one thinks their first draft is superb and the whole world cannot wait to see it, other readers are likely to disagree. If someone else really thinks the early drafts are great and the manuscript does not come back all marked up and dripping with red ink, then that person probably did not read it closely. Published articles usually are far different than how they started out. One should not worry; be patient, work at it, and the manuscript will eventually be presentable and acceptable.

To help remember ideas, we recommend keeping a file of thoughts, rough notes, miscellaneous references that one stumbles across, and creating outlines to help organize. The main point is to write things down whenever they occur to you, on whatever paper available.

Beyond these general comments on writing, we recommend courses and references on writing. Day (1988) and the CBE manual (1983) are particularly useful. (Day is easy and often entertaining reading, in spite of the subject.)

SUGGESTED EXERCISES FOR OBSERVING BEHAVIOR

Four-hour Quantification of Behavior

This project is the most straightforward and least likely to encounter unexpected problems. One can use any species that may be reasonably active (i.e., shows a diversity of behavior) and observe it during an active period (not during periods when it is likely to be sleeping, resting, or inactive during the entire period). Almost all animals will be inactive during some of the observation time, and one cannot always predict when they will be inactive, but one should try to choose a 4-hour block of time when the animal is likely to be most active. The animal should be acting "on its own," that is, not in response to the observer or another human being, although it may be responding to other animals of its own or other species. The animal should be in an environment that is not too simple and restricted; that is, it should be able to show a reasonable diversity of behavior. Mice that are alone in simple cages, for example, cannot do much. Mice that are housed in more complex surroundings with other mice, exercise wheels, and a variety of structures on which to move, on the other hand, have more opportunity for expressing behavior. Single cattle, horses, or other livestock in a small, confined area also cannot be expected to do much, whereas those in fields with others of their species may show more activities and movements. Also situations should be avoided in which the animal is likely to be doing the same thing the entire time, such as trying to escape from confinement, or, as with many species of fish confined to an aquarium, simply moving back and forth.

Animals that might make good subjects include mice and other pets, wild or pet birds, many domestic livestock species (a flock of bantam chickens or pigeons is particularly good for choosing one animal and following it), and some fish or other aquarium species such as crabs or crayfish. Many arthropods (insects, crustaceans, spiders, and others, including some that might be readily available as pests) also make good subjects. Fruit flies are excellent subjects if one can get close enough or have some kind of magnifying ability to see them; one or a few pairs can be set up in a small container with plants, food, etc., and the one to be followed can be identified by using a mutant, for example, one with white eyes.

Once the animals, time, and place have been chosen, spend at least three 4-hour blocks of time with it. The first period of 4 hours or longer is simply to familiarize oneself with the animal's behavior and to devise a scheme for watching and recording the behavior, perhaps by speaking into a tape recorder or devising a system of shorthand notation. It is necessary to recognize and name different movements and behavior—something that may be easy with some behaviors and difficult with others.

One must be conscious of and alert to all movements the animal makes and not just those which are familiar. Watch for subtle movements and activities, such as scratching a part of the body, different postures, etc., that might otherwise be considered insignificant. If the animal makes sounds, a system will be needed for describing the different sounds and calls. If preparing a report, describe the initial observations, problems experienced, and how you decided to identify and record the behaviors, including a key to symbols for each behavior pattern.

Next (perhaps the next day at the same time period) watch the animal and record all behavior for one continuous 4-hour period. The observations should be recorded on a form or sheet that has been devised to maintain a running tally of everything the animal does and when. This can be done with a sheet prepared beforehand that has each minute on a separate line, or a log of everything that happens may be kept; that is, every time the animal changes behavior, write down the time and behavior. Some form of shorthand will almost be necessary because most animals can do a lot of things faster than one can write them down in longhand.

After the continuous 4-hour observation is completed, conduct another session, preferably the next day during the same 4-hour period, in which the behavior is only randomly sampled. (One may, of course, watch the animal between sampling periods, but only the behavior for the predetermined sampling segments of time is recorded.) The way to do this is to divide the 4 hours into equal chunks of time and then randomly choose (with a random numbers table, by drawing slips out of a hat, flipping coins, or any other suitable randomizing procedure) a number of segments amounting to a total observation time of 1 hour (one fourth of the total time available). For example, one could choose 60 1-minute segments, 12 5-minute segments, or some other choice (each segment should be no longer than 5 minutes). Alternatively, one can spot sample by choosing a large number of instants (at least 360) or, for example, 10-second segments, during which one does not follow the behavior for a period of time but simply records what the animal is doing at the instant of sampling. The particular times of sampling, regardless of length of sampling segments, are to be randomly chosen before the observation begins.

After both the continuous and sampling 4-hour observations are finished, quantitatively compare the results of the two ways of observing; that is, from the data, figure out how much actual time the animal spent doing different things and what the proportion or percent of time spent in each activity was. If one has had statistics and wants to make comparisons statistically, this is to be encouraged. The behaviors are to be quantified by measuring the amount of time or number of instances each behavior is shown, but the results of the two methods (continuous and sampling) may be compared by subjectively describing how one thinks they compare. Did both methods give similar percentages, or were they quite different? Use graphs and tables as appropriate.

Species Ethogram

As described in Chapter 3, determine as complete an ethogram for a species as possible, based on one's observations plus those from the literature. This does not involve sampling or quantitative measurements necessarily but, rather, simply compiling a qualitative list and descriptions of the various acts that the species engages in.

Determination of Home Range or Territory

This project has the potential of being quite interesting, but it also may end up being more time consuming, difficult, and vulnerable to unexpected problems and not getting data. It may, however, offer a little more adventure. Persons wanting to do this project must know something about home range and will have to do some reading on the subject in general before getting started (see discussion and references in Chapter 7).

Subjects can be wild animals, domestic animals, or pets—as long as they are reasonably unconfined (or confined in an area larger than their natural home range), and they must be unaffected by observation of them, so they are choosing where they go on their own. Individuals may be followed by tracking, marking, or observing at a distance. It may be possible to do this project with animals in a large aquarium or terrarium if the animals do not occupy the whole area. Sufficient time and number of observation periods must be allowed to gain a reasonable picture of the space (or volume, if in water) that the animal is occupying. This will preferably involve sampling the animal's paths or whereabouts over more than a week. One should plan to spend at least 12 hours (total time—not necessarily all at once) observing or tracking the animal. This does not count time for capturing, marking, or otherwise finding the animal or setting up aquaria initially.

Conduct Markov Chain or Other Multiple-behavior Analysis

See the discussion and references in Chapter 3 and further references for a guide to performing multiple-behavior analysis.

Plot Numbers of Behaviors Against Observation Time

See Figure 3-4 and the related discussion.

Prepare a List of Species Communication Behaviors

This is similar to the species ethogram but is confined to communication behavior. Compile a list and description of all communication signals that the animals use with others of their own species. Use both personal observations and literature sources. Consider that not all communication signals are visual or auditory.

Modification of Behavior

This project requires some knowledge and understanding of learning before it can be done (see Chapters 2, 19, and 20). In addition, one may need to read other sources, such as an introductory psychology or learning text. Familiar categories of learning include classical and operant conditioning, but there are also many others, including imprinting (if one can hatch some chicks or get other suitable young animals) and taste aversions.

To document the acquisition of modified behavior, one needs data before, during, and after (or at the end). For before and after this can include photographs, motion pictures, videotape, or simple written descriptions. For documenting the acquisition, one needs to keep careful records and present the data in tables, graphs of learning curves, or whatever is most appropriate in the particular situation.

This should be more than simply teaching a dog to do cute tricks. Such tricks are completely legitimate instances of behavioral modification and may indeed be used. However, one should carefully document the trial-by-trial acquisition of the modified behavior. Include a carefully recorded description of the progress (including problems and lack of progress) that was involved.

Behavior Experiment

If one chooses this route for a project, inferential statistics and at least a minimum experimental design will be needed. Statistical tests would involve, for example, t test, ANOVA, or an appropriate nonparametric test (Conover 1980, Zar 1984). One also should have some reasonable hypothesis about behavior—either a new one or one someone else has proposed. If one has not had a course in statistics and knows little or nothing about at least simple experimental design and statistics, guidance and instructions will be required. Otherwise one of the other projects might be preferable.

LITERATURE CITED

Abele, L.G., and S. Gilchrist. 1977. Homosexual rape and sexual selection in acanthocephalan worms. Science 197:81-83.

Able, K.P. 1974. Environmental influences on the orientation of free-flying nocturnal bird migrants. Animal Behaviour 22:224-238.

Able, K.P., and V.P. Bingman. 1987. The development of orientation and navigation behavior in birds. Quarterly Review of Biology 62:1-29.

Adams, J.L. 1974. Conceptual blockbusting. W.H. Freeman, San Francisco.

Aird, S.D., and M.E. Aird. 1990. Rain-collecting behavior in a great basin rattlesnake (Crotalus viridis lutosus). Bulletin of the Chicago Herpetological Society 25:217.

Albers, H.E., C.F. Ferris, S.E. Leeman, and B.D. Goldman. 1984. Avian pancreatic polypeptide phase shifts hamster circadian rhythms when microinjected into the suprachiasmatic region. Science 223:833-835.

Alcock, J. 1973. Cues used in searching for food by red-winged blackbirds (Agelaius phoeniceus). Behaviour 46:174-188.

Alcock, J. 1979. Multiple mating in Calopteryx maculata (Odonata: Calopterygidae) and the advantage of non-contact guarding by males. Journal of Natural History 13:439-446.

Alcock, J., E.M. Barrows, G. Gordh, L.J. Hubbard, L.L. Kirkendall, D. Pyle, T.L. Ponder, and F.G. Zalom. 1978. The ecology and evolution of male reproductive behaviour in the bees and wasps. Journal of the Linnean Society of London, Zoology 64:293-326.

Aldis, O. 1975. Play fighting. Academic Press, New York.

Alexander, R.D. 1961. Aggressiveness, territoriality, and sexual behaviour in field crickets. Behaviour 17:130-223.

Alexander, R.D. 1974. The evolution of social behaior. Annual Review of Ecology and Systematics 5:325-383.

Alexander, R.D. 1975. Natural selection and specialized chorusing behavior in acoustical insects. In: D. Pimental, editor. Insects, science, and society. Academic Press, New York.

Alexander, R.D. 1979. Darwinism and human affairs. University of Washington Press, Seattle, Wash.

Alexander, R.D., and D.W. Tinkle. 1981. Natural selection and social behavior: recent research and new theory. Chiron Press, New York.

Alexander, R.D., and P.W. Sherman. 1977. Local mate competition and parental investment in social insects. Science 196:494-500.

Alexander, R.D., J.L. Hoogland, R.D. Howard, K.M. Noonan, and P.W. Sherman. 1979. Sexual dimorphisms and breeding systems in pinnipeds, ungulates, primates and humans. In: N.A. Chagnon and W. Irons, editors. Evolutionary biology and human social behavior: an anthropological perspective. Wadsworth, Belmont, Calif.

Alkon, D.L. 1983. Learning in a marine snail. Scientific American 249:70-84.

Alkon, D.L. 1984. Calcium-mediated reduction of ionic currents: a biophysical memory trace. Science 226:1037-1045.

Alkon, D.L. 1987. Memory traces in the brain. Cambridge University Press, New York.

Allee, W.C., A.E. Emerson, O. Park, T. Park, and K.P. Schmidt. 1949. Principles of animal ecology. W.B. Saunders, Philadelphia.

Allison, T., and H. van Twyver. 1970. The evolution of sleep. Natural History 79:56-65.

Allman, J., F. Miezin, and E. McGuinness. 1985. Stimulus specific responses from beyond the classical receptive field: neurophysiological mechanisms for local-global comparisons in visual neurons. Annual Review of Neuroscience 8:407-430.

Alloway, T.M. 1972. Retention of learning through metamorphosis in the grain beetle, Tenebrio molitor. American Zoologist 12:471-477.

Alloway, T.M. 1973. Learning in insects except Apoidea. In: W.C. Corning, J.A. Dyal, and A.O.D. Willows, editors. Invertebrate learning. Vol. 2. Plenum Press, New York.

Allport, S. 1986. Explorers of the black box. W.W. Norton, New York.

Altmann, J. 1974. Observational study of behavior: sampling methods. Behaviour 49:227-267.

Altmann, J. and K. Sudarshan. 1975. Postnatal development of locomotion in the laboratory rat. Animal Behaviour 23:896-920.

Altmann, S.A. 1965. Sociobiology of rhesus monkeys. II. Stochastics of social communication. Journal of Theoretical Biology 8:490-522.

Altmann, S.A., and J. Altmann. 1977. On the analysis of rates of behaviour. Animal Behaviour 25:364-372.

Alvarez, F., F. Braza, and A. Norzagaray. 1976. The use of the rump patch in the fallow deer (D. dama). Behaviour 56:298-308.

Anand, B.K., and J.R. Brobeck. 1951. Hypothalamic control of food intake in rats and cats. Yale Journal of Biology and Medicine 24:123-140.

Anctil, M. 1984. Commentary on the endocrine system. Science 224:240.

Anderson, D.J. 1982. The home range: a new nonparametric estimation technique. Ecology 63(1):103-112.

Anderson, J.A., and E. Rosenfeld, editors. 1988. Neurocomputing: foundations of research. MIT Press, Cambridge, Mass.

Andersson, M. 1980. Why are there so many threat displays? Journal of Theoretical Biology 86:773-781.

Andersson, M. 1982. Female choice selects for extreme tail length in a widowbird. Nature 299:818-820.

Andersson, M., and C.G. Wiklund. 1978. Clumping versus spacing out: experiments on nest predation in fieldfares (Turdus pilaris). Animal Behaviour 26:1207-1212.

Andersson, M., and J. Krebs. 1978. On the evolution of hoarding behaviour. Animal Behaviour 26(3):707-711.

Andersson, M., F. Gotmark, and C.G. Wiklund. 1981. Food information in the black-headed gull, Larus ridibundus. Behavioral Ecology and Sociobiology 9:199-202.

Andrew, R.J. 1974. Arousal and the causation of behavior. Behaviour 51:135-165.

Animal Behavior Society, Ethical and Animal Care Committees. 1986. Guidelines for the use of animals in research. Animal Behaviour 34:315-318.

Anokhin, P.K. 1964. Systemogenesis as a

general regulator of brain development. In: W.A. Himwich and H.E. Himwich, editors. The developing brain. Elsevier North-Holland, New York.

Antin, J., J. Gibbs, J. Holt. R.C. Young, and G.P. Smith. 1975. Cholecystokinin elicits the complete behavioral sequence of satiety in rats. Journal of Comparative and Physiological Psychology 89:784-790.

Aoki, C., and P. Siekevitz. 1988. Plasticity in brain development. Scientific American 259:56-64.

Appleby, M.C. 1983. The probability of linearity in hierarchies. Animal Behaviour 31:600-608.

Apps, P.J., A. Rasa, and H.W. Vlijoen. 1988. Quantitative chromatographic profiling of odours associated with dominance in male laboratory mice. Aggressive Behavior 14:451-461.

Archibald, G.W. 1976 a. The unison call of cranes as a useful taxonomic tool. Ph.D. thesis. Cornell University, Ithaca, N.Y.

Archibald, G.W. 1976 b. Crane taxonomy as revealed by the unison call. Proceedings of the International Crane Workshop 1:225-251.

Arditi, R., and B. Dacorogna. 1988. Optimal foraging on arbitrary food distributions and the definition of habitat patches. American Naturalist 131:837-846.

Aronson, L.R., E. Toback, D.S. Lehrman, and J.S. Rosenblatt, editors. 1970. Development and evolution of behavior. W.H. Freeman, San Francisco.

Aschoff, J. 1979. Circadian rhythms: influences of internal and external factors on the period measured in constant conditions. Zeitschrift für Tierpsychologie 49:225-249.

Aschoff, S.D., and G.A. Groos, editors. 1980. Vertebrate circadian systems. Structure and physiology. Springer-Verlag, New York.

Atchley, W.R. 1985. Genetics of growth predict patterns of brain-size evolution. Science 229:668-671.

Austad, S.N. 1983. A game theoretical interpretation of male combat in the bowl and doily spider (Frontinella pyramitela). Animal Behaviour 31:59-73.

Austad, S.N. 1984. A classification of alternative reproductive behaviors and methods for field-testing ESS models. American Zoologist 24:309-319.

Axelrod, R., and D. Dion. 1988. The further evolution of cooperation. Science 242:1385-1390.

Axelrod, R., and W.D. Hamilton. 1981. The evolution of cooperation. Science 211:1390-1396.

Baars, B.J. 1986. The cognitive revolution in psychology. Guilford, New York.

Baerends, G.P. 1976. The functional organization of behaviour. Animal Behaviour 24:726-738.

Baerends, G.P. 1985. Do the dummy experiments with sticklebacks support the IRM-concept? Behaviour 93:258-277.

Baerends, G.P., R. Brouwer, and H.T. Waterbolk. 1955. Ethological studies on Lebistes reticulatus (Peters). I. An analysis of the male courtship pattern. Behaviour 8:249-334.

Bakeman, R., and J.M. Gottman. 1986. Observing interaction: an introduction to sequential analysis. Cambridge University Press, New York.

Baker, M.C., K.J. Spitler-Nabors, and D.C. Bradley. 1981. Early experience determines song dialect responsiveness of female sparrows. Science 214:819-821.

Baker, R.R. 1978. The evolutionary ecology of animal migration. Holmes & Meier, New York.

Baker, R.R. 1980 a. The mystery of migration. Viking Press, New York.

Baker, R.R. 1980 b. Goal orientation by blindfolded humans after long-distance displacement: possible involvement of a magnetic sense. Science 210:555-557.

Baker, R.R. 1982. Migration. Paths through time and space. Hodder and Stoughton, London.

Baker, R.R. 1983. Insect territoriality. Annual Review of Entomology 28:65-89.

Baker, R.R. 1984 a. Bird navigation. The solution of a mystery? Hodder and Stoughton, London.

Baker, R.R. 1984 b. The dilemma: when and how to go or stay. In: R.I. Vane-Wright and P.R. Ackery, editors: The biology of butterflies. Symposium of the Royal Entomological Society of London, No. 11. Academic Press, New York.

Baker, R.R. 1987. Human navigation and magnetoreception: the Manchester experiments do replicate. Animal Behaviour 35:691-704.

Baker, R.R., and G.A. Parker. 1979. The evolution of bird coloration. Philosophical Transactions of the Royal Society of London. B287:63-130.

Balaban, E. 1988. Cultural and genetic variation in swamp sparrows (Melospiza georgiana). Behaviour 105:250-291.

Balaban, E., M.A. Teillet, and N. Le Douarin. 1988. Application of the quail-chick chimera system to the study of brain development and behavior. Science 241:1339-1342.

Balda, R.P. 1980. Recovery of cached seeds by a captive Nucifraga caryoca-

tactes. Zeitschrift für Tierpsychologie 52:331-346.

Balda, R.P., A.C. Kamil, and K. Grim. 1986. Revisits to emptied cache sites by Clark's nutcrackers (Nucifraga columbiana). Animal Behaviour 34:1289-1298.

Baldaccini, N.E., S. Benvenuti, V. Fiaschi, and F. Papi. 1975. Pigeon navigation: effects of wind deflection at home cage on homing behaviour. Journal of Comparative Physiology 99:177-186.

Ball, G.F., and J.C. Wingfield. 1987. Changes in plasma levels of luteinizing hormone and sex steroid hormones in relation to multiple-broodedness and nest-site density in male starlings. Physiological Zoology 60:191-199.

Balsbaugh, E.U., Jr. 1988. Mimicry and the chrysomelidae. North Dakota Agricultural Experiment Station Publication No. 1582.

Barchas, J.D., H. Akil, G.R. Elliott, R.B. Holman, and S.J. Watson. 1978. Behavioral neurochemistry: neuroregulators and behavioral states. Science 200:964-973.

Barchas, P.R., and S.D. Mendoza. 1984. Emergent hierarchical relationships in rhesus macaques: an application of Chase's model. In: P.R. Barchas, editor. Social hierarchies: essays toward a sociophysiological perspective. Greenwood Press, Westport, Conn.

Barclay, J.H., and T.J. Cade. 1983. Restoration of the peregrine falcon in the eastern United States. Bird Conservation 1:3-37.

Barker, J.L., and T.G. Smith, Jr., editors. 1980. The role of peptides in neuronal function. Marcel Dekker, New York.

Barlow, G.W. 1968. Ethological units of behavior. In: D. Ingle, editor. The central nervous sytem and fish behavior. University of Chicago Press, Chicago.

Barlow, G.W. 1974. Hexagonal territories. Animal Behaviour 22:876-878.

Barlow, G.W. 1977. Modal action patterns. In: T.A. Sebeok, editor. How animals communicate. Indiana University Press, Bloomington, Ind.

Barlow, G.W., and J. Silverberg, editors. 1980. Sociobiology: beyond nature/nurture? AAAS Symposium 35. Westview Press, Boulder, Colo.

Barnard, C.J., and T. Burk. 1979. Dominance hierarchies and the evolution of "individual recognition." Journal of Theoretical Biology 81:65-73.

Barnes, D.M. 1987. Broad attack launched on the nervous system. Science 238:1651-1653.

Barnes, D.M. 1987. Neural models yield

data on learning. Science 236:1628-1629.

Barnes, D.M. 1988. The biological tangle of drug addiction. Science 241:415-417.

Barnes, D.M. 1988. Meeting on the mind. Science 239:142-144.

Baron, R.A., D. Byrne, and B. Kantowitz. 1978. Psychology: understanding behavior. W.B. Saunders, Philadelphia.

Bartels, M. 1929. Sinnesphysiologische und psychologische Untersuchungen an der trichterspinne *Agelena labyrinthica* (Cl.). Zeitschrift für Vergleichende Physiologie 10:527-593.

Bateman, A.J. 1948. Intrasexual selection in *Drosophila*. Heredity 2:349-368.

Bateson, P. 1979. How do sensitive periods arise and what are they for? Animal Behaviour 27:470-486.

Bateson, P., and B. D'Udine. 1986. Exploration in two inbred strains of mice and their hybrids: additive and interactive models of gene expression. Animal Behaviour 34:1026-1032.

Bateson, P.P.G. 1982. Preferences for cousins in Japanese quail. Nature 295:236-237.

Bateson, P.P.G., and P.H. Klopfer, editors. 1985. Perspectives in ethology. Vol. 6: Mechanisms. Plenum Press, New York.

Bayer, R.D. 1982. How important are bird colonies as information centers? Auk 99:31-40.

Baylis, J.R. 1976 a. A quantitative study of long-term courtship: I. ethological isolation between sympatric populations of the midas cichlid, *Cichlasoma citrinellum*, and the arrow cichlid, *C. zaliosum*. Behaviour 59:59-69.

Baylis, J.R. 1976 b. A quantitative study of long-term courtship: II. a comparative study of the dynamics of courtship in two new world cichlid fishes. Behaviour 59:117-161.

Bayliss, R., N.L. Bishop, and R.C. Fowler. 1985. Pineal gland calcification and defective sense of direction. British Medical Journal 291:1758-1759.

Bays, S.M. 1962. Training possibilities of *Araneus diadematus*. Experientia 18:423.

Beach, F.A. 1950. The snark was a boojum. American Psychologist 5:115-124.

Beach, F.A. 1955. The descent of instinct. Psychological Review 62:401-410.

Beach, F.A. 1976. Sexual attractivity, proceptivity, and receptivity in female mammals. Hormones and Behavior 7:105-308.

Beaugrand, J.P., and R. Zayan. 1985. An experimental model of aggressive dominance in *Xiphophorus helleri* (Pisces,

Poeciliidae). Behavioural Processes 10:1-52.

Bednarz, J.C. 1988. Cooperative hunting in Harris' hawks *(Parabuteo unicinctus)*. Science 239:1525-1527.

Beecher, M.D., and I.M. Beecher. 1979. Sociobiology of bank swallows: reproductive strategy of the male. Science 205:1282-1285.

Beecher, M.D., I.M. Beecher, and S. Hahn. 1981. Parent-offspring recognition in bank swallows *(Riparia riparia)*: II. Development and acoustic basis. Animal Behaviour 29: 95-101.

Beehler, B.M., and M.S. Foster. 1988. Hotshots, hotspots, and female preference in the organization of lek mating systems. American Naturalist 203-219.

Beer, C.G. 1975. Was professor Lehrman an ethologist? Animal Behaviour 23:957-964.

Beitinger, T.L., J.J. Magnuson, W.H. Neill, and W.R. Shaffer. 1975. Behavioural thermoregulation and activity patterns in the green sunfish, *Lepomis cyanellus*. Animal Behaviour 23:222-229.

Bekoff, A. 1976. Ontogeny of leg motor output in the chick embryo: a neural analysis. Brain Research 106:271-291.

Bekoff, A. 1978. A neuroethological approach to the study of the ontogeny of coordinated behavior. In: G.M. Burghardt and M. Bekoff, editors. The development of behavior: comparative and evloutionary aspects. Garland Publishing, New York.

Bekoff, A., M.P. Nusbaum, A.L. Sabichi, and M. Clifford. 1987. Neural control of limb coordination. I. Comparison of hatching and walking motor output patterns in normal and deafferented chicks. Journal of Neuroscience 7:2320-2330.

Bekoff, M. 1976. Animal play: problems and perspectives. In: P.P.G. Bateson and P.H. Klopfer, editors. Perspectives in ethology: Vol. 2. Plenum Press, New York.

Bekoff, M. 1977. Quantitative studies in three areas of classical ethology: social dominance, behavioral taxonomy, and behavioral variability. In: B.A. Hazlett, editor. Quantitative methods in the study of animal behavior. Academic Press, New York.

Bell, G. 1982. The masterpiece of nature: the evolution and genetics of sexuality. University of California Press, Berkeley, Calif.

Bellrose, F.C. 1958. Celestial orientation in wild mallards. Bird Banding 29:75-90.

Belovsky, G.E. 1978. Diet optimization in a generalist herbivore: the moose. Theoretical Population Biology 14:105-134.

Belwood, J.J., and G.K. Morris. 1987. Bat predation and its influence on calling behavior in neotropical katydids. Science 238:64-67.

Belzer, W.R. 1970. The control of protein ingestion in the black blowfly, *Phormia regina* (Meigen). Ph.D. thesis. University of Pennsylvania, Philadelphia.

Benderly, B.L. 1980. The great ape debate. Science 80 1(5):61-65.

Bentley, D.R. 1971. Genetic control of an insect network. Science 174:1139-1141.

Bentley, D.R., and R.R. Hoy. 1970. Development of motor patterns in crickets. Science 170:1409-1411.

Bentley, D.R., and R.R. Hoy. 1972. Genetic control of the neuronal network generating cricket *(Teleogryllus gryllus)* song patterns. Animal Behaviour 20:478-492.

Bentley, D.R., and R.R. Hoy. 1974. The neurobiology of cricket song. Scientific American 231(2):34-44.

Benzer, S. 1973. Genetic dissection of behavior. Scientific American 229(6):24-37.

Bequaert, J. 1922. Ants in their diverse relations to the plant world. Bulletin of the American Museum of Natural History 45:333-585.

Berger, J. 1983. Induced abortion and social factors in wild horses. Nature 303:59-61.

Berger, J. 1986. Wild horses of the great basin: social competition and population size. University of Chicago Press, Chicago.

Berndt, R., and H. Sternberg. 1969. Alters und geschlechtsunterschiede in der dispersion des trauerschnappers *(Ficedula hypoleuca)*. J. Ornithologie 110:22-26.

Bernstein, I.S. 1981. Dominance: the baby and the bathwater. Behavioral and Brain Sciences 4:419-457.

Berthold, P. 1986. The control of migration in European warblers. In: H. Ouellet. Acta XIX Congressus Internationalis Ornithologici. Vol. 1, pages 215-248. University of Ottawa Press, Ottawa, Canada.

Berthold, P., and U. Querner. 1981. Genetic basis of migratory behavior in European warblers. Science 212:77-79.

Bertram, B.C.R. 1976. Kin selection in lions and in evolution. In: P.P.G. Bateson and R.A. Hinde, editors. Growing points in ethology. Cambridge University Press, Cambridge, England.

Bertram, B.C.R. 1983. Cooperation and competition in lions. Nature 302:356.

Biological Sciences Curriculum Study

(BSCS). 1976. Investigating behavior. W.B. Saunders, Philadelphia.

Bishop, J.A., and L.M. Cook. 1975. Moths, melanism and clean air. Scientific American 232:90-99.

Bitterman, M.E. 1975. The comparative analysis of learning. Science 188:699-709.

Bitterman, M.E. 1979. Historical introduction. In: M.E. Bitterman et al, editors. Animal learning, survey and analysis. Plenum Press and NATO, New York.

Bitterman, M.E. 1988. Creative deception. Science 239:1360.

Bitterman, M.E., V.M. LoLordo, J.B. Overmier, and M.E. Rashotte. 1979. Animal learning. Plenum Press, New York.

Black, I.B., J.E. Adler, C.F. Dreyfus, W.F. Friedman, E.F. LaGamma, and A.H. Roach. 1987. Biochemistry of information storage in the nervous system. Science 236:1263-1268.

Black, C.H., and R.H. Wiley. 1977. Spatial variation in behavior in relation to territoriality in dwarf cichlids *Apistogramma ramirezi*. Zeitschrift für Tierpsychologie 45:288-297.

Bligh, J., and K.G. Johnson. 1973. Glossary of terms for thermal physiology. Journal of Applied Physiology 35:941-961.

Block, R.A., and J.V. McConnell. 1967. Classically conditioned discrimination in the planarian, *Dugesia dorotocephala*. Nature 215:1465-1466.

Block, G.D., and S.F. Wallace. 1982. Localization of a circadian pacemaker in the eye of a mollusc, *Bulla*. Science 217:155-157.

Bloedel, J.R. 1987. The cerebellum and memory storage. Science 238:1728-1729.

Blois, C., and A. Cloarec. 1985. Influence of experience on prey selection by *Anax imperator* larvae (Aeschnidae-Odonata). Zeitschrift für Tierpsychologie 68:303-312.

Bloom, F.E. 1981. Neuropeptides. Scientific American 245:148-168.

Bloom, F.E., editor. 1980. Peptides: integrators of cell and tissue function. Raven Press, New York.

Blough, D.S. 1977. Visual search in the pigeon: hunt and peck method. Science 196:1013-1014.

Blum, M.S., and N.A. Blum, editors. 1979. Sexual selection and reproductive competition in insects. Academic Press, New York.

Blum, V. 1986. Vertebrate reproduction. Springer-Verlag, Berlin and New York. (Translated by A.C. Whittle.)

Boake, C.R.B. 1986. A method for testing adaptive hypotheses of mate choice. American Naturalist 127:654-666.

Boakes, R. 1984. From Darwin to behaviourism. Cambridge University Press, Cambridge, England.

Bodian, D. 1972. Neuron junctions: a revolutionary decade. Anatomical Record 174:73-82.

Bodmer, W.F. 1972. Race and IQ: the genetic background. In: K. Richardson et al, editors. Race, culture and intelligence. Penguin Books, New York.

Boice, R. 1973. Domestication. Psychological Bulletin 80:215-230.

Bolles, R.C. 1975. Theory of motivation. Harper & Row, New York.

Bolles, R.C. 1979. Learning theory. Holt, Rinehart & Winston, New York.

Bolles, R.C. 1980. Some functionalistic thoughts about regulation. In: F.M. Toates and T.R. Halliday, editors. Analysis of motivational processes. Academic Press, New York.

Bond, A.B., G.W. Barlow, and W. Rogers. 1985. Two modal action patterns with a continuous temporal distribution. Zeitschrift für Tierpsychologie, 68:326-334.

Bond, R.R. 1957. Ecological distribution of breeding birds in the upland forests of southern Wisconsin. Ecological Monographs 27:351-384.

Bonner, J.T. 1980. The evolution of culture in animals. Princeton University Press, Princeton, N.J.

Boorman, S.A., and P.R. Levitt. 1980. The genetics of altruism. Academic Press, New York.

Borbely, A. 1986. Secrets of sleep. Basic Books, New York.

Borgia, G. 1979. Sexual selection and the evolution of mating systems. In: M.S. and N.A. Blum, editors. Sexual selection and reproductive competition in insects. Academic Press, New York.

Borgia, G. 1986 a. Satin bowerbird parasites: a test of the bright male hypothesis. Behavioral Ecology and Sociobiology 19:355-358.

Borgia, G. 1986 b. Sexual selection in bowerbirds. Scientific American 154(6):92-100.

Borgia, G. 1987. A critical review of sexual selection models. In: J.W. Bradbury and M.B. Andersson, editors. Sexual selection: testing the alternatives. John Wiley & Sons, Chichester, N.J.

Boring, E.G. 1957. A history of experimental psychology, ed 2. Appleton-Century-Crofts, New York.

Bouchard, T.J., Jr. 1983. Do environmental similarities explain the similarity in intelligence of identical twins reared apart? Intelligence 7:175-184.

Bouchard, T.J., Jr. and M. McGue. 1981. Familial studies of intelligence: a review. Science 212:1055-1059.

Bouchard, T.J., Jr., L. Heston, E. Eckert, M. Keys, and S. Resnick. 1981. The Minnesota study of twins reared apart: project description and sample results in the development domain. In: L. Gedda, P. Parisi, and W. Nance, editors: Twins Research 3: Part B. Intelligence, Personality and Development. Alan R. Liss, New York.

Boudreau, J.C., and C. Tsuchitani. 1973. Sensory neurophysiology. Van Nostrand Reinhold, New York.

Bourke, A.F.G. 1988. Worker reproduction in the higher eusocial Hymenoptera. Quarterly Review of Biology 63:291-311.

Boycott, B.B. 1965. Learning in the octopus. Scientific American 212:42-50.

Boyd, R., and J.B. Silk. 1983. A method for assigning cardinal dominance ranks. Animal Behaviour 31:45-58.

Boyd, R., and J.P. Lorberbaum. 1987. No pure strategy is evolutionarily stable in the repeated prisoner's dilemma game. Nature 327:58-59.

Bradbury, J.W. 1981. The evolution of leks. In: R.D. Alexander and D.W. Tinkle, editors. Natural selection and social behavior: recent research and new theory. Chiron Press, New York.

Bradbury, J.W. 1985. Contrasts between insects and vertebrates in the evolution of male display, female choice, and lek mating. In: B. Holldobler and M. Lindauer, editors. Experimental behavioral ecology and sociobiology. Sinauer Associates, Sunderland, Mass.

Bradbury, J.W., and M.B. Andersson, editors. 1987. Sexual selection: testing the alternatives. John Wiley & Sons, Chichester, N.J.

Bradbury, J.W., and R.M. Gibson. 1983. Leks and mate choice. In: P. Bateson, editor. Mate choice. Cambridge University Press, Cambridge, England.

Brandes, Ch., B. Frisch, and R. Menzel. 1988. Time-course of memory formation differs in honey bee lines selected for good and poor learning. Animal Behaviour 36:981-985.

Brandon, R.N. and R.M. Burian, editors. 1984. Genes, organisms, populations: controversies over the units of selection. MIT Press, Cambridge, Mass.

Bray, O., J. Kennelly, and J. Guarino. 1975. Fertility of eggs produced on territories of vasectomized red-winged blackbirds. Wilson Bulletin 87:187-195.

Breed, M.D. 1983. Cockroach mating systems. In: D.T. Gwynne and G.K. Morris, editors. Orthopteran mating systems: sexual competition in a diverse group of insects. Westview Press, Boulder, Colo.

Breed, M.D., C.M. Hinkle, and W.J. Bell. 1975. Agonistic behavior in the German cockroach, *Blattella germanica*. Zeitschrift für Tierpsychologie 39:24-32.

Brenowitz, W.A. 1981. "Territorial song" as a flocking signal in red-winged blackbirds. Animal Behaviour 29(2):641-642

Brines, M.L., and J.L. Gould. 1979. Bees have rules. Science 206:571-573.

Brockmann, H.J., A. Grafen, and R. Dawkins. 1979. Evolutionarily stable nesting strategy in a digger wasp. Journal of Theoretical Biology 77:473-496.

Brockmann, H.J., and C.J. Barnard. 1979. Kleptoparasitism in birds. Animal Behaviour 27:487-514.

Brockmann, H.J., and J.P. Hailman. 1976. Fish cleaning symbiosis: notes on juvenile angelfishes (Pomacanthus, Chaetodontidae) and comparisons with other species. Zeitschrift für Tierpsychologie 42:129-138.

Brodie, E.D., Jr. 1978. Biting and vocalization as antipredator mechanisms in terrestrial salamanders. Copeia 1978: 127-129.

Brower, L. 1985. New perspectives on the migration biology of the monarch butterfly, *Danaus plexippus* L. In: M.A. Rankin, editor. Migration: mechanisms and adaptive significance. Contributions in Marine Science, 27(suppl):748-785.

Brower, L.P., J.V.Z. Brower, and E.P. Cranston. 1965. Courtship behavior of the queen butterfly, *Danaus gilippus*. Zoologica 50:1-39.

Brown, C.R. 1977. Brown-headed cowbird courting a purple martin. Auk 94:395.

Brown, C.R. 1984. Laying eggs in a neighbor's nest: benefit and cost of colonial nesting in swallows. Science 224:518-519.

Brown, C.R. 1986. Cliff swallow colonies as information centers. Science 234:83-85.

Brown, C.R. 1986. Ectoparasitism as a cost of coloniality in cliff swallows (*Hirundo pyrrhonota*). Ecology 67(5):1206-1218.

Brown, C.R. 1988 a. Enhanced foraging efficiency through information centers: a benefit of coloniality in cliff swallows. Ecology 69:602-613.

Brown, C.R. 1988 b. Social foraging in cliff swallows: local enhancement, risk sensitivity, competition and the avoidance of predators. Animal Behaviour 36:780-792.

Brown, C.R., and J.L. Hoogland. 1986. Risk in mobbing for solitary and colonial

swallows. Animal Behaviour 34:1319-1323.

Brown, C.R., and M.B. Brown. 1986. Cliff swallow colonies as information centers. Science 234:83-85.

Brown, C.R., and M.B. Brown. 1986. Ectoparasitism as a cost of coloniality in cliff swallows (*Hirundo pyrrhonota*). Ecology 67:1206-1218.

Brown, C.R., and M.B. Brown. 1987. Group-living in cliff swallows as an advantage in avoiding predators. Behavioral Ecology and Sociobiology 21:97-107.

Brown, C.R., and M.B. Brown. 1988 a. A new form of reproductive parasitism in cliff swallows. Nature 331:66-68.

Brown, C.R., and M.B. Brown. 1988 b. The costs and benefits of egg destruction by conspecifics in colonial cliff swallows. Auk 105:737-748.

Brown, C.R., and M.B. Brown. 1989. Behavioural dynamics of intraspecific brood parasitism in colonial cliff swallows. Animal Behaviour 37:777-796.

Brown, E.D., and C.J. Veltman. 1987. Ethogram of the Australian magpie (*Gymnorhina tibicen*) in comparison to other Cracticidae and Corvus species. Ethology 76:309-333.

Brown, J.A., and P.W. Colgan. 1985. Interspecific differences in the ontogeny of feeding behaviour in two species of centrarchid fish. Zeitschrift für Tierpsychologie, 70:70-80.

Brown, J.L. 1963. Ecogeographic variation and introgression in an avian visual signal: the crest of the Stellar's jay, *Cyanocitta stelleri*. Evolution 17:23-39.

Brown, J.L. 1964. The evolution of diversity in avian territorial systems. Wilson Bulletin 76:160-169.

Brown, J.L. 1964. The integration of agonistic behavior in the Steller's jay *Cyanocitta stelleri* (Gmelin). University of California Publications in Zoology 60:223-328.

Brown, J.L. 1975. The evolution of behavior. W.W. Norton, New York.

Brown, J.L. 1987. Helping and communal breeding in birds: ecology and evolution. Princeton University Press, Princeton, N.J.

Brown, K.T. 1974. Physiology of the retina. In: V.B. Mountcastle, editor. Medical physiology, ed 13. The C.V. Mosby Co., St. Louis.

Brown, L., and J.F. Downhower. 1988. Analyses in behavioral ecology, a manual for lab and field. Sinauer Associates, Sunderland, Mass.

Brown, T.H., P.F. Chapman, E.W. Kairiss, and C.L. Keenan. 1988. Long-term

synaptic potentiation. Science 242:724-728.

Bruce, H.M. 1960. A block to pregnancy in the house mouse caused by the proximity of strange males. Journal of Reproduction and Fertility 1:96-103.

Bruggers, R.L., and C.C.H. Elliott. 1989. *Quelea quelea*: Africa's bird pest. Oxford University Press, New York.

Buckle, G.R., and L. Greenberg. 1981. Nestmate recognition in sweat bees (*Lasioglossum zephyrum*): does an individual recognize its own odour or only of its nestmates? Animal Behaviour 29:802-809.

Buechner, H.K. 1974. Implications of social behavior in the management of Uganda kob. In: V. Geist and F. Walther, editors. The behaviour of ungulates and its relation to management. International Union for Conservation of Nature and Natural Resources, Morges, Switzerland.

Bullock, T.H. 1982. Afterthoughts on animal minds. In: D.R. Griffin, editor: Animal mind-human mind. Springer-Verlag, New York.

Bullock, T.H. 1984. Comparative neuroscience holds promise for quiet revolutions. Science 225:473478.

Bullock, T.H., and W. Heiligenberg, editors. 1986. Electroreception. John Wiley & Sons, New York.

Bullock, T.H., R. Orkland, and A. Grinnell. 1977. Introduction to nervous systems. W.H. Freeman, San Francisco.

Bunch, K.G., and D.F. Tomback. 1986. Bolus recovery by gray jays: an experimental analysis. Animal Behaviour 34:754-762.

Burghardt, G.M. (in press) Human-bear bonding in research on black bear behavior. In: H. Davis and D. Balfour, editors: The inevitable bond. Cambridge University Press, Cambridge, England.

Burghardt, G.M. 1970 a. Intraspecific geographical variation in chemical food preferences of newborn garter snakes (*Thamnophis sirtalis*). Behaviour 36: 246-257.

Burghardt, G.M. 1970 b. Defining "communication." In: J.W. Johnston, Jr., D.G. Moulton, and A. Turk, editors. Communication by chemical signals. Appleton-Century-Crofts, New York.

Burghardt, G.M. 1977 a. Ontogeny of communication. In: T.A. Sebeok, editor. How animals communicate. Indiana University Press, Bloomington, Ind.

Burghardt, G.M. 1977 b. Learning processes in reptiles. In: C. Gans and D. Tinkle, editors. Biology of the reptilia. Vol. 7. Ecology and behavior. Academic Press, London.

Burghardt, G.M. 1982. Comparison matters: curiosity, bears, surplus energy and why reptiles don't play. Behavioral and Brain Sciences 5:159-160.

Burghardt, G.M. 1984. On the origins of play. In: P.K. Smith, Play in animals and humans. Basil Blackwell, New York.

Burghardt, G.M. 1985 a. Animal awareness. American Psychologist 40:905-9l9.

Burghardt, G.M., editor. 1985 b. Foundations of comparative ethology. van Nostrand Rheinhold, New York.

Burghardt, G.M. 1986. Book review of Dewsbury (1984 b). Ethology 73:78-88.

Burghardt, G.M. 1988 a. Developmental creationism. Behavioral and Brain Sciences 11:632.

Burghardt, G.M. 1988 b. Precosity, play, and the ectotherm-endotherm transition. In: E.M. Blass, editor. Handbook of Behavioral Neurobiology. Vol. 9. Plenum Press, New York.

Burghardt, G.M. 1988 c. Anecdotes and critical anthropomorphism. Behavioral and Brain Sciences 11:249.

Burghardt, G.M. 1991. Cognitive ethology and critical anthropomorphism: a snake with two heads and hognose snakes that play dead. In: C.A. Ristau and P. Marler, editors. Cognitive ethology: the minds of other animals. Lawrence Erlbaum, Hillsdale, N.J.

Burghardt, G.M., and H.A. Herzog, Jr. 1980. Beyond conspecifics: is Brer Rabbit our brother? Bioscience 30:763-768.

Burghardt, G.M., and H.A. Herzog, Jr. 1989. Animals, evolution, and ethics. In: R.J. Hoage, editor. 1989. Perceptions of animals in American culture. Smithsonian Institution Press, Washington, D.C.

Burghardt, G.M., and M. Bekoff, editors. 1978. The development of behavior: comparative and evolutionary aspects. Garland Publishing, New York.

Burk, T. 1979. An analysis of social behaviour in crickets. Ph.D. thesis, University of Oxford, Oxford, England.

Burk, T. 1982. Evolutionary significance of predation on sexually signalling males. Florida Entomologist 65:90-104.

Burk, T. 1983 a. Behavioral ecology of mating in the Caribbean fruit fly, Anastrepha suspensa (Loew) (Diptera: Tephritidae). Florida Entomologist 66:330-344.

Burk, T. 1983 b. Male agression and female choice in a field cricket (Teleogryllus oceanicus): the importance of courtship song. In: D.T. Gwynne and G.K. Morris, editors. Orthopteran mating systems: sexual competition in a diverse group of insects. Westview Press, Boulder, Colo.

Burk, T. 1984. Male-male interactions in Caribbean fruit flies, Anastrepha suspensa (Loew) (Diptera: Tephritidae): territorial fights and signalling stimulation. Florida Entomologist 67:542-547.

Burk, T. 1986. Sexual selection, feminism, and the behavior of biologists: changes in the study of animal behavior, 1953-1985. Creighton University Faculty Journal 5:1-16.

Burk, T. 1988 a. Acoustic signals, arms races, and the evolution of honest signalling. Florida Entomologist 71:400-409.

Burk, T. 1988 b. Insect behavioral ecology: some future paths. Annual Review of Entomology 33:319-335.

Burk, T., and J.C. Webb. 1983. Effect of male size on calling propensity, song parameters, and mating success in Caribbean fruit flies, Anastrepha suspensa (Loew) (Diptera: Tephritidae). Annals of the Entomological Society of America 76:678-682.

Burkhardt, R.W., Jr. 1981. On the emergence of ethology as a scientific discipline. Conspectus of History 1(7):62-81.

Burley, N., G. Krantzberg, and P. Radman. 1982. Influence of colour-banding on the conspecific preferences of zebra finches. Animal Behaviour 30:444-455.

Burtt, E.H., Jr. 1977. Some factors in the timing of parent-chick recognition in swallows. Animal Behaviour 25:231-239.

Busnel, M.C. 1967. Rivalite acoustique et hierarchie chez l' Ephippiger (Insecte, Orthoptere, Tettigoniidea). Zeitschrift für Vergleichende Physiologie 54:232-245.

Butler, S.R., and E.A. Rowe. 1976. A data acquisition and retrieval system for studies of animal social behaviour. Behaviour 57:281-287.

Byers, J.A. 1977. Terrain preferences in the play behavior of Siberian ibex kids (Capra ibex sibirica). Zeitschrift für Tierpsychologie 45:199-209.

Bygott, J.D., B.C.R. Bertram, and J.P. Hanby. 1979. Male lions in large coalitions gain reproductive advantages. Nature 282:839-841.

Cade, T.J. 1988. The breeding of peregrines and other falcons in captivity: an historical summary. In: T.J. Cade, J.H. Enderson, C.G. Thelander, and C.M. White, editors. Peregrine falcon populations, their management and recovery. The Peregrine Fund, Boise, Idaho.

Cade, T.J., and G.L. Maclean. 1967. Transport of water by adult sandgrouse to their young. Condor 69:323-343.

Cade, T.J., and S.A. Temple. 1977. The Cornell University falcon programme. In: R.D. Chancellor, editor. Proceedings of the ICBP World Conference on Birds of Prey. International Council for Bird Preservation, London.

Cade, T.J., J.H. Enderson, C.G. Thelander, and C.M. White. 1988. Peregrine falcon populations: their management and recovery. The Peregrine Fund, Boise, Idaho.

Cade, W. 1975. Acoustically orienting parasitoids: fly phonotaxis to cricket song. Science 190:1312-1313.

Cade, W. 1980. Alternative male reproductive behaviors. Florida Entomologist 63:30-45.

Cade, W.H. 1979. The evolution of alternative male reproductive strategies in field crickets. In: M.S. and N.A. Blum, editors. Sexual selection and reproductive competition in insects. Academic Press, New York.

Cade, W.H. 1984. Genetic variation underlying sexual behavior and reproduction. American Zoologist 24:355-366.

Caldwell, P.C., and R.D. Keynes. 1957. The utilization of phosphate bond energy for sodium extrusion from giant axons. Journal of Physiology 137:12P-13P.

Caldwell, R.L. 1979. Cavity occupation and defensive behaviour in the stomatopod Gonodactylus festai: evidence for chemically mediated indivual recognition. Animal Behaviour 27:194-201.

Calvert, W.H., L.E. Hedrick, and L.P. Brower. 1979. Mortality of the monarch butterfly (Danaus plexippus L.): avian predation at five overwintering sites in Mexico. Science 204:847-851.

Camhi, J.M. 1977. Behavioral switching in cockroaches: transformations of tactile reflexes during righting behavior. Journal of Comparative and Physiological Psychology 113:283-301.

Camhi, J.M. 1984. Neuroethology. Nerve cells and the natural behavior of animals. Sinauer Associates, Sunderland, Mass.

Campbell, B., editor. 1972. Sexual selection and the descent of man. Aldine Publishing, Hawthorne, N.Y.

Campbell, N.A. 1989. Biology. ed 2. Benjamin/Cummings, Menlo Park, Calif.

Candland, D.K. 1971. The ontogeny of emotional behavior. In: H. Moltz, editor. The ontogeny of vertebrate behavior. Academic Press, New York.

Candland, D.K. 1979. A profound Quaker meeting. Contemporary Psychology 24:965-967.

Capen, D.E., editor. 1981. The use of multivariate statistics in studies of wildlife habitat. U.S. Department of Agriculture Forest Service, General Technical Report RM-87, Fort Collins, Colo.

Caraco, T., S. Martindale, and H.R. Pulliam. 1980 b. Flocking: advantages and disadvantages. Nature 285:400-401.

Caraco, T., S. Martindale, and T.S. Whitham. 1980 a. An empirical demonstration of risk-sensitive foraging preferences. Animal Behaviour 28:820-830.

Carew, T.J., E.A. Marcus, T.G. Nolen, C.H. Rankin, and M. Stopfer. 1990. The development of learning and memory in *Aplysia*. In: J.L. McGaugh, N.M. Weinberger, and G. Lynch, editors. Brain organization and memory: cells, systems, and circuits. Oxford University Press, New York.

Carew, T.J., R.D. Hawkins, and E.R. Kandel. 1983. Differential classical conditioning of a defensive withdrawal reflex in *Aplysia californica*. Science 219:397-400.

Carey, F.G., and K.D. Lawson. 1973. Temperature regulation in free-swimming bluefin tuna. Comparative Biochemistry and Physiology A44:375-392.

Carlin, N.F., R. Halpern, Bert Holldobler, and P. Schwartz. 1987. Early learning and the recognition of conspecific cocoons by carpenter ants (*Camponotus* sp.). Ethology 75:306-316.

Carlson, A.D., and J. Copeland. 1978. Behavioral plasticity in the flash communication systems of fireflies. American Scientist 66:340-346.

Carlson, J.R. 1977. The imaginal ecdysis of the cricket *Teleogryllus ocianicus*. 1. Organization of motor programs and roles of central and sensory control. Journal of Comparative Physiology 115:299-317.

Carlson, J.R., and D. Bentley. 1977. Ecdysis: neural orchestration of a complex behavioral performance. Science 195:1006-1008.

Carlson, N.R. 1980. Physiology of behavior, ed 2. Allyn & Bacon, Boston.

Carmichael, L. 1963. The onset and early development of behavior. In: L. Carmichael, editor. Manual of child psychology. John Wiley & Sons, New York.

Caro, T.M. 1979. Relations between kitten behaviour and adult predation. Zeitschrift für Tierpsychologie 51:158-168.

Caro, T.M. 1980. Predatory behaviour in domestic cat mothers. Behaviour 74:128-148.

Caro, T.M. 1981. Predatory behaviour and social play in kittens. Behaviour 76:1-24.

Caro, T.M. 1988. Adaptive significance of play: are we getting closer? Trends in Ecology and Evolution 3(2):50-54.

Carr, A. 1965. The navigation of the green turtle. Scientific American 212(5):79-86.

Carthy, J.D. 1956. Animal navigation. Charles Scribner's Sons, New York.

Carthy, J.D. 1968. The pectines of scorpions. Zoological Society of London, Symposia 23:251-261.

Carthy, J.D., and G.E. Newell, editors. 1968. Invertebrate receptors. Zoological Society of London, Symposia No. 23.

Caryl, P.G. 1979. Communication by agonistic displays: what can games theory contribute to ethology? Behaviour 68:136-169.

Cassidy, J. 1979. Half a century on the concepts of innateness and instinct: survey, synthesis, and philosophical implications. Zeitschrift für Tierpsychologie 50:364-386.

Casy, A.F. and R.T. Parfitt. 1987. Opioid analgesics: chemistry and receptors. Plenum Press, New York.

Catchpole, C.K. 1980. Sexual selection and the evolution of complex songs among European warblers of the genus *Acrocephalus*. Behaviour 74:149-166.

Cavalli-Sforza, L.L., and M.W. Feldman. 1981. A quantitative approach. Princeton University Press, Princeton, N.J.

Chamberlain, T.J., P. Halick, and R.W. Gerard. 1963. Fixation of experience in the rat spinal cord. Journal of Neurophysiology 26:662-673.

Chamove, A.S., L.A. Rosenblum, and H.F. Harlow. 1973. Monkeys *(Macaca mulatta)* raised only with peers: a pilot study. Animal Behaviour 21:316-325.

Chance, M.R.A. 1959. What makes monkeys sociable? New Scientist 5:520-522.

Chappell, M.A. 1978. Behavioral factors in the altitudinal zonation of chipmunks *(Eutamias)* Ecology 59:565-579.

Charnov, E.L. 1976. Optimal foraging, the marginal value theorem. Theoretical Population Biology 9:129-136.

Charnov, E.L. 1982. The theory of sex allocation. Princeton University Press, Princeton, N.J.

Chase, I.D. 1974. Models of hierarchy formation in animal societies. Behavioral Science 19: 374-382.

Chase, I.D. 1982 a. Behavioral sequences during dominance hierarchy formation in chickens. Science 216:439-440.

Chase, I.D. 1982 b. Dynamics of hierarchy formation: the sequential development of dominance relationships. Behaviour 80:218-240.

Chase, I.D. 1985. The sequential analysis of aggressive acts during hierarchy formation: an application of the "jigsaw puzzle" approach. Animal Behaviour 33:86-100.

Chase, I.D., and S. Rohwer. 1987. Two methods for quantifying the development of dominance hierarchies in large groups with applications to Harris' sparrows. Animal Behaviour 35:1113-1128.

Chase, M.H., and F.R. Morales. 1983. Subthreshold excitatory activity and motoneuron discharge during REM periods of active sleep. Science 221:1195-1198.

Chen, D.M., J.S. Collins, and T.H. Goldsmith. 1984. The ultraviolet receptor of bird retinas. Science 225:337-340.

Chen, W.Y., L.C. Aranda, and J.V. Luco. 1970. Learning and long- and short-term memory in cockroaches. Animal Behaviour 18:725-732.

Cheney, D.L., P.C. Lee, and R.M. Seyfarth. 1981. Behavioral correlates of non-random mortality among free-ranging adult female vervet monkeys. Behavioral Ecology and Sociobiology 9:153-161.

Chepko, B.D. 1971. A preliminary study of the effects of play deprivation on young goats. Zeitschrift für Tierpsychologie 28:517-576.

Christian, J.J. 1960. Adrenocortical and gonadal responses of female mice to increased population density. Proceedings of the Society for Experimental Biology and Medicine 104:330-332.

Christian, J.J., and D.E. Davis. 1964. Endocrines, behavior, and population. Science 146:1550-1560.

Clark, G.A., Jr. 1975. Additional records of passerine terrestrial gaits. Wilson Bulletin 87:384-389.

Clarke, S.E., P.W. Colgan, and N.P. Lester. 1984. Courtship sequences and ethological isolation in two species of sunfish *(Lepomis* spp) and their hybrids. Behaviour 91:93-113.

Clayton, D. 1976. The effects of pre-test conditions on social facilitation of drinking in ducks. Animal Behaviour 24:125-134.

Clayton, N.S. 1988. Song tutor choice in zebra finches and bengalese finches: the relative importance of visual and vocal cues. Behaviour 104:281-299.

Cloarec, A. 1980. Post-moult behaviour in the water-stick insect *Ranatra linearis*. Behaviour 73:304-324.

Clutton-Brock, J. 1981. Domesticated animals. University of Texas Press, Austin, Texas.

Clutton-Brock, T. 1982. The red deer of Rhum. Natural History 91(11):42-47.

Clutton-Brock, T.H., and P.H. Harvey. 1977. Primate ecology and social organization. Journal of Zoology 183:1-39.

Clutton-Brock, T.H., editor. 1988. Reproductive success. University of Chicago Press, Chicago.

Clutton-Brock, T.H., F.E. Guinness, and S.D. Albon. 1982. Red deer: behavior and ecology of two sexes. University of Chicago Press, Chicago.

Clutton-Brock, T.H., S.D. Albon, and F.E. Guinness. 1989. Fitness costs of gestation and lactation in wild mammals. Nature 337:260-262.

Cochran, W.W. 1987. Orientation and other migratory behaviours of a Swainson's thrush followed for 1500 km. Animal Behaviour 35:927-929.

Cochran, W.W., G.G. Montgomery, and R.R. Graber. 1967. Migratory flights of *Hylocichla* thrushes in spring: a radiotelemetry study. Living Bird 6:213-225.

Cody, M.L. 1968. On the methods of resource division in grassland bird communities. American Naturalist 102:107-147.

Cody, M.L. 1974. Optimization in ecology. Science 183:1156-1164.

Cody, M.L., editor. 1985. Habitat selection in birds. Academic Press, Orlando, Fla.

Coghill, G.E. 1929. Anatomy and the problem of behavior. Cambridge University Press, Cambridge, England.

Cohem, D.B. 1979. Sleep and dreaming: origins, nature, and function. Pergamon Press, New York.

Cole, L.C. 1954. The population consequences of life history phenomena. Quarterly Review of Biology 29:103-137.

Cole, L.C. 1957. Biological clock in the unicorn. Science 125:874-876.

Colgan, P.W., editor. 1978. Quantitative ethology. John Wiley & Sons, New York.

Colgan, P.W., W.A. Nowell, and N.W. Stokes. 1981. Spatial aspects of nest defence by pumpkinseed sunfish *(Lepomis gibbosus)*: stimulus features and an application of catastrophe theory. Animal Behaviour 29:433-442.

Collias, N.E. 1960. An ecological and functional classification of animal sounds. In: W.E. Lanyon and W.N. Tavolga, editors. Animal sounds and communication. American Institute of Biological Sciences 7:368-391.

Collias, N.E., and E.C. Collias, editors. 1976. External construction by animals: benchmark papers in animal behavior. Vol. 4. Dowden, Hutchinson, and Ross, Stroudsburg, Pa.

Collias, N.E., and E.C. Collias. 1984. Nest building and bird behavior. Prince-

ton University Press, Princeton, N.J.

Collins, B.J., and J.D. Davis. 1978. Long term inhibition of intake by mannitol. Physiology and Behavior 21:957-965.

Colwell, R.K. 1981. Group selection is implicated in the evolution of female-biased sex ratios. Nature 290:401-404.

Conner, R.N., and C.S. Adkisson. 1977. Principal component analysis of woodpecker nesting habitat. Wilson Bulletin 89:122-129.

Conner, W.E., and W.M. Masters. 1978. Infrared video viewing. Science 199:1004.

Conover, W.J. 1980. Practical nonparametric statistics, ed 2. John Wiley & Sons, New York.

Cook, A. 1977. Mucus trail following by the slug *Limax grossui* Lupu. Animal Behaviour 25:774-781.

Cook, A. 1979. Homing by the slug *Limax pseudoflavus*. Animal Behaviour 27:545-552.

Cooper, W.E. Jr. and D. Crews. 1987. Hormonal induction of secondary sexual coloration and rejection behaviour in female keeled earless lizards, *Holbrookia propinqua*. Animal Behaviour 35:1177-1187.

Corben, C.J., G.J. Ingram, and M.J. Tyler. 1974. Electrophysiological correlates of meaning. Science 186:944-947.

Corning, W.C., and R. von Burg. 1973. Protozoa. In: W.C. Corning, J.A. Dyal, and A.O.D. Willows, editors. Invertebrate learning. Vol. 3. Plenum Press, New York.

Corning, W.C., and S. Kelly. 1973. Platyhelminthes: the turbellarians. In: W.C. Corning. J.A. Dyal, and A.O.D. Willows, editors. Invertebrate learning. Vol. 3. Plenum Press, New York.

Corning, W.C., J.A. Dyal, and A.O.D. Willows, editors. 1973, 1975. Invertebrate learning, 3 vols. Plenum Press, New York.

Cornsweet, T.N. 1962. The staircase method in psychophysics. American Journal of Psychology 75:485-491.

Corrent, G., D.J. McAdoo, and A. Eskin. 1978. Serotonin shifts the phase of the circadian rhythm from the *Aplysia* eye. Science 202:977-979.

Coss, R.G., and A. Globus. 1978. Spine stems on tectal interneurons in jewel fish are shortened by social stimulation. Science 200:787-790.

Coulson, J.C. 1966. The influence of the pair bond and age on the breeding biology of the kittiwake gull *Rissa tridactyla*. Journal of Animal Ecology 35:269-279.

Cousteau, J. 1973. Attack and defense. World Publishing, New York.

Cowan, W.M., editor. 1981. Studies in developmental neurobiology. Oxford University Press, New York.

Cox, C.R. 1981. Agonistic encounters among male elephant seals: frequency, context, and the role of female preference. American Zoologist 21:197-209.

Cox, C.R., and B.J. LeBoeuf. 1977. Female incitation of male competition: a mechanism in sexual selection. American Naturalist 111:317-335.

Craig, J.V. 1981. Domestic animal behavior: causes and implications for animal care and management. Prentice-Hall, Englewood Cliffs, N.J.

Crandall, L.S. 1964. Management of wild animals in captivity. University of Chicago Press, Chicago.

Cregier, S.E. 1981. Alleviating road transit stress on horses. Animal Regulation Studies 3:223-227.

Crews, D. 1975. Psychobiology of reptilian reproduction. Science 189:1059-1065.

Crews, D. 1979. The hormonal control of behavior in a lizard. Scientific American 241:180-187.

Crews, D. 1982. On the origin of sexual behavior. Psychoneuroendocrinology 7:259-270.

Crews, D. 1983. Alternative reproductive tactics in reptiles. Bioscience 33:562-566.

Crews, D. 1987 a. Diversity and evolution of behavioral controlling mechanisms. In: D. Crews, editor. Psychobiology of reproductive behavior: an evolutionary perspective. Prentice-Hall, Englewood Cliffs, N.J.

Crews, D., editor. 1987 b. Psychobiology of reproductive behavior: an evolutionary perspective. Prentice-Hall, Englewood Cliffs, N.J.

Crews, D., and R. Silver. 1985. Reproductive physiology and behavior interactions in nonmammalian vertebrates. In: N. Adler, D. Pfaff, and R.W. Goy, editors. Handbook of behavioral neurobiology 7:101-182.

Crews, D., and W.R. Garstka. 1982. The ecological physiology of a garter snake. Scientific American 247(5):158-168.

Crick, F.H.C. 1979. Thinking about the brain. Scientific American 241(3):219-232.

Crook, J.H. 1964. The evolution of social organization and visual communication in the weaver birds (Ploceinae). Behaviour. Supplement 10.

Crossley, S.A. 1988. Failure to confirm rhythms in *Drosophila* courtship song. Animal Behaviour, 36:1098-1109.

Crow, J.F. 1986. Basic concepts in population, quantitative, and evolutionary genetics. W.H. Freeman, New York.

Cruze, W.W. 1935. Maturation and learning in chicks. Journal of Comparative Psychology 19:371-409.

Cullen, E. 1957. Adaptations in the kittiwake to cliff-nesting. Ibis 99:275-302.

Culliton, B.J. 1987. Take two pets and call me in the morning. Science 237:1560-1561.

Curio, E. 1976. The ethology of predation. Springer-Verlag, New York.

Curio, E., U. Ernst, and W. Vieth. 1978 a. Cultural transmission of enemy recognition: one function of mobbing. Science 202:899-901.

Curio, E., U. Ernst, and W. Vieth. 1978 b. The adaptive significance of avian mobbing. II. Cultural transmission of enemy recognition in blackbirds: effectiveness and some constraints. Zeitschrift für Tierpsychologie 48:184-202.

Currie, R.W. 1987. The biology and behaviour of drones. Bee World 68:129-143.

Cusson, M., and J.N. McNeil. 1989. Involvement of juvenile hormone in the regulation of pheromone release activities in a moth. Science 243:210-212.

Czeisler, C.A., J.S. Allan, S.H. Strogatz, J.M. Ronda, R. Sanchez, C.D. Rios, W.O. Freitag, G.S. Richardson, R.E. Kronauer. 1986. Bright light resets the human circadian pacemaker independent of the timing of the sleep-wake cycle. Science 233:667-671.

Czeisler, C.A., M.C. Moore-Ede, and R.M. Coleman. 1982. Rotating shift work schedules that disrupt sleep are improved by applying circadian principles. Science 217:460-462.

Dahl, F. 1884. Das Gehor- und Geruchsorgan der Spinnen. Arch. Mikroskop. Anat. 24:1-10.

Dal Molin, C. 1979. An external scent as the basis of a rare-male mating advantage in Drosophila melanogaster. American Naturalist 113:951-954.

Dale, N., S. Schacher, E.R. Kandel. 1988. Long-term facilitation in Aplysia involves increases in transmitter release. Science 239:282-285.

Dall, W.H. 1881. Intelligence in a snail. American Naturalist 15:976-978.

Daly, M. 1979. Why don't male mammals lactate? Journal of Theoretical Biology 78:325-245.

Daly, M., and M. Wilson. 1983. Sex, evolution, and behavior, ed 2. Willard Grant Press, Boston.

Daly, M., and M. Wilson. 1988. Evolutionary social psychology and family homicide. Science 242:519-524.

Darling, F.F. 1938. Bird flocks and the breeding cycle. Cambridge University Press, Cambridge, England.

Darwin, C. 1859. The origin of species. John Murray, London.

Darwin, C. 1871. The descent of man, and selection in relation to sex. D. Appleton, New York. (Reprinted 1971, John Murray, London.)

Darwin, C. 1873. Expression of the emotions in man and animals. D. Appleton, New York.

Darwin, C. 1889. The various contrivances by which orchids are fertilised by insects. ed 2. D. Appleton, New York.

Davidson, D.W. 1978. Experimental tests of the optimal diet in two social insects. Behavioral Ecology and Sociobiology 4:35-41.

Davidson, J.M., and R.J. Davidson, editors. 1980. The psychobiology of consciousness. Plenum Press, New York.

Davies, N.B. 1978. Territorial defence in the speckled wood butterfly (Pararge aegea): the resident always wins. Animal Behaviour 26:138-147.

Davies, N.B. 1983. Polyandry, cloaca-pecking and sperm competition in dunnocks. Nature 302:334-336.

Davies, N.B. 1985. Cooperation and conflict among dunnocks, Prunella modularis, in a variable mating system. Animal Behaviour 33:628-648.

Davies, N.B., A.F.G. Bourke, and M. de L. Brooke. 1989. Cuckoos and parasitic ants: interspecific brood parasitism as an evolutionary arms race. Trends in Ecology and Evolution 4:274-278.

Davies, N.B., and A. Lundberg. 1984. Food distribution and a variable mating system in the dunnock, Prunella modularis. Journal of Animal Ecology 53:895-913.

Davies, N.B., and A.I. Houston. 1981. Owners and satellites: the economics of territory defence in the pied wagtail, Motacilla alba. Journal of Animal Ecology 50:157-180.

Davies, N.B., and A.I. Houston. 1984. Territory economics. In: J.R. Krebs and N.B. Davies, editors. Behavioural ecology: an evolutionary approach, ed 2. Sinauer Associates, Sunderland, Mass.

Davies, N.B., and T.R. Halliday. 1977. Optimal mate selection in Bufo bufo. Nature 269: 56-58.

Davies, N.B., and T.R. Halliday. 1979. Competitive mate searching in male common toads, Bufo bufo. Animal Behaviour 27:1253-1267.

Davis, J.M. 1975. Socially induced flight reactions in pigeons. Animal Behaviour 23: 597-601.

Davis, W.H., and H.B. Hitchcock. 1965. Biology and migration of the bat Myotis lucifugus in New England. Journal of Mammalogy 46:296-313.

Davis, W.J. 1973. Development of locomotor patterns in the absence of peripheral sense organs and muscles. Proceedings of the National Academy of Sciences of the United States of America 70:954-958.

Dawkins, M.S. 1971. Perceptual changes in chicks: another look at the "search image" concept. Animal Behaviour 19:566-574.

Dawkins, M.S. 1977. Do hens suffer in battery cages? Environmental preferences and welfare. Animal Behaviour 25:1034-1046.

Dawkins, M.S. 1980. Animal suffering: the science of animal welfare. Chapman and Hall, New York.

Dawkins, M.S. 1986. Unravelling animal behaviour. Longman Group Limited, Essex, England.

Dawkins, R. 1972. A cheap method of recording behavioural events, for direct computer access. Behaviour 40:162-173.

Dawkins, R. 1976. The selfish gene. Oxford University Press, New York.

Dawkins, R. 1978. Replicator selection and the extended phenotype. Zeitschrift für Tierpsychologie 47:61-76.

Dawkins, R. 1979. Twelve misunderstandings of kin selection. Zeitschrift für Tierpsychologie 51:184-200.

Dawkins, R. 1980. Good strategy or evolutionarily stable strategy? In: G.W. Barlow and J. Silverberg, editors. Sociobiology: beyond nature/nurture? Westview Press, Boulder, Colo.

Dawkins, R. 1982. The extended phenotype. W.H. Freeman, San Francisco.

Dawkins, R. 1987. The blind watchmaker. W.W. Norton, New York.

Dawkins, R. 1989. The selfish gene, ed 2. Oxford University Press, New York.

Dawkins, R., and H.J. Brockmann. 1980. Do digger wasps commit the Concorde fallacy? Animal Behaviour 28:892-896.

Dawkins, R., and J.R. Krebs. 1978. Animal signals: information or manipulation? In: J.R. Krebs and N.B. Davies, editors. Behavioural ecology, an evolutionary approach. Blackwell Scientific Publications, London.

Dawkins, R., and J.R. Krebs. 1979. Arms races within and between species. Proceedings of the Royal Society of London B 205:489-511.

Dawkins, R., and M. Dawkins. 1976. Hierarchical organization and postural facilitation: rules for grooming in flies. Animal Behaviour 24:739-755.

Dawkins, R., and T.R. Carlisle. 1976. Parental investment, mate desertion and a fallacy. Nature 262:131-133.

de Groot, P. 1980. Information transfer in

a socially roosting weaver bird (*Quelea quelea*; Ploceinae): an experimental study. Animal Behaviour 28:1249-1254.

de Robertis, E. 1960. Some observations on the ultrastructure and morphogenesis of photoreceptors. Journal of General Physiology 43(Suppl.):1-13.

de Waal, F.B.M. 1982. Chimpanzee politics. Harper & Row, New York.

de Vos, G.J. 1979. Adaptations of arena behavior in black grouse *(Tetrao tetrix)* and other grouse species (Tetraoninae). Behaviour 68:277-314.

de Vries, T. 1973. The Galapagos hawk. Ph.D. thesis. University of Amsterdam.

Deese, J. 1978. Thought into speech. American Scientist 66:314-321.

Deguchi, T. 1981. Rhodopsin-like photosensitivity of isolated chicken pineal gland. Nature 209:706-707.

del Rio, C.M., and B.R. Stevens 1989. Physiological constraint on feeding behavior: intestinal membrane disaccharides of the starling. Science 243:794-796.

Demarest, J. 1985. Book review of Dewsbury 1984 a. Animal Behaviour 33:689-691.

Dement, W., and N. Kleitman. 1957. Cyclic variations of EEG during sleep and their relations to eye movements, body motility, and dreaming. Electroencephalography and Clinical Neurophysiology 9:673.

Demong, N.J., and S.T. Emlen. 1978. Radar tracking of experimentally released migrant birds. Bird Banding 49:342-359.

Denenberg, V.H. 1963. Early experience and emotional development. Scientific American 208:138-146.

Denenberg, V.H. 1972 a. Biobehavioral bases of development. In: V.H. Denenberg, editor. The development of behavior. Sinauer Associates, Sunderland, Mass.

Denenberg, V.H. 1972 b. Readings in the development of behavior. Sinauer Associates, Sunderland, Mass.

Denny, M. 1980. Locomotion: the cost of gastropod crawling. Science 208:1288-1290.

Dethier, V.G. 1962. To know a fly. Holden-Day, Oakland, Calif.

Dethier, V.G. 1971. A surfeit of stimuli: a paucity of receptors. American Scientist 59:706-715.

Dethier, V.G. 1976. The hungry fly. Harvard University Press, Cambridge, Mass.

Deutsch, J.A., W.G. Young, and T.J. Kalogeris. 1978. The stomach signals satiety. Science 201:165-167.

Dewdney, A.K. 1988. Computer recreations: random walks that lead to fractal crowds. Scientific American 259:116-119.

Dewsbury, D.A. 1978. What is (was?) the "fixed action pattern?" Animal Behaviour 26:310-311.

Dewsbury, D.A. 1984 a. Comparative psychology in the twentieth century. Hutchinson Ross, Stroudsburg, Pa.

Dewsbury, D.A. 1984 b. Foundations of comparative psychology. van Nostrand Rheinhold, New York.

Dewsbury, D.A. 1985. Leaders in the study of animal behavior: autobiographical perspectives. Bucknell University Press, Lewisburg, Penn.

Diamond, J. 1985. Everything *else* you've always wanted to know about sex... Discover 6(4):70-82.

Diamond, J. 1988. Survival of the sexiest. Discover 8(5):74-81.

Diamond, J.M. 1987. A darwinian theory of divorce. Nature 329:765-766.

Diamond, S., editor. 1974. The roots of psychology. Basic Books, New York.

Dickemann, M. 1985. Human sociobiology: the first decade. New Scientist 108:38-42.

Dilger, W.C. 1962. The behavior of lovebirds. Scientific American 206(1):88-98.

Dill, P.A. 1977. Development of behaviour in alevins of Atlantic salmon, *Salmo salar*, and rainbow trout, *S. gairdneri*. Animal Behaviour 25:116-121.

Dingle, H. 1969. A statistical and information analysis of aggressive communication in the mantis shrimp *Gonodactylus bredini manning*. Animal Behaviour 17:561-575.

Dingle, H. 1972. Aggressive behavior in stomatopods and the use of information theory in the analysis of animal communication. In: H.E. Winn and B.L. Olla, editors. Behavior of marine animals. Vol. 1: invertebrates. Plenum Press, New York.

Dissanayake, E. 1974. A hypothesis of the evolution of art from play. Leonardo 7:211-217.

Dodds, W.J., and F.B. Orlans, editors. 1982. Scientific perspectives on animal welfare. Academic Press, New York.

Dominey, W.J. 1981. Anti-predator function of bluegill sunfish nesting colonies. Nature 290:586-588.

Domjan, M., and B. Burkhard. 1982. The principles of learning and behavior. Brooks/Cole, Monterey, Calif.

Donnelley, S., and K. Nolan. 1990. Animals, science, and ethics. Hastings Center Report, Special Supplement, May/June, Publications Department, The Hastings Center, Briarcliff Manor, NY 10510.

Douglas, M.M. 1986. The lives of butterflies. University of Michigan Press, Ann Arbor, Mich.

Dowling, J.E., 1987. The retina. Belknap. Harvard University Press, Cambridge, Mass.

Downes, J.A. 1978. Feeding and mating in the insectivorous Ceratopogoninae (Diptera). Memoirs of the Entomological Society of Canada 104:1-62.

Downhower, J.F., and K.B. Armitage. 1971. The yellow-bellied marmot and the evolution of polygamy. American Naturalist 105:355-370.

Drickamer, L.C. 1972. Experience and selection behavior in the food habits of *Peromyscus*: use of olfaction. Behaviour 41:269-287.

Drucker-Colin, R.R., and J.L. McGaugh, editors. 1977. Neurobiology of sleep and memory. Academic Press, New York.

Duncan, P. 1985. Time-Budgets of camargue horses: III. Environmental influences. Behaviour 92:189-209.

Dunstone, N., and R.J. O'Connor. 1979. Optimal foraging in an amphibious mammal. Animal Behaviour 27:1182-1201.

Durant, J.R. 1985. The science of sentiment: the problem of the cerebral localization of emotion. In: P.P.G. Bateson and P.H. Klopfer. Perspectives in ethology. Vol. 6. Plenum Press, New York.

Durant, J.R. 1986. The making of ethology: the Association for the Study of Animal Behaviour, 1936-1986.

Durden, K. 1972. Gifts of an eagle. Simon & Schuster, New York.

Dussourd, D.E., and T. Eisner. 1987. Vein-cutting behavior: insect counterploy to the latex defense of plants. Science 237:898-901.

Dwyer, T.J. 1975. Time budget of breeding gadwalls. Wilson Bulletin 87:335-343.

Dyal, J.A. 1973. Behavior modification in annelids. In: W.C. Corning, J.A. Dyal, and A.O.D. Willows, editors. Invertebrate learning. Vol. 1. Plenum Press, New York.

Dyer, F.C., and J.L. Gould. 1983. Honey bee navigation. American Scientist 71:587-597.

Eaton, R.L. 1970. The predatory sequence, with emphasis on killing behavior and its ontogeny, in the cheetah *(Acinonyx jubatus* Schreber). Zeitschrift für Tierpsychologie 27:492-504.

Eaton, R.L. 1974. The cheetah: the biology, ecology, and behavior of an endangered species. Van Nostrand Reinhold, New York.

Eccles, J.C. 1970. Facing reality: philosophical adventures by a brain scientist. Springer, New York.

Eccles, J.C. 1977. The understanding of the brain, ed 2. McGraw-Hill, New York.

Eckert, R. 1963. Electrical interaction of paired ganglion cells in the leech. Journal of General Physiology 46:575-587.

Edelman, G.M., W.E. Gall, and W.M. Cowan, editors. 1985. Molecular bases of neural development. John Wiley & Sons, New York.

Ehrman, L., and J. Probber. 1978. Rare *Drosophila* males: the mysterious matter of choice. American Scientist 66:216-222.

Ehrman, L., and P.A. Parsons. 1976. The genetics of behavior. Sinauer Associates, Sunderland, Mass.

Eibl-Eibesfeldt, I. 1951. Beobachtungen zur Fortpflanzungsbiologie und Jungendentwicklung des Eichhornchens. Zeitschrift für Tierpsychologie 8:370-400.

Eibl-Eibesfeldt, I. 1956. Angebornes und Erworbenes in der technik des Beutetotens (Versuche am Iltis, *Putorius putorius* L.). Zeitschrift für Saugetierkunde 21:135-137.

Eibl-Eibesfeldt, I. 1958. Versuche uber den Nestbau erfahrungsloser Ratten. (Wiss. Film B757.) Inst. Wiss. Film, Gottingen.

Eibl-Eibesfeldt, I. 1961. The fighting behavior of animals. Scientific American 205112-122.

Eibl-Eibesfeldt, I. 1963. Angebornes und Erworbenes im Verhalten einiger Sauger. Zeitschrift für Tierpsychologie 20:705-754.

Eibl-Eibesfeldt, I. 1970. Ethology. The biology of behavior. Holt, Rinehart and Winston, New York.

Eibl-Eibesfeldt, I. 1972. Ethology: the biology of behavior, ed 2. Holt, Rinehart & Winston, New York. (Translated by E. Klinghammer.)

Eisenberg, J.F. 1967. A comparative study in rodent ethology with emphasis on the evolution of social behavior. Proceedings of the United States National Museum 122:1-51.

Eisenberg, J.F., and P. Leyhausen. 1972. The phylogenesis of predatory behavior in mammals. Zeitschrift für Tierpsychologie 30:59-93.

Eisner, T., and D.J. Aneshansley. 1982. Spray aiming in bombardier beetles: jet deflection by the coanda effect. Science 215:83-85.

Eisner, T., K. Hicks, M. Eisner, and D.S. Robson. 1978. "Wolf-in-sheep's clothing" strategy of a predaceous insect larva. Science 199:790-794.

Ekman, J. 1987. Exposure and time use in willow tit flocks: the cost of subordination. Animal Behaviour 35:445-452.

Elliot, R.D. 1977. Hanging behavior in common ravens. Auk 94:777-778.

Ellis, D.H. 1979. Development of behavior in the golden eagle. Wildlife Monographs 70:1-94.

Ellis, M.J. 1973. Why people play. Prentice-Hall, Englewood Cliffs, N.J.

Ely, C.R. 1987. An inexpensive device for recording animal behavior. Wildlife Society Bulletin 15:264-265.

Elzinga, R.J. 1978. Fundamentals of entomology. Prentice-Hall, Englewood Cliffs, N.J.

Emlen, J.M. 1966. The role of time and energy in food preference. American Naturalist 100:611-617.

Emlen, J.M., and M.G. Emlen. 1975. Optimal choice in diet: test of a hypothesis. American Naturalist 109:427-435.

Emlen, S. 1976. Lek organization and mating strategies in the bullfrog. Behavioral Ecology and Sociobiology 1:283-313.

Emlen, S.T. 1972. An experimental analysis of the parameter of bird song eliciting species recognition. Behaviour 41:130-171.

Emlen, S.T. 1975. The stellar-orientation system of a migratory bird. Scientific American 233(2):102-111.

Emlen, S.T., and N.J. Demong. 1975. Adaptive significance of synchronized breeding in a colonial bird: a new hypothesis. Science 188:1029-1031.

Emlen, S.T., and J.T. Emlen. 1966. A technique for recording migratory orientation of captive birds. Auk 83:361-367.

Emlen, S.T., and L.W. Oring. 1977. Ecology, sexual selection, and the evolution of mating systems. Science 197:215-233.

Emlen, S.T., N.J. Demong, and D.J. Emlen. 1989. Experimental induction of infanticide in female wattled jacanas. Auk 106:1-7.

Enright, J.T. 1980. The timing of sleep and wakefulness. Springer-Verlag, New York.

Ens, B.J., and J.D. Goss-Custard. 1984. Interference among oystercatchers, *Haematopus ostralegus*, feeding on mussels, *Mytilus edulis*, on the Exe estuary. Journal of Animal Ecology 53:217-232.

Eoff, M. 1977. Artificial selection in *Drosophila simulans* males for increased and decreased sexual isolation from *D.*

melanogaster females. American Naturalist 3:259-277.

Epstein, R., R.P. Lanza, and B.F. Skinner. 1980. Symbolic communication between two pigeons. Science 207:543-545.

Epstein, R., R.P. Lanza, and B.F. Skinner. 1981. "Self-awareness" in the pigeon. Science 212:695-696.

Ericsson, K.A., and W.G. Chase, 1982. Exceptional memory. American Scientist 70:607-617.

Erwin, R.M. 1978. Coloniality in terns: the role of social feeding. Condor 80:211-215.

Erwin, R.M. 1983. Feeding habitats of nesting wading birds: spatial use and social influences. Auk 100:960-970.

Esch, H. 1967. The evolution of bee language. Scientific American 216(4):96-104.

Eskin, A. 1971. Some properties of the system controlling the circadian activity rhythm of sparrows. In: M. Menaker, editor. Biochronometry. National Academy of Sciences, Washington, D.C.

Estes, R.D. 1976. The significance of breeding synchrony in the wildebeest. East African Wildlife Journal 14:135-152.

Etienne, A.S., R. Maurer, F. Saucy, and E. Teroni. 1986. Short-distance homing in the golden hamster after a passive outward journey. Animal Behaviour 34:696-715.

Evans, D.L. 1980. Multivariate analysis of weather and fall migration of saw-whet owls at Duluth, Minnesota, M.S. thesis. North Dakota State University, Fargo, N.D.

Evans, H.E. 1977. Extrinsic versus intrinsic factors in the evolution of insect sociality. Bioscience 27:613-617.

Evans, H.E. 1984. Insect biology. Addison-Wesley, Reading, Mass.

Evans, R.L. 1976. The making of psychology. Alfred A. Knopf, New York.

Evans, R.M. 1983. Do secondary roosts function as information centers in black-billed gulls? Wilson Bulletin 95:461-462.

Ewert, J.P. 1980. Neuroethology, an introduction to the neurophysiological fundamentals of behavior. Springer-Verlag, New York.

Ewert, J.P. 1985. Concepts in vertebrate neuroethology. Animal Behaviour 33:1-29.

Ewing, A.W. 1988. Cycles in the courtship song of male *Drosophila melanogaster* have not been detected. Animal Behaviour, 36:1091-1097.

Ewing, A.W., and V. Evans. 1973. Studies on the behaviour of cyprinodont fish. I.

The agonistic and sexual behaviour of *Aphyosemion biviltatum* (Lonnberg 1895). Behaviour 46:264-278.

Faegri, K., and L. van der Pijl. 1966. The principles of pollination ecology. Pergamon Press, New York.

Fagen, R.M. 1976. Exercise, play, and physical training in animals. In: P.P.G. Bateson and P.H. Klopfer, editors. Perspectives in ethology. Vol. 2. Plenum Press, New York.

Fagen, R.M. 1978. Information measures: statistical confidence limits and inference. Journal of Theoretical Biology 73:61-79.

Fagen, R.M. 1981. Animal play behavior. Oxford University Press, New York.

Fagen, R.M., and R.N. Goldman. 1977. Behavioural catlogue analysis methods. Animal Behaviour 25:261-274.

Falconer, D.S. 1989. Introduction to quantitative genetics, ed 3. John Wiley & Sons, New York.

Farish, D.J. 1972. The evolutionary implications of qualitative variation in the grooming behaviour of the hymenoptera (Insecta). Animal Behaviour 20:662-676.

Farkas, S.R., and H.H. Shorey. 1976. Anemotaxis and odour trail following by the terrestrial snail *Helix aspersa*. Animal Behaviour 24:686-689.

Feder, M.E., and G.V. Lauder. 1986. Predator-prey relationships. Perspectives and approaches from the study of lower vertebrates. University of Chicago Press, Chicago.

Feist, J.M. 1982. Bats away! American Heritage 33:93-95.

Fenton, M.B. 1982. Echolocation, insect hearing, and feeding ecology of insectivorous bats. In: T.H. Kunz, editor. Ecology of bats. Plenum Press, New York.

Fenton, M.B. 1983. Just bats. University of Toronto Press, Toronto.

Fenton, M.B., and J.H. Fullard. 1979. The influence of moth hearing on bat echolocation strategies. Journal of Comparative Physiology 132:77-86.

Fenton, M.B., P. Racey, and J.M.V. Rayner, editors. 1987. Recent advances in the study of bats. Cambridge University Press, New York.

Fentress, J.C., editor. 1976. Simpler networks and behavior. Sinauer Associates, Sunderland, Mass.

Ferguson, N.B.L., and R.E. Keesey. 1975. Effect of a quinine-adulterated diet upon body weight maintenance in male rats with vertromedical hypothalamic lesions. Journal of Comparative and Physiological Psychology 89:478-488.

Ferguson, W. 1968. Abnormal behavior in domestic birds. In: M.W. Fox, editor. Abnormal behavior in animals. W.B. Saunders, Philadelphia.

Fernald, R.D., and N.R. Hirata. 1979. The ontogeny of social behavior and body coloration in the African cichlid fish *Haplochromis burtoni*, Zeitschrift für Tierpsychologie 50:180-187.

Fernald, R.D., and P. Heinecke. 1974. A computer compatible multi-purpose event recorder. Behaviour 48:269-275.

Ferron, J. 1976. Comfort behavior of the red squirrel *(Tamiasciurus hudsonicus)*. Zeitschrift für Tierpsychologie 42:66-85.

Ferron, J. 1981. Comparative ontogeny of behaviour in four species of squirrels (Sciuridae). Zeitschrift für Tierpsychologie 55:193-216.

Ficken, R.W., M.S. Ficken, and J.P. Hailman. 1978. Differential aggression in genetically different morphs of the white-throated sparrow *(Zonotrichia albicollis)*. Zeitschrift für Tierpsychologie 46:43-57.

Fillion, T.J., and E.M. Blass. 1986. Infantile experience with suckling odors determines adult sexual behavior in male rats. Science 231:729-731.

Finger, T.E., and W.L. Silver. 1987. Neurobiology of taste and smell. John Wiley & Sons, New York.

Fischer, G.L. 1975. The behaviour of chickens. In: E.S.E. Hafez, editor. The behaviour of domestic animals. Williams & Wilkins, Baltimore.

Fisher, J. 1954. Evolution and bird sociality. In: Huxley, J., A.C. Hardy, and E.B. Ford, editors. Evolution as a process. George Allen and Unwin, London.

Fisher, R.A. 1930. The genetical theory of natural selection. Clarendon Press, Oxford.

Fisher, R.A. 1958. The genetical theory of natural selection, ed 2. Dover, New York.

Fisler, G.F. 1977. Interspecific hierarchy at an artificial food source. Animal Behaviour 25:240-244.

Fitzgerald, J.W., and G.E. Woolfenden. 1984. The helpful shall inherit the scrub. Natural History 93(5):55-63.

Fitzpatrick, J.W. 1981. Search strategies of tyrant flycatchers. Animal Behaviour 29:810-821.

Fivizzani, A.J., and L.W. Oring. 1986. Plasma steroid hormones in relation to behavioral sex role reversal in the spotted sandpiper, *Actitis macularia*. Biology of Reproduction 35:1195-1201.

Fivizzani, A.J., M.A. Colwell, and L.W. Oring. 1986. Plasma steroid hormone levels in free-living Wilson's phalar-

opes, *Phalaropus tricolor*. General and Comparative Endocrinology 62:137-144.

Fjerdingstad, E., editor. 1971. Chemical transfer of learned information. North-Holland, Amsterdam.

Fleming, T.H. 1981. Winter roosting and feeding behaviour of pied wagtails *Motacilla alba* near Oxford, England. Ibis 123:463-476.

Flugel, J.C. 1933. A hundred years of psychology. G. Duckworth, London.

Follett, B.K., C.G. Scanes, and F.J. Cunningham. 1972. A radioimmunassay for avian luteinizing hormone. Journal of Endocrinology 52:359-378.

Foltz, D.W., and J.L. Hoogland. 1981. Analysis of the mating system in the black-tailed prairie dog *(Cynomys ludovicianus)* by likelihood of paternity. Journal of Mammalogy 62: 706-712.

Forrest, T.G. 1980. Phonotaxis in mole crickets: its reproductive significance. Florida Entomologist 63:42-53.

Forster, L. 1982. Vision and prey-catching strategies in jumping spiders. American Scientist 70:165-175.

Fouts, R.S. 1972. The use of guidance in teaching sign language to a chimpanzee. Journal of Comparative and Physiological Psychology 80:515-522.

Fouts, R.S., and R.L. Mellgren. 1976. Language, signs, and cognition in the chimpanzee. Sign Language Studies 13:319-346.

Fowler, M.E. 1978. Restraint and handling of wild and domestic animals. Iowa State University Press, Ames.

Fox, L.R., and P.A. Morrow. 1981. Specialization: species property or local phenomenon? Science 211:887-893.

Fox, M.W. 1967. The place and future of animal behavior studies in veterinary medicine. Journal of the American Veterinary Medicine Association 151: 609-615.

Fox, M.W. 1969. Ontogeny of prey-killing behavior in the Canidae. Behaviour 35:259-272.

Fox, M.W. 1980. Returning to Eden: animal rights and human responsibility. Viking Press, New York.

Fox, M.W., editor. 1968. Abnormal behavior in animals. W.B. Saunders, Philadelphia.

Fraenkel, G.S., and D.L. Gunn. 1961. The orientation of animals. Dover, New York.

Frankel, R.B., R.P. Blakemore, and R.S. Wolfe. 1979. Magnetite in freshwater magnetotactic bacteria. Science 203: 1355-1356.

Franks, N.R. 1985. Reproduction, foraging efficiency and worker polymorphism in army ants. In: B. Hölldobler and M. Lindauer, editors. Experimental behavioral ecology and sociobiology. Fortschritte der Zoologie (31:91-107). G. Fischer Verlag, New York.

Fraser, A.F. 1968. Behavior disorders in domestic animals. In: M.W. Fox, editor. Abnormal behavior in animals. W.B. Saunders, Philadelphia.

Fraser, D. 1979. Aquatic feeding by a woodchuck. Canadian Field-Naturalist 93:309-310.

Fraser, D.F., and F.A. Huntingford. 1986. Feeding and avoiding predation hazard: the behavioral response of the prey. Ethology 73:56-68.

Freemon, F.R. 1972. Sleep research. Charles C. Thomas, Springfield, Ill.

Fretwell, S.D. 1972. Populations in a seasonal environment. Princeton University Press, Princeton, N.J.

Fretwell, S.D., and H.L. Lucas, Jr. 1970. On territorial behavior and other factors infulencing habitat distribution in birds. I. Theoretical development. Acta Biotheoretica 19:16-36.

Fricke, H.W. 1979. Mating system, resource and defense and sex change in the anemonefish *Amphiprion akallopisos*. Zeitschrift für Tierpsychologie 50:313-326.

Frisch, K. von. 1914. Demonstration von Versucher zum Nachweis des Farbensinnes bei angeblich total farbenblinder Tieren. Verhandl. d. Deutsch. Zool. Ges. in Freiburg, Berlin.

Frisch, K. von. 1962. Dialects in the language of the bees. Scientific American 207(2):78-87.

Frisch, K. von. 1967. The dance language and orientation of bees. Belknap/Harvard University Press, Cambridge, Mass.

Frisch, K. von. 1971. Bees, their vision, chemical senses, and language. Cornell University Press, Ithaca, N.Y.

Frisch, K. von. 1974. Animal architecture. Harcourt Brace Jovanovich, New York.

Frohman, L.A., and L.L. Bernardis. 1968. Growth hormone and insulin levels in weanling rats with vertromedial hypothalamic lesions. Endocrinology 82:1125-1132.

Fulker, D.W. 1966. Mating speed in male *Drosophila melanogaster*: a psychogenetic analysis. Science 153:203-205.

Futuyma, D.J. 1986. Evolutionary biology, ed 2. Sinauer Associates, Sunderland, Mass.

Fuzessery, Z.M., and G.D. Pollak. 1984. Neural mechanisms of sound localization in an echolocating bat. Science 225:725-727.

Gadgil, M. 1972. The function of communal roosts: relevance of mixed roosts. Ibis 114:531-533.

Gaioni, S.J., and C.S. Evans. 1986. Mallard duckling response to distress calls with reduced variability: a constraint on stereotypy in a "fixed action pattern." Ethology 72:1-14

Gaither, N.S., and B.E. Stein. 1979. Reptiles and mammals use similar sensory organizations in the midbrain. Science 205:595-597.

Galef, B.G. Jr., and S.W. Wigmore. 1983. Transfer of information concerning distant foods: a laboratory investigation of the "information-centre" hypothesis. Animal Behaviour 31:748-758.

Gallup, G.G., Jr., and J.W. Beckstead. 1988. Attitudes toward animal research. American Psychologist 43:474-476.

Gamboa, G.J. 1978. Intraspecific defense: advantage of social cooperation among paper wasp foundresses. Science 199:1463-1465.

Gamow, R.I., and J.F. Harris. 1973. The infrared receptors of snakes. Scientific American 228(5):94-100.

Garcia, J., W.G. Hankins, and K.W. Rusiniak, 1974. Behavioral regulation of the milieu interne in man and rat. Science 185:824-831.

Gardner, B.T., and R.A. Gardner. 1969. Teaching sign language to a chimpanzee. Science 165:664-672.

Gardner, B.T., and R.A. Gardner. 1978. Comparative psychology and language acquisition. In: K. Salzinger and F. Denmark, editors. Psychology: the state of the art. Annals of the New York Academy of Sciences 309:37-76.

Garson, P.J., W.K. Pleszczynska, and C.H. Holm. 1981. The "polygyny threshold" model: a reassessment. Canadian Journal of Zoology 59:902-911.

Garstka, W., B. Camazine, and D. Crews. 1982. Interactions of behavior and physiology during the annual reproductive cycle of the red-sided garter snake *(Thamnophis sirtalis parietalis)*. Herpetologica 38:104-123.

Gauthreaux, S.A., Jr., editor. 1980. Animal migration, orientation, and navigation. Academic Press, New York.

Gautier, J.Y. 1974. Etude comparee de la distribution spatiale et temporelle des adultes de *Blaberus atropos* et *B. colosseus* (Dictyopteres) dans cinq grottes de l'ile de Trinidad. Rev. Comp. Animal 9:237-258.

Geist, V. 1971. Mountain sheep: a study in behavior and evolution. University of Chicago Press, Chicago.

Geist, V. 1972. An ecological and behavioural explanation of mammalian characteristics and their implication to therapsid evolution. Zeitschrift für Saugetierkunde 37:1-15.

Geist, V., and F. Walther, editors. 1974. Behaviour of ungulates and its relation to management. Vols. 1 and 2. International Union for Conservation of Nature and Natural Resources, Morges, Switzerland.

Gentry, R.L. 1973. Thermoregulatory behavior of eared seals. Behaviour 46:73-93.

Gersuni, G.V. 1971. Sensory processes at the neuronal and behavioral levels. Academic Press, New York.

Geschwind, N., and A.M. Galaburda, editors. 1984. Cerebral dominance: the biological foundations. Harvard University Press, Cambridge, Mass.

Getty, T. 1981 a. Analysis of central-place space-use patterns: the elastic disc revisited. Ecology 62(4):907-914.

Getty, T. 1981 b. Territorial behavior of eastern chipmunks *(Tamias striatus)*: encounter avoidance and spatial time-sharing. Ecology 62(4):915-921.

Getty, T. 1981 c. Structure and dynamics of chipmunk home range. Journal of Mammalogy 62(4):726-737.

Getz, W.M., and K.B. Smith. 1983. Genetic kin recognition: honey bees discriminate between full and half sisters. Nature 302:147-148.

Gherardi, F., F. Tarducci, and M. Vannini. 1988. Locomotor activity in the freshwater crab *Potamon fluviatile*: the analysis of temporal patterns by radio-telemetry. Ethology 77:300-316.

Gibb, J.A. 1957. Food requirements and other observations on captive tits. Bird Study 4:207-215.

Gibson, R.M., and J.W. Bradbury. 1986. Male and female mating strategies on sage grouse leks. In: D.I. Rubenstein and R.W. Wrangham, editors. Ecological aspects of social evolution. Princeton University Press, Princeton, N.J.

Gilbert, L.E. 1976. Postmating female odor in *Heliconius* butterflies: a male-contributed antiaphrodisiac? Science 193:419-420.

Gilbert, L.E. 1982. The coevolution of a butterfly and a vine. Scientific American 247(2):110-121.

Gill, F.B., and L.L. Wolf. 1975. Economics of feeding territoriality in the golden-winged sunbird. Ecology 56:333-345.

Ginsburg, B., and W.C. Allee. 1942. Some effects of conditioning on social dominance and subordination in inbred strains of mice. Physiological Zoology 15:485-506.

Giraldeau, L., and L. Lefebvre. 1987. Scrounging prevents cultural transmission of food-finding behaviour in pigeons. Animal Behaviour 35:387-394.

Gittleman, J.L. 1989. Carnivore group living: comparative trends. In: J.L. Gittleman, editor. Carnivore behavior, ecology, and evolution. Comstock, Ithaca, N.Y.

Gloor, P., A. Olivier, L.F. Quesney, F. Andermann, and S. Horowitz. 1982. The role of the limbic system in experiential phenomena of temporal lobe epilepsy. Annals of Neurology 12:129-144.

Gold, R.M. 1973. Hypothalamic obesity: the myth of the ventromedial nucleus. Science 82:488-490.

Goldman-Rakic, P.S. 1988. Topography of cognition: parallel distributed networks in primate association cortex. Annual Review of Neuroscience 11:137-156.

Goldstein, M.C. 1971. Stratification, polyandry, and family structure in central Tibet. Southwest Journal of Anthropology 27:64-74.

Gomendio, M. 1988. The development of different types of play in gazelles: implications for the nature and functions of play. Animal Behaviour 36:825-836.

Goodall, J. 1986. The chimpanzees of Gombe: patterns of behavior. Harvard University Press, Cambridge, Mass.

Gorner, P. 1958. Die optische und kinasthetische Orientierung der trichterspinne Agelena labyrinthica (Cl.). Zeitschrift für Vergleichende Physiologie 41:111-153.

Gossow, H. 1970. Vergleichende Verhaltensstudien an Marderartigen I. Uber LautauBerungen und zum Beuteverhalten. Zeitschrift für Tierpsychologie 27:405-480.

Gotmark, F., D.W. Winkler, and M. Andersson. 1986. Flock-feeding on fish schools increases individual success in gulls. Nature 319:589-591.

Gottfried, B.M., and E.C. Franks. 1975. Habitat use and flock activity of dark-eyed juncos in winter. Wilson Bulletin 87(3):374-383.

Gottlieb, G. 1970. Conceptions of prenatal behavior. In: L.R. Aronson, E. Tobach, D.S. Lehrman, and J.S. Rosenblatt, editors. Development and evolution of behavior. W.H. Freeman, San Francisco.

Gottlieb, G. 1971. Development of species identification in birds. University of Chicago Press, Chicago.

Gould, J.L. 1974. Genetics and molecular ethology. Zeitschrift für Tierpsychologie 36:267-292.

Gould, J.L. 1975. Honey bee recruitment: the dance-language controversy. Science 189:685-693.

Gould, J.L. 1980. Sun compensation by bees. Science 207:545-547.

Gould, J.L. 1984. Processing of sun-azimuth information by bees. Animal Behaviour 32.

Gould, J.L. 1985. How bees remember flower shapes. Science 227:1492-1494.

Gould, J.L. 1986. Pattern learning by honey bees. Animal Behaviour 34:990-997.

Gould, J.L. 1987 a. Honey bees store learned flower-landing behaviour according to time of day. Animal Behaviour 35:1579-1581.

Gould, J.L. 1987 b. Landmark learning by honeybees. Animal Behaviour 35:26-34.

Gould, J.L. 1988. A mirror-image "ambiguity" in honey bee pattern matching. Animal Behaviour 36:487-492.

Gould, J.L., and C. Gould. 1989. The honey bee. W.H. Freeman, San Francisco.

Gould, J.L., and K.P. Able. 1981. Human homing: an elusive phenomenon. Science 212:1061-1063.

Gould, J.L., J.L. Kirschvink, and K.S. Deffeyes. 1978. Bees have magnetic remanence. Science 201:1026-1028.

Gould, S.J. 1974. Racist arguments and IQ. Natural History 83:24-29.

Gould, S.J. 1976. Biological potential vs. biological determinism. Natural History 85(5):12-22.

Gould, S.J. 1980 a. Is a new and general theory of evolution emerging? Paleobiology 6(1):119-130.

Gould, S.J. 1980 b. Sociobiology and the theory of natural selection. In: G.W. Barlow and J. Silverberg, editors. Sociobiology: beyond nature/nurture? Westview Press, Boulder, Colo.

Gould, S.J. 1985. The flamingo's smile. W.W. Norton, New York.

Gould, S.J., and E.S. Vrba. 1982. Exaptation—a missing term in the science of form. Paleobiology 8(1)4-15.

Gould, S.J., and R.C. Lewontin. 1979. The spandrels of San Marco and the Panglossian paradigm: a critique of the adaptationist programme. Proceedings of the Royal Society of London. B205:581-98.

Gouzoules, H., S. Gouzoules, and L. Fedigan. 1982. Behavioural dominance and reproductive success in female Japanese monkeys (M. fuscata). Animal Behaviour 30:1138-1151.

Gowaty, P.A. 1982. Sexual terms in sociobiology: emotionally evocative and, paradoxically, jargon. Animal Behaviour 30:630-631.

Grafen, A. 1982. How not to measure inclusive fitness. Nature 298:425-426.

Grafen, A. 1984. Natural selection, kin selection and group selection. In: J.R. Krebs and N.B. Davies, editors. Behavioural Ecology: an evolutionary approach. Blackwell Scientific Publications. Oxford.

Grafen, A. 1987. Measuring sexual selection: why bother? In: J.W. Bradbury and M.B. Andersson, editors. Sexual selection: testing the alternatives. John Wiley & Sons, Chichester, N.J.

Gray, J.A.B. 1959. Initiation of impulses at receptors. In: J. Field, editor. Handbook of physiology. Section I, Neurophysiology. Vol 1. Waverly Press, Baltimore.

Green, G.W. 1964. The control of spontaneous locomotor activity in Phormia regina Meigen. I. Locomotor activity patterns in intact flies. Journal of Insect Physiology 10:711-726.

Greenberg, L. 1979. Genetic component of bee odor in kin recognition. Science 206:1095-1097.

Greenberg, N. 1976. Thermoregulatory aspects of behavior in the blue spiny lizard Sceloporus cyanogenys (Sauria, Iguanidae). Behaviour 59:1-21.

Greene, E. 1987. Individuals in an osprey colony discriminate between high and low quality information. Nature 329:239-241.

Greene, E., L.J. Orsak, D.W. Whitman. 1987. A tephritid fly mimics the territorial displays of its jumping spider predators. Science 236:310-312.

Greene, H.W., and G.M. Burghardt. 1978. Behavior and phylogeny: constriction in ancient and modern snakes. Science 200:74-77.

Greene, H.W., and R.W. McDiarmid. 1981. Coral snake mimicry: does it occur? Science 21:1207-1212.

Greenfield, M.D. 1981. Moth sex pheromones: an evolutionary perspective. Florida Entomologist 64:4-17.

Greenfield, M.D., and K.C. Shaw. 1983. Adaptive significance of chorusing with special reference to the Orthoptera. In: D.T. Gwynne and G.K. Morris, editors. Orthopteran mating systems: sexual competition in a diverse group of insects. Westview Press, Boulder, Colo.

Greenough, W.T. 1975. Experiential modification of the developing brain. American Scientist 63:37-46.

Greenwald, O.E. 1978. Kinematics and time relations of prey capture by gopher snakes. Copeia 1978(2):263-268.

Greenwood, P.J. 1980. Mating systems, philopatry, and dispersal in birds and mammals. Animal Behaviour 28:1140-1162.

Greenwood, P.J., P.H. Harvey, and M. Slatkin, editors. 1985. Evolution: essays in honour of John Maynard Smith.

Cambridge University Press, Cambridge, England.

Grier, J.W. 1968. Pre-attack behavior of the red-tailed hawk. M.S. thesis. University of Wisconsin, Madison, Wis.

Grier, J.W. 1971. Pre-attack posture of the red-tailed hawk. Wilson Bulletin 83:115-123.

Grier, J.W. 1973. Techniques and results of artificial insemination with eagles. Raptor Research 7:1-12.

Grier, J.W. 1975. Avian spread-winged sunbathing in thermoregulation and drying. Ph.D. thesis. Cornell University, Ithaca, N.Y.

Griffin, D.R. 1955. Bird navigation. In: A. Wolfson, editor. Recent studies in avian biology. University of Illinois Press, Urbana, Ill.

Griffin, D.R. 1958. Listening in the dark. Yale University Press, New Haven, Conn. (Reprinted 1974, Dover, New York.)

Griffin, D.R. 1959. Echoes of bats and men. Doubleday & Co., New York.

Griffin, D.R. 1970. Migrations and homing of bats. In: W.A. Winsatt, editor. Biology of bats. Academic Press, New York.

Griffin, D.R. 1976 a. The question of animal awareness. Rockefeller University Press, New York.

Griffin, D.R. 1976 b. A possible window on the minds of animals. American Scientist 64:530-535.

Griffin, D.R. 1977. Anthropomorphism. Bioscience 27:445-446.

Griffin, D.R. 1981. The question of animal awareness, ed 2. Rockefeller University Press, New York.

Griffin, D.R., editor. 1982. Animal mind-human mind. Springer-Verlag, New York.

Griffin, D.R. 1984 a. Animal thinking. Harvard University Press, Cambridge, Mass.

Griffin, D.R. 1984 b. Animal thinking. American Scientist 72:456-464.

Griffin, D.R., and R.J. Hock. 1949. Airplane observations of homing birds. Ecology 30:176-198.

Griffin, D.R., J. Friend, and F. Webster. 1965. Target discrimination by the echolocation of bats. Journal of Experimental Zoology 158:155-168.

Griffiths, M. 1988. The platypus. Scientific American 258:84-91.

Grillner, S., and P. Wallen. 1985. Central pattern generators for locomotion, with special reference to vertebrates. Annual Review of Neuroscience 8:233-261.

Grinvald, A. 1985. Real-time optical mapping of neuronal activity: from single growth cones to the intact mammalian

Brain. Annual Review of Neuroscience 8:263-305.

Grohmann, J. 1939. Modifikation oder Funktionsreifung? Ein Beitrag zur Karung der wechselseitigen Beziehungen zwischen Instinkthandlung und Erfahrung. Zeitschrift für Tierpsychologie 2:132-144.

Groos, K. 1898. The play of animals. D. Appleton, New York. (Translated by E.L. Baldwin.)

Gross, M.R. 1985. Disruptive selection for alternative life histories in salmon. Nature 313:47-48.

Gross, M.R., and A.M. MacMillan. 1981. Predation and the evolution of colonial nesting in bluegill sunfish (Lepomis macrochirus). Behavioral Ecology and Sociobiology 8:163-174.

Gross, M.R., and R. Shine. 1981. Parental care and mode of fertilization in ectothermic vertebrates. Evolution 35:775-793.

Gross, M.R., and R.C. Sargent. 1985. The evolution of male and female parental care in fishes. American Zoologist 25:807-822.

Gross, M.R., R.M. Coleman, and R.M. McDowall. 1988. Aquatic productivity and the evolution of diadromous fish migration. Science 239:1291-1293.

Grossman, S.P. 1975. Role of the hypothalamus in the regulation of food and water intake. Psychological Review 82:200-224.

Grubb, T.C., Jr. 1974. Olfactory navigation to the nesting burrow in Leach's petrel (Oceanodroma leucorrhoa). Animal Behaviour 22:192-202.

Grubb, T.C., Jr. 1977. Why ospreys hover. Wilson Bulletin 89:149-150.

Guhl, A.M. 1962. The behaviour of chickens. In: E.S.E. Hafez, editor. The behaviour of domestic animals. Balliere, Tindall, and Co., London.

Guilford, T., and M.S. Dawkins. 1987. Search images not proven: a reappraisal of recent evidence. Animal Behaviour 35:1838-1845.

Gurin, J. 1980. Chemical feelings. Science 80 1(1):28-33.

Gustafsson, L. 1987. Interspecific competition lowers fitness in collared flycatchers Ficedula albicollis: an experimental demonstration. Ecology 68:291-296.

Guthrie, E.R., and G.P. Horton. 1946. Cats in a puzzle box. Rinehart, New York.

Guthrie, R.D. 1970. Evolution of human threat display organs. Evolutionary Biology 4:257- 302.

Gottinger, H.R. 1985. Consequences of

domestication on the song structures in the canary. Behaviour 94:254-278.

Gwinner, E. 1986. Circannual rhythms. Springer-Verlag, New York.

Gwynne, D.T. 1981. Sexual difference theory: Mormon crickets show role reversal in mate choice. Science 213:779-780.

Gwynne, D.T. 1982. Mate selection by female ketydids (Orthoptera: Tettigoniidae, Conocephalus nigropleurum). Animal Behaviour 30:734-738.

Gwynne, D.T. 1983. Coy conquistadors of the sagebrush. Natural History 92(10):70-75.

Gwynne, D.T. 1988. Courtship feeding and the fitness of female katydids (Orthoptera: Tettigoniidae). Evolution 42:545-555.

Gyger, M., and P. Marler. 1988. Food calling in the domestic fowl, Gallus gallus: the role of external referents and deception.

Hadidian, J. 1980. Yawning in an old world monkey, Macaca nigra (primates: Cercopithecidae). Behaviour 75:133-147.

Hadley, M.E. 1988. Endocrinology, ed 2. Prentice Hall, Englewood Cliffs, N.J.

Hafez, E.S.E., editor. 1975. Behavior of domestic animals. Williams & Wilkins, Baltimore.

Hage, S.R., and J. Mellen. 1984 Research methods for studying animal behavior in a zoo setting. Minnesota and Washington Park Zoos. (Minnesota Zoo, Education Department, Apple Valley, MN 55124).

Hagedorn, M., and W. Heiligenberg. 1985. Court and spark: electric signals in the courtship and mating of gymnotoid fish. Animal Behaviour 33:254-265.

Haig, S.M., and L.W. Oring. 1988. Distribution and dispersal in the piping plover. Auk 105:630-638.

Hailman, J.P. 1967. The ontogeny of an instinct: the pecking response in chicks of the laughing gull (Larus atricilla L.) and related species. Behaviour Supplement 15.

Hailman, J.P. 1969. How an instinct is learned. Scientific American 221(6):98-106.

Hailman, J.P. 1977. Optical signals. Indiana University Press, Bloomington, Ind.

Hailman, J.P. 1978 a. The question of animal awareness: evolutionary continuity of mental experience. Auk 95:614-615.

Hailman, J.P. 1978 b. Rape among mallards: technical comments. Science 201:280-281.

Hailman, J.P. 1978 c. The behavior of communicating (review). Auk 95:771-774.

Hailman, J.P. 1982. Evolution and behavior: an iconoclastic view. In: H.C. Plotkin, editor. Learning, development, and culture. John Wiley & Sons, New York.

Hailman, J.P. 1988. Operationalism, optimality and optimism: suitabilities versus adaptations of organisms. In M.W. Ho and S.W. Fox, editors: Evolutionary processes and metaphors. John Wiley & Sons, New York.

Hailman, J.P., and J.J.I. Dzelzkalns. 1974. Mallard tail-wagging: punctuation for animal communication? American Naturalist 108:236-238.

Hailman, J.P., M.S. Ficken, and R.W. Ficken. 1987. Constraints on the structure of combinatorial "chick-a-dee" calls. Ethology 75:62-80.

Hainsworth, F.R. 1986. Why hummingbirds hover: a commentary. Auk 103:832-833.

Hainsworth, F.R., and L.L. Wolf. 1978. Regulation of metabolism during torpor in "temperate" zone hummingbirds. Auk 95:197-199.

Hainsworth, F.R., and L.L. Wolf. 1979. Feeding: an ecological approach. Advances in the Study of Behavior 9:53-96.

Halberg, F. 1973. Laboratory techniques and rhythmometry. In: J.N. Mills, editor. Biological aspects of circadian rhythms. Plenum Press, New York.

Haldane, J.B.S., and Spurway, H. 1954. A statistical analysis of communication in *Apis mellifera* and a comparison with communication in other animals. Insectes Soc. 1:247-283.

Haley, T.J., and R.S. Snider, editors. 1964. Responses of the nervous system to ionizing radiation. Second International Symposium of University of California at Los Angeles. Little, Brown, & Co., Boston.

Halgren, E., R.D. Walter, A.G. Cherlow, and P.H. Crandall. Mental phenomena evoked by electrical stimulation of the human hippocampal formation and amygdala. Brain 101:83-117.

Halliday, T. 1982. Sexual strategy. University of Chicago Press, Chicago.

Halliday, T., and S.J. Arnold. 1987. Multiple mating by females: a perspective from quantitative genetics. Animal Behaviour 35:939-941.

Halliday, T.R. 1975. An observational and experimental study of sexual behaviour in the smooth newt, *Triturus vulgaris* (Amphibia: Salamandridae). Animal Behaviour 23:291-322.

Halliday, T.R. 1976. The libidinous newt: an analysis of variations in the sexual behaviour of the male smooth new, *Triturus vulgaris*. Animal Behaviour 24:398-414.

Halliday, T.R. 1978. Sexual selection and mate choice. In: J.R. Krebs and N.B. Davies, editors. Behavioural ecology: an evolutionary approach. Blackwell Scientific Publications, Oxford.

Halliday, T.R. 1987. Physiological constraints on sexual selection. In: J.W. Bradbury and M.B. Andersson, editors. Sexual selection: testing the alternatives. John Wiley & Sons, Chichester, N.J.

Halliday, T.R., and P.J.B. Slater, editors. 1983. Genes, development and learning. Animal Behaviour. Vol. 3. W.H. Freeman, New York.

Hamburger, V., E. Wenger, and R. Oppenheim. 1966. Motility in the chick embryo in the absence of sensory input. Journal of Experimental Zoology 162:133-160.

Hamerstrom, F. 1957. The influence of a hawk's appetite on mobbing. Condor 59:192-194.

Hamilton, W.D. 1963. The evolution of altruistic behavior. American Naturalist 97:354-356.

Hamilton, W.D. 1964. The genetical evolution of social behavior. I and II. Journal of Theoretical Biology 7:1-52.

Hamilton, W.D. 1970. Selfish and spiteful behaviour in an evolutionary model. Nature 1218-1220.

Hamilton, W.D. 1971. Geometry for the selfish herd. Journal of Theoretical Biology 31:295-311.

Hamilton, W.D. 1972. Altruism and related phenomena, mainly in social insects. Annual Review of Ecology and Systematics 3:193-232.

Hamilton, W.D. 1980. Sex vs. non-sex vs. parasite. Oikos 35:282-290.

Hamilton, W.D., and M. Zuk. 1982. Heritable true fitness and bright birds: a role for parasites? Science 218:384-387.

Hamilton, W.D., and R.M. May. 1977. Dispersal in stable habitats. Nature 269:578-581.

Hamilton, W.J. III. 1973. Life's color code. McGraw-Hill, New York.

Hanby, J.P., and J.D. Bygott. 1987. Emigration of subadult lions. Animal Behaviour 35:161-169.

Hanken, J., and P.W. Sherman. 1981. Multiple paternity in Belding's ground squirrels. Science 212:351-353.

Hansell, M.H. 1984. Animal architecture and building behaviour. Longman, London.

Hansen, E.W. 1966. The development of maternal and infant behavior in the rhesus monkey. Behaviour 27:107-149.

Harcourt, A.H. 1978. Activity periods and patterns of social interaction: a neglected problem. Behaviour 66:121-135.

Harcourt, A.H. 1989. Environment, competition and reproductive performance of female monkeys. Trends in Ecology and Evolution 4:101-105.

Harden Jones, F.R. 1968. Fish migration. Edward Arnold, London.

Harden Jones, F.R. 1986. Book review of B.A. McKeown (1984). Animal Behaviour 33:1046.

Hardy, J.W. 1976. Comparative breeding behavior and ecology of the bushy-crested and Nelson san blas jays. Wilson Bulletin 88:96-120.

Harlow, H.F. 1959. Love in infant monkeys. Scientific American 200:68-74.

Harlow, H.F., and R.R. Zimmerman. 1959. Affectional responses in the infant monkey. Science 130:421-432.

Harosi, F.I., and Y. Hashimoto. 1983. Ultraviolet visual pigment in a vertebrate: a tetrachromatic cone system in the dace. Science 222:1021-1023.

Harris, M.P., U.N. Safriel, M. De L. Brooke, and C.K. Britton. 1987. The pair bond and divorce among oystercatchers *Haematopus ostralegus* on Skokholm Island, Wales. Ibis 129:45-57.

Hart, B.L. 1985. The behaviour of domestic animals. W.H. Freeman, New York.

Harth, E., K.P. Unnikrishnan, and A.S. Pandya. 1987. The inversion of sensory processing by feedback pathways: a model of visual cognitive functions. Science 237:184-187.

Hartzler, J.E. 1972. An analysis of sage grouse lek behavior. Ph.D. dissertation, University of Montana, Missoula, Mont.

Hasler, A.D. 1985. Book review of B.A. McKeown (1984). Zeitschrift für Tierpsychologie 69:168-169.

Hasler, A.D. 1986. Book review of R.J.F. Smith (1985). Zeitschrift für Tierpsychologie 70:168-169.

Hasler, A.D., A.T. Scholz, and R.M. Horrall. 1978. Olfactory imprinting and homing in salmon. American Scientist 66:347-355.

Hausfater, G. 1975. Dominance and reproduction in baboons *(Papio cynocephalus):* a quantitative analysis. Contributions to Primatology 7:1-150.

Hausfater, G. 1976. Predatory behavior of yellow baboons. Behaviour 56:44-68.

Hazlett, B.A., and Bossert, W.H. 1965. A statistical analysis of the aggressive communications systems of some hermit crabs. Animal Behaviour 13:357-373.

Hazlett, B.A., editor. 1977. Quantitative

methods in the study of animal behavior. Academic Press, New York.

Heath, J.E. 1965. Temperature regulation and diurnal activity in horned lizards. University of California Publications in Zoology 64:97-136.

Hebb, D.O. 1958. Textbook of psychology. W.B. Saunders, Philadelphia.

Hediger, H. 1964. Wild animals in captivity. Dover, New York.

Hediger, H. 1968. The psychology and behaviour of animals in zoos and circuses. Dover, New York.

Hedrick, A.V. 1986. Female preferences for male calling bout duration in a field cricket. Behavioral Ecology and Sociobiology 19:73-77.

Hedrick, A.V. 1988. Female choice and the heritability of attractive male traits: an empirical study. American Naturalist 132:267-276.

Hegner, R.E., and J.C. Wingfield. 1986 a. Behavioral and endocrine correlates of multiple brooding in the semicolonial house sparrow *Passer domesticus*. I. Males. Hormones and Behavior 20:294-312.

Hegner, R.E., and J.C. Wingfield. 1986 b. Behavioral and endocrine correlates of multiple brooding in the semicolonial house sparrow *Passer domesticus*. II. Females. Hormones and Behavior 20:313-326.

Hegner, R.E., and J.C. Wingfield. 1987 a. Effects of experimental manipulation of testosterone levels on parental investment and breeding success in male house sparrows. Auk 104:462-469.

Hegner, R.E., and J.C. Wingfield. 1987 b. Effects of brood-size manipulations on parental investment, breeding success, and reproductive endocrinology of house sparrows. Auk 104:470-480.

Heid, P., H.R. Guttinger, and E. Prove. 1985. The influence of castration and testosterone replacement on the song architecture of canaries *(Serinus canaria)*. Zeitschrift für Tierpsychologie 69:224-236.

Heinrich, B. 1975. Thermoregulation in bumblebees. II. Energetics of warm-up and free flight. Journal of Comparative and Physiological Psychology 96:155-166.

Heinrich, B. 1979. Bumblebee economics. Harvard University Press, Cambridge, Mass.

Heinrich, B., and G.A. Bartholomew. 1979. The ecology of the African dung beetle. Scientific American 241(5):146-156.

Held, R., and A. Hein. 1963. Movement produced stimulation in the development of visually guided behavior. Journal of Comparative and Physiological Psychology 56:872-876.

Hendrichs, H. 1975. Changes in a population of dikdik *Madoqua (Rhynchotragus) kirki* (Gunther 1880). Zeitschrift für Tierpsychologie 38:55-69.

Hendrichs, J.P. 1986. Sexual selection in wild and sterile Caribbean fruit flies, *Anastrepha suspensa* (Loew) (Diptera: Tephritidae). M.Sc. thesis, University of Florida, Gainesville, Fla.

Hennessy, D.F., and D.H. Owings. 1988. Rattlesnakes create a context for localizing their search for potential prey. Ethology 77:317-329.

Hensler, G.L., S.S. Klugman, and M.R. Fuller. 1986. Portable microcomputers for field collection of animal behavior data. Wildlife Society Bulletin 14:189-192.

Herbers, J.M. 1981. Time resources and laziness in animals. Oecologia 49:252-262.

Herter, K. 1962. Der Temperatursinn der Tiere. Ziensen Verlag, Wittenberg, Germany.

Herzog, H.A., Jr. 1988. The moral status of mice. American Psychologist 43:473-474.

Hess, E.H. 1956. Space perception in the chick. Scientific American 195:71-80.

Hess, E.H. 1962. Ethology: an approach toward the complete analysis of behavior. In: New directions in psychology. Holt, Rinehart & Winston, New York.

Hess, E.H. 1973 a. Imprinting. Van Nostrand Reinhold, New York.

Hess, E.H. 1973 b. Comparative sensory processes. In: D.A. Dewsbury and D.A. Rethlingshafer, editors. Comparative psychology. McGraw-Hill, New York.

Hess, E.H., and S.B. Petrovich, editor. 1977. Imprinting. Dowden, Hutchinson, & Ross, Stroudsburg, Pa.

Hetherington, A.W., and S.W. Ranson. 1940. Hypothalamic lesions and adiposity in the rat. Anatomical Record 78:149.

Hetherington, A.W., and S.W. Ranson. 1942. Effect of early hypophysectomy on hypothalamic obesity. Endocrinology 31:30-34.

Heymer, A. 1977. Ethologisches Wörterbuch/Ethological Dictionary/Vocabulaire Ethologique. Paul Parey, Berlin.

Hibbard, L.S., J.S. McGlone, D.W. Davis, R.A. Hawkins. 1987. Three-dimensional representation and analysis of brain energy metabolism. Science 236:1641-1646.

Hickey, J.J., editor. 1969. Peregrine falcon populations: their biology and decline. University of Wisconsin Press, Madison, Wis.

Hickman, C.P., Jr., L.S. Roberts, and F.M. Hickman. 1988. Integrated principles of zoology, ed 8. Times Mirror/Mosby College Publishing, St. Louis.

Hill, J.E., and J.D. Smith. 1984. Bats, a natural history. University of Texas Press, Austin, Tex.

Hinde, R.A. 1956 a. Ethological models and the concept of drive. British Journal of Philosophy and Science 6:321.

Hinde, R.A. 1956 b. The biological significance of the territories of birds. Ibis 98: 340-369.

Hinde, R.A. 1970. Animal Behaviour, ed 2. McGraw-Hill, New York.

Hinde, R.A. 1977. Mother-infant separation and the nature of inter-individual relationships: experiments with rhesus monkeys. Proceedings of the Royal Society of London B 196:29-50.

Hinde, R.A. 1981. Animal signals: ethological and games-theory approaches are not incompatible. Animal Behaviour 29:535-542.

Hinde, R.A. 1985. Was "the Expression of the Emotions" a misleading phrase? Animal Behaviour 33:985-992.

Hinde, R.A., and J. Fisher. 1951. Further observations on the opening of milk bottles by birds. British Birds 44(12):393-396.

Hindmarsh, A.M. 1984. Vocal mimicry in starlings. Behaviour 90:302-324.

Hindmarsh, A.M. 1986. The functional significance of vocal mimicry in song. Behaviour 99:87-100.

Hobson, J.A., and R.W. McCarley. 1977. The brain as a dream state generator: an activation-synthesis hypothesis of the dream process. American Journal of Psychology 134(12):1335-1348.

Hobson, J.A., T. Spagna, and R. Malenka. 1978. Ethology of sleep studied with time-lapse photography: postural immobility and sleep-cycle phase in humans. Science 201:1251-1253.

Hodges, C.M. 1981. Optimal foraging in bumblebees: hunting by expectation. Animal Behaviour 29:1166-1171.

Hodos, W., and C.B.G. Campbell. 1969. Scala naturae: why there is no theory in comparative psychology. Psychological Review 76:337-350.

Hoebel, B.G., and P. Teitelbaum. 1966. Weight regulation in normal and hypothalamic hyperphagic rats. Journal of Comparative and Physiological Psychology 61:189-193.

Holden, C. 1981. Human-animal relationship under scrutiny. Science 214:418-420.

Hölldobler, B. 1967. Zur Physiologie der Gast-Wirt-Beziehungen (Myrmecophilie) bei Ameisen: I, das Gastverhältnis der *Atemeles* und *Lomechusa*-Larven (Col. Staphlinidae) zu *Formica* (Hym. Formicidae). Zeitschrift für Vergleichende Physiologie 56:1-21.

Hölldobler, B. 1969. Orientierungsmechanismen des Ameisengastes *Atemeles* (Coleoptera, Staphlinidae) bei der Wirtssuche. Verh. Deutsch. Zool. Gesell. 33:580-585.

Hölldobler, B. 1970. Zur Physiologic der Gast-Wirt-Beziehungen (Myrmecophilie) bei Ameisen: II, das Gastverhaltnis des imaginalen *Atemeles pubicollis* Bris. (Col. Staphylinidae) zu *Myrmica* und *Formica* (Hym. Formicidae). Zeitschrift für Vergleichende Physiologie 66:215-250.

Hölldobler, B. 1971 a. Recruitment behavior in *Camponotus socius* (Hym. Formicidae). Zeitschrift für Vergleichende Physiologie 75:123-142.

Hölldobler, B. 1971 b. Communication between ants and their guests. Scientific American 224(3):85-93.

Hölldobler, B. 1976. Tournaments and slavery in a desert ant. Science 192:912-914.

Holling, C.S. 1959. The components of predation as revealed by a study of small-mammal predation of the European pine saw-fly. Canadian Entomologist 91:293-320.

Holling, C.S. 1966. The functional response of invertebrate predators to prey density. Memoirs of the Entomological Society of Canada 48:1-86.

Holmes, W.G., and P.W. Sherman. 1983. Kin recognition in animals. American Scientist 71:46-55.

Holmes, W.N., and J.G. Phillips. 1976. The adrenal cortex of birds. In: I. Chester-Jones and I.W. Henderson, editors. General, comparative and clinical endocrinology of the adrenal cortex. Academic Press, New York.

Holt, E.B. 1931. Animal drive and the learning process. Holt, Rinehart & Winston, New York.

Hoogland, J.L. 1979 a. Aggression, ectoparasitism, and other possible costs of prairie dog (Sciuridae, *Cynomys* spp.) coloniality. Behaviour 69:1-35.

Hoogland, J.L. 1979 b. The effect of colony size on individual alertness of prairie dogs (Sciuridae: *Cynomys* spp.). Animal Behaviour 27:394-407.

Hoogland, J.L. 1982. Prairie dogs avoid extreme inbreeding. Science 215:1639-1641.

Hoogland, J.L. 1983. Nepotism and alarm calling in the black-tailed prairie dog (*Cynomys ludovicianus*). Animal Behaviour 31:472-479.

Hoogland, J.L. 1985. Infanticide in prairie dogs: lactating females kill offspring of close kin. Science 230:1037-1040.

Hopfield, J.J. and D.W. Tank. 1986. Computing with neural circuits: a model. Science 233:625-633.

Hopkins, C.D. 1974. Electric communication in the reproductive behavior of *Sternopygus macrurus* (Gymnotoidei). Zeitschrift für Tierpsychologie 35:518-535.

Hopkins, C.D. 1988. Neuroethology of electric communication. Annual Review of Neuroscience 11:497-535.

Hopson, J. 1980. Growl, bark, whine, and hiss. Science 80 1:81-85.

Horn, H.S. 1968. The adaptive significance of colonial nesting in the Brewer's blackbird (*Euphagus cyanocephalus*). Ecology 49:682-694.

Horn, H.S. 1983. Some theories about dispersal. In: I.R. Swingland and P.J. Greenwood, editors. The ecology of animal movement. Clarendon Press, Oxford, England.

Horridge, G.A. 1965. Intracellular action potentials associated with the beating of the cilia in ctenophore comb plate cells. Nature 205:602.

Hotta, Y., and S. Benzer. 1970. Genetic dissection of the *Drosophila* nervous system by means of mosaics. Proceedings of the National Academy of Sciences of the United States of America 67:1156-1163.

Hotta, Y., and S. Benzer. 1972. Mapping of behavior in *Drosophila* mosaics. Nature 240:527-535.

Howard, H.E. 1920. Territory in bird life. John Murray, London.

Howard, R.D. 1978. The evolution of mating strategies in bullfrogs, *Rana catesbiana*. Evolution 32:850-871.

Howse, P.E. 1970. Termites: a study in social behaviour. Hutchinson University Library, London.

Hoyt, D.F., and Taylor, C.R. 1981. Gait and the energetics of locomotion in horses. Nature 191:239-240.

Hrdy, S.B. 1977 a. Infanticide as a primate reproductive strategy. American Scientist 65:40-49.

Hrdy, S.B. 1977 b. The langurs of Abu: female and male strategies of reproduction. Harvard University Press, Cambridge, Mass.

Hrdy, S.B. 1979. Infanticide among animals: a review, classification, and examination of the reproductive strategies of females. Ethology and Sociobiology 1:13-40.

Hrdy, S.B. 1981. The woman that never evolved. Harvard University Press, Cambridge, Mass.

Hubel, D.H. 1963. The visual cortex of the brain. Scientific American 209:54-62.

Hubel, D.H. 1979. The brain. Scientific American 241(3):45-53.

Hubel, D.H., and T.N. Wiesel. 1979. Brain mechanisms of vision. Scientific American 241(3):150-162.

Huber, P. 1810. Recherches sur les Moeurs des Fourmis Indigenes. J.J. Paschoud, Paris.

Huey, R.B., and E.R. Pianka. 1977. Natural selection for juvenile lizards mimicking noxious beetles. Science 195:201-203.

Hughes, G.M. 1957. The coordination of insect movements. II. The effect of limb amputation and the cutting of commissures in the cockroach (*Blatta orientalis*). Journal of Experimental Biology 34:306-333.

Hunsaker, D. 1962. Ethological isolating mechanisms in the *Sceloporus torquatus* group of lizards. Evolution 16:62-74.

Hunter, M.L., Jr. 1980. Microhabitat selection for singing and other behaviour in great tits, *Parus major:* some visual and acoustical considerations. Animal Behaviour 28:468-475.

Hunter, M.L. Jr., and J.R. Krebs. 1979. Geographical variation in the song of the great tit (*Parus major*) in relation to ecological factors. Journal of Animal Ecology. 48:759-785.

Huntingford, F., and A. Turner. 1987. Animal conflict. Chapman and Hall, London.

Huntingford, F.A. 1976. The relationship between anti-predator behaviour and aggression among conspecifics in the three-spined stickleback, *Gasterosteus aculeatus*. Animal Behaviour 24:245-260.

Hutchinson, G.E. 1981. Random adaptation and imitation in human evolution. American Scientist 69:161-165.

Hutt, C. 1966. Exploration and play in children. In: P.A. Jewell and C. Loizos, editors. Play, exploration, and territory in mammals. Academic Press, London.

Huxley, J.S. 1914. The courtship habits of the great crested grebe (*Podiceps cristatus*), with an addition to the theory of sexual selection. Proceedings of the Zoological Society of London 2:491-562.

Huxley, J.S. 1938. The present standing of the theory of sexual selection. In: G.R. de Beer, editor. Evolution. Clarendon Press, Oxford.

Hyatt, G.W. and M. Salmon. 1978. Combat in the fiddler crabs *Uca pugilator* and *U. pugnax:* a quantitative analysis. Behaviour 65:182-211.

Immelmann, K. 1972. Sexual and other

long-term aspects of imprinting in birds and other species. Advances in the Study of Behavior 4:147-174.

Immelmann, K. 1980. Introduction to ethology. Plenum Press, New York.

Immelmann, K., J.P. Hailman, J.R. Baylis. 1982. Reputed band attractiveness and sex manipulation in zebra finches. Science 215:422.

Ingle, D., and D. Crews. 1985. Vertebrate neuroethology: definitions and paradigms. Annual Review of Neuroscience 8:457-494.

Inglis, I.R., and N.J.K. Ferguson. 1986. Starlings search for food rather than eat freely-available, identical food. Animal Behaviour 34:614-617.

Ioale, P. 1980. Further investigations on the homing behaviour of pigeons subjected to reverse wind direction at home loft. Monit. Zool. Ital. 14:77-87.

Ioale, P., F. Papi, V. Fiaschi, and N.E. Baldaccini. 1978. Pigeon navigation: effects upon homing behaviour by reversing wind direction at the loft. Journal of Comparative Physiology 128:285-295.

Irwin, R.E. 1988. The evolutionary importance of behavioural development: the ontogeny and phylogeny of bird song. Animal Behaviour 36:814-824.

Isack, H.A., and H.U. Reyer. 1989. Honeyguides and honey gatherers: Interspecific communication in a symbiotic relationship. Science 243:1343-1346.

Ito, Y. 1989. The evolutionary biology of sterile soldiers in aphids. Trends in Ecology and Evolution 4:69-73.

Iversen, L.L. 1979. The chemistry of the brain. Scientific American 241(3):134-149.

Jacobs, C.H., N.E. Collias, and J.T. Fujimoto. 1978. Nest colour as a factor in nest selection by female village weaverbirds. Animal Behaviour 26:463-469.

Jacobs, J. 1981. How heritable is innate behaviour? Zeitschrift für Tierpsychologie 55:1-18.

Jacobs, M.E. 1955. Studies on territorialism and sexual selection in dragonflies. Ecology 36:566-586.

Jaeger, R.G., R.G. Joseph, and D.E. Barnard. 1981. Foraging tactics of a terrestrial salamander: sustained yield in territories. Animal Behaviour 29:1100-1105.

Jaenike, J. 1978. An hypothesis to account for the maintenance of sex within populations. Evolutionary Theory 3:191-194.

James, F.C. 1971. Ordinations of habitat relationships among breeding birds. Wilson Bulletin 83:215-236.

James, W. 1890. Principles of psychology. Macmillan, New York.

Janzen, D.H. 1966. Coevolution of mutualism between ants and acacias in Central America. Evolution 20:249-275.

Janzen, D.H. 1967. Interaction of the bull's-horn acacia (Acacia cornigera L.) with an ant inhabitant (Pseudomyrmex furruginea F. Smith) in eastern Mexico. University of Kansas Science Bulletin 47:315-558.

Jaynes, J. 1969. The historical origins of "ethology" and "comparative psychology." Animal Behaviour 17:601-606.

Jennings, H.S. 1905. Modifiability in behavior. I. Behavior of sea anemones. Journal of Experimental Zoology 2:447-473.

Jennings, H.S. 1906. Behavior of the lower organisms. Columbia University Press, New York.

Jennings, H.S. 1907. Behaviour of starfish Asterias forreri. University of California Publications in Zoology 4:53-185.

Jennings, T., and S.M. Evans. 1980. Influence of position in the flock and flock size on vigilance in the starling, Sturnus vulgaris. Animal Behaviour 28:634-635.

Jensen, D.D. 1967. Polythetic operationism and the phylogeny of learning. In: W.C. Corning and S.C. Ratner, editors. Chemistry of learning. Plenum Press, New York.

Jeronen, E., R. Iosmetsa, R. Hissa, and A. Pyornila. 1976. Effect of acute temperature stress on the plasma catecholamine, corticosterone and metabolite levels in the pigeon. Comparative Biochemistry and Physiology 55C:17-22.

Jewell, P.A. 1966. The concept of home range in mammals. Zoological Society of London, Symposia 18:85-109.

Jewell, P.A., and C. Loizos, editors. 1966. Play, exploration and territory in mammals. Zoological Society of London, Symposia 18:1-280.

John, E.R., Y. Tang, A.B. Brill, R. Young, and K. Ono. 1986. Double-labeled metabolic maps of memory. Science 233:1167-1175.

Johnsgard, P.A. 1983. Cranes of the World. Indiana University Press, Bloomington, Ind.

Johnson, C.H., and J.W. Hastings. 1986. The elusive mechanism of the circadian clock. American Scientist 74:29-36.

Johnson, D.H. 1980. The comparison of usage and availability measurements for evaluating resource preference. Ecology 61:65-71.

Johnson, D.H. 1981. The use and misuse of statistics in wildlife habitat studies.

In: D.E. Capen, editor. The use of multivariate statistics in studies of wildlife habitat. U.S. Department of Agriculture Forest General Technical Report RM-87, Fort Collins, Colo.

Johnson, V.R., Jr. 1977. Individual recognition in the banded shrimp Stenopus hispidus (Olivier). Animal Behaviour 25:418-428.

Johnston, R.E., and T. Schmidt. 1979. Responses of hamsters to scent marks of different ages. Behavioral and Neural Biology 26:64-75.

Jordan, R.H., and G.M Burghardt. 1986. Employing an ethogram to detect reactivity of black bears (Ursus americanus) to the presence of humans. Ethology 73:89-115.

Jukes, T.H. 1980. Silent nucleotide substitutions and the molecular evolutionary clock. Science 210:973-978.

Kacelnik, A. 1979. The foraging efficiency of great tits (Parus major L.) in relation to light intensity. Animal Behaviour 27:237-241.

Kalat, J.W. 1977. Biological significance of food aversion learning. In: N.W. Milgram, L. Krames, and T.M. Alloway, editors. Food aversion learning. Plenum Press, New York.

Kamil, A.C., and T.D. Sargent, editors. 1981. Foraging behavior: ecological, ethological, and psychological approaches. Garland Publishing, New York.

Kamil, A.C., J.R. Krebs, and H.R. Pulliam. 1987. Foraging behavior. Plenum Press, New York.

Kamil, A.C., S.I. Yoerg, and K.C. Clements. 1988. Rules to leave by: patch departure in foraging blue jays. Animal Behaviour 36:843-853.

Kandel, E.R. 1976. Cellular basis of behavior. W.H. Freeman, San Francisco.

Kandel, E.R. 1979 a. Behavioral biology of Aplysia. W.H. Freeman, San Francisco.

Kandel, E.R. 1979 b. Small systems of neurons. Scientific American 241(3):66-76.

Kandel, E.R., and J.H. Schwartz, editors. 1985. Principles of neural science. Elsevier, New York.

Karplus, I. 1979. The tactile communication between Cryptocentrus steinitzi (Pisces, Gobiidae) and Alpheus purpurilenticularis (Crustacea, Alpheidae). Zeitschrift für Tierpsychologie 49:173-196.

Karplus, I., M. Tsurnamal, R. Szlep, and D. Algom. 1979. Film analysis of the tactile communication between Cryptocentrus steinitzi (Pisces, Gobiidae) and Alpheus purpurilenicularis (Crustacea,

Alpheidae), Zeitschrift für Tierpsychologie 49:337-351.

Kavaliers, M., M. Hirst, G.C. Teskey. 1983. A functional role for an opiate system in snail thermal behavior. Science 220:99-101.

Kavanagh, M.W. 1987. The efficiency of sound production in two cricket species, *Gryllotalpa australis* and *Teleogryllus commodus* (Orthoptera: Grylloidea). Journal of Experimental Biology 130: 107-119.

Keeler, C.E., and H.O. King. 1942. Multiple effects of coat color genes in the Norway rat, with special reference to temperament and domestication. Journal of Comparative and Physiological Psychology 34:241-250.

Keeton, W.T. 1974. The orientational and navigational basis of homing in birds. Advances in the Study of Behavior 5:47-132.

Keeton, W.T. 1974 a. The mystery of pigeon homing. Scientific American 231(6):96-107.

Keeton, W.T. 1974 b. The orientational and navigational basis of homing in birds. Advances in the Study of Behavior 5:47-132.

Keiper, R.R., and H.H. Sambraus. 1986. The stability of equine dominance hierarchies and effects of kinship, proximity, and foaling status on hierarchy rank. Applied Animal Behaviour Science 16:121-130.

Kelley, D. 1980. Auditory and vocal nuclei in the frog brain concentrate sex hormones. Science 207:553-555.

Kelley, D.B. 1988. Sexually dimorphic behaviors. Annual Review of Neuroscience 11:225-251.

Kendrick, K.M., and B.A. Baldwin. 1987. Cells in temporal cortex of conscious sheep can respond preferentially to the sight of faces. Science 236:448-450.

Kennedy, C.E.J., J.A. Endler, S.L. Poynton, and H. McMinn. 1987. Parasite load predicts mate choice in guppies. Behavioral Ecology and Sociobiology 21:291-295.

Kennedy, G.C. 1950. The hypothalamic control of food intake in rats. Proceedings of the Royal Society of London 137:535-548.

Kennedy, R.J. 1969. Sunbathing behaviour in birds. Br. Birds 62:249-258.

Kenward, R.E. 1978. Hawks and doves: factors affecting success and selection in goshawk attacks on wood-pigeons. Journal of Animal Ecology 47:449-460.

Kessel, B. 1953. Distribution and migration of the European starling in North America. Condor 55:49-67.

Kessel, E.L. 1955. Mating activities of balloon flies. Systematic Zoology 4:97-104.

Kevan, P.G. 1975. Sun-tracking solar furnaces in high arctic flowers: significance for pollination and insects. Science 189:723-726.

Kevan, P.G. 1976. Sir Thomas More on imprinting: observations from the sixteenth century. Animal Behaviour 24:16-17.

Kiester, E., Jr. 1980. Images of the night. Science 80 1:36-43.

Kihlstrom, J.F. 1987. Response to letters. Science 238:1638

Kihlstrom, J.F. 1987. The cognitive unconscious. Science 237:1445-1452.

Kiis, A., and A.P. Moller. 1986. A field test of food information transfer in communally roosting greenfinches *Carduelis chloris*. Animal behaviour 34:1251-1255.

Kiley, M. 1972. The vocalizations of ungulates, their causation and function. Zeitschrift für Tierpsychologie 31:171-222.

Kiley-Worthington, M. 1976. The tail movements of ungulates, canids, and felids with particular reference to their causation and function as displays. Behaviour 56:69-115.

Kiley-Worthington, M. 1977. Behavioural problems of farm animals. Oriel Press, Boston.

King, A.P., M.J. West, and D.H. Eastzer. 1986. Female cowbird song perception: evidence for different developmental programs within the same subspecies. Ethology 72:89-98.

King, J.A., D. Maas, and R.G. Weisman. 1964. Geographic variation in nest size among species of *Peromyscus*. Evolution 18:230-234.

Kingett, P.D., and D.M. Lambert. 1981. Does mate choice occur in *Drosophila melanogaster?* Nature 293:492.

Kirkpatrick, M. 1982. Sexual selection and the evolution of female choice. Evolution 36:1-12.

Kirkpatrick, M., and C.D. Jenkins. 1989. Genetic segregation and the maintenance of sexual reproduction. Nature 339:300-301.

Kirschvink, J.L., D.S. Jones, and B.J. MacFadden, editors. 1985. Magnetite biomineralization and magnetoreception in organisms: a new biomagnetism. Geobiology 5. Plenum Press, New York.

Kitcher, P. 1985. Vaulting ambition. MIT Press: Cambridge, Mass.

Klein, D.B. 1970. A history of scientific psychology. Basic Books, New York.

Klopfer, P.H. 1963. Behavioral aspects of habitat selection: the role of early experience. Wilson Bulletin 75(1):15-22.

Klopfer, P.H. 1965. Behavioral aspects of habitat selection: a preliminary report on stereotypy in foliage preferences of birds. Wilson Bulletin 77(4):376-381.

Klopfer, P.H. 1969. Habitats and territories. Basic Books, New York.

Klopfer, P.H. 1974. An introduction to animal behavior: ethology's first century, revised ed. Prentice-Hall, Englewood Cliffs, N.J.

Klopfer, P.H. 1985. On central controls for aggression. In: P.P.G. Bateson and P.H. Klopfer. Perspectives in ethology. Vol. 6. Plenum Press, New York.

Klopfer, P.H., and J.P. Hailman. 1965. Habitat selection in birds. Advances in the Study of Behavior 1:279-303.

Knight, R.L., and S.A. Temple. 1986 a. Methodological problems in studies of avian nest defence. Animal Behaviour, 34:561-566.

Knight, R.L., and S.A. Temple. 1986 b. Why does intensity of avian nest defense increase during the nesting cycle? Auk 103:318-327.

Knowlton, N. 1974. A note on the evolution of gamete dimorphism. Journal of Theoretical Biology 46: 283-285.

Knowlton, N. 1979. Reproductive synchrony, parental investment and the evolutionary dynamics of sexual selection. Animal Behaviour 27:1022-1033.

Knowlton, N., and G.A. Parker. 1979. An evolutionary stable strategy approach to indiscriminate spite. Nature 279:419-421.

Knowlton, N., and S.R. Greenwell. 1984. Male sperm competition avoidance mechanisms: the influence of female interests. In: R.L. Smith, editor. Sperm competition and the evolution of animal mating systems. Academic Press, New York.

Knudson, E.I., and M. Konishi. 1978. A neural map of auditory space in the owl. Science 200:795-797.

Kobler, J.B., S.F. Isbey, and J.H. Casseday. 1987. Auditory pathways to the frontal cortex of the mustache bat, *Pteronotus parnellii*. Science 236:824-826 (plus cover illustration).

Kolata, G. 1985. Genes and biological clocks. Science 230:1151-1152.

Konishi, M. 1965. The role of auditory feedback in the control of vocalization in the white-crowned sparrow. Zeitschrift für Tierpsychologie 22:770-783.

Konishi, M. 1986. Centrally synthesized maps of sensory space. Trends in Neuroscience 9:163-168.

Konishi, M., and E.I. Knudsen. 1979. The oilbird: hearing and echolocation. Science 204:425-427.

Kosslyn, S.M. 1988. Aspects of a cognitive

neuroscience of mental imagery. Science 240:1621-1626.

Kovach, J.K. 1980. Mendelian units of inheritance control color preferences in quail chicks *(Coturnix coturnix japonica)*. Science 207:549-551.

Kovach, J.K., and G.C. Wilson. 1981. Behaviour and pleiotropy: generalization of gene effects in the colour preferences of Japanese quail chicks *(C. Coturnix japonica)*. Animal Behaviour 29:746-759.

Krajewski, C. 1989. Phylogenetic relationships among cranes (Gruiformes, Gruidae) based on DNA hybridization. Auk 106:603-618.

Kraly, F.S., W.S. Carty, S. Resnick, and G.P. Smith. 1978. Effect of cholecystokinin on meal size and intermeal interval in the sham-feeding rat. Journal of Comparative and Physiological Psychology 92:697-707.

Kramer, B. 1978. Spontaneous discharge rhythms and social signalling in the weakly electric fish *Pollimyrus isidori* (Cuvier et Valenciennes) (Mormyridae, Teleostei). Behavioral Ecology and Sociobiology 4:61-74.

Kramer, D.L., and W. Nowell. 1980. Central place foraging in the eastern chipmunk, *Tamias striatus*. Animal Behaviour 28:772-778.

Kramer, G. 1953. Die Sonnenorientiering der Vogel. Verh. Deut. Zool. Ges. Freiburg 1952:72-84.

Kramer, G. 1957. Experiments in bird orientation and their interpretation. Ibis 99:196-227.

Krasne, F.B. 1973. Learning in Crustacea. In: W.C. Corning, J.A. Dayal, and A.O.D. Willows, editors. Invertebrate learning. Vol. 2. Plenum Press, New York.

Krebs, C.J. 1985. Ecology. Harper & Row, New York.

Krebs, H.A. 1975. The August Krogh principle: "For many problems there is an animal on which it can be most conveniently studied." Journal of Experimental Zoology 194:221-226.

Krebs, J., and R.M. May. 1976. Social insects and the evolution of altruism. Nature 260:9-10.

Krebs, J.R. 1971. Territory and breeding density in the great tit, *Parus major* L. Ecology 52:2-22.

Krebs, J.R. 1973. Behavioral aspects of predation. In: P.P.G. Bateson and P.H. Klopfer, editors. Perspectives in ethology. Vol. 1. Plenum Press, New York.

Krebs, J.R. 1974. Colonial nesting and social feeding as strategies for exploiting food resources in the great blue heron *(Ardea herodias)*. Behaviour 51:100-131.

Krebs, J.R. 1977. Song and territory in the great tit *Parus major*. In: B. Stonehouse and C.M. Perrins, editors. Evolutionary ecology. Macmillan, London.

Krebs, J.R. 1978. Optimal foraging: decision rules for predators. In: J.R. Krebs and N.B. Davies, editors. Behavioural ecology, an evolutionary approoach. Blackwell Scientific Publications, London.

Krebs, J.R., and D.E. Kroodsma. 1980. Repertoires and geographical variation in bird song. Advances in the Study of Behavior 11:143-177.

Krebs, J.R., and N.B. Davies, editors. 1978. Behavioural ecology, an evolutionary approach. Blackwell Scientific Publications, London.

Krebs, J.R., and N.B. Davies. 1987. An introduction to behavioural ecology, ed 2. Sinauer Associates, Sunderland, Mass.

Krebs, J.R., and R. Dawkins. 1984. Animal signals: mind-reading and manipulation. In: J.R. Krebs and N.B. Davies, editors. Behavioural ecology: an evolutionary approach, ed 2. Sinauer Associates, Sunderland, Mass.

Krebs, J.R., and R.H. McCleery. 1984. Optimization in behavioural ecology. In: J.R. Krebs and N.B. Davies, editors. Behavioural ecology. An evolutionary approach, ed 2. Sinauer Associates, Sunderland, Mass.

Krebs, J.R., J.C. Ryan, and E.L. Charnov. 1974. Hunting by expectation or optimal foraging? A study of patch use by chickadees. Animal Behaviour 22:953-964.

Krebs, J.R., J.T. Erichsen, M.L. Webber, and E.L. Charnov. 1977. Optimal prey selection in the great tit *(Parus major)*. Animal Behaviour 25:30-38.

Krebs, J.R., M.H. MacRoberts, and J.M. Cullen. 1972. Flocking and feeding in the great tit *Parus major*—an experimental study. Ibis 114:507-530.

Krebs, J.R., R. Ashcroft, and M. Webber. 1978. Song repertoires and territory defense in the great tit. Nature 211:539-542.

Kreithen, M.L., and W.T. Keeton. 1974. Detection of polarized light by the homing pigeon, *Columbia livia*. Journal of Comparative and Physiological Psychology 89:83-92.

Krieger, D.T. 1983. Brain peptides: what, where, and why? Science 222:975-985.

Krieger, D.T. 1984. Response to Anctil (1984) concerning invertebrate endocrine systems. Science 224:240.

Kruuk, H. 1972. The spotted hyena. University of Chicago Press, Chicago.

Kucharski, D., and W.G. Hall. 1987. New routes to early memories. Science 238:786-788.

Kummer, H. 1968. Social organization of hamadryas baboons. University of Chicago Press, Chicago.

Kummer, H. 1971. Primate societies. Aldine, Atherton, Chicago.

Kung, C., S.Y. Chang, Y. Satow, J. Van Houten, and H. Hansma. 1975. Genetic dissection of behavior in *Paramecium*. Science 188:898-904.

Kuo, Z.Y. 1939. Studies in the physiology of the embryonic nervous system. II. Experimental evidence on the controversy over the reflex theory in development. Journal of Comparative Neurology 70:437-459.

Kuo, Z.Y. 1967. The dynamics of behavior development. Random House, New York.

Kurz, E.M., D.R. Sengelaub, and A.P. Arnold. 1986. Androgens regulate the dendritic length of mammalian motorneurons in adulthood. Science 232:395-398.

Kushlan, J.A. 1978. Nonrigorous foraging by robbing egrets. Ecology 59:649-653.

Kuterbach, D.A., and B. Walcott. 1982. Iron-containing cells in the honey bee *(Apis mellifera)*. Science 218:695-697.

Kutsch, W. 1974. The influence of the wing sense organs on the flight motor pattern in maturing adult locusts. Journal of Comparative and Physiological Psychology 88:413-424.

Lack, D. 1933. Habitat selection in birds with special reference to the effects of afforestation on the Breckland avifauna. Journal of Animal Ecology 2:239-262.

Lack, D. 1943. The life of the robin. H.F. & G. Witherby, London.

Lack, D. 1966. Population studies of birds. Oxford University Press, Oxford, England.

Lack, D. 1968. Ecological adaptations for breeding in birds. Methuen, London.

Lade, B.I., and W.H. Thorpe. 1964. Dove songs as innately coded patterns of specific behaviour. Nature 202:366-368.

Lahue, R., and W.C. Corning. 1975. Synthesis: a comparative look at vertebrates. In: W.C. Corning, J.A. Dyal, and A.O.D. Willows, editors. Invertebrate learning. Vol. 3. Plenum Press, New York.

Laidlaw, H.H., Jr., and R.E. Page, Jr. 1984. Polyandry in honey bees *(Apis mellifera* L.): sperm utilization and intracolony genetic relationships. Genetics 108:985-997.

Lall, A.B., H.H. Seliger, W.H. Biggley,

and J.E. Lloyd. 1980. Ecology of colors of firefly bioluminescence. Science 210:560-562.

Lamb, T. 1987. Call site selection in a hybrid population of treefrogs. Animal Behaviour 35:1140-1144.

Landau, H.G. 1951 a. On dominance relations and the structure of animal societies: I. The effect of inherent characteristics. Bulletin of Mathematical Biophysics 13:1-19.

Landau, H.G. 1951 b. On dominance relations and the structure of animal societies: II. Some effects of possible social factors. Bulletin of Mathematical Biophysics 13:245-266.

Landau, H.G. 1968. Models of social structure. Bulletin of Mathematical Biophysics 30:215-224.

Lande, R. 1981. Models of speciation by sexual selection of polygenic traits. Proceedings of the National Academy of Sciences of the United States of America 78:3721-3725.

Lang, J.W. 1987. Crocodilian thermal selection. In: G.J.W. Webb, S.C. Manolis, and P.J. Whitehead, editors. Wildlife management: crocodiles and alligators. Surrey Beatty and Sons, Northern Territory, Australia.

Lang, J.W. 1988. Crocodilian behaviour: implications for management. In: G.J.W. Webb, S.C. Manolis, and P. J. Whitehead, editors. Wildlife management: crocodiles and alligators. Surrey Beatty and Sons, Northern Territory, Australia.

Lashley, K. 1950. In search of the engram. Society for Experimental Biology, Symposia 4:454-482.

Lauder, G.V. 1986. Homology, analogy, and the evolution of behavior. In: M.H. Nitecki, and J.A. Kitchell, editors. Evolution of animal behavior: paleontological and field approaches. Oxford University Press, New York.

Laufer, H., and R.G.H. Downer, editors. 1988. Endocrinology of selected invertebrate types. Invertebrate endocrinology. Vol. 2. Liss, New York.

Laven, H. 1940. Beiträge zur Biologie des Sandregenpfeifers (Charadrius hiaticula L.). J. Ornithologie 88:183-288.

Laverty, T.M., and R.C. Plowright. 1988. Flower handling by bumblebees: a comparison of specialists and generalists. Animal Behaviour 36:733-740.

Lawrence, E.S. 1985. Evidence for search image in blackbirds Turdus merula L.: long-term learning. Animal Behaviour 33:1301-1309.

Lawrence, E.S. 1985. Evidence for search image in blackbirds (Turdus merula L.):

short-term learning. Animal Behaviour 33:929-937.

Layzer, D. 1974. Heritability analyses of IQ scores: science or numerology? Science 183:1259-1266.

Le Magnen, J. 1986. Hunger. Cambridge University Press, Cambridge, England.

Le Moli, F., and A. Mori. 1984. The effect of early experience on the development of "aggressive" behaviour in Formica lugubris Zett. (Hymenoptera: Formicidae). Zeitschrift für Tierpsychologie 65:241-249.

Leahey, T.H. 1987. A history of psychology. Prentice-Hall, Englewood Cliffs, N.J.

LeBouef, B.J., and R.S. Peterson. 1969. Social status and mating activity in elephant seals. Science 163:91-93.

Lee, P.C. 1984. Ecological constraints on the social development of vervet monkeys. Behaviour 91:245-258.

Lefebvre, Louis. 1986. Cultural diffusion of a novel food-finding behaviour in urban pigeons: an experimental field test. Ethology 71:295-304.

Leger, D.W. and D.H. Owings. 1978. Responses to alarm calls by California ground squirrels: effects of call structure and maternal status. Behavioral Ecology and Sociobiology 3:177-186.

Leger, D.W., D.H. Owings, and D.L. Gelfand. 1980. Single-note vocalizations of California ground squirrels: graded signals and situation-specificity of predator and socially evoked calls. Zeitschrift für Tierpsychologie 52:227-246.

Leger, D.W., D.H. Owings, and L.M. Boal. 1979. Contextual information and differential responses to alarm whistles in California ground squirrels. Zeitschrift für Tierpsychologie 49:142-155.

Leger, D.W., S.D. Berney-Key, and P.W. Sherman. 1984. Vocalizations of Belding's ground squirrels (Spermophilus beldingi). Animal Behaviour 32:753-764.

Lehner, P.N. 1979. Handbook of ethological methods. Garland Publishing, New York.

Lehrman, D.S. 1953. A critique of Konrad Lorenz's theory of instinctive behavior. Quarterly Review of Biology 28:337-363.

Lehrman, D.S. 1964. The reproductive behavior of the ring dove. Scientific American 211:48-54.

Leinaas, H.P. 1983. Synchronized moulting controlled by communication in group-living Collembola. Science 219:193-195

LeMagnen, J. 1971. Advances in studies on the physiological control and regulation of food intake. In: E. Stellar and

J.M. Sprague, editors. Progress in physiological psychology. Vol 4. Academic Press, New York.

Lenington, S. 1980. Female choice and polygyny in redwinged blackbirds. Animal Behaviour 28: 347-361.

Lent, C.M., and M.H. Dickinson. 1988. The neurobiology of feeding in leeches. Scientific American 258:98-103.

Leonard, J.L., and K. Lukowiak. 1984. An ethogram of the sea slug, Navanax inermis (Gastropoda, Opisthobranchia). Zeitschrift für Tierpsychologie, 65:327-345.

Leonard, J.L., and K. Lukowiak. 1985. The standard ethogram: a two-edged sword? Zeitschrift für Tierpsychologie, 68:335-337.

Leonard, J.L., and K. Lukowiak. 1986. The behavior of Aplysia californica Cooper (Gastropoda; Opisthobranchia): I. Ethogram. Behaviour 98:320-360.

Leonard, M.L., and J. Picman. 1986. Why are nesting marsh wrens and yellow-headed blackbirds spatially segregated? Auk 103:135-140.

Leopold, A. 1933. Game management. Charles Scribner's Sons, New York.

Lerwill, C.J., and P. Makings. 1971. The agonistic behavior of the golden hamster Mesocricetus auratus (Waterhouse). Animal Behaviour 19:714-721.

Leshner, A.I. 1978. An introduction to behavioural endocrinology. Oxford University Press, New York.

Levi, H.W. 1986. Spiders: webs, behavior, and evolution. New Biological Books 63:123-124.

Levine, J.D., and R.J. Wyman. 1973. Neurophysiology of flight in a mutant Drosophila. Proceedings of the National Academy of Sciences 70:1050-1054.

Levine, L. 1958. Studies on sexual selection in mice. American Naturalist 92:21-26.

Levinson, B.M. 1968. Interpersonal relationships between pet and human being. In: M.W. Fox, editor. Abnormal behavior in animals. W.B. Saunders, Philadelphia.

Levitzki, Alexander. 1988. From epinephrine to cyclic AMP. Science 241:800-806.

Lewin, R. 1981 a. Lamarck will not lie down. Science 213:316-321.

Lewin, R. 1981 b. Seeds of change in embryonic development. Science 214:42-44.

Lewin, R. 1984. The continuing tale of a small worm. Science 225:153-156

Lewin, R. 1987. Do animals read minds, tell lies? Science. 238:1350-1351.

Lewin, R. 1988. Response to Bitterman. Science 239:1360.

Lewis, A.C. 1986. Memory constraints and flower choice in *Pieris rapae*. Science 232:863-865.

Lewontin, R.C. 1970. Race and intelligence. Science and Public Affairs, the Bulletin of the Atomic Scientists. March:2-8.

Lewontin, R.C. 1970. The units of selection. Annual Review of Ecology and Systematics 1:1-18.

Lewontin, R.C., S. Rose, and L.J. Kamin. 1984. Not in our genes. Pantheon Books, New York.

Lewy, A.J., R.L. Sack, L.S. Miller, and T.M. Hoban. 1987. Antidepressant and circadian phase-shifting effects of light. Science 235:352-354.

Leyhausen, P. 1956. Verhaltensstudien an Katzen. Zeitschrift für Tierpsychologie Beiheft 2.

Leyhausen, P. 1965. Uber die Funktion der relativen Stimmungshierarchie. Dargestellt am Beispiel der phylogenetischen und ontogenetischen Entwicklung des Beutefangs von Raubtieren. Zeitschrift für Tierpsychologie 22:412-494.

Lidicker, W.Z., Jr., and R.L. Caldwell. 1982. Dispersal and migration. Benchmark Papers in Ecology/ 11. Hutchinson Ross, Stroudsburg, Pa.

Ligon, J.D., and D.J. Martin. 1974. Pinon seed assessment by the pinon jay, *Gymnorhinus cyanocephalus*. Animal Behaviour 22:421-429.

Lin, N., and C.D. Michener. 1972. Evolution of sociality in insects. Quarterly Review of Biology 47: 131-159.

Lindauer, M. 1952. Ein Beitrag zur Frage der arbeitsteilung in Bienenstaat. Zeitschrift für Vergleichende Physiologie 34:299-345.

Lindauer, M. 1961. Communication among social bees. Harvard University Press, Cambridge, Mass.

Lindauer, M. 1971 a. Communication among social bees. Harvard University Press, Cambridge, Mass.

Lindauer, M. 1971 b. The functional significance of the honeybee waggle dance. American Naturalist 105:89-96.

Lissman, H.W. 1958. On the function and evolution of electric organs in fish. Journal of Experimental Biology 35:156-191.

Lissman, H.W. 1963. Electric location by fishes. Scientific American 208(3):50-59.

Liu, C.M., and T.A. Yin. 1974. Caloric compensation to gastric loads in rats with hypothalamic hyperphagia. Physiological Behaviour 13:231-238.

Livingstone, M., and D. Hubel. 1988. Segregation of form, color, movement, and depth: anatomy, physiology, and perception. Science 240:740-749.

Lloyd, J.E. 1966. Studies on the flash communication system in *Photinus* fireflies. Miscellaneous Publications of the Museum of Zoology of the University of Michigan 130:1-95.

Lloyd, J.E. 1975. Aggressive mimicry in *Photuris* fireflies: signal repertoires by femmes fatales. Science 187:452-453.

Lloyd, J.E. 1980. Male *Photuris* fireflies mimic sexual signals of their females' prey. Science 210:669-671.

Lloyd, J.E. 1981. Mimicry in the sexual signals of fireflies. Scientific American 245(1):139-145.

Lockard, R.B. 1971. Reflections on the fall of comparative psychology: is there a lesson for us all? American Psychologist 26:168-179.

Lockley, R.M. 1967. Animal navigation. Hart Publishing, New York.

Loewenstein, W.R. 1971. Mechano-electric transduction in the Pacinian corpuscle. Initiation of sensory impulses in mechanorecepters. In: W.R. Loewenstein, editor. Handbook of sensory physiology. Vol. 1. Springer-Verlag, New York.

Loffredo, C.A., and G. Borgia. 1986. Male courtship vocalizations as cues for mate choice in the satin bowerbird *(Ptilonorhynchus violaceus)*. Auk 103:189-195.

Loiselle, P.V., and G.W. Barlow. 1978. Do fishes lek like birds? In: E.S. Reese and F.J. Lighter, editors. Contrasts in behavior: adaptations in the aquatic and terrestrial environments. John Wiley & Sons, New York.

Loizos, C. 1966. Play in mammals. In: P.A. Jewell and C. Loizos, editors. Play, exploration, and territory in mammals. Academic Press, London.

Lorenz, K.Z. 1932. A consideration of methods of identification of species-specific instinctive behaviour patterns in birds. J. Ornithologie 80:50-98. (Translated by R. Martin in Lorenz, K.Z. 1970. Studies in animal and human behavior. Harvard University Press, Cambridge, Mass.)

Lorenz, K.Z. 1935. Der Kumpan in der Umwelt des Vogels. J. Ornithologie 83:137-213.

Lorenz, K.Z. 1941. Vergleichende Bewegungsstudien an Anatinen. Suppl. J. Ornithologie 89:194-294.

Lorenz, K.Z. 1952. King Solomon's ring. Thomas Y. Crowell, New York.

Lorenz, K.Z. 1958. The evolution of behavior. Scientific American 199(6):67-78.

Lorenz, K.Z. 1966. On aggression. Harcourt, Brace, and World, New York.

Lorenz, K.Z. 1970. Studies in animal and human behaviour. Vol. I. Harvard University Press, Cambridge, Mass. (Translation of earlier papers.)

Lorenz, K.Z. 1979. The year of the greylag goose. Harcourt, Brace, Jovanovich, New York.

Lorenz, K.Z. 1981. The foundations of ethology. Springer-Verlag, New York.

Lorenz, K.Z., and N. Tinbergen. 1938. Taxis und Instinkthandlung in der Eirollbewegung der Graugans I. Zeitschrift für Tierpsychologie 2:1-29.

Losey, George Jr. 1978. Information theory and communication. In: P.W. Colgan, editor. Quantitative ethology. Wiley-Interscience, New York.

Lott, D.F. 1979. Dominance relations and breeding rate in mature male American bison. Zeitschrift für Tierpsychologie 49:418-432.

Lott, D.F. 1981. Sexual behavior and intersexual strategies in American bison. Zeitschrift für Tierpsychologie 56:97-114.

Louw, G. 1979. Biological "strategies." Science 203:955.

Low, W.A., R.L. Tweedie, C.B.H. Edwards, R.M. Hodder, K.W.J. Malafant, and R.B. Cunningham. 1981 a. The influence of environment on daily maintenance behaviour of free-ranging shorthorn cows in central Australia. I. General introduction and descriptive analysis of day-long activities. Applied Animal Ethology 7:11-26.

Low, W.A., R.L. Tweedie, C.B.H. Edwards, R.M. Hodder, K.W.J. Malafant, and R.B. Cunningham. 1981 b. The influence of environment on daily maintenance behaviour of free-ranging shorthorn cows in central Australia. II. Multivariate analysis of duration and incidence of activities. Applied Animal Ethology 7:27-38.

Luce, R.D., 1986. Response Times. Clarendon. Oxford University Press, New York.

Ludlow, A.R. 1976. The behaviour of a model animal. Behaviour 58:131-172.

Lumsden, C.J., and E.O. Wilson. 1983. Promethean fire: reflections on the origin of mind. Harvard University Press, Cambridge, Mass.

Lustick, S., B. Battersby, and M. Kelty. 1978. Behavioral thermoregulation: orientation toward the sun in herring gulls. Science 200:81-83.

Luttenberger, V.F. 1975. Zum problem des gahnens bei reptilien. Zeitschrift für Tierpsychologie 37:113-137.

Lyman, C.P., J.S. Willis, A. Malan, and L.C.H. Wang. 1982. Hibernation and torpor in mammals and birds. Academic Press, New York.

Lynch, G., and M. Baudry. 1984. The biochemistry of memory: a new and specific hypothesis. Science 224:1057-1063.

MacArthur, R.H. 1958. Population ecology of some warblers of northeastern coniferous forests. Ecology 39:599-619.

MacArthur, R.H., and E.R. Pianka. 1966. On optimal use of a patchy environment. American Naturalist 100:603-609.

Mackintosh, N.J. 1974. The psychology of animal learning. Academic Press, New York.

Mackintosh, N.J. 1983. General principles of learning. In: T.R. Halliday and P.J.B. Slater, editors. Genes, development and learning. Animal Behaviour. Vol. 3. W.H. Freeman, New York.

MacLean, P. 1981. Training the brain. Interview on options in education series. National Public Radio Education Services, Washington, D.C. (Cassette tape available from N.P.R.)

MacNally, R.C. 1981. An analysis of factors affecting metabolic rates of two species of *Ranidella* (Anura). Comparative Biochemistry and Physiology 69A:731-737.

MacNally, R.C., and D. Young. 1981. Song energetics of the bladder cicada, *Cystosoma saundersii*. Journal of Experimental Biology 90:185-196.

MacNichol, E.F., Jr. 1964. Three-pigment color vision. Scientific American 211(6):48-56.

Maier, N.R.F., and T.C. Schneirla. 1935. Principles of animal psychology. McGraw-Hill, New York.

Mainardi, D., M. Mainardi, and A. Pasquali. 1986. Genetic and experiential features in a case of cultural transmission in the house mouse *(Mus musculus)*. Ethology 72:191-198.

Majerus, M.E.N. 1986. The genetics and evolution of female choice. Trends in Ecology and Evolution 1:1,3-7.

Major, P.F. 1978. Predator-prey interactions in two schooling fishes, *Caranx ignobilis* and *Stolephorus purpureus*. Animal Behaviour 26:760-777.

Maki, W.S. 1979. Discrimination learning without short-term memory: dissociation of memory processes in pigeons. Science 204:83-85.

Mandelbrot, B.B. 1982. The fractal geometry of nature. W.H. Freeman, New York.

Manning, A. 1967. An introduction to animal behavior. Addison-Wesley, Reading, Mass.

Manning, J.T. 1985. Choosy females and correlates of male age. Journal of Theoretical Biology 116: 349-354.

Marcus, E.A., T.G. Nolen, C.H. Rankin, and T.J. Carew. 1988. Behavioral dissociation of dishabituation, sensitization, and inhibition in *Aplysia*. Science 241:210-213.

Margolis, R.L., S.K. Mariscal, J.D. Gordon, J. Dollinger, and J.L. Gould. 1987. The ontogeny of the pecking response of laughing gull chicks. Animal Behaviour 35:191-202.

Markl, H. 1985. Manipulation, modulation, information, cognition: some of the riddles of communication. In: B. Holldobler and M. Lindauer, editors. Experimental behavioral ecology and sociobiology. Fortschritte der Zoologie 31:163-194. G. Fischer Verlag, New York.

Marler, P. 1952. Variations in the song of the chaffinch, *Fringilla coelebs*. Ibis 94:458-472.

Marler, P. 1957. Specific distinctiveness in the communication signals of birds. Behaviour 11:13-39.

Marler, P. 1959. Developments in the study of animal communication. In: P.R. Bell, editor. Darwin's biological work: some aspects reconsidered. John Wiley & Sons, New York.

Marler, P., and H.S. Terrace, editors. 1984. The biology of learning. Springer-Verlag, Berlin.

Marler, P., and J.G. Vandenbergh, editors. 1979. Social behavior and communication. Handbook of behavioral neurobiology. Vol. 3. Plenum Press, New York.

Marler, P., and M. Tamura. 1964. Culturally transmitted patterns of vocal behavior in sparrows. Science 146:1486.

Marler, P., and P. Mundinger. 1971. Vocal learning in birds. In: H. Holtz, editor. The ontogeny of vertebrate behavior. Academic Press, New York.

Marler, P., and S. Peters. 1981. Sparrows learn adult song and more from memory. Science 213:780-782.

Marler, P., and S. Peters. 1988 a. Sensitive periods for song acquisition from tape recordings and live tutors in the swamp sparrow, *Melospiza georgiana*. Ethology 77:76-84.

Marler, P., and S. Peters. 1988 b. The role of song phonology and syntax in vocal learning preferences in the song sparrow, *Melospiza melodia*. Ethology 77:125-149.

Marler, P., A. Dufty and R. Pickert. 1986 a. Vocal communication in the domestic chicken: I. Does a sender communicate information about the quality of a food referent to a receiver? Animal Behaviour 34:188-193.

Marler, P., A. Dufty, and R. Pickert. 1986

b. Vocal communication in the domestic chicken II. Is a sender sensitive to the presence and nature of a receiver? Animal Behaviour 34:194-198.

Marrocco, R.T. 1986. The neurobiology of perception. In: J.E. Ledoux, and W. Hirst. Mind and brain. Dialogues in cognitive neuroscience. Cambridge University Press, Cambridge, England.

Martin, C.R. 1985. Endocrine physiology. Oxford University Press, New York.

Martin, P. 1982. The energetic costs of play: definition and estimation. Animal behaviour 30:292.

Martin, P. 1984. The time and energy costs of play behaviour in the cat. Zeitschrift für Tierpsychologie 64:298-312.

Martin, P., and P. Bateson. 1986. Measuring behavior. Cambridge University Press, New York.

Martin, R.R., and M. Menaker. 1988. A mutation of the circadian system in golden hamsters. Science 241:1225-1227.

Marx, J.L. 1980. Ape-language controversy flares up. Science 207:1330-1333.

Marx, J.L. 1981. Genes that control development. Science 213:1485-1488.

Mason, P.R. 1975. Chemo-klino-kinesis in planarian food location. Animal Behaviour 23:460-469.

Mason, W.A. 1976. Windows on other minds. Science 194:930-931.

Mast, S.O. 1911. Light and the behavior of organisms. John Wiley & Sons, New York.

Mather, M.H., and B.D. Roitberg. 1987. A sheep in wolf's clothing: tephritid flies mimic spider predators. Science 236:308-310.

Matsubara, J.A. 1981. Neural correlates of a nonjammable electrolocation system. Science 211:722-725.

Matthews, G.V.T. 1951 a. The sensory basis of bird navigation. Journal of the Institute for Navigation 4:260-275.

Matthews, G.V.T. 1951 b. The experimental investigation of navigation in homing pigeons. Journal of Experimental Biology 28:508-536.

Matthews, G.V.T. 1953. Navigation in the Manx shearwater. Journal of Experimental Biology 30:370-396.

Matthews, G.V.T. 1955. Bird navigation. Cambridge University Press, Cambridge, England.

Matthews, R.W., and J.R. Matthews. 1978. Insect behavior. John Wiley & Sons, New York.

Maturana, H.R., J.Y. Lettvin, W.S. McCulloch, and W.H. Pitts. 1960. Anatomy and physiology of vision in the frog *(Rana pipiens)*. Journal of General Physiology 43:129-175.

Mayer, J. 1955. Regulation of energy intake and body weight. The glucostatic theory and the lipostatic hypothesis. Annals of the Academy of Science 63:15-43.

Maynard Smith, J. 1964. Group selection and kin selection. Nature 201:1145-1147.

Maynard Smith, J. 1968. Evolution in sexual and asexual populations. American Naturalist 102:469-473.

Maynard Smith, J. 1971. What use is sex? Journal of Theoretical Biology 30:319-335.

Maynard Smith, J. 1972. Game theory and the evolution of fighting. In: J. Maynard Smith, editor. On evolution. Edinburgh University Press, Edinburgh, England.

Maynard Smith, J. 1976. Evolution and the theory of games. American Scientist 64:41-45.

Maynard Smith, J. 1976. Group selection. Quarterly Review of Biology 51:277-283.

Maynard Smith, J. 1978. The evolution of sex. Cambridge University Press, Cambridge, England.

Maynard Smith, J. 1982. Evolution and the Theory of Games. Cambridge University Press, Cambridge, England.

Maynard Smith, J., and G.A. Parker. 1976. The logic of asymmetric contests. Animal Behaviour 24:159-175.

Maynard Smith, J., and G.R. Price. 1973. The logic of animal conflict. Nature 246:15-18.

Mayr, E. 1963. Animal species and evolution. Belknap/Harvard University Press, Cambridge, Mass.

Mayr, E. 1974. Behavior programs and evolutionary strategies. American Scientist 62:650-658.

Mayr, E., and W.B. Provine, editors. 1980. The evolutionary synthesis. Harvard University Press, Cambridge, Mass.

McArdle, W.D., F.I. Katch, and V.L. Katch. 1981. Exercise physiology. Lea & Febiger, Philadelphia.

McCann, T.S. 1981. Aggression and sexual activity of male southern elephant seals, *Mirounga leonina*. Journal of Zoology, London 195:295-310.

McCarley, R.W., and J.A. Hobson. 1977. The neurobiological origins of psychoanalytic dream theory. American Journal of Psychiatry 134(11):1211-1221.

McConnell, J.V., A.L. Jacobson, and D.P. Kimble. 1959. The effects of regeneration upon retention of a conditioned response in the planarian. Journal of

Comparative and Physiological Psychology 52:1-5.

McConnell, J.V., and A.L. Jacobson. 1973. Learning in invertebrates. In: D.A. Dewsbury and D.A. Rethlingshafer, editors. Comparative psychology. McGraw-Hill, New York.

McConnell, J.V., and J. Shelby. 1970. Memory transfer in invertebrates. In: G. Ungar, editor. Molecular mechanisms in memory and learning. Plenum Press, New York.

McCosker, J.E. 1977. Fright posture of the Plesiopid fish *Calloplesiops altivelis*: an example of batesian mimicry. Science 197:400-401.

McCracken, G.F. 1984. Communal nursing in Mexican free-tailed bat maternity colonies. Science 223:1090-1091.

McCracken, G.F., and M.K. Gustin. 1987. Batmom's daily nightmare. Natural History 96(10):66-73.

McEwen, B.S. 1976. Interactions between hormones and nerve tissue. Scientific American 235(1):48-58.

McFarland, D., and R. Sibly. 1972. "Unitary drives" revisited. Animal Behaviour 20:548-563.

McGrew, R.E. 1985. Encyclopedia of medical history. McGraw-Hill, New York.

McGue, M., and T.J. Bouchard, Jr. 1984. Information processing abilities in twins reared apart. Intelligence 8:239-250.

McGue, M., and T.J. Bouchard, Jr. 1987. Intelligence. In: P. McGuffin, R.M. Murray, and A.M. Reveley, editors. Psychiatric genetics. Churchill Livingstone, London.

McGue, M., and T.J. Bouchard, Jr. 1989. Genetic and environmental determinants of information processing and special mental abilities: a twin analysis. In: R.J. Sternberg, editor. Advances in the psychology of human intelligence. Vol. 5:7-45.

McKean, K. 1985. Of two minds: selling the right brain. Discover April:30- 41.

McKeown, B.A. 1984. Fish migration. Timber Press, Portland, Ore.

McKinney, F. 1975. The behaviour of ducks. In: E.S.E. Hafez, editor. The behaviour of domestic animals, ed 3. Williams & Wilkins, Baltimore.

McLain, D.K. 1980. Female choice and the adaptive significance of prolonged copulation in *Nezara viridula* (Hemiptera: Pentatomidae). Psyche 87:325-336.

McNaughton, S.J. 1976. Serengeti migratory wildebeest: facilitation of energy flow by grazing. Science 191:92-94.

McQuade, D.B., E.H. Williams, and H.B. Eichenbaum. 1986. Cues used for lo-

calizing food by the gray squirrel *(Sciurus carolinensis)*. Ethology 72:22-30.

Mech, L.D. 1966. The wolves of Isle Royale. Fauna of the National Parks of the U.S.—Fauna Series 7. U.S. Government Printing Office, Washington, D.C.

Mech, L.D. 1970. The wolf: the ecology and behavior of an endangered species. Natural History Press, Garden City, N.Y.

Mech, L.D. 1977. Wolf-pack buffer zones as prey reservoirs. Science 198:320-321.

Meddis, R. 1975. On the function of sleep. Animal Behaviour 23:676-691.

Medin, D.L., W.A. Roberts, and R.T. Davis. 1976. Processes of animal memory. Lawrence Erlbaum Associates, Hillsdale, N.J.

Meire, P.M., and A. Ervynck. 1986. Are oystercatchers *(Haematopus ostralegus)* selecting the most profitable mussels *(Mytilus edulis)?* Animal Behaviour 43:1427-1435.

Meisel, R.L., and I.L. Ward. 1981. Fetal female rats are masculinized by male littermates located caudally in the uterus. Science 213:239-242.

Melzack, R., and T.H. Scott. 1957. The effects of early experience on the response to pain. Journal of Comparative and Physiological Psychology 50:155-161.

Menaken, M. 1971. Biochronometry. National Academy of Sciences. Washington, D.C.

Menzel, R., and A. Mercer. 1987. Neurobiology and behavior of honeybees. Springer-Verlag, New York.

Menzel, R., and J. Erber. 1978. Learning and memory in bees. Scientific American 239(1):102-108.

Mertens, R. 1956. Das Problem der Mimikry bei Korallenschlangen. Zool. Jahrb. Abt. Syst. Oekol. Geogr. Tiere. 84:541-576.

Metcalf, R.A. 1980. Sex ratios, parent offspring conflict, and local competition for mates in the social wasps *Polistes metricus* and *Polistes variatus*. American Naturalist 116:642-654.

Metcalfe, N.B., F.A. Huntingford, and J.E. Thorpe. 1988. Predation risk impairs diet selection in juvenile salmon. Animal Behaviour 35:931-933.

Metzgar, L.H. 1967. An experimental comparison of screech owl predation on resident and transient white-footed mice *(Peromyscus leucopus)*. Journal of Mammalogy 48:387-391.

Meyer-Holzapfel, M. 1968. Abnormal behavior in zoo animals. In: M.W. Fox, editor. Abnormal behavior in animals. W.B. Saunders, Philadelphia.

Meyerriecks, A.J. 1972. Man and birds: evolution and behavior. The Bobbs-Merrill, Indianapolis.

Meyers, B. 1971. Early experience and problem-solving behavior. In: H. Moltz, editor. The ontogeny of vertebrate behavior. Academic Press, New York.

Meylan, A. 1988. Spongivory in hawksbill turtles: a diet of glass. Science 239:393-395.

Michelsen, A. 1979. Insect ears as mechanical systems. American Scientist 67:696-706.

Michelsen, A., and B.B. Andersen. 1989. Honeybees can be recruited by a mechanical model of a dancing bee. Naturwissenschaften 76:277-280.

Miles, L.E.M., D.M. Raynal, and M.A. Wilson. 1977. Blind man living in normal society has circadian rhythms of 24.9 hours. Science 198:421-423.

Milgram, N.W., L. Krames, and T.M. Alloway, editors. 1977. Food aversion learning. Plenum Press, New York.

Milinski, M. 1988. Games fish play: making decisions as a social forager. Trends in Ecology and Evolution 3:325-330.

Milinski, M., and R. Heller. 1978. Influence of a predator on the optimal foraging behaviour of sticklebacks (Gasterosteus aculeatus). Nature 275:642-644.

Miller, D.B. 1977. Social display of mallard ducks. (Anas platyrhynchos): effects of domestication. Journal of Comparative and Physiological Psychology 91:221-232.

Miller, N.E., and M.L. Kessen. 1952. Reward efforts of food via stomach fistula compared with those of food via mouth. Journal of Comparative and Physiological Psychology 45:555-564.

Miller, R.S. 1967. Pattern and process in competition. Advances in Ecological Research 4:1-74.

Miller, R.S. 1969. Competition and species diversity. Brookhaven Symposia in Biology 22:63-70.

Miller, R.S. 1985. Why hummingbirds hover. Auk 102:722-726.

Miller, R.S. 1986. Response to F.R. Hainsworth. Auk 103:834.

Mills, J.N. 1973. Biological aspects of circadian rhythms. Plenum Press, New York.

Milne, L., and M. Milne. 1964. The senses of animals and men. Atheneum, New York.

Mimura, K. 1986. Development of visual pattern discrimination in the fly depends on light experience. Science 232:83-85.

Miura, S. 1984. Social behavior and territoriality in male sika deer (Cervus nippon Temminck 1838) during the rut.

Zeitschrift für Tierpsychologie, 64:33-73.

Mock, D.W. 1977. Maintenance behavior and communication in the brown pelican (Review of Schreiber 1977). Wilson Bulletin 89:639-641.

Mock, D.W. 1980. Behavioral mechanisms in ecology. American Scientist 70:325.

Mock, D.W. 1984. Siblicidal aggression and resource monopolization in birds. Science 225:731-733.

Mock, D.W. 1985. Knockouts in the nest. Natural History 94(5):54-61.

Mock, D.W., and G.A. Parker. 1986. Advantages and disadvantages of egret and heron brood reduction. Evolution 40:459-470.

Moehlman, P.D. 1979. Jackal helpers and pup survival. Nature 277:382-383.

Moericke, V. 1950. Ueber das Farbensehen der Pfirsichblattlaus, Myxodes persical Sulz. Zeitschrift für Tierpsychologie 7:265-274.

Moermond, T.C. 1979. The influence of habitat structure on anolis foraging behavior. Behaviour 70:147-167.

Mohl, B. 1968. Auditory sensitivity of the common seal in air and water. Journal of Auditory Research 8:27-38.

Moller, A.P. 1988. False alarm calls as a means of resource usurpation in the great tit Parus major. Ethology 79:25-30.

Moller, P., and J. Serrier. 1986. Species recognition in mormyrid weakly electric fish. Animal Behaviour 34:333-339.

Molnar, R.E. 1977. Analogies in the evolution of combat and display structures in ornithopods and ungulates. Evolutionary Theory 3:165-190.

Moltz, H. 1965. Contemporary instinct theory and the fixed action pattern. Psychological Review 72:27-47.

Moltz, H., editor. 1971. The ontogeny of vertebrate behavior. Academic Press, New York.

Montagu, A., editor. 1975. Race and IQ. Oxford University Press, New York.

Montgomery, G. 1990. A brain reborn. Discover 11:48-53.

Montgomery, G.G., and M.E. Sunquist. 1975. Impact of sloths on neotropical forest energy flow and nutrient cycling. In: F.B. Golley and E. Medina, editors. Tropical ecological systems: trends in terrestrial and aquatic research. Springer-Verlag, New York.

Mook, D.C., and E.M. Blas. 1968. Quinine-aversion thresholds and finickiness in hyperphagic rats. Journal of Comparative and Physiological Psychology 65:202-207.

Moore, A.J., M.D. Breed, and M.J. Moor.

1987. The guard honey bee: ontogeny and behavioural variability of workers performing a specialized task. Animal Behaviour 35:1159-1167.

Moore, B.R., and S. Stuttard. 1979. Dr. Guthrie and Felis domesticus or: tripping over the cat. Science 205:1031-1033.

Moore, M.C., J.M. Whittier, A.J. Billy, and D. Crews. 1985. Male-like behaviour in an all-female lizard: relationship to ovarian cycle. Animal Behaviour 33:284-289.

Moore, R.G. 1978. Seasonal and daily activity patterns and thermoregulation in the southwestern speckled rattlesnake (Crotalus mitchelli pyrrhus) and the Colorado desert sidewinder (Crotaluscerastes laterorepens). Copeia 3:439-442.

Moore, R.Y. 1978. Central neural control of circadian rhythms. Frontiers in Neuroendocrinology 5:185-206.

Moore-Ede, M.C., F.M. Sulzman, and C.A. Fuller. 1982. The clocks that time us. Harvard University Press, Cambridge, Mass.

Moranto, G., and S. Brownlee. 1984. Why sex? Discover 5(2):24-28.

Morgan, C.L. 1894. Introduction to comparative psychology. Scribner, New York.

Morrell, G.M., and J.R. Turner. 1970. Experiments on mimicry. I. The response of wild birds to artificial prey. Behaviour 36:116-130.

Morris, D. 1957. "Typical intensity" and its relation to the problem of ritualization. Behaviour 11:1-12.

Morris, D. 1964. The response of animals to a restricted environment. Symp. Zool. Soc. London 13:99-118. (Reprinted in D. Morris. 1970. Patterns of reproductive behaviour. McGraw-Hill, New York.)

Morris, D., editor. 1970. Patterns of reproductive behavior. McGraw-Hill, New York.

Morris, R., and D. Morris. 1965. Men and snakes. McGraw-Hill, New York.

Morrison, A.R. 1983. A window on the sleeping brain. Scientific American 248:94-102.

Morrison, G.R. 1974. Alterations in palatability of nutrients for the rat as a result of prior tasting. Journal of Comparative and Physiological Psychology 86:56-61.

Morton, E.S. 1977. On the occurrence and significance of motivational-structural rules in some bird and mammal sounds. American Naturalist 111:855-869.

Motley, M.T. 1985. Slips of the tongue. Scientific American 253:116-127.

Mountcastle, V.B., editor. 1980. Medical

physiology, ed 14. Vol. 1. The C.V. Mosby Co., St. Louis.

Moynihan, M.H. 1962. The organization and probable evolution of some mixed species flocks of neotropical birds. Smithsonian Miscellaneous Collections 143(7):1-140.

Moynihan, M.H. 1970. Control, suppression, decay, disappearance and replacement of displays. Journal of Theoretical Biology 29:85-112.

Moynihan, M.H., and A.F. Rodaniche. 1977. Communication, crypsis, and mimicry among cephalopods. In: T.A. Sebeok, editor. How animals communicate. Indiana University Press, Bloomington, Ind.

Mrosovosky, N., and T. Powley. 1977. Set points for body weight and fat. Behavioral Biology 20:205-233.

Mueller, H.C. 1974. The development of prey recognition and predatory behavior in the American kestrel *Falco sparverius*. Behaviour 49:313-324.

Mueller, H.C. 1977. Prey selection in the American kestrel: experiments with two species of prey. American Naturalist 3:25-29.

Mueller, H.C., and D.D. Berger. 1973. The daily rhythm of hawk migration at Cedar Grove, Wisconsin. Auk 90(3):591-596.

Mueller, H.C., and P.G. Parker. 1980. Naive ducklings show different cardiac response to hawk than to goose models. Behaviour 74:1-2.

Muller, H.J. 1932. Some genetic aspects of sex. American Naturalist 66:118-138.

Muller-Schwarze, D. 1968. Play deprivation in deer. Behaviour 31:144-162.

Muller-Schwarze, D. 1971. Pheromones in black-tailed deer *(Odocoileus hemionus columbianus)*. Animal Behaviour 19:141-152.

Muller-Schwarze, D., and R.M. Silverstein, editors. 1980. Chemical signals, vertebrates and aquatic invertebrates. Plenum Press, New York.

Munn, C.A. 1984. Birds of different feather also flock together. Natural History 93(11):34-42.

Munn, C.A. 1986. Birds that "cry wolf." Nature 319:143-145.

Muntz, W.R.A. 1962 a. Microelectrode recordings from the diencephalon of the frog *(Rana pipiens)*, and a blue-sensitive system. Journal of Neurophysiology 25:699-711.

Muntz, W.R.A. 1962 b. Effectiveness of different colors of light in releasing the positive phototactic behavior of frogs, and a possible function of the retinal

projection to the diencephalon. Journal of Neurophysiology 25:712-720.

Muntz, W.R.A. 1963 a. The development of phototaxis in the frog *(Rana temporaria)*. Journal of Experimental Biology 40:371-379.

Muntz, W.R.A. 1963 b Phototaxis and green rods in urodeles. Nature 199:620.

Murata, M., K. Miyagawa-Kohshima, K. Nakanishi, and Y. Naya. 1986. Characterization of compounds that induce symbiosis between sea anemone and anemone fish. Science 234:585-587.

Murton, R.K., A.J. Isaacson, and N.J. Westwood. 1971. The significance of gregarious feeding behaviour and adrenal stress in a population of woodpigeons *Columba palumbus*. Journal of Zoology, London 165:53-84.

Mykytowycz, R., and E.R. Hesterman. 1975. An experimental study of aggression in captive European rabbits, *Oryctolagus cuniculus* (L.). Behaviour 52:104-123.

Myles, T.G., and W.L. Nutting. 1988. Termite social evolution: a re-examination of Bartz's hypothesis and assumptions. Quarterly Review of Biology 63:1-23.

Myrberg, A.A., Jr. 1972. Social dominance and territoriality in the bicolor damselfish, *Eupomacentrus partitus* (Poey) (Pisces: Pomacentridae). Behavior 41:207-231.

Mysterud, I., and H. Dunker. 1979. Mammal ear mimicry: a hypothesis on the behavioural function of owl "horns." Animal Behaviour 27:315-317.

Nagel, W.A. 1894. Experimentelle sinnesphysiologische Untersuchungen an Coelenteraten. Arch. Ges. Physiol. 57:493-552.

Nash, J. 1970. Developmental psychology. Prentice-Hall, Englewood Cliffs, N.J.

Nauta, W.J.H., and M. Feirtag. 1979. The organization of the brain. Scientific American 241(3)88-111.

Nebes, D., and R.W. Sperry. 1971. Hemispheric deconnection syndrome with cerebral birth injury in the dominant arm area. Neuropsychologia 9:247-259.

Needham, J. 1959. A history of embryology. Abelard-Schuman, New York.

Nei, M. 1987. Molecular evolutionary genetics. Columbia University Press, New York.

Neill, S.R.St.J., and J.M. Cullen. 1974. Experiments on whether schooling by their prey affects the hunting behaviour of cephalopod and fish predators. Journal of Zoology, London 172: 549-569.

Neisser, U. 1966. Cognitive psychology. Appleton-Century-Crofts, New York.

Nelissen, M.H.J. 1986. The effect of tied

rank numbers on linearity of dominance hierarchies. Behavioural Processes 12:159-168.

Nelson, M.E., and L.D. Mech. 1981. Deer social organization and wolf predation in northeastern Minnesota. Wildlife Monographs 77:1-53.

Neuchterlein, G.L. 1981. "Information parasitism" in mixed colonies of western grebes and Forster's terns. Animal Behaviour, 29:985-989.

Newton, I., editor. 1989. Lifetime reproduction in birds. Academic Press, New York.

Nicolai, V.J. 1976. Evolutive neuerungen in der Balz von haustaubenrassen als ergebnis menschlicher zuchtwahl. Zeitschrift für Tierpsychologie 40:225-243.

Nisbet, I.C.T. 1973. Courtship feeding, egg-size, and breeding success in common terns. Nature 241:141-142.

Nitecki, M.H., and J.A. Kitchell, editors. 1986. Evolution of animal behavior: paleontological and field approaches. Oxford University Press, New York.

Noble, G.K. 1936. Courtship and sexual selection of the flicker *(Colaptes auratus luteus)*. Auk. 53:269-282.

Noble, G.K. 1939. The role of dominance in the social life of birds. Auk 56:263-273.

Nolen, T.G., and R.R. Hoy. 1984. Initiation of behavior by single neurons: the role of behavioral context. Science 226:992-994.

Norberg, U.M., and J.M.V. Rayner. 1987. Ecological morphology and flight in bats (Mammalia; Chiroptera): wing adaptations, flight performance, foraging strategy and echolocation. Philosophical Transactions of the Royal Society of London B 316:335-427.

Nordeen, E.J., K.W. Nordeen, D.R. Sengelaub, A.P. Arnold. 1985. Androgens prevent normally occurring cell death in a sexually dimorphic spinal nucleus. Science 229:671-672.

Norman, C. 1986. Mexico acts to protect overwintering monarchs. Science 233:1252-1253.

Norris, D.O., and R.E. Jones, editors. 1987. Hormones and Reproduction in fishes, amphibians, and reptiles. Plenum Press, New York.

Norton-Griffiths, M.N. 1969. The organisation, control and development of parental feeding in the oystercatcher *(Haematopus ostralegus)*. Behaviour 34:55-114.

Nottebohm, F. 1981. A brain for all seasons: cyclical anatomical changes in song control nuclei of the canary brain. Science 214:1368-1370.

Nottebohm, F. 1984. Birdsong as a model

in which to study brain processes related to learning. Condor 86:227-236.

Nottebohm, F., editor. 1985. Hope for a new neurology. Annals of the New York Academy of Sciences, Vol. 457.

Nottebohm, F. 1989. From bird song to neurogenesis. Scientific American 260:74-79.

Nottebohm, F., M.E. Nottebohm, and L. Crane. 1986. Developmental and seasonal changes in canary song and their relation to changes in the anatomy of song-control nuclei. Behavioral and Neural Biology 46:445-471.

Nottebohm, F., M.E. Nottebohm, L.A. Crane, and J.C. Wingfield. 1987. Seasonal changes in gonadal hormone levels of adult male canaries and their relation to song. Behavioral and Neural Biology 47:197-211.

Novin, D., W. Wyrwicka, and G.A. Bray, editors. 1976. Hunger: basic mechanisms and clinical implications. Raven Press, New York.

Nuechterlein, G.L. 1981. "Information parasitism" in mixed colonies of western grebes and forster's terns. Animal Behaviour 29:985-989.

Nuechterlein, G.L., and R.W. Storer. 1985. Aggressive behavior and interspecific killing by flying steamer-ducks in Argentina. Condor 87:87-91.

O'Brien, D.F. 1982. The chemistry of vision. Science 218:961-965.

Ohguchi, O. 1981. Prey density and selection against oddity by three-spined sticklebacks. Supplements to Journal of Comparative Ethology, Vol. 23. Verlag Paul Parey, Berlin.

Ohmart, R.D., and R.C. Lasiewski. 1971. Roadrunners: energy conservation by hypothermia and absorption of sunlight. Science 172:67-69.

Okanoya, K., and R.J. Dooling. 1988. Hearing in the swamp sparrow, *Melospiza georgiana*, and the song sparrow, *Melospiza melodia*. Animal Behaviour 36:726-732.

Olds, J., and P. Milner. 1954. Positive reinforcement produced by electrical stimulation of septal area and other regions of the rat brain. Journal of Comparative and Physiological Psychology 47:419-427.

Oppenheim, R.W. 1972. Prehatching and hatching behaviour in birds: a comparative study of altricial and precocial species. Animal Behaviour 20:644-655.

Orchard, I. 1984. The role of biogenic amines in the regulation of peptidergic neurosecretory cells. In: A.B., Borkovec, and T.J. Kelly, editors. Insect neurochemistry and neurophysiology. Plenum Press, New York.

Orians, G.H. 1969. On the evolution of mating systems in birds and mammals. American Naturalist 103:589-603.

Orians, G.H., and N.E. Pearson. 1979. On the theory of central place foraging. In: D.J. Horn, G.R. Stairs, and R.D. Mitchell, editors. Analysis of ecological systems. Ohio State University Press, Columbus, Ohio.

Oring, L.W. 1982. Avian mating systems. Avian Biology 6:1-92.

Oring, L.W. 1986. Avian polyandry. Current Ornithology 3:309-351.

Oring, L.W., and D.B. Lank. 1986. Polyandry in spotted sandpipers: the impact of environment and experience. In: D. Rubenstein and R.W. Wrangham, editors. Ecological aspects of social evolution. Princeton University Press, Princeton, N.J.

Oring, L.W., A.J. Fivizzani, M.A. Colwell, and M.E. El Halawani. 1988. Hormonal changes associated with natural and manipulated incubation in the sex-role reversed Wilson's phalarope. General and Comparative Endocrinology 72:1-10.

Oring, L.W., A.J. Fivizzani, M.E. El Halawani, and A. Goldsmith. 1986. Seasonal changes in prolactin and luteinizing hormone in the polyandrous spotted sandpiper, *Actitis macularia*. General and Comparative Endocrinology 62:394-403.

Ortega, J.C. and M. Bekoff. 1987. Avian play: comparative evolutionary and developmental trends. Auk 104:338-341.

Ostrom, J.H. 1979. Bird flight: how did it begin? American Scientist 67:46-56.

Ostrom, J.H. 1980. The evidence for endothermy in dinosaurs. In: D.K. Thomas and E.C. Olson, editors. A cold look at the warm-blooded dinosaurs. Westview Press, Boulder, Colo.

Ostrom, J.H. 1986. Social and unsocial behavior in dinosaurs. In: M.H. Nitecki, and J.A. Kitchell, editors. Evolution of animal behavior: paleontological and field approaches. Oxford University Press, New York.

Otis, L.S., and J.A. Cerf. 1963. Conditioned avoidance learning in two fish species. Psychological Reports 12:679-682.

Otte, D. 1972. Simple versus elaborate behavior in grasshoppers: an analysis of communication in the genus *Syrbula*. Behaviour 42:291-322.

Otte, D. 1974. Effects and functions in the evolution of signaling systems. Annual Review of Ecology and Systematics 5:385-417.

Owings, D.H. and D.W. Leger. 1980. Chatter vocalizations of California

ground squirrels: predator-and social-role specificity. Zeitschrift für Tierpsychologie 54:163-184.

Packer, C. 1977. Reciprocal altruism in olive baboons. Nature 265:441-443.

Packer, C. 1986. The ecology of sociality in felids. In: D.I. Rubenstein and R.W. Wrangham, editors. Ecological aspects of social evolution. Princeton University Press, Princeton, N.J.

Packer, C., and A.E. Pusey. 1982. Cooperation and competition within coalitions of male lions: kin selection or game theory? Nature 296:740-742.

Packer, C. and A.E. Pusey. 1983. Male takeovers and female reproductive parameters: a simulation of oestrous synchrony in lions (*Panthera leo*). Animal Behaviour 31:334-340.

Packer, C., and L. Ruttan. 1988. The evolution of cooperative hunting. American Naturalist 132:159-198.

Packer, C., L. Herbst, A.E. Pusey, J.D. Bygott, J.P. Hanby, S.J. Cairns, and M.B. Mulder. 1988. Reproductive success in lions. In: T.H. Clutton-Brock, editor. Reproductive success: studies of individual variation in contrasting breeding systems. University of Chicago Press, Chicago.

Page, T.L. 1982. Transplantation of the cockroach circadian pacemaker. Science 216:73-75.

Pajunen, V.I. 1966. The influence of population density on the territorial behaviour of *Leucorrhinia rubicunda* (Odon., Libellulidae). Ann. Zool. Fenn. 3:40-52.

Pakes, S.P. (chairman), et al. (16 other persons), Committee on Care and Use of Laboratory Animals of the Institute of Laboratory Animal Resources Commission on Life Sciences of the National Research Council. 1985. Guide for the care and use of laboratory animals. U.S. Dept. of Health and Human Services. Public Health Service, National Institutes of Health, NIH Publication No. 85-23.

Palameta, B., and L. Lefebvre. 1985. The social transmission of a food-finding technique in pigeons: what is learned? Animal Behaviour 33:892-896

Palka, J., and K.S. Babu. 1967. Toward the physiological analysis of defensive responses of scorpions. Zeitschrift für Vergleichende Physiologie 55:286-298.

Palm, C.E., et al. 1970. Vertebrate pests: problems and control. National Academy of Sciences, Washington, D.C.

Panaman, R. 1981. Behaviour and ecology of free-ranging female farm cats (*Felis catus* L.). Zeitschrift für Tierpsychologie 56:59-73.

Pantin, C.F.A. 1952. The elementary nervous system. Proceedings of the Royal Society of London 140:147-168.

Papi, F. 1982. Olfaction and homing in pigeons: ten years of experiments. In: F.P. and H.G. Wallraff, editors. Avian Navigation. Springer-Verlag, Berlin.

Papi, F., L. Fiore, V. Fiaschi, and S. Benvenuti. 1971. The influence of olfactory nerve section on the homing capacity of carrier pigeons. Monit. Zool. Ital. 5:265-267.

Papi, F., L. Fiore, V. Fiaschi, and S. Benvenuti. 1972. Olfaction and homing in pigeons. Monit. Zool. Ital. 6:85-95.

Park, T. 1954. Experimental studies of interspecific competition, II: Temperature, humidity and competition in two species of Tribolium. Physiological Zoology 27:177-238.

Parker, G.A. 1970. Sperm competition and its evolutionary consequences in the insects. Biological Reviews 45:525-567.

Parker, G.A. 1974. Assessment strategy and the evolution of fighting behaviour. Journal of Theoretical Biology 47:223-243.

Parker, G.A. 1978 a. Evolution of competitive mate searching. Annual Review of Entomology 23:173-196.

Parker, G.A. 1978 b. Searching for mates. In: J.R. Krebs and N.B. Davies, editors. Behavioural ecology. Blackwell Scientific Publications, London.

Parker, G.A. 1983. Mate quality and mating decisions. In: P. Bateson, editor. Mate choice. Cambridge University Press, Cambridge, England.

Parker, G.A. 1984 a. Evolutionarily stable strategies. In: J.R. Krebs and N.B. Davies, editors. Behavioural ecology: an evolutionary approach. Blackwell Scientific Publications, Oxford, England.

Parker, G.A. 1984 b. Sperm competition and the evolution of animal mating strategies. In: R.L. Smith, editor. Sperm competition and the evolution of animal mating systems. Academic Press, New York.

Parker, G.A., R.R. Baker, and V.G.F. Smith. 1972. The origin and evolution of gamete dimorphism and the male-female phenomenon. Journal of Theoretical Biology 36:529-553.

Parker, G.H. 1896. The reactions of Metridium to food and other substances. Bulletin of the Museum, Harvard 29:107-119.

Parker, G.H. 1919. The elementary nervous system. J.B. Lippincott, Philadelphia.

Parsons, J. 1971. Cannibalism in herring gulls. British Birds 64:528-537.

Partridge, B.L. 1982. The structure and function of fish schools. Scientific American 246(6):114-123.

Partridge, B.L., and T.J. Pitcher. 1979. Evidence against a hydrodynamic function for fish schools. Nature 279:418-419.

Partridge, L. 1974. Habitat selection in titmice. Nature 247:573-574.

Partridge, L. 1976 a. Individual differences in feeding efficiencies and feeding preferences of captive great tits. Animal Behaviour 24:230-240.

Partridge, L. 1976 b. Field and laboratory observations on the foraging and feeding techniques of blue tits (Parus caeruleus) and coal tits (P. ater) in relation to their habitats. Animal Behaviour 24:534-544.

Partridge, L. 1980. Mate choice increases a component of offspring fitness in fruit flies. Nature 283:290-291.

Pasmussen, T., and W. Penfield. 1947. Further studies of sensory and motor cerebral cortex of man. Federation Proceedings 6:452.

Paton, J.A., and F.N. Nottebohm. 1981. Neurons generated in the adult brain are recruited into functional circuits. Science 225:1046-1048.

Patterson, F. 1978. The gestures of a gorilla: sign language acquisition in another pongid species. Brain and Language 5:72-97.

Paul, D.B. 1985. Textbook treatments of the genetics of intelligence. Quarterly Review of Biology 60:317-326.

Pavlov, I.P. 1906, 1927. Conditioned reflex. Oxford University Press, Oxford, England. (Translated by G.V. Aurep.)

Payne, R., editor. 1983. Communication and behavior of whales. Westview Press, Boulder, Colo.

Pearl, R. 1903. The movements and reactions of freshwater planaria. Quarterly Journal of Microscopical Science 46:509-714.

Pearson, O.P. 1966. The prey of carnivores during one cycle of mouse abundance. Journal of Animal Ecology 35:217-233.

Peck, J.W. 1978. Rats defend different body weights depending on palatability and accessibility of their food. Journal of Comparative and Physiological Psychology 92:555-570.

Peckham, G.W., and E.G. Peckham. 1894. The sense of sight in spiders with some observations on the color sense. Transactions of the Wisconsin Academy of Science 10:231-261.

Peek, F. 1972. An experimental study of the territorial function of vocal and visual display in the male red-winged blackbird (Ageliaus phoeniceus). Animal Behaviour 20:112-118.

Peitgen, H.O., and P.H. Richter. 1986. Images of complex dynamical systems. Springer-Verlag, New York.

Pelkwijk, J., J. Ter, and N. Tinbergen. 1937. Eine reizbiologische Analyse einiger Verhaltensweisen von Gasterosteus aculeatus L. Zeitschrift für Tierpsychologie 1:193-204.

Penfield, W.W., and H. Jasper. 1954. Epilepsy and the functional anatomy of the human brain. Little, Brown & Co., Boston.

Pengelley, E.T. 1974. Circannual clocks. Academic Press, New York.

Pengelley, E.T., and S.J. Asmundson. 1971. Annual biological clocks. Scientific American 224(4):72-79.

Pennisi, E. 1986. Not just another pretty face. Discover 7(3):68-78.

Pennycuick, C.J., and J.A. Rudnai. 1970. A method of identifying individual lions Panthera leo with an analysis of the reliability of identification. Journal of Zoology, London 160:497-508.

Pepperberg, I.M. 1981. Functional vocalizations by an African grey parrot. Zeitschrift für Tierpsychologie 55:139-160.

Pepperberg, I.M., and D.M. Neapolitan. 1988. Second language acquisition: a framework for studying the importance of input and interaction in exceptional song acquisition. Ethology 77:150-168.

Perdeck, A.C. 1958. Two types of orientation in migrating starlings, Sturnus vulgaris L., and chaffinches, Fringilla coelebs L., as revealed by displacement experiments. Ardea 46:1-37.

Peters, H. 1932. Experimente uber die Orientierung der Kreuzspinne Epeira diademata Cl. im Netz. Zool. Jahrb. 51:239-288.

Peters, M. 1971. Sensory mechanisms of homing in salmonids: a comment. Behaviour 39:18-19.

Peters, R.P., and L.D. Mech. 1975. Scent marking in wolves. American Scientist 63:628-637.

Peterson, S.R., and R.S. Ellarson. 1977. Food habits of oldsquaws wintering on Lake Michigan. Wilson Bulletin 89(1):81-91.

Petrie, M. 1983. Female moorhens compete for small fat males. Science 220:413-415.

Pettingill, O.S., Jr. 1970. Ornithology in laboratory and field. Burgess, Minneapolis, Minn.

Pfaff, D.W., editor. 1982. The physiological mechanisms of motivation. Springer-Verlag, New York.

Phelps, M.E., D.E. Kuhl, and H.C. Mazziotta. 1981. Metabolic mapping of the brain's response to visual stimulation: studies in humans. Science 211:1445-1448.

Pianka, E.R. 1970. On r- and K-selection. American Naturalist 104:592-597.

Picton, T.W., S.A. Hillyard, H.I Krausz, and R. Galambos. 1973. Human auditory evoked potentials. I. Evaluation of components. Electroencephalography and Clinical Neurophysiology 36:179-190.

Pierotti, R. 1982. Habitat selection and its effect on reproductive output in the herring gull in Newfoundland. Ecology 63(3):854-868.

Pietrewicz, A.T., and A.C. Kamil. 1979. Search image formation in the blue jay (Cyanocitta cristata). Science 204:1332-1333.

Pietsch, T.W., and D.B. Grobecker. 1978. The compleat angler: aggressive mimicry in an antennariid anglefish. Science 201:369-370.

Pitcher, T.J., editor. 1986. The behavior of teleost fishes. Johns Hopkins University Press, Baltimore, Md.

Pittendrigh, C.S. 1958. Perspectives in the study of biological clocks. In: A.A. Buzzati-Traverso, editor. Perspectives in marine biology. Scripps Institution of Oceanography, La Jolla, Calif.

Pleszczynska, W.K. 1978. Microgeographic prediction of polygyny in the lark bunting. Science 201:935-937.

Plomin, R., J.C. DeFries, and G.E. McClearn. 1990. Behavioral genetics: a primer. W.H. Freeman, New York.

Polis, G.A. 1981. The evolution and dynamics of intraspecific predation. Annual Review of Ecology and Systematics 12:225-251.

Pollak, G., D. Marsh, R. Bodenhamer, and A. Souther. 1977. Echo-detecting characteristics of neurons in inferior colliculus of unanesthetized bats. Science 196:675-678.

Polyak, S. 1941. The retina. University of Chicago Press, Chicago.

Polyak, S. 1957. In: H. Kluver, editor. The vertebrate visual system. University of Chicago Press, Chicago.

Poole, T.B. 1966. Aggressive play in polecats. In: P.A. Jewell and C. Loizos, editors. Play, exploration, and territory in mammals. Academic Press, London.

Popper, A.N., and S. Combs. 1980. Auditory mechanisms in teleost fishes. American Scientist 68:429-440.

Porter, J.P. 1904. A preliminary study of the psychology of the English sparrow. American Journal of Psychology 15:313-346.

Porter, R.H., M. Wyrick, and J. Pankey. 1978. Sibling recognition in spiny mice (Acomys cahirinus). Behavioral Ecology and Sociobiology 3:61-68.

Posner, M.I., S.E. Petersen, P.T. Fox, & M.E. Raichle. 1988. Localization of cognitive operations in the human brain. Science 240:1627-1631.

Potts, G.W. 1973. The ethology of Labroides dimidiatus (Cuv, and Val) (Labridae, Pisces) on Aldabra. Animal Behaviour 21:250-291.

Powell, R.A. 1978. A comparison of fisher and weasel hunting behavior. Carnivore 1:28-34.

Powley, T.L. 1977. The ventromedial hypothalamic syndrome, satiety, and a cephalic hypothesis. Psychological Review 84:89-126.

Powley, T.L., and C.A. Opsahl. 1974. Ventromedial hypothalamic obesity by subdiagraphramatic vagotomy. American Journal of Physiology 226:25-33.

Powley, T.L., and R.E. Keesey. 1970. Relationship of body weight to the lateral hypothalamic syndrome. Journal of Comparative and Physiological Psychology 70:25-36.

Pratt, H.M. 1980. Directions and timing of great blue heron foraging flights from a California colony: implications for social facilitation of food finding. Wilson Bulletin 92(4):489-496.

Premack, D. 1972. Teaching language to the ape. Scientific American 227:92-99.

Premack, D., and G. Woodruff. 1978. Chimpanzee problem-solving: a test for comprehension. Science 202:532-535.

Preston, J.L. 1978. Communication systems and social interactions in a goby-shrimp symbiosis. Animal Behaviour 26:791-802.

Prestwich, K.N., and T.J. Walker. 1981. Energetics of singing in crickets: effect of temperature in three trilling species (Orthoptera: Gryllidae). Journal of Comparative Physiology 143:199-212.

Preyer, W. 1885. Specielle physiologie des embryo. Grieben, Leipzig.

Price, E.O. 1984. Behavioral aspects of animal domestication. Quarterly Review of Biology 59:1-32.

Price, E.O., and A.W. Stokes, 1975. Animal behavior in laboratory and field. W.H. Freeman, San Francisco.

Pringle, J.W.S. 1957. Insect flight. Cambridge University Press, Cambridge, England.

Prioli, C.A. 1982. The Fu-Go project. American Heritage 33:88-92.

Prokopy, R.J. 1980. Mating behavior of frugivorous Tephritidae in nature. In: Proceedings of the International Symposium on Fruit Fly Problems. XVI International Congress of Entomology, Kyoto.

Provine, R.R. 1979. "Wing-flapping" develops in wingless chicks. Behavioral and Neural Biology 27:233-237.

Provine, R.R. 1986. Yawning as a stereotyped action pattern and releasing stimulus. Ethology 72:109-122.

Provine, R.R., H.B. Hamernik, and B.C. Curchack. 1987. Yawning: relation to sleeping and stretching in humans. Ethology 76:152-160.

Pruett-Jones, M.A., and S.G. Pruett-Jones. 1985. Food caching in the tropical frugivore, MacGregor's bowerbird (Amblyornis macgregoriae). Auk 102:334-341.

Pusey, A., and C. Packer. 1983. Once and future kings. Natural History 92(8):54-63.

Pyke, G.H. 1979. The economics of territory size and time budget in the golden-winged sunbird. American Naturalist 114:131-145.

Pyke, G.H. 1981 a. Why hummingbirds hover and honeyeaters perch. Animal Behaviour 29:861-867.

Pyke, G.H. 1981 b. Honeyeaters foraging: a test of optimal foraging theory. Animal Behaviour 29:878-888.

Pyke, G.H. 1981 c. Optimal foraging in hummingbirds: rule of movement between inflorescences. Animal Behaviour 29:889-896.

Pyke, G.H., H.R. Pulliam, and E.I. Charnov. 1977. Optimal foraging: a selective review of theory and tests. Quarterly Review of Biology 52(2):137-154.

Pyrcraft, W.P. 1912. The infancy of animals. Hutchinson Publishing, London.

Queller, D.C., and J.E. Strassmann. 1988. Reproductive success and group nesting in the paper wasp, Polistes annularis. In: T.H. Clutton-Brock, editor. Reproductive success: studies of individual variation in contrasting breeding systems. University of Chicago Press, Chicago.

Raabe, M. 1986. Insect reproduction: regulation of successive steps. Advances in Insect Physiology 19:29-154.

Raber, H. 1950. Das Verhalten gefangenen Waldohreulen (Asio otus otus) and Waldkauze (Strix aluco aluco) zur Beute. Behaviour 2:1-95.

Radcliffe, C.W., K. Estep, T. Boyer, and D. Chiszar. 1986. Stimulus control of predatory behavior in red spitting cobras (Naja mossambica pallida) and prairie rattlesnakes (Crotalus v. viridis). Animal Behaviour 34:804-814.

Rado, R., N. Levi, H. Hauser, J. Witcher, N. Adler, N. Intrator, Z. Wollberg, and J. Terkel. 1987. Seismic signalling as a means of communication in a subterranean mammal. Animal Behaviour 35:1249-1251.

Rakic, P. 1985. Limits of neurogenesis in primates. Science 227:1054-1056.

Ralph, M.R., and M. Menaker. 1988. A mutation of the circadian system in golden hamsters. Science 241:1225-1227.

Randall, J.A. 1978. Behavioral mechanisms of habitat segregation between sympatric species of Microtus: habitat preference and interspecific dominance. Behavioral Ecology and Sociobiology 3:187-202.

Randall, J.A. 1981. Comparison of sand-bathing and grooming in two species of kangaroo rat. Animal Behaviour 29:1213-1219.

Rasa, O.E.A. 1971. The causal factors and function of "yawning" in Microspathadon chrysurus (Pisces, Pomacentridae). Behaviour 39:39-57.

Ratner, S.C., and K.R. Miller. 1959. Classical conditioning in earthworms, Lumbricus terrestris. Journal of Comparative and Physiological Psychology 52:102-105.

Ratner, S.C., and R. Boice. 1975. Effects of domestication on behaviour. In: E.S.E. Hafez, editor. The behavior of domestic animals, ed 3. Williams & Wilkins, Baltimore.

Raven, P.H. 1977. A suggestion concerning the Cretaceous rise to dominance of the angiosperms. Evolution 31:451-452.

Raven, P.H., and G.B. Johnson. 1989. Biology. Times Mirror/Mosby College Publishing, St. Louis.

Ray, T.S., and C.C. Andrews. 1980. Ant-butterflies: butterflies that follow army ants to feed on antbird droppings. Science 210:1147-1148.

Raynaud, A., and C. Pieau. 1985. Embryonic development of the genital system [in reptiles]. In: C. Gans and F. Billett, editors. Biology of the reptilia. Vol. 15, Development B. John Wiley & Sons, New York.

Razran, G.A. 1971. Mind in evolution: an east-west synthesis. Houghton Mifflin, Boston.

Read, A.F. 1987. Comparative evidence supports the Hamilton and Zuk hypothesis on parasites and sexual selection. Nature 328:68-70.

Read, A.F., and P.H. Harvey. 1989. Reassessment of comparative evidence for Hamilton and Zuk theory on the evolution of secondary sexual characters. Nature 339:618-620.

Real, L., and T. Caraco. 1986. Risk and foraging in stochastic environments.

Annual Review of Ecology and Systematics 17:371-390.

Rechtschaffen, A., M.A. Gilliland, B.M. Bergmann, and J.B. Winter. 1983. Physiological correlates of prolonged sleep deprivation in rats. Science 221:182-184.

Regnier, F.E., and E.O. Wilson. 1971. Chemical communication and "propaganda" in slave-maker ants. Science 172:267-269.

Reichert, S.E. 1985. Why do some spiders cooperate? Agelena consociata, a case study. Florida Entomologist 68:105-116.

Reinoso-Suarez, F., and C. Ajmone-Marsan, editors. 1984. Cortical integration: basic archicortical, and cortical association levels of neural integration. Raven Press, New York.

Renner, M. 1960. The contribution of the honey bee to the study of time-sense and astronomical orientation. Cold Spring Harbor Symp. Quant. Biol. 25:361-367.

Reppert, S.M., and W.J. Schwartz. 1983. Maternal coordination of the fetal biological clock in utero. Science 220:969-971.

Reyer, H.-U. 1984. Investment and relatedness: a cost/benefit analysis of breeding and helping in the pied kingfisher (Ceryle rudis). Animal Behaviour 32:1163-1178.

Richards, O.W. 1927. The specific characters of the British bumblebees (Hymenoptera). Transactions of the Entomological Society of London 75:233-268.

Richards, S.M. 1974. The concept of dominance and methods of assessment. Animal Behaviour 22:914-930.

Richardson, W.J. 1978. Timing and amount of bird migration in relation to weather: a review. Oikos 30:224-272.

Ridgway, S.H., R.J. Harrison, and P.L. Joyce. 1975. Sleep and cardiac rhythm in the gray seal. Science 187:553-555.

Ridley, M. 1978. Paternal care. Animal Behaviour 26:904-932.

Ridley, M. 1983. The explanation of organic diversity. Clarendon Press, Oxford, England.

Ridley, M. 1986. The number of males in a primate troup. Animal Behaviour 34:1848-1858.

Rijnsdorp, A., S. Daan, and C. Dijkstra. 1981. Hunting in the kestrel Falco tinnunculus, and the adaptive significance of daily habits. Oecologia 50:391-406.

Robertson, J.G.M. 1986. Male territoriality, fighting and assessment of fighting ability in the Australian frog, Uperoleia rugosa. Animal Behaviour 34:763-772.

Robinson, D.N. 1981. The psychobiology

of consciousness. American Scientist 69:463-464.

Robinson, J.G. 1984. Syntactic structures in the vocalizations of wedge-capped capuchin monkeys, Cebus olivaceus. Behaviour 90:46-79.

Robson, E.A. 1961. Some observations on the swimming behaviour of the anemone Stomphia coccinea. Journal of Experimental Biology 38:343-363.

Rodda, G.H. 1985. Navigation in juvenile alligators. Zeitschrift für Tierpsychologie 68:65-77.

Rodgers, J.A., Jr. 1978. Breeding behavior of the Louisiana heron. Wilson Bulletin 90:45-59.

Roeder, K.D. 1948. Organization of the ascending giant fiber system in the cockroach (Periplaneta americana L.) Journal of Experimental Zoology 108:243-262.

Roeder, K.D. 1962. The behaviour of free flying moths in the presence of artificial ultrasonic pulses. Animal Behaviour 10:300-304.

Roeder, K.D. 1963. Echoes of ultrasonic pulses from flying moths. Biological Bulletin 124:200-210.

Roeder, K.D. 1967. Nerve cells and insect behavior. Harvard University Press, Cambridge, Mass.

Roeder, K.D., and A.E. Treat. 1961. The detection and evasion of bats by moths. American Scientist 49:135-148.

Roggenbuck, M.E., and T.A. Jenssen. 1986. The ontogeny of display behaviour in Sceloporus undulatus (Sauria: Iguanidae). Ethology 71:153-165.

Roitblat, T.G., B. and H.S. Terrace, editors. 1984. Animal cognition. Lawrence Erlbaum, N.J.

Rollin, B.E. 1981. Animal rights and human morality. Prometheus Books, Buffalo, N.Y.

Rolls, E.T. 1978. Neurophysiology of feeding. Trends in Neuroscience 1:1-3.

Romanes, G.J. 1884. Mental evolution in animals. Keegan, Paul, Trench & Co., London.

Romanes, G.J. 1889. Mental evolution in man. D. Appleton, New York.

Roper, T.J. 1983. Learning as a biological phenomenon. In: T.R. Halliday and P.J.B. Slater, editors. Genes, development and learning. Animal Behaviour. Vol. 3. W.H. Freeman, New York.

Rose, G.J., R. Zelick, and A.S. Rand. 1988. Auditory processing of temporal information in a Neotropical frog is independent of signal intensity. Ethology 77:330-336.

Rose, G.J., R. Zelick, and A.S. Rand. 1988. Auditory processing of temporal information in a neotropical frog is independent of signal intensity. Ethology 77:330-336.

Rosenthal, G.A., D.L. Dahlman, and D.H. Jansen. 1978. L-canaline detoxification: a seed predator's biochemical mechanism. Science 202:528-529

Rosenzweig, M.R. 1984. Experience, memory, and the brain. American Psychologist 39:365-376.

Rosenzweig, M.R., and E.L. Bennett, editors. 1976. Neural mechanisms of learning and memory. MIT Press, Cambridge, Mass.

Rosenzweig, M.R., E.L. Bennett, and M.C. Diamond. 1972. Brain changes in response to experience. Scientific American 226(2):22-29.

Roskaft, E., and S. Rohwer. 1987. An experimental study of the function of the red epaulettes and the black body colour of male red-winged blackbirds. Animal Behaviour 35:1070-1077.

Roskaft, E., T. Jarvi, M. Bakken, C. Bech, and R.E. Reinertsen. 1986. The relationship between social status and resting metabolic rate in great tits *(Parus major)* and pied flycatchers *(Ficedula hypoteuca)*. Animal Behaviour 34:838-842.

Ross, D.M. 1965. The behavior of sessile coelenterates in relation to some conditioning experiments. Animal Behaviour (Suppl)1:43-55.

Ross, W.N., B.M. Salzberg, L.B. Cohen, and H.V. Davila. 1974. A large change in dye absorption during the action potential. Biophysical Journal 14:983-986.

Rossler, V.E. 1978. Ubertragung von verhaltensweisen durch transplantation von anlagen neuroanatomischer strukturen bei amphibienlarven. Zeitschrift für Tierpsychologie 46:1-13.

Roth, G. 1986. Neural mechanisms of prey recognition: an example in amphibians. In: M.E. Feder, and G.V. Lauder, editors. Predator-prey relationships. University of Chicago Press, Chicago.

Rothenbuhler, N. 1964. Behavior genetics of nest cleaning in honey bees. IV. Responses of F1 and backcross generations to disease-killed brood. American Zoologist 4:111-123.

Rothstein, S.I., and R.C. Fleischer. 1987. Brown-headed cowbirds learn flight whistles after the juvenile period. Auk 104:512-516.

Rothstein, S.I., D.A. Yokel, and R.C. Fleischer. 1988. The agonistic and sexual functions of vocalizations of male brown-headed cowbirds, *Molothrus ater*. Animal Behaviour 36:73-86.

Rovner, J.S., G.A. Higashi, and R.F. Foelix. 1973. Maternal behavior in wolf spiders: the role of abdominal hairs. Science 182:1153-1155.

Rowe, M.P., R.G. Coss, and D.H. Owings. 1986. Rattlesnake rattles and burrowing owl hisses: a case of acoustic batesian mimicry. Ethology 72:53-71.

Rowell, T.E. 1974. The concept of social dominance. Behavioral Biology 11:131-154.

Rowell, T.E., R.A. Hinde, and Y. Spencer-Booth. 1964. "Aunt"-infant interaction in captive rhesus monkeys. Animal Behaviour 12:219-226.

Rumbaugh, D.M., editor. 1977. Language learning by a chimpanzee: the LANA Project. Academic Press, New York.

Rumelhart, D.E., and J.L. McClelland. 1986. Parallel distributed processing. Explorations in the microstructure of cognition. Two volumes. MIT Press, Cambridge, Mass.

Runfeldt, S., and J.C. Wingfield. 1985. Experimentally prolonged sexual activity in female sparrows delays termination of reproductive activity in their untreated mates. Animal Behaviour 33:403-410.

Rusak, B., and G. Groos. 1982. Suprachiasmatic stimulation phase shifts rodent circadian rhythms. Science 215:1407-1409.

Rushforth, N.B. 1973. Behavioral modifications in coelenterates. In: W.C. Corning, J.A. Dyal, and A.O.D. Willows, editors. Invertebrate learning. Vol. 1. Plenum Press, New York.

Ryan, M.J. 1985. The Tungara frog. University of Chicago Press, Chicago.

Ryan, M.J. 1988. Energy, calling, and selection. American Zoologist 28:885-898.

Ryan, M.J., M.D. Tuttle, and L.K. Taft. 1981. The costs and benefits of frog chorusing behavior. Behavioral Ecology and Sociobiology 8:273-278.

Sackett, G.P. 1968. Abnormal behavior in laboratory reared rhesus monkeys. In: M.W. Fox, editor. Abnormal behavior in animals. W.B. Saunders, Philadelphia.

Sade, D.S. 1965. Some aspects of parent-offspring and sibling relations in a group of rhesus monkeys, with a discussion of grooming. American Journal of Physical Anthropology 23:1-17.

Sagan, C. 1977. The dragons of Eden. Random House, New York.

Sahlins, M. 1976. The use and abuse of biology: an anthropological critique of sociobiology. University of Michigan Press, Ann Arbor, Mich.

Sale, P.F. 1971. Apparent effect of prior experience on a habitat preference exhibited by the reef fish, *Dascyllus aru-anus* (Pisces: Pomacentridae). Animal Behaviour 19:251-256.

Sales, G., and D. Pye. 1974. Ultrasonic communication by animals. Chapman & Hall, London.

Sanders, G.D. 1975. The cephalopods. In: W.C. Corning, J.A. Dyal, and A.O.D. Willows, editors. Invertebrate learning. Vol. 3. Plenum Press, New York.

Sanger, V.L., and A.H. Hamdy. A strange fright-flight behavior pattern (hysteria) in hens. American Veterinary Medicine Association 140:455-459.

Sargant, W.W. 1957. Battle for the mind, a physiology of conversion and brainwashing. Doubleday, New York.

Sasvari, L. 1985. Different observational learning capacity in juvenile and adult individuals of congeneric bird species. Zeitschrift für Tierpsychologie 69:293-304.

Sauer, E.G.F. 1954. Die Entwicklung der Lautau Berungen vom Ei ab schalldicht gehaltener Dorngrasmucken *(Sylvia c. communis* Latham) im Vergleich mit spater isolierten und mit wildlebenden Artgenossen. Zeitschrift für Tierpsychologie 11:10-93.

Saunders, D.S. 1977. An introduction to biological rhythms. John Wiley & Sons, New York.

Savage-Rumbaugh, E.S., D.M. Rumbaugh, and S. Boysen. 1980. Do apes use language? American Scientist 68:49-61.

Savage-Rumbaugh, S. 1988. A new look at ape language: comprehension of vocal speech and syntax. Nebraska Symposium on Motivation 35:201-255.

Scanes, C.G., G.F. Merrill, R. Ford, P. Mauger, and C. Horowitz. 1980. Effects of stress (hypoglycaemia, endotoxin, and ether) on the peripheral circulating concentration of corticosterone in the domestic fowl *(Gallus domesticus)*. Comparative Biochemistry and Physiology 66C:183-186.

Schacher, S., V.F. Castellucci, E.R. Kandel. 1988. cAMP evokes long-term facilitation in *Aplysia* sensory neurons that requires new protein synthesis. Science 240:1667-1669.

Schacter, D.L. Stranger behind the engram. 1982. Erlbaum, Hillsdale, N.J.

Schaeffer, S.W., C.J. Brown, and W.W. Anderson. 1984. Does mate choice affect fitness? Genetics 107:94.

Schaller, G.B. 1967. The deer and the tiger. University of Chicago Press, Chicago.

Schaller, G.B. 1972. The Serengeti lion. University of Chicago Press, Chicago.

Schally, A.V., T.W. Redding, H.W. Lucien, and J. Meyer. 1967. Enterogasterone inhibits eating by fasted mice. Science 157:210-211.

Scharrer, B. 1935. Uber das Hanströmsche Organ X bei Opisthobranchiern. Pubbl. Staz. Zool. Napoli 15:132-142.

Scharrer, B. 1987. Neurosecretion: beginnings and new directions in neuropeptide research. Annual Review of Neuroscience 10:1-17.

Scharrer, E. 1928. Die Lichtempfindlichkeit blinder Elritzen (Untersuchungen uber das Zwischenhirn der Fische). Zeitschrift für Vergleichende Physiologie 7:1-38.

Scharrer, E. 1952. The general significance of the neurosecretory cell. Scientia 46:177-183.

Scharrer, E., and B. Scharrer. 1963. Neuroendocrinology. Columbia University Press, New York.

Scheller, R.H., J.F. Jackson, L.B. McAllister, B.S. Rothman, E. Mayeri, and R. Axel. 1983. A single gene encodes multiple neuropeptides mediating a stereotyped behavior. Cell 32:7-22.

Scheller, R.H., J.F. Jackson, L.B. McAllister, J.H. Schwartz, E.R. Kandel, and R. Axel. 1982. A family of genes that codes for ELH, a neuropeptide eliciting a stereotyped pattern of behavior in Aplysia. Cell 28:709-719.

Schenkel, R. 1956. Zur Deutung der Balzleistungen einiger Phasianiden und Tetraoniden. Ornithol. Beobacht. 53:182-201.

Schenkel, R. 1966. Play, exploration and territoriality in the wild lion. In: P.A. Jewell and C. Loizos, editors. Play, exploration, and territory in mammals. Academic Press, Inc., London.

Scherzinger, W. 1970. Zum Aktionssystem des Sperlingskauzes (Glaucidium passerinum L.) Zoologica 118:1-120.

Schiller, C.H. 1957. Instinctive behavior. International Universities Press, New York.

Schjelderup-Ebbe, T. 1935. Social behavior of birds. In: C.A. Murchison, editor. A handbook of social pscyhology. Clark University Press, Worcester, Mass.

Schleidt, M. 1988. A universal time constant operating in human short-term behaviour repetitions. Ethology 77:67-75.

Schleidt, W.M. 1961. Reaktionen von truthuhnern auf fliegende Raubvogel und Versuche zur Analyse ihrer AAM's. Zeitschrift für Tierpsychologie 18:534-560.

Schleidt, W.M. 1974. How "fixed" is the fixed action pattern? Zeitschrift für Tierpsychologie 36:184-211.

Schleidt, W.M. 1986. Book review of Dewsbury 1984 a. Ethology 72:352.

Schleidt, W.M., G. Yakalis, M. Donnelly, and J. McGarry. 1984. A proposal for a standard ethogram, exemplified by an ethogram of the bluebreasted quail (Coturnix chinensis). Zeitschrift für Tierpsychologie, 64:193-220.

Schleidt, W.M., M. Schleidt, and M. Magg. 1960. Storung der Mutter-Kind-Beiziehung bei Truthuhern durch Gehorverlust. Behaviour 16:254-260.

Schmid-Hempel, P. 1986. The influence of reward sequence on flight directionality in bees. Animal Behaviour 34:831-837.

Schmidt, J.P. 1968. Psychosomatics in veterinary medicine. In: M.W. Fox, editor. Abnormal behavior in animals. W.B. Saunders, Philadelphia.

Schmidt, R.H. 1990. Why do we debate animal rights? Wildlife Society Bulletin 18:459-461.

Schmidt-Koenig, K. 1975. Migration and homing in animals. Springer-Verlag, New York.

Schmidt-Koenig, K. 1987. Bird navigation: has olfactory orientation solved the problem? Quarterly Review of Biology 62:31-47.

Schmidt-Koenig, K., and C. Walcott. 1978. Tracks of pigeons homing with frosted lenses. Animal Behaviour 26:480-486.

Schmidt-Koenig, K., and J. Kiepenheuer. 1986. On the relevance of site-specific stimuli for pigeon homing. In: H. Ouellet. Acta XIX Congressus Internationalis Ornithologici, Vol. I, pages 317-322. University of Ottawa Press.

Schmidt-Koenig, K., and W.T. Keeton. 1978. Animal migration, navigation, and homing. Springer-Verlag, New York.

Schmidt-Nielsen, K. 1984. Scaling; why is animal size so important? Cambridge University Press, Cambridge, England.

Schneider, D. 1974. The sex-attractant receptor of moths. Scientific American 231(1):28-35.

Schneider, S.S., J.A. Stamps, and N.E. Gary. 1986. The vibration dance of the honey bee. I. Communication regulating foraging on two time scales. II. The effects of foraging success on daily patterns of vibration activity. Animal Behaviour 34:377-391.

Schneirla, T.C. 1929. Learning and orientation in ants. Comparative Psychology Monographs 6(4):1-143.

Schneirla, T.C. 1959. An evolutionary and developmental theory of biphasic processes underlying approach and withdrawal. Nebraska Symposium on Motivation 7:1-42.

Schneirla, T.C. 1971. Army ants, a study

in social organization. W.H. Freeman, San Francisco.

Schnell, G.D. 1970 a. A phenetic study of the suborder Lari (Aves). I. Methods and results of principal components analyses. Systematic Zoology 19:35-57.

Schnell, G.D. 1970 b. A phenetic study of the suborder Lari (Aves). II. Phenograms, discussion, and conclusions. Systematic Zoology 19:264-302.

Schoener, T.W. 1971. Theory of feeding strategies. Annual Review of Ecology and Systematics 2:369-404.

Schoener, T.W. 1981. An empirically based estimate of home range. Theoretical Population Biology 20:281-325.

Schoener, T.W., and A. Schoener. 1982. Intraspecific variation in home-range size in some anolis lizards. Ecology 63(3):809-823.

Scholz, K.P. and J.H. Byrne. 1987. Long-term sensitization in Aplysia: biophysical correlates in tail sensory neurons. Science 235:685-687.

Schone, H. 1984. Spatial orientation. The spatial control of behavior in animals and man. Princeton University Press, Princeton, N.J. (Translated by C. Strausfed.)

Schoner, G., and J.A.S. Kelso. 1988. Dynamic pattern generation in behavioral and neural systems. Science 239:1513-1519.

Schreiber, R.W. 1977. Maintenance behavior and communication in the brown pelican. Ornithological Monographs 22:1-78.

Schuz, E. 1971. Grundriss der Vogelzugskunde. Verlag, Paul Parey, Berlin.

Schwagmeyer, P.L. 1988. Scramble-competition polygyny in an asocial mammal: male mobility and mating success. American Naturalist 131:885-892.

Schwagmeyer, P.L., and S.J. Wootner. 1985. Mating competition in an asocial ground squirrel, Spermophilus tridecemlineatus. Behavioral Ecology and Sociobiology 17:291-296.

Schwartzbaum, J.S., and H.P. Ward. 1958. An osmotic factor in the regulation of food intake in the rat. Journal of Comparative and Physiological Psychology 51:555-560.

Sclafani, A., and C.N. Berner. 1977. Hyperphagia and obesity produced by parasagittal and coronal hypothalamic knife cuts. Journal of Comparative and Physiological Psychology 91:1000-1018.

Sclafani, A., and D. Springer. 1976. Dietary obesity: similarities to hypothalamic and human obesity syndromes. Physiology and Behavior 17:461-471.

Searcy, W.A., and K. Yasukawa. 1983.

Sexual selection and red-winged blackbirds. American Scientist 71:166-174.

Searcy, W.A., and M. Andersson. 1986. Sexual selection and the evolution of song. Annual Review of Ecology and Systematics 17:507-533.

Searcy, W.A., and P. Marler. 1981. A test for responsiveness to song structure and programming in female sparrows. Science 213:926-928.

Sebeok, T.A., editor. 1977. How animals communicate. Indiana University Press, Bloomington, Ind.

Seeley, T.D. 1985. Honeybee ecology: a study of adaptation in social life. Princeton University Press, Princeton, N.J.

Seely, M.K., and W.J. Hamilton, III. 1976. Fog catchment sand trenches constructed by tenebrionid beetles, *Lepidochora*, from the Namib Desert. Science 193:484-486.

Segal, N.L. 1985. Monozygotic and dizygotic twins: a comparative analysis of mental ability profiles. Child Development 56:1051-1058.

Seger, J., and R. Trivers. 1986. Asymmetry in the evolution of female mating preferences. Nature 319:771-773.

Selander, R.K. 1972. Sexual selection and dimorphism in birds. In: B. Campbell, editor. Sexual selection and the descent of man. Aldine, Chicago.

Seligman, M.E.P. 1975. Helplessness: on depression, development, and death. W.H. Freeman, San Francisco.

Selye, H. 1956. The stress of life. McGraw-Hill, New York.

Semon, R. 1904. die Mneme als erhaltendes Prinzip im Wechsel des organischen Geschehens. Wilhelm Engelmann, Leipzig.

Seyfarth, R.M. 1981. Do monkeys rank each other? Behavior and Brain Sciences 4:447-448.

Seyfarth, R.M., and D.L. Cheney. 1982. The ontogeny of vervet monkey alarm calling behavior: a preliminary report. Zeitschrift für Tierpsychologie 54:37-56.

Seyfarth, R.M., D.L. Cheney, and P. Marler. 1980. Monkey responses to three different alarm calls: evidence of predator classification and semantic communication. Science 210:801-803.

Shaffery, J.P., N.J. Ball, and C.J. Amlaner, Jr. 1985. Manipulating daytime sleep in herring gulls *(Larus argentatus)*. Animal Behaviour, 33:566-572.

Shannon, C.E., and W. Weaver. 1949. The mathematical theory of communication. University of Illinois Press, Urbana, Ill.

Shapiro, C.M., R. Bortz, and D. Mitchell. 1981. Slow-wave sleep: a recovery period after exercise. Science 214:1253-1254.

Shashoua, V.E. 1985. The role of extracellular proteins in learning and memory. American Scientist 73:364-370.

Shaw, K.C. 1968. An analysis of the phonoresponse of males of the true katydid, *Pterophylla camellifolia* (Fabricius) (Orthoptera: Tettigoniidae). Behaviour 31:203-260.

Shear, W.A. editor. 1986. Spiders: webs, behavior, and evolution. Standford University Press, Standford, Calif.

Shephard, R.S. 1982. Physiology and biochemistry of exercise. Praeger, New York.

Shepher, J. 1971. Mate selection among second generation kibbutz adolescents and adults: incest avoidance and negative imprinting. Archives Sex. Behav. 1:293-307.

Sherman, P.W. 1977. Nepotism and the evolution of alarm calls. Science 197:1246-1253.

Sherman, P.W. 1980. The limits of ground squirrel nepotism. In: G.W. Barlow and J. Silverberg, editors. Sociobiology: beyond nature/nurture? Westview Press, Boulder, Colo.

Sherman, P.W. 1981. Reproductive competition and infanticide in Belding's ground squirrels and other animals. In: R.D. Alexander and D.W. Tinkle, editors. Natural selection and social behavior: recent research and new theory. Chiron Press, New York.

Sherman, P.W. 1985. Alarm calls of Belding's ground squirrels to aerial predators: nepotism or self-preservation? Behavioral Ecology and Sociobiology 17:313-323.

Sherry, D.F. 1981. Parental care and the development of thermoregulation in red junglefowl. Behaviour 76:250-279.

Sherry, D.F., J.R. Krebs, and R.J. Cowie. 1981. Memory for the location of stored food in marsh tits. Animal Behaviour 29:1260-1266.

Sherwood, N.M., J.C. Wingfield, G.F. Ball, and A.M. Dufty. 1988. Identity of gonadotropin-releasing hormone in passerine birds: comparison of GnRH in song sparrow *(Melospiza melodia)* and starling *(Sturnus vulgaris)* with five vertebrate GnRHs. General and Comparative Endocrinology 69:341-351.

Shettleworth, S.J. 1983. Memory in food-hoarding birds. Scientific American 248(3):102-110.

Shettleworth, S.J. 1984. Learning and behavioural ecology. In: J.R. Krebs and N.B. Davies. Behavioural ecology. An evolutionary approach, ed 2. Sinauer Associates, Sunderland, Mass.

Sibatani, A. 1980. The Japanese brain. Science 80 1(8):22-27.

Sidman, R.L., S.H. Appel, and J.L. Fuller. 1965. Neurological mutants of the mouse. Science 150:513-516.

Sidtis, J.J., B.T. Volpe, J.D. Holtzman, D.H. Wilson, and M.S. Gazzaniga. 1981. Cognitive interaction after staged callosal section: evidence for transfer of semantic activation. Science 212:344-346.

Siegel, H.S. 1980. Physiological stress in birds. Bioscience 30:529-534.

Sih, A. 1980. Optimal behavior: can foragers balance two conflicting demands? Science 210:1041-1043.

Sijnowski, T.J., C. Koch, and P.S. Churchland. 1988. Computational neuroscience. Science 241:1299-1306.

Silberglied, R.E. 1979. Communication in the ultraviolet. Annual Review of Ecology and Systematics 10:373-398.

Silk, J.B. 1987. Social behavior in evolutionary perspective. In: B.B. Smuts, D.L. Cheney, R.M. Seyfarth, R.W. Wrangham, and T.T. Struhsaker, editors. Primate societies. University of Chicago Press, Chicago.

Silver, R. 1977. Effects of the anti-androgen cyproterone acetate on reproduction in male and female ring doves. Hormones and Behavior 9:371-379.

Silver, R. 1978. The parental behavior of ring doves. American Scientist 66:209-215.

Silver, R., and C. Barbiere. 1977. Display of courtship and incubation behavior during the reproductive cycle of the male ring dove *(Streptopelia risoria)*. Hormones and Behavior 8:8-21.

Simmons, J.A. 1979. Perception of echo phase information in bat sonar. Science 204:1336-1338.

Simmons, J.A., and R.A. Stein. 1980. Acoustic imaging in bat sonar: echolocation signals and the evolution of echolocation. Journal of Comparative Physiology 135:61-84.

Simmons, J.A., M.B. Fenton, and M.J. O'Farrell. 1979. Echolocation and pursuit of prey by bats. Science 203:14-21.

Simmons, L.W. 1986. Intermale competition and mating success in the field cricket, *Gryllus bimaculatus* (De Geer). Animal Behaviour 34:567-579.

Simmons, L.W. 1987. Female choice contributes to offspring fitness in the field cricket, *Gryllus bimaculatus* (De Geer). Behavioral Ecology and Sociobiology 21:313-321.

Simon, C. 1979. Debut of the seventeen-year-old cicada. Natural History 88(5):38-45.

Simon, D., and R.H. Barth. 1977. Sexual

behavior in the cockroach genera *Periplaneta* and *Blatta*. I. Descriptive aspects. Zeitschrift für Tierpsychologie 44:80-107.

Simpson, G.G. 1970. Uniformitarianism: an inquiry into principle, theory, and method in geohistory and biohistory. In: M.K. Hecht and W.C. Steere, editors. Essays in evolution and genetics in honor of Theodosius Dobzhansky. Appleton-Century-Crofts, New York.

Simpson, M.J.A., and A.E. Simpson. 1977. One-zero and scan methods for sampling behaviour. Animal Behaviour 25:726-731.

Simpson, S.J., M.S.J. Simmonds, A.R. Wheatley, and E.A. Bernays. 1988. The control of meal termination in the locust. Animal Behaviour 36:1216-1227.

Sims, E.A.H. 1976. Experimental obesity, dietary-induced thermogenesis, and their clinical implications. Clinical Endocrinology and Metabolism 5:377-395.

Singer, A. 1978. Cranes of the world. Audubon 80(2):17-24.

Sivinski, J. 1978. Intrasexual aggression in the stick insect *Diapheromera veliei* and *D.* covilleae, and sexual dimorphism in the Phasmatodea. Psyche 85:395-406.

Sivinski, J. 1984. Effect of sexual experience on male mating success in a lek forming tephritid *Anastrepha suspensa* (Loew). Florida Entomologist 67:126-130.

Skinner, B.F. 1938. The behavior of organisms. Appleton-Century-Crofts, New York.

Skinner, B.F. 1960. Pigeons in a pelican. American Psychologist 15:28-37.

Skinner, B.F. 1974. About behaviorism. Random House, New York.

Skinner, B.F. 1976. Particulars of my life. Alfred A. Knopf, New York.

Skinner, B.F. 1979. The shaping of a behaviorist: part two of an autobiography. Alfred A. Knopf, New York.

Skinner, B.F. 1981. Selection by consequences. Science 212:501-504.

Skinner, B.F. 1989. The origins of cognitive thought. American Psychologist 44:13-18.

Slater, P.J.B. 1986. Individual differences and dominance hierarchies. Animal Behaviour 34:1264-1265.

Sluckin, W. 1979. Fear in animals and man. Van Nostrand Reinhold, New York.

Small, W.S. 1901. Experimental study of the mental processes of the rat. American Journal of Psychology 12:206-239.

Smartt, R.A. 1978. A comparison of ecological and morphological overlap in a *Peromyscus* community. Ecology 59(2):216-220.

Smith, D.C. 1985. Home range and territory in the striped plateau lizard *(Sceloporus virgatus)*. Animal Behaviour 33:417-427.

Smith, D.G. 1972. The role of the epaulets in the red-winged blackbird *(Agelaius phoeniceus)* social system. Behaviour 41:251-268.

Smith, D.G. 1976. An experimental analysis of the function of red-winged blackbird song. Behaviour 56:136-156.

Smith, D.G. 1979. Male singing ability and territory integrity in male red-winged blackbirds *(Agelaius phoeniceus)*. Behaviour 68:193-206.

Smith, E.O., editor. 1977. Social play in primates. Academic Press, New York.

Smith, J.N.M. 1974. The food searching paths of two European thrushes. I: Description and analysis of search paths. Behaviour 48:276-302.

Smith, J.N.M., and H.P. Sweatman. 1974. Food searching behaviour of titmice in patchy environments. Ecology 55:1216-1232.

Smith, J.N.M., and R. Dawkins. 1971. The hunting behaviour of individual great tits in relation to spatial variations in their food density. Animal Behaviour 19:695-706.

Smith, N.G. 1968. The advantages of being parasitized. Nature 219:690-694.

Smith, P.K., editor. 1986. Play in animals and humans. Basil Blackwell, New York.

Smith, R.J.F. 1985. The control of fish migration. Springer-Verlag, Berlin.

Smith, R.L. 1979. Repeated copulation and sperm precedence: paternity assurance for a male brooding water bug. Science 205:1029-1031.

Smith, S. 1983. Ideas of the great psychologists. Barnes & Noble Books, Philadelphia.

Smith, W.J. 1977. The behavior of communicating. Harvard University Press, Cambridge, Mass.

Smuts, B. 1987. What are friends for? Natural History 96(2):36-45.

Smuts, B.B. 1985. Sex and friendship in baboons. Aldine Press, Hawthorne, N.Y.

Sneath, P.H.A., and R.R. Sokal. 1973. Numerical taxonomy. W.H. Freeman, San Francisco.

Snodgrass, R.E. 1935. Principles of insect morphology. McGraw-Hill, New York.

Snowdon, C.T. 1969. Motivation, regulation, and the control of meal parameters with oral and intragastric feeding. Journal of Comparative and Physiological Psychology 69:91-100.

Snowdon, C.T. 1975. Production of satiety with small intraduodenal infusions in the rat. Journal of Comparative and Physiological Psychology 88:231-238.

Snyder, N.F.R., and J.D. Taapken. 1977. Puerto Rican parrots and nest predation by pearly-eyed thrashers. In: S.A. Temple, editor. Endangered birds. University of Wisconsin Press, Madison, Wis.

Snyder, S.H. 1980. Brain peptides as neurotransmitters. Science 209:976-983.

Solso, R.L. 1988. Cognitive psychology. Allyn and Bacon, Boston.

Southwick, E.E., and G. Heldmaier. 1987. Temperature control in honey bee colonies. Bioscience 37:395-399.

Sparks, J. 1982. The discovery of animal behaviour. Little, Brown and Co., Boston.

Spear, N.E., and B.A. Campbell. 1979. Ontogeny of learning and memory. Halsted Press, New York.

Spear, N.E., and R.R. Miller, editors. 1981. Information processing in animals. Lawrence Erlbaum Associates, Hillsdale, N.J.

Spencer, H. 1855. Principles of psychology. D. Appleton, New York.

Spencer, H. 1892. The principles of ethics. Vol. I. D. Appleton, New York.

Spencer, H. 1896. Principles of psychology, ed 2. D. Appleton, New York.

Sperry, R.W. 1970. Perception in the absence of the neocortical commissures. In: Perception and its disorders. Research Publications of the Association of Research on Nervous and Mental Disease. Vol. 48.

Sperry, R.W. 1974. Lateral specialization in the surgically separated hemispheres. In: F.O. Schmitt and F.G. Worden, editors. The neurosciences: third study program. MIT Press, Cambridge, Mass.

Sperry, R.W. 1982. Some effects of disconnecting the cerebral hemispheres. Science 217:1223-1226.

Spetch, M.L., and C.A. Edwards. 1988. Pigeons, *Columba livia*, use of global and local cues for spatial memory. Animal Behaviour 36:293-296.

Spiess, E.B., and J.F. Kruckeberg. 1980. Minority advantage of certain eye color mutants of *Drosophila melanogaster*. II. A behavioral basis. American Naturalist 115:307-327.

Spieth, H.T. 1974. Courtship behavior in *Drosophila*. Annual Review of Entomology 19:385-405.

Sprague, R.H., and J.J. Anisko. 1973. Elimination patterns in the laboratory beagle. Behaviour 47:257-267.

Springett, B.P. 1968. Aspects of the relationship between burying beetles, *Necrophorus* spp. and the mite *Poecil-*

ochirus necrophori Vitz. Journal of Animal Ecology 37:417-424.

Squire, L.R. 1987. Memory and the brain. Oxford University Press, New York.

Stallcup, J.A., and G.E. Woolfenden. 1978. Family status and contributions to breeding by Florida scrub jays. Animal Behaviour 26:1144-1156.

Stanley, S.M. 1974. Relative growth of the titanothere horn: a new approach to an old problem. Evolution 28:447-457.

Stapanian, M.A., and C.C. Smith. 1978. A model for seed scatterhoarding: coevolution of fox squirrels and black walnuts. Ecology 59:884-896.

Stefansson, V. 1944. The friendly arctic. Macmillan, New York.

Steinberg, J.B. 1977. Information theory as an ethological tool. In: Hazlett, B.A., editor. Quantitative methods in the study of animal behavior. Academic Press, New York.

Steinberg, J.B., and R.C. Conant. 1974. An informational analysis of the intermale behaviour of the grasshopper *Chortophaga viridifasciata*. Animal Behaviour 22:617-627.

Stellar, E. 1954. The physiology of motivation. Psychological Review 61:5-22.

Stenseth, N.C. 1983. Causes and consequences of dispersal in small animals. In: I.R. Swingland and P.J. Greenwood, editors. The ecology of animal movement. Clarendon Press, Oxford, England.

Stenseth, N.C. 1988. Book review of Kamil et al. (1984). Science 240:1212-1213.

Stephens, D.W., and J.R. Krebs. 1986. Foraging theory. Princeton University Press, Princeton, N.J.

Stephenson, G.R., and T.W. Roberts. 1977. The SSR system 7: a general encoding system with computerized transcription. Behavioral Research Methods and Instrumentation 9:434-441.

Sternberg, R.J. 1986. Inside intelligence. American Scientist 74:137-143.

Stetson, M.H., and M. Watson-Whitmyre. 1976. Nucleus suprachiasmaticus: the biological clock in the hamster? Science 191:197-199.

Stevens, C.F. 1966. Neurophysiology: a primer. John Wiley & Sons, New York.

Stinson, C.H. 1979. On the selective advantage of fratricide in raptors. Evolution 33: 1219-1225.

Stockman, E.R., H.E. Albers, and M.J. Baum. 1985. Activity in the ferret: oestradiol effects and circadian rhythms. Animal Behaviour, 33:150-154.

Stoddard, P.K. 1988. The "bugs" call of the cliff swallow: a rare food signal in a colonially nesting bird species. Condor 90:714-715.

Stoddart, D.M. 1976. Mammalian odours and pheromones. Camelot Press, Ltd., London.

Storm, R.M. 1967. Animal orientation and navigation. Oregon State University Press, Corvallis, Ore.

Stowe, M.K., J.H. Tumlinson, and R.R. Heath. 1987. Chemical mimicry: bolas spiders emit components of moth prey species sex pheromones. Science 236:964-967.

Strassmann, J.E. 1988. Worker reproduction in social insects. Trends in Ecology and Evolution 3:286-287.

Street, P. 1976. Animal migration and navigation. Charles Scribner's Sons, New York.

Stricker, E.M., and J.G. Verbalis. 1988. Hormones and behavior: the biology of thirst and sodium appetite. American Scientist 76:261-267.

Sulak, K.J. 1975. Cleaning behaviour in the centrarchid fishes, *Lepomis macrochirus* and *Micropterus salmoides*. Animal Behaviour 23:331-334.

Suomalainen, E. 1969. Evolution in parthenogenetic Curculionidae. Evolutionary Biology 3:261-296.

Suomi, S.J., and H.F. Harlow. 1972. Social rehabilitation of isolate-reared monkeys. Developmental Psychology 6:487-496.

Sutherland, N.S., and N.J. Mackintosh. 1971. Mechanisms of animal discrimination learning. Academic Press, New York.

Sutherland, W.J. 1985. Measures of sexual selection. Oxford Surveys in Evolutionary Biology 2:90-101.

Sutherland, W.J., and D. Moss. 1985. The inactivity of animals: influence of stochasticity and prey size. Behaviour 92:1-8.

Swenson, E.A. 1929. The active simple movements of the albino rat fetus: the order of their appearance, their qualities, and their significance. Anatomical Record 42:40.

Taigen, T.L., K.D. Wells, and R.L. Marsh. 1985. The enzymatic basis of high metaboloic rates in calling frogs. Physiological Zoology 58:719-726.

Takahashi, J.S., and M. Zatz. 1982. Regulation of circadian rhythmicity. Science 217:1104-1111.

Takei, Y., and H. Kobayashi. 1980. Angiotensin and drinking behavior in birds. In: Y. Tanabe, K. Tanaka, and T. Ookawa, editors. Biological rhythms in birds. Neural and endocrine aspects. Japan Scientific Societies Press, Tokyo/Springer-Verlag, New York.

Tallamy, D.W. 1984. Insect parental care. Bioscience 34:20-24.

Tamm, G.R., and J.S. Cobb. 1978. Behavior and the crustacean molt cycle: changes in aggression of *Homarus americanus*. Science 200: 79-81.

Tank, D.W. and J.J. Hopfield. 1987. Collective computation in neuronlike circuits. Scientific American 257(6):104-114.

Tavolga, W.N., editor. 1976. Sound reception in fishes. Benchmark papers in animal behavior. Vol. 7. Dowden, Hutchinson, & Ross, Stroudsburg, Pa.

Tavolga, W.N., and J. Wodinsky. 1963. Auditory capacities in fishes: pure tone thresholds in nine species of marine teleosts. Bulletin American Museum Natural History 126:177, 179-239. (Reprinted in Tavolga 1976.)

Taylor, E.L. 1962. The place of animal behavior studies in veterinary science. Veterinary Record 74:521-524.

Tegner, M.J., and P.K. Dayton. 1976. Sea urchin recruitment patterns and implications of commercial fishing. Science 196:324-326.

Teitelbaum, P., and A.N. Epstein. 1962. The lateral hypothalamic syndrome: recovery of feeding and drinking after lateral hypothalamic lesions. Psychological Review 69:74-90.

Temple, S.A. 1974. Plasma testosterone titers during the annual cycle of starlings, *Sturnus vulgaris*. General and Comparative Endocrinology 22:470-479.

Temple, S.A. 1977 a. Plant-animal mutualism: coevolution with dodo leads to near extinction of plant. Science 197:885-886.

Temple, S.A., editor. 1977 b. Endangered birds: management techniques for preserving threatened species. University of Wisconsin Press, Madison, Wis.

Terman, G.W., Y. Shavit, J.W. Lewis, J.T. Cannon, J.C. Liebeskind. 1984. Intrinsic mechanisms of pain inhibition: activation by stress. Science 226:1270-1277.

Terrace, H.S., L.A. Petitto, R.J. Sanders, and T.G. Bever. 1979. Can an ape create a sentence? Science 206:891-902.

Thiessen, D.D., K. Owen, and M. Whitsett. 1970. Chromosome mapping of behavioral activities. In: G. Lindzey and D.D. Thiessen, editors. Contributions to behavior genetic analysis: the mouse as a prototype. Appleton-Century-Crofts, New York.

Thomas, D.W., and J. Mayer. 1978. Meal size as a determinant of food intake in normal and hypothalamic obese rats. Physiology and Behavior 21:113-117.

Thomas, R.D.K., and E.C. Olson. 1980. A cold look at the warm-blooded dinosaurs. American Association for the Advancement of Science (AAAS) Selected Symposium 28. Westview Press, Boulder, Colo.

Thompson, C.R., and R.M. Church. 1980. An explanation of the language of a chimpanzee. Science 208:313-314.

Thompson, J.N. 1982. Interaction and coevolution. John Wiley & Sons, New York.

Thompson, R.F. 1986. The neurobiology of learning and memory. Science 233:941-947.

Thompson, R.F., and W.A. Spencer. 1966. Habituation: a model phenomenon for the study of neuronal substrates of behavior. Psychological Review 73:16-43.

Thorndike, E.L. 1911. Animal intelligence: experimental studies. Macmillan, Inc., New York.

Thornhill, R. 1980. Competitive, charming males and choosy females: was Darwin correct? Florida Entomologist 63:5-30.

Thornhill, R. 1980. Rape in *Panorpa* scorpionflies and a general rape hypothesis. Animal Behaviour 28:52-59.

Thornhill, R. 1980. Sexual selection in the black-tipped hangingfly. Scientific American 242(6):162-172.

Thornhill, R. 1981. *Panorpa* (Mecoptera: Panorpidae) scorpionflies: systems for understanding resource-defense polygyny and alternative male reproductive efforts. Annual Review of Evolution and Systematics 12:355-386.

Thornhill, R. 1986. Relative parental contribution of the sexes to their offspring and the operation of sexual selection. In: M.H. Nitecki, and J.A. Kitchell, editors. Evolution of animal behavior: paleontological and field approaches. Oxford University Press, New York.

Thornhill, R., and J. Alcock. 1983. The evolution of insect mating systems. Harvard University Press, Cambridge, Mass.

Thornhill, R., and N.W. Thornhill. 1983. Human rape: an evolutionary analysis. Ethology and Sociobiology 4:137-173.

Thorpe, W.H. 1948. The modern concept of instinctive behaviour. Bulletin of Animal Behaviour 7:2-12.

Thorpe, W.H. 1951. The definition of terms used in animal behaviour studies. Bulletin of Animal Behaviour 9:34-40.

Thorpe, W.H. 1951. The learning abilities of birds. Ibis 93(1):251.

Thorpe, W.H. 1958. The learning of song patterns by birds, with special reference to the song of the chaffinch, *Fringilla coelebs*. Ibis 100:535-570.

Thorpe, W.H. 1963. Learning and instinct in animals, ed 2. Methuen, London.

Thorpe, W.H. 1979. The origins and rise of ethology. Heinemann Educational Books, London.

Tinbergen, L. 1960. The natural control of insects in pinewoods. I. Factors influencing the intensity of predation by song birds. Arch. Neerl. Zool. 13:265-343.

Tinbergen, N. 1942. An objectivistic study of the innate behavior of animals. Bibliotheca Biotheoretica 1:39-98.

Tinbergen, N. 1948. Social releasers and the experimental method required for their study. Wilson Bulletin 60:6-51.

Tinbergen, N. 1951. The study of instinct. Oxford University Press, New York.

Tinbergen, N. 1952. The curious behavior of the stickleback. Scientific American 187(6):22-26.

Tinbergen, N. 1953. The herring gull's world. William Collins Sons, London.

Tinbergen, N. 1954. The origin and evolution of courtship and threat display. In: J. Huxley, A.C. Hardy, and E.B. Ford, editors. Evolution as a process. George Allen and Unwin, London.

Tinbergen, N. 1958 (reprinted 1968). Curious naturalists. Natural History Library, Anchor Books, Doubleday, Garden City, N.Y.

Tinbergen, N. 1959. Comparative studies of the behavior of gulls (Laridae): a progress report. Behaviour 15:1-70.

Tinbergen, N. 1960. The evolution of behavior in gulls. Scientific American 203(6):118-130.

Tinbergen, N. 1965. Animal behavior. Time-Life Books, New York.

Toates, F.M., and T.R. Halliday. 1980. Analysis of motivational processes. Academic Press, New York.

Tobias, J.M. 1952. Some optically detectable consequences of activity in nerve. Cold Spring Harbor Symposia on Quantitative Biology 17:15-25.

Tolman, E.C. 1932. Purposive behavior in animals and men. Century, New York.

Topoff, H., and J. Mirenda. 1980. Army ants do not eat and run: influence of food supply on emigration behaviour in *Neiv amyrmex nigrescens*. Animal Behaviour 28:1040-1045.

Treat, A.E. 1975. Mites of moths and butterflies. Cornell University Press, Ithaca, N.Y.

Trivers, R. 1985. Social evolution. Benjamin/Cummings, Menlo Park, Calif.

Trivers, R.L. 1971. The evolution of reciprocal altruism. Quarterly Review of Biology 46:35-57.

Trivers, R.L. 1972. Parental investment and sexual selection. In: B. Campbell, editor. Sexual selection and the descent of man 1871-1971. Aldine, Chicago.

Trivers, R.L. 1974. Parent-offspring conflict. American Zoologist 14:249-264.

Trivers, R.L., and H. Hare. 1976. Haplodiploidy and the evolution of the social insects. Science 191:249-263.

Trumler, E. 1959. Das "Rossigkeitsgesicht" und ahnliches Ausdrucksverhalten bei Einhufern. Zeitschrift für Tierpsychologie 16:478-488.

Tsingalia, H.M., and T.E. Rowell. 1984. The behaviour of adult male blue monkeys. Zeitschrift für Tierpsychologie, 64:253-268.

Tucker, D.W. 1959. A new solution to the Atlantic eel problem. Nature 183:495-501.

Tuffery, A.A. 1987. Laboratory animals: an introduction for new experimenters. John Wiley & Sons, Chinchester, N.J.

Turner, E.R.A. 1961. Survival values of different methods of camouflage as shown in a model population. Proceedings of the Zoological Society of London 136:273-284.

Tuttle, M.D. 1988. America's neighborhood bats. University of Texas Press, Austin, Tex.

Tuttle, M.D., and M.J. Ryan. 1981. Bat predation and the evolution of frog vocalizations in the neotropics. Science 214:677-678.

Tyler, M.J., and D.B. Carter. 1981. Oral birth of the young of the gastric brooding frog *Rheobatrachus silus*. Animal Behaviour 29:280-282.

Tyler, S. 1979. Time-sampling: a matter of convention. Animal Behaviour 27:801-810.

Tzeng, Ovid J.L., William S.Y. Wang. 1983. The first two r's. The way different languages reduce speech to script affect how visual information is processed in the brain. American Scientist 71:238-243.

Ubico, S.R., G.O. Maupin, K.A. Fagerstone, and R.G. McLean. 1988. A plague epizootic in the white-tailed prairie dogs *(Cynomys leucurus)* of Meeteetse, Wyoming. Journal of Wildlife Diseases 24:399-406.

Ueda, T. 1979. Plasticity of the reproductive behaviour in a dragonfly, *Sympetrum parvulum* Barteneff, with reference to the social relationship of males and the density of territories. Research on Population Ecology 21:135-152.

Uemura, H., H. Kobayashi, Y. Okawara, and K. Yamaguchi. 1983. Neuropeptides and drinking in birds. In: S. Mikami, K. Homma, and M. Wada, editors. Avian endocrinology. Environmental and ecological perspectives. Japan Scientific Societies Press, Tokyo/Springer-Verlag, New York.

Ullman, S. 1986. Artificial intelligence and the brain: computational studies of the visual system. Annual Review of Neuroscience 9:1-26.

Underwood, H. 1977. Circadian organization in lizards: the role of the pineal organ. Science 195:587-589.

Ungar, G., editor. 1970. Molecular mechanisms in memory and learning. Plenum Press, New York.

Universities Federation for Animal Welfare. 1972. The U.F.A.W. handbook on the care and management of laboratory animals. Churchill Livingstone, London.

Urquhart, F.A. 1976. Found at last: the monarch's winter home. National Geographic 150:160-173.

Urquhart, F.A. 1987. The monarch butterfly: international traveler. Nelson-Hall, Chicago.

Urquhart, F.A., and N.R. Urquhart. 1979. Breeding areas and overnight roosting locations in the northern range of the monarch butterfly (Danaus plexippus plexippus) with a summary of associated migratory routes. Canadian Field-Naturalist 93:41-47.

Ursin, H. 1985. The instrumental effects of emotional behavior. In: P.P.G. Bateson and P.H. Klopfer. Perspectives in ethology. Vol. 6. Plenum Press, New York.

Valenstein, E.S., V.C. Cox, and J.W. Kakolewski. 1970. Reexamination of the role of the hypothalamus in motivation. Psychological Review 77:16-31.

van Abeelen, J.H.F. 1979. Genetic analysis of locomotor activity in immature mice from two inbred strains. Behavioral Biology 27:214-217.

van den Assem, J., and P. Sevenster, editors. 1985. First international symposium on stickleback behaviour. Behaviour 93:1-265.

van den Berghe, E.P. 1988. Piracy as an alternative reproductive tactic for males. Nature 334:697-698.

van Der Kloot, W.G. 1956. Brains and cocoons. Scientific American 194(4):131-140.

van der Pijl, L., and C.H. Dodson. 1966. Orchid flowers. University of Miami Press, Coral Gables, Fla.

van Lawick, H., and J. van Lawick-Goodall. 1970. Innocent killers. Ballantine Books, New York.

van Lawick-Goodall, J. 1968. The behaviour of free-living chimpanzees in the

Gombe Stream Reserve. Animal Behaviour Monographs 1:165-311.

van Lawick-Goodall, J., and H. van Lawick. 1965. New discoveries among Africa's chimpanzees. National Geographic 128:802-831.

van Oye, P. 1920. Over het geheugen by fr flatwormen en andere biologische waarnemingen bji deze dieren. Natuurwet. Tydschr. 2:1.

van Rhijn, J.G. 1977. The patterning of preening and other comfort behaviour in a herring gull. Behaviour 63:71-109.

van Tets, G.F. 1965. A comparative study of some social communication patterns in the Pelecaniformes. Ornithological Monographs 2:1-88.

van Tienhoven, A. 1968. Reproductive physiology of vertebrates. W.B. Saunders, Philadelphia.

Vance, D.R., and R.L. Westemeier. 1979. Interactions of pheasants and prairie chickens in Illinois. Wildlife Society Bulletin 7(4):221-225.

Verbeek, N.A.M. 1985. Behavioural interactions between avian predators and their avian prey: play behaviour or mobbing? Zeitschrift für Tierpsychologie 67:204-214.

Verner, J., and M.F. Willson. 1966. The influence of habitats on mating systems of North American passerine birds. Ecology 47:143-147.

Vieth, W., E. Curio, and U. Ernst. 1980. The adaptive significance of avian mobbing. III. Cultural transmission of enemy recognition in blackbirds: cross-species tutoring and properties of learning. Animal Behaviour 28:1217-1229.

Vilberg, T.R., and W.W. Beatty. 1975. Behavioral changes following VMH lesions in rats with controlled insulin levels. Pharmacology, Biochemistry and Behavior 3:377-384.

Vincent, L.E., and M. Bekoff. 1978. Quantitative analyses of the ontogeny of predatory behaviour in coyotes, Canis latrans. Animal Behaviour 26:225-231.

Visscher, P.K., and T.D. Seeley. 1982. Foraging strategy of honeybee colonies in a temperate deciduous forest. Ecology 63:1790-1801.

Visser, M. 1981. Prediction of switching and counterswitching based on optimal foraging. Zeitschrift für Tierpsychologie 55:129-138.

Voith, V.L., and P.L. Borchelt, editors. 1982. Symposium on animal behavior. The veterinary clinics of North America/small animal practice. Vol. 12, No. 4. W.B. Saunders, Philadelphia.

von Frisch: see under Frisch

von Holst, E., and U. von Saint Paul. 1960, 1963. On the functional organi-

zation of drives. Animal Behaviour 11:1-20. (Translated from Naturwiss. 18:409-422.)

vom Saal, F.S., W.M. Grant, C.W. McMullen, K.S. Laves. 1983. High fetal estrogen concentrations: correlation with increased adult sexual activity and decreased aggression in male mice. Science 220:1306-1309.

von Uexkull, J. 1909. Umwelt und Innerwelt der tiere. Springer-Verlag, Berlin.

von Uexkull, J. 1934. A stroll through the worlds of animals and men. In: C.H. Schiller, editor. 1957. Instinctive behavior. International Universities Press, New York.

Waage, J.K. 1979. Dual function of the damselfly penis: sperm removal and transfer. Science 203:916-918.

Waage, J.K. 1984. Sperm competition and the evolution of odonate mating systems. In: R.L. Smith, editor. Sperm competition and the evoution of animal mating systems. Academic Press, New York.

Waage, J.K. 1986. Evidence for widespread sperm displacement ability among Zygoptera (Odonata) and the means for predicting its presence. Journal of the Linnean Society of London, Biology 28:285-300.

Wade, M.J. 1987. Measuring sexual selection. In: J.W. Bradbury and M.B. Andersson, editors. Sexual selection: testing the alternatives. John Wiley & Sons, Chichester, N.J.

Wade, M.J., and S.J. Arnold. 1980. The intensity of sexual selection in relation to male sexual behaviour, female choice, and sperm precedence. Animal Behaviour 28:446-461.

Wade, N. 1980. Does man alone have language? Apes reply in riddles, and a horse says neigh. Science 208:1349-1351.

Walcott, C. 1969. A spider's vibration receptor: its anatomy and physiology. American Zoologist 99:133-144.

Walcott, C. 1986. Homing in pigeons: are differences in experimental results due to different home-loft environments? In: H. Ouellet. Acta XIX Congressus Internationalis Ornithologici. Vol. I. University of Ottawa Press, Ottawa, Canada.

Walcott, C., and R.P. Green. 1974. Orientation of homing pigeons altered by a change in the direction of an applied magnetic field. Science 184:180-182.

Walcott, C., J.L. Gould, and J.L. Kirschvink. 1979. Pigeons have magnets. Science 205:1027-1028.

Waldrop, M. Mitchell. 1988. A landmark in speech recognition. Science 240:1615.

Waldrop, M. Mitchell. 1988. Toward a unified theory of cognition. Science 241:27-29

Waldvogel, J.A., J.B. Phillips, and A.I. Brown. 1988. Changes in the short-term deflector loft effect are linked to the sun compass of homing pigeons. Animal Behaviour 36:150-158.

Walker, E.P., F. Warnick, K.I. Lange, H.E. Uible, S.E. Hamlet, M.A. Davis, and P.F. Wright. 1975. Mammals of the World, Vol. II. Johns Hopkins University Press, Baltimore.

Walker, W.F. 1980. Sperm utilization strategies in nonsocial insects. American Naturalist 115: 780-799.

Walkinshaw, L.H. 1973. Cranes of the world. Winchester Press, Tulsa, Okla.

Wallace, R.L. 1980. Ecology of sessile rotifers. Hydrobiologia 73:181-193.

Wallin, A. 1988. The genetics of foraging behaviour: artificial selection for food choice in larvae of the fruitfly, Drosophila melanogaster. Animal Behaviour, 36:106-114.

Wallraff, H.G. 1966. Uber die Heimfindeleistungen von Brieftauben nach Haltung in verschiedenartig ab geschirmten Volieren. Zeitschrift für Vergleichende Physiologie 52:215-259.

Wallraff, H.G. 1970. Uber die Flugrichtungen verfrachteter Brieftauben in Abhangigkeit vom Heimatort und vom Ort der Freilassung. Zeitschrift für Tierpsychologie 27:303-35l.

Wallraff, H.G. 1974. Das Navigationssystem der Vogel. Oldenbourg, Munchen Wien.

Wallraff, H.G., and U. Sinsch. 1988. The role of "outward-journey information" in homing experiments with pigeons: new data on ontogeny of navigation and general survey. Ethology 77:10-27.

Walsh, C., and C.L. Cepko. 1988. Clonally related cortical cells show several migration patterns. Science 241:1342-1345.

Walter, H.E. 1908. The reactions of planaria to light. Journal of Experimental Zoology 5:35-163.

Walters, J.F., and R.M. Seyfarth. 1986. Conflict and cooperation. In: B.B. Smuts, D.L. Cheney, R.M. Seyfarth, R.W. Wrangham, and T.T. Struhsaker, editors. Primate societies. University of Chicago Press, Chicago.

Waltz, E.C. 1987. A test of the information-centre hypothesis in two colonies of common terns, Sterna hirundo. Animal Behaviour 35:48-59.

Waltz, E.C., and L.L. Wolf. 1984. By Jove! Why do alternative mating tactics assume so many different forms? American Zoologist 24:333-343.

Wampler, R.S., and C.T. Snowdon. 1979. Development of VMH obesity in vagotomized rats. Physiology and Behavior 22:85-93.

Ward, P. 1965. Feeding ecology of the black-faced dioch (Quelea quelea) in Nigeria. Ibis 107:173-214.

Ward, P., and A. Zahavi. 1973. The importance of certain assemblages of birds as "information-centres" for food-finding. Ibis 115:517-534.

Ward, P.I. 1988. Sexual selection, natural selection, and body size in Gammarus pulex (Amphipoda). American Naturalist 131:348-359.

Ward, P.S. 1983. Genetic relatedness and colony organisation in a species complex of pomerine ants. II. Patterns of sex ratio investment. Behavioral Ecology and Sociobiology 12:301- 307.

Waring, G.H. 1983. Horse behavior. Noyes Publications, Park Ridge, N.J.

Warren, J.M. 1973. Learning in vertebrates. In: D.A. Dewsbury and D.A. Rethlingshafer, editors. Comparative psychology. McGraw-Hill, New York.

Wasserman, P.D. 1989. Neural computing: theory and practice. Van Nostrand and Reinhold, New York.

Waterman, T.H. 1989. Animal navigation. Scientific American Library, New York.

Watkins, L.R., and D.J. Mayer. 1982. Organization of endogenous opiate and nonopiate pain control systems. Science 216:1185-1192.

Watson, D.D., and F.H.C. Crick. 1953. Molecular structure of nucleic acids. Nature 171:737-738.

Watson, J.B. 1913. Psychology as the behaviorist views it. Psychological Review 20:158-177.

Watson, J.B. 1930. Behaviorism. W.W. Norton, New York.

Watson, R.I. 1971. The great psychologists, ed 3. J.B. Lippincott, Philadelphia.

Weatherhead, P.J. 1983. Two principal strategies in avian communal roosts. American Naturalist 121:237-243.

Weatherhead, P.J. 1985. The birds' communal connection. Natural History 94(2):34-41.

Weatherhead, P.J. 1987. Field tests of information transfer in communally roosting birds. Animal Behaviour 35:614-615.

Wecker, S.C. 1963. The role of early experience in habitat selection by the prairie deermouse, Peromyscus maniculatus bairdi. Ecological Monographs 33:307-325.

Wecker, S.C. 1964. Habitat selection. Scientific American 211(4):109-116.

Weeks, H.P., Jr. 1977. Abnormal nest building in the eastern phoebe. Auk 94:367-369.

Wehner, R. 1976. Polarized-light navigation by insects. Scientific American 235(1):106-114.

Weichel, K., G. Schwager, P. Heid, H.R. Guttinger, and A. Pesch. 1986. Sex differences in plasma steroid concentrations and singing behaviour during ontogeny in canaries (Serinus canaria). Ethology 73:281-294.

Weihs, D. 1973. Hydrodynamics of fish schooling. Nature 241:290-291.

Weiss, B.A., and T.C. Schneirla. 1967. Intersituational transfer in the ant Formica schaufussi as tested in a two-phase single choice point maze. Behaviour 28:269-279.

Wells, K.D., and T.L. Taigen. 1986. The effect of social interactions on calling energetics in the gray treefrog (Hyla versicolor). Behavioral Ecology and Sociobiology 19:9-18.

Wells, M.J. 1958. Factors affecting reactions to Mysis by newly hatched Sepia. Behaviour 13:96-111.

Wells, M.J., and S.K.L. Buckley. 1972. Snails and trails. Animal Behaviour 20:345-355.

Wenner, A.M. 1964. Sound communication in honeybees. Scientific American 210(4):116-124.

Wenner, A.M. 1971. The bee language controversy. Educational Programs Improvement Corporation, Boulder, Colo.

Werner, E.E., and D.J. Hall. 1976. Niche shifts in sunfishes: experimental evidence and significance. Science 191:404-406.

West, M.J., and A.P. King. 1985. Social guidance of vocal learning by female cowbirds: validating its functional significance. Zeitschrift für Tierpsychologie 70:225-235.

West, M.J., A.P. King, and D.H. Eastzer. 1981. The cowbird: reflections on development from an unlikely source. American Scientist 69:56-66.

West-Eberhard, M.J. 1975. The evolution of social behavior by kin selection. Quarterly Review of Biology 50:1-33.

West-Eberhard, M.J. 1978 a. Polygyny and the evolution of social behavior in wasps. Journal of the Kansas Entomological Society 51:832-856.

West-Eberhard, M.J. 1978 b. Temporary queens in Metapolybia wasps: nonreproductive helpers without altruism? Science 200:441-443.

Westerman, R.A. 1963. A study of the habituation of responses to light in the planarian Dugesia dorotocephala. Worm Runner's Digest 5:6-11.

Wever, R.A. 1979. The circadian system

in man. Springer-Verlag. New York.

Whalen, D.H., and A.M. Liberman. 1987. Speech perception takes precedence over nonspeech perception. Science 237:169-171.

White, D.H., and D. James. 1978. Differential use of fresh water environments by wintering waterfowl of coastal Texas. Wilson Bulletin 90(1):99-111.

White, G. 1788-1789. The natural history of Selbourne. Penguin Books, London (1977 printing).

White, M.J.D. 1970. Heterozygosity and genetic polymorphism in parthenogenetic animals. In: M.K. Hecht and W.C. Steere, editors. Essays in evolution and genetics. Appleton-Century-Crofts, New York.

Whitman, D.W. 1987. Thermoregulation and daily activity patterns in a black desert grasshopper, *Taeniopoda eques*. Animal Behaviour 35:1814-1826.

Whitney, C.L., and J.R. Krebs. 1975. Mate selection in Pacific tree frogs. Nature 255:325-326.

Whittaker, R.H., and P.P. Feeney. 1971. Allelochemics: chemical interactions between species. Science 171:757-770.

Whittier, J.M., R.T. Mason, and D. Crews. 1987. Plasma steroid hormone levels of female red-sided garter snakes, *Thamnophis sirtalis parietalis*: relationship to mating and gestation. General and Comparative Endocrinology 67:33-43.

Wickler, W. 1968. Mimicry in plants and animals. McGraw-Hill, New York. (Translated by R.D. Martin.)

Wicksten, M.K. 1980. Decorator crabs. Scientific American 242(2):146-154.

Wiens, J.A. 1966. On group selection and Wynne-Edwards' hypothesis. American Scientist 54:273-287.

Wiens, J.A., S.G. Martink, W.R. Holthaus, and F.A. Iwen. 1970. Metronome timing in behavioral ecology studies. Ecology 51:350-352.

Wiesel, T.N., and D.H. Hubel. 1963. Effects of visual deprivation on morphology of cells in the cat's lateral geniculate body. Journal of Neurophysiology 26:978-993.

Wiesel, T.N., and D.H. Hubel. 1965. Comparison of the effects of unilateral and bilateral eye closure on cortical unit responses in kittens. Journal of Neurophysiology 28:1029-1040.

Wiggins, D.A., and R.D. Morris. 1986. Criteria for female choice of mates: courtship feeding and paternal care in the common tern. American Naturalist 128:126-129.

Wilcox, R.S. 1979. Sex discrimination in

Gerris remigis: role of a surface wave signal. Science 206:1325-1327.

Wilczek, M. 1967. The distribution and neuroanatomy of the labellar sense organs of the blowfly *Phormia regina* Meigen. Journal of Morphology 122:175-201.

Wiley, R.H. 1974. Evolution of social organization and life history patterns among grouse (Aves: Tetraonidae). Quarterly Review of Biology 49:209-227.

Wiley, R.H. 1983. The evolution of communication: information and manipulation. In: T.R. Halliday and P.J.B. Slater, editors. Communication. Animal Behaviour. Vol 2. W.H. Freeman, New York.

Wiley, R.H., Jr. 1978. The lek mating system of the sage grouse. Scientific American 238(5):114-125.

Wilkinson, G.S. 1984. Reciprocal food sharing in the vampire bat. Nature 308:181-184.

Wilkinson, G.S. 1990. Food sharing in vampire bats. Scientific American 262(2):76-82.

Williams, G.C. 1966. Adaptation and natural selection. Princeton University Press, Princeton, N.J.

Williams, G.C. 1975. Sex and evolution. Princeton University Press, Princeton, N.J.

Williams, K.S., and L.E. Gilbert. 1981. Insects as selective agents on plant vegetative morphology: egg mimicry reduces egg laying by butterflies. Science 212:467-469.

Willows, A.O.D. 1973. Learning in gastropod mollusks. In: W.C. Corning, J.A. Dyal, and A.O.D. Willows, editors. Invertebrate learning. Vol. 2. Plenum Press, New York.

Willows, A.O.D., and W.C. Corning. 1975. The echinoderms. In: W.C. Corning, J.A. Dyal, and A.O.D. Willows, editors. Invertebrate learning. Vol. 3. Plenum Press, New York.

Wilson, D.S. 1975. A theory of group selection. Proceedings National Academy of Science of the United States of America 72:143-146.

Wilson, D.S. 1980. The natural selection of populations and communities. Benjamin/Cummings, Menlo Park, Calif.

Wilson, D.S. 1983. The group selection controversy: history and current status. Annual Review of Ecology and Systematics 14:159-187.

Wilson, E.O. 1962. Chemical communication among workers of the fire ant *Solenopsis saevissima* (Fr. Smith). 1. The organization of mass-foraging. 2. An information analysis of the odour trail. 3.

The experimental induction of social responses. Animal Behaviour 10:134-164.

Wilson, E.O. 1965. Chemical communication in the social insects. Science 149:1064-1071.

Wilson, E.O. 1971. The insect societies. Belknap/Harvard University Press, Cambridge, Mass.

Wilson, E.O. 1972. Animal communication. Scientific American 227(3):52-60.

Wilson, E.O. 1975 a. *Leptothorax duloticus* and the beginnings of slavery in ants. Evolution 29:108-119.

Wilson, E.O. 1975 b. Sociobiology: the new synthesis. Belknap/Harvard University Press, Cambridge, Mass.

Wilson, E.O. 1976. Academic vigilantism and the political significance of sociobiology. Bioscience 26:183, 187-190.

Wilson, E.O. 1980. A consideration of the genetic foundation of human social behavior. In: G.W. Barlow and J. Silverberg, editors. Sociobiolgy: beyond nature/nurture? Westview Press, Boudler, Colo.

Wilson, E.O., and B. Hölldobler. 1990. The ants. Harvard University Press, Cambridge, Mass.

Wilson, E.O., and W.H. Bossert. 1963. Chemical communication among animals. Recent Progress in Hormone Research 19:673-716.

Wilson, S. 1973. The development of social behavior in the vole *(Microtus agrestis)*. Journal of the Linnean Society, Zoology 52:45-62.

Wiltschko, R., and W. Wiltschko. 1985. Pigeon homing: change in navigational strategy during ontogeny. Animal Behaviour 33:583-590.

Wiltschko, R., and W. Wiltschko. 1989. Pigeon homing: olfactory orientation a paradox. Behavioral Ecology and Sociobiology 24:163-173.

Wiltschko, W. 1972. The influence of magnetic total intensity and inclination on directions preferred by migrating European robins *(Erithacus rubecula)*. In: S.R. Galler, et al., editors. Animal orientation and navigation. Science and Technical Information Office, NASA Special Publications, Washington, D.C.

Wiltschko, W., and R. Wiltschko. 1988. Magnetic orientation in birds. Current Ornithology 5:67-121.

Wiltschko, W., R. Wiltschko, and M. Jahnel. 1987. The orientation behaviour of anosmic pigeons in Frankfurt, Germany. Animal Behaviour 35:1324-1333.

Wiltschko, W., R. Wiltschko, W.T. Keeton, and A.I. Brown. 1987. Pigeon homing: the orientation of young birds that had been prevented from seeing the sun. Ethology 76:27-32.

Wimsatt, W.A. 1970, 1977. Biology of bats. 3 volumes. Academic Press, New York.

Windle, W.F. 1944. Genesis of somatic motor function in mammalian embryos: a synthesizing article. Physiological Zoology 17:247-260.

Wingfield, J.C. 1984. Influence of weather on reproduction. Journal of Experimental Zoology 232:589-594.

Wingfield, J.C. and B. Silverin. 1986. Effects of corticosterone on territorial behavior of free-living male song sparrows *Melospiza melodia*. Hormones and Behavior 20:405-417.

Wingfield, J.C., and D.S. Farner. 1976. Avian endocrinology field investigations and methods. Condor 78:570-573.

Wingfield, J.C. and D.S. Farner. 1978. The endocrinology of a natural breeding population of the white-crowned sparrow (*Zonotrichia leucophrys pugetensis*). Physiological Zoology 51:188-205.

Wingfield, J.C., J.P. Smith, and D.S. Farner. 1982. Endocrine responses of white-crowned sparrows, *Zonotrichia leucophyrs*, to environmental stress. Condor 84:399-409.

Wingfield, J.C., M.C. Moore, and D.S. Farner. 1983. Endocrine responses to inclement weather in naturally breeding populations of white-crowned sparrows (*Zonotrichia leucophrys pugetensis*). Auk 100:56-62.

Winkler, H. 1973. Nahrungserwerb und Konkurrenz des Blutspechts, *Picoides (Dendrocopos) syriacus*. Oecologia 12: 193-208.

Winston, M.L. 1987. The biology of the honey bee. Harvard University Press, Cambridge, Mass.

Witelson, Sandra F. 1985. The brain connection: the corpus callosum is larger in left-handers. Science 229:665-668.

Wittenberger, J.F. 1981. Animal social behavior. Duxbury Press, Boston.

Wolf, L.L. 1975. "Prostitution" behavior in a tropical hummingbird. Condor 77:140-144.

Wolgin, D.L., J. Cytrawa, and P. Teitelbaum. 1976. The role of activation in the regulation of food intake. In: D. Novin, W. Wyrwicka, and G. Bray, editors. Hunger: basic mechanisms and clinical implications. Raven Press, New York.

Wood, D.S. 1979. Phenetic relationships within the family Gruidae. Wilson Bulletin 91:384-399.

Woodcock, A., and M. Davis. 1978. Catastrophe theory. Avon Books, New York.

Woolfenden, G.E., and J.W. Fitzpatrick. 1978. The inheritance of territory in group-breeding birds. Bioscience 28: 104-108.

Woolfenden, G.E., and J.W. Fitzpatrick. 1984. The Florida scrub jay: demography of a cooperative-breeding bird. Princeton, University Press, Princeton, N.J.

Worden, A.N. 1968. Nutritional factors and abnormal behavior. In: M.W. Fox, editor. Abnormal behavior in animals. W.B. Saunders, Philadelphia.

Wrangham, R.W. 1980. Female choice of least costly males; a possible factor in the evolution of leks. Zeitschrift für Tierpsychologie 54:357-367.

Wrangham, R.W., and D.I. Rubenstein. 1986. Social evolution in birds and mammals. In: D.I. Rubenstein and R.W. Wrangham, editors. Ecological Aspects of Social Evolution. Princeton University Press: Princeton, N.J.

Wurtman, R.J., and J.J. Wurtman. 1989. Carbohydrates and depression. Scientific American 259:68-75.

Wynne-Edwards, V.C. 1962. Animal dispersion in relation to social behavior. Hafner, New York.

Wynne-Edwards, V.C. 1965. Self-regulating systems in populations of animals. Science 147:1543-1548.

Wynne-Edwards. V.C. 1977. Intrinsic population control: an introduction. Inst. Biol. Symp. on Population Control by Social Behavior, London.

Yalden, D.W., and P.A. Morris. 1975. The lives of bats. Quadrangle/New York Times Book, New York.

Yasukawa, K. 1981. Song and territory defense in the red-winged blackbird. Auk 98:185-187.

Yerkes, R.M. 1912. The intelligence of earthworms. Journal of Animal Behaviour 2:332-352.

Young, E., editor. 1973. The capture and care of wild animals. South Africa Nature Foundation/World Wildlife Fund. Ralph Curtis Books, Hollywood, Fla.

Young, J.Z. 1965. The organization of a memory system. Proceedings of the Royal Society of London, Biology 163:285-320.

Zach, R. 1978. Selection and dropping of whelks by northwestern crows. Behaviour 67:134-148.

Zach, R. 1979. Shell dropping: decision-making and optimal foraging in northwestern crows. Behaviour 68:106-117.

Zack, S. 1978. Head grooming behavior in the praying mantis. Animal Behaviour 26:1107-1119.

Zahavi, A. 1971. The function of pre-roost gatherings and communal roosts. Ibis 113:106-109.

Zahavi, A. 1979. Why shouting? American Naturalist 113(1):155-156.

Zar, J.H. 1984. Biostatistical analysis. Prentice-Hall, Englewood Cliffs, N.J.

Zeigler, H.P. 1976. Feeding behavior of the pigeon. Advances in the Study of Behavior 7:285-389.

Zeigler, H.P., and H.S. Karten. 1974. Central trigeminal structures and the lateral hypothalamic syndrome. Science 186:636-637.

Zimen, E. 1976. On the regulation of pack size in wolves. Zeitschrift für Tierpsychologie 40:300-341.

Zimmerman, N.H., and M. Menaker. 1979. The pineal gland: a pacemaker within the circadian system of the house sparrow. Proceedings of the National Academy of Sciences of the United States of America 76:999-1003.

Zoeger, J., J.R. Dunn, and M. Fuller. 1981. Magnetic material in the head of the common Pacific dolphin. Science 213:892-894.

Zuckerman, L., editor. 1980. Great zoos of the world. Westview Press, Boulder, Colo.

Zuk, M. 1984. A charming resistance to parasites. Natural History 93(4):28-34.

Zuk, M. 1988. Parasite load, body size, and age of wild-caught male field crickets (Orthoptera: Gryllidae): effects on sexual selection. Evolution 42:969-976.

Zuk, M. 1989. Validity of sexual selection in birds. Nature 340:104-105.

Zwislocki, J.J. 1981. Sound analysis in the ear: a history of discoveries. American Scientist 69:184-191.

INDEX

Page numbers in italics indicate illustrations; *t* indicates tables.

873